CHINA NATIONAL
BOTANICAL GARDEN

中国
二十一世纪的
园林之母

第八卷

CHINA

Mother of Gardens, in the Twenty-first Century

Volume 8

马金双　主编

Editor in Chief: MA Jinshuang

中国林业出版社
China Forestry Publishing House

内容提要

　　《中国——二十一世纪的园林之母》为系列丛书，记载今日中国观赏植物研究与历史以及相关的人物与机构，其宗旨是总结中国观赏植物资源及其现状，弘扬园林之母对世界植物学，乃至园林学和园艺学的贡献。全书拟分卷出版，本书为第八卷，共2章：第1章，蓝果树科的珙桐——中国鸽子树；第2章，韩尔礼采集植物志。

图书在版编目（CIP）数据

　　中国——二十一世纪的园林之母. 第八卷 /（美）马
金双主编. -- 北京：中国林业出版社, 2025. 3.
ISBN 978-7-5219-3119-8

　　Ⅰ. S68

　　中国国家版本馆 CIP 数据核字第202513XB34号

责任编辑：张　华　贾麦娥
装帧设计：刘临川

出版发行：中国林业出版社
　　　　　（100009，北京市西城区刘海胡同7号，电话010-83143566）
电子邮箱：43634711@qq.com
网址：https://www.cfph.net
印刷：北京博海升彩色印刷有限公司
版次：2025年3月第1版
印次：2025年3月第1次印刷
开本：889mm×1194mm　1/16
印张：39.5
字数：1320千字
定价：498.00元

编写说明

《中国——二十一世纪的园林之母》为系列丛书，由多位作者集体创作，完成的内容组成一卷即出版一卷。

《中国——二十一世纪的园林之母》记载中国观赏植物资源以及有关的人物与机构，其顺序为植物分类群在前，人物与机构于后。收录的类群以中国具有观赏价值和潜在观赏价值的种类为主；其系统排列为先蕨类植物后种子植物（即裸子植物和被子植物），并采用最新的分类系统（蕨类植物：CHRISTENHUSZ et al., 2011；裸子植物：CHRISTENHUSZ et al., 2011；被子植物：APG IV, 2016）。人物与机构的排列基本上以汉语拼音顺序记载，其内容则侧重于历史上为中国观赏植物做出重要贡献的主要人物以及以研究与收藏中国观赏植物为主的重要机构。植物分类群的记载包括隶属简介、分类历史与系统、分类群（含学名以及模式信息）介绍、识别特征、地理分布和资源引种以及传播历史等。人物侧重于其主要经历、与中国观赏植物和机构的关系及其主要成就；而机构则侧重于基本信息、自然地理概况、历史变迁、现状以及收藏的具有特色的中国观赏植物资源及其影响等。

本丛书不设具体的收载文字与照片限制，这不仅仅是因为植物类群不一、人物和机构不同，更考虑到其多样性以及其影响。特别是通过这样的工作能够使作者们充分发挥潜力并提高研究水平，不仅仅是记载相关的历史渊源与文化传承，更重要的是借以提高对观赏植物资源开发利用和保护的科学认知。

欢迎海内外同仁与同行加入编写行列。在21世纪的今天，我们携手总结中国观赏植物概况，不仅仅是充分展示今日园林之母的成就，弘扬中华民族对世界植物学乃至园林学和园艺学的贡献；而且希望通过这样的工作，锻炼、培养一批有志于该领域的人才，继承传统并发扬光大。

本丛书第一卷和第二卷于2022年秋天出版，并得到业界和读者的广泛认可。2023年再次推出第三、第四和第五卷。2024年继续完成第六卷、第七卷出版工作。特别感谢各位作者的真诚奉献，使得丛书能够在4年时间内完成7卷本的顺利出版！感谢各位照片拍摄者和提供者，使得丛书能够图文并茂并增加可读性。特别感谢国家植物园（北园）领导的大力支持、有关部门的通力协助以及有关课题组与相关人员的大力支持；感谢中国林业出版社编辑们的全力合作与辛苦付出，使得本书顺利面世。

因时间紧张，加之水平有限，错误与不当之处，诚挚地欢迎各位批评指正。

编者

2025年春

前言

中国是世界著名的文明古国，同时也是世界公认的园林之母！数千年的农耕历史不仅为中国积累了丰富的栽培与利用植物的宝贵经验，而且大自然还赋予了中国得天独厚的自然条件，因而孕育了独特而又丰富的植物资源。多重因素叠加，使得中国成为举世公认的植物大国！中国高等植物总数超过欧洲和北美洲的总和，高居北半球之首，而且名列世界前茅。然而，园林之母也好，植物大国也罢，我们究竟有多少具有观赏价值或者潜在观赏价值（尚未开发利用）的植物，要比较准确或者可靠地回答这个问题，则是摆在业界面前比较困难的挑战。特别是，中国观赏植物在世界园林历史上的作用与影响，我们还有哪些经验教训值得总结，更值得我们深思。

百余年来，经过几代人的艰苦奋斗，先后完成《中国植物志》（1959—2004）中文版和英文版（*Flora of China*，1994—2013）两版国家级植物志和近百部省（自治区、直辖市）植物志，特别是近年来不断地深入研究使得数据更加准确，这使得我们有可能进一步探讨中国观赏植物的资源现状，并总结这些物种及其在海内外的传播与利用，辅之以学科有关的重要人物与主要机构介绍。这在21世纪的今天，作为园林之母的中国显得格外重要。一方面，我们要清楚自己的家底，总结其开发与利用的经验教训，以便进一步保护与利用；另一方面，我们要激发民族的自豪感与优越感，进而鼓励业界更好地深入研究并探讨，充分扩展我们的思路与视野，真正引领世界行业发展。

改革开放40多年来，国人的生活水准有了极大的改善与提高，国民大众的生活不仅仅满足于温饱而更进一步向小康迈进，尤其是在休闲娱乐、亲近自然、欣赏园林之美等层面不断提出更高要求。作为专业人士，一方面，我们应该尽职尽责做好本职工作，充分展示园林之母对世界植物学乃至园林学和园艺学的贡献；另一方面，我们要开阔自己的视野，以园林之母主人公姿态引领时代的需求，总结丰富的中国观赏植物资源，以科学的方式展示给海内外读者。中国是一个14亿人口的大国，要将植物知识和园林文化融合发展，讲好中国植物故事，彰显中华文化和生物多样性魅力以及提高国民素质，科学普及工作可谓任重道远。

基于此，我们组织业界有关专家与学者，对中国观赏植物以及具有潜在观赏价值的植物资源进行了总结，充分记载中国观赏植物的资源现状及其海内外引种传播历史和对世界园林界的贡献。与此同时，对海内外业界有关采集并研究中国观赏植物比较突出的人物与事迹，相关机构的概况等进行了介绍；并借此机会，致敬业界的前辈，同时激励民族的后人。

国家植物园（北园），期待业界的同仁与同事参与，我们共同谱写二十一世纪园林之母新篇章。

贺 然 魏 钰 马金双

2025 年

目录

Summary

Explanation

Preface

园林之母
China

01

-ONE-

蓝果树科的珙桐——中国鸽子树

Davidia involucrata of Nyssaceae —— Chinese Dove tree

傅承新　李文昊
（浙江大学）

FU Chengxin　LI Wenhao
(Zhejiang University)

cxfu@zju.edu.cn

摘　要：本文基于保护遗传学的研究成果，通过翔实的文献和图片，梳理了"活化石"中国鸽子树——珙桐的发现与相关研究历程，介绍了珙桐的分类地位、物种分化及系统位置的确立过程、种群现状以及珙桐的海内外传播、引种栽培、园林利用等方面的内容。

关键词：中国鸽子树　珙桐　活化石

Abstract: Based on the research findings of conservation genetics, this paper systematically reviews the discovery and research history of the "living fossil" Chinese dove tree (*Davidia involucrata*) through comprehensive literature and illustrative images. The taxonomic status of *Davidia*, the establishment process of species differentiation and phylogenetic position, current population status, as well as its domestic and international dissemination, introduction and cultivation, and horticultural utilization, have been introduced.

Keywords: Chinese dove tree, *Davidia involucrata*, Living fossil

傅承新，李文昊，2025，第1章，蓝果树科的珙桐——中国鸽子树；中国——二十一世纪的园林之母，第八卷：001-033页.

1 珙桐的分类及系统位置

1.1　珙桐属 *Davidia* Baillon

Baillon in Adansonia 10: 114. 1871; Wangerin in Engl., Pflanzenr. 41: (IV. 220a):17. 1910; Qin et Phengklai in Flora of China. 13:301. 2007.

Typus: *Davidia involucrata* Baillon

本属仅有1种，中国西南部特产。

1.2　珙桐

Davidia involucrata Baillon, in Adansonia 10:115. 1871; Wangerin in Engl., Pflanzenr. 41 (IV. 220a): 18, f. 4. 1910; Fang, Icon. Pl Omeiens. 1: Pl. 16. 1942. ——*D. tibetana* David in Nuov. Arch. Mus. Hist. Nat. Paris II. 5: 1884. 1882, nom. nud.; Qin et Phengklai in Flora of China. 13:301. 2007.

珙桐（图1）最初是1871年由拜伦（Henri E Baillon, 1827—1895）基于法国博物学家戴维（Armand David, 中文名谭卫道, 1826—1900）采集的标本（图

图1　珙桐开花时形态（2024年4月11日傅承新摄于杭州植物园）

图2 珙桐模式标本（A：模式；B：等模式，现存于巴黎自然博物馆，P）（A：引自http://plants.jstor.org/stable/10.5555/al.ap.specimen.p00545534；B：引自http://plants.jstor.org/stable/10.5555/al.ap.specimen.p00545535）

2）描述的，他把它放在使君子科（Combretaceae），靠近蓝果树属（*Nyssa*）的位置。

1897年，德国植物学家哈姆斯（Hermann Harms，1870—1942）以珙桐属（*Davidia*）为模式属建立了珙桐亚科（Davidioideae）隶属于山茱萸科（Cornaceae）（Harms，1997）。Engler（1909）在他的系统中把蓝果树、珙桐等由山茱萸科分出，成为一个独立的科：蓝果树科（Nyssaceae）。Wangerin（1910）把珙桐亚科重组为蓝果树科的一个亚科。后来，Takhtajan（1954）在他的分类系统中将珙桐亚科上升为珙桐科（Davidiaceae），Li（1955）也提出建立珙桐科；他们的依据是珙桐子房室数目和蓝果树科其他植物不同而支持单独设科。方文培和宋滋圃（1975）也赞同这一点。但由于《中国植物志》和*Flora of China*采用的是恩格勒系统，因而仍将珙桐属置于蓝果树科（方文培等，1983；Qin & Phengklai，2007）。

在上述观点基础上，后来的形态与解剖学研究揭示蓝果树（*Nyssa sinensis*）和喜树（*Camptotheca acuminata*）的花序和花的形态、导管穿孔类型等与珙桐（*Davidia involucrata*）不同，其厚珠心胚珠、具绒毡层等胚胎学特征也差异较大，认为珙桐的导管分子在系统演化上比喜树和蓝果树更进化；此外，珙桐的花粉、化学成分和血清学研究也揭示它和蓝果树科其他植物具有较大差异；在化学成分方面，喜树含抗癌性较强的喜树碱，珙桐则没有；在染色体数目上，喜树、蓝果树的染色体数目均为2n=44，而珙桐与其变种光叶珙桐的染色体数目相同，为2n=42；上述研究结论均支持建立珙桐科（胡进耀等，2003；陈树思，1999；邹广权等，2022；向桂琼和卢馥荪，1989；贺军辉，1991）。

自20世纪90年代末被子植物分子系统学系统（APG）提出以来，APG系统Ⅰ、Ⅱ、Ⅲ版

（1998，2003，2009）基于少量DNA片段的研究仍然支持将喜树属、蓝果树属和珙桐属划入山茱萸科。Xiang等（2011）基于cpDNA 6个片段揭示了珙桐与蓝果树属和喜树属（*Camptotheca*）为姐妹关系（图3），支持珙桐为蓝果树科成员。因此，APG Ⅳ（2016）综合多方面数据，接受了珙桐属与蓝果树属、喜树属的近缘关系，支持置于蓝果树科。Yang & Ji（2017）通过对喜树属、珙桐属和蓝果树属的叶绿体基因组测序分析，也揭示喜树属、珙桐属和蓝果树属与山茱萸科植物的分化，支持成立蓝果树科。

Chen等（2020）发表了珙桐全基因组序列。Ren等（2024）进一步研究揭示它与喜树在分化之前共享了最近的全基因组复制，分化在约6000万年前。

综上所述，珙桐的系统位置已经确定：山茱萸目蓝果树科珙桐属。

1.3　种下分类

1.3.1　珙桐（原变种）

Davidia involucrata Baill. var. ***involucrata***

产湖北西部（兴山、长阳、五峰、恩施、利川、建始、巴东、宣恩、鹤峰、神农架、十堰）、湖南西部（石门、慈利、桑植）、重庆（南川、巫山、巫溪、酉阳）、四川（平武、峨边、马边、峨眉山、洪雅、筠连、荥经、天全、宝兴、美姑、雷波）、贵州北部（江口、毕节）和云南北部（大关、绥江、镇雄），以及甘肃（文县）和陕西南部（平利、岚皋、镇坪）。在四川西部的宝兴、天全、峨眉、马边、峨边等县常见；生于海拔1 500～2 200m的润湿的常绿阔叶与落叶阔叶混交林中。模式标本采自四川宝兴，*A. David 1869*（P，P00545534）。

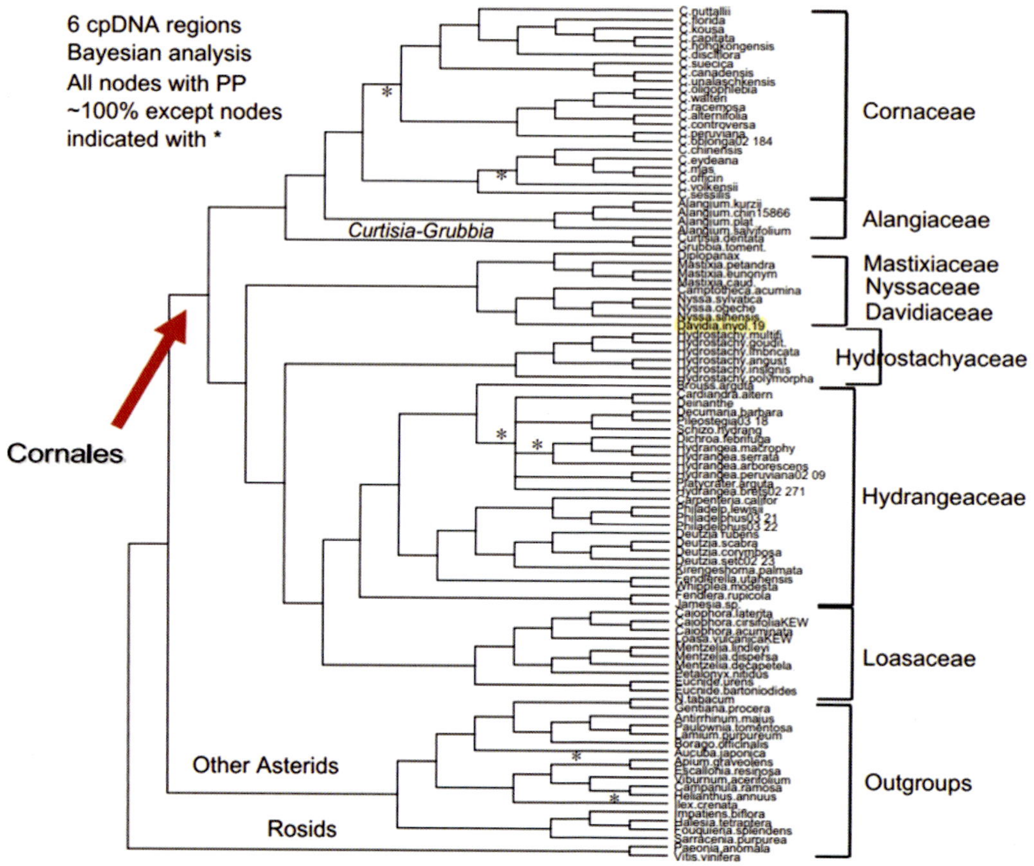

图3　基于6个cpDNA片段揭示的珙桐系统位置（Xiang et al., 2011）

01

1.3.2 光叶珙桐（变种）

Davidia involucrata Baill. var. ***vilmoriniana***
(Dode) Wangerin in Engl., Pflanzenr. 41（Ⅳ. 220a):
17. 1910. ——*D. vilmoriniana* Dode in Rev. Hort. II.
8. 406. 1908。

模式标本采自四川巫山（今重庆巫山），*A.*

Henry. 1889, 5577 (K). —— *D. laeta* Dode op. cit.
407. 1908（图4）.

与原变种的区别在于本变种叶下面常无毛或幼时
叶脉上被很稀疏的短柔毛及粗毛，有时下面被白霜。

产湖北西部、四川、重庆和贵州等地；生于
1 500～2 000m的森林中，常与原变种珙桐混生
（Qin & Phengklai, 2007）。

图4　光叶珙桐（现存K，傅承新摄于邱园标本馆K）

1.4　种下分类研究

1869年，法国传教士和博物学家戴维在四川省西部宝兴首次发现珙桐，法国植物学家拜伦于1871年首次命名并发表了珙桐的植物学描述，以发现者戴维的名字作为属名 *Davidia*。光叶珙桐最初是基于奥古斯丁·亨利（Augustine Henry，中文名韩尔礼）1889年采于四川巫山（今重庆巫山）的标本，多德（L A Dode）把它作为珙桐属的新种发表的（Dode, 1908），随后，Wangerin（1910）认为光叶珙桐其他特征与珙桐无区别，因此将光叶珙桐降为珙桐的变种处理。此后，从形态学上，对珙桐的种内分类问题再没有新的讨论，因此，光叶珙桐其变种的地位沿用至今。

21世纪以来，国内学者对珙桐种内的遗传多样性、分化及与光叶珙桐的关系从形态学、分子生物学、生态学、比较基因组学角度进行了研究。李雪萍等（2012）对52个珙桐个体和60个光叶珙桐个体进行的比较研究表明，除了叶背被毛外，其他性状没有差异。Ma等（2015）基于22个居群的cpDNA片段和nSSR标记均揭示珙桐种内（不论叶背有毛无毛）具有东西两个主要谱系和4个次级谱系（图5）。Chen等（2015）基于32个居群的6个cpDNA片段分析也揭示存在东西两个主要谱系，并发现谱系内存在着进一步分化。最近，随着珙桐全基因组的发表，Ren等（2024）基于东西两大谱系的22个居群的77个个体的全基因组重测序分析已完全揭示了珙桐种内的遗传分化，即存在东西两大主要谱系（分化时间大约在300万前），而西部谱系由2个次级支系组成（即西部谱系和西南部谱系，分化时间约在30万年前）组成，东部谱系也具有2个次级支系。研究表明谱系之间存在频繁的杂交事件（图6）。

宋帅帅（2023）对珙桐叶背的毛被进行了深入研究，揭示珙桐叶片下表皮从无毛—有少量被毛—有密被毛是一个连续的过程，且多数采样点珙桐与

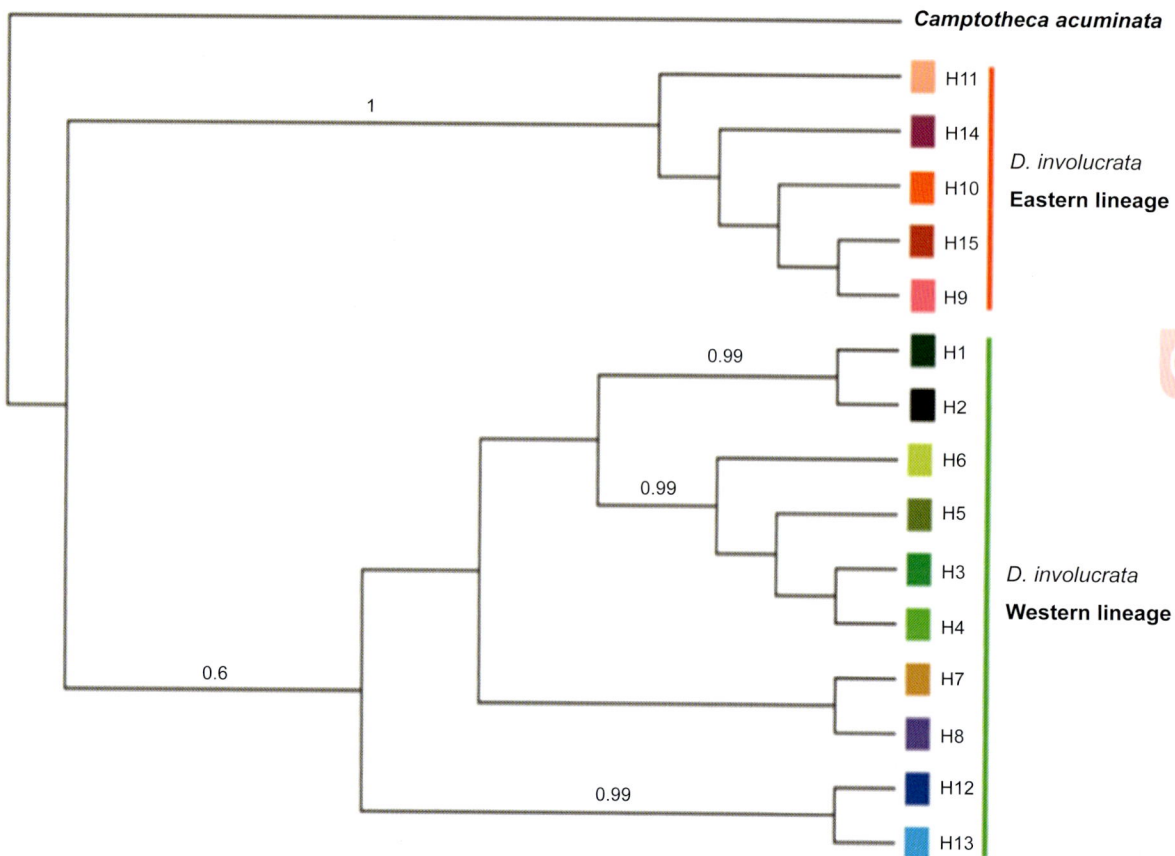

图5　基于cpDNA的珙桐亲缘地理研究揭示的种内东西两个谱系（Ma et al., 2015）

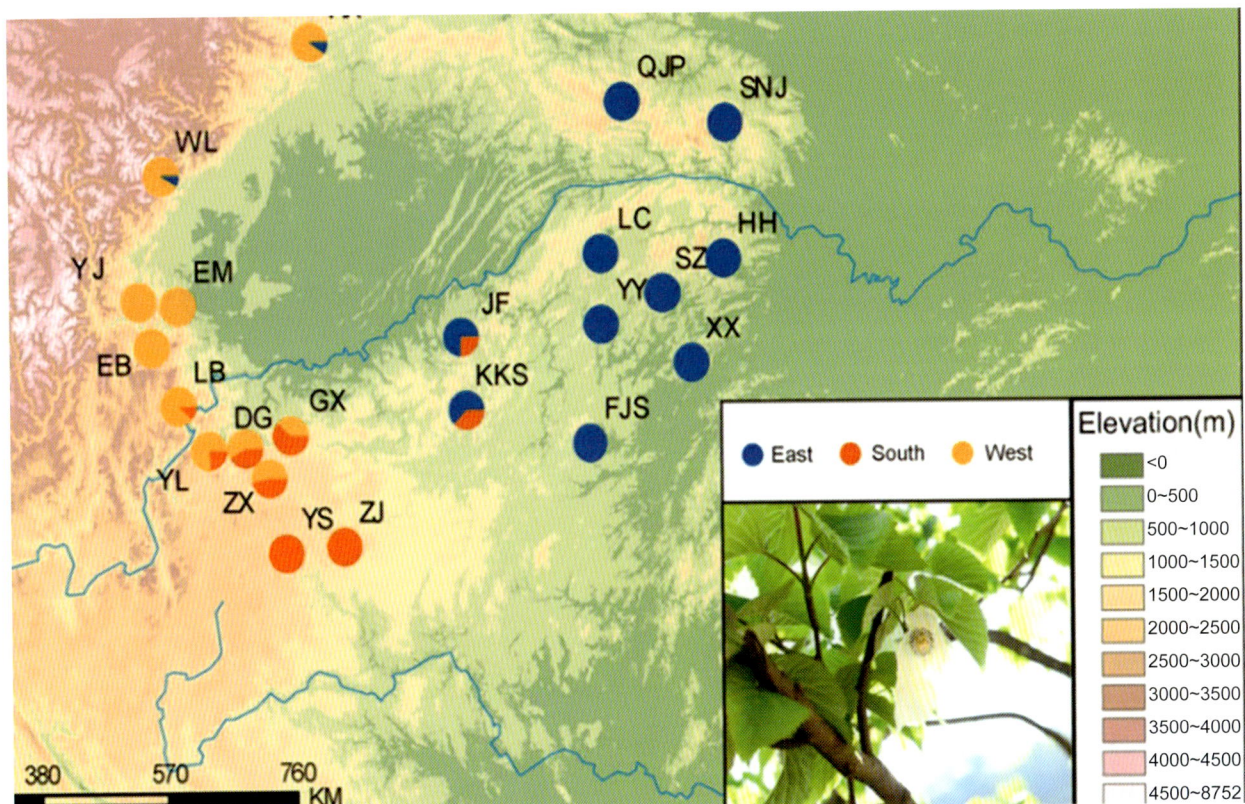

图6 基于基因组重测序分析揭示的珙桐种内隐分化（Ren et al., 2024）

光叶珙桐是混生的；发现珙桐叶片被毛密度与叶片虫食率呈显著负相关，即珙桐叶片被毛能帮助叶片抵御植食性昆虫的取食。因此，笔者认为珙桐科仅珙桐属珙桐一个种，无种下分类单位。

2 地史背景与演化

珙桐最早的化石发现于早第三纪。化石记录表明，珙桐在古新世、上新世已在北美广泛分布，上新世珙桐种群也曾广泛分布于东亚。Kokawa（1965）第一次在日本中部川崎发现了内果皮化石，Tsukagoshi等（1997）研究发现珙桐内果皮化石在日本中部岐阜和西南部早更新世地层，认为珙桐应该在上新世就已存在日本，到早更新世灭绝。Manchester（2002）研究了北美的珙桐化石，发现珙桐属在北美中纬度的古新世植被中起了突出的作用，在北达科他州、蒙大拿州和怀俄明州等八个地方存在珙桐叶和果实的化石记录。类似的树叶和果实化石记录在俄罗斯东部的古新世沉积层中也被发现，确定了现存的 *Davidia* 是在第三纪古新世起源的。Paclyutkin（2009）在俄罗斯远东的涅日诺也发现珙桐属的化石。

珙桐是距今6 000万年前新生代第三纪古热带植物区系的孑遗种，但早期应该存在多个种。日本记录过 *Davidia kabutoiwana* Ozaki（Tsukagoshi et al., 1997），北美记录过 *D. palaeoivolucrata* Pavlyutkin以及俄罗斯远东涅日诺记录过 *D. nezhinoensis*

Paclyutkin(Paclyutkin, 2009)。

在距今6 500万年的早第三纪，地球的气候变得温暖湿润。当时的地球正处于一个巨变的时期，蕨类植物已经大大减少，裸子植物中苏铁类和银杏类只剩下了很少的成员，松柏类也远不如中生代时那样繁盛，而木本被子植物则获得了大量的发展。珙桐是在这样的地球环境中发展起来，并和其他同时代的植物通过自身扩散遍布于当时地球的古热带地区。而大约从渐新世早期（距今约340万年前）开始，地球气温逐渐变冷，这些物种开始逐步向南退缩到较低纬度的地区（即现在的长江流域一带），到了晚第三纪至第四纪期间，进一步南退到东亚、北美和欧洲西南部三块区域。至末次冰期盛期（约距今2万年），地球气温降至最低点。此时的欧洲、北美洲以及亚洲的日本多被冰雪覆盖，导致其发生了大规模的灭绝（Eyde, 1997; Manchester et al., 2009）。中国所属的东亚地区，尤其是长江流域在第四纪冰期仅形成了很有限的冰川覆盖；其次，中国南方地区众多的高山和峡谷缓冲了剧烈的气候波动，为物种原地存留提供了长期稳定的生存环境；再次，在冰川时期，由于珙桐种子具有休眠期长的特点，相当于被冰藏起来了，直到冰川退却，地球温度逐渐回暖，种子会再生根发芽。在这些因素的共同作用下，中国成为保存远古植物种类最为丰富和多样的国家，即被称为"第三纪植物的避难所"。珙桐作为其中的代表物种，因此得以幸存并且繁衍至今（Peng et al., 2003）。

3 珙桐的发现历史

珙桐由法国传教士和博物学家戴维（图7A）于1869年在西康穆坪（今四川宝兴）首次发现的。1871年，巴黎自然历史博物馆分类学家拜伦根据这份标本，建立了单种新属珙桐属（Davidia），其属名就是为了纪念戴维。宝兴是珙桐的模式标本产地，被誉为"中国鸽子花的故乡"。珙桐发表之后到1888年，被法国植物学家佛朗切特（Adrien Franchet）写进其著作《戴维采集植物志》（Plantae Davidianae）第二卷，该卷的副标题是"藏东植物"，记载了宝兴植物402种，并附有1页珙桐的手绘彩色解剖图（图7B），素描和彩色的渲染让珙桐花枝显得栩栩如生，在夸张的苞片上，人们甚至可以看到细致清晰的脉络。当时，珙桐被发现这件事引起欧洲园艺界的轰动！很多园林公司纷纷派人前往中国采集标本和树种。此后直到1897年才由另一位法国神父保罗·法格斯（Paul Guillaume Farges, 1844—1912）在川西北采集到珙桐种子并进行了异地繁育研究。而1900年英国"植物猎人"及探险家威尔逊（Ernest Wilson, 1876—1930）第一次到达中国，在湖北长阳采集到大量珙桐种子，才实现在欧洲和北美广泛种植（Mcclintock, 1991）。

在此，需要重点介绍一下第一个发现珙桐的法国神父戴维（Armand David, 中文名谭卫道，1826—1900），他是19世纪法国杰出的动物学家、植物学家和博物学家。1826年，他出生在法国比利牛斯山区一个叫埃斯布莱特的小镇。少年时期的戴维聪慧好学，活泼好动，迷醉于家乡的自然山水。受当医生的父亲影响，戴维非常喜爱动植物。青年时期，他立志要做一名博物学家，到世界各地去考察动植物状况。同时，戴维也是一个虔诚的天主教徒，有为上帝献身的精神。1846年，他进入巴黎遣使会学习。1850年，24岁的戴维当上神父。1860年，35岁的戴维在巴黎结识斯坦尼斯拉斯·艾尼昂·朱利安（Stanislas Aignan Julien，1797—1873；中文名儒莲，法国汉学家），燃起了他对中国的兴趣，并向朱利安学习中文。在巴黎，他也认识了不少著名的动植物学家。1862年2月，戴维受教会派遣，以英国皇家动物学会和法国科学院及巴黎自然历史博物馆通讯员、传教士的身份，与其他传教士一起前往中国。戴维历时12年在中国进行传教与自然考察，历尽艰险，他用他的眼睛让全世界了解了中国的丰富自然资源，为东西方交流做出了杰出贡献。在中国的12年间，戴维三次深入中国内陆（从北京周边到秦陕山地，从内蒙古草原到闽西山地）进行宗教活动和动植物资源的研究、考察。1869年年初，戴维来到成都传教，听说了穆坪这个动植物宝库，随即要求去穆坪邓池沟天主堂传教。从成都到穆坪的路程在当时是非常艰难的，戴维请当地年轻教徒为向导，沿着临邛古道，穿过雅安的碧峰峡和上里古镇，翻越了一座林木茂密的高山，整整走了8天才来到穆坪。在穆坪一个叫邓池沟的地方有一个天主堂，戴维也就成为该天主堂的一个神父。凭着他多年野外考察积累的经验感觉到，这里正是他长久以来苦苦追寻的梦想之地，可能是一个动植物的王国。他在日记里写道："如我所预料的那样，这里虽然离成都不算遥远，但由于崇山峻岭的阻隔，是一个封闭的、原始的、摆脱了中国当局管辖的部落。""这里的高山和河谷都被原始森林所覆盖，使得当地的野生动物得以生存和延续下来"（David，1949）。当年在西康穆坪进行的科学考察和传教活动是他在中国最有意义的一段日子。他也因此取得了非凡的科研成就，并将永远为学界铭记。他的卓越贡献表现在以下几方面：①1869年，在西康穆坪发现了第三纪孑遗植物珙桐。当时，他将珙桐的种子、标本和照片寄回巴黎，虽然当时并没有促成珙桐在欧洲的成功种植繁衍，但催生了

图7　A：发现珙桐的法国传教士戴维（谭卫道）（引自 Wikimedia Commons）；B：珙桐花纵剖图（引自 Franchet, 1888）

该种的命名和发表，促进了后来威尔逊的中国行（胡启明，2022）。②将麋鹿引到欧洲。1862年到达中国后，他先是在北京传教和考察。自然资源丰富的中国没有让戴维失望，很快他见到了北京南海子皇家猎苑中饲养的、举世无双的"四不像"（即麋鹿，学名 Elaphurus davidianus）。当时他购得两套骨骼和角寄回了法国，这种中国湿地特有的鹿种，被西方称为"戴维鹿"。由于戴维的发现，使后来欧洲从中国盗买走了几十头麋鹿到西方喂养。然而由于战乱、洪涝等原因，后来麋鹿从中国彻底绝迹。幸运的是那些在欧洲的麋鹿保存了下来。1985年，英国第十四世贝福特公爵将18头麋鹿送回了其原产地中国，并成功繁育至今天的数千头，成为世界濒危动物回归故土成功繁育的典范。说起来这也有戴维的功劳。③发现大熊猫并介绍给了欧洲。1869年，他在西康穆坪发现珙桐之前，先发现了第三纪孑遗动物：大熊猫（学名 Ailuropoda melanoleuca），这在博物学家中也是非常值得纪念的。1869年2月，戴维到达西康穆坪后不久的3月11日，在一个姓李的庄主猎人家里（也是教徒），无意间瞥见了墙上挂着的黑白兽皮，他不由得看呆了。主人见他如此，便告诉他这是白熊皮，附近山上都有。戴维给这个神秘动物临时起了个名字叫"白熊"。他怀着欣喜在日记里写道："这可能会成为科学上的新种"（David，1949）。3月23日，当地猎人捕获并杀死一头幼年"白熊"，并将其卖给了戴维。戴维将之制成标本，送回巴黎，并称之为"黑白熊"。不久他又获得一只活的母熊，本想运回法国，但因为饲养不当而死去。戴维只好将其制成标本寄回了巴黎自然历史博物馆，成为该馆镇馆之宝一直到今天。法国国家博物馆主任米勒-爱德华（Henri Milne-Edwards，1800—1885）认真研究了这个他从来没见过的动物的皮和骨骼，认定它不是熊，除了四肢、耳朵和眼圈是黑色以外，其余部分都呈白色。头很大，嘴短圆，不像熊嘴那么尖长，世界上所有的熊都没有它漂亮。他确信这是一个新的、唯独中国才有的物种，他在1870年发表了题为《论西藏东部的几种哺乳动物》的文章，认为"就其外貌而言，大熊猫的确与熊很相似，但其骨骼特征和牙齿系统的区别十分明

显，而与小熊猫和浣熊相近。他是一新属新种，我已经将它命名为 Ailuropoda melanoleuca"（Milne-Edwards，1870）。

1869—1872年，戴维在宝兴发现或命名的动植物模式标本达数十种。不仅大熊猫是戴维神父发现的，声名赫赫的金丝猴也是戴维神父首先记载史册的，还有牛羚、鬣羚、短尾猴、亚洲黑熊、绿尾虹雉、雉鹑、朱雀以及许多种鹛类。他还对宝兴的植物进行了探究，除了发现珙桐外，还发现了铁筷子、水月华等多种植物。

1874年，由于健康状况每况愈下，他不得不离开生活、工作和考察十二年的中国，从上海乘船返回法国。1877年，戴维撰写的《中国之鸟类》两卷出版，记载他在中国发现的772种鸟类，其中有60余种是前人没有记载的。1884—1888年，《戴维采集植物志》出版（Franchet，1888），记载了珙桐、杜鹃花科52种，木兰属3种，冷杉属4种以及一些蔷薇科植物。此外，经戴维亲自鉴定的哺乳动物共有200余种，其中63种是新种，包括扭角羚、金丝猴、绿尾虹雉都是戴维在穆坪地区发现并亲手制作标本后，送到法国自然历史博物馆珍藏的。

1872年，戴维入选法国科学院院士。法国社会科学学会授予他大师称号、法国地理学会授予他金质奖章。1900年，戴维因病去世，享年74岁。戴维去世后，安葬在巴黎市中心一个叫盟帕拉斯的公墓，作为一名动植物学家，戴维为科学的神圣殿堂增添了一束瑰丽的光芒，永为世人怀念。戴维还是第一个到我国西南高山区收集生物标本的西方人。他在西南山区的一系列发现为西方学术界对中国的动植物的认识展开了新的视野，他们注意到了中国西南山区丰富的动植物区系和大量的珍稀特有物种。他的工作也促使了不少西方学者前来这一地区考察与采集。

爱尔兰植物学家韩尔礼（Augustine Henry）先生，在湖北宜昌巴东海关工作七年（1882—1889）（许亦华，2024）。1885年后，他常利用闲暇时间在宜昌及周边地区采集植物，成为非植物学专业出身的植物学家，且影响力足够大。以至于当他被调离宜昌海关之后，伦敦邱园园长威廉西塞尔

顿-戴尔（William Thiselton - Dyer）爵士仍先后三次写信向他寻求珙桐树种子。他的努力对后继者威尔逊的影响是深刻而巨大的。

1888年，韩尔礼在宜昌海关工作的第六年，为了扩大采集范围，他和他的团队进入了湖北西部恩施土家族苗族自治州巴东县（与原四川今重庆巫山接壤处）。5月17日，在经过一个名叫绿葱坡的村庄时，他们突然发现，在一块大岩石的脚下，有一棵正值花期的珙桐树，"仿佛有几千条幽灵般的白色手帕悬挂在枝头随风飘动"，十分引人注目。这年秋季，韩尔礼专门派出两名植物采集者重回绿葱坡采集该树的果实，并将果实和植物标本一起寄往伦敦邱园。这是继法国人戴维之后，西方植物猎人第一次在中国湖北西部—四川东部，现重庆巫山发现珙桐树。可惜邱园收到种子后只顾着研究，却忘记了播种。

当时韩尔礼无意中的发现使他情不自禁地赞叹道："珙桐太美了！它简直就是无价之宝，虽然我只看到了一棵，但是毫无疑问，在这个区域肯定还有很多的珙桐"。他把这个情况写信告诉了英国皇家植物邱园，建议邱园派人到这里寻找。这就是为什么戴维在四川穆坪首次发现了珙桐，而威尔逊的首次中国行却来到湖北宜昌的原因。

这里还必须提一下英国当年著名的园林公司——维奇（Veitch）苗圃公司，其老板维奇（John Veitch，1752—1839）爵士。有一年，维奇苗圃公司举办名贵苗木展，韩尔礼先生应邀观展，韩尔

礼在中国工作7年，在鄂西地区采集过许多植物种子和标本。他向维奇着重介绍了鄂西山区生长的珙桐，认为是园艺界的极品，维奇对此非常感兴趣，当即开出巨额赏金，招募前往中国寻宝的"植物猎人"。这个人选既需要丰富的专业经验，还要具备强健的体魄和顽强的意志。当年，年轻力壮且专业知识丰富的威尔逊，成为不二人选。维奇苗圃公司给他提供了500英镑出差费和100英镑的年薪。这份年薪，相当于当时英国中产家庭3年多的收入（央视纪录片，2016）。

1899年4月，被英国维奇苗圃公司选中的23岁的"植物猎人"威尔逊（图8），被派往中国寻觅园林观赏植物，当时他要完成的一项重要任务，就是引种戴维发现的珙桐树，并要求他"不可在任何其他事务上浪费时间、精力或金钱"即不遗余力找到珙桐（王康，2022）。威尔逊于1900年2月到达湖北宜昌，路上花去了近一年时间。在那个年代，那可真正是远渡重洋了。威尔逊进入宜昌之前先要找到韩尔礼先生，需要他提供多方面的引导，尤其是要获取他宜昌海关工作时曾经发现的珙桐树的确切位置。在寻找韩尔礼的过程中，适逢中国大清王朝的末世时光，庚子国难爆发，作为洋人的威尔逊要在那样的年代进入中国，尤其要从当年长江最繁华的宜昌南津关码头登陆，确实是要冒生命危险的。稍有中国历史常识的人都知道，那时扶清灭洋的义和团运动风起云涌，对于一个欧洲人来说，在中国有多凶

However, all went well, and we drank in the beauties of this extraordinary tree. The distinctive beauty of Davidia is in the two snow-white connate bracts which subtend the flower proper. These are always unequal in size, the larger usually 6 inches long by 3 inches broad, and the smaller $3\frac{1}{2}$ inches by $2\frac{1}{2}$ inches; they range up to 8 inches by 4 inches and 5 inches by 3 inches. At first greenish, they become pure white as the flowers mature and change to brown with age. The flowers and their attendant bracts are pendulous on fairly long stalks, and when stirred by the slightest breeze they resemble huge Butterflies hovering amongst the trees. The bracts are somewhat boat-shaped and flimsy in texture, and the leaves often hide them considerably, but so freely are they borne that the tree looks, from a short distance, as if flecked with snow. On dull days and in the early morning and evening the bracts are most conspicuous. The fruit superficially resembles a small walnut, but the inner shell is absolutely unbreakable. To my mind *Davidia involucrata* is at once the most interesting and beautiful of all trees of the north-temperate flora.

图8　威尔逊在中国西部考察途中（引自胡启明，2022）及其笔下的珙桐（引自 Wilson，1913）

险可想而知。除了动荡的社会环境之外，长江三峡所处之地恶劣的交通也让威尔逊的中国行处处涉险，步步惊心。

年轻的威尔逊，从未走出过国门，既没有经验，也没有方向，韩尔礼发现的那棵珙桐树在当时几乎是他心中的唯一目标，他历尽艰险，执意先找到韩尔礼，就是想搞清楚韩尔礼发现的那棵珙桐树的具体位置。1899年9月，威尔逊经香港—越南老街进入云南，最后在云南思茅找到了在海关任职的韩尔礼先生，韩尔礼给他绘制了发现珙桐的大致地点和线路草图，还传授了湖北西部和四川东部（今重庆）一带的植被情况、风土人情，也提供了在宜昌可以提供帮助和雇佣人力的联系人，这些信息使威尔逊受益匪浅。因此可以说，韩尔礼先生是威尔逊进入中国寻找珙桐树的第一个引路人（王康，2022）。当时拿着韩尔礼先生给他手绘的地图和寥寥几句前辈的手写指引，再回到香港经上海沿长江溯流而上到湖北宜昌寻找珙桐（胡启明 译，2022）。

1900年4月，当他历尽各种磨难，抵达恩施巴东绿葱坡，找到了韩尔礼先生所说的那棵珙桐树所在地。遗憾的是，上帝跟他开了一个大玩笑，那棵珙桐树已被乡民在一年前砍伐去做了房梁。他万万没想到千辛万苦得到的是这样的结果（央视纪录片，2016）。

短暂的沮丧之后，威尔逊并没有放弃，他重新振作，折回宜昌继续在大山里寻找植物，同时还考虑到四川穆坪去寻找珙桐。

1900年4月，威尔逊一行在恩施扑空后，带领随从和挑夫一行，从绿葱坡向宜昌折返，行经至长阳（今天的长阳乐园）二墩岩山脚的时候，他留意到此处山体比较特别，决定在此逗留探寻，碰一碰运气。

长阳榔坪乐园二墩岩高山上一整片山林的所有者，是土家庄主康远德。据"欧美珙桐原产地博物馆"常务副馆长李茂清考证[1]，康远德的祖辈是湖北松滋人。康熙年间，松滋发大水，康家逃荒来到长阳乐园。当地一名姓覃的大地主，将康家人收留后，安置在一片荒林中搭棚栖身，康家就在这里安了家。经过几代人的辛勤劳作，康家在山上开辟出一座山林和一大片良田，当地人都称这片土地叫康家湾。传到康远德手上的时候，此处已经营成一座颇具规模的山庄。1900年，康远德时年45岁，带着近30名家眷，生活在山庄中，农忙时组织耕作，农闲时赶山打猎消遣。

1900年4月底，威尔逊来到了康家湾山庄。主人康远德（图9）热情款待了这名远道来访的异域探险家。一个是阅历丰富的"植物猎人"，一个是久居乡野的山庄主人，在翻译的帮助下，两人相谈甚欢。威尔逊向康远德打听珙桐树，也许是因为叫法不同，康远德不明白这名"洋人"要找的是什么，只得带他进山寻找。据威尔逊的日记记载，当年的5月19日，康远德带他到一片名为"大湾"的山林打野鸡，他在林间穿梭时，不小心被藤蔓绊倒，情急之下他抓住身旁的灌木枝才没有让自己摔下山谷，正当他想抱怨上天时，抬头却发现眼前的树枝叶茂密，正开着大朵的白色鲜花，成双成对，好似白鸽站在枝头，又好似一群人摇曳着手帕在向他打招呼，那正是他数月以来艰辛寻找的珙桐，他不禁大喊"珙桐"（Wilson，2013）！康远德这才知道，威尔逊要找的其实是后山上常见的"木梨子树"（当地俗称）。威尔逊找到了梦寐以求的珙桐树后，与康远德约定，秋冬时节将再访康家湾，来收集珙桐种子。康远德答应会帮他收集珙桐果实，按他教的方法阴干储存。当年11月底，威尔逊如约而至，带走了14 875粒珙桐种子和一些植物标本，并将这些宝贝运回了英国。目前我们认为，威尔逊采集到种子的那片珙桐居群是光叶珙桐，与韩尔礼当年采到的是同一个变种。

在西方负有盛名的"植物猎人"英国人威尔逊的笔下，珙桐则有这样的风姿："我们对这种树

---

[1] 自（https://news.qq.com/rain/a/20230519A03BA100?suid=&media_id=）湖北头条2023.5.19，极目新闻记者刘俊华的"风靡欧美的'中国鸽子花'树，牵出一段跨国友谊"（2024年11月进入）。

图9　威尔逊的植物采集团队1900年在湖北，左1为气度不凡的康远德，带威尔逊（右1）在山林里打猎（引自Wilson, 1913）

木的美丽感触很深……珙桐特别漂亮之处，在于它那两片承托花序的雪白苞片……每当微风吹过，就像一大群蝴蝶飞翔于大树间；在稍远一些的地方观看，被绿叶部分遮掩的白花，又使全树像被雪花装点。在阴天、清晨和黄昏时分，它纯白的花苞尤为动人……在我的思想中，珙桐是北温带植物中最有趣、最美丽的树种。"（Wilson，1913；图8右）。

不久，威尔逊得到一个消息：在巴黎的一个植物园里，已经成功栽种了一棵珙桐树，后来确认这是另一来源的珙桐，种子是一位法国传教士保罗法格斯（Paul Guillaume Farges）收集的37颗种子中培育而来。

令威尔逊沮丧的是，他寄回英国的种子当年未能发芽。但是，英国的植物家们并没有气馁，尝试了各种方法。到1902年4月，威尔逊从宜昌返回英国后，十分欣喜地发现，两年前发回维奇苗圃公司的种子播种一年后有13 000颗已经发芽。随后，他和助手在苗圃里盆栽培育了这13 000株珙桐树苗，几乎每一棵树苗最后都成活了，成为欧洲园艺界的奢侈品和抢手货。1911年5月，引自湖北长阳的珙桐树后代，9年后第一次在英国维奇苗圃公司开花，到秋天结出了小梨子一般的果实，他们也发现了珙桐种子有二次休眠特性（胡启明，2022）。维奇苗圃公司因此获得了英国皇家园艺学会颁发的最高荣誉：一级勋章（袁明华，2022）。

在此，我们还需提到，1903年秋天，威尔逊从乐山出发经雅安，去了一趟穆坪（今宝兴），专程采集珙桐种子。据考证，威尔逊先后五次到雅安，足迹遍布宝兴、汉源、荥经、雨城等，在雅安及周边发现大量有观赏价值的物种，如珙桐、报春花、杜鹃花、百合等，并引种到欧洲。威尔逊在宝兴发现命名了西南蔷薇、扁刺栲等28个新种，发现新物种的数量仅次于戴维和韩尔礼。

威尔逊第三次到访康家湾，已经是1908年的秋天。这次他带着一部当时最先进的德国产相机，为他的中国之行留下了大量照片。

根据威尔逊拍的照片和写的日记推测，1908年秋天，威尔逊邀请康远德陪他西进四川，采集珍稀植物种子和标本。直到冬天，威尔逊和挑夫、助手、翻译一行16人，又返回住进了康远德家里。从腊月二十四，一直停留到次年正月十五，过完春节才离开。在康家湾的山水之间，威尔逊与康远德、英国助手和康远德的儿子等人，都有照片留影。

民国后，康远德的山庄因屡遭匪患，家族日渐没落。近年来，随着中国鸽子树在欧美越来越受追捧，有关威尔逊寻找珙桐的故事又被挖掘出来。威尔逊五到中国、三访长阳乐园康家湾的史料，陆续在哈佛大学网站上披露[2]。

2021年5月19日，长阳椰坪康家湾迎来一个重大纪念日子："欧美珙桐原产地乐园博物馆"成立（图10）。该馆收集陈列了许多有关珙桐被发现

图10　位于湖北长阳康家湾的欧美珙桐原产地乐园博物馆外貌和馆匾（李茂清 摄）

---

2 自（https://news.qq.com/rain/a/20230519A03BA100?suid=&media_id=）湖北头条2023.5.19，极目新闻记者刘俊华的"风靡欧美的'中国鸽子花'树，牵出一段跨国友谊"（2024年11月进入）。

以及传播到欧美这段往事的实物和资料。人们可能知道，5月19日是《徐霞客游记》的开篇日，已被浙江宁海政府设立为"徐霞客开游节"。2011年开始，经国务院批准，升级为"中国旅游日"。但绝大多数人可能不知道，5月19日也是威尔逊发现珙桐树的日子，当然也就是乐园成为"欧美珙桐原产地"的起始日。当然，更重要的是，5月19日是中国原产地鸽子花和走向世界的鸽子花共同绽放的日子。

该馆总策划、总设计、总撰稿为李茂清先生（曾在长阳县委办公室工作），为家乡珙桐树的宣传推广呕心沥血几十年。目前，乐园所在的方圆40余平方千米（含核心珙桐居群）已全部纳入湖北长阳清江国家森林公园自然保护区管理。落户在康家湾的"欧美珙桐树原产地乐园博物馆"，是征用了康远德玄孙女康祖梅的老屋修缮而成（从博物馆到大湾康家老屋场遗址还得走两三个小时）（图11）。生命有她自己的钟声，作为孑遗植物和世界顶级观赏植物，乐园珙桐树每年5月19日这天会迎来属于自己节日的钟声。中国鸽子花，就是飞向世界的和平鸽，向世界传递着中国人民的美好期许，也是全世界人民共同的心声。

图11 中国珙桐故乡——长阳康远德第五代孙女康祖梅故居（威尔逊当年寻找珙桐曾居住过）（李茂清 摄）

4 中国面积最大的野生珙桐分布区——四川雅安

我国著名的植物分类学家和园艺学家俞德浚院士（1908—1986），1932—1934年，在任四川北碚中国西部科学院生物研究所植物部主任期间，两次到川西考察。1934年，俞德浚发表《四川植物采集记二、峨眉与峨边之初春采集》，讲述了在峨边牛心山发现并采集珙桐等标本的情景"林中最易引人注目者，当推珙桐之花，新叶初放，花已盛开，花序头状，深紫色，外有两个乳白色大形苞片，细枝下垂如钟状，迎风招展，极为美丽"。

我国著名植物生态学家曲仲湘（1905—1990），1936年，任中国西部科学院生物研究所植物部主任（继俞德浚之后），率调查队赴天全进行植物采集和森林调查，在大河头发现了正处于盛花期的珙桐。目前，在中国数字植物标本馆，能查阅到曲仲湘采集于天全的珙桐标本（图12）。

图12 曲仲湘1936年在四川天全采集的珙桐标本（引自中国数字植物标本馆，https://www.cvh.ac.cn/spms/detail.php?id=bf796d71；2025年3月13日进入）

据1939年的《国立四川大学校刊》记载："三年级学生李彩琪等十余人，趁本年暑期机会，专赴雅安、天全一带采集植物标本，由该系讲师戴蕃晋参与指导。闻采得动植物标本，多属珍贵之品……"通过中国数字植物标本馆，能查到戴蕃晋1939年9月在天全采集制作的珙桐标本，此号标本带有珙桐的一颗果实。

同在1939年，中国科学社生物研究所的植物学家姚仲吾在荥经采集到珙桐标本，其中一号标本现藏于中国科学院植物研究所标本馆。69年后的2008年，华中农业大学、湖北民族学院组成联合调查组，对雅安荥经大相岭省级自然保护区龙苍沟的珙桐进行了种群生态学调查测算——境内珙桐资源分布不低于50万亩，集中成片部分10万亩，认定此为中国面积最大的野生珙桐分布区之一（左凌仁，2011；图13）。

图13　四川荥经最大珙桐群落花期外貌（引自左凌仁，2011）

5 珙桐——和平之鸽、友谊之树

自从戴维1869年发现珙桐，1871年拜伦命名珙桐，1897年，法国传教士法格斯采集到37枚种子（仅1枚在法国发芽），到1900年威尔逊采集到大批种子引种欧洲的100多年来，珙桐已成为世界闻名的观赏树、友谊树。

珙桐树形优雅，茎干斑驳，略显淡黄色的叶片碧绿青翠，枝叶婆娑，尤其是珙桐的花朵，硕大洁白，苞片犹如鸽子翅膀，整树花开，远远望去，恰如群鸽，时而展翅飞翔，时而振羽起舞，时而休憩假寐，时而啄食觅物，翩翩姿态，令人着迷，让人感叹，极具视觉冲击力，因而具有极高观赏价值。尤其是珙桐别具一格的形象，就像耀眼的光芒照耀大地，当时一下子就引起了西方世界的极大关注。科学家称它为"北温带植物群最有趣和最漂亮的木本植物"。德国慕尼黑植物园、瑞士日内瓦植物园、英国皇家植物园邱园、美国白宫、哈佛大学阿诺德树木园等植物园和风景名胜区纷纷引种，很快扩展到了欧美许多国家（图14至图19）。

目前，欧洲和北美是珙桐在海外的主要迁地引种区域。珙桐与大熊猫一样已成为世界和平的使者。珙桐在国内开始引种栽培和引起重视始于20世纪60年代，当时北京、南京等地开展了珙桐

的引种试验，其栽培技术获得突破。从此珙桐在国内成为与大熊猫齐名的中国植物的形象代言，即植物界的"大熊猫"。

2008年12月23日，17棵来自"5·12"汶川特大地震重灾区北川县禹里乡珙桐苗圃的珙桐树苗，与赠台大熊猫"团团""圆圆"一起，搭乘台湾长荣航空公司的专机飞往台湾，成为两岸人民相互扶持的见证。

2011年9月29日，珙桐的种子和其他三种珍稀濒危植物（普陀鹅耳枥、望天树、大树杜鹃）的种子，随着"天宫一号"空间站一同进入了太空。

2016年6月18日，从北川羌族自治县精心挑选的珙桐树，带着中国人民的深情厚谊，漂洋过海到塞尔维亚，由彭丽媛女士与塞尔维亚总统夫人德拉吉察女士，共同种植在贝尔格莱德植物园中。

2023年，在成都举行的第31届世界大学生夏季运动会，其组委会执行主席、四川省省长黄强

图14　A：伦敦皇家植物园邱园栽培的珙桐（傅承新摄于2024年5月13日）；B：英国剑桥大学植物园栽培的珙桐（傅承新摄于2024年5月12日）

01

图15　A：伦敦金斯顿公园栽培的珙桐（2024年5月 Dr. Mark Fan 摄）；B：伦敦 Postman 公园栽培的珙桐（傅承新摄于2024年5月，花已谢）

图16　A：巴黎蒙索公园栽培的珙桐（引自公园官网 https://www.pariscotejardin.fr/2015/04/ l-arbre-aux-mouchoirs-du-parc-monceau- ）；B：美国俄勒冈州立大学校园栽培的珙桐（引自大学官网 https://landscapeplants.oregonstate.edu/plants/davidia-involucrata ）

图17 A：在威尔逊家乡奇平卡姆登镇的威尔逊纪念植物园中的珙桐；B：在威尔逊家乡农家院子里的珙桐（傅承新2024年5月19日摄）

图18 哈佛大学阿诺德树木园的引自法国扦插苗、已栽培120多年的珙桐（A、C为2024年5月16日，哈佛大学阿诺德树木园Dr. Micheal Dosmann摄；B为马金双摄）

图19　美国纽约布鲁克林植物园栽培的珙桐（马金双　摄）

在致辞中讲到，"象征坚强、团结、和平的中国珙桐树'鸽子花'今夜绽放"，并与大家分享了历久弥新的珙桐故事。当各参赛代表团旗帜及标识牌进入会场时，一棵距今6 000万年、生在中国、被誉为植物界"活化石"的珙桐显现在舞台上，大运会的引导员以象征着和平的珙桐树鸽子花为造型，将鸽子花的白色苞片幻化成一对对飞翔的翅膀，愿全世界的青年朋友乘愿而来，乘梦飞翔（图20）（成都方志，2023）。

最初，珙桐作为西方园艺和博物爱好者追求的神秘物种进入大众的视野，但是如今的珙桐已经超越了原有的传统印象，作为国宝的珙桐成为象征和平和友谊的大使漂洋过海，散播世界。

2015年8月央视纪录频道首播的三集纪录片《中国威尔逊》呈现了威尔逊从1900年起在中国西部寻找珙桐及采集中国植物标本的故事，使珙桐"中国植物大熊猫"走向世界园林的和平之路。

同样，在国内珙桐也逐渐被引种到各地作为观赏植物。而种植在城市中的珙桐，也在用它独有的美丽身姿和传奇故事，激发着我们对脚下这一片乡土的热爱。国家植物园（北园，原北京市植物园）栽培的珙桐已能正常开花（图21）。这是中国大陆地区陆地栽培的最北位置（北纬40°）。杭州植物园里有三件镇园之宝——天目玉兰、夏蜡梅以及珙桐（图22），其中的珙桐是1982年从湖北引种的（郭馨怡，2021）。珙桐的原产地海拔高达1 200m，且生性喜阴，要在杭州生长并不容易，经过多年的精心养护，这些珙桐从2010年开始能正常开花，并且一年比一年开得好。

图20 2023年成都大学生运动会开幕式上的珙桐元素（成都方志，2023）

图21　国家植物园（北园）栽培的珙桐（王康 摄）

图22　杭州植物园种植的珙桐（傅承新摄于2024年4月10日）

6 珙桐——从中国西南山区走向世界园林

1911年，威尔逊采回的珙桐种子育成的幼苗苗壮成长，在英国绽放出第一朵鸽子花。后来，全世界广泛引种珙桐，大量栽植，成为世界十大观赏植物之一，被西方园林人士称为"北半球最漂亮的树木"。在法国巴黎，珙桐广植于皇宫、教堂、公园等重要场所，甚至美国的白宫前也种植这种树冠开阔、满树盛开具有巨大白色苞片花朵的风景树。

据不完全统计（BGCI数据库、中国各植物园调查反馈等）[3]，100多年来，至少有23个国家的178个城市或大学植物园、树木园、城市花园、森林公园等引种栽培珙桐（表1），海外引种栽培主要分布在欧美国家，最多的国家是英国和美国，在乔木中栽种的个体数可能仅次于银杏和水杉。

国内引种较多的省份主要在其野生分布地的四川和湖北，引种栽培最多个体的是陕西秦岭国家植物园（约2万株），引种最北的是北京地区、最东部引种的是台湾、上海及宁波的植物园，最南部引种的是位于广州的华南国家植物园和深圳中国科学院仙湖植物园（表2）。每当4～5月开花季节，会有很多人慕名而来欣赏这北半球最有历史故事的"手帕树""鸽子树"。然而由于它生长在高山凉爽生境和花期较短、难以育种的习性，使它还主要局限于在植物园、树木园中作为"标本"种植。今后，利用分子育种技术培育新的、具有不同苞片颜色的、能广泛适应低海拔城市园林种植的品种是现代园艺学的方向。

表1 珙桐在国外的引种栽培记录

国家	引种栽培点	国家	引种栽培点
爱尔兰	都柏林基尔马库拉格国家植物园	美国	北卡罗来纳州布恩树木园
爱尔兰	格拉斯尼文国家植物园	美国	亚特兰大植物园
奥地利	维也纳大学植物园	美国	肯塔基州贝克树木园
澳大利亚	墨尔本皇家植物园	美国	北卡罗来纳州夏洛特树木园
澳大利亚	堪培拉国家树木园	美国	佐治亚州巴富维植物园
澳大利亚	丹顿农山脉植物园	美国	纽约州贝亚德卡廷树木园
澳大利亚	南澳植物园	美国	佐治亚州立植物园
澳大利亚	悉尼蓝山植物园	美国	加利福尼亚州立大学萨克拉门托树木园
澳大利亚	塔斯马尼亚树木园	美国	新泽西州威洛伍德树木园
比利时	维斯佩拉尔树木园	美国	西雅图华盛顿大学植物园
比利时	佛兰德老福特萨克森树木园	美国	宾夕法尼亚州泰勒树木园
比利时	卡尔姆特豪特树木园	美国	加利福尼亚州立大学伯克利植物园
丹麦	赫斯霍尔姆农业大学树木园	美国	哈佛大学阿诺德树木园
德国	哥廷根实验植物园	挪威	卑尔根博物馆花园
德国	塔朗特森林植物园	挪威	卑尔根植物园
德国	哥廷根森林植物园和树木园	瑞典	贝尔尤斯植物园
德国	埃森格鲁加公园	瑞士	佛里堡大学植物园
德国	波恩大学植物园	瑞士	日内瓦雅尔丹植物保护区

3 自 BGCI 数据（https://www.bgci.org/resources/bgci-databases/plantsearch/；2024 年 11 月进入）。

国家	引种栽培点	国家	引种栽培点
德国	德累斯顿植物园	瑞士	伯尔尼大学植物园
德国	耶拿植物园	瑞士	苏黎世大学植物园
德国	佛莱堡大学植物园	斯洛文尼亚	卢布热那亚大学植物园
德国	基尔大学植物园	斯洛文尼亚	马里博尔大学植物园
德国	乌尔姆大学植物园	西班牙	坎塔布里亚职业培训中心
德国	佛莱堡贡特施塔尔树木园	新西兰	普克库拉花园
德国	巴伐利亚拜罗伊特植物园	新西兰	东木山坡国家植物园
法国	巴黎植物园	新西兰	但尼丁植物园
法国	里昂市政植物园	新西兰	奥克兰植物园
法国	南特植物园	以色列	耶路撒冷大学植物园
法国	南锡大学植物园	意大利	托斯卡尼植物园
法国	卢瓦雷省巴雷斯国家树木园	意大利	比萨大学植物园
法国	滨海塞纳省植物园	意大利	锡纳耶大学穆塞奥植物园
芬兰	赫尔辛基大学植物园	意大利	马尔凯卡梅拉科尔蒂尼植物园
荷兰	阿姆斯特丹霍图斯植物园	意大利	威尼托帕多瓦大学植物园
荷兰	瓦格宁根贝尔马利树木园	意大利	佩鲁贾大学植物园
荷兰	奥登博斯树木园	意大利	贝加莫植物园
荷兰	代尔夫特霍图斯花园	意大利	奥罗帕植物园
荷兰	多尔恩骏宝树木园	英国	牛津大学植物园和树木园
荷兰	乌特勒支大学植物园	英国	德文郡佩恩顿动物园
加拿大	安大略省尼亚加拉公园植物园	英国	北约克郡皇家园艺学会花园
加拿大	不列颠哥伦比亚省米尔纳花园和林地	英国	埃塞克斯，切姆斯福德皇家园艺学会花园
加拿大	温哥华范达森植物园	英国	北德文郡皇家园艺学会花园
加拿大	不列颠哥伦比亚大学植物园	英国	萨里郡皇家园艺学会花园
捷克	赞帕克树木园	英国	爱丁堡皇家植物园
美国	北卡罗来纳州杜克植物园	英国	伦敦皇家植物园邱园
美国	圣约瑟夫大学树木园	英国	西萨塞克斯郡皇家邱植物园
美国	俄亥俄州道斯树木园	英国	南约克郡谢菲尔德植物园
美国	罗得岛州纽波特树木园	英国	南约克郡温特沃斯卡斯特尔花园
美国	麻省波莉·希尔树木园	英国	汉普郡哈罗德·希利尔树木园
美国	斯沃斯莫尔学院树木园	英国	南威勒尔半岛海角植物园
美国	密苏里植物园	英国	斯特兰拉尔洛根植物园
美国	新罕布什尔州奥本山公墓	英国	伦敦切尔西药用植物园
美国	俄亥俄州艾姆格山树木园	英国	皮布尔斯郡道克植物园
美国	弗吉尼亚诺福克植物园	英国	格拉斯哥植物园
美国	北卡罗来纳州立植物园	英国	卡尔·奥斯基茨植物园
美国	俄勒冈州霍伊特树木园	英国	伯明翰植物园
美国	亚拉巴马州亨茨维尔植物园	英国	剑桥大学植物园
美国	弗吉尼亚路易斯·金特植物园	英国	肯特郡百吉百利松树园
美国	弗吉尼亚梅多伍德植物园	英国	班摩尔植物园
美国	俄亥俄州辛辛那提动植物园	英国	柴郡塔顿植物园协会树木园
美国	康涅狄格州树木园	英国	贝德福德郡英伦群岛树木园
美国	纽约州康奈尔大学植物园	英国	约克郡约克树木园
美国	特拉华河谷学院树木园	英国	班戈特雷博思植物园

（续）

国家	引种栽培点	国家	引种栽培点
美国	俄亥俄州奥尔登树木园	英国	温特沃斯城堡花园
美国	北卡罗来纳州教堂山金银花农场	英国	康沃尔郡特鲁罗特雷戈斯南庄园
美国	得克萨斯州福特沃斯植物园	美国	纽约布鲁克林植物园
美国	新泽西州弗里林海森树木园	美国	马里兰州布鲁克赛德花园
美国	加利福尼亚州旧金山金门植物园	美国	费城天普大学安布勒树木园
美国	弗吉尼亚绿春花园	美国	肯塔基州尤戴尔植物园
美国	纽约布鲁克林格林伍德	日本	京都植物园
美国	新泽西州格林伍德花园	日本	东京都立植物园
美国	加利福尼亚州立大学圣克鲁兹分校树木园	韩国	济州观光植物园

表2 珙桐在中国的引种栽培记录

国家	引种栽培点	国家	引种栽培点
中国	国家植物园（北园）	中国	恩施华中药用植物园
中国	北京市十三陵林场	中国	湖南南岳树木园
中国	北京松山国家级自然保护区	中国	湖南省植物园
中国	北京市园林绿化科学研究院	中国	长沙中南林业科技大学校园
中国	西安周至秦岭国家植物园	中国	贵阳药用植物园
中国	中国科学院武汉植物园	中国	贵州植物园
中国	郑州植物园	中国	江西庐山植物园
中国	南京中山植物园	中国	上海植物园
中国	宜昌三峡植物园和卷桥河湿地引种园	中国	上海松江天马山森林公园
中国	雅安四川农业大学校园	中国	宁波植物园
中国	都江堰华西亚高山植物园	中国	杭州植物园
中国	都江堰市壹街区同心广场	中国	中国科学院华南国家植物园
中国	泸州植物园	中国	深圳中国科学院仙湖植物园
中国	成都植物园	中国	中国科学院昆明植物园
中国	内江四川农业特色植物研究院	中国	台湾屏东辜严倬云植物保种中心
中国	绵阳太乙仙山植物园	中国	中国科学院西双版纳热带植物园

7 珙桐的天然分布和保护

7.1 珙桐的天然分布

珙桐天然分布于中国亚热带西部中高山地区，呈间断零星分布。从行政区域角度看，分布在中国的8个省（直辖市），贵州、湖南、湖北、四川、重庆、云南、甘肃及陕西（贺金生，林洁，1995），其水平分布呈马蹄状。垂直分布从600m（湖南石门县壶瓶山）到2400m的四川天全县和3200m

云南高黎贡山（张清华 等，2000; 贺金生，林洁，1995）。

珙桐喜欢生长在海拔1 500~2 200m的温暖润湿的落叶阔叶混交林中，多生于空气阴湿处，在干燥多风、日光直射之处生长不良，不耐瘠薄，不耐干旱。幼苗生长缓慢，喜阴湿，成年树趋于喜光。其分布区的气候为凉爽湿润型，湿潮多雨，夏凉冬温，年平均气温8.9~15℃，1月平均气温0.43~3.60℃，7月平均气温18.4~22.5℃，年降水量600~2 600.9mm。其土壤多为山地黄壤和山地黄棕壤，pH 4.5~6.0，喜中性或微酸性腐殖质深厚的土壤；生长地多含有大量砾石碎片的坡积物，基岩为砂岩、板岩和页岩。常出现在深切的山间溪沟两侧，山坡沟谷地段，坡度约在30°以上（贺金生，林洁，1995; 张家勋 等，1995）。You等（2014）研究表明珙桐林以温带植被为主，群落通常以落叶阔叶物种为主，有时与一些常绿物种混合。

然而，由于珙桐分布区局限，自身种群繁衍相对困难。特别是人为乱砍滥伐现象严重，使得珙桐主要局限分布在人烟稀少的高山峡谷地带。

直到1984年，珙桐才被列入《中国珍稀濒危保护植物名录》第一批八个重点保护的一级濒危物种。1999年《国家重点保护野生植物名录》第一批发布，珙桐被正式列入国家重点保护对象。

近几十年来，在中国西部陆续发现一些大的珙桐野生居群、零星分布点和几百年以上的古老个体（图23、图24）。

湖北长阳土家族自治县系欧美珙桐原产地，在崩尖子国家级自然保护区和二墩岩省级自然保护区各有一处野生珙桐居群，其中，二墩岩省级自然保护区，珙桐生长面积达十多平方千米，百年以上的珙桐有25株。目前，已建立了100亩珙桐拯救繁育基地。

在威尔逊当年寻觅珙桐的乐园康家湾，已开辟乐园珙桐观赏处，修建了"中国珙桐故乡纪念广场""欧美珙桐原产地乐园博物馆"和威尔逊、康远德纪念亭（黄善君，2022）。

20世纪80年代，在四川大凉山发现大片珙桐野生居群，达2万多株，其中最大一株高近20m，胸径超过1m，估计其树龄在几百年。

图23　贵州阔宽水国家级自然保护区野生珙桐林（林汉扬2018年摄于贵州）

图24 A：四川峨眉山万年寺的野生珙桐（引自新乐山 2021-04-20）；B：峨眉山珙桐幼果期（林汉扬 2018年摄）

图25 八大公山的"珙桐王"（李文昊 2024年5月 摄）

在湖南桑植八大公山国家级自然保护区内的天平山，有一株古老的珙桐，树龄已有400多年，树高接近30m，胸径1.3m，冠幅达18m，被称为"世界珙桐王"（图25）。它的主干早已经折断，但又长出四根树干，茎干上遍布苔藓。即便历经沧桑，仍然焕发着生机，每当开花时，葱绿的树上，仿佛无数的白鸽展翅欲飞。在"珙桐王"的生长之地，还分布着上千亩野生珙桐林。

图26 四川荣经珙桐（A：古老的"独木成林"珙桐；B：珙桐大树）（李文昊 摄）

2008年，四川荣经县龙苍沟乡发现近10万亩野生珙桐居群，其密集程度之高、面积之大，实属罕见（图26）。在洪雅县瓦屋山，也发现有30万亩天然珙桐林，被誉为"中国鸽子花的故乡"。最近，在贵州赫章县结构乡大山、毛姑等村已先后发现2 000多株珙桐。2022年乌蒙山国家级自然保护区发现13万亩珙桐居群。

通过分子生物学、地理信息学结合生态信息系统等研究手段，基于全球气候预测模型预测：到2030年珙桐的适宜分布区将发生改变，适宜分布区有可能减少20%，珙桐现在的分布区将更加碎片化（张清华 等，2000）。

珙桐的保护价值：珙桐为古老、珍贵、稀有植物，是我国特有的单属植物，系第三纪古热带植物区系的孑遗种，也是全世界著名的观赏植物，有"中国鸽子树"之美称，它除了对研究古植物区系、古地理和植物系统进化均具有重要的科学价值外，其自身的木材有很高的经济价值，因材质沉重，是建筑的上等用材，可制作家具和作雕刻材料；其球状花序上的花是良好的蜜源，种子和果皮能榨出味道清香的油，树皮可提取栲胶或制备活性炭；其叶片中含有的槲皮素、山柰酚等黄酮类化合物具有抗肿瘤、抗衰老、增强免疫力等药理作用（黄江 等，2005）。

7.2　珙桐的生存策略和生存威胁因素

珙桐的化石发现于北美洲、欧洲等许多地区，中全新世和末次冰盛期（Last Glacial Maximum，LGM）的研究结果表明，中国中南部和西南部是它长期稳定的避难所（Tang et al., 2017）。

珙桐由于生长在年降水量1 200mm的中国西南地区，每年5月的花期正是雨季，而经过6 000万年的演化，珙桐形成了独特的白色宽大的苞叶，目的是为了让花序免遭雨淋，避免花粉受损；同时，宽大、柔软的白色苞叶还能吸引昆虫前来采蜜并完成授粉（Sun et al., 2008）。这也是珙桐能够生存至今的重要因素之一。

在冰川时期，珙桐种子因为具有长时间休眠的能力，相当于被冷冰藏起来了，直到冰川融化，等到温度适宜了，种子再萌发、生根发芽，这也是它能够度过末次冰期生存至今的另一因素。

但是人类的活动和珙桐本身的经济价值，使珙桐一直以来面临着生存威胁。人类肆意采挖野生苗、采伐树木、摘取还未成熟的果实以及各种项目开发破坏了生态环境等，使得野外较难发现自然生长的珙桐树苗，对其种群维持造成了一定的影响。近年来，随着珙桐知名度的提高，挖掘野生苗，过早、过多地采集珙桐果实现象常有发生。研究表明，人类对珙桐果实的采摘，会导致许多

图27 A：珙桐的头状花序；B：核果（引自杭州日报客户端2022.4.15）；C：去掉部分外、中果皮后，可见包有内果皮的种子（李文昊2024年1月17日摄于张家界）

地区珙桐种群严重缺乏实生苗（吴刚 等，2000）。

除了外在原因外，珙桐在生长发育和遗传上也存在许多内在不利因素，导致其种群复壮的困难。研究发现，珙桐对生长环境要求苛刻，其原生境多为温凉、湿润、多雨、多雾的山地环境（不能忍受38℃以上的高温）。幼龄时期，喜欢荫蔽的环境；中龄后，则更加喜光，光线不足，生长会受到抑制。研究还发现，珙桐通过种子萌发实生苗和根萌蘖（从根上长出新芽）两种繁殖形式进行自然更新，在自然界有一定困难。珙桐的核果呈褐色，外果皮薄，有很薄的果肉（中果皮），内果皮橄榄形具骨质沟纹，即果核，内有种子（图27）。种子成熟后会经历2～3年的长休眠期，用种子繁殖一般需要15～18年才能开花结实，20～25年才进入盛果期，即便结果，其大小年现象也非常明显，且早期落果严重，有"千花一果"之说，但珙桐的寿命可以达到800年以上。

而根萌蘖繁殖的方式虽然是居群不断扩大生存空间、增强抵御不良环境的能力，从而增加了生存机会，但却限制了珙桐的扩展范围，不利于扩散。还有一点，观察发现珙桐的种子具有致密厚实的木质外皮，阻挡着种胚对外界水分的吸收，因而很不容易发芽。一个果实中多数情况仅有1～3枚种子发育成熟，而且大多数胚存在败育现象，

降低了种子的发芽率。由于上述原因以及种子苗生长对环境要求高，导致居群内幼苗、幼树死亡率较高，使珙桐自然更新能力缺乏（罗世家 等，2009）。

7.3 珙桐的保护生物学研究

近40年来，珙桐作为植物界的"大熊猫"日益被人们认识到。在许多野生珙桐产地相继建立了省级和国家级自然保护区，实施就地保护，如早期建立的卧龙国家级自然保护区和峨眉山自然保护区，1978年建立的梵净山国家级自然保护区，2012年建立的湖北长阳乐园二墩岩省级自然保护区，2013年建立的贵州纳雍珙桐省级自然保护区，湖南省桑植县八大公山国家级自然保护区，四川宝兴邓池沟自然教育基地，云南省袁家湾珙桐市级自然保护区，四川荥经大相岭省级自然保护区龙苍沟等几十处保护地。但是，Tang 等（2017）研究指出随着全球气候变暖，珙桐的适宜分布区会收缩，使得珙桐的生境变得高度脆弱。另一方面，研究发现在现有保护区外的珙桐分布居群高达53%，应优先保护，尤其四川盆地西南山区边缘和华中山区（即三峡周边地区）急需新建以珙桐为主的自然保护区。各地应通过禁止乱砍滥伐、

采摘果实，对生境加强保护，确保珙桐异交结实、种子萌发、幼苗生长，以最大限度保留珙桐的遗传多样性，实现就地复壮。

在迁地保护方面，珙桐特有的生物学特性使得迁地保护、引种栽培存在一定难度。从威尔逊引种欧洲开始的100多年来，在国外，主要在欧美的植物园、公园里获得存活并开花，大面积栽种观赏成功的不多。国内也同样，20世纪80年代以来，也只是在一些植物园里引种成功（胡进耀 等，2002），还没有成为城市的观赏树种。然而，人们通过不断研究实践，已掌握了种子育苗技术，突破人工栽培的难点，为珙桐的迁地保护和利用积累了宝贵经验，正在开展珙桐对干热气候的适应性驯化。

珙桐种子一般需2~3年休眠时间才能发芽出苗。经过多次实验已发明了种子雪藏变温处理等人工干预技术促使萌动。具体方法是将收集的种子洗净后用湿沙混合，埋藏于雪地里，使其寒冷时冰冻，升温时融化，达到温差较大幅度变化。这样经2~3个月时间，促成坚硬的内果皮松动、开裂，到春天再行播种，可大大加快其发芽速度。也可以通过冰箱变温处理，使种胚能提前发芽。如今，许多长江流域的植物园和保护区都培育出了大量的珙桐树苗。北京植物园培育的珙桐树已正常开花，连同在波士顿哈佛大学阿诺德树木园栽培成功的，是该种目前迁地最高纬度的地区。

近年，随着基因组学的快速发展，从植物比较基因组研究来探讨濒危物种的保护已成为保护生物学的热点。Chen 等（2020）使用单分子实时长读取和染色体构象捕获（Hi-C）技术报道了珙桐的染色体尺度基因组，该种基因组大小为1 169Mb，包含42 554个基因，比近亲喜树（397Mb和31 825个基因）要大而多。认为这两个属种在分化之前共享了最近的全基因组复制。珙桐基因组重复序列在两个属种分化后的大量增加是基因组增大的主要因素；研究揭示珙桐花序苞片中光合作用相关基因几乎缺失或很少表达，而化学防御的相关基因则显著增加，凸显了苞片保护花朵和吸引传粉者的重要作用；也揭示在第四纪气候变化过程中，珙桐的有效居群规模持续减

小，指出在未来气候持续变暖的背景下，对这一濒危物种的保护应充分考虑气候敏感性。

7.4 珙桐的繁育措施

首先，增加保护地建设。有的分布区虽已建自然保护区，但无严格保护措施，在其他分布区也应设置保护点，应制定具体的保护管理措施，积极开展引种栽培和繁殖试验，进行人工造林，扩大其分布区。Tang 等（2017）研究认为四川盆地西南山区边缘和华中山区（即三峡周边地区）最急需新建自然保护区。

其次，开展栽培驯化的规范性技术建设（张征云 等，2003）。种子采集，首先对秋天采集果实进行破碎处理，包括①将收集到的果实放在捣臼中，用木槌反复敲打，将捣碎的果实用清水洗净，拌上草木灰后即可播种；②将成熟的果实堆放在阴暗潮湿的露地，任其自然冷冻，促进生理转化，烂掉果肉，内果皮壳冻脆，堆置3个月（平均气温6~7℃），可使种子发芽率达到92%左右，而平均气温降到-2.5℃时则发芽率可高达98%以上；③最简便易行的方法是将成熟的果实倒入有盖的粪便桶，桶中注满新鲜人尿，全部淹没果实，经6~7周，果实全部腐烂后用清水冲洗，再将种子拌上草木灰后即可播种，此法在湖北省长阳土家族自治县已广泛应用，出苗率均在95%以上。

第三，在幼苗阶段搭棚庇荫，保持苗床湿润。苗木移栽宜在落叶后或翌春芽苞萌动前进行。起苗时不可伤根皮和顶芽，并适当修剪，栽植时要求穴大底平，苗正根展，并灌足定根水。

对于无性繁殖，扦插宜用一年生枝条（长15~20cm，直径0.3cm以上），每个插穗上至少要有2个节间、2个芽，切口平滑不伤皮。3月上旬，在土壤刚化冻、芽萌动前扦插。插前细致整地，施足基肥，使土壤疏松、水分充足。行距20cm，株距10cm，以直插为主。插后应适当覆土，在温暖、湿润条件下，上面第一个芽子微露地表即可。插穗成活后，要适时灌溉、松土、除草、施肥和防治病虫。扦插育苗成苗率在60%左右。扦插前若用ABT生根粉对插穗进行处理，成苗率可提高

10%~20%。

参考文献

陈树思, 1999. 豆科和蓝果树科植物导管分子比较研究 [D]. 武汉: 华中师范大学.

成都方志, 2023. 珙桐花开成都大运会闭幕式, 背后有何深意? [N]. 8 月 9 日.

E. H. 威尔逊, 2022. 中国——园林之母 [M]. 胡启明译. 北京: 北京大学出版社.

方文培, 宋滋圃, 1975. 中国植物志增补资料 2. 珙桐科 [J]. 四川大学学报(自然科学版), 1: 63-68.

方文培, 张泽荣, 宋滋圃, 等, 1983. 中国植物志 [M]. 北京: 科学出版社, 52(2): 157-158.

郭馨怡, 2021. 杭州植物园: 植物界的"大熊猫"珙桐 [N]. 中国花卉报. 9 月 2 日第 2 版. 花卉园艺.

贺军辉, 1991. 蓝果树科 3 种代表植物的染色体观察 [J]. 湖南林业科技, 2: 40-41.

贺金生, 林洁, 1995. 中国珍稀特有植物珙桐的现状及其保护 [J]. 生物多样性, 3(4): 213-221.

胡进耀, 苏智先, 黎云祥, 2003. 珙桐生物学研究进展 [J]. 中国野生植物资源, 22(4): 15-19.

黄江, 刘荣, 王从周, 等. 2005. 珙桐科植物化学成分研究进展(综述)[J]. 亚热带植物科学, 34(2): 70-75.

黄善君. 2022. 长阳榔坪乐园是欧美珙桐种源地 [N]. 湖北日报. https://roll.sohu.com/a/549708360_121372103.

李雪萍, 朱文琰, 贺春玲, 等, 2012. 珍稀濒危植物珙桐与其变种光叶珙桐的差异初探 [J]. 中国农学通报, 28(19): 1-5.

罗世家, 包满珠, 赵善雄, 等, 2009. 大相岭龙苍沟珙桐种群空间分布格局研究 [J]. 生物数学学报, 3: 6.

宋帅帅, 2023. 珙桐的分类学、叶片被毛功能与叶片防御策略研究 [D]. 拉萨: 西藏大学.

王康, 2022. 威尔逊与园林之母——中国 [M] // 马金双, 贺然, 魏钰. 中国——二十一世纪的园林之母: 第 1 卷. 北京: 中国林业出版社: 553-575.

吴刚, 肖寒, 李静, 等, 2000. 珍稀濒危植物珙桐的生存与人为活动的关系 [J]. 应用生态学报, 11(4): 493-496.

向桂琼, 卢馥荪, 1989. 中国特有植物珙桐化学成分研究 [J]. 植物学报, 31(7): 540-543.

许亦华, 2024. 韩尔礼的植物学之路 [M]// 马金双. 中国——二十一世纪的园林之母: 第 7 卷. 北京: 中国林业出版社: 339-361.

央视纪录片, 2016.《中国威尔逊》第一集 https://www.bilibili.com/video/BV1wW411R7wo/?vd_source=d8c505ff615e7b74331f1cc114b601b9.

俞德浚, 1934. 四川植物采集记二、峨眉与峨边之初春采集 [J]. 中国植物学杂志 (1): 325-344.

袁明华, 2022. 中国鸽子树探源——献给"欧美珙桐原产地乐园博物馆"开馆日 [N]. 中华读书报, 5-19.

邹广权, 袁俊吕, 黄富忠, 2022. 珙桐系统分类及遗传变异 [J]. 林业科技通讯, 7:34-38.

张清华, 郭泉水, 徐德应, 等, 2000. 气候变化对中国珍稀濒危树种——珙桐地理分布的影响研究 [J]. 林业科学, 36(2): 47-52.

张家勋, 李俊清, 周宝顺, 等, 1955. 珙桐的天然分布和人工引种分析 [J]. 北京林业大学学报, 17(1): 25-29.

张征云, 苏智先, 申爱英, 2003. 中国特有植物珙桐的生物学特性、濒危原因及保护 [J]. 淮阴师范学院学报(自然科学版), 2: 66-69.

左凌仁, 2011. 荥经发现了中国面积最大的野生珙桐林 [J]. 中国自然地理, 8: 136-145.

APG I, 1998. An ordinal classification for the families of flowering plants[J]. Annals of the Missouri Botanical Garden, 85: 531-553.

APG II, 2003. An update of the Angiosperm Phylogeny Group classification for the orders and families of flowering plants: APG II [J]. Botanical Journal of the Linnean Society, 141: 399-436.

APG III, 2009. An update of the Angiosperm Phylogeny Group classification for the orders and families of flowering plants: APG III [J]. Botanical Journal of the Linnean Society, 161: 105-121.

APG IV, 2016. An update of the Angiosperm Phylogeny Group classification for the orders and families of flowering plants: APG IV [J]. Botanical Journal of the Linnean Society, 181(1): 1-20.

BAILLON H. B, 1871. Nyssaceae[J]. Adansonia, 10: 114-115.

CHEN J M, ZHAO S Y, LIAO Y Y, et al., 2015. Chloroplast DNA phylogeographic analysis reveals significant spatial genetic structure of the relictual tree *Davidia involucrata* (Davidiaceae)[J]. Conserv Genet, 16:583-593.

CHEN Y, MA T, ZHANG L, et al., 2020. Genomic analyses of a "living fossil": The endangered dove-tree[J]. Mol Ecol Resour, 20:756-769.

DAVID A, COX (Ed.), 1949. Abbé David's Diary[M]. Cambridge Boston: Havard University Press.

DODE L A, 1908. *Davidia vilmoriniana*[J]. Revue horticole (Paris), 80: 406.

ENGLER A, 1909. Nyssaceae[M]. Syllabus der Pflanzenfamilen ed. 6.179.

EYDE R H, 1997. Fossil record and ecology of *Nyssa* (Cornaceae)[J]. The Botanical Review, 63: 97-123.

FRANCHET M, A, 1888. Plantae Davidianae Vol. 2[M]. Paris: G. Masson, Editeur.

HARMS H, 1998. Die Gattiingen der Cornaceen. Ber. Deuts. Bot. Gesell[J]. Band xv: 21-29.

LI H L, 1955. Davidiaceae[J]. Lloydia, 17: 330.

KOKAWA S, 1965. Fossil endocarp of *Davidia* in Japan[J]. J. Biol. Osaka City, 16: 45-61.

MA Q, DU Y J, CHEN N, et al., 2015. Phylogeography of *Davidia involucrata* (Davidiaceae) Inferred from cpDNA Haplotypes and nSSR Data[J]. Systematic Botany, 40(3): 796-810.

MANCHESTER S R, 2002. Leaves and Fruits of *Davidia* (Cornales) from the Paleocene of North America[J].

Systematic Botany, 27(2): 368-382.

MANCHESTER S R, CHEN Z D, LU A M, et al., 2009. Eastern Asian endemic seed plant genera and their paleogeographic history throughout the Northern Hemisphere[J]. Journal of Systematics and Evolution, 47: 1-42.

MCCLINTOCK E, 1991. *Davidia involucrata* and *Acer pentaphyllun*[J]. Pacific Horticulture, 52(3): 57-61.

MILNE EDWARDS H, 1870. Note sur quelques mammifères du Thibet oriental[M]. Comptes rendus hebdomadaires des séances de l'Académie des Sciences, 70: 341-342.

PACLYUTKIN B I, 2009. Leaf and fruit remains of *Davidia* (Cornales) from the Nezhino Flora (Miocene of Primoye)[J]. Paleontological Journal, 3: 339-344.

PENG Y L, HU Y Q, SUN H G, 2003. Allozyme analysis of *Davidia involucrata* var. *vilmoriniana* and its biogeography significance[J]. Acta Botanica Yunnanica, 25: 55-62.

QIN H N, CHAMLONG P, 2007. Flora of China. 13[M]. Beijing: Science Press.

REN Y M, ZHANG L S, YANG X C, et al., 2024. Cryptic divergences and repeated hybridizations within the endangered "living fossil" dove tree (*Davidia involucrata*) revealed by whole genome[J]. Plant Diversity, 46(2):169-180.

SUN J F, GONG Y B, SUSANNE S, et al., 2008. Multifunctional Bracts in the Dove Tree *Davidia involucrata* (Nyssaceae: Cornales): Rain Protection and Pollinator Attraction[J]. The American Naturalist, 171(1): 119-124.

TAKHTAJAN A L, 1954. Davidiaceae[M]. Proiskh. Pokruitosem. Rast. 89.

TANG C Q, DONG Y F, SONIA H M, et al., 2017. Potential effects of climate change on geographic distribution of the Tertiary relict tree species *Davidia involucrata* in China[J]. Scientific Reports, 7: 43822. | DOI: 10.1038/srep43822.

TSUKAGOSHI M, YASUO O, TADASHI H, 1997. Fossil endocarp of *Davidia* from the early Pleistocene sediments of the Tokai Group in Gifu Prefecture, central Japan[J]. Bulletin of the Osaka Museum, 51:13-23.

WANGERIN W, 1910. Nyssaceae[M]// Engler A, ed. Das Pflanzenreich. Heft 41, IV. Leipzig: Engelmann: 1-20.

WILSON E H, 1913. A Naturalist in Western China with vasculum, camera and gun[M]. Methuen & co. ltd.

XIANG Q Y, et al., 2011. Resolving and dating the phylogeny of Cornales – Effects of taxon sampling, data partitions, and fossil calibrations[J]. Molecular Phylogenetics and Evolution, 59: 123-138.

YANG Z Y, JI Y H, 2017. Comparative and Phylogenetic Analyses of the Complete Chloroplast Genomes of Three Arcto-Tertiary Relicts: *Camptotheca acuminata*, *Davidia involucrata*, and *Nyssa sinensis*[J]. Front. Plant Sci, 7: 1538.

YOU H, FUJIWARA K, LIU Y, 2014. A Preliminary Vegetation-Ecological Study of *Davidia involucrata* Forest[J]. Natural Science, 6: 1012-1029.

致谢

感谢马金双博士在编写过程中给予的细致帮助和宝贵意见，也感谢美国哈佛大学阿诺德树木园Michael Dosmann博士，英国邱园Mike Fay教授以及国家植物园王康研究员、四川成都生物研究所印开蒲研究员、湖北长阳李茂清先生提供部分照片。

作者简介

傅承新（男，浙江杭州人，1954年生），浙江大学教授。1982年2月毕业于哈尔滨师范大学生物学专业，获学士学位；1985年2月研究生毕业于哈尔滨师范大学植物分类专业，获理学硕士。1985年入职浙江农业大学基础部植物教研室，1998年并入新浙江大学生命科学学院，建立了浙江大学植物系统进化与生物多样性研究室，1999—2004年期间，先后到美国纽约植物园植物分子系统学实验室、德国图宾根大学进化生物学系高访（累计一年）。曾任浙江大学生物系主任、生命科学学院副院长、浙江大学植物研究所副所长、教育部濒危野生动植物保护生物学重点实验室副主任等职，历任中国植物学会理事、常务理事，浙江省植物学会理事长、名誉理事长等职。长期从事植物系统分类与进化以及生物多样性研究，对世界菝葜科系统发育与分类有全面、深入的研究；对世界玄参属、苍术属的系统发育等类群做了深入的工作；对第三纪孑遗植物、珍稀濒危植物，如银杏、珙桐、连香树、黄山梅等，进行了大量的比较基因组、群体遗传、亲缘地理及保护生物学研究；对一些道地药用植物（玄参、白术、元胡等）的种质资源、亲缘地理、栽培起源及道地性等方面有独特的研究思路。主持国家基金重点项目、面上项目、中美生物多样性国际合作项目多项，主持科技部重大调查专项1项。在国内外重要主流、核心刊物发表论文100余篇；获教育部、浙江省自然科学奖、科技进步奖4项。

李文昊（男，湖南长沙人，1997年生），浙江大学生态学在读博士生；本科毕业于浙江大学生物科学专业。主要研究中国特有孑遗植物珙桐的进化生态学和保护生物学，在Journal of Ecology, Journal of Systematics and Evolution等期刊发表论文数篇，曾获第八届长三角植物科学学术研讨会青年论坛报告二等奖。

園林之母
China

02

-TWO-

韩尔礼采集植物志

Plantae Henryanae

叶文* 译
（厦门大学）

YE Wen*
(Xiamen University)

———————

* 邮箱：wenye@xmu.edu.cn

摘　要:《韩尔礼采集植物志》是奥古斯丁·亨利博士（韩尔礼，1857—1930）于1885—1900年间在中国采集的158 620份植物标本（馆藏）的详细记录。这些标本被送往英国皇家植物园邱园，当时该园的分类学家正在编纂中国首部大型植物志——《中国植物索引》。韩尔礼的复份标本随后被分发至全球各地的植物标本馆。韩尔礼曾任职于清朝海关总税务司署，其职责是对中药材（主要为植物性药材货物）征收关税，这激发了他对植物学的早期兴趣。

他的采集成果堪称有史以来从中国送往欧洲的最重要的植物标本之一，涵盖了多样生境中的丰富新分类群。当前这部编目记录了5 364个分类群，其中1 188个被证实为科学上的新发现，且仍为当前的植物命名系统所接受。每个采集号都附有详细信息，包括采集地点、发表刊物以及韩尔礼记录的俗名。中国已被证明是欧洲、北美、澳大利亚和新西兰温带地区优质园林植物主要来源地。文中亦探讨了这些地区早期从中国引种植物的细节，并着重提及了其他西方植物采集者的贡献，例如威尔逊、赖神甫、谭卫道、傅礼士、骆克和金登－沃德等。

关键词: 韩尔礼　植物标本　采集名录　中国植物志　中国植物索引　邱园　中药材

叶文，2025，第2章，韩尔礼采集植物志；中国——二十一世纪的园林之母，第八卷：034-042页.

1 引言

哈佛大学阿诺德树木园（Arnold Arboretum）主任查尔斯·斯普拉格·萨金特（Charles Sprague Sargent, 1841—1927）在《威尔逊采集植物志》（*Plantae Wilsonianae*, 1911—1917年出版）的序言中写道:"尽管法国传教士谭卫道（又译阿尔芒·大卫、谭微道、阿曼德·大卫、阿尔芒·，Père Armand David）早在1870年就已在四川进行了重要的植物采集活动，但世人对中国中西部植物区系的非凡壮丽和丰富仍几乎一无所知，直到韩尔礼（又译奥古斯丁·亨利，Augustine Henry）——一位在1882—1889年间驻扎在宜昌的大清皇家海关税务总司属官员——将他在湖北西部采集的腊叶植物标本寄往英国。首批标本在1886年抵达英国，经检视，在这些标本中发现了许多新属和大量新种"（Sargent, 1911）。

韩尔礼于1857年7月2日出生，在爱尔兰蒂龙郡（County Tyrone）的库克斯敦（Cookstown）和德里郡（County Derry）的泰阿尼（Tyanee）长大。他在库克斯敦学院接受教育，之后进入高威女王学院（现为爱尔兰国立高威大学）学习自然科学和哲学，并于1887年以一等荣誉学位（文学士）毕业。次年，他从贝尔法斯特女王学院（现为贝尔法斯特女王大学）获得了文学硕士学位。

1879年，在伦敦的一所教学医院工作一年后，韩尔礼重返贝尔法斯特女王学院。期间，他的一位昔日教授告知他，另一位爱尔兰人罗伯特·赫德爵士（Sir Robert Hart, 1835—1911）所主持的中国海关总税务司署有一个职位空缺。当时赫德爵士正为监督第三届巴黎博览会中国展区的工作而返回欧洲，并借机造访了母校。在此期间，他透露希望招募一位受过良好教育的青年——若可能，最好具备医学知识。尽管韩尔礼当时的医学资质尚不够，但他迅速在爱丁堡获得了相关学位，并于1881年夏天被海关总税务司署录用（O'Brien, 2011）。

1863年，慈禧太后（1835—1908）任命赫德为中国海关总税务司。由此，他成为受雇于清政府的最高级别的外国人，并在任职期间招募了许多爱尔兰青年进入海关体系。在赫德领导下，海关总税务司署不仅负责国内关税管理，还统辖邮政服务、港口运营、缉私行动，以及对中国沿海和长江沿岸地区的测绘、警务和照明工作。该机构还组织中国参与了近30场世界博览会。至1895年，这一国际性文职机构已雇用超过700名西方人（主要为欧美籍），巅峰时期甚至贡献了清廷一半的财政收入。海关总税务司署的站点遍布中国，对于韩尔礼这样对植物学感兴趣的人而言，这为深入大清帝国最偏远的角落采集植物提供了绝佳机会。

正如萨金特教授所指出的，在韩尔礼抵达中国时，人们对华中地区的植物区系几乎一无所知，而英国皇家植物园邱园（Royal Botanic Gardens, Kew）的植物标本馆中的大部分腊叶标本都来自中国东部沿海地区。韩尔礼于1881年7月抵达上海，并在接下来的几个月里接受了培训，为即将代表海关履行的各种职责做准备。1882年3月，他被派驻湖北长江沿岸的通商口岸宜昌（时称Ichang）。

当时，宜昌的主要出口商品是从四川和西藏山区通过长江三峡运下来的药材。这些货物在离港之前，需要缴纳税金或征税。韩尔礼的主要职责之一便是管理这些货物的税务。这类利润丰厚的贸易商品主要包括药用大黄（*Rheum officinale*）根、党参（*Codonopsis tangshen*）根、杜仲（*Eucommia ulmoides*）树皮、厚朴（*Magnolia officinalis*）树皮以及许多其他在中医中广泛使用的植物。作为一名医生，韩尔礼对这些制品非常感兴趣。尽管他能够获知这些植物的本地俗名，却始终无法确定它们的植物学双名。

正因如此，1885年3月，韩尔礼联系了英国著名植物学家、英国皇家植物园邱园馆长约瑟夫·道尔顿·胡克爵士（Sir Joseph Dalton Hooker, 1817—1911），提出若邱园的分类学家能协助鉴定植物，他将采集并寄送标本。对于当时处于大英帝国植物学核心地位的胡克来说，这是一个无法拒绝的提议。当时，邱园正主导着殖民地农业扩张，而胡克深知这位驻扎在中国偏远地区的无名爱尔兰人所提供标本的潜在价值。胡克大力鼓励韩尔礼的工作，要求他为每份采集物仔细编号，且每种植物编号的标本需采集并编号10份而非1份。此后十五年间，韩尔礼的足迹遍布福建、广西、广东、海南、河北、湖北、湖南、江西、四川、台湾、云南和浙江等地，甚至从中国云南南部跨境至越南北部短暂采集，累计采集了15 862份编号标本。

据邱园标本馆与图书馆馆长威廉·博廷·赫姆斯利（William Botting Hemsley, 1843—1924）估算，由于每份编号标本均包含复份及10份样本，至1900年12月韩尔礼离开中国时，他已经采集了158 050份腊叶标本（如果包括他在中国各地的所有采集编号，实际更精确的数字应是158 620份）。第一套标本由邱园保存，副本则分发至全球其他标本馆。今天，这些标本分别保存在全球的多个标本馆中，包括A（美国哈佛大学）、B（德国柏林植物园）、BM（英国自然历史博物馆）、BRSL（波兰弗罗茨瓦夫大学）、CAL（印度植物研究中心）、CGE（英国剑桥大学）、DBN（爱尔兰都柏林植物园）、E（英国爱丁堡植物园）、F（美国费尔德自然史博物馆）、FHO（英国牛津大学）、G（瑞士日内瓦植物园）、GH（美国哈佛大学）、HK（渔农自然护理署香港植物标本室）、K（英国邱园）、L（荷兰国家植物标本馆）、LE（俄罗斯科马洛夫植物研究所）、MANCH（英国曼彻斯特大学）、MO（美国密苏里植物园）、MPU（法国蒙特利埃大学）、NMW（英国威尔斯卡迪夫大学）、NY（美国纽约植物园）、P（法国国立自然历史博物馆）、PE（中国科学院植物研究所）、US（美国史密森尼学会）、W（奥地利维也纳自然历史博物馆）和WRSL（波兰华沙大学）。

在《威尔逊采集植物志》的序言中，查尔斯·斯普拉格·萨金特进一步指出，韩尔礼"仅仅采集植物标本和一些百合鳞茎，并未尝试将他的非凡发现引入西方园林"。这一说法并不准确，萨金特并不知晓韩尔礼曾向邱园寄送过大量种子。但不幸的是，来自温带地区的种子在并不适合的温室条件下萌芽并培育，很快死亡。邱园入库记录中关于韩尔礼在中国采集的种子培育幼苗的情况令人失望。如果邱园的培育更加成功，韩尔礼可能会继续寄送更多的种子。在当时，植物园并非寄送种子的最佳地点，这也解释了为何法国传教士植物采集者往往将种子寄送给像维尔莫兰（Vilmorin）这样的商业苗圃，而不是巴黎植物园（Jardin des Plantes）。

19世纪末至20世纪初，几位法国传教士积极在华采集植物，其中代表性的人物有谭卫道（1826—1900）、赖神甫（又译让·玛丽·德拉维，Père Jean-Marie Delavay, 1834—1895）、让·安德烈·苏利埃神父（Père Jean-André Soulié, 1858—1905）和保罗·吉约姆·法尔热神父（Père Paul Guillaume Farges, 1844—1912）。他们所采集的标本被送往巴黎自然历史博物馆，由著名的法国分类学家阿德里安·雷内·弗朗谢（Adrien René Franchet, 1834—1900）命名和描述。1899—1900年间，弗朗谢发表了《德拉维植物采集志，德拉维神父在云南采集的中国植物》（*Plantae Delavayanae, Plants de Chine recueillies au Yun-nan par l'abbé Delavay*），详细记录了赖神甫在云南采集的大量植物（虽未完全收录），并首次描述了超过140个新分类单元。

西方植物采集者中鲜有人能将其全部采集成果以单一著作的形式出版。前述《威尔逊采集植物志》仅列出了 E. H. 威尔逊为哈佛大学阿诺德树木园（1906—1909年、1910—1911年、1918年）远征采集的木本植物，但未包括他早期为英国维奇苗圃公司（Veitch Nursery）进行的两次远征（1899—1902年和1903—1905年）中采集的植物，也不包括他采集的草本或球茎类植物。这本《韩尔礼采集植物志》可能是迄今为止由一位采集者在中国所采集的植物的最完整目录。

韩尔礼在中国开始采集工作的时机恰逢其时——邱园标本馆正准备启动编写《中国植物索引》（*Index Florae Sinensis*）。这是中国植物区系的首部主要研究成果，由威廉·博廷·赫姆斯利和美国植物学家弗朗西斯·布莱克威尔·福布斯（Francis Blackwell Forbes, 1839—1908）合著，并于1886—1905年间发表在《林奈学会杂志（植物学）》[*Journal of the Linnean Society（Botany）*]上。韩尔礼的采集成果占据了《中国植物名录》的大量篇幅，正是他的采集工作为如今我们对中国丰富的植物区系的了解奠定了基础。

与职业植物采集者，如欧内斯特·亨利·威尔逊（Ernest Henry Wilson, 1876—1930）、乔治·福雷斯特（傅礼士，George Forrest, 1873—1932）、约瑟夫·洛克（骆克，Joseph Rock, 1884—1962）和弗兰克·金登-沃德上尉（Captain Frank Kingdon Ward, 1885—1958）不同——这些人主要专注为英国、爱尔兰和北美的富有的赞助人采集具有园艺潜力的观赏植物，韩尔礼始终秉持科学的态度。他致力于向邱园提供尽可能广泛的植物材料，以供收录于《中国植物索引》。为了扩大采集范围，他雇用了中国采集者，亲自培训他们。这些采集者很快熟练掌握了从中国不同栖息地采集和压制植物标本的技能，采集范围从四川康定的高山到云南南部的西双版纳热带雨林。韩尔礼的标本收藏涵盖了被子植物、裸子植物、蕨类植物、蕨类近缘植物、苔藓、地衣和藻类。鉴于他对经济植物学的兴趣，其标本收藏还包含了许多中国的药用植物和重要的粮食作物。1893年，他在上海出版了《中国经济植物学笔记》（*Notes on Economic Botany of China*），这是他的一部早期著作。在书中，他详细记录了中国的重要经济植物，列出它们的中文俗名并匹配了对应的学名。例如，他提到杜仲（*Eucommia ulmoides*），指出其树皮是一种昂贵的药材。邱园植物学家丹尼尔·奥利弗（Daniel Oliver, 1830—1916）于1890年根据韩尔礼采集的标本将杜仲描述为一个新属新种。后来，杜仲又被确立为一个全新的科——杜仲科（Eucommiaceae）（Henry, 1893）。

韩尔礼的《中国经济植物学笔记》深刻影响了美国农业部（USDA）植物引进部主任大卫·费尔柴尔德（David Fairchild, 1869—1954）。费尔柴尔德在1900年访问中国期间联系了韩尔礼，询问有关中国经济植物的情况。韩尔礼寄给他一本书，并告诉他"不要浪费邮费，而是派一位采集者到中国来"。基于此，费尔柴尔德派遣了荷兰出生的弗兰克·N. 梅耶（Frank N. Meyer, 1875—1918）前往中国。在1905—1918年间的四次考察中，梅耶向美国引入了约2 500种经济植物，其中包括著名的中国作物如大豆、豆芽、荸荠、谷物、水果、蔬菜以及防风林树种等。1922年，费尔柴尔德又派遣另一位采集者约瑟夫·洛克前往中国为美国农业部进行采集。

韩尔礼在中国的工作对后世的影响难以估量。他不断致信英国皇家植物园邱园和阿诺德树木园的相关负责人，建议派遣专业植物采集者到中国，以便将更多中国美丽的植物引入欧洲和北美的花园中。尽管当时邱园在英国没有足够的资源开展这项工作，但他们成功说服了著名的维奇苗圃（Veitch Nursery）派遣一位采集者。而他们选中的人正是前文提到的英国年轻人，欧内斯特·亨利·威尔逊。威尔逊在1899年前往中国之前，还在邱园研究了韩尔礼的植物标本。

威尔逊在抵达香港时收到维奇苗圃的指示，前往南云南的思茅（现普洱）与韩尔礼会面，并向他获取珙桐（*Davidia involucrata*，鸽子树、手帕树）的具体位置。1888年5月，韩尔礼曾在四川巫山南部采集到这种植物。在思茅，韩尔礼指示威尔逊重新雇佣他先前在宜昌训练过的植物采集者，并告诉他珙桐的确切分布地点。威尔逊为维奇苗圃两次赴中国采集，后来受雇于阿诺德树木园，又在中国中部和西部进行了两次采集探险，并于1918年短暂前往台湾。

同样也是受韩尔礼的影响，苏格兰植物采集者乔治·福雷斯特于1904年前往云南，为阿瑟·基尔平·布利（Arthur Kilpin Bulley, 1861—1942）的蜜蜂苗圃（Bees Nurseries）工作。布利的指派也为弗兰克·金登-沃德上尉开启了后者的植物采集生涯。金登-沃德是所有植物采集者中采集时间最长的一位。通过这些人的引种，再加上法国传教士植物采集者的引种，改变了欧洲和北美温带地区的园林景观。以爱尔兰为例，种植于大型花园的植物大约40%原产于中国。

韩尔礼在1900年12月31日离开中国，后来在法国南锡国立林业学校接受林业培训。1907年，他被任命为剑桥大学新成立的林业学院的讲师（后为教授）。1913年4月，韩尔礼回到家乡爱尔兰，担任皇家科学学院（现为都柏林大学学院）新设立的林业教授之职。在爱尔兰，韩尔礼被誉为"爱尔兰林业之父"。当爱尔兰创建林业时，韩尔礼是创业委员会最具影响力的顾问，如今爱尔兰的林业非常成功。

1930年3月23日，韩尔礼在都柏林病逝，他被认为是爱尔兰有史以来最伟大的探险植物学家。在他去世时，其私人标本馆包含超过1万份标本，其中许多是由其他伟大的收藏家在世界各地收集的。他的遗孀爱丽丝·H.（埃尔西）亨利 [Alice H. (Elsie) Henry, 1881—1956] 花了8年时间整编了这些标本，这些标本随后被赠予了都柏林格拉斯尼文国家植物园（National Botanic Gardens, Glasnevin, Dublin）。在那里，这些收藏被称为"韩尔礼林业标本馆（The Augustine Henry Forestry Herbarium）"，与他在中国的采集一同保存在爱尔兰国家植物标本馆（Irish National Herbarium, DBN）中。格拉斯内文国家植物园的园长托马斯·J.沃尔什博士（Dr. Thomas J. Walsh）于1957年1月出版了韩尔礼林业标本馆的目录（Walsh, 1957）。

我对韩尔礼植物采集工作的兴趣实际上始于1999年。当时我受命担任爱尔兰花园植物协会（Irish Garden Plant Society）下属一个小组委员会的主席，负责为2002年5月的英国皇家园艺学会（Royal Horticultural Society, RHS）切尔西花展（Chelsea Flower Show）准备一个展览。一份简单的韩尔礼在中国"发现"的植物清单逐渐扩充为一份详细的目录，现在以这部作品的形式出版。当我说"发现"时，我完全尊重中国当地居民，远早于韩尔礼来到中国之前，他们就熟知这些植物的俗名。我所说的"发现"和"发现过程"，是指他让植物学界注意到了这些植物，并首次赋予它们植物学双名法的名称。

在英国皇家园艺学会切尔西花展的展品获奖之后，我于2002年9～10月以及2004年9～10月组织并领导了两次由格拉斯尼文国家植物园（Glasnevin National Botanic Gardens）发起的格拉斯尼文中国中部探险（Glasnevin Central China Expedition, GCCE 2002, 2004），前往湖北和四川。这些活动是与中国科学院（Chinese Academy of Sciences）和武汉植物园（Wuhan Botanical Gardens）的资深工作人员合作进行的。在获得许可证后，我们在2002年三峡大坝竣工之前收集了种子和标本。两年后，我们再次返回，通过船闸航行，见证了已竣工的大坝及其蓄水过程。2004年的行程还包括一次前往台湾南部的采集探险（格拉斯尼文台湾探险——Glasnevin Taiwan Expedition, GTE），我们拜访了许多韩尔礼早期的采集地点和

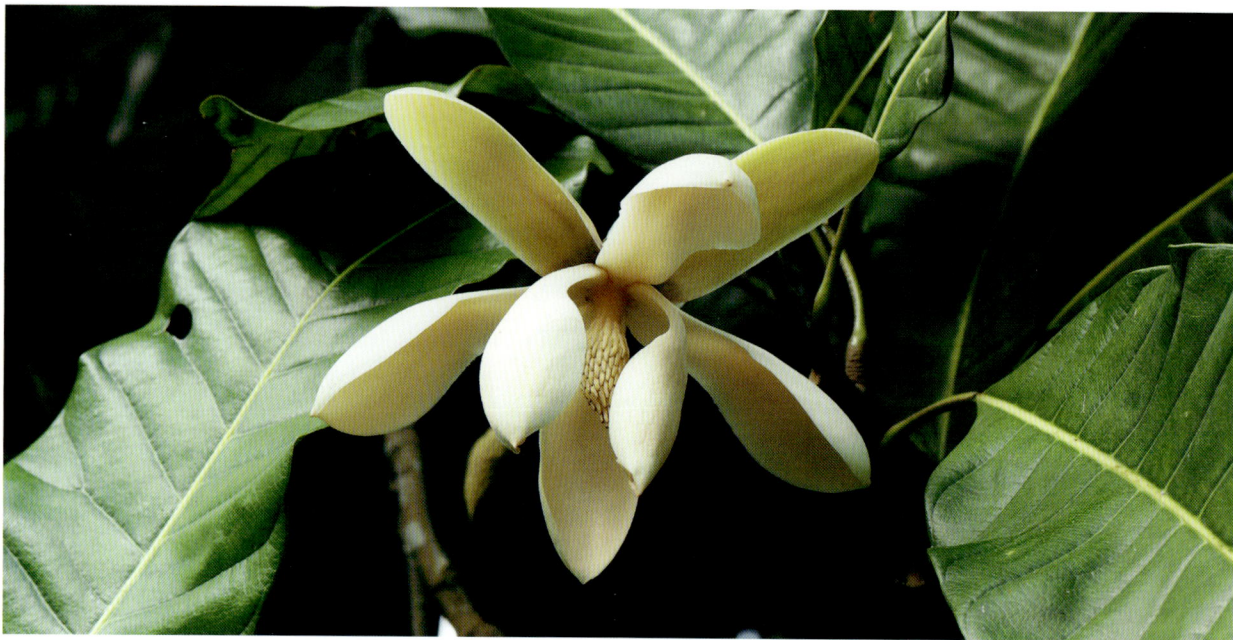

大叶木兰（*Lirianthe henryi*，叶文 摄）

基地。2005年，我带领一个团队前往云南南部，与昆明植物园（Kunming Botanical Garden）合作，参观了韩尔礼在中国的最后几个基地，特别是蒙自（Mengzi）和思茅（Simao）。在两次格拉斯尼文中国中部探险中（2002年、2004年）采集的许多植物，如今生长在都柏林格拉斯尼文国家植物园及其姐妹园和乡间附属园区——基尔马库拉国家植物园（National Botanic Gardens, Kilmacurragh），这些采集的细节信息也收录在本部编目中。

我认为补充介绍本书中列出的许多分类群在西方园林的早期引入历史也很重要。中国一直是欧洲、北美洲、澳大利亚和新西兰温带地区优良园林植物的重要来源。正如前文所述，韩尔礼的"发现"后来被E. H. 威尔逊引种，而德拉维神父的发现则由乔治·福雷斯特引种。不过，本书中还列出了许多其他历史上的采集者及其从中国不同地区引种的植物。

目录中每种分类单元后的注释，特别是中国植物名称和植物的当地用途，部分基于韩尔礼和E. H. 威尔逊的观察，同时也结合了多个其他来源的信息，其中大多数在讨论的具体物种文本中已被引用。

在整本《韩尔礼采集植物志》中，我选择采用目前通用的汉语拼音罗马化系统来表示中国地名，替换了旧的罗马化方式——威妥玛式（Wade-Giles）或极少数情况下，邮政式拼音（Post Office Romanisation）。因此，用宜昌代替"Ichang"，用巴东代替"Patung"，用高雄代替"Takow"。有关韩尔礼在中国采集地名的更详细更新列表，请参阅《追寻韩尔礼及其中国植物采集者的足迹》（*In the Footsteps of Augustine Henry and his Chinese Plant Collectors*）第336-337页的附录2；O'Brien, 2011）。

需要指出的是，韩尔礼在中国大陆和海南岛共采集了13 826份编号标本（每个编号有多份标本）。然而，由于他的第一任妻子卡罗琳生病和去世引发的混乱，以及1895年5月日本侵略台湾的双重影响，他在台湾采集的标本（编号1-2036号）是一个完全独立的新编号序列，独立于中国其他地区的采集。在本书中，标本根据韩尔礼在中国各个基地采集的时间顺序排列，因此从湖北和四川开始，然后是海南，再接着是台湾（如前文所述，采用全新的编号序列），最后是他的最后一个基地云南。

在编写本编目时，我参考了引用韩尔礼标本的出版物，而没有查看他实际的标本馆藏标本。这也意味着参考了一些未发表的来源，主要包括两份《邱园鉴定清单》（*Kew Lists of Determinations*）。其中第24卷聚焦于韩尔礼在中国中部和海南的采集，存于都柏林国家植物园的档案中；第36卷列出了韩尔礼在云南的采集，保存在邱园皇家植物园的图书馆和档案中。虽然我在引用时未注明作者姓名，但这些清单很可能是由邱园的植物学家威廉·博廷·赫姆斯利和丹尼尔·奥利弗教授编写的。在更新植物名称和植物科属信息时，我使用了邱园皇家植物园的在线资源——世界植物在线（Plants of the World Online, POWO）[1]。

目前的编目几乎完整记录了韩尔礼在中国的采集，只有600多个编号的标本我无法追溯到，不过其中许多可能是已列出的标本的副本。在本编目中共有植物的308个科1 776个属5 099个种210个变种30个亚种8个型1个亚型15个杂交种（包括1个属间杂交种）和1个栽培种，使韩尔礼的记录采集总计达5 364个分类类群。其中以下为科学新发现：1 046个种109个变种28个亚种和5个新型，总计1 188个为当前植物命名法所接受的新分类单元。这成为中国史上成果最丰富的个人植物采集之一。

西默斯·奥布莱恩
爱尔兰基尔马库拉国家植物园
2024年11月

2 作者简介

西默斯·奥布莱恩（Seamus O'Brien）是位于威克洛郡的基尔马库拉（Kilmacurragh）国家植物园的首席园艺师（Head Gardener）。该植物园是都柏林格拉斯尼文爱尔兰国家植物园的乡村庄园及附属园区。他是爱尔兰研究中国植物区系以及植物采集历史的权威。在基尔马库拉，奥布莱恩管理着一批来自世界温带地区的重要植物收藏，尤其是来自中国、喜马拉雅地区、南美洲、新西兰、澳大利亚和塔斯马尼亚的植物。植物园拥有大量濒危和受威胁的植物类群，并以植物保护为重要内容，包括通过全球性的植物考察以及与爱丁堡皇家植物园国际针叶树保护计划（International Conifer Conservation Programme）等合作项目获取的植物材料。

2.1 教育背景与职业生涯

西默斯，1970年出生于威克洛郡的巴尔廷格拉斯（Baltinglass），1989年在那里的斯库尔康莱斯（Scoil Conglais）完成中学教育。1990—1993年，他在爱尔兰格拉斯尼文国家植物园接受了正规的园艺培训，并于2006年获得伦敦邱园皇家植物园颁发的国际植物园管理文凭。

西默斯曾管理过几座拥有重要的历史植物收藏的著名爱尔兰植物园，包括位于克里郡瓦伦蒂亚岛（Valentia Island）的格兰利姆（Glanleam）植物园（1993—1997年），该园以其南半球的乔木和灌木而闻名；还有位于都柏林克隆西拉（Clonsilla）的比奇公园（Beech Park）（1997—2000年），当时它因拥有不列颠及爱尔兰最大的草本植物和高山植物收藏之一而备受赞誉。2000年2月，他回到格拉斯尼文国家植物园工作，并于2006年5月调任至基尔马库拉国家植物园从事管理工作。

西默斯游历全球，到原生生境研究植物，其中最值得关注的包括澳大利亚（包括塔斯马尼亚）、亚速尔群岛、不丹、中国（香港、湖北、四川、云南、台湾和西藏）、美国（加利福尼亚）、智利、印度（阿萨姆邦、拉达克、曼尼普尔邦、那加兰邦、西孟加拉邦）、缅甸、尼泊尔、新西兰、肯尼亚、乌干达和南非。

1 https://powo.science.kew.org/。

西默斯重访了爱尔兰植物探险家韩尔礼（Augustine Henry）博士（许亦华，2024）在中国的采集路线，并出版了广受国际赞誉和奖项的《追寻韩尔礼及其中国植物采集者的足迹》（In the Footsteps of Augustine Henry and his Chinese Plant Collectors）（2011年）（叶文，马金双，2014）。他还根据最近在印度东北部的旅行经历出版了《追寻约瑟夫·道尔顿·胡克的足迹：锡金探险之旅》（In the Footsteps of Joseph Dalton Hooker: A Sikkim Adventure, O'Brien, 2018）。西默斯目前正在撰写英国植物采集者和探险家弗兰克·金登-沃德上尉（1885—1958）的传记。他也为《柯蒂斯植物学杂志》《爱尔兰园林》《爱尔兰园林植物协会通讯》以及《国际树木学协会年鉴》等多部出版物撰稿。

西默斯在国际上讲授园林历史和以植物学为主题的旅行。他是英国皇家园艺学会木本植物委员会成员，同时也是英国皇家园艺学会早春花展的评委。他钟情于杜鹃花，担任皇家园艺学会杜鹃花山茶木兰组爱尔兰分会主席，还是埃克斯伯里植物园信托基金会（Exbury Garden Trust）的受托人。埃克斯伯里植物园归罗斯柴尔德家族所有，位于英格兰南安普敦附近，占地200英亩。这里的杜鹃花属收藏世界闻名。该植物园由莱昂内尔·德·罗斯柴尔德（1882—1942）创建，园内收藏了乔治·福雷斯特（傅礼士，1873—1932）（王涛 等，2023）和弗兰克·金登-沃德上尉（黄伟，2024）从中国采集的植物。他还是安斯利植物园（Annesley Garden）和卡斯尔韦兰（Castlewellan）北爱尔兰国家树木园咨询小组的成员。

2.2 奖项与荣誉

2018年，西默斯荣获爱尔兰皇家园艺学会金质荣誉奖章，并于2019年成为该学会的荣誉会员。2021年，他被探险家俱乐部（纽约）授予会员资格。2022年，他因在保护基尔马库拉著名杜鹃花属收藏方面的工作，以及通过书籍和出版物分享知识的意愿，荣获皇家园艺学会（英国）的洛德杜鹃花杯。2023年，他当选为林奈学会会员（Fellows of the Linnean Society of London, FLS）。

参考文献

黄伟，2024. 传奇一生的弗兰克·金敦-沃德 [M]// 马金双. 中国——二十一世纪的园林之母：第六卷. 北京：中国林业出版社：391-425.

王涛，2023. 植物猎人——傅礼士 [M]// 马金双. 中国——二十一世纪的园林之母：第五卷. 北京：中国林业出版社：455-513.

许亦华，2024. 韩尔礼的植物学之路 [M]// 马金双. 中国——二十一世纪的园林之母：第七卷. 北京：中国林业出版社：339-361.

叶文，马金双，2014(2012). 重叠的脚印——两个爱尔兰青年相距百年的中国之旅 [J]. Journal of Fairylake Botanical Garden, 11(3-4): 56-58.

HENRY A, 1893. Notes on Economic Botany of China 16 [M]. Shanghai: The Presbyterian Mission Press.

O'BRIEN S, 2011. In the Footsteps of Augustine Henry and his Chinese Plant Collectors 19 [M]. Woodbridge, UK: Garden Art Press.

O'BRIEN S, 2018. In the Footsteps of Joseph Dalton Hooker: A Sikkim adventure [M]. Royal Botanic Gardens, Kew: Kew Publishing.

SARGENT C S, 1911. Plantae Wilsonianiae [M]. Preface. V. Cambridge, The University Press.

WALSH T, 1957. The Augustine Henry Forestry Herbarium – A Catalogue of Specimens [M]. An Roinn Talmhaíochta (Department of Agriculture), Dublin.

译者简介

叶文（女，福建厦门人，1982年生），2004年获厦门大学生物科学专业学士学位，2007年获厦门大学生态学专业硕士学位，2010年获华东师范大学植物学专业博士学位，2011—2013年浙江大学生命科学学院博士后；2013—2021年任中国科学院华南植物园助理研究员/陈焕镛副研究员；2021年至今任厦门大学生命科学学院副教授。主要从事苔藓植物学研究，尤其关注苔藓植物的多样性与进化。研究领域涉及植物分类学、植物系统发育和进化、苔藓植物与环境因子的相关性等。参与《中国植物分类学纪事》和《解译法规》（《国际藻类、菌物和植物命名法规》读者指南）的编写与翻译。

说明

本文为《韩尔礼采集植物志》英文的第一部分（引言）和最后一部分（作者简介的部分内容，不包括出版物）的中文翻译版。英文正文（第66-621页）以及作者简介的出版物部分（第622-624页）请参见英文原文。另外，英文正文部分以属的字母顺序列出韩尔礼采集的具体类群（Abelia-Ziziphus），所以没有列入本卷的索引。特此说明。

Plantae Henryanae

A catalogue of the conifers, gymnosperms, flowering plants, ferns, fern allies, bryophytes, lichens and algae collected by Dr. Augustine Henry in China during the years between 1885 and 1900.

Seamus O'Brien

(National Botanic Gardens of Ireland, Kilmacurragh, Ireland)

Email: seamus.obrien@opw.ie

Abstract: *Plantae Henryanae* is a catalogue of the 158,620 botanical (herbarium) collections made by Dr Augustine Henry (1857—1930) in China between 1885 and 1900. Sent to the Royal Botanic Gardens, Kew when taxonomists there were compiling the *Index Florae* Sinensis, the first major flora of China, his duplicate specimens were further distributed to herbaria worldwide. Henry worked as an official in the Chinese Imperial Maritime Customs Service, his duty being to place levies on items of *materia medica*, mainly plant-derived medicinal goods, and this stimulated an early interest in botany. His collections proved to be one of the most important ever sent to Europe from China and were rich in new taxa from a diverse range of habitats. The current catalogue documents 5,364 taxa, of which 1,188 proved new to science and are still recognised in current botanical nomenclature. Individual collection numbers bear details of where they were collected, published, and their colloquial names as recorded by Augustine Henry. China has proved an incredibly rich source of good garden plants for temperate regions of Europe, North America, Australia and New Zealand. The details of early introductions of taxa from China to these regions is discussed, highlighting the work of other Western collectors like Ernest Henry Wilson, Père Jean Marie Delavay, Père Armand David, George Forrest, Joseph Rock and Frank Kingdon Ward, for example.

Keywords: Augustine Henry, Botanical collections, Catalogue, Flora of China, Index Florae Sinensis, Kew, Materia medica

Seamus O'Brien, 2025, Part 2, PLANTAE HENRYANAE; CHINA-Mother of Gardens, in the Twenty-first Century, Volume 8: p043-619.

Introduction

In his preface to *Plantae Wilsonianae* (published between 1911—1917) Charles Sprague Sargent (1841—1927), Director of Harvard University's Arnold Arboretum wrote 'Although important collections of plants had been made in Szech'uan by the French missionary Armand David as early as 1870, the world knew little of the remarkable beauty and richness of the flora of west central China until Augustine Henry, an officer of the Chinese Imperial Maritime Customs Service stationed at Yichang (then Ichang) from 1882 to 1889, sent to England the dried plants which he had collected in western Hubei (then Hupeh). An examination of these collections, the first of which reached England in 1886, disclosed many new genera and a great number of new species.' (Sargent, 1911)

Born on July 2 1857, Augustine Henry grew up in Cookstown, County Tyrone and at Tyanee, County Derry in north-west Ireland and was educated at Cookstown Academy, before moving to Queen's College, Galway (now the National University of Ireland, Galway), to study natural science and philosophy, graduating with a first-class degree (B.A.) in 1887. The following year he obtained a Master of Arts (M.A.) postgraduate degree from Queen's College (now Queen's University), Belfast.

In 1879, having spent a year in one of London's teaching hospitals, Henry returned to Queen's College, Belfast where one of his former Professors told him of a vacancy in the Chinese Imperial Maritime Customs Service under another Irishman, Sir Robert Hart (1835—1911). Hart had returned to Europe to oversee the Chinese exhibit at the Third Paris Exhibition and while visiting his former College, he let it be known that he was looking to recruit a well-educated man – if possible, one with a knowledge of medicine. Henry's medical qualifications at that time were not sufficient, but he obtained his degree quickly in Edinburgh and was accepted into the Customs Service in the summer of 1881. (O'Brien, 2011)

In 1863, the Empress Dowager Cixi (1835—1908) appointed Sir Robert Hart Inspector General of the Customs Service. As such, he was the highest ranking foreigner employed by the Chinese government, and, over the years, he was to recruit many young Irishmen into the service. Under Hart, the Customs Service was responsible for domestic customs administration, postal services, harbour management, anti-smuggling operations and for mapping, policing and lighting China's coast and areas along the Yangtze River. The service also organised China's representation at nearly 30 world fairs. By 1895, more than 700 Westerners (mainly European and American) were employed in this international Civil Service and at one time, the Chinese Customs Service provided half the revenue of the Emperor. The service had stations throughout China and to those like Augustine Henry, who were interested in botany, it provided an unrivalled opportunity to collect plants in the most remote corners of the Chinese empire.

As pointed out by Professor Sargent, at the time of Henry's arrival in China, virtually nothing was known of the flora of central China and the bulk of dried specimens in the herbarium at the Royal Botanic Gardens, Kew came from eastern coastal regions. Henry arrived in Shanghai in July 1881 and spent the following months there training for the various duties he would soon carry out on behalf of the Customs Service. In March 1882 he was posted to Yichang a treaty port on the Yangtze River in Hubei province.

The principal exports from Yichang at the time were items of *materia medica* from the mountains of Sichuan and Tibet (Xizang) that were transported down the Yangtze River through the famous Three Gorges. One of Henry's customs duties was to place a tax or levy on these before they were dispatched from the port. This highly-lucrative trade consisted largely of medicinal rhubarb root (*Rheum officinale*), roots of *Codonopsis tangshen*, the medicinal bark *Eucommia ulmoides* and *Magnolia officinalis,* and many other plants widely used in traditional Chinese medicine. Henry, a medical doctor, took great interest in these products, and while he could get the local colloquial names of these plants, he found it impossible to get their botanical binomials.

It was for that reason he contacted the great British botanist, Sir Joseph Dalton Hooker (1817—1911), Director of the Royal Botanic Gardens, Kew, in London, in March 1885, offering to collect and forward specimens if taxonomists at Kew could identify them for him. For Hooker, then at the centre of imperial botany, it was an offer too good to refuse. Kew was then a key player in the expansion of colonial agriculture and Hooker knew the potential of receiving material from this obscure Irishman based in rural China. Joseph Hooker greatly encouraged Henry in his work, asking him to carefully number his collections and to collect not one, but ten specimens of each plant gathered and numbered. Over the course of the next fifteen years Henry was to collect 15,862 numbered specimens across Fujian, Guangxi, Guangdong, Hainan, Hebei, Hubei, Hunan, Jiangxi, Sichuan, Taiwan, Yunnan and Zhejiang, even crossing the border from southern Yunnan to briefly collect in northern Vietnam. William Botting Hemsley (1843—1924), Keeper of the Kew Herbarium and Library, estimated that since Henry had collected duplicates and ten specimens of each collection number, that by the

time of his departure from China in December 1900, he had collected 158,050 dried plant specimens in China (more accurately 158,620 specimens when all his collection numbers from across China are included). The first set of specimens were retained by Kew while duplicates were set to other herbaria worldwide. Today these are housed in herbaria across the world including A, B, BM, BRSL, CAL, CGE, DBN, E, F, FHO, G, GH, HK, K, L, LE, MANCH, MO, MPU, NMW, NY, P, PE, US, W and WRSL.

In his preface to *Plantae Wilsonianae*, Charles Sprague Sargent further stated that Augustine Henry 'collected only herbarium specimens and a few lily bulbs, and took no steps to introduce into Western gardens his remarkable discoveries.' That statement is not at all accurate, Sargent was unaware that Henry had sent large volume of seeds to Kew where unfortunately taxa from temperate regions were germinated and grown-on in unsuitable stove house conditions and soon died. The fate of seedlings raised from Henry's Chinese collection as recorded in the Kew Inwards book makes for disappointing reading and had Kew had better success Henry might have continued sending further large volumes of seeds. Botanic Gardens at this time were not the best places to send seeds and it was for that reason that French missionary collectors sent their seed collections from China to commercial nurseries like Messrs. Vilmorin rather than the Jardin des Plantes in Paris.

Several French missionaries were actively collecting in China in the late 19th and early 20th centuries, particularly Père Armand David (1826—1900), Père Jean-Marie Delavay (1834—1895), Père Jean-André Soulié (1858—1905) and Père Paul Guillaume Farges (1844—1912). Their collections were sent to the Paris Museum of Natural History where they were named and described by the famous French taxonomist Adrien René Franchet (1834—1900). In 1899—1890 Franchet published *Plantae Delavayanae, Plants de Chine recueillies au Yun-nan par l'abbé Delavay* (Plants from China collected in Yunnan by Father Delavay), listing many (but not all) of Delavay's collections and describing over 140 new taxa for the first time.

Few of the Western plant collectors have had their entire collections published in a single work. The aforementioned *Plantae Wilsonianae* enumerates the woody plants collected by E. H. Wilson on expeditions for the Arnold Arboretum of Harvard University (1906—1909, 1910—1911, 1918) but includes comparatively few of his earlier collections (and none of his herbaceous or bulbous material) made on two successive expeditions (1899—1902 and 1903—1905) for the Veitch nursery firm in England. This current *Plantae Henryanae* is probably the most complete catalogue of plants made by a collector in China.

By coincidence, Augustine Henry had chosen a very opportune time to begin collecting work in China since the Kew Herbarium was about to embark on the *Index Florae Sinensis*, the first ever major flora of China, co-authored by William Botting Hemsley and the American botanist, Francis Blackwell Forbes (1839—1908). Published in the *Journal of the Linnean Society* (Botany) between 1886 and 1905, Augustine Henry's collections were to dominate its pages and it was his collections that were to lay the basis of our present knowledge of the vast flora of China.

Unlike professional plant collectors such as Ernest Henry Wilson (1876—1930), George Forrest (1873—1932), Joseph Rock (1884—1962) and Captain Frank Kingdon Ward (1885—1958), who were primarily interested in collecting ornamental plants with horticultural potential for wealthy patrons in Britain, Ireland and North America, Augustine Henry maintained a scientific approach to his work. He was keen to supply as wide a range of material to Kew for inclusion in the *Index Florae Sinensis*. To expand his work he employed Chinese collectors, men he trained himself and who soon became highly skilled in collecting and pressing plants from a range of habitats across China, from the alpine slopes above Kangding in Sichuan to the tropical rainforest of Xishuangbanna in southern Yunnan. Henry's herbarium is wide-ranging in its scope, containing angiosperms, gymnosperms, ferns, fern-allies, bryophytes, lichens and algae, and given his interest in economic botany it is not surprising that it also contains many of China's medicinal plants and important food crops. In 1893 his *Notes on Economic Botany of China* was published in Shanghai, an early work in which he highlighted important economic Chinese plants, giving their Chinese colloquial names and matching them with their scientific equivalents. Thus, for example, he highlighted *tu-chung* or *du zhong*, whose bark he stated was an expensive drug, it was described (from Henry's collections) as a new genus and species, *Eucommia ulmoides* (later placed in an entirely new family, Eucommiaceae) by the Kew botanist Daniel Oliver (1830—1916) in 1890. (Henry, 1893)

Notes on Economic Botany of China made a major impression on David Fairchild (1869—1954), Head of the Plant Introduction Department of the United States Department of Agriculture (USDA). In 1900, during a visit to China, Fairchild contacted Henry to enquire about economic plants in China. Henry sent him a copy of the publication and told him 'not to waste money on postage, but to send a collector to China.' It was based on this advice that Fairchild sent the Dutch-born Frank N. Meyer (1875—1918) to China, and over the course of four expeditions between 1905 and 1918, Meyer sent about 2,500 economic

plant introductions to America. This included notable Chinese crops like soybean, bean sprouts, water chestnuts, cereals, fruits, vegetables and windbreak trees. In 1922, Fairchild sent another collector, Joseph Rock, to China to collect for the USDA.

It is hard not to underestimate the influence of Augustine Henry's work in China. He constantly wrote to authorities at the Royal Botanic Gardens, Kew and to the Arnold Arboretum to send professional plant collectors to China to introduce its many beautiful plants into European and North American gardens. While in England Kew didn't have the resources to carry out this work, they did manage to persuade the famous Veitch Nursery to send a collector and the man they chose was the aforementioned Ernest Henry Wilson, a young Englishman, who before departing for China in 1899, studied Henry's herbarium specimens at Kew.

His instructions from the Veitch Nursery (which he received on arrival in Hong Kong) was to travel to meet Augustine Henry, then based in Simao in southern Yunnan, and to get from him the location of *Davidia involucrata*, the pigeon, dove or handkerchief tree, which Henry had previously collected in South Wushan, Sichuan in May 1888. At Simao, Augustine Henry instructed E. H. Wilson to re-hire the men at Yichang he had previously trained as botanical collectors and told him exactly where to find *Davidia*. Wilson travelled twice in China for the Veitch Nursery and was later employed by the Arnold Arboretum, collecting over two further expeditions for that garden in central and western China and briefly in Taiwan in 1918.

Similarly, in 1904, as a result of Henry's influence, the Scottish plant collector, George Forrest (1873—1932), travelled to Yunnan on behalf of Arthur Kilpin Bulley (1861—1942) of Bees Nurseries and Bulley was also to launch the plant collecting career of Captain Frank Kingdon Ward, the longest serving of all plant hunters. Their introductions, and the introductions of the French missionary plant collectors, transformed gardens in temperate regions of Europe and North America, in Ireland for example, approximately 40% of the plants grown in major gardens originate in China.

Augustine Henry left China on December 31 1900, later training as a forester at the French National School of Forestry in Nancy, and in 1907 was appointed Reader (later Professor) of Forestry in the newly established School of Forestry at the University of Cambridge. In April 1913 he returned to his native Ireland to take the newly established Chair of Forestry at The Royal College of Science for Ireland (now University College, Dublin). In Ireland he is remembered as 'the Founding Father of Irish Forestry', having been the most influential advisor to a committee formed to establish Ireland's now highly-successful forestry industry.

Augustine Henry died in Dublin following a short illness on 23 March 1930 and is regarded as Ireland's greatest ever exploring botanist. At the time of his death his private herbarium contained over ten thousand specimens, many of which had been gathered across the world by other great collectors. His widow, Alice H. (Elsie) Henry (1881-1956), assembled and arranged them, a task that took her eight years to complete, and these were presented to the National Botanic Gardens, Glasnevin, Dublin where the collection is known as 'The Augustine Henry Forestry Herbarium' and resides in the Irish National Herbarium (DBN), alongside his Chinese collections. A catalogue for his Forestry Herbarium was published by Glasnevin's Director, Dr Thomas J. Walsh in January 1957. (Walsh, 1957)

My interest in Augustine Henry's plant collections really began in 1999 when I was tasked with chairing a subcommittee of the Irish Garden Plant Society to bring an exhibit to the Royal Horticultural Society's Chelsea Flower Show, in London, in May 2002. A simple list of Henry's 'discoveries' in China grew into a substantial catalogue, now published in this work. I say 'discoveries' in complete respect to the inhabitants of China, who knew these plants by their colloquial names long before Augustine Henry arrived in China. By 'discoveries and discovering' I mean he brought these plants to the attention of botanical science where they were given their botanical binomials for the first time.

Following the award-winning exhibit at the RHS Chelsea Flower Show, in September-October 2002 and again in September-October 2004, I organised and led two expeditions (the Glasnevin Central China Expedition, GCCE 2002, 2004) from the National Botanic Gardens, Glasnevin to Hubei and Sichuan. These were carried out in collaboration with the Chinese Academy of Sciences and senior staff from the Wuhan Botanical Gardens. With permits we collected seeds and herbarium specimens before the dam on the Yangtze was completed in 2002, returning two years to navigate through ship locks and see the completed dam and flooding process. The 2004 visit also included a collecting expedition (the Glasnevin Taiwan Expedition – GTE) to southern Taiwan where we visited many of Henry's former collecting areas and bases. In 2005 I led a group to southern Yunnan, collaborating with Kunming Botanical Gardens and visiting Henry's final bases in China, particularly Mengzi and Simao. Many of the plants raised on the two Glasnevin Central China Expeditions (2002, 2004),

grow at the National Botanic Gardens, Glasnevin in Dublin and its sister garden and rural annex, the National Botanic Gardens, Kilmacurragh, and details of these collections also appear in this catalogue.

I have also thought it important to add details of the earlier introduction of many of the taxa listed to Western gardens. China has proved an incredibly rich source of good garden plants for temperate regions of Europe, North America, Australia and New Zealand, the discoveries of Augustine Henry, as previously stated, were later introduced by E. H. Wilson, while those of Père Jean Marie Delavay were later introduced by George Forrest, though the catalogue lists many other historic collectors and their introductions from across various regions of China.

The notes after each taxon, particularly Chinese plant names and local uses of plants, are partly based on observations made by Augustine Henry and also by Ernest Henry Wilson, but are also compiled for multiple other sources, most of which are cited in the text for the individual species discussed.

Throughout *Plantae Henryanae* I have opted to use the current Pinyin Romanisation system for Chinese place names, replacing the older Romanisations – Wade-Giles, or rarely, Post Office – as applicable. Thus Yichang appears instead of Ichang, Badong instead of Patung or Kaohsiung instead of Takow. For a more detailed list of updated place name of Henry's bases in China see appendix 2 of *In the Footsteps of Augustine Henry and his Chinese Plant Collectors* pp. 336-337. (2011).

It is important to point out that Henry collected 13,826 numbered specimens (each number in a multiple of ten sheets) on mainland China and Hainan, but because of the confusion caused by the illness and death of his first wife, Caroline, and the Japanese invasion of Taiwan in May 1895, his collections from Taiwan (1-2,036 numbers), run in an entirely new sequence, separate to his other Chinese collections. Within the catalogue, the specimens are listed in order of the timeline of Henry's bases in China, so beginning with Hubei and Sichuan, followed by Hainan, then followed by Taiwan (running, as previously stated in an entirely new sequence of numbers) and finally, his last base, Yunnan.

In compiling this catalogue I have consulted published works citing Henry's collections rather than viewing his actual herbarium specimens. This has also meant referring to some un-published sources, principally in two 'Kew Lists of Determinations' (volume 24, focusing on Henry's collections from central China and Hainan, is held in the archives at the National Botanic Gardens, Glasnevin, Dublin and volume 36, listing Henry's Yunnan collections, is preserved in the library and archives at the Royal Botanic

02

Gardens, Kew). While I have cited the authors as anonymous, these lists are most likely the work of Kew botanists William Botting Hemsley and Professor Daniel Oliver. In bringing plant names and plant families up to date I have used the Royal Botanic Gardens, Kew's Plants of the World Online (POWO) https://powo.science.kew.org/.

The current catalogue is an almost complete record of Augustine Henry's collections in China, there are just over 600 numbers that I've been unable to trace, though many of these are probably duplicate of specimens already listed. Within the catalogue are 308 plant families, 1,776 genera, 5,099 species, 210 varieties, 30 subspecies, 8 forms, 1 subforma, 15 hybrids (including 1 bigeneric hybrid) and 1 cultivar, bringing Augustine Henry's documented collection to a total of 5,364 taxa. Of these, the following proved new to science: 1,046 species, 109 varieties, 28 subspecies and 5 new forms, giving a total of 1,188 new taxa that are recognised in current botanical nomenclature, one of the richest collections ever made by an individual in China.

A portrait of Augustine Henry by Celia Harrison, 1929. The leaves in the background are of *Rhododendron augustinii*, one of his most famous finds. Copyright: National Botanic Gardens of Ireland, Glasnevin.

02

Augustine Henry in the deer park at the National Botanic Gardens, Kilmacurragh in the summer of 1904 with *Alnus glutinosa*. Henry studied forestry on his return to Europe from China. Copyright: National Botanic Gardens of Ireland, Kilmacurragh.

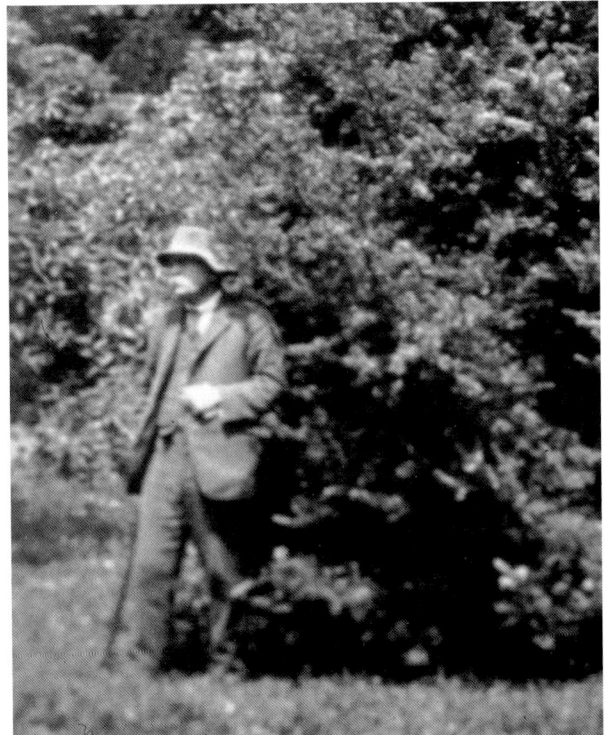

Henry's interest in trees remained a passion all his life. He is seen here with *Podocarpus hallii* at the National Botanic Gardens, Kilmacurragh, in 1929.

Yi people at Iwu, southern Yunnan in the 1890s. Augustine Henry wrote a dictionary of the language of this ethnic group. Copyright: National Botanic Gardens of Ireland, Glasnevin.

Ernest Henry Wilson who collected in China for the Veitch Nursery in England and later for the Arnold Arboretum in America. He introduced many of Henry's earlier collections and always credited Henry for helping launch his career as a plant hunter in China. Copyright: Roy Briggs.

The Scottish plant collector George Forrest was sent to Yunnan as a result of Augustine Henry's advice to his employer Arthur Kilpin Bulley. He is seen here at the China Inland Mission, Dali, Yunnan. Copyright: Royal Botanic Gardens, Kew.

The plant collector Captain Frank Kingdon Ward was sent to north-west Yunnan in 1911 as a result of Henry's advice to his employer Arthur Kilpin Bulley. Henry had enormous influence on the later botanical exploration of central and western China. Copyright: Royal Botanic Gardens, Kew.

02

Augustine Henry's Chinese collectors at Yichang, Hubei. They were later re-hired by a young Ernest Henry Wilson when he arrived there in 1900. Copyright: Arnold Arboretum of Harvard University.

Kew's senior herbarium staff, the men who described Augustine Henry's collections from China. Seated left to right are: John Gilbert Baker, a noted authority on ferns, William Botting Hemsley who named most of Henry's newly discovered species, Otto Stapf, then an assistant in the herbarium and Professor Daniel Oliver, Keeper of the Herbarium. Copyright: Royal Botanic Gardens, Kew.

Yichang (then Ichang), Augustine Henry's first base in China. Boats line the bund and the Yangtze River is a kilometre wide having just exited the famous Three Gorges. Photographed in 1906 by E. H. Wilson. Copyright: Arnold Arboretum of Harvard University.

The foreign compound outside the eastern city walls at Yichang where Henry lived during the 1880s. Photographed in 1906 by E. H. Wilson. Copyright: Arnold Arboretum of Harvard University.

Following his meeting with Augustine Henry in August 1903, David Fairchild, Head of the plant introduction department of the United States Department of Agriculture (ASDA) decided to send plant collectors to China. The first he was to employ was Frank Nicholas Meyer. Copyright: National Agricultural Library, United States Department of Agriculture.

A 2,000 year old tree of *Ginkgo biloba* by a temple at Chongqing in central China. The photograph was published by Augustine Henry in *The Trees of Great Britain and Ireland* which he co-authored with Henry Elwes. Copyright: National Botanic Gardens of Ireland, Glasnevin.

The Wu Gorge, one of the three famous Three Gorges on the Yangtze where Augustine Henry, and later E. H. Wilson, carried out their collecting work. Copyright: Seamus O'Brien.

Members of the 2002 Glasnevin Central China Expedition in the bole of a 1,000 year old *Ginkgo biloba*, near Bai Sha, Hubei. The author is on the right of the second row with Professor Ding Zhaohua from Wuhan Botanical Gardens. Copyright: Seamus O'Brien.

Davidia involucrata, one of China's best-known and best-loved trees in Western gardens. E. H. Wilson travelled to southern Yunnan in 1899 to meet Augustine Henry to get details of where to find this tree. Copyright: Seamus O'Brien.

Lilium henryi, one of Augustine Henry's earliest collections and a popular plant in European and North American gardens. He found it on Moji Shan, near Yichang. Copyright: Seamus O'Brien.

Rhododendron augustinii, one of the most beautiful of all Chinese rhododendron species, it commemorates Augustine Henry from whose specimens it was described as new to science. Copyright: Seamus O'Brien.

A catalogue of the conifers, gymnosperms, flowering plants, ferns, fern allies, bryophytes, lichens and algae collected by Dr. Augustine Henry in China during the years between 1885 and 1900.

Abelia chinensis R. Brown (CAPRIFOLIACEAE)

Forbes & Hemsley in Journ. Linn. Soc. Bot. 23: 358. (1888), Rehder in Sargent Pl. Wilson. 1: 121. (1913), Zheng in Journ. Wuhan Bot. Res. 2(1): 190. (1984).

Hubei: Yichang, 35. Liantuo, 2688 (February 1887), 2087.

A deciduous shrub to 1.5 metres tall, bearing clusters of white fragrant flowers from terminal leaf-axils in summer and autumn. Discovered by Clarke Abel in 1816—1817 in China where it is widely distributed. *Abelia chinensis* was collected by the Glasnevin Central China Expedition (GCCE 597) near Henji Town in Xingshan, Hubei in October 2004. In that region it formed lax, low spreading shrubs and was abundant on hillsides where it filled the air with a powerful fragrance.

Abelia engleriana (Graebn.) Rehder (CAPRIFOLIACEAE) New species

in Sargent Pl. Wilson. 1: 120. (1913), Bean in Trees & Shrubs 1: 137. (1989), O' Brien in Augustine Henry, An Irish Plant Collector in China 9. (2002).

Abelia uniflora Forbes & Hemsley non R. Br. in Journ. Linn. Soc. Bot. 23: 358. (1888) *Linnea engleriana* Graebner in Bot. Jahrb. Syst. 29: 132. (1900).

Hubei: Badong, 1737, 5563b. Jianshi, 5573, **Sichuan:** South Wushan, 5563 (type), 5563a., 5573.

A very handsome deciduous shrub to almost 2 metres tall, bearing in summer a multitude of rose-coloured funnel shape flowers. Introduced to cultivation by Ernest Henry Wilson when collecting for the Arnold Arboretum. Plants cultivated in the old Chinese shrubbery at Glasnevin are Wilson originals.

Abelia graebneriana Rehder (CAPRIFOLIACEAE) New species

in Sargent Pl. Wilson. 1: 118. (1913), O' Brien in Augustine Henry, An Irish Plant Collector in China 9. (2002).

Abelia uniflora Forbes & Hemsley non R. Br. in Journ. Linn. Soc. Bot. 23: 359. (1888).

Hubei: Yichang, 846. Badong, 1768 (collected July 1886), 1808. Liantuo, 1965. Antelope Glen, 3436.

A deciduous shrub to 2.5 metres tall in its native habitat and bearing funnel shaped pink flowers with a yellow throat in June. In Ireland this species is cultivated at the National Botanic Gardens, Glasnevin. Introduced to cultivation by Wilson in 1910.

Abelia macrotera (Graebn. & Buchw.) Rehder (CAPRIFOLIACEAE) New species

in Sargent Pl. Wilson. 1: 126. (1913), Zheng in Journ. Wuhan Bot. Res. 2.(1): 191. (1984).

Linnaea macrotera Graebner & Buchwald in Bot. Jahrb. 29: 131. (1900).

Hubei: Liantuo, 1893, 6398 (type).

Abelia parvifolia Hemsley (CAPRIFOLIACEAE) New species

in Journ. Linn. Soc. Bot. 23: 359. (1888), Bretschneider in Hist. Eur. Bot. Disc. China 2: 785. (1898), Rehder in Sargent Pl. Wilson. 1: 121. (1913), Morley in Glasra 3: 78. (1979), Zheng in Journ. Wuhan Bot. Res. 2.(1): 191. (1984).

Linnaea parvifolia (Hemsl.) Graebner in Bot. Jahrb. 29: 129. (1900) *Abelia longituba* Rehder in Sarg. Pl. Wilson. 1: 126. (1913).

Hubei: Yichang, 664, 1356 (holotype of *Abelia longituba* Rehder), 2337. On the banks of the Yangtze opposite Yichang, 4225. Donghu, 6398.

A deciduous shrub to 2 metres tall, producing cymes of pink to pale-purple tubular flowers in the upper leaf axils from April to August. Distributed in Fujian, Gansu, Guizhou, Hubei, Shaanxi, Sichuan and Yunnan.

Abelia sp. (CAPRIFOLIACEAE)

Anonymous in Kew List of Determinations (volume 24).

Hubei: Yichang, 693.

Abelia uniflora R. Br. (CAPRIFOLIACEAE)

Forbes & Hemsley in Journ. Linn. Soc. Bot. 23: 359. (1888).

Hubei: Antelope Glen, 3642.

Distributed in Fujian, Hubei and Sichuan.

Abelmoschus manihot (L.) Medic. (MALVACEAE)

Hibiscus manihot L. Rehder & E. H. Wilson in Sargent Pl. Wilson 2: 374. (1916).

Hubei: Yichang, 4183. **Yunnan:** Mengzi at 1,600 metres, 9263.

An annual or perennial herb to 2 metres tall, the whole plant is sparsely hirsute. The large yellow flowers are held solitary in leaf axils or at the tips of branches during June and July. Distributed in India and China, a common plant at low altitudes in Hubei, Sichuan and Yunnan.

Abelmoschus manihot (L.) Medik. var. ***pungens*** (Roxb.) Hochr. (MALVACEAE)

Hibiscus pungens L. Anonymous in Kew List of Determinations (volume 34).

Yunnan: Simao, 12543.

Widespread over India, Nepal, China and the Philippines.

Abelmoschus moschatus Medicik. (MALVACEAE)

Hibiscus abelmoschus L. Henry in Trans. Asiat. Soc. Jap. 24 suppl: 21. (1896) (List Pl. Formosa).

Hainan: Environs of Haikou, 7963, 8437. Without locality, 8579. Environs of Kiungchow, 8694. Ling-men (in the interior of the island), 8781. **Taiwan:** Wanchin, 806, 808. Oluanpi, 1265. Kaohsiung Plain, without number.

A tall annual herb to 2 metres, native of the old world tropics, often cultivated or escaped.

Abelmoschus sp. (MALVACEAE)

Hibiscus aff. *tetraphylla* Roxb. ex Hornem. Anonymous in Kew List of Determinations (volume 34).

Yunnan: Simao, 12347.

Abies fargesii Franch. (PINACEAE) New species

Diels in Bot. Jahrb. 29: 218. (1900), Rehder & E. H. Wilson in Sargent Pl. Wilson 2: 48. (1916), Walsh in Aug. Henry. Forst. Herb. 7. (1957).

Abies chensiensis Zheng non Van Teigh in Journ. Wuhan Bot. Res. 2.(1): 8. (1984).

Hubei: Fang Xian, 6881.

The Sichuan fir, a rare tree in Hubei where it grows to about 35 metres tall with girths of up to 4.5 metres. *Abies fargesii* was based on material collected in the 1890s by the French missionary and plant collector, Père Paul Farges, but had been previously discovered by Henry in September 1888. Wilson introduced it from the same region as where Henry had discovered it and stated it was the most common fir there and remnants of old growth forest were still to be found. He introduced it to cultivation through Messrs Veitch in 1901 and it has remained rare in cultivation. The finest specimen in the British Isles and Ireland is a 20 metre tall specimen with a diameter of 59 cm at Castlewellan, County Down, Ireland.

Abildgaardia ovata (Burm. f.) Kral. (CYPERACEAE)

Fimbristylis monostachya (L.) Hassk Henry in Trans. Asiat. Soc. Jap. 24 suppl: 104. (1896) (List Pl. Formosa), C. B. Clarke in Journ. Linn. Soc. Bot. 36: 239. (1903).

Taiwan: Kaohsiung, 732, 759, 1029.

Abrus precatorius L. (FABACEAE)

Henry in Trans. Asiat. Soc. Jap. 24 suppl: 34. (1896) (List Pl. Formosa) Li in Woody Flora of Taiwan 332. fig. 115. (1963).

Taiwan: Kaohsiung, without number.

A pantropic climbing shrub. Common in Taiwan and known as *chi-mu-chu* according to Henry.

Abrus pulchellus Wallich ex Thwaites (FABACEAE)

Anonymous in Kew List of Determinations (volume 34).

Yunnan: Simao, 12470, 13459.

Native to the provinces of Guangxi and Yunnan in south-west China and also distributed in Sri Lanka.

Abutilon gebauerianum Handel-Mazzetti (MALVACEAE) New species

in Symb. Sin. 7: 607. (1933).

Abutilon sinense Oliv. var. *yunnanense* Hochr. in Annuaire Conserv. Jard. Bot. Geneve 21: 447. (1920).

Yunnan: Simao, 11584, 11584b., (type of *Abutilon sinense* Oliv. var. *yunnanense* Hochr.)

Abutilon indicum (L.) Sweet (MALVACEAE)

Henry in Trans. Asiat. Soc. Jap. 24 suppl: 21. (1896) (List Pl. Formosa).

Abutilon asiaticum (L.) G. Don Henry in Trans. Asiat. Soc. Jap. 24 suppl: 20. (1896) (List Pl. Formosa).

Taiwan: Kaohsiung, 292. Oluanpi, 404. Lambay Isle, 1132 (collected August 15th 1893). Wanchin, without number. **Yunnan:** Mengzi, 11316. Simao, 13262.

A erect suffrutescent herb to 1.5 metre tall, common in southern China, including Taiwan, where it grows in thickets or along the roadside. Native to the Old World tropics. This species was collected by the Glasnev in Taiwan Expedition (GTE 47) by the lighthouse at Oluanpi (South Cape of Henry and Wilson) in October 2004.

Abutilon sinense Oliver (MALVACEAE) New species

in Hooker's Icon. Pl. t. 1750. (1888), Henry in Notes on Economic Botany of China 65. (1893), Bretschneider in Hist. Eur. Bot. Disc. China. 2: 780. (1898), Gurke & Diels in Bot. Jahrb. 29: 469. (1900), Henry in Flora & Sylva 1: 218. (1903), Rehder & E. H. Wilson in Sargent Pl. Wilson 2: 373. (1916), Zheng in Journ. Wuhan. Bot. Res. 2.(1): 148. (1984).

Hubei: Xiling (Yichang) Gorge, 3454. Liantuo, 3822. **Yunnan:** Lunan, 10584.

The *yeh ch'ing ma,* a very ornamental shrub to 3 metres tall with golden yellow flowers. Discovered by Henry in

1887, *Abutilon sinense* was later collected by Wilson on the cliffs of the Wu Gorge in 1908 and in Yunnan by George Forrest in April 1906. Henry stated it deserved introduction into Europe as an ornamental plant, seeds from his 3822 were taken from the herbarium specimen at Kew on October 29th 1887, obviously nothing ever became of these, as it has yet to be introduced to cultivation. Endemic to China and distributed in the provinces of Guangxi, Guizhou, Hubei, Sichuan and Yunnan.

Abutilon theophrasti Mediik. (MALVACEAE)

Zheng in Journ. Wuhan Bot. Res. 2.(1): 148. (1984).

Abutilon avincenna Henry non Gaertn. in Journ. China Br. Roy. Asiat. Soc. 22: 257. (1887) (Chinese Names of Plants); in Kew Bull. Misc. Inf. 250. (1891); in Notes Econ. Bot. China 11. (1893).

Hubei: Yichang 292, without locality, 292a.

According to Henry this was the source of abutilon hemp. It was widely cultivated in Hubei and Sichuan, from where large quantities of the *ch'ing ma* or 'hemp' passed through the customs at Yichang and eventually made its way to London where it was known as 'China jute'. This species does not appear to have been collected by Wilson on either of his expeditions for the Arnold Arboretum. An annual summer crop once widely cultivated in western and central China up to an altitude of 1,000 metres. Cultivated at the National Botanic Gardens, Glasnevin.

Acacia confusa Merr. (FABACEAE)

Li in Woody Flora of Taiwan 332. fig. 116. (1963).

Acacia richii Henry non A. Gray in Trans. Asiat. Soc. Jap. 24 suppl: 39. (1896) (List Pl. Formosa).

Taiwan: Oluanpi, 774. Kaohsiung, without number.

An evergreen tree to 15 metres tall with trunks up to one metre across. Native to the northern Philippines and Taiwan. According to Henry the wood of this species is excellent and was used in the construction of junk-frames, rudders, shafts of sugar-mills, etc., and was known locally as *song-si*. The same tree was also collected in Taiwan in 1882 by the Irish collectors, William Hancock and Thomas Watters in the area of Tamsui. *Acacia confusa* is widely cultivated in Hawaii and has became naturalised there. This species was collected by the Glasnevin Taiwan Expedition (GTE 114) on Wanshoushan (Ape's Hill of Henry and Robert Swinhoe) in October 2004. It is cultivated in Ireland at the Talbot Botanic Gardens, Malahide Castle. The form grown there remains a dwarfed shrub to about 70 cm tall.

Acaciella glauca (L.) L. Rico (FABACEAE)

Leucaena glauca (L.) Benth. Henry in Trans. Asiat. Soc. Jap. 24 suppl: 38. (1896) (List Pl. Formosa).

Taiwan: Kaohsiung, cultivated 709.

A small tree, widely distributed in tropical and subtropical regions. Thought to be native to the southern USA, this tree is abundant in southern Taiwan. *Acaciella glauca* was collected by the Glasnevin Taiwan Expedition (GTE 18) by the lighthouse at Oluanpi (South Cape of Henry and Wilson) in October 2004.

Acalypha angatensis Blanco (EUPHORBIACEAE)

Acalypha grandis Henry non Benth. in Trans. Asiat. Soc. Jap. 24 suppl: 84. (1896) (List Pl. Formosa), *Acalypha akoensis* Hayata Li in Woody Flora of Taiwan 411. (1963).

Taiwan: Wanchin, 5, 72, 95.

A shrub found at low altitude in the central and southern regions of Taiwan where it is endemic. First collected there by Henry in 1893.

Acalypha australis L. (EUPHORBIACEAE)

Forbes & Hemsley in Journ. Linn. Soc. Bot. 26: 437. (1894), Pax in Engler, Pflanzenr. 147: 35. (1924).

Acalypha gemina var. *genuina* Henry non Muell.-Arg. in Journ. China Br. Roy. Asiat. Soc. 22: 248. (1887) (Chinese Names of Plants).

Hubei: Yichang, 379, 2799a., 4891. Badong, 2799, 2820.

Acalypha brachystachya Hornem. (EUPHORBIACEAE)

Forbes & Hemsley in Journ. Linn. Soc. Bot. 26: 437. (1894), Pax in Engler, Pflanzenr. 147: 210. (1924).

Sichuan: North Wushan, 7061.

Acalypha indica L. (EUPHORBIACEAE)

Henry in Trans. Asiat. Soc. Jap. 24 suppl: 84. (1896) (List Pl. Formosa).

Taiwan: Kaohsiung, without number.

An annual herb to 80 cm tall with a massive geographical range across tropical Africa, Madagascar, India, Sri Lanka, Thailand, Singapore, Java, the Philippines, China (Taiwan), Japan (south) and the Pacific Islands. In Taiwan this species is mainly found by roadsides and waste land at low altitude.

Acalypha kerrii Craib (EUPHORBIACEAE)

Acalypha acmophylla Hemsley in Journ. Linn. Soc. Bot. 26: 436. (1894), Bretschneider in Hist. Eur. Bot. Disc. China 2: 791. (1898), Hutchinson in Sargent Pl. Wilson 2: 523. (1916), Pax in Engler, Pflanzenr. 147: 137. (1924), Zheng in Journ. Wuhan. Bot. Res. 2.(1): 110. (1984).

Hubei: Yichang, 627, 1188. Liantuo, 3824. Antelope Glen, 3586.

A small shrub to a metre tall, common on the cliffs of the Xiling (Yichang) Gorge during Henry's time in central China. First found by Henry, described from material later collected in Thailand by the Irish botanist Dr Arthur Francis George Kerr (1877—1942).

Acalypha sp. (EUPHORBIACEAE)

Anonymous in Kew List of Determinations (volume

34).

Yunnan: Mengzi, 13776.

Acanthaceae (ACANTHACEAE)

Henry in Trans. Asiat. Soc. Jap. 24 suppl: 70. (1896) (List Pl. Formosa).

Taiwan: Wanchin, 1651. Oluanpi, 1214, 1224, 1283. **Yunnan:** Mengzi, 9022, 10039. Simao, 12915.

Acanthus ilicifolius L. (ACANTHACEAE)

Forbes & Hemsley in Journ. Linn. Soc. Bot. 26: 242. (1890).

Hainan: Environs of Haikou, without number.

Acer albopurpurascens Hayata (SAPINDACEAE) New species

Li in Woody Flora of Taiwan 486. (1963).

Acer oblongum Henry non Wall. ex DC. in Trans. Asiat. Soc. Jap. 24 suppl: 29. (1896) (List Pl. Formosa), *Acer lanceolatum* Rehder non Molliard. in Sargent, Trees & Shrubs. 1: 180. (1905).

Taiwan: Oluanpi, 1257.

A handsome evergreen tree to 15 metres tall, endemic to Taiwan where it was first collected by Henry. Wilson later collected this species at Nantuo (Wilson 10,082).

Acer campbellii Hook. f. & Thoms. ex Hiern (SAPINDACEAE) Rehder in Sargent, Trees and Shrubs 1: 179. (1905).

Yunnan: Mengzi, 9179.

A tree to 30 metres tall in its native habitat. Described from a specimen collected by Griffith in the Sikkim Himalaya and named for Archibald Campbell, a botanical explorer and friend of Sir Joseph Hooker. Native to the Himalaya and western China, Myanmar and northern Vietnam. In Sikkim and Nepal this maple often grows alongside *Magnolia campbellii*.

Acer campbellii Hook. f. & Thoms. ex Hiern subsp. *flabellatum* (Rehder) A. E. Murray (SAPINDACEAE) New subspecies

van Gelderen et al. in Maples of the World 127. (1994).

Acer flabellatum Rehder in Sargent, Trees & Shrubs 1: 161. fig. lxxxi. (1905), Rehder in Fedde, Repert. Spec. Nov. Regni Veg. 1: 7. (1906), Zheng in Journ. Wuhan Bot. Res. 2.(1): 130. (1984), O' Brien in Augustine Henry, An Irish Plant Collector in China 9. (2002). *Acer campbellii* Pax non Hook. f. in Engler, Pflanzenr. 163. 21. (1902), Henry in Gard. Chron. ser. 3, 33: 100. (1903). *Acer campbellii* Hook. f. & Thoms. var. *yunnanense* Rehder in Sargent, Trees & Shrubs 1: 179. (1905); in Fedde Repert. Spec. Nov. Regni Veg. 1:6. (1906).

Hubei: Fang Xian, 6900 (type of *Acer flabellatum* Rehder). **Yunnan:** Mengzi at 2,600 metres, 10495, (type of *Acer campbellii* Hook. f. & Thoms. var. *yunnanense* Rehder).

A small tree to about 6 metres tall, allied to the Himalayan *Acer campbelli*. W. J. Bean states it was probably introduced to cultivation by Wilson in 1907. Native to Hubei, Sichuan and Yunnan and also found in northern Laos and Vietnam. In Ireland this species is cultivated at the John F. Kennedy Arboretum, Mount Usher and Killarney National Park.

Acer cappadocicum Gleditsch subsp. *sinicum* (Rehder) Hand-Mazz. (SAPINDACEAE) New subspecies

van Gelderen in Maples of the World 217. (1994), *Acer cappadocicum* Gleditch var. *indicum* Rehder in Sargent. Pl. Wilson. 1: 86. (1913).

Acer laetum C. A. Meyer var. *cultratum* (Wall.) Pax in Engler, Pflanzenr. 163: 48. (1902) Henry in Gard. Chron. ser. 3, 33: 100. (1903) Veitch in Journ. Roy. Hort. Soc. 29: 354. (1904—1905) *Acer laetum* Zheng non C. A. Meyer in Journ. Wuhan Bot. Res. 2.(1): 131. (1984).

Hubei: South Badong, 5347, 5480. **Yunnan:** Mengzi, 10872, 10877.

A deciduous tree to 30 metres tall, this variety differs from the type (itself from the Himalaya) in its smaller leaves which have longer and narrower lobes. Discovered by Henry in May 1888 and later based on Wilson's material. Wilson also introduced this tree to cultivation through Veitch's Coombe Wood nursery where seedlings were raised in the spring of 1902. Forrest also sent back seeds while based in Yunnan. Native to the Himalaya and China (Hubei, Sichuan and Yunnan), seeds of Wilson's 1009 from western Sichuan reached Glasnevin in 1909. The tallest specimen in Britain or Ireland grows at the Royal Botanic Gardens, Kew. Raised from Wilson's 1358, by 2001 it was 14 metres tall with a girth of 57 cm.

Acer caudatum Wall. var. *multiserratum* (Maxim.) A. E. Murray (SAPINDACEAE)

Zheng in Journ. Wuhan Bot. Res. 2.(1): 128. (1984).

Acer erosum Pax in Hooker's Icon. Pl. 19: t. 1897. (1899), Bretschneider in Hist. Eur. Bot. Disc. China 2: 781. (1898), Pax in Engler, Pflanzenr. 163: 69. (1902), Henry in Gard. Chron. ser. 3, 33: 100. (1903). *Acer multiserratum* Maximowicz in Act. Hort. Petrop. 19: t. 1897. (1889). *Acer caudatum* Wallich var. *prattii* Rehder in Sargent, Trees & Shrubs 1: 178. (1905); in Fedde, Repert. Spec. Nov. Regni. Veg. 1: 7. (1906).

Hubei: Fang Xian, 6788, (type of *Acer erosum* Pax) 6937, (isotype of *Acer erosum* Pax). **Sichuan:** Henry's Chinese collector with Antwerp Pratt, no locality stated but may be any of the following, Leshan, Emei Shan, Wa Shan or Kangding, 8802.

Seeds of *Acer caudatum* var. *multiserratum* Wilson 1161 from Wa Shan in Sichuan were received at Glasnevin in 1909 from the Arnold Arboretum.

Acer cordatum Pax (SAPINDACEAE) New species

in Hooker's Icon. Pl. 19: t. 1897. (1889), Bretschneider in Hist. Eur. Bot. Disc. China 2: 781. (1898), Pax in Engler, Pflanzenreich. Heft 8(iv. 163) 33, fig. 6. (1902), Henry in Gard. Chron. ser. 3, 33: 62. (1903), Rehder in Sargent, Trees & Shrubs 1: 180. (1905), Zheng in Journ. Wuhan Bot. Res. 2.(1): 128. (1984).

Hubei: Jianshi, 7721 (type).

A small tree to about 13 metres tall in its native range of Fujian, Hubei and Zhejiang where it grows in mountainous areas between 500 to 1,200 metres. The original type specimen was destroyed in Berlin during the Second World War.

Acer davidii Franch. (SAPINDACEAE)

Henry in Notes Econ. Bot. China 48. (1893), Rehder in Sargent, Trees & Shrubs 1: 167. fig. lxxxiii. (1905), Zheng in Journ. Wuhan Bot. Res. 2.(1): 129. (1984).

Acer davidii Franch. var. *glabrescens* Pax in Hooker's Icon. Pl. 19: t. 1897. (1889); in Engler, Pflanzenr. 163: 36. (1902), Rehder in Sargent, Trees & Shrubs 1: 167. (1905), Zheng in Journ. Wuhan Bot. Res. 2.(1): 129. (1984). *Acer davidii* Franch. var. *tomentellum* Graf. Pax in Engler, Pflanzenr. 163: 36. (1902).

Hubei: Badong, 3716, 5356. Jianshi, 5827. Fang Xian, 6910. **Sichuan:** South Wushan, 5643, 7199. North Wushan, 7085 (type of *Acer davidii* Franch. var. *glabrescens* Pax).

David's snake-bark maple, discovered in Baoxing in western Sichuan in April 1869 by Père Armand David and later introduced to cultivation by Charles Maries from Yichang in 1879 for Messrs Veitch. Henry stated that from the bark of this tree string and sandals were made, he called it the *ch'ing-hsia-ma*. According to Wilson, David's maple was abundant in the region of Yichang. Wilsons seed collection of this species (Wilson 649 from Fang and Wilson 1005 from Sichuan.) arrived at Glasnevin in 1908 and 1909 respectively from the Arnold Arboretum. Forrest later collected David's maple in the Mekong-Salween divide in October 1918. It was collected by the Glasnevin Central China Expedition (GCCE 796) on Tianthu Shan in Changyang, Hubei in October 2004.

Acer erianthum Schwerin (SAPINDACEAE) New species

in Mitt. Deutsch. Dendrol. Ges. 10: 59. (1901), Pax in Engler, Pflanzenr. Heft 8 (iv. 163), 22. (1902), Rehder in Sargent, Ttrees & Shrubs 159. fig. lxxx. (1905), de Jong et al. in Maples of the World 131. (1994), O'Brien in Augustine Henry, An Irish Plant Collector in China 9. (2002).

Sichuan: Henry's Chinese collector with Pratt, no locality stated but may be Leshan, Emei Shan, Wa Shan or Kangding, 8989 (isotype).

A very handsome slow growing maple to about 7 metres tall, native to the mountains of Sichuan, Guangxi and Hubei at altitudes between 2,000 and 3,000 metres. Introduced to cultivation by Wilson in 1907, it has remained very rare in gardens and is cultivated in Ireland at Birr Castle, the John F. Kennedy Arboretum, Mount Congreve and Glasnevin. The tallest specimen in Britain and Ireland grows at Borde Hill in the UK where it has reached 8 metres tall with a diameter of 0.9 metres.

Acer henryi Pax (SAPINDACEAE) New species

in Hooker's. Icon. Pl. 19: t. 1896. (1889), Bretschneider in Hist. Eur. Bot. Disc. China 2: 781. (1898), Pax in Engler, Pflanzenr. 5: 30. (1902), Henry in Gard. Chron. ser. 3, 33: 100. (1903), Veitch in Journ. Roy. Hort. Soc. 29: 353. (1904—1905), Rehder in Sargent, Trees & Shrubs 1: 181. (1905), Henry in Elwes and Henry, The Trees Gt. Brit. & Irel. 3: 640. (1908), Schneider in Illus. Handb. Laubholzk. 2: 210. (1912), Besant in Gard. Chron. ser. 3, 98: 335. (1935), Morley in Glasra 3: 78. (1979), Zheng in Journ. Wuhan Bot. Res. 2.(1): 130. (1984), Nelson in Pim, The Wood & the Trees 229. (1984), Lancaster in Travels in China 398. (1993), van Gelderen et al. in Maples of the World 172. (1994), Nelson in A Heritage of Beauty 309. (2000), O'Brien in Augustine Henry, An Irish Plant Collector in China 11. (2002), O'Brien in Ir. Garden 13(5): 53. (2004).

Crula henryi (Pax) Nieuwland in Amer. Midland Nat. 2: 142. (1911).

Hubei: Xingshan, 5644a. Without locality, 5644b. **Sichuan:** South Wushan, 5655.

A deciduous tree to about 10 metres tall belonging to the same group as *Acer nikoense* and *Acer cissifolium* though different from other trifoliate maples in the entire margins of leaflets and the stalkless flowers. Introduced to cultivation by Wilson through Veitch's Coombe Wood nursery in 1903, by 1908 plants at Veitch's Coombe Wood nursery were 3 metres (10 feet) tall. On returning to Ireland the Henry's grew this species in their Ranelagh (in the south Dublin suburbs) garden. There is an old Wilson-Veitch tree in the arboretum at Glasnevin that is very handsome in spring when in flower and turns a good shade of amber-red in autumn. *Acer henryi* was reintroduced to cultivation by the Sino-American Expedition from the Shennongjia Forest District in 1980, from this collection there are several fruiting trees in the U.S. National Arboretum, Washington, DC. In Britain and Ireland Henry's maple has remained a rare tree being confined mostly to botanical collections, the tallest specimen recorded is a tree in the superb collection at Caerhays Castle in Cornwall which measured 9 metres tall with a diameter of 0.3 metres in 1984. A tree at Powis Castle, Powis measured 7.5 metres tall with a diameter of 75 cm in 2002 and had a spread of 21.5 metres.

On his six-month expedition in 1888 Henry stated he found no less than sixteen species of *Acer*, of which nine turned out to be new to science. Not only that, but Henry found *Dipteronia*, a distinct new genus, differing from *Acer* in having the fruit winged all around and the large pinnate leaves.

Acer laetum C. A. Meyer (SAPINDACEAE)

Rehder in Sargent, Trees & Shrubs 1: 177. (1905).

Acer truncatum Pax non Bunge in Engler, Pflanzenr. 163: 46. (1902).

Hubei: South Badong, 5410.

Acer laevigatum Wall. (SAPINDACEAE)

Henry in Gard. Chron. ser. 3, 33: 62. (1903), Rehder in Sargent, Trees & Shrubs 1: 180. (1905).

Sichuan: South Wushan, 5538.

An evergreen tree to 15 metres tall, native from the central Himalaya to eastern China and bearing oblong, unlobed, pinnately veined leaves. A tender species best suited to the milder coastal gardens of the British Isles and Ireland.

Acer maximowiczianum Miquel (SAPINDACEAE)

Acer nikoense Bean non Maxim in Trees & Shrubs 1: 214. (1989).

Hubei: Fang Xian, 6599.

A deciduous tree to 15 metres tall, native to Japan and central China though not common in either region. Introduced to cultivation by Maries from Japan in 1881 through Messrs Veitch. Still better known in cultivation as *Acer nikoense*, Miquel's *Acer maximowiczianum* holds priority since Mamimowicz incorrectly diagnosed a species of *Parthenocissus* and named it *Negundo nokoense*. The tallest specimen in Britain and Ireland is a tree (planted in 1904) at Tortworth Court, Glostershire, UK which measured 12 metres tall with a diameter of 58 cm in 1993.

Acer oblongum Wall. ex DC. (SAPINDACEAE)

Pax in Engler, Pflanzenr. 5: 31. (1902), Henry in Journ. China Br. Roy. Asiat. Soc. 22: 236. (1887) (Chinese Names of Plants); in Notes Econ. Bot. China 55. (1893); in Gard. Chron. ser. 3, 33: 62. (1903), Rehder in Sargent, Trees & Shrubs 1: 179. (1905), Zheng in Journ. Wuhan Bot. Res. 2.(1): 132. (1984).

Acer oblongum Wall. ex DC. var. *trilobum* Henry in Gard. Chron. ser. 3. 33: 62. (1903), Rehder in Sargent, Trees & Shrubs 1: 179. (1905) Bean in Trees & Shrubs 1: 215. (1989).

Hubei: Yichang, 1685, Liantuo, 3199, 3795. Without locality, 3199a., (type of *Acer oblongum* Wall. ex DC. var. *trilobum* Henry). **Sichuan:** South Wushan, 7226.

A semi-evergreen tree with a wide distribution across the Himalaya and into China as far east as Taiwan. A common maple on the plains around Yichang according to Henry who stated it was known colloquially as *chi huo shu*. The Himalayan form is tender, Wilson introduced a hardier form from Hubei in 1901.

Acer oblongum Wall. ex DC. var. *concolor* Pax

(SAPINDACEAE) New variety

in Hooker's. Icon. Pl. 19: t. 1897. (1889); in Engler, Pflanzenr. 163: 32. (1902), Rehder in Sargent, Trees & Shrubs 1: 180. (1905), van Gelderen in Maples of the World 201. (1994).

Acer oblongum Wall. var. *glaucum* Graf. Pax in Engler, Pflanzenr. 5: 31. (1902).

Hubei: Changyang, 7677. **Yunnan:** Mengzi, 10957.

Leaves glaucous on both sides. Introduced to cultivation by Wilson from South Wushan in Sichuan (Wilson was en route to the region where Henry had found *Davidia* in 1888 when he collected the seeds of this maple), through Messrs Veitch's Coombe Wood nursery were seedlings were raised in April 1901.

Acer oblongum Wall. ex DC. var. *latialatum* Pax

(SAPINDACEAE) New variety

in Engler, Pflanzenr. 4, Fam. 163: 31. (1902), Rehder in Sargent, Trees & Shrubs 1: 179. (1905), Zheng in Journ. Wuhan Bot. Res. 2.(1): 132. (1984), van Gelderen in Maples of the World 201. (1994).

Hubei: Donghu, 6392 (type).

Wings of fruit broad and almost semicircular.

Acer oliverianum Pax (SAPINDACEAE) New species

in Hooker's Icon. Pl. 19: t. 1897. (1889), Bretschneider in Hist. Eur. Bot. Disc. China 2: 781. (1898), Pax in Engler, Pflanzenreich Heft. 8. (iv. 163), 21. (1902), Henry in Gard. Chron. ser. 3, 33: 100. (1903), Rehder in Sargent, Trees & Shrubs 1: 153. fig. lxxvii. (1905), Henry in Elwes and Henry, The Trees Gt. Brit. & Irel. 3: 638. (1908), Schneider in Illus. Handb. Laubholzk. 2: 206. (1912), Zheng in Journ. Wuhan Bot. Res. 2.(1): 133. (1984), Nelson in Pim, The Wood & the Trees 229. (1984), Bean in Trees & Shrubs 1: 216. (1989), van Gelderen in Maples of the World 135. (1994), Nelson in A Heritage of Beauty 309. (2000), O' Brien in Augustine Henry, An Irish Plant Collector in China 11. (2002).

Hubei: South Badong, 6187, Xingshan, 6512 (type).

A deciduous tree to 9 metres tall, leaves five-lobed and long and pointed, glabrous except for down along the veins. Introduced to cultivation by Wilson in 1901 during his first expedition for Messrs Veitch. Forrest also collected this species in north-west Yunnan in May 1925 (Forrest 26,503). The specific epithet commemorates Professor Daniel Oliver (1830—1916), keeper of the Kew herbarium. According to

02

Bean it is allied to *Acer sinense* but has smaller more finely and evenly toothed leaves. The original type specimen was destroyed in Berlin during World War Two. *Acer oliverianum* is extremly rare in cultivation, the tallest specimen in the British Isles and Ireland grows at Endsleigh, Devon, in 1990 this tree measured 14 metres tall with a diameter of 42 cm.

Acer pectinatum Wall. subsp. *maximowiczii* (Pax) A. E. Murray (SAPINDACEAE) New subspecies

van Gelderen et al. in Maples of the World 155. (1994), O' Brien in Augustine Henry, An Irish Plant Collector in China 11. (2002).

Acer maximowiczii Pax in Hooker's Icon. Pl. 19: t. 1897. (1889) Bretschneider in Hist. Eur. Bot. Disc. China 2: 781. (1898) Pax in Engler, Pflanzenr. 163: 70. (1902) Henry in Gard. Chron. ser. 3, 33: 100. (1903) Schneider in Illus. Handb. Laubholzk. 2: 239. (1912) Zheng in Journ. Wuhan Bot. Res. 2.(1): 131. (1984) Bean in Trees & Shrubs 1: 209. (1989) *Acer urophyllum* Maxim. Rehder in Sargent, Trees & Shrubs 1: 169. fig. lxxxiv. (1905).

Hubei: Fang Xian, 6783, 6857 (type of *Acer maximowiczii* Pax).

A handsome, small deciduous tree, young stems purple-red eventually taking on a snakebark appearance. Leaves three to five-lobed. Introduced to cultivation by Wilson as *Acer laxiflorum* (Wilson 4100). James Veitch writing in the *Journal of the Royal Horticultural Society* (28: 60. 1903—1904) stated that during Henry's six month plant collecting expedition in 1888 he found sixteen distinct species of maple, of which nine were new. The specific epithet commemorates the Russian botanist and plant hunter, Carl Maximowicz (1827—1891). The original type specimen was destroyed in Berlin during World War 2. The tallest specimen in the the British Isles and Ireland grows at Caerhays Castle, Cornwall, in 1984 this tree measured 15 metres tall with a diameter of 32 cm.

Acer shenkanense W. P. Fang ex C. C. Fu (SAPINDACEAE) New Species

Acer laetum C. A. Meyer var. *tricaudatum* Rehder in Sargent, Trees & Shrubs 1: 161. (1905), *Acer cappadocicum* Gleditsch ssp. *sinicum* van Gelderen non (Rehder) Handel-Mazzetti in Maples of the World 218. (1994).

Hubei: Badong, 234a., (type of *Acer laetum* C. A. Meyer var. *tricaudatum* Rehder).

A tree to 15 metres tall, in this form of the var. *sinicum*, leaves have only three lobes. Introduced to cultivation by Wilson in 1901 through Veitch's Coombe Wood nursery. Endemic to Hubei and rare in cultivation. The tallest specimen in Britain or Ireland grows at Westonbirt National Arboretum in Gloustershire where in 2002 it measured

22 metres tall with a diameter of 40 cm. First collected by Henry.

Acer sikkimense Miq. var. *serrulatum* Pax (SAPINDACEAE)

in Engler, Pflanzenr. 163: 34. (1902), Henry in Gard. Chron. ser. 3, 33: 63. (1903), Rehder in Sargent, Trees & Shrubs 1: 180. (1905); in Journ. Arnold Arb. 14: 346. (1933).

Yunnan: Mengzi, 10640.

Relatively new in Western cultivation.

Acer sinense Pax (SAPINDACEAE) New species

in Hooker's Icon. Pl. 19: t. 1897. (1889), Pax in Engler, Pflanzenr. Heft 8 (iv. 163): 22. (1902), Henry in Gard. Chron. ser. 3, 33: 100. (1903), Rehder in Sargent, Trees & Shrubs 1: 155. fig. lxxviii. (1905), Schneider in Illus. Handb. Laubholzk. 2: 205. (1912), Rehder in Sargent, Trees & Shrubs 2: 26. (1913), Zheng in Journ. Wuhan Bot. Res. 2.(1): 133. (1984), O' Brien in Augustine Henry, An Irish Plant Collector in China 12. (2002).

Acer sinense Pax var. *concolor* Pax in Engler, Pflanzenr. 163: 22. (1902), Veitch in Journ. Roy. Hort. Soc. 29: 348. (1904—1905), van Gelderen et al. in Maples of the World 128. (1994), *Acer palmatum* Henry non Thunb. in Gard. Chron. ser. 3, 33: 100. (1903), *Acer campbellii* Hook. f. & Thomson ex Hiern subsp. *sinensis* (Pax) de Jong in Maples of the World 128. (1994).

Hubei: Jianshi, 5831 (type). **Sichuan:** South Wushan, 5641, (isoparatype of *Acer sinense* Pax). North Wushan, 7081 (isotype of *Acer sinense* Pax var. *concolor*). **Yunnan:** Mountains near Mengzi at 1,650 metres, 10244.

A deciduous tree to about 9 metres tall, leaves 5 lobed, to 15 cm long, lobes long and drawn out at points and irregularly toothed. Native to Hubei and Sichuan. Rare in gardens, in Ireland this species is cultivated at Mount Usher, the John F. Kennedy Arboretum and the National Botanic Gardens, Glasnevin.

Acer spp. (SAPINDACEAE)

Anonymous in Kew List of Determinations (volume 24 & 34.).

Hubei: South Badong, 7330. **Yunnan:** Mengzi, 10218.

Acer stachyophyllum Hiern. (SAPINDACEAE)

Pax in Engler, Pflanzenr. 163: 34. (1902), de Jong, et. al. in Maples of the World 167. (1994).

Acer tetramerum Pax in in Hooker's Icon. Pl. 19: t. 1897. (1989), Bretschneider in Hist. Eur. Bot. Disc. China 2: 781. (1898), Pax in Engler, Pflanzenr. 163: 73. (1902), Rehder in Sargent, Trees & Shrubs 1: 171. fig. lxxxv. (1905), Schneider in Illus. Handb. Laubholzk. 2: 245. (1912), Bean in Trees & Shrubs 1: 236. (1989), O' Brien in Augustine Henry, An Irish Plant Collector in China 12. (2002). *Acer tetramerum* Pax var. *elobulatum* Rehder in Sargent Pl. Wilson. 1: 95. (1913), Veitch in Journ. Roy. Hort. Soc. 29:

353. (1904—1905), Zheng in Journ.Wuhan Bot. Res. 2.(1): 133. (1984).

Hubei: South Badong, 5313, (type of *Acer tetramerum* Pax) **Sichuan:** Henry's Chinese collector with Pratt, 8799 (syntype of *Acer tetramerum* Pax var. *elobulatum* Rehder).

A deciduous tree to 9 metres tall. Discovered by William Griffith in Bhutan, it was found again by Henry in central China where it meets its most easterly limits. Wilson sent seeds to Messrs Veitch in 1901. A handsome maple with reddish-purple stems and striate young bark, it is cultivated in several major Irish collections. The tallest specimen in the British Isles and Ireland grows at Westonbirt in Gloucestershire where in 2002 it measured 11 metres tall with a diameter of 51 cm.

Acer stachyophyllum Hiern. subsp. *betulifolium* (Maxim) de Jong (SAPINDACEAE) New subspecies

in Maples of the World 168. (1995), O' Brien in Augustine Henry, An Irish Plant Collector in China 12. (2002).

Acer betulifolium Maxim. in Trudy. Imp. S. Petersburgsk. Bot. Sada. 11: 108. (1890), Bretschneider in Hist. Eur. Bot. Disc. China 2: 781. (1898), Pax in Engler, Pflanzenr. 163: 10. (1902), *Acer tetramerum* Rehder non Pax in Sargent, Trees & Shrubs 1: 181. (1905). *Acer tetramerum* Pax var. *betulifolium* (Maxim.) Rehder in Sargent Pl. Wilson. 1: 95. (1913), Zheng in Journ. Wuhan Bot. Res. 2.(1): 128. (1984).

Hubei: Badong, 515 (September 1885), (isotype of *Acer betulifolium* Maxim.) 1885, (syntype of *Acer betulifolium* Maxim).

Wilson sent seeds of this tree from Sichuan in 1910 while collecting for the Arnold Arboretum under his 4102.

Acer sterculiaceum Wall. (SAPINDACEAE)

Acer villosum Wall. Forbes & Hemsley in Journ. Linn. Soc. Bot. 23: 142. (1886).

Hubei: Badong, 525 (September 1885).

Acer sterculiaceum Wall. subsp. *franchetii* (Pax) A. E. Murray (SAPINDACEAE) New subspecies

van Gelderen et al. in Maples of the World 210. (1994).

Acer franchetii Pax in Hooker's Icon. Pl. 19: t. 1897. (1889), Bretschneider in Hist. Eur. Bot. Disc. China 2: 781. (1898), Pax in Engler, Pflanzenr. Heft 8 (iv. 163) 71. (1902), Henry in Gard. Chron. ser. 3, 33: 100. (1903), Veitch in Journ. Roy. Hort. Soc. 29: 353. (1904—1905), Rehder in Sargent, Trees & Shrubs. 1: 175. fig. lxxxvii. (1905), Schneider in Ill. Handb. Laubholzk. 2: 240. (1912), Zheng in Journ. Wuhan Bot. Res. 2.(1): 130. (1984), O' Brien in Augustine Henry, An Irish Plant Collector in China 9. (2002). *Acer schoenermarkiae* Pax in Engler, Pflanzenreich. 8. (iv. 163): 71. fig. 13. (1902), Henry in Gard. Chron. ser. 3, 33:

100. (1903), Rehder in Sargent, Trees & Shrubs 1: 181. (1905).

Hubei: Xingshan, 6456, (type of *Acer franchetii* Pax). Fang Xian, 6894. **Yunnan:** Mengzi at 2,300 metres, 10497 (type of *Acer schoenermarkiae* Pax).

A deciduous tree to about 12 metres tall, leaves three lobed, outward pointing and triangular. Introduced to cultivation by Wilson through Veitch's Coombe Wood Nursery in 1901. In China this subspecies is native to the mountains of Guangxi, Guizhou, Hubei, Sichuan (Emei Shan), south-east Tibet (Xizang) and Yunnan between 1,600 and 3,000 metres. In Ireland it is cultivated at Dunloe Castle and the John F. Kennedy Arboretum. The tallest specimen in Britain and Ireland grows at the Savill Garden, Windsor, in 2000 it measured 10 metres tall with a diameter of 31 cm.

Acer sutchuenense Franch. (SAPINDACEAE) New species

Henry in Gard. Chron. ser. 3, 33: 100. (1903), J. H. Veitch in Journ. Roy. Hort. Soc. 29: 353. (1904—1905), van Gelderen in Maples of the World 207. (1994).

Sichuan: Wushan, without number.

Discovered by Henry in 1888, this species was described by Franchet from material later collected by Farges in the 1890s and was introduced to cultivation by Wilson from Hubei. Native to Hubei and Sichuan.

Acer tataricum L. (SAPINDACEAE)

Anonymous in Kew List of Determinations (volume 24).

Hubei: Badong, 236 (June 1885). **Hunan:** Shimen, 7557.

Acer tenellum Pax (SAPINDACEAE) New species

in Hooker's Icon. Pl. 19: t. 1897. (1889); in Engler, Pflanzenr. 163: 53. (1902), Henry in Gard. Chron. ser. 3, 33: 100. (1903).

Sichuan: South Wushan, 5612 (type of *Acer tenellum* Pax).

A distinct species with three lobed leaves with long slender petioles. Introduced to cultivation by Wilson in 1901. In Ireland this rare tree is cultivated at Mount Usher and the John F. Kennedy Arboretum. A tree at Glasnevin is the tallest in the British Isles and Ireland and was 16 metres tall with a diameter of 84 cm when measured in 2003.

Acer wilsonii Rehder (SAPINDACEAE) New species

in Sargent, Trees & Shrubs 1: 157. fig. lxxix. (1905), Rehder in Fedde, Repert. Spec. Nov. Regni. Veg. 1: 6. (1906), Schneider in Illl. Handb. Laubholzk. 2: 204. (1912), Bean in Trees & Shrubs 1: 240. (1989), Nelson in A Heritage of Beauty 309. (2000).

Acer campbellii Hook. f. & Thomson ex Hiern subsp. *wilsonii* (Rehder) de Jong in Maples of the World 129. (1994)

Yunnan: Simao, 12044.

02

Wilson's maple, a deciduous tree to 13 metres tall, leaves to 10 cm across and three lobed. Discovered in Yunnan by Henry and later collected in Hubei by Wilson, for whom it was named. Wilson introduced it in 1907 while collecting for the Arnold Arboretum under his 233. Native to Guangdong, Hubei, Yunnan and Zhejiang, it is a rare tree in cultivation where it normally makes a medium sized shrub. The tallest specimen in the British Isles and Ireland grows at Birr Castle, County Offaly, in 2000 this tree measured 9.5 metres tall with a diameter of 27 cm.

Achillea spp. (ASTERACEAE)

Anonymous in Kew List of Determinations (volume 24 & 34).

Hubei: Badong, 122. **Yunnan:** Simao, 12222.

Achillea wilsoniana (W. Heim.) Handel-Mazzetti (ASTERACEAE) New species

Forbes & Hemsley in Journ. Linn. Soc. Bot. 23: 437. (1888).

Achillea siberica Forbes & Hemsley non Ledeb. in Journ. Linn. Soc. Bot. 23: 437. (1888).

Hubei: Badong, 5032. Fang Xian, 6587. Jianshi, 7433.

Discovered by Henry in May of 1888 and based on collections later made by Wilson.

Achudemia hilliana (Handel-Mazzetti) L. F. Fu & Y. G. Wei (URTICACEAE) New species

Pilea hilliana Handel-Mazzetti in Symb. Sin. 7: 129. (1929), *Pilea stipulosa* Forbes & Hemsley non (Miq.) Miq. in Journ. Linn. Soc. Bot. 26: 478. (1899).

Yunnan: Mengzi, 10294, 10295 (type of *Pilea hilliana* Handel-Mazzetti), 11294. In mountains near Mengzi, 11344.

Native to China and Vietnam.

Achudemia japonica Maxim. (URTICACEAE)

Forbes & Hemsley in Journ. Linn. Soc. Bot. 480. (1890).

Hubei: In the Lung tung chi Glen near Yichang, (October 1886), 2896. Liantuo, 4461. **Hainan:** Environs of Kiungchow, 8745.

Native to Russia, China (including Taiwan), Korea and Japan.

Achudemia sp. (URTICACEAE)

Anonymous in Kew List of Determinations (volume 34).

Yunnan: Mengzi, 9771.

Achyranthes aspera L. (AMARANTHACEAE)

Forbes & Hemsley in Journ. Linn. Soc. Bot. 26: 321. (1891), Henry in Trans. Asiat. Soc. Jap. 24 suppl: 75. (1896) (List Pl. Formosa).

Sichuan: South Wushan, 7265. **Taiwan:** Wanchin, 438. Oluanpi, 627. Tamsui, 1749, 1752. Kaohsiung, without number.

According to Henry this was a medicinal plant known as *cho-pi-ts'ao*.

Achyranthes bidentata Blume (AMARANTHACEAE)

Henry in Journ. China Br. Roy. Asiat. Soc. 22: 261. (1887) (Chinese Names of Plants), Forbes & Hemsley in Journ. Linn. Soc. Bot. 26: 322. (1891), Henry in Trans. Asiat. Soc. Jap. 24 suppl: 75. (1896) (List Pl. Formosa).

Hubei: Yichang, 985, 2200, 2222, 3227, 3244. Lung Tung Chi Glen, (October 1886) 2925. Liantuo, 3075, without locality 3244a. San You Dong Glen, 4371. **Taiwan:** Wanchin, 1544.

A perennial herb to 80 cm tall, native to much of tropical Asia. Henry stated it was known colloquially in the Three Gorges area as *niu hsi* (niu xi i.e. ox's knee) the plant is used in Chinese herbal medicine to invigorate the blood and to stregnthen the sinews and bones. In Taiwan this species inhabits the edges of forests and roadsides in the central and northern parts of the island.

Achyranthes sp. (AMARANTHACEAE)

Anonymous in Kew List of Determinations (volume 34).

Yunnan: Mengzi, 10318. Simao 12492.

Acilepis aspera (Buch.-Ham.) H. Rob. (ASTERACEAE)

Vernonia aspera Buch.-Ham. Anonymous in Kew List of Determinations (volume 34).

Yunnan: Mengzi, 9131. Simao, 12707.

Native to Myanmar, western China, Thailand, Laos and Vietnam.

Acilepis clivorum (Hance) H. Rob. (ASTERACEAE)

Vernonia clivorum Hance Anonymous in Kew List of Determinations (volume 34).

Yunnan: Simao, 11692, 12714.

Native to Myanmar and China (Guangdong and Yunnan).

Acilepis nantcianensis (Pamp.) H. Rob. (ASTERACEAE)

Vernonia bracteata Forbes & Hemsley non Wallich in Journ. Linn. Soc. Bot. 23: 400. (1888).

Hubei: Yichang, 296, 296a., 958, 2306a. Badong, 521 (September 1885). Liantuo, 2176. Xingshan, 4529. San You Dong Glen, 2306. **Sichuan:** South Wushan, 7187.

A very handsome perennial to 1.5 metres tall, a common species in the Three Gorges region, especially in the mountains to the north of the Yangtze. Cultivated at the National Botanic Gardens, Glasnevin from collections made by the Glasnevin Central China Expedition in Xingshan (GCCE 506) and on Tianthu Shan in Changyang (GCCE 821). Distributed in Hubei and Sichuan.

Acilepis saligna (DC.) H. Rob. (ASTERACEAE)

Vernonia saligna DC. Anonymous in Kew List of Determinations (volume 34).

Yunnan: Mengzi, 9841.

Distributed in India, Bangladesh, Nepal, Myanmar, China, Thailand and Vietnam.

Acilepis silhetensis (DC.) H. Rob. (ASTERACEAE)

Vernonia bracteata Wall. ex C. B. Clarke Anonymous in Kew List of Determinations (volume 34).

Yunnan: Simao, 12498.

Acilepis squarrosa D. Don (ASTERACEAE)

Vernonia teres Wall. Anonymous in Kew List of Determinations (volume 34).

Yunnan: Mengzi, 9504.

Native to India, the Himalaya, Myanmar, China (Yunnan), Thailand, Cambodia and Vietnam.

Acmella calva (de Candolle) R. K. Jansen. (ASTERACEAE)

Spilanthes callimorpha A. H. Moore in Proc. Amer. Acad. Arts 42: 536. (1907), Fedde in Repert. Nov. Spec. Regni Veg. 5: 323. (1908).

Yunnan: Near Simao, 12260 (isotype of *Spilanthes callimorpha* A. H. Moore), 12260a., (holotype of *Spilanthes callimorpha* A. H. Moore).

Aconitum carmichaelii Debx. (RANUNCULACEAE)

Handel-Mazzetti in Acta Hort. Gotob. 13: 117. (1939), Fletcher & Lauener in Notes Roy. Bot. Gard. Edinburgh 25: 200. (1939—1949).

Aconitum fischeri Henry non Reichb. in Journ. China Br. Roy. Asiat. Soc. 22: 280. (1887) (Chinese Names of Plants); in Notes Econ. Bot. China 30. (1893).

Hubei: Badong, 121, 2855,. Liantuo, 3041. Without locality, 3041a., 3041b.

This species occured wild near Yichang in mountain meadow according to Henry. The specific epithet commemorates Dr. J. R. Carmichael of the hospital of the London Mission Society. According to Bretschneider, Carmichael arrived at Guangzhou (formerly Canton) in 1862. He also married his eldest sister. Henry stated Carmichael's aconite was known colloquially in the Three Gorges region as *wu tu* (it is also the *fu zi*) and the root is used as a drug. According to Henry large quantities of these roots were exported from Sichuan. In China the tubers have been used topically as analgesic, anti-neuralgic, anti-inflammitory and antipyretic. The most common poisons used on the spear-tips of indigenous tribes in both the Old and New World are derived from various species of *Aconitum*. In the National Museum of Scotland is a crossbow and arrow that was once used by the Lisu ethnic group who inhabited the Salween valley in western Yunnan. These were donated to the museum by Forrest who traversed this region in 1905, Forrest stated the arrows of split bamboo were coated with *Aconitum* poison (10 grams of *Aconitum* tincture is enough to kill a human). *Aconitum carmichaelii*

was collected by the Glasnevin Central China Expedition (GCCE 298) on the hills above Badong in September 2002.

Aconitum hemsleyanum E. Pritzel (RANUNCULACEAE)
New species

in Bot. Jahrb. Syst. 29: 329. (1900), Handel-Mazzetti in Acta Hort. Gotob. 13: 123. (1939), Fletcher & Lauener in Notes Roy. Bot. Gard. Edinburgh 25: 202. (1939—1949), O' Brien in Augustine Henry, An Irish Plant Collector in China 12. (2002).

Aconitum volubile Henry non Pall. ex Koelle in Journ. China Br. Roy. Asiat. Soc. 22: 272. (1887) (Chinese Names of Plants); in Notes Econ. Bot. China 31. (1893).

Hubei: Badong, 2458, 4997. South Badong 7336. Liantuo, 2596, 4477. Fang Xian, 6773, 6773a., 6646 (isotype).

The *ch'uan wu tu*, a climbing, herbaceous perennial with large showy, rich voilet-blue flowers. Wilson sent roots from central China in 1901, these failed to grow but a second consignment sent in 1903 fared better and plants flowered at Veitch's Coombe Wood nursery in the summer of 1905. In *Veitch's Novelties for 1908—1909* plants sold for three shillings and six pence each. Named in honour of the keeper of the Kew herbarium, William Botting Hemsley, a personal friend of Henry. While traveling on the 2004 Glasnevin Central China Expedition we found this to be locally abundant in open glades in Shennongjia Forest District. An important medicinal plant in China.

Aconitum henryi E. Pritzel (RANUNCULACEAE)
New species

in Bot. Jahrb. Syst. 29: 329. (1900), Fletcher & Lauener in Notes Roy. Bot. Gard. Edinburgh 25: 202. (1939—1949).

Aconitum henryi E. Pritzel var. *compositum* Hand.-Mazz. in Acta Horti Gothob. 13: 130. (1939).

Hubei: Xingshan, 6979. North Badong, 7012 (isosyntype), 7012a. (isosyntype). Sichuan: North Wushan, 7017 (type of *Aconitum henryi* E. Pritzel var. *compositum* Hand.-Mazz.).

A handsome lax-habited monkshood carrying 1 metre-tall spikes of mid-blue flowers in autumn. This species was also later collected by Farges in Sichuan. Material was also collected by the Glasnevin Central China Expedition (GCCE 779) in October 2004 on Tianthu Shan, Changyang in Hubei.

Aconitum ichangense (Fin. & Gagnep.) Handel-Mazzetti RANUNCULACEAE) New Species

in Acta. Hort. Gotob. 13: 111. (1939).

Aconitum semigaleatum Pall. var. *ichangense* Fin. & Gagnep. in Bull. Soc. Bot. Fr. li. 511. (1904).

Hubei: Xingshan, 6976, (type of *Aconitum semigaleatum* Pall. var. *ichangense* Fin. & Gagnep.).

Aconitum scaposum Franch. (RANUNCULACEAE)

Aconitum jucundum Diels Handel-Mazzetti in Acta. Hort. Gotob. 13: 78. (1939). *Aconitum vaginatum* E. Pritz. Handel-Mazzetti in Acta. Hort. Gotob. 13: 77. (1939), Fletcher & Lauener in Notes Roy. Bot. Gard. Edinburgh 25: 205. (1939—1949).

Hubei: Yichang, 1678. Badong, 4900. Xingshan, 6547. Fang Xian, 6547a., 6547b., 6828.

Aconitum sinomontanum Nakai (RANUNCULACEAE)

Housed in the Harvard University Herbarium (GH).

Aconitum excelsum Handel-Mazzetti non Rchb. in Acta. Hort. Gotob. 13: 82. (1939), Fletcher & Lauener in Notes Roy. Bot. Gard. Edinburgh 25: 201. (1939—1949).

Hubei: Jianshi, 5904. Donghu, 6426. Xingshan, 6426a. Without locality, 5904a.

A magnificent monkshood to fully 3 metres tall. While travelling on the Glasnevin Central China Expedition through Shennongjia Forest District in northwest Hubei in September 2004 we met with several colonies of this species, most of which were cultivated for their medicinal tubers.

Aconitum spp. (RANUNCULACEAE)

Anonymous in Kew List of Determinations (volume 24).

Hubei: Badong, 876. Liantuo, 2075, 3064. Xingshan, 6464. Fang Xian, 6671, 6867, 6870, 6967. **Sichuan:** South Wushan, 7632.

Aconitum sp. no. 1. (RANUNCULACEAE)

Aconitum, sect. lycoctonum Henry in Notes Econ. Bot. China 32. (1893).

Hubei: Donghu, 6426. Badong, 4901, 4668. South Badong, 7307. Jianshu, 7423. **Sichuan:** South Wushan, 7244.

An important medicinal plant which Henry stated was known as the *hao-erh-wu*.

Aconitum sp. no 2. (RANUNCULACEAE)

Aconitum uncinatum Henry non L. in Notes Econ. Bot. China 31. (1893).

Hubei: Badong, 534 (September 1885).

Aconitum vilmorinianum Komar. (RANUNCULACEAE)

Handel-Mazzetti in Acta Hort. Gotob. 13: 131. (1939), Fletcher & Lauener in Notes Roy. Bot. Gard. Edinburgh 25: 205. (1939—1949).

Yunnan: Mengzi, 9825.

A climbing perennial with twining stems to 4 metres long. The purple-blue flowers are carried during August, September and October. Distributed in Guizhou, Sichuan and Yunnan. The root has medicinal uses. Described from plants that first flowered at the Vilmorin Nursery, France, in 1908, presumably raised from seeds sent by one of the many French missionaries based in western China at the time.

Acorus calamus L. (ACORACEAE)

Henry in Journ. China Br. Roy. Asiat. Soc. 235. (1887) (Chinese Names of Plants), Brown in Journ. Linn. Soc. Bot. 26: 187. (1903).

Hubei: Liantuo, 4506.

The sweet flag, a reed-like aquatic perennial, indigenous to most of the north temperate zones (Europe, Asia and North America) and in the Asian tropics from India to New Guinea. In China the rhizomes are used to treat indigestion, flatulence, to stimulate appetite and to relieve stomache ache. *Acorus calamus* was found in Tutankhamen's tomb in Egypt. According to Henry this species was known colloquially at Yichang as *ch'ang pu*. To the Hani (an ethnic group) of Xishuangbanna in southern Yunnan this species is called *lanxal* and is used as an antiseptic.

Acorus gramineus Ait. (ACORACEAE)

Henry in Trans. Asiat. Soc. Jap. 24 suppl: 101. (1896) (List Pl. Formosa), Brown in Journ. Linn. Soc. Bot. 26: 187. (1903), Engler, in Engler, Pflanzenr. iv. 23B: 312. (1905).

Hubei: Yichang, 727, 7857. Liantuo, 1904. Antelope Glen, 3370. Changyang, 5246. **Taiwan:** Wanchin, 142. **Yunnan:** In Pingbian in forests on the Daweishan Range at 1,650 metres, 10840. Simao 11691.

A grass-like herb with aromatic creeping rhizomes which are used for medicinal purposes. This species was known as the *shui chang p'u* around Yichang according to Henry. Native to India, China (including Taiwan), Korea, Japan, the Philippines and Indonesia.

Acroglochin persicarioides (Poir.) Moq. (CHENOPODIACEAE)

Acroglochin chenopoidioides Schrad. Henry in Journ. China Br. Roy. Asiat. Soc. 22: 248. (1887) (Chinese Names of Plants), Forbes & Hemsley in Journ. Linn. Soc. Bot. 26: 323. (1891).

Hubei: Badong, 423 (June 1885), 2382, 4906, 5053, 5143. Fang Xian, 6774. South Badong, 7348. Liantuo, 4470. **Sichuan:** South Wushan, 7045.

Known colloquially in the Three Gorges region as *yeh han ts'ai* according to Henry, the leaves and shoots of this species are eaten as a vegetable.

Acronychia pedunculata (L.) Miq. (RUTACEAE)

Acronychia laurifolia Blume Anonymous in Kew List of Determinations (volume 24 & 34).

Hainan: Hummocks near Haikou, 8091. Environs of Haikou, 8409. 8220. At Nodoa in the interior of the island, 8458. Without locality, 8552. **Yunnan:** Simao, 12263.

A shrub or small to large tree (depending on habitat), a widespread species native to the Himalaya east to China (Taiwan) and south to Sri Lanka, Java and New Guinea.

Acrorumohra diffracta (Baker) H. Ito (DRYOPTERIDACEAE)
New species

Monachosorum subdigitatum Christ non. (Bl.) Kuhn. in Bull. Herb. Boiss. 6: 868. (1898). *Nephrodium diffractum* Baker in Kew Bull. Misc. Inf. 230. (1898), Christ in Bull. Herb. Boiss. 7: 17. (1899).

Yunnan: In Pingbian in woods on the Daweishan Range at 2,600 metres, 9028, (type of *Nephrodium diffractum* Baker).

A beautiful fern to about 40 cm tall, distributed across Indochina, mainland China and Taiwan.

Actaea brachycarpa (P. K. Hsiao) J. Compton (RANUNCULACEAE)

Compton in Taxon 47: 593. (1998).

Hubei: Yichang, 6073a.

Actaea dahurica (Turcz. Ex Fisch. & C. A. Mey) Franch. (RANUNCULACEAE)

Cimicifuga dahurica (Turcz. Ex Fisch. & C. A. Mey) Maxim. Anonymous in Kew List of Determinations (volume 24).

Hubei: Badong, 1875, 2247. **Sichuan:** South Wushan 7261.

Actaea japonica Thunb. (RANUNCULACEAE)

Cimicifuga japonica (Thunb.) Spreng. Henry in Notes Econ. Bot. China 59. (1893).

Hubei: Badong, 4763, 4853.

According to Henry this was the *chu-ken-ch'i*, the rhizome of which is a drug.

Actaea purpurea (P. K. Hsiao) J. Compton (RANUNCULACEAE)

Compton in Taxon 47: 593. (1998).

Hubei: South Badong, 6063 (May 1888).

Actaea simplex (DC.) Prantl. (RANUNCULACEAE)

Compton in Taxon 47: 593. (1998).

Cimicifuga simplex (DC.) Wormsk. ex Turcz. Yang & Compton in Bot. Bull. Acad. Sin. 36: 356. (1995).

Hubei: Yichang, 415. Badong, 4867, 4903, 4990, 6073. Fang at 1,900 to 3,000 metres, 6073b., 6812.

Distributed through Siberia, Mongolia, Korea, southwest, central and northern China and Japan.

Actaea sp. (RANUNCULACEAE)

Cimicifuga sp. Anonymous in Kew List of Determinations (volume 24).

Hubei: Badong, 4935.

Actaea spicata L. (RANUNCULACEAE)

Diels in Bot. Jahrb. 29: 326. (1900).

Hubei: Badong, 4038. Changyang, 5257. Donghu (July 1888) 5257a. South Badong (April 30th 1888) 5257c. **Sichuan:** South Wushan (May 26th 1888) 5257b. Henry's Chinese collector with Pratt, no locality stated but may be any of the following, Leshan, Emei Shan, Wa Shan or Kangding 8789.

Actinidia arguta (Sieb. & Zucc.) Planch. ex Miq. (ACTINIDIACEAE)

Actinidia rufa Pl. var. *typica* Dunn in Journ. Linn. Soc. Bot. 39: 402. (1911). *Actinidia purpurea* Rehder in Sargent Pl. Wilson.2: 378. (1916); in Journ. Arnold Arb. 15: 96. (1934), Hand.-Mazz. in Symb. Sin. 7: 390. (1931), Nakai in Tokyo Bot. Mag. 47: 253. (1933).

Sichuan: South Wushan, 5622 (type of *Actinidia rufa* Pl. var. *typica* Dunn). **Yunnan:** Mengzi at 2,100 metres, 11008. Mengzi, 9694.

A climbing shrub to 10 metres tall, a widespread species found across eastern Siberia, China, Korea and Japan. Seeds of Wilson's 1314, collected to the south-east of Kangding in October 1908 reached Glasnevin through the Arnold Arboretum.

Actinidia callosa Lindl. (ACTINIDIACEAE)

Henry in Trans. Asiat. Soc. Jap. 24 suppl: 20. (1896) (List Pl. Formosa), Dunn in J. Linn Soc. Bot. 39: 405. (1911), Rehder in Sargent Pl. Wilson. 2: 382. (1916), Li in Journ. Arnold Arb. 33: 46. (1952).

Actinidia callosa Lindl. var. *formosana* Finet & Gagnep. Dunn in Journ. Linn. Soc. Bot. 39: 406. (1911). *Actinidia callosa* Lindl. var. *typica* Dunn in Journ. Linn. Soc. Bot. 39: 405. (1911). *Actinidia lutea* Stapf in Bot. Mag. sub. t. 9140, in adnot. (1928).

Taiwan: Tamsui, 1388. Wanchin, 1650. **Yunnan:** South of the Red River from Manpan, 10056 (isotype of *Actinidia lutea* Stapf), 10056a. South of the Red River from Manpan at 2,300 metres 10056b (isotype of *Actinidia lutea* Stapf). Mengzi at 1,600 metres 10824, 10824a.

A deciduous climbing shrub to almost 10 metres tall, bearing in summer fragrant creamy-white flowers to 4 cm wide followed by small red spotted fruits in autumn. First known in the Himalaya, it was described by Lindley from a specimen collected in northern India. *Actinidia callosa* was found by Griffith in Bhutan, Wallich in Nepal and Hooker in the Sikkim Himalaya. Henry found in China and it was also later collected there by Forrest. This species is also found in Indochina and Malaysia.

Actinidia callosa Lindl. var. *henryi* Maxim. (ACTINIDIACEAE) New variety

in Acta. Hort. Petrop. 11: 36. (1890), Leveille in Fl. Kouy-Tcheou 413. (1915), Rehder in Sargent Pl. Wilson 2: 382. (1916), Zheng in Journ. Wuhan Bot. Res. 2.(1): 149. (1984).

Actinidia curvidens Dunn in Kew Bull. Misc. Inf. 1. (1906), Hand.-Mazz. in Symb. Sin. 7: 390. (1931).

Hubei: Antelope Glen, 3471, 3494 (co-type of *Actinidia curvidens* Dunn), 4377 (co-type of *Actinidia* *curvidens* Dunn). In a ravine on Moji Shan (the Dome), 3564. Yichang, 3955 (type). Badong, 724 (May 1885). Jianshi, 5797 (March 1888), 6010. Without locality, 4377a. **Sichuan:** South Wushan, 7243.

A climber to 7 metres tall with ovoid to elongated fruits. Judging by Wilson's collections this was a common plant in western Hubei at the turn of the 20th century. He collected it around Yichang and Fang Xian, and in Sichuan on Emei Shan.

Actinidia callosa Lindl. var. *pubescens* Dunn (ACTINIDIACEAE)

Li in Journ. Arnold Arb. 33: 49. (1952).

Actinidia callosa forma. Dunn in Journ. Linn. Soc. Bot. 26: 406. (1911).

Yunnan: Mengzi, 10780.

Actinidia chinensis Planch. (ACTINIDIACEAE)

Henry in Notes Econ. Bot. China 46. (1893); in Flora & Sylva 1: 218. (1903), Dunn in Journ. Linn. Soc. Bot. 39: 408. (1911), Rehder in Sargent Pl. Wilson 2: 385. (1916), Besant in Gard. Chron. ser. 3, 98: 335. (1935), Li in Journ. Arnold Arb. 33: 55. (1952), Zheng in Journ. Wuhan Bot. Res. 2.(1): 149. (1984), Nelson in Pim, The Wood & the Trees 229. (1984), Nelson in A Heritage of Beauty 309. (2000), O' Brien in Augustine Henry, An Irish Plant Collector in China 12. (2002).

Hubei: Yichang, 1166. Badong, 1754, Liantuo, 2076. Jianshi, 5834, 5834a. Without locality, 5834b. **Sichuan:** South Wushan, 5574.

The *yang-t'sao*, the Yichang or Chinese gooseberry, a vigorous climbing shrub first collected by Robert Fortune in 1847. The commercial possibilities of the Yichang gooseberry were promoted by Henry who sent the first (pickled) fruits to Europe. Writing in *Flora & Sylva* in 1903 he stated:

Actinidia chinensis produces in the wild state excellent fruit about the size of a big plum. This climber would be perfectly hardy in this country, and the fruit would be a great acquisition, I think.

In the past it has been claimed that both Henry's and Wilson's collections were the kiwi fruit, now much cultivated in New Zealand. It is not, (though the fruits are similar) the kiwi is derived from *Actinidia chinensis* var. *deliciosa* and neither Henry or Wilson for that matter have any associations with that variety. In Chinese herbal medicine the dried roots of this species are used to treat rheumatic arthritis and cancer. This species was collected by the Glasnevin Central China Expedition (GCCE 252 & GCCE 267) on the hills above Badong in September 2002. Widely distributed in western, central, eastern and southern China.

Actinidia henryi Dunn (ACTINIDACEAE) New species

in Kew Bull. Misc. Inf. 1. (1906,) exclude specimens A. Henry 11307, 13335.; in Journ. Linn. Soc. Bot. 39: 407. (1911), Besant in Gard. Chron. ser. 3, 98: 335. (1935), Li in Journ. Arnold Arb. 33: 12. (1952), Bean in Trees & Shrubs 1: 244. (1989), Nelson in A Heritage of Beauty 309. (2000).

Yunnan: In Pingbian, on the Daweishan Range at 1,650 metres, 10381 (isotype). Mengzi, 10381a., (type).

A vigorous climber, young growth ribbed and covered in dense reddish bristles. Small white flowers are produced in the leaf axils in late spring followerd by small cylindrical fruits in autumn. *Actinidia henryi* was introduced to cultivation by Wilson while collecting for the Arnold Arboretum and Augustine and Alice Henry grew this in their garden at Ranelagh in the south Dublin city suburbs in the early 20th century.

Actinidia kolomikta (Rupr. & Maxim) Maxim. (ACTINIDIACEAE)

Rehder in Sargent Pl. Wilson 2: 381. (1916).

Actinidia gagnepainii Nakai in Bot. Mag. Tokyo 47: 258. (1933) *Actinidia kolomikta* (Rupr. & Maxim) Maxim. var. *gagnepainii* (Nakai) Li in Journ. Arnold Arb. 33: 19. (1952)

Hubei: South Badong, 7334. Changleping 7637. **Sichuan:** Henry's Chinese collector with Pratt, no locality stated but may be Leshan, Emei Shan, Wa Shan or Kangding, 8806, 8994.

Actinidia latifolia (Gardn. & Champ.) Merr. (ACTINIDIACEAE)

Li in Journ. Arnold Arb. 33: 50. (1952), Li in Woody Flora of Taiwan 571. (1963).

Actinidia championi Henry non Benth. in Trans. Asiat. Soc. Jap. 24 suppl: 20. (1896) (List Pl. Formosa).

Taiwan: Wanchin, 825.

A climbing shrub to 7 metres tall, native to western and southern China (including Taiwan), Vietnam, Malaysia and Cambodia. In Taiwan this species grows in thickets at low altitude.

Actinidia latifolia (Gardn. & Champ.) Merr. var. *mollis* (Dunn) Handel-Mazzetti (ACTINIDIACEAE) New variety

in Symb. Sin. 7: 391. (1931).

Actinidia championii Benth. var. *mollis* Dunn in Journ. Linn. Soc. Bot. 39: 407. (1911).

Yunnan: Simao, 12041, (isotype of *Actinidia championii* Benth. var. *mollis* Dunn).

Endemic to southern Yunnan.

Actinidia melanandra Franch. (ACTINIDIACEAE) New species

Rehder in Sargent Pl. Wilson. 2: 378. (1916), Li in Journ. Arnold Arb. 33: 27: (1952), Zheng in Journ. Wuhan Bot. Res. 2.(1): 149. (1984).

Actinidia rufa Pl. var. *parviflora* Dunn in Journ. Linn. Soc. Bot. 39: 403. (1911) in part.

Hubei: Jianshi, 5938. Fang at 2,250 metres, 6794. Without locality, 5938a., (type of *Actinidia rufa* Pl. var. *parviflora* Dunn), 5938b.

Described by Franchet from material gathered by the French missionary, Farges, this vigorous deciduous climber had been discovered by Henry several years previously. It was introduced to cultivation by Wilson while collecting for the Arnold Arboretum.

Actinidia polygama (Sieb. & Zucc.) Planch. ex Maxim (ACTINIDIACEAE)

Rehder in Sargent Pl. Wilson. 2: 380. (1916).

Actinidia rufa Pl. var. *parviflora* Dunn in Journ. Linn. Soc. Bot. 39: 403. (1911), in part. *Actinidia lecomptei* Nakai in Bot. Mag. Tokyo 47: 253. (1933). *Actinidia polygama* (Sieb. & Zucc.) Maxim. var. *lecomptei* (Nakai) Li in Journ. Arnold Arb. 33: 22. (1952).

Hubei: Badong, 1788. Jianshi, 5922. Fang Xian, 6644. Zigui, 5922b. **Sichuan:** South Wushan, 5764, 5922a.

A vigorous climber to about 7 metres tall. Fragrant white flowers are produced in spring, followed by soft translucent fruits in autumn. The fruits are a handsome feature but with a disagreeable taste. Nakai based his *Actinidia lecomptei* on collections made by David, Farges and Henry in central China.

Actinidia rubricaulis Dunn (ACTINIDIACEAE) New species

in Kew Bull. Misc. Inf. 2. (1906); in J. Linn. Soc. Bot. 39: 405. (1911), Li in Journ. Arnold Arb. 33: 35. (1952), O' Brien in Augustine Henry, An Irish Plant Collector in China 12. (2002).

Yunnan: To the south of the Red River in mountain forest on Fengchunling at 2,300 metres, 10696 (type). In forests near Mengzi 11334 (collected 15th November 1898).

In Ireland this species is cultivated at the National Botanic Gardens, Glasnevin.

Actinidia rudis Dunn (ACTINIDIACEAE) New species

in Journ. Linn. Soc. Bot. 39: 408. (1911), Li in Journ. Arnold Arb. 33: 8. (1952).

Actinidia henryi Dunn in Kew Bull. Misc. Inf. 1. (1906), in part, quoad A. Henry 11307, 13335.

Yunnan: In mountain forest to the south of Mengzi at 2,000 metres, 11307, (isotype). Yuanjiang 13325.

Alongside A. Henry 10381, Stephen Troyte Dunn originally cited the two fruiting specimens, A. Henry 11307 and 13335 as his *Actinidia henryi*. Subsequently he established *Actinidia rudis* on the same two collections.

Actinidia spp. (ACTINIDIACEAE)

Anonymous in Kew List of Determinations (volume 24).

Hubei: Yichang, 3468. Badong 4075. Changleping, 6289. Fang Xian, 6617. **Sichuan:** South Wushan, 5719.

Actinidia tetramera Maxim. (ACTINIDIACEAE)

Rehder in Sargent Pl. Wilson. 2: 381. (1916), Zheng in Journ. Wuhan Bot. Res. 2.(1): 150. (1984).

Hubei: Fang Xian, 6821.

Allied to both *Actinidia polygama* (Sieb. & Zucc) Maxim and *Actinidia kolomikta* (Maxim. & Rupr.) Maxim.

Actinidia trichogyna Franch. (ACTINIDIACEAE) New species

Rehder in Sargent Pl. Wilson. 2: 384. (1916), Li in Journ. Arnold Arb. 33: 43. (1952).

Sichuan: South Wushan, 7135.

Franchet based this species on Farges' 370 from Sichuan during the early 1890s. Henry had discovered it in May of 1888.

Actinodaphne cupularis (Hemsl.) Gamble (LAURACEAE) New species

in Sargent Pl. Wilson. 2: 75. (1916), Allen in Ann. Missouri Bot. Gard. 25: 405. (1938), Zheng in Journ. Wuhan Bot. Res. 2.(1): 48. (1984).

Litsea cupularis Hemsley in Journ. Linn. Soc. Bot. 26: 380. (1891), Bretschneider in Hist. Eur. Bot. Disc. China 2: 790. (1898). *Actinodaphne lancifolia* Zheng non (Sieb. & Zucc.), Meissn. in Journ. Wuhan Bot. Res. 2.(1): 48. (1984).

Hubei: Antelope Glen, 3473 (syntype), 4382. Yichang 3240 (syntype), 4320 (syntype). Antelope Glen 4370 (isosyntype). Liantuo 4584 (syntype). South Donghu 7711 (syntype). Changyang 7750 (syntype). Without locality 3473a, 3473b. **Sichuan:** North Wushan 7122 (syntype).

Actinodaphne forrestii (C. K. Allen) Kosterm. (LAURACEAE) New species

Actinodaphne reticulata Liou Ho non Meissn. in Laurac. Chine & Indochine 158. (1934), *Actinodaphne reticulata* Meissn. var. *forrestii* C. K. Allen in Ann. Missouri Bot. Gard. 25: 412. (1938).

Yunnan: Mengzi, 11436, (syntype of *Actinodaphne reticulata* Meissn. var. *forrestii* C. K. Allen). Yuanjiang, 13323, (type of *Actinodaphne reticulata* Meissn. var. *forrestii* C. K. Allen, collected December 28th 1899).

Allen based her description of this tree on material collected by Henry and Forrest in Yunnan. Endemic to China and distributed in the montane forests of Guangxi, Guizhou and Yunnan. Henry described his 13323 from Yuanjiang as a 15 metre tall tree with white flowers.

Actinodaphne henryi Gamble (LAURACEAE) New species

in Kew Bull. Misc. Inform. 265. (1913), Liou Ho in Laurac. Chine & Indoch. 157. (1934), Allen in Ann. Missouri Bot. Gard. 25: 407. (1938).

Yunnan: Simao at 1,650 metres, 11799, (type). Simao at 1,300 metres 11799a., (isotype).

02

Endemic to Yunnan.

Actinodaphne longipes Kosterm. (LAURACEAE)

Actinodaphne reticulata Meissn. var. *glabra* Meissn. Liou Ho in Laurac. Chine & Indochine 158. (1934).

Yunnan: South of the Red River, 9805.

Actinodaphne pilosa (Lour.) Merr. (LAURACEAE)

Allen in Ann. Missouri Bot. Gard. 25: 404. (1938).

Yunnan: Simao, 13588.

Distributed in southern China, Laos and Vietnam.

Actinodaphne spp. No. 1. (LAURACEAE)

Anonymous in Kew List of Determinations (vol. 24).

Hubei: Badong, 5171. **Hainan:** Ling-men (in the interior of the island), 8676.

Actinodaphne spp. No. 2. (LAURACEAE)

Henry in Trans. Asiat. Soc. Jap. 24 suppl: 79. (1896) (List Pl. Formosa).

Taiwan: Wanchin, 1599.

Actinoschoenus yunnanensis (C.B.Clarke) Y. C. Tang (CYPERACEAE) New species

Fimbristylis yunnanensis C. B. Clarke in Journ. Linn. Soc. Bot. 36: 247. (1903).

Yunnan: Mengzi at 1,650 metres, 11150 (in part).

Distributed in China (Yunnan), Thailand and Vietnam.

Actinostemma sp. (CUCURBITACEAE)

Anonymous in Kew List of Determinations (volume 34).

Yunnan: Mengzi, 10478. Simao 12277, 13542.

Actinostemma tenerum Griff. (CUCURBITACEAE)

Actinostemma lobatum (Maxim.) Maxim. ex Fr. & Sav. Henry in Trans. Asiat. Soc. Jap. 24 suppl: 46. (1896), (List Pl. Formosa).

Taiwan: Wanchin, on the plain, 1666.

A slender pilose vine, distributed over much of India, China, Indochina, Korea and Japan.

Adenocaulon bicolor Hook. (ASTERACEAE)

Forbes & Hemsley in Journ. Linn. Soc. Bot. 23: 432. (1888), Henry in Notes Econ. Bot. China 29. (1893).

Hubei: Badong, 4800. Fang Xian, 6633. South Badong, 7477. **Sichuan:** North Wushan, 6633a. **Hunan:** Shimen, 7555. **Yunnan:** Mengzi, 9199.

Henry stated that in the mountainous areas of Hubei this species was called *hu-lu-yeh*.

Adenophora capillaris Hemsley (CAMPANULACEAE) New species

in Journ. Linn. Soc. Bot. 26: 10. (1889), Bretschneider in Hist. Eur. Bot. Disc. China 2: 786. (1898), Nelson in A Heritage of Beauty 310. (2000).

Hubei: Badong, 954, 4799. Xingshan, 6476. Fang Xian, 6667.

James Veitch in *Hortus Veitchii* described this species

as 'a hardy herbaceous perennial about 2 feet (60cm.) high, with loose, graceful panicles of blue campanula-like flowers.' Introduced from Henry's type locality (Badong) by Wilson in 1901, plants first flowered in Veitch's Coombe Wood nursery in the summer of 1905. While travelling on the Glasnevin Central China Expedition through Shennongjia Forest District in the autumn of 2004 we found this to be a relatively common inhabitant of damp limestone cliffs where it carried a profusion of tiny blue bell-shaped flowers on pendelous stems, a beautiful plant deserving reintroduction to cultivation.

Adenophora petiolata Pax & K. Hoffm. subsp. **hunanensis** (Nannf.) D. Y. Hong & S. Ge (CAMPANULACEAE)

Adenophora polymorpha Henry non Ledeb. in Journ. China Br. Roy. Asiat. Soc. 22: 269. (1887) (Chinese Names of Plants), Forbes & Hemsley in Journ. Linn. Soc. Bot. 26: 11. (1889).

Hubei: Yichang, 975a. Badong, 1052 (December 1885), 2409, 2879 (October 1886), 3156. 4866, 4880, 4992. Liantuo, 2685. Xingshan, 6519.

A very beautiful perennial to 90 cm tall. This species was collected on both the 2002 and 2004 Glasnevin Central China Expeditions. According to Henry, in the Three Gorges region it was known colloquially as *sha shen*. *Adenophora petiolata* subsp. *hunanensis* was introduced to cultivation by Wilson and first flowered at the Royal Botanic Gardens, Kew in September 1903 (as *Adenophora polymorpha* Ledeb.) A common plant in Hubei.

Adenophora rupincola Hemsley (CAMPANULACEAE) New species

in Journ. Linn. Soc. Bot. 26: 13. (1889), Bretschneider in Hist. Eur. Bot. Disc. China 2: 786. (1898).

Adenophora pubescens Hemsley in Journ. Linn. Soc. Bot. 26: 12. (1889), Bretschneider in Hist. Eur. Bot. Disc. China 2: 786. (1898).

Hubei: Badong, 1 (collected September 1885, received at Kew 12th March 1886). On the banks of the Yangtze at Yichang, 4360 (type). Liantuo, 2178, (type of *Adenophora pubescens* Hemsley), 4476.

Adenophora sp. No. 1. (CAMPANULACEAE)

Anonymous in Kew List of Determinations (volume 24).

Hubei: Badong, 532, 536 (September 1885), 5153. Yichang, 975, 2157. Liantuo, 2119. Fang Xian, 6636, 6677. **Sichuan:** South Wushan, 7202.

Adenophora sp. No. 2. (CAMPANULACEAE)

Adenophora verticillata Henry non Fisch. in Trans. Asiat. Soc. Jap. 24 suppl: 56. (1896) (List Pl. Formosa).

Taiwan: Oluanpi, 988, 1291, 1364.

Adenophora sp. No. 3. (CAMPANULACEAE)

Adenophora latifolia Anonymous non Fisch. in Kew List of Determinations (volume 34).

Yunnan: Mengzi, 9273, 9680.

Adenostemma lavenia (L.) Kuntze (ASTERACEAE)

Adenostemma viscosum Henry non J. R. Forst. & G. Forst. in Trans. Asiat. Soc. Jap. 24 suppl: 51. (1896) (List Pl. Formosa).

Taiwan: Kaohsiung, without number. **Yunnan:** Mengzi, 9829. Simao, 12538.

Adenostemma sp. (ASTERACEAE)

Adenostemma grandiflorum Anonymous non Benth. in Kew List of Determinations (volume 24).

Hainan: Hummocks near Haikou, 8102, Environs of Haikou, 8167. Environs of Kiungchow, 8297.

Adenostemma viscosum J. R. Forst. & G. Forst. (ASTERACEAE)

Forbes & Hemsley in Journ. Linn. Soc. Bot. 23: 403. (1888).

Hubei: Yichang, 73, 2359. Changyang, 7484.

Adiantum capillus-junonis Ruprecht (ADIANTACEAE)

Christ in Bull. Soc. Bot. France 5: 61. (1905), Diels in Bot. Jahrb. 29: 200. (1900), Henry in Trans. Asiat. Soc. Jap. 24 suppl: 110. (1896) (List Pl. Formosa).

Hubei: On Tsui Fu Shan near Yichang (May 1888), 1620 (in part). On a stone ditch at Pingshanba near Yichang, 1680. **Taiwan:** Wanshoushan, without number.

Distributed through China and Japan.

Adiantum capillus-veneris L. (ADIANTACEAE)

Baker in Journ. Bot. 26: 225. (1888), Henry in Trans. Asiat. Soc. Jap. 24 suppl: 110. (1896) (List Pl. Formosa), Christ in Bull. Herb. Boiss. 7: 8. (1899), Diels in Bot. Jahrb. 29: 201. (1900).

Hubei: Liantuo, 2000. Antelope Glen, 3449 (in part). In caves in the San You Dong Glen in the Xiling Gorge, 5959. **Taiwan:** Tamsui, 1409. Kaohsiung, without number. **Yunnan:** Mengzi at 1,650 metres, 10264.

According to Wilson there was a grotto-like cave full of stalagmites in the San You Dong Glen, (a rugged valley with cliffs 160 metres sheer at the mouth of the Xiling Gorge) and its roof was one mass of maiden-hair fern. These were detached with a piece of limestone and sold along the lower reaches of the Yangtze as far as Shanghai where they were known as Ichang fern-stones. Widely dispersed through tropical and temperate parts of the world.

Adiantum caudatum L. (ADIANTACEAE)

Henry in Trans. Asiat. Soc. Jap. 24 suppl: 110. (1896) (List Pl. Formosa).

Taiwan: Wanchin, 1546. Kaohsiung, without number. **Yunnan:** Simao, 13119.

Native to Africa, the Himalaya, India, Myanmar, central to southern China (including Taiwan), Indochina, Malay Peninsula, the Philippines, Malay Isl. and New Guinea. A widespread species in Asia.

Adiantum edgeworthii Hook. (ADIANTACEAE)

Anonymous in Kew List of Determinations (volume 34).

Yunnan: Simao, 12821.

A handsome little fern distributed across the Himalaya, India, Myanmar, China, Japan and the Philippines. The specific epithet was coined by Hooker to commemorate the Irish amateur botanist, M. Pakenham Edgeworth from Edgeworthstown in County Longford. Edgeworth served as a civil servant in India and made several notable discoveries there.

Adiantum flabellulatum L. (ADIANTACEAE)

Henry in Trans. Asiat. Soc. Jap. 24 suppl: 110. (1896) (List Pl. Formosa).

Taiwan: Tamsui, 1420.

Distributed over much of India, Sri Lanka, south-east China, Indochina, southern Japan and Malesia.

Adiantum hispidulum Sw. (ADIANTACEAE)

Henry in Trans. Asiat. Soc. Jap. 24 suppl: 110. (1896) (List Pl. Formosa).

Taiwan: Wanchin, 1488. **Yunnan:** Yuanjiang, 13224.

Distributed through east Africa, southern India, China (including Taiwan), the Malay Peninsula, the Philippines, Polynesia, Java and New Zealand.

Adiantum myriosorum Baker (ADIANTACEAE) New species

in Kew Bull. Misc. Inf. 230. (1898).

Yunnan: On the mountains to the south of Mengzi at 1,950 metres, 9266 (type).

A beautiful fern, (similar to *Adiantum pedatum*), distributed across central and southern China and Myanmar. This species was collected in the Shennongjia Forest District, Hubei by the Sino-American Expedition in 1980.

Adiantum pedatum L. (ADIANTACEAE)

Baker in Journ. Bot. 26: 225. (1888).

Hubei: Badong, 3688. Xingshan at 1,400 metres, 6557.

This species was collected by the Sino-American Expedition in Shennongjia in 1980, it was also collected by the Glasnevin Central China Expedition in Dragon Gate River Forest Park in Xingshan, Hubei in October 2004.

Adiantum philippense L. (ADIANTACEAE)

Adiantum lunatum Henry non Burm. f. in Trans. Asiat. Soc. Jap. 24 suppl: 110. (1896) (List Pl. Formosa). *Adiantum lunulatum* Burm. f. Anonymous in Kew List of Determinations (volume 34).

Taiwan: Kaohsiung, without number. Wanchin, without

02

number. **Yunnan:** South of the Red River from Manmei at 1,650 metres, 9468.

Distributed through tropical and tropical regions of the Old World.

Adiantum soboliferum Wall. ex Hook. (ADIANTACEAE)

Adiantum balansae Baker Christ in Bull. Soc. Bot. France 5: 61. (1905).

Hubei: Yichang, 1934.

Adina pilulifera (Lam.) Franch. (RUBIACEAE)

Adina pilulifera (Lam.) Franch. var. *tonkinense* (Pitard.) Merr. Li in Journ. Arnold Arb. 25: 317. (1944).

Yunnan: In Pingbian in forests on the Daweishan Range at 1,640 metres, 13466.

Adina racemosa (Sieb. & Zucc.) Miq. (RUBIACEAE)

Henry in Journ. China Br. Roy. Asiat. Soc. 22: 271. (1887) (Chinese Names of Plants), Forbes & Hemsley in Journ. Linn. Soc. Bot. 23:370. (1888), Li in Woody Flora of Taiwan 846. fig. 340. (1963).

Adina mollifolia Hutchinson in Sargent Pl. Wilson. 3: 391. (1916), Li in Journ. Arnold Arb. 25: 317. (1944). *Sinoadina racemosa* (Sieb. & Zucc.) Ridst. Zheng in Journ. Wuhan Bot. Res. 2.(1): 187. (1984).

Hubei: Yichang, 1557, 2236. In a cave in the San You Dong Glen, 2893. Liantuo, 6381. **Taiwan:** Oluanpi, without number. Kaohsiung, 710. Hengchun, 710c. **Yunnan:** Mountains to the west of Simao at 1,320 metres, 12853. Mountains to the west of Simao at 1,640 metres, 11888, (type of *Adina mollifolia* Hutchinson). Near Tonguan at 1,640 metres, 13265.

A small tree, common on the hills near Yichang. Henry stated it was known there as *shui tung kua*.

Adina rubella Hance (RUBIACEAE)

Forbes & Hemsley in Journ. Linn. Soc. Bot. 23: 371. (1888), Zheng in Journ. Wuhan Bot. Res. 2.(1): 188. (1984).

Hubei: Yichang, 50. Liantuo, 3197.

A bush to about 1.5 metres tall, discovered by Theo. Sampson near Macao in June 1864. Common along the tributaries of the Yangtze in the Three Gorges region. This species was abundant just above river-level in the lesser Three Gorges between Wushan and Dachang, a region visited by the 2002 Glasnevin expedition to central China. These colonies are now submerged beneath the waters of the new Three Gorges reservoir. Plants presently cultivated at the National Botanic Gardens, Kilmacurragh were raised from seeds collected by the Glasnevin Central China Expedition (2004) at Gaojiayan town, between Yichang and Changyang in October 2004 (GCCE 745). The dried roots, leaves, flowers and fruits of this species are currently used in Chinese herbal medicine to treat cold and fever, mumps and sore-throat.

Adina spp. (RUBIACEAE)

Anonymous in Kew List of Determinations (volume 24 & 34).

Hainan: At Nodoa in the interior of the island, 8464. Without locality, 8570. **Yunnan:** Mengzi, 9012. Simao, 13279.

Adinandra cf. *griffithii* Dyer (PENTAPYLACACEAE)

Anonymous in Kew List of Determinations (volume 34).

Yunnan: Simao, 12028.

Adinandra formosana Hayata (PENTAPYLACACEAE)

in Journ. Coll. Sci. Tokyo 22: 45. (1906), Li in Woody Flora of Taiwan 576. (1963).

Adinandra sp. Henry in Trans. Asiat. Soc. Jap. 24 suppl: 19. (1896) (List Pl. Formosa). *Adinandra millettii* Henry non (Hook. & Arn.) Benth & Hook. f in Trans. Asiat. Soc. Jap. 24 suppl: 19. (1896) (List Pl. Formosa). *Adinandra millettii* (Hook. & Arn.) Benth & Hook. f. var. *formosana* (Hayata) Kobuski in Journ. Arnold Arb. 28: 14. (1947).

Taiwan: Wanchin, 514, 1583. Oluanpi, 981, 1985, 2058.

A small tree, endemic to Taiwan where it grows at low to medium altitude. It appears to have been first found in Taiwan by Richard Oldham, Hayata based the species on Henry's collections.

Adinandra formosana Hayata var. *obtusissima* (Hayata) Keng (PENTAPYLACACEAE) New variety

Adinandra sp. Henry in Trans. Asiat. Soc. Jap. 24: suppl: 19. (1896) (List Pl. Formosa), *Adinandra millettii* Henry non (Hook. & Arn.) Benth & Hook. f. in Trans. Asiat. Soc. Jap. 24 suppl: 19. (1896) (List Pl. Formosa) *Adinandra millettii* (Hook. & Arn.) Benth. & Hook. f. var. *obtusissima* (Hayata) Kobuski in Journ. Arnold. Arb. 28: 577. (1947).

Taiwan: Oluanpi, 1985.

A small tree, endemic to Taiwan where it is known only from the southern part of the island.

Adinandra sp. (PENTAPYLACACEAE)

Adinandra millettii (Hook. & Arn.) Benth & Hook. f Anonymous in Kew List of Determinations (volume 24).

Hainan: Without locality, 8564.

Aegiceras corniculatum (L.) Blanco (PRIMULACEAE)

Aegiceras majus Gaertn. Anonymous in Kew List of Determinations (volume 34).

Yunnan: Mengzi, 9516.

Distributed in India, Sri Lanka, southern China, Vietnam and the Philippines.

Aeginetia indica L. (OROBANCHACEAE)

Henry in Trans. Asiat. Soc. Jap. 24 supp: 68. (1896) (List Pl. Formosa) Beck-Mannagetta in Engler, Pflanzenr. 261: 19. (1930).

Taiwan: Kaohsiung, without number. Wanchin, 2.

Aeginetia sp. (OROBANCHACEAE)

Anonymous in Kew List of Determinations (volume 34).

Yunnan: Mengzi, 9258.

Aegopodium henryi Diels (APIACEAE) New species

in Engl. Bot. Jahrb. 29: 497. (1900), Wolff in Engler, Pflanzenr. iv. 228: 330. (1927).

Hubei: Badong, 4946 (type).

Aerva sanguinolenta (L.) Bl. (AMARANTHACEAE)

Aerva scandens (Roxb.) Wall. Henry in Trans. Asiat. Soc. Jap. 24 suppl: 75. (1896) (List Pl. Formosa).

Taiwan: Wanchin, 1619.

A climbing undershrub, native to the tropics of Africa, Malaysia, southern China (including Taiwan) and the Philippines. In Taiwan this taxon is mainly found on the southern part of the island.

Aerva sp. (AMARANTHACEAE)

Anonymous in Kew List of Determinations (volume 34).

Yunnan: Mengzi, 9389, 10846. Simao, 11634.

Aeschynanthus acuminatus Wall. ex A. DC. (GESNERIACEAE)

Henry in Trans. Asiat. Soc. Jap. 24 suppl: 68. (1896) (List Pl. Formosa).

Taiwan: Oluanpi, 1363. Yunnan: Mengzi, 10449.

A widespread species with a large range over Bhutan, Nepal, Myanmar, China, Thailand, Laos and Vietnam.

Aeschynanthus buxifolius Hemsley (GESNERIACEAE) New species

in Journ. Linn. Soc. Bot. 35: 515. (1903).

Yunnan: Mengzi at 2,300 metres, 11217.

An evergreen shrub to 30 cm tall, carrying from June to September purple-red flowers in pairs in the leaf-axils of the upper parts of stems. The species was based on Hancock's material from the Red River valley and Henry's Mengzi collection. Native to southern China and Vietnam.

Aeschynanthus humilis Dunn (GESNERIACEAE) New species

in Journ. Linn. Soc. Bot. 35: 516. (1903).

Yunnan: Epiphytic on trees in a forest near Simao at 1,800 metres, 13204.

China (Southern Yunnan) and Indochina.

Aeschynanthus sp. (GESNERIACEAE)

Anonymous in Kew List of Determinations (volume 34).

Yunnan: Mengzi, 9221, 11386. Simao, 11984, 12184, 12247, 12533, 12637, 12857, 12868, 12995.

Aeschynanthus superba C. B. Clarke (GESNERIACEAE)

Anonymous in Kew List of Determinations (volume 34).

Yunnan: Simao, 12811.

Aeschynomene indica L. (FABACEAE)

Forbes & Hemsley in Journ. Linn. Soc. Bot. 23: 170. (1887), Henry in Trans. Asiat. Soc. Jap. 24 suppl: 32. (1896) (List Pl. Formosa).

Hubei: In glens near Yichang, 1642. Hainan: Environs of Kiungchow, 8777. Taiwan: Kaohsiung, without number.

An erect annual herb to 90 cm tall, pantropical in distribution and often found growing near paddy fields where it is used as a green manure.

Aeschynomene sp. (FABACEAE)

Anonymous in Kew List of Determinations (volume 24).

Hainan: On Liang Shan, 16 km (10 miles) east of Haikou, 8201.

Aesculus chinensis Bunge var. *wilsonii* (Rehder) Turland & N. H. Xia (SAPINDACEAE) New variety

Aesculus wilsonii Rehder in Sargent Pl. Wilson. 1: 498. (1913). *Actinotinus sinensis* Oliver in Hooker's Icon. Pl. 18: t. 1740. (1888), Forbes & Hemsley in Journ. Linn. Soc. Bot. 23: 357. (1888). *Aesculus chinensis* Henry non Bunge in Notes Econ. Bot. China 53. (1893), Bretschneider in Hist. Eur. Bot. Disc. China 2: 784. (1898), Henry in Elwes & Henry in the Trees Gt. Brit. & Irel. 2: 427. (1907), Schneider in Illus. Handb. Laubholzk. 2: 249. fig. 173. (1912), Morley in Glasra 3: 78. (1979), Pim in The Wood & the Trees 32. (1984).

Hubei: Badong, 4058 (type of *Actinotinus sinensis* Oliver). Sichuan: South Wushan, 5892, 7203.

Henry called this fine tree the *hou-pan-li* or 'monkey-chestnut'. Initially confused with *Aesculus chinensis* Bunge, Wilson's horse-chestnut was discovered about 1887 by Henry's Badong collector. There are magnificent trees on Emei Shan in Sichuan and in the Sanxia Botanical Gardens near Yichang in Hubei.

In May of 1887 Henry's Badong collector decided to play a practical joke on Henry and taxonomists at Kew by inserting the inflorescence of *Viburnum plicatum* var. *tomentosum* into the terminal bud of the as-then-unknown, *Aesculus wilsonii*. The specimen reached Kew where with great excitment Daniel Oliver declared it a hitherto unknown genus, *Actinotinus sinensis*. This proved to be one of the most bizarre specimens to have been sent to Kew, a tree with large digitate horsechestnut-like leaves, yet with flowers clearly belonging to the Viburnaceae. Henry's Badong man described *Actinotinus* as 'a tree 10 feet or more tall on the high mountains and stated it was very rare there. The hoax was only noticed after the genus was published and Henry heard of it from Charles Ford of Hong Kong Botanic Gardens while he was recuperating from malaria following a three month stay in Hainan.

Other hoaxes from this mischievous collector included A. Henry 6261 from Changleping in Hubei which was noted in the Kew List as; "Flowers of a *Viburnum* stuck into a branch of a *Rhus*. - (a fraud)." Henry's 7566 (from Shimen in Hunan province) proved to be a species of *Alangium* with the flowers of a *Lespedeza* inserted. Hemsley described *Rhododendron aucubifolium* from the foliage of *Daphniphyllum macropodum* with the flowers of *Rhododendron stamineum*! Henry's 5787 and 6432 were of the same combination. Whatever sort of pranks he may have carried out and despite the embarrassment he caused Henry, this Badong man was a first-rate collector and his specimens were always of a very high standard.

Aesculus chinensis var. *wilsonii* was collected by the Glasnevin Central China Expedition (GCCE 336) in the Forest Reserve at Sanxia Botanic Gardens near Yichang in October 2004. In *The Trees of Great Britain and Ireland* Augustine states that he sucessfully introduced this species to cultivation through the Royal Botanic Gardens, Kew in 1888, but all plants had died within a year or so. The 'Kew Inwards' book, which records donations of seeds from Henry between 1885 and 1900 makes dismal reading, Henry sent hundreds of packets of seeds yet the vast majority of these appear to have died. The French missionaries had the same problem with the Jardin des Plantes in Paris and it wasn't until the nursery firms, Messrs. Lemoine, Vilmorin and Messrs. Veitch became involved in the introduction of hardy Chinese plants that any rate of sucess was achieved. *Aesculus chinensis* var. *wilsonii* was later successfully introduced in 1908 by Wilson (for whom it was named) and first flowered at Caerhays Castle, Cornwall in 1934.

Aganosma cymosa (Roxb.) G. Don (APOCYNACEAE)
Tsiang in Sunyatsenia 2: 154. (1934)

Ichnocarpus cymosus (Roxb.) Mottet Anonymous in Kew List of Determinations (volume 34).

Yunnan: In woods near Mengzi at 1,600 metres, 10806. Simao, 11980. In forests near Simao at 1,300 metres, 12407a.

Aganosma schlechteriana H. Levl. (APOCYNACEAE)
Aganosma schlecteriana H. Leveille var. *leptanthum* Tsiang in Sunyatsenia 4: 34. (1939).

Yunnan: Near Simao at 1,900 metres, 11862 (syntype of *Aganosma schlecteriana* H. Leveille var. *breviloba* Tsiang). In forests near Simao at 1,300 metres, 12047 (holotype of *Aganosma schlecteriana* H. Leveille var. *leptanthum* Tsiang).

Agapetes mannii Hemsley (ERICACEAE)
In the Harvard University Herbarium (A.).

Yunnan: On rocky mountains near Mengzi at 2,300 metres, 11218, 11219.

An epiphytic shrub to 1.5 metres tall with a swollen turnip-like base. Distributed in India, Myanmar, Thailand and China (Yunnan). *Agapetes mannii* was introduced to cultivation from the Khasia hills by Gustav Mann through the Royal Botanic Gardens, Kew in 1889.

Agastache rugosa (Fisch.& Mey.) Kuntze (LAMIACEAE)
Lophanthus rugosus Fisch. & Mey. Henry in Journ. China Br. Roy. Asiat. Soc. 22: 243. (1887) (Chinese Names of Plants), Forbes & Hemsley in Journ. Linn. Soc. Bot. 26: 288. (1890), Dunn in Notes Roy. Bot. Gard. Edinburgh 6: 165. (1915).

Hubei: Badong, 424 (June 1885), 4779, 5066. **Sichuan:** South Wuhan, 7228. **Yunnan:** In forests on the slopes of Fengchunling at 2,300 metres, 11215.

According to Henry this species was known colloquially in the Three Gorges region as *ho hsiang* (*huo xiang*). This species is currently used in Chinese herbal medicine to supress abdominal pain, headache, diarrhea, fever and to stop vomiting and control morning sickness. Cultivated in the order beds at the NBG, Glasnevin from seeds collected in October 2004 by the Glasnevin Central China Expedition.

Agave angustifolia Haw (ASPARAGACEAE)
Agave rigida Mill. Henry in Trans. Asiat. Soc. Jap. 24 suppl: 97. (1896) (List Pl. Formosa).

Taiwan: Kaohsiung, 1016.

Mexico and Central America.

Ageratum conyzoides L. (ASTERACEAE)
Henry in Trans. Asiat. Soc. Jap. 24 suppl: 51. (1896) (List Pl. Formosa).

Hainan: Without locality, 8377. **Taiwan:** Wanchin, 424. Kaohsiung 1198a. **Yunnan:** Mengzi, 9094.

A widespread pantropical weed, annual or a sub-shrub where conditions are favourable.

Aglaia elaeagnoidea Benth. (MELIACEAE)
Aglaia sp. Henry in Trans. Asiat. Soc. Jap. 24 suppl: 26. (1896) (List. Pl. Formosa). *Aglaia formosana* (Hayatya) Hayata Li in Woody Flora of Taiwan 395. (1963).

Taiwan: Oluanpi, 602, 673.

Tropical and subtropical Asia to Vanuatu. This species was collected by the Glasnevin Taiwan Expedition (GTE 50) by the lighthouse at Oluanpi (South Cape of Henry and Wilson) in October 2004.

Aglaia lawii (Wight) Saldanha ex Ramamoorthy (MELIACEAE)
Aglaia yunnanensis Li in J. Arn. Arb. 25: 305. (1944) *Aglaia attenuata* H. L. Li in Journ. Arnold Arbor. 25: 303. (1944).

Yunnan: Mountains to the south of Simao at 1,400

metres, 12170, (syntype of *Aglaia attenuata* H. L. Li). Simao at 1,450 metres, 12228, (syntype of *Aglaia attenuata* H. L. Li). In forests near Simao at 1,650 metres, 12228a.

Aglaia odorata Lour. (MELIACEAE)

Henry in Trans. Asiat. Soc. Jap. 24 suppl: 26. (1896) (List Pl. Formosa).

Hainan: Environs of Haikou, 8163. Ling-men (in the interior of the island) 8673. **Taiwan:** Kaohsiung, without number. Wanchin, without number.

Known as *shu-lan* or *ch'iu-lan* according to Henry. The flowers were used on the mainland for scenting tea. According to Robert Fortune the following plants were used to scent tea - rose, plum, orange, *Jasminum sambac* Ait; *Jasminum lanceolaria* Roxb; *Aglaia odorata* Lour; *Osmanthus fragrans* Lour; *Gardenia jasminoides* J. Ellis. Henry stated that in the returns of the Imperial Chinese Maritime Customs Service the dried flowers of this species were known as *lan-hua-mi*.

Aglaia rimosa (Blanco) Merr. (MELIACEAE)

Agalia roxburghiana Henry non W. & A. in Trans. Asiat. Soc. Jap. 24 suppl: 26: (1896) (List Pl. Formosa. *Agalia* sp. Henry in Trans. Asiat. Soc. Jap. 24 suppl: 26. (1896) (List Pl. Formosa). *Aglaia elliptifolia* Merr. Li in Woody Flora of Taiwan 393. (1963).

Taiwan: Oluanpi, 260, 1298.

An evergreen shrub or small tree, native from China (Taiwan) to Papuasia.

Aglaia sp. (MELIACEAE)

Anonymous in Kew List of Determinations (volume 34).

Yunnan: Simao, 12996.

Aglaonema sp. (ARACEAE)

Anonymous in Kew List of Determinations (volume 34).

Yunnan: Simao, 12287.

Agrimonia pilosa Ledeb. (ROSACEAE)

Agrimonia eupatoria Henry non L. in Journ. China Br. Roy. Asiat. Soc. 22: 270. (1887) (Chinese Names of Plants).

Hubei: Yichang, 402. Badong, 282 (July 1885), 1834, 4839.

A hirsute perennial to 50 cm tall, native to eastern Europe, Siberia, China, Korea and Japan. Known colloquially in the Three Gorges region as *she ke ta*, according to Henry the roots of this plant entered into the composition of wine-ferment. It is currently used in Chinese herbal medicine to relieve dysentry, to arrest bleeding and to counteract toxicity.

× *Agropogon lutosus* (Poir.) P. Fourn. (POACEAE)

Polypogon × littoralis Sm. Henry in Journ. China Br. Roy. Asiat. Soc. 22: 263. (1887) (Chinese Names of Plants),

Rendle in Journ. Linn. Soc. Bot. 36: 386. (1904).

Hubei: Yichang, 190. **Yunnan:** Near Mengzi in a dry rice field at 1,500 metres, 10246. Near Mengzi, on rocks in a streamlet at 1,500 metres, 10246a. By a streamlet near Mengzi at 1,500 metres, 10556.

Agrostis stolonifera × Polypogon monspliensis. Henry stated it was known colloquially at Yichang as *pang t'ou ts'ao*.

Agrostis canina L. (POACEAE)

Rendle in Journ. Linn. Soc. Bot. 36: 389. (1904).

Hubei: Yichang, 245. Hills behind Yichang, 3464.

Agrostis micrantha Steud. (POACEAE)

Rendle in Journ. Linn. Soc. Bot. Bot. 36: 390. (1904).

Agrostis milioides C. C. Mez in Repert. Spec. Nov. Regni Veg. 17: 302. (1921). *Agrostis perennans* Rendle non (Walter) Tuck in Journ. Linn. Soc. Bot. 36: 390. (1903). *Agrostis myriantha* Hook. f. Rendle in Journ. Linn. Soc. Bot. 36: 390. (1904).

Hubei: Badong, 4036, 4697, 4698, (type of *Agrostis milioides* C. C. Mez), 4722, 4723. **Yunnan:** In forests near Mengzi at 1,750 metres, 11257.

Agrostis perennans (Walter) Tuck. (POACEAE)

Rendle in Journ. Linn. Soc. Bot. 36: 390. (1904).

Hubei: Badong, 4034.

Agrostophyllum inocephalum (Schauer) Ames. (ORCHIDACEAE)

Agrostophyllum formosanum Rolfe in Ann. Bot. 9: 157. (1895), Henry in Trans. Asiat. Soc. Jap. 24 suppl: 92. (1896) (List Pl. Formosa), Bretschneider in Hist. Eur. Bot. Disc. China 2: 792. (1898), Rolfe in Journ. Linn. Soc. Bot. 36: 21. (1904—1906).

Taiwan: Oluanpi, 1350 (type of *Agrostophyllum formosanum* Rolfe).

A terrestrial orchid to 40 cm tall, distributed throughout the Philippines and China (Taiwan). In Taiwan this species inhabits evergreen forests on the south of the island.

Aidia canthioides (Champ ex Benth.) Masam. (RUBIACEAE)

Randia dumetorum Henry non (Retz.) Poir. in Trans. Asiat. Soc. Jap. 24 suppl: 50. (1896) (List Pl. Formosa). *Randia canthioides* Champ ex Benth. Li in Woody Flora of Taiwan 869. (1963).

Taiwan: Wanchin, 1687.

An erect shrub, native to southern China (including Taiwan), Vietnam and Japan. Distributed at low altitude in Taiwan.

Aidia henryi (E. Pritzel) T. Yamaz. (RUBIACEAE) New species

Randia henryi E. Pritzel in Bot. Jahrb. 29: 581. (1901).

Sichuan: Henry's Chinese collector with Pratt, no

locality stated but most likely Kangding, 8924.

Wilson later collected this shrub near Kangding, the species extends into Hubei province.

Aidia yunnanensis (Hutchinson) T. Yamaz. (RUBIACEAE) New species

in Journ. Jap. Bot. 45: 339. (1970)

Randia yunnanensis Hutchinson in Sargent Pl. Wilson. 3: 400. (1917).

Yunnan: Mountains to the south of Simao at 1,640 metres, 11750, (type of *Randia yunnanensis* Hutchinson). Simao, 11750a., 11750b., 11750c., 11750d.

Southern Yunnan to Thailand.

Ailanthus altissima (Mill.) Swingle var. *sutchuenensis* (Dode) Rehder & E. H. Wilson (SIMAROUBACEAE) New variety

in Sargent Pl. Wilson. 3. 449. (1917), Zheng in Journ. Wuhan Bot. Res. 2 (1): 109. (1984), Bean in Trees & Shrubs 1: 266. (1989).

Ailanthus glandulosa Henry non Desf. in Journ. China Br. Royal Asiat. Soc. 22: 401. (1887); in Notes Econ. Bot. China 38. (1893) (Chinese Names of Plants). *Ailanthus sutchuenensis* Dode in Bull. Soc. Dend. France 192. fig. A. (1907); in Fedde, Repert. Nov. Spec. Regni Veg. 6: 8. (1908—1909). *Ailanthus cacodendron* (Ehrhart.) Schinz. & Thellung. var. *sutchuenensis* (Dode) Rehder & E. H. Wilson in Sargent Pl. Wilson. 2: 153. (1916).

Hubei: Yichang, 223, 1591, 1591a. Liantuo, 3886.

A common tree around Yichang where Henry said it was known as the *ch'ao ch'un shu* or stinking *Cedrela* tree. Dode based his *Ailanthus sutchuensis* on Henry's Hubei collections and on material later gathered in Sichuan by Farges. According to Henry the bark of this tree was used in the treatment of dysentry.

Ainsliaea apteroides (C. C. Chang) Y. C. Tseng (ASTERACEAE) New species

in Acta Phytotax. Sin. 31: 363. (1993).

Ainsliaea pteropoda DC. var. *apteroides* Chang in Sinensia 4: 227. (1934), S. Y. Hu in Quart. Journ. Taiwan Mus. 18: 99. (1965).

Yunnan: Mengzi, 9907, (isotype). Milê, 9907a., (isotype). Simao, 12763.

Distributed in India, Nepal, Bhutan and China [Tibet (Xizang) and Yunnan].

Ainsliaea bonatii Beauverd (ASTERACEAE)

Tseng in Acta Phytotax. Sin. 31: 362. (1993).

Yunnan: Milê, 9955.

Endemic to China and distributed in the provinces of Guizhou, Sichuan and Yunnan,

Ainsliaea elegans Hemsley (ASTERACEAE) New species

in Hooker's Icon. Pl. 28: t. 2747. (1902), Tseng in Acta Phytotax. Sin. 31: 364. (1993).

Yunnan: Mengzi at 2,300 to 2,650 metres, 9108 (type). Distributed in China (Guizhou, Yunnan) and Vietnam.

Ainsliaea fragrans Champ. ex Benth. (ASTERACEAE)

Anonymous in Kew List of Determinations (volume 24).

Hubei: Changyang, 7674.

Ainsliaea glabra Hemsl. var. *sutchuenensis* (Franch.) S. E. Freire (ASTERACEAE)

Ainsliaea tenuicaulis Mattf. in Notizbl. Bot. Gart. Berlin 11: 108. (1931), Tseng in Acta Phytotax. Sin. 31: 366. (1993).

Ainsliaea glabra Hemsl. var. *tenuicaulis* (Mattf.) Chang in Bull. Fan. Mem. Inst. Biol. 5: 319. (1934), S. Y. Hu in Quart. Journ. Taiwan Mus. 18: 96. (1965).

Hubei: Changyang, 7599.

Ainsliaea gracilis Franchet (ASTERACEAE)

In the Harvard University Herbarium (GH).

Hubei: South Badong, 6163. Fang Xian, 6637c.

Ainsliaea latifolia (D. Don) Sch.-Bip. (ASTERACEAE)

Ainsliaea pteropoda DC. Anonymous in Kew List of Determinations (volume 24).

Ainsliaea scabrida Dunn in Journ. Linn. Soc. Bot. 35: 510. (1903).

Hubei: without locality, 7079a. **Sichuan:** North Wushan, 7079. **Yunnan:** On mountain slopes near Mengzi at 1,950 metres, 9851, (isotype of *Ainsliaea scabrida* Dunn, collected May 1st 1898).

Ainsliaea pertyoides Franchet (ASTERACEAE)

In the Harvard University Herbarium (A.).

Yunnan: In the mountains to the west of Simao at 1,650 metres, 9882a., 9882b.

Native to India and China (Guizhou, Sichuan and Yunnan).

Ainsliaea pertyoides Franch. subsp. *albo-tomentosa* (Beauverd) T. G. Gao (ASTERACEAE)

In the Harvard University Herbarium (A.).

Yunnan: Mengzi, 9926.

Endemic to China, distributed in Guizhou, Sichuan and Yunnan.

Ainsliaea ramosa Hemsley (ASTERACEAE) New species

in Journ. Linn. Soc. Bot. 23: 471. (1888), Bretschneider in Hist. Eur. Bot. Disc. China 2: 786. (1898).

Hubei: Xiling (Yichang) Gorge, 4422 (type). **Yunnan:** Mengzi, 10891.

Ainsliaea spp. (ASTERACEAE)

Anonymous in Kew List of Determinations (volume 24 & 34).

Hubei: Badong, 4701. Jianshi, 5932, 5983, 7461.

Fang Xian, 6639, 6620. Without locality, 6637a. **Yunnan:** Mengzi, 9294.

Ajania myrianthum (Franch.) Ling ex C. Shih (ASTERACEAE)

Tanacetum pallasianum Trautv. & Mey. forma Anonymous in Kew List of Determinations (volume 24).

Hubei: Fang Xian, 6931.

Ajuga ciliata Bunge var. *glabrescens* Hemsley (LAMIACEAE) New variety

in Journ. Linn. Soc. Bot. 26: 315. (1890).

Ajuga ciliata Dunn non Bunge in Notes Roy. Bot. Gard. Edinburgh 6: 194. (1915).

Hubei: On the mountains above Jianshi at 1650 metres, 6038. Zigui, 6038a., Changleping, 6285.

Ajuga decumbens Thunb. (LAMIACEAE)

Henry in Journ. China Br. Roy. Asiat. Soc. 22: 248. (1887) (Chinese Names of Plants), Forbes & Hemsley in Journ. Linn. Soc. Bot. 26: 315. (1890), Dunn in Notes Roy. Bot. Gard. Edinburgh 6: 193. (1915), C. Y. Wu in Flora Yunnanica 1: 522. (1977).

Ajuga remota Dunn non Benth. in Notes Roy. Bot. Gard. Edinburgh 6: 195. (1915).

Hubei: Yichang, 770, 814, 1213. Liantuo, 3837. **Yunnan:** Mengzi at 1,450 metres, 10861.

Henry stated this species was known colloquially at Yichang as *hsia ku ts'ao*. In Chinese herbal medicine this species is currently used to treat cases of tonsillitis, bronchitis and pneumonia.

Ajuga macrosperma Wall. ex Benth. (LAMIACEAE)

Dunn in Notes Roy. Bot. Gard. Edinburgh 6: 193. (1915).

Yunnan: South of the Red River from Manpan at 1,300 metres, 10364. Mengzi, 10900. Simao, 11684, 11684a., 11684b., 11684c.

Ajuga multiflora Bunge (LAMIACEAE)

Ajuga genevensis Forbes & Hemsley non L. in Journ. Linn. Soc. Bot. 26: 315.(1890), Dunn in Notes Roy. Bot. Gard. Edinburgh 6: 195. (1890).

Hubei: In a ravine on Moji Shan, 3364.

Ajuga nipponensis Makino (LAMIACEAE)

Ajuga genovensis Anonymous non L. in Kew List of Determinations (volume 34).

Yunnan: Lunan, 10851.

Akebia longeracemosa Matsum. (LARDIZABALACEAE) New species

Akebia sp Henry in Trans. Asiat. Soc. Jap. 24 suppl: 17. (1896) (List Pl. Formosa). *Akebia quinata* (Hout.) Denes. var. *longeracemosa* (Matsum) Rehder & E. H. Wilson in Sargent Pl. Wilson. 1: 349. (1913).

Taiwan: Wanshoushan, 319. Wanchin, 1829 (May 12th 1894).

An evergreen climbing vine, native to southern China, including Taiwan, where it was first collected by Henry.

Akebia quinata (Thunb. ex Houtte) Decne. (LARDIZABALACEAE)

Henry in Journ. China Br. Roy. Asiat. Soc. 22: 262. (1887) (Chinese Names of Plants; in Notes Econ. Bot. China 47. (1893). Rehder & Wilson in Sargent Pl. Wilson. 1: 347. (1913). Zheng in Journ. Wuhan Bot. Res. 2.(1): 41. (1984).

Hubei: Yichang, 1223. Liantuo, 3805, 3807.

A vigorous evergreen climber, well-known in gardens and valued for its handsome foliage and pendant racemes of fragrant chocolate-purple flowers. Native to Japan, Korea and China, it was introduced to cultivation by Robert Fortune in 1845 from Zhoushan (the Island of Chusan where Fortune also procured plants of the Chusan palm, *Trachycarpus fortunei*) off the coast of Zhejiang province in eastern China. According to both Henry and Wilson the fleshy fruits of this species were eaten in parts of Hubei where it shared with various species of *Holboellia* and *Akebia trifoliata*, the colloquail name of *pa-yuen-cha*. In current Chinese herbal medicine the fruits of this species and *Akebia trifoliata* are used as a diuretic, as a painkiller and in cases of insect and snake-bite. *Akebia quinata* was collected by the Glasnevin Central China Expedition (GCCE 512) near Zhenzi town, Xingshan in September 2004, there it grew on the edge of *Castanea seguinii* woodland.

Akebia trifoliata (Thunb.) Koidz. (LARDIZABALACEAE)

Zheng in Journ. Wuhan Bot. Res. 2.(1): 41. (1984)

Akebia lobata Decne. Henry in Notes Econ. Bot. China 47. (1893), Rehder & E. H. Wilson in Sargent Pl. Wilson. 1: 348. (1913).

Hubei: Badong, 3752, Changleping, 7655.

A deciduous, rampantly vigorous climbing shrub, carrying pendant racemes of dark purple flowers in April occassionaly followed by sausage-shaped fruits. Distributed throughout the reaches of the Yangtze River and also in the provinces of Gansu, Hebei, Henan, Shaanxi and Shandong. Introduced to Kew in 1897. In the almost frost-free climate at Glanleam on Valentia Island off the coast of County Kerry there is a remarkable plant that has scaled its way over five trees of *Acer pseudoplatanus* on sea cliffs and continues to vigorously colonise neighbouring trees.

Akebia trifoliata (Thunb.) Koidz. subsp. *australis* (Diels) T. Shimizu (LARDIZABALACEAE)

Akebia trifoliata (Thunb.) Koidz. var. *australis* Diels Rehder & E. H. Wilson in Sargent Pl. Wilson. 1: 348. (1913), Zheng in Journ. Wuhan Bot. Res. 2.(1): 41. (1984). *Akebia lobata* Decne var. *australis* Diels Rehder & E. H. Wilson in Sargent Pl. Wilson. 1: 348. (1913).

02

Hubei: Badong, 1415. Antelope Glen, 3382. On the road to Moji Shan 1548 (June 1886), 4342. Liantuo, 7856. Without locality, 7856a. **Sichuan:** South Wushan, 5636. **Yunnan:** In mountains near Mengzi at 1,300 metres, 10679.

In China this variety along with *Akebia quinata*, *Clematis armandii* and *Clematis montana* are used medicinally and are the *mu tong* - literally 'wood with holes from one end to the other', the stems are the source of a medicine used to promote urination. This variety was first found in China by the French missionary, Père Armand David, Diels based his description on David's and Henry's collections.

Alangium chinense (Lour.) Rehder (ALANGIACEAE)

Rehder in Sargent Pl. Wilson. 2: 552. (1916), Zheng in Journ. Wuhan Bot. Res. 2.(1): 158. (1984).

Marlea begoniaefolia Roxb. Henry in Journ. China Br. Roy. Asiat. Soc. 22: 262. (1887) (Chinese Names of Plants), Forbes & Hemsley in Journ. Linn. Soc. Bot. 23: 344. (1887) in part.

Hubei: Yichang, 224, 1399. San You Dong Glen near Yichang (July 1886), 1503. Liantuo, 2017. Jianshi, 7425. **Hainan:** Near the Kiungchow Pagoda, 7976. **Yunnan:** Mengzi at 1,500 metres, 10142a., 10142b. In the Red River valley near Manpan at 700 metres, 10647. Simao at 1,500 metres, 13030.

A widespread shrub or small tree to about 4 metres common on roadsides and in thickets in Hubei and widely dispersed across east Asia. *Alangium chinense* was collected at the Sanxia (Three Gorges) Botanical Gardens in an area of low altitude secondary forest by the Glasnevin Central China Expedition (2002) where it grew with *Aesculus wilsonii*. In Ireland this species is cultivated at Airfield House, Dundrum, Dublin. According to Henry this species was known colloquially in the Three Gorges region as *pa chio shu*.

Alangium platanifolium (Sieb. & Zucc.) Harms (ALANGIACEAE) New record for China

Rehder in Sargent Pl. Wilson. 2: 552. (1916), Zheng in Journ. Wuhan Bot. Res. 2.(1): 159. (1984), Bean in Trees & Shrubs 1: 270. (1989), O' Brien in Augustine Henry, An Irish Plant Collector in China 12. (2002),

Marlea platanifolia Sieb. & Zucc. Forbes & Hemsley in Journ. Linn. Soc. Bot. 23: 344. (1887) Henry in Journ. China Br. Roy. Asiat. Soc. 22: 280. (1887) (Chinese Names of Plants)

Hubei: Badong, 1731, 2439, 4770. Jianshi, 5813, 7345. Fang Xian, 5813a. (September 1888).

A bush or small tree to about 5 metres tall, introduced to cultivation by Maries from Japan in 1879. This species appears perfectly hardy at the National Botanic Gardens,

Glasnevin where it grows in the central section of the Chinese slope. Henry stated in Badong it was known colloquially as *ya chio pan shu*. *Alangium platanifolium* was collected by the Glasnevin Central China Expedition near Zhenzi town, Xingshan (GCCE 552) in September 2004), at Dragon Gate Forest River Park, also in Xingshan (GCCE 616) in September 2004 and near the San Yuo Dong Glen (Glen of the three pilgrims) near Yichang (GCCE 710) in October 2004.

Alangium sinicum (Nakai) S.Y. Hu, Sponberg & Z. Cheng (ALANGIACEAE)

Zheng in Journ. Wuhan Bot. Res. 2.(1): 159. (1984).

Marlea begoniaefolia Henry non Roxb. in Journ. China Br. Roy. Asiat. Soc. 22: 262. (1887) (Chinese Names of Plants), Forbes & Hemsley in Journ. Linn. Soc. Bot. 23: 344. (1887) in part. *Alangium chinense* Rehder non (Lour.) Rehd. in Sargent Pl. Wilson 2: 552. (1916).

Hubei: Jianshi, 5855. Without locality, 5855a., (March 1888), 5855b. Xingshan, 7666. Donghu, 6416.

The presence of this taxon in the Shennongjia Forest District was confimed by the 1980 Sino-American Expedition to that region.

Albizia chinensis (Osbeck) Merrill (FABACEAE)

Albizia stipulata (DC.) Boiv. Anonymous in Kew List of Determinations (volume 24).

Hainan: Hummocks near Haikou, 8107. At Nodoa in the interior of the island, 8445. Ling-men (in the interior of the island), 8617.

Albizia corniculata (Lour.) Druce (FABACEAE)

Albizia millettii Benth. Anonymous in Kew List of Determinations (volume 24 & 34).

Hainan: Environs of Haikou, 8396. **Yunnan:** Simao, 11916.

Albizia julibrissin Durazz. (FABACEAE)

Forbes & Hemsley in Journ. Linn. Soc. Bot. 23: 216. (1887), Ricker in J. Wash. Acad. Sci. 8: 245. (1918), Zheng in Journ. Wuhan Bot. Res. 2.(1): 95. (1984).

Hubei: In a temple courtyard beside the Customs House at Yichang (June 4th 1886), 1590. Badong, 2870, 6185. **Yunnan:** Mengzi, 10194, 13630. Simao, 12990.

The *yeh ho* or 'night fold', referring to the leaflets which fold together at night. A handsome deciduous tree, well known in the west though thriving better in continental Europe than in Ireland and the U.K. *Albizia julibrissin* grows well against the brick walls of the alpine yard at Glasnevin where it flowers during warm sunny summers, thought is worth travelling to the Jardin des Plantes, Paris, to see it at its best. According to Wilson it made trees from 7 to 16 metres tall in Hubei and Sichuan, but was rare. In China it is also called the *he huan pi* and the bark is used as a drug in cases of insomnia and irritability. From trees on the Bund, Shanghai (GCCE 889) seedlings were raised at the National Botanic Gardens, Glasnevin in April 2005. Introduced to cultivation in 1745.

Albizia kalkora (Roxb.) Prain (FABACEAE)
Ricker in J. Wash. Acad. Sci. 8: 243. (1918), Zheng in Journ. Wuhan Bot. Res. 2.(1): 96. (1984).

Albizia lebbek Forbes & Hemsley non (L.) Benth. in Journ. Linn. Soc. Bot. 23: 216. (1887). *Albizia henryi* Ricker in J. Wash. Acad. Sci. 8: 243. (1918).

Hubei: At Shih Liu Hung near Yichang (June 1886), 1605. Yichang, 1959, 2870a., 2977. Zigui, 6203. Liantuo, 6203 (March 1888). Changyang, 6203a. **Yunnan:** In woods near Mengzi at 1,300 metres, 9374, 10683, (isotype of *Albizia henryi* Ricker).

A flat-topped tree to 24 metres tall with girths of up to 3 metres, native to tropical and temperate Asia, though often planted. A characteristic feature of the vegetation of the more arid valleys.

Albizia lucidior (Steud.) J. C. Nielson ex H. Hara (FABACEAE)
Albizia bracteata Dunn in Journ. Linn. Soc. Bot. 35: 493. (1903), Ricker in J. Wash. Acad. Sci. 8: 242. (1918).

Yunnan: On the Mengzi plain at 1,450 metres, 9997, (isosyntype of *Albizia bracteata* Dunn). Mountains to the east of Simao at 1,650 metres 9997a., (syntype of *Albizia bracteata* Dunn), 9997d.

Albizia odoratissima (L. f.) Benth. (FABACEAE)
Ricker in J. Wash. Acad. Sci. 8: 245. (1918).

Hainan: Without locality, 8485. **Yunnan:** Mengzi, 9910, 10811a., 11274.

Albizia procera (Roxb.) Benth. (FABACEAE)
Li in Woody Flora of Taiwan 335. fig. 117. (1963) Ricker in J. Wash. Acad. Sci. 8: 245. (1918).

Albizia sp. Henry in Trans. Asiat. Soc. Jap. 24 suppl: 39. (1896) (List Pl. Formosa).

Hainan: Ling-men (in the interior of the island), 8668. **Taiwan:** Wanchin, 1574, 1613.

A tree to 20 metres tall, native to Indo-Malaysia and Taiwan where it grows in open dry areas at low altitude in the eastern and southern parts of the island. Henry stated this species was known as *yen-ch'ai.*

Albizia retusa Benth. (FABACEAE)
Henry in Trans. Asiat. Soc. Jap. 24 suppl: 39. (1896) (List Pl. Formosa), Ricker in J. Wash. Acad. Sci. 8: 244. (1918).

Taiwan: Oluanpi, 992.

A small tree, native to Java, the Philippines, Japan (south) and Micronesia. In Taiwan this species is found in coastal areas of the southern tip of the island.

Alcea rosea L. (MALVACEAE)
Althaea rosea (L.) Cav. Henry in Journ. China Br. Roy. Asiat. Soc. 22: 259. (1887) (Chinese Names of Plants).

Hubei: Badong, 373 (June 1885). Liantuo 1900. **Yunnan:** Mengzi, 10609.

The hollyhock, Henry's Badong collector called this species the *hung mou tan.* Long cultivated in China but thought to be a hybrid between *A. setosa* and *A. pallida*, both of which come from Turkey, it is likely therefore that the hollyhock travelled via the silkroad to China. An old Chinese name for the hollyhock is *jung k'uei* or 'mallow of the tribes of the west' indicating its foreign origins. An infusion of the roots is used as a soothing medicine for coughs, peptic ulcers and inflammation of the mouth, throat and stomach. Leaf infusions are used specifically for dry cough associated with inflammation of the respiratory tract.

Alchornea davidii Franch. (EUPHORBIACEAE)
Henry in Journ. China Br. Roy. Asiat. Soc. 22: 234. (1887) (Chinese Names of Plants), Forbes & Hemsley in Journ. Linn. Soc. Bot. 26: 438. (1894), Pax in Engler, Pflanzenr. iv. 147. vii: 249. (1914).

Hubei: Near Yichang by a cave, 1187.

A shrubby weed to about 2 metres tall, common at low altitudes in the glens around Yichang. According to Henry this plant was known colloquially at Yichang as *cha pao yeh.* It was discovered by Père Armand David in southern Shaanxi in April 1873.

Alchornea rugosa (Lour.) Müell.-Arg. (EUPHORBIACEAE)
Forbes & Hemsley in Journ. Linn. Soc. Bot. 26: 438. (1894).

Alchornea hainensis Pax & Hoffm. var. *glabrescens* Pax & Hoffm. in Engler, Pflanzenr. iv. 147. vii: 242. (1914).

Hainan: Hummocks near Haikou, 8104, 8119. Without locality, 8536, 8543. Environs of Kiungchow, 8778.

Alchornea sp. (EUPHORBIACEAE)
Anonymous in Kew List of Determinations (volume 34).

Yunnan: In the Red River Valley near Manhao, 9538. Mengzi, 9558, 10525.

Alchornea tiliifolia (Benth.) Müell.-Arg. (EUPHORBIACEAE)
Pax in Engler, Pflanzenr. iv. 147. vii: 250. (1914).

Yunnan: Simao, 12131, 12131a., 12131c.

Native to India, Bangladesh, Myanmar, Thailand and southern China.

Alectra arvensis (Benth.) Merr. (SCROPHULARIACEAE)
Alectra indica Benth. Anonymous in Kew List of Determinations (volume 34).

Yunnan: Mengzi, 11283.

Aletris spicata (Thunb.) Franch. (MELANTHIACEAE)
Aletris japonica Lamb. Henry in Journ. China Br. Roy.

Asiat. Soc. 22: 272. (1887) (Chinese Names of Plants), Wright in Journ. Linn. Soc. Bot. 36: 76. (1903).

Hubei: Yichang, 269. Badong, 1748.

According to Henry this species was known colloquially in the Three Gorges region as *t'ai yang ts'ao*.

Aleurites moluccana (L.) Willd. (EUPHORBIACEAE)

Aleurites triloba Forst. Forbes & Hemsley in Journ. Linn. Soc. Bot. 26; 343. (1894).

Hainan: Environs of Haikou, 8019. Without locality, 8336.

The candlenut tree, an evergreen tree to 20 metres tall, native to south-east Asia and widely cultivated in the tropics.

Aleurites sp. (EUPHORBIACEAE)

Anonymous in Kew List of Determinations (volume 34).

Yunnan: Mengzi, 9547.

Alisma lanceolatum Withering (ALISMATACEAE)

Alisma plantago L. var. *angustifolium* Kunth. Wright in Journ. Linn. Soc. Bot. 36: 189. (1903).

Hubei: Badong, 1852, 5117.

Alisma plantago-aquatica L. (ALISMATACEAE)

Anonymous in Kew List of Determinations (volume 24).

Hubei: Badong, 3160. Yichang, 4301.

Alkekengi officinarum L. (SOLANACEAE)

Physalis alkekengi L. Henry in Journ. China Br. Roy. Asiat. Soc. 22: 273. (1887) (Chinese Names of Plants), Forbes & Hemsley in Journ. Linn. Soc. Bot. 26: 173. (1890).

Hubei: Jianshi, 6043. Changyang, 6225. **Sichuan:** South Wushan, 5758.

Henry stated that in the Three Gorges region this species was locally known as *t'ien p'ao tzu*.

Allium chinense G. Don (AMARYLLIDACEAE)

Wright in Journ. Linn. Soc. Bot. 36: 121. (1903), Henry in Journ. China Br. Roy. Asiat. Soc. 22: 238. (1887) (Chinese Names of Plants).

Allium thunbergii Henry non G. Don G. Don in Journ. China Br. Roy. Asiat. Soc. 22: 240. (1887) (Chinese Names of Plants), *Allium bakeri* Regel Wright in Journ. Linn. Soc. Bot. 36: 120. (1903).

Hubei: Liantuo, 3071, 4413. Fang Xian, 6926, 6739. Donghu, 7493.

Chinese onion or Rakkyo, a perennial species, grown in both China and Japan for its small shallot-like bulbs. Native to central and eastern China and very handsome when in flower. It is cultivated on the terraces of a number of monasteries on Emei Shan as an ornamental plant. According to Henry this species was known colloquially in the regions of Yichang, Badong and Liantou as *yeh ch'iao t'ou tzu*. Widely cultivated in China, Indochina and India.

Allium funküfolium Handel-Mazzetti (AMARYLLIDACEAE)
New species

Anz. Akad. Wiss. Wien, Math.-Naturwiss. Kl. 57: 175 (1920), *Allium victorialis* Wright non L. in Journ. Linn. Soc. Bot. 36: 126. (1903).

Hubei: Xingshan, 5590f. (isotype).

Allium henryi C. H. Wright (AMARYLLIDACEAE)
New species

in Kew Bull. Misc. Inform. 119. (1895), Bretschneider in Hist. Eur. Bot. Disc. China 2: 793. (1898), Wright in Journ. Linn. Soc. Bot. 36: 122. (1903), Airy Shaw in Notes Roy. Bot. Gard. Edinburgh 16: 143. (1928—1932), Morley in Glasra 3: 78. (1979).

Hubei: In grass on the mountains near Fang Xian at 2,100 metres, 6924 (type).

A beautiful little onion, carrying globular heads of blue flowers on pendelous stems to 15 cm tall. When travelling on the 2004 Glasnevin Central China Expedition our group met with this aristocratic little onion on the tallest mountain peaks in Shennongjia Forest District in September of that year. There it transformed the limestone summits into a blue haze, it has yet to be introduced to cultivation but would be worth any effort taken. *Allium henryi* was collected by the Sino-American Expedition to the Shennongjia Forest District in 1980.

Allium macrostemon Bunge (AMARYLLIDACEAE)

Airy-Shaw in Notes Roy. Bot. Gard. Edinburgh 16: 136. (1928—1932)

Allium thunbergii Henry non G. Don in Journ. China Br. Roy. Asiat. Soc. 22: 240. (1887) (Chinese Names of Plants). *Allium bakeri* Wright non Regel in Journ. Linn. Soc. Bot. 36: 120. (1903).

Hubei: Yichang, 563.

Allium paepalanthoides Airy-Shaw (AMARYLLIDACEAE)
New species

in Notes Roy. Bot. Gard. Edinburgh 16: 142. (1928—1932)

Allium thunbergii Henry non G. Don in Journ. China Br. Roy. Asiat. Soc. 22: 240. (1887) (Chinese Names of Plants), *Allium bakeri* Wright non Regel in Journ. Linn. Soc. Bot. 36: 120. (1903).

Sichuan: North Wushan at 2,100 metres, on a cliff wall, 7038 (type).

Allium ramosum L. (AMARYLLIDACEAE)

Henry in Journ. China Br. Roy. Asiat. Soc. 22: 240. (1887)

Allium odorum L. Wright in Journ. Linn. Soc. Bot. 36: 123. (1903).

Hubei: Yichang, 443. Badong, 2884, 4656, 5109,

6929a. Fang Xian, 6929. **Hunan:** Shimen, 7537, 7940.

Chinese chives, according to Henry this species was cultivated around Yichang where it was known colloquially as *chiu ts'ai*. A perennial, clump forming species with umbels of white flowers on 45 cm tall stems. In China, these Chives are blanched like seakale and eaten as a vegetable rather than as a flavouring. The blanched leaves are called *kau wong* and are harvested between August and November and are often eaten as a delicacy with *chow mien* (fried noodles). It is not certain to where this species is indigenous but it is found in the Khasi hills in Assam, north-east India. The Hani (an ethnic group) of Xishuangbanna in southern Yunnan call this species *sailqiv* and use the roots to treat bronchitis. Seeds were taken from Henry's 6929 at Kew on April 10th 1889 and sent for propagation.

Allium sp. (AMARYLLIDACEAE)
Anonymous in Kew List of Determinations (volume 34).

Yunnan: Mengzi, 9817.

Allium tenuissimum L. (AMARYLLIDACEAE)
Wright in Journ. Linn. Soc. Bot. 36: 125. (1903).
Hubei: Fang Xian, 6860.

Allium thunbergii G. Don (AMARYLLIDACEAE)
Allium odorum Henry non L. in Trans. Asiat. Soc. Jap. 24 suppl: 97. (1896) (List Pl. Formosa), Wright in Journ. Linn. Soc. Bot. 36: 123. (1903).

Taiwan: Kaohsiung Plain, 1856.

A perennial bulb, native to Japan, Korea and China (Taiwan) (where it is uncommon). Henry's collection locality (the Kaohsiung Plain) is now covered by the city of Kaohsiung.

Allium victoralis L. (AMARYLLIDACEAE)
Wright in Journ. Linn. Soc. Bot. 36: 126. (1903).
Hubei: Changleping, 5590b. South Badong, 5590c. Zigui, 5590d. Donghu, 5590e. **Sichuan:** South Wushan, 5590.

Allium wallichii Kunth (AMARYLLIDACEAE)
Allium bulleyanum Diels var. *tchongchanense* (Levl.) Airy Shaw in Notes Roy. Bot. Gard. Edinburgh 16: 139. (1928—1932).

Yunnan: On rocky hills near Mengzi, 9156. On the mountains to the east of Mengzi at 1,950 metres, 9156a.

Joseph Rock also later collected this handsome pink to violet flowered *Allium* west of the Mekong in Yunnan in October 1922. Native to India, Bhutan, Nepal, Myanmar and China.

Allmania nodiflora (L.) R. Br. ex Wight (AMARANTHACEAE)
Anonymous in Kew List of Determinations (volume 24).

Hainan: On Liang Shan, 16 km (10 miles) east of Haikou, 8211.

Allophylus cobbe (L.) Raeusch. (SAPINDACEAE)
Anonymous in Kew List of Determinations (volume 34).

Yunnan: Mengzi, 11459.

Allophylus timorensis (DC.) Blume (SAPINDACEAE)
Schmidelia cobbe Anonymous non DC. in Kew List of Determinations (volume 24).

Hainan: On hummocks near Haikou, 8117. 32 km (20 miles) west of Haikou 8128. Environs of Kiungchow, 8691.

An evergreen shrub to 3 metres tall, widespread across Malaysia and China where it is found in Taiwan and Hainan.

Alloteropsis semialata (R. Br.) Hitchc. (POACEAE)
Axonopus semialatus (R. Br.) Hook. f. Rendle in Journ. Linn. Soc. Bot. 36: 334. (1904).

Guangdong: without locality, 39.

Alniaria alnifolia (Sieb. & Zucc.) Rushforth (ROSACEAE)
Sorbus alnifolia (Sieb. & Zucc.) K. Koch Rehder in Sargent Pl. Wilson 2: 270. (1915).

Hubei: Fang Xian, 6791.

A deciduous tree to over 20 metres tall, native to Japan, Korea and northern and central China. An uncommon tree in cultivation and deserving to be more widely grown on account of its beautiful orange-scarlet autumn colour. Wilson reintroduced it to cultivation from Hubei in 1901 and again many years later from Korea.

Alniaria folgneri (Schneid.) Rushforth (ROSACEAE) New species
Pyrus aria Forbes & Hemsley non (L.) Ehrh. in Journ. Linn. Soc. Bot. 254. (1887) Henry in Notes Econ. Bot. China 47. (1893). *Micromeles folgneri* Schneider in Bull. Herb. Boissier ser. 2, 6: 318. (1906); in Illus. Handb. Laubhilzk. 1: 704. fig. 386. (1906); in Fedde, Repert. Nov. Spec. Regni. Veg. 3: 152. (1906), Bean in Kew Bull. Misc. Inf. 175. (1910), Rehder in Moller's Deutch. Gartn.-Zeit. xxvii. 136. (1912). *Pyrus folgneri* (Schneid.) Leveille apud Bean in Trees & Shrubs 2: 283. (1914), Leveille in Fl. Kouy-Tcheou 349. (1915). *Sorbus folgneri* (Schneid.) Rehder in Sargent Pl. Wilson. 2: 271. (1915), O' Brien in Augustine Henry, An Irish Plant Collector in China 36. (2002).

Hubei: Badong on high hills, 2814, 4065. Changyang, 5273. Jianshi, 5804. **Sichuan:** South Wushan, 7075.

According to Henry this handsome tree varies in height from 3 to 10 metres in height in its native habitat and was the *shih-hui* tree of the Hubei mountains. Introduced to cultivation by Wilson through Messrs Veitch's Coombe Wood nursery in 1901. It has remained very rare in cultivation, a tree at Caerhays Castle Cornwall had reached a height of 18 metres by 1984.

Alniphyllum fortunei (Hemsl.) Makino (STYRACEAE)
Perkins in Engler, Pflanzenr. 241: 91. (1907) Rehder in

Sargent Pl. Wilson. 1: 294. (1913).

Alniphyllum macranthum Perkins in Bot. Jahrb. Syst. 31: 488. (1902).

Yunnan: In mountain forests on the Daweishan Range in Pingbian at 1,600 metres, 10593. In mountain forest to the south-east of Simao at 1,600 metres 11608, (isotype of *Alniphyllum macranthum* Perkins). Simao at 1,500 to 1,600 metres, 11957, 11957a.

A deciduous tree to 15 metres tall with dark grey bark mottled grey-white. In April and May white, sometimes pink-tinged bell-shaped flowers are borne in upright racemes. Distributed through China (including Taiwan), India, Laos and Vietnam.

Alniphyllum pterospermum Matsum. (STYRACACEAE) New species

in Tokyo Bot. Mag. 171: 67. (1901), Li in Woody Flora of Taiwan 748. fig. 301.(1963).

Halesia fortunei Henry non Hemsl. in Trans. Asiat. Soc. Jap. 24 suppl: 59. (1896) (List Pl. Formosa).

Taiwan: Wanchin, 430 (co-type), 1578. Oluanpi, 2057.

A shrub or small tree to 10 metres, native to southern China, including Taiwan.

Alnus cremastogyne Burkill (BETULACEAE) New species

in Journ. Linn. Soc. Bot. 26: 499. (1899), Diels in Bot. Jahrb. 29: 282. (1900), Winkler in Engler. Pflanzenr. iv.-61. 127. fig. 28. (1904); in Bot. Jahrb. xxxvi. Biebl. lxxxii. 33. (1905), Callier apud Schneider in Illus. Handb. Laubholzk. 2: 891. (1912), Bean in Kew Bull. Misc. Inf. xxvi. 164. (1913).

Sichuan: Kangding, Henry's Chinese collector with Pratt, 8890 (type).

A tree to 30 metres tall, common in western Sichuan. Wilson who recorded seeing trees with girts of 4.5 metres near Wa Shan stated it was very common in the valley of the Min River and was planted there for fuel. On the Chengdu plain it was planted by the sides of fields, (*Metasequoia glyptostroboides* is largely used for this purpose at present) streams and irrigation canals. Henry stated it was known colloquially as *ching-shu* in the same region. *Alnus cremastogyne* was introduced to cultivation by Wilson in 1908 (see his photograph of this tree in Kew Bull. Misc. Inf. facing page 164. 1913). In Ireland this tree grows at the John F. Kennedy Arboretum in County Wexford. It is allied to *Alnus lanata* Duthie from Sichuan which grows in the arboretum at Glasnevin.

Alnus formosana (Burkill.) Makino (BETULACEAE) New species

Schneider in Sargent Pl. Wilson. 2: 501. (1916)

Henry in Elwes and Henry The Trees Gt. Brit. & Irel. 4: 953. (1909) Li in Woody Flora of Taiwan 80. fig. 24. (1963)

Alnus maritima Henry non (Marshall) Muhl. ex Nuttall in Trans. Asiat. Soc. Jap. 24 suppl: 90. (1896) (List Pl. Formosa). *Alnus maritima* (Marshall) Muhl. ex Nuttall var. *formosana* Burkill in Journ. Linn. Soc. Bot. 500. (1899). *Alnus japonica* Henry non (Thunb.) Steud. in Elwes and Henry The Trees Gt. Brit. & Irel. 4: 953. (1909) Li in Woody Flora of Taiwan 80. fig. 24. (1963).

Taiwan: Tamsui, 1394 (type of *Alnus maritima* Nutt. var. *formosana* Burkill), 1732.

Alnus henryi C. K. Schneider (BETULACEAE) New species

in Sargent Pl. Wilson. 2: 495. (1916), Chang et al. in Flora of Taiwan 2: 43. (1976), O' Brien in Augustine Henry, an Irish Plant Collector in China 12. (2002).

Taiwan: Tamsui, 1389 (type).

Cultivated at Mount Usher, County Wicklow.

Alnus nepalensis D. Don (BETULACEAE)

Burkill in Journ. Linn. Soc. Bot. 26: 500. (1899), Winkler in Engler, Pflanzenr. 61: 108. (1904), Schneider in Sargent Pl. Wilson. 2: 502. (1916).

Yunnan: On mountains near Mengzi at 1,500 to 1,800 metres, forming woods, 9223.

A deciduous tree to about 17 metres tall (up to 30 metres under favourable conditions). Native to northern India, Nepal, south-east Tibet (Xizang) and Yunnan. There are magnificent trees by the entrance to the Royal Botanic Gardens, Kathmandu, Nepal. In Nepal it is called the *utis* and is widely planted to control soil erosion. Wilson observed this species in south-west Yunnan in 1899 while visiting Henry and stated it was abundant in that region and often formed pure stands. Forrest also collected *Alnus nepalensis* in the Mekong-Salween divide in October 1918 (Forrest 17,204). To the Hani (an ethnic group) of Xishuangbanna in southern Yunnan this species is called *haqnyuq* and its bark is used to treat cases of dysentry, diarrhoea, rheumatism, sprains and fractures. *Alnus nepalensis* is an autumn flowered species (September to November) and is a beautiful sight when laden with creamy-yellow catkins. It is cultivated at Ness Botanic Gardens, the former home of A. K. Bulley now managed by the University of Liverpool.

Alocasia cucullata (Lour.) G. Don (ARACEAE)

Henry in Notes Econ. Bot. China 35. (1893), Brown in Journ. Linn. Soc. Bot. 36: 183. (1903).

Hainan: Lingmen (in the interior of the island), 8626.

Native to India, Thailand, China and Japan. Cultivated for its edible roots. On Hainan, according to Henry, this species was known as *fan-yu*.

Alocasia hainanica N. E. Brown (ARACEAE) New species

in Journ. Linn. Soc. Bot. 36: 183. (1903).

Alocasia hainanensis Krause in Engler, Pflanzenr. iv. 23Ba: 91. (1912).

Hainan: Environs of Kiungchow, 8679, (type of *Alocasia hainanensis* Krause).

Alocasia odora (Lodd.) Spach. (ARACEAE)

Alocasia macrorrhiza Henry non (L.) Schott. Henry in Trans. Asiat. Soc. Jap. 24 suppl: 101. (1896) (List Pl. Formosa).

Taiwan: Oluanpi, 310. Wanshoushan, 1915. Tamsui, 2014.

A handsome perennial to about 1.5 metres tall and carrying large peltate leaves. Native to Bhutan, India, Indonesia, Malaysia, Thailand, China (including Taiwan) and the Philippines. According to Henry this species was known as *shan-yu*, or *ku-p'o-yu*. Common in tropical rainforest in Taiwan. This species was collected by the Glasnevin Taiwan Expedition (GTE 27) by the lighthouse at Oluanpi (South Cape of Henry and Wilson) in October 2004.

Alocasia sp. (ARACEAE)

Anonymous in Kew List of Determinations (volume 34).

Yunnan: Mengzi, 11847. Simao, 12510, 13019.

Aloe vera (L.) Burm. f. (ASPHODELIACEAE)

Aloe chinensis (Haw.) Baker Henry in Trans. Asiat. Soc. Jap. 24 suppl: 97. (1896) (List Pl. Formosa). *Aloe vera* L. var. *chinensis* (Haw.) Berger in Engler, Pflanzenr. 38: 230. (1908).

Taiwan: Wanchin, 486.

Henry called this the *lu hui*, stating that the dried concentrate of the leaf is used to treat constipation. *Aloe vera* is an ancient cultigen from North Africa.

Alopecurus aequalis Sobol. (POACEAE)

Rendle in Journ. Linn. Soc. Bot. 36: 384. (1904).

Hubei: Yichang, 3475. Badong, 4733.

Alopecurus japonicus Steud. (POACEAE)

Rendle in Journ. Linn. Soc. Bot. 36: 385. (1904).

Hubei: Yichang, 7869.

Alphonsea mollis Dunn (ANNONACEAE) New species

in Journ. Linn. Soc. Bot. 35: 485. (1903) Merrill & Chun in Sunyatsenia 2: 230. (1935).

Yunnan: In Yulo forest to the south of Simao (Xishuangbanna) at 1,300 metres, 11923, (isotype). Simao, 12923.

Endemic to China, distributed in Guangxi, Hainan and Yunnan.

Alpinia blepharocalyx Schumann (ZINGIBERACEAE) New species

in Engler, Pflanzenr. 20.(iv. 20): 334. (1904).

Yunnan: Simao at 1,400 metres, 11962, (type).

A widespread species, native to India, Bangladesh, Myanmar, southern China, Thailand, Laos and Vietnam.

Alpinia calcarata (Haw.) Roscoe (ZINGIBERACEAE)

Alpinia aff. *calcarata* Anonymous in Kew List of Determinations (volume 24).

Hainan: Lingmen (in the interior of the island), 8630.

Alpinia galanga (L.) Willd. (ZINGIBERACEAE)

Henry in Trans. Asiat. Soc. Jap. 24 suppl: 94.(1896) (List Pl. Formosa), Wright in Journ. Linn. Soc. Bot. 36: 71. (1903).

Hainan: Lingmen (in the interior of the island), 8672, 8552. **Taiwan:** Wanchin, 804.

Greater galangal, the *nan-ch'iang*, an ingredient of herbal medicine in China. Native to the Himalaya, Ceylon, Indochina, China, Java, Borneo and the Philippines. The rhizomes when cut fresh are used to flavour foods and curries. In the 13th century Marco Polo noted it was grown in southern China and exports of it (as had been the case for centuries before) were brought to Europe through India to the Red Sea and along the Mesopotamian trade routes. It is used to dye wool yellow and is effective in stimulating the appetite of sick elephants.

Alpinia hainanensis K. Schum. (ZINGIBERACEAE) New species

in Engler, Pflanzenr. 20: 335. (1904) Wu in Novon 7(4): 441. (1997).

Alpinia speciosa (Wendl.) Schum. var. *longiramosa* Schumann non Gagnep. in Engler, Pflanzenr. 20: 339. (1904). *Alpinia henryi* K. Schum. in Engler, Pflanzenr. 20: 335. (1904), O' Brien in Augustine Henry, An Irish Plant Collector in China 12. (2002).

Hainan: Environs of Kiungchow, 8773, (type), Environs of Haikou 8402, (type of *Alpinia henryi* K. Schum.).

Alpinia japonica (Thunb.) Miq. (ZINGIBERACEA)

Wright in Journ. Linn. Soc. Bot. 36: 72. (1904).

Hubei: South Donghu, 7710.

Native to China and Japan.

Alpinia platychilus K. Schum. (ZINGIBERACEAE) New species

in Engler, Pflanzenr. 20.(iv. 20): 334. (1904).

Yunnan: Simao at 1,300 to 1,600 metres, 12227 (type). Endemic to Yunnan.

Alpinia sp. No. 1. (ZINGIBERACEAE)

Anonymous in Kew List of Determinations (volume 24).

Hainan: Without locality, 8552. Ling-men (in the interior of the island), 8631.

Alpinia sp. No. 2. (ZINGIBERACEAE)

Alpinia sp. Henry in Trans. Asiat. Soc. Jap. 24 suppl:

95. (1896) (List Pl. Formosa).

Taiwan: Tamsui, 2079.

Alpinia spp. (ZINGIBERACEAE)

Anonymous in Kew List of Determinations (volume 34).

Yunnan: Mengzi, 9410, 10838, 11494, 11504, 11506. Simao, 12127, 12709, 13127.

Alpinia zerumbet (Persoon) B. L. Burtt & R. M. Smith (ZINGIBERACEAE)

Alpinia nutans Henry non (L.) Rosc. in Trans. Asiat. Soc. Jap. 24 suppl: 94. (1896) (List Pl. Formosa), Forbes & Hemsley in Journ. Linn. Soc. Bot. 36: 72. (1903).

Hainan: Without locality, 8594. Environs of Kiungchow, 8731. **Taiwan:** Wanchin, 427. Kaohsiung, 782. Oluanpi, 1331.

A perennial to 2 metres or more tall, native to the Old World tropics, and, according to Henry, was known as *yueh-t'a*. Henry also stated its fibre was occasionally used in making hats and mats. This species was collected by the Glasnevin Taiwan Expedition (GTE 40) by the lighthouse at Oluanpi (South Cape of Henry and Wilson) and on Wanshoushan (Ape's Hill of Henry and Swinhoe) (GTE 106) in October 2004.

Alpuda mutica L. (POACEAE)

Alpuda mutica L. var. *aristata* (L.) Hack. Rendle in Journ. Linn. Soc. Bot. 36: 379. (1904).

Yunnan: In a ravine near Mengzi at 1,700 metres, 9616.

Alseodaphne andersonii (King ex Hook. f.) H. W. Li & J. Li (LAURACEAE)

Alseodaphne keenanii Gamble Liou Ho in Laurac. Chine & Indoch. 43. (1934).

Yunnan: Simao, A 12152.

A widespread species found over much of India, Myanmar, China [Yunnan, Tibet (Xizang)], Thailand, Laos and Vietnam.

Alsomitra sp. (CUCURBITACEAE)

Anonymous in Kew List of Determinations (volume 34).

Yunnan: Simao, 13375.

Alsophila costularis Baker (CYATHEACEAE) New species

in Kew Bull. Misc. Inf. 8. (1906), Ching & Ling in acta Phytotax. Sin. 22: 195. (1984), Xia in Acta. Phytotax. Sin. 27: 6. (1989).

Cyathea yunnanensis Domin in Acta Bot. Bohem. 3: 172. (1930) nom. nov. for *Alsphilla costularis*.

Yunnan: In forests near Simao at 1,950 metres, 13136, (type of *Alsophila costularis* Baker).

Alsophila glabra (Blume) Hook. (CYATHEACEAE)

Anonymous in Kew List of Determinations (volume 34).

Yunnan: Simao, 11961.

Belongs to a different species than cited in the Kew List.

Alsophila henryi Baker (CYATHEACEAE) New species

in Kew Bull. Misc. Inf. 227. (1898)

Cyathea henryi (Bak.) Copel. in Phillip. J. Sci. Bot. 4: 38. (1909), Holttum in Kew Bull. 19: 478. (1965). *Alsophila gigantea* Wall. ex Hook. var. *polynervata* (Miou) Xia in Acta Phytotax. Sin. 27: 11. (1989).

Yunnan: In forests near Mengzi at 1,300 metres, 11451 (type of *Alsophila henryi* Baker). Mengzi, 11541c., (syntype of *Alsophila henryi* Baker)

Alsophila latebrosa Wall. ex Hook. (CYATHEACEAE)

Henry in Trans. Asiat. Soc. Jap. 24 suppl: 109. (1896) (List Pl. Formosa), Christ in Bull. Herb. Boiss. 7: 18. (1899).

Taiwan: Oluanpi, 907. **Yunnan:** In mountain forest near Mengzi at 1,750 metres, fronds 3 metres long, 11531.

Alstonia sebusi (Van Heurk & Müll.Arg) Monach. (APOCYNACEAE) New species

Alstonia henryi Tsiang in Sunyatsenia 6: 112. (1941).

Yunnan: Simao, 11932 (type of *Alstonia henryi* Tsiang).

Alternanthera sessilis (L.) DC. (AMARANTHACEAE)

Henry in Journ. China Br. Roy. Asiat. Soc. 22: 263. (1887) (Chinese Names of Plants), Forbes & Hemsley in Journ. Linn. Soc. Bot. 26: 322. (1891).

Alternanthera nodiflora R. Br. Henry in Trans. Asiat. Soc. Jap. 24 suppl: 75. (1896) (List Pl. Formosa).

Hubei: Yichang, 301, 2791, 2791a. **Hainan:** On Liang Shan, 16 km (10 miles) east of Haikou, 8204. **Taiwan:** Kaohsiung, 275, 810. Wanshoushan, 255. Oluanpi, 2062. Wanchin, 447, 810.

A prostrate annual herb, widespread in the tropics and subtropics. Widely spread in warm regions of both hemispheres, in the Three Gorges region Henry stated it was known colloquially as *pa ko ts'ao*. In Taiwan this species is a common weed of open fields.

Alternanthera sp. (AMARANTHACEAE)

Anonymous in Kew List of Determinations (volume 34).

Yunnan: Mengzi, 9846.

Alysicarpus bupleurifolius (L.) DC. (FABACEAE)

Henry in Trans. Asiat. Soc. Jap. 24 suppl: 34. (1896) (List Pl. Formosa).

Taiwan: Kaohsiung, without number.

A slender erect herb to 45 cm tall, native to much of

India, south-east Asia and China (including Taiwan).

Alysicarpus vaginalis (L.) DC. (FABACEAE)

Henry in Trans. Asiat. Soc. Jap. 24 suppl: 34. (1896) (List Pl. Formosa).

Taiwan: Kaohsiung, 706, 1157, 1158.

An erect herb to 1.5 metres tall, native to the tropics of the Old World. Common in Taiwan throughout the island by roadsides and in pasture.

Alyxia sp. (APOCYNACEAE)

Anonymous in Kew List of Determinations (volume 34).

Yunnan: Mengzi, 10881.

Amalocalyx microlobus Pierre (APOCYNACEAE)

Amalocalyx yunnanensis Tsiang in Bull. Fan. Mem. Biol. Bot. 9: 19. (1939).

Yunnan: Mountains to the east of Simao at 1,300 metres, 11756, (isosyntype of *Amalocalyx yunnanensis* Tsiang). Mountains to the west of Simao at 1,650 metres, 11865, (isosyntype of *Amalocalyx yunnanensis* Tsiang).

Native to Myanmar, Thailand, Vietnam and China (Yunnan).

Amana edulis (Miq.) Honda (LILIACEAE)

Tulipa edulis (Miq.) Baker Henry in Notes Econ. Bot. China 26. (1893), Wright in Journ. Linn. Soc. Bot. 36: 138. (1903).

Hubei: Yichang, 3323, 3323a.

Henry stated that the bulbs of *Amana edulis* were an article of conciderable export in 19th century China and were the source of the drug *kuang-ku*. Common on the banks of the Yichang (Xiling) Gorge during the late 19th century.

Amaranthus blitum L. (AMARANTHACEAE)

Forbes & Hemsley in Journ. Linn. Soc. Bot. 26: 319. (1891), Henry in Notes Econ. Bot. China 52. (1893).

Hubei: Yichang, 147.

A widespread weed of cultivation in warm countries, according to Henry it was called the *han-ts'ai* in Hubei.

Amaranthus caudatus L. (AMARANTHACEAE)

Forbes & Hemsley in Journ. Linn. Soc. Bot. 26: 319. (1891), Henry in Notes Econ. Bot. China 52. (1893).

Hubei: Changyang, 7517.

Love-lies-bleeding, a widespread species in hot climates and often an escape from cultivation. Called the *lung-hsu-han* in the Three Gorges region. Native to the Andes and once an important food crop for the Aztecs.

Amaranthus cruentus L. (AMARANTHACEAE)

Amaranthus paniculatus L. Henry in Journ. China Br. Roy. Asiat. Soc. 22: 249. (1887) (Chinese Names of Plants), Forbes & Hemsley Journ. Linn. Soc. Bot. 26: 320. (1891,) Henry in Notes Econ. Bot. China 52. (1893).

Hubei: Badong, 2390, 4808. Yichang, 2732. Liantuo, 4590.

Commonly cultivated in China and widely spread in tropical Asia. In the Three Gorges region Henry stated it was known colloquially as *jua ku* or *ya-ku-han*.

Amaranthus spinosus L. (AMARANTHACEAE)

Henry in Journ. China Br. Roy. Asiat. Soc. 22: 248. (1887) (Chinese Names of Plants) (1887), Forbes & Hemsley Journ. Linn. Soc. Bot. 26: 320. (1891), Henry in Notes Econ. Bot. China 52. (1893); in Trans. Asiat. Soc. Jap. 24 suppl: 75. (1896) (List Pl. Formosa).

Hubei: Badong, 665. **Taiwan:** Oluanpi, 333, 662.

An annual herb to about 70 cm tall, native to tropical America but now widely naturalised around the world. A weed of waste places in Taiwan where according to Henry it is the *tz'u-han*. Henry also stated it was known colloquially at Badong as *tz'u han*.

Amaranthus tricolor L. (AMARANTHACEAE)

Amaranthus gangeticus L. Forbes & Hemsley in Journ. Linn. Soc. Bot. 26: 319. (1891), Henry in Notes Econ. Bot. China 52. (1893); in Trans. Asiat. Soc. Jap. 24 suppl: 75. (1896) (List Pl. Formosa).

Hubei: Yichang, 3085. **Hainan:** Environs of Kiungchow, 8680. **Taiwan:** Oluanpi, 206.

Chinese spinach, widespread in warm countries in south-east Asia, Japan, China and India. The young shoots and leaves are eaten. Needs a very warm climate and a rich soil to succeed and may be harvested after 6 to 8 weeks from sowing. Over 50 species of amaranths in temperate and tropical regions are eaten. An ancient vegetable in south-east Asia, amaranths have been hailed as a valuable crop for the future with the potential of helping to feed the planet's ever increasing population. The fruits of amaranths resemble grains in flavour and flourish in arid and saline habits providing a nutritious harvest grown on otherwise useless land.

Amaranthus viridis L. (AMARANTHACEAE)

Henry in Trans. Asiat. Soc. Jap. 24 suppl: 75. (1896) (List Pl. Formosa).

Hainan: Environs of Haikou, 8426. **Taiwan:** Kaohsiung, 274.

An annual to 80 cm tall, native to tropical America and now a weed in warmer areas of the world.

Amelanchier sinica (Schneid.) Chun (ROSACEAE)
New species

in Chin. Econ. Trees 168. f. 62. (1921), Zheng in Journ. Wuhan Bot. Res. 2.(1): 66. (1984).

Amelanchier asiatica (Sieb & Zucc.) Endl. var. *sinica* Schneider in Illus. Handb. Laubholz. I. 736. (1906); in Repert. Nov. Spec. 3: 181. (1907), E. H. Wilson in Sargent

Pl. Wilson. 1: 195. (1913).

Sichuan: South Wushan, 5521 (sosyntype of *Amelanchier asiatica* (Sieb & Zucc.) Endl. var. *sinica* Schneider). **Hubei:** Changleping, 6278.

A small deciduous tree to 5 metres (rarely up to 15 metres) tall and carrying in May many-flowered racemes of white fragrant flowers. According to Wilson who later collected this species himself stated that this was one of the commonest and most beautiful of the small trees in the thickets and woods of western Hubei. Introduced to cultivation in 1920.

Amentotaxus formosana Li (TAXACEAE)

Podocarpus argotaenia Henry non Hance in Trans. Asiat. Soc. Jap. 24 suppl: 91. (1896) (List Pl. Formosa), Masters in Journ. Linn. Soc. Bot. 26: 547. (1902). *Podocarpus latifolius* Masters non Wallich in Journ. Linn. Soc. Bot. 26: 547. (1902). *Podocarpus macrophyllus* Masters non (Thunb.) Sweet in Journ. Linn. Soc. Bot. 26: 548. (1902).

Taiwan: Oluanpi, 2075.

A small tree to 10 metrestall with trunks up to 30 cm in diameter. Endemic to Taiwan where it is confined to the southeastern part of the island between 1,200 and 1,300 metres. The current *Flora of Taiwan* describes it as being very scarce.

Amethystea caerulea L. (LAMIACEAE)

Forbes & Hemsley in Journ. Linn. Soc. Bot. 26: 310. (1890), Dunn in Notes Roy. Bot. Gard. Edinburgh. 6: 190. (1915).

Hubei: Liantuo, 4488.

Amischotolype glabrata Hassk. (COMMELINACEAE)

Forrestia hispida Henry non A. Rich. in Trans. Asiat. Soc. Jap. 24 suppl: 99. (1896) (List Pl. Formosa). *Forrestia chinensis* N. E. Brown in Journ. Linn. Soc. Bot. 36: 158. (1903).

Hainan: Ling-men, 8670, (isotype of *Forrestia chinensis* N. E. Brown). **Taiwan:** Wanchin, 149 (type of *Forrestia chinensis* N. E. Brown), 1618a.

species was collected by the Glasnevin Taiwan Expedition (GTE 70) in the nature reserve of the Hong Chuaw Tropical Botanic Gardens in Kenting National Park. A common understorey perennial on the floor of the tropical rainforests of southern Taiwan.

Amischotolype sp. (COMMELINACEAE)

Forrestia sp. Anonymous in Kew List of Determinations (volume 34).

Yunnan: Mengzi, 10884. Simao, 12204, 12400, 13377.

Ammannia baccifera L. (LYTHRACEAE)

Forbes & Hemsley in Journ. Linn. Soc. Bot. 23: 302. (1887), Henry in Trans. Asiat. Soc. Jap. 24 suppl: 44. (1896) (List Pl. Formosa), Koehne in Engler, Pflanzenr. 216: 285. (1903).

Hubei: Yichang, 410. Changyang 7739. **Taiwan:** Kaohsiung, in rice fields, 1191.

An erect much branched annual to 30 cm tall, native to Europe, Africa, China (including Taiwan), Japan, south east Asia and Australia. A weed of paddy fields in Taiwan.

Ammannia multiflora Roxb. (LYTHRACEAE)

Koehne in Engler, Pflanzenr. 216: 285. (1903).

Hainan: Without locality, 8370.

An annual weed, widespread across the tropics and subtropics of Africa, Asia and Australia.

Ammannia senegalensis Lam. (LYTHRACEAE)

Anonymous in Kew List of Determinations (volume 24).

Hainan: Environs of Haikou, 8365.

Ammannia sp. No. 1. (LYTHRACEAE)

Ammannia pentandra Roxb. Anonymous in Kew List of Determinations (volume 24).

Hubei: Yichang, 2369.

Ammannia sp. No. 2. (LYTHRACEAE)

Ammannia senegalensis Forbes & Hemsley non Lam. in Journ. Linn. Soc. Bot. 23: 304. (1887).

Hubei: Yichang, 2754.

Ammannia verticillata Lam. (LYTHRACEAE)

Ammannia salicifolia Monti. Anonymous in Kew List of Determinations (volume 34).

Yunnan: Mengzi, 9493.

Amomum sp. (ZINGIBERACEAE)

Anonymous in Kew List of Determinations (volume 24).

Hubei: Badong, 861. Yichang, 4141.

Amorphophallus henryi N. E. Brown (ARACEAE)
New species

in Journ. Linn. Soc. Bot. 36: 181. (1903), Engler in Engler, Pflanzenr. iv. 23C: 93. (1911), Liu & Huang in Fl. Taiwan 5: 801. (1978).

Amorphophallus sp. nova. Henry in Trans. Asiat. Soc. Jap. 24 suppl: 100. (1896) (List Pl. Formosa).

Taiwan: Wanshoushan, in rocky situations, 776, (type).

A tuberous perennial to 60 cm tall and native to the area west of Taiwan's central range of mountains in the southern half of the island where it is found in shaded places between 20 to 660 metres. Endemic to Taiwan, Henry stated it was known as *shih-su* or *shan-su*. This species is cultivated at both Glasnevin and Kew and commercially available in the USA. *Amorphophallus henryi* was collected by the Glasnevin Taiwan Expedition (GTE 111) in Henry's type locality at Wanshoushan, in October 2004. It is still a common plant on the floor of tropical rainforest on

Wanshoushan.

Augustine Henry had the following to say about this curious aroid;

Of peculiar plants, a new species of *Amorphophallus*, found on Ape's Hill, is perhaps the most curious. From a tuber arises a leafless spadex, expanded above into a hollow organ, covered in bristles, dull red or purple in colour, a gruesome sight, at which I have seen dogs take fright. In the following year, it sends up its leaves.

Amorphophallus hirtus N. E. Brown (ARACEAE)
New species

in Journ. Linn. Soc. Bot. 36: 181. (1903), Engler in Engler, Pflanzenr. iv. 23C: 67. (1911).

Amorphophallus sp. nova. Henry in Trans. Asiat. Soc. Jap. 24 suppl: 100. (1896) (List Pl. Formosa).

Taiwan: Wanshoushan, 1914, (type).

A tuberous perennial to 1 metre tall (though specimens up to 2.5 metres tall have been recorded on Wanshoushan), native to the area of Taiwan south of the Tropic of Cancer between altitudes of 90 to 800 metres. A spectacular aroid, on Wanshoushan this species is commonly seen in the tropical forests that surround the National Sun Yat Sen University. Endemic to Taiwan.

Amorphophallus konjac K. Koch (ARACEAE)
Brown in Journ. Linn. Soc. Bot. 36: 182. (1903).

Amorphophallus rivieri Durieu Henry in Notes Econ. Bot. China 35. (1893).

Hubei: Yichang, wild in the Antelope Gorge, 4223. **Sichuan:** South Wushan, cultivated, 5544, 5544a.

The Devil's tounge, snake palm or umbrella arum. According to Henry this was the *mo-yu*, the plant was cultivated in garden plots for its tubers, when the root is deprived of its acrid qualities by washing it with water it is made into what was sold at Yichang as *mo-yii tou-fu* (i.e. similar to tou fu, the extract of soja beans). Henry's collections from Sichuan were from cultivated plants, those from the Antelope Gorge near Yichang were from wild plants. It is still a valuable crop plant in central China, the Glasnevin Central China Expedition 2002 met with it in the mountainous districts above the Three Gorges where it grew between rows of maize. It is exported from China to Japan. Cultivated at the National Botanic Gardens, Glasnevin from material collected by the Glasnevin Central China Expedition (GCCE 354) near the village of Bai Sha in Badong in October 2004.

Amorphophallus sp. No. 1. (ARACEAE)
Henry in Trans. Asiat. Soc. Jap. 24 suppl: 100. (1896) (List Pl. Formosa).

Taiwan: Wanchin, without number.

Henry describes this un-numbered collection as a large species with tubers 8 inches in diameter. Alas, he also states that the specimens obtained were unluckily destroyed.

Amorphophallus sp. No. 2. (ARACEAE)
Anonymous in Kew List of Determinations (volume 34).

Yunnan: Simao, 13057. Mengzi 13710, 13804.

Ampelopsis aconitifolia Bunge (VITACEAE)
Zheng in Journ. Wuhan Bot. Res. 2.(1): 141. (1984).

Hubei: On Tsui Fu Shan near Yichang (May 1888), 1622. In the glens near Yichang, 2031. Without locality, 3632a.

Accordiing to W. J. Bean, of vines with compound leaves and deeply cut leaflets this is the hardiest and most luxuriant in growth. Introduced to cultivation by the Emil Bretschneider through the Arnold Arboretum from the Beijing area in 1881. This handsome climber was collected by the Glasnevin Central China Expedition near the San You Dong Glen, Yichang (GCCE 716) in October 2004.

Ampelopsis delavayana Planch. ex Franch. (VITACEAE)
Zheng in Journ. Wuhan Bot. Res. 2.(1): 142. (1984).

Hubei: Yichang, 2765 (February 1887), 4106. Xiling (Yichang) Gorge, 3620.

Introduced by Wilson in 1900 when collecting for Messrs Veitch.

Ampelopsis delavayana Planch. ex Franch. var. *glabra* Diels & Gilg (VITACEAE) New variety
in Bot. Jahrb. 29: 465. (1900), Zheng in Journ. Wuhan Bot. Res. 2.(1): 141. (1984).

Hubei: Yichang, 3632 (isotype of *Ampelopsis aconitifolia* Bunge var. *glabra* Diels & Gilg.).

Ampelopsis delavayana Planch. ex Franch. var. *setulosa* (Diels & Gilg.) C. L. Li (VITACEAE) New variety
Ampelopsis aconitifolia Bunge f. *setulosa* Diels & Gilg in Bot. Jahrb. 29: 465. (1900).

Hubei: Xingshan, 6479 (syntype of *Ampelopsis aconitifolia* Bunge f. *setulosa* Diels).

Ampelopsis glandulosa (Wall.) Momiy. (VITACEAE)
Ampelopsis brevipedunculata Maxim. & Tratuv. var. *vestita* Rehder in Journ. Arnold Arb. 2: 176. (1922), Zheng in Journ. Wuhan Bot. Res. 2.(1): 142. (1984).

Hubei: Jianshi, 7519.

Ampelopsis glandulosa (Wall.) Momiy. var. *brevipedunculata* (Maxim. & Tratuv.) Monmiy. (VITACEAE)
Ampelopsis brevipedunculata Maxim. & Tratuv. Zheng in Journ. Wuhan Bot. Res. 2.(1): 142. (1984).

Hubei: On the road to the Dome, (June 1886), 1552. On Tsui Fu Shan near Yichang (May 1888), 1617. Xiling (Yichang) Gorge, 3624 (October 1887). Yichang, 6239.

Ampelopsis glandulosa (Wall.) Momiy. var. *hancei* (Planch.) Rehd. (VITACEAE)
Ampelopsis brevipedunculata Maxim. & Tratuv. var.

02

hancei (Planch.) Rehd. Zheng in Journ. Wuhan Bot. Res. 2.(1): 142. (1984).

Hubei: On the hills behind the Monastery Valley, 1660.

Ampelopsis glandulosa (Wall.) Momiy. var. *heterophylla* (Thunb.) Momiy. (VITACEAE)

Vitis heterophylla Thunb. Henry in Trans. Asiat. Soc. Jap. 24 suppl: 28: (1896) (List Pl. Formosa).

Hubei: Yichang, 352, 580 (in part). Liantuo, 4394, 4632. Changyang, 7768. South Donghu, 7581, 7582. **Hainan:** On Liang Shan, 16 km (10 miles) east of Haikou, 8199. Environs of Kiungchow, 8280. Without locality, 8475. **Taiwan:** Oluanpi, 283, 661. Wanchin, 454.

Ampelopsis japonica (Thunb.) Makino (VITACEAE)

Zheng in Journ. Wuhan Bot. Res. 2.(1): 142. (1984).

Vitis serianaefolia (Bunge) Maxim. Forbes & Hemsley in Journ. Linn. Soc. Bot. 23: 136. (1886), Henry in Journ. China Br. Roy. Asiat. Soc. 22: 258. (1887) (Chinese Names of Plants). *Ampelopsis serianaefolia* (Thunb.) Regel Diels & Gilg. in Bot. Jahrb. 29: 466. (1900). *Ampelopsis mirabilis* Diels & Gilg in Bot. Jahrb. Syst. 29: 465. (1900), Schneider in Illus. Handb.Laubholzk. 2: 320. (1912), Zheng in Journ. Wuhan Bot. Res. 2.(1): 143. (1984).

Hubei: Yichang, 341, 745, 990, 3638 (type of *Ampelopsis mirabilis* Diels & Gilg., October 1887). Yichang, on the hills behind the San You Dong Glen (July 1886), 1513. On Tsui Fu Shan near Yichang (May 1888), 1607. Liantuo, 3004.

Native to China and Korea and only cultivated in Japan. This species has tuberous roots like a *Dahlia* which are used in Chinese herbal medicine to treat boils, scalds and burns. According to Henry this species was known colloquially in the Three Gorges region as *mao erh luan*.

Amphicarpaea edgeworthii Benth. (FABACEAE)

Anonymous in Kew List of Determinations (volume 34). Henry in Journ. China Br. Roy. Asiat. Soc. 22: 274. (1887) (Chinese Names of Plants). *Amphicarpaea edgeworthii* Benth. var. *japonica* Oliv. Forbes & Hemsley in Journ. Linn. Soc. Bot. 23: 188. (1887).

Hubei: Yichang, 84, 2713. Antelope Glen, 4380. Badong, 4955. **Sichuan:** South Wushan, 7221. **Yunnan:** Milê 10331. Yuanjiang, 13237.

A widespread species with a large range over Russia, India, China, Korea, Vietnam and Japan. According to Henry this species was known colloquially in the Three Gorges region as *yeh mao pien tou*.

Amphicarpaea ferruginea Benth. (FABACEAE)

Shuteria ferruginea (Benth.) Baker Anonymous in Kew List of Determinations (volume 34).

Yunnan: Mengzi, 10291.

Anacardiaceae (ANACARDIACEAE)

Anonymous in Kew List of Determinations (volume 34).

Yunnan: Simao, 12904.

Anaphalis margaritacea (L.) Benth. & Hook. f. (ASTERACEAE)

Forbes & Hemsley in Journ. Linn. Soc. Bot. 23: 425. (1888).

Hubei: Badong, 364 (June 1885), 1041 (December 1885), 2402, 2543, 2861, 4994, 5005, 5039. Yichang, 927, 1054, 2056. Liantuo, 4634. Without locality 2543a. **Yunnan:** In Pingbian on the Daweishan Range at 1,950 metres, 9052.

Anaphalis sinica Hance (ASTERACEAE)

Anaphalis pterocaulon Maxim. Forbes & Hemsley in Journ. Linn. Soc. Bot. 23: 426. (1888).

Hubei: Yichang, 163, 163a., 2300. Antelope Glen, 3339. Badong, 278 (July 1885), 1704, 1811, 2518, 4662, 4795, 4965, 4993, 5048.

Anaphalis sp. No. 1. (ASTERACEAE)

Anaphalis araneosa DC. Anonymous in Kew List of Determinations (volume 24).

Hubei: San You Dong Glen, 1696. Yichang, 2715. Antelope Glen near Yichang, 3134. Badong, 2576, 2394, 2460, 2477. Liantuo, 3033.

Anaphalis sp. No. 2. (ASTERACEAE)

Anonymous in Kew List of Determinations (volume 34).

Yunnan: Mengzi, 9308 (in part).

Anaphalis viridis Cummins (ASTERACEAE) New species

in Kew Bull. Misc. Inf. 19. (1911).

Sichuan: Henry's Chinese collector with Pratt, no locality stated but may be Leshan, Emei Shan or Kangding, 8922.

Ancylostemon humilis W. T. Wang (GESNERIACEAE)

Didissandra saxatilis Hemsl. var. *microcalyx* Hemsley in Journ. Linn. Soc. Bot. 26: 227. (1890). *Ancylostemon saxatilis* (Hemsl.) Craib in Notes Roy. Bot. Gard. Edinburgh 11: 266. (1918—1919) in part.

Sichuan: South Wushan, on vertical cliffs, without number.

Andropogon chinensis (Nees) Merr. (POACEAE)

Andropogon apricus Trin. var. *indicus* Hack. Rendle in Journ. Linn. Soc. Bot. 36: 370. (1904).

Yunnan: On barren clay hills near Mengzi at 1,300 metres, 10995.

Andropogon sp. (POACEAE)

Anonymous in Kew List of Determinations (volume 34).

Yunnan: Simao, 12457, 13732, 13735.

Androsace axillaris (Franch.) Franch. (PRIMULACEAE)

Pax & Knuth in Engler, Pflanzenr. iv. 236: 178. (1905).

Yunnan: Mengzi, 10868.

Discovered by Père Delavay before 1885.

Androsace henryi Oliver (PRIMULACEAE) New species

in Hooker's. Icon. Pl. 18: t. 1973. (1891), Bretschneider in Hist. Eur. Bot. Disc. China 2: 786. (1898), Pax & Knuth in Engler, Pflanzenr. iv. 236: 176. (1905), J. H. Veitch in Hortus Veitchii 414. (1906), W. I. in Gard. Chron. ser. 3, 51: 354. (1912), Fedde in Repert. Spec. Nov. Regni Veg. 19: 21. (1924), Handel-Mazetti in Notes Roy. Bot. Gard. Edinburgh 15: 270. (1925—1927), Morley in Glasra 3: 78. (1979), Hu in Acta Phytotax. Sin. 24: 116. (1986), Nelson in A Heritage of Beauty 310. (2000).

Hubei: South Badong, 4868, 5364, 5364a. Changleping, 6364c. Without locality, 5364b., 5364d.

James Veitch in *Hortus Veitchii* described this plant as "a somewhat large species for the genus with reniform crenate leaves on hairy petioles 4 to 6 inches (10~15 cm.) in length, and umbels of pure white flowers freely produced." Introduced to cultivation by Wilson. This species was later collected in Sichuan by Wilson and Harry Smith, in the same province on Emei Shan by Ernst Faber, on the Yunnan-Myanmar border by Forrest and Handel-Mazzettii and in Myanmar by Frank Kingdon Ward. It is no longer in cultivation.

Androsace umbellata (Lour.) Merr. (PRIMULACEAE)

Androsace saxifragaefolia Bunge Henry in Journ. China Br. Roy. Asiat. Soc. 22: 275. (1887) (Chinese Names of Plants), Forbes & Hemsley in Journ. Linn. Soc. Bot. 26: 45. (1889), Henry in Trans. Asiat. Soc. Jap. 24 suppl: 57. (1896) (List Pl. Formosa), Pax & Knuth in Engler, Pflanzenr. iv. 236: 179. (1905).

Hubei: Yichang, 805. Badong, 1280. Henry's Chinese collector with Pratt, no locality stated but may be any of the following, Leshan, Emei Shan, Wa Shan or Kangding, 8853. **Taiwan:** Wanchin, without number.

According to Henry this species was known colloquially in the Three Gorges region as *t'ung ch'ien ts'ao.*

Anemone davidii Franch. (RANUNCULACEAE)

Ziman, S., Keener, C. S., Kodata, Y., Bulakh, E., Tsarenko, O., & Dutton, B. E. in J. Jpn. Bot. 79: (2004).

Sichuan: South Wushan, 5581.

Anemone flaccida Fr. Schmidt (RANUNCULACEAE)

Anonymous in Anonymous in Kew List of Determinations (volume 24).

Hubei: Changyang, 5235.

Anemone hupehensis (Lemoine) Lemoine (RANUNCULACEAE) New species

Anemone japonica Henry non (Thunb.) Sieb. & Zucc. in Journ. China Br. Roy. Asiat. Soc. 22: 259. (1887) (Chinese Names of Plants), Diels in Bot. Jahrb. 29: 331. 205. (1900).

Hubei: Yichang, 3, 2086.

According to Henry this species was a common sight in the glens above the Three Gorges in autumn. He stated that in the same region it was colloquially known as *yeh mien hua. Anemone hupehensis* is still one of the commonest wild flowers of Hubei and forms a charming picture when seen in masses in open glades in the mountains of the western part of the province. Several collections of this species were made by both the 2002 and 2004 Glasnevin Central China Expeditions and plants derived from these trips grow in the central section of the Chinese slope and in the order beds at Glasnevin. *Anemone hupehensis* was described by Lemoine from plants which flowered in his nursery in 1909 having been presumably raised from seeds sent to France by one of the French missionaries. It was one of Henry's earliest finds having been collected by him in the autumn of 1885.

Anemone hupehensis V. Lemoine var. ***japonica*** (Thunb.) Bowles & Stearn (RANUNCULACEAE)

Anemone japonica (Thunb.) Sieb. & Zucc. Anonymous in Kew List of Determinations (volume 34).

Yunnan: Mengzi, 9313.

Native to China and Japan.

Anemone rivularis Buch.-Ham. ex DC. (RANUNCULACEAE)

Anonymous in Kew List of Determinations (volume 34).

Yunnan: Mengzi, 9435.

A very beautiful perennial to 1.5 metres tall, native to India, Sri Lanka, Bhutan, Nepal, and China. A spectacular plant when seen in masses and forming extensive colonies in the meadows of south-east Tibet (Xizang).

Anemone sp. (RANUNCULACEAE)

Anonymous in Kew List of Determinations (volume 24).

Hubei: Xingshan, 6978.

Angelica setchuenensis Diels (APIACEAE)

Angelica henryi H. Wolff in Repert. Spec. Nov. Regni Veg. 28: 109. (1930).

Hubei: Badong, 4961 (type of *Angelica henryi* H. Wolff).

Angelica sinensis (Oliv.) Diels (APIACEAE) New species

Angelica polymorpha Maxim. var. *sinensis* Oliver in Hooker's Icon. Pl. t. 1999. (1891), Henry in Notes Econ. Bot. China 23. (1893), Bretschneider in Hist. Eur. Bot. Disc. China 2: 784. (1898).

Hubei: Fang Xian, 6897. **Sichuan:** South Wushan,

7143 (type of *Angelica polymorpha* Maxim. var. *sinensis* Oliver).

Henry discovered this species in 1888, though he gathered his specimens from cultivated material. The roots of *Angelica sinensis* are the source of *tang-kui* (*dang gui* i.e. 'state of return'), one of the most important drugs used in everyday Chinese medicine. Henry stated it was much esteemed in the treatment of womens diseases. *Dang gui* is widely available in western and central China, it is sold by vendors on Emei Shan in Sichuan and is also available in health shops in Dublin and other European cities. The roots are used as a tonic and are also used in cases of abdominal pain and treating traumatic injuries. There are root specimens in the National Herbarium at Glasnevin.

Angelica spp. (APIACEAE)

Anonymous in Kew List of Determinations (volume 24).

Hubei: Yichang, 2188. Badong, 3166, 4872. South Badong, 7363. South Donghu, 7634. Changyang, 7516. Without locality, 2421a., 6629a. **Sichuan:** South Wushan, 7471. North Wushan, 7634a., 7634b., 7436c.

Angiopteris caudatiformis Hieron. (MARATTIACEAE)
New species

in Hedwigia 61: 278. (1919).

Yunnan: Mengzi, 9399d.

Angiopteris evecta (Forst.) Hoffm. (MARATTIACEAE)

Henry in Trans. Asiat. Soc. Jap. 24 suppl: 117. (1896) (List Pl. Formosa).

Taiwan: Wanchin, 1499.

Angiopteris latipinna (Ching) Z. R. He, W. M. Chu & Christenh. (MARATTIACEAE)

Archangiopteris henryi Christ & Giesenhagen in Flora Jena 86: 77. (1899), Christ in Bull. Herb. Boiss. 7: 19. (1899), Hemsley in Journ. Linn. Soc. Bot. 34: 178. (1900), E. H. Wilson in A Naturalist in Western China 13. (1913).

Yunnan: In Pingbian on the Daweishan Range at 1,650 metres, 11544 (type). Mengzi 11560a.

Endemic to Yunnan.

Anisadenia pubescens Griff. (LINACEAE)

Oliver in Hooker's Ic. 26: t. 2593. (1901).

Yunnan: In Pingbian on the Daweishan Range, 9046.

Distributed in India and China [Tibet (Xizang) and Yunnan].

Anisochilus pallidus Wall. ex Benth. (LAMIACEAE)

Dunn in Notes Roy. Bot. Gard. Edinburgh 6: 145. (1915).

Yunnan: In forests to the south of Simao at 1,300 metres, 12675.

A widespread species found over much of India, Myanmar, Laos, Vietnam and China (Yunnan).

Anisomeles indica (L.) Kuntze (LAMIACEAE)

Anisomeles ovata R. Br. Forbes & Hemsley in Journ. Linn. Soc. Bot. 26: 299. (1890), Henry in Trans. Asiat. Soc. Jap. 24 suppl: 73. (1896) (List Pl. Formosa), Dunn in Notes Roy. Bot. Gard. Edinburgh 6: 178. (1915).

Hubei: Liantuo, on the banks of the Yangtze, 4442. **Yunnan:** Mengzi, 10149, 10149a. **Taiwan:** Kaohsiung, without number. Oluanpi, 223, 642.

Anisopappus chinensis (L.) Hook & Arn. (ASTERACEAE)

Inula yunnanensis Anthony in Notes Roy. Bot. Gard. Edinburgh 18: 189. (1933—1935).

Yunnan: On grassy hills near Mengzi at 1,650 metres, 9145 (syntype of *Inula yunnanensis* Anthony).

Anisopappus chinensis was also collected in Yunnan by Forrest and Rock.

Anneslea fragrans Wall. (THEACEAE)

Kobuski in Journ. Arnold Arb. 33: (1952).

Symplocos neriifolia Henry non Sieb. & Zucc. in Trans. Asiat. Soc. Jap. 24 suppl: 58. (1896) (List Pl. Formosa). *Anneslea fragrans* Wallich var. *lanceolata* Hayata Li in Woody Flora of Taiwan 579. fig. 228. (1963).

Taiwan: Oluanpi, 1359. **Yunnan:** Yuanjiang at 1,300 to 1,650 metres, 11591. Mountains to the east of Simao at 1,300 to 1,650 metres 11591a., 11591c.

An evergreen shrub or small tree to 15 metres tall. Distributed in Nepal, Myanmar, China, Vietnam, Laos, Thailand and Cambodia. The genus *Anneslea* in honour of George Annesley, Lord Mountnorris, who collected plants in north Africa and southern Europe while Viscount Valentia.

Annona squamosa L. (ANNONACEAE)

Henry in Trans. Asiat. Soc. Jap. 24 suppl: 16. (1896) (List Pl. Formosa).

Hainan: 32 km (20 miles) west of Haikou, 8154. Environs of Haikou 8400. **Taiwan:** Kaohsiung, without number.

The sugar apple, an American native, cultivated and introduced to Taiwan (according to Henry) by Dutch colonists, it was known colloquially in Taiwan as *shih-chia-kuo*.

Annonaceae (ANNONACEAE)

Anonymous in Kew List of Determinations (volume 34).

Yunnan: Simao, 12154.

Anodendron affine (Hook. & Arn.) Druce (APOCYNACEAE)

Li in Woody Flora of Taiwan 782. fig. 314. (1963).

Anodendron laeve (Champ. ex Benth.) Maxim. Henry in Trans. Asiat. Soc. Jap. 24 suppl: 60. (1896) (List Pl. Formosa).

Taiwan: Kaohsiung, without number. Wanchin, 19. Oluanpi, 262.

A climbing shrub, native to southern China and Japan. In Taiwan this species is found at low and medium altitude throughout the island.

Anodendron benthamianum Hemsley (APOCYNACEAE)

Henry in Trans. Asiat. Soc. Jap. 24 suppl: 60. (1896) (List Pl. Formosa).

Taiwan: Wanchin, 185, 801. Oluanpi, 683.

A climbing shrub, endemic to Taiwan where it is widely distributed and locally common. Discovered by the Kew collector, Richard Oldham.

Anodendron sp. (APOCYNACEAE)

Henry in annotation in the Index Florae Sinensis (copy at N.B.G., Glasnevin).

Taiwan: Oluanpi, 1308.

Anoectochilus formosanus Hayata (ORCHIDACEAE)

Anoectochilus roxburghii Henry non (Wall.) Lindl. in Trans. Asiat. Soc. Jap. 24 suppl: 92. (1896) (List Pl. Formosa), Rolfe in Journ. Linn. Soc. Bot. 36: 42. (1903).

Taiwan: Wanchin, 1626.

A terrestrial orchid to 15 cm tall, endemic to Taiwan where it is common in forests below 1,500 metres.

Anotis sp. (RUBIACEAE)

Anonymous in Kew List of Determinations (volume 34).

Yunnan: Simao, 13512, 13536.

Anthogonium gracile Wallich (ORCHIDACEAE)

Rolfe in Journ. Linn. Soc. Bot. 36: 21. (1903).

Yunnan: On grassy mountains near Mengzi at 1,600 metres, 11099. Simao 12555.

Anthoxanthum hookeri (Griseb.) Rendle (POACEAE)

Rendle in Journ. Linn. Soc. Bot. 36: 380. (1904).

Yunnan: On grassy mountains near Mengzi at 1,600 metres, 10517.

Distributed from India to western and central China.

Anthriscus sylvestris (L.) Hoffm. (APIACEAE)

Diels in Bot. Jahrb. 29: 492. (1900).

Hubei: Fang Xian, 6922. **Sichuan:** South Wushan, 5609.

Cow parsley, distributed across Europe, north Africa and Asia.

Anticlea sibiricus (L.) Kunth (MELANTHIACEAE)

Zigadenus sibiricus (L.) A. Gray Wright in Journ. Linn. Soc. Bot. 36: 148. (1903).

Hubei: Fang Xian, 6889. **Sichuan:** North Wushan, without number.

Antidesma ambiguum Pax & K. Hoffman (PHYLLANTHACEAE) New species

in Pflanzenr. (Engler) 4. Fam. 147: 127. (1922).

Yunnan: Mengzi at 1,650 metres, 10767, (isosyntype). Native to China (Yunnan) and Vietnam.

Antidesma bunius (L.) Spreng. (PHYLLANTHACEAE)

Forbes & Hemsley in Journ. Linn. Soc. Bot. 26; 430. (1894), Pax & Hoffmann in Engler, Pflanzenr. iv. 147. xv: 160. (1922).

Hainan: Environs of Kiungchow, 8316 (in part).

Antidesma costulatum Pax & Hoffmann (PHYLLANTHACEAE) New species

in Engler, Pflanzenr. iv. 147. xv: 129. (1922).

Yunnan: Mengzi, 13690.

Endemic to China and distributed in Sichuan and Yunnan.

Antidesma fordii Hemsley (PHYLLANTHACEAE)

Antidesma yunnanense Pax & K. Hoffmann in Pflanzenr. (Engler) 4. Fam. 147: 157. (1922).

Yunnan: Mengzi at 1,650 metres, 10768 (isosyntype of *Antidesma yunnanense* Pax & K. Hoffmann). Simao at 1,400 metres 12130 (isosyntype of *Antidesma yunnanense* Pax & K. Hoffmann).

Distributed in southern China, Laos and Vietnam.

Antidesma ghaesembilla Gaertn. (PHYLLANTHACEAE)

Forbes & Hemsley in Journ. Linn. Soc. Bot. 26: 431. (1894), Pax & Hoffmann in Engler, Pflanzenr. iv. 147. xv: 155. (1922).

Hainan: Hummocks near Haikou, 8099, 8100. Environs of Haikou, 8371. Without locality, 8488.

Antidesma japonicum Sieb. & Zucc. (PHYLLANTHACEAE)

Antidesma japonicum Henry in Trans. Asiat. Soc. Jap. 24 suppl: 83. (1896) (List Pl. Formosa).

Taiwan: Wanchin, 176, 539, 874.

Antidesma montanum Blume (PHYLLANTHACEAE)

Pax & Hoffmann in Engler, Pflanzenr. iv. 147. xv: 158. (1922).

Antidesma henryi Hemsley in Journ. Linn. Soc. Bot. 26: 432. (1894), Pax & K. Hoffmann in Pflanzenr. (Engler) 4. Fam. 147(15): 132. (1922). *Antidesma apiculatum* Hemsley in Journ. Linn. Soc. Bot. 26: 430. (1894), Pax & Hoffmann in Engler, Pflanzenr. iv. 147. xv: 127. (1922). *Antidesma calvescens* Pax & Hoffmann in Engler, pflanzenr. iv. 147. xv: 118. (1922). *Antidesma diandrum* Engler non (Roxb.) Roth. Pax & Hoffmann in Pflanzenr. iv. 147. xv: 143. (1922). *Antidesma paxii* F. P. Metcalf in Lingnan Sci. J. 10: 485. (1931). *Antidesma* sp. Henry in Trans. Asiat. Soc. Jap. 24 suppl: 83. (1896) (List Pl. Formosa). *Antidesma pentandrum* (Blanco) Merrill var. *barbatum* (Presl.) Merrill Pax & Hoffmann in Engler, Pflanzenr. iv. 147. xv: 125. (1922).

Hainan: Without locality, 36, 8512 (type of *Antidesma apiculatum* Hemsley), 8562 (type of *Antidesma henryi* Hemsley). **Taiwan:** Kaohsiung, 1144, 1885. **Yunnan:** Mountains to the west of Simao at 1,500 metres, (syntype of *Antidesma calvescens* Pax & Hoffmann). South of the

Red River from Manmei, 13667a., (syntype of *Antidesma henryi* Pax & K. Hoffmann). South of the Red River 13688. Simao, 12735, 13032, 13032a. South of the Red River from Manmei, 13667, (syntype of *Antidesma paxii* F. P. Metcalf).

Distributed in Myanmar, China, Cambodia, Laos and Vietnam. Collected by the Glasnevin Taiwan Expedition (GTE 36) by the lighthouse at Oluanpi (South Cape of Henry and Wilson) in October 2004.

Antidesma sp. No. 1. (PHYLLANTHACEAE)

Anonymous in Kew List of Determinations (volume 24).

Hainan: Without locality, 8471.

Antidesma sp. No. 2. (PHYLLANTHACEAE)

Henry in Trans. Asiat. Soc. Jap. 24 suppl: 83. (1896) (List Pl. Formosa).

Taiwan: Kaohsiung, 780. Wanchin, 915.

Antidesma sp. No. 3. (PHYLLANTHACEAE)

Anonymous in Kew List of Determinations (volume 34).

Yunnan: At Hsienkei on the Red River near the Vietnamese border, 9561. Mengzi 10667. Simao 12029.

Antidesma venosum E. Mey ex Tul. (PHYLLANTHACEAE)

Antidesma microphyllum Hemsley Pax & Hoffmann in Engler, Pflanzenr. iv. 147. xv: 139. (1922).

Yunnan: In the Red River Valley near Manhao, 9530b.

Distributed in southern China, Laos, Thailand and Vietnam.

Antiotrema dunnianum (Diels) Hand.-Mazz. (BORAGINACEAE)

Cynoglossum dunnianum Diels in Notes Roy. Bot. Gard. Edinburgh 5: 169. (1912). *Henryettana mirabilis* Brand in Repert. Spec. Nov. Regni Veg. 26: 171. (1929) (genus novum).

Yunnan: On grassy mountains near Mengzi at 1,800 metres, 10600 (isotype of *Henryettana mirabilis* Brand). Mengzi, 10600a.

An upgight perennial herb to 30 cm tall, producing from March to July masses of blue-purple tubular flowers. This handsome perennial was discovered by William Hancock near Mengzi in 1893, Henry stated it grew as extensive colonies on the Mengzi plain. Endemic to China and distributed in Guangxi, Guizhou, Sichuan and Yunnan.

Antrophyum formosanum G. Hieronymus (PTERIDACEAE) New species

in Hedwigia 57: 210. (1915).

Taiwan: Kaohsiung, without number.

Native to southern China (including Taiwan) and Japan (south).

Antrophyum henryi G. Hieronymus (PTERIDACEAE) New species

in Hedwigia 57: 208. (1916).

Yunnan: In Pingbian on the Daweishan Range at 1,900 metres, on trees, 11517a.

Distributed in India, southern China and Thailand.

Antrophyum obovatum Baker (PTERIDACEAE) New species

in Kew Bull. Misc. Inf. 233. (1898).

Antrophyum latifolium Christ non Blume in Bull. Herb. Boiss. 6: 868., non Bl. (1898), *Antrophyum petiolatum* Baker in Kew Bull. Misc. Inf. 14. (1906).

Yunnan: On rocks in mountain forest near Mengzi at 1,950 metres, 9153a., (type). On the mountains to the east of Mengzi at 1,950 metres, 9153 (type of *Antrophyum petiolatum* Baker).

Distributed in India, Bhutan, Nepal, Myanmar, China, Thailand, Vietnam and Japan.

Antrophyum plantagineum Kaulf. (PTERIDACEAE)

Henry in Trans. Asiat. Soc. Jap. 24 suppl: 116.(1896) (List Pl. Formosa).

Taiwan: Wanshoushan, without locality.

Wet tropical Asia to the Pacific.

Antrophyum reticulatum (G. Forst.) Kaulf. (PTERIDACEAE)

Anonymous in Kew List of Determinations (volume 34).

Yunnan: Mengzi, 11562.

Antrophyum vittarioides Baker (PTERIDACEAE)

Christ in Bull. Herb. Boiss. 6: 868. (1898).

Antrophyum stenophyllum Baker in Kew Bull. Misc. Inf. 233. (1898).

Yunnan: On rocks by Hsien t'an above Hsinkei on the Red River, 9607 (type of *Antrophyum stenophyllum* Baker, collected 19th June 1896).

Aphanamixis polystachya (Wall.) R. N. Parker (MELIACEAE)

Amoora rohituka (Roxb.) Wight. & Arn. Henry in Notes Econ. Bot. China 61. (1893).

Hainan: Environs of Kiungchow, 8233, 8253, 8404.

A small tree, native to India, Myanmar, Indochina, western and southern China, the Malay Peninsula to the Philippines and south to Borneo, Sumatra, Java, New Guinea and Australia. This is the *t'ung-yu-shu*, according to Henry the oil obtained from this species was used for lighting purposes.

Aphyllorchis caudata Rolfe ex Downie (ORCHIDACEAE)

Anonymous in Kew List of Determinations (volume 34).

Yunnan: Simao, 12969.

Apiaceae (APIACEAE)

Umbellifera dubia Forbes & Hemsley in Journ. Linn. Soc. Bot. 23: 337. (1887). Umbellifera, undetermined.

Henry in Trans. Asiat. Soc. Jap. 24 suppl: 47. (1896) (List Pl. Formosa).

Hubei: Badong, 1742, 2389, 4079, 4977, 4978, 6072, 6254. South Badong, 7386. Antelope Glen, 3257. Changleping, 6254. **Hunan:** Shimen, 7563. **Sichuan:** Henry's Chinese collector with Pratt, no locality but may be Leshan, Emei Shan, Wa Shan or Kangding, 8943. **Taiwan:** Wanchin, 385. Oluanpi, 600, 1356. Kaohsiung, 1812. **Yunnan:** Mengzi, 13789, 13791.

Apios americana Medik. (FABACEAE)
Apios tuberosa Moench Anonymous in Kew List of Determinations (volume 34).

Yunnan: Mengzi, 9337.

Apios carnea (Wall.) Benth. ex Baker (FABACEAE)
Anonymous in Kew List of Determinations (volume 34).

Yunnan: Simao, 12586.

Native to India, Nepal, western China, Thailand and Vietnam.

Apios delavayi Franch. var. *gracillima* (Dunn)B. Pan bis (FABACEAE) New variety
Apios gracillima Dunn in Journ. Linn. Soc. Bot. 35: 488. (1903).

Yunnan: Mengzi at 1,900 metres, 9828 (isotype, collected September 14th 1896).

Endemic to Yunnan.

Apios fortunei Maxim. (FABACEAE)
Anonymous in Kew List of Determinations (volume 24).

Hubei: Yichang, 4270. Donghu, 6394. South Badong, 7303.

A tuberous yellow-flowered climber to 3 metres tall. Native to both Japan and China.

Apios macrantha Oliver (FABACEAE) New species
in Hooker's Icon. Pl. t. 1946. (1890).

Sichuan: Emei Shan, Henry's Chinese collector with Pratt, 8984.

Apium graveolens L. (APIACEAE)
Henry in Trans. Asiat. Soc. Jap. 24 suppl: 47. (1896) (List Pl. Formosa), Wolff in Engler, Pflanzenr. iv. 228: 28. (1927).

Hainan: Without locality, 17. **Taiwan:** Kaohsiung, without number.

Wild celery, a biennial species native to temperate Europe and Asia and in the mountains of the tropics. It occurs both wild and cultivated in China, in China celery is derived from an Asian plant, the crop grown in Europe and north America may be traced back to wild European plants. Herklots in Vegetables in South-East Asia 128. (1972) states that strong flavoured varieties (cultivars) have been bred in

China, both in Hainan and in Taiwan.

Apluda mutica L. (POACEAE)
Henry in Trans. Asiat. Soc. Jap. 24 suppl: 108. (1896) (List Pl. Formosa), Rendle in Journ. Linn. Soc. Bot. 36: 378. (1904).

Taiwan: Kaohsiung, 743, 1114.

A widespread grass, distributed from India to Australia.

Apocynaceae (APOCYNACEAE)
Anonymous in Kew List of Determinations (volume 24 & 34). Asclepiadaceae, Anonymous in Kew List of Determinations (volume 24 & 34).

Hainan: On hummocks near Haikou, 8125. Without locality, 8534. **Yunnan:** Simao, 12908, 13164, 13194, 13490, 13510, 13756.

Aporosa octandra (Buch.-Ham. Ex D. Don)Vickery var. *chinensis* (Champ) Schot (PHYLLANTHACEAE)
Aporosa leptostachya Benth Forbes & Hemsley in Journ. Linn. Soc. Bot. 26: 429. (1894). *Aporosa microcalyx* Hassk. var. *chinensis* (Champ) Muell.-Arg. Pax & Hoffmann in Engler, Pflanzenr. iv. 147. xv: 102. (1922).

Hainan: Without locality, 8495, 8519.

Aporosa yunnanensis (Pax & K. Hoffm.) F. P. Metcalf. (PHYLLANTHACEAE) New species
Aporosa wallichii J. D. Hooker var. *yunnanensis* Pax & K. Hoffmann in Pflanzenr. (Engler) 4. Fam. 147(15): 90. (1922).

Yunnan: Simao, 11638, (isotype of *Aporosa wallichii* J. D. Hooker var. *yunnanensis* Pax & K. Hoffmann). Simao at 1,650 metres 11638b., (isosyntype of *Aporosa wallichii* J. D. Hooker var. *yunnanensis* Pax & K. Hoffmann). Mountains to the south of Simao at 1,400 metres 11638e., (isosyntype of *Aporosa wallichii* J. D. Hooker var. *yunnanensis* Pax & K. Hoffmann). Simao at 990 metres 12828, (isosyntype of *Aporosa wallichii* J. D. Hooker var. *yunnanensis* Pax & K. Hoffmann).

Aquilaria sinensis (Lour.) Merr. (THYMELAEACEAE)
Aquilaria grandiflora Benth. Forbes & Hemsley in Journ. Linn. Soc. Bot. 26: 402. (1894).

Hainan: Environs of Kiungchow, 8768.

Aquilegia oxysepala Trautv. & Mey. (RANUNCULACEAE)
Aquilegia vulgaris Anonymous non L. in Kew List of Determinations (volume 24).

Hubei: Badong, 4736. Changleping, 6255. Donghu, 6427. Fang Xian, 6427a., 6427b. Xingshan, 6427c.

A. oxysepala was discovered by Nicolai Przewalski in Gansu on 7th August 1872. Seeds were taken from Henry's 6427 in the Kew herbarium on April 15th 1889 for propagation.

Aquilegia sp. (RANUNCULACEAE)
Anonymous in Kew List of Determinations (volume

02

24).

Hubei: Badong, 877.

Arabis hirsuta (L.) Scop. (BRASSICACEAE)

Anonymous in Kew List of Determinations (volume 24 & 34).

Hubei: Badong, 4028. **Yunnan:** Mengzi, 10709.

Distributed from Russia to Kazakstan, China and Korea.

Arabis sp. (BRASSICACEAE)

Anonymous in Kew List of Determinations (volume 24).

Hubei: Yichang, 3734. South Badong, 5408. Fang Xian, 6882.

Either *Arabis paniculata* Franch. or *Arabis pendula* L.

Arachis hypogaea L. (FABACEAE)

Henry in Journ. China Br. Roy. Asiat. Soc. 22: 244. (1887) (Chinese Names of Plants), Henry in Trans. Asiat. Soc. Jap. 24 suppl: 32. (1896) (List Pl. Formosa).

Hubei: Yichang, 2318. **Taiwan:** Kaohsiung, without number.

The ground nut, monkey nut or peanut, a tropical annual from southern Bolivia and northern Argentina. China is a leading producer of this crop and according to Henry it was known colloquially at Yichang as *hua sheng*. In the late 19th century the peanut was cultivated on the sandy plains around Kaohsiung (Takow) and according to Henry it was used as food while oil extracted from the seeds was used for lighting.

Arachniodes amabilis (Blume) Tindale (POLYPODIACEAE)

Aspidium amabile Blume Henry in Trans. Asiat. Soc. Jap. 24 suppl: 113. (1896) (List Pl. Formosa).

Taiwan: Wanchin, 1636.

A widespread fern, distributed from Nepal to the Philippines.

Arachniodes aristata (G. Forst) Tindale (POLYPODIACEAE)

Aspidium aristatum Sw. Baker in Journ. Bot. 26: 228. (1888), Henry in Trans. Asiat. Soc. Jap. 24 suppl: 113. (1896) (List Pl. Formosa). *Polystichum aristatum* (G. Forst.) Presl. Baker in Journ. Bot. 26: 228. (1888).

Hubei: Liantuo, 3048, 4408. Changleping, 7819. **Sichuan:** North Wushan, 7108. **Taiwan:** Tamsui, 1416, 1425, 1437. **Yunnan:** Mengzi, 11535.

Distributed in India, Sri Lanka, China, Korea and Japan.

Arachniodes carvifolia (Kunze) Ching (POLYPODIACEAE)

Polystichum coniifolium C. Presl. Christ in Bull. Soc. Bot. France 5: 32. (1905).

Hubei: Badong, 1717, 3678.

Arachniodes coniifolia (T. Moore) Ching (POLYPODIACEAE)

Polystichum speciosum (Don.) J. Smith Christ in Notul.

Syst. (Paris) 1: 36. (1909).

Yunnan: Mengzi at 1,650 metres, 10461.

Native to Nepal, Myanmar and southern China.

Arachniodes henryi (H. Christ) Ching (POLYPODIACEAE)

New species

in Acta Bot. Sin. 10: 258. (1962).

Polystichum henryi H. Christ in Lecompte, Notul. Syst. (Paris) 1: 36. (1909).

Yunnan: Simao, 13351.

Endemic to China and distributed across Sichuan and Yunnan.

Arachniodes standishii (T. Moore) Ohwi. (POLYPODIACEAE)

Aspidium laserpitiifolium Mett. Anonymous in Kew List of Determinations (volume 24).

Hubei: Badong, 1718.

Aralia bipinnata Blanco (ARALIACEAE)

Li in Woody Flora of Taiwan. 665. fig. 271. (1963).

Aralia sp. Henry in Trans. Asiat. Soc. Jap. 24 suppl: 47. (1896) (List Pl. Formosa).

Taiwan: Wanchin, 15.

A small prickly tree, native to western New Guinea, the Philippines, Japan and China (Taiwan). In Taiwan this species is commonly encountered on the edges of forest at low altitude.

Aralia chinensis L. (ARALIACEAE)

Zheng in Journ. Wuhan Bot. Res. 2.(1): 160. (1984).

Aralia spinosa Henry non L. in Journ. China Br. Roy. Asiat. Soc. 22: 279. (1887) (Chinese Names of Plants).

Hubei: Liantuo, 2104, 2585. Badong 2535, 2585.

A small spiny tree to 7 metres tall, known colloquially in the Three Gorges region as *tz'u pang t'ou*. In Chinese herbal medicine the stems of this species is currently used as a diuretic, to promote blood circulation and as a painkiller. Allied to *Aralia spinosa* is the remarkably handsome *Aralia echinocaulis* Hand.-Mazz. This species was introduced to cultivation by the Glasnevin Central Expedition (GCCE 259) from Badong in October 2002. It grows in the central section of the Chinese slope at Glasnevin and at the Talbot Botanic Gardens, Malahide Castle in County Dublin.

Aralia cordata Thunb. (ARALIACEAE)

Henry in Notes Econ. Bot. China 23. (1893).

Hubei: Fang Xian, 6785.

A perennial to 1.5 metres tall, found in woods and mountain ravines throughout Japan, Sakhalin, Korea and China. Henry stated that in the Three Gorges region this species was treated as a vegetable, being popular in soups and was known as *t'u-tan-kui*. It is still grown as a vegetable in parts of Asia, the young shoots are blanched in spring like seakale when about 60 cm tall. It is said to resemble asparagus in taste. Henry's found this species at Fang in

1888 and it was a first record for China. Scores of other plants were also first records but for the sake of space I have not recorded them all here.

Aralia decaisneana Hance (ARALIACEAE)

Aralia chinensis Anonymous non L. in Kew List of Determinations (volume 24).

Hainan: Without locality, 8776.

A sparingly prickly shrub to 3 metres tall, endemic to China.

Aralia foliolosa (Wall.) Seem. ex Clarke (ARALIACEAE)

Wen in Acta Phytotax. Sin. 38: 7. (2000).

Yunnan: Simao, 12711.

Distributed in India, China (Hunan and Yunnan) and Vietnam.

Aralia henryi Harms (ARALIACEAE) New species

in Bot. Jahrb. Syst. 23: 12. (1896), Morley in Glasra 3: 79. (1979), Nelson in A Heritage of Beauty 310. (2000).

Aralia pilosa Franchet in Journ. de Bot. 302. (1896), Bretschneider in Hist. Eur. Bot. Disc. China 2: 748. (1898).

Hubei: Fang Xian, 6655 (isolectotype).

Aralia searelliana Dunn (ARALIACEAE) New species

in Journ. Linn. Soc. Bot. 35: 498. (1903).

Yunnan: Between Pu'er and Simao and in the forests south of Simao at 1,300 metres, 13426 (type).

Distributed in Yunnan, Myanmar and Vietnam.

Aralia spp. (ARALIACEAE)

Anonymous in Kew List of Determinations (volume 24 & 34).

Hubei: Badong, 555 (September 1885). South Badong, 5396, 6088. Liantuo, 5396b. Jianshi, 5396c., 5396h. Xinghan, 5396f., 5396g., 5396e. On the borders of Fang and Xingshan, 5396j. **Sichuan:** South Wushan, 5396d. North Wushan, 5396i. **Yunnan:** Mengzi, 13646.

Aralia thomsonii Seem. ex C. B. Clarke (ARALIACEAE)

Wen in Acta Phytotax. Sin. 38: 6. (2000).

Yunnan: Mengzi, 9479a.

Distributed in India, southern China and Vietnam.

Aralia verticillata (Dunn) J. Wen (ARALIACEAE) New species

Pentapanax verticillatus Dunn in Journ. Linn. Soc. Bot. 35: 498. (1903).

Yunnan: Mountains to the south-west of Mengzi at 2,300 metres, 9284 (type of *Pentapanax verticillatus* Dunn).

China (Guangxi and Yunnan) to northern Vietnam.

Archidendron lucidum (Benth.) I. Nielsen (FABACEAE)

Pithecellobium lucidum Benth. Henry in Trans. Asiat. Soc. Jap. 24 suppl: 40. (1896) (List Pl. Formosa), Li in Woody Flora of Taiwan 357. fig. 127. (1963).

Taiwan: Wanchin, 433, 1657.

A small evergreen tree, native to Japan (south) and China (Hong Kong, Hainan and Taiwan).

Archidendron robinsonii (Gagnep.) I. C. Nielson (FABACEAE)

Cylindrokelupha robinsonii (Gagnep.) Kosterm. Wu in Acta Phytotax. Sin. 19: 218. (1981).

Yunnan: Mengzi, 9373.

Distributed in western China and Vietnam.

Arctium lappa L. (ASTERACEAE)

Henry in Journ. China Br. Roy. Asiat. Soc. 22: 261. (1887) (Chinese Names of Plants), Forbes & Hemsley in Journ. Linn. Soc. Bot. 23: 460. (1888), Henry in Notes Econ. Bot. China 7. (1893).

Hubei: In a ravine on Moji Shan, 4277. Liantuo, 4507. Badong, 286 (July 1885), 4835. **Yunnan:** Simao, 11879.

Burdock, a European native. Henry's specimens were collected from both wild and cultivated plants. He stated it was cultivated for its seeds which were used as a drug known as *ta-li-tzu*. This species was known colloquially in the Three Gorges region as *niu p'ang tzu*. Burdock is a root crop in Japan and it is cultivated in other parts of the world for its edible leaves. In China the fruits are used with the root of *Platycodon grandiflorus* to treat inflammation of the throat and with the fruits of the *lian qiao* (*Forsythia suspensa*) for pain and swelling of the throat.

Ardisia conspersa E. Walker (PRIMULACEAE)

Ardisia undulata Mez, non C. B. Clarke, in Pflanzenr. (Engler) Myrsin. 4. Fam. 236: 146. (1902).

Yunnan: Mengzi at 1,600 metres, 10779 (isotype of *Ardisia undulata* Mez).

An evergreen shrub to 2 metres tall, unbranched except for the lateral flowering shoots which are borne on upper part of the plant. Flowering occurs in June, followed by attractive umbels of red globose fruits. Distributed in China (southern Yunnan) and Vietnam.

Ardisia cornudentata Mez (PRIMULACEAE) New species

in Engler, Pflanzenr. 4. Fam. 236: 104. (1902), Li in Woody Flora of Taiwan 710. (1963).

Ardisia sp. Henry in Trans. Asiat. Soc. Jap. 24 suppl: 58. (1896) (List Pl. Formosa).

Taiwan: Wanchin, 54 (isosyntype)1074a., (isosyntype), 975, 1074.

A shrub to one metre tall, endemic to Taiwan where it is common from low to high altitude. This species was collected by the Glasnevin Taiwan Expedition (GTE 53) by the lighthouse at Oluanpi (South Cape of Henry and Wilson) in October 2004.

Ardisia corymbifera Mez (PRIMULACEAE) New species

in Pflanzenr. (Engler) 4. Fam. 236: 149. (1902), Pipoly

02

& Chen in Novon 5: 360. (1995).

Yunnan: Simao at 1,300 metres, 12000a., (isosyntype). Distributed in southern China, Vietnam and Japan.

Ardisia crenata Sims. (PRIMULACEAE)

Forbes & Hemsley in Journ. Linn. Soc. Bot. 26: 63. (1889), Mez in Engler, Pflanzenr. iv. 236: 144. (1902), Zheng in Journ. Wuhan Bot. Res. 2.(1): 172. (1984).

Ardisia crispa Mez non (Thunb.) A. DC. in Engler, Pflanzenr. iv. 236: 144. (1902), Rehder in Sargent Pl. Wilson. 2: 581. (1916).

Hubei: Xiling (Yichang) Gorge, 3265 (February 1887). Liantuo, 4409. Hainan: Environs of Haikou, 8417. Lingmen (in the interior of the island) 8650. Yunnan: Mengzi, 10832, 9333, 9852, 11328, 11356. Simao 12088.

An evergreen shrub to 2 metres tall with a flat tuber-like root, flowering occurs in May and June followed by cymose umbels of long-lasting red, fleshy fruits from October to December. To the Hani (an ethnic group) of Xishuangbanna in southern Yunnan this species is called *hmqmaqlavqssev* and the roots and leaves are used to treat laryngitis, bronchitis and joint pain. Distributed in India, China, the Indochina Peninsula, Indonesia, Myanmar, Japan, the Philippines and the Malay Peninsula.

Ardisia crispa (Thunb.) A. DC. (PRIMULACEAE)

Mez in Engler, Pflanzenr. iv. 236: 144. (1902), Rehder in Sargent Pl. Wilson. 2: 581. (1916).

Ardisia henryi Hemsley in Journ. Linn. Soc. Bot. 26: 65. (1889), Bretschneider in Hist. Eur. Bot. Disc. China 2: 787. (1898), Diels in Bot. Jahrb. 29: 518. (1900), Mez in Engler. Pflanzenr. iv.-236. 149. (1902), Rehder in Sargent Pl. Wilson 2: 582. (1916), Morley in Glasra 3: 78. (1979), Zheng in Journ. Wuhan Bot. Res. 2.(1): 172. (1984), Nelson in A Heritage of Beauty 310. (2000), O' Brien in Augustine Henry, An Irish Plant Collector in China 12. (2002). *Tinus henryi* Kuntze in Rev. Gen. 2: 974. (1891). *Ardisia crispa* (Thunb.) A. DC. var. *amplifolia* E. Walker in J. Wash. Acad. Sci. 29: 259. (1939).

Hubei: Yichang, 3972, 3455. Changyang, 7747. Xiling (Yichang) Gorge, 3455 (type of *Ardisia henryi* Hemsley), 4314. Fang Xian, 6733. Yunnan: Mengzi at 1,650 metres, 9653, [type of *Ardisia crispa* (Thunb.) A. DC. var. *amplifolia* E. Walker]. Mengzi at 1,600 metres, 9791, 9791a. Simao, 12088c.

Distributed in Vietnam, China, Korea and Japan.

Ardisia faberi Hemsley (PRIMULACEAE) New species

in Journ. Linn. Soc. Bot. 26: 64. (1889), Mez in Engler, Pflanzenr. iv. 236: 153. (1902).

Hubei: Yichang, 3304. Changyang, 6237. Yunnan: Mengzi, 11338.

Hemsley based this species on material gathered on Emei Shan by Faber (Faber, 795) and Henry's Yichang specimen, (A. Henry 3304). Native to southern and central China.

Ardisia humilis Vahl. (PRIMULACEAE)

Ardisia hainanensis Mez in Engler, Pflanzenreich, Myrsinac. iv. 236: 138. (1902).

Hainan: Environs of Haikou, 8017 (syntype of *Ardisia hainanensis* Mez).

Ardisia hypargyrea C. Y. Wu & C. Chen (PRIMULACEAE) New species

in Fl. Yunnan. 340. pl. 79. (1977).

Ardisia salicifolia E. Walker in Bull. Fan. Mem. Inst. Biol. Bot. 9: 156. (1939) non A. de Candolle (1844).

Yunnan: In Pingbian, in forests on the Daweishan Range at 1,650 metres, 11032 (isotype of *Ardisia salicifolia* E. Walker).

Native to western China and Vietnam.

Ardisia japonica (Thunb.) Blume (PRIMULACEAE)

Henry in Journ. China Br. Roy. Asiat. Soc. 22: 234. (1887) (Chinese Names of Plants), Forbes & Hemsley in Journ. Linn. Soc. Bot. 26: 65. (1889), Mez in Engler, Pflanzenr. iv. 236: 151. (1902), Rehder in Sargent Pl. Wilson 2: 582. (1916), Li in Woody Flora of Taiwan 714. (1963).

Hubei: Yichang, 98, 2275. Changyang, 7683. Taiwan: Wanchin, 461.

A decumbent suffrutescent undershrub, native to China (including Taiwan) and Japan. According to Henry this small shrub was known colloquially as *ch'a kou*. It is currently used in Chinese herbal medicine as a diuretic and to promote blood circulation.

Ardisia nigropilosa Pit. (PRIMULACEAE) New species

Ardisia stellata E. Walker in Bull. Fan Mem. Inst. Biol. Bot. 9: 155. (1939).

Yunnan: In the mountains at Yuanjiang at 1,650 metres, 13226 (isotype of *Ardisia stellata* E. Walker).

Distributed in China (Yunnan) and Vietnam.

Ardisia obtusa Mez (PRIMULACEAE) New species

in Engler, Pflanzenreich, Myrsinac. iv. 236: 104. (1902), Pipoly & Mez in Novon 5: 359. (1995).

Hainan: Near the Kiungchow Pagoda, 7990 (type).

Ardisia polysticta Miq. (PRIMULACEAE)

Li in Woody Flora of Taiwan 710. (1963).

Ardisia crenata Henry non Sims. in Trans. Asiat. Soc. Jap. 24 suppl: 57. (1896) (List Pl. Formosa), *Ardisia radians* Hemsley & Mezz in Engler, Pflanzenreich. Myrsin, 146. (1902). *Ardisia patens* Mez in Engler, Pflanzenr. Mrysin. 236: 149. (1902). *Ardisia maculosa* Mez in Engler, Pflanzenr. 9. (iv. 236): 146. (1902). *Ardisia corymbifera* Mez in Pflanzenr. (Engler) 4. Fam. 236: 149. (1902), Pipoly

& Chen in Novon 5: 360. (1995).

Taiwan: Wanchin, 85, 555. Oluanpi, 909. Yunnan: Simao at 1,600 metres, 11724, 12000. Simao at 1,650 metres 12088b., (isotype of *Ardisia patens* Mez). Simao at 2,000 metres, 12088d., (syntype of *Ardisia patens* Mez). Simao at 1,950 metres 12088e., (holotype of *Ardisia maculosa* Mez). Simao 11624 c., (syntype of *Ardisia radians* Hemsley & Mezz), 13578. Mengzi, 9226a., (syntype of *Ardisia radians* Hemsley & Mezz), 11400, 13671.

A shrub to 3 metres tall, widely distributed across India, Myanmar, southern mainland China (including Hainan and Taiwan), Thailand and Vietnam.

Ardisia purpurea Reinw. ex Blume (PRIMULACEAE)

Ardisia lanceolata Roxb. Anonymous in Kew List of Determinations (volume 34).

Yunnan: Mengzi, 13773.

Belongs to another taxon.

Ardisia quinquegona Blume (PRIMULACEAE)

Mez in Engler, Pflanzenreich, Myrsinac. iv. 236: 108. (1902), Li in Woody Flora of Taiwan 709. (1963).

Ardisia sieboldii Henry non Miq. in Trans. Asiat. Soc. Jap. 24 suppl: 57. (1896) (List Pl. Formosa).

Hainan: Without locality, 8500. Taiwan: Oluanpi, 651, 698.

A shrub or small tree to 6 metres tall, native to Malaysia, southern China (including Taiwan) and Japan (south). A common inhabitant of forests in Taiwan.

Ardisia replicata Walker (PRIMULACEAE) New species

in Bull. Fan. Mem. Inst. Biol. 9: 169. fig. 21. (1939).

Yunnan: In Pingbian on the Daweishan Range at 1,666 to 2,000 metres, 11444 (type).

Native to China (Yunnan) and Vietnam.

Ardisia scalarinervis E. Walker (PRIMULACEAE) New species

in J. Wash. Acad. Sci. 21: 477. (1931).

Yunnan: Simao at 1,450 metres, 12021 (isotype).

Endemic to Yunnan.

Ardisia sieboldii Miq. (PRIMULACEAE)

Henry in Trans. Asiat. Soc. Jap. 24 suppl: 57. (1896) (List Pl. Formosa), Li in Woody Flora of Taiwan 709. fig. 291. (1963).

Taiwan: Wanchin, 414. Oluanpi, 667. Tamsui, 1733. Wanshoushan, 1881.

A shrub or small tree, native to southern Japan and Taiwan of China where it is common at low altitude throughout the island.

Ardisia spp. (PRIMULACEAE)

Anonymous in Kew List of Determinations (volume 24 & 34).

Hubei: Liantuo, 6351. Yunnan: In the Red River Valley near Manhao, 9536. Mengzi, 11028, 11463. Simao, 11629, 12693, 13407.

Ardisia thyrsiflora D. Don (PRIMULACEAE)

Pipoly & Chen in Novon 5: 359. (1995).

Ardisia tenera Mez in Engler, Pflanzenr. 4. Fam. 236: 104. (1902). *Ardisia yunnanensis* Mez in part, misapplied by Walker in Philip. Journ. Sci. 73: 64. (1940.) *Ardisia austroasiatica* Walker in Journ. Arnold Arb. 23: 347. (1942). *Ardisia yunnanensis* Mez in Engler, Pflanzenr. Myrsin. 4. Fam. 236: 107. (1902), Walker in Journ. Arnold Arb. 23: 348. (1942). *Ardisia penduliflora* Mez in Engler, Pflanzenr. iv. 236: 150. (1902).

Hubei: Liantuo, 6365 (syntype of *Ardisia penduliflora* Mez). Yunnan: Simao at 1,450 metres, 12094, (isosyntype of *Ardisia tenera* Mez). Simao at 1,500 metres, 12123a. In forests near Simao at 1,650 metres 13095 (type of *Ardisia yunnanensis* Mez). Simao, 11994.

Distributed in India, Nepal, Myanmar, China and Vietnam.

Ardisia villosa Roxb. (PRIMULACEAE)

Ardisia vestita Wall. in Engler, Pflanzenreich, Myrsinac. iv. 236: 408. (1902).

Hainan: Hummocks near Haikou, 8103. Environs of Haikou, 8388. Yunnan: Simao, 12175.

Distributed in southern China and Thailand.

Areca catechu L. (ARECACEAE)

Henry in Trans. Asiat. Soc. Jap. 24 suppl: 99. (1896) (List Pl. Formosa), Wright in Journ. Linn. Soc. Bot. 36: 167. (1903).

Hainan: Environs of Haikou, 8406. Ling-men (in the interior of the island), 8655. Taiwan: Kaohsiung Plain, without number.

According to Henry the areca or betelnut palm was known in Chinese as *pin-lang* (*bing-lang),* the nut together with the husk as *ta-fu-tze* and the husk is *ta-fu-p'i.* Fans made from the leaf of this palm were known as *lao-hsio-shan.* Wilson stated that betel-chewing was not in general vogue with the Chinese who valued it more as a medicinal plant. In current Chinese medicine it is used to kill parasites like tapeworm, pinworm, roundworm and flukes.

Arecaceae (ARECACEAE)

Palmae Anonymous in Kew List of Determinations (volume 34).

Yunnan: Simao, 11808.

Arenaria serpyllifolia L. (CARYOPHYLLACEAE)

Henry in Journ. China Br. Roy. Asiat. Soc. 22: 279. (1887) (Chinese Names of Plants).

Hubei: Yichang, 904, 1194.

According to Henry this species was known

colloquially in the Three Gorges district as *wo pu shih ts'ao.*

Arenga tremula (Blanco) Becc. (ARECACEAE)

Arenga engleri Henry non Becc. in Trans. Asiat. Soc. Jap. 24 suppl: 99. (1896) (List Pl. Formosa), Wright in Journ. Linn. Soc. Bot. 36: 167. (1903), Li in Woody Flora of Taiwan 917. fig. 365. (1963).

Taiwan: Wanshoushan, 798.

A large shrub to 3.5 metres tall, native to tropical Asia. This species is a common and very beautiful under-storey palm in the tropical forests of southern Taiwan. According to Henry this palm was known as the *ts'ung*, a name given on mainland China to *Trachycarpus fortunei*. *Arenga tremula* was collected by the Glasnevin Taiwan Expedition (GTE 98) on Wanshoushan (Ape's Hill of Henry, Swinhoe and Wilson) where it is an abundant under-storey species in dense tropical forest.

Argentina lineata (Trevir.) Soják. (ROSACEAE)

Potentilla fulgens Wall. ex Hook. Anonymous in Kew List of Determinations (volume 34).

Yunnan: Fengchunling, 10723.

Native to India, Nepal, Bhutan, Myanmar, central and western China and Vietnam.

Argostemma verticillata Wall. ex Roxb. (RUBIACEAE)

Anonymous in Kew List of Determinations (volume 34).

Yunnan: Simao, 13122.

Argyreia argentea Roxb.) Sweet (CONVOLVULACEAE)

Anonymous in Kew List of Determinations (volume 34).

Yunnan: Simao, 12511.

Argyreia bella (C. B. Clarke) Raizada (CONVOLVULACEAE)

Lettsomia bella C. B. Clarke Anonymous in Kew List of Determinations (volume 34).

Yunnan: Mengzi, 9904.

Argyreia capitiformis (Poir.) Ooststr. (CONVOLVULACEAE)

Lettsomia strigosa Roxb. Anonymous in Kew List of Determinations (volume 34).

Yunnan: Mengzi, 9777. Simao, 12656.

Argyreia convolvuloides (Prain) Rattanakr. & Traiperm (CONVOLVULACEAE)

Blinkworthia convolvuloides Prain Anonymous in Kew List of Determinations (volume 34).

Yunnan: Mengzi, 11178.

Native to south-west China and Myanmar.

Argyreia formosana Ishigami ex Yamaz. (CONVOLVULACEAE)

Argyreia tiliifolia Henry non Wight in Trans. Asiat. Soc. Jap. 24 suppl: 82. (1896) (List Pl. Formosa).

Taiwan: Oluanpi, 199a. Wanchin, 1610, 1671.

Argyreia henryi (Craib) Craib (CONVOLVULACEAE) New species

in Kew Bull. Misc. Inf. 9. (1914).

Ipomoea henryi Craib in Kew Bull. Misc. Inf. 423. (1911).

Yunnan: Simao at 1,650 metres, 13387 (isosyntype of *Ipomoea henryi* Craib).

Distributed in Yunnan (Xishuangbanna) and Thailand.

Argyreia hirsutissima (C. B. Clarke) Thoth. (CONVOLVULACEAE)

Lettsomia hirsutissima C. B. Clarke Anonymous in Kew List of Determinations (volume 34).

Yunnan: Simao, 13444.

Argyreia obtusifolia Lour. (CONVOLVULACEAE)

Lettsomia chalmersii Hance Anonymous in Kew List of Determinations (volume 24).

Hainan: Environs of Kiungchow, 8306.

Argyreia sp. No. 1. (CONVOLVULACEAE)

Henry in annotation in the *Index Florae Sinensis* (copy at National Botanic Gardens, Glasnevin).

Taiwan: Kaohsiung, 1699.

Argyreia sp. No. 2. (CONVOLVULACEAE)

Lettsomia sp. Anonymous in Kew List of Determinations (volume 34).

Yunnan: In woods on Fengchunling at 2,100 metres, 9701.

Argyreia sp. nova (CONVOLVULACEAE)

Lettsomia sp. nova. Henry in Trans. Asiat. Soc. Jap. 24 suppl: 63. (1896) (List Pl. Formosa).

Taiwan: Wanchin, 775.

Arisaema amurense Maxim. (ARACEAE)

Arisaema amurense Maxim. var. *magnidens* N. E. Bown in Journ. Linn. Soc. Bot. 36: 176. (1903).

Hubei: Liantuo, 5393a. Badong, 5393 (isotype of *Arisaema amurense* Maxim. var. *magnidens* N. E. Brown). Without locality, 6363a., (syntype of *Arisaema amurense* Maxim. var. *magnidens* N. E. Brown), 6730a.

A striking aroid to 40 cm tall, plants currently cultivated at the National Botanic Gardens, Glasnevin were raised from seeds collected in October 2004 on Tianthu Shan, Changyang by the Glasnevin Central China Expedition (GCCE 774). On Tianthu Shan in Changyang it grew on the forest floor with *Paris* and *Cardiocrinum*. The last western botanist to collect in the Changyang region before the Glasnevin group was British plant hunter, Ernest Wilson.

Arisaema asperatum N. E. Brown (ARACEAE) New species

in Journ. Linn. Soc. Bot 36: 176. (1903), Engler in Engler, Pflanzenr. iv. 23F: 213. (1920).

Hubei: Badong, 3776 (type)

Henry's type specimen was cultivated in Badong where the root was used as a drug.

Arisaema bockii Engl. (ARACEAE)

Arisaema sazensoo (Buerg ex Bl.) Makino var. *henryanum* Engler in Pflanzenr. 4, Fam. 23: 205. (1920). *Arisaema sikokianum* Franch & Sav. var. *henryanum* (Engl.) H. Li Acta Phytotax. Sin. 15(2): 108. (1977).

Hubei: South Badong, 5394a., [isotype of *Arisaema sazensoo* (Buerg ex Bl.) Makino. var *henryanum* Engler].

This very beautiful aroid was collected by the Glasnevin Central China Expedition on Tianthu Shan, Changyang (GCCE 791) in October 2004. Cultivated in the woodland garden at the National Botanic Gardens, Glasnevin. A spectacular plant.

Arisaema consanguineum Schott. (ARACEAE)

Brown in Journ. Linn. Soc. Bot. 176. (1903).

Arisaema consanguineum Schott f. *latisectum* Engl. Engler in Engler, Pflanzenr. iv. 23F: 176. (1920).

Hubei: Badong, 3780, 4653, 5303a., 5397a, 5397b. South Badong, 5303, 9397. Liantuo, 4546. **Yunnan:** Mengzi at 1,900 metres, 10133a.

The highly poisonous tubers are used as a source of medicine.

Arisaema franchetianum Engler (ARACEAE)

Engler in Engler, Pflanzenr. iv. 23F: 184. (1920).

Yunnan: Mengzi, 10969. Simao, 13048.

A handsome and sinister-looking cobra lily with deep purple, white-striped sphates. Distributed in Guangxi, Guizhou, Sichuan and Yunnan (and N. Myanmar). The poisonous tubers are used as a source of medicine.

Arisaema heterophyllum Blume (ARACEAE)

Henry in Journ. China Br. Roy. Asiat. Soc. 22: 274. (1887) (Chinese Names of Plants); in Notes Econ. Bot. China 26. (1893), Brown in Journ. Linn. Soc. Bot. 36: 178. (1903).

Arisaema amurense Maxim. var. *sazensoo* Brown non Engl. in Journ. Linn. Soc. Bot. 36: 175. (1903). *Arisaema ambiguum* Engler in Engler, Pflanzenr. 23F: 187. (1920). *Arisaema thunbergii* Brown non Blume in Journ. Linn. Soc. Bot. 36: 180. (1903). *Arisaema multisectum* Engler in Pflanzenr. 4. Fam. 23: 186. (1920), in part.

Hubei: Yichang, 212. Liantuo, 3030 (syntype of *Arisaema ambiguum* Engler). Xiling (Yichang) Gorge, 3574. South Badong, 5508. Jianshi, (June 23rd 1888), 5370a., (type of *Arisaema multisectum*, Engler). Badong, 5370, (syntype of *Arisaema multisectum* Engler).

In the Three Gorges region this species was colloquially known as *tu chio lien*, according to Henry it was the most common aroid in the plains near Yichang. Its roots were used in medicine.

Arisaema japonicum Blume (ARACEAE)

Henry in Notes Econ. Bot. China 34. (1893), Takeda in Bull. Misc. Inf. 216. (1912), Brown in Journ. Linn. Soc. Bot. 36: 178. (1903).

Hubei: South Badong, 5371. **Sichuan:** Without locality, 5536.

The *tian nan xing* or 'star of the southern heavens, this species is used in Chinese herbal medicine in cases of deep rooted sores, ulcers and carbuncles. When used with the rhizome of *tian mu* (*Gastrodia elata*) it is used to treat cases of dizziness, vertigo, seizures and muscle spasm. The Hani (an ethnic group) of Xishuangbanna in southern Yunnan call this species *taimavq* and use it to treat gastritis.

Arisaema lobatum Engl. (ARACEAE)

Arisaema pictum N. E. Brown in Journ. Linn. Soc. Bot. 36: 179. (1903). *Arisaema lobatum* Engl. var. *rosthornianum* Engler in Engler in Engler, Pflanzenr. iv. 23F: 202. (1920). *Arisaema lobatum* Engl. var. *latisectum* Engler in Pflanzenr. 73. (iv. 23f.): 202. (1920).

Hubei: South Badong, 5381(syntype of *Arisaema pictum* N. E. Br.). **Sichuan:** South Wushan, 5381a., 5381c., (type of *Arisaema lobatum* Engl. var. *latisectum* Engler). Without locality, 5381b.

Arisaema prazeri Hook. f. (ARACEAE)

Arisaema prazeri Hook. f. var. *variegatum* Engler in Engler, Pflanzenr. 73.(iv. 23f.): 160. (1920).

Yunnan: Mengzi at 1,560 metres, 9583a., (syntype of *Arisaema prazeri* Hook. f. var. *variegatum* Engler). Simao at 1,500 metres, 13081.

Distributed in Myanmar and China (Yunnan).

Arisaema sikokianum Franch. & Sav. (ARACEAE)

Arisaema sazensoo (Buerg.) Makino var. *serrato-dentatum* Engl. Engler in Engler, Pflanzenr. iv. 23F: 205. (1920).

Hubei: Liantuo, 4571, 5394b. South Badong, 5394.

Arisaema spp. (ARACEAE)

Anonymous in Kew List of Determinations (volume 24 & 34).

Hubei: San You Dong Glen, 4371 (in part). Fang Xian, 6824. **Yunnan:** Mengzi, 10992. Simao, 12196, 12899, 13054.

Aristolochia contorta Bunge (ARISTOLOCHIACEAE)

Forbes & Hemsley in Journ. Linn. Soc. Bot. 26: 361. (1891).

Hubei: Liantuo, 4487.

Aristolochia cucurbitifolia Hayata (ARISTOLOCHIACEAE)

New species

in Ic. Pl. Formosa 5: 137. (1915), Ma in Acta Phytotax. Sin. 27: 355. (1989).

Taiwan: Kaohsiung, 719.

A scandent perennial herb, endemic to Taiwan where it is found in thickets and open forest at low altitudes in the central and southern parts of the island.

Aristolochia debilis Sieb. & Zucc. (ARISTOLOCHIACEAE)

Forbes & Hemsley in Journ. Linn. Soc. Bot. 26: 361. (1891), Henry in Notes Econ. Bot. China 19. (1893).

Hubei: Liantuo, 4649.

Yields the 'green putchuk' used as an antidote to snake-bite in China. Rare at Yichang according to Ernest Wilson.

Aristolochia gentilis Franch (ARISTOLOCHIACEAE)

Ma in Acta Phytotax. Sin. 27: 339. (1989).

Yunnan: Mengzi, 11222.

Endemic to China and distributed in Yunnan and Sichuan.

Aristolochia heterophylla Hemsley (ARISTOLOCHIACEAE)
New species

in Journ. Linn. Soc. Bot. 26: 361. (1891), Henry in Notes on Econ. Bot. China 19. (1893), Bretschneider in Hist. Eur. Bot. Disc. China 2: 790. (1898), Diels in Bot. Jahrb. 29: 310. (1900), J. H. Veitch in Hortus Veitchii 357. (1906), Bean in Trees & Shrubs. 1: 207. (pro parte) (1914), Rehder & E. H. Wilson in Sargent Pl. Wilson. 3: 323. (1917), Ma in Acta Phytotax. Sin. 27: 356. (1989); 32: 291. (1994), Nelson in A Heritage of Beauty 310. (2000), O' Brien in Augustine Henry, An Irish Plant Collector in China 12. (2002).

Aristolochia sp. Henry in Journ. China Br. Roy. Asiat. Soc. 22: 259. (1887) (Chinese Names of Plants).

Hubei: Antelope Glen, 3493, Donghu, 6417. Fang Xian, 6417a. Xingshan, 6490. Badong, 4665. **Sichuan:** South Wushan, 5620.

The roots of this species are an important ingredient of traditional Chinese herbal medicine. Henry stated that in the Three Gorges area it was known colloquially as *ch'ing mu hsiang*. *Aristolochia heterophylla* was introduced to cultivation by Wilson through Veitch's Coombe Wood nursery and in Ireland it grows at Mount Usher and Glasnevin. Several species of *Aristolochia* are used medicinally in various parts of the world. Their original use (in Europe) seems to stem from the doctrine of signatures, the flowers are reminiscent of the shape of a human foetus prior to birth hence the Greek *aristos* (best) *lochia* (childbirth). In China other species of the genus used in medicine include *A. fangchi* and *A. kaempferi*. On the Indian subcontinent *A. indica* is used as a contraceptive. In Iran *A. rotunda* is used as a tonic. In North America, the Virginian snake root, *A. serpentaria* is used as a snake-bite remedy. In South America *A. cymbifera* and *A. ringens* are used in cases of ulcers, colic, fever and snake bite.

Aristolochia mollissima Hance (ARISTOLOCHIACEAE)

Forbes & Hemsley in Journ. Linn. Soc. Bot. 26: 362. (1891), Rehder & E. H. Wilson in Sargent Pl. Wilson. 3: 324. (1917).

Aristolochia aff. *kaempferi* W. Henry in Journ. China Br. Roy. Asiat. Soc. 22: 248. (1887) (Chinese Names of Plants).

Hubei: Yichang, 779, 4144, 4144a.

A slender climber to 1 metre tall, common on the hills around Yichang according to Henry who also stated it was known colloquially as *hsun ku feng*. It was discovered by W. J. Clarke near Shanghai on May 13th 1877.

Aristolochia sp. No. 1. (ARISTOLOCHIACEAE)

Anonymous in Kew List of Determinations (volume 24).

Hubei: Badong, 642 (July 1885).

Aristolochia sp. No. 2. (ARISTOLOCHIACEAE)

Henry in Trans. Asiat. Soc. Jap. 24 suppl: 77. (1896) (List Pl. Formosa).

Taiwan: Wanchin, 402 (cultivated as a medicinal plant).

Aristolochia sp. No. 3. (ARISTOLOCHIACEAE)

Anonymous in Kew List of Determinations (volume 36).

Yunnan: Mengzi, 10712, 13778.

Artabotrys hexapetalus (L. f.) Bhand. (ANNONACEAE)

Artabotrys odoratissimus R. Br. Anonymous in Kew List of Determinations (volume 24).

Hainan: At Nodoa in the interior of the island, 8468.

A scandent shrub to 4 metres tall, climbing by means of hook-like, modified peduncles and carrying carrying highly fragrant yellow to ivory flowers. Native to southern India and Sri Lanka and widely naturalised in the Old World tropics.

Artemisia annua L. (ASTERACEAE)

Henry in Journ. China Br. Roy. Asiat. Soc. 22: 243. (1887) (Chinese Names of Plants), Forbes & Hemsley in Journ. Linn. Soc. Bot. 23: 441. (1888).

Hubei: Yichang, 2221, 4322. South Donghu, 7699. Changyang, 7510, 7742. **Hainan:** 8 km (5 miles) south of Kuingchow, 8175.

The Chinese or sweet wormwood which Henry stated was known colloquially at Yichang as *ch'ou hao qing hao*. A common weed distributed from Eastern Europe to Asia and naturalised in North America. In China this species is cultivated on a commercial scale, it is highly effective against malaria, both as a preventative and in treating the disease (it is said to be effective against drug resistant strains of malaria).

Artemisia anomala S. Moore (ASTERACEAE)

Anonymous in Kew List of Determinations (volume 24).

Hubei: Zigui, 6195.

According to Bretschneider *Artemisia anomala* was discovered in 1873 by Dr. George Shearer on the banks of the Yangtze near Jiujing in Jiangxi province in 1873.

Artemisia capillaris Thunb. (ASTERACEAE)

Henry in Trans. Asiat. Soc. Jap. 24 suppl: 54. (1896) (List Pl. Formosa).

Taiwan: Wanchin, 4, 1597. Tamsui, 1448.

Artemisia dubia Wall. ex Besser var. *septentrionalis* Pamp. (ASTERACEAE)

In the Herbarium of Missouri Botanic Gardens (MO).

Hainan: On Liang Shan, 16 km (10 miles) east of Haikou, 8197.

Artemisia aff. *glauca* Pall. ex. Willd. (ASTERACEAE)

Anonymous in Kew List of Determinations (volume 24).

Hubei: Yichang, 19, 354a.

Artemisia cf. *japonica* Thunb. (ASTERACEAE)

Anonymous in Kew List of Determinations (volume 24).

Hubei: Badong, 329 (June 1885).

Artemisia gmelinii Weber ex Stechmann (ASTERACEAE)

Artemisia sacrorum Led. Forbes & Hemsley in Journ. Linn. Soc. Bot. 23: 444. (1888).

Hubei: Yichang, 3235.

Artemisia japonica Thunb. (ASTERACEAE)

Henry in Journ. China Br. Roy. Asiat. Soc. 22: 243. (1887) (Chinese Names of Plants), Forbes & Hemsley in Journ. Linn. Soc. Bot. 23: 443. (1888).

Hubei: Yichang, 343, 3096, 3236, 3286. Badong 280 (July 1885), 1042 (December 1885), 3143, 5031, 5055. **Hainan:** At Nodoa in the interior of the island, 8451. Lingmen (in the interior of the island), 8636.

According to Henry, in the Three Gorges region this species was known colloquially as *ch'i t'ou hao.*

Artemisia lactiflora Wall. ex DC. (ASTERACEAE)

Henry in Journ. China Br. Roy. Asiat. Soc. 22: 271. (1887) (Chinese Names of Plants), Forbes & Hemsley in Journ. Linn. Soc. Bot. 23: 444. (1888), Irving in The Garden 64: 294. (1903).

Hubei: Yichang, 151, 152, 1119 (cultivated as a vegetable). Badong, 2441. South Badong, 7355. Liantuo, 3028, 4465. Fang Xian, 6590.

A vigorous showy perennial to over a metre tall producing terminal panicles of milky-white flowers. Introduced to cultivation by Wilson who sent seeds to Veitch's Coombe Wood nursery where plants first flowered in 1903. According to Henry this species was known in the Three Gorges region as the *ssu chi ts'ai,* Henry also states it was cultivated in Badong as a vegetable In Veitch's *Novelties for 1908—1909* plants were available for one shilling each. This species was collected by the Glasnevin Central China Expedition (GCCE 2) on Emei Shan (Mount Omei) in September 2002.

Artemisia lavandulifolia DC. (ASTERACEAE)

Forbes & Hemsley in Journ. Linn. Soc. Bot. 23: 446. (1888).

Artemisia vulgaris Henry non L. in Journ. China Br. Roy. Asiat. Soc. 22: 243. (1887) (Chinese Names of Plants), Forbes & Hemsley in Journ. Linn. Soc. Bot. 23: 446. (1888).

Hubei: Yichang, 64, 71, 633, 2780. Liantuo, 2069. Badong, 2403, 4984, 5145. Xiling (Yichang) Gorge, 4315.

According to Henry this species was known colloquially at Yichang as *peh hao tzu.*

Artemisia lavandulifolia DC. subforma. *minutiflora* Pamp. (ASTERACEAE)

In the Herbarium of Missouri Botanic Gardens (MO).

Hubei: Yichang, 4233.

Artemisia mongolica (Fisch. ex Besser) Nakai (ASTERACEAE)

Artemisia vulgaris Henry non L. in Trans. Asiat. Soc. Jap. 24 suppl. 55. (1896) (List Pl. Formosa).

Taiwan: Kaohsiung, without number. Oluanpi, 371.

Artemisia scoparia Waldst. & Kit. (ASTERACEAE)

Forbes & Hemsley in Journ. Linn. Soc. Bot. 23: 445. (1888), Henry in Trans. Asiat. Soc. Jap. 24 suppl: 55. (1896) (List Pl. Formosa).

Hubei: Yichang, 357. Badong 2549, 3696. South Donghu, 7698. **Taiwan:** Oluanpi, 240, 1755.

Artemisia sp. (ASTERACEAE)

Anonymous in Kew List of Determinations (volume 24).

Hubei: Yichang, 2775, 2906. Badong 2401, 3153.

Artemisia sp. **nov.** (ASTERACEAE)

Journ. Linn. Soc. Bot. 23: 446. (1888).

Hubei: Badong, 4677.

Arthraxon hispidus (Thunb.) Makino (POACEAE)

Arthraxon ciliaris Beauv. subsp. *langsdorfii* (Trin.) Hack. Rendle in Journ. Linn. Soc. Bot. 36: 360. (1904), *Arthraxon ciliaris* Beauv. subsp. *submuticus* Hack. Rendle in Journ. Linn. Soc. Bot. 36: 360. (1904).

Hubei: Badong, 4806. **Yunnan:** Mengzi at 1,600 metres, 9137a.

Arthraxon lanceolatus (Roxb.) Hochst. (POACEAE)

Rendle in Journ. Linn. Soc. Bot. 36: 361. (1904).

Hubei: Yichang, 2282. **Yunnan:** Mengzi at 1,600 metres, 9137.

Distributed in India, Pakistan, Nepal and western China.

Artocarpus lakoocha Roxb. (MORACEAE)

Artocarpus sp. Anonymous in Kew List of Determinations (volume 34).

Yunnan: Simao, 11746, 13015.

Artocarpus nitidus subsp. *lingnanensis* (Merr.) F. M. Jarrett (MORACEAE)

Artocarpus integrifolius Forbes & Hemsley non L. f. in Journ. Linn. Soc. Bot. 26: 470. (1899).

Hainan: Environs of Kiungchow, 8255.

Henry stated this was a 3 metre tall tree with large edible fruits.

Aruncus dioicus (Walter) Fernald (ROSACEAE)

Astilbe polyandra Hemsley in Journ. Linn. Soc. Bot. 23: 265. (1887), Henry in Journ. China Br. Roy. Asiat. Soc. 22: 270. (1887) (Chinese Names of Plants), Bretschneider in Hist. Eur. Bot. Disc. China 2: 783. (1898).

Hubei: Badong, 1816 (syntype of *Astilbe polyandra* Hemsley), 4052, 4794. South Badong, 6106, 7276. Changleping, 6251. Jianshi, 5882.

Arundina graminifolia (D. Don) Hochr. (ORCHIDACEAE)

Arundina chinensis Blume Henry in Trans. Asiat. Soc. Jap. 24 suppl. 92. (1896) (List Pl. Formosa), Rolfe in Journ. Linn. Soc. Bot. 36: 27. (1903).

Fujian: Fuzhou, 2078. **Sichuan:** Henry's Chinese collector with Pratt, no locality stated but may be Leshan, Emei Shan, Wa Shan or Kangding, 8915. **Taiwan:** Siko Hills, Tamsui, 1386, 1386a. **Yunnan:** Near Mengzi on grassy mountain slopes at 1,750 metres, 9988. Simao, 12256.

A graceful terrestrial orchid to 150 cm tall with grass-like foliage and carrying in June and July, a terminal inflorescence of pink flowers. This widespread species is distributed from the Himalaya, across China, Indochina and Malaysia. A. Henry 2078 was collected by Hurst in Fujian.

Arundinella hirta (Thunb.) C. Tanaka (POACEAE)

Arundinella anomala Steud. Rendle in Journ. Linn. Soc. Bot. 36: 341. (1904). *Arundinella anomala* Steud. var. *depauperata* Rendle in Journ. Linn. Soc. Bot. 36: 341. (1904).

Hubei: Yichang, 431, 1054 (type of *Arundinella anomala* Steud. var. *depauperata* Rendle), 2265, 4284. On Pingshanba near Yichang, 2272 (August 1886). Xiling (Yichang) Gorge, 4290. Badong, 4845.

Arundinella nepalensis Trin. (POACEAE)

Rendle in Journ. Linn. Soc. Bot. 36: 341. (1904).

Hubei: On Tsui Fu Shan near Yichang (May 1888), 1630. Yichang 2264. **Yunnan:** Mengzi at 1,500 metres, 9139.

Arundinella setosa Trin. (POACEAE)

Rendle in Journ. Linn. Soc. Bot. 36: 342. (1904).

Taiwan: Kaohsiung, 771. Wanshoushan, 1033, 1107. **Yunnan:** Mengzi, 10518, 10518a. In mountains near Simao at 1,600 metres, 13193.

Widespread across south-east Asia.

Arundinella sp. (POACEAE)

Anonymous in Kew List of Determinations (volume 34).

Yunnan: Simao, 12741.

Arundo donax L. (POACEAE)

Rendle in Journ. Linn. Soc. Bot. 36: 408. (1904).

Yunnan: Mengzi at 1,500 metres, 9660a.

Arundo formosana Hack. (POACEAE) New species

in Bull. Herb. Boiss. vii. 724. (1899), Rendle in Journ. Linn. Soc. Bot. 36: 408. (1904).

Taiwan: Kaohsiung, 1153. Wanshoushan, 1153a. Wanchin, 1153b.

Native to China (Taiwan), Japan (south) and the Philippines.

Arundo formosana Hack. var. *gracilis* Hack. (POACEAE) New variety

in Bull. Herb. Boiss. vii. 724. (1899), Rendle in Journ. Linn. Soc. Bot. 36: 408. (1904).

Taiwan: Tamsui, 1462.

Asarum cf. *caudigerum* Hance (ARISTOLOCHIACEAE)

Anonymous in Kew List of Determinations (volume 34).

Yunnan: Mengzi, 11452.

Asarum himalaicum Hook. f. & Thoms. ex Klotzsch (ARISTOLOCHIACEAE)

Forbes & Hemsley in Journ. Linn. Soc. Bot. 26: 359. (1891).

Hubei: South Badong, 5433.

According to Henry this species was known as *hsi-hsin* and the root was employed as a valuable drug.

Asarum maximum Hemsley (ARISTOLOCHIACEAE) New species

in Gard. Chron. ser. 3, 7: 422. (1890); in Journ. Linn. Soc. Bot. 26: 359. (1891), Henry in Notes on Economic Botany of China 29. (1893), J. D. Hooker in Curtis's Bot. Mag. 122: t. 7456. (1896), Bretschneider in Hist. Eur. Bot. Disc. China 2: 790. (1898), Morley in Glasra 3: 78. (1979), Cheng & Yang in Journ. Arnold Arb. 64: 589. (1983), Nelson in A Heritage of Beauty 310. (2000), O'Brien in Augustine Henry, An Irish Plant Collector in China 12. (2002).

Asarum blumei Henry non Duch. in Journ. China Br. Roy. Asiat. Soc. 22: 247. (1887) (Chinese Names of Plants) I. C. 22: 257. (1887).

Hubei: Yichang, 3369 (type).

According to Henry's field notes this plant occured in glens around the Yichang (Xiling) Gorge, on the sides of cliffs and always some distance up. He stated it was known as *ma-ti-hsiang* i.e. horse-hoof fragrance and its root was used in medicine. By 1893 *Asarum maximum* was in cultivation in the public gardens at Shanghai. It was introduced to cultivation by Henry through the Royal Botanic Gardens, Kew where it first flowered in 1895. This is one of the most beautiful members of a remarkable genus and was exhibited amongst many other plants in the silver-gilt award winning exhibit based on Henrys botanical work in China by the Irish Garden Plant Society and Bord Glas at the Chelsea Flower Show, London in May 2002.

Asarum pulchellum Hemsley (ARISTOLOCHIACEAE)
New species

in Gard. Chron. ser. 3, 7: 422. (1890); in Journ. Linn. Soc. Bot. 26: 360. (1891), Bretschneider in Hist. Eur. Bot. Disc. China 2: 790. (1898), Cheng & Yang in Journ. Arnold Arb. 64: 569. (1983), O' Brien in Augustine Henry, an Irish Plant collector in China 12. (2002).

Hubei: Yichang, amongst stones in one spot in the Antelope glen, 7800 (isotype).

Henry discovered this species in the Antelope Glen, the same glen is located about 8 kms from the city of Yichang on the northern bank of the Yangtze, given this proximity it was an area regularly visited by Henry. *Asarum pulchellum* was introduced to cultivation by Henry through the Royal Botanic Gardens, Kew. Native to the provinces of Anhui, Guizhou, Hubei, Jiangxi and Sichuan where it grows in montane woodland and is most abundant between 700 and 1,700 metres alt. *Asarum pulchellum* was collected by the Glasnevin Central China Expedition in October 2004 (GCCE 615) in Dragon Gate River Forest Park, Xingshan where it grew in a humus-rich cleft on a damp cliff face in shade. This species is cultivated in several Irish gardens.

Asarum sieboldii Miq. (ARISTOLOCHIACEAE)

Henry in Journ. China Br. Roy. Asiat. Soc. 22: 247. (1887) (Chinese Names of Plants), Forbes & Hemsley in Journ. Linn. Soc. Bot. 26: 360. (1891), Henry in Notes Econ. Bot. China 4. (1893).

Hubei: South Badong, 5392a.

The root of this species was employed as a drug and was known according to Henry as *hsi-hsin*.

Asclepias curassavica L. (APOCYNACEAE)

Henry in Trans. Asiat. Soc. Jap. 24 suppl: 60. (1896) (List Pl. Formosa).

Sichuan: North Wushan, 7127. **Taiwan:** Kaohsiung, without number. **Yunnan:** Mengzi, 10788.

The blood flower or Indian root, an annual or short lived evergreen subshrub to 1.5 metres tall, a native of tropical America and now a pantropic weed.

Asparagus cochinchinensis (Lour.) Merr. (ASPARAGACEAE)

Asparagus lucidus Lindl. Henry in Journ. China Br. Roy. Asiat. Soc. 22: 273. (1887) (Chinese Names of Plants), Wright in Journ. Linn. Soc. Bot. 36: 102. (1903), Henry in Trans. Asiat. Soc. Jap. 24 suppl: 97. (1896) (List Pl. Formosa).

Hubei: Yichang, 192, 1124. Badong, 1009. Liantuo, 2093. **Hainan:** Environs of Haikou, 8021, 8164, 8335.

A tuberous perennial, native to China, Vietnam, Laos, Thailand, Japan and the Philippines. Henry stated that in the Three Gorges region this species was locally known as *t'ien ken tzu ts'ao* and *t'ien-men-tung* in Taiwan. The dried roots are currently used in Chinese herbal medicine to treat both dry cough and constipation. *Asparagus cochinchinensis* was collected by the Glasnevin Central China Expedition (GCCE 770) on Tianthu Shan, Changyang in October 2004.

Asparagus filicinus Buch.-Ham. ex D. Don (ASPARAGACEAE)
Wright in Journ. Linn. Soc. Bot. 36: 102. (1903).

Hubei: Xingshan, 5712a. South Badong, 5712b. **Sichuan:** South Wushan, 5712.

Asparagus sp. (ASPARAGACEAE)
Anonymous in Kew List of Determinations (volume 34).

Yunnan: Mengzi, 9865, 10782, 11319. Simao, 12181.

Aspidistra sp. (ASPARAGACEAE)
Anonymous in Kew List of Determinations (volume 34).

Yunnan: Mengzi, 13739.

Aspidium parasiticum (L.) Sw. (DRYOPTERIDACEAE)
Christ in Bull. Herb. Boiss. 7: 17. (1899).

Yunnan: Lunan, 11538.

Belongs to another taxon.

Aspidium pennigerum H. Christ (DRYOPTERIDACEAE)
Christ in Bull. Herb. Boiss. 7: 17. (1899).

Yunnan: In dry places in the Simao forests at 1,650 metres, 11809.

Belongs to another taxon.

Aspidium sp. No. 1. (DRYOPTERIDACEAE)
Anonymous in Kew List of Determinations (vol. 24).

Hunan: Shimen, 7539.

Aspidium sp. No. 2. (DRYOPTERIDACEAE)
Henry in Trans. Asiat. Soc. Jap. 24 suppl: 113. (1896) (List Pl. Formosa).

Taiwan: Tamsui, 1440.

Aspidocarya uvifera Hook. f. & Thoms. (MENISPERMACEAE)
Diels in Engler, Pflanzenr. iv. 94: 127. (1910), Forman in Kew Bull. 39: 101. (1984).

Yunnan: Mountains to the west of Simao at 1,640 metres, 11867a.

A slender woody climber, discovered in the Sikkim Himalaya by Joseph Hooker who stated in his field notes that the fruits were 'tolerably good'. The fruits are said to taste of green grapes. Distributed in India and China (Yunnan).

Aspidopterys glabriuscula (Wall.) Juss. var. *brevicuspis* Niedenzu (MALPIGHIACEAE)

Niedenzu in Engler, Pflanzenr. iv. 141: 30. (1928).

Aspidopterys floribunda Hutchinson in Kew Bull. Misc. Inf. 94. (1917), in part

Yunnan: Mengzi, 10455, (isotype of *Aspidopterys floribunda* Hutchinson), 10455b., (isotype of *Aspidopterys floribunda* Hutchinson).

New record of the genus for China.

Aspidopterys henryi Hutch. (MALPIGHIACEAE) New species

Bull. Misc. Inf. Kew 94. 1917.

Aspidopterys floribunda Hutchinson in Kew Bull. Misc. Inf. 94. (1917), in part. *Aspidopterys glabriuscula* (Wall.) Juss. var. *subrotunda* Niedenzu Niedenzu in Engler, Pflanzenr. iv. 141: 30. (1928).

Yunnan: In Pingbian, in forests on the Daweishan Range at 1,650 metres, 11055a., (isotype of *Aspidopterys floribunda* Hutchinson).

Aspidopterys tomentosa (Blume) Juss. var. *obcordata* (Hemsl.) Niedenzu (MALPIGHIACEAE)

New variety. in Engler, Pflanzenr. iv. 141: 21. (1928).

Aspidopterys obcordata Hemsley in Hooker's Ic. Pl. 27: t. 2673. (1901).

Yunnan: Simao at 1,640 metres, 12894 (type of *Aspidopterys obcordata* Hemsley).

Asplenium adiantoides (L.) C. Chr. (ASPLENIACEAE)

Asplenium falcatum Henry non Lam. in Trans. Asiat. Soc. Jap. 24 suppl: 112. (1896) (List Pl. Formosa).

Taiwan: Wanshoushan, 796.

A widespread fern, distributed from Madagascar to Australia. Grows on rocks or tree trunks.

Asplenium aethiopicum (Burm. f.) Becherer (ASPLENIACEAE)

Asplenium furcatum Thunb. Anonymous in Kew List of Determinations (volume 34).

Yunnan: Simao, 13607.

Asplenium aff. *alatum* Humb. & Bonpl. ex Willd. (ASPLENIACEAE)

Christ in Bull. Soc. Bot. France 5: 47. (1905).

Yunnan: Mengzi, 10350.

Asplenium bulbiferum G. Forst. (ASPLENIACEAE)

Anonymous in Kew List of Determinations (volume 34).

Yunnan: Simao, 12636.

A New Zealand species. This collection number belongs to another taxon.

Asplenium bullatum Wall. ex Mett. (ASPLENIACEAE)

Christ in Bull. Herb. Boiss. 7: 9. (1899).

Yunnan: In mountain forest to the east of Mengzi at 2,200 metres, 10108a., (type of *Asplenium grandifrons* Christ.).

Asplenium cancellatum Alston (ASPLENIACEAE)

Asplenium dimidiatum Sw. Christ in Bull. Herb. Boiss. 7: 9. (1899).

Yunnan: On a cliff in woods near Mengzi at 1,650 metres, 11542.

A New World species, this collection number belongs elsewhere.

Asplenium caudatum G. Forst. (ASPLENIACEAE)

Anonymous in Kew List of Determinations (volume 34).

Yunnan: Mengzi, 10591.

Asplenium delavayi (Franch.) Copel. (ASPLENIACEAE) New species

Christ in Bull. Soc. Bot. France 5: 47. (1905).

Scolopendrium delavayi Franchet Christ in Bull. Herb. Boiss. 7: 11. (1899).

Hubei: Yichang, 1061. **Yunnan:** At Yuanjiang, in a shady ravine at 1,100 metres, 11580.

Though named for Père Delavay (who later collected this species in Yunnan in May 1894), Henry had first found it near Yichang about 1886. Distributed through India, Myanmar, central and western China and Vietnam.

Asplenium ensiforme Wall. ex Hook. & Grev. (ASPLENIACEAE)

Anonymous in Kew List of Determinations (volume 34).

Yunnan: Mengzi, 10045, 11515.

Distributed across the Himalaya, southern India, Sri Lanka, China and Japan.

Asplenium exiguum Bedd. subsp. *yunnanense* (Franch.) Fraser-Jenk., Pangtey & Khullar (ASPLENIACEAE)

Asplenium yunnanense Franchet Ching & Wu in Acta Phytotax. Sin. 23: 13. (1985).

Yunnan: Mengzi, 10106, 10107, 13392, 13606.

Distributed across India, Myanmar, China and Vietnam. This species was also collected near Mengzi by William Hancock.

Asplenium griffithianum Hook. (ASPLENIACEAE)

Anonymous in Kew List of Determinations (volume 34).

Yunnan: Mengzi, 10467.

Distributed across Nepal, India, Bhutan, China,

Vietnam and Japan.

Asplenium heterophyllum C. Presl. (ASPLENIACEAE)

Anonymous in Kew List of Determinations (volume 34).

Yunnan: Simao, 13568.

This collection belongs to an allied taxon.

Asplenium holosorum Christ (ASPLENIACEAE) New species

in Bull. Herb. Boiss. 7: 10. (1899).

Yunnan: In Pingbian on the Daweishan Range in forest at 1,650 metres, 11543.

Asplenium incisum Thunb. (ASPLENIACEAE)

Baker in Journ. Bot. 26: 227. (1888), Diels in Bot. Jahrb. 29: 198. (1900), Christ in Bull. Soc. Bot. France 5: 54. (1905).

Hubei: Badong, 3674. Liantuo, 2146, 4431, 4435.

Distributed through Russia, India, Nepal, China, Korea and Japan. This species was collected by the Sino-American Expedition in Shennongjia Forest Disctrict in 1980.

Asplenium laciniatum D. Don (ASPLENIACEAE)

Christ in Bull. Herb. Boiss. 7: 10. (1899).

Asplenium varians Wall. ex Hook. & Grev. Baker in Journ. Bot. 27: 176. (1889), Diels in Bot. Jahrb. 29: 198. (1900).

Hubei: Jianshi, 5964. **Yunnan:** In Pingbian on the Daweishan Range, in mountain forest at 1,650 metres, 9056. Simao 12471. On mountains to the east of Mengzi at 1,950 metres, 9743d.

Native to India, Sri Lanka, Nepal, Bhutan, Myanmar, and China including Taiwan.

Asplenium laserpitiifolium Lam. (ASPLENIACEAE)

Henry in Trans. Asiat. Soc. Jap. 24 suppl: 112. (1896) (List Pl. Formosa).

Taiwan: Oluanpi, 1249.

Native to Malaysia, Indonesia, the Philippines, China (Taiwan) and Japan (south).

Asplenium latifolium Bory (ASPLENIACEAE)

Anonymous in Kew List of Determinations (volume 34).

Yunnan: Simao, 13130, 13182.

This collection belongs to an allied species.

Asplenium macrocarpum Desv. (ASPLENIACEAE)

Anonymous in Kew List of Determinations (volume 34).

Yunnan: Mengzi, 10109, 13310.

This collection belongs to an allied species.

Asplenium macrophyllum Sw. (ASPLENIACEAE)

Henry in Trans. Asiat. Soc. Jap. 24 suppl: 112. (1896) (List Pl. Formosa).

Taiwan: Oluanpi, 1255.

Asplenium microtum Maxon (ASPLENIACEAE) New species

in Contr. U. S. Natl. Herb. 12: 411. (1909); in Notul. Syst. (Paris) 1: 57. (1909).

Yunnan: Mountains to the south-west of Mengzi at 1,950 metres, 10344 (isotype).

Asplenium nidus L. (ASPLENIACEAE)

Henry in Trans. Asiat. Soc. Jap. 24 suppl: 112. (1896) (List Pl. Formosa).

Taiwan: Tamsui, 1380. Wanchin, 1496.

The bird's nest fern, a large epiphytic fern, distributed through Africa to Polynesia. A common species in Taiwan.

Asplenium normale D. Don (ASPLENIACEAE)

Anonymous in Kew List of Determinations (volume 34).

Yunnan: Mengzi, 10104.

A widespread species, distributed from Africa to Polynesia.

Asplenium paradoxum Blume (ASPLENIACEAE)

Anonymous in Kew List of Determinations (volume 34).

Yunnan: Mengzi, 10590.

Asplenium prolongatum Hook. (ASPLENIACEAE)

Christ in Bull. Soc. Bot. France 5: 54. (1905).

Asplenium rutaefolium Baker non Kze. Baker in Journ. Bot. 26: 227. (1888), Diels in Bot. Jahrb. 29: 198. (1900).

Hubei: Antelope Glen, 3291.

Distributed in India, Nepal, Sri Lanka, China, Vietnam, Japan and Korea.

Asplenium pubicostatum Ching & Z. Y. Liu (ASPLENIACEAE)

Zhang in Acta Phytotax. Sin. 34: 187. (1996).

Asplenium nigripes Christ non (Fee) Hook. in Bull. Soc. Bot. France 5: 48. (1905).

Hubei: Yichang, 1716.

Asplenium pulcherrimum (Baker) Ching (ASPLENIACEAE)

Asplenium billetii Christ. Christ. in Bull. Herb. Boiss. 7: 10. (1899).

Yunnan: On a cliff in woods near Mengzi at 1,650 metres, 11532b.

Asplenium ritoense Hayata (ASPLENIACEAE)

Asplenium davallioides Hook. Henry in Trans. Asiat. Soc. Jap. 24 suppl: 112. (1896) (List Pl. Formosa).

Taiwan: Tamsui, 1421.

Native to Korea, Japan, southern China (including Taiwan).

Asplenium sarelii Hook (ASPLENIACEAE)

Asplenium saulii Baker Baker in Journ. Bot. 26: 227. (1888) Diels in Bot. Jahrb. 29: 198. (1900).

Hubei: Yichang, 219. Badong, 3789.

02

Asplenium sarelii Hook. subsp. ***pekinense*** (Hance) Fraser-Jenk., Pangtey & Khullar (ASPLENIACEAE)

Asplenium pekinense Hance Christ in Bull. Soc. Bot. France 5: 54. (1905).

Hubei: Yichang, 2169. Xingshan, 7924. **Sichuan:** North Wushan, 7060.

A widespread species found over much of India, China, Korea and Japan.

Asplenium sp. (ASPLENIACEAE)

Anonymous in Kew List of Determinations (volume 34).

Yunnan: Mengzi, 9416, 9228.

Asplenium sp. nova (ASPLENIACEAE)

Henry in Trans. Asiat. Soc. Jap. 24 suppl: 112. (1896) (List Pl. Formosa).

Taiwan: Oluanpi, 1248.

Asplenium trapezoideum Ching (ASPLENIACEAE) New species

in Bull. Fan. Mem. Inst. Biol. 2(10): 209. pl. 26. (1931).

Yunnan: In forests near Mengzi at 1,650 metres, 11540.

Endemic to western China.

Asplenium trichomanes L. (ASPLENIACEAE)

Baker in Journ. Bot. 26: 226. (1888), Diels in Bot. Jahrb. 29: 198. (1900).

Hubei: Badong, 2567, 3675, 5144, 6205a. South Badong, 6146. Changleping, 6205. Changyang, 7650.

Distributed through North and South America, Europe, Africa, Asia and Australia.

Asplenium yoshinagae Makino subsp. ***indicum*** (Sledge) Fraser-Jenk. (ASPLENIACEAE)

Asplenium planicaule Wall. ex Mett. Anonymous in Kew List of Determinations (volume 34).

Yunnan: Mengzi, 9749.

Distributed from India to Borneo.

Aster ageratoides Turcz. var. ***heterophyllus*** Maxim. (ASTERACEAE)

In the Harbarium of Harvard University (GH.).

Sichuan: South Wushan, 7269.

Aster ageratoides Turcz. var. ***lasiocladus*** (Hayata) Hand.-Mazz. (ASTERACEAE)

Aster trinervius Henry non Roxb. in Trans. Asiat. Soc. Jap. 24 suppl: 52. (1896) (List Pl. Formosa).

Taiwan: Wanchin, 1632.

Aster alatipes Hemsley (ASTERACEAE) New species

in Journ. Linn. Soc. Bot. 23: 407. (1888), Bretschneider in Hist. Eur. Bot. Disc. China 2: 785. (1898).

Hubei: Liantuo, 4496 (type).

Aster baccharoides Steetz. (ASTERACEAE)

Henry in Trans. Asiat. Soc. Jap. 24 suppl: 52. (1896)

(List Pl. Formosa).

Hubei: Xingshan at 1,400 metres, 6565. **Taiwan:** Tamsui, 1445.

Aster glehnii F. Schmidt (ASTERACEAE)

Forbes & Hemsley in Journ. Linn. Soc. Bot. 23: 411. (1888).

Hubei: Yichang, 39, 386. Liantuo, 2656. **Sichuan:** South Wushan, 7198.

Aster hispidus Thunb. (ASTERACEAE)

Forbes & Hemsley in Journ. Linn. Soc. Bot. 23: 411. (1888).

Hubei: Yichang, 794, 795, 977. Badong, 937, 946 (November 1885), 1040 (December 1885), 3175, 3702, 4788, 5125. Liantuo, 3189. Antelope Glen, 3338.

Aster incisus Fisch. (ASTERACEAE)

Boltonia incisa (Fisch.) Bentham Anonymous in Kew List of Determinations (volume 24).

Hubei: Yichang, 2968.

Aster indicus L. (ASTERACEAE)

Forbes & Hemsley in Journ. Linn. Soc. Bot. 23: 413. (1888), Henry in Journ. China Br. Roy. Asiat. Soc. 22: 235. (1887) (Chinese Names of Plants); in Trans. Asiat. Soc. Jap. 24 suppl: 52. (1896) (List Pl. Formosa).

Hubei: Yichang, 70, 89, 216, 294, 384, 647, 835, 991, 1113, 1342, 1343, 2099, 2138. Without locality, 70a., 647a. Badong, 1051 (December 1885), 4983. San You Dong Glen, (July 1886) 1526. **Taiwan:** Wanchin, 533. **Yunnan:** Mengzi, 9754.

According to Henry this species was known colloquially at Yichang as *ch'ai hao*. Distributed in India, China, Korea, Japan, Thailand, Loas and Vietnam.

Aster indicus L. subsp. ***stenolepis*** (Handel-Mazzetti) Soejima & Igari (ASTERACEAE)

In the Herbarium of the Royal Botanic Gardens, Kew (K.).

Aster indicus Forbes & Hemsley non L. in Journ. Linn. Soc. Bot. 23: 413. (1888), Henry in Journ. China Br. Roy. Asiat. Soc. 22: 235. (1887) (Chinese Names of Plants). *Boltonia indica* Henry non (L.) Benth. Henry in Journ. China Br. Roy. Asiat. Soc. 22: 260. (1887) (Chinese Names of Plants).

Hubei: Badong, 2474. **Sichuan:** South Wushan, 7204.

Aster limosus Hemsley (ASTERACEAE) New species

in Journ. Linn. Soc. Bot. 23: 413. (1888).

Hubei: Yichang, growing in dirty water fields, 102 (type).

Aster moupinensis (Franch.) Hand.-Mazz. (ASTERACEAE)

Aster henryi Hemsley in Journ. Linn. Soc. Bot. 23: 411. (1888), Morley in Glasra 3: 78. (1979).

Hubei: On the banks of the Yangtze at Yichang, 133

(isosyntype of *Aster henryi* Hemsley). Yichang 3262 (type of *Aster henryi* Hemsley).

Aster nigromontanus Dunn (ASTERACEAE) New species
in Journ. Linn. Soc. Bot. 35: 501. (1903).

Yunnan: Obtained by Henry's Chinese collector, Ho, on the summit of Great Black Mountain at 2,650 metres, 11302.

Aster panduratus Nees ex Walp. (ASTERACEAE)
Aster fordii Hemsley in Journ. Linn. Soc. Bot. 23: 410. (1888).

Hubei: Yichang, 984, 3229, 4235. In the Kan Chi Glen near Yichang (October 1886), 2910. Badong, on high mountains, 2494 (type of *Aster fordii* Hemsley).

Aster pekinensis (Hance) F. H. Chen (ASTERACEAE)
Aster holophyllus Hemsl. Forbes & Hemsley in Journ. Linn. Soc. Bot. 23: 412. (1888).

Hubei: Yichang, 891.

Aster procerus Hemsley (ASTERACEAE) New species
in Journ. Linn. Soc. Bot. 23: 415. (1888).

Hubei: Antelope Glen, 4278 (type).

Aster scaber Thunb. (ASTERACEAE)
Forbes & Hemsley in Journ. Linn. Soc. Bot. 23: 415. (1888).

Hubei: Badong, 552 (September 1885), 938, 2566, 4859. Fang Xian, 6580.

Aster spp. (ASTERACEAE)
Anonymous in Kew List of Determinations (volume 24 & 34). *Heteropappus* sp. Anonymous in Kew List of Determinations (volume 24).

Hubei: Yichang, 148, 472, 946a., 4108. San You Dong Glen, 1699. Badong, 866, 3174 (September 1885), 1034 (December 1885), 2525, 3161, 4781, 4966, 5101, 5008, 5038, 5078, 5152. Liantuo, 2653, 2994, 3032. Changyang, 7686. **Yunnan:** Mengzi, 9445, 9445c., 10943, 13783, 13784.

Aster trinervius Roxb. ex D. Don (ASTERACEAE)
Forbes & Hemsley in Journ. Linn. Soc. Bot. 23: 416. (1888).

Hubei: Badong, 114, 951 (November 1885), 2446, 3164, 4798, 4932, 4934, 4963, 5049, 5063, 5076, 5077, 5188. Yichang, 987, 2172, 2789, 2790, 2890, 2932, 2942. In a cave on Moji Shan (the Dome), 3093. Liantuo, 2657, 3055, 4432, 4493. Xingshan, 6492, 7001. Jianshi, 7437. **Yunnan:** Mengzi, 9348, 10322, 10323, 10324. Simao, 12617. South of the Red River near Manmei at 1,433 metres, 9632.

Aster verticillatus (Reinw.) Brouillet, Semple & Y. L. Chen (ASTERACEAE)
Rhynchospermum verticillatum Reinw. Anonymous in Kew List of Determinations (volume 34).

Yunnan: Mengzi, 9873.

Distributed across Nepal, India, Bhutan, Myanmar, China and Japan.

Asteraceae. No. 1. (ASTERACEAE)
Anonymous in Kew List of Determinations (volume 24).

Hubei: Jianshi, 7431. Badong, 4016. Yichang, 2173.

Asteraceae No. 2 (ASTERACEAE)
Atractylis sp. Anonymous in Kew List of Determinations (volume 24).

Hubei: Fang Xian, 6605. **Sichuan:** North Wushan 6605a.

Asteropyrum peltatum (Franch.) Drummond & Hutchinson (RANUNCULACEAE)
Isopyrum peltatum Franch. Diels in Bot. Jahrb. 29: 325. (1900).

Hubei: South Badong, 5630.

Discovered by Père Armand David in Baoxing, Sichuan in May 1869.

Astilbe chinensis (Maxim.) Franch. & Savat. (SAXIFRAGACEAE)
Anonymous in Kew List of Determinations (volume 34).

Yunnan: Mengzi, 11018.

Astilbe rubra Hook. f. & Thoms. (SAXIFRAGACEAE)
Pan in Acta Phytotax. Sin. 23: 435. (1985).

Astilbe chinensis Henry non (Maxim.) Franch. & Sav. in Journ. China Br. Roy. Asiat. Soc. 22: 270. (1887) (Chinese Names of Plants), Forbes & Hemsley in Journ. Linn. Soc. Bot. 23: 265. (1887). *Astilbe chinensis* (Maxim.) Franch. & Sav. var. *davidii* Henry non Franchet in Gard. Chron. ser. 3, 32: 95. (1902), O' Brien in Augustine Henry, An Irish Plant Collector in China 15. (2002).

Hubei: Yichang, 174. Badong, 1853. South Badong, 7278. Jianshi, 7446.

According to Henry this species was known colloquially in the Three Gorges region as the red *sheng ma*.

Astilbe sp. (SAXIFRAGACEAE)
Anonymous in Kew List of Determinations (volume 24).

Hubei: Yichang, 176.

Astilbe thunbergii Miq. (SAXIFRAGACEAE)
Forbes & Hemsley in Journ. Linn. Soc. Bot. 23: 266. (1887), Henry in Journ. China Br. Roy. Asiat. Soc. 22: 270. (1887) (Chinese Names of Plants).

Hubei: Badong, 1775, 4706, 4734, 6048. Xingshan, 6467. **Hunan:** Shimen, 7548. **Sichuan:** South Wushan, 7727.

Astraeus hygrometricus (Pers.) Morgan (GEASTRACEAE)
Geaster hygrometricus Pers. Anonymous in Kew List

of Determinations (volume 24).

Hubei: South Badong, 5353. (Fungus).

This collection belong to an allied species.

Astragalus henryi Oliver (FABACEAE) New species

in Hooker's Icon. Pl. 20: t. 1959. (1891), Henry in Notes Econ. Bot. China 8. & 17. (1893), Bretschneider in Hist. Eur. Bot. Disc. China 2: 781. (1898), Forbes & Hemsley in Journ. Linn. Soc. Bot. 26: 459. (1904), Morley in Glasra 3: 78. (1979), Zheng in Journ. Wuhan Bot. Res. 2.(1): 96. (1984).

Hubei: Fang Xian, 6902 (type).

Henry stated that the root of *Astragalus henryi* was the source of the important drug *hoan-ch'i*, (in Sichuan it was the *huang-ch'i*) which is still widely used in China. In the 'Kew Inwards' book for 1888—1889 is a note by Professor Daniel Oliver that over 160 tons of roots were exported annually from Hankow (now part of Wuhan). Seeds were collected from Henry's 6902 with a note 'Any chance of germinating?' at Kew in 1889. There are samples of this drug in the herbarium at Glasnevin from material purchased by the Glasnevin Central China Expedition (2002) in Chengdu, Sichuan.

Astragalus khasianus Benth. ex Bunge (FABACEAE)

Astragalus englerianus Ulbr. in Eng. Bot. Jahrb. xxxvi. Beibl. 82, 60. (1905), Simpson in Notes Roy. Bot. Gard. Edinburgh viii. 259. (1913—1915).

Yunnan: Mengzi at 1,800 metres, 9783a.

Bhutan to western China.

Astragalus sinicus L. (FABACEAE)

Forbes & Hemsley in Journ. Linn. Soc. Bot. 23: 166. (1887), Simpson in Notes Roy. Bot. Gard. Edinburgh 8: 250. (1913—1915).

Astragalus sinicus L. var. *macrocalyx* Ulbrich Simpson in Notes Roy. Bot. Gard. Edinburgh 8: 251. (1913—1915).

Hubei: Yichang, 253, 657, 1137. Antelope Glen, 3459. Badong, 3771, 4067. South Badong, 5504, 5398. Liantuo, 6374. **Yunnan:** Mengzi at 1,380 metres, 9356. Mengzi at 1,500 metres, 9356a.

A biennial species to 20 cm tall, China to Vietnam and Korea.

Astragalus sutchuenensis Franch. (FABACEAE)

Simpson in Notes Roy. Bot. Gard. Edinburgh 8: 254. (1913—1915).

Hubei: Yichang, 763, 764, 1352. **Sichuan:** Emei Shan, Henry's Chinese collector with Pratt, without number.

Introduced to cultivation by Wilson for Messrs Veitch & Sons.

Astragalus wushanicus N. D. Simpson (FABACEAE) New species

in Notes Roy. Bot. Gard. Edinburgh. 8: 248. (1913—1915).

Sichuan: North Wushan, 7071 (type)

A plant with pink flowers growing on the cliffs, rare.

Astronia ferruginea Elmer (MELASTOMATACEAE)

Li in Woody Flora of Taiwan 651. fig. 258. (1963).

Astronia sp. Henry in Trans. Asiat. Soc. Jap. 24 suppl: 44. (1896) (List Pl. Formosa).

Taiwan: Oluanpi, 658, 658a., 1288.

Taiwan of China to the Philippines.

Asyneuma fulgens (Wall.) Briq. (CAMPANULACEAE)

Campanula fulgens Wall. Anonymous in Kew List of Determinations (volume 34).

Yunnan: Mengzi, 9980.

Asystasiella sp. (ACANTHACEAE)

Asystasia sp. Anonymous in Kew List of Determinations (volume 34).

Yunnan: Yuanjiang, 13330.

Atalantia buxifolia (Poir.) Oliv. Ex Benth. (RUTACEAE)

Henry in Trans. Asiat. Soc. Jap. 24 suppl: 25. (1896) (List Pl. Formosa).

Hainan: Environs of Haikou, 8047. **Taiwan:** Kaohsiung, without number. Wanchin, 1623.

A widely distributed shrub with a range from southern China, to Malaysia and the Philippines to Japan. In Taiwan this small shrub is mainly found along the sea-shore on the Hengchun Peninsula on the southern part of the island.

Atalantia henryi (Swingle) C. C. Huang (RUTACEAE) New species

Atalantia racemosa Wight & Arn. var. *henryi* Swingle in Journ. Arnold Arbor. 21: 127. (1940).

Yunnan: Simao at 1,300 metres, 12930 (type of *Atalantia racemosa* Wight & Arn. var. *henryi* Swingle).

Atalantia sp. (RUTACEAE)

Henry in Trans. Asiat. Soc. Jap. 24 suppl: 25. (1896) (List Pl. Formosa).

Taiwan: Wanchin, 843.

Athyriopsis japonica (Thunb.) Ching (ATHYRIACEAE)

Asplenium japonicum Thunb. Henry in Trans. Asiat. Soc. Jap. 24 suppl: 112. (1896) (List Pl. Formosa).

Taiwan: Tamsui, 1411, 1415.

A widespread fern, native to China (including Taiwan), Korea and Japan.

Athyrium arisanense (Hayata) Tagawa (ASPLENIACEAE)

Asplenium sinense Baker in Kew Bull. Misc. Inf. 9. (1906). *Athyrium mengtziense* Hieron in Hedwigia 59: 319. (1918). *Athyrium roseum* H. Christ in Bull. Herb. Boiss. 6: 961. (1898), Zhang in Acta Phytotax. Sin. 34: 190. (1996).

Yunnan: In mountain forest to the north of Mengzi, 10101 (syntype of *Asplenium sinense* Baker). Mengzi at 1,500 metres, 9918 (type of *Athyrium mengtziense* Hieron., syntype of *Athyrium roseum* H. Christ)

Native to China (including Taiwan) and Japan.

Athyrium atkinsonii Bedd. (ASPLENIACEAE)

Asplenium atkinsonii (Bedd.) C. B. Clarke Christ. in Bull. Herb. Boiss. 7: 13. (1899).

Yunnan: In forest on Fengchunling at 1,650 metres, 11525.

Distributed across Nepal, India, Bhutan, Myanmar, China, Korea and Japan.

Athyrium brevisorum (Wall.) Bedd. (ASPLENIACEAE)

Athyrium fissum Christ in Notul. Syst. (Paris) 1: 47. (1909).

Yunnan: Simao, 13157. In the mountains near Simao at 1,650 metres, 13107 (syntype of *Athyrium fissum* Christ).

Distributed from Pakistan to Nepal, Myanmar and China (Yunnan).

Athyrium clarkei Bedd. (ASPLENIACEAE)

Asplenium clarkei (Bedd.) Atk. ex C. B. Clarke Anonymous in Kew List of Determinations (volume 34).

Yunnan: Mengzi, 11523, 11524.

Distributed in Nepal, India, Myanmar and south-west China.

Athyrium davidii (Franch.) Christ (ASPLENIACEAE)

Polypodium davidii Franch. Christ in Bull. Herb. Boiss. 6: 872. (1898).

Yunnan: Milê, 9061.

Athyrium dissitifolium (Baker) C. Chr. (ASPLENIACEAE)

Phegopteris incrassata Christ in Bull. Herb. Boiss. 6: 963. (1898).

Yunnan: Mengzi, 10105 (type of *Phegopteris incrassata* Christ).

Native to India, Nepal, Myanmar, China, Thailand and Vietnam.

Athyrium drepopterum (Kunze) A. Br. (ASPLENIACEAE)

Christ in Notul. Syst. (Paris) 1: 46. (1909).

Yunnan: Simao, 13078.

Distributed across India, Myanmar, China and the Philippines.

Athyrium drepopterum (Kunze) A. Br. var. *decompositum* Christ (ASPLENIACEAE)

New variety. in Notul. Syst. (Paris) 1: 46. (1909).

Yunnan: Simao, 13078c.

Athyrium eburneum (J. Sm. Ex Mett.) J. Sm. (ASPLENIACEAE)

Asplenium niponicum Mett. var. *elatius* Christ in Bull. Soc. Bot. France 5: 48. (1905).

Hubei: Liantou, 2145.

Athyrium filix-femina (L.) Roth. (ASPLENIACEAE)

Asplenium filix-femina (L.) Bernh. Anonymous in Kew List of Determinations (volume 24).

Hubei: Badong, 1727. Liantuo, 2148.

These collections belong to an allied species.

Athyrium nephrodioides (Baker) Christ (ASPLENIACEAE)

New species

Asplenium nephrodioides Baker in Journ. Bot. 25: 170. (1887), Baker in Ann. Bot. 5: 306. (1890), Bretschneider in Hist. Eur. Bot. Disc. China 2: 794. (1898).

Hubei: Badong, 1858 (type).

Athyrium nigripes (Blume) T. Moore (ASPLENIACEAE)

Asplenium nigripes (Blume) Hook. Christ in Bull. Soc. Bot. France 5: 48. (1905).

Christ in Bull. Herb. Boiss. 7: 13. (1899).

Hubei: Liantuo, 4603. Badong, 5036, 5046, 5088 (in part), 5091, 5163, 5199, 6175. Jianshi, 5060, 5809. Fang Xian, 6601. Changyang, 7645, 7546. **Hunan:** Shimen, 7538. **Yunnan:** In Pingbian on the Daweishan Range in forests at 1,650 metres, 10101a. In forests near Mengzi at 1,700 metres, 11522. In forests to the east of Mengzi at 1,900 metres, 11522a.

Athyrium nigripes (Blume) T. Moore var. *elongatum* Christ (ASPLENIACEAE) New variety

in Bull. Herb. Boiss. 7: 13. (1899).

Yunnan: In Pingbian on the Daweishan Range at 2,200 to 2,600 metres, 10101c. In mountain forest near Mengzi at 1,950 metres, 11522b.

Athyrium niponicum (Mett.) Hance (ASPLENIACEAE)

Asplenium niponicum Mett. Christ in Bull. Soc. Bot. France 5: 48. (1905). *Dryopteris yunnanensis* (Baker) Copel Christ in Notul. Syst. (Paris) 1: 324. (1909). *Aspidium yunnanensis* (Baker) Christ in Notul. Syst. (Paris) 1: 324. (1909). *Aspidium apiifolium* Christ non Schkuhr in Bull. Herb. Boiss. 7: 16. (1899). *Nephrodium yunnanense* Baker in Kew Bull. Misc. Inf. 11 (1906).

Hubei: Yichang, 2133. Badong, 5067. Liantuo, 2584, 2988, 3059. Liantuo, on the banks of the Yangtze, 4443. **Yunnan:** In Pingbian on the Daweishan Range in forests at 1,650 metres, 10340b., 10341. Yuanjiang, 13232.

Distributed across China, Korea and Japan.

Athyrium puncticaule (Blume) T. Moore (ASPLENIACEAE)

Asplenium anisopteron H. Christ. in Bull. Herb. Boissier 6: 962. (1898).

Yunnan: Mengzi, 10109b., (type of *Asplenium anisopteron* H. Christ.).

Athyrium sheareri (Baker) Ching (ASPLENIACEAE)

Nephrodium sheareri Baker Anonymous in Kew List of Determinations (volume 24).

Hubei: Yichang, 2973. Liantuo, 4429.

According to Bretschneider this species was discovered by Dr. George Shearer on the banks of the Yangtze near Jiujing in Jiangxi province in 1873. Distributed in China and

Japan.

Athyrium sp. (ATHYRIACEAE)

Anonymous in Kew List of Determinations (volume 24).

Hubei: Badong, 2834.

Atractylodes lancea (Thunb.) DC. (ASTERACEAE)

Atractylis ovata Thunb. Henry in Notes Econ. Bot. China 6. (1893).

Sichuan: North Wushan, 7105.

The source of a drug called *ts'ang shu*, according to Henry.

Atrichum henryi (E.S.Salmon) E. B. Bartram (POLYTRICHACEAE)

Catharinea henryi E.S.Salmon in Journ. Bot. 40: 1. (1902).

Yunnan: In forests near Simao at 1,650, 13608 (type of *Catharinea henryi* E. S. Salmon). (Moss).

Atrichum obtusulum (C. Mull.) Jaeg. (POLYTRICHACEAE)

Salmon in Journ. Linn. Soc. Bot. 34: 460. (1900).

Sichuan: South Wushan, 5888 (in part, June 1888), Fang Xian, 7482a., (September 1888).

Previous to Henry's collections this species of moss was only known from the Himalaya.

Atropanthe sinensis (Hemsl.) Pascher (SOLANACEAE) New species

Scopolia sinensis Hemsley in Journ. Linn. Soc. Bot. 26: 176. (1890), Nelson in A Heritage of Beauty 319. (2000).

Hubei: Badong, 1801. Jianshi, 5923, 5923a. **Sichuan:** North Wushan, 7062, 7449 (type of *Scopolia sinensis* Hemsley). Without locality, 5923b.

An herbaceous perennial to 1.5 metres tall with obovate leaves and solitary axillary greenish-purple flowers surrounded by a large green foliaceous involucre. In central China, during Henry's time there, this species was much valued for the sake of a cardiac drug extracted from it. Seeds of Henry's 7449 were removed the herbarium sheet at Kew and sent for propagation. In south-east Tibet (Xizang), the closely related *Anisodus stramoniifolius* is highly valued as a vetinerary medicine. While ascending the Showa Range in 2001 our porters gathered the roots of this plant beneath the pass, to be later used in treating ailments of their yak herds. *Atropanthe sinensis* was introduced to cultivation by Wilson through Messrs Veitch who first flowered plants in their Coombe Wood nursery in 1902.

Atylosia sp. (FABACEAE)

Anonymous in Kew List of Determinations (volume 34).

Yunnan: Simao, 13283.

Aucuba chinensis Benth. (GARRYACEAE)

Henry in Trans. Asiat. Soc. Jap. 24 suppl: 48. (1896) (List Pl. Formosa), Wagnerin in Engler, Pflanzenr. iv. 229. 40. (1910), Rehder in Sargent Pl. Wilson. 2: 572. (1916), Li in Woody Flora of Taiwan 675. fig. 282. (1963).

Hubei: Yichang, 4388. **Taiwan:** Wanchin, 140. **Yunnan:** At Fengchunling in forest at 2,300 to 2,500 metres, 10122, 10123. In forests to the east of Mengzi, 10123a.

A dichotomously branched shrub to 3 metres tall, endemic to China.. A variable shrub in regard to foliage, this species was introduced to cultivation by Wilson in 1901. It is better suited to the warmer counties of the British Isles and Ireland. *Aucuba chinensis* was collected by the Glasnevin Central China Expedition (GCCE 627) in Dragon Gate River Forest Park in October 2004.

Aucuba chinensis Benth. var. *angustata* F. T. Wang (GARRYACEAE) New variety

Aucuba chinensis Benth. f. *angustifolia* Rehder in Sargent Pl. Wilson. 2: 573. (1916), Zheng in Journ. Wuhan Bot. Res. 2.(1): 163. (1984), O' Brien in Augustine Henry, An Irish Plant Collector in China 15. (2002).

Hubei: South Badong, 5383. Without locality, 5383a. **Yunnan:** In the mountains near Yuanjiang at 2,300 metres, 10123b., (type), 13300.

Leaves long and narrow, up to 17 cm long by 3 cm wide. In Ireland this form is cultivated on Garinish Island (Ilnacullin), County Cork and Mount Usher in County Wicklow.

Aucuba obcordata (Rehd.) Fu (GARRYACEAE) New species

in Flora Hubeiensis 3: 257. (1998).

Aucuba chinensis Wagnerin non Benth. in Engler, Pflanzenr. iv. 229. 40. (1910.) *Aucuba chinensis* Benth. forma *obcordata* Rehder in Sargent Pl. Wilson. 2: 572. (1916), Zheng in Journ. Wuhan Bot. Res. 2.(1): 163. (1984).

Hubei: Antelope Glen, 3353 , 3353a., 3353d. Moji Shan, 3353c.

A very handsome and distinct foliage plant worthy of introduction to our westerly gardens, discovered by Henry in 1887 and later collected by Wilson (Wilson-Veitch 101, collected in April 1901 forms the type) in the same area as Henry's original collection.

Auricularia polytricha (Mont.) Sacc. (AURICULARIACEAE)

Hirneola polytricha (Mont.) Fr. Anonymous in Kew List of Determinations (volume 24).

Hubei: South Badong, 5472.

This edible fungus was cultivated in Hubei on felled saplings of *Quercus variabilis* (though Henry in his field notes at Kew states this particular collection was cultivated on the bark of a wild cherry - *Prunus* sp.). He stated that young saplings about 15 cm thick were cut down and were allowed to lie on the ground for a year. They were then cut into staves about 3 metres long and built into a shed-like structure and alllowed to rot for a year. Spores of this species were then established and a crop soon after was harvested during the second year (in the mountains of Badong this mushroom was cultivated between 1,000 to 1,700 metres above sea level). The ear shaped fungus was esteemed by the Chinese as a delicacy, it was exported all over China and was a crop of great commercial importance. Wilson also came across this method of cultivation in the high mountains near Xingshan, for his trouble in sampling this delicacy he was rewarded with a stomach ache. For a note on this method of cultivation see Henry in *Notes on Economic Botany of China* 15. (1893).

Avena fatua L. (POACEAE)

Avena strigosa Pilger non Schreb. in Engl. Bot. Jahrb. 29: 225. (1900). *Avena fatua* L. var. *glabrata* Stapf Rendle in Journ. Linn. Soc. Bot. 36: 401. (1904).

Hubei: Yichang, wild at sides of wheat fields, 3474.

Avena nuda L. (POACEAE)

Rendle in Journ. Linn. Soc. Bot. 36: 401. (1904).

Hubei: South Badong, 1080 (seeds only), 6098. In the Lung wang tung valley, 3573.

Cultivated oats, the *yen meh* of the Chinese, W. B. Hemsley (RBG, Kew) requested some grains from Henry for cultivation. The Chinese prefer this species to *A. fatua*, which the Tibetan tribespeople grow (A. Henry).

Avicennia officinalis L. (ACANTHACEAE)

Henry in Trans. Asiat. Soc. Jap. 24 suppl: 72. (1896) (List Pl. Formosa).

Taiwan: Kaohsiung Lagoon, in tidal mangrove swamp, without number.

A native of tropical Asia and the Pacific Islands. Once a common tree the edge of the Kaohsiung lagoon according to Henry and associated with mangrove plants in brackish water. The last of the mangrove forests at Kaohsiung dissappeared in the late 1960s.

Ayenia integrifolia (Lace) Christenh & Byng (MALVACEAE) New species

Byttneria integrifolia Lace in Kew Bull. Misc. Inf. 9: 396. (1915).

Yunnan: Simao, 13370 (syntype of *Byttneria integrifolia* Lace).

Azanza lampas (Cav.) Alef. (MALVACEAE)

Thespesia lampas (Cav.) Dalz. ex Dalz. & A. Gibson Anonymous in Kew List of Determinations (volume 34).

Yunnan: Simao, 13281, 12367.

An evergreen shrub to 3 metres tall, widely distributed through east Africa and southern and south-east Asia.

Azolla pinnata R. Br. (AZOLLACEAE)

Baker in Journ. Bot. 26: 231. (1888), Diels in Bot.

Jahrb. 29: 209. (1900).

Hubei: Yichang, 3977.

A small aquatic fern, floating on ponds, paddy fields and irrigation channels. Distributed in Africa, Asia, Australia and the Pacific Islands.

Baccaurea ramiflora Lour. (PHYLLANTHACEAE)

Baccaurea sapida (Roxb.) Müell.-Arg. Pax & Hoffmann in Engler, Pflanzenr. iv. 147. xv: 52. (1922)

Yunnan: Simao, 11781a., 11783, 11783a., 11783b.

Distributed in India, Myanmar, southern China, Thailand, Cambodia, Laos and Vietnam.

Bacopa monnieri (L.) Wettst. (PLANTAGINACEAE)

Herpestis monniera (L.) Pennell. Henry in Trans. Asiat. Soc. Jap. 24 suppl: 67. (1896) (List Pl. Formosa).

Hainan: Environs of Haikou, 8331. **Taiwan:** Kaohsiung, 725.

A semi-aquatic, succulent mat-forming perennial. Pantropic in distribution.

Baeckea frutescens L. (MYRTACEAE)

Merrill & Perry in Journ. Arnold Arb. 19: 194. (1938).

Hainan: At Nodoa in the interior of the island, 8461.

A widespread species ranging from India, southern China, Indochina, the Malay Peninsula, Sumatra, Borneo and New Guinea.

Balanophora harlandii Hook. f. (BALANOPHORACEAE)

Balanophora henryi Hemsley in Journ. Linn. Soc. Bot. 26: 410. (1894), Bretschneider in Hist. Eur. Bot. Disc. China 2: 791. (1898). *Balania henryi* (Hemsl.) van Teighem in Ann. Soc. Nat. Paris. Bot. ser. 9. 202. (1907); in Fedde, Repert. Nov. Spec. Regni Veg. 6: 271. (1909). *Balanophora minor* Hemsley in Journ. Linn. Soc. Bot. 410. t. 9. fig. 2-3. (1894), Bretschneider in Hist. Eur. Bot. Disc. China 2: 791. (1898).

Hubei: Changyang and Yichang, without number. Fang Xian, without number. **Sichuan:** South Wushan, 7191 (holotype of *Balanophora henryi* Hemsley).

India (Assam) to China (Taiwan).

Balanophora involucrata Hook. f. & Thoms. (BALANOPHORACEAE)

Forbes & Hemsley in Journ. Linn. Soc. Bot. 26: 410. (1894)

Balania involucrata (Hook. f. & Thoms.) van Teighem in Ann. Soc. Nat. Paris. Bot. ser. 9. 143 & 205. (1907); in Fedde, Repert. Nov. Spec. Regni Veg. 6: 271. (1909).

Hubei: Fang Xian, 6825, 6851, 6880. **Sichuan:** Henry's Chinese collector with Pratt, no locality stated but may be Leshan, Emei Shan, Wa Shan or Kangding, 8888.

Pakistan to China.

Balanophora laxiflora Hemsley (BALANOPHORACEAE) New species

in Journ. Linn. Soc. Bot. 410. t. 9. fig. 2-3. (1894),

Bretschneider in Hist. Eur. Bot. Disc. China 2: 791. (1898).

Sichuan: North Wushan, 7112.

A parasitic dioecious perennial, native to China and Indochina.

Balanophora polyandra Griff. (BALANOPHORACEAE)

Forbes & Hemsley in Journ. Linn. Soc. Bot. 26: 411. (1894).

Hubei: Fang Xian, 6689, 6689a., 7934.

Himalaya to southern China.

Baliospermum calycinum Müell.-Arg. (EUPHORBIACEAE)

Baliospermum effusum Pax & K. Hoffmann in Pflanzenr. (Engler) 4. Fam. 147(4): 27. (1912).

Yunnan: Mountains to the west of Simao at 1,650 metres, 12000b., (isosyntype *Baliospermum effusum* Pax & K. Hoffmann). Simao at 1,500 metres, 12053, 12053b., (isosyntype of *Baliospermum effusum* Pax & K. Hoffmann).

Nepal to southern China and Indochina.

Baliospermum sp. (EUPHORBIACEAE)

Anonymous in Kew List of Determinations (volume 34).

Yunnan: Mengzi, 13777.

Bambusa flexuosa Munro (POACEAE)

Rendle in Journ. Linn. Soc. Bot. 36: 445. (1904).

Hainan: Environs of Haikou, 7996.

Bambusa oldhamii Munro (POACEAE)

Rendle in Journ. Linn. Soc. Bot. 36: 448. (1904).

Bambusa tuldoides Rendle non Munro in Journ. Linn. Soc. Bot. 36: 447. (1904).

Taiwan: Oluanpi, 686. Kaohsiung 1955.

A vigorous bamboo, culms to 12 metres tall and 12 cm across, this species was discovered by the Kew collector Richard Oldham in 1864. Native to southern China, though introduced to Taiwan where it grows throughout the island. Bamboo is one of the most important of economic plants in China being used in all sorts of situations from house building, as household utensils, furniture, rafts, ropes, bridges, boats, irrigation-wheels, umbrellas, hats, shoes, sandals and in the past for building sedan-chairs, for example. The young shoots of many species are edible and bamboo is commonly planted around houses and villages, anyone who has passed through the Chendu plain (one of the most densly populated agricultural areas in the world) in Sichuan cannot have but marvelled at the many little villages sheltered by the towering culms of various species of bamboo. A favourite Taoist saying is ' No man can live without a bamboo tree in the immediate vicinity of his house but he can live without meat.' The general Chinese name for bamboo is *chu*.

Barleria noctiflora L.f. (ACANTHACEAE)

Barleria cristata Lam. Forbes & Hemsley in Journ.

Linn. Soc. Bot. 26: 242. (1890).

Hainan: Haikou, without number. **Yunnan:** Milê district, 9940.

A much branched subshrub to 2 meters tall, carrying from August to October clusters of blue-purple funnel-shaped flowers in the axils of leaves. Distributed from Nepal to China and the Indochina Peninsula.

Barnardia japonica (Thunb.) Schult. & Schult. f. (ASPARAGACEAE)

Scilla chinensis Benth. Wright in Journ. Linn. Soc. Bot. 36: 127. (1903).

Hubei: Yichang, 126, 2927, 3264. Liantuo, 4411.

Barringtonia asiatica (L.) Kurz. (LECYTHIDACEAE)

Li in Woody Flora of Taiwan 629. fig. 248. (1963).

Barringtonia speciosa J. R. Forst. & G. Forst. Henry in Trans. Asiat. Soc. Jap. 24 suppl: 43. (1896) (List Pl. Formosa).

Taiwan: Oluanpi, 327.

An evergreen medium-sized tree, native to the tropics of the Old World. The large buoyant fruits of this species are dispersed by floating. Henry described it as a remarkable tree on the sea-shore at Oluanpi, with large leaves and great quadrangular fruits. The indigenous inhabitants of Taiwan named it ramudan. *Barringtonia* comprises of 39 species largely found in tropical Africa and all are nocturnal bloomers and the large silky-white flowers of this species are pollinatd by bats and flying foxes. In Polynesia the toxic fruits are grated and used as bait to stun and catch fish. *Barringtonia asiatica* was collected by the Glasnevin Taiwan Expedition (GTE 1) at Banana Bay Coastal Forest on the Hengchun Peninsula and by the lighthouse at Oluanpi (GCCE 19) (South Cape of Henry and Wilson) in October 2004.

Barringtonia racemosa (L.) Spreng. (LECYTHIDACEAE)

Henry in Trans. Asiat. Soc. Jap. 24 suppl: 43. (1896) (List Pl. Formosa) Li in Woody Flora of Taiwan 631. (1963).

Taiwan: Oluanpi, 1006.

A small tree, native to the tropics of the Old World, it grows as scattered trees in coastal regions of the northern and southern parts of Taiwan.

Bartramia halleriana Hedw. (BARTRAMIACEAE)

Salmon in Journ. Linn. Soc. Bot. 34: 459. (1900).

Sichuan: North Wushan, on rocks at 1,900 to 2,600 metres, 7151. (September 1888). (Moss).

Basella alba L. (CHENOPODIACEAE)

Basella rubra L. Forbes & Hemsley in Journ. Linn. Soc. Bot. 26: 331. (1891) Henry in Trans. Asiat. Soc. Jap. 24 suppl: 76. (1896) (List Pl. Formosa).

Hainan: Without locality, 8129. **Taiwan:** Oluanpi, 346.

Ceylon spinach, widely cultivated in the tropics of

02

Africa and Asia and often existing as an escape from cultivation. It is thought to have originated in Asia but has been long grown in China for the plants fleshy berries which yield a dye used as rouge by women and by officials for seal impressions. The Hani (an ethnic group) of Xishuangbanna in southern Yunnan call this species *savciqqilni* and use the stems to treat fractures and sprains. The leaves are used as spinach (see Herklots, *Vegetables in South-East Asia* 152. 1972).

Basilicum polystachyon (L.) Moench. (LAMIACEAE)

Moschosma polystachyon (L.) Benth. Forbes & Hemsley in Journ. Linn. Soc. Bot. 26: 269. (1890), Dunn in Notes Roy. Bot. Gard. Edinburgh 6: 135. (1915).

Hainan: 8 km (5 miles) south of Kuingchow, 8181.

Bauhinia brachycarpa Wallich ex Bentham (FABACEAE)

Bauhinia faberi Oliv. var. *microphylla* Oliver ex Craib in Sargent Pl. Wilson. 2: 89. (1916). *Bauhinia brachycarpa* Wall. ex Benth. var. *microphylla* (Oliver ex Craib) K. Larsen & S. S. Larsen, Zheng in Journ. Wuhan Bot. Res. 2.(1): 96. (1984).

Hubei: On the banks of the Yangtze at Badong, 7179 (isotype of *Bauhinia faberi* Oliv. var. *microphylla* Oliver ex Craib). **Yunnan:** Simao, 13264.

Distributed in India, Myanmar, China and Thailand.

Bauhinia bryoniflora Franchet (FABACEAE)

Anonymous in Kew List of Determinations (volume 34).

Yunnan: Mengzi, 10289.

Bauhinia diptera Blume ex Miq. (FABACEAE)

Anonymous in Kew List of Determinations (volume 34).

Yunnan: Simao, 13098.

This collection belongs to an allied species.

Bauhinia sp. (FABACEAE)

Anonymous in Kew List of Determinations (volume 34).

Yunnan: Simao, 11894, 12059, 13259, 13357, 13434.

Bauhinia variegata L. (FABACEAE)

Anonymous in Kew List of Determinations (volume 34).

Yunnan: Near the Red River at Manhao, 9574. Mengzi, 10596.

The genus *Bauhinia* commemorates the French botanists and brothers, Johannes Bauhin (1541—1613) and Caspar Bauhin (1560—1624) because of the two-lobed leaves. *Bauhinia variegata* has a large range from the foothills of the Himalaya, India, Myanmar and China where it is often grown by Buddhist temples and monasteries. It was collected by the Glasnevin Taiwan Expedition (GTE 61) by the lighthouse at Oluanpi (South Cape of Henry and

Wilson) in Taiwan in October 2004. *Bauhinia variegata* is commonly known as the poor man's orchid and carries showy fragrant magenta flowers during winter and spring. It is one of the parents of the spectacular Hong Kong orchid tree, *Bauhinia × blakeana*.

Beaumontia grandiflora Wall. (APOCYNACEAE)

Tsiang in Sunyatsenia 2: 160. (1934).

Yunnan: In mountain forest to the south of Simao (Xishuangbanna), 11950.

Distributed in Nepal, India, Bhutan, Bangladesh, Myanmar, southern China, Laos, Thailand and Vietnam. Henry's collections proved to be a new record for China.

Beaumontia sp. (APOCYNACEAE)

Anonymous in Kew List of Determinations (volume 34).

Yunnan: Simao, 13289.

Beesia calthifolia (Maxim. ex Oliv.) Ulbr. (RANUNCULACEAE)

Cimicifuga calthifolia Maxim. ex Oliver Anonymous in Kew List of Determinations (volume 24).

Hubei: Badong, 4669. South Badong, 5432. Jianshi, 5790. Zigui, 5790a. Xingshan, 5790b. **Sichuan:** South Wushan, 5708, 7220.

Beesia calthifolia was discovered by the Russian explorer, Grigori Nicolaevich Potanin in Gansu on July 15th 1885.

Begonia acetosella Craib (BEGONIACEAE)

Begonia tetragona Irmsch in Mitt. Inst. Bot. Hamburg. 10: 515. (1939).

Yunnan: Mengzi, 10137a., (type of *Begonia tetragona* Irmsch.).

E. Himalaya to China (Yunnan) and Indochina.

Begonia acetosella Craib var. *hirtifolia* E. Irmscher (BEGONIACEAE) New variety

in Mitt. Inst. Allg. Bot. Hamburg. 10: 515. (1939).

Yunnan: Simao at 1,372 metres, 12251a., (isotype).

North-east India, Myanmar and China (Yunnan).

Begonia augustinei Hemsley (BEGONIACEAE) New species

in Gard. Chron. ser. 3, 286. (1900), Nelson in Pim, The Wood & the Trees 221. (1984), Nelson in A Heritage of Beauty 310. (2000).

Yunnan: Simao, 12333.

Described by W. B. Hemsley from plants raised at the Royal Botanic Gardens, Kew from seeds sent there by Henry. Distributed in China (Guangxi, Yunnan) and Laos.

Begonia cathayana Hemsley (BEGONIACEAE) New species

in Bot. Mag.134. t. 8202. (1908), Nelson in Pim, The Wood & the Trees 221. (1984), Nelson in A Heritage of

Beauty 310. (2000).

Yunnan: Near Mengzi at 1,650 metres, 9198 (syntype), 13516 (syntype).

Introduced to cultivation by Henry through the Royal Botanic Gardens, Kew. Endemic to China and distributed in the provinces of Guangxi and Yunnan.

Begonia grandis subsp. *sinensis* (A.DC.) Imrsch. (BEGONIACEAE)

Begonia sinensis A. DC. Forbes & Hemsley in Journ. Linn. Soc. Bot. 23: 323. (1887), Henry in Notes on Economic Botany of China 60. (1893), J. D. Hooker in Bot. Mag. 125: t. 7673. (1899), Nelson in Pim, The Wood & the Trees 221. (1984), Nelson in A Heritage of Beauty 310. (2000).

Hubei: Liantuo, 2110. Changyang 6234, 6236. In glens on way to Moji Shan, 2228. Yichang, 2707. Without locality, 6234a., 6236a. **Yunnan:** Simao, without number.

The *kou-tzu-ch'i*, introduced to cultivation by Henry through the Royal Botanic Gardens, Kew in 1898, plants first flowered there in the autumn of the following year. A widespread subspecies in China as can be seen from the locations of Henry's collections.

Begonia hemsleyana Hook. f. (BEGONIACEAE) New species

in Bot. Mag. t. 7685. (1899), Nelson in Pim, The Wood & the Trees 220. (1984), Nelson in A Heritage of Beauty 310. (2000).

Yunnan: Simao, 12290.

Introduced to cultivation by Henry through the Royal Botanic Gardens, Kew from seeds collected near Mengzi in 1898. Henry described it as a pretty plant, 30 to 45cm tall. At Kew it flowered freely and continuously in a warm glasshouse from April 1899 onwards and formed a mass of foliage comparable to a well grown *Helleborus*. Hooker named this species in honour of William Botting Hemsley because of the services he rendered to botanical science in his researches into the flora of China and his enormous contribution to the *Index Florae Sinensis*, the first major flora of China. Distributed in south-west China and Indochina.

Begonia henryi Hemsley (BEGONIACEAE) New species

in Journ. Linn. Soc. Bot. 23: 322. (1887), Bretschneider in Hist. Eur. Bot. Disc. China 2: 784. (1898), Morley in Glasra 3: 78. (1979).

Hubei: Yichang, 277. Liantuo 2109. San You Dong Glen, 2305.

Begonia labordei H. Lev. (BEGONIACEAE)

Begonia harrowiana Diels in Notes Roy. Bot. Gard. Edinburgh 5: 166. (1912).

Yunnan: Mengzi, 9677a., (syntype of *Begonia harrowiana* Diels).

North-east India to southern China and Indochina, in Yunnan it was also collected by Delavay and Forrest.

Begonia lacerata E. Irmscher (BEGONIACEAE) New species

in Mitt. Inst. Allg. Bot. Hamb. 10: 535. (1939).

Yunnan: Mengzi at 1,524 metres, 11458 (isotype, collected 10th of January 1898).

Endemic to south-east Yunnan.

Begonia modestiflora Kurz (BEGONIACEAE) New species

Begonia yunnanensis H. Léveille in Fedde, Repert. Spec. Nov. Regni Veg. 6: 20. (1909).

Yunnan: Simao, 12403c., (isotype of *Begonia yunnanensis* H. Léveille, collected 17th September 1898).

Indian (Assam) to China (Yunnan) and Indochina.

Begonia palmata D. Don (BEGONIACEAE)

Begonia laciniata Roxb. Henry in Trans. Asiat. Soc. Jap. 24 suppl: 46. (1896) (List Pl. Formosa), Nelson in Pim, The Wood & the Trees 221. (1984). *Begonia edulis* Levl. var. *henryi* Leveille in Fedde, Repert. Nov. Spec. Regni Veg. 6: 20. (1909).

Taiwan: Wanchin, 110. **Yunnan:** Mengzi, 9205. Simao, 12250 (type of *Begonia edulis* Levl. var. *henryi* Leveille).

Distributed in Nepal, India, Bangladesh, Bhutan, Myanmar, China and Vietnam. A succulent, tuberous perennial to 50 cm tall. Introduced to cultivation by Henry through the Royal Botanic Gardens, Kew.

Begonia ravenii Peng & Chen (BEGONIACEAE)

Begonia sinensis Henry non ADC. in Trans. Asiat. Soc. Jap. 24 suppl: 46. (1896) (List Pl. Formosa).

Taiwan: Tamsui, 1396, 1459.

Endemic to Taiwan.

Begonia sp. No. 1. (BEGONIACEAE)

Henry in Trans. Asiat. Soc. Jap. 24 suppl: 46. (1896) (List Pl. Formosa).

Taiwan: Oluanpi, 930, 995, 1275.

Begonia sp. No. 2. (cf. Faber 491) (BEGONIACEAE)

Anonymous in Kew List of Determinations (volume 24).

Sichuan: Wa Shan, Henry's Chinese collector with Pratt, 8946.

Begonia spp. (BEGONIACEAE)

Anonymous in Kew List of Determinations (volume 34).

Yunnan: Mengzi, 9186, 9204, 10241, 10243, 10368, 10538, 11258, 11420. Simao, 11628, 12505, 12645, 13266, 13347, 13482, 13508, 13537, 13600.

02

Beilschmiedia erythrophloia Hayata (LAURACEAE)
New species

Li in Woody Flora of Taiwan 200. fig. 74. (1963).

Beilschmeidia sp. Henry in Trans. Asiat. Soc. Jap. 24 suppl: 78. (1896) (List Pl. Formosa).

Taiwan: Oluanpi, 1001, 1260, 1264.

A large evergreen tree to 25 metres tall, native to Vietnam, China (Hainan and Taiwan) and Japan (south) where it is distributed in broad-leaved forest. First collected there by Henry in 1893.

Beilschmiedia robusta C. K. Allen (LAURACEAE)
New species

J. Arnold. Arbor. 23: 447. (1942).

Yunnan: Simao, 12777.

Distributed in southwest China and Vietnam.

Beilschmiedia sp. (LAURACEAE)

Anonymous in Kew List of Determinations (volume 34).

Yunnan: Simao, 12356, 11703.

Benincasa hispida (Thunb.) Cogn. (CUCURBITACEAE)

Benincasa cerifera Savi. Forbes & Hemsley in Journ. Linn. Soc. Bot. 23: 315. (1887).

Hubei: South Donghu, 7713. **Hainan:** Environs of Haikou, cultivated, 8002.

Henry called this the *dong gua ren*, the Chinese waxgourd, winter gourd or Chinese preserving melon, a hispid climbing annual, native of Java and perhaps Japan but cultivated in China for thousands of years. The cylindrical fruits can weigh almost 40 kg. In cultivation it needs a warm wet season and an extemly rich growing medium. The dried seeds are currently used in Chinese herbal medicine as a diuretic.

Benkara evenosa (Hutchinson) Ridsdale (RUBIACEAE)
New species

Randia evenosa Hutchinson in Sargent Pl. Wilson. 3: 400. (1916).

Yunnan: In woods near Mengzi at 1,520 metres, 10363, (type of *Randia evenosa* Hutchinson), 10363a., (isotype of *Randia evenosa* Hutchinson).

Benkara scandens (Thunb.) Ridsdale (RUBIACEAE)

Randia accedens Hance Anonymous in Kew List of Determinations (volume 24).

Hainan: Environs of Haikou, 8044.

Benkara sinensis (Lour.) Ridsdale (RUBIACEAE)

Randia sinensis (Lour.) Roem & Schult. Henry in Trans. Asiat. Soc. Jap. 24 suppl: 50. (1896) (List Pl. Formosa), Li in Woody Flora of Taiwan 871. (1963).

Taiwan: Wanchin, 93, 172. Wanshoushan, 753, 1891. Oluanpi, 2049.

An erect shrub, native to southern China, including Taiwan, where it is found on the southern part of the island.

Berberis bealei Fortune (BERBERIDACEAE)

Berberis fortunei Henry non Lindl. in Journ. China Br. Roy. Asiat. Soc. 22: 258. (1887) (Chinese Names of Plants). *Mahonia bealei* (Fort.) Carr. Takeda in Notes Roy. Bot. Gard. Edinburgh 6: 226. (1917), Ahrendt in Journ. Linn. Soc. Bot. 57: 319. (1961), Zheng in Journ, Wuhan Bot. Res. 2.(1): 43. (1984).

Hubei: Antelope Glen, 3250, 3283. Badong, 1450. Liantuo, 3913.

A shrub to 2 metres tall, discovered by Robert Fortune in 1848 and kept in the garden of T. C. Beale at Shanghai before being sent on to the nursery of Standish and Noble at Sunningdale and put into the trade in 1858. Henry stated that Berberis (formerly *Mahonia*) *bealei* was known colloquially in the Three Gorges region as *mao erh t'ou*. It was also collected on Emei Shan (Mount Omei or Omei Shan) in Sichuan by the Irish plant collector and missionary, Fr. Hugh Scallan. Fr. Scallan (Pater Hugo) was the first Irish collector to climb this floristically rich mountain (in about 1900). It is one of the best scented of the hardy *Berberis* species and often confused in cultivation with *Berberis japonica* Thunb. This species was collected by the Glasnevin Central China Expedition (GCCE 621) in Dragon Gate River Forest Park in Xingshan in October 2004.

Berberis brevipaniculata Schneider (BERBERIDACEAE)
New species

in Bull. Herb. Boissier. (2)8: 263. (1908) Ahrendt in Journ. Linn. Soc. Bot. 57: 204. (1961).

Hubei: Badong, 4675 (type).

Berberis dasystachya Maxim. (BERBERIDACEAE)

Schneider in Sargent Pl. Wilson. 3: 442. (1917), Ahrendt in Journ. Linn. Soc. Bot. 57: 184. (1961), Zheng in Journ. Wuhan Bot. Res. 2.(1): 42. (1984).

Hubei: Fang at 1,900 to 3,000 metres, 6816.

Discoverey by Przewalski in Gansu in 1873, this species was also collected in the same province in 1914 by Reginald Farrer. The Irish missionary Hugh Scallan collected *Berberis dasystachya* in northern Hubei in June 1897. Introduced to cultivation by Farrer.

Berberis eurybracteata (Fedde) Laferr. (BERBERIDACEAE)
New species

Mahonia confusa Sprague in Kew. Bull. Misc. Inform. 339. (1912), Schneider in Sargent Pl. Wilson. 3: 443. (1917), Takeda in Notes Roy. Bot. Gard. Edinburgh 6: 234. (1917), O' Brien in Augustine Henry, An Irish Plant Collector in China 26. (2002). *Mahonia fortunei* Schneider non (Lindl.) Fedde in Sargent Pl. Wilson. 1: 380. (1913). *Mahonia zemanii* Schneider in Sargent Pl. Wilson. 1: 378. (1913).

Hubei: Antelope Glen, 3117 (type), 3351 (isosyntype

of *Mahonia confusa* Sprague), 3177a., 3351a.

Discovered by Henry near Yichang in February 1887, this species was later collected on Emei Shan (Mount Omei) by the Irish missionary, Fr. Hugh Scallan (Pater Hugo). A century after Pater Hugo climbed the mountain's 22,000 steps to the summit, the Glasnevin expedition (2002) found it locally common on that mountain's mid-forested slopes while botanising the region. It is cultivated in Ireland at the National Botanic Gardens, Glasnevin and in the walled garden of Beech Park House, Clonsilla. The Beech Park plant was a gift from Roy Lancaster who introduced this species into cultivation almost a century after Henry discovered it.

Berberis ferdinandi-coburgii Schneider (BERBERIDACEAE) New species

in Sargent Pl. Wilson. 1: 364. (1913), Ahrendt in Journ. Linn. Soc. Bot. 57: 70. (1961), O' Brien in Augustine Henry, an Irish Plant Collector in China 15. (2002).

Hubei: Badong, 1458. **Yunnan:** In woods near Mengzi at 2,400 metres, 10257 (type). In mountain forest to the east of Simao at 2,000 metres, 11617, 11617a.

The specific epithet commemorates King Ferninand 1. of Bulgaria. In Ireland this species is cultivated at the Talbot Botanic Gardens, Malahide Castle and Mount Usher.

Berberis ganpinensis H. Lév. (BERBERIDACEAE)

Mahonia confusa Sprague var. *bournei* Ahrendt in Journ. Linn. Soc. Bot. 57: 316. (1961).

Hubei: Liantuo, 2689.

Berberis henryi Laferr. (BERBERIDACEAE) New species

in Bot. Zhurn. (Moscow & Leningrad) 82(9): 97. (1997) *Mahonia conferta* Takeda in Notes Roy. Bot. Gard. Edinburgh 6: 230. (1917), Ahrendt in Journ. Linn. Soc. Bot. 57: 305. (1961).

Yunnan: In forests on Fengchunling at 2,150 metres, 10180a., (type of *Mahonia conferta* Takeda).

An evergreen shrub to 2 metres tall, with pinnate leaves to 20 cm long and carrying up to 40 shortly petioled leaflets. During June, July and August clustered upright racemes of yellow flowers make a striking display. Distributed in Guizhou and Yunnan. A beautiful plant.

Berberis henryana Schneider (BERBERIDACEAE) New species

in Bull. Herb. Boissier. ser. 2, 5: 664. (1905), Ahrendt in Journ. Linn. Soc. Bot. 57: 192. (1961), Zheng in Journ. Wuhan Bot. Res. 2.(1): 42. (1984), O' Brien in Augustine Henry, An Irish Plant Collector in China 15. (2002).

Berberis feddeana Schneider in Bull. Herb. Boissier. (2)5: 665. (1905), Ahrendt in Journ. Linn. Soc. Bot. 57: 184. (1961).

Hubei: South Badong, 4937, 5470, 5470a. Changleping, 5470b., (type) 5470c. **Sichuan:** South Wushan, 5470d.

In Ireland this very handsome shrub is cultivated at the John F. Kennedy Arboretum, Kilmacurragh and the National Botanic Gardens, Glasnevin.

Berberis julianae Schneider (BERBERIDACEAE) New species

in Sargent Pl. Wilson. 1: 360. (1913).

Berberis julianae Schneider var. *oblongifolia* Ahrendt in Journ. Linn. Soc. Bot. 57: 69. (1961). *Berberis zanlanscianensis* Pamp. Ahrendt in Journ. Linn. Soc. Bot. 57: 67. (1961).

Berberis ferdinandi-coburgii Schneider in Sargent Pl. Wilson. 1: 364. (1913) in part.

Hubei: Badong, 3170 (syntype *Berberis julianae* Schneider var. *oblongifolia* Ahrendt). **Yunnan:** I-men district, 10618.

Described from the collections of Wilson, but first collected by Henry.

Berberis napaulensis (DC.) Spreng. (BERBERIDACEAE)

Mahonia flavida Schneider in Sargent Pl. Wilson. 1: 382. (1913), Takeda in Notes Roy. Bot. Gard. Edinburgh. 6: 227. (1917), Ahrendt in Journ. Linn. Soc. Bot. 57: 313. (1961).

Yunnan: In Pingbian on the Daweishan Range at 1,650 metres, 10180 (holotype of *Mahonia flavida* Schneider, collected 20th February 1897).

Berberis nivea (C. K. Schneid.) Laferr. (BERBERIDACEAE) New species

in Bot. Zhurn. (Moscow & Leningrad) 82(9): 98. (1997).

Mahonia nivea Schneider in Bot. Gaz. 63: 521. (1917). *Mahonia hypoleuca* H. Takeda in Notes Roy. Bot. Gard. Edinburgh 6: 238. (1917), Ahrehd in Journ. Linn. Soc. Bot. 57: 330. (1961).

Yunnan: Mountains south-west of Mengzi at 1,950 metres, 9863 (holotype of *Mahonia hypoleuca* H. Takeda, holotype of *Mahonia nivea* Schneider).

Known only from the type specimen which is without flower or fruit.

Berberis oiwakensis (Hayata) Laferr. (BERBERIDACEAE)

Mahonia lomariifolia Takeda in Notes Roy. Bot. Gard. Edinburgh 6: 231. (1917), Ahrendt in Journ. Linn. Soc. Bot. 57: 311. (1961), O' Brien in Augustine Henry, An Irish Plant Collector in China 26. (2002).

Yunnan: In mountain forest near Milê, 10309 (syntype of *Mahonia lomariifolia* Takeda, collected 1st of November 1897).

Berberis polydonta (Fedde) Laferr. (BERBERIDACEAE)
New species

in Bot. Zhurn. (Moscow & Leningrad) 82(9): 98. (1997).

Berberis veitchiorum Hemsley & E. H. Wilson in Kew Bull. Misc. Inf. 152. (1906). *Mahonia veitchiorum* (Hemsl. & E. H. Wilson) Schneider in Sargent Pl. Wilson. 1: 383. (1913), Ahrendt in Journ. Linn. Soc. Bot. 57: 301. (1961).

Sichuan: Henry's Chinese collector with Pratt, no locality stated but may be Leshan, Emei Shan, Wa Shan or Kangding, 8993 (syntype of *Berberis veitchiorum* Hemsley & E. H. Wilson).

Discovered by Henry's Chinese collector in the autumn of 1889 and based on a specimen collected two years later by Bock and von Rosthorn (c. 1891).

Berberis sanguinea Franch. (BERBERIDACEAE)
Ahrendt in Journ. Linn. Soc. Bot. 57: 57. (1961).

Sichuan: South Wushan, 5753.

An evergreen shrub to 2.5 metres tall, flowers golden yellow. Discovered by Père Armand David in Baoxing (Moupin) in Sichuan in April 1869 and introduced by him through the gardens of the Paris Museum. It was later reintroduced by Père Paul Farges who sent seeds to Maurice de Vilmorin. Endemic to Sichuan.

Berberis sp. (BERBERIDACEAE)
Berberis insignis Henry non Hook. f et Thoms. in Journ. China. Br. Roy. Asiat. Soc. 22: 272. (1887) (Chinese Names of Plants).

Hubei: Badong, 703 (July 1885).

The *ta wang tzu* according to Henry.

Berberis subacuminata C. K. Schneider (BERBERIDACEAE)
New species

in Sargent Pl. Wilson. 1: 364. (1913), Ahrendt in Journ. Linn. Soc. Bot. 57: 62. (1961).

Yunnan: Yuanjiang at 2,000 metres, 13267 (type).

2 metres tall shrub, flowers yellow. China (Yunnan) and Vietnam.

Berberis triacanthophora Fedde (BERBERIDACEAE)
New species

in Bot. Jahrb. 36: Beibl. 82: 43. (1905), Ahrendt in Journ. Linn. Soc. Bot. 57: 51. (1961), Zheng in Journ. Wuhan Bot. Res. 2.(1): 43. (1984).

Hubei: Changyang, 5681 (type). Jianshi, 5681a. Without number, 5681b.

A handsome evergreen shrub to 2 metres tall, pale yellow flowers are carried on gracefully arching branches. Introduced to cultivation by Wilson.

Berberis veitchii Schneid. (BERBERIDACEAE)
Ahrendt in Journ. Linn. Soc. Bot. 57: 51. (1961).

Hubei: Badong, without number.

Discovered by Delavay in 1882 and later described as a new species by Schneider from a collection made by Wilson in Hubei in 1900. Wilson introduced this handsome shrub in 1900 through Messrs Veitch who first flowered it in 1904.

Berchemia flavescens (Wall.) Brong. (RHAMNACEAE)
Anonymous in Kew List of Determinations (volume 34).

Yunnan: Mengzi, 10771.

Distributed in India, Nepal, Bhutan and China.

Berchemia floribunda (Wall.) Brongn. (RHAMNACEAE)
Rehder & E. H. Wilson in Sargent Pl. Wilson. 2: 213. (1916).

Berchemia racemosa Sieb. & Zucc. Forbes & Hemsley in Journ. Linn. Soc. Bot. 23: 127. (1886), Henry in Journ. China Br. Roy. Asiat. Soc. 22: 250. (1887) (Chinese Names of Plants); in Notes Econ. Bot. China 64. (1893). *Berchemia floribunda* (Wall.) Brong. var. *megalophylla* Schneider in Sargent Pl. Wilson. 2: 213. (1916). *Berchemia giraldiana* Schneider in Sargent Pl. Wilson. 2: 213. (1916), Zheng in Journ. Wuhan Bot. Res. 2.(1): 137. (1984).

Hubei: Badong, 326 (June 1885). Liantuo, 1883, 2979. Xiling (Yichang) Gorge, 3621. Jianshi, 5995. Donghu, 6391. **Yunnan:** In woods near Mengzi at 2,200 metres, 10803. Simao at 1,800 metres, 11747a., [syntype of *Berchemia floribunda* (Wall.) Brong. var. *megalophylla* Schneider].

A deciduous climbing shrub to 6 metres tall, widely distributed in Nepal, India, Bhutan, China, Vietnam and Japan. According to Henry this species was known colloquially at Badong as *kou erh ch'a*.

Berchemia lineata (L.) DC. (RHAMNACEAE)
Anonymous in Kew List of Determinations (volume 24).

Hainan: Environs of Haikou, 8051, 8091.

Native to southern China (including Hainan and Taiwan) and Japan (south). A very variable species.

Berchemia polyphylla Wall. ex Laws. var. **leioclada** (Hand.-Mazz.) Hand.- Mazz. (RHAMNACEAE) New variety

Zheng in Journ. Wuhan Bot. Res. 2.(1): 137. (1984).

Berchemia lineata Forbes & Hemsley non (L.) DC. in Journ. Linn. Soc. Bot. 23: 127. (1886), Henry in Journ. China Br. Roy. Asiat. Soc. 22: 250. (1887); in Notes Econ. Bot. China 64. (1893).

Hubei: Yichang, 52. Badong, 1795 (February 1887), 2551. Liantuo, 3201. Without locality, 3201a.

According to Henry this species was known colloquially in the Three Gorges region as *lu mi k'o-tzu*, the leaves were once used as a substitute for tea (*Camellia sinensis*) Discovered by Henry in 1885, Handel-Mazzetti based his description on a collection made by Wilson in July

1907.

Berchemia sp. (RHAMNACEAE)

Anonymous in Kew List of Determinations (volume 24).

Hubei: Liantuo, 3015.

Berchemia yunnanensis Franch. (RHAMNACEAE)

Schneider in Sargent Pl. Wilson. 2: 216. (1916).

Berchemia lineata Forbes & Hemsley non (L.) DC. in Journ. Linn. Soc. Bot. 23: 127. (1886).

Sichuan: South Wushan, 5283. **Yunnan:** Simao at 2,000 metres, 10376a.

Discovered by Delavay in May 1883.

Bergera koenigii L. (RUTACEAE)

Murraya koenigii (L.) Spreng. Anonymous in Kew List of Determinations (volume 34).

Yunnan: Mengzi, 9651. Simao 13038.

A tree to about 10 metres tall and native to much of tropical Asia. Known as curry leaf, the pungent foliage of this species is scented of curry.

Bergia ammannioides Roxb. (ELATINACEAE)

Bergia glandulosa Henry non Turcz. in Trans. Asiat. Soc. Jap. 24 suppl: 19. (1896) (List Pl. Formosa).

Taiwan: Kaohsiung, 246, 1187. Oluanpi, 278, 1722, in rice fields.

An erect herb to 30 cm tall, native to much of tropical Asia and Australia. A weed of paddy fields in Taiwan.

Berneuxia thibetica Decne. (DIAPENSIACEAE)

Shortia thibetica (Decne.) (Franchet) Anonymous in Kew List of Determinations (volume 24).

Sichuan: Henry's Chinese collector with Pratt, no locality stated but may be any of the following, Leshan, Emei Shan, Wa Shan or Kangding, 8876.

Discovered by Père Armand David in Baoxing, Sichuan in May 1869.

Beta vulgaris L. (AMARANTHACEAE)

Anonymous in Kew List of Determinations (volume 24).

Hubei: Yichang, 4334, 4335 (cultivated).

This species has provided a wide array of vegetables such as sugar beet, beetroot, mangolds, chard and perpetual spinach. *Beta vulgaris* L. var. *cicla* L., better known as chard appears in the current (2001) *Flora Hubeiensis* and Henry's specimens may belong here.

Betula alnoides Hamilton apud D. Don (BETULACEAE)

Schneider in Sargent Pl. Wilson. 2: 467. (1916).

Betula alnoides Ham. apud. D. Don var. *acuminata* (Wall.) Wink. Winkler in Engler, Pflanzenr. 61: 89. (1904).

Yunnan: In forests on the Daweishan Range in Pinbian County at 1,500 metres, 10437. In forests on the Daweishan Range in Pinbian County at 1,800 metres, 10437a.

Mountains to the south of Simao at 1,500 metres, 12869, 12869a., 12869b. Mengzi at 1,400 metres, 11387.

A tree to over 20 metres tall, native to the Himalaya, south-east Tibet (Xizang) and western China. The bark of this species, like the North American cherry birch (*Betula lenta*) smells of wintergreen when crushed.

Betula delavayi Franch. (BETULACEAE) New species

Skvortsov in Harvard Papers in Botany 3(1): 109. (1998).

Betula fargesii Burkill non Franch. in Journ. Linn. Soc. Bot. 26: 498. (1899), Henry in The Garden 61: 4. (1902), Winkler in Engler, Pflanzenr. 61: 66. (1904), Schneider in Sargent Pl. Wilson 2: 478. (1916), Zheng in Journ. Wuhan Bot. Res. 2(1): 18. (1984).

Hubei: Fang Xian, 6879.

A small slender tree to 12 metres tall. Franchet based this species on a specimen collected by Delavay in western Sichuan on the 24th of May 1889, Henry had discovered the tree some nine months previously in September 1888. Wilson later collected it at Kangding and (both in Sichuan) and it was also collectd by the Sino-American Expedition in Shennongjia Forest District in 1980 (the same region where Henry had made his collection in 1888).

Betula luminifera Winkler (BETULACEAE) New species

Schneider in Sargent Pl. Wilson. 2: 455. (1916), Zheng in Journ. Wuhan Bot. Res. 2.(1): 19. (1984).

Betula cylindrostachya Henry non Wall. in Journ. China Br. Roy. Asiat. Soc. 22: 245. (1887) (Chinese Names of Plants), *Betula alnoides* Hamilton ex D. Don var. *pyrifolia* (Franch.) Burkill in Journ. Linn. Soc. Bot. 26: 497. (1899).

Hubei: Badong, 1414, 1443. Liantuo, 3942. Changyang, 5238. Jianshi, 7402. Without locality, 7402b. **Sichuan:** South Wushan, 5667.

Discovered by Henry in 1886 and described by Franchet from material later collected in Sichuan by Farges. In Hubei and eastern Sichuan this tree is capable of growing to 20 metres tall though it normally is found as a smaller tree (7 to 15 metres) and is usually found at low altitudes. A pioneer species of rapid growth, the bark on mature trees is orange-red or a dull reddish brown. According to Henry this species was known colloquially at Yichang as *hua ko shu*. *Betula luminifera* was introduced to cultivation by Wilson in 1901 from Hubei through Messrs Veitch. It has remained a great rarity in the gardens of the British Isles and Ireland, the largest recorded specimen in these islands grows at Hergest Croft, UK (planted 1920) and was 19 metres tall with a trunk diameter of 35 cm when measured in 1995.

02

Betula utilis D. Don subsp. ***albosinensis*** (Burkill) Ashburner and McAll. (BETULACEAE) New subspecies

Betula albosinensis Burkill Schneider in Sargent Pl. Wilson. 2: 457. (1916).

Betula utilis Burkill non D. Don in Journ. Linn. Soc. Bot. 26: 499. (1899), Winkler in Engler, Pflanzenr. 61: 61. (1904). *Betula utilis* D. Don var. *sinensis* (Franch.) Winkler Henry in Elwes & Henry, Trees Gr. Brit. & Irel. iv. 981. (1909).

Hubei: Fang at 2,500 to 2,900 metres, 6798, 6798a.

A tree to about 26 metres tall with girths of up to 3.5 metres in its native western Hubei, where it occasionally forms pure woods. This magnificent tree was discovered by Henry in 1888 in present day Shennongjia Forest District (The *Index Florae Sinensis* states Fang but the given the altitudes quoted by Burkill the area definately appears to be the highest peaks of Shennongjia where it is still a common tree). *Betula utilis* subsp. *albosinensis* was described from material later collected by Farges in Sichuan during the early 1890s. The bark, which is bright orange-red peels off in thin sheets. Introduced by Wilson for Messrs Veitch in 1901.

Biancaea decapetala (Roth.) O. Deg. (FABACEAE)

Caesalpinia sepiaria Roxb. Forbes & Hemsley in Journ. Linn. Soc. Bot. 23: 206. (1887), Henry in Journ. China Br. Roy. Asiat. Soc. 22: 272. (1887) (Chinese Names of Plants).

Hubei: Yichang, 338, 1088. **Yunnan:** Mengzi, 10222.

Widely spread in tropical Asia. In Hubei it forms semi-scandent, thorny shrubs from 1 to 7 metres in height. A handsome species with yellow, fragrant flowers. According to Henry this species was known in the Three Gorges region as *yeh-tsao-chio* or *tao kua tz'u*.

Biancaea sappan (L.) Tod. (FABACEAE)

Caesalpinia sappan L. Anonymous in Kew List of Determinations (volume 24 & 34).

Hainan: Environs of Kiungchow, 8268. **Yunnan:** By the Red River, 9812.

The sappanwood, bakam or pattangi, a small tree to 7 metres tall and native to India, Sri Lanka, Myanmar and Vietnam. The heartwood yields a red and black dye which in south-east Asia is used to dye wool and calico.

Bidens bipinnata L. (ASTERACEAE)

Henry in Journ. China Br. Roy. Asiat. Soc. 22: 266. (1887) (Chinese Names of Plants), Forbes & Hemsley in Journ. Linn. Soc. Bot. 23: 434. (1888).

Hubei: Yichang, 1158, 2794. North Badong, 7011.

Henry stated that both this species and *Bidens pilosa* were known colloquially in the Three Gorges region as *p'o p'o chen*. In Chinese herbal medicine this species is currently used to treat upper respiratory tract infection and rheumatic arthritis.

Bidens parviflora Willd. (ASTERACEAE)

Forbes & Hemsley in Journ. Linn. Soc. Bot. 23: 435. (1888).

Hubei: Badong, 2875, 4791.

Bidens pilosa L. (ASTERACEAE)

Forbes & Hemsley in Journ. Linn. Soc. Bot. 23: 435. (1888), Henry in Trans. Asiat. Soc. Jap. 24 suppl: 54. (1896) (List Pl. Formosa).

Hubei: Yichang, 388, 1111. In the Lung tung chi Glen near Yichang (October 1886), 2895. **Hainan:** Environs of Kiungchow, 8269. Without locality, 8514. Environs of Kiungchow, 8769. **Taiwan:** Wanchin, 108. Oluanpi, 373.

Generally spread in the tropics and extending into some temperate regions

Bidens sp. (ASTERACEAE)

Anonymous in Kew List of Determinations (volume 24).

Hubei: Badong, 2491.

Bidens tripartita L. (ASTERACEAE)

Forbes & Hemsley in Journ. Linn. Soc. Bot. 23: 436. (1888).

Hubei: Liantuo, 2658.

Biophytum sensitivum (L.) DC. (OXALIDACEAE)

Henry in Trans. Asiat. Soc. Jap. 24 suppl: 24. (1896) (List Pl. Formosa), Knuth in Engler, Pflanzenr. 10: 393. (1930).

Oxalis stricta Diels non L. in Bot. Jahrb. 29: 420. (1900)

Hubei: Yichang, 4203, 4759. **Taiwan:** Wanchin, 1549. **Yunnan:** Mengzi, 13769.

Bischofia javanica Blume (PHYLLANTHACEAE)

Henry in Trans. Asiat. Soc. Jap. 24 suppl: 83. (1896) (List Pl. Formosa), Forbes & Hemsley in Journ. Linn. Soc. Bot. 26: 428. (1894), Li in Woody Flora of Taiwan 417. fig. 156. (1963).

Hainan: 32 km (20 miles) west of Haikou, 8127. **Taiwan:** Oluanpi, without number. Wanchin, 914.

A widely dispersed large deciduous tree with a range from India, Malaysia, southern China, Japan (south) to Polynesia and Australia. The *ka-ang* tree of Taiwan, whose wood according to Henry was used in furniture making. The Hani (an ethnic group) of Xishuangbanna in southern Yunnan call this species *siqpanq* and use its roots, bark and leaves to treat rheumatism, mumps and skin ulcers.

Bischofia polycarpa (H. Lev.) Airy-Shaw (PHYLLANTHACEAE) New species

Airy Shaw in Kew Bull. 27: 271. (1972), Zheng in Journ. Wuhan Bot. Res. 2.(1): 111. (1984).

Bischoffia javanica Forbes & Hemsley non Blume in Journ. Linn. Soc. Bot. 26: 428. (1894), Pax & Hoffmann in engler, Pflanzenr. iv. 147. xv: 313. (1922).

Hubei: Beside a house in the Lung Wang Tung Glen, 1558. Yichang 1558a., 3097.

This species was found in Hubei by Henry in Hubei in 1887 and about the same time by Charles Ford in Guangdong. Leveille, who misidentified many of the collections of the Roman Catholic missionaries in China based his *Celtis polycarpa* on a collection made by E. E. Maire (in Cavalerie 4473) in 1918 in Yunnan. A rare tree in Hubei where it grows at low altitude, and, according to Henry, was known as *wu-yang*. It makes a tree to 20 metres with girths up to 2 metres. The wood is soft and of no value.

Bischofia sp. (PHYLLANTHACEAE)
Anonymous in Kew List of Determinations (volume 34).

Yunnan: Mengzi, 9909.

Bistorta amplexicaulis (D. Don) Greene (POLYGONACEAE)
Polygonum amplexicaule D. Don var. *sinense* Hemsley in Hooker's Icon. Pl. t. 1743 (1888). *Polygonum amplexicaule* D. Don Forbes & Hemsley in Journ. Linn. Soc. Bot. 26: 333. (1891).

Hubei: Badong, 1818, 2521 (type of *Polygonum amplexicaule* D. Don var. *sinense* Hemsley), 4061, 4976, 4998. Fang Xian, 6743.

Bistorta officinalis Delarbre (POLYGONACEAE)
Polygonum bistorta L. Forbes & Hemsley in Journ. Linn. Soc. Bot. 26: 334. (1891).

Hubei: Xingshan, 6956. North Badong, 7163.

Found throughout Eurasia and Morocco, the dried rhizome of this species is used in Chinese herbal medicine to treat diarrhea, acute respiratory infection with cough and venomous snake-bite.

Bistorta suffulta (Maxim.) Greene ex H. Gross (POLYGONACEAE)
Polygonum suffultum Maxim. Forbes & Hemsley in Journ. Linn. Soc. Bot. 26: 350. (1891).

Hubei: Badong, 1454, 3727, 5317.

Blachia pentzii (Muell.-Arg.) Benth. (EUPHORBIACEAE)
Forbes & Hemsley in Journ. Linn. Soc. Bot. 26: 435. (1894).

Hainan: Environs of Kiungchow, 8726.

Blainvillea acmella (L.) Philipson (ASTERACEAE)
Spilanthes acmella (L.) Murray Henry in Trans. Asiat. Soc. Jap. 24 suppl: 54. (1896) (List Pl. Formosa).

Taiwan: Oluanpi, 219, 655. Wanchin, 812. **Yunnan:** Simao, 12706.

The toothache plant, a weedy annual to less than 0.5 metres tall. Native to south America, this plant is thought to be an ancient cultigen of Peruvian origin. A popular salad plant in South America, it is cultivated throughout most of the tropical world. In South America it is used in herbal medicine to treat inflammation of the mouth and gums, toothache, digestive complaints, intestinal worms, colds, flu, respiratory tract infection, ear infection, headache, cold sores and herpes. When chewed the flower heads produce a remarkable tingling sensation, with the effect of a local anaesthetic.

Blechnum eburneum Christ (BLECHNACEAE)
Christ in Bull. Soc. Bot. France 5: 64. (1905) quoted as A. Henry 7316 in error.
Lomaria spicant Baker non (L.) Desv. in Journ. Bot. 26: 226. (1888).

Hubei: Antelope Glen, 3316.

This species was collected by the Sino-American Expedition in Shennongjia Forest District in 1980.

Blechnum insignis (Hook.) C. M. Kuo (BLECHNACEAE)
Brainea insignis (Hook.) J. Sm. Christ in Bull. Herb. Boiss. 7: 9. (1899).

Yunnan: In mountain forest near Simao at 1,650 metres, woody stem to 2 metres high, plant has exact habit of *Cycas,* 11938.

A small tree fern to about a metre tall.

Blechnum orientale L. (BLECHNACEAE)
Henry in Trans. Asiat. Soc. Jap. 24 suppl: 111. (1896) (List Pl. Formosa).

Taiwan: Oluanpi, 502. Wanchin, 1500. Kaohsiung Plain, 2039.

Widely distributed through tropical Asia, Australia and the Pacific Islands.

Bletilla striata (Thunb.) Reichb. f. (ORCHIDACEAE)
Bletia hyacinthina (Sm.) Aiton Henry in Journ. China Br. Roy. Asiat. Soc. 22: 265. (1887) (Chinese Names of Plants), Rolfe in Journ. Linn. Soc. Bot. 36: 19. (1903). *Bletia hyacinthina* (Sm.) Aiton var. *gebina* (Lindl.) Blume Rolfe in Journ. Linn. Soc. Bot. 19. (1903).

Hubei: Yichang, 492. Liantuo, 1909. Jianshi, 5940. **Yunnan:** Mengzi at 1,900 metres, 11104, 11104a., 11104b., 11104c.

According to Henry this species was known colloqially in the Three Gorges region as *peh chi*, this handsome terrestrial orchid is the source of the drug *bai ji* which is used to stop bleeding. Native to south-west, south-east and northern China and Japan where it grows in forests and open pasture between 110 to 3,200 metres.

Blumea aromatica DC. (ASTERACEAE)
Randeria in Blumea 10: 230. (1960).

Yunnan: Mengzi, 10127, 10480. Simao, 11677, 12784, 12854, 12953.

An undershrub to 3 metres tall, native to the Himalaya, Myanmar, China, Indochina and Thailand. The genus *Blumea* was named to commemorate the Dutch botanist C. L. Blume and is a widespread group of herbs and undershrubs characteristic of the 'weedy flora' of tropical and subtropical south east Asia. It is distributed from Africa through south-east Asia and New Guinea to Australia and the Pascific Islands extending as far cast as Hawaii (it was introduced to Hawaii). The genus has its centre of origin in south-east Asia.

Blumea axillaris (Lam.) DC. (ASTERACEAE)

Blumea mollis (D. Don) Merr. Randeria in Blumea 10: 261. (1960).

Yunnan: Mengzi, 10897. Simao, 11971, 11971a., 12955.

A very widespread herb to 90 cm with habitats from Africa to New Caladonia.

Blumea balsamifera (L.) DC. (ASTERACEAE)

Henry in Trans. Asiat. Soc. Jap. 24 suppl: 52. (1896) (List Pl. Formosa), Randeria in Blumea 10: 237. (1960).

Taiwan: Wanchin, 418. Oluanpi, 612. **Yunnan:** Simao, 12332, 12951, 12952.

An evergreen shrub, sometimes a herb to 4 metres tall, of widespread distribution from India to the Philippines. The most fragrant of all *Blumea* species, all parts of the plant smell of camphor. According to Augustine a type of camphor was commercially extracted from this species and was exported by the Chinese and Burmese. Henry stated the refined product was known in China as *ngai-pien*. To the Hani (an ethnic group) of Xishuangbanna in southern Yunnan this species is called *hhaoqsaqlaqma yaoheeqjiaq* and is used to treat sprains and contusions.

Blumea fistulosa (Roxb.) Kurz (ASTERACEAE)

Randeria in Blumea 10: 256. (1960).

Yunnan: Mengzi, 10356a.

A widespread herb to 1.5 metres, native from India to Indochina.

Blumea flava DC. (ASTERACEAE)

Laggera flava (DC.) Benth.& Hook. f. Anonymous in Kew List of Determinations (volume 34).

Yunnan: Simao, 11595.

Distributed in India, China, Laos, Cambodia and Vietnam.

Blumea hieraciifolia (Spreng.) DC. (ASTERACEAE)

Randeria in Blumea 10: 246. (1960)

Blumea sericans (Kurz.) Hook. f. Randeria in Blumea 10: 246. (1960).

Hainan: Without locality, 8248, 8505. Ling-men (in the interior of the island), 8613. **Yunnan:** Mengzi, 10799, 10799a. Simao, 11672.

The *wu yang* according to Henry.

Blumea lacera (Burm. f.) DC. (ASTERACEAE)

Henry in Trans. Asiat. Soc. Jap. 24 suppl: 53. (1896) (List Pl. Formosa), Randeria in Blumea 10: 164. (1960).

Hainan: 8 km (5 miles) south of Kuingchow, 8177. **Taiwan:** Wanchin, 103. Kaohsiung, 230, 1198. Tamsui, 1738. **Yunnan:** Mengzi, 10863.

The most widespread species in the genus with a range from Africa, across south-east Asia to northern Australia.

Blumea lanceolaria (Roxb.) Druce (ASTERACEAE)

Randeria in Blumea 10: 218. (1960).

Blumea myriocephala DC. Henry in Trans. Asiat. Soc. Jap. 24 suppl: 53. (1896) (List Pl. Formosa).

Taiwan: Wanchin, 181. Wanshoushan, 721. **Yunnan:** Simao, 11730.

A shrub to 2 metres tall, native from southern and eastern India to the Philippines. An extremely variable species.

Blumea macrostachya DC. (ASTERACEAE)

Blumea hieraciifolia (Spreng.) DC. var. *macrostachya* (DC.) Hook. f. Randeria in Blumea 10: 248. (1960).

Yunnan: In the mountains to the west of Simao at 1,300 metres, a large climber on trees, 12844.

Blumea martiniana Vaniot (ASTERACEAE)

Randeria in Blumea 10: 233. (1960).

Blumea henryi Dunn in Journ. Linn. Soc. Bot. 35: 503. (1903).

Yunnan: Mengzi, 10405 (syntype of *Blumea henryi* Dunn). In a dark ravine near Simao at 1,350 metres 10405b., (syntype of *Blumea henryi* Dunn).

An undershrub to 2.5 metres tall, distributed from north-east India to China and Vietnam.

Blumea megacephala (Randeria) C. C. Chang & Y. Q. Tseng (ASTERACEAE) New species

in Acta Phytotax. Sin. 12: 310. (1974).

Blumea riparia (Blume) DC. var. *megacephala* A. J. Randeria in Blumea 10: 215. (1960).

Yunnan: Mountains to the west of Simao at 1,650 metres, 10979 [holotype of *Blumea riparia* (Blume) DC. var. *megacephala* A. J. Randeria].

Distributed from Indian (Assam) to Japan and Vietnam.

Blumea membranacea DC. (ASTERACEAE)

Randeria in Blumea 10: 269. (1960).

Yunnan: Simao, 12954.

Blumea oxyodonta DC. (ASTERACEAE)

Henry in Trans. Asiat. Soc. Jap. 24 suppl: 53. (1896) (List Pl. Formosa), Randeria in Blumea 10: 280. (1960).

Taiwan: Wanchin, 1727. Oluanpi, 674. **Yunnan:** Simao, 11915.

A common annual weed of rice fields in parts of south-east Asia.

Blumea repanda (Roxb.) Hand.-Mazz. (ASTERACEAE)

Blumea procera DC. Randeria in Blumea 10: 212. (1960).

Yunnan: Mengzi, 10481. In the mountains to the west of Simao at 1,640 metres, 11587.

An undershrub to 3.5 metres tall, native from eastern Pakistan, Nepal, north-east India, Myanmar and western China where it grows in evergreen forest to 1,500 metres in altitude.

Blumea riparia (Blume) DC. (ASTERACEAE)

Randeria in Blumea 10: 213. (1960).

Yunnan: Simao, 11811.

A scandent undershrub to 2.5 metres in length, of widespread distribution from India to New Guinea and the Solomon Islands. Described from material gathered by Blume in Java.

Blumea sinuata (Lour.) Merr. (ASTERACEAE)

Blumea laciniata (Roxb.) DC Randeria in Blumea 10: 258. (1960).

Taiwan: Kaohsiung, 229, 1810. **Yunnan:** Mengzi, 9306, 9656, 10817. Simao, 13284.

Blumea sp. No. 1. (ASTERACEAE)

Henry in Trans. Asiat. Soc. Jap. 24 suppl: 53. (1896) (List Pl. Formosa).

Taiwan: Kaohsiung, 1196, 1942.

Blumea spp. No. 2. (ASTERACEAE)

Anonymous in Kew List of Determinations (vol. 34).

Yunnan: Mengzi, 13787.

Blumea tenuifolia C. Y. Wu (ASTERACEAE)

Blumea gracilis Dunn in Journ. Linn. Soc. Bot. 35: 502. (1903).

Yunnan: In wet places in open parts of a river-ravine near Simao, 12526 (isotype *Blumea gracilis* Dunn). Near Simao, 12783.

Distributed in India, Myanmar, southern China, Thailand, Cambodia, Laos and Vietnam.

Blumea virens DC. (ASTERACEAE)

Anonymous in Kew List of Determinations (volume 34).

Yunnan: Mengzi, 10188.

Boea spp. (GESNERIACEAE)

Anonymous in Kew List of Determinations (volume 34).

Yunnan: Mengzi, 9753, 11236. In Pingbian, in forests on the Daweishan Range at 1,650 metres, 11039. Simao, 12574, 13394.

Boehmeria clidemioides Miq. var. *diffusa* (Wedd.) Hand.-Mazz. (URTICACEAE)

Henry in Journ. China Br. Roy. Asiat. Soc. 22: 242. (1887) (Chinese Names of Plants).

Boehmeria diffusa Wedd. Henry in Journ. China Br. Roy. Asiat. Soc. 22: 242. (1887) (Chinese Names of Plants), Forbes & Hemsley in Journ. Linn. Soc. Bot. 26: 484. (1899).

Hubei: Yichang, 2924, 4274. Liantuo, 3215, 2671, 4622. **Yunnan:** Mengzi, 10736. Simao, 12262

A widespread variety found over much of India, Nepal, Bhutan, Myanmar, Laos and Vietnam. According to Henry this species was known colloquially in the Three Gorges region as *shui ch'u ma*.

Boehmeria clidemioides Miq. var. *umbrosa* Hand.-Mazz. (URTICACEAE)

In the Kew Herbarium (K.).

Yunnan: Simao, 12795

Boehmeria densiflora Hook. & Arn. (URTICACEAE)

Henry in Trans. Asiat. Soc. Jap. 24 suppl: 89. (1896) (List Pl. Formosa), Forbes & Hemsley in Journ. Linn. Soc. Bot. 26: 484. (1899), Li in woody Flora of Taiwan 131. (1963).

Taiwan: Wanchin, 37, 184. Oluanpi, 213, 688.

A shrub to 2 metres with a wide geographical range from southern China to the Philippines. Common in Taiwan from sea level to 1,600 metres.

Boehmeria depauperata Wedd. (URTICACEAE)

In the Kew Herbarium (K.).

Yunnan: Simao, 12896.

Boehmeria diffusa Wedd. (URTICACEAE)

Henry in Trans. Asiat. Soc. Jap. 24 suppl: 89. (1896) (List Pl. Formosa).

Taiwan: Tamsui, 1471.

Boehmeria hamiltoniana Wedd. (URTICACEAE)

In the Kew Herbarium (K.).

Yunnan: Simao, 12353.

Boehmeria japonica (L. f.) Miq. (URTICACEAE)

Boehmeria grandifolia Wedd. Forbes & Hemsley in Journ. Linn. Soc. Bot. 26: 485. (1899).

Yunnan: Mengzi, 9792e.

Boehmeria japonica (L. f.) Miq var. *tenera* (Blume) Friis & Wilmot-Dear (URTICACEAE)

Boehmeria gracilis C. H. Wright in Journ. Linn. Soc. Bot. 26: 485. (1899). *Boehmeria spicata* Forbes & Hemsley non (Thunb.) Thunb. in Journ. Linn. Soc. Bot. 26: 488. (1899).

Hubei: Badong, 1808, 4728, 4958, 4692 (type of *Boehmeria gracilis* C. H. Wright). Changleping, 6258.

Boehmeria nivea (L.) Gaud. (URTICACEAE)

Henry in Journ. China Br. Roy. Asiat. Soc. 22: 242. (1887) (Chinese Names of Plants); in Kew Bull. Misc. Inf. 250. (1891); in Notes Econ. Bot. China 65. (1893), Forbes & Hemsley in Journ. Linn. Soc. Bot. 26: 486. (1899).

Hubei: Yichang, 15, 2159. Badong, 4878 (the cultivated form of China grass). **Yunnan:** Mengzi, 11247. Simao, 12542.

Ramie, a perennial to about 1.5 metres tall. Henry called it the *ch'u ma* from which rhea fibre, China grass or grasscloth (in Chinese *hsia pu*) was made. It was extensively cultivated in Hubei and Sichuan and other Chinese provinces. Wilson records this species as being wild and abundant around Yichang to about 1,300 metres in altitude. In 1910 the exports from Hankow (Wuhan) amounted to 120,034 piculs. Ramie or China grass has never become an important fibre outside of China due to agronomic problems in its cultivation and because of undesirable chemicals in the fibres. Ramie has been gathered in Asia for fibre extraction for over 7,000 years and a number of Egyptian mummies from 3,300 B.C. were found to be bound with ramie strips.

Boehmeria nivea (L.) Gaudich. var. *tenacissima* (Gaudich.) Miq. (URTICACEAE)

Boehmeria nivea Henry non (L.) Gaudich. in Trans. Asiat. Soc. Jap. 24 suppl: 89. (1896) (List Pl. Formosa), Forbes & Hemsley in Journ. Linn. Soc. Bot. 26: 486. (1899). *Boehmeria frutescens* Thunb. var. *viridula* Yamamoto Li in Woody Flora of Taiwan 130. (1963).

Taiwan: Wanchin, 173, 483.

According to Henry this variety was known as *tui*, and in mandarin, *ch'u*. The 19th century indigenous inhabitants of Taiwan made fibre from various wild *Boehmeria*. Native to India, China and Japan to Java. A common shub at low altitude in Taiwan, it was collected by the Glasnevin Taiwan Expedition (GTE 118) on Wanshoushan, in October 2004.

Boehmeria ourantha Miq. (URTICACEAE)

In the Kew Herbarium (K.).

Yunnan: Simao, 12303.

Boehmeria penduliflora Wedd. ex D. G. Long (URTICACEAE)

Boehmeria macrophylla D. Don Forbes & Hemsley in Journ. Linn. Soc. Bot. 26: 485. (1899).

Yunnan: Mengzi, 9063, 9063a., 9063b., 9063c.

Boehmeria pilosciuscula (Blume) Hasskarl. (URTICACEAE)

Li in Woody Flora of Taiwan 133. (1963). *Boehmeria platyphylla* D. Don. Henry in Trans. Asiat. Soc. Jap. 24 suppl: 89. (1896) (List Pl. Formosa). *Boehmeria platyphylla* D. Don. *clidemioides* Forbes & Hemsley non (Miq.) Wedd. in Journ. Linn. Soc. Bot. 26: 487. (1899).

Taiwan: Wanchin, 33. **Yunnan:** Simao, 12370, 12696.

A procumbent undershrub, native to India and southern China to Indonesia. Native from the tropics of India to southern China and Java.

Boehmeria platanifolia (Franch. & Sav.) C. H. Wright (URTICACEAE)

Forbes & Hemsley in Journ. Linn. Soc. Bot. 26: 486. (1899).

Hubei: Liantuo, 2049. South Badong, 6150. **Sichuan:** North Wushan, 7033.

Boehmeria sieboldiana Blume (URTICACEAE)

Boehmeria formosana Hayata Li in Woody Flora of Taiwan 131. fig. 41. (1963).

Boehmeria sp. Henry in Trans. Asiat. Soc. Jap. 24 suppl: 89. (1896). *Boehmeria platyphylla* D. Don var. *stricta* C. H. Wright in Journ. Linn. Soc. Bot. 26: 487. (1899).

Hubei: San You Dong Glen, 1692. **Taiwan:** Wanchin, 187. Oluanpi, 1293 (isosyntype of *Boehmeria platyphylla* D. Don var. *stricta* C. H. Wright).

Native from China to southern Japan and Korea.

Boehmeria sieboldiana Blume var. *fuzhouensis* (W. T. Wang) Friis & Wilmot-Dear (URTICACEAE)

In the Kew Herbarium (K.).

Yunnan: Simao, 12540.

Boehmeria sp. (URTICACEAE)

Anonymous in Kew List of Determinations (volume 24 & 34).

Hubei: Badong, 376, 396 (June 1885), 933 (November 1885).

Boehmeria virgata (G. Forst.) Guill. (URTICACEAE)

In the Kew Herbarium (K.).

Yunnan: Simao, 12371.

Boehmeria virgata (G. Forst.) Guill. var. *macrostachya* (Wight) Friis & Wilmot-Dear (URTICACEAE)

Boehmeria platyphylla D. Don var. *macrostachya* (Wight) Wedd. Forbes & Hemsley in Journ. Linn. Soc. Bot. 26: 487. (1899).

Yunnan: Mengzi, 9792, 9792a., 9792b.

Distributed in Nepal, India, Sri Lanka, Bhutan, Myanmar China, Laos and Vietnam.

Boehmeria zollingeriana Wedd. (URTICACEAE)

In the Kew Herbarium (K.).

Yunnan: Simao, 13189.

Boehmeria zollingeriana Wedd. var. *blinii* (H. Lèvl.) C. J. Chen (URTICACEAE)

Boehmeria sp. Henry in Trans. Asiat. Soc. Jap. 24 suppl: 89. (1896) (List Pl. Formosa). *Boehmeria densiflora* Forbes & Hemsley non Hook. & Arn. in Journ. Linn. Soc. Bot. 484. (1899). *Boehmeria blinii* Levl. var. *podocarpa* W. T. Wang Li in Woody Flora of Taiwan 131. (1963).

Taiwan: Wanchin, 468. Without locality, 468a., 468d. Wanshoushan, 468e. **Yunnan:** Simao, 12541.

02

Boenninghausenia albiflora (Hook.) Reichenb. ex Meiss. (RUTACEAE)

Anonymous in Kew List of Determinations (volume 24 & 34).

Hubei: Xiling (Yichang) Gorge, 2415. Antelope Glen, 4263.

A slender, glabrous foetid perennial or a deciduous subshrub to 90 cm tall bearing loose panicles of small white flowers on the current years growth during August. This widespread species grows over much of India, Myanmar, China and Japan. It is best cultivated against a warm wall and Farrer regarded it as a good plant for the rock garden. To the Hani (an ethnic group) of southern Yunnan this is the *beqpucaol* or the 'white tiger plant', and the stem and leaves are used to treat cases of influenza, laryngitis, bronchitis and malaria. A widespread species found growing over a large range from the Himalaya to China and Japan.

Boerhavia repens L. (NYCTAGINACEAE)

Forbes & Hemsley in Journ. Linn. Soc. Bot. 26: 317. (1891), Henry in Trans. Asiat. Soc. Jap. 24 suppl: 74. (1896) (List Pl. Formosa).

Hainan: Environs of Haikou, 8069. Environs of Kiungchow, 8308. **Taiwan:** Kaohsiung, without number.

A loosely branched herb to 50 cm tall, native to much of tropical Africa, Asia, Australia and Polynesia.

Boerhavia sp. (NYCTAGINACEAE)

Anonymous in Kew List of Determinations (volume 34).

Yunnan: Simao, 13261.

Bolbitis appendiculata (Willd.) JK. Iwats. (POLYPODIACEAE)

Acrostichum appendiculatum Willd. Henry in Trans. Asiat. Soc. Jap. 24 suppl: 116. (1896) (List Pl. Formosa).

Taiwan: Wanchin, 81.

A widely distributed fern, native to India, Malaysia, Thailand, Indonesia and China (Taiwan) where it grows on rocks in mountain streams.

Bolbitis heteroclita (C. Presl.) Ching (POLYPODIACEAE)

Gymnopteris flagellifera (Wall. ex Hook. & Grev.) Bedd. Christ in Bull. Herb. Boiss. 6: 866. (1898). *Nephrodium cuspidatum* C. Presl. Anonymous in Kew List of Determinations (volume 34).

Yunnan: In Pingbian, in mountain forest on the Daweishan Range at 1,600 metres, 10762. Mengzi, 10825. Simao, 12907, 13064.

Bolbitis sinense (Baker) K. Iwats. (POLYPODIACEAE) New species

Acrostichum sinense Baker in Kew Bull. Misc. Inf. 14. (1906).

Yunnan: On shaded cliffs, in forests near Simao, 12494 (isotype of *Acrostichum sinense* Baker).

Nepal to New Guinea.

Bolbostemma biglandulosum (Hemsl.) Franquet (CUCURBITACEAE) New species

Actinostemma biglandulosum Hemsley in Hooker's Icon. Pl. 17: t. 2622. (1899); 18: t. 2645. (1900), Nelson in A Heritage of Beauty 309. (2000).

Yunnan: Mengzi at 1,950 metres, 9390 (isosyntype of *Actinostemma biglandulosum* Hemsley). Mengzi at 1,200 metres 9390a., (type of *Actinostemma biglandulosum* Hemsley, collected 4th November 1896).

Discovered by the Irish collector, William Hancock in 1894 and introduced to cultivation from Mengzi by Henry through the Royal Botanic Gardens, Kew. A note with Henry's 9390a. stated 'ripe seed', it is possible that these were collected from this particular herbarium specimen and it was by this means it was raised at the Royal Botanic Gardens, Kew and Edinburgh. It is no longer in cultivation. (Nelson, 2000)

Bombax ceiba L. (BOMBACEAE)

Bombax malabaricum DC. Henry in Trans. Asiat. Soc. Jap. 24 suppl: 21. (1896) (List Pl. Formosa).

Taiwan: Kaohsiung, without number. **Yunnan:** Mengzi, 11088. Simao, 12901.

The red silk cotton tree, a large imposing tree to 25 metres tall. In spring this tree presents a magnificent sight when it is covered with masses of short-lived, waxy red flowers. Native from India to Malaysia, the Philippines and south to Australia. Naturalised in Taiwan where it is cultivated as an ornamental tree. The flowers are used in Chinese medicine and the fruit capsules furnish a cotton that in 19th century China, was used for stuffing pillows. In Myanmar the flowers are relished as a kind of vegetable curry.

Bombax insigne Wall. (BOMBACACEAE)

Bombax tenebrosum Dunn in Journ. Linn. Soc. Bot. 35: 486. (1903).

Yunnan: In a very dry part of the forest west of Simao at 1,650 metres, 12666 (type of *Bombax tenebrosum* Dunn).

Henry stated he only ever saw one tree himself and then only when it was in bud, its remarkable appearance induced him to send back Chinese collectors later in the season to gather flowers and fruit.

Bonnaya antipoda (L.) Druce (LINDERNIACEAE)

Bonnaya verbenifolia (Colsm.) Spreng. Forbes & Hemsley in Journ. Linn. Soc. Bot. 26: 192. (1890).

Hubei: Liantuo, 3205. **Hainan:** Environs of Kiungchow, 8281. 8335.

Bonnaya ciliata (Colsm.) Spreng. (LINDERNIACEAE)

Bonnaya brachiata Link & Otto Henry in Trans. Asiat.

Soc. Jap. 24 suppl: 68. (1896) (List Pl. Formosa).

Taiwan: Wanchin, 879. Kaohsiung, 1869. **Yunnan:** Simao, 12396.

Bonnaya* cf. *tenuifolia (Colsm.) Spreng. (LINDERNIACEAE)

Anonymous in Kew List of Determinations (volume 24).

Hainan: 8 km (5 miles) south of Kuingchow, 8178.

Bonnaya veronicifolia (Retz.) Spreng. (LINDERNIACEAE)

Henry in Trans. Asiat. Soc. Jap. 24 suppl: 68. (1896) (List Pl. Formosa).

Taiwan: Wanchin, 884. Kaohsiung, 1909, 2019.

Boottia sp. (HYDROCHARITACEAE)

Anonymous in Kew List of Determinations (volume 34).

Yunnan: Simao, 12391.

Boraginaceae (BORAGINACEAE)

Anonymous in Kew List of Determinations (volume 34).

Yunnan: Mengzi, 13793.

Bosmania membranacea (D. Don) Testo (POLYPODIACEAE)

Polypodium membranaceum D. Don Christ in Bull. Herb. Boiss. 6: 874. (1898), Takeda in Notes Roy. Bot. Gard. Edinburgh 8: 310. (1913—1915).

Yunnan: South of the Red River from Manmei at 1,650 metres, 9469, 9469a. In Pingbian, in forests on the Daweishan Range at 1,950 metres, 11488. Simao, 9469d., 11488c.

Distributed across India, Sri Lanka, China, Laos, Vietnam, Thailand and the Philippines.

Bothriochloa intermedia (R. Br.) A. Camus (POACEAE)

Andropogon vachellii Henry non Nees in Journ. China Br. Roy. Asiat. Soc. 22: 241. (1887) (Chinese Names of Plants), *Andropogon intermedius* R. Br. Rendle in Journ. Linn. Soc. Bot. 36: 373. (1904).

Hubei: Yichang, 2373, 2374, 2766, 4096. **Taiwan:** Kaohsiung, 1039, 1094, 1152. **Yunnan:** Mengzi at 1,500 metres, 9614.

Native to tropical Africa, Asia, Australia and the Pacific Islands. A common grass around Yichang where according to Henry it was colloquially known as *ch'ou ken tzu ts'ao*. A fodder grass in Taiwan.

Bothriochloa ischaemum (L.) Keng (POACEAE)

Andropogon ischaemum L. Rendle in Journ. Linn. Soc. Bot. 36: 374. (1904).

Taiwan: Kaohsiung, 761.

An annual grass with a large geographical range over Europe, North Africa, the northwestern Himalaya to China including Taiwan.

Bothriospermum secundum Maxim. (BORAGINACEAE)

Bothriospermum kusnezowii Forbes & Hemsley non Bunge in Journ. Linn. Soc. Bot. 26: 151. (1890).

Hubei: Yichang, 773, 1383.

Bothriospermum sp. (BORAGINACEAE)

Anonymous in Kew List of Determinations (volume 34).

Yunnan: Mengzi, 11411.

Bothriospermum zeylanicum (J. Jacq.) Druce (BORAGINACEAE)

Bothriospermum tenellum (Horn.) Fisch & Mey. Forbes & Hemsley in Journ. Linn. Soc. Bot. 26: 152. (1890).

Fisch & Mey. in Trans. Asiat. Soc. Jap. 24 suppl: 63. (1896) (List Pl. Formosa).

Hubei: Yichang, 651, 841. Near Yichang, on the hills behind the San You Dong Glen (July 1886), 1508, 1511. **Taiwan:** Oluanpi, 679.

Botrychium lanuginosum (Wall.) Hook. & Grev. (OPHIOGLOSSACEAE)

Anonymous in Kew List of Determinations (volume 34).

Yunnan: Mengzi, 9213.

Distributed through India, Sri Lanka, Indonesia, China and the Philippines.

Botrychium ternatum (Thunb.) Sw. (BOTRYCHIACEAE)

Christ in Bull. Herb. Boiss. 7: 21. (1899), Diels in Bot. Jahrb. 29: 209. (1900).

Hubei: Xingshan, 6944. **Yunnan:** In Pingbian on the Daweishan Range in forests at 1,650 metres, 11951.

A widespread fern, distributed in North America, Asia and Australia.

Botrychium virginianum (L.) Sw. (BOTRYCHIACEAE)

Henry in Notes Econ. Bot. China 60.(1893) Diels in Bot. Jahrb. 29: 209. (1900).

Hubei: Badong, 2568, 5047. Jianshi, 5799. **Sichuan:** North Wushan, 7078.

The rattlesnake fern, distributed in North and South America and temperate Eurasia. According to Henry this 'flowering fern' is the *chueh-ch'i*, source of the drug *leng-shui-ch'i*.

Bougainvillea spectabilis Willd. (NYCTAGINACEAE)

Henry in Trans. Asiat. Soc. Jap. 24 suppl: 74. (1896) (List Pl. Formosa).

Taiwan: Kaohsiung, 735.

Cultivated, a subtropical climber from Brazil.

Brachycorythis galeandra (Reichb. f.) Summerh. (ORCHIDACEAE)

Habenaria galeandra (Reichb. f.) Benth. Henry in Trans. Asiat. Soc. Jap. 24 suppl: 93. (1896) (List Pl. Formosa). *Platanthera obcordata* Rolfe non Lindl. ex Wall.

in Journ. Linn. Soc. Bot. 36: 56. (1903).

Taiwan: Wanchin, 850.

A terrestrial orchid to 35 cm tall, native to north-east India, Myanmar, Vietnam, Thailand and China.

Brachycorythis henryi (Schltr.) Summerh. (ORCHIDACEAE) New species

in Kew Bull. Misc. Inf. 10: 235. (1955).

Platanthera iantha Rolfe non Wight in Journ. Linn. Soc. Bot. 36: 55. (1903). *Phyllomphax henryi* Schlecter in Repert. Spec. Nov. Regni. Veg. Beih. 4: 45. (1919).

Yunnan: On the slopes of hills at Mengzi at 1,800 metres, 11111 (isotype of *Phyllomphax henryi* Schlecter). Simao, 12418.

Distributed in south-west China, Myanmar and Thailand. Rare in Yunnan according to Henry.

Brachypodium sylvaticum (Huds.) P. Beauv. (POACEAE)

Rendle in Journ. Linn. Soc. Bot. 36: 431. (1904).

Hubei: Badong, 4721. Fang Xian at 2,250 metres, 6745.

Brachypodium sylvaticum P. Beauv. var. *breviglume* Keng (POACEAE)

Rendle in Journ. Linn. Soc. Bot. 431. (1904).

Brachypodium sylvaticum P. Beauv. var. *breviglume* Keng in Acta. Bot. Yunnanica 4(3): 277. (1982).

Yunnan: In woods near Mengzi at 1,500 metres, 11255 (type *Brachypodium sylvaticum* P. Beauv. var. *breviglume* Keng).

Brachypterum scandens (Roxb.) Wight & Arn. ex Miq. (FABACEAE)

Derris scandens (Roxb.) Benth. Anonymous in Kew List of Determinations (volume 34).

Yunnan: Mengzi, 10770.

Brachythecium wichurae (Broth.) Par. (BRACHYTHECIACEAE)

Salmon in Journ. Linn. Soc. Bot. 34: 470. (1900).

Hubei: Glens near Yichang, 7912. (Moss).

Brandisia discolor Hook. f. & Thoms. (OROBANCHACEAE)

Rehder in Sargent Pl. Wilson. 1: 573. (1914).

Yunnan: In forests near Simao at 1,300 metres, 12605b.

Distributed in India, Myanmar, China (Yunnan), Thailand, Laos and Vietnam.

Brandisia glabrescens Rehder (OROBANCHACEAE) New species

in Sargent Pl. Wilson. 1: 574. (1913).

Yunnan: In forests near Mengzi at 2,000 metres, 9176a., (type). South of the Red River from Manmei at 2,300 metres, 9716. In forests near Mengzi at 2,300 metres, 9716a. Native to China (Yunnan) and Vietnam.

Brandisia hancei Hook. f. (OROBANCHACEAE)

Forbes & Hemsley in Journ. Linn. Soc. Bot. 26: 179. (1890), Rehder in Sargent Pl. Wilson. 1: 573. (1913), Zheng in Journ. Wuhan. Bot. Res. 2.(1): 187. (1984), Nelson in Pim, The Wood & the Trees 219. (1984).

Brandisia laetevirens Rehder in Sargent Pl. Wilson. 1: 573. (1913).

Hubei: Yichang, 313, 976. Xiling (Yichang) Gorge, 1099, 1150. Liantuo, 3007, 3213. **Yunnan:** Mengzi at 1,600 metres, 9013. In mountains to the east of Simao at 1,300 metres, 12605 (type of *Brandisia laetevirens* Rehder).

Discovered by Charles Maries in the Yichang (Xiling) Gorge in 1879 while collecting for James Veitch & Sons, Chelsea. Henry sent seeds of this species to Kew where they were received on November 27th 1886. Endemic to China.

Brandisia racemosa Hemsl. (OROBANCHACEAE)

Rehder in Sargent Pl. Wilson. 1: 574. (1913).

Yunnan: Mengzi at 1600 metres, hanging down from cliffs, stems to 1 metre, flowers scarlet, 9973.

According to Alfred Rehder this is the most beautiful member of the genus. Wilson introduced *Brandisia racemosa* to cultivation, but due to its parasitic nature it was never sucessful as its proper host plant was unknown. It was discovered by the Irish plant collector and Imperial Maritime Customs officer, William Hancock in 1894. Endemic to China and distributed in the provinces of Guizhou and Yunnan.

Brassaiopsis ciliata Dunn (ARALIACEAE) New species

in Journ. Linn. Soc. Bot. 35: 499. (1903).

Yunnan: In mountain forest near Mengzi at 1,650 metres, 9180a., (isotype).

Distributed in western China and Vietnam.

Brassaiopsis ficifolia Dunn (ARALIACEAE) New species

in Journ. Linn. Soc. Bot. 35: 500. (1903).

Euaraliopsis ficifolia (Dunn) Hutch in Gen. Fl. Pl. 2: 624. (1967).

Yunnan: In mountain forest to the east of Simao at 1,650 metres, 11650, (isosyntype). In forests near Simao, 12653. In forests to the south of Simao at 1,650 metres, 12653a., (isosyntype). Common along rocky banks of streamlets in shade of forests near Simao, 12653b., (type).

North-east India to China (Guangxi and Yunnan) and northern Indochina.

Brassaiopsis glomerulata (Blume) Regel. (ARALIACEAE)

Brassaiopsis speciosa Decne. & Planch. Anonymous in Kew List of Determinations (volume 34).

Yunnan: Simao, 13621, 13456.

A widespread species, distributed in Nepal, India, western China, Cambodia and Vietnam.

Brassaiopsis hainla (Buch.-Ham.) Seem. (ARALIACEAE)

Anonymous in Kew List of Determinations (volume 34).

Yunnan: Simao, 11882. Yuanjiang at 1,800 metres, 13294.

Distributed in Nepal, Bhutan, China [Tibet (Xizang) and Yunnan].

Brassaiopsis producta (Dunn) C. B. Shang (ARALIACEAE) New species

Heptapleurum productum Dunn in Journ. Linn. Soc. Bot. 35: 499. (1903).

Yunnan: In mountain forest near Mengzi at 1,650 metres, 9350 (isotype of *Heptapleurum productum* Dunn). Mengzi, 11382 (isotype of *Heptapleurum productum* Dunn).

Endemic to China and distributed in the provinces of Guangxi, Guizhou and Yunnan.

Brassaiopsis sp. (ARALIACEAE)

Anonymous in Kew List of Determinations (volume 34).

Yunnan: Simao, 12846.

Brassica juncea (L.) Czern. & Cross. ex Czern. (BRASSICACEAE)

Diels in Bot. Jahrb. 29: 357. (1900).

Hubei: Badong, 4661, 4957.

According to Henry this was the *aai ts'oi* or Chinese mustard, long grown in China as an oil seed and has been developed in that country into an enormous range of different vegetables. The centre of diversity for this species is China and its cultivation dates back to ancient times, *Brassica juncea* is of hybrid origin between *Brassica rapa* x *Brassica nigra* (black mustard) thus doubling the chromosome number to form a new species.

Brassica rapa L. (BRASSICACEAE)

Brassica campestris L. Schulz in Engler, Pflanzenr. iv. 105: 45. (1896). *Brassica juncea* Henry non (L.) Czern. & Cross. Henry in Trans. Asiat. Soc. Jap. 24 suppl: 17. (1896) (List Pl. Formosa). *Brassica napus* L. var. *chinensis* (L.) O. E. Schulz. Schulz in Engler, Pflanzenr. iv. 105: 45. (1919).

Hubei: Yichang, 819. Badong, 3710. Liantuo, 3797. **Taiwan:** Wanchin, 1720.

Pak Choi or Chinese white cabbage. This well known Chinese vegetable has been grown in China since the 5th century AD. Given suitable growing conditions crops can be harvested as soon as five weeks from sowing. Cultivated throughout most of the country.

Brassica sp. (BRASSICACEAE)

Anonymous in Kew List of Determinations (volume 24).

Hubei: Badong, 4008 (cultivated).

Brassicaceae (BRASICACEAE)

Anonymous in Kew List of Determinations (volume 34).

Yunnan: Simao, 12945.

Bredia oldhamii Hook. f. (MELASTOMACEAE)

Henry in Trans. Asiat. Soc. Jap. 24 suppl: 44. (1896) (List Pl. Formosa), Li in Woody Flora of Taiwan 654. fig. 261. (1963).

Taiwan: Wanchin, 520. Oluanpi, 1222, 2017, 2071.

A small to medium sized shrub, endemic to Taiwan where it grows throughout the island in broad-leaved forest at low altitude. Discovered near Tamsui by the Kew collector, Richard Oldham.

Bretschneidera sinensis Hemsley (AKANIACEAE) New genus & species

in Hooker's Icon. Pl. 28: t. 2708. (1901).

Yunnan: Mengzi, 10540 (isosyntype). Simao, 11050 (type). Simao at 1,650 metres, 11651 (isosyntype).

A very beautiful deciduous tree to 20 metres tall with large pinnate leaves to 80 cm long and bearing racemes of flesh pink bell-shaped flowers in April and May. Monotypic and distributed in north-east India, China, Vietnam and Thailand. This new genus commemorates a correspondent of Henry, the Latvian botanist Emil Bretschneider (1833—1901). It is extremly rare in the wild and is a nationally protected species.

Breynia fruticosa (L.) Müll.Arg. (PHYLLANTHACEAE)

Breynia fruticosa (L.) Hook. f. (comb. nov. superf.) Forbes & Hemsley in Journ. Linn. Soc. Bot. 26: 427. (1894).

Hainan: Environs of Haikou, 8045. Environs of Kiungchow, 8229, 8775. Without locality, 8530.

Breynia garrettii (Craib) Chakrab. & N. P. Balakr. (PHYLLANTHACEAE)

Sauropus yunnanensis Pax & Hoffmann in Engler, Pflanzenr. 81: (iv. 147. xv.): 220. (1921). *Sauropus garrettii* Craib Li in Acta Phytotax. Sin. 25: 136. (1987).

Yunnan: Mengzi, 9359, 9359b., (type of *Sauropus yunnanensis* Pax & Hoffmann). Simao at 1,300 metres, 13144 (isotype of *Sauropus yunnanensis* Pax & Hoffmann).

Southern China and Indochina.

Breynia macrantha (Hassk.) Chakrab. & N. P. Balakr. (PHYLLANTHACEAE)

Sauropus macranthus Hassk. Li in Acta Phytotax. Sin. 25: 133. (1987). *Sauropus grandifolius* Pax & Hoffmann in Engler, Pflanzenr. 81 (iv. 147. xv.): 222. (1921).

Yunnan: Simao at 1,300 metres, 11765a., (type of *Sauropus grandifolius* Pax & Hoffmann).

Asia to Australia.

Breynia officinalis Hemsl. (PHYLLANTHACEAE)

Henry in Trans. Asiat. Soc. Jap. 24 suppl: 83. (1896) (List Pl. Formosa), Li in Woody Flora of Taiwan 418. (1963).

Taiwan: Oluanpi, 241, 1305. Wanchin, 505. Kaohsiung, 748, 749.

According to the Irish collector, Thomas Watters (who was based in Taiwan as British Consul and sent specimens fron Tamsui to Kew) this shrub was known as *shan-ch'i-chin*. The roots and leaves of this species were used to make a wash, which was said to be very effective in removing blisters caused by paint and varnish. It was discovered in Taiwan by the Kew collector, Richard Oldham in 1864. Native to China (Fujian and Taiwan) and Japan.

Breynia sp. (PHYLLANTHACEAE)

Sauropus sp. Anonymous in Kew List of Determinations (volume 34).

Yunnan: Mengzi, 10307, 10673, 10989.

Breynia vitis-idaea (Burm. f.) C. E. C. Fisch. (PHYLLANTHACEAE)

Breynia rhamnoides (Retz.) Muell.-Arg. Forbes & Hemsley in Journ. Linn. Soc. Bot. 26: 428. (1894).

Hubei: In a glen near Pingshanba (near Yichang, May 1888), 1589. Antelope Glen, 3578, 4258. San You Dong Glen (July 1886), 1531, 1633. South Badong 7304.

Bridelia affinis Craib (PHYLLANTHACEAE) New species

Bridelia henryana Jablonsky in Pflanzenr. (Engler) 65: 62. (1915).

Yunnan: Mengzi at 1,550 metres, 9666 (type of *Bridelia henryana* Jablonsky).

This tree was discovered by Henry in 1897, it was described by Craib from material later collected by Kerr in Thailand. Distributed in China in Hainan and Yunnan, and also in Thailand.

Bridelia balansae Tutcher (PHYLLANTHACEAE)

Forbes & Hemsley in Journ. Linn. Soc. Bot. 419. (1905), Li in Woody Flora of Taiwan 419. fig. 158. (1963).

Taiwan: Wanchin, 472. Oluanpi, 935.

A small tree to 7 metres tall, native to Indochina, southern China (including Taiwan) and Japan (south). Distributed at low altitude throughout Taiwan.

Bridelia glauca Blume (PHYLLANTHACEAE)

Bridelia pubescens Kurz Jablonsky in Engler, Pflanzenr. iv. 147. viii: 73. (1915).

Yunnan: By the Red River, 13027.

Distributed across tropical and subtropical Asia.

Bridelia sp. (PHYLLANTHACEAE)

Anonymous in Kew List of Determinations (volume 34).

Yunnan: Mengzi, 10129. Simao, 12575.

Bridelia stipularis (L.) Blume (PHYLLANTHACEAE)

Jablonsky in Engler, Pflanzenr. iv. 147. viii: 55. (1915).

Yunnan: South of the Red River, 10595.

Bridelia tomentosa Blume (PHYLLANTHACEAE)

Henry in Trans. Asiat. Soc. Jap. 24 suppl: 82. (1896) (List Pl. Formosa), Jablonsky in Engler, Pflanzenr. iv. viii: 58. (1915).

Bridelia monoica Li non (Lour.) Merr. in Woody Flora of Taiwan 419. (1963)

Taiwan: Wanchin, 44. Oluanpi, 199, 242. Kaohsiung, 757, 1083.

A shrub or small tree of wide distribution in India, southern China and the Philippines to New Guinea. Known colloquially as *t'u-mi-shu* according to Henry. This species was collected by the Glasnevin Taiwan Expedition (GTE 107) on Wanshoushan, in October 2004.

Briggsia sp. (GESNERIACEAE)

Near *Didymocarpus* Anonymous in Kew List of Determinations (volume 24).

Hubei: Liantuo, 2108.

Bromus japonicus Houtt. (POACEAE)

Bromus japonicus Thunb. (nom. illeg.) Rendle in Journ. Linn. Soc. Bot. 36: 430. (1904).

Hubei: On the hills near Yichang, 3523. Badong, 4033, 4732.

Bromus ramosus Huds. (POACEAE)

Bromus asper Murr. Rendle in Journ. Linn. Soc. Bot. 36: 430. (1904).

Hubei: Antelope Glen, 3560. Xingshan at 2,600 to 2,900 metres, 6994.

Broussonetia kaempferi Siebold (MORACEAE)

Forbes & Hemsley in Journ. Linn. Soc. Bot. 26: 455. (1894), Henry in Trans. Asiat. Soc. Jap. 24 suppl: 86. (1896) (List Pl. Formosa), Schneider in Sargent Pl. Wilson. 3: 304. (1917).

Hubei: In a glen near Pingshanba (near Yichang, May 1886), 1586. Yichang 3668. Changyang (April 1888), 5226. In a ravine on Moji Shan, 3668a. Jianshi, 5972. Changleping, 6257. **Yunnan:** In Pingbian on the Daweishan Range at 1,800 metres, 10959, 13644. In Pingbian on the Daweishan Range at 1,700 metres, 10959a. In Yulo forest to the south of Simao, 12019.

Widespread in China and also native to Japan.

Broussonetia papyrifera (L.) L' Heritier ex Vent. (MORACEAE)

Henry in Journ. China Br. Roy. Asiat. Soc. 22: 250. (1887) (Chinese Names of Plants), Forbes & Hemsley in Journ. Linn. Soc. Bot. 26: 455. (1894), Schneider in Sargent Pl. Wilson. 3: 304. (1917), Zheng in Journ. Wuhan Bot. Res. 2.(1): 31. (1984).

Hubei: Badong, 1426, 3740. Yichang, 3666. **Hainan:** Environs of Kiungchow, 8714. **Taiwan:** Kaohsiung, without number. Wanchin, without number. **Yunnan:** Milê district, 10569. Mountains to the north of Mengzi at 1,600 metres, 10569a. Forests to the east of Simao at 1,900 metres, 10569b.

Henry called this the *kou shu*, or paper mulberry a tree to 13 metres tall, abundant in the warmer parts of China. At Yichang in the late 19th century, the bark of the paper mulberry was used for making string and paper (the paper being called *kou p'i chih*; literally 'bark paper'). Paper making in China dates from the dawn of Christianity, previous to this period silk and cloth provided a medium on which to write and before this the early annals were recorded on tablets of bamboo. Paper money first originated in Sichuan province in the reign of the first emperor of the Sung Dynasty (A.D. 960). Marco Polo mentions the Kublai Khan's mint at Peking (Beijing) stating:

He makes them take the bark of a certain tree, in fact, of the mulberry tree, the leaves of which are the food of the silkworms, these trees being so numerous that whole districts are full of them. What they take is a certain fine white bast or skin which lies between the wood of the tree and the thick outer bark and this they make into something resembling sheets of paper.

Marco Polo was really referring to, *Broussonetia* a close relative of the mulberry and somewhat similar in appearance. It is extremely common in China and a weed around Chengdu in Sichuan. In cultivation this species normally forms a large bush, at Cambridge Botanic Gardens there is a remarkable double-stemmed tree, when last measured (1991) it was 15 metres tall, its first stem had a trunk diameter of 37 cm, the second was 34 cm. Native to Indo-Malaysia, China, Japan and the Pacific Islands.

Broussonetia sp. (MORACEAE)

Anonymous in Kew List of Determinations (volume 34).

Yunnan: Simao, 11611.

Brucea javanica (L.) Merr. (SIMAROUBACEAE)

Brucea sumatrana Roxb. Henry in Trans. Asiat. Soc. Jap. 24 suppl: 26. (1896) (List Pl. Formosa). *Rhus javanica* L. Rehder & E. H. Wilson in Sargent Pl. Wilson. 2: 178. (1916).

Hainan: Environs of Haikou, 7959. At Nodoa in the interior of the island, 8466. **Taiwan:** Kaohsiung, without number. **Yunnan:** In mountains near Mengzi at 1,600 metres, 11034.

A paleotropical shrub or small tree, ranging from India to southern China, Malaysia to Australia. Found along riversides in thickets in southern Taiwan. Henry stated it was known in China as the *ya dan zi* or 'crow's gall bladder seed', the fruits are used to treat chronic or reoccuring dysentry-like disorders and are also used to treat warts and corns.

Bruguiera gymnorhiza (L.) Lam. ex Savigny (RHIZOPORACEAE)

Bruguiera cylindrica Henry non (L.) Blume in Trans. Asiat. Soc. Jap. 24 suppl: 42. (1896) (List Pl. Formosa).

Taiwan: Kaohsiung Lagoon in tidal mangrove swamp, without number.

A small or medium butressed tree, native to tropical Africa, Madagascar and south east Asia. Until recent years this species was an inhabitant of the mangrove swamps at Kaohsiung.These have now been destroyed due to urban development and the species no longer part of the flora of Taiwan.

Bryum pseudotriqeutrum (Hedw.) Gaertn. (BRYACEAE)

Salmon in Journ. Linn. Soc. Bot. 34: 458. (1900).

Sichuan: South Wushan, 5593 (May 1888) (Moss).

Buchanania arborescens (Blume) Blume (ANACARDIACEAE)

Henry in Trans. Asiat. Soc. Jap. 24 suppl: 30. (1896) (List Pl. Formosa), Li in Woody Flora of Taiwan 445. fig. 172. (1963) Li in Flora of Taiwan 2: 581. (1993).

Taiwan: Kaohsiung, without number. Wanchin, 315. Oluanpi, 620.

A common tree resembling the mango in foliage hense the common name, *shan-shuain*. Distributed from Indo-Malaysia to the Philippines. A sea-shore tree in China (southern Taiwan).

Buchanania lanzan Spreng. (ANACARDIACEAE)

Buchanania latifolia Roxb. Anonymous in Kew List of Determinations (volume 34).

Yunnan: Simao, 13158.

Distributed in Nepal, India, Myanmar, China (Hainan and Yunnan), Cambodia, Laos and Vietnam.

Buchnera cruciata Buch.-Ham. ex D. Don (OROBANCHACEAE)

Forbes & Hemsley in Journ. Linn. Soc. Bot. 26: 201. (1900).

Hubei: Yichang, 2244, 4231. **Hainan:** Environs of Haikou, 8360. Ling-men (in the interior of the island) 8634. **Yunnan:** Mengzi, 9960.

Buckleya henryi Diels (SANTALACEAE) New species in Bot. Jahrb. Syst. 29: 306. (1900).

Buckleya lanceolata Forbes & Hemsley non (Sieb. & Zucc.) Miq. in Journ. Linn. Soc. Bot. 26: 409. (1894).

Hubei: Fang Xian, 6694. Xingshan at 1,400 metres 6539 (isosyntype).

Buddleja albiflora Hemsley (BUDDLEJACEAE) New species

in Journ. Linn. Soc. Bot. 26: 118. (1889), Bretschneider in Hist. Eur. Bot. Disc. China 2: 788. (1898), J. H. Veitch in Journ. Roy. Hort. Soc. 28: 65. (1903—1904), E. H. Wilson in Flora and Sylva. 3: 335. (1905), J. H. Veitch in Hortus Veitchii 358. (1906), Schneider in Illus. Handb. Laubholzk. 2: 845. fig. 530. (1912), Rehder & E. H. Wilson in Sargent Pl. Wilson. 1: 569. (1913), Zheng in Journ. Wuhan Bot. Res. 2(1): 182. (1984).

Buddleia sp. nova. Henry in Journ. China Br. Roy. Asiat. Soc. 22: 266. (1887) (Chinese Names of Plants).

Hubei: Yichang, 156a. Badong, 1871, 2351 (in part), 4689. Zigui, 6193. **Yunnan:** Mengzi, 10915.

Because of inaccurate information supplied by Henry's Badong collector, this species was described by Hemsley as a tree 7 to 9 metres tall with white flowers, whereas Wilson stated it rarely exceeded 4 metres in height and the flowers are lilac coloured. Unlike the four angled *Buddleja davidii*, this species is distinguished by its round stems and stamens inserted immediately below the mouth of the corolla tube. Henry stated that in the Three Gorges region it was known colloquially as *peh hui shu*. Introduced to cultivation by Wilson through Veitch's Coombe Wood nursery in 1900. In Veitch's *Novelties for 1908—1909* this species was selling for one shilling and six pence each. In Ireland it is cultivated at the John F. Kennedy Arboretum and the National Botanic Gardens, Glasnevin.

Buddleja asiatica Lour. (BUDDLEJACEAE)

Forbes & Hemsley in Journ. Linn. Soc. Bot. 26: 119. (1889), Henry in Trans. Asiat. Soc. Jap. 24 suppl: 61. (1896) (List Pl. Formosa), Rehder & E. H. Wilson in Sargent Pl. Wilson. 1: 566. (1913), Zheng in Journ. Wuhan. Bot. Res. 2.(1): 182. (1984), Li in Woody Flora of Taiwan 772. fig. 309. (1963).

Hubei: Yichang, 3285. San You Dong Glen, 3456. **Hainan:** Environs of Kiungchow, 8759. **Taiwan:** Wanchin, 50. Oluanpi, 200. **Yunnan:** Vacinity of Mengzi, in ravines at 1,500 to 1,600 metres, 10443, 10443b., 10433c. Che Yuan 10443a. Simao at 1,600 metres, 10443d., 11679.

A shrub to 8 metres tall, widely distributed from India to Malaysia to China, including Taiwan where it grows at low altitude on mountain slopes. According to Henry this species was known as *peh pu ch'iang* or 'white vitex'.

Buddleja davidii Franch. (BUDDLEJACEAE)

Rehder & E. H. Wilson in Sargent Pl. Wilson. 1: 567. (1913), Besant in Gard. Chron. ser. 3, 98: 335. (1935), Morley in Glasra 3: 78. (1979), Nelson in Pim, The Wood & the Trees 220. (1984), Zheng in Journ. Wuhan Bot. Res. 2.(1): 182. (1984), Bean in Trees & Shrubs 1: 450. (1989), O'

Brien in Augustine Henry, An Irish Plant collector in China 15. (2002).

Buddlea variabilis Hemsley in Journ. Linn. Soc. Bot. 26: 120. (1889), Hook. f. in Bot. Mag. cxxiv. t. 7609. (1898), Bean in The Garden 54: 6. (1898), Bretschneider in Hist. Eur. Bot. Disc. China 2: 788. (1898), Anon in Gard. Chron. ser. 3, 24: 138. fig. 36. (1898).

Hubei: Yichang, 156 (syntype of *Buddleia variabilis* Hemsley), 1069, 2060. Badong, 2351 (in part). South Badong, 7281. Antelope Glen, 4166. Without locality, 3110a. On the floor of a cave on Moji Shan, 3110b. On way to Moji Shan, 4166a.

Discovered by Père Armand David near Baoxing in western Sichuan in August 1869 and first introduced to cultivation by Henry when he sent seeds to the Royal Botanic Gardens, Kew on March 18th 1889. It was later reintroduced by the French missionary Père Soulie through the garden of the Natural History Museum in Paris in 1893 where it flowered for the first time in 1894. The plants raised at Kew proved to be far better forms than sent to the Jardin des Plants in Paris (Henry described the flowers on wild plants as rose, orange-coloured inside, the plants raised from his seeds were rosy purple). *Buddleja davidii* was growing in Ireland in 1898 (though it probably grew at Glasnevin previous to this date) in the garden of William Gumpleton at Belgrove in County Cork. In a hand written list 'Henry Plants in our Garden' (Alice Henry, c. 1920's) this species was included (as *B. variabilis*) in the list of plants grown in the garden at Ranelagh, Dublin. Still a common plant in the mountains above the gorges.

Buddleja davidii Franch. var. **magnifica** (E. H. Wilson) Rehder & E. H. Wilson (BUDDLEJACEAE)

New variety. in Sargent Pl. Wilson. 1: 567. (1913), Zheng in Journ. Wuhan Bot. Res.2. (1): 183. (1984).

Buddleia variabilis Hemsley in Journ. Linn. Soc. Bot. 26: 120. (1889), Hook. f. in Bot. Mag. cxxiv. t. 7609. (1898), Bean in The Garden 54: 6. (1898), Bretschneider in Hist. Eur. Bot. Disc. China 2: 788. (1898), Anon in Gard. Chron. ser. 3, 24: 138. fig. 36. (1898).

Hubei: North Badong, 7008.

Introduced to cultivation by Wilson.

Buddleja forrestii Diels (BUDDLEJACEAE) New species

Buddleia henryi Rehder & E. H. Wilson in Sargent Pl. Wilson. 1: 571. (1913), *Buddleia henryi* Rehder & E. H. Wilson var. *glabrescens* Marquand in Kew Bull. Misc. Inf. 192. (1930).

Yunnan: In woods and ravines on the Daweishan Range in Pingbian at 1,600 metres, 9025 (type of *Buddleia henryi* Rehder & E. H. Wilson), 9025b. South of the Red

River from Manmei at 1,800 metres 9025a., (isolectotype of *Buddleia henryi* Rehder & E. H. Wilson var. *glabrescens* Marquand).

Discovered by Henry in 1896 and described from material later collected by the Scottish plant hunter, George Forrest.

Buddleja lindleyana Fortune (BUDDLEJACEAE)

Forbes & Hemsley in Journ. Linn. Soc. Bot. 26: 120. (1889), Zheng in Journ. Wuhan Bot. Res. 2.(1): 183. (1984).

Buddleja lindleyana Fortune var. *sinuato-dentata* Hemsley in Journ. Linn. Soc. Bot. 26: 121. (1889), Rehder & E. H. Wilson in Sargent Pl. Wilson. 1: 564. (1913).

Hubei: Yichang, 624 (type of *Buddleja lindleyana* Fortune var. *sinuato-dentata* Hemsley), 3979. At Shih Pan Shan (off the Xiling (Yichang) Gorge), 4190.

Lindley's butterfly bush was originally discovered by Robert Fortune on the archipelago of Zhoushan (formerly Chusan) off the coast Zhejiang province in eastern China and introduced by him through his sponsors, the Horticultural Society (now the RHS) in 1843. Fortune named this beautiful shrub for Dr. John Lindley, one of the leading botanists of his day. Henry stated that in China it was called *tsui yu ts'ao* or 'fish-stupefying herb' since the crushed flowers when thrown into water will stupefy fish. At Glasnevin this shrub makes vigorous growth and puts on a handsome floral display in early autumn. A plant that deserves to be better known in cultivation, it has a quiet charm matched by few other species in this genus. *Buddleja lindleyana* was collected by the Glasnevin Central China Expedition in September 2004 near Yichang (GCCE 450) and near Wudu village, Hubei (GCCE 891).

Buddleja macrostachya Benth. ex Wall. (BUDDLEJACEAE)
New to China

Rehder & E. H. Wilson in Sargent Pl. Wilson. 1: 572. (1913).

Buddleia cylindrostachya Kraenzlin in Bot. Jahrb. Syst. 1, Biebl. 3: 35. (1913).

Yunnan: At Fengchunling at 1,600 metres, 10251a., (isolectotype of *Buddleia cylindrostachya* Kraenzlin). Mountains to the west of Simao at 1,600 metres, 10251c., 10251d., (isotype of *Buddleia cylindrostachya* Kraenzlin). Mengzi, 10251 (isotype of *Buddleia cylindrostachya* Kraenzlin, collected 30th January 1897).

Lectotypified in Meded. Landbonwhogeschool Wagengingen 79.(6): 35. (1979).

Buddleja officinalis Maxim. (BUDDLEJACEAE)

Henry in Journ. China Br. Roy. Asiat. Soc. 22: 259. (1887) (Chinese Names of Plants), Forbes & Hemsley in Journ. Linn. Soc. Bot. 26: 120. (1889), Oliver in Hooker's Icon. Pl. t. 1972. (1891), Henry in Notes Econ. Bot. China 22. (1893), Rehder & E. H. Wilson in Sargent Pl. Wilson. 1: 565. (1913), Zheng in Journ. Wuhan Bot. Res. 2.(1): 183. (1984).

Hubei: Yichang, 1117, 1291, 3110, 7852. San You Dong Glen, (July 1886) 1527. Badong, 1447. In a ravine on Moji Shan, 3363. Xiling (Yichang) Gorge, 7884.

Discovered in Gansu about 1874 by the Russian explorer Pavel Jakovlevich Piasetski in Shaanxi and Gansu. Wilson, who later collected this species for Messrs Veitch at Yichang in March 1900 stated that *Buddleja officinalis* was a common shrub in rocky places to an altitude of about 1,000 metres. He also described it as being very floriferous, fragrant and ornamental. According to Henry, his collector at Liantuo (a small hamlet in the Yichang (Xiling) Gorge) stated the dried flowers were the source of a drug known as *meng-hua*, the same name was applied at Yichang to the flowers of *Edgeworthia chrysantha* Lindl. Henry stated that 20 tons of *meng-hua* were annually exported fron Hankow (now part of the city of Wuhan on the eastern plains of Hubei). The *Buddleja* flowers were obtained from Shaanxi and Gansu while the flowers of *Edgeworthia* were collected from shrubs cultivated at Yichang. These are still used in China to treat red, swollen, painful eyes, visual obstruction or sensitivity to light.

Buddleja paniculata Wallich (BUDDLEJACEAE)

Buddleja lavandulacea Kränzl. in Bot. Jahrb. Syst. 50 (2-3, Biebl. 111): 45. (1913).

Yunnan: Mengzi, 10178 (isotype of *Buddleja lavandulacea* Kränzl., collected 27th January 1897).

A shrub or small tree to 6 metres, distributed in northern Nepal, India, Bhutan, Myanmar and China.

Buddleja spp. (BUDDLEJACEAE)

Anonymous in Kew List of Determinations (volume 24 & 34).

Hubei: Yichang, 815. **Yunnan:** Simao, 11619.

Buddleja yunnanensis L. F. Gagnepain (BUDDLEJACEAE)
New species

in Lecompte, Not. Syst. 2: 192. (1912), Rehder & E. H. Wilson in Sargent Pl. Wilson. 1: 564. (1913).

Yunnan: In forests near Simao at 1,500 metres, 12214.

A shrub to two metres tall, flowers lilac, distributed from north-east India to China (Yunnan).

Buglossoides arvensis (L.) I. M. Johnst. (BORAGINACEAE)

Lithospermum arvense L. Henry in Journ. China Br. Roy. Asiat. Soc. 22: 258. (1887) (Chinese Names of Plants), Forbes & Hemsley in Journ. Linn. Soc. Bot. 26: 154. (1890).

Hubei: Yichang, 1265, 1282.

According to Henry this species was known colloquially in the Three Gorges region as *me ka kung*.

Bulbophyllum andersonii (Hook. f.) J. J. Sm. (ORCHIDACEAE)

Cirrhopetalum henryi Rolfe in Journ. Linn. Soc. Bot. 36: 15. (1903).

Yunnan: Amoung rocks on mountains near Mengzi at 1,600 metres, 11264 (type of *Cirrhopetalum henryi* Rolfe).

Bulbophyllum griffithii (Lindley) Rchb. f. (ORCHIDACEAE)

Bulbophyllum calodictyon Schlecter in Repert Spec. Nov. Regni Veg. Beih. 4: 70. (1919).

Yunnan: Simao, 13609 (isotype of *Bulbophyllum calodictyon* Schlecter).

Bulbophyllum hirundinis (Gagnep.) Seidenf. (ORCHIDACEAE)

Cirrhopetalum aurantiacum W. W. Smith in Notes Roy. Bot. Gard. Edinburgh 13: 195. (1931).

Yunnan: Epiphytic on trees on the mountain range to the west of Simao at 1,950 metres, 13087 (type of *Cirrhopetalum aurantiacum* W. W. Smith).

An epiphytic or epiphylitic orchid, native to Vietnam and southern China including Taiwan.

Bulbophyllum odoratissimum (Sm.) Lindl. ex Wall. (ORCHIDACEAE)

Bulbophyllum congestum Rolfe in Kew Bull. Misc. Inf. 198. (1896).

Yunnan: Simao, 12286a. Simao, epiphytic on trees, flowers yellow, 12291 (isotype *Bulbophyllum congestum* Rolfe, collected 20th June 1898).

Bulbophyllum radiatum Lindl. (ORCHIDACEAE)

Forbes in Journ. Linn. Soc. Bot. 36: 13. (1903).

Hubei: Liantuo, 4548.

This species does not appear in the current Flora Hubeiensis, it may be iether *B. omerandrum* Hayata or *B. kwangtungense* Schltr.

Bulbophyllum retusiusculum Rchb. f. (ORCHIDACEAE)

Cirrhopetalum retusiusculum (Rchb. f.) Hook. f. Rolfe in Journ. Linn. Soc. Bot. 36: 15. (1903).

Yunnan: In Pingbian on the Daweishan Range at 2,300 metres, epiphytic on trees, 11106a.

Bulbophyllum sp. (ORCHIDACEAE)

Cirrhopetalum sp. Anonymous in Kew List of Determinations (volume 34).

Yunnan: Simao, 12067, 12729, 13052.

Bulbophyllum wallichii Rchb. F. (ORCHIDACEAE)

Cirrhopetalum wallichii Lindl. Rolfe in Journ. Linn. Soc. Bot. 36: 16. (1903).

Yunnan: In mountain forest 40 miles (64 kms.) southeast of Mengzi in Pingbian on the Daweishan Range at 2,300 metres, epiphytic on trees, 11106.

Distributed in Nepal, Bhutan, Myanmar, China, Thailand, Laos and Vietnam.

Bulbophyllum yunnanense Rolfe (ORCHIDACEAE)

New species

in Journ. Linn. Soc. Bot. 36: 14. (1903).

Yunnan: Mountains to the north of Mengzi at 2,300 metres, 11370.

Rolfe based his description of this species on material gathered near Mengzi by both William Hancock and Augustine Henry. Nepal to China (Yunnan).

Bulboschoenus maritimus (L.) Palla (CYPERACEAE)

Scirpus maritimus L. Henry in Trans. Asiat. Soc. Jap. 24 suppl: 105. (1896) (List Pl. Formosa), C. B. Clarke in Journ. Linn. Soc. Bot. 36: 251. (1903).

Taiwan: Kaohsiung, 1818.

Bulbostylis barbata (Rottb.) C. B. Clark (CYPERACEAE)

C. B. Clarke in Journ. Linn. Soc. Bot. 36: 247. (1903).

Fimbristylis barbata (Rottb.) Benth. Henry in Trans. Asiat. Soc. Jap. 24 suppl: 105. (1896) (List Pl. Formosa)

Hainan: Environs of Haikou, 8241. **Taiwan:** Kaohsiung, 1866.

A densely tufted annual, culms to 30 cm tall. Pantropic in distribution.

Bulbostylis sp. (CYPERACEAE)

Bulbostylis capillaris Kunth Anonymous in Kew List of Determinations (volume 24).

Hubei: South Badong, 7404.

Bupleurum candollei Wall. ex DC. (APIACEAE)

Anonymous in Kew List of Determinations (volume 34).

Yunnan: Mengzi, 11168.

Distributed in Nepal, India, Bhutan, Myanmar and western China.

Bupleurum falcatum L. (APIACEAE)

Anonymous in Kew List of Determinations (volume 34).

Yunnan: Mengzi, 10745.

A European species, this collection belongs to an allied species.

Bupleurum hamiltonii N. P. Balkr. (APIACEAE)

Bupleurum tenue Buch.-Ham ex Don var. *genuinum* Wolff in Engler, Pflanzenr. iv. 228: 145. (1910).

Yunnan: Mengzi, 9200. Simao, 13493.

A widespread species, distributed in Pakistan, Nepal, India, Bhutan, China, Thailand and Vietnam.

Bupleurum longicaule Wall. ex DC. (APIACEAE)

Bupleurum longicaule Wall. ex DC. var. *strictum* C. B. Clarke Wollf in Engler, Pflanzenr. iv. 228: 123. (1910).

Hubei: Fang Xian, 6930.

Bupleurum longiradiatum Turcz. var. ***porphyranthum*** Shan & Li (APIACEAE)

Bupleurum longiradiatum Henry non Turcz. in Notes

Econ. Bot. China 25. (1893). *Bupleurum longiradiatum* Turcz. var. *genuinum* Wolff in Engler, Pflanzenr. iv. 228: 55. (1910).

Hubei: Donghu, 6420. Without locality, 6420a., b.

Bupleurum scorzonerifolium Willd. (APIACEAE)

Bupleurum falcatum L. var. *scorzoneraefolium* (Wall.) Willd. Forbes & Hemsley in Journ. Linn. Soc. Bot. 23: 327. (1887). *Bupleurum falcatum* Henry non L. in Notes Econ. Bot. China 25. (1893).

Hubei: Yichang, 16, 81, 3238. Liantuo, 3020. 6721. Without locality, 3238a.

A common plant in the Three Gorges region according to Henry who also stated it was the source of the drug *ch'ai-hu.*

Burmannia sp. (BURMANNIACEAE)

Anonymous in Kew List of Determinations (volume 34).

Yunnan: Mengzi, 9444. Simao, 12043.

Burretiodendron esquirolii (Levl.) Rehder (MALVACEAE) New genus

in Journ. Arnold Arb. 17: 48. (1936).

Yunnan: Near the Red River at Manhao (near the Vietnamese border), 9572, 9573.

This genus was first found by Henry in 1896, it was later collected by J. Cavalerie in November 1905 and by J. Esquirol in August 1911, both collections being made in the adjacent province of Guizhou. Endemic to western China.

Buxus henryi Mayr (BUXACEAE) New species

in Fremland Wald und Parkbuume. 451. (1906), Dummer in Gard. Chron. ser. 3, 52: 423. fig. 182. (1912), Rehder & E. H. Wilson in Sargent Pl. Wilson. 2: 168. (1916).

Buxus sempervirens Forbes & Hemsley non L. in Journ. Linn. Soc. Bot. 26: 418. (1894)

Hubei: Antelope Glen, 3387. South Badong, 7438.

A much branched shrub to 1.75 metres tall. For a good description of this species see Dummer in the *Gardeners' Chronicle* ser. 3, 52: 423. (1912).

Buxus ichangensis Hatus. (BUXACEAE) New species

Buxus sempervirens Forbes & Hemsley non L. in Journ. Linn. Soc. Bot. 26: 418. (1894). *Buxus harlandii* Rehder & E. H. Wilson non Hance in Sargent Pl. Wilson. 2: 166. (1916), Walsh. in Aug. Henry Forst. Herb. 95. (1957).

Hubei: Xiling (Yichang) Gorge, a fluviatile shrub, 3313 (misquoted by Hatus. as 3318), 3313a.

In both Henry's and Wilson's time in western Hubei this was an abundant dwarf shrub in the gorges and glens near Yichang where it grew in rock crevices and amoung stones in the bed and banks of streams where during the summer floods it is submerged. It is found at low altitudes where the summers are mild.

Buxus liukiuensis (Makino) Makino (BUXACEAE)

Buxus sempervirens Henry non L. in Trans. Asiat. Soc. Jap. 24 suppl: 82. (1896) (List Pl. Formosa). *Buxus microphylla* Sieb. & Zucc. var. *sinica* Rehder & E. H. Wilson in Sargent Pl. Wilson. 2: 165. (1916), in part. *Buxus microphylla* Sieb. & Zucc. var. *intermedia* Li non (Hatus.) H. L. Li in Woody Flora of Taiwan 440. fig. 170. (1963).

Taiwan: Wanchin, without number. Wanshoushan, 1177.

Buxus rugulosa Hatus (BUXACEAE) New species

In J. Dept. Agric. Kyushu Imp. Univ. 6: 303 (1942).

Buxus microphylla Sieb. & Zucc. var. *sinica* Rehder & E. H. Wilson in Sargent Pl. Wilson. 2: 165. (1916). Pro parte, quoad pl. ex Yunnan.

Yunnan: Forests near Mengzi at 2,800 metres, 11157.

Buxus sinica (Rehder & E. H. Wilson) M. Cheng (BUXACEAE) New species

Buxus sempervirens Forbes & Hemsley non L. in Journ. Linn. Soc. Bot. 26: 418. (1894). *Buxus sinica* (Rehder & E. H. Wilson) Ching var. *aemulans* Zheng non (Rehder & E. H. Wilson) M. Cheng in Journ. Wuhan Bot. Res. 2.(1): 115. (1984). *Buxus microphylla* Sieb. & Zucc. var. *sinica* Rehder & E. H. Wilson in Sargent Pl. Wilson. 2: 165. (1916), Walsh in Aug. Henry Forst. Herb. Cat. 96. (1957).

Hubei: Badong, 4863, 5455. Fang Xian, 6886. North Badong, 7159.

Buxus sinica (Rehder & E. H. Wilson) M. Cheng var. *aemulans* (Rehder & E. H. Wilson) P. Bruckn. & T. L. Ming (BUXACEAE) New variety

Buxus sempervirens Forbes & Hemsley non L. in Journ. Linn. Soc. Bot. 26: 418. (1894). *Buxus microphylla* Sieb. & Zucc. var. *aemulans* Rehder & E. H. Wilson in Sargent Pl. Wilson. 2: 169. (1914). *Buxus sinica* (Rehder & E. H. Wilson) M. Cheng var. *aemulans* (Rehder & E. H. Wilson) M. Cheng Zheng in Journ. Wuhan Bot. Res. 2.(1): 115. (1984).

Hubei: Yichang, 3293. Changyang, 7807 (type). Without locality, 3293a.

According to Wilson this box was a favourite garden shrub with the Chinese.

Buxus sp. (BUXACEAE)

Buxus sempervirens Forbes & Hemsley non L. in Journ. Linn. Soc. Bot. 26: 418. (1894).

Hubei: Jianshi, 5956, 5957. **Sichuan:** Henry's Chinese collector with Pratt, no locality stated but may be Leshan, Emei Shan, Wa Shan or Kangding, 8894.

Caesalpinia cf. *parviflora* Prain ex King (FABACEAE)

Anonymous in Kew List of Determinations (volume 34).

Yunnan: Mengzi, 9789.

Caesalpinia pulcherrima (L.) Sw. (FABACEAE)

Henry in Trans. Asiat. Soc. Jap. 24 suppl: 38. (1896) (List Pl. Formosa).

Hainan: On Liang Shan, 16 km (10 miles) east of Haikou, 8207. **Taiwan:** Wanchin, without number. Kaohsiung 330.

Cultivated in Taiwan. According to Henry this species was known coloqually there as *mei-a-huei* or 'butterfly flower'.

Cajanus cajan (L.) Huth (FABACEAE)

Cajanus indicus Spreng. Henry in Trans. Asiat. Soc. Jap. 24 suppl: 37. (1896) (List Pl. Formosa).

Taiwan: Kaohsiung, without number. **Yunnan:** Mengzi, 10357.

The Catjang pea, red gram, dahl, dhal or pigeon pea, an erect shrub to 3 metres tall and cultivated in Taiwan where it was also collected by the Veitchian collector, Charles Maries. A perennial pea forming a short lived shrub to 3 metres tall. Henry stated it was known as *shu-tou*. Now naturalised in Taiwan, this species may have originated in India and was cultivated by the Egyptians 4,000 years ago. Its cultivation spread to China in about 500 AD. The young pods and seeds are edible when cooked. Widely dispersed in the Old World tropics.

Cajanus grandiflorus (Bentham ex Baker) Maesen. (FABACEAE)

Atylosia grandiflora Bentham ex Baker Anonymous in Kew List of Determinations (volume 34).

Yunnan: Simao, 12558, 13369.

Cajanus scarabaeoides (L.) du Petit-Thouars (FABACEAE)

Atylosia scarabaeoides (L.) Benth. Henry in Trans. Asiat. Soc. Jap. 24 suppl: 37. (1896) (List Pl. Formosa).

Taiwan: Wanchin, 1539.

Calamagrostis arundinacea (L.) Roth. (POACEAE)

Deyeuxia sylvatica (Scrad.) Kunth. Rendle in Journ. Linn. Soc. Bot. 36: 395. (1904). *Deyeuxia sylvatica* (Scrad.) Kunth. var. *brachytricha* (Steud.) Hack Rendle in Journ. Linn. Soc. Bot. 36: 396. (1904). *Deyeuxia sylvatica* (Scrad.) Kunth. var. *ligulata* Rendle in Journ. Linn. Soc. Bot. 36: 398. (1904). *Deyeuxia henryi* Rendle in Journ. Linn. Soc. Bot. 36: 393. (1904). *Deyeuxia hupehensis* Rendle in Journ. Linn. Soc. Bot. 36: 394. (1904).

Hubei: Badong, 2564, 5052, 5021 (type of *Deyeuxia sylvatica* (Scrad.) Kunth. var. *ligulata* Rendle). Jianshi, 7411. Yichang, 2745. Fang at 2,250 metres, 6726 (type of *Deyeuxia hupehensis* Rendle), 6727, 6724 (type of *Deyeuxia henryi* Rendle). **Sichuan:** North Wushan, without number.

Calamagrostis effusiflora (Rendle.) P. C. Kuo & S. L. Lu ex J. L. Yang (POACEAE)

Deyeuxia sylvatica (Scrad.) Kunth. var. *laxiflora* Rendle in Journ. Linn. Soc. Bot. 36: 397. (1904).

Hubei: Yichang, 978 (type of *Deyeuxia sylvatica* (Scrad.) Kunth. var. *laxiflora* Rendle).

Calamagrostis epigejos (L.) Roth. (POACEAE)

Rendle in Journ. Linn. Soc. Bot. 36: 391. (1904).

Hubei: Yichang, 3967. Badong, 4772. Changleping, 6213. South Badong, 7371.

Calamagrostis sp. (POACEAE)

Deyeuxia sp. Anonymous in Kew List of Determinations (volume 24).

Hubei: Yichang, 571, 850.

Calamus formosanus Becc. (ARACEAE)

Becc. in Records Bot. Survey India, ii. 211. (1902), Wright in Journ. Linn. Soc. Bot. 36: 170. (1903), Li in Woody Flora of Taiwan 917. fig. 366. (1963).

Calamus sp. Henry in Trans. Asiat. Soc. Jap. 24 suppl: 100. (1896) (List Pl. Formosa).

Taiwan: Wanchin, 522. Oluanpi, 587, 687, 1354.

An eleagant palm forming a climbing vine to 50 metres long. Endemic to Taiwan and found there in broadleaved forest at low to medium altitudes. According to Henry's Chinese collectors this rattan palm was called *t'u teng*.

Calamus henryanus Beccari (ARECACEAE) New species

in Rec. Bot. Survey India. 2: 199. (1902), Wright in Journ. Linn. Soc. Bot. 36: 170. (1903).

Yunnan: In mountain forest on the Daweishan Range at 1,400 metres, 12239 (type, collected July 20th 1898).

Distributed across southwest China, Myanmar and Indochina.

Calamus quiquesetinervius Burret (ARACEAE)

Calamus margaritae Henry non Hance in Trans. Asiat. Soc. Jap. 24 suppl: 100. (1896) (List Pl. Formosa), Wright in Journ. Linn. Soc. Bot. 36: 170. (1903). *Daemonorops margaritae* Li non (Hance) Beccari in Woody Flora of Taiwan 918. fig. 367. (1936).

Taiwan: Wanchin, 521, 521a. Oluanpi, 910.

A climbing palm with stems reaching 70 metres long. Endemic to Taiwan and found in forests at 300 to 1,000 metres altitude throughout the island. According to Henry's Chinese collector, this rattan palm was called *cheng-t'eng*.

Calamus sp. (ARECACEAE)

Anonymous in Kew List of Determinations (volume 34).

Yunnan: Mengzi, 10475.

Calamus tetradactylus Hance (ARACEAE)

Wright in Journ. Linn. Soc. Bot. 36: 170. (1903).

Hainan: Environs of Haikou, 8213.

Calanthe alpina Hook. f. ex Lindl. (ORCHIDACEAE)

Calanthe buccinifera Rolfe in Journ. Linn. Soc. Bot.

29: 318. (1892), Bretschneider in Hist. Eur. Bot. Disc. China 2: 792. (1898), Rolfe in Journ. Linn. Soc. Bot. 36: 25. (1903).

Hubei: South Badong at 1,900 metres, 6064. North Badong, 7161 (isosyntype of *Calanthe buccinifera* Rolfe).

Calanthe arcuata Rolfe (ORCHIDACEAE) New species

in Kew Bull. Misc. Inf. 196. (1896), Bretschneider in Hist. Eur. Bot. Disc. China 2: 792. (1898), Rolfe in Journ. Linn. Soc. Bot. 36: 25. (1903).

Hubei: Xingshan, 6514 (isotype).

A terrestrial or sometimes epiphytic orchid, widespread across Nepal to China (including Taiwan).

Calanthe davidii Franch. (ORCHIDACEAE)

Calanthe ensifolia Rolfe in Kew Bull. Misc. Inf. 197. (1896), Bretschneider in Hist. Eur. Bot. Disc. China 2: 792. (1898), Rolfe in Journ. Linn. Soc. Bot. 36: 25. (1903).

Hubei: Jianshi, 6005 (syntype of *Calanthe ensifolia* Rolfe).

A handsome terrestrial orchid, this species is distributed in India, the Himalaya, China and Japan. The specific epithet commemorates its discoverer, Père Armand David.

Calanthe formosana Rolfe (ORCHIDACEAE)

in Ann. of Bot. 9: 157. (1895), Henry in Trans. Asiat. Soc. Jap. 24 suppl: 92. (1896) (List Pl. Formosa), Bretschneider in Hist. Eur. Bot. Disc. China 2: 792. (1898), Rolfe in Journ. Linn. Soc. Bot. 36: 25. (1903).

Taiwan: Oluanpi, 1347 (type of *Calanthe formosana* Rolfe).

A handsome terrestrial orchid to 70 cm tall, native to southern China (Hainan, Hong Kong, Taiwan), Java, Sumatra and Malaysia.

Calanthe henryi Rolfe (ORCHIDACEAE) New species

in Kew Bull. Misc. Inf. 197. (1896), Bretschneider in Hist. Eur. Bot. Disc. China 2: 792. (1898), Rolfe in Journ. Linn. Soc. Bot. 36: 25. (1903), Morley in Glasra 3: 78. (1979).

Hubei: Changyang, 5253, 5253a., 5958a. Jianshi, (June 19th 1888) 5253b. **Sichuan:** South Wushan, (May 26th 1888) 5253d.

Calanthe herbacea Lindl. (ORCHIDACEAE)

Rolfe in Journ. Linn. Soc. Bot. 36: 26. (1903).

Yunnan: In mountain forest on the Daweishan Range in Pingbian at 1,600 metres, 11130.

Distributed from Sikkim to China and Vietnam.

Calanthe lamellosa Rolfe (ORCHIDACEAE) New species

in Kew Bull. Misc. Inf. 197. (1896), Bretschneider in Hist. Eur. Bot. Disc. China 2: 792. (1898), Rolfe in Journ. Linn. Soc. Bot. 36: 26: (1903).

Calanthe yunnanensis Rolfe in Journ. Linn. Soc. Bot. 36: 27. (1903).

Hubei: Jianshi, 5958 (isotype of *Calanthe lamellosa* Rolfe). **Yunnan:** In mountain forest to the north of Mengzi at 2,800 metres, 11107 (type of *Calanthe yunnanensis* Rolfe).

Henry described the flower of his Yunnan collection as creamy green, with a lilac lip and with the fragrance of grated cocoa-nut. This species is distributed in north-east India, Bhutan, Nepal, Myanmar and China.

Calanthe sp. No. 1. (ORCHIDACEAE)

Anonymous in Kew List of Determinations (volume 24).

Hubei: South Badong, 6108. Fang Xian, 6833.

Calanthe sp. No. 2. (ORCHIDACEAE)

Calanthe sp. Rolfe in Journ. Linn. Soc. Bot. 36: 27. (1903).

Yunnan: In mountain forest to the east of Mengzi at 2,300 metres, 11359.

In fruit only and therefore indeterminable.

Calanthe sp. No. 3. (ORCHIDACEAE)

Anonymous in Kew List of Determinations (volume 34).

Yunnan: Mengzi, 11819. Simao, 12179, 12464.

Calanthe tricarinata Lindl. (ORCHIDACEAE)

Rolfe in Journ. Linn. Soc. Bot. 36: 26. (1903).

Yunnan: In mountain forest near Mengzi at 1,800 metres, 11297.

A common terrestrial orchid, widespread across the Himalaya, north-east India, southern and western China and Japan, where it grows in forests between 1,000 to 3,500 metres above sea level. In its native habitat this species flowers from late April to early July. The pseudobulbs are used in Chinese medicine.

Calanthe triplicata (Willem.) Ames (ORCHIDACEAE)

Calanthe veratrifolia R. Br. Henry in Trans. Asiat. Soc. Jap. 24 suppl: 92. (1896) (List Pl. Formosa), Forbes & Hemsley in Journ. Linn. Soc. Bot. 26. (1904).

Taiwan: Oluanpi, 1327.

A terrestrial orchid to 60 cm tall, widely distributed from India to China, southern Japan, Malaysia to Australia and the Pacific Islands.

Calathodes palmata Hook.f. & Thoms. (RANUNCULACEAE)

Hemsley in Hooker's Icon. Pl. t. 1935. (1890).

Trollius palmatus (Hook. f. & Thoms.) Baill. Diels in Bot. Jahrb. 29: 324. (1900).

Hubei: South Badong, 6177. Xingshan, 6977.

Callerya cinerea (Benth.) Schot. (FABACEAE)

Millettia cinerea Benth. Anonymous in Kew List of Determinations (volume 34).

Yunnan: Simao, 12992.

Native to India, Bangladesh, Nepal, Bhutan, Myanmar and China (Sichuan, Tibet (Xizang), and Yunnan).

Callerya dielsiana (Harms ex Diels) L. K. Phan ex Z. Wei & Pedley (FABACEAE) New species

Millettia dielsiana Harms ex Diels Dunn in Journ. Linn. Soc. Bot. 41: 160. (1912).

Hubei: In the San You Dong Glen (May 1886), 1584. In the Hsiao Pingshanba Glen (October 1886), 2917. Liantuo, 1938, 3074. Badong, 2561. **Yunnan:** South of the Red River from Mengzi, 9633. Mengzi, 10813.

Harms based this species on a collection made by Bock and von Rosthorn in 1891, Henry had discovered it in the spring of 1887. Distributed in China, Myanmar, Vietnam and Laos.

Callerya oosperma (Dunn) Z. Wei & Pedley (FABACEAE) New species

Millettia oosperma Dunn in Journ. Linn. Soc. Bot. 41: 157. (1912).

Hainan: Environs of Kiungchow, 8235. **Yunnan:** Mengzi at 1,830 metres, 10265 (syntype). In Pingbian on the Daweishan Range at 1,300 metres, 10670 (syntype). Simao, 11670c., (syntype). Simao at 1,450 metres, 12992a., (type).

A large climber, distributed in China and Vietnam.

Callicarpa arborea Roxb. (LAMIACEAE)

Anonymous in Kew List of Determinations (volume 34).

Yunnan: Mengzi, 9551, 9551a. Simao, 12093.

Distributed in the Himalaya, India, Bangladesh, Myanmar, western China, Thailand, Cambodia, Laos and Vietnam.

Callicarpa bodinieri H. Levl. (LAMIACEAE) New species

Zheng in Journ. Wuhan Bot. Res. 2.(1): 184. (1984), Bean in Trees & Shrubs 1: 468. (1989).

Callicarpa giraldiana Hesse Rehder in Sargent Pl. Wilson. 3: 366. (1917), in part, omit A. Henry 3107, 5864. *Callicarpa giraldiana* Hesse. var. *subcanescens* Rehder in Sargent Pl. Wilson. 3: 368. (1917).

Hubei: Moji Shan, 3107. Jianshi, 5864 (type of *Callicarpa giraldiana* Hesse var. *subcanescens* Rehder).

An erect shrub to 2.5 metres tall, well known in cultivation for the handsome masses of globose, glossy lilac fruits it carries in autumn. Native to the central and southern provinces of China and Indochina, *Callicarpa bodinieri* was discovered by Henry in 1887 (Bodinier's collection dates to October 28th 1897).

Callicarpa candicans (Burm. f.) Hochr. (LAMIACEAE)

Callicarpa cana L. Forbes & Hemsley in Journ. Linn. Soc. Bot. 26: 252. (1890).

Hainan: Haikou, without number.

Callicarpa dichotoma (Lour.) K. Koch (LAMIACEAE)

Li in Woody Flora of Taiwan 823. (1963)

Callicarpa purpurea Juss. Forbes & Hemsley in Journ. Linn. Soc. Bot. 26: 254. (1890). *Callicarpa* sp. Henry in Trans. Asiat. Soc. Jap. 24 suppl: 70. (1896) (List Pl. Formosa).

Hubei: On the road to Moji Shan (June 1886), 1551. Yichang, 3107b. **Taiwan:** Wanchin, 435.

A native of China, Vietnam and Japan. Introduced to cultivation by Robert Fortune in 1857.

Callicarpa giraldii Hesse ex Rehder (LAMIACEAE) New species

Callicarpa giraldiana Hesse Rehder in Sargent Pl. Wilson. 3: 366. (1917), in part. *Callicarpa bodinieri* Zheng non H. Levl. in Journ. Wuhan Bot. Res. 2.(1): 184. (1984).

Hubei: Yichang, 3107a. Jianshi, 5992. Badong, 6127. South Badong, 7312. Changyang, 7497. **Yunnan:** Near Simao at 1,400 metres, 12119, 12119a., 12119c.

First introduced through the nursery of H. A. Hesse, Germany by the missionary, G. Giraldi from Shaanxi at the end of the 19th century. Perhaps the finest of the callicarpas, this species was collected by the Glasnevin Central China Expedition (GCCE 678) in the Red Dragon Reserve in Xingshan in October 2004.

Callicarpa gracilipes Rehder (LAMIACEAE) New species

in Sargent Pl. Wilson. 3: 371. (1917).

Hubei: Changyang, 7690 (holotype).

Callicarpa japonica Thunb. (LAMIACEAE)

Anonymous in Kew List of Determinations (volume 24).

Hubei: Liantuo, 1946. Yichang, 2155.

Callicarpa longifolia Lam. (LAMIACEAE)

Forbes & Hemsley in Journ. Linn. Soc. Bot. 26: 253. (1890).

Hainan: Haikou, without number.

Callicarpa macrophylla Vahl. (LAMIACEAE)

Forbes & Hemsley in Journ. Linn. Soc. Bot. 26: 254. (1890).

Hainan: Environs of Kiunchow, without number. **Yunnan:** Mengzi, 9262.

Distributed in the Himalaya, India, Sri Lanka, Myanmar, China, Thailand and Vietnam.

Callicarpa membranacea C. H. Chang (LAMIACEAE)

Callicarpa japonica Thunb. var. *angustata* Rehder in Sargent Pl. Wilson. 3: 369. (1917), Zheng in Journ. Wuhan Bot. Res. 2.(1): 185. (1984).

Hubei: Xingshan, 6515. Fang Xian, 6679. **Hunan:** Shimen, 7554.

02

Callicarpa nudiflora Hook. & Arn. (LAMIACEAE)

Callicarpa reevesii Wall. Forbes & Hemsley in Journ. Linn. Soc. Bot. 26: 254. (1890).

Hainan: Environs of Haikou, without number.

Callicarpa pedunculata R. Br. (LAMIACEAE)

Callicarpa formosana Rolfe Forbes & Hemsley in Journ. Linn. Soc. Bot. 26: 252. (1890), Li in Woody Flora of Taiwan 821. fig. 330. (1963). *Callicarpa* sp. Henry in Trans. Asiat. Soc. Jap. 24 suppl: 70. (1896) (List Pl. Formosa).

Taiwan: Wanchin, 78. Oluanpi, 287, 616. Kaohsiung 741. Kaohsiung Plain 1947. Wanshoushan, 1048..

A small shrub, native from Asia to E. Australia. Common in Taiwan at low altitude.

Callicarpa pilosissima Maxim. (LAMIACEAE)

Li in Woody Flora of Taiwan 819. (1963).

Callicarpa sp. nova. Henry in Trans. Asiat. Soc. Jap. 24 suppl: 70. (1896) (List Pl. Formosa). *Callicarpa pilosissima* Maxim. var. *henryi* Yamamoto in Journ. Soc. Trop. Agr. 6: 554. (1934).

Taiwan: Wanchin, 120. Oluanpi, 267.

A small shrub, endemic to Taiwan where it is scattered in forests throughout the island. Discovered by the Kew collector, Richard Oldham.

Callicarpa rubella Lindl. f. ***angusta*** C. P'ei (LAMIACEAE) New form

in Mem. Sci. Soc. China 1: 40. (1932).

Yunnan: Mengzi, 9412 (type), 9412a., (syntype).

Callicarpa spp. (LAMIACEAE)

Anonymous in Kew List of Determinations (volume 24).

Hubei: Yichang, 1185, 3999. South Badong, 7296. Liantuo, 3934. **Sichuan:** South Wushan, 7236.

Calocedrus macrolepis Kurz (CUPRESSACEAE)

Lancaster in Travels in China 182. (1993).

Libocedrus macrolepis (Kurz) Bentham & Hooker Henry in Elwes and Henry The Trees Gt. Brit. & Irel. 3: 488. (1907), E. H. Wilson in Journ. Arnold Arb. 7: 62. (1926), Walsh in Aug. Henry Forst. Herb. 21. (1957).

Yunnan: Tonguan at 1,600 metres, 11566. Simao at 1,400 metres, 11566a., 11566b.

An evergreen tree to 35 metres tall. Native to Myanmar, Thailand, Indochina and to southern Yunnan where it was discovered in May 1868 by Dr. John Anderson while travelling to Yunnan on Sladen's expedition from India. A handsome pyramidal tree, the bark of the trunk was described by Henry as being remarkably white. A rare tree in Yunnan where it is confined to low altitudes and is often planted in the courtyards of temple grounds. In the *Trees of Great Britain and Ireland* Henry states it was; introduced by E. H. Wilson who collected seeds when paying me a visit in the autumn of 1899. Young plants raised at Coombe Wood have beautiful, glaucous, large flat (sprays of) foliage, they may be seen at the temperate house at Kew.

This handsome tree is rare in cultivation and suited only to the mildest counties of the British Isles and Ireland.

Calophyllum inophyllum L. (CLUSIACEAE)

Henry in Trans. Asiat. Soc. Jap. 24 suppl: 19. (1896) (List Pl. Formosa), Anonymous in Kew List of Determinations (volume 24).

Hainan: Environs of Kiungchow, 8225. **Taiwan:** Tamsui, 1694. Oluanpi, 903.

A large tree to 20 metres tall, native to India, Madagascar, Malaya, the Philippines, Polynesia, Melenesia and Australia. In Taiwan this tree is mostly found on the southern part of the island. The generic epithet is derived from Greek and means 'beautiful leaf' and it is indeed a handsome foliage plant. The flowers begin to open at 3 to 4 a.m. and wither the next day, these are followed by golf ball sized fruits which are eaten and dispersed by bats. The timber is excellent and called Borneo mahogany and is used for furniture and boat building. Plants currently cultivated at the National Botanic Gardens, Glasnevin were raised from collections made by the Glasnevin Taiwan Expedition (GTE 37) in the tropical rainforest that surrounds the lighthouse at Oluanpi (South Cape of Henry and Wilson) in October 2004.

Calotropis gigantea (L.) W. T. Aiton (APOCYNACEAE)

Anonymous in Kew List of Determinations (volume 34).

Yunnan: Mengzi at 1,500 metres, 9690.

Caltha palustris L. (RANUNCULACEAE)

Diels in Bot. Jahrb. 29: 324. (1900).

Hubei: Badong, 3719.

Calystegia hederacea Wall. (CONVOLVULACEAE)

Forbes & Hemsley in Journ. Linn. Soc. Bot. 26: 164. (1890).

Hubei: Yichang, 1215.

Calystegia sepium (L.) R. Br. (CONVOLVULACEAE)

Henry in Journ. China Br. Roy. Asiat. Soc. 22: 275. (1887) (Chinese Names of Plants), Forbes & Hemsley in Journ. Linn. Soc. Bot. 26: 164. (1890), Henry in Notes Econ. Bot. China 48. (1893).

Hubei: Yichang, 547, 1567. Badong, 722, 1839. At side of rice fields in the Monastery Valley, 1647. Liantuo, 1975, 4475. At the side of rice fields near Yichang, 4090.

A native of north temperate and subtropical regions, Australia and New Zealand. Henry stated that in the Three Gorges region this species was known colloquially as *t'u erh miao* though he also gives the name *fan-t'eng*. A common plant in the gorges area.

Camellia caudata Wall. var. ***gracilis*** (Hemsl.) Yamam ex H. Keng (THEACEAE) New variety

Camellia gracilis Hemsley in Ann. of Bot. 9: 146. (1895), Forbes & Hemsley in Journ. Linn. Soc. Bot. 146. (1892), Bretschneider in Hist. Eur. Bot. Disc. China 2: 780. (1898). *Camellia caudata* Li non Wall. in Woody Flora of Taiwan 581. (1963).

Taiwan: Wanchin, mountains, 1612 (isotype of *Camellia gracilis* Hemsley). **Yunnan:** Mengzi, 9417, 11170, 11467.

Native to south-east China including Hainan and Taiwan.

Camellia cuspidata (Kochs.) Veitch (THEACEAE) New species

Zheng in Journ. Wuhan Bot. Res. 2.(1): 150. (1984).

Thea cuspidata Kochs in Bot. Jahrb. Syst. 27: 586. (1900), Rehder in Sargent Pl. Wilson 2: 390. (1916).

Hubei: Liantuo, 3024 (February 1887), 3216. Badong, 5151, 5165, 5360, 7922. On the Yangtze above Liantuo, 7917. Yichang, 7864. Changleping, 6323. Jianshi, 7442. **Sichuan:** North Wushan, 7026 (isotype of *Thea cuspidata* Kochs).

An evergreen shrub to almost 2 metres tall, young emerging leaves are copper tinted and the white flowers are a charming picture in spring. Introduced to cultivation by Wilson in 1900 through Veitch's Coombe Wood nursery. It is one of the parents of the very beautiful *Camellia* 'Cornish Snow'.

Camellia euryoides Lindl. (THEACEAE)

Henry in Journ. China Br. Roy. Asiat. Soc. 22: 257. (1887) (Chinese Names of Plants), Zheng in Journ. Wuhan Bot. Res. 2.(1): 150. (1984).

Hubei: South Donghu, 7715.

Henry stated that in Badong this species was known colloquially as *lu kuo ch'ing*.

Camellia euryoides Lindl. var. ***nokoensis*** (Hayata) Tang S. Liu (THEACEAE) New variety

Camellia euryoides Henry non Lindl. in Trans. Asiat. Soc. Jap. 24 suppl: 20. (1896) (List Pl. Formosa), Rehder & E. H. Wilson in Sargent Pl. Wilson. 2: 390. (1916). *Camellia nokoensis* Hayata Li in Woody Flora of Taiwan 582. (1963).

Taiwan: Wanchin, 90.

Native to Hunan, Jiangsu, Sichuan and Taiwan where it is known only from Mount Nengkaoshan (from where Kaneshira and Sasaki collected the type specimen), Nantou and Hengchun (where Henry gathered his specimen). Henry's material was the earliest collected.

Camellia fraterna Hance (THEACEAE)

Zheng in Journ. Wuhan Bot. Res. 2.(1): 150. (1984).

Thea fraterna (Hance) O. Kuntze. Rehder in Sargent Pl. Wilson 2: 390. (1916)

Hubei: Xiling (Yichang) Gorge, 3374. In a ravine on Moji Shan, 3374a.

An evergreen shrub to 3 metres tall producing fragrant axillary white flowers in March.

Camellia grijsii Hance (THEACEAE)

Zheng in Journ. Wuhan Bot. Res. 2.(1): 151. (1984).

Thea grisjii Kochs. Rehder & E. H. Wilson in Sargent Pl. Wilson 2: 394. (1916).

Hubei: Xiling (Yichang) Gorge, 3335, 4192. Liantuo, 3796. Badong, 7921.

A handsome shrub to 1.5 metres bearing faint-pink or ivory-white flowers in spring. Introduced to cultivation by Wilson in 1901 while collecting for Messrs Veitch.

Camellia mairei (H. Levl.) Melch. var. ***velutina*** Sealy (THEACEAE) New variety

in Rev. Genus, Camellia 174. f. 82D-F. (1958).

Yunnan: Simao, 12875.

Endemic to Yunnnan.

Camellia oleifera C. Abel (THEACEAE)

Camellia sasanqua Henry non Thunb. in Notes Econ. Bot. China 47. (1893).

Hubei: Badong, 4850.

Henry called this the *yu-ch'a*, stating it was cultivated in Hubei and Sichuan for the tea-oil extracted from its seeds. An evergreen shrub to about 7 metres tall, flowering from November to Ferbuary according to Bean though I have seen it on Emei Shan in Sichuan carrying both flowers and fruits in September. This species was collected by the Glasnevin Central China Expedition (GCCE 47) on Emei Shan (Mount Omei) in September 2002.

Camellia reticulata Lindl. (THEACEAE)

Anonymous in Kew List of Determinations (volume 34).

Camellia pitardii Cohen Stuart var. *yunnanica* Sealy in Kew Bull. 219. (1949).

Yunnan: Mountains to the north of Mengzi, 9109. Mengzi, 10427. Chuyuan, 10589. Yimen, 10509a., (type of *Camellia pitardii* Cohen Stuart var. *yunnanica* Sealy).

An evergreen shrub or sometimes a small tree to 15 metres tall in its native habitat. Distributed in western Guizhou, south-west Sichuan and Yunnan. The parent of many cultivars.

Camellia rosiflora Hook. (THEACEAE)

Anonymous in Kew List of Determinations (volume 34).

Yunnan: Mengzi, 10424, 10895.

Camellia sasanqua Thunb. (THEACEAE)

Anonymous in Kew List of Determinations (volume 24).

Hainan: Environs of Kiungchow, 8249.

A Japanese plant.

Camellia sinensis (L.) Kuntze (THEACEAE)

Sealy in Rev. Genus Camellia 161. (1958), Zheng in Journ. Wuhan Bot. Res. 2(1): 151. (1984), O' Brien in Augustine Henry, An Irish Plant Collector in China 15. (2002).

Camellia thea Link. Augustine Henry in Journ. China Br. Roy. Asiat. Soc. 22: 234. (1887) (Chinese Names of Plants). *Thea sinensis* L. Rehder & E. H. Wilson in Sargent Pl. Wilson. 2: 391. (1916). *Camellia sinensis* (L.) O. Kuntze f. *macrophylla* (Sieb.) Kitamura Sealy in Rev. Genus Camellia 116. (1958).

Hubei: Liantuo, 1917, 2214 (February 1887), 2978. Badong, 2499. Changleping, 7822 (March 1888). **Yunnan:** In virgin forest south of the Red River from Manmei at 2,300 metres, 9722 (from a truly wild plant). In forest at Fengchunling at 2,300 metres, 9722a. In forests near Mengzi at 1,600 metres, 10377. Yibang, (near the Laos border), 13183 (cultivated, the source of Pu'er tea).

During Henry's stay in central China, the tea plant was sparingly cultivated in western Hubei and in eastern Sichuan. On the Chengdu plain in western Sichuan plantations supplied large areas of Tibet (Xizang). In his *Chinese Names of Plants* Henry states it was colloquially known as *ch'a*. Tea is known to have been cultivated in Sichuan in the early Han Dynasty (202 BC.-25 AD.) but did not become a common beverage until the seventh century AD. Europeans first met with tea when the Portuguese began to explore the east coast of China in the sixteenth century though the beverage was to be later introduced to England by Dutch traders in the 17th century. The first tea offered in a London coffee-house appeared in 1657. By the 18th century tea had become an important trade item for which China held the monopoly. For the international market most teas are manufactured into black tea, the Chinese preferring green tea or in some cases brick tea, though *Camellia sinensis* is not the only tea source in China.

The tea plant was long thought to have been native to Assam in northern India and to have been introduced to China from there until in 1896 when Henry's Chinese collector, Ho brought him from the virgin forests of south-west Yunnan, undoubted wild tea (see Kew Bull 100 1897). Wild tea can grow to 10 metres tall, though cultivated plants are generally kept lower than 1.5 metres to facilitate harvesting.

Hankow, a port down the Yangtze River from Yichang (and now part of Wuhan) was the greatest tea mart of 19th century China. Brick tea (still popular in parts of China) was an important export to 19th century Tibet (Xizang).

This was used in the preparation of the Tibetan national beverage, butter tea. Transporting this tea to Lhasa from the tea growing centres in China was a major ordeal and from Kangding in Sichuan (the "gateway to Tibet" as Wilson once called it) the journey took porters over three months, crossing dangerous mountain passes and often carrying enormous loads up to 40 miles a day. Thousands of men were once employed in this brutally hard work. Brick tea in its crudest form has been compared to "crow's nests compressed into cakes." Good quality brick tea is worth sourcing however and in modern day Chengdu it is possible to buy century old brick tea, apparently the older the product the better the flavour.

Another type of tea grown to the south of Simao at Yibang in Xishuangbanna in south-west Yunnan and famous throughout China is Pu'er tea (Pu'er is a city to the north of Simao from where the product was traded, Henry collected in all these regions). This tea is highly regarded in China as a medicine and was also esteemed by Tibetan Lamaseries.

Camellia sp. (THEACEAE)

Henry in Trans. Asiat. Soc. Jap. 24 suppl: 20. (1896) (List Pl. Formosa).

Taiwan: Wanchin, 123, 503, 832.

Camellia spp. (THEACEAE)

Anonymous in Kew List of Determinations (vol. 24 & 34.)

Hubei: Badong, 1423, 4964, 5161, 5178. South Badong, 7285, 7315. Liantuo, 2678. Yichang, 1246. Changyang, 5261. Changleping, 7827. **Sichuan:** South Wushan, 7195. **Yunnan:** Mengzi, 10975. Fengchunling, 11052. Simao, 12796, 12814, 13088, 13551, 13558.

Camellia stuartiana Sealy (THEACEAE) New species in Kew Bull. 220. (1949).

Yunnan: Yuanjiang, 13278 (type).

Endemic to Yunnan.

Camellia yunnanensis (Pitard ex Diels) Cohen-Stuart (THEACEAE)

Sealy in Rev. Genus Camellia 163. (1950).

Camellia henryana Cohen-Stuart in Meded. Proefstat. Thee xl. 132. fig. 13. (1916), Cohen Stuart in Bull. Jard. Bot. Buitenzorg. ser. 3. i. 290. t. 30. fig. 15. (1919), Melchior in Engler, Pflanzenfam. ed. 2, 21: 130. (1925), Nakai in Journ. Jap. Bot. 16: 694. (1940), Sealy in Camellias and Magnolias Rep. Conf. R. H. S. 88. (1950); in Rev. Genus Camellia 161. (1958). *Thea henryana* (Cohen-Stuart) Hu in Journ. Arnold Arb. 5: 238. (1924).

Yunnan: Mengzi at 1,800 metres, 10908, (isolectotype of *Camellia henryana* Cohen-Stuart). Mengzi at 1,650 metres, 12883, (syntype). Fengchunling at 1,800 metres, 10908b., (syntype of *Camellia henryana* Cohen-Stuart).

Simao, 10908c., 13593.

An evergreen shrub or small tree to 7 metres tall, bearing solitary white, fragrant flowers in November and December. Distributed in Sichuan and Yunnan. Discovered by Delavay in October 1889 and also later collected by Forrest and Rock.

Camellia yunnanensis (Pitard ex Diels) Cohen Stuart var. *camellioides* (Hu) T. L. Ming (THEACEAE)

Camellia skogiana C. X. Ye in Acta Sci. Nat. Univ. Sunyatseni 35: 127. (1966).

Yunnan: Yuanjiang, 13318 (isotype of *Camellia skogiana* C. X. Ye).

Endemic to Yunnan.

Camonea pilosa (Houtt.) A. R. Simões & Staples (CONVOLVULACEAE)

Ipomoea cymosa (Desr.) Roem. & Schult. Anonymous in Kew List of Determinations (volume 34).

Yunnan: Mengzi, 9786.

Widespread in Asia.

Campanula dimorphantha Schweinf. (CAMPANULACEAE)

Campanula canescens Wall. ex A. DC. (nom. Illeg.) Forbes & Hemsley in Journ. Linn. Soc. Bot. 26: 9. (1889).

Hubei: Liantuo, 3895.

Campanula pallida Wall. (CAMPANULACEAE)

Campanula colorata Wall. Anonymous in Kew List of Determinations (volume 34).

Yunnan: Mengzi, 9421.

Distributed from Afghanistan to western China and Laos.

Campanulaceae (CAMPANULACEAE)

Anonymous in Kew List of Determinations (volume 34).

Yunnan: Simao, 13040.

Campeiostachys dahurica (Turcz. ex Griseb.) B. R. Baum, J. L. Lang & C. Yen (POACEAE)

Elymus dahuricus Turcz. ex Griseb. Rendle in Journ. Linn. Soc. Bot. 36: 433. (1904).

Hubei: Badong, 4709.

Campsis grandiflora (Thunb.) K. Schum. (BIGNONIACEAE)

Zheng in Journ. Wuhan Bot. Res. 2.(1): 187. (1984). *Tecoma grandiflora* (Thunb.) Loisel. Henry in Journ. China Br. Roy. Asiat. Soc. 22: 278. (1887) (Chinese Names of Plants), Forbes & Hemsley in Journ. Linn. Soc. Bot. 26: 235. (1890). *Campsis chinensis* (Lam.) Voss. Rehder in Sargent Pl. Wilson. 1: 303. (1913).

Hubei: Yichang, among rocks, 1535.

A very beautiful deciduous climber to 10 metres tall. Introduced from China in 1800, Henry stated it was known colloquially in the Three Gorges region as *tzu wei*.

Camptotheca acuminata Decne. (NYSSACEAE)

E. H. Wilson in Sargent Pl. Wilson. 2: 254. (1916), Zheng in Journ. Wuhan Bot. Res. 2.(1): 158. (1984).

Hubei: Changyang, 7606. **Yunnan:** Simao at 1,600 to 2,000 metres, 13091, 13433.

The happy tree or cancer tree, a handsome tree of rapid growth to about 25 metres tall and with recorded girths of up to 2 metres. Found in western and central China, though Wilson recorded it as being rare in Hubei. The genus was discovered by Père Armand David on the Lushan range in Jiangxi in 1868. In Chinese herbal medicine the fruits of this species are used to treat patients suffering from cancer of the digestive tract and leukemia. The active compound, which shows such promise in treating cancer is Camptothecine and cultivars with higher yields of this compound are being developed. Endemic to China it is now being grown commercially as a crop plant in India, Japan and the USA. The parts originally used were the stem, bark and seeds but it is now mainly the young leaves and trees are clipped for repeated harvests. In China it is the *xi shu* or happy tree and has for centuries been used to treat colds and diseases of the spleen, liver, stomach and gall bladder. *Camptotheca acuminata* is cultivated at the National Botanic Gardens, Glasnevin from collections made by the Glasnevin Central China Expedition (GCCE 216) between Yichang and Badong in October 2002 and near Changyang (GCCE 741) in October 2004

Campylotropis argentea Schindler (FABACEAE)

New species in Repert. Nov. Spec. Nov. Regni Veg. 11: 426. (1912).

Yunnan: In the mountains near Mengzi at 1,640 metres, 10384 (type).

Endemic to Yunnan.

Campylotropis capillipes (Franch.) Schindl. subsp. *prainii* (Collett & Hemsley) Iokawa & H. Ohashi (FABACEAE)

Lespedeza prainii Collett & Hemsley Anonymous in Kew List of Determinations (volume 34). *Campylotropis prainii* (Collett & Hemsley) Schildler Schindler in Sargent Pl. Wilson. 2: 115. (1916).

Yunnan: In woods near Mengzi at 1,500 metres, 9803, 9803b. Simao, 13366.

Distributed across southwest China and northern Indochina.

Campylotropis delavayi (Franch.) Shindl. (FABACEAE)

Lespedeza delavayi Franchet Anonymous in Kew List of Determinations (volume 34).

Yunnan: Mengzi, 9718.

Endemic to China and distributed in Sichuan and Yunnan.

Campylotropis diversifolia (Hemsley) Schindler (FABACEAE) New species

in Fedde Repert. Regni. Veg. Sp. Nov. xi. 342. (1912), Schindler in Sargent Pl. Wilson. 2: 115. (1916).

Lespedeza diversifolia Hemsley in Hooker's Icon. Pl. t. 2625. (1901); in Journ. Linn. Soc. Bot. 34: 475. (1900).

Yunnan: Mengzi at 2,300 metres, 9243a., (type of *Lespedeza diversifolia* Hemsley). Mengzi 9243b., (syntype of *Lespedeza diversifolia* Hemsley). Mountains to the south of Mengzi, 9243.

Endemic to Yunnan.

Campylotropis fulva Schindler (FABACEAE) New species

in Fedde, Repert. nov. Spec. Regni Veg. 11: 426. (1912).

Yunnan: Mengzi, 9689a.

Endemic to Yunnan. Known only from Henry's type specimen.

Campylotropis grandifolia Schindller (FABACEAE) New species

in Fedde, Rep. Sp. Nov. Regni. Veg. 11: 346. (1912).

Yunnan: Milê, 9888 (type), 9890.

Endemic to south-east Yunnan and known only from Henry's type specimen.

Campylotropis harmsii Schindler (FABACEAE) New species

in Fedde, Repert. Nov. Spec. Regni Veg. 11: 342. (1912).

Yunnan: In woods near Simao at 1,300 metres, 9803d., (type).

Distributed in China (Yunnan) and Thailand.

Campylotropis henryi (Schindl.) Schindler (FABACEAE) New species

in Fedde, Repert. Spec. Nov. Regni. Veg. 11: 347. (1912).

Lespedeza henryi Schindler in Fedde, Repert. Spec. Nov. Regni. Veg. 9: 517. (1911).

Yunnan: Yuanjiang at 750 metres, 13212.

Distributed in western China, Laos, Cambodia and Thailand.

Campylotropis hirtella (Franchet) Schindl. (FABACEAE)

Schindler in Sargent Pl. Wilson. 2: 115. (1916).

Yunnan: Mengzi at 1,500 metres, 9689.

Distributed in north-east India and western China.

Campylotropis latifolia (Dunn) Schindler (FABACEAE) New species

in Fedde, Repert. Nov. Spec. Regni Veg. 11: 428. (1912).

Lespedeza latifolia Dunn in Journ. Linn. Soc. Bot. 35: 488. (1903)

Yunnan: Milê district, 9899.

Endemic to Yunnan.

Campylotropis macrocarpa (Bunge) Rehder (FABACEAE)

Zheng in Journ. Wuhan Bot. Res. 2.(1): 97. (1984).

Lespedeza macrocarpa Bunge Forbes & Hemsley in Journ. Linn. Soc. Bot. 23: 182. (1887), Henry in Journ. China Br. Roy. Asiat. Soc. 22: 257. (1887) (Chinese Names of Plants), *Lespedeza ichangensis* Schindler in Fedde, Repert. Spec. Nov. Regni Veg. 9: 515. (1911). *Campylotropis ichangensis* (Schindl.) Schindler Fl. Ill. Prim. Sin. 544. (1955).

Hubei: Yichang, 45, 400 (in part), 400b., 645, 744, 1058, 2174. On road to Moji Shan, (June 1886), 1547 (type of *Lespedeza ichangensis* Schindler). Xiling (Yichang) Gorge, 2920. Xingshan, 6508. Changyang, 7511.

According to Henry this species was known colloquially in the Three Gorges region as *ma hu shao*.

Campylotropis macrocarpa (Bunge) Rehder var. **hupehensis** (Pamp.) Iokawa & H. Ohashi (FABACEAE)

Lespedeza juncea Henry non (l.) Pers. in Trans. Asiat. Soc. Jap. 24 suppl: 34. (1896) (List Pl. Formosa).

Taiwan: Tamsui, 1479.

Campylotropis parviflora (Kurz.) Schindler (FABACEAE)

Schindler in Sargent Pl. Wilson. 2: 115. (1916).

Yunnan: In forests near Simao at 1,500 metres, 12172.

Distributed in India (Assam), Myanmar, China (Yunnan), Vietnam and Thailand.

Campylotropis pinetorum (Kurz) Schindler subsp. **velutina** (Dunn) H. Ohashi New subspecies (FABACEAE)

Schindler in Fedde, Repert. Nov. Spec. Regni Veg. 11: 429. (1912).

Lespedeza velutina Dunn in Hooker's Icon. Pl. 7: t. 2700. (1901).

Yunnan: Near Manpan in the Red River valley at 950 metres, 10447 (syntype of *Lespedeza velutina* Dunn). Mengzi in forests at 1,150 metres, 10447a. Simao, 11590a., 11590b.

Endemic to southern China.

Campylotropis polyantha (Franch.) Schindl. (FABACEAE)

Schindl. in Sargent Pl. Wilson. 2: 114. (1916), in part. *Campylotropis reticulata* P. L. Ricker in J. Wash. Acad. Sci. 36: 40. (1946).

Yunnan:. Mengzi, 9626a., (syntype of *Campylotropis reticulata* P. L. Ricker).

Campylotropis polyantha (Franch.) Schindl. var. **neglecta** (Schind.) H. Ohashi New variety (FABACEAE)

Campylotropis neglecta Schindler in Repert. Spec. Nov. Regni Veg. 11: 340. (1912). *Campylotropis reticulata* P. L.

02

Ricker in J. Wash. Acad. Sci. 36: 40. (1946). *Campylotropis polyantha* Schindler non (Franch.) Schindl. in Sargent Pl. Wilson. 2: 114. (1916), in part, exclude A. Henry 9626a.

Yunnan: On grassy mountains near Mengzi at 1,400 metres, 9626. On grassy mountains near Mengzi at 1,350 to 1,500 metres, 9626b., (isotype of *Campylotropis neglecta* Schindler).

Native to western China.

Campylotropis trigonoclada (Franch.) Schindl. (FABACEAE)

Schindler in Fedde, Repert. Nov. Spec. Regni Veg. 11: 430. (1912), Schindler in Sargent Pl. Wilson. 2: 114. (1916).

Lespedeza angulicaulis Harms in Fedde, Repert. Nov. Spec. Regni Veg. 9: 522. (1911).

Yunnan: Mengzi at 1,500 metres, 9135 (type of *Lespedeza angulicaulis* Harms), 9622. Milê, 9135a., (isotype of *Lespedeza angulicaulis* Harms).

Endemic to western China.

Campylotropis yunnanensis (Franch.) Schindler (FABACEAE)

Schindler in Sargent Pl. Wilson. 2: 114. (1916).

Yunnan: In woods near Mengzi at 1,380 metres, 9134. In woods near Mengzi at 1,400 metres, 9702, 9702a.

Forrest also later collected this species on dry scrubby hills in western Yunnan (Forrest 16,884) and described it as a shrub to 1.5 metres tall with white flowers tipped and margined pale-blue. Endemic to western China.

Canarium pimela K. D. Koenig (BURSERACEAE)

Anonymous in Kew List of Determinations (volume 24). Anonymous in Kew List of Determinations (volume 34).

Hainan: Without locality, 8731. **Yunnan:** Simao, 13171.

Canavalia ensiformis (L.) DC. (FABACEAE)

Henry in Trans. Asiat. Soc. Jap. 24 suppl: 36. (1896) (List Pl. Formosa).

Hainan: Near the Kiungchow Pagoda, 7983. Environs of Haikou, 8046. **Taiwan:** Wanchin, 1670. **Yunnan:** Simao, 12424.

The Jack Bean, a twining annual herb to 160 cm tall, native to the West Indies, Mexico, Peru and Brazil. Widely grown in Asia as a food crop, as a green manure prior to planting tobacco and for providing shade to pineapples. An ancient crop, it was domesticated by American Indians and has been found at various archaeological sites in south-western USA dating mostly from about 1,300 A.D. Cultivated in Taiwan and locally known there as *tao-to* according to Henry.

Canavalia rosea (Sw.) DC. (FABACEAE)

Canavalia obtusifolia DC. Henry in Trans. Asiat. Soc. Jap. 24 suppl: 36. (1896) (List Pl. Formosa).

Taiwan: Kaohsiung, without number.

A common sea-coast climber in Taiwan with large purplish-pink flowers.

Canavalia sp. (FABACEAE)

Henry in Trans. Asiat. Soc. Jap. 24 suppl: 36. (1896) (List Pl. Formosa).

Taiwan: Wanchin, 1719.

Canna indica L. (CANNACEAE)

Henry in Trans. Asiat. Soc. Jap. 24 suppl: 95. (1896) (List Pl. Formosa).

Canna indica L. var. *orientalis* Baker Wright in Journ. Linn. Soc. Bot. 36: 73. (1903). *Canna chinensis* Willd. Kranzlin in Engler, Pflanzenr. iv. 47: 46. (1912).

Sichuan: Kangding, Henry's Chinese collector with Pratt, 8828. **Hainan:** Environs of Haikou, 8030. Ling-men, 8656. **Taiwan:** Wanchin, without number. Oluanpi, without number.

Native to tropical America, naturalised and cultivated in the tropics throughout the world.

Cannabis sativa L. (CANNABIDACEAE)

Henry in Journ. China Br. Roy. Asiat. Soc. 22: 258. (1887) (Chinese Names of Plants); in Chinese Imp. Marit. Customs, 2nd special series, no. 16. pp. 5 & 7. (1891); in Kew Bull. Misc. Inf. 250. (1891); in Notes Econ. Bot. China 11. (1893), Forbes & Hemsley in Journ. Linn. Soc. Bot. 26: 453. (1894).

Hubei: Yichang, 2024. Liantuo, 2431. **Yunnan:** At Fengchunling near the Red River at 2,200 metres, 11209.

Hemp, a stately annual to 1.5 metres tall, native to Asia and cultivated for its fibre. It is often seen as an escape. According to a Neolithic Chinese legend, the gods gave the Yang Shao culture (c. 4,000 B.C.) one plant to fulfill all their needs. That plant was *ma*, hemp or marijuana and the assertion to its usefullness is quite correct. The Chinese were the first race to cultivate this plant and it was from hemp fibre that the first true paper was manufactured in China. Though they were aware of the plant's psychoactive properties, early Chinese were more interested in its medicinal virtues. The Chinese grew their plants in tightly spaced rows to discourage branching and improve fibre production, wheras in India, where the hallucinogenic properties of *Cannabis sativa* ssp. *indica* were first exploited, the Bengalis developed cultivation techniques that maximised the production of psychoactive compounds. It is one of the oldest cultivated plants, grown for both its oil bearing seeds and the fibre extracted from its stems. In his *Memorandum on the Jute and Hemp of China* Henry states that this is the *ta ma* of books; *hsiao ma* in northern China and *huo ma* in southern China. Henry stated it was known

02

in the Three Gorges region as *t'ang ma* and was grown at Yichang both for its fibre and the oil from its seeds. In Chinese the character *ma* signifies plants producing textile fabrics and also includes certain plants whose seeds are used for their oil. (The castor oil plant, *Ricinus communis* is sometimes known as *ta ma*, i.e. the large ma, from its structure). *Cannabis sativa* is a spring crop, sown in February the crop is harvested in early June as flowering commence It is a common weed of arable land in parts of south-east Tibet (Xizang) and a street weed in Kunming, Yunnan. In Chinese herbal medicine the seeds are used to treat constipation of the elderly and to promote the healing of sores.

Canthium horridum Blume (RUBIACEAE)

Canthium parvifolium Hemsley non Roxb. in Kew List of Determinations (volume 24).

Hainan: Environs of Haikou, 8337.

Canthium sp. (RUBIACEAE)

Anonymous in Kew List of Determinations (volume 24).

Hainan: Without locality, 8576.

Capillipedium assimilis (Steud.) A. Camus (POACEAE)

Andropogon assimilis Steud. Rendle in Journ. Linn. Soc. Bot. 36: 370. (1904).

Hubei: In a ravine on Moji Shan, 4177. Yichang, 4220. **Yunnan:** Mengzi at 1,530 to 1, 600 metres, 9138. Mengzi at 1,430 to 1,500 metres, 9615, 9617. Simao at 1,670 metres, 13414.

Native to India, Myanmar, Thailand and China.

Capillipedium parviflorum (R. Br.) Stapf. (POACEAE)

Andropogon micranthus Kunth. Rendle in Journ. Linn. Soc. Bot. 36: 374. (1904).

Hubei: Yichang, 26, 2373 (in part), 4285. Changleping, 7815. **Yunnan:** Grassy mountains near Mengzi at 1,750 metres, 9615a.

Widely distributed in the tropics of the Old World including Ethiopia, China, Malaysia, Japan and Australia.

Capparis acutifolia Sweet (CAPPARIDACEAE)

Capparis membranacea Gard. & Champ. varietates. Henry in Trans. Asiat. Soc. Jap. 24 suppl: 18. (1896) (List Pl. Formosa). *Capparis membranacea* Gard. & Champ. var. *angustissima* Hemsley in Ann. of Bot. 9: 145. (1895), Henry in Trans. Asiat. Soc. Jap. 24 suppl: 18. (1896) (List Pl. Formosa), Forbes & Hemsley in Journ. Linn. Soc. Bot. 26: 145. (1892), Bretschnieder in Hist. Eur. Bot. Disc. China 2: 780. (1898), Li in Woody Flora of Taiwan 236. (1963).

Taiwan: Wanchin, 410, 471 (type of *Capparis membranacea* Gard. & Champ. var. *angustissima* Hemsley), 844, 1005. **Yunnan:** Mengzi, 9124.

A small shrub, native to Vietnam, southern China (including Taiwan) and Japan (south).

Capparis formosana Hemsley (CAPPARIDACEAE) New species

in Ann. of Bot. 9: 145. (1895), Henry in Trans. Asiat. Soc. Jap. 24 suppl: 18. (1896) (List Pl. Formosa), Bretschnieder in Hist. Eur. Bot. Disc. China 2: 780. (1898), Li in Woody Flora of Taiwan 235. (1963).

Capparis sikkimensis Kurz. subsp. *formosana* (Hemsl.) Jacobs in Blumea 12: 497. (1965).

Taiwan: Wanchin, 501 (type of *Capparis formosana* Hemsley). Wanshoushan, 160. Without locality, 501a., 501b. (Syntype of *Capparis formosana* Hemsley), 501c., 501d. Oluanpi, 2069.

A climbing shrub with stems up to 30 cm. in diameter and reaching 6 metres tall. Native to China (Guangdong, Hainan, Taiwan), northern Vietnam and Japan (south).

Capparis henryi Matsum. (CAPPARIDACEAE) New species

in Bot. Mag. Tokyo 13: 33. (1899); Matsum. & Hayata in Journ. Coll. Sci. Univ. Tokyo 22: 26. t. 3. (1906), Hayata in Ic. Pl. Formos. 33: 3. (1911), Li in Woody Flora of Taiwan 235. fig. 86. (1963).

Capparis sp. Henry in Trans. Asiat. Soc. Jap. 24 suppl: 18. (1896) (List Pl. Formosa), *Capparis micrantha* DC. var. *henryi* (Matsum.) Jacobs in Blumea 12: 470. (1965).

Taiwan: Wanchin, 570.

A shrub to 2 metres tall, native to Myanmar, Thailand and Indochina to Malaysia. Mainly found in the southern part of Chinese Taiwan, this species was also collected by Wilson on Taiwan's Hengchun Peninsula.

Capparis membranifolia Kurz (CAPPARIDACEAE)

Capparis membranacea Gard. & Champ. varietates. Henry in Trans. Asiat. Soc. Jap. 24 suppl: 18. (1896) (List Pl. Formosa). *Capparis viminea* Hook.f. & Thoms. (nom. Illeg.) Li in Woody Flora of Taiwan 236. (1963).

Taiwan: Wanchin, 405.

A small shrub with a wide geographical range from India to southern China. In Chinese Taiwan it grows on the southern part of the island.

Capparis multiflora Hook. f. & Thoms. (CAPPARIDACEAE)

Jacobs in Blumea 12: 475. (1965).

Yunnan: In Pingbian in forests on the Daweishan Range at 1,640 metres, 11146.

A shrub or small tree, native from the Himalaya to Myanmar (where it was collected by Kingdon Ward) and in China in Yunnan province.

Capparis sepiaria L. (CAPPARIDACEAE)

Jacobs in Blumea 12: 489. (1965).

Hainan: 8 km (5 miles) south of Kuingchow, 8176.

Vietnam: Haiphong, 9518.

A widespread tropical with an enormous range from Africa to Australia.

Capparis sp. **nov.** (CAPPARIDACEAE)

Anonymous in Kew List of Determinations (volume 34).

Yunnan: Simao, 12830.

Capparis urophylla F. Chun (CAPPARIDACEAE) New species

Jacobs in Blumea 12: 501. (1965).

Yunnan: Mengzi, 10215.

The type specimen was collected by Z. S. Chung (T. S. Tsoong) but this shrub had previously been discovered near Simao by Henry and was also later collected at the watershed of the Black River and Maokai by Joseph Rock. Distributed in south-west China and Laos.

Capparis viburnifolia Gagn. (CAPPARIDACEAE)

Jacobs in Blumea 12: 505. (1965).

Yunnan: Simao, 12380.

Gagnepain described this liana from collections made by Poilane in Vietnam in June 1909, the species appears to have been first collected by Bons d'Anty near Menghet in southern Yunnan and was collected shortly after by Henry.

Capparis yunnanensis Craib & W. W. Smith (CAPPARIDACEAE) New species

in Notes Roy. Bot. Gard. Edinburgh 9: 91. (1916).

Capparis sikkimensis Kurz. ssp. *yunnanensis* (Craib & W. W. Sm.) M. Jacobs in Blumea 12: 496. (1965).

Yunnan: In forests to the south-west of Simao at 1,300 metres, 12986 (type).

An evergreen shrub or climber to 6 metres tall, native to northern Myanmar, Vietnam and China (Yunnan). A spectacular plant.

Capparis zeylanica L. (CAPPARIDACEAE)

Capparis swinhoei Hance Anonymous in Kew List of Determinations (volume 24).

Hainan: Environs of Haikou, 8015.

Caprifoliaceae (CAPRIFOLIACEAE)

Anonymous in Kew List of Determinations (volume 34).

Yunnan: Mengzi, 11335.

Capsella bursa-pastoris (L.) Medik. (BRASSICACEAE)

Henry in Journ. China Br. Roy. Asiat. Soc. 22: 272. (1887) (Chinese Names of Plants), Henry in Trans. Asiat. Soc. Jap. 24 suppl: 17. (1896) (List Pl. Formosa).

Hubei: Yichang, 799. Xiling (Yichang) Gorge, 1102. Badong, 1439, 4695. Taiwan: Oluanpi, 1992.

Shepherd's purse, a common and variable annual weed found throughout the world. Henry stated that in the Three Gorges region this species was known as *ti mi ts'ai*.

In Shanghai it is grown and sold as a vegetable and used in salads and soups and the flowers are also said to be edible. In China because of its high vitamin A content it is used as an eye medicine, it is richer in vitamin C than oranges. In Chinese herbal medicine it is used to lower blood pressure, to stop bleeding (eg. nose-bleed), to treat heavy periods, diarrhea and as a diuretic.

Capsicum frutescens L. (SOLANACEAE)

Henry in Notes Econ. Bot. China 39. (1893).

Capsicum minimum Roxb. Henry in Trans. Asiat. Soc. Jap. 24 suppl: 65. (1896) (List Pl. Formosa).

Hainan: Environs of Kiungchow, 8734. Taiwan: Wanchin, without number. Oluanpi, 356.

Tobasco, chili, goat or spur pepper, native to tropical South America and cultivated in Peru since ancient times. Widely cultivated in China, Henry stated that this and othe species of chilli or cayenne peppers were known as *la-chiao*. Henry went on to say that this species was known in Taiwan to the Pingpu or Pepohuans (the local indigenous tribe) as *shan-sin-i, sin-i* being a corruption of chili. China is the world's largest producer of chilis and green peppers. *Capsicum chinense* Jacq. is not native to China but the western Amazon basin.

Caragana sinica (Buc' hoz) Rehd. (FABACEAE)

Zheng in Journ. Wuhan Bot. Res. 2.(1): 97. (1984).

Hubei: Liantuo, 3812. South Badong, 5378.

A deciduous shrub to 5 metres tall. According to Wilson this shrub was frequently cultivated to the north and south of Yichang and the flowers when cooked with eggs were eaten. Introduced from China in 1773, the bruised bark is said to smell of liquorice.

Carallia sp. (RHIZOPHORACEAE)

Anonymous in Kew List of Determinations (volume 34).

Yunnan: Simao, 12045.

Cardamine circaeoides Hook. f. & Thoms. (BRASSICACEAE)

Cardamine insignis O.E. Schulz in Bot. Jahrb. 32: 439. (1903). *Cardamine violifolia* O. E. Schulz var. *diversifolia* O. E. Schulz in Bot. Jahrb. Syst. 32(4): 440. (1903).

Hubei: Xiling (Yichang) Gorge, 3298, (type of *Cardamine violifolia* O. E. Schulz var. *diversifolia* O. E. Schulz). Yunnan: In mountains near Mengzi, 11349, 10659. Simao, 13616. Simao at 2,000 metres, 13090, (type of *Cardamine insignis* O.E. Schulz).

Distributed in India, Myanmar, China, Laos, Thailand and Vietnam.

Cardamine fragariifolia Schulz (BRASSICACEAE) New species

in Bot. Jahrb. Syst. 32: 446. (1903).

Hubei: Jianshi, 5803 (isotype).

Cardamine hirsuta L. (BRASSICACEAE)

Henry in Journ. China Br. Roy. Asiat. Soc. 22: 239. (1887) (Chinese Names of Plants), Diels in Bot. Jahrb. 29: 358. (1900).

Hubei: Yichang, 1313. Xiling (Yichang) Gorge, 3331, 3297. Changyang, 7740, 7794, 7854. **Yunnan:** Mengzi, 9277.

A common weed around Yichang where Henry stated it was colloquially known as *yeh ch'in ts'ai.*

Cardamine impatiens L. (BRASSICACEAE)

Henry in Journ. China Br. Roy. Asiat. Soc. 22: 271. (1887) (Chinese Names of Plants), Diels in Bot. Jahrb. 29: 35. (1900).

Hubei: South Badong, 350 (June 1885), 5367. Badong, 4037. Changyang, 7855. **Yunnan:** Mengzi, 10513.

An annual, sometimes perennial herb to about 50 cm tall, native to China, Korea and Japan. Henry stated it was kocally known in the Three Gorges region as *shui ts'ai hua.*

Cardamine macrophylla Willd. (BRASSICACEAE)

Henry in Journ. China Br. Roy. Asiat. Soc. 22: 239. (1887) (Chinese Names of Plants), Henry in Notes Econ. Bot. China 60. (1893).

Cardamine urbaniana Schulz in Bot. Jahrb. 32: 396. (1903).

Hubei: Badong, 4007. Without locality, 5635a. **Sichuan:** South Wushan, 5580, 5635 (type of *Cardamine urbaniana* Schulz). Without locality, Henry's Chinese collector with Pratt, no locality stated but may be Leshan, Emei Shan, Wa Shan or Kangding, 8790.

Henry stated that this species was known colloquially around Badong as *chia ch'in ts'ai* and in the China in gereral it was the *wu-kung-ch'i.*

Cardamine spp. (BRASSICACEAE)

Anonymous in Kew List of Determinations (volume 24).

Hubei: Badong, 1440, 1472, 3779, 5462. Changyang, 5260, 7791. Fang Xian, 6729. **Sichuan:** South Wushan, 5572, 5594.

Cardamine stenoloba Hemsley (BRASSICACEAE)
New species

in Journ. Linn. Soc. Bot. 29: 303. (1892), Bretschneider in Hist. Eur. Bot. Disc. China 2: 779. (1898).

Sichuan: Henry's Chinese collector with Pratt, no locality stated but may be Leshan, Emei Shan, Wa Shan or Kangding, 8794 (quoted by Hemsley as 8724).

Cardiocrinum cathayanum (E. H. Wilson) Stearn (LILIACEAE)

in Gard. Chron. ser. 3, 124: 4. (1948).

Lilium giganteum Henry non Wallich in Journ. China

Br. Roy. Asiat. Soc. 22: 265. (1887) (Chinese Names of Plants); in Journ. Roy. Hort. Soc. 26: 348. (1901—1902) Wright in Journ. Linn. Soc. Bot. 36: 130. (1903) *Lilium cathayanum* E. H. Wilson in Lilies of Eastern Asia 99. (1925).

Hubei: Badong, 3141, 4085.

A stout-stemmed bulbous perennial to 150 cm tall, producing in June and July a terminal raceme of 10 to 15 large bell-shaped white flowers striped purple on the inner central part of trhe corolla. Distributed in Anhui, Fujian, Hubei, Hunan, Jiangsu, Jiangxi and Zhejiang. According to Henry, his Badong collector called this species *peh wa.* He also stated that large lilies with edible bulbs were named *peh ho* at Badong and Yichang and at Badong the roots of *Cardiocrinum cathayanum* were used in surgery. It was discovered in 1868 by Père David. On May 5th 1888 a consignment of Henry's seeds including this species (as *Lilium giganteum*) arrived at Kew, the seeds were sent to the temperate pits for propagation.

Cardiocrinum giganteum (Wall.) Makino var. *yunnanense* (Leichtlin ex Elwes) Stearn (LILIACEAE)

New variety in Gard. Chron. ser. 3. 124: 4. (1948), Nelson in A Heritage of Beauty 310. (2000).

Lilium giganteum Henry non Wallich in Journ. Chin. Branch Roy. Asiat. Soc. 22: 265. (1887); in Journ. Roy. Hort. Soc. 26: 348. (1901—1902); in Flora & Sylva 1: 217. (1903), Wright in Journ. Linn. Soc. Bot. 36: 130. (1903). *Lilium giganteum* Wall. var. *yunnanense* Leichtlin apud Elwes in Gard. Chron. ser. 3, lx. fig. 18. (1916), E. H. Wilson in Lilies of Eastern Asia 96. (1925).

Hubei: Liantuo, 2681, 4545, 6197a. South Badong, 5376. Zigui, 6197. Changyang, 6197a. **Yunnan:** Near Mengzi at 1,900 to 2,940 metres, 10805.

This variety differs from the type plant from the Himalayan region in its outer morphological features, the dark stems, bronze spring foliage, the (horizontally held) flowers habit of opening from the top of the stem downwards and the shape of the seed capsules. It is also shorter, rarely more than 2 metres tall. The type i.e. *Cardiocrinum giganteum* was discovered by Dr. Nathaniel Wallich in Nepal in 1825 and as stated by E. H. Wilson, was introduced by seeds sent by the Irish collector, Colonel Edward Madden to Glasnevin in the 1840s. The seeds and seedlings were liberally distributed by Dr. David Moore and the giant Himalayan lily first flowered in cultivation in 1851 at Lamorran Rectory near Truro in Cornwall (and the following year at Glasnevin). In 1852 Thomas Lobb sent bulbs to Messr. Veitch from Nepal. The native range of *Cardiocrinum giganteum* extends from the Himalaya as far as Tibet (Xizang) and parts of north-west Yunnan. I have walked though valleys in the same area filled with hundreds of this magnificent lily in flower, their scent wafting through the ancient forests of those wonderful valleys. Henry's var. *yunnanense* extends from Yunnan and Sichuan to Hubei and the north-western region of Myanmar that joins Yunnan. Once a common plant in Hubei, Henry made the following remark in Flora & Sylva;

In open glades rather high up *Lilium giganteum* may be spied miles away across the valley with its gorgeous turret of flowers.

This variety was collected by the Glasnevin Central China Expedition (GCCE 337) in a valley near Yesanguan in Badong, Hubei in the autumn of 2002 and on Tianthu Shan, Changyang (GCCE 790) in October 2004. This variety has long been grown in Irish gardens and flowers regularly in the gardens at the National Botanic Gardens, Kilmacurragh.

Cardiospermum halicacabum L. (SAPINDACEAE)
Henry in Trans. Asiat. Soc. Jap. 24 suppl: 28. (1896) (List Pl. Formosa).

Hubei: Yichang, 749. **Hainan:** Near the Kiungchow Pagoda, 7978. **Taiwan:** Kaohsiung, without number.

Carduus crispus L. (ASTERACEAE)
Anonymous in Kew List of Determinations (volume 24 & 34).

Hubei: Jianshi, 6044. Fang Xian, 6647. **Yunnan:** Mengzi, 10025.

Carex alliiformis C.B. Clarke (CYPERACEAE) New species
in Journ. Linn. Soc. Bot.36: 270. (1903), Kukenthal in Engler, Pflanzenr. 38: 618. (1909).

Hubei: Antelope Glen, 3409. Antelope Glen, 3483.

A handsome species found in central and southern China (including Taiwan), Vietnamand Japan. Clarke based this species on material gathered by Henry and Wilson. The specific epithet refers to the spikes of fruit that suggest lilliputian ropes of onions.

Carex alopecuroides D. Don ex Tilloch & Taylor (CYPERACEAE)
Clarke in Journ. Linn. Soc. Bot. 36: 271. (1903).

Hubei: Badong, 3744.

Carex ascotreta C. B. Clarke ex Franch. (CYPERACEAE)
Carex davidii Franch. var. *ascocetra* (C. B. Clarke) Kukenth. Kukenthal in Engler, Pflanzenr. 38: 479. (1909). *Carex ichangensis* C. B. Clarke in Journ. Linn. Soc. Bot. 36: 290. (1903), Morley in Glasra 3: 78. (1979).

Hubei: Yichang, 7888, 7893, 7894, 7860 (isosyntype of *Carex ichangensis* C. B. Clarke).

Carex baccans Nees (CYPERACEAE)
Henry in Trans. Asiat. Soc. Jap. 24 suppl: 106. (1896) (List Pl. Formosa), C. B. Clarke in Journ. Linn. Soc. Bot.

36: 274. (1903).

Taiwan: Tamsui, 1402.

Native to the Himalaya, India, Sri Lanka, southern China and south-east Asia.

Carex basiflora C. B. Clarke (CYPERACEAE) New species
in Journ. Linn. Soc. Bot. 36: 274. (1903).

Carex brevicuspis Clarke. var. *basiflora* (C. B. Clarke) Kukenth in Engler, Pflanzenr. 38: 630. (1909).

Hubei: Badong, 3748.

Clarke's *Carex basiflora* was based on material gathered by Henry at Badong and in north-central China (Gansu or Shaanxi) by the Irish missionary Hugh Scallan (Pater Hugo).

Carex biwensis Franch. (CYPERACEAE)
C. B. Clarke in Journ. Linn. Soc. Bot. 36: 275. (1903).

Hubei: Yichang, 7865.

Carex breviculmis R. Br. (CYPERACEAE)
C. B. Clarke in Journ. Linn. Soc. Bot.36: 276. (1903).

Hubei: Badong, 3732.

Distributed across the Himalaya, China, Korea, Japan, southeast Asia, Australia and New Zealand.

Carex brunnea Thunb. (CYPERACEAE)
Henry in Trans. Asiat. Soc. Jap. 24 suppl: 106. (1896) (List Pl. Formosa), C. B. Clarke in Journ. Linn. Soc. Bot. 36: 278. (1904—1906), Kukenthal in Engler, Pflanzenr. 38: 599. (1909).

Hubei: Antelope Glen, 3341. Yichang, 4267. Badong, 5020, 5156, 5180. **Taiwan:** Tamsui, 1401. Wanshoushan, 2009. **Yunnan:** Mengzi at 1,600 metres, 11401.

Native to Madagascar, the Himalaya, India, Sri Lanka, China, Japan, south-east Asia, Australia and the Pacific Islands.

Carex chinensis Retz. (CYPERACEAE)
C. B. Clarke in Journ. Linn. Soc. Bot. 36: 280. (1903), Kukenthal in Engler, Pflanzenr. 38: 625. (1909).

Hubei: Yichang, 3530.

Carex chinensis Retz. var. longkiensis (Franch.) Kükenth. (CYPERACEAE)
Kükenthal in Engler, Pflanzenr. 38: 625. (1909).

Hubei: South Badong, 5466.

Carex condensata Nees. (CYPERACEAE)
C. B. Clarke in Journ. Linn. Soc. Bot. 36: 280. (1903).

Yunnan: Mengzi at 1950 metres, 9576, 9576a., 9576b.

Carex continua C. B. Clarke (CYPERACEAE)
C. B. Clarke in Journ. Linn. Soc. Bot. 36: 261. (1903), Kukenthal in Engler, Pflanzenr. 38: 281. (1909).

Yunnan: Mengzi, 10161.

Carex cruciata Wahlenb. (CYPERACEAE)
C. B. Clarke in Journ. Linn. Soc. Bot. 36: 281. (1903),

Kukenthal in Engler, Pflanzenr. 38: 265. (1909).

Carex valida Henry non Nees. in Trans. Asiat. Soc. Jap. 24 suppl: 106. (1896) (List Pl. Formosa).

Hubei: At Shih Pan Shan (off the Xiling (Yichang) Gorge), 4209. Liantuo, 4628. Changyang, 7753. **Taiwan:** Wanchin, 840. **Yunnan:** Mengzi at 1,600 metres, 9971. Mengzi at 1,300 to 1,600 metres 11151, 11440. Simao at 1,400 metres, 12404.

A widespread species, native to the Himalaya, India, southern China, Japan (south), south-east Asia and Australia.

Carex davidii Franch. (CYPERACEAE)

C. B. Clarke in Journ. Linn. Soc. Bot. 36: 282. (1903), Kukenthal in Engler, Pflanzenr. 38: 479. (1909).

Hubei: Badong, 5320. Yichang, 7899.

Discovered by Père Armand David in southern Shaanxi in April 1873.

Carex dispalata Boott (CYPERACEAE)

Carex pollens C. B. Clarke var. *angustior* C. B. Clarke in Journ. Linn. Soc. Bot. 36: 305. (1903).

Hubei: Yichang, 3356, 7858. Changyang, 7101.

Carex doniana Spreng. (CYPERACEAE)

Carex alopecuroides D. Don var. *chlorostachya* (D. Don) C. B. Clarke Clarke in Journ. Linn. Soc. Bot. 36: 271. (1903).

Hubei: Badong, 5655. **Sichuan:** South Wushan, 5500.

A widespread species, found over much of the Himalaya, India, central and southern China, Korea, Japan (south) and southeast Asia.

Carex fargesii Franch. (CYPERACEAE) New species

Carex immanis C. B. Clarke in Journ. Linn. Soc. Bot. 36: 290. (1903). *Carex prescottiana* Boott var. *fargesii* (Franch.) Kukenth. Kukenthal in Englr, Pflanzenr. 38: 356. (1909).

Hubei: Badong, 6113 (type of *Carex immanis* C. B. Clarke).

Discovered by Henry in 1888 and based on material later collected by Père Farges in Sichuan in 1892.

Carex filicina Nees (CYPERACEAE)

C. B. Clarke in Journ. Linn. Soc. Bot. 36: 285. (1903), Kukenthal in Engler, Pflanzenr. 38: 274. (1909).

Hubei: Liantuo, 2669. **Yunnan:** Mengzi at 2,200 to 2,650 metres, 10210, 10210a., 10210b.

Native to the Himalaya, India, Sri Lanka, China and southeast Asia.

Carex gibba Wahlenb. (CYPERACEAE)

C. B. Clarke in Journ. Linn. Soc. 36: 287. (1903,) Kukenthal in Engler, Pflanzenr. 38: 238. (1909).

Hubei: Badong, 1746. Xiling (Yichang) Gorge, 3479. Liantuo, 3862. Jianshi, 5879. **Hunan:** Shimen, 7544.

Carex henryi (C. B. Clarke) T. Koyama (CYPERACEAE) New species

in Nouv. Arch. Mus. 3. ser. 8, 243. (1896).

Carex longicruris Nees. var. *henryi* C. B. Clarke in Journ. Linn. Soc. Bot. 36: 295. (1903), Kukenthal in Engler, Pflanzenr. 38: 603. (1909).

Hubei: Liantuo, 4495. Badong, 5185, 5186. Changyang, 7502. Antelope Glen, 4266 (syntype of *Carex longicruris* Nees. var. *henryi* C. B. Clarke).

Carex lanceolata Boott (CYPERACEAE)

Carex lanceolata Boott. var. *macrosandra* Franchet in Nouv. Arch. Mus. 9: 169. (1897). *Carex pediformis* C. A. Mey. var. *macrosandra* (Franch.) Clarke C. B. Clarke in Journ. Linn. Soc. Bot. 36: 303. (1903).

Hubei: Yichang, 1288, 7887, 7890. Changleping, 7846.

Carex lanceolata Boott var. *subpediformis* Kukenthal (CYPERACEAE) New variety

in Engler, Pflanzenr. 38: 493. (1909).

Carex pediformis C. B. Clarke non C. A. Mey. in Journ. Linn. Soc. Bot. 36: 303. (1903).

Hubei: Badong, 5321.

Carex lancifolia C. B. Clarke (CYPERACEAE) New species

in Journ. Linn. Soc. Bot. 36: 293. (1903), Kukenthal in Engler, Pflanzenr. 38: 496. (1909).

Hubei: South Badong, 5368, 5467 (syntype).

Carex lehmannii Drejer (CYPERACEAE)

C. B. Clarke in Journ. Linn. Soc. Bot. 36: 293. (1903), Kukenthal in Engler, Pflanzenr. 38: 387. (1909).

Hubei: Fang Xian at 2,900 metres, 6837.

Carex ligulata Nees (CYPERACEAE)

C. B. Clarke in Journ. Linn. Soc. Bot. 36: 294. (1903).

Carex hebecarpa C. A. Mey var. *ligulata* (Nees) Kukenth. Kukenthal in Engler, Pflanzenr. 38: 745. (1909).

Hubei: Badong, 1864, 3689. Antelope Glen, 3491. Liantuo, 3909, 3948. **Sichuan:** South Wushan, 5748.

Carex longipes D. Don (CYPERACEAE)

C. B. Clarke in Journ. Linn. Soc. Bot. 36: 295. (1903), Kukenthal in Engler, Pflanzenr. 38: 603. (1909).

Hubei: Badong, 4713.

Carex maculata Boott (CYPERACEAE)

C. B. Clarke in Journ. Linn. Soc. Bot. 36: 297. (1903).

Hubei: Yichang, 3650.

A widespread species, distributed in India, Sri Lanka, China, Japan, southeast Asia, Australia and the Pacific Islands.

Carex manciformis Clarke ex Franchet (CYPERACEAE) New species

in Nouv. Arch. Mus. 3rd ser. 10: 62. (1898). C. B. Clarke in Journ. Linn. Soc. Bot. 36: 297. (1903), Kukenthal

02

in Engler, Pflanzenr. 38: 628. (1909).

Hubei: San You Dong Glen, 3390. Changyang, 5222. Yichang, 7889.

Carex moupinensis Franch. (CYPERACEAE)

C. B. Clarke in Journ. Linn. Soc. Bot. 36: 297. (1903), Kukenthal in Engler, Pflanzenr. 38: 289. (1909).

Hubei: Yichang, 1234.

Discovered by Père Armand David near Baoxing (Moupin) in western Sichuan in April 1869.

Carex nemostachys Steud. (CYPERACEAE)

Clarke in Journ. Linn. Soc. Bot. 36: 300. (1903).

Hubei: Yichang, 2931, 3531.

Native to Myanmar, Indochina, China and Japan.

Carex nubigena D. Don ex Tilloch & Taylor (CYPERACEAE)

C. B. Clarke in Journ. Linn. Soc. Bot. 36: 300. (1903), Kukenthal in Engler, Pflanzenr. 38: 145. (1909).

Hubei: Badong at 1,900 metres, 6103. Fang Xian, 6928.

Native to the Himalaya, India, Sri Lanka, China, Japan and southeast Asia.

Carex pediformis C. A. Mey. (CYPERACEAE)

C. B. Clarke in Journ. Linn. Soc. Bot. 36: 303. (1903).

Hubei: Badong, 3747, 4921. Changleping, 6326. Xiling (Yichang) Gorge, 7886. Yichang, 7897.

Carex perakensis C. B. Clarke (CYPERACEAE)

Carex prainii C. B. Clarke (nom. illeg.) in Journ. Linn. Soc. Bot. 36: 305. (1904).

Yunnan: On Fengchunling at 3,100 metres, 10839 (isosyntype of *Carex prainii* C. B. Clarke).

Carex phacota Sprengel (CYPERACEAE)

Henry in Trans. Asiat. Soc. Jap. 24 suppl: 106. (1896) (List Pl. Formosa), C. B. Clarke in Journ. Linn. Soc. Bot. 36: 303. (1903).

Taiwan: Wanchin, 540. **Yunnan:** In the mountains near Simao at 1,600 metres, 11759.

Native to the Himalaya, Sri Lanka, Thailand, China, Korea and Japan.

Carex pruinosa Boott. (CYPERACEAE)

C. B. Clarke in Journ. Linn. Soc. Bot. 36: 306. (1903).

Hubei: On the hills behind Yichang, 3501.

Carex repanda C. B. Clarke (CYPERACEAE)

C. B. Clarke in Journ. Linn. Soc. Bot. 36: 308. (1903).

Carex wightiana Nees var. *repanda* (C. B. Clarke) Kukenth. Kukenthal in Engler, Pflanzenr. 38: 288. (1909).

Yunnan: Mountains near Simao at 1,600 metres, 9971a. On mountains near Simao at 1,650 metres, 11719.

Carex rochebrunei Franch. & Sav. (CYPERACEAE)

Carex remota L. var. *rochebrunei* (Franch. & Sav.) C. B. Clarke C. B. Clarke in Journ. Linn. Soc. Bot. 36: 307. (1904).

Sichuan: South Wushan, 5722.

Native to India, China, Japan, Sumatra and Java.

Carex rubro-brunnea C. B. Clarke (CYPERACEAE)

Clarke in Journ. Linn. Soc. Bot. 36: 309. (1903).

Yunnan: Simao at 1,400 metres, 11722.

Carex rubro-brunnea C. B. Clarke var. *taliensis* (Franch.) Kukenthal (CYPERACEAE)

in Engler, Pflanzenr. 38: 344. (1909).

Carex taliensis Franch. Clarke in Journ. Linn. Soc. Bot. 36: 313. (1903).

Hubei: Badong, 1725. South Badong, 5431. **Yunnan:** Mountains near Simao, 11722 (in part).

Discovered by Delavay on the Cangshan. Franchet chose the specific epithet *taliensis* from Tali now the city of Dali in north-west Yunnan. Dali was Delavay's base while exploring the Cangshan range.

Carex sampsonii Hance (CYPERACEAE) New species

Carex laticeps C. B. Clarke ex Franch. in Nouv. Arch. Mus. 3me. ser. 9: 178. (1897), C. B. Clarke in Journ. Linn. Soc. Bot. 36: 293. (1903), Kukenthal in Engler, Pflanzenr. 38: 635. (1909).

Hubei: Yichang, 3332.

Carex sendaica Franch. (CYPERACEAE)

Carex longistolon C. B. Clarke ex Franch in Nouv. Arch. Mus. d'Hist. Nat. 8: 243. (1896), C. B. Clarke in Journ. Linn. Soc. Bot. 36: 296. (1903), Bretschneider in Hist. Eur. Bot. Disc. China 2: 793. (1898).

Hubei: Yichang, 86, 87, 2776.

Carex siderosticta Hance (CYPERACEAE)

C. B. Clarke in Journ. Linn. Soc. Bot. 36: 310. (1903).

Hubei: Yichang, 3758.

Carex simulans C. B. Clarke (CYPERACEAE) New species

in Journ. Linn. Soc. Bot. 36: 310. (1903), Kukenthal in Engler, Pflanzenr. 38: 629. (1909).

Hubei: Badong, 3712. Yichang, 7892.

Carex sparsinux C. B. Clarke ex Franch. (CYPERACEAE)

Carex filipes Franch. et Sav. var. *sparsinux* (C. B. Clarke ex Franch.) Kukenth. Kukenthal in Engler, Pflanzenr. 38: 639. (1909). *Carex filipes* Clarke non Franch. & Sav. in Journ. Linn. Soc. Bot. 36: 285. (1903).

Hubei: Badong, 3745. Changleping, 6271.

Carex spp. (CYPERACEAE)

Anonymous in Kew List of Determinations (volume 24 & 34).

Hubei: Badong, 5458. Changyang, 7801. Yichang, 7891. **Yunnan:** Mengzi, 13730.

Carex stipitinux C. B. Clarke ex Franch. (CYPERACEAE) New species

in Bull. Soc. Philom. Paris, 5me ser., 7: 31 (1895), Bretschneider in Hist. Eur. Bot. Disc. China 2: 793 (1898),

02

Clarke in Journ. Linn. Soc. Bot. 36: 312. (1903), Franch in Nouv. Arch. Mus. 3me ser. 8: 246. (1896), Diels in Engl. Bot. Jahrb. 29: 230. (1901).

Carex brunnea Thunb. var. *stipitinux* (C. B. Clarke) Kukenth. Kukenthal in Engler, Pflanzenr. 38: 602. (1909).

Hubei: Badong, 4852. Jianshi, 5838, 5971. Fang Xian, 6610.

Discovered by Henry's Badong collector in the summer of 1887, this species was also later collected by Père Paul Farges in Sichuan. Allied to *Carex brunnea* Thunb.

Carex stramentitia Boott ex Boeckeler (CYPERACEAE)

Anonymous in Kew List of Determinations (volume 34).

Yunnan: Simao, 11769.

Carex thibetica Franch. (CYPERACEAE)

C. B. Clarke in Journ. Linn. Soc. Bot. 36: 314. (1903).

Hubei: Liantuo, 3833. Changyang, 5245. Changleping, 6332.

Discovered by Père Armand David near Baoxing in western Sichuan in April 1869.

Carex thomsonii Boott (CYPERACEAE)

C. B. Clarke in Journ. Linn. Soc. Bot. 36: 314. (1903), Kukenthalin Engler, Pflanzenr. 38: 143. (1909).

Hubei: Yichang, 3392.

Carex transversa Boott (CYPERACEAE)

C. B. Clarke in Journ. Linn. Soc. Bot. 36: 314. (1904).

Carex brownii Tuckerm. var. *transversa* (Boott.) Kukenth ex Matsum. Kukenthal in Engler, Pflanzenr. 38: 614. (1909).

Zhejiang: Punto Isle, 24.

This number is quite seperate to Henry's Hubei no. 24.

Carex tristachya Thunb. (CYPERACEAE)

C. B. Clarke in Journ. Linn. Soc. Bot. 36: 315. (1903), Kukenthal in Engler, Pflanzenr. 38: 471. (1909).

Hubei: Yichang, 7898. Antelope Glen near Yichang, 7901.

Carex tumida Boott (CYPERACEAE)

C. B. Clarke in Journ. Linn. Soc. Bot. 36: 316. (1903), Kukenthal in Engler, Pflanzenr. 38: 615. (1909).

Yunnan: Mengzi at 1,800 metres, 10833.

Distributed from Bhutan to New Guinea.

Carex unisexualis C. B. Clarke (CYPERACEAE) New species

in Journ. Linn. Soc. Bot. 36: 316. (1903).

Carex fluviatilis Boott var. *unisexualis* (C. B. Clarke) Kukenth. Kukenthal in Engler, Pflanzenr. 38: 144. (1909).

Hubei: Yichang, 594.

Carex wahuensis C. A. Mey. subsp. *robusta* (Franch. & Sav.) T. Koyama (CYPERACEAE)

Carex wilfordii C. B. Clarke in Journ. Linn. Soc. Bot.

36: 318. (1903).

Zhejiang: Shrub Island, 47.

Carica papaya L. (CARICACEAE)

Anonymous in Kew List of Determinations (volume 24).

Hainan: Environs of Kiungchow, 8247.

The papaya or paw-paw, a musky flavoured tropical fruit native to Central America. A short lived, soft wooded single stemmed tree with large palmately lobed leaves clustered on the uppermost part of the stem. The flowers and fruits are borne in the leaf axils, male and female flowers are borne on seperate trees. The fruits are large and fleshy and ripen to bright orange and are ased as a meat tenderiser, to reduce cloudiness in beer and to shrinkproof wool and silk.

Carlemannia sp. (CARLEMANNIACEAE)

Anonymous in Kew List of Determinations (volume 34).

Yunnan: In Pingbian, in forests on the Daweishan Range at 1,650 metres, 11045. Mengzi, 11474. Simao 12461.

Carlemannia tetragona Hook. f. (CARLEMANNIACEAE)

Carlemannia henryi H. Léveille in Fedde, Repert. Nov. Spec. Regni Veg. 13: 178. (1914).

Yunnan: Simao, 12545 (syntype of *Carlemannia henryi* H. Léveille).

Carpesium abrotanoides L. (ASTERACEAE)

Forbes & Hemsley in Journ. Linn. Soc. Bot. 23: 430. (1888).

Hubei: Badong, 10 (collected September 1885, received Kew March 12th 1886), 2461. Yichang, 2201, 2737. **Yunnan:** Mengzi, 10866.

Distributed in Myanmar, China, Korea, Vietnam and Japan.

Carpesium cernuum L. (ASTERACEAE)

Forbes & Hemsley in Journ. Linn. Soc. Bot. 23: 430. (1888).

Hubei: Badong, 108, 2462, 4811. Yichang, 1661, 2185. Fang Xian, 6578. **Yunnan:** Mengzi, 11021.

Carpesium minus Hemsl. (ASTERACEAE)

Forbes & Hemsley in Journ. Linn. Soc. Bot. 23: 431. (1888).

Hubei: South Donghu, 7716. **Sichuan:** North Wushan, 7110.

Discovered by the Rev'd. Dr. Ernst Faber on the lower slopes of Emei Shan in 1887.

Carpesium nepalense Less var. *lanatum* (Hook f. & Thoms. ex C. B. Clarke) Kitam. (ASTERACEAE)

Carpesium cernuum L. var. *lanatum* Hook f. & Thoms. ex C. B. Clarke Forbes & Hemsley in Journ. Linn. Soc. Bot. 23: 431. (1888).

Hunan: Shimen, 7944.

Carpesium sp. (ASTERACEAE)

Anonymous in Kew List of Determinations (volume 24).

Hubei: Badong, 2449. Fang Xian, 6619. **Sichuan:** South Wushan, 7235, 7272.

Carpesium trachelifolium Less. var. *foliis* lanceolatis. (ASTERACEAE)

Forbes & Hemsley in Journ. Linn. Soc. Bot. 23: 431. (1888).

Hubei: Badong, 2387, 2466, 4793, 4987, 5193. Yichang, 6496. Fang Xian, 6847.

Carpinus cordata Blume var. *chinensis* Franch. (BETULACEAE) New variety

Burkill in Journ. Linn. Soc. Bot. 26: 501. (1899), Winkler in Engler, Pflanzenr. 61: 27. (1904), Schneider in Sargent Pl. Wilson 2: 437. (1916).

Carpinus cordata Henry non Blume in Notes Econ. Bot. 54. (1893).

Hubei: Jianshi, 5886. **Sichuan:** North Wushan, 5886a.

Discovered by Henry about 1887 and described by Franchet from material later collected by Farges in Sichuan in the early 1890s. Henry stated this was the *wo-erh-li*. Introduced to cultivation by Wilson in 1901.

Carpinus fargesiana H. J. P. Winkl. var. *ovalifolia* (H. J. P. Winkl.) Holstein & Weigend (BETULACEAE) New variety

Carpinus polyneura Burkill non Franchet in Journ. Linn. Soc. Bot. 26: 501. (1899), (in part). *Carpinus turczaninowii* Henry non Hance in Notes Econ. Bot. China 55. (1893), Winkler in Engler, Pflanzenr. 61: 38. (1904). *Carpinus turczaninovii* Hance var. *ovalifolia* Winkler in Bot. Jahrb. L. suppl. 505. (1914), Schneider in Sargent Pl. Wilson 2: 427. (1916), Henry in Elwes & Henry, Trees Gr. Brit. & Irel. 3: 530. (1908), Nelson in Pim, The Wood & the Trees 221. (1984), Bean in Trees & Shrubs 1: 510. (1989), Nelson in A Heritage of Beauty 310. (2000).

Sichuan: North Wushan, 7020. South Wushan, 7219.

The *chien-feng-kan*, a tree to 10 metres tall, introduced to cultivation by Henry through the Royal Botanic Gardens, Kew. The tallest specimen in the British Isles and Ireland grows at Burford Lodge, Surrey, when measured in 2000 it was 11 metres tall with a trunk diameter of 45 cm at 1.2 metres from its base.

Carpinus henryana (Winkler) Winkler (BETULACEAE) New species

in Bot. Jahrb. L. suppl. 507. fig. 7. (1914), Schneider in Sargent Pl. Wilson 2: 429. (1916), Bean in Trees & Shrubs 3: 72. (1936), Bean in Trees & Shrubs 1: 507. (1989), Nelson in A Heritage of Beauty 310. (2000), O' Brien in Augustine

Henry, an Irish Plant Collector in China 15. (2002).

Carpinus polyneura Burkill non Franchet in Journ. Linn. Soc. Bot. 26: 501. (1899), (in part). *Carpinus tschonoskii* Maxim. var. *henryana* H. J. P. Wink. in Engler, Pflanzenr. 4, Fam. 61: 36. (1904).

Sichuan: North Wushan, 7063 (type).

A deciduous tree to about 20 metres tall, introduced to cultivation by Wilson through the Arnold Arboretum in 1907, it was in the Kew collection by 1912. In Ireland Henry's hornbeam is cultivated at Dunloe Castle, Kilmacurragh and the John F. Kennedy Arboretum. The tallest tree in the British Isles and Ireland grows at the Royal Botanic Gardens, Kew. The same tree is from Wilson's 4443 (living plants collected near Fang in 1907) and was planted out at Kew in 1912. When measured in 2001 this tree was 11 metres tall with a trunk diameter of 25 cm.

Carpinus londoniana H. J. P. Winkler (BETULACEAE) New species

in Pflanzenr. (Engler) iv.-61: 32. (1904), Winkler in Bot. Jahrb. L. suppl. 492. (1914), Schneider in Ill. Handb. Laubholzk. 2: 894. fig. 558. (1912), Schneider Sargent Pl. Wilson. 2: 438. (1916), Hu in Sunyatsenia 1: 111. (1933).

Yunnan: In forests near Simao at 1,300 metres, 11640 (type). In the mountains to the east of Simao at 1,300 metres, 11640a., 11640b.

Joseph Rock also collected this species when he visited Henry's type locality at Simao in March 1922. Distributed in Myamar, China, Thailand, Laos and Vietnam.

Carpinus polyneura Franch. (BETULACEAE) New species

Burkill in Journ. Linn. Soc. Bot. 26: 501. (1889), Winkler in Engler, Pflanzenr. 61: 38. (1904), Schneider in Sargent Pl. Wilson 2: 430. (1916).

Carpinus faginea Burkill non Lindl. in Journ. Linn. Soc. Bot. 26: 501. (1899).

Hubei: Liantuo, 4472. **Sichuan:** South Wushan, 5520, 5520a., 7020a.

Discovered by Henry in 1887 and described by Franchet from material later collected by Farges in Sichuan in the 1890s.

Carpinus pubescens Burkill (BETULACEAE) New species

in Journ. Linn. Soc. Bot. 26: 502. (1899), Winkler in Engler. Pflanzenr. iv.-61, 37. (1904); in Bot. Jahrb. L. suppl. 501. (1914), (in part), Schneider in Ill. Handb. Laubholzk. 2: 895. (1914), (in part); in Sargent Pl. Wilson. 2: 442. (1916), Hu in Sunyatsenia 1: 119. (1933).

Yunnan: In old woods near Feng-tien in Milê district, 9929 (type). Mengzi, 13775.

Distributed in western China and Vietnam.

02

Carpinus sp. (BETULACEAE)
Carpinus viminea Henry non Wallich in Notes Econ. Bot. China 55. (1893).

Hubei: Jianshi, 7429.

Henry called this species the *lei-kung-li* or *lei-kung-yu*.

Carpinus viminea Lindl. ex Wall. (BETULACEAE)
Carpinus laxiflora (Sieb. & Zucc.) Blume var. *macrostachya* Oliver in Hooker's Icon. Pl. xx. t. 1989. (1891), Bretschneider in Hist. Eur. Bot. Disc. China 2: 791. (1898), Burkill in Journ. Linn. Soc. Bot. 26: 501. (1899), Schneider in Sargent Pl. Wilson.2: 425. (1916), Hu in Sunyatsenia 1: 111. (1933), Morley in Glasra 3: 78. (1979), Zheng in Journ. Wuhan Bot. Res. 2.(1): 20. (1984), Bean in Trees & Shrubs 1: 509. (1989), Nelson in A Heritage of Beauty 310. (2000), O' Brien in Augustine Henry, An Irish Plant Collector in China 15. (2002), *Carpinus laxiflora* Henry non (Sieb. & Zucc.) Blume in Notes Econ. Bot. China 55. (1893); in Elwes & Henry, Trees Gr. Brit. & Irel. 3: 530. (1908), *Carpinus laxiflora* (Sieb. & Zucc.) Blume var. *davidii* Franch. Winkler in Engler, Pflanzenr. 61: 33. (1904).

Hubei: North Badong, 7013 [type of *Carpinus laxiflora* (Sieb. & Zucc.) Blume. var. *macrostachya* Oliver].

Carthamus tinctorius L. (ASTERACEAE)
Forbes & Hemsley in Journ. Linn. Soc. Bot. 23: 470. (1888).

Sichuan: Henry's Chinese collector with Pratt, no locality stated but may be Leshan, Emei Shan, Wa shan or Kangding, 8917.

Safflower, the *hung hwa* or *hong hua* i.e 'red flower' (though the flowers are orange-yellow), a thistle-like annual bearing handsome orange flowers above toothed prickly leaves. This species was introduced to China from western Asia in the third century A.D. and was extensively cultivated in 19th century China and is now more or less naturalised in some parts. According to Wilson it was esteemed for dyeing costly silk fabrics. It is thought that the safflower was originally domesticated in the eastern Mediterranean region (the earliest archaelogical records are from Mesopotamia) though today it is only known in cultivation. In Chinese herbal medicine safflower is used to treat inflammation, heart disorder and fever. It was originally grown for its flowers which contain the deep orange pigment carthamin and it is much grown in developing countries for safflower oil used as a cooking oil and in paints and varnishes.

Carum sp. (APIACEAE)
Anonymous in Kew List of Determinations (volume 34).

Yunnan: Simao, 12481.

Caryodaphnopsis henryi Airy Shaw (LAURACEAE)
New species
in Kew Bull. Misc. Inf. 75. (1940).

Nothaphoebe tonkinensis Lecompte f. *brevipedicellata* Liou Ho in Laur. Chine & Indochine 77. (1934).

Yunnan: At Fengchunling, south of Red River, in mountain forest at 2100 metres, 10692 (type of *Caryodaphnopsis henryi* Airy Shaw, isotype of *Nothaphoebe tonkinensis* Lecompte f. *brevipedicellata* Liou Ho).

Endemic to SE Yunnan.

Caryopteris incana (Thunb. ex Houtt.) Miq. (LAMIACEAE)
Rehder in Sargent Pl. Wilson. 3: 378. (1917), Zheng in Wuhan Bot. Res. 2.(1): 185. (1984).

Caryopteris mastacanthus Schauer. Forbes & Hemsley in Journ. Linn. Soc. Bot. 26: 263. (1890), Henry in Trans. Asiat. Soc. Jap. 24 suppl: 72. (1896) (List Pl. Formosa).

Hubei: Yichang, 161, 2192, 2782. Liantuo, 2597, 4581.
Taiwan: Wanshoushan, without number.

A deciduous shrub to over a metre tall, bearing masses of violet-blue flowers in hemispherical cymes during the month of October. Native to China and Japan and introduced to cultivation by Robert Fortune who found it wild near Guangzhou (Canton). It was then lost to cultivation for many years until reintroduced by Charles Maries from Japan. In current Chinese herbal medicine this species is used to induce diaphoresis and to relieve pain.

Caryopteris sp. (LAMIACEAE)
Anonymous in Kew List of Determinations (volume 24).

Hubei: Yichang, 997.

Caryota mitis Lour. (ARECACEAE)
Wright in Journ. Linn. Soc. Bot. 36: 167. (1904).
Hainan: Kuanglang, 8373.

The Burmese fishtail palm, a handsome palm distributed from Myanmar to the Malay Peninsula, Java and the Philippines.

Caryota sp. (ARECACEAE)
Anonymous in Kew List of Determinations (volume 34).

Yunnan: Mengzi, 11838.

Cascabela thevetia (L.) Lippold (APOCYNACEAE)
Thevetia neriifolia A. Juss ex A. DC. Anonymous in Kew List of Determinations (volume 34).

Yunnan: Simao, 12747.

Yellow oleander, a shrub or small tree to 8 metres tall, native to tropical America. Naturalised in many of the warmer parts of China, widely planted in southern Yunnan.

Casearia membranacea Hance (SALICACEAE)
Li in Woody Flora of Taiwan 603. fig. 237. (1963).
Casearia sp. Henry in Trans. Asiat. Soc. Jap. 24 suppl: 45. (1896) (List Pl. Formosa).

Taiwan: Wanchin, 440, 1586.

A small evergreen tree, native to Hainan and Taiwan.

Casearia spp. (SALICACEAE)

Anonymous in Kew List of Determinations (volume 24 & 34).

Hainan: Environs of Haikou, 7961, 8325, 8398. Hummocks near Haikou, 8084. On Liang Shan, 16 km (10 miles) east of Haikou, 8202. Environs of Kiungchow, 8221. **Yunnan:** Simao, 12056.

Cassia sp. (FABACEAE)

Anonymous in Kew List of Determinations (volume 34).

Yunnan: Simao, 11725.

Cassiope selaginoides Hook. f. & Thoms. (ERICACEAE)

Rehder & E. H. Wilson in Sargent Pl. Wilson. 1: 551. (1913).

Sichuan: Kangding, Henry's Chinese collector with Pratt, 8871.

A dwarf evergreen shrub to about 20 cm tall carrying stalkless overlapping leaves in four rows along the stems of plants. The handsome white bell-shaped flowers are carried in late summer and autumn. Discovered by Sir Joseph Hooker in the Sikkim Himalaya in 1849, this species grows at very high altitudes and in places is a very conspicuous feature of the Himalayan flora. I have seen it cover enormous tracts of high alpine moorland just below scree on the mountains of south-east Tibet (Xizang) where it grows with dwarf *Rhododendron* and willows (*Salix* spp.).

Cassytha filiformis L. (LAURACEAE)

Henry in Trans. Asiat. Soc. Jap. 24 suppl: 80. (1896) (List Pl. Formosa).

Taiwan: Wanchin, 1567.

A twining holoparasitic chamaephyte, native to the tropics of Africa, Asia and Australia. In Chinese Taiwan this species was known as *wu-ken-ts'ao* according to Henry.

Castanea henryi (Skan) Rehder & E. H. Wilson (FAGACEAE) New species

in Sargent Pl. Wilson. 3. 196. (1917), Zheng in Journ. Wuhan Bot. Res. 2.(1): 22. (1984), Li and Zhang in Journ. Wuhan Bot. Res. 16: 242. (1998), Nelson in A Heritage of Beauty 311. (2000), O' Brien in Augustine Henry, An Irish Plant Collector in China 15. (2002).

Castanea vulgaris Henry non Lam. in Journ. China Br. Roy. Asiat. Soc. 22: 254. (1887) (Chinese Names of Plants). *Castanea sativa* Skan non Mill. in Journ. Linn. Soc. Bot. 26: 525. (pro parte) (1899). *Castanea sativa* Mill. var. *acuminatissima* Seemen in Bot. Jahrb. 29: 287. (1900). *Castanopsis henryi* Skan in Journ. Linn. Soc. Bot. 26: 524. (1899), Seemen in Bot. Jahrb. xxix. 287. (1900), Koidzumi in Tokyo Bot. Mag. 30: 101. (1916), Morley in Glasra 3: 78. (1979).

Hubei: Badong, 125 (in part, co-type of *Castanopsis*

henryi Skan), 2878 [isotype of (Skan) Rehder & E. H. Wilson]. Without locality, 5793a., (co-type of *Castanea sativa* Mill. *acuminatissima* Seemen). Jianshi, 5793 (May 1888), 5800 [co-type of *Castanopsis henryi* Skan, isotype of *Castanea henryi* (Skan) Rehder & E. H. Wilson].

A tree to about 25 metres tall with trunk girths of up to 2.5 metres in the wild, but extremely slow growing in cultivation, probably as a result of our rather dull, insular summers in the British Isles and Ireland. *Castanea henryi* was introduced to cultivation by Wilson in 1900 through Veitch's Coombe Wood nursery. In Ireland this species is cultivated at the John F. Kennedy Arboretum in Co Wexford.

Castanea mollissima Blume (FAGACEAE)

Rehder & E. H. Wilson in Sargent Pl. Wilson. 3: 192. (1917), Walsh in A. Henry. Forst. Herb. Cat. 62. (1957), Zheng in Journ. Wuhan Bot. Res. 2.(1): 22. (1984).

Castanea sativa Skan non Mill. in Journ. Linn. Soc. Bot. 26: 525. (1899).

Hubei: Yichang, 493, 892, 1570. Liantuo, 1924. Badong, 124. South Badong, 6306. **Yunnan:** Mountains to the south of Mengzi at 1,600 metres, 10701.

The Chinese chestnut, a tree to about 20 metres tall with trunks to 2 metres in girth. Introduced to cultivation by Professor Charles Sprague Sargent (the founding director of the Arnold Arboretum), who purchased seeds in a Beijing market in 1903 (though a tree at Buckingham Palace is said to predate this introduction. China is the world's largest producer of chestnuts and it is almost entirely this species and its numerous cultivars that supply this trade. The tallest specimen in Britain and Ireland grows at the Royal Botanic Gardens, Kew, when measured in 2001 it was 11 metres tall with a trunk diameter of 35 cm. This species was collected by the Glasnevin Central China Expedition (GCCE 250) on the hills above Badong in September 2002.

Castanea seguinii Dode (FAGACEAE)

Rehder & E. H. Wilson in Sargent Pl. Wilson. 3: 194. (1917), Walsh in Aug. Henry. Forst. Herb. Cat. 62. (1957), Zheng in Journ. Wuhan Bot. Res. 2.(1): 23. (1984).

Castanea vulgaris Lam. Henry in Journ. China Br. Roy. Asiat. Soc. 22: 254. (1887) (Chinese Names of Plants).

Hubei: Badong, 2867, 6046. Changyang, 5800a.

A bush or small tree to 10 metres tall, according to Wilson the fruits were collected and eaten by the peasantry of Hubei and according to Henry it was known as *mao pan-li*. The tallest specimen in the British Isles and Ireland grows at Borde Hill, Sussex, planted in 1930, when measured in1995 this tree was found to be 18 metres tall with a trunk diameter of 67 cm. This species was collected by the Glasnevin Central China Expedition (GCCE 519) in September 2004 near Zhenzi town in Xingshan where it formed woods and in Dragon Gate River Forest Park (GCCE 652) in October of the same year. The fruits of this species are far superior in taste to the more widely grown *Castanea mollissima.*

Castanopsis calathiformis (Skan.) Rehder & E. H. Wilson (FAGACEAE) New species

in Sargent Pl. Wilson. 3: 204. (1917), Walsh in Aug. Henry Forst. Herb. 62. (1957).

Quercus calathiformis Skan in Journ. Linn. Soc. Bot. 26: 508. (1899). *Synaedrys calathiformis* (Skan) Koidzumi in Tokyo Bot. Mag. 30: 188. (1916).

Yunnan: On the Daweishan Range in Pingbian at 1,500 metres, 9070, 9070a., (co-type of *Quercus calathiformis* Skan), 9070b., (co-type of *Quercus calathiformis* Skan), 13682. In mountains near Simao at 1,600 metres, 11696, 11696a., 11696b., 11696c.

Distributed in India (Assam), Myanmar, China [Tibet (Xizang), Yunnan], Laos, Thailand and Vietnam.

Castanopsis carlesii (Hemsl.) Hayata (FAGACEAE)

Quercus cuspidata Henry non (Thunb. ex Murray) Schottky in Trans. Asiat. Soc. Jap. 24 suppl: 90. (1896) (List Pl. Formosa). *Castanopsis cuspidata* Rehder & E. H. Wilson non (Thunb. ex Murray) Schottky in Sargent Pl. Wilson. 3: 204. (1917). *Castanopsis carlesii* (Hemsl.) Hayata var. *sessilis* Nakai Li in Woody Flora of Taiwan 88. (1963).

Taiwan: Oluanpi, 1313.

Castanopsis ceratacantha Rehder & E. H. Wilson (FAGACEAE) New species

in Sargent Pl. Wilson. 3: 199. (1917).

Yunnan: In forests near Simao at 1,500 metres, 12831.

Castanopsis delavayi Franch. (FAGACEAE)

Skan in J. Linn. Soc. Bot. 26: 523. (1899).

Yunnan: Mengzi, 9457, 9457a., 9457b., 9457c.

Endemic to western and southern China.

Castanopsis echinocarpa Miq. (FAGACEAE)

Castanopsis tribuloides A. de Candolle var. *echidnocarpa* (Hook. f. & Thoms. ex Miq.) King apud Hook. f. Rehder & E. H. Wilson in Sargent Pl. Wilson. 3: 203. (1917).

Yunnan: In woods near Simao at 1,500 to 1,600 metres, 11565, 11565a., 11565c., 12328a. In the valley of the Red River near Manpan at 1150 metres, 12847.

Himalaya to China.

Castanopsis fargesii Franch. (FAGACEAE) New species

Render & E. H. Wilson in Sargent Pl. Wilson. 3: 198. (1917), Zheng in Journ. Wuhan Bot. Res. 2.(1): 23. (1984).

Sichuan: South Wushan, 5551. **Yunnan:** South of the Red River from Manmei at 2,000 metres, 9202a.

A handsome evergreen tree to 30 metres tall with trunks up to 4 metres in girth. According to Henry this species was known colloquially in Hubei as *tsze-li-tzu-shu.* Discovered by Henry in May of 1888 and described by Franchet from material later collected by Farges in Sichuan during the early 1890s.

Castanopsis ferox (Roxb.) Spach. (FAGACEAE)

Castanopsis tribuloides A. de Candolle var. *ferox (*Roxb.) King apud Hooker. f. Rehder & E. H. Wilson in Sargent Pl. Wilson. 3: 201. (1917).

Yunnan: In forests near Simao at 1,500 metres, 12831a.

Distributed in India (Sikkim), Bangladesh, Myanmar, western China, Thailand, Laos and Vietnam.

Castanopsis fissa (Champ. ex Benth.) Rehder & E. H. Wilson (FAGACEAE)

Quercus fissa Champ. Skan in Journ. Linn. Soc. Bot. 26: 512. (1899).

Hainan: Without locality, without number.

Castanopsis formosana (Skan) Hayata (FAGACEAE) New species

in Icon. Pl. Form. 3: 189. (1913) Li in Woody Flora of Taiwan 85. (1963).

Quercus sp. Henry in Trans. Asiat. Soc. Jap. 24 suppl: 90. (1896) (List Pl. Formosa). *Castanopsis chinensis* Henry non Hance in Trans. Asiat. Soc. Jap. 24 suppl: 90. (1896) (List Pl. Formosa). *Castanopsis tribuloides* A.D.C. var. *formosana* Skan in Journ. Linn. Soc. Bot. 524. (1899).

Taiwan: Wanchin, 60 (syntype of *Quercus formosana* Skan), 556, 1641. Oluanpi, 1371 (isosyntype of *Quercus formosana* Skan), 1710 (syntype of *Quercus formosana* Skan).

A large evergreen tree which Henry stated was known as the *kou-li,* the wood was once popular for making carts in Taiwan. Also native to Hainan, in Taiwan it is found in forests in the central and southern parts of the island.

Castanopsis hystrix Hook. f. & Thoms. ex A. de Candolle (FAGACEAE)

Forbes & Hemsley in Journ. Linn. Soc. Bot. 26: 524. (1899), Rehder & E. H. Wilson in Sargent Pl. Wilson. 3: 198. (1917).

Yunnan: In forests in Mengzi at 1,600 metres, 11313. In woods near Simao at 1,500 metres, 11738, 11738a., 11738b.

Distributed in Nepal, India, Bhutan, Myanmar, China, Laos, Vietnam and Cambodia.

Castanopsis indica (Roxb. ex Lindl.) A. de Candolle (FAGACEAE)

Rehder & E. H. Wilson in Sargent Pl. Wilson. 3: 202. (1917).

Yunnan: Near Menglieh at 1,300 metres, 12834.

The common Indian chestnut, a medium-sized evergreen tree to about 15 metres tall, native to Nepal, India, Bangladesh, the Himalaya, Myanmar, western China, Laos, Thailand and Vietnam.

Castanopsis orthacantha Franchet (FAGACEAE)

Castanopsis concolor Rehder & E. H. Wilson in Sargent Pl. Wilson. 3: 203. (1917).

Yunnan: Yuanday at 2,500 metres, 13242 (type of *Castanopsis concolor* Rehder & E. H. Wilson). Milê district, 10369. Yuanjiang at 2,300 metres, 13342.

Endemic to western China.

Castanopsis platyacantha Rehder & E. H. Wilson (FAGACEAE) New species

in Sargent Pl. Wilson. 3: 200. (1917).

Yunnan: In forests near Mengzi at 2,300 metres, 10610.

Endemic to western China.

Castanopsis sclerophylla (Lindl. & Paxton) Schottky (FAGACEAE)

Rehder & E. H. Wilson in Sargent Pl. Wilson. 3: 201. (1917), Zheng in Journ. Wuhan Bot. Res. 2.(1): 23. (1984).

Quercus sclerophylla Lindl. & Paxton Henry in Journ. China Br. Roy. Asiat. Soc. 22: 242. (1887) (Chinese Names of Plants), Skan in Journ. Linn. Soc. Bot. 26: 520. (1889), Henry in Notes Econ. Bot. China 53. (1893).

Hubei: Yichang, 2964, 3218, 3218a. Liantuo, 3870. South Donghu, 7596, 7707.

A common evergreen tree in Hubei where it is found from river level to 1,500 metres in altitude. Both Henry and Wilson stated that the tree was known as *chu li* in Hubei and the fruits were gathered there and crushed by local peasants into a paste known as *tou-fu,* the paste resembled bean-curd.

Castanopsis sp. (FAGACEAE)

Anonymous in Kew List of Determinations (volume 34).

Yunnan: Simao, 13103, 13159.

Castanopsis tribuloides (Sm.) A. de Candolle (FAGACEAE)

Rehder & E. H. Wilson in Sargent Pl. Wilson. 3: 203. (1917).

Yunnan: In the valley of the Red River near Manpan at 1150 metres, 10847.

A tree of moderate height, native to the Himalaya, Myanmar, western China and Thailand.

Catalpa fargesii Bureau (BIGNONIACEAE) New species

Rehder in Sargent Pl. Wilson. 1: 305. (1913), Bean in Trees & Shrubs 1: 537. (1989), Lancaster in Travels in China 136. (1993).

Sichuan: South Wushan, 5856a.

A tree to about 15 metres tall, discovered by Henry in May of 1888 and described from material later collected by Farges in Sichuan. Introduced to cultivation by Wilson in 1901 through Veitch's Coombe Wood nursery and again in 1907 when collecting for the Arnold Arboretum. A tree from the latter seed collection grows at the Royal Botanic Gardens, Kew. When measured in 1997 it was 18 metres tall with a trunk diameter of 56 cm, the tallest in Britain and Ireland in fact.

Catalpa fargesii Bureau f. *duclouxii* (Dode) Gilmour. (BIGNONIACEAE)

Catalpa duclouxii Dode Rehder & E. H. Wilson in Sargent Pl. Wilson. 3: 433. (1917).

Hubei: Jianshi, 5856.

A deciduous tree to 25 metres tall producing panicles of light red bell-shaped flowers with purple-brown spotted throats in early summer. Distributed in Guizhou, Hubei, Sichuan and Yunnan. A beautiful tree commemorating two French missionaries, Père Paul Farges and Père Francois Ducloux, though it had been earlier collected by Delavay in Yunnan. Introduced to cultivation by Wilson in 1907.

Catalpa ovata G. Don (BIGNONIACEAE)

Catalpa kaempferi (DC.) Sieb. & Zucc. Henry in Journ. China Br. Roy. Asiat. Soc. 22: 240. (1887); in Notes Econ. Bot. China 43. (1893), Forbes & Hemsley in Journ. Linn. Soc. Bot. 26: 235. (1890), Henry in Elwes and Henry The Trees Gt. Brit. & Irel. 6: 1487. (1912), *Catalpa henryi* Dode in Bull. Soc. Dendr. France 2: 199. (1907); in Fedde, Repert. Nov. Spec. Regni Veg. 6: 7. (1908—1909), Schneider in Ill. Handb. Laubholzk. 2: 625. (1912).

Hubei: Yichang, 1391, 1391a., 1684.

The Chinese bean tree, endemic to China and cultivated for centuries in Japan from where it was introduced to Europe by Siebold in 1849. Because of its long pods Henry stated this tree was known colloquially as *ch'iu shu* around Yichang. Still a common tree in Hubei and widely planted along roadsides. Collected by the Glasnevin Central; China Expedition in Xingshan (GCCE 461) in September 2004 and in the Dragon Gate River Forest Park (GCCE 630), also in Xingshan, in October 2004.

Catharanthus roseus (L.) G. Don (APOCYNACEAE)

Vinca rosea L. Henry in Trans. Asiat. Soc. Jap. 24 suppl: 59. 1896 (List Pl. Formosa).

Hainan: Environs of Haikou, 8188. **Taiwan:** Wanchin, 446. Kaohsiung Spit, without number.

The Madagascar periwinkle, a short lived perennial to about 0.5 metres tall, native to Madagascar and widely naturalised in the tropics, Henry stated it was known to the Chinese as *ssu-shih-chun,* the isolated alkaloids are used to treat breast, lung and uterine cancer.

Cathcartia oliveriana (Franchet & Prain) Grey-Wilson (PAPAVERACEAE) New species

Gen. Meconopsis 376. (2014)

Meconopsis oliveriana Franchet & Prain in Journ. Asiat. Soc. Bengal xliv. 312. (1895), Prain in Ann. of Bot. 20: 365. (1906), Fedde in Engler, Pflanzenr. 104: 270. (1909).

Hubei: Fang Xian, 6863. A very beautiful annual to 90 cm tall, carrying solitary golden-yellow flowers in the axils of leaves on the upper part of stems from May to July. Distributed through the provinces of Henan, Hubei, Shaanxi, and Sichuan, where it grows on grassy slopes and thickets at altitudes between 1,700 to 2,400 metres. Allied to *Cathcartia chelidoniifolia*, this species has never reached cultivation, efforts were made at Kew when seeds were collected from Henry's 6863 in the herbarium on May 4th 1889, these were sent for propagation but obviously failed. *Cathcartia oliveriana* is said to prefer a shady, moist habitat with deep acid soil. In Hubei it is restricted to Badong and Shennongjia. The specific epithet commemorates the Kew botanist, Professor Daniel Oliver (1830—1916).

Cathormion umbellatum (Vahl.) Kosterm. (FABACEAE)

Pithecellobium umbellatum (Vahl.) Benth. Anonymous in Kew List of Determinations (volume 34).

Yunnan: Simao, 11875.

Catunaregam spinosa (Thunb.) Tirveng. (RUBIACEAE)

Randia dumetorum (Retz.) Poir. Anonymous in Kew List of Determinations (volume 24).

Hainan: Environs of Haikou, 8037, 8266.

Caulophyllum robustum Maxim. (BERBERIDACEAE)

Henry in Notes Econ. Bot. China 51. (1893); in Flora & Sylva 1: 217. (1903).

Hubei: Badong, 4915. South Badong, 5435. **Sichuan:** South Wushan. 5435a.

This species shares with *Clematis armandii* the name *wei-ling-hsien*. Henry's Badong collection proved to be a new record for the genus in China.

Causonis japonica (Thunb.) Raf. (VITACEAE)

Li in Woody Flora of Taiwan 523. (1963).

Vitis japonica Thunb. Forbes & Hemsley in Journ. Linn. Soc. Bot. 23: 134. (1886), Henry in Journ. China Br. Roy. Asiat. Soc. 22: 280. (1887) (Chinese Names of Plants); in Trans. Asiat. Soc. Jap. 24 suppl: 28. (1896) (List Pl. Formosa).

Hubei: Yichang, 455. South Badong, 6119. **Taiwan:** Wanchin, 463.

A glabrous or pubescent vine, native to Indochina, southern China and the Philippines. Henry stated that in the Three Gorges region this species was known colloquially as *wu chao t'eng*. The dried plant is used in Chinese herbal medicine to promote blood circulation, to arrest bleeding, in cases of snake-bite and as a diuretic.

Causonis mollis (Wall. ex M. A. Lawson) G. Parmar & J. Wen (VITACEAE)

Vitis mollis Wall. ex M. A. Lawson Henry in Notes Econ. Bot. China 52. (1893).

Hubei: Changyang, 6232. South Badong, 7290.

In the Three Gorges region this wild vine was called the *mao-wu-yueh-wu*.

Causonis trifolia (L.) Mabb. & J. Wen (VITACEAE)

Vitis carnosa (Lam.) Wall. Anonymous in Kew List of Determinations (volume 24).

Hainan: Environs of Haikou, 7951.

Cautleya gracilis (Sm.) Dandy (ZINGIBERACEAE)

Cautleya aff. *lutea* Royle var. *robusta* Schumann in Engler, Pflanzenr. 20: 125. (1904).

Yunnan: Mengzi, 9480a.

Cayaponia laciniosa (L.) C. Jeffrey (CUCURBITACEAE)

Bryonia laciniosa L Henry in Trans. Asiat. Soc. Jap. 24 suppl: 55. (1896) (List Pl. Formosa).

Taiwan: Oluanpi, 324.

A Jamaican species, this collection belongs elsewhere.

Cayratia albiflora C. L. Li (VITACEAE)

Cayratia oligocarpa (Levl. & Vant.) Gagn var. *glabra* (Gagn.) Rehder Zheng in Journ. Wuhan Bot. Res. 2.(1): 143. (1984).

Hubei: Xingshan, 6486. Fang Xian, 6592.

Celastrus angulatus Maxim. (CELASTRACEAE)

Rehder & E. H. Wilson in Sargent Pl. Wilson. 2: 346. (1916), Schneider in Illus. Handb. Laubholzk. 2: 184. (1912), Ding Hou in Ann. Missouri Bot. Gard. 42: 239. (1955), Morley in Glasra 3: 78. (1979), Zheng in Journ. Wuhan Bot. Res. 2.(1): 121. (1984), Nelson in A Heritage of Beauty 311. (2000).

Celastrus latifolius Hemsley in Journ. Linn. Soc. Bot. 23: 123. (1886), Oliver in Hooker's Ic. Pl. xxiii. t. 2206. (1892), J. H. Veitch in Hortus Veitchii 359. (1906).

Hubei: Yichang, 485, 1774, 3405, 3404a., 5925a. Liantuo, 2084, 3883. Jianshi, 5925 (type of *Celastrus latifolius* Hemsley).

A scandent shrub to 6 metres tall with large handsome leaves. At Yichang the leaves and roots of this shrub were dried in the sun and pounded into a powder and this was said to have been effacious in killing insects that infested turnips and other root crops. *Celastrus angulatus* was introduced by Wilson from Hubei in 1900 through Messrs Veitch. This is the largest leaved species of any *Celastrus*. Cultivated at the National Botanic Gardens, Glasnevin.

Celastrus gemmatus Loesener (CELASTRACEAE) New species

in Engl., Bot. Jahrb. 30: 468. (1902), Rehder & E. H.

02

Wilson in Sargent Pl. Wilson 2: 352. (1916), Ding Hou in Ann. Missouri Bot. Gard. 42: 258. (1955).

Celastrus articulatus Forbes & Hemsley non Thunb. in Journ. Linn. Soc. Bot. 23: 122. (1886). *Celastrus orbicularis* Thunb. f. *microphylla* Loesener in Engl. Bot. Jahrb. 30: 469. (1901).

Hubei: North Donghu, 7614. **Yunnan:** South of the Red River from Manmei at 2,000 metres, 9679a. In forests near Mengzi at 1,800 to 2,000 metres, 9782a., (isosyntype). Mengzi, 9872, 9872a., (type). In forests near Mengzi at 1,800 to 2,000 metres, 10531 (paratype), 11471 (paratype).

A scandent shrub to 7 metres tall, widely distributed in China.

Celastrus glaucophyllus Rehder & E. H. Wilson (CELASTRACEAE) New species

Ding Hou in Ann. Missouri Bot. Garden 42: 253. (1955).

Yunnan: Mengzi, 9679.

Discovered by Henry in 1896, Rehder & E. H. Wilson based this species on Wilson's 952 from Baoxing in western Sichuan (collected in June 1908).

Celastrus hindsii Benth. (CELASTRACEAE)

Henry in Journ. China Br. Roy. Asiat. Soc. 22: 259. (1887) (Chinese Names of Plants), Rehder & E. H. Wilson in Sargent Pl. Wilson. 2: 357. (1916), Loesener in Bot. Jahrb. 30: 466. (1932), Ding Hou in Ann. Missouri Bot. Gard. 42: 249. (1955), Zheng in Journ. Wuhan Bot. Res. 2.(1): 122. (1984).

Celastrus hindsii Benth. var. *henryi* Loesener in Bot. Jahrb. 29: 444. (1900); 30: 467. (1902).

Hubei: Antelope Glen, 3495 (paratype of *Celastrus hindsii* Benth. var. *henryi* Loesener). Yichang 3495a., (paratype of *Celastrus hindsii* Benth. var. *henryi* Loesener), 3241 (lectotype of *Celastrus hindsii* Benth. var. *henryi* Loesener). Liantuo, 3856. **Hainan:** Without locality, 8556. **Yunnan:** In mountains near Mengzi at 2,300 metres, 10559. Simao, 11972c.

A scandent shrub to 19 metres tall, widely distributed from India to Myanmar, China, Indonesia, Indochina and Thailand. In the Three Gorges region Henry stated it is colloquially known as *mien t'eng* and went on to say that the fruits of this species were eaten by wild antelope in the Three Gorges region.

Celastrus hookeri Prain (CELASTRACEAE)

Zheng in Journ. Wuhan Bot. Res. 2.(1): 122. (1984).

Celastrus stylosus Ding Hou non Wallich in Ann. Missouri Bot. Gard. 42: 272. (1955).

Sichuan: South Wushan, 5559.

Celastrus hypoleucus (Oliv.) Warburg apud Loesener (CELASTRACEAE) New species

in Bot. Jahrb. 29: 445. (1900), Rehder & E. H. Wilson in Sargent Pl. Wilson 2: 346. (1916), Ding Hou in Ann. Missouri Bot. Gard. 42: 257. (1955), Zheng in Journ. Wuhan Bot. Res. 2.(1): 122. (1984), Nelson in A Heritage of Beauty 311. (2000), O' Brien in Augustine Henry, An Irish Plant Collector in China 15. (2002).

Erythospermum hypoleucum Oliver in Hooker's Icon. Pl. 9: t. 1899. (1889). *Celastrus hypoglaucus* Hemsley in Ann. Bot. 9: 151. (1895), Bretschneider in Hist. Eur. Bot. Disc. China 2: 781. (1898), J. H. Veitch in Hortus Veitchii 359. (1906). *Celastrus hypoleucus* (Oliv.) Warb. apud Loes. f. *angustior* Loes in Bot. Jahrb. xxx. 445. (1902).

Hubei: South Badong, 5887a., 2837 (type of *Celastrus hypoglaucus* Hemsley). Fang at 1,900 to 3,000 metres, 6811, 6771 [type of *Celastrus hypoleucus* (Oliv.) Warb. apud. Loes. forma *angustior* Loesener]. **Sichuan:** South Wushan, 5887 (type of *Erythrospermum hypoleucum* Oliver). **Hunan:** Shimen, 7553 (isotype of *Erythrospermum hypoleucum* Oliver).

A scandent shrub to 5 metres tall. A common species in the mountains of north-west Hubei where according to Wilson the long terminal pendant cymose racemes of long-stalked orange-yellow fruits are strikingly beautiful in autumn. Introduced to cultivation by Wilson. Cultivated at the National Botanic Gardens, Glasnevin.

Celastrus kusanoi Hayata (CELASTRACEAE) New species

Rehder & E. H. Wilson in Sargent Pl. Wilson. 2: 356. (1916), Ding Hou in Ann. Missouri Bot. Gard. 42: 268. (1955), Li in Woody Flora of Taiwan 469. (1963).

Celastrus articulatus Thunb. Henry in Trans. Asiat. Soc. Jap. 24 suppl: 27. (1896) (List Pl. Formosa).

Taiwan: Wanshoushan, 1893.

A scandent shrub to 18 metres tall, native to southern China, including Hainan and Taiwan, it was first collected in Taiwan by Henry in 1893.

Celastrus monospermus Roxb. (CELASTRACEAE)

Loesener in Bot. Jahrb. 30: 467. (1930), Rehder & E. H. Wilson in Sargent Pl. Wilson. 2: 357. (1916), Ding Hou in Ann. Missouri Bot. Gard. 42: 244. (1955).

Yunnan: In forests on the Daweishan Range in Pingbian at 1,500 to 1,600 metres, 10446. In forests on the Daweishan Range in Pingbian at 1,640 metres, 10955. In forests on the Daweishan Range in Pingbian at 1,600 metres, 11399. Mountains to the west of Simao at 1,500 to 1,600 metres, 11972, 11972b.

A scandent shrub to 10 metres tall, widely distributed through Pakistan, India, Myanmar, Indochina and China.

Celastrus orbiculatus Thunb. (CELASTRACEAE)

Ding Hou in Ann. Missouri Bot. Gard. 42: 260. (1955).

Celastrus articulatus Thunb. Forbes & Hemsley in Journ. Linn. Soc. Bot. 23: 122. (1886), Henry in Journ. China Br. Roy. Asiat. Soc. 22: 245. (1887) (Chinese Names of Plants). *Celastrus orbicularis* Thunb. f. *microphylla* Loesener in Engl. Bot. Jahrb. 30: 469. (1901).

Hubei: Yichang, 456, 588, 1386, 1568, 2962. Liantuo, 3827 (type of *Celastrus orbicularis* Thunb. f. *microphylla* Loesener). In a ravine on Moji Shan, 3563.

A scandent shrub to 12 metres tall, widely distributed in northern and central Japan, Korea, the Russian Far East and China. A beautiful plant introduced to cultivation by Charles Sprague Sargent who sent seeds to Kew in 1870. Bretschneider later sent seeds from the region of Beijing in 1883. Henry stated that at Yichang it was known colloquially as *huang tou pan*.

Celastrus paniculatus Willdenow (CELASTRACEAE)

Ding Hou in Ann. Missouri Bot. Gard. 42: 229. (1955).

Yunnan: Simao, 11572.

A scandent shrub to 6 metres tall, widely distributed in India, Myanmar, Thailand, Indochina, China, Malaya, Indonesia and the Philippines.

Celastrus paniculatus Willdenow. subsp. *multiflorus* (Roxb.) D. Hou (CELASTRACEAE)

Ding Hou in Ann. Missouri Bot. Gard. 42: 231. (1955).

Celastrus paniculatus Rehder & E. H. Wilson non Willd. in Sargent Pl. Wilson. 2: 355. (1916). *Celastrus dependens* Rehder & E. H. Wilson in Sargent Pl. Wilson. 2: 355. (1916).

Yunnan: In forests near Simao at 1,500 to 1,600 metres, 11993, 11993a., 12122, 12122a., 12122b, 12572, 12572b. Simao, 11972a.

Celastrus rosthornianus Loes. (CELASTRACEAE) New species

Rehder & E. H. Wilson in Sargent Pl. Wilson. 2: 351. (1916), Ding Hou in Ann. Missouri Bot. Gard. 42: 265. (1955), Zheng in Journ. Wuhan Bot. Res. 2.(1): 122. (1984).

Celastrus articulatus Forbes & Hemsley non Thunb. in Journ. Linn. Soc. Bot. 23: 122. (1886).

Hubei: Antelope Glen, 3115. **Sichuan:** South Wushan, 5640, 5734.

A deciduous scandent shrub to about 7 metres tall, in autumn bearing heavy crops of orange-yellow fruits. Discovered by Henry in 1887, it was described from material later collected by A. von Rosthorn. In Emil Bretschneider's *A History of European Botanical Discoveries in China* 2: 1093 (1898) Bretschneider (from information supplied by Henry) stated that von Rosthorn was formerly employed by the Chinese Imperial Maratime Customs Service and he made his collection of plants through Chinese collectors in Sichuan. By 1898 von Rosthorn was based in Beijing (then Peking), where he was attached to the Austrian Legation. In a letter to Sir William Thistleton-Dyer, director of the Royal Botanic Gardens, Kew (2nd March 1895) Henry stated he had been corrisponding with von Rosthorn (who was then based in Shanghai) and from him learned that he was based in Sichuan during 1891 and 1892. During this time, von Rosthorn and a Mr. Bock, the Consul General for Sweden and Norway, employed Chinese collectors who worked on the Daba Shan range separating Hubei from Sichuan (near where Henry himself collected), Nanchuan to the south-east of Chongqing and the region of the upper Min river in western Sichuan. von Rosthorn sent his collections to Professor Baron Ettingshausen in Austria, a total of about 149 species, though a larger collection was sent to the University of Christiana in Mr. Bock's native Norway (it is this collection that is mentioned in the *Gardeners' Chronicle* ser. 3, 15: 177. 1894). *Celastrus rosthornianus* commemorates this little known collector and was introduced to cultivation by Wilson in 1910 through the Arnold Arboretum. It is cultivated at the National Botanic Gardens, Glasnevin.

Celastrus rosthornianus Loes. var. *loeseneri* (Rehder & E. H. Wilson) C. Y. Wu & Y. C. Ho (CELASTRACEAE) New variety

Celastrus loeseneri Rehder & E. H. Wilson in Sargent Pl. Wilson 2: 350. (1916), O' Brien in Augustine Henry, An Irish Plant Collector in China 16. (2002).

Hubei: Yichang, 5986a. Jianshi, 5909, 5986.

Celastrus spp. (CELASTRACEAE)

Anonymous in Kew List of Determinations (volume 24 & 34).

Hubei: Badong, 1734. **Yunnan:** Mengzi, 10615. Simao 11662, 11925.

Celastrus stylosus Wall. (CELASTRACEAE)

Loesener in Bot. Jahrb. 30: 469. (1902), Ding Hou in Ann. Missouri Bot. Gard. 42: 272. (1955).

Yunnan: Mengzi at 2,000 metres, 9679b. Mengzi, 10522, 11267. Simao, 13089.

A scandent shrub to 4 metres tall, native to China and Indochina.

Celastrus vaniotii (H. Lévl.) Rehd. (CELASTRACEAE)

Ding Hou in Ann. Missouri Bot. Gard. 42: 256. (1955), Zheng in Journ. Wuhan Bot. Res. 2.(1): 123. (1984).

Celastrus latifolius Hemsley non. Hemsl. in Journ. Linn. Soc. Bot. 23: 123. (1886). *Celastrus spiciformis* Rehder & E. H. Wilson in Sargent Pl. Wilson 2: 348. (1916). *Celastrus spiciformis* Rehd. & E. H. Wils. var. *laevis* Rehder & E. H. Wilson in Sargent Pl. Wilson. 2: 349. (1916).

Hubei: Jianshi, 5935 (May 1888). **Yunnan:** In mountains to the north of Mengzi at 2,300 metres, 11006.

According to W. J. Bean this species is not in cultivation, he also goes on to state it was discovered by Wilson, however Henry's specimen dates from 1888.

Celosia argentea L. (AMARANTHACEAE)

Henry in Journ. China Br. Roy. Asiat. Soc. 22: 236. (1887) (Chinese Names of Plants), Forbes & Hemsley in Journ. Linn. Soc. Bot. 26: 318. (1891), Henry in Trans. Asiat. Soc. Jap. 24 suppl. 75. (1896) (List Pl. Formosa).

Hubei: Yichang, 4, 2336. **Hainan:** Environs of Kiungchow, 8237. At Nodoa in the interior of the island, 8454. Environs of Kiungchow, 8696. **Taiwan:** Kaohsiung, without number. Wanchin, 866.

The silver cockscomb, known colloquially at Yichang as *chi kung hua* (*qing xian zi*) according to Henry. A handsome annual herb to 60 cm tall bearing silvery-white plume like flowers, widely spread in tropical and subtropical regions of Africa and Asia. The seeds of this species are used with the seeds of *Senna tora* and the flowers of *Buddleja officinalis* to treat cases of impaired or blurred vision. This species is currently cultivated in the order beds at the National Botanic Gardens, Glasnevin from material collected by the Glasnevin Central China Expedition (GCCE 610) near Henji town in Xingshan.

Celtis biondii Pamp. (CANNABACEAE) New species

Schneider in Sargent Pl. Wilson. 3: 272. (1917), Zheng in Journ. Wuhan Bot. Res.2.(1): 29. (1984), Bean in Trees & Shrubs 1: 568. (1989).

Celtis sinensis Forbes & Hemsley non Pers. in Journ. Linn. Soc. Bot. 26: 450. (1894). *Celtis bungeana* Forbes & Hemsley non Blume in Journ. Linn. Soc. Bot. 26: 449. (1894). *Celtis biondii* Pamp. var. *cavaleriei* (Lev.) Schneid. Schneider in Sargent Pl. Wilson. 3: 273. (1917). *Celtis leveillei* Nakai. var. *holophylla* Nakai in Journ. Arnold. Arb. 5: 74. (1924).

Hubei: On the summit of Moji Shan, at 700 metres alt., 3100 (February 1887). Badong, 3150 (syntype of *Celtis leveillei* Nakai. var. *holophylla* Nakai). Changyang, 5276. **Sichuan:** South Wushan, 5735. **Yunnan:** Milê district, 9938 (collected November 1st 1896). Pu'er fu at 1,300 metres, 13509.

A deciduous tree to 15 metres tall. Pampanini based this species on specimens gathered in Hubei by Silvestri in 1907, Henry had discovered it about 1887. It is a relatively common tree in Hubei and eastern Sichuan where Henry stated it was known as *p'o shu*. Introduced to cultivation in 1902 by Wilson.

Celtis bungeana Blume (CANNABACEAE)

Forbes & Hemsley in Journ. Linn. Soc. Bot. 26: 449.

(1894), Schneider in Sargent Pl. Wilson. 3: 269. (1917), Zheng in Journ. Wuhan Bot. Res. 2.(1): 29. (1984), Bean in Trees & Shrubs 1: 568. (1989).

Celtis sinensis Forbes & Hemsley non Pers. in Journ. Linn. Soc. Bot. 26: 450. (1894). *Celtis amphibola* Schneider in Sargent Pl. Wilson. 3: 279. (1917).

Hubei: Yichang, 4214. Xingshan, 6483. **Yunnan:** Near Mengzi in a wood in the great gully at 1,500 metres, 9323 (in part, type *Celtis amphibola* Schneider, collected October 18th 1896).

A tree to 15 metres tall with trunks from 1 to 2 metres in girth. Introduced to cultivation through the Royal Botanic Gardens, Kew from seeds sent there by Bretschneider from the hills north of Beijing. It is extremely rare in cultivation, the tallest known tree grows at Kew and was 14 metres tall with a girth of 43 cm in 1991.

Celtis julianae Schneider (CANNABACEAE) New species

Zheng in Journ. Wuhan Bot. Res. 2.(1): 29. (1984).

Hubei: Changyang, 7755.

A handsome tree to 26 metres tall with a girth of 3 metres. Discovered by Henry in April of 1888, it was later collected by Wilson. The specific epithet commemorates Mrs. C. K. Schneider.

Celtis labilis Schneider (CANNABACEAE) New species

in Sargent Pl. Wilson. 3: 267. (1917), Zheng in Journ. Wuhan Bot. Res. 2.(1): 29. (1984).

Celtis sinensis Forbes & Hemsley non Pers. in in Journ. Linn. Soc. Bot. 26: 450. (1894).

Hubei: Yichang, 3404 (October 1887), 7866 (March 1888).

A tree to 16 metres tall with girths of 1 to 2 metres in its native habitat. Discovered by Henry in October 1887 and introduced to cultivation by Wilson through the Arnold Arboretum twenty years later. Wilson's 444 from Changyang reached Glasnevin in the spring of 1908 (as it did to Kew) from the Arnold Arboretum. At Borde Hill in Sussex a tree planted in 1934 was 8 metres tall with a trunk diameter of 37 cm when measured in 1995.

Celtis philippensis Blanco (CANNABACEAE)

Forbes & Hemsley in Journ. Linn. Soc. Bot. 26: 450. (1894).

Hainan: Without locality, 8263.

An evergreen tree, widespread across tropical Africa, Madagascar, India, Myanmar, Indochina, China (Hainan), south through Malesia, north-east Australia and the Solomon Islands.

Celtis sinensis Pers. (CANNABACEAE)

Henry in Journ. China Br. Roy. Asiat. Soc. 22: 267. (1887) (Chinese Names of Plants); in Notes Econ. Bot.

China 27. (1893), (Forbes & Hemsley in Journ. Linn. Soc. Bot. 26: 450. (1894), Henry in Trans. Asiat. Soc. Jap. 24 suppl: 85. (1896) (List Pl. Formosa), Li in Woody Flora of Taiwan 107. (1963).

Celtis bodinieri Leville Schneider in Sargent Pl. Wilson. 3: 276. (1917). *Celtis cercidifolia* Schneider in Sarg. Pl. Wilson. 3: 276. (1917), Zheng in Journ. Wuhan Bot. Res. 2.(1): 29. (1984). *Celtis nervosa* Hemsl. Henry in Trans. Asiat. Soc. Jap. 24 suppl: 85. (1896) (List Pl. Formosa), Schneider in Sargent Pl. Wilson. 3: 280. (1917), Walsh in Aug. Henry Forst. Herb. 73. (1957).

Hubei: Yichang, 2262 (type of *Celtis cercidifolia* Schneider), 7872. Liantuo, 2997, 3851, 4489, 4528. **Hainan:** Without locality, 8573. **Taiwan:** Oluanpi, 1892. Kaohsiung, 2035. **Yunnan:** In Yulo mountains to the south of Simao at 1,600 metres, 12881.

According to Henry this tree was called the *as p'o shu* at Badong and *ch'ing* at Yichang. It is a fast growing species, a tree in the Kew collection raised from seed sown in 1976 measured 11 metres tall with a trunk diameter of 27 cm at 1 metre above ground level when measured in 2001.

Celtis spp. (CANNABACEAE)

Anonymous in Kew List of Determinations (volume 24 & 34).

Hubei: Liantuo, 4575. **Yunnan:** Simao, 13535.

Celtis tetrandra Roxb. (CANNABACEAE)

Celtis sinensis Henry non Pers. in Trans. Asiat. Soc. Jap. 24 suppl: 85. (1896) (List Pl. Formosa). *Celtis salvatiana* Schneider in Sargent Pl. Wilson. 3: 279. (1916). *Celtis yunnanensis* Schneider in Sargent Pl. Wilson. 3: 279. (1917). *Celtis formosana* Hayata Schneider in Sargent Pl. Wilson. 3: 280. (1917), Li in Woody Flora of Taiwan 106. fig. 31. (1963).

Taiwan: Wanchin, 1615. **Yunnan:** On the bank of the Red River near Manpan, 10848 (type of *Celtis salvatiana* Schneider). Milê 9323a., (type of *Celtis yunnanensis* Schneider).

Distributed in the Himalaya, India, Bangladesh, China, Thailand and Vietnam.

Cenchrus alopecuroides (L.) Thunb. (POACEAE)

Pennisetum cenchroides Henry non Rich. in Journ. China Br. Roy. Asiat. Soc. 22: 254. (1887) (Chinese Names of Plants), *Pennisetum compressum* R. Br. Rendle in Journ. Linn. Soc. Bot. 36: 338. (1904).

Hubei: Yichang, 97, 2364. Badong, 2540. **Hainan:** Environs of Haikou, 7950. **Yunnan:** Mengzi at 1,400 metres, 10170. Mengzi at 1,500 metres, 10174.

A handsome perennial grass to about 80 cm tall, widespread and abundant across Myanmar, China, Japan, Malesia, Australia and Polynesia. An abundant species in

Hubei, several collections were made by the Glasnevin Central China Expedition in the autumn of 2004 including a particularly good black-flowered form that grew near the San You Dong Glen above Yichang (GCCE 722).

Cenchrus flaccidus (Griseb.) Morrone (POACEAE)

Pennisetum flaccidum Griseb. Rendle in Journ. Linn. Soc. Bot. 36: 339. (1904).

Yunnan: Mengzi, 9619, 9619a.

Centella asiatica (L.) Urban (APIACEAE)

Hydrocotyle asiatica L. Henry in Trans. Asiat. Soc. Jap. 24 suppl: 47. (1896) (List Pl. Formosa).

Taiwan: Wanchin, 453. Kaohsiung, 1918.

A perennial to 10 cm tall, widely distributed in tropical and subtropical regions.

Centipeda minima (L.) A. Br. & Asch. (ASTERACEAE)

Myriogyne minuta Less. Henry in Trans. Asiat. Soc. Jap. 24 suppl: 54. (1896) (List Pl. Formosa).

Taiwan: Oluanpi, 277. Kaohsiung Plain, 1199, 1833.

Centotheca lappacea (L.) Desv. (POACEAE)

Rendle in Journ. Linn. Soc. Bot. 36: 419. (1904).

Centhotheca lappacea Desv. var. *inermis* Rendle in Journ. Linn. Soc. Bot. 36: 420. (1904).

Taiwan: Wanchin, 99, 417 (type of *Centhotheca lappacea* Desv. var. *inermis* Rendle).

Widely distributed in the Northern Hemisphere.

Centranthera cochinchinensis (Lour.) Merr. (OROBANCHACEAE)

Centranthera brunoniana Thwaites Forbes & Hemsley in Journ. Linn. Soc. Bot. 26: 201. (1890).

Hubei: Yichang, 2245.

Centranthera grandiflora Benth. (OROBANCHACEAE)

Anonymous in Kew List of Determinations (volume 34).

Yunnan: Mengzi, 9019.

Distributed in the Himalaya, Myanmar, China and Vietnam.

Centranthera hispida R. Br. (OROBANCHACEAE)

Anonymous in Kew List of Determinations (volume 24 & 34).

Hubei: Yichang, 140, 2936. **Yunnan:** Simao, 12430.

These collections belong to an allied species.

Ceodes umbellifera J. R. Forst. & G. Forst. (NYCTAGINACEAE)

Pisonia brunoniana Henry non Endl. in Trans. Asiat. Soc. Jap. 24 suppl: 74. (1896) (List Pl. Formosa). *Pisonia umbellifera* (J. R. Forst. & G. Forst.) Seem. Li in Woody Flora of Taiwan 153. (1963).

Taiwan: Oluanpi, without number. Wanchin, 329.

A medium-sized tree to 14 metres tall, native to tropical Asia, Australia, Hawaii and Polynesia. Distributed in damp coastal forests of eastern and southern Taiwan, China.

02

Cephalanthera erecta (Thunb.) Blume (ORCHIDACEAE)

Diels in Bot. Jahrb. 29: 268. (1900), Rolfe in Journ. Linn. Soc. Bot. 36: 48. (1903).

Hubei: Badong, 1429, 3761, 5464. **Sichuan:** South Wushan, 5464a.

Cephalanthera falcata (Thunb.) Blume (ORCHIDACEAE)

Rolfe in Journ. Linn. Soc. Bot. 36: 48. (1903).

Sichuan: South Wushan, 5621.

Cephalanthus occidentalis L. (RUBIACEAE)

Anonymous in Kew List of Determinations (volume 24).

Hainan: Environs of Haikou, 8013. Ling-men (in the interior of the island), 8601.

Native to the USA and Mexico, naturalised in warmer parts of Asia.

Cephalostachyum chinense (Rend.) D. Z. Li & H. Q. Yang (POACEAE)

in Acta Phytotax. Sin. 19(2): 212. (1981).

Schizostachyum chinensis Rendle in Journ. Linn. Soc. Bot. 36: 448. (1904), McClure in Lignan Sci. Journ. 14: 596. fig. 38. (1935).

Yunnan: On the Daweishan Range in Pingbian in forests on cliffs at 2,600 metres, 10420 (type of *Schizostachyum chinensis* Rendle).

Endemic to Yunnan where it grows in evergreen broad-leaved forest between 1,500 and 2,600 metres. The culms of this bamboo are popular in weaving and the native people of south-east Yunnan use it to make *gao-sheng*, a rocket used in celebration festivals. The province of Yunnan is the centre of disibution for bamboo species.

Cephalotaxus fortunei Hook. (CEPHALOTAXACEAE)

Henry in Journ. China Br. Roy. Asiat. Soc. 22: 269. (1887) (Chinese Names of Plants); in Notes Econ. Bot. China 36. (1893), Masters in Journ. Linn. Soc. Bot. 26: 545. (1902), Pilger in Engler, Pflanzenr. 5: 103. (1903), Henry in Elwes and Henry, The Trees Gt. Brit. & Irel. 6: 1470. (1912), Rehder & E. H. Wilson in Sargent Pl. Wilson 2: 4. (1916), E. H. Wilson in Journ. Arnold Arb. 7: 39. (1926), Zheng in Journ. Wuhan Bot. Res. 2.(1): 8. (1984).

Hubei: Liantuo, 1925, 3879, 7186. On the banks of the Yangtze at Badong, 7172. **Sichuan:** North Wushan, 7018. **Yunnan:** Mountains to the north of Mengzi at 2,300 metres, 9100.

According to Wilson this was the *lo-han-shu* of the Chinese (Henry called it the *san chien sha*) and is frequently associated with wayside shrines, tombs and temples. It makes a small tree to 10 metres tall in the Three Gorges region and is most common on limestone cliffs. *Cephalotaxus fortunei* was introduced to cultivation by Robert Fortune in 1849. The finest specimens in the British

Isles and Ireland are found in the Irish province of Leinster, one of these, a remarkable tree, grows at Belvedere House in County Westmeath and when measured in 2000 was 11 metres tall with a trunk diameter of 28 cm. Another tree at Knocklofty House in County Tipperary measured 7.5 metres tall with a trunk diameter of 77 cm in 2000.

Cephalotaxus oliveri Masters (CEPHALOTAXACEAE)

in Bull. Herb. Boiss. 6: 270. (1898), Masters in Journ. Linn. Soc. Bot. 26: 545. (1902); in Gard. Chron. ser. 3, xxxiii. 227. fig. 93. (1903); in Journ. Bot. xli. 269. (1903) Pritzel in Bot. Jahrb. xxix. 214. (1900), Pilger in Engler, Pflanzenr. iv.-5. 104. (Taxaceae) (1903), Patachke in Bot. Jahrb. xlviii. 629. (1913, Rehder & E. H. Wilson in Sargent Pl. Wilson. 1: 6. (1913), Fu in acta Phytotax. Sin. 22: 279. (1984).

Hubei: Changleping, 7832, 7833 (syntype). Changyang, 7479. Without locality, 7832a.

A striking evergreen tree, discovered by Faber on Emei Shan (Mount Omei) in Sichuan in the autumn of 1887, it was found again in April of 1888 by Henry. Dr. Masters used Henry's material in the description of this new tree. Introduced to cultivation by Wilson in 1900 through Veitch's Coombe Wood nursery.

Cephalotaxus sinensis (Rehder & E. H. Wilson) H. L. Li (CEPHALOTAXACEAE)

Zheng in Journ. Wuhan Bot. Res. 2.(1): 8. (1984).

Cephalotaxus drupacea Masters non Sieb. & Zucc. in Journ. Linn. Soc. Bot. 26: 545. (1902), Pilger in Engler, Pflanzenr. 5: 100. (1903), Henry in Elwes and Henry, The Trees Gt. Brit. & Irel. 6: 1469. (1912), Rehder & E. H. Wilson in Sargent Pl. Wilson. 2: 3. (1916), Walsh in Aug. Henry Forst. Herb. 111. (1957).

Hubei: Yichang, 107, 5030. Changleping, 7831.

Henry called this species the 'cow's-tail pine', he also stated it grew to a large size with trees sometimes gaining a circumference of 20 feet. In central China the wood of this tree was used in the manufacture of coffins. When collecting seeds of *Cephalotaxus fortunei* in 1849, Robert Fortune's Chinese collectors mixed these two species, though when seedlings were raised they were noticed to be distinct.

Ceratopteris thalictroides (L.) Brongn. (PTERIDACEAE)

Henry in Trans. Asiat. Soc. Jap. 24 suppl: 111. (1896) (List Pl. Formosa).

Taiwan: Oluanpi, 270. Kaohsiung Plain, 1924.

A widespread fern in the tropics of southeast Asia, the Pacific Islands, Madagascar, the West Indies, Central and South America. Common in paddy fields and marshes.

Ceratosanthes palmata (L.) Urb. (CUCURBITACEAE)

Trichosanthes palmata L. Anonymous in Kew List of Determinations (volume 34).

02

Yunnan: Mengzi, 10787, 11062.

A New World species, Henry's collection belongs elsewhere.

Ceratostigma griffithii C. B. Clarke (PLUMBAGINACEAE)

E. H. Wilson in Sargent Pl. Wilson. 2: 586. (1916).

Yunnan: In ravines near Mengzi at 1,500 metres, 9586.

A much branched shrub to about 60 cm tall bearing terminal clusters of bright blue flower in August. Native to the eastern Himalaya and Yunnan province. It was introduced from Yunnan by Forrest but has proved tender in Britain and Ireland.

Cerbera manghas L. (APOCYNACEAE)

Li in Woody Flora of Taiwan 784. fig. 315. (1963).

Cerbera odollam Henry non Gaertn. in Trans. Asiat. Soc. Jap. 24 suppl: 59. (1896) (List Pl. Formosa).

Hainan: On Liang Shan, 16 km (10 miles) east of Haikou, 8205. **Taiwan:** Oluanpi, 342.

A tropical tree to about 20 metres tall (though usually smaller), native to the coastal regions of Polynesia westwards to south-east Asia and the eastern shores of Africa. The delightfully fragrant white blooms are carried almost all year round and are followed by highly toxic, oily rounded fruits. The tree's poisonous nature is mirrored in the generic epithet *Cerbera* which is derived from Cerberus, the name of a three-headed dog from Greek mythology that guarded the entrance to Hades, the underworld. The fruits are lightweight and waterproof and are distributed on ocean currents. A sea-shore tree in Chinese Taiwan, mainly on the southern part of the island.

Cercis chinensis Bunge (FABACEAE)

Henry in Journ. China Br. Roy. Asiat. Soc. 22: 256. (1887) (Chinese Names of Plants), Forbes & Hemsley in Journ. Linn. Soc. Bot. 23: 213. (1887), Zheng in Journ. Wuhan Bot. Res. 2.(1): 97. (1984).

Hubei: Yichang, 1027, 1239. Badong, 3715. Liantuo, 1951, 7662. North Badong, 7662a. **Sichuan:** North Wushan, 7662b.

The Chinese redbud, in a note accompanying Dr. Henry's specimens he stated that this was a valuable timber tree, fifty feet high, with a trunk sometimes twelve feet in girth. In Henry's time in western Hubei this was a very common tree. The flowers vary from peach coloured, to pale pink, to red-pink. The colloquial name *lo-chiang shu* signifies that the flowers have the appearance of an open basket. The largest tree in the genus, according to Bean it needs a warmer summer than we normally get in the British Isles and Ireland to flower successfully. Collected by the Glasnevin Central China Expedition (GCCE 865) near the Dragon Gate River Forest Park in Xingshan, Hubei in October 2004.

Cercis racemosa Oliver (FABACEAE) New species

in Hooker's Icon. Pl. 19: t. 1894. (1889), Bretschneider in Hist. Eur. Bot. Disc. China 2: 782. (1898), Schneider in Illus. Handb. Laubholzk. 2: 4. (1912), Bean in Trees & Shrubs 1: 335. (1921), Nelson in Pim, The Wood & the Trees 222. (1984), Nelson in A Heritage of Beauty 311. (2000), O' Brien in Augustine Henry, An Irish Plant Collector in China 16. (2002).

Sichuan: South Wushan at Ma-huang-p'o at 1,800 metres, 5602 (type). **Hubei:** Fang Xian, without number.

According to Wilson this species was common in the woods of central Fang Xian, but was rare in western Hubei. That great authority on temperate woody plants, W. J. Bean stated that "Wilson told me that this is one of the very best and most beautiful flowering trees he has introduced". It is a low growing tree, sometimes reaching 12 metres tall with trunks from 1 to 2 metres in diameter. The flowers are silvery-rose coloured, are carried on naked wood and produced in great profusion. The species rarely fruits and Wilson stated that its introduction to cultivation was only accomplished in 1907 by his third expedition to China. (His first two expeditions were for the nurserymen, Messrs Veitch, the third and fourth were for the Arnold Arboretum of Harvard University) In Ireland this tree is cultivated at the National Botanic Gardens, Glasnevin, the John F. Kennedy Arboretum and Mount Congreve. The tallest known specimen in cultivation grows at the Hillier Arboretum, Hampshire. When measured in 1997 this tree was 12 metres tall with a trunk diameter of 30 cm.

Ceropegia dolichophylla Schlecter (APOCYNACEAE) New species

in Notes Roy. Bot. Gard. Edinburgh 8: 17. (1913—1915).

Yunnan: Mengzi at 1,950 metres, 9490.

A scandent plant to 0.5 metres tall, the flowers are deep purplish-maroon and green. According to George Forrest who collected this species on the Cangshan Range, the fleshy roots were eaten by hill tribes in the area. Distributed from north-east India to China (Guangxi and Yunnan).

Ceropegia dryophila Schneider (APOCYNACEAE) New species

in Sargent Pl. Wilson. 3: 349. (1916).

Hubei: Changyang, 6222.

A low climber to 1.25 metres tall with dark red flowers. Wilson later collected this species at Badong.

Ceropegia salicifolia H. Huber (APOCYNACEAE) New species

in Mem. Soc. Brot. 12: 51. (1957).

Yunnan: In forests to the north-west of Simao at 1,650 metres, 9729 (isotype).

Chaenomeles cathayensis (Hemsl.) C. K. Schneider (ROSACEAE) New species

in Illus. Handb. Laubholzk. 1: 730. fig. 405. (1906), Bean in Trees & Shrubs 1: 452. (1921), Nelson in A Heritage of Beauty 311. (2000).

Pyrus cathayensis Hemsley in Journ. Linn. Soc. Bot. 23: 256. (1887), Henry in Journ. China Br. Roy. Asiat. Soc. 22: 259. (1887) (Chinese Names of Plants), *Cydonia cathayensis* (Hemsl.) Hemsley in Hooker's Icon. Pl. xxvii. t. 2657—2658. (1900).

Hubei: Liantuo 1916. Yichang, 3396. Changyang, 5263 [type of *Pyrus cathayensis* Hemsley, lectotype of *Chaenomeles cathayensis* (Hemsl.) Schneid.].

A deciduous open-branched shrub to 5 metres tall, carrying in February clusters of two to three pale pink flowers on short spurred growths followed by sweetly fragrant egg-shaped fruits to 15 cm long. Henry referred to this species as the 'Chinese quince' and stated it was known colloquially in the Three Gorges region as *mu kua*. Though described as a new species from his herbarium matrial it was already in cultivation in Europe at the time.

Chaenomeles japonica (Thunb.) Lindl. ex Spach. (ROSACEAE)

Pyrus japonica Thunb. Anonymous in Kew List of Determinations (volume 24). *Cydonia japonica* (Thunb.) Pers. Anonymous in Kew List of Determinations (volume 34).

Hubei: Badong, 550 (September 1885). **Yunnan:** Mengzi, 10730.

Introduced, native to Japan.

Chaenomeles speciosa (Sweet) Nakai (ROSACEAE)

Chaenomeles lagenaria (Loisel) Koidz. Rehder in Sargent Pl. Wilson. 2: 269. (1916).

Hubei: Changyang, 5249.

A widespreading deciduous shrub to 3 metres tall, though usually only seen half that height in cultivation. Flowers scarlet to blood-red followed by fragrant apple or pear-shaped fruits. A native of China and Myanmar, this beautiful shrub has been cultivated in Japan for centuries and was introduced to cultivation by Sir Joseph Banks through the Royal Botanic Gardens, Kew in 1796. *Chaenomeles speciosa* 'Phylis Moore' is an especially attractive selection with double salmon-rose flowers. It was raised at Knap Hill nursery in the UK and named for the wife of Sir Frederick Moore, one time Keeper of the National (formerly Royal) Botanic Gardens, Glasnevin.

Chaetomitrium sp. (SYMPHYODONTACEAE)

Anonymous in Kew List of Determinations (volume 24).

Hubei: South Badong, 6173. (Moss).

Chamabainia cuspidata Wight (URTICACEAE)

Forbes & Hemsley in Journ. Linn. Soc. Bot. 26: 489. (1899).

Yunnan: Mengzi, 10208.

A polymorphic species distributed across India, Bhutan, China and southeast Asia.

Chamaecrista mimosoides (L.) Greene (FABACEAE)

Cassia mimosoides L. Forbes & Hemsley in Journ. Linn. Soc. Bot. 23: 211. (1887), Henry in Trans. Asiat. Soc. Jap. 24 suppl: 38. (1896) (List Pl. Formosa).

Hubei: Yichang, 666, 2285. In glens near Yichang, 1644. Liantuo, 2078. **Hainan:** Environs of Haikou, 8318. Ling-men (in the interior of the island), 8637. **Taiwan:** Oluanpi, 1376. Wanchin, 1548. **Yunnan:** Mengzi, 10760.

Almost universal in tropical countries and extending into some temperate regions.

Champereia manillana (Blume) Merr. (SANTALACEAE)

Li in Woody Flora of Taiwan 142. (1963).

Champereia griffithiana Henry non Planch. in Trans. Asiat. Soc. Jap. 24 suppl: 81. (1896) (List Pl. Formosa).

Taiwan: Kaohsiung, without locality. Wanchin, 129, 141. Oluanpi, 303.

A small evergreen tree to 7 metres tall, native to Malaysia. In Chinese Taiwan this species is distributed in coastal forests on the southern part of the island. Known coloquially as *shan-kan* according to Henry. *Champereia manillana* was collected by the Glasnevin Taiwan Expedition (GTE 97) on Wanshoushan (Ape's Hill of Henry and Wilson) in October 2004.

Chara vulgaris L. (CHARACEAE)

Anonymous in Kew List of Determinations (volume 24).

Hubei: Yichang, 7878.

Chassalia curviflora (Wall.) Thwaites (RUBIACEAE)

Hutchinson in Sargent Pl. Wilson. 3: 416. (1917).

Yunnan: Near Manpan in the Red River Valley at 830 metres, 10691. In Pingbian in forests on the Daweishan Range at 1,640 metres, 10758. Near Simao, 11966b., 11966c., 11966d., 11966e., 11966f., 12283.

Cheirostylis chinensis Rolfe (ORCHIDACEAE) New species

in Ann. of Bot. 9: 158. (1895), Henry in Trans. Asiat. Soc. Jap. 24 suppl: 92. (1896) (List Pl. Formosa), Bretschneider in Hist. Eur. Bot. Disc. China 2: 792. (1898).

Taiwan: Summit of Wanshoushan, 320 (syntype).

A small terrestrial orchid, native to China, Indochina and the Philippines.

Cheirostylis yunnanensis Rolfe (ORCHIDACEAE)

Rolfe in Kew Bull. Misc. Inf. 201. (1896), Rolfe in Journ. Linn. Soc. Bot. 36: 43. (1903).

Yunnan: In woods near Mengzi at 1,500 metres, on stones, 10423.

Discovered by the Irish plant collector and Imperial Maritime Customs officer, William Hancock near Mengzi in 1894, recollected by Henry two years later. Native from Sikkim to China and Vietnam.

Chelidonium majus L. (PAPAVERACEAE)

Anonymous in Kew List of Determinations (volume 24).

Hubei: On the banks of the Yangtze at Badong, 7177.

Greater celandine, a short lived perennial, though self-seeding widely and valuable in dry situations in the wooodland garden where other perennials would fail. Native to Europe, Asia and North Africa and naturalised in North America. The yellow sap has been traditionally used to treat warts and ringworm.

Chelonopsis deflexa (Benth.) Diels (LAMIACEAE)

in Engl. Bot. Jahrb. xxix. 554. (1900).

Chelonopsis benthamiana Hemsl. Forbes & Hemsley in Journ. Linn. Soc. Bot. 26: 298. (1890).

Hubei: Jianshi, 7325.

Chengiodendron marginatum (Champ ex Benth.) C. B. Shang, X. R. Wang, Yi F. Duang & Yong F. Li (OLEACEAE)

Linociera sp. Henry in Trans. Asiat. Soc. Jap. 24 suppl: 59. (1896) (List Pl. Formosa). *Osmanthus marginatus* (Champ ex Benth.) Hemsl Green in Notes Roy. Bot. Gard. Edinburgh 22: 470. (1958), Li in woody Flora of Taiwan 769. (1963).

Taiwan: Oluanpi, 1252.

An evergreen shrub or small tree to 10 metres tall, widely spread from southern China to Japan (south).

Chengiodendron matsumuranum (Hayata) C. B. Shang, X. R. Wang, Yi F. Duang & Yong F. Li (OLEACEAE) New species

Linociera sp. Henry in Trans. Asiat. Soc. Jap. 24 suppl: 59. (1896) (List Pl. Formosa). *Osmanthus matsumuranus* Hayata Green in Notes Roy. Bot. Gard. Edinburgh 22: 470. (1958), Li in Woody Flora of Taiwan 767. fig. 308. (1963).

Taiwan: Oluanpi, 824. Wanchin, 543, 1535, 1686.

A large evergreen shrub or a tree to 20 metres tall with a range from Khasia in India to southern China and Indochina. In Chinese Taiwan this species is distrributed in broad-leaved forest where it was first collected by Augustine Henry.

Cheniella tenuiflora (Wall. ex C. B. Clarke) R. Clark & Mackinder (FABACEAE)

Bauhinia glauca Henry non (Benth.) Wall. ex Benth. in Journ. China Br. Roy. Asiat. Soc. 22: 241. (1887) (Chinese Names of Plants), Forbes & Hemsley in Journ. Linn. Soc.

Bot. 23: 212. (1887), in part. *Bauhinia hupehana* Craib in Sargent Pl. Wilson. 2: 89. (1916), Zheng in Journ. Wuhan Bot. Res. 2.(1): 96. (1984). *Bauhinia caterviflora* H. Y. Chen in Journ. Arnold. Arb. 19: 129. (1938). *Bauhinia pernervosa* H. Y. Chen in Journ. Arnold. Arb. 29: 132. (1938).

Hubei: Badong, 706 (July 1885). Yichang, 2938, 2938a., 3551a. Antelope Glen, 3551. **Yunnan:** Mengzi, 10763. Mountains to the south of Simao at 1,300 metres, 12344 (type of *Bauhinia caterviflora* H. Y. Chen). In Pingbian on the Daweishan Range, 10763a. (collected 5th January 1898), 10763b., (type of *Bauhinia pernervosa* H. Y. Chen).

Diostributed from the eastern Himalaya to Indochina. A climbing species with fragrant flowers varying from white to rose-pink, collected by Henry in 1885 and later collected by Wilson. According to Henry this species was known colloquially in the Three Gorges region as *chueh yueh t'eng.*

Chenopodium acuminatum Willd. (AMARANTHACEAE)

Henry in Trans. Asiat. Soc. Jap. 24 suppl: 75. (1896) (List Pl. Formosa).

Taiwan: Kaohsiung, 1061.

An annual herb to 80 cm tall, distributed across European Russia to Japan and the Philippines.

Chenopodium album L. (AMARANTHACEAE)

Henry in Journ. China Br. Roy. Asiat. Soc. 22: 246. (1887) (Chinese Names of Plants), Forbes & Hemsley in Journ. Linn. Soc. Bot. 26: 323. (1891), Henry in Trans. Asiat. Soc. Jap. 24 suppl: 75. (1896) (List Pl. Formosa).

Hubei: Yichang, 56, 619, 2352. Liantuo, 3207. Badong, 952 (November 1885), 2381, 2531, 4909, 4947, 4956. **Taiwan:** Oluanpi, 629.

Fat hen, an almost cosmopolitan annual in tropical and temperate regions. According to Henry this species was known colloquially in the Three Gorges region as *hui t'ien han.* Once grown as a vegetable in eastern Europe.

Chenopodium ficifolium Sm. (AMARANTHACEAE)

Henry in Trans. Asiat. Soc. Jap. 24 suppl: 75. (1896) (List Pl. Formosa).

Taiwan: Kaohsiung, without locality. Oluanpi, 332.

Chenopodium sp. (AMARANTHACEAE)

Anonymous in Kew List of Determinations (volume 34).

Yunnan: Mengzi, 9322.

Chiloschista yunnanensis Schlecter (ORCHIDACEAE) New species

in Repert. Spec. Nov. Regni. Veg. Beih. 4: 74. (1919).

Yunnan: Simao, 11792a., (isotype).

Endemic to China and distributed across Sichuan, south-east Tibet (Xizang) and Yunnan.

Chimaphila japonica Miq. (ERICACEAE)

Anonymous in Kew List of Determinations (volume 24).

Hubei: Jianshi, 5906.

This species was collected by the Sino-American Expedition in the Shennongjia Forest District in 1980.

Chimonanthus nitens Oliver (CALYCANTHACEAE)
New species

in Hooker's. Icon. Pl. 16: t. 1600. (1887), Henry in Journ. China Br. Roy. Asiat. Soc. 22: 252. (1887) (Chinese Names of Plants), Bretschneider in Hist. Eur. Bot. Disc. China 2: 779. (1898), Diels in Bot. Jahrb. 29: 345. (1900), Morley in Glasra 3: 78. (1979), Bean in Trees & Shrubs 1: 605. (1989), Zheng in Journ. Wuhan Bot. Res. 2.(1): 48. (1984).

Meratia nitens (Oliv.) Rehder & E. H. Wilson in Sargent Pl. Wilson.1: 420. (1913).

Hubei: In the Hsiao Pingshanba Glen near Yichang (October 1886), 2915 (type). Antelope Glen, 3433. Xiling (Yichang) Gorge, 4387.

Henry discovered this species in the glens off the Yichang (Xiling) Gorge. He described it as a small evergreen shrub carrying white flowers in October. Distributed through Anhui, Guangxi, Fujian, Hubei, Hunan, Jiangsu, Jiangxi, Shaanxi, Yunnan and Zhejiang. It is cultivated at the Strybing Arboretum in Golden Gate Park, San Francisco (2004) and in a few British collections. It is a more slender shrub than *Chimonanthus praecox*.

Chimonanthus praecox (L.) Link. (CALYCANTHACEAE)

Zheng in Journ. Wuhan Bot. Res. 2.(1): 48. (1984).

Chimonanthus fragrans (Loisel.) Lindl. Henry in Journ. China Br. Roy. Asiat. Soc. 22: 252. (1887) (Chinese Names of Plants). *Meratia praecox* (L.) Rehder & E. H. Wilson in Sargent Pl. Wilson. 1: 419. (1913).

Hubei: In a ravine on the way towards Moji Shan, 3288. Yichang, 3565. Without locality, 3288a., 3288b.

Wintersweet, a deciduous shrub to 2 metres tall, producing intensely fragrant, pendant, almost transparent flowers from November to March. Common in Henry's time on the cliffs and in the glens and gorges around Yichang, according to Wilson it was called the *la-mei-haw*. It has been long cultivated in China. This species was first described by Linnaeus as *Calycanthus praecox* and was introduced from China in 1776. Wintersweet belongs to the family Calycanthaceae, itself comprising of three genera, *Calycanthus*, *Chimonanthus* and *Idiospermum* all of which are represented in cultivation but by no means common. At Glasnevin the cultivars 'Luteus' (syn. 'Concolor') and 'Grandiflorus' (the latter arose in cultivation in China) have been long grown against the warmest walls (in the Three Gorges district this shrub grows on limestone cliffs and receives a summer baking in a similar manner). *Chimonanthus praecox* was collected by the Glasnevin Central China Expedition (GCCE 839) near Dragon Gate River Forest Park in October 2004. Several cultivars of *Chimonanthus praecox* are cultivated in Chinese gardens and would be worth importing into the West.

Chionanthus henryanus (H. L. Li) P. S. Green (OLEACEAE)
New species

in Kew Bull. 50. 326. (1995).

Linociera henryi H. L. Li in Journ. Arnold Arbor. 25: 313. (1944).

Yunnan: Mountains to the east of Simao at 1,400 metres, 12042 (holotype of *Linociera henryi* H. L. Li). In mountain forest to the east of Simao at 1,400 metres, 12236. In mountain forest to the east of Simao at 1,300 metres, 12236a.

Native to Myanmar and China (Yunnan).

Chionanthus ramiflorus Roxb. (OLEACEAE)

Linociera sp. Henry in Trans. Asiat. Soc. Jap. 24 suppl: 59. (1896) (List Pl. Formosa). *Linociera ramiflora* (Roxb.) Wall. Li in Woody Flora of Taiwan 765. fig. 307. (1963).

Taiwan: Wanchin, 413.

A small evergreen tree from 3 to 12 metres tall, native from Malaysia, the Philippines, and China (Hainan and Taiwan, where it grows on the southern part of the island).

Chionanthus retusus Lindley & Paxton (OLEACEAE)

Rehder in Sargent Pl. Wilson. 2: 611. (1916).

Yunnan: In woods near Mengzi at 1,500 metres, 10532, 10532c. By a riverside at Lunan, 10532a.

The Chinese fringe tree, a deciduous tree to 20 metres tall, carrying cymose panicles of white flowers in April and May. Distributed in China, Korea and Japan. There is an old tree supplied by Messrs. Veitch by the pond at Glasnevin.

Chionanthus sp. (OLEACEAE)

Linociera aff. *cambodiana* Hance Anonymous in Kew List of Determinations (volume 24).

Hainan: At Nodoa in the interior of the island, 8457.

Chloranthus angustifolius Oliver (CHLORANTHACEAE)
New species

in Hooker's Icon. Pl. 16: t. 1580. (1887), Henry in Journ. China Br. Roy. Asiat. Soc. 22: 247. (1887) (Chinese Names of Plants), Forbes & Hemsley in Journ. Linn. Soc. Bot. 26: 367. (1891), Bretschneider in Hist. Eur. Bot. Disc. China 2: 790. (1898).

Hubei: Yichang, 658, 797, 1195 (isotype). Liantuo, 3859. San You Dong Glen, 3403a.

According to Henry this species was known colloquially at Yichang as *ch'i-hsin*.

02

Chloranthus fortunei (A. Gray) Solms-Laub. (CHLORANTHACEAE)

Henry in Journ. China Br. Roy. Asiat. Soc. 22: 247. (1887) (Chinese Names of Plants), Forbes & Hemsley in Journ. Linn. Soc. Bot. 367. (1891).

Hubei: Yichang, 272, 1321.

Chloranthus henryi Hemsley (CHLORANTHACEAE) New species

in Journ. Linn. Soc. Bot. 26: 367. (1891), Bretschneider in Hist. Eur. Bot. Disc. China 2: 790. (1898), Pe'i in Sinensia 6: 679. f. 6. (1935), Anonymous in Icon. Corm. Sin. 1: 347. f. 694. (1972), K. F. Wu in Fl. Reip. Pop. Sin. 20: 92. t. 29. (1982), Kong in Acta Phytotax. Sin. 38: 355. (2000).

Hubei: San You Dong Glen, 3403. Antelope Glen, 3447 (lectotype). Yichang, 3447a., 5402a., 3447. Liantuo, 3920. Badong, 4072. South Badong, 5402. **Sichuan:** South Wushan, 5402b., (isosyntype), 7719 (isosyntype).

This species was collected again by the 2004 Glasnevin Expedition (GCCE 548) near Zhenzi town in Xingshan in September 2004.

Chloranthus oldhamii Solms-Laub. (CHLORANTHACEAE)

Henry in Trans. Asiat. Soc. Jap. 24 suppl: 78. (1896) (List Pl. Formosa), Kong in Acta Phytotax. Sin. 38: 363. (2000).

Taiwan: Oluanpi, 649, 649a.

An erect perennial to 50 cm tall, native to southeastern China including Taiwan. Discovered near Tamsui by the Kew collector, Richard Oldham in 1864.

Chloranthus serratus (Thunb.) Roem. & Schult. (CHLORANTHACEAE)

Forbes & Hemsley in Journ. Linn. Soc. Bot. 26: 369. (1891).

Hubei: Liantuo, 2673. Xingshan, 7636. Changleping, 7636a.

The dried stems and leaves of this species are currently used in Chinese herbal medicine to relax mucles and tendons, it is also administered to victims of snake bite and those suffering from traumatic injury.

Chloranthus spp. (CHLORANTHACEAE)

Anonymous in Kew List of Determinations (volume 24 & 34).

Hubei: Badong, 4071. **Yunnan:** Mengzi, 9962, 10834. Simao, 12183, 12687.

Chloris barbata Sw. (POACEAE)

Henry in Trans. Asiat. Soc. Jap. 24 suppl: 109. (1896) (List Pl. Formosa), Rendle in Journ. Linn. Soc. Bot. 36: 403. (1904).

Taiwan: Kaohsiung, on cliffs and near sea level, 726, 1023.

Native to tropical south-east Asia, introduced elsewhere. This handsome grass was collected by the Glasnevin Taiwan Expedition (GTE 62) by the lighthouse at Oluanpi (South Cape of Henry and Wilson) in October 2004.

Chlorophytum laxum R. Br. (ASPARAGACEAE)

Wright in Journ. Linn. Soc. Bot. 36: 119. (1903).

Hainan: Environs of Haikou, 8364.

Chlorophytum sp. (ASPARAGACEAE)

Anonymous in Kew List of Determinations (volume 34).

Yunnan: Mengzi, 9501. Simao 12169.

Choerospondias axillaris (Roxb.) B. L. Burtt & A. W. Hill (ANACARDIACEAE)

Spondias axillaris Roxb. Rehder & E. H. Wilson in Sargent Pl. Wilson. 2: 172. (1916).

Yunnan: In forests near Simao at 1,600 metres, 11690.

A deciduous tree to about 20 metres tall bearing alternately arranged leaves to 30 cm long. These are divided into 6 to 8 pairs of leaflets and one terminal leaflet. Greeny-white flowers are borne in panicles at the ends of branches between April and May. Native to the Himalaya and western China. It is also widely planted in the Kathmandu valley in Nepal and there are good trees in the Royal Botanic Gardens, Gadawari.

Choerospondias axillaris (Roxb.) Burtt & Hill. var. ***pubinervis*** (Rehder & E. H. Wilson) B. L. Burtt & A. W. Hill (ANACARDIACEAE) New variety

Spondias axillaris Roxb. var. *pubinervis* Rehder & E. H. Wilson in Sargent Pl. Wilson 2: 173. (1916).

Sichuan: South Wushan, 5535.

A tree to 25 metres tall, bearing yellow edible fruits and found at low altitudes in Sichuan (up to 1,000 metres in alt.). According to Henry this species was known to the Chinese as *hsuan tsao*.

Christia obcordata (Poir.) Bakh. f. ex van Meeuwen (FABACEAE)

Lourea obcordata (Poir) Desv. Henry in Trans. Asiat. Soc. Jap. 24 suppl: 34. (1896) (List Pl. Formosa).

Hainan: Environs of Haikou, 8323. **Taiwan:** Kaohsiung, 2002.

A spreading herb with stems to 60 cm long, native to tropical Asia and Australia.

Christia vespertilionis (L. f.) Bakh. F. (FABACEAE)

Lourea vespertilionis (L. f.) Desv. Anonymous in Kew List of Determinations (volume 24).

Hainan: Near the Kiungchow Pagoda, 7989, 8249.

Chroesthes lanceolata (T. Anderson) B. Hansen (ACANTHACEAE)

Asystasia silvicola W. W. Smith in Notes Roy. Bot.

Gard. Edinburgh 9: 170. (1916).

Yunnan: In Yulo forests to the south of Simao at 1,300 metres, 12934 (isosyntype of *Asystasia silvicola* W. W. Smith). In the mountains to the east of Simao at 1,450 metres, 11600a., (isosyntype of *Asystasia silvicola* W. W. Smith).

Chrysanthemum indicum L. (ASTERACEAE)

Forbes & Hemsley in Journ. Linn. Soc. Bot. 23: 437. (1888), Henry in Gard. Chron. ser. 3, 8: 565. (1890).

Hubei: Yichang, on the banks of the Yangtze at 300 metres, 887. On the way to Moji Shan, 3081. Badong, 115, 5071. Fang Xian, 6765. At Yin-Yu-Ho in Fang Xian, in montane coniferous woodland at 2,750 metres, collected August 28th 1888, 6765a. **Sichuan:** South Wushan, 7252. **Yunnan:** Milê district, 9941. Simao 12629.

The *chu hua*, one of the parents of the florist's *Chrysanthemum* and cultivated in China since time immemorial, it is found in all parts of China. and it is thought that the florist's *Chrysanthemum* is the result of crossing *Chrysanthemum indicum* with *Chrysanthemum* × *morifolium*. The hardy, small flowered garden types are said to be attributed to *C. indicum* while the larger flowered indoor types stem from *C.* × *morifolium*. It is through thousands of years of artificial selection in China that we grow the many showy cultivars today. *Chrysanthemum* culture is mentioned in several ancient Chinese texts, such as the Book of Rites and the Book of Odes, revised by Confucius in the fifth century B.C.

Phillip Miller cultivated this species in the Chelsea Physic Garden in 1764 only 13 years after its discovery by Osbeck at Macao in south-east China. Chinese chrysanthemums reached Europe in 1689 and were first cultivated in Dutch gardens. According to Henry *Chrysanthemum indicum* grows on exposed limestone and flowers in November. Such was the habitat also noted by members of both the 2002 and 2004 Glasnevin Central China Expeditions, material of GCCE 368 grew as a single isolated plant on an exposed limestone ridge in south Badong and was collected in October 2002. North of the Yangtze where its habitat is less disturbed man this species is abundant, GCCE 510 was collected near Zhenzi town in Xingshan in September 2004 and GCCE 832 was collected near Dragon Gate River Forest Park also in Xingshan in October 2004. Seedlings from both of the Xingshan collections grow at Glasnevin. *Chryanthemum indicum* is particularly abundant on the upper slopes of Wudang Shan in north-west Hubei.

Henry also stated that his 6765a. was gathered by him on August 28th 1888, on rocks in a mountain coniferous wood at 2,850 metres above sea level at Yin-Yu-Ho in Fang

district. At that altitude the plant was graceful, delicate and small in stature. It had quite a different aspect from the weedy plants, which occurred along the ditches of fields, and along the banks of the Yangtze at altitudes under 300 metres. The dried flowers of certain cultivated kinds are used in traditional Chinese medicine in cases of headache, boils and inflammation of the eye. Dried *Chrysanthemum indicum* formed a considerable article of trade, e.g. 22 tons were annually exported from Hankow (now part of Wuhan) in the 19th century. *Ye ju hua* is an infusion of this herb that is used to cure hypertension and to lower blood pressure.

Chrysanthemum indicum var. *aromaticum* is endemic to Shennongjia Forest District where it grows on the tallest most exposed peaks and is an altogether more refined plant than the type and worthy of introduction to cultivation. Despite the specific epithet *Chrysanthemum indicum* is not native to India (Linnaeus was a little confused as to the position of China).

Chrysanthemum × *morifolium* (Ramat.) Hemsl. (ASTERACEAE)

Chrysanthemum × *sinense* Sabine Henry in Journ. China Br. Roy. Asiat. Soc. 22: 241. (1887) (Chinese Names of Plants), Forbes & Hemsley in Journ. Linn. Soc. Bot. 23: 438. (1888).

Hubei: Badong, 960.

Known colloquially in the Three Gorges region as *chu hua* according to Henry.

Chrysanthemum × *morifolium* (Ramat.) Hemsl Double-flowered. (ASTERACEAE)

Chrysanthemum × *sinense* Sabine Double-flowered. Henry in Gard. Chron. ser. 3, 8: 564. (1890).

Hubei: Badong, 2812. Liantuo, 3051.

Henry stated that these double-flowered specimens were collected by his local collector and probably came from cultivated plants.

Chrysanthemum vestitum (Hemsley) Stapf (ASTERACEAE) New species

Chrysanthemum sinense Sabine var. *vestitum* Hemsley in Journ. Linn. Soc. Bot. 23: 439. (1888), Bretschneider in Hist. Eur. Bot. Disc. China 2: 785. (1898).

Hubei: Moji Shan, at 300 to 500 metres, 6 miles south of Yichang, (on the opposite bank of the Yangtze to the foreign compound), 1115, 3102, 3365 (type).

Of his number 1115 Dr. Henry wrote, 'common in winter in the mountains;' and of 3102, 'I have only seen this in the neighbourhood of Moji Shan (the Dome),' a peak 2000 feet high a little south-west of Yichang.

Chrysojasminum floridum (Bunge) Banfi (OLEACEAE)

Jasminum floridum Bunge Henry in Journ. China Br. Roy. Asiat. Soc. 22: 261. (1887) (Chinese Names of Plants),

Rehder in Sargent Pl. Wilson 2: 614. (1916), Kobuski in Journ. Arnold Arb. 13: 147. (1932), Zheng in Journ. Wuhan Bot. Res. 2.(1): 179. (1984), Bean in Trees & Shrubs 2: 464. (1992).

Hubei: Yichang, 351, 1082, 2700, without number. Changleping, 6288 (March 1888).

An evergreen rambling shrub resembling *Jasminum humile* and like it, carrying yellow flowers. Native to northern and central China and introduced to cultivation by Lord Ilchester during the 1850s. According to Henry this species was known colloquially in the Three Gorges region as *niu she tzu.*

Chrysojasminum subhumile (W. W. Smith) Banfi & Galasso (OLEACEAE) New species

Jasminum subhumile W. W. Smith in Notes Roy. Bot. Gard. Edinburgh 12: 209. (1920). *Jasminum heterophyllum* Roxb. var. *glabricymosum* W. W. Smith Kobuski in Journ. Arnold Arb. 13: 149. (1932).

Yunnan: In woods near Mengzi at 1,475 to 1,525 metres, 9107. In woods near Mengzi at 1,550 metres, 9107a. In woods near Mengzi at 1,650 metres, 9107b.

Native to India, Nepal, Myanmar and China (Sichuan and Yunnan). Described from a collection made by George Forrest in April 1910.

Chrysopogon aciculatus (Retz.) Trin. (POACEAE)

Henry in Trans. Asiat. Soc. Jap. 24 suppl: 108. (1896) (List Pl. Formosa), Rendle in Journ. Linn. Soc. Bot. 36: 368. (1904).

Taiwan: Kaohsiung, 1034.

A widespread over the tropics of Asia.

Chrysosplenium cavaleriei H. Lév. & Vaniot (SAXIFRAGACEAE)

Chrysosplenium nepalense D. Don var. *vegetum* K. Hara in J. Fac. Sci. Univ. Tokyo Sect. 3. 7: 30. (1957).

Hubei: South Badong, 5430 (isotype of *Chrysosplenium nepalense* D. Don var. *vegetum* K. Hara).

Chrysosplenium lanuginosum Hook. f. & Thoms. (SAXIFRAGACEAE)

Chrysosplenium henryi Franchet in Bull. Soc. Philom. 112. (1890), Bretschneider in Hist. Eur. Bot. Disc. China 2: 783. (1898). *Chrysosplenium lanuginosum* Hook. f. & Thoms. var. *ciliatum* (Franch.) J. T. Pan in Acta Phytotax. Sin. 24: 96. (1986). *Chrysosplenium ciliatum* Franchet in Bull. Soc. Philom. 102. (1890); in Nouv. Arch. du Mus. 108. (1890), Bretschneider in Hist. Eur. Bot. Disc. China 2: 783. (1898).

Hubei: South Badong, 5270 (type of *Chrysosplenium henryi* Franchet), 5429 (isotype of *Chrysoplenium ciliatum* Franchet).

Chrysosplenium macrophyllum Oliver (SAXIFRAGACEAE) New species

in Hooker's. Icon. Pl. t. 18: 1744. (1888), Franch in Nuov. Arch. Mus. Hist. Nat. Paris ser. 3, 2: 102. (1890), Diels in Bot. Jahrb. 29: 367. (1900), Bretschneider in Hist. Eur. Bot. Disc. China 2: 783. (1898), Hand.-Mazz. in Symb. Sin. 7(2): 428. (1931), Hara in Journ. Fac. Sci. Univ. Tokyo 7: 86 f. 17D. (1957).

Chrysosplenium barbeyi A. Terracciano in Bull. Soc. Bot. Geneve. ser. 2, 7: 150. f. 1. (1915).

Hubei: Liantuo, 3846, 6315. South Badong, 5443 (isotype of *Chrysosplenium barbeyi* Terr.).

Chrysosplenium microspermum Franchet (SAXIFRAGACEAE) New species

in Nouv. Arch. Mus. Hist. Nat. ser. 3, 2: 109. (1890), Bretschneider in Hist. Eur. Bot. Disc. China 2: 783. (1898), Hara in Journ. Fac. Sci. Tokyo Bot. 7: 77. f. 15C. (1957); 2: 37. (1983).

Sichuan: South Wushan, 5582 (isotype).

Chrysosplenium cf. *nepalense* D. Don (SAXIFRAGACEAE)

Anonymous in Kew List of Determinations (volume 34).

Yunnan: Mengzi, 10534.

Chukrasia tabularis A. Juss. (MELIACEAE)

Anonymous in Kew List of Determinations (volume 24).

Hainan: 32 km (20 miles) west of Haikou, 8149.

Cibotium barometz (L.) J. Sm. (DICKSONIACEAE)

Dicksonia barometz (L.) Link Henry in Trans. Asiat. Soc. Jap. 24 suppl: 113. (1896) (List Pl. Formosa).

Taiwan: Tamsui, 1385 (in part). **Yunnan:** Mengzi, 9418. Simao, 11739.

A widespread species distributed in India, southeast Asia to Indonesia and the Philippines. Rare in Taiwan. Henry called it the *gou ji* or 'dog spine', this low growing tree fern is used as a drug in China, which is used as a tonic and to stregnthen the sinews and bones.

Cinnadenia liyuyingii (H. Liou) de Kok & Sengun (LAURACEAE) New species

Litsea liyuyingii H. Liou in Bull. Soc. Bot. France 80: 566. fig. 1. (1933), Allen in Ann. Missouri Bot. Gard. 25: 380. (1938).

Yunnan: Simao at 1,450 metres, 12839 (type).

Distributed across southern Yunnan of China and northern Indochina.

Cinnamomum bejolghota (Buch.-Ham.) Sweet (LAURACEAE)

Cinnamomum iners Liou Ho non (Reinw. ex Nees & T. Nees) Blume in Laurac. Chine & Indoch. 30. (1934). *Cinnamomum obtusifolium* (Roxb.) Nees Allen in Journ.

Arnold Arb. 20: 61. (1939).

Yunnan: Mengzi, 10440.

Used in the manufacture of incense, Henry stated it was known as *kue pi*.

Cinnamomum bodinieri H. Lév. (LAURACEAE)

Cinnamomum parthenoxylum Forbes & Hemsley non (Jack) Meisn. in Journ. Linn. Soc. Bot. 26: 372. (1891). *Cinnamomum hupehanum* Gamble in Sargent Pl. Wilson 2: 69. (1916).

Hubei: Liantuo, 3936.

Cinnamomum heyneanum Nees (LAURACEAE)

Zheng in Journ. Wuhan Bot. Res. 2.(1): 49. (1984).

Cinnamomum pedunculatum Nees var. *angustifolium* Hemsley in Journ. Linn. Soc. Bot. 373. (1891,) Liou Ho in Laurac. Chine & Indoch. 37. (1934). *Cinnamomum burmani* (Nees) Bl. var. *angustifolium* (Hemsl.) C. K. Allen in Journ. Arnold Arb. 20: 51. (1939), Zheng in Journ. Wuhan Bot. Res. 2.(1): 49. (1984).

Hubei: Yichang, 1193 (syntype of *Cinnamomum pedunculatum* Nees var. *angustifolium* Hemsley), 1293 (isotype of *Cinnamomum pedunculatum* Nees var. *angustifolium* Hemsley), 1353 (syntype of *Cinnamomum pedunculatum* Nees var. *angustifolium* Hemsley), 2759 (isotype of *Cinnamomum pedunculatum* Nees var. *angustifolium* Hemsley), 3466. Liantuo 3881. **Yunnan:** South of the Red River from Mengzi, 10888.

Cinnamomum iners (Reinw. ex Nees & T. Nees) Blume (LAURACEAE)

Liou Ho in Laurac. Chine & Indoch. 30. (1934).

Yunnan: Simao, 11785.

Native to the provinces of Guangxi, Tibet (Xizang) and Yunnan, also distributed in India, Sri Lanka and Myanmar.

Cinnamomum loureiroi Nees (LAURACEAE)

Liou Ho in Laurac. Chine & Indoch. 39. (1934).

Yunnan: South of the Red River, 12819.

Cinnamomum nominale (Hayata) Hayata (LAURACEAE) New species

Li in Woody Flora of Taiwan 206. (1963).

Cinnamomum camphora Henry non (L.) Presl. Henry in Trans. Asiat. Soc. Jap. 24 suppl: 78. (1896) (List Pl. Formosa).

Taiwan: Oluanpi, 251.

Cinnamomum officinarum Nees (LAURACEAE)

Cinnamomum camphora (L.) Presl. Henry in Journ. China Br. Roy. Asiat. Soc. 22: 235. (1887) (Chinese Names of Plants), Forbes & Hemsley in Journ. Linn. Soc. Bot. 26: 371. (1891), Henry in Trans. Asiat. Soc. Jap. 24 suppl: 78. (1896) (List Pl. Formosa), Li in Woody Flora of Taiwan 206. fig. 75. (1963). *Cinnamomum simondii* Lecompte Allen in Journ. Arnold Arb. 20: 45. (1939).

Hubei: Yichang, 965. **Taiwan:** Wanchin, 421. Oluanpi, 640, 648. **Yunnan:** Simao, 11960b.

A tall, aromatic evergreen tree to more than 50 metres tall with a massive trunk and spreading crown with a dense mass of foliage. Native to tropical Asia and Malaysia. According to Henry the wood of this tree was widely used in central China but camphor was not extracted. He went on to say that this was the *chang shu* (camphor tree of the Chinese) and was rare in Hubei. According to Wilson who later visited Chinese Taiwan, the camphor tree is found in submontane forest in Taiwan from sea-level to 1,400 metres alt. It occurs as a scattered tree of evergreen forests, growing with various Lauraceae, Fagaceae, *Ficus*, miscellaneous undergrowth shrubs, tree ferns and lianas. Often its trunks are clothed with epiphytes, especially ferns and orchids. The tree reaches its western limit of distribution in central China where it is confined to low altitudes. When Wilson visited Taiwan he stated that that island held the monopoly on the Asian camphor trade. The native tribes of head-hunters however made life for early Chinese settlers working in the industry extremly dangerous. Originally the area where most of these camphor trees grew was in territory occupied by these native peoples, following much warfare the Chinese forced the native tribes higher into the mountains and by the 1920s, Japan, who then occupied the island, had brought these tribes under control. The camphor tree has been known in China since ancient times for its excellent wood. The main centres of production today are Japan and China (Taiwan), the timber is distilled to yield an essential oil used to treat colds, influenza, fever, pneumonia, inflammation and diarrhoea. It has also been used in making incense, disinfectants, celluloid and (in the past) to make smokeless gunpowder. Old-fashioned mothballs were also made from camphor (these are now made of a synthetic product). Externally the oil is used to treat rheumatic conditions of the muscles. The wood is highly aromatic and is used in cabinet making and high-class furniture. At Glanleam Gardens, on Valentia Island (off the coast of Co Kerry) grows a fine 18 metre tall tree, planted by Sir Peter Fitzgerald, the 19th Knight of Kerry during the 1870s.

Cinnamomum osmophloeum Kanehira (LAURACEAE) New species

Li in Woody Flora of Taiwan 206. (1963).

Cinnamomum pedunculatum Henry non (Thunb.) Nees in Trans. Asiat. Soc. Jap. 24 suppl: 79. (1896) (List Pl. Formosa).

Taiwan: Wanchin, 451.

A medium-sized tree with aromatic bark. Endemic to Taiwan where it was first collected by Henry. According to Henry the fragrant root-bark of this species was known as *t'u-jou-kuei.*

Cinnamomum sp. (LAURACEAE)

Henry in Trans. Asiat. Soc. Jap. 24 suppl: 78. (1896) (List Pl. Formosa).

Taiwan: Oluanpi, 1330. **Yunnan:** Mengzi, 10054. Simao, 11598.

Cinnamomum wilsonii Gamble (LAURACEAE) New species.

Gamble in Sargent Pl. Wilson. 2: 66. (1916), Zheng in Journ. Wuhan Bot. Res.2. (1): 49. (1984).

Cinnamomum sp. Henry in Journ. China Br. Roy. Asiat. Soc. 22: 252. (1887) (Chinese Names of Plants). *Cinnamomum tamala* Forbes & Hemsley non (Buch.-Ham.) T. Nees & C. H. Eberm. in Journ. Linn. Soc. Bot. 26: 373. (1891), Henry in Notes Econ. Bot. China 21. (1893).

Hubei: Badong, 713 (June 1885). Changyang, 7669. **Sichuan:** South Wushan, 5497.

The *zigui-p'i shu*, discovered by Henry in June 1885. According to Wilson, for whom the species was named, it was common at the beginning of the 20th century in western Hubei and Sichuan. According to Henry the bark of this species was used locally as a medicine. Its pungent twigs and bark were used as a condiment, tonic and stimulant. Material in the Irish National Herbarium at Glasnevin was collected by the Glasnevin Central China Expedition (GCCE 680) in the Red Dragon Reserve in Xingshan in October 2004 (GCCE 680).

Cipadessa baccifera (Roxb. ex Roth.) Miq. (MELIACEAE)

Cipadessa fruticosa Blume Anonymous in Kew List of Determinations (volume 34). *Cipadessa baccifera* (Roth.) Miq. var. *sinensis* Rehder & E. H. Wilson in Sargent Pl. Wilson. 2: 159. (1914).

Yunnan: Mengzi, 10677. South of the Red River from Manmei, 9461 (in part). Mountains to the west of Simao at 1,500 metres, 9461c.

The Hani (an ethnic group) of Xishuangbanna in southern Yunnan call this small tree *manqya'laiq* and the roots and leaves are used to treat constipation, gastroenteritis and malaria.

Circaea alpina L. (ONAGRACEAE)

Anonymous in Kew List of Determinations (volume 24).

Hubei: Zigui, 6086. Fang Xian, 6906. **Sichuan:** Henry's Chinese collector with Pratt, no locality stated but may be Leshan, Emei Shan, Wa Shan or Kangding, 8950.

Circaea canadensis (L.) Hill var. *quadrisulcata* (Maxim.) Bufford (ONAGRACEAE)

Circaea quadrisulcata (Maxim.) Franch. & Savat. Anonymous in Kew List of Determinations (volume 34).

Hubei: Yichang, 5106. Jianshi, 6041. South Badong,

7279. Xingshan at 1,400 metres, 6552. **Yunnan:** Mengzi, 9733.

Circaea cordata Royle (ONAGRACEAE)

Forbes & Hemsley in Journ. Linn. Soc. Bot. 26: 310. (1887).

Hubei: Badong, 867 (September 1885), 4943, 5102. South Badong, 7380. Fang Xian, 6563. Without locality 6573a., 6573b. **Sichuan:** South Wushan, 7215.

A perennial rhizomatous herb to 1 metre tall, forming large colonies. A widespread species distributed from north-east Afghanistan eastwards along the south face of the Himalaya to Assam and from there into northwest Myanmar, China and north-west Vietnam.

Circaea mollis Sieb. & Zucc. (ONAGRACEAE)

Anonymous in Kew List of Determinations (volume 24).

Hubei: Liantuo, 4640.

Native to south-east Russia, India (including Assam), Myanmar, China, Vietnam, Laos and Cambodia.

Circulifolium microdendron (Mont.) S. Olsson, Enroth & D. Quandt (NECKERACEAE)

Homalia glossophylla (Mitt.) Jaeg. Salmon in Journ. Linn. Soc. Bot. 34: 467. (1900).

Taiwan: On the summit of Wanshoushan, on rocks in dark places, 717. Kaohsiung, 752, 1931, 2023. **Yunnan:** Simao, 12812. (Moss).

Cirsium arvense (L.) Scop. (ASTERACEAE)

Cnicus arvensis (L.) Roth. Anonymous in Kew List of Determinations (volume 24).

Hubei: Badong, 710 (July 1885).

Cirsium arvense (L.) Scop. var. *integrifolium* Wimm. & Grab. (ASTERACEAE)

Cnicus segetum (Bunge) Maxim. Henry in Journ. China Br. Roy. Asiat. Soc. 22: 278. (1887) (Chinese Names of Plants), Forbes & Hemsley in Journ. Linn. Soc. Bot. 23: 462. (1888).

Hubei: Badong, 6 (collected September 1885, received at Kew on March 12th 1886). Yichang, 1368. Liantuo, 1892, 2025.

According to Henry this species was known colloquially in the Three Gorges region as *hsiao tzu kai.*

Cirsium chinensis Gardner & Champ. (ASTERACEAE)

Cnicus chinensis (Gardner & Champ.) Benth ex Maxim. Henry in Journ. China Br. Roy. Asiat. Soc. 22: 278. (1887) (Chinese Names of Plants); in Trans. Asiat. Soc. Jap. 24 suppl: 55. (1896) (List Pl. Formosa), Forbes & Hemsley in Journ. Linn. Soc. Bot. 23: 461. (1888).

Hubei: Yichang, 119, 429, 883, 890, 2738. Badong, 2469, 3148, 5006, 5069, 5108. South Badong, 4869. Liantuo, 2634. Fang Xian, 6725. Changyang, 7504.

02

Sichuan: South Wushan, 7478. **Taiwan:** Wanchin, 1498.

According to Henry this species was known colloquially in the Three Gorges region as *ma tzu kai.*

Cirsium fargesii (Franchet) Diels (ASTERACEAE) New species

Diels in Bot. Jahrb. 29: 627. (1900).

Hubei: Yichang 6189. **Yunnan:** Mengzi, 9083.

First described by Franchet (as *Cnicus fargesii*) from Henry's collection from Yichang, and a later collection made by Farges in Sichuan. This very beautiful biennial herb was introduced to cultivation by the Glasnevin Central China Expedition (GCCE 465) from Xingshan in October 2004.

Cirsium henryi (Franchet) Diels (ASTERACEAE) New species

in Bot. Jahrb. 29: 627. (1901), Shih in Acta Phytotax. Sin. 22: 389. (1984).

Cnicus henryi Franchet in Journ. de Bot. (Morot) 11: 21. (1897) Bretschneider in Hist. Eur. Bot. Disc. China 2: 786. (1898) Morley in Glasra 3: 79. (1979)

Hubei: Fang Xian, 6764 (isotype of *Cnicus henryi* Franchet).

A very beautiful biennial species with a handsome, lace-like finely decorative calyx. Herbarium material of this species was collected by the Sino-American Expedition to the Shennongjia Forest District in 1980 and it was introduced to cultivation by members of the Glasnevin Central China Expedition (GCCE 473) from Xingshan in September 2004 and first flowered in the order beds in front of the Great Palm House at the National Botanic Gardens, Glasnevin.

Cirsium japonicum DC. (ASTERACEAE)

Cnicus japonicus (DC.) Maxim. Henry in Journ. China Br. Roy. Asiat. Soc. 22: 278. (1887), Forbes & Hemsley in Journ. Linn. Soc. Bot. 23: 461. (1888).

Hubei: Yichang, 1370. South Badong, 6054. Without locality, 5858. **Hainan:** Environs of Haikou, 8074.

According to Henry in the Three Gorges region this species was known colloquially as *ta tzu kai* or *da ji,* a drug made from it is used to stop bleeding.

Cirsium sp. No. 1. (ASTERACEAE)

Cnicus sp. Anonymous in Kew List of Determinations (volume 24).

Hubei: Badong, 4899. Yichang 7379.

Cirsium sp. No. 2. (ASTERACEAE)

Cnicus sp. Anonymous in Kew List of Determinations (volume 34).

Yunnan: Mengzi, 9662.

Cissampelopsis corifolia C. Jeffrey & Y. L. Chen (ASTERACEAE) New species

in Kew Bull. 39: 342. (1985).

Yunnan: Mengzi, 9178a., 9178b., (holotype), 9178c. In Pingbian in forests on the Daweishan Range at 1,640 metres, 13473.

A large scandent herb or subshrub to 7 metres, tall climbing on trees and shrubs in forests and thickets at 1,500 to 2,800 metres. Native to Sikkim, Myanmar, China [Tibet (Xizang), Yunnan] and Indochina.

Cissampelopsis spelaeicola (Vanihot) C. Jeffrey & Y. L. Chen (ASTERACEAE)

Jeffrey & Chen in Kew Bull. 39: 346. (1984).

Yunnan: Mountains to the south-west of Mengzi at 1,900 metres, 9274. On a wooded cliff near Mengzi at 1,650 metres 9274a., 9274b.

Distributed in China and Vietnam.

Cissampelopsis volubilis (Blume) Miquel (ASTERACEAE)

Jeffrey & Chen in Kew Bull. 39: 344. (1985).

Senecio hoi Dunn in Journ. Linn. Soc. Bot. 35: 506. (1903).

Yunnan: In forests near Mengzi at 1,650 to 1900 metres, 10392 (syntype of *Senecio hoi* Dunn), 10392a.

A scandent subshrub to 3 metres tall, widely distributed through north-east India, Myanmar, China, Thailand, Indochina and Malesia where it is found climbing on trees in forests between 800 to 2,000 metres. Dunn's *Senecio hoi* commemorated 'Ho' one of Henry's Yunnan collectors. Henry was particularly fond of Ho and regarded him as his best collector.

Cissampelos pareira L. (MENISPERMACEAE)

Diels in Engler, Pflanzenr. iv. 94: 286. (1910).

Yunnan: In the Red River Valley near Manhao, 9532.

Cissus discolor Blume (VITACEAE)

Vitis discolor (Blume) Dalzell Anonymous in Kew List of Determinations (volume 34).

Yunnan: Simao, 12163, 12406.

Native to China, southeast Asia and Malaysia.

Cissus repens Lamarck (VITACEAE)

Li in Woody Flora of Taiwan 524. (1963).

Vitis repens (Lam.) Wight & Arnott Henry in Trans. Asiat. Soc. Jap. 24 suppl: 28: (1896) (List Pl. Formosa).

Hainan: Environs of Haikou, 8036. 32 km (20 miles) west of Haikou, 8148. **Taiwan:** Kaohsiung, without number. **Yunnan:** Mengzi, 9217, 11306.

A vine to 10 metres tall, native from India to southern China, the Philippines, Malaysia and Australia. Common at low altitudes in Chinese Taiwan, this species was collected by the Glasnevin Taiwan Expedition (GTE 115) on Wanshoushan (Ape's Hill of Henry and Wilson) in October 2004.

Cissus sp. (VITACEAE)

Vitis cf. *repens* (Lam.) Wight & Arnott Anonymous in Kew List of Determinations (volume 24).

02

Hainan: Without locality, 8567.

Citrullus lanatus (Thunb.) Matsum. & Nakai (CUCURBITACEAE)

Citrullus vulgaris Schrad. Henry in Trans. Asiat. Soc. Jap. 24 suppl: 46. (1896) (List Pl. Formosa).

Hainan: Environs of Haikou, 8381. **Taiwan:** Wanchin, 380.

The *sai kwa* or water-melon, an ancient crop known since prehistoric times. An African native, according to Bretschneider the water-melon reached China in the 10th century A.D. and was introduced by a man called Hun Hao who lived between A.D. 1090 to 1155. China is the world's largest producer of this fruit (6,530,000 metric tons in 1995).

Citrus × *aurantium* L. (RUTACEAE)

Swingle in Sargent Pl. Wilson. 2: 147. (1916).

Citrus × *sinensis* (L.) Osbeck Swingle in Sargent Pl. Wilson. 2: 148. (1916).

Taiwan: Wanchin, 150. **Yunnan:** In mountains near Simao at 1,400 metres, 11605.

The Seville or bitter orange, the *ch'eng* of Mandarin Chinese (according to Wilson), a tree to 10 metres tall, the fruits are used mainly in the production of preserves, marmalade and orange liqueurs. Henry stated that his Chinese Taiwan collection was from a wild plant in the mountains, though it was most likely a relict or an escape from cultivation. This species is thought to be a natural hybrid between *Citrus maxima* and *Citrus reticulata*. Oranges were carried from the orient along the silk route to the Persian empire and from there were taken by the Moors to Spain (where the fruit was greatly improved, Valencia oranges are now grown in many warm temperate and subtropical parts of the world). On his second voyage, Columbus introduced *Citrus* × *aurantium*, *Citrus* × *limon* and *Citrus medica* to the New World where they were successfully established. The Spanish later introduced them to the new world in the 1500s. *Citrus* × *aurantium* grows well in subtropical and tropical areas. In the lowland tropics it develops a good flavour but the skin remains yellowish green even when fully ripe and lacks the glowing, orange tones of those produced in cooler Mediterranean regions of the world. Oranges remained a delicacy reserved for the rich and affluent until the 19th century. After Brazil and the USA, China is the world's third largest producer of oranges. The roots of this orange are resistant to *Phytopthera* rot and young plants are used as rootstock on which other *Citrus* are grafted.

Citrus cavaleriei H. Lévl. ex Cavalerie (RUTACEAE)

Citrus ichangensis Swingle in Journ. Agric. Research. I. 1, fig. 1-7. t. 1. (1913), Zheng in Journ. Wuhan Bot. Res. 2.(1): 104. (1984).

Hubei: Antelope Glen, 3423. Changyang, 7695.

The Ichang (or Yichang) lemon, a shrub or small tree from 1 to 10 metres tall. Discovered by Henry in October 1887 and later collected by Wilson, named for the French missionary Père Julien Cavalerie who was based in China between 1894 and 1927. This species occurs wild in Hubei but was cultivated around Yichang for its coarse lemon-like fruits and was known by European residents as the Ichang lemon. Wilson collected cultivated plants in the Xiling Gorge, in the San You Dong Glen (near Yichang), Liantuo and in the area of Xingshan. He did find wild plants in Sichuan and Henry's specimen also came from a wild plant. From Yichang it was shipped as far as Hankow, now part of Wuhan in eastern Hubei. The Ichang lemon was collected by the Glasnevin Central China Expedition (GCCE 660) in the Red Dragon Reserve in Xingshan in October 2004. This species grows further north and at higher altitudes than any other other species of Citrus and as a consequence is hardy in Ireland, there are plants at Glanleam, the former seat of the Knights of Kerry, on Valentia Island off the coast of Co Kerry.

Citrus × *limon* (L.) Osbeck (RUTACEAE)

Citrus × *limonia* Osbeck Swingle in Sargent Pl. Wilson. 2: 146. (1914).

Yunnan: On mountains near Mengzi at 1,445 metres, 10445, collected March 21st 1897.

The lemon, a small evergreen tree to 6 metres tall and thought to be of hybrid origin (*C. maxima* × *C. medica* × *C. reticulata*). The lemon was first recorded in China about 1,000 A.D.

Citrus maxima (Burm.) Merr. (RUTACEAE)

Citrus decumana L. Henry in Trans. Asiat. Soc. Jap. 24 suppl: 26. (1896) (List Pl. Formosa). *Citrus grandis* (L.) Osbeck Swingle in Sargent Pl. Wilson. 2: 144. (1914).

Hubei: Badong, 4743. **Taiwan:** Kaohsiung, 318. Pummelo, cultivated plants.

Citrus medica L. (RUTACEAE)

Anonymous in Kew List of Determinations (volume 24).

Sichuan: North Wushan, 7130.

The citron, a small thorny tree to 2 metres tall. Although the citron is native to south-east Asia (some authorities state it is endemic to India) it was known to the ancient Greeks and it is said to have been brought from northern India to the Mediterranean by Alexander the Great. When devising the generic name *Citrus*, Linnaeus used Pliny's name citron as the basis for this group of fruits. The Hani (an ethnic group) of Xishuangbanna in southern Yunnan call this species *siqlyulqyul* and use the fruits to treat vomiting. Seeds from Henry's 7130 were removed in the Kew herbarium and sent for propagation on May 4th

1889.

Citrus medica L. var. *sarcodactylis* (Nooten) Swingle (RUTACEAE)

Swingle in Sargent Pl. Wilson. 2: 141. (1916).

Yunnan: At Pan-ya-huan near Mengzi, 10445a., collected March 21st 1897.

The *fo shou* or Buddha's hand citrus, a small evergreen tree. Henrys specimen was collected from a 1 metre tall shrub with flowers dark pink on the reverse of the petals. Wilson who collected it from cultivated plants in Sichuan stated that the fruits were valued because of their delicious odour and because they were used in medicine. He also stated that it originated in India and was carried to China by Buddist monks. Most likely this particular collection was cultivated in the grounds of a monastery, the Buddha's hand citrus is believed to bring good health and to symbolise wealth. Because the fruit resembles the classic prayer position of Buddha's hands, the long fruit fingers connote Buddhism. Probably the strangest and most exotic looking fruit in Henry's collections. Robert Fortune introduced it through the Horticultural Society in 1844. In China it forms the basis of a drug used to alleviate pain and is a traditional spring festival fruit.

Citrus reticulata Blanco (RUTACEAE)

Citrus deliciosa Ten var. *kinokuni* Zheng non Hort in Journ. Wuhan Bot. Res. 2(1): 104. (1984).

Hubei: Antelope Glen, 3308.

The Mandarin orange or *chen pi*, originated in China and still an important crop in the river-side areas of the Three Gorges region.

Cladonia chlorophaea (Flörke ex Sommerf.) Spreng ex Asahina (CLADONIACEAE)

Wei in An Enumeration of Lichens in China 66. (1991).
Cladonia pyxidata (L.) Hoffm. var. *chlorophaea* Flörke ex Sommerf. Muell.-Arg. in Bull. Herb. Boissier 1: 235. (1893).

Hubei: Xingshan, 6962. (Lichen).

Cladonia coccifera (L.) Willd. (CLADONIACEAE)

Muell.-Arg. in Bull. Herb. Boissier 1: 235. (1893), Wei in An Enumeration of Lichens in China 67. (1991).

Hubei: Fang Xian, 6772. (Lichen).

Cladonia coniocraea (Flörke) Spreng. forma *phyllostrata* (Flörke) Vain. (CLADONIACEAE)

Wei in An Enumeration of Lichens in China 67. (1991).
Cladonia ochrochlora Flörke f. *phyllostrota* Flörke Mueller in Bull. Herb. Boissier 1: 235. (1893).

Hubei: Xingshan, 6943. (Lichen).

Cladonia floerkeana (Fr.) Flörke f. *carcata* (Ach.) Thoms. (CLADONIACEAE)

Muell.-Arg. in Bull. Herb. Boissier 1: 235. (1893), Wei in An Enumeration of Lichens in China 70. (1991).

Hubei: Xingshan, 6973. (Lichen).

Cladonia pyxidata (L.) Hoffm. (CLADONIACEAE)

Muell.-Arg. in Bull. Herb. Boissier 1: 235. (1893), Wei in An Enumeration of Lichens in China 77. (1991).

Hubei: Xingshan, 6971. (Lichen).

Cladonia rangiferina (L.) Weber ex F. H. Wigg. (CLADONIACEAE)

Muell.-Arg. in Bull. Herb. Boissier 1: 235. (1893).

Hubei: Fang Xian, 6947. (Lichen).

Cladonia squamosissima (Muell.-Arg.) Ahti (CLADONIACEAE) New species

in Ann. Bot. Fennici 17: 233. (1980), Wei in An Enumeration of Lichens in China 80. (1991).
Cladonia gracilis Hoffm. var. *squamosissima* Muell.-Arg. in Flora 74: 372. (1891), Muell.-Arg. in Bull. Herb. Boissier 1: 235. (1893).

Hubei: Xingshan, 6965, 6959 (type of *Cladonia gracilis* Hoffm. var. *squamossima* Muell.-Arg). (Lichen).

Cladrastis delavayi (Franch.) Prain (FABACEAE)

Cladrastis sinensis Hemsl. Henry in Elwes and Henry The Trees Gt. Brit. & Irel. 2: 450. (1907), Takeda in Sargent Pl. Wilson. 2: 97. (1916), Bean in Trees & Shrubs 1: 623. (1989).

Yunnan: Mengzi, 10784.

A medium sized tree to 21 metres tall with widespreading branches forming an oval head. Branched panicles of flowers are freely produced and vary from white to rosy-pink. Discovered in Yunnan in July 1888 by Delavay, it was later collected in Hubei by Wilson and in Yunnan province by Forrest and Henry. Wilson introduced it to cultivation from Fang Xian (north of Yichang) through Messrs Veitch's Coombe Wood nursery in 1901. According to Henry plants from Wilson's 1901 introduction were 1.5 metres (5 feet) tall in Veitch's Coomb Wood nursery by 1906. The tallest specimen in Britain and Ireland grows at Newcastle House, Glamorgan. UK, when measured in 1990 it was 13 metres tall with a trunk diameter of 28 cm. At the Royal Botanic Gardens, Kew it was 9 metres tall with a trunk diameter of 48 cm in 2001. Bhutan to China.

Claoxylon hainanense Pax & Hoffm. (EUPHORBIACEAE) New species

in Engler, Pflanzenr. iv. 147. vii: 128. (1914), Forbes & Hemsley in Journ. Linn. Soc. Bot. 26: 444. (1894).
Baliospermum sp. Forbes & Hemsley in Journ. Linn. Soc. Bot. 26: 444. (1894).

Hainan: Environs of Kiungchow, 8742.

Claoxylon indicum (Reinw.ex Blume) Hassk. (EUPHORBIACEAE)

Forbes & Hemsley in Journ. Linn. Soc. Bot. 26; 436.

(1894), Pax in Engler, Pflanzenr. iv. 147. vii: 108. (1914).

Hainan: Environs of Haikou, 8018, 8767. **Yunnan:** Simao, 12206.

Distributed in India, China, Vietnam and the Philippines.

Clausena anisata (Willd.) Hook. f. ex Benth. (RUTACEAE)

Clausena suffruticosa (Roxb.) Wight & Arn. Rehder & E. H. Wilson in Sargent Pl. Wilson. 2: 140. (1914), Zheng in Journ. Wuhan Bot. Res. 2.(1): 105. (1984). *Clausena dentata* (Willd.) M. Roem. var. *robusta* Tanaka in Lingnan Sci. J. 7: 346. (1931).

Hubei: Antelope Glen near Yichang, 3127. Yichang, 4122. **Yunnan:** Simao at 1,450 metres, 13028 [type of *Clausena dentata* (Willd.) M. Roem. var. *robusta* Tanaka].

A shrub to 3 metres tall, rather rare in the Three Gorges region of Hubei where it is found in glens and ravines at low altitudes. The fruits are orange-yellow before ripening to black and are edible.

Clausena excavata Burm. f. (RUTACEAE)

Henry in Trans. Asiat. Soc. Jap. 24 suppl: 25. (1896) (List Pl. Formosa), Li in Woody Flora of Taiwan 367. fig. 132. (1963).

Hainan: Environs of Haikou, 8023, 8324. Hummocks near Haikou, 8090. **Taiwan:** Wanchin, 401. Oluanpi, 599. **Yunnan:** Simao, 11914.

A tropical tree to 13 metres tall with a wide geographical range from India to southern Asia, Malaysia and New Guinea. Found in the central and southern parts of Chinese Taiwan.

Clausena lansium (Lour.) Skeels (RUTACEAE)

Clausena wampi (Blanco) Oliv. Henry in Trans. Asiat. Soc. Jap. 24 suppl: 25. (1896) (List Pl. Formosa). *Clausena punctata* (Sonn.) Rehder & E. H. Wilson in Sargent Pl. Wilson. 2: 140. (1914).

Hainan: Environs of Haikou, 7997. **Taiwan:** Wanchin, 42, 493.

The wampi, a tree to 12 metres tall, cultivated in southern China for its popular lime-like fruits which can be sour or sweet depending on the cultivar grown. Henry stated that in Chinese Taiwan this plant was known as *kou-tzu-huang*, on the eastern part of the mainland the name *huang-p'i* was applied.

Clausena sp. (RUTACEAE)

Anonymous in Kew List of Determinations (volume 24).

Hainan: Ling-men, 8653.

Clavaria emerici Berk. (CLAVARIACEAE)

Anonymous in Kew List of Determinations (volume 24).

Sichuan: North Wushan, 7023. (Fungus).

Clavulina coralloides (L.) Schröt. (CLAVULINACEAE)

Clavaria cristata Jungh. Anonymous in Kew List of Determinations (volume 24). *Clavaria coralloides* L. Anonymous in Kew List of Determinations (volume 24).

Hubei: Fang Xian, 6650, 7629. (Fungus).

Cleidion javanicum Blume (EUPHORBIACEAE)

Pax in Engler, Pflanzenr. iv. 147. vii: 290. (1914).

Yunnan: Simao, 11787a.

In China this species is distributed in Yunnan and Tibet (Xizang).

Cleidion sp. (EUPHORBIACEAE)

Anonymous in Kew List of Determinations (volume 34).

Yunnan: Simao, 12142.

Cleisostoma fuerstenbergianum Kraenzl. (ORCHIDACEAE)

Sarcanthus flagellaris Schlecter in Repert. Spec. Nov. Regni. Veg. Beih. 4: 76. (1919).

Yunnan: Simao, 12319, 12319a., 12383 (type of *Sarcanthus flagellaris* Schlecter).

Native to China, Thailand, Cambodia, Laos and Vietnam.

Cleisostoma racemiferum (Lindl.) Garay (ORCHIDACEAE)

Sarcanthus pallidus Lindl. W. W. Smith in Notes Roy. Bot. Gard. Edinburgh 13: 219. (1921).

Yunnan: Epiphytic (or epithylithic in this case) on cliffs on the mountain range to the east of Simao at 1,950 metres, 13080.

Distributed in India, Bhutan, Myanmar, China (Yunnan), Thailand, Laos and Vietnam.

Clematis acuminata DC. (RANUNCULACEAE)

Anonymous in Kew List of Determinations (volume 34).

Yunnan: Tonguan, 11568.

Distributed in the Himalaya (Nepal to Bhutan and Assam) and in Tibet (Xizang) and Yunnan in China.

Clematis acutangula Hook. f. & Thoms. (RANUNCULACEAE)

Anonymous in Kew List of Determinations (volume 34).

Yunnan: Mengzi, 9292. Simao, 12582. Yuanjiang, 13231.

Clematis apiifolia DC. var. *argentilucida* (H. Lévl. & Vantioy) W. T. Wang (RANUNCULACEAE)

Clematis obtusidentata (Rehd. & Wils.) H. Eichler. Zheng in Journ. Wuhan Bot. Res. 2.(1): 37. (1984). *Clematis apiifolia* DC. var. *obtusidentata* Rehder & E. H. Wilson in Sargent Pl. Wilson. 1: 336. (1913).

Hubei: Moji Shan, (June 1886), 1556.

A vigorous deciduous climber to 5 metres tall, allied to *Clematis apiifolia* DC. and native to central and southern

China.

Clematis armandii Franch. (RANUNCULACEAE)

Henry in Notes Econ. Bot. China 51. (1893), Rehder & E. H. Wilson in Sargent Pl. Wilson. 1: 326. (1913), Zheng in Journ. Wuhan Bot. Res. 2.(1): 37. (1984), Wang in Acta Phytotax. Sin. 41:133. (2003).

Clematis hedysarifolia DC. var. *armandii* Franch. Henry in Journ. China Br. Roy. Asiat. Soc. 22: 279. (1887) (Chinese Names of Plants).

Hubei: Badong, 502, 1468, 1469, 3720. Yichang, 1388, 3377 (October 1887). Changyang, 7784. Liantuo, 3817. Without locality, 5223a., 5223b., 5223c. **Sichuan:** South Wushan, 7904. **Yunnan:** Mengzi, 9349.

A vigorous climber to 10 metres tall. 'The *wei-ling-hsien,* a beautiful species occuring in the mountains and glens near Yichang,' was Henry's statement about this popular evergreen climber which is an important medicinal plant in China (the stems are sold with other medicinal plants by vendors on Emei Shan, the holy Buddhist mountain in western Sichuan). To the Hani (an ethnic group) of southern Yunnan this species is called *eelpuqcyuq* and is used to treat menstrual disorders. A widespread species in western and central China, it is also found in northern Myanmar and northern Vietnam. In its native habitat this species inhabits woods and thickets in warm-temperate and subtropical regions. Discovered by Père Armand David near Baoxing probably about 1879, introduced to cultivation by Wilson in 1900 when collecting for Messrs Veitch, plants first flowered in their Coombe Wood nursery in April 1905.

Clematis armandii Franch. var. *farquhariana* (Rehder & E. H. Wilson) W. T. Wang (RANUNCULACEAE) New variety

Wang in Acta Phytotax. Sin. 41: 136. (2003).

Hubei: South Badong, 5388. Jianshi, 5998.

Originally described as *Clematis armandii* f. *farquhariana* by Rehder & E. H. Wilson (in Sargent Pl. Wilson. 1: 327. 1913) on a collection made by Wilson at Changyang in Hubei in April 1907, Henry had discovered the plant two decades previously. Wang raised it to varietal level in 1998 (see Acta Phytotax. Sin. 36: 158. (1998) The variety differs from the type in its larger, pale pink coloured flowers. Endemic to China where it is found in western Hubei, western and eastern Hunan and eastern Sichuan.

Clematis brevicaudata DC. (RANUNCULACEAE)

Rehder & E. H. Wilson in Sargent Pl. Wilson. 1: 340. (1913), Zheng in Journ. Wuhan Bot. Res. 2.(1): 37. (1984).

Clematis benthamiana Henry non Hemsl. in Journ. China Br. Roy. Asiat. Soc. 22: 240. (1887) (Chinese Names of Plants), Rehder & E. H. Wilson in Sargent Pl. Wilson. 1: 330. (1913), Zheng in Journ. Wuhan Bot. Res. 2(1): 37.

(1984).

Hubei: Antelope Glen, 4341.

Clematis chinensis Osbeck (RANUNCULACEAE)

Rehder & E. H. Wilson in Sargent Pl. Wilson. 1: 329. (1913), Zheng in Journ. Wuhan Bot. Res. 2.(1): 37. (1984), Wang in Acta Phytotax. Sin. 39: 314. (2001) 41: 138. (2002); in Acta Phytotax. Sin. 41: 138. (2003).

Hubei: Near Moji Shan, 1639. On Tsui Fu Shan near Yichang (May 1888), 1601. In glens near Yichang, 2033. On the banks of the Yangtze at Yichang, 4347, 4348. Changleping, 6220. Without locality 4328a., 4348a. **Taiwan:** Kaohsiung, without number. Wanchin, 189.

A strong growing semi-evergreen climber to 3 metres tall, small, starry, fragrant white flowers are carried in large cymose panicles and appear in August to October. A widespread species found in central, southern and eastern China (including Taiwan), Vietnam and Japan. The Chinese name for this species, according to Wilson was *ching-lung-hsu,* he introduced it through Veitch's Coombe Wood nursery in 1900.

Clematis chrysocoma Franch. (RANUNCULACEAE)

Wang in Acta Phytotax. Sin. 41: 220. (2003).

Yunnan: Mengzi, 9835, 9835a., 9835b.

A subshrub to 2 metres tall, though in its native habitat it is more often only 20 to 40 cm tall. Solitary flowers are carried in the lower leaf axils on a long pedicel above densely pubescent foliage. Flower colour is variable, ranging from rose-pink to rose-purple, sometimes white. Native to western China in the provinces of Guizhou, Yunnan and Sichuan, discovered by Delavay on the 1st of August 1884 and introduced to Kew by Maurice de Vilmorin in 1910 (Vilmorin had originally received seeds from Delavay). A beautiful plant, though probably best suited to the warmer counties of the British Isles and Ireland.

Clematis connata DC. (RANUNCULACEAE)

Clematis gracilis Edgew. Anonymous in Kew List of Determinations (volume 34).

Yunnan: Mengzi, 9362.

Distributed in the Himalaya (Kashmir to Bhutan) and in western China [Guizhou, western Sichuan, south-east Tibet (Xizang) and Yunnan].

Clematis fasciculiflora Franch. (RANUNCULACEAE)

Rehder & E. H. Wilson in Sargent Pl. Wilson. 1: 331. (1913), Wang in Acta Phytotax. Sin. 41: 231. (2003).

Yunnan: On grassy hills to the west of Mengzi at 1,600 to 1,800 metres, climbing over low shrubs, 10114. In the mountains to the west of Simao at 1,600 metres, 10114a. In forests near Yuanjiang at 1,600 metres, 11575. Mengzi at 1,300 to 1,600 metres, 13627.

An evergreen climber to 6 metres tall, native to south-west China (Guangxi, Guizhou, Sichuan and Yunnan), northern Myanmar and northern Vietnam. Creamy-yellow bell shaped flowers are carried in whorls of three to seven from the axils of leaves in November to April. Discovered by Delavay in March 1885 and introduced to cultivation by Forrest through the Royal Botanic Gardens, Edinburgh in 1910.

Clematis finetiana H. Lévl. & Vantiot (RANUNCULACEAE)

Zheng in Journ. Wuhan Bot. Res. 2.(1): 40. (1984), Wang in Acta Phytotax. Sin. 41: 125. (2003).

Clematis pavoliniana Pamp. Rehder & E. H. Wilson in Sargent Pl. Wilson. 1: 328. (1913).

Hubei: Yichang, 2744, 3499, 3529. Jianshi, 7523.

An evergreen climber to 5 metres tall, bearing lateral clusters of strongly fragrant starry-white flowers in April to June. According to Wilson, it was very common in the glens and ravines around Yichang. The species seems to have been first collected near Ningbo in Zhejiang by Robert Fortune and later at Longki in Yunnan by Père Delavay. Wilson introduced it in 1908 while collecting for the Arnold Arboretum. This species resembles *Clematis armandii* in foliage and is endemic to China where it is found in woodlands and open ravines at 1,200 metres in the provinces of Anhui, Hubei, Sichuan and southern Zhejiang.

Clematis florida Thunb. (RANUNCULACEAE)

Henry in Gard. Chron. ser. 3, 51. fig 20. (1902), Rehder & E. H. Wilson in Sargent Pl. Wilson. 1: 325. (1913), Zheng in Journ. Wuhan. Bot. Res. 2.(1): 38. (1984), Bean in Trees & Shrubs 1: 644. (1989).

Hubei: Yichang, 3516, 3516a.

A handsome semi-evergreen climber to 3 metres tall, bearing and July creamy-white flowers to almost 10cm across during the months of June. First noticed in Japan (where it has long been cultivated) by Thunberg, Henry found the first truly wild plants in Hubei (not Wilson as often stated). Wilson claimed it was extremly rare in central China and knew it only from the vacinity of Yichang. Introduced to cultivation from Japan in 1776. In China it is also native to Guangdong, Guangxi, Hunan and Jiangxi, it is naturalised in Japan.

Clematis formosana Kunze (RANUNCULACEAE)

Wang in Acta Phytotax. Sin. 41: 19. (2003).

Taiwan: Wanchin, 31.

A deciduous herbaceous climber, flowers white, 3 to 5 flowered lateral cymes. Endemic to Taiwan where it was first collected on Wanshoushan, by the Irish plant collector and Consul, George M. H. Playfair in 1888. It is still a common plant in the tropical forests on Wanshoushan where it was collected by the Glasnevin Taiwan Expedition (GTE 100) in October 2004.

Clematis fulvicoma Rehder & E. H. Wilson (RANUNCULACEAE) New species

in Sargent Pl. Wilson. 1: 327.(1913).

Yunnan: Mountains to the south of Mengzi, 9377 (type).

A vigorous climber to 10 metres tall, flowers dark violet, to 6 cm across and carried in lateral cymes. Native to southern Yunnan of China where it was first found by Henry, it is also native to India (Assam), Myanmar, northern Laos and northern Thailand.

Clematis gouriana Roxb. ex DC. (RANUNCULACEAE)

Rehder & E. H. Wilson in Sargent Pl. Wilson. 1: 339. (1913), Zheng in Journ. Wuhan Bot. Res. 2.(1): 38. (1984), Wang in Acta Phytotax Sin. 41: 37. (2003).

Hubei: Badong, 2464. Yichang 2946, 3090. On the banks of the Yangtze at Yichang 4329. Antelope glen near Yichang 3128, 3132. Without locality 4329a., 4329b., 4329c. **Yunnan:** Mengzi at 1,300 to 1,600 metres, 9104. Mengzi at 1,500 to 1,600 metres 10148. Simao at 2,000 metres 13432.

An extremly widespread deciduous climber, native to Pakistan, Nepal, Bhutan, India (including Sikkim), Sri Lanka, Bangladesh, Myanmar, China, Laos, Thailand, northern Vietnam and the Philippines. Fragrant creamy-yellow flowers are produced in dense clusters during August and September. Introduced to cultivation by Wilson from Hubei in 1901. The Veitch Catalogue *New Hardy Plants from Western China* (Autumn 1912) had this climber available for one shilling and six pence.

Clematis grandidentata (Rehder & E. H. Wilson) W. T. Wang (RANUNCULACEAE) New species

Wang in Acta Phytotax. Sin. 38: 309. (2000); in Acta Phytotax. Sin. 41: 23. (2003).

Clematis grata Henry non Wallich in Journ. China Br. Roy. Asiat. Soc. 22: 260. (1887) (Chinese Names of Plants); in Notes Econ. Bot. China 8. (1893). *Clematis grata* Wall. var. *grandidentata* Rehder & E. H. Wilson in Sargent Pl. Wilson.1: 338. (1913).

Hubei: Yichang, 733. On floor of a cave on Moji Shan, near Yichang, 3092. Badong, 1738, 2483. Liantuo, 2015, 2619, 4588. Jianshi, 5647a. Without locality, 1062a., 2015a. **Sichuan:** South Wushan, 5578, 5647.

A vigorous deciduous climber to 10 metres tall bearing in May or June panicles of pure white flowers in axillary or terminal clusters. A widespread species in south-west to eastern China, first found near Yichang by Henry. Colloquially known in the Three Gorges region as *mu t'ung*, according to Henry the root was used as a drug. Allied to *Clematis grata*, it was introduced to cultivation by Wilson in 1904. According to Wilson this mountainous species was

02

abundant in central China in the early 20th century and was highly valued in Chinese medicine as the source of *mu-t'ung* used as an emetic purgative and vesicant. Seeeds of his 1100 and 1233 from Sichuan reached Glasnevin in 1909.

Clematis grata Wall. (RANUNCULACEAE)

Henry in Trans. Asiat. Soc. Jap. 24 suppl: 14. (1896) (List Pl. Formosa).

Taiwan: Oluanpi, 997.

A vigorous deciduous climber to 5 metres tall, flowers cream, fragrant and produced in dense panicles from July to September. Distributed from Afghanistan, across the Himalaya into southern Tibet (Xizang) and Yunnan to Fujian and Taiwan. Introduced to cultivation in 1830.

Clematis gratopsis W. T. Wang (RANUNCULACEAE)
New species

Zheng in Journ. Wuhan Bot. Res. 2.(1): 38. (1984).

Clematis grata Henry non Wall. in Trans. Asiat. Soc. Jap. 24 suppl: 14. (1896) (List Pl. Formosa). *Clematis grata* Wall. var. *lobulata* Rehd. & Wils. in Sargent Pl. Wilson. 1: 337. (1913).

Hubei: Yichang, 2721 (February 1887), 4330 (type of *Clematis grata* Wall. var. *lobulata* Rehder & E. H. Wilson). **Sichuan:** South Wushan, 7230 (paratype of *Clematis grata* Wall. var. *lobulata* Rehder & E. H. Wilson), 7231, 7261. **Taiwan:** Oluanpi, 904, 904a.

This species was first found by Henry near Yichang and was later collected near Badong by Wilson. Native to the provinces of Gansu, Hubei, Sichuan, Hunan and Shaanxi. Seeds of Wilson's 665 from Badong reached Glasnevin in 1908.

Clematis grewiiflora DC. (RANUNCULACEAE)

In the Harvard University Herbarium (A.).

Yunnan: In the mountains to the south of Mengzi at 1,950 metres, 9383. Mengzi at 1,950 metres, 9381a. Near Yuanyang, on Fengchunling at 1,250 metres, 9383b. In the mountains to the east of Mengzi at 2,300 metres, 9383c. Simao at 1,950 metres, 9383c., 9383d. In the mountains to the south-east of Simao at 1,650 metres, 9383e.

A vigorous deciduous climber to 6 metres tall, the small urn-shaped flowers may be solitary or carried in three-flowered cymes. Native to the Himalaya and south-west China where it grows in subtropical and warm-temperate broad-leaved forest between 750 and 1,800 metres. Flowering fom November to February. The entire plant including the flowers is covered in a dense yellow-brown down.

Clematis hayatae Kudô & Masam. (RANUNCULACEAE)
New species

Clematis henryi Oliver in Hooker's Icon. Pl. 19: t. 1819. (1889), nom. illeg., Bretschneider in Hist. Eur. Bot.

Disc. China 2: 778. (1898), Finet & Gagnepain in Bull. Soc. Bot. France, L. 540. (1903); Contrib. Fl. As. Or. I. 25. (1905), Rehder & E. H. Wilson in Sargent Pl. Wilson. 1: 432. (1916), Morley in Glasra 3: 79. (1979), Zheng in Journ. Wuhan Bot. Res. 2.(1): 39. (1984), Grey-Wilson in Clematis, The Genus 147. (2000), O' Brien in Augustine Henry, An Irish Plant Collector in China 16. (2002).

Hubei: Yichang, 266, 3280a., 3280b. Antelope Glen (February 1887), 3280 (type). South Badong, 7351. **Yunnan:** In mountain forests near Mengzi at 2,000 metres, 9864. Mountains to the west of Simao at 1,600 metres, 9864a. In the mountains to the west of Simao at 1,640 metres, 9864b.

Henry 266 was the very first herbarium specimen ever collected by Augustine Henry and was first described as a new species, *Clematis henryi* (now *C. hayatae*), which later led to confustion with the white-flowered cultivar *Clematis* 'Henryi'. E. H. Wilson later collected *C. hayatae* around Yichang and Badong, stating it was not a common plant and that it was confined to low-level glens, ravines and rocky places. *C. hayatae* is a winter and early spring flowered species (November to February) and while its evergreen foliage is very handsome and distinct its creamy-white flowers are of no great horticultural value, being small and often hidden by leaves. Henry stated that in the Three Gorges region it was known colloquially as *mu t'ung*.

Clematis heracleifolia DC. (RANUNCULACEAE)

Zheng in Journ. Wuhan Bot. Res. 2.(1): 39. (1984).

Clematis heracleifolia DC. var. *ichangensis* Rehder & E. H. Wilson in Sargent Pl. Wilson. 1: 321.(1913), O' Brien in Augustine Henry, An Irish Plant Collector in China 16. (2002).

Hubei: Yichang, 2179, 3053, 4359. Liantuo, 4478. Changyang, 7494.

A semi-shrubby bush to about 80 cm tall bearing in late summer dense short terminal clusters of tubular scented deep blue flowers. Native to central and eastern China, introduced to cultivation in 1837.

Clematis javana DC. (RANUNCULACEAE)

Henry in Trans. Asiat. Soc. Jap. 24 suppl: 14. (1896) (List Pl. Formosa), W. T. Wang in Acta Phytotax. Sin. 39: 3. (2000).

Taiwan: Wanchin, 537.

A vigorous climber to 6 metres tall, carrying masses of white flowers in lateral cymes. A widespread species native to Chian (Taiwan), Japan, the Philippines, Timor, Indonesia and Papua New Guinea. In the Philippines the juice extracted from the leaves of this species are used to treat wounds.

Clematis lasiandra Maxim. (RANUNCULACEAE)

Rehder & E. H. Wilson in Sargent Pl. Wilson. 1: 322. (1913), Zheng in Journ. Wuhan Bot. Res. 2.(1): 39. (1984).

Hubei: Badong, 2859, 3694, 5099. Liantuo, 3006, 4460, 4589. Badong, 4898, 4959. Fang Xian, 6713. South Donghu, 7583. **Sichuan:** Henry's Chinese collector with Pratt, no locality stated but may be Leshan, Emei Shan, Wa Shan or Kangding, 8803.

A vigorous deciduous climber to 5 metres tall bearing axillary cymes of three bell-shaped flowers in October, varying in colour on different plants from white to voilet-purple, rosy-purple or slaty-purple. Native to southern and central China (including Taiwan) and southern Japan. A common species in Hubei, *Clematis lasiandra* was introduced to cultivation by Wilson in 1900. Seeds of Wilson's 679a. from Yichang and his 1315a. from Kangding in Sichuan were received at Glasnevin in 1908 and 1909 respectively. It was met with by members of the Glasnevin Central China Expedition (GCCE 361) near the village of Bai Sha in the mountains of south Badong in western Hubei in the autumn of 2002, nearby it grew a remarkable 1,000 year old tree of *Ginkgo biloba*. It was again collected two years later by the 2004 Glasnevin Central China Expedition (GCCE 582) near Henji town in Xingshan in October of that year.

Clematis leschenaultiana DC. (RANUNCULACEAE)

Zheng in Journ. Wuhan Bot. Res. 2.(1): 39. (1984).

Hubei: Antelope Glen, 3284 (February 1887). **Yunnan:** Mountains to the south of Mengzi at 1,800 metres, 9360. In Pingbian, in forests on the Daweishan Range at 1,950 metres, 11477.

A vigorous evergreen climber to 8 metres tall, flowers bell shaped, greenish-yellow to 3 cm long with up to 15 flowers in lateral cymes. Native to southern and south-east China (including Taiwan), Vietnam, the Philippines, Sumatra, Java, Bali and Lambok. One of a few species in the genus to occur in both the northern and southern hemisphere. Flowers appear in January to March in the Northern Hemisphere.

Clematis longistyla Handel-Mazzetti (RANUNCULACEAE) New species

in Acta Hort. Gotob. 13: 131. (1939).

Clematis florida Henry non Thunb. in Gard. Chron. ser. 3. 32: 51. fig. 20. (1902), in part.

Hubei: Yichang, 791 (March 1886), 1398 (Sept. 1886), 3516 (December 1887), 3516a.

Clematis meyeniana Walp. (RANUNCULACEAE)

Wang in Acta Phytotax. Sin. 41: 120. (2003).

Yunnan: Simao at 1,650 metres, 9349b. Mengla, 12270. In the mountains to the south of Simao at 1,450 metres, 12270a.

A vigorous climber to 7 metres tall, small, white starry flowers are carried in lateral cymes of 3 to 5 flowers which join to create lax masses in April to July. Native to subtropical southern and south-east China, including the islands of Hainan, Hong Kong and Taiwan, also native to Laos, Vietnam and southern Japan. According to Grey-Wilson this species thrives in the temperate house at the Royal Botanic Gardens, Kew. A. Henry 12267 is the most southerly collection made by Henry in Yunnan, Mengla lies less than 20 km from the Laos border.

Clematis meyeniana Walp. var. *granulata* Finet & Gagnepain (RANUNCULACEAE)

New variety. in Bull. Soc. Bot. France 50: 530. (1903), Wang in Acta Phytotax. Sin. 41: 124. (2003).

Hainan: Hummocks near Haikou, 8087 (syntype). Environs of Kiungchow, 8682.

Close to the type but with the leaflets more finely rugose on the uppersurface. Native to southern China (Guangdong, Guangxi, Hainan and south-east Yunnan), also Laos and northern Vietnam.

Clematis montana Buch. -Ham. ex DC. (RANUNCULACEAE)

Henry in Journ. China Br. Roy. Asiat. Soc. 22: 245. (1887) (Chinese Names of Plants), Rehder & E. H. Wilson in Sargent Pl. Wilson. 1: 333. (1913), Zheng in Journ. Wuhan Bot. Res. 2.(1): 39. (1984), O' Brien in Augustine Henry, An Irish Plant Collector in China 16. (2002), Wang in Acta Phytotax. Sin. 41: 206. (2003).

Hubei: Badong, 204, 207 (both June 1885). Fang Xian, 6887. **Sichuan:** Henry's Chinese collector with Pratt, no locality stated but may be Leshan, Emei Shan, Wa Shan or Kangding, 8800.

A vigorous deciduous climber to 12 metres, native to the Himalaya, Myanmar and China (including central Taiwan), well known in cultivation and known colloquially at Yichang as *ta huai t'ung* according to Henry. Introduced to cultivation from the Himalaya by Lady Amherst in 1831, most of the forms in cultivation today are of Chinese origin.

Clematis montana Buch.-Ham. ex DC. var. *rubens* E. H. Wilson (RANUNCULACEAE) New variety

Rehder & E. H. Wilson in Sargent Pl. Wilson. 1: 333. (1913), Zheng in Journ. Wuhan Bot. Res. 2.(1): 39. (1984), Grey-Wilson in Clematis, The Genus 85. (2000).

Hubei: Yichang, 1240. Badong, 5437. Sichuan: South Wushan, 5437a.

Discovered by Henry in 1886 and introduced to cultivation in 1900 by Wilson through Messrs Veitch who first flowered this variety in September 1903. The variety was described from one of these seedlings and no type specimen was ever designated. The foliage of this variety is purplish and more downy than the type and the rosy red flowers appear in June. In *Veitch's Novelties for 1908—1909* plants were available for two shillings and six pence to seven shillings and six pence. It was awarded a First Class Certificate when shown at a Royal Horticultural Society Show in 1905 from one of the Coombe Wood plants.

Clematis montana Buch.-Ham. var. *wilsonii* Sprague (RANUNCULACEAE) New variety

Rehder & E. H. Wilson in Sargent Pl. Wilson. 1: 333. (1913), Wang in Acta Phytotax. Sin. 41: 215. (2003).

Yunnan: Mountains to the north of Mengzi at 2,300 metres, 10748.

This variety was described from a cultivated plant grown from one of Wilson's 1907 central Chinese collections. Henry had first collected the variety in Yunnan. Native to Sichuan, western Hubei and Yunnan, it bears large scented white flowers in July and August. Wilson stated that this was the commonest form of *Clematis montana* to be seen in western Sichuan.

Clematis oreophila Hance (RANUNCULACEAE)

Anonymous in Kew List of Determinations (volume 24).

Hubei: Badong, 714.

Clematis parviloba Gard. & Champ. (RANUNCULACEAE)

Henry in Trans. Asiat. Soc. Jap. 24 suppl: 14. (1896) (List Pl. Formosa).

Taiwan: Wanchin, 846.

A deciduous climber, white flowers are carried in lateral cymes from April to July. Also native to Fujian, Guangdong, Guangxi, Hong Kong, Jiangxi, Sichuan and southern Zhejiang where it grows in forests, ravines and thickets between 500 and 3,200 metres.

Clematis pashanensis (M. C. Chang) W. T. Wang (RANUNCULACEAE)

Wang in Acta Phytotax Sin. 41: 142. (2003).

Clematis benthamiana Henry non Hemsl. in Journ. China Br. Roy. Asiat. Soc. 22: 240. (1887) (Chinese Names of Plants), Rehder & E. H. Wilson in Sargent Pl. Wilson. 1: 330. (1913), Zheng in Journ. Wuhan Bot. Res. 2(1): 37. (1984).

Hubei: Yichang, 309a., 1062, 1497, 2773, 4368. San You Dong Glen near Yichang (July 1886), 1497, 1518, 4339. Badong 2881. Without locality, 4368a., 4368b.

This species which is endemic to central and eastern China was first found by Henry near Yichang and later at Badong by Wilson. It was known colloquially around Badong as *chiu lung hsu,* according to Henry.

Clematis peterae Hand.-Mazz. (RANUNCULACEAE)

Zheng in Journ. Wuhan Bot. Res. 2.(1): 38. (1984), Wang in Acta Phytotax. Sin. 41: 30. (2003).

Clematis gouriana Roxb. ex DC. var. *finetii* Rehder & E. H. Wilson in Sargent Pl. Wilson. 1: 339. (1913), non Levl. & Vant.

Hubei: Badong, 2420 Xingshan, 6461.

A deciduous climber to 5 metres tall bearing clusters of small white flowers in terminal cymes during May and July. A widespread species in China, it was collected in Gansu by Potanin and Rock, Hubei by Henry and Wilson, Sichuan by Farges and Soulie and in Yunnan by Delavay, Monbeig, Schneider, Forrest and Ducloux.

Clematis peterae Hand.-Mazz. var. *trichocarpa* W. T. Wang (RANUNCULACEAE)

Wang in Acta Phytotax. Sin. 41: 33. (2003).

Clematis gouriana Rehder & E. H. Wilson non Roxb. ex DC. in Sargent Pl. Wilson. 1: 339. (1913).

Yunnan: Milê district, 10306.

A deciduous climber to 5 metres tall, differing from the type by having pubescent achenes. A widespread plant in China where it was first found by Potanin.

Clematis pogonandra Maxim. (RANUNCULACEAE)

Rehder & E. H. Wilson in Sargent Pl. Wilson. 1: 319. (1913), Zheng in Journ. Wuhan Bot. Res. 2.(1): 40. (1984).

Clematis prattii Hemsley in Kew Bull. Misc. Inf. 82. (1892), Bretschneider in Hist. Eur. Bot. Disc. China 2: 778. (1898), Finet & Gagnepain in Bull. Soc. Bot. France L. 550. (1903); Contrig. Fl. As. Or. I. 35. (1905), Hemsley in Kew Bull. Misc. Inf. 148. (1906). *Clematis otophora* Zheng non Franch. ex Finet & Gagn. in Journ. Wuhan. Bot. Res. 2.(1): 38. (1984).

Hubei: Badong, 4920 (syntype of *Clematis prattii* Hemsley). **Sichuan:** North Wushan, 6704 (syntype of *Clematis prattii* Hemsley). Fang at 1,900 to 3,000 metres 6817.

An evergreen climber to 2.5 metres tall, flowers (carried in May to June) small and solitary, yellow, flushed purple-brown and carried in the axils of leaves. Native to southern Gansu, western Hubei, southern Shaanxi and western Sichuan where it grows on forest margins, in thickets and in ravines between 2,200 to 3,400 metres. Discovered by the Russian explorer, Grigori Nicolaevich Potanin in Gansu on July 15th 1885.

Clematis pterantha Dunn (RANUNCULACEAE) New variety

in Hooker's Icon. Pl. t. 2713. (1901), Wang in Acta Phytotax. Sin. 31: 224. (1993).

Clematis ranunculoides Franch. var. *pterantha* (Dunn) M. Y. Fang in Fl. Reip. Pop. Sin. 27: 119. (1980).

Yunnan: Mountain forest to the west of Simao at 1,650 metres, 12452 (type of *Clematis pterantha* Dunn).

The flowers are borne in larger lateral cymes than the type and are white or pale rose coloured and the stems and

02

leaves are glabrous. According to Grey-Wilson this species is restricted to Pu'er, a few miles north of where Henry discovered the variety.

Clematis puberula Hook. f. & Thoms. var. *ganpiniana* (H. Lévl. & Vantiot) W. T. Wang (RANUNCULACEAE)

Wang in Acta Phytotax. Sin. 38: 407. (2000); in Acta Phytotax. Sin. 41: 57. (2003).

Clematis parviloba Hook. f. & Thoms. var. *glabrescens* Finet & Gagnepain in Bull. Soc. Bot. France 50: 834. (1903), Levl. in Fl. Kouy-tcheou 333. (1915). *Clematis brevicaudata* DC. var. *lissocarpa* Rehder & E. H. Wilson in Sargent Pl. Wilson.1: 340. (1913).

Hubei: Xingshan, 6462 (paratype of *Clematis brevicaudata* DC. var. *lissocarpa* Rehder & E. H. Wilson). In a ravine on Moji Shan, 4338 (syntype of *Clematis parviloba* Gardn. & Champ. var. *glabrescens* Finet & Gagnepain).

Endemic to China. The type specimen was collected by Ganpin in Guizhou on the 4th of August 1897, Henry had discovered the variety over a decade previous to this.

Clematis puberula Hook. f. & Thoms. var. *tenuisepala* (Maxim.) W. T. Wang (RANUNCULACEAE)

Wang in Acta Phytotax. Sin. 38: 406. (2000); in Acta Phytotax. Sin. 41: 56. (2003).

Clematis puberula Henry non Hook. f. & Thoms. in Journ. China Br. Roy. Asiat. Soc. 22: 245. (1887) (Chinese Names of Plants). *Clematis brevicaudata* DC. var. *filipes* Rehder & E. H. Wilson in Sargent Pl. Wilson. 1: 340. (1913). *Clematis ganpiniana* (Levl. & Vant.) Tamura var. *tenuisepala* (Maxim.) C. T. Ting Zheng in Journ. Wuhan Bot. Res. 2.(1): 37. (1984).

Hubei: Yichang, 2347. On the banks of the Yangtze at Yichang, 4361. Antelope Glen, 4340. Liantuo, 4583 (type of *Clematis brevicaudata* DC. var. *filipes* Rehder & E. H. Wilson, May 1888).

According to Henry's Badong collector this variety was colloquially as *huai t'ung*.

Clematis quinquefoliolata Hutchchinson (RANUNCULACEAE) New species

in Gard. Chron. ser. 3, 41: 3. (1907), Fedde in Repert. Nov. Spec. Regni Veg. 6: 317. (1908), Rehder & E. H. Wilson in Sargent Pl. Wilson.1: 328. (1913), Bean in Trees & Shrubs 3: 98. (1936), Zheng in Journ. Wuhan. Bot. Res. 2.(1): 40. (1984), Nelson in Pim, The Wood & the Trees, 230. (1984), Bean in Trees & Shrubs 1: 657. (1989), Grey-Wilson in Clematis, The Genus 105. (2000), Nelson in A Heritage of Beauty 311. (2000), Wang in Acta Phytotax. Sin. 41: 128. (2003).

Hubei: At Shih Pan Shan [off the Xiling (Yichang) Gorge], 4185. Yichang, 4332.

An evergreen climber to 5 metres tall, producing lateral cymes of three, five or seven, white flowers to 5 cm across in early autumn. Discovered by Henry in 1887 and introduced from Zigui in Hubei by E. H. Wilson in 1900 for Messrs Veitch who flowered it in their Coombe Wood nursery in September 1906. Native to north-east Guizhou, Hubei, north-west Hunan, eastern Sichuan and northen Yunnan where it grows on woodland verges and in thickets at 1,000 to 1,800 metres and flowering in July to September. Allied to *Clematis armandii* but differing from that species by its pinnate leaves and autumn flowered habit.

Clematis ranunculoides Franch. (RANUNCULACEAE)

Clematis acutangula Hook. f. & Thoms. ssp. *ranunculoides* (Franch.) W. T. Wang in Acta Phytotax. Sin. 31: 233. (1993). *Clematis acutangula* Hook. f. & Thoms. f. *major* W. T. Wang in Acta Phytotax. Sin. 31: 222. (1993).

Yunnan: On grassy hills near Mengzi at 1,500 metres, 9326. Mengzi, 9174.

An herbaceous perennial from 30 to 60 cm tall, though if allowed to climb through shrubs or up a wall it may become to 3 metres tall. Flowers bell-shaped with recurved petals, appearing from September to November, rose-pink, to 20 mm are carried in terminal clusters. Native to north-west Guangxi, western Guizhou, south-west Sichuan and Yunnan where it grows in grass and in thickets at 500 to 3,000 metres. I have seen good plants of this growing in the grounds of the Yu Feng Temple near the base of Yu Long Xue Shan, at Lijiang in north-west Yunnan. Discovered by Père Delavay in September 1883, it was introduced in 1906 from the Cangshan Range above Dali by George Forrest.

Clematis rubifolia C. H. Wright (RANUNCULACEAE)

In the Harvard University Herbarium (A.).

Yunnan: Mengzi at 1,650 metres, 9844.

An ally of *Clematis leschenaultiana*. Flowers are greenish-yellow and appear between November to January. Discovered by the Irish plant collector and Chinese Maritime Customs official, William Hancock near Mengzi in 1894. Native to western Guangxi, southern Guizhou and Yunnan where it grows in thickets and forest margins between 800 to 2,000 metres.

Clematis siamensis Drumm. & Craib. (RANUNCULACEAE) New species

Clematis acuminata DC. var. *multiflora* H. F. Comber in Notes Roy. Bot. Gard. Edinburgh 18: 234. (1934). *Clematis acuminata* DC. var. *hirtella* Hand.-Mazz. in Acta Horti. Gothob. 13: 194. (1940).

Yunnan: In mountain forest to the east of Simao at 1,500 metres, 11586 (isotype of *Clematis acuminata* DC. var. *hirtella* Hand.-Mazz.). In the mountains to the east of Simao at 1,650 metres, 11586a., (syntype of *Clematis* acuminata DC. var. *multiflora* H. F. Comber). In the mountains to the west of Simao at 1,640 metres 11586b., (syntype of *Clematis acuminata* DC. var. *multiflora* H. F. Comber), 11586c., (syntype of *Clematis acuminata* DC. var. *multiflora* H. F. Comber). Simao at 1,650 metres 13282 (syntype of *Clematis acuminata* DC. var. *multiflora* H. F. Comber).

This species was discovered by Henry in 1897 and was described from collections later made by the Irish collector, Kerr in northern Thailand.

Clematis smilacifolia Wall. (RANUNCULACEAE)

In the Harvard University Herbarium (A.).

Yunnan: In woods near Simao at 1,950 metres, 12766.

A vigorous climber to 10 metres tall, flowers appear in September to November in the northern hemisphere, these are violet-purple and are carried in a many flowered terminal panicle. Native to eastern Nepal, India (Assam), Myanmar, Thailand, Vietnam, Malaysia and in China in southern Guangdong, Guangxi, southern Guizhou, Hainan, south-east Tibet (Xizang) and Yunnan, growing in subtropical forest at 1,000 to 2,300 metres.

Clematis smilacifolia Wall. var. *peltata* (W. T. Wang) W. T. Wang. (RANUNCULACEAE)

In the Harvard University Herbarium (A.).

Yunnan: In the mountains to the south-east of Simao at 1,640 metres, 12766a. In mountain forest to the south-east of Simao at 1,650 metres, 13561.

Clematis sp. (Connata Group) (RANUNCULACEAE)

Clematis gracilis Edgew. Anonymous in Kew List of Determinations (volume 24).

Hubei: Without locality, 3284a.

Clematis gracilis Edgeworth is synonomous with *Clematis connata* DC., a species from the Himalaya and western China, not central China. I have not traced this particular collection nor have I checked in the Kew Herbarium.

Clematis spp. (RANUNCULACEAE)

Anonymous in Kew List of Determinations (volume 24).

Hubei: Yichang, 1398. Liantuo, 4608. **Sichuan:** Henry's Chinese collector with Pratt, no locality stated but may be Leshan, Emei Shan, Wa Shan or Kangding, 8807.

Clematis subumbellata Kurz (RANUNCULACEAE)

Wang in Acta Phytotax. Sin. 38: 310. (2000); in Acta Phytotax. Sin. 41: 48. (2003).

Yunnan: On the Red River near Manpan, 10919. Yuanjiang, 10919a.

Henry described his Red River collection as a large climber on trees. Allied to *Clematis gouriana*, this species is native to China (southern Yunnan), Laos, northern Myanmar,

and northern Vietnam where it grows in thickets and on forest margins between 400 to 1,900 metres. The flowering period of this species is from December to February.

Clematis tashiroi Maxim. (RANUNCULACEAE)

Wang in Acta Phytotax. Sin. 38: 419. (2000).

Clematis parviloba Henry non Gard. & Champ. in Trans. Asiat. Soc. Jap. 24 suppl: 14. (1896) (List Pl. Formosa)

Taiwan: Oluanpi, 1320.

A woody climber, dark-purple flowers appearing between August to October. Native to Vietnam, China (Taiwan) and Japan where it grows on stream margins and beaches between 700 to 2,800 metres.

Clematis terniflora DC. (RANUNCULACEAE)

Clematis recta L. ssp. *terniflora* (DC.) Kuntze Anonymous in Kew List of Determinations (volume 24).

Hubei: Yichang, 308, 440, 4367. Liantuo, 2128.

A vigorous climber to 10 metres tall, starry white flowers are borne in lateral cymes in June to September and said to be scented of hawthorn (*Crataegus monogyna*). Native to central and eastern China, Japan and Korea. Introduced to cultivation from Japan by von Siebold in 1830 through Ghent Botanic Garden in Belgium. Grey-Wilson likens this species as the eastern equivalent of *Clematis flammula*.

Clematis terniflora DC. var. *garanbiensis* (Hayata) M. C. Chang (RANUNCULACEAE) New variety

Yang et al. in Flora of Taiwan 2: 541. (1996).

Clematis recta Henry non L. in Trans. Asiat. Soc. Jap. 24 suppl: 14. (1896) (List Pl. Formosa).

Taiwan: Oluanpi, 1289.

A scandent perennial vine, endemic to coastal areas of southern Taiwan. Discovered by Henry in 1893.

Clematis tsaii W. T. Wang (RANUNCULACEAE) New species

Wang in Acta Phytotax. Sin. 41: 29. (2003).

Yunnan: In mountain forest to the west of Simao at 1,650 metres, 12444.

A climber to about 5 metres tall, allied to *Clematis grata* Wall. and *Clematis chingii* W. T. Wang. Native to south-east Tibet (Xizang) and Yunnan where it grows in thickets and the verges of forests between 1,500 and 2,000 metres. The type specimen was collected by H. T. Tsai on the 18th of September 1933 in Yunnan near Fugong. Henry had found the species over three decades previous to this. The flowering period for this species is between September and October.

Clematis uncinata Champ. ex Benth. (RANUNCULACEAE)

Rehder & E. H. Wilson in Sargent Pl. Wilson. 1: 327. (1913), Morley in Glasra 3: 79. (1979), Bean in Trees &

Shrubs 1: 661. (1989), Zheng in Journ. Wuhan Bot. Res. 2.(1): 40. (1984), Grey-Wilson in Clematis, The Genus 112. (2000), Wang in Acta Phytotax. Sin. 41: 163. (2003).

Clematis leiocarpa Oliver in Hooker's Icon. Pl. 16: t. 1533. (1886), Bretschneider in Hist. Eur. Bot. Disc. China 2: 778. (1898). *Clematis chinensis* Henry non Retz. in Trans. Asiat. Soc. Jap. 24 suppl: 14. (1896) (List Pl. Formosa).

Hubei: Yichang, 309b., 714a., (syntype of *Clematis leiocarpa* Oliver)., 1173, 4097, 4107, 4385, 3622a. Liantuo, 3922. On road to Moji Shan (June 1886), 1553. Xiling (Yichang) Gorge, 3622. South Badong, 5510. Jianshi, 6008. Changleping, 6212, 6287. **Taiwan:** Wanchin, 465. **Yunnan:** Vacinity of Mengzi between 1,600 and 2,000 metres, 9431, 9431a., 9431b., 9431c., 9431d.

An evergreen climber from 4 to 12 metres tall bearing numerous creamy fragrant white flowers to 10 cm across in leafless cymes in summer. Native to central and southern China (including Taiwan), northern Vietnam and southern Japan. *Clematis uncinata* was discovered by Champion in a ravine near Mount Parker in Hong Kong in 1848 and was introduced to cultivation by Wilson in 1901. This species is cultivated at the National Botanic Gardens, Kilmacurragh from collections made by the Glasnevin Central China Expedition in Badong in October 2002.

Clematis uncinata Champ. ex Benth. var. *coriacea* Pamp. (RANUNCULACEAE) New variety

Wang in Acta Phytotax. Sin. 38: 312. (2000); in Acta Phytotax. Sin. 41: 166. (2003).

Clematis leiocarpa Oliver in Hooker's Icon. Pl. 16: t. 1533. (1886), in part. Bretschneider in Hist. Eur. Bot. Disc. China 2: 778. (1898).

Hubei: Yichang, 288, 309. Near Yichang by the Pingshanba creek (May 1886), 1596. Badong, 1413.

While describing *Clematis leiocarpa*, Daniel Oliver cited two specimens. Wang Wen-Tsai from the Institute of Botany, Beijing (an authority on *Clematis*) studied both of these at Kew and has determined they represent two taxa. The flowering specimen, Henry 714a. is *Clematis uncinata* var. *uncinata* whereas the other syntype, A. Henry 309, a fruiting specimen is Pampanini's *Clematis uncinata* var. *coriacea*. Pampanini based this variety on specimens collected by the Italian missionary C. Sylvestri at Zan-lan-sian in Hubei in 1913, Henry had found the plant some 28 years earlier. This variety differs from the type in its more leathery, sometimes bipinnate leaves. Distributed through the provinces of southern Gansu, western Hubei, north-west Hunan, southern Shaanxi and north-east Sichuan.

Clematis urophylla Franch. (RANUNCULACEAE)

Rehder & E. H. Wilson in Sargent Pl. Wilson. 1: 323. (1913).

Yunnan: In mountains near Mengzi, 11347.

An evergreen climber carrying pendant cymes of up to 3 bell-shaped white flowers. A beautiful plant both in terms of foliage and flower, native to south-west China where it grows on forest margins and thickets at 500 to 2,400 metres. Discovered by the French missionary, Père Paul Perny in Guizhou, it was recently introduced to cultivation (as a conservatory climber) by Martyn Rix. *Clematis urophylla* received an Award of Merit in 1998. For an illustration of this species see Grey-Wilson in *Clematis, The Genus* 152. (2000).

Clematis wissmanniana Handel-Mazzetti (RANUNCULACEAE) New species

in Acta. Hort. Gotob. 13: 212. (1939), Wang in Acta Phytotax. Sin. 41: 47. (2003).

Yunnan: Mengzi, 9346a. On rocky mountains near Mengzi at 1,800 metres, 9346b. Yuanjiang, 13219 (holotype). Simao, at 1525 metres 13346.

A deciduous climber bearing cymes of small starry white flowers, Endemic to Yunnan where it grows in woods and thickets at 1,200 to 1,800 metres and flowering in September and October. Discovered by Henry near Yuanjiang and collected four decades later in the same region by Wissmann for whom the species was named.

Clematis yuanjiangensis W. T. Wang (RANUNCULACEAE) New species

in Acta Phytotax. Sin. 31: 224. (1993).

Yunnan: Near Yuanjiang at 1,650 metres, 13229 (holotype).

An herbaceous climber with thin four furrowed stems, allied to *Clematis ranunculoides*, but with larger flowers and with widely spreading pinkish sepals. Native to southern Yunnan where it grows in thickets at 1,600 metres. A very pretty plant, it also grows in the Stone Forest to the south of Kunming. For an illustration of this species see Martyn Rix's photograph in Grey-Wilson's Clematis, The Genus 135. fig. 99. (2000).

Clematis yunnanensis Franch. (RANUNCULACEAE)

In the Harvard University Herbarium (A.).

Yunnan: Near Yuanyang, on Fengchunling in bamboo forest at 2,600 metres, 10843.

An evergreen bushy climber, carrying in October to December lateral cymes of up to 7 bell-shaped flowers. Native to western Guangxi, south-west Sichuan, eastern Tibet (Xizang) and Yunnan (where it was discovered by Père Jean Marie Delavay in December 1884) where it grows on forest margins and in scrub between 1,600 and 3,000 metres. A handsome plant recently introduced to cultivation.

Clematoclethra scandens (Franch.) Maxim. ssp. ***hemsleyi*** (Baill.) Y. C. Tang & Q. Y. Xing (ACTINIDIACEAE)
in Acta Phytotax. Sin. 27: 90. (1989)

Clematoclethra hemsleyi Baill. Rehder in Sargent Pl. Wilson. 2: 389. (1913), Zheng in Journ. Wuhan Bot. Res. 2.(1): 150. (1984), Rehder in Sargent Pl. Wilson 2: 389. (1916). *Clematoclethra henryi* Franchet ex Komarov in Acta. Hort. Petrop. 29: 94. (pro synon.) (1908).

Hubei: Fang at 1,900 to 3,000 metres, 6818, 6885.

Cleome gynandra L. (CLEOMACEAE)

Gynandropsis pentaphylla (L.) DC. Henry in Journ. China Br. Roy. Asiat. Soc. 22: 266. (1887) (Chinese Names of Plants); in Trans. Asiat. Soc. Jap. 24 suppl: 18. (1896) (List Pl. Formosa).

Hubei: Yichang, 169, 1505. **Taiwan:** Wanchin, without number.

An annual herb to 130 cm tall, widely distributed in tropical and subtropical regions According to Henry this was a common weed at Yichang where it was known as *peh hua ts'ai.*

Cleome viscosa L. (CLEOMACEAE)

Polanisia viscosa (L.) DC. Henry in Trans. Asiat. Soc. Jap. 24 suppl: 18. (1896) (List Pl. Formosa).

Hainan: Environs of Haikou, 8065.

Tickweed, a pantropical annual to 1.5 metres tall, a common weed on tropical seashores.

Clerodendrum bungei Steud. (LAMIACEAE)

Zheng in Journ. Wuhan Bot. Res. 2.(1): 185. (1984).

Clerodendron foetidum Bunge Henry in Journ. China Br. Roy. Asiat. Soc. 22: 241. (1887), Forbes & Hemsley in Journ. Linn. Soc. Bot. 26: 259. (1890), Rehder in Sargent Pl. Wilson.3: 375. (1917).

Hubei: Yichang, 189, 1393, 2307. **Yunnan:** Mengzi, 10299.

A shrub to 2 metres tall bearing large heart shaped leaves, in August and September terminal rounded corymbs of purple-red blossoms to 12 cm across. A common plant around Yichang according to Henry who also stated it was known colloquially as *ch'ou mou tan.* Introduced to cultivation by Robert Fortune in 1844, this species is best suited to sheltered gardens and does well in the maritime Irish climate. *Clerodendrum bungei* was collected by the Glasnevin Central China Expedition (GCCE 1) on Emei Shan in September 2002.

Clerodendrum chinense (Osbeck) Mabb. (LAMIACEAE)

Clerodendrum fragrans (Vent.) R. Br. Forbes & Hemsley in Journ. Linn. Soc. Bot. 26: 260. (1890).

Hainan: Twenty miles west of Haikou and near Kiunchow, without number.

Clerodendrum colebrookianum Walp. (LAMIACEAE)

Clerodendrum colebrookianum Walp. var. *henryanum* Moldenke in Phytologia 52: 330. (1983).

Yunnan: Simao, 12421a., (type of *Clerodendrum colebrookianum* Walp. var. *henryanum* Moldenke, collected September 1898), 12421.

Clerodendrum cyrtophyllum Turcz. (LAMIACEAE)

Forbes & Hemsley in Journ. Linn. Soc. Bot. 26: 259. (1890), Rehder in Sargent Pl. Wilson. 3: 377. (1917), Li in Woody Flora of Taiwan 826. (1963).

Clerodendrum fragrans Henry non (Vent.) R. Br. in Trans. Asiat. Soc. Jap. 24 suppl: 71. (1896) (List Pl. Formosa).

Hainan: Haikou, without number. **Taiwan:** Wanchin, 1, 393, 1470, 562. Oluanpi, 1295. Wanshoushan, 1873. Tamsui, 1743.

Native to China, Korea and Indochina. A widespread species in Chinese Taiwan.

Clerodendrum japonicum (Thunb.) Sweet (LAMIACEAE)

Rehder in Sargent Pl. Wilson. 3: 377. (1916).

Yunnan: Simao at 1,200 metres, 12060.

The Japanese glorybower, a shrub to 2 metres tall, producing from April to November terminal panicles of spectacular orange-red flowers. Distributed in India (including Sikkim), Bhutan, Bangladesh, Laos, Vietnam, the Malay Peninsula and Japan. The Hani (an ethnic group) of Xishuangbanna in southern Yunnan call this species *haqhmldiaqceiq* and the roots are used to treat cases of rheumatism, joint pain, back ache and contusions.

Clerodendrum laevifolium Blume (LAMIACEAE)

Clerodendrum henryi C. P'ei in Mem. Sci. Soc. China 1: 40. (1932), C. Y. Wu in Flora Yunnanica 1: 463. (1977). *Clerodendrum longilimbum* C. P'ei in Mem. Sci. Soc. China 1(3): 151. (1932), C. Y. Wu in Flora Yunnanica 1: 463. (1977).

Yunnan: Tonguan, 11585b., (type of *Clerodendrum henryi* C. P'ei). Mengzi 9302. In the mountains to the south of Simao at 1,450 metres, 11585a., (isotype of *Clerodendrum longilimbum* C. P'ei).

Clerodendrum macrostachya (Turcz.) Leerat. & Chantar. (LAMIACEAE)

Clerodendrum subscaposum Hemsley in Hookers Icon. Pl. 27: t. 2675. (1901).

Yunnan: In Pingbian on the Daweishan Range at 1,950 metres, 9181 (isotype of *Clerodendrum subscaposum* Hemsley, collected September 23rd 1896).

Clerodendrum mandarinorum Diels (LAMIACEAE)

Rehder in Sargent Pl. Wilson. 3: 375. (1916).

Yunnan: In Pingbian on the Daweishan Range at 1,500 metres, 9026, 9026a.

Distributed in western China and Vietnam.

Clerodendrum paniculatum L. (LAMIACEAE)

Forbes & Hemsley in Journ. Linn. Soc. Bot. 26: 261. (1890), Henry in Notes Econ. Bot. China 61. (1893); in Trans. Asiat. Soc. Jap. 24 suppl: 71. (1896) (List Pl. Formosa).

Hainan: Haikou, without number. **Taiwan:** Kaohsiung, 159. Oluanpi, 617.

The *chu-t'ung* according to Henry, a spectacular flowering shrub to 3 metres tall bearing upright racemes of fiery orange-red flowers. Native from China to Malaysia and common in Chinese Taiwan at low altitude. This species was collected by the Glasnevin Taiwan Expedition (GTE 96) on Wanshoushan in October 2004. The rare *Clerodendrum paniculatum* L. forma *albiflorum* (Hemsl.) Hsieh was also collected by the Glasnevin Taiwan Expedition (GTE 51) by the lighthouse at Oluanpi.

Clerodendrum peii Moldenke (LAMIACEAE) New species

Clerodendrum longipetiolatum C. P'ei in Mem. Sci. Soc. China 1(3): 159. (1932), non Gurcke (1894).

Yunnan: Yuanjiang, 13511 (type of *Clerodendrum longipetiolatum* C. P'ei).

Southern Yunnan, northern Vietnam.

Clerodendrum sp. (LAMIACEAE)

Anonymous in Kew List of Determinations (volume 34).

Yunnan: Mengzi, 11390.

Clerodendrum trichotomum Thunb. (LAMIACEAE)

Henry in Journ. China Br. Roy. Asiat. Soc. 22: 241. (1887) (Chinese Names of Plants), Forbes & Hemsley in Journ. Linn. Soc. Bot. 26: 262. (1890).

Clerodendron sp. Henry in Trans. Asiat. Soc. Jap. 24 suppl: 71. (1896) (List Pl. Formosa). *Clerodendrum trichotomum* Thunb. var. *fargesii* (Dode) Rehd. Rehder in Sargent Pl. Wilson. 3: 376. (1916), Zheng in Journ. Wuhan Bot. Res. 2.(1): 186. (1984), Li in Woody Flora of Taiwan 829. (1963).

Hubei: Badong, 2434, 2839. South Badong, 7327. Liantuo, 2616, 2620, 4646. Xingshan, 4520. Donghu, 6422. North Badong, 6422a. Jianshi, 7327. **Taiwan:** Wanchin, 28, 422. Oluanpi, 291, 961.

A large deciduous shrub or small tree to 7 metres tall bearing small white flowers in July to September. It is the bright blue fruits surrounded by a persistent crimson calyx that brought this handsome plant to the attention of horticulturists. Native to India, Japan E. Asia, Korea and China, the leaves for some have an unpleasant odour when crushed, other claim it smells of roast beef! According to Henry this shrub was known colloqually in the Three Gorges

region as *ch'ou mou tan shu*. Cultivated at the National Botanic Gardens, Glasnevin and Kilmacurragh, from collections made by the Glasnevin Central China Expedition (GCCE 95) on Emei Shan.

Clethra delavayi Franch. (CLETHRACEAE)

Clethra canescens Forbes & Hemsley non Reinw. ex Blume in Journ. Linn. Soc. Bot. 26: 33. (1889). *Clethra esquirolii* H. Lévl. Rehder & E. H. Wilson in Sargent Pl. Wilson. 1: 502. (1913), Hu in Jour. Arnold Arb. xli. 185. (1960), Zheng in Journ. Wuhan Bot. Res. 2.(1): 166. (1984).

Hubei: Yichang, 2838 (February 1887).

A deciduous tree to 12 metres tall, native to India, Bhutan and China and discovered by Delavay in Yunnan in 1884 and introduced to cultivation by Forrest from the same province in 1913. A highly variable species and a beautiful plant when carrying masses of lily of the valley like flowers. There is a fine tree on Garinish Island (Ilnacullin) in Bantry Bay, County Cork.

Clethra fargesii Franch. (CLETHRACEAE) New species

Rehder & E. H. Wilson in Sargent Pl. Wilson. 1: 502. (1913), Rehder & E. H. Wilson in Sargent Pl. Wilson. 1: 502. (1913), Hu in Jour. Arnold Arb. xli. 185. (1960), Zheng in Journ. Wuhan Bot. Res. 2.(1): 166. (1984).

Clethra barbinervis Henry non Sieb. & Zucc. in Journ. China Br. Roy. Asiat. Soc. 22: 256. (1887) (Chinese Names of Plants). *Clethra canescens* Forbes & Hemsley non Reinw. ex Blume in Journ. Linn. Soc. Bot. 26: 33. (1889).

Hubei: Yichang, 177. Liantuo 4492. Xingshan, 4531. Badong, 4694. Donghu, 6407 (March 1888). Jianshi 5818. **Sichuan:** South Wushan, 7270.

A handsome shrub to 4 metres tall, bearing fragrant terminal clusters of pure white flowers on slender racemes to 25 cm long. *Clethra fargesii* is common in open woodlands and thickets in the middle Yangtze region according to Hu. It was first found in central China by Henry in 1885 (one of his earliest discoveries in fact) and later by Farges on whose 108 from Sichuan the species was based. In the Three Gorges region Henry stated it was known as *shan lui*. Introduced to cultivation by Wilson while collecting for Messrs Veitch in 1900.

Clethra sp. (CLETHRACEAE)

Anonymous in Kew List of Determinations (volume 24).

Hubei: Badong, 1783, 1829.

Cleyera japonica Thunb. (PENTAPHYLACACEAE)

Ternstroemia japonica (Thunb.) Thunb. Anonymous in Kew List of Determinations (volume 34). *Eurya japonica* Henry non Thunb. in Trans. Asiat. Soc. Jap. 24 suppl: 19. (1896) (List Pl. Formosa). *Cleyera ochnacea* DC. Henry in Trans. Asiat. Soc. Jap. 24 suppl: 19. (1896) (List Pl. Formosa). *Eurya chinensis* Kobuski non R. Brown in Ann. Missouri. Bot. Gard. 25: 319. (1938), for A. Henry 1465. *Cleyera japonica* Thunb. var. *morii* (Yamamoto) Masamune Li in Woody Flora of Taiwan 586. fig. 230. (1963).

Taiwan: Tamsui, 1465, 1468. **Yunnan:** In Pingbian on the Daweishan Range at 1,950 metres, 9183.

Native to India, Nepal, Myanmar, China, Japan, Indochina and the Philippines.

Climacium japonicum Lindb. (CLIMACIACEAE)
Salmon in Journ. Linn. Soc. Bot. 34: 466. (1900).

Hubei: Jianshi, 5934 (June 1888). **Sichuan:** South Wushan, 7059 (September 1888). (Moss).

Clinopodium chinense (Benth.) O. Ktze. (LAMIACEAE)
Calamintha chinensis Benth. Forbes & Hemsley in Journ. Linn. Soc. Bot. 26: 283. (1890).

Hubei: Liantuo, 2598, 5187. **Yunnan:** Mengzi at 1,650 metres, 9321.

Distributed in China and Japan.

Clinopodium discolor (Diels) C. Y. Wu & Hsuan ex H. W. Li (LAMIACEAE) New species
Calamintha clinopodium Benth. var. *discolor* (Diels) Dunn Dunn in Notes Roy. Bot. Gard. Edinburgh 6: 159. (1915).

Yunnan: Mengzi, 9766.

Discovered by Henry in 1896 and described from material later collected by Forrest (Forrest 4527, type) from the Cangshan Range above Dali. Wilson had also collected this species previous to Forrest. Native to Yunnan and Tibet (Xizang).

Clinopodium gracile (Benth.) Kunze (LAMIACEAE)
Calamintha gracilis Benth. Forbes & Hemsley in Journ. Linn. Soc. Bot. 26: 283. (1890), Dunn in Notes Roy. Bot. Gard. Edinburgh 6: 158. (1915).

Hubei: Yichang, 252, 770a., 1002, 1198, 1957. South Badong, 5337. Jianshi, 5844.

This species is currently used in Chinese herbal medicine to treat sore-throat.

Clinopodium megalanthum (Diels) C. Y. Wu & Hsuan ex H. W. Li (LAMIACEAE)
Calamintha clinopodium Dunn non Benth. in Notes Roy. Bot. Gard. Edinburgh 6: 158. (1915).

Yunnan: Mengzi at 1,640 metres, 9231a.

Clinopodium polycephalum (Vaniot) C. Y. Wu & S. J. Hsuan (LAMIACEAE)
Calamintha clinopodium Benth. var. *pratensis* Dunn in Notes Roy. Bot. Gard. Edinburgh 6: 159. (1915).

Hubei: Yichang, 1333, 1830a., 4916.

Clinopodium sp. **nov.** (LAMIACEAE)
Forbes & Hemsley in Journ. Linn. Soc. Bot. 26: 284.

(1890).

Hubei: Yichang, 316.

Clinopodium umbrosum (M. Bieb.) K. Koch (LAMIACEAE)
Calamintha umbrosa (M. Bieb.) Rchb. Forbes & Hemsley in Journ. Linn. Soc. Bot. 26: 284. (1890), Dunn in Kew Bull. Misc. Inf. 200. (1917). *Calamintha clinopodium* Benth. var. *umbrosa* Dunn in Notes Roy. Bot. Gard. Edinburgh 6: 160. (1915).

Hubei: Yichang, 757, 943, 1013, 1017, 2778, 3661. **Yunnan:** Mengzi at 1,650 metres, 11824.

Clintonia udensis Trautv. & C. A. Mey. (CONVALLARIACEAE)
Wright in Journ. Linn. Soc. Bot. 36: 144. (1903).

Hubei: South Badong, 6085. Fang Xian, 6085a. Zigui 6085b. **Sichuan:** Kangding, Henry's Chinese collector with Pratt, 8832.

This species was collected by the Sino-American expedition in Shennongjia Forest District in 1980.

Clitoria ternatea L. (FABACEAE)
Henry in Trans. Asiat. Soc. Jap. 24 suppl: 34. (1896) (List Pl. Formosa).

Taiwan: Kaohsiung, without number. **Yunnan:** In Pingbian in forests on the Daweishan Range at 1,640 metres, 11147. Simao, 12242.

A tall, striking scandent vine, native to the tropics of the Old World. This species was collected by the Glasnevin Taiwan Expedition (GTE 95) on Wanshoushan, in October 2004 and plants currently growing in the Great Palm House at Glasnevin were raised from that collection.

Cnidium monnieri (L.) Cuss. (APIACEAE)
Selinum monnieri L. Anonymous in Kew List of Determinations (volume 24).

Hainan: Environs of Kiungchow, 8272.

Cnidium sp. (APIACEAE)
Anonymous in Kew List of Determinations (volume 24).

Hubei: Without locality, 828a.

Coccinia grandis (L.) Viogt. (CUCURBITACEAE)
Coccinia indica Wight & Arn. Anonymous in Kew List of Determinations (volume 24).

Hainan: Environs of Haikou, 8374.

The ivy or scarlet-fruited gourd, a vigorous perennial with stems to 20 metres long. A widespread species, distributed across tropical Africa and Asia to northern Australia. Introduced in tropical America. The small bright-red beaked fruits are very handsome.

Cocculus sp. (MENISPERMACEAE)
Anonymous in Kew List of Determinations (volume 34).

02

Yunnan: Fengchunling, 11660.

Cochlianthus gracilis Benth. (FABACEAE)

Anonymous in Kew List of Determinations (volume 34).

Yunnan: Mengzi, 11760.

Distributed from Nepal to western China.

Codariocalyx gyroides (Roxb. ex Link) Hassk. (FABACEAE)

Desmodium gyroides (Roxb. ex Link) DC. Henry in Trans. Asiat. Soc. Jap. 24 suppl: 33. (1896) (List Pl. Formosa).

Taiwan: Oluanpi, 1232. Yunnan: Simao, 12361.

Distributed in India, Myanmar, China and Indochina.

Codariocalyx motorius (Houtt.) H. Ohashi (FABACEAE)

Desmodium gyrans (L. f.) DC. Henry in Trans. Asiat. Soc. Jap. 24 suppl: 33. (1896) (List Pl. Formosa).

Taiwan: Wanchin, 852. Yunnan: On the Daweishan Range in Pingbian at 1,500 metres, 9071.

An erect herb to 90 cm tall, native to tropical and subtropical Asia and Australia.

Codonacanthus pauciflorus (Nees) Nees (ACANTHACEAE)

Henry in Trans. Asiat. Soc. Jap. 24 suppl: 69. (1896) (List Pl. Formosa).

Taiwan: Oluanpi, 217, 1986. Wanchin, 1523.

Codonopsis henryi Oliver (CAMPANULACEAE) New species

in Hooker's Icon. Pl. 20: t. 1967. (1891), Bretschneider in Hist. Eur. Bot. Disc. China 2: 786. (1898), Morley in Glasra 3: 79. (1979).

Hubei: Fang Xian, 6651 (type), 6651a., (isotype)

Codonopsis javanica (Blume) Hook. f. & Thoms. (CAMPANULACEAE)

Codonopsis cordifolia Komarow in Acta Horti Petrop.29: 108. (1908), Komarow in Fedde, Repert. Spec. Nov. 8: 417. (1910).

Yunnan: In the mountains above the Red River at Manmei at 1,950 metres, 9634, (isotype of *Codonopsis cordifolia* Komarow). Mengzi at 1,150 metres 9634a., collected September 5th 1896.

A widespread herbaceous climber with a large geographical range from Bhutan to Japan.

Codonopsis javanica (Blume) Hook. f. & Thoms. subsp. ***japonica*** (Makino) Lammers (CAMPANULACEAE)

Campanumoea javanica Forbes & Hemsley non Blume in Journ. Linn. Soc. Bot. 26: 8. (1889).

Hubei: Yichang, 1687, 3109 (in part), 3211a. Liantuo, 3211, 4437. Sichuan: South Wushan, 7249.

Codonopsis lanceolata (Sieb. & Zucc.) Benth & Hook. f. ex Trautv. (CAMPANULACEAE)

Henry in Journ. China Br. Roy. Asiat. Soc. 22: 260.

(1887) (Chinese Names of Plants), Forbes & Hemsley in Journ. Linn. Soc. Bot. 26: 5. (1889), Henry in Notes Econ. Bot. China 18. (1893).

Hubei: Yichang, 2342. Xingshan, 6527. Sichuan: South Wushan, 7248.

A handsome perennial vine to 2 metres tall. *Codonopsis lanceolata* is currently cultivated in the order beds in front of the Great Palm House at Glasnevin from material collected by the Glasnevin Central China Expedition on Tianthu Shan, Changyang in October 2004. According to Henry the roots of this species were occasionally used as a substitute for *tang shen*, (*Codonopsis tangshen*), it was known colloquially at Yichang as *nai shu*.

Codonopsis micrantha Chipp (CAMPANULACEAE)

Anonymous in Kew List of Determinations (volume 34).

Yunnan: Mengzi, without number.

Endemic to China and distributed in Guizhou, Sichuan and Yunnan.

Codonopsis pilosula (Franch.) Nannf. subsp. ***tangshen*** (Oliver) D. Y. Hong (CAMPANULACEAE) New subspecies

Codonopsis tangshen Oliver in Hooker's Icon. Pl. 20: t. 1969. (1891), Henry in Notes on Economic Botany of China 19. (1893), Bretschneider in Hist. Eur. Bot. Disc. China 2: 786. (1898), Hemsley in Curtis's Bot. Mag. 132: t. 8090. (1906), Morley in Glasra 3: 79. (1979), Nelson in Pim, The Wood & the Trees 230. (1984), Nelson in A Heritage of Beauty 311. (2000), O' Brien in Augustine Henry, An Irish Plant Collector in China 16. (2002).

Hubei: Xingshan, 6468 (isotype of *Codonopsis tangshen* Oliver) 6468a. Fang at 2,600 metres, 6468b.

Tang-shen, an important drug is yielded in Hupeh and Shensi by *Codonopsis tangshen* Oliver, a new species, a beautiful climber found by me in the mountains of Fang district, Hupeh,

(Henry in *Notes on Economic Botany of China* 18. 1893)

Still one of the most important drugs in Chinese herbal medicine, *tang-shen* is credited with tonic and aphrodisiac properties and was once used as a substitute for the very costly ginseng. From Hankow (Wuhan) alone, in Henry's and Wilson's time in central China over 500 tons were exported annually. The roots of wild plants were dug up and dried in the sun by medicine gatherers and sent all over China. An inferior kind was derived from the roots of *Codonopsis lanceolata*. Introduced to cultivation by Henry and later reintroduced by Wilson through Veitch's Coombe Wood nursery.

Codonopsis sp. (CAMPANULACEAE)

Campanumoea sp. Anonymous in Kew List of

Determinations (volume 24).

Hubei: Badong, 859 (September 1885). Xiling (Yichang) Gorge, 3342.

Codonopsis tubulosa Komarow (CAMPANULACEAE)
New species

in Acta Horti Petrop. 29: 112. fig. 3. (1908), Komarow in Fedde, Repert. Spec. Nov. Regni. Veg. 8: 418. (1910).

Yunnan: On grassy hills near Mengzi at 2,300 metres, 10167 (type).

Native to western China and Myanmar.

Coelogyne articulata (Lindl.) Rchb. f. (ORCHIDACEAE)

Pholidota lugardii Rolfe Rolfe in Journ. Linn. Soc. Bot. 36: 24. (1903), Kranzlin in Engler, Pflanzenr. 50: 156. (1907).

Yunnan: On a bank near Mengzi at 1,650 metres, 11126.

Coelogyne brunnea Lindl. (ORCHIDACEAE)
Coelogyne fuscescens Lindl. var. *brunnea* (Lindl.) Lindl Kranzlin in Engler, Pflanzenr. 50: 43. (1907).

Yunnan: In forests near Simao at 1,300 metres, 12964.

Distributed in the subtropical forests of Myanmar, Yunnan, Laos, Thailand and Vietnam.

Coelogyne corymbosa Lindl. (ORCHIDACEAE)
Rolfe in Journ. Linn. Soc. Bot. 36: 22. (1903).

Yunnan: South of the Red River near Mengting at 2,600 metres, epiphytic on trees, 11114.

A handsome epiphyte, previously only known from the Himalaya. In Myanmar, China [Tibet (Xizang) and Yunnan] it grown on trees in forests at 1,300 to 2,900 metres above sea level.

Coelogyne fimbriata Lindl. (ORCHIDACEAE)
Anonymous in Kew List of Determinations (volume 34).

Yunnan: Simao, 13598.

Native to India, China, Thailand, Laos, Vietnam and Cambodia.

Coelogyne gardneriana Lind. (ORCHIDACEAE)
Neogyna gardneriana (Lind.) Reichb. f. Kranzlin in Engler, Pflanzenr. 50: 129. (1907).

Yunnan: Mengzi, 13635.

Coelogyne imbricata (Hook.) Rchb. f. (ORCHIDACEAE)
Pholidota imbricata Lindl. Anonymous in Kew List of Determinations (volume 34).

Yunnan: Simao, 12641, 12973, 13488.

Widely distributed in Asia, in China this species is native to Sichuan, Tibet (Xizang) and Yunnan.

Coelogyne kouytcheensis (Gagnep.) M. W. Chase & Schuit. (ORCHIDACEAE)
Pholidota yunnanensis Rolfe in Journ. Linn. Soc.

Bot. 36: 24. (1903), Kranzlin in Engler, Pflanzenr. 50: 149. (1907).

Yunnan: Near Mengzi on exposed rocky mountains at 2,100 metres, 11144 (isotype of *Pholidota yunnanensis* Rolfe).

Native to southern China and Vietnam.

Coelogyne leucantha W. W. Sm. (ORCHIDACEAE)
Coelogyne leucantha W. W. Sm. var. *heterophylla* T. Tang & F.T. Wang in Acta Phytotax. Sin. 1: 78. (1951).

Yunnan: South of the Red River from Mengzi at 1,950 metres, epiphytic on trees, 13703.

Coelogyne ovalis Lindl. (ORCHIDACEAE)
Anonymous in Kew List of Determinations (volume 34).

Yunnan: Simao, 13539.

Distributed in the forests of India, Nepal, Myanmar and western China.

Coelogyne spp. (ORCHIDACEAE)
Anonymous in Kew List of Determinations (volume 34). *Pholidota* sp. Anonymous in Kew List of Determinations (volume 34).

Yunnan: Mengzi, 9627, 11849. Simao, 11807, 12612, 12974, 13085, 13586, 13619, 13620.

Coffea sp. (RUBIACEAE)
Anonymous in Kew List of Determinations (volume 34).

Yunnan: Mengzi, 9590.

Coix lacryma-jobi L. (POACEAE)
Henry in Journ, China Br. Roy. Asiat. Soc. 22: 272. (1887) (Chinese Names of Plants), Rendle in Journ. Linn. Soc. Bot. 36: 345. (1904).

Coix lacryma L. Henry in Trans. Asiat. Soc. Jap. 24 suppl: 107. (1896) (List Pl. Formosa)

Hubei: Liantuo, 2635a. Without locality, 7567. **Taiwan:** Wanchin, cultivated, 51. **Yunnan:** Mengzi, 10137.

Job's tears, Henry stated that in the Three Gorges region this species was known colloquially as *ta wan tzu*, its seeds constituted the drug *yi yi ren* which promotes urination.

Coldenia procumbens L. (BORAGINACEAE)
Henry in Trans. Asiat. Soc. Jap. 24 suppl: 62. (1896) (List Pl. Formosa), Johnston in Journ. Arnold Arb. 32: 13. (1951).

Taiwan: Kaohsiung, without number.

A prostrate, annual xerophytic herb which according to Johnston can develop and flourish on sunbaked soils of rice paddies subjected to droughts, great heat and intense light. Found as a weed of dried out paddies in India, the east Indies, Indochina and China (Hainan and Taiwan).

Colebrookea oppositifolia Smith (LAMIACEAE)

Dunn in Notes Roy. Bot. Gard. Edinburgh 6: 147. (1915).

Yunnan: Near Mengzi at 1,850 metres, 10921. Shiping at 990 to 1,300 metres, 11593, 11593a., 11593b. Simao at 1,450 metres, 12740.

The Hani (an ethnic group) in southern Yunnan, call this species *Anlbiqmal* and used it in herbal remedies to reduce swelling and pain due to fractures. It is used with other herbs to treat weak kidneys, male impotence and to improve nervous function.

Coleus bracteatus Dunn (LAMIACEAE) New species

in Notes Roy. Bot. Gard. Edinburgh 8: 158. (1913), Dunn in Notes Roy. Bot. Gard. Edinburgh vi. 145. (1911-17).

Yunnan: In the mountains to the west of Simao at 1,650 metres, 12537. On cliffs to the west of Pu'er at 1,450 metres, 13498 (isosyntype).

China (Yunnan) and Indochina.

Coleus parishii (Hook. f.) A. J. Paton (LAMIACEAE)

Coleus esquirolii (Levl.) Dunn Dunn in Notes Roy. Bot. Gard. Edinburgh 8: 158. (1913).

Labiatae, undetermined. Henry in Trans. Asiat. Soc. Jap. 24 suppl: 74. (1896) (List Pl. Formosa).

Taiwan: Wanchin, 1674. **Yunnan:** In Pingbian, on the Daweishan Range in a ravine amoung dry stones at 1,300 to 1,950 metres, 9222.

Colocasia esculenta (L.) Schott. (ARACEAE)

Colocasia antiquorum Schott. Henry in Notes Econ. Bot. China 35. (1893), Brown in Journ. Linn. Soc. Bot. 36: 183. (1903).

Sichuan: Henry's Chinese collector with Pratt, no locality stated but may be Leshan, Emei Shan, Wa Shan or Kangding, 8830.

Taro or *yu-t'ou*, commonly cultivated in China and an important root crop in the tropics. An ancient crop that originated about 7,000 years ago, it is no longer found in the wild but may have came from the wet rice growing regions of India. From India, it reached Egypt about the time of Christ eventually finding its way to France and Spain. From Spain it was taken to the New World shortly after its discovery. It has been cultivated in China since about 200 A.D. It is the staple food of the rainforest inhabitants of New Guinea and a subsistance plant in New Caledonia and Fiji. Over 200 varieties (cultivars) have been selected over the past 2,000 years and the plant over that period has been almost exclusively propagated by vegetative means. Taro requires the same growing conditions as rice and will grow in soil too wet for the sweet potato. A handsome plant adding an exotic feel to the intensly cultivated warmer arable regions of China.

Colona sp. (MALVACEAE)

Columbia sp. Anonymous in Kew List of Determinations (volume 34).

Yunnan: By the Red River at Manpan, 11175.

Colquhounia elegans Wall. var. ***tenuiflora*** (Hook. f.) Prain (LAMIACEAE)

in Journ. Asiat. Soc. Bengal lxii. 38. (1899), Dunn in Notes Roy. Bot. Gard. Edinburgh 5: 179. (1915), Rehder in Sargent Pl. Wilson. 3: 380. (1917).

Yunnan: Mountains to the east of Simao at 1,200 metres, 12607. Simao at 1,650 metres, 12607a., 12607b.

Distributed in Myanmar, Yunnan, Thailand, Cambodia, Laos and Vietnam.

Colquhounia seguinii Vaniot (LAMIACEAE)

Rehder in Sargent Pl. Wilson. 3: 380. (1917).

Colquhounia coccinea Forbes & Hemsley non Wallich in Journ. Linn. Soc. Bot. 26: 299. (1890). *Colquhounia elegans* Wall. var. *pauciflora* Prain in Journ. Asiat. Soc. Bengal. lxii. 38. (1893), Dunn in Notes Roy. Bot. Gard. Edinburgh 6: 179. (1915).

Hubei: On cliffs near Yichang, stems from 3 to 4 metres long, hanging down from the face of rocks, flowers red., 3334, 3334a., (type of *Colquhounia elegans* Wall. var. *pauciflora* Prain). **Yunnan:** Mengzi, 9089, 9089a., 9089b. Shiping 11582.

Native to Myanmar and China.

Colquhounia sp. (LAMIACEAE)

Anonymous in Kew List of Determinations (volume 34).

Yunnan: Mengzi, 9088.

Colubrina asiatica (L.) Brongn. (RHAMNACEAE)

Henry in Trans. Asiat. Soc. Jap. 24 suppl: 27. (1896) (List Pl. Formosa).

Taiwan: Oluanpi, 2016. **Yunnan:** Fengchunling, 11191.

A scandent shrub, native from Africa to India, Malaysia, Philippines, Australia and Polynesia. In China (Taiwan) this species is found along the coast of the Hengchun Peninsula.

Coluria henryi Batalin (ROSACEAE) New species

in Acta Hort. Petrop. xiii. 94. (1893), Bretschneider in Hist. Eur. Bot. Disc. China 2: 783. (1898), Diels in Bot. Jahrb. 29: 404. (1900), Evans in Notes Roy. Bot. Gard. Edinburgh15: 50. (1925—1927).

Geum sp. nov. Forbes & Hemsley in Journ. Linn. Soc. Bot. 29: 239. (1887).

Hubei: South Badong, 5400 (type).

Comastoma henryi (Hemsl.) Holub (GENTIANACEAE) New species

Gentiana henryi Hemsley in Journ. Linn. Soc. Bot. 26:

128. (1890), Bretschneider in Hist. Eur. Bot. Disc. China 2: 788. (1898), Morley in Glasra 3: 79. (1979).

Hubei: Fang Xian, 6936, 6936a., (type).

Combretum indicum (L.) DeFlipps (COMBRETACEAE)

Quisqualis indica L. Exell in Sunyatsenia 1: 89. (1933), Li in Woody Flora of Taiwan 840. fig. 338. (1963).

Hainan: Environs of Haikou, 8070. **Taiwan:** Wanchin, 558. Oluanpi, 576.

The Rangoon climber, a deciduous climbing shrub to 8 metres tall. A very beautiful and ornamental plant, this species carries masses of deep red flowers in pendelous terminal spikes. Worth growing in any region of the tropics, there is a particularly good plant near the central fountain at Hong Kong Botanic Gardens.

Combretum latifolium Blume (COMBRETACEAE)

Exell in Sunyatsenia 1: 88. (1934).

Yunnan: Simao, 13013.

A widespread species, distributed in Asia from India, Sri Lanka, across Myanmar, into China, Laos, Thailand, Vietnam and the Philippines.

Combretum wallichii DC. (COMBRETACEAE)

Exell in Sunyatsenia 1: 87. (1934).

Yunnan: In woods near Mengzi at 1,650 metres, 9242a. Simao 12509, 12546, 12546, 12546a., 12546b.

Native to India, Bangladesh, Nepal, Myanmar and western China.

Commelina auriculata Blume (COMMELINACEAE)

Commelina undulata Henry non R. Br. in Trans. Asiat. Soc. Jap. 24 suppl: 98. (1896) (List Pl. Formosa).

Taiwan: Kaohsiung, without number.

Native to China (Taiwan), Japan, Indonesia and western Oceania.

Commelina benghalensis L. (COMMELINACEAE)

Henry in Journ. China Br. Roy. Asiat. Soc. 22: 242. (1887), Brown in Journ. Linn. Soc. Bot. 36: 155. (1903).

Hubei: Yichang, 2353.

According to Henry this species was known as *chu yeh ts'ai* at Yichang.

Commelina communis L. (COMMELINACEAE)

Henry in Journ. China Br. Roy. Asiat. Soc. 22: 242. (1887) (Chinese Names of Plants), Brown in Journ. Linn. Soc. Bot. 36: 155. (1903).

Hubei: Yichang, 76. Badong, 1800, 4927.

The common dayflower, which Henry stated was colloquially known as *chu yeh ts'ai* at Yichang and *shui chu yeh* at Badong. Distributed throughout the north temperate zone. The dried aerial parts of this species are currently used in Chinese herbal medicine to treat upper respiratory infection with high fever and sore throat and acute urinary infection. *Commelina communis* was collected by the

Glasnevin Central China Expedition (GCCE 694) in October 2004 near Xiakou (Daxiakou) on the Xiang Xi River, a tributary of the Yangtze that meets the Yangtze in the Xiling (Yichang) Gorge.

Commelina diffusa Burm. f. (COMMELINACEAE)

Commelina nudiflora Henry non L. in Trans. Asiat. Soc. Jap. 24 suppl: 98. (1896) (List Pl. Formosa), Brown in Journ. Linn. Soc. Bot. 36: 156. (1903).

Taiwan: Oluanpi, 357. Kaohsiung Plain, 2041.

A widespread perennial found over much of India, China, south-east Asia, Japan, Korea, the Philippines, Indonesia, Australia and the Pacific Islands.

Commelina paludosa Blume (COMMELINACEAE)

Commelina obliqua Buch.-Ham. ex DC. Henry in Trans. Asiat. Soc. Jap. 24 suppl: 98. (1896) (List Pl. Formosa), Brown in Journ. Linn. Soc. Bot. 36: 157. (1903).

Taiwan: Tamsui, 1447.

A widely distributed perennial, native to the Himalaya, India, southern China, Malaysia and Indonesia.

Commelina spp. (COMMELINACEAE)

Anonymous in Kew List of Determinations (volume 34).

Yunnan: Mengzi, 9211, 11228. Simao, 12285.

Commersonia bartramia (L.) Merr. (MALVACEAE)

Commersonia platyphylla Andr. Anonymous in Kew List of Determinations (volume 24).

Hainan: Without locality, 8560.

Coniogramme fraxinea (D. Don) Fee ex Diels (PTERIDACEAE)

Gymnogramma javanica Blume Anonymous in Kew List of Determinations (volume 24).

Hubei: Xingshan at 1,400 metres, 6548. Donghu, 7101a. **Sichuan:** North Wushan, 7107.

Coniogramme japonica (Thunb.) Diels (PTERIDACEAE)

Gymnogramma japonica (Thunb.) Desv. Baker in Journ. Bot. 27: 178. (1889).

Hubei: Changleping, 7821. Without locality, 7101a.

Coniogramme robusta Christ (PTERIDACEAE)

Gymnogramma javanica Blume var. *robusta* Christ Christ in Bull. Soc. Bot. France 5: 55.(1905).

Hubei: Liantuo, 2204.

Conioselinum anthriscoides (H. Boissieu) Pimenov & Kljuykov (APIACEAE) New species

Ligusticum sinense Oliver in Hooker's Icon. Pl. 20(3): t. 1958. (1891), Henry in Notes Econ. Bot. China 20. (1893), Bretschneider in Hist. Eur. Bot. Disc. China 2: 784. (1898), Shan in Sinensia 12: 176. (1941), Leute in Ann. Naturhist. Mus. Wien 74: 490. (1970), H. T. Chang in Fl. Reip. Pop. Sin. 55(2): 252. (1985), Pan in Acta Phytotax. Sin. 29: 529. (1991).

Hubei: Fang Xian, 6759 (isosyntype of *Ligusticum sinense* Oliver). Xingshan, 6759b. **Sichuan:** South Wushan, 6759a., (isosyntype of *Ligusticum sinense* Oliver).

Chinese lovage root, an important medicinal plant and stated by Henry to be the source of the drug *kao-pen* (*gao ben*). In current Chinese herbal medicine it is used to allieviate acute lower-back pain, headache, toothache and when used with the fruits of the *wu zhu yu* (*Tetradium ruticarpum)* it is used in cases of abdominal pains and hernia-like conditions.

Conioselinum tenuisectum (H. Boissieu) Pimenov & Kljuykov (APIACEAE)

Ligusticum tenuisectum H. Boissieu in Bull. Herb. Boiss. 3: 843. (1903), Pu in Acta Phytotax. Sin. 29: 540. (1991).

Hubei: Fang Xian, 6921.

Connarus henryi G. Schellenberg (CONNARACEAE) New species

in Engler, Pflanzenr. iv. 127: 228. (1938).

Yunnan: South of the Red River from Manpan, 9571a., (isotype).

Connarus yunnanensis G. Schellenberg (CONNARACEAE) New species

in Engler, Pflanzenr. iv. 127: 228. (1938).

Yunnan: South of the Red River from Manpan, 9571a., (isotype).

Convolvulus sp. (CONVOLVULACEAE)

Henry in annotation in the *Index Florae Sinensis* (copy at NBG, Glasnevin).

Taiwan: Kaohsiung, 1145.

Coptis chinensis Franchet (RANUNCULACEAE) New species

in Journ. de Bot. 321. (1897), Bretschneider in Hist. Eur. Bot. Disc. China 2: 779. (1898).

Coptis teeta Henry non Wallich in Journ. China Br. Roy. Asiat. Soc. 22: 245. (1887) (Chinese Names of Plants).

Hubei: North Badong, 6984.

The Chinese goldthread or *huang-lian*, a small perennial, the rhizomes of which are an all-around medicine, said to promote the flow of milk in nursing mothers, is used as a tonic to treat bacterial diarrhoea, gastroenteritis, conjunctivitis, external sores, ulcers (of the mouth and tounge) and as a blood-purifier. It is also used to reduce high fevers and in cases of irritability, delerium, sore-throat, insomnia and in extreme cases, coma. It is still grown on a large scale in the mountains of central China, in 2002 while visiting the *Metasequoia* valley in western Hubei members of the Glasnevin team visited a plantation where this crop was cultivated beneath an enclosure of raised (to 1.5 metre) poles draped with branches of *Cryptomeria japonica* to provide the deep shade necessary for this crop to thrive. *Coptis* is used as a medicine in various parts of the world, eg. *C. chinensis* and *C. deltoidea* in China. *C. japonica* in Japan. *C. trifolia* in North America and *C. teeta* in India. The active ingredient is Berberine.

Corallodiscus lanuginosus (Wall. ex R. Br.) B. L. Burtt. (GESNERIACEAE)

Didissandra lanuginosa (Wall. ex R. Br.) C. B. Clarke Forbes & Hemsley in Journ. Linn. Soc. Bot. 26: 226. (1890), *Didissandra mengtzeana* Craib in Notes Roy. Bot. Gard. Edinburgh 11: 247. (1921).

Hubei: San You Dong Glen (July 1886) on the vertical face of rocks, 1533. Liantuo, 4565. In a ravine on Moji Shan, 4167. South Badong, 4702, 6131. **Yunnan:** On rocks near Mengzi at 1,500 to 1,650 metres, 9190, (type of *Didissandra mengtzeana* Craib).

Distributed in India, the Himalaya, China and Thailand.

Corchoropsis sp. (MALVACEAE)

Corchoropsis crenata Sieb. & Zucc. Anonymous non Sieb. & Zucc. in Kew List of Determinations (volume 24).

Hubei: Yichang, 446, 2231, 2350. Badong, 1049 (December 1885), 2490, 4952.

Corchorus acutangulus Lam. (MALVACEAE)

Anonymous in Kew List of Determinations (volume 24).

Hubei: Yichang, 2182, 4236.

Corchorus aestuans L. (MALVACEAE)

Corchorus acutangulus Lam. Henry in Trans. Asiat. Soc. Jap. 24 suppl: 23. (1896) (List Pl. Formosa).

Hainan: Environs of Haikou, 8367. Environs of Kiungchow, 8755. **Taiwan:** Tamsui, 1475. **Yunnan:** Mengzi, 9672.

An undershrub to 50 cm tall, widely distributed over tropical Asia, Africa and the West Indies.

Corchorus capsularis L. (MALVACEAE)

Henry in Trans. Asiat. Soc. Jap. 24 suppl: 23. (1896) (List Pl. Formosa).

Hainan: Environs of Haikou, 8363. **Taiwan:** Kaohsiung Plain, without number.

Cultivated. The white jute plant. According to Henry the fibre was known in Taiwan as *ma-p'e* and was made into ropes and coarse sacking, the 'Hemp bags' of the Imperial Maritime Customs Service' returns. On mainland China the jute plant was known as *lu-ma* or *huang-ma*. According to Henry the word *ma* has a very extended significance, being given (1.) to vegetable fibres of all kinds (ramie, jute, flax, hemp, pine-apple fibre, etc.) (2.) to oil producing plants like sesame and *Ricinus*. (3.) to medicinal herbs with a certain kind of foliage.

02

Corchorus olitorius L. (MALVACEAE)

Henry in Trans. Asiat. Soc. Jap. 24 suppl: 23. (1896) (List Pl. Formosa).

Taiwan: Kaohsiung, without number. Wanchin, 439. Tamsui, 1461.

The jute mallow. A common weed in Taiwan sometimes termed *shan-ma*, according to Henry's Chinese collector at Wanchin, the young leaves were sometimes used as a vegetable and was named *tou-lu*. Native to India but widely naturalised in the tropics.

Cordia dichotoma G. Forst. (BORAGINACEAE)

Johnston in Journ. Arnold Arb. 32: 8. (1951), Li in Woody Flora of Taiwan 810. fig. 327. (1963).

Cordia myxa Henry non L. in Trans. Asiat. Soc. Jap. 24 suppl: 62. (1896) (List Pl. Formosa).

Taiwan: Kaohsiung, without number.

According to Henry this small tree was known as *p'o-ku-tze,* and colloquially in Chinese Taiwan as *p'o-tze*, it also bore edible fruits. Native from India to Indochina, southern China and the Philippines.

Cordia furcans I. M. Johnston (BORAGINACEAE) New species

in Journ. Arnold Arb. 32: 5. (1951).

Yunnan: Yuanjiang at 1,650 metres, 13251 (type). At Namban Ho near Simao at 1,650 metres, 13386.

Distributed in India, Myanmar, China, Vietnam and Thailand.

Cordyline fruticosa (L.) A. Chev. (AGAVACEAE)

Cordyline terminalis (L.) Kunth. Henry in Trans. Asiat. Soc. Jap. 24 suppl: 97. (1896) (List Pl. Formosa), Wright in Journ. Linn. Soc. Bot. 36: 118. (1903).

Taiwan: Oluanpi, without number. Wanchin, 813.

A shrub to 3 metres tall, native to the eastern Himalaya, southern China and Australia. According to S. S. Ying (in *Flora of Taiwan* 5: 75. 2000) this species was introduced to Chinese Taiwan in 1910. This is clearly incorrect since Henry's specimens date from 1893—1894. Henry stated it was known colloquially as *hung-chu.*

Coriandrum sativum L. (APIACEAE)

Anonymous in Kew List of Determinations (volume 24).

Hubei: Yichang, 839. Badong, 4823. Changyang, 7506.

Coriander, an ancient spice, seeds over 3,000 years old have been found in Egyptian tombs belonging to the 21st dynasty. According to Bretschneider, the first time it is mentioned in Chinese agricultural references is during the 5th century A.D. Native to southern Europe and Asia.

Coriaria sinica Maxim. (CORIARIACEAE)

Rehder & E. H. Wilson in Sargent Pl. Wilson. 2: 170.

(1916), Zheng in Journ. Wuhan Bot. Res. 2.(1): 115. (1984).

Coriaria napalensis Forbes & Hemsley non Wallich in Journ. Linn. Soc. Bot. 23: 149. (1886), Henry in Journ. China Br. Roy. Asiat. Soc. 22: 267. (1887) (Chinese Names of Plants).

Hubei: Yichang, 483, 1289. **Yunnan:** Mengzi, 10361.

A deciduous shrub to 6 metres tall, resembling the Himalayan *Coriaria napalensis* Wall. Common in Hubei where it was known as *ma-sang*, according to Wilson the shoots are poisonous to cattle. Introduced to cultivation by Wilson in 1907 while collecting for the Arnold Arboretum. A handsome sight in early summer when seen carrying masses of red flowers in short racemes. *Coriaria sinica* is cultivated at the National Botanic Gardens Glasnevin from collections made by the Glasnevin Central China Expedition (GCCE 242) in Badong in October 2004.

Coriolopsis floccosa (Jungh.) Ryvarden (POLYPORACEAE)

Polystictus sordidus (Berk.) Fr. Anonymous in Kew List of Determinations (volume 24).

Hubei: Changyang, on trees, 7930. (Fungus).

Coriolus azureus (Fr.) G. Cunn. (POLYPORACEAE)

Polystictus azureus Fr. Anonymous in Kew List of Determinations (volume 24).

Hubei: Fang Xian, 6618. (Fungus).

Cornus capitata Wallich (CORNACEAE)

Forbes & Hemsley in Journ. Linn. Soc. Bot. 23: 345. (1887), Henry in Notes Econ. Bot. China 46. (1893), Wangerin in Engler, Pflanzenr. iv. 229. 90. (1910), Rehder in Sargent Pl. Wilson 2: 578. (1916), Zheng in Journ. Wuhan Bot. Res. 2.(1): 163. (1984).

Hubei: Badong, 208 (June 1885), 483, 1289. **Yunnan:** Mengzi, 9176, 9176a., 9176b.

A very beautiful evergreen tree to 15 metres tall, distributed through the Himalaya and China. Introduced to cultivation in 1825, this species thrives in the Irish climate and according to W. J. Bean the old tree at Fota, County Cork dates from the original introduction. There is a magnificent specimen near the pond at the National Botanic Gardens, Glasnevin. Henry stated that in the Three Gorges region this fine tree was called the *shih-p'ao-tzu.*

Cornus chinensis Wangerin (CORNACEAE) New species

in Repert. Spec. Nov. Regni Veg. 6: 100. (1908), Wangerin in Engler, Pflanzenr. iv. 229. 80. (1910), Schneider in Illus. Handb. Laubholzk. 2: 452. (1912), Rehder in Sargent Pl. Wilson 2: 577. (1916), Zheng in Journ. Wuhan Bot. Res. 2.(1): 163. (1984), Bean in Trees & Shrubs 1: 706. (1989).

Cornus officinalis Henry non Sieb. & Zucc. in Notes Econ. Bot. China 41. (1893).

Hubei: Xingshan at 1,400, 6560. Fang Xian, 5733 (isosyntype). **Sichuan:** South Wushan, 6707.

A fine tree with exceedingly handsome foliage, it is an eastern ally of *Cornus mas*. Source of the drug *shan-chu-yu*. Introduced to cultivation by Captain Frank Kingdon Ward from Assam.

Cornus controversa Hemsl. (CORNACEAE)

Wagnerin in Engler, Pflanzenr. iv. 229. 49. (1910), Rehder in Sargent Pl. Wilson. 2: 573. (1916).

Hubei: Badong, 725, 3773. **Yunnan:** In the Red River valley near Manpan at 600 metres, 10747a.

Distributed in Nepal, India, Bhutan, Myanmar, China, Korea and Japan. In Xishuangbanna, in southern Yunnan, *Cornus controversa* is employed in local medicine and acts as a cough suppressant and has asthma relieving properties.

Cornus kousa Büerger ex Hance subsp. *chinensis* (Osborn) Q. Y. Xiang (CORNACEAE) New subspecies

Cornus kousa Henry non Büerger ex Hance in Notes Econ. Bot. China 47. (1893), Wangerin in Engler, Pflanzenr. iv. 229. 88. (1910), Rehder in Sargent Pl. Wilson.2: 577. (1916). *Cornus kousa* Büerger ex Hance var. *chinensis* Osborn Zheng in Journ. Wuhan Bot. Res. 2.(1): 164. (1984).

Hubei: Badong, 4073. North Badong, 5672b. Jianshi, 5672c. Without locality, 5672a. **Sichuan:** South Wushan, 5672.

One of the finest of the Chinese dogwoods in terms of flower, foliage and habit. Henry stated it was called the *shih-tsao*. Wilson recorded this species as being common around Yichang and his 223 collected from the same area in September 1907 has formed a fine tree at the western end of the pond at Glasnevin where in most years it carries an abundant supply of flowers. It was from Wilson's 223 that Osborn based *Cornus kousa* var. *chinensis* (see Gard. Chron. ser. 3, 72: 310. 1922) though Henry had found this subspecies some 20 years earlier in 1887. Giraldi also found this fine tree in Shaanxi in July 1897, Silvestri collected it in Hubei in 1906 and Frank Meyer collected material in Zhejiang in 1915. When travelling with other members of the Glasnevin Central China Expedition (2002) we noted this tree as being locally common at high altitude in South Badong, where it grew with *Rosa multiflora* var. *cathayensis*, *Populus lasiocarpa* and *Liriodendron chinense*. Material was collected by the 2004 Glasnevin Expedition on Tianthu Shan in Changyang (GCCE 755).

Cornus macrophylla Wall. (CORNACEAE)

Forbes & Hemsley in Journ. Linn. Soc. Bot. 23: 345. (1887), Henry in Notes Econ. Bot. China 49. (1893), Rehder in Sargent Pl. Wilson 2: 575. (1916), Zheng in Journ. Wuhan Bot. Res. 2.(1): 164. (1984).

Cornus brachypoda Rehder non C. A. Mey in Sargent, Trees & Shrubs 1: 81. (1905), Wangerin in Engler, Pflanzenr. iv. 229. 64. (1910). *Cornus alba* Zheng non L. in Journ. Wuhan Bot. Res. 2.(1): 163. (1984).

Hubei: Badong, 3772. Liantuo, 4555. South Badong, 5506. Changleping, 6266, 6300. Fang Xian, 6696. Jianshi, 7434. Without locality, 6300a. **Sichuan:** Henry's Chinese collector with Pratt, no locality stated but may be any of the following, Leshan, Emei Shan, Wa Shan or Kangding, 8970. **Yunnan:** Mengzi at 2,100 metres, 10747.

Native to India, Afghanistan, Pakistan, Bhutan, Nepal, Myanmar, China and Japan.

Cornus oblonga Wall. (CORNACEAE)

Wagnerin in Engler, Pflanzenr. iv. 229. (1910), Rehder in Sargent Pl. Wilson. 2: 579. (1916), Bean in Trees & Shrubs 1: 707. (1989), Rehder in Sargent Pl. Wilson. 2: 579. (1916).

Cornus oblonga Wall. f. *pilosa* H. L. Li in Journ. Arnold Arb. 25: 311. (1944)

Sichuan: North Wushan, 7031. **Yunnan:** Mengzi at 1,800 to 2,800 metres, 11161, 11397. Milê district, 9930.

An evergreen shrub to 7 metres tall bearing scented white flowers in terminal pyramidal panicles. Previously only known from the Himalaya, Henry's collections in Sichuan and Yunnan were new records for China. Loudon stated it was introduced from the Himalaya in 1818, it was later reintroduced by Wilson through Veitch's Coombe Wood nursery. Native to Sri Lanka, Pakistan, Nepal, Bhutan, Myanmar, China, Thailand and Vietnam.

Cornus quinquenervis Franch. (CORNACEAE)

Cornus paucinervis Hance Forbes & Hemsley in Journ. Linn. Soc. Bot. 23: 346. (1887), Wangerin in Engler, Pflanzenr. iv. 229. 72. (1910), Rehder in Sargent Pl. Wilson. 2: 576. (1916).

Hubei: Yichang, 218, 471, 674, 1683. Badong, 471a. **Yunnan:** Mengzi at 1,600 metres, 10800, 10800a.

Endemic to China.

Cornus walteri Wangerin (CORNACEAE) New species

in Repert. Spec. Nov. Regni. Veg. 6: 99. (1908), Wangerin in Engler, Pflanzenr. iv. 229. 72. (1910), Schneider in Illus. Handb. Laubholzk. 2: 250. (1913), Rehder in Sargent Pl. Wilson 2: 576. (1916), Zheng in Journ. Wuhan Bot. Res. 2.(1): 165. (1984).

Cornus henryi Hemsley ex Wagnerin in Engler, Pflanzenr. iv.-229, 90. (1910).

Hubei: Liantuo, 3891, 6382.

A deciduous tree to 15 metres tall with black-grey bark often splitting longtidinally, native to central and western China. Introduced to cultivation by Wilson in 1907 through the Arnold Arboretum. The large terminal cymes of creamy-white flowers which appear in June are particularly attractive.

Cornus wilsoniana Wangerin (CORNACEAE) New species

Rehder in Sargent Pl. Wilson. 2: 579. (1916).

Cornus fordii Hemsley in Kew Bull. Misc. Inf. 8: 334. (1909), Wangerin in Engler, Pflanzenr. iv. 229. 90. (1910).

Hubei: Changyang, 7751 (syntype of *Cornus fordii* Hemsley).

Discovered by Henry in 1888, this species was based on material later collected by Wilson in western Hubei.

Corydalis bungeana Turcz. (PAPAVERACEAE)

Anonymous in Kew List of Determinations (volume 24).

Hubei: Xingshan, 6992.

Corydalis cheilanthifolia Hemsley (PAPAVERACEAE) New species

in Journ. Linn. Soc. Bot. 29: 302. (1892), Bretschneider in Hist. Eur. Bot. Disc. China 2: 779. (1898), Diels in Bot. Jahrb. 29: 354. (1900), Henry in Gard. Chron. ser. 3, 32: 288. (1902), Morley in Glasra 3: 79. (1979), Nelson in A Heritage of Beauty 311. (2000), O' Brien in Augustine Henry, An Irish Plant Collector in China 16. (2002).

Hubei: South Badong, 3723, 5399 (type).

Henry discovered this charming plant amoungst stones on the banks of streams in the higher mountains above the gorges. Introduced to cultivation by Wilson through Veitch's Coombe Wood nursery. Cultivated in several Irish gardens.

Corydalis edulis Maxim. (PAPAVERACEAE)

Diels in Bot. Jahrb. 29: 355. (1900).

Corydalis micropoda Franch. Diels in Bot. Jahrb. 29: 355. (1900).

Hubei: Yichang, 807.

Discovered by Pavel Jakovlevich Piasetski, an interpreter and photographer on a Russian expedition sent by the Tsar to China in 1874 to explore trade routes to China.

Corydalis hemsleyana Franchet & Prain (PAPAVERACEAE) New species

in Journ. Asait. Soc. Bengal lxv. 2: 29. (1897), Fedde in Repert. Spec. Nov. Regni Veg. 21: 49. (1925).

Corydalis incisa Henry non (Thunb.) Pers. in Journ. China Br. Roy. Asiat. Soc. 22: 246. (1887) (Chinese Names of Plants), Diels in Bot. Jahrb. 29: 355. (1900).

Hubei: Badong, 1459, 3729.

According to Henry this species was known colloquially at Yichang as *yeh huang lien*.

Corydalis ochotensis Turcz. (PAPAVERACEAE)

Fedde in Repert. Spec. Nov. Regni Veg. 21: 48. (1925).

Hubei: Fang Xian, 6711.

Corydalis ophiocarpa Hook.f. & Thoms. (PAPAVERACEAE)

Diels in Bot. Jahrb. 29: 355. (1900), Henry in Gard.

Chron. ser. 3, 32: 288. (1902), Nelson in Pim, The Wood & the Trees 221. (1984); in A Heritage of Beauty 311. (2000), O' Brien in Augustine Henry, An Irish Plant Collector in China 16. (2002).

Hubei: Jianshi, 5822. Fang Xian, 6884.

Introduced to cultivation by Henry through the Royal Botanic Gardens, Kew (from seeds taken from A. Henry 6884). Henry stated 'It has tall lax foliage, and perhaps is not worth cultivating as an ornamental plant.' This statement is partly true, however if plants are kept cut back (and one or two left to self-seed) is is worth cultivating for its handsome finely dissected bronze-silver foliage.

Corydalis pallida (Thunb.) Pers. (PAPAVERACEAE)

Diels in Bot. Jahrb. 29: 355. (1900).

Hubei: Yichang, 1260.

Native to China and Japan.

Corydalis racemosa (Thunb.) Pers. (PAPAVERACEAE)

Henry in Trans. Asiat. Soc. Jap. 24 suppl: 17. (1896) (List Pl. Formosa).

Hubei: Yichang, 3541. **Taiwan:** Tamsui, 1781.

A soft glabrous biennial herb, bearing yellow flowers in spring. Native to China (including Taiwan) and Japan.

Corydalis saxicola Bunting (PAPAVERACEAE) New species

Corydalis thalictrifolia Franchet in Journ. de Botanique. viii. 291. (1894), Bretschneider in Hist. Eur. Bot. Disc. China 2: 799. (1898), Diels in Bot. Jahrb. 29: 355. (1900), Henry in Gard. Chron. ser. 3, 32: 288. fig. 94. (1902), Besant in Gard. Chon. ser. 3, 98: 335. (1935), Pim in The Wood & the Trees 35. (1984).

Corydalis sp. nova Henry in Journ. China Br. Roy. Asiat. Soc. 22: 245. (1887) (Chinese Names of Plants).

Hubei: On rocky ledges of cliffs in the Xiling (Yichang) Gorge, 743, 3463 (type).

According to Henry this species, known as *ai-huang-lien*, was a common plant on the ledges of cliffs in the gorges and spring it carried conspicuous yellow flowers. The large woody rhizomes of this species were used as a drug (as a substitute for *huang-lien*, *Coptis chinensis*) the Chinese name *ai-huang-lien* meaning cliff coptis. In the gorges it flowered in March and April. *Corydalis saxicola* was introduced to cultivation by Wilson while collecting for Messrs Veitch. Frederick Burbidge, writing in the Gardener's Chronicle stated it was cultivated with other rarities in the enclosed garden at Glasnevin at the beginning of the 20th century. Henry grew it in his greenhouse in Ranelagh in the south Dublin city suburbs where it flowered throughout the winter months. *Corydalis saxicola* was collected by the Glasnevin Central China Expedition (GCCE 702) in the San You Dong Glen above Yichang in October

2004.

Corydalis sheareri S. Moore (PAPAVERACEAE)

Diels in Bot. Jahrb. 29: 355. (1900).

Hubei: Yichang, 1310, 1316. Antelope Glen, 3385. Changyang, 7790, 7797.

According to Bretschneider *Corydalis sheareri* was first found by Robert Fortune and was described from material later collected by Dr. George Shearer of the London Missionary Society who was based at Hankow, now part of Wuhan, a city to the east of Yichang.

Corydalis spp. (PAPAVERACEAE)

Anonymous in Kew List of Determinations (volume 24 & 34).

Hubei: Yichang, 1192, 1319 (Henry's note; plant has a stinking odour), 7779. Badong, 1712, 3788. South Badong, 5440, 6314. Liantuo, 1982. Sichuan: South Wushan, 5291. **Yunnan:** Mengzi, 9831, 11369.

Corydalis taliensis Franch. (PAPAVERACEAE)

Anonymous in Kew List of Determinations (volume 34).

Yunnan: Mengzi, 9375. Lunan, 10578.

An annual species to 45 cm tall, bearing racemes of purple flowers from March to November. Named for the city of Tali, now Dali in north-west Yunnan. Endemic to Yunnan, this species is cultivated in a number of Irish gardens.

Corydalis temulifolia Franchet (PAPAVERACEAE) New species

Diels in Bot. Jahrb. 29: 355. (1900), Henry in Gard. Chron. ser. 3, 32: 288. (1902), F. Fedde in Repert. Spec. Nov. Regni Veg. 20: 288. (1924).

Sichuan: South Wushan, 5284, 5284a.

Franchet based the species on Farges' 501 from Sichuan and Henry's 5284a., Henry had first found this species in May of 1888, Fages later gathered his material during the early 1890s. Henry stated it had large pink flowers and leaves of great size. He also stated that Wilson missed collecting seeds of this species.

Corydalis wilfordii Regel (PAPAVERACEAE)

Diels in Bot. Jahrb. 29: 355. (1900).

Hubei: Yichang, 660.

Corylopsis henryi Hemsley (HAMAMELIDACEAE) New species

Hooker's Icon. Pl. 29: t. 2820, fig. 5 & 6. (1906), Fedde in Repert. Nov. Spec. Regni Veg. 5: 267. (1908), Morley & Chao in Journ. Arnold Arb. 58: 319. fig. 5. (1977).

Hubei: Badong, 1444.

A shrub or small tree, discovered by Henry at Badong and later collected by Maire in Yunnan.

Corylopsis sinensis Hemsley (HAMAMELIDACEAE) New species

in Gard. Chron. ser. 3, 39: 18. (1906), Hemsley in Hooker's Icon. Pl. xxix. t. 2820. fig. 17-20. (1906); in Fedde. Rep. Spec. Nov. Regni. Veg. iv. 363. (1907), Schneider in Illus. Handb. Laubhilzk. 2: 955. fig. 587. (1912), Rehder & E. H. Wilson in Sargent Pl. Wilson. 1: 424. (1913), Morley & Chao in Journ. Arnold Arb. 58: 406. (1977), Zheng in Journ. Wuhan Bot. Res. 2.(1): 64. (1984), O' Brien in Augustine Henry, An Irish Plant Collector in China 16. (2002).

Corylopsis spicata Forbes & Hemsley non Sieb. & Zucc. in Journ. Linn. Soc. Bot. 23: 290. (1887).

Hubei: Jianshi, 5854.

A shrub to 4 metres tall, bearing in spring masses of pendant racemes of fragrant primrose-yellow flowers. Introduced to cultivation by Wilson in 1901 through Veitch's Coombe Wood nursery where it first flowered in the spring of 1905. In Ireland this species is cultivated at Birr Castle and the National Botanic Gardens, Glasnevin.

Corylus chinensis Franch. (BETULACEAE)

Schneider in Sargent Pl. Wilson. 2: 444. (1916).

Corylus colurna L. var. *chinensis* (Franch.) Burkill in Journ. Linn. Soc. Bot. 26: 503. (1899), Winkler in Engler, Pflanzenr. 61: 50. (1904), Henry in Elwes and Henry The Trees Gt. Brit. & Irel. 3: 521. (1908).

Sichuan: North Wushan, 7111. **Hubei:** South Badong, 7533.

A magnificent tree varying from 25 to 40 metres tall with girths from 2 to 5 metres. According to Wilson who later collected in the same localities as Henry, it occurs sparingly on the mountains of western Hubei but is remarkably rare in Sichuan. It was introduced to cultivation by Wilson in 1901 through Messrs Veitch. There is a very good tree at Mount Usher, County Wicklow.

Corylus ferox Wall. var. *tibetica* (Batal.) Franch. (BETULACEAE) New variety

Burkill in Journ. Linn. Soc. Bot. 26: 503. (1899).

Corylus tibetica Batalin Schneider in Sargent Pl. Wilson 2: 443. (1916), Zheng in Journ. Wuhan Bot. Res. 2.(1): 116. (1984).

Hubei: Badong, 608 (July 1885). Fang Xian, 6778, 6778a.

This species was first collected by Henry in 1885 and was described from a collection made by Potanin in the same year. Maurice de Vilmorin introduced it through his French nursery in 1898 and it was later reintroduced by Wilson in 1901. There is an exceptionally large tree at the National Botanic Gardens, Glasnevin.

Corylus heterophylla Fisch ex Trautv. var.
sutchuenensis Franch. (BETULACEAE) New variety

Schneider in Sargent Pl. Wilson. 2: 445. (1916), Zheng in Journ. Wuhan Bot. Res. 2.(1): 20. (1984).

Corylus heterophylla Henry non Fisch. ex Trautv. in Journ. China Br. Roy. Asiat. Soc. 22: 269. (1887) (Chinese Names of Plants). *Corylus heterophylla* Fisch. ex Trautv. var. *crista-galli* Burkhill in J. Linn. Soc . Bot 26: 504. (1899), Winkler in Engler, Pflanzenr. 61: 50. (1904).

Hubei: Badong, 1449, 2854, 2866. Donghu, 6413. **Hunan:** Shimen, 7942 (type of *Corylus heterophylla* Fisch. ex Trautv. var. *crista-galli* Burkhill).

Discovered by Henry in 1886, described from material later collected by Père Paul Farges near Kangding in Sichuan in the early 1890s. Henry stated that in the Three Gorges region it was known colloquially as *shan peh kuo*.

Corylus yunnanensis (Franch.) A. Camus (BETULACEAE)

Corylus heterophylla Fisch. ex Bess. var. *yunnanensis* Franch. Burkill in Journ. Linn. Soc. Bot. 26: 504. (1899), Schneider in Sargent Pl. Wilson. 2: 451. (1916).

Yunnan: Near Mengzi on the ascent of Mount Benvenu at 3,200 metres, 9682 (collected September 13th 1896). Milê district at 1,800 metres, 9894.

Discovered by Delavay on the Cangshan range above Dali in Yunnan. Distributed in western and central China.

Coryphopteris viscosa (Baker) Holttum. (ASPLENIACEAE)

Nephrodium calcaratum (Blume) Hook. Anonymous in Kew List of Determinations (volume 34).

Yunnan: Simao, 12827, 12890.

Cotinus coggygria Scop. ANACARDIACEAE)

Rhus cotinus L. Forbes & Hemsley in Journ. Linn. Soc. Bot. 23: 146. (1886), Henry in Journ. China Br. Roy. Asiat. Soc. 22: 257. (1887) (Chinese Names of Plants); in Notes Econ. Bot. China 42. (1893). *Cotinus coggygria* Scop. var. *pubescens* Engler Rehder & E. H. Wilson in Sargent Pl. Wilson. 2: 175. (1914), Zheng in Journ. Wuhan Bot. Res. 2.(1): 116. (1984).

Hubei: Yichang, 735, 1148, 1196. On Tsui Fu Shan near Yichang (May 1888), 1627.

Henry stated that in the Three Gorges region where this shrub is a common inhabitant of the cliffs above the Yangtze it was known colloquially as *lu mu*. Collected by the Glasnevin Central China Expedition (GCCE 600) near Henji town in Xingshan in September 2004. Spectacular on the dark limestone cliffs of the Three Gorges when it takes on fiery autumnal hues.

Cotoneaster acutifolius Turcz. (ROSACEAE)

Schneider in Illus. Handb. Laubholzk. 1: 752. (1906).

Cotoneaster sp. Zheng in Journ. Wuhan Bot. Res. 2.(1): 69. (1984).

Hubei: Jianshi, 6003, 7426. Fang Xian, 6649. Xingshan, 6750. Yichang, 7611. **Sichuan:** South Wushan, 5587.

Cotoneaster bullatus Bois (ROSACEAE)

Schneider in Illus. Handb. Laubholzk. 1: 747. (1906).

Sichuan: Henry's Chinese collector with Pratt, no locality stated but may be any of the following, Leshan, Emei Shan, Wa Shan or Kangding, 8954.

A handsome deciduous shrub to 4 metres tall, first cultivated in France in 1898 by Maurice de Vilmorin at Les Barres, it rapidly became a popular plant on account of its beautiful clusters of fruits. Reintroduced by Wilson from central China in 1900.

Cotoneaster dammeri C. K. Schneider (ROSACEAE) New species

in Illus. Handb. Laubholzk. I. 760. (1904), Zheng in Journ. Wuhan Bot. Res. 2.(1): 67. (1984), Nelson in Pim, The Wood & the Trees 230. (1984), Bean in Trees & Shrubs 1: 738. (1989), Nelson in A Heritage of Beauty 311. (2000), O' Brien in Augustine Henry, An Irish Plant Collector in China 16. (2002), O' Brien in Ir. Garden 13(5): 53. (2004).

Hubei: On Huang Po Shan in Changyang, 4930 (type).

A prostrate, evergreen shrub much use in modern day landscape design, especially for ground cover. Introduced to cultivation by Wilson in 1900.

Cotoneaster divaricatus Rehder & E. H. Wilson (ROSACEAE) New species

in Sargent Pl. Wilson. 1: 157. (1913), Bean in Trees & Shrubs 1: 408. (1921), Nelson in A Heritage of Beauty 311. (2000), O' Brien in Augustine Henry, An Irish Plant Collector in China 16. (2002).

Cotoneaster integerrimus Forbes & Hemsley non Medik. in Journ. Linn. Soc. Bot. 23: 260. (1887)

Sichuan: South Wushan, 5701.

A shrub to 2 metres tall and one of the finest Chinese cotoneasters in terms of fruit. Discovered by Henry in May of 1888, later introduced to cultivation by E. H. Wilson in 1904 through Veitch's Coombe Wood nursery.

Cotoneaster gracilis Rehder & E. H. Wilson (ROSACEAE) New species

in Sargent Pl. Wilson. 1: 167. (1912).

Cotoneaster integerrimus Forbes & Hemsley non Medik. in Journ. Linn. Soc. Bot. 23: 260. (1887).

Hubei: Liantuo, 7918.

Discovered by Henry in the spring of 1889, later collected by Wilson in Hubei and Sichuan and by William Purdom in Shaanxi.

Cotoneaster harrovianus E. H. Wilson (ROSACEAE) New species

Rehder & E. H. Wilson in Sargent Pl. Wilson. 1: 173.

(1911), Bean in Trees & Shrubs 1: 410. (1921), Nelson in A Heritage of Beauty 311. (2000), O' Brien in Augustine Henry, An Irish Plant Collector in China 17. (2002).

Yunnan: Mountains to the north of Mengzi at 1,600 metres, 10785.

Wilson described this species (naming it for George Harrow, Foreman at Veitch's Coombe Wood nursery) from plants raised by the Veitch firm from seeds (Wilson Veitch 1315) he had collected about 10 miles to the south-west of Mengzi in November 1899 while visiting Henry to secure details of the location of *Davidia.* Henry had collected his herbarium specimen two years previously. It is cultivated in Ireland at the John F. Kennedy Arboretum and the National Botanic Gardens, Glasnevin. Endemic to south-east Yunnan.

Cotoneaster henryanus (Schneid.) Rehder & E. H. Wilson (ROSACEAE) New species

in Sargent Pl. Wilson. 1: 174. (1913), Besant in Gard. Chron. ser. 3, 98: 335. (1935), Kingdon Ward in Berried Treasure 77. (1954), Morley in Glasra 3: 79. (1979), Nelson in Pim, The Wood & the Trees 230. (1984), Bean in Trees & Shrubs 1: 745. (1989), Nelson in A Heritage of Beauty 311. (2000), O' Brien in Augustine Henry, An Irish Plant Collector in China 17. (2002).

Cotoneaster rugosus E. Pritz. var. *henryana* Schneid. in Illus. Handb.Laub. 1: 758. (1906).

Sichuan: South Wushan, 5752 (type).

An evergreen shrub to 2 metres tall with a graceful, lax pendelous habit and a handsome sight when carrying a heavy crop of fruits. Introduced to cultivation by Wilson in 1901 through Messrs Veitch and collected by him in Xingshan and Badong, (Hubei) and in Sichuan in South Wushan. Henry and his wife Alice grew this species in their garden at Ranelagh in the south Dublin suburbs in the early 20th century. It was available from Messrs Veitch in 1908 for five shillings. *Cotoneaster henryanus* was collected by the Glasnevin Central China Expedition (GCCE 613) in Dragon Gate River Forest Park in Xingshan in October 2004. It appears to be rare in its native habitat and many of plants in cultivation are anything but the real species.

Cotoneaster horizontalis Decne. var. *perpusillus* Schneider (ROSACEAE) New variety

in Illus. Handb. Laubholzk. 1: 745. fig. 419. (1906), Rehder & E. H. Wilson in Sargent Pl. Wilson. 1: 155. (1913), Zheng in Journ. Wuhan Bot. Res. 2.(1): 68. (1984).

Cotoneaster microphyllus Henry non Wallich ex Lind. in Journ. China Br. Roy. Asiat. Soc. 22: 267. (1887) (Chinese Names of Plants), Forbes & Hemsley in Journ. Linn. Soc. Bot. 23: 261. (1887).

Hubei: Badong, 2858.

A prostrate shrub differing from the type in its smaller leaves and abundant on open moorland in parts of Hubei. According to Henry this prostrate shrub was known colloquially in the mountainous areas above the Three Gorges as *p'u ti wu kung.* Introduced to cultivation by Wilson in 1908 when collecting for the Arnold Arboretum. The typical *Cotoneaster horizontalis* was described from plants grown in the garden of the Paris Museum and raised from seeds sent by Père Armand David about 1870. There is a good plant of the variety in the old Chinese shrubbery at Glasnevin. *Cotoneaster horizontalis* var. *perpusilla* is still a relatively common plant on the higher mountains in the Three Gorges region, it was collected by the Glasnevin Central China Expedition (GCCE 579) in Xingshan in September 2004.

Cotoneaster moupinensis Franch. (ROSACEAE)

Cotoneaster foveolatus Rehder & E. H. Wilson in Sargent Pl. Wilson. 1: 162. (1912), Zheng in Journ. Wuhan Bot. Res. 2.(1): 67. (1984).

Hubei: Badong, 3760. Changleping, 6328 (March 1888).

Cotoneaster pannosus Franchet (ROSACEAE)

Cotoneaster amoenus E. H. Wilson, Rehder & E. H. Wilson in Sargent Pl. Wilson. 1: 165. (1913), Bean in Trees & Shrubs 1: 735. (1989).

Yunnan: Mountains to the north of Mengzi, 11341.

Cotoneaster pannosus was first introduced by Delavay through Paris in 1888. It was reintroduced by Wilson through Veitch's nursery from seeds collected by Wilson in mountains to the south-west of Mengzi in the autumn of 1899 while visiting Henry for advice and guidance on his (Wilson's) future exploration of the flora of central China. The seedlings first flowered in the summer of 1902. (Wilson-Veitch 1889).

Cotoneaster sp. No. 1. Wall. (ROSACEAE)

Cotoneaster rotundifolius Anonymous non Wall. ex Lindl. in Kew List of Determinations (volume 24).

Hubei: Yichang, 1242.

Cotoneaster sp. No. 2. (ROSACEAE)

Zheng in Journ. Wuhan Bot. Res. 2.(1): 69. (1984).

Hubei: Badong, 910, 4054. South Badong, 7274. Fang Xian, 6787.

Cotoneaster sp. No. 3. (ROSACEAE)

Anonymous in Kew List of Determinations (volume 34).

Yunnan: Milê, 9951

Cotoneaster sylvestrii Pamp. (ROSACEAE)

Cotoneaster hupehensis Rehder & E. H. Wilson in Sargent Pl. Wilson. 1: 169. (1912), Bean in Trees & Shrubs 3: 115. (1936), Nelson in A Heritage of Beauty 311. (2000), O' Brien in Augustine Henry, An Irish Plant Collector in

China 17. (2002).

Cotoneaster integerrimus Forbes & Hemsley non Medik. in Journ. Linn. Soc. Bot. 23: 260. (1887).

Hubei: Badong, 2851. Jianshi, 7468. **Sichuan:** South Wushan, 5525, 7201.

Cotoneaster zabelii C. K. Schneid. (ROSACEAE)

Rehder & E. H. Wilson in Sargent Pl. Wilson. 1: 166. (1912), Zheng in Journ. Wuhan Bot. Res. 2.(1): 69. (1984).

Cotoneaster integerrimus Forbes & Hemsley non Medik. in Journ. Linn. Soc. Bot. 23: 260. (1887).

Hubei: South Badong, 5463. Without locality, 5463a.

A deciduous shrub to 3 metres tall , introduced to cultivation by Wilson in 1907 while collecting for the Arnold Arboretum.

Craibiodendron henryi W. W. Smith (ERICACEAE) New species

in Notes Roy. Bot. Gard. Edinburgh 5: 158. (1912), Judd in Journ. Arnold Arbor. 459. (1986).

Yunnan: In mountains near Simao at 1,600 to 1,950 metres, 9505a. Mengzi, 10459 (paratype). In forests near Simao at 1,950 metres, 13137 (lectotype), 13260.

A tree or shrub to 15 metres tall. Also native to India, northern Myanmar and northern Thailand where it is found in warm-temperate and subtropical forests between 1,500 to 2,850 metres altitude. According to Judd (1986) A. Henry 9505 may represent a hybrid between *Craibiodendron henryi* and *Craibiodendron stellatum* both species occur in southern Yunnan. The genus commemorates William Craib (1882-1933), of the Royal Botanic Gardens, Kew.

Craibiodendron stellatum (Laness.) W. W. Sm (ERICACEAE)

Judd in Journ. Arnold Arb. 67: 464. (1986).

Craibiodendron shanicum (W. W. Sm. Smith in Notes Roy. Bot. Gard. Edinburgh 5: 158. (1912).

Yunnan: On grassy mountains to the north of Mengzi at 1,650 metres, 9505.

Native to Myanmar, China, Thailand, Laos, Cambodia and Vietnam.

Craibiodendron yunnanense W. W. Sm. (ERICACEAE) New species

Judd in Journ. Arnold Arb. 67: 453. (1986).

Yunnan: In Pingbian on the Daweishan Range at 1,950 metres, 9182.

A large evergreen shrub or a small tree to 6 metres tall, this species is particularly striking when its newly emerged copper-red growths are appearing. The type was based on material collected by E. E. Maire in 1912, Henry discovered the tree in 1896. For an illustration of this species see Lancaster in *Travels in China* 166. (1993). Distributed in Myanmar and western China.

Craigia yunnanensis W. W. Sm. & W. E. Evans (MALVACEAE)

Burretiodendron yunnanense Kosterm. in Reinwardtia 6: 8. (1961).

Yunnan: Near Mengzi, in a gully on the northern side of So-Tu-To, 11442 (holotype of *Burretiodendron yunnanense* Kosterm., collected 26th December 1897).

Craniotome furcata (Link) O. Kuntze (LAMIACEAE)

Craniotome versicolor Reichb. Dunn in Notes Roy. Bot. Gard. Edinburgh 6: 178. (1915).

Yunnan: In Pingbian, on the Daweishan Range at 2,300 metres, 9162, 11251.

Distributed in India, the Himalaya (Bhutan and Nepal), Myanmar, western China, Laos and Vietnam.

Craspedolobium unijugum (Gagnepain) Z. Wei & Pedley (FABACEAE) New genus

Craspedolobium schochii Harms in Fedde, Repert. Spec. Nov. Regni Veg. 17: 135. (1921).

Yunnan: Mengzi at 2,000 metres, 9241a., 9241c.

Crataegus cuneata Sieb. & Zucc. (ROSACEAE)

Forbes & Hemsley in Journ. Linn. Soc. Bot. 23: 259. (1887), Henry in Journ. China Br. Roy. Asiat. Soc. 22: 244. (1887) (Chinese Names of Plants), Zheng in Journ. Wuhan Bot. Res. 2.(1): 69. (1984).

Hubei: Yichang, 591, 767, 1378, 2241. Without locality, 591a.

The *hou chua tzu* according to Henry. Wilson described this as a common plant at Yichang where grows as shrubs to 30 to 75 cm tall. It is also known as the *shan zha* and the fruits are used in herbal medicine. Also native to Japan, it is rare in cultivation and its leaves are still used as a tea substitute in China.

Crataegus hupehensis Sargent (ROSACEAE) New species

in Sargent Pl. Wilson. 1: 178. (1912), Zheng in Journ. Wuhan Bot. Res. 2.(1): 69. (1984).

Crataegus pinnatifida Henry non Bunge in Journ. China Br. Roy. Asiat. Soc. 22: 244 & 269. (1887) (Chinese Names of Plants), Forbes & Hemsley in Journ. Linn. Soc. Bot. 259. (1887), Henry in Notes Econ. Bot. China 7. (1893).

Hubei: Yichang, 1675, 2062. Jianshi, 7522 (syntype). By a cave in the San You Dong Glen (October 1886), 2894. Without locality, 2894a.

This small tree was discovered by Henry in 1886 and was also later collected by Wilson at Yichang and Liantuo. In the Three Gorges region Henry stated it was colloquially known as *shan cha* or *yo shan cha* and was cultivated in parts of Hubei for its large edible fruits. *Crataegus hupehensis* was collected by the Glasnevin Central China Expedition (GCCE 703) in the San You Dong Glen near

Yichang in October 2004. The last Western plant hunters to collect in the San You Dong Glen (Glen of the Three Pilgrims) before the Glasnevin group were Augustine Henry, Antwerp Pratt and Ernest Wilson.

Crataegus scabrifolia (Franch.) Rehder (ROSACEAE)
Rehder in Journ. Arnold Arb. 12: 71. (1931)

Crataegus henryi Dunn in Journ. Linn. Soc. Bot. 35: 494. (1903), Schneider in Ill. Handb. Laubholzk. 1: 770. f. 435. (1906), Diels in Notes Roy. Bot. Gard. Edinburgh. 7: 236. (1912), Sargent in Sargent Pl. Wilson. 1: 181. (1913), Leveille in Cat. Pl. Yun-Nan 229. (1917), Bean in Trees & Shrubs 3: 121. (1936), Nelson in A Heritage of Beauty 311. (2000), O' Brien in Augustine Henry, An Irish Plant Collector in China 17. (2002), Johnson in Champ. Trees of Brit. & Irel. 48. (2003).

Yunnan: Mengzi at 1,650 to 2,000 metres, 9426 (type of *Crataegus henryi* Dunn, July 18th 1896), 9426a., 9426b.

A handsome hawthorn to 9 metres tall, remarkable for its large red fruits and toothed, not lobed leaves. This tree was first collected near Dali in north-west Yunnan by Delavay in June 1882, Forrest gathered specimens on the Cangshan range behind Dali. In Ireland it is cultivated at Birr Castle, County Offaly. Endemic to China and distributed in the western provinces.

Crateva religiosa Forst. f. (CAPPARACEAE)
Henry in Trans. Asiat. Soc. Jap. 24 suppl: 18. (1896) (List Pl. Formosa), Li in Woody Flora of Taiwan 236. fig. 87. (1963).

Taiwan: Wanshoushan, 794. Kaohsiung, 794a. **Yunnan:** In Red River valley, 10597.

The sacred garlic pear is a very beautiful deciduous flowering tree to about 15 metres tall, native to tropical Africa, south-east Asia, southern China to northern Australia and the Pacific islands. Typical of the family to which it belongs (Capparidaceae) the blooms are spectacular, with creamy four-petalled flowers and long whiskery purple stamens. The petals deepen to an orange hue as they age and are garlic scented. The fruits are about the size of a tennis ball and dangle on long stalks. Inside is a mass of fleshy poisonous pulp, in which horseshoe shaped seeds are imbedded. *Crateva religiosa* is held sacred in India and is traditionally planted around tombs and temples. Found in all parts of Chinese Taiwan at low altitude.

Cratoxylum cochinchinense (Lour.) Blume (HYPERICACEAE)
Cratoxylum polyanthum Korth. Anonymous in Kew List of Determinations (volume 24).

Hainan: Near the Kiungchow Pagoda, 7986, Environs of Haikou 8062. Without locality 8499.

Cratoxylum formosum (Jack) Benth & Hook. f. ex Dyer (HYPERICACEAE)
Cratoxylum prunifolium Dyer Anonymous in Kew List of Determinations (volume 36).

Yunnan: South of the Red River, 10687.

Crepidiastrum denticulatum (Houtt.) Pak & Kawano (ASTERACEAE)
Lactuca denticulata Houtt. (Maxim.) Henry in Journ. China Br. Roy. Asiat. Soc. 22: 250. (1887) (Chinese Names of Plants), Forbes & Hemsley in Journ. Linn. Soc. Bot. 23: 480. (1888), Henry in Trans. Asiat. Soc. Jap. 24 suppl: 56. (1896) (List Pl. Formosa).

Hubei: Yichang, 343, 392, 769, 3108, 1958 (in part). Liantuo, 1891, 3108a. In a cave on Moji Shan, 3108. San You Dong Glen, 3108b., 3108c. Badong, 3108d. **Taiwan:** Oluanpi, without number. Tamsui, 1739. **Yunnan:** Mengzi, 9661.

According to Henry this species was known colloquially at Yichang as *ku ching ts'ao*, the leaves and young shoots of this species are used as a vegetable. Distributed in Russia, Mongolia, China, Korea and Japan.

Crepidium biauritum (Lindl.) Szlach. (ORCHIDACEAE)
Microstylis brevicaulis Schlecter in Repert. Spec. Nov. Regni Veg. Beih. 4: 62. (1919).

Yunnan: Simao, 12284. Simao at 1,300 metres, 12284a., (type of *Microstylis brevicaulis* Schlecter).

Crepidium szemaoensis (Tang & F. T. Wang) Nuammee, Seelanan, Suddee & H. A. Pedersen (ORCHIDACEAE) New species
in Thai Forest Bull., Bot. 44: 36 (2016).

Malaxis szemaoensis Tang & F. T. Wang in Acta Phytotax. Sin. 1(1): 75. (1951).

Yunnan: Mountains to the north-west of Simao at 1,650 metres, 13128.

Crepis chloroclada Coll. & Hemsl. (ASTERACEAE)
Anonymous in Kew List of Determinations (volume 34).

Yunnan: Mengzi, 10165.

Crepis napifera (Franch.) Babc. (ASTERACEAE)
Lactuca napifera Franch. Anonymous in Kew List of Determinations (volume 34).

Yunnan: Milê, 9952. Simao 12669.

Endemic to China and distributed in the provinces of Guizhou, Sichuan and Yunnan.

Crepis phoenix Dunn (ASTERACEAE) New species
in Journ. Linn. Soc. Bot. 35: 511. (1903)

Yunnan: In grass mountains near Mengzi at 1950 metres, 10290.

This species was based on material gathered near

Mengzi by William Hancock and Augustine Henry and near Kunming by Père Ducloux. Six of the eight specimens sent to Kew by these collectors had apparently sprung up after older stems had been destroyed by fire, hense Dunn coined the specific epithet *phoenix*. Endemic to Yunnan.

Crepis sp. (ASTERACEAE)

Anonymous in Kew List of Determinations (volume 34).

Yunnan: Mengzi, 10048, 13798. Simao 11788.

Crepis subscaposa Coll. & Hemsl. (ASTERACEAE)

Anonymous in Kew List of Determinations (volume 34).

Yunnan: Simao, 13004.

Native to Myammar, China (Yunnan) and Indochina.

Crinum asiaticum L. (AMARYLLIDACEAE)

Henry in Trans. Asiat. Soc. Jap. 24 suppl: 95. (1896) (List Pl. Formosa).

Taiwan: Kaohsiung, without number. Wanchin, without number. Oluanpi, without number.

A handsome, large bulbous perennial bearing white fragrant flowers, native from India to southern China and Japan. The Hani (an ethnic group) of Xishuangbanna in southern Yunnan call this species *qaqpavqyaivma* and use the leaves and bulbs to treat sprains and contusions and infected stings and bites.

Crossandra infundibuliformis (L.) Nees (ACANTHACEAE)

Crossandra undulifolia Salisbury Henry in Trans. Asiat. Soc. Jap. 24 suppl: 69. (1896) (List Pl. Formosa).

Taiwan: Wanchin, cultivated in gardens, 1723.

Crotalaria acicularis Buch.-Ham. ex Benth. (FABACEAE)

Crotalaria sp. Henry in Trans. Asiat. Soc. Jap. 24 suppl: 30. (1896) (List Pl. Formosa).

Taiwan: Wanchin, 1520. **Yunnan:** Mengzi, 13718.

Widespread across India, southern China, the Philippines and Java.

Crotalaria albida B. Heyne ex Roth (FABACEAE)

Henry in Trans. Asiat. Soc. Jap. 24 suppl: 30. (1896) (List Pl. Formosa).

Hubei: Yichang, 2180. **Taiwan:** Wanshoushan, 711. Wanchin, 448, 1551. **Yunnan:** Milê district, 10574. Simao, 12553.

An erect perennial herb to 70 cm tall, native to India, Malaysia, the Philippines and southern China. The Hani (an ethnic group) of Xishuangbanna in southern Yunnan call this species *qullubaivssaq* and use it to treat urinary track infections, kidney infections, bronchitis, chest infections and asthma.

Crotalaria calycina Schrank (FABACEAE)

Henry in Trans. Asiat. Soc. Jap. 24 suppl: 30. (1896)

(List Pl. Formosa).

Hainan: Environs of Kiungchow, 8310. **Taiwan:** Kaohsiung, 877. Oluanpi, 1230. **Yunnan:** Mengzi at 1,380 metres, 9814.

An annual herb to 70 cm tall, native to much of tropical Asia, a garden escape in Chinese Taiwan.

Crotalaria chinensis L. (FABACEAE)

Crotalaria sp. Henry in Trans. Asiat. Soc. Jap. 24 suppl: 30. (1896) (List Pl. Formosa).

Hainan: Environs of Kiungchow, 8771. **Taiwan:** Wanchin, 1524. Oluanpi, 1233.

A herb to 15 cm tall, a widespread species from eastern India to southern China, Malaysia and the Philippines.

Crotalaria khasiana Balansa ex Thoth. & A. A. Ansari (FABACEAE)

Crotalaria capitata Benth. ex Baker Anonymous in Kew List of Determinations (volume 34).

Yunnan: Mengzi, 10040. Yuanjiang, 11577. Simao, 12466.

Distributed in India, the Himalaya and western China.

Crotalaria kurzii Baker ex Kurz (FABACEAE)

Crotalaria kurzii Baker ex Kurz Anonymous in Kew List of Determinations (volume 34).

Yunnan: Mengzi, 10971. Simao, 12589, 11602, 12703, 12791.

Crotalaria lachnophora A. Rich. (FABACEAE)

Crotalaria elata Welw. ex Baker Anonymous in Kew List of Determinations (volume 34).

Yunnan: South of the Red River from Manmei, 10064.

Crotalaria lejoloba Bartl. (FABACEAE)

Crotalaria ferruginea Graham ex Benth. Forbes & Hemsley in Journ. Linn. Soc. Bot. 23: 151. (1886), Henry in Journ. China Br. Roy. Asiat. Soc. 22: 245. (1887) (Chinese Names of Plants).

Hubei: Yichang, 397, 1659, 2123. Near Yichang, on the hills behind the San You Dong Glen, (July 1886), 1507. **Yunnan:** Mengzi, 9508. Simao, 12271.

Of widespread distribution from India to China, Malaysia and the Philippines. According to Henry this species was known colloquially at Yichang as *yeh hua-sheng*.

Crotalaria linifolia L.f. (FABACEAE)

Henry in Trans. Asiat. Soc. Jap. 24 suppl: 30. (1896) (List Pl. Formosa).

Hainan: Without locality, 8313, (in part). **Taiwan:** Kaohsiung, 1140. **Yunnan:** Mengzi, 9423. Simao, 12372.

An erect herb to 60 cm tall, native to India and southern China.

Crotalaria medicaginea Lam. (FABACEAE)

Anonymous in Kew List of Determinations (volume

34).

Yunnan: Mengzi, 9736.

An erect perennial to 30 cm tall, widespread across India to China to the Philippines and Australia.

Crotalaria pallida Aiton (FABACEAE)

Crotalaria striata DC. Henry in Trans. Asiat. Soc. Jap. 24 suppl: 30. (1896) (List Pl. Formosa).

Taiwan: Wanchin, 397, 1683. **Yunnan:** Simao, 12381.

Crotalaria sericea Burm. f. (FABACEAE)

Crotalaria assamica Benth. Anonymous in Kew List of Determinations (volume 34).

Yunnan: Mengzi, 9017. Simao, 11594.

A subshrub to 1.5 metres tall, carrying terminal spikes of up to 30 yellow flowers from August to October. Widely distributed in India to the Philippines, throughout Yunnan except in alpine areas.

Crotalaria sessiliflora L. (FABACEAE)

Forbes & Hemsley in Journ. Linn. Soc. Bot. 23: 152. (1886)

Crotalaria brevipes Champ. ex Benth. Anonymous in Kew List of Determinations (volume 24).

Hubei: Yichang, 411, 2328, 2328a.

In Chinese herbal medicine this species is currently used to correct cases of dizziness, deafness, blurred vision and to act as a tonic to the kidneys.

Crotalaria similis Hemsley (FABACEAE) New species

in Ann. of Bot. 152. (1895) Henry in Trans. Asiat. Soc. Jap. 24 suppl: 80. (1896) (list Pl. Formosa), Forbes & Hemsley in Journ. Linn. Soc. Bot. 152. (1892).

Taiwan: Oluanpi, 252.

A small decumbent herb to 20 cm tall, endemic to Taiwan where it is confined to the Hengchun Peninsula (where it was discovered by Henry).

Crotalaria spp. (FABACEAE)

Anonymous in Kew List of Determinations (volume 34).

Yunnan: Mengzi, 10155. Simao, 12465, 12859.

Crotalaria sp. **nov**. (FABACEAE)

Anonymous in Kew List of Determinations (volume 24).

Hainan: Environs of Kiungchow, 8684.

Crotalaria tetragona Roxb. ex Andrews (FABACEAE)

Anonymous in Kew List of Determinations (volume 34).

Yunnan: South of the Red River from Manmei, 10181. Simao, 12596.

Native to India, the Himalaya, Myanmar, western China and Vietnam.

Crotalaria uncinella Lam. subsp. *elliptica* (Roxb.) Polhill (FABACEAE)

Crotalaria elliptica Roxb. Henry in Trans. Asiat. Soc.

Jap. 24 suppl: 30. (1896) (List Pl. Formosa).

Taiwan: Oluanpi, 1237.

A perennial herb to 60 cm tall, native to the Andaman Islands to Malaysia

Crotalaria verrucosa L. (FABACEAE)

Henry in Trans. Asiat. Soc. Jap. 24 suppl: 30. (1896) (List Pl. Formosa).

Taiwan: Kaohsiung, without number.

An erect, stout herb to 70 cm tall, native to southern China (including Hong Kong and Taiwan).

Croton cascarilloides Raeush. (EUPHORBIACEAE)

Li in Woody Flora of Taiwan 421. fig. 160. (1963).

Croton cumingii Muell.-Arg. Henry in Trans. Asiat. Soc. Jap. 24 suppl: 84. (1896) (List Pl. Formosa).

Taiwan: Kaohsiung, without number. Wanchin, without number. Oluanpi, 218.

A wide ranging species distributed from Malaysia through to southern China to the Philippines. Common in coastal forests in southern and eastern Taiwan of China.

Croton spp. (EUPHORBIACEAE)

Anonymous in Kew List of Determinations (volume 34).

Yunnan: Mengzi, 9522. Simao, 12095, 12106, 12425, 12917, 13165, 13365.

Croton tiglium L. (EUPHORBIACEAE)

Forbes & Hemsley in Journ. Linn. Soc. Bot. 26: 435. (1894), Henry in Trans. Asiat. Soc. Jap. 24 suppl: 84. (1896) (List Pl. Formosa).

Hubei: Liantuo, cultivated 4402. **Taiwan:** Wanchin, 495.

A small tree to about 5 metres tall, the leaves remain on the tree throughout winter and become highly coloured before they fall in spring. Native to tropical and subtropical Asia, according to Henry it was known colloqually in the Three Gorges region as the *pa-tou* tree, its seeds are used medicinally. *Croton tiglium* is naturalised in Chinese Taiwan, according to Henry it was used for killing fish.

Cryptocarya concinna Hance (LAURACEAE)

Cryptocarya konishii Hayata Li in Woody Flora of Taiwan 209. (1963).

Symplocos sp. Henry in Trans. Asiat. Soc. Jap. 24 suppl: 58. (1896) (List Pl. Formosa). *Cryptocarya lenticellata* Liou Ho non H. Lec. in Laurac. Chine & Indochine 99. (1934).

Taiwan: Oluanpi, 2068.

Cryptocarya densiflora Blume (LAURACEAE)

Liou Ho in Laurac. Chine & Indochine 98. (1934).

Yunnan: Simao, 11645.

Distributed in China, Laos, Vietnam and the Philippines to Queensland, Australia.

Cryptogramma brunoniana Wall. ex Hook. & Grev. (PTERIDACEAE)

Baker in Journ. Bot 27: 176. (1889).

Hubei: Fang Xian, 6948.

A handsome little fern, distributed in Siberia, the Himalaya and China (including Taiwan). Though the locality given by Henry for this species is Fang, by his discription in *The Garden* 61: 4. (1902) it is clear that this collect is from present-day Shennongjia Forest District.

Cryptolepis buchananii R. Br. ex. Roem. & Schult. (APOCYNACEAE)

Anonymous in Kew List of Determinations (volume 34).

Yunnan: Mengzi, 9862. Simao, 12576, 13546.

Distributed in Pakistan, Nepal, India, Sri Lanka, China, Thailand, Laos and Vietnam.

Cryptolepis sinensis (Lour) Merr. (APOCYNACEAE)

Li in Woody Flora of Taiwan 797. fig. 321. (1963).

Cryptolepis elegans Wall. ex G. Don Henry in Trans. Asiat. Soc. Jap. 24 suppl: 60. (1896) (List Pl. Formosa).

Taiwan: Kaohsiung, without number. Wanchin, 626. Oluanpi, 1706.

A slender vine, native to south-east Asia and Malaysia. In China (Taiwan) this species is found on the southern coastal part of the island.

Cryptomeria japonica (Thunb. ex L. f.) D. Don (CUPRESSACEAE)

Henry in Journ. China Br. Roy. Asiat. Soc. 22: 269. (1887) (Chinese Names of Plants); in Notes Econ. Bot. China 35. (1893), Masters in Journ. Linn. Soc. Bot. 26: 544. (1902), Henry in Elwes & Henry in the Trees Gt. Brit. & Irel. 1: 132. (1906), Rehder & E. H. Wilson in Sargent Pl. Wilson. 2: 52. (1914), Wilson in Journ. Arnold Arb. 7: 59. (1926).

Hubei: Yichang, 1340. **Yunnan:** By a temple near Mengzi at 2,000 metres, an enormous tree with a girth of 5.5 metres (18 feet), 9667. Planted by a village near Mengzi, 9667a.

The *kung ch'io sung* or peacock pine. Henry stated that the tree did not occur wild in the Three Gorges region but were brought as potted plants from Sichuan. Wilson's Chinese collections were all from cultivated plants. Native to central and southern Japan and long cultivated in China.

Cryptotaenia japonica Hassk. (APIACEAE)

Cryptotaenia canadensis Forbes & Hemsley non (L.) DC. in Journ. Linn. Soc. Bot. 23: 329. (1887), Henry in Journ. China Br. Roy. Asiat. Soc. 22: 279. (1887) (Chinese Names of Plants).

Hubei: Badong, 1874, 4724, 4829. Antelope Glen, 3634. Liantuo, 2080, 4218. **Yunnan:** Mengzi, 11353.

According to Henry this species was known colloquially in the Three Gorges area as *wo chio pan*. In China (Taiwan) and Japan this annual is grown as a green and either sold as bunches of seedlings or bunches of leaves. These are cooked for a minute or two or used in soups and salads. Distributed in China, Vietnam, Korea and Japan.

Cryptotaenia sp. (APIACEAE)

Anonymous in Kew List of Determinations (volume 24).

Hubei: Badong, 5419.

Cucumis leiospermus (Wight & Arn.) Ghebret. & Thulin (CUCURBITACEAE)

Melothria leiosperma (Wight & Arn.) Cogn. Cogniaux in Engler, Pflanzenr. iv. 275. i: 125. (1916).

Yunnan: Mengzi, 9784.

Cucumis maderaspatanus L. (CUCURBITACEAE)

Mukia scabrella Henry non (L. f.) Arn. in Trans. Asiat. Soc. Jap. 24 suppl: 46. (1896) (List Pl. Formosa). *Melothria maderaspatana* (L.) Cogn. Cogniaux in Engler, Pflanzenr. iv. 275. i: 126. (1916).

Hainan: Environs of Kiungchow, 8230. **Taiwan:** Kaohsiung, without number. Oluanpi, without number.

A slender hirsute vine, distributed through tropical Africa, India to China and Malaysia and Australia.

Cucumis sp. (CUCURBITACEAE)

Mukia sp. Anonymous in Kew List of Determinations (volume 34).

Yunnan: Simao, 12177.

Cucumis trigonus Roxb. (CUCURBITACEAE)

Forbes & Hemsley in Journ. Linn. Soc. Bot. 23: 317. (1887). *Cucumis trigonus* Roxb. in Kew List of Determinations (volume 24).

Hubei: Yichang, 295. **Hainan:** Without locality, 8510.

The gourd, probably cultivated. According to Henry this species was known colloquially at Yichang as *yeh k'u kua*.

Cucurbita maxima Duch. ex Lam. (CUCURBITACEAE)

Henry in Notes Econ. Bot. China 52. (1893).

Hubei: Fang Xian, 6104. **Hainan:** Environs of Haikou, cultivated, 8372.

Winter squash, a cultigen found across Chile, Argentina, Bolivia and Uruguay and cultivated in the mountainous areas of Hubei where according to Henry it was known as the *peh-kua* or *nan kua* (southern gourd), *kua* in China is a generic term for cucurbitaceous plants.

Cucurbitaceae (CUCURBITACEAE)

Henry in Trans. Asiat. Soc. Jap. 24 suppl: 46. (1896) (List Pl. Formosa).

Taiwan: Wanshoushan, 779, 1861. **Yunnan:** Mengzi, 9837, 13785, 13801. Simao, 12512, 12732.

Cullen corylifolium (L.) Medik. (FABACEAE)

Psoralea corylifolia L. Anonymous in Kew List of Determinations (volume 34).

Yunnan: Mengzi, 9436.

An annual suffruticose herb to about 1.5 metres tall, distributed across India, Myanmar, China (Sichuan and Yunnan) the Malay Peninsula and Indonesia.

Cunninghamia lanceolata (Lamb.) Hook. (CUPRESSACEAE)

E. H. Wilson in Journ. Arnold Arb. 7: 57. (1926), Zheng in Journ. Wuhan Bot. Res. 2.(1): 11. (1984), Walsh in Aug. Henry Forst. Herb. 6. (1957).

Cunninghamia sinensis A. Rich. Henry in Journ. China Br. Roy. Asiat. Soc. 22: 269. (1887) (Chinese Names of Plants); in Notes Econ. Bot. China 36. (1893), Masters in Journ. Linn. Soc. Bot. 26: 548. (1902), Henry in Elwes and Henry The Trees Gt. Brit. & Irel. 3: 494. (1908).

Hubei: Yichang, 1086. **Yunnan:** In Pingbian, on the Daweishan Range at 1,600 metres, 9148a. **Taiwan:** Tamsui, without number.

The *sha shu* or Chinese fir, according to Henry this tree was found chiefly in the mountain valleys near the Xiling Gorge where it was known colloquially as the *sha* and its timber was chiefly used for planking and roofing purposes. According to Wilson was esteemed as the most useful of all Chinese timber trees and grew to 35 metres tall with a straight mast-like stem. The wood is fragrant, soft and easily worked and used in general construction and as masts for boats. It was the principal coffin wood of central and western China, its fragrance being concidered to act as a preservative. In parts of western Sichuan, trees that had been engulfed by an earthquake in the 16th century were being mined in the early 19th century and furnished the most valuable coffin timber. The same logs were known as *hsiang-mu* (fragrant wood) or *yin-chen-mu* (long buried wood), from them planks of enormous sizes were cut, coffins made of this well preserved timber commanded astronomical prices.

Those who could afford such coffins had them lacquered jet-black using the varish obtained from the sap of the varnish tree, *Rhus verniciflua*. The most expensive coffins were those hewn from a single log of timber or even more expensive again were those made from *yin chen* (long buried timber) such a coffin costed up to 1,000 ounces of silver in 19th century China. Introduced to Taiwan from mainland China in the early 1800s. The native *Cunninghamia konishii* Hayata is endemic to Taiwan.

Of the many conifers that Henry collected in central China it is worth noting the absence of the *pai-kuo*, *Ginkgo biloba* L. Wilson's 2109, collected at Yichang was from a tree cultivated by a temple as is often the case in China. He stated that the maidenhair tree was commonly cultivated in the region though he never met with wild specimens. In October 2002 a team from the National Botanic Gardens, Glasnevin visited Badong in Hubei province with staff from Wuhan Botanical Gardens, a region where Henry himself had botanised and received material from his Chinese collectors. In this region the expedition gathered herbarium material and fruits from two trees; the first tree was a 1,000-year-old specimen and annually carried over 250 kilos of fruit. The second tree, a mere 800 year old, had belonged to the same family of farmers over its entire life-span. It is not known if these trees are truly wild (both were female) or remnants of cultivation. Yet it is a mystery that Henry never collected specimens at Yichang of such a celebrated tree. Neither Henry nor Wilson collected the Chusan palm, *Trachycarpus fortunei* in the Three Gorges region. Though now common there, it must have colonised the area since both men made their famous collections.

Cunninghamia lanceolata was introduced to Kew by William Kerr in 1804. At Mount Usher in County Wicklow a tree planted in 1880 is 20 metres tall with a trunk diameter of 93 cm. The Chinese fir was collected by the Glasnevin Central; China Expedition (GCCE 341) in October 2002 in Badong and near Zhenzi town in Xingshan (GCCE 492) in September 2004.

Cupressus funebris Endl. (CUPRESSACEAE)

Henry in Journ. China Br. Roy. Asiat. Soc. 22: 265. (1887) (Chinese Names of Plants), Henry in Notes Econ. Bot. China 37. (1893), Skan in Journ. Linn. Soc. Bot. 26: 540. (1902), Henry in Elwes and Henry The Trees Gt. Brit. & Irel. 5: 1162. (1910).

Hubei: Glens about Yichang, 1070. Liantuo, 1898.

During Henry's stay in central China the Chinese weeping cypress was abundant around Yichang where it formed pure woods with trees to 20 metres tall .According to Henry it provided a home to Reeve's pheasant. *Cupressus funebris* is still a relatively common tree in the Three Gorges region. Like other conifers that are evergreen and so symbolised eternal life, it was often planted around tombs, shrines and temples. The Chinese call it the *peh-mu-shu* (white wood tree), its wood is highly regarded in China and in the 19th century it was much used in the hulls and decks of boats and for pillars due to its durability. It was first noticed by the members of Lord Macartney's embassy to China in 1793 in the vale of the tombs where it was planted around graves and was introduced to cultivation by Robert Fortune in 1849. Fortunes description of this tree is worth repeating;

A specimen of one of the most beautiful trees in China.

It has the leaves of a *Thuja*; the trunk is tall and straight, 50 to 60 feet, the branches are horizontal, inclining upwards at the points, and the branchlets are long and pendulous. It forms a striking feature in the landscape owing to its weeping habit. If one could imagine an evergreen weeping-willow with a straight trunk, he would have a good idea of the form and beauty of the tree. - Fortune MSS. in Herb. Kew.

The mourning cypress is rare in gardens and the finest specimen in cultivation grows at Kilmacurragh in County Wicklow (a satellite garden of the National Botanic Gardens, Glasnevin). When measured in 1990 this tree was 15.5 metres tall with the trunk diameter of both limbs reaching 32 cm and 29 cm.

Curculigo spp. (HYPOXIDACEAE)

Anonymous in Kew List of Determinations (volume 34).

Yunnan: Mengzi, 10281, 11076, 11077. In Pingbian, in forests on the Daweishan Range at 1,950 metres, 11484. Simao, 11982.

Curcuma longa L. (ZINGIBERACEAE)

Wright in Journ. Linn. Soc. Bot. 36: 69. (1903).

Taiwan: Wanchin, 89. Kaohsiung Plain, 1606.

Turmeric or *ch'iang-huang*, a stemless, leafy ginger-like perennial with handsome yellow and white flowers carried in oblong spikes. The fleshy rhizomes are bright orange-yellow inside. An ancient cultigen thought to have originated in southwest India, it is one of the most widely cultivated spices in the tropics and was an important export in late 19th century Chinese Taiwan. In the region of the Min river in central China's Sichuan province, a yellow vegetable dye was obtained from the roots of this plant (in parts of China it is still used to dye wool, cotton and silk). The rhizome is the source of the drug *yu-jin* which is used to invigorate the blood. It is also used as an anti-inflammatory (one of its traditional uses is in the treatment of arthritis) and has a cholesterol lowering effect. Studies on animals have shown that turmeric can prevent or inhibit the development of certain cancer cells. It has long been used as a natural antibiotic agent. As a culinary spice it is well known and after being peeled and grated the rhizomes form the basis of curry powder.

Curcuma spp. (ZINGIBERACEAE)

Anonymous in Kew List of Determinations (volume 34).

Yunnan: Mengzi, 11315. Simao, 12261.

Cuscuta chinensis Lam. (CONVOLVULACEAE)

Cuscuta reflexa Roxb. Henry in Journ. China. Br. Roy. Asiat. Soc. 22: 280. (1887) (Chinese Names of Plants).

Hubei: Badong, 2475, 3185. Liantuo, 2632.

Chinese dodder, stated by Henry to be known colloquially in the Three Gorges region as *wu niang t'eng*. The dried stems and leaves of this dodder are currently used in Chinese herbal medicine to relieve diarrhea. *Cuscuta chinensis* was collected by the Glasnevin Central China Expedition (GCCE 223) between Yichang and Badong in October 2002 and two years later by the 2004 Glasnevin Expedition (GCCE 585) near Henji town in Xingshan.

Cuscuta japonica Choisy (CONVOLVULACEAE)

Forbes & Hemsley in Journ. Linn. Soc. Bot. 26: 168. (1890) Henry in Notes Econ. Bot. China 9. (1893).

Hubei: Badong, 358.

Japanese dodder, according to Henry the seeds were the source of a drug called *t'u-tsu-tzu* or *tu si zi,* they are currently used as a tonic and cases of threatened miscarriage in Chinese herbal medicine.

Cuscuta sp. (CONVOLVULACEAE)

Anonymous in Kew List of Determinations (volume 34).

Yunnan: Mengzi, 9693. Simao, 12738.

Cyananthus sp. (CAMPANULACEAE)

Anonymous in Kew List of Determinations (volume 34).

Yunnan: Yuanjiang, 13257.

Cyanotis arachnoidea C. B. Clarke (COMMELINACEAE)

Henry in Trans. Asiat. Soc. Jap. 24 suppl: 99. (1896) (List Pl. Formosa), Brown in Journ. Linn. Soc. Bot. 36: 157. (1903).

Taiwan: Wanchin, 900.

A widespread perennial found over much of tropical Africa, India, Sri Lanka, Vietnam. Laos, Cambodia and southern China.

Cyanotis cristata Roem. & Schult. (COMMELINACEAE)

Cyanotis cristata Roem. & Schult. var. *griffithii* C. B. Clarke Brown in Journ. Linn. Soc. Bot. 36: 158. (1903).

Hainan: Environs of Haikou, 8413.

Cyanotis spp. (COMMELINACEAE)

Anonymous in Kew List of Determinations (volume 34).

Yunnan: Mengzi, 9002, 9984, 10965, 13745. In the mountains to the east of Mengzi at 2,200 metres, 9060. Simao, 12268.

Cyanthillium cinereum (L.) H. Rob. (ASTERACEAE)

Blumea chinensis (L.) DC. Henry in Trans. Asiat. Soc. Jap. 24 suppl: 53. (1896) (List Pl. Formosa). *Vernonia cinerea* (L.) Less. Henry in Trans. Asiat. Soc. Jap. 24 suppl: 51. (1896) (List Pl. Formosa).

Hainan: Near the Kiungchow Pagoda, 7982. Environs of Haikou, 8245. Without locality, 8504. **Taiwan:** Wanchin, 18, 109. Kaohsiung, 8, 1051, 1138, 1785. **Yunnan:** Mengzi at 1,500 metres, 9691, 9886.

Distributed in India, China, Laos, Thailand, Laos, Thailand, Vietnam and Japan.

Cyanthillium patulum (Aiton) H. Rob. (ASTERACEAE)

Vernonia chinensis (L.) Less. Henry in Trans. Asiat. Soc. Jap. 24 suppl: 51. (1896) (List Pl. Formosa), Anonymous in Kew List of Determinations (volume 24 & 34).

Hainan: Without locality, 8140. **Taiwan:** Kaohsiung, 1193. **Yunnan:** Mengzi, 11195.

Native to India, China, Thailand, Laos, Vietnam and the Philippines.

Cyathula prostrata (L.) Blume (AMARANTHACEAE)

Henry in Trans. Asiat. Soc. Jap. 24 suppl: 75. (1896) (List Pl. Formosa), Forbes & Hemsley in Journ. Linn. Soc. Bot. 26: 321. (1903).

Hainan: Without locality, 8398. **Taiwan:** Oluanpi, 207.

Dispersed throughout the tropics of the Old World.

Cyathus montagnei Tul. & C. Tul. (NIDULARIACEAE)

Anonymous in Kew List of Determinations (volume 24).

Sichuan: South Wushan, 7132. (Fungus).

Cycas balansae Warburg (CYCADACEAE)

Cycas siamensis Thistleton-Dyer non Miq. in Journ. Linn. Soc. Bot. 26: 559. (1902).

Yunnan: In forests near Simao at 1,300 to 1,600 metres, 13637.

Chinese plants of *Cycas balansae* have been misidentified as *Cycas siamensis*, a species occuring in Myanmar, Indochina and Thailand.

Cyclea gracillima Diels (MENISPERMACEAE) New species

in Engler, Pflanzenr. 46(iv. 94): 319. (1910).

Menospermaceae, undetermined. Henry in Trans. Asiat. Soc. Jap. 24 suppl: 17. (1896) (List Pl. Formosa). *Paracyclea gracillima* (Diels) Yamamoto in Trop. Agr. 12: 245. f. 1. (1940), Li in Woody Flora of Taiwan 179. (1963).

Taiwan: Wanshoushan, 1166, 1864 (isotype of *Cyclea gracillima* Diels).

A scandent shrub to 3 metres tall, endemic to Taiwan where it grows on the margins of low-altitude forests on the southern part of the island.

Cyclea polypetala Dunn (MENISPERMACEAE) New species

in Journ. Linn. Soc. Bot. 35: 485. (1903), Diels in Engler, Pflanzenr. iv. 94: 317. (1910).

Yunnan: In Pingbian, on the Daweishan Range at 1,500 metres, 11460. In forests near Simao at 1,250 to 1,600 metres, 12072. Near Simao, between 500 and 1,300 metres, 12072a., 12072b. In mountains to the south-west of Simao

at 1,600 metres, 12072c., (isotype).

China and Indochina.

Cyclea racemosa Oliver (MENISPERMACEAE) New species

in Hooker's. Icon. Pl. t 1938. (1890), Bretschneider in Hist. Eur. Bot. Disc. China 2: 779. (1898), Diels in Engler, Pflanzenr. iv. 94: 318. (1910), Zheng in Journ. Wuhan Bot. Res. 2.(1): 44. (1984).

Hubei: Yichang, 2030, 3628, 4113. Liantuo, 3925. Without locality, 5539a., 5539b. **Sichuan:** South Wushan, 5539. **Yunnan:** Simao, 12528.

Endemic to China.

Cyclea sp. (MENISPERMACEAE)

Anonymous in Kew List of Determinations (volume 34).

Yunnan: Simao, 11979.

Cyclea sutchuenensis Gagnep. (MENISPERMACEAE)

Diels in Engler, Pflanzenr. iv. 94: 319. (1910).

Yunnan: South of the Red River on Fengchunling, 10660.

China and Vietnam.

Cyclea tonkinensis Gagnep. (MENISPERMACEAE)

Diels in Engler, Pflanzenr. iv. 94: 318. (1910).

Yunnan: Mountains to the south and south-west of Mengzi at 1,500 metres, 9406a., 9406b.

Distributed in western China and Vietnam.

Cyclocarya paliurus (Batalin) Iljinskaja (JUGLANDACEAE) New species

Pterocarya paliurus Batalin Skan in Journ. Linn. Soc. Bot. 26: 494. (1899), Rehder & E. H. Wilson in Sargent Pl. Wilson. 3: 182. (1917), Zheng in Journ. Wuhan Bot. Res. 2.(1): 18. (1984).

Hubei: Fang Xian, 6598. Changyang, 7694.

Discovered by Henry in 1888 this species was based on material later collected by Père Paul Farges. It is a relatively common tree in moist woods in Hubei and Henry stated it was known colloquially in Hubei as the *chin-chien-li*. Introduced to cultivation by Wilson through Messrs Veitch's Coombe Wood nursery in 1901.

Cyclocodon axillaris (Oliver) W. J. de Wilde & Duyfjes (CAMPANULACEAE)

Campanumoea axillaris Oliver Henry in Trans. Asiat. Soc. Jap. 24 suppl: 56. (1896) (List Pl. Formosa). Anonymous in Kew List of Determinations (volume 34).

Taiwan: Tamsui, 1458. **Yunnan:** Mengzi, 11194. Simao, 12590.

Cymbidium aloifolium (L.) Sw. (ORCHIDACEAE)

Cymbidium pendulum (Roxb.) Sw. Anonymous in Kew List of Determinations (volume 34).

Yunnan: Simao, 12968.

02

Cymbidium elegans Lindl. (ORCHIDACEAE)

Cymbidium longifolium D. Don Rolfe in Journ. Linn. Soc. Bot. 36: 31. (1903).

Yunnan: Mountains to the south-west of Mengzi at 1,650 metres, 11000. Mengzi, 11100, 11372. On cliffs in a forest near Mengzi at 1,900 metres, 11371. Simao, 13531.

Cymbidium faberi Rolfe (ORCHIDACEAE) New species

in Kew Bull. Misc. Inform. 198. (1896), Rolfe in Journ. Linn. Soc. Bot. 36: 30. (1903).

Eulophia yunnanensis Rolfe in Journ. Linn. Soc. Bot. 36: 29. (1903).

Sichuan: South Wushan, 5515 (syntype). **Yunnan:** On rocky mountains in Mengzi at 1,900 metres, 11125 (type of *Eulophia yunnanensis* Rolfe).

Described by Rolfe from collections made by Ernst Faber in Zhejiang in 1886 and from material collected by Henry in Sichuan in May 1888. A handsome terrestrial species known in China as the 'nine-joint orchid'. Native to Nepal, Myanmar, south-west, central and eastern China including Taiwan. The entire plant is used in Chinese herbal medicine and the fragrant flowers appear in March to May.

Cymbidium hookerianum Rchb. f. (ORCHIDACEAE)

Cymbidium grandiflorum Griff. Rolfe in Journ. Linn. Soc. Bot. 36: 30. (1903).

Yunnan: Growing epiphytically on a tree on a wooded cliff near Mengzi at 1,650 metres, 11097. Mengzi, 11858.

Cymbidium iridioides D. Don (ORCHIDACEAE)

Cymbidium giganteum Wall. ex Lind. Anonymous in Kew List of Determinations (volume 34).

Yunnan: Simao, 12965.

Cymbidium lancifolium Hook. (ORCHIDACEAE)

Anonymous in Kew List of Determinations (volume 34).

Yunnan: Simao, 12723, 12975.

A handsome species to about 30 cm tall and bearing white flowers with purple-red blotched lips from April to July, widespread through the Himalaya, southern China, southern Japan, southeast Asia and Indonesia to Papua New Guinea where it grows as an epiphyte on trees and rocks.

Cymbidium lowianum (Rchb. f.) Rchb. f. (ORCHIDACEAE)

Anonymous in Kew List of Determinations (volume 34).

Yunnan: Simao, 12211.

A very beautiful epiphytic orchid carrying pendelous scapes of yellow-green flowers with a conspicuous v-shaped red blotch of the lip from February to May. Distributed from India (Assam) to China (Guangdong and Yunnan), Indochina, Myanmar and Thailand.

Cymbidium sp. No. 1. (ORCHIDACEAE)

Henry in Trans. Asiat. Soc. Jap. 24 suppl: 92. (1896) (List Pl. Formosa), Rolfe in Journ. Linn. Soc. Bot. 36: 32. (1903).

Taiwan: Oluanpi, 1352.

According to Rolfe this collection is allied to *Cymbidium aloifolium* (L.) Sw. (*C. pendulum* (Roxb.) Sw., but with leaves less than half as broad. The specimen is in fruit and the two remaining flowers have lost their lips.

Cymbidium sp. No. 2. (ORCHIDACEAE)

Anonymous in Kew List of Determinations (volume 34).

Yunnan: Simao, 12722, 12971.

Cymbopogon caesius (Hook. & Arn.) Stapf (POACEAE)

Rendle in Journ. Linn. Soc. Bot. 36: 377. (1904).

Cymbopogon schoenanthus (L.) Spreng. var. *caesius* Hack. Rendle in Journ. Linn. Soc. Bot. 36: 377. (1904).

Hainan: Environs of Kiungchow, 8286.

Cymbopogon goeringii (Steud.) A. Camus (POACEAE)

Cymbopogon nardus Rendle ssp. *marginatus* var. *goeringii* Hack. Rendle in Journ. Linn. Soc. Bot. 36: 376. (1904).

Taiwan: Oluanpi, 1309.

Cymbopogon nardus (L.) Rendle (POACEAE)

Rendle in Journ. Linn. Soc. Bot. 36: 376. (1904).

Taiwan: Kaohsiung, 1021.

Cymbopogon tortilis (J. Presl.) A. Camus (POACEAE)

Cymbopogon nardus (L.) Rendle subsp. *hamatulus* (Hook. & Arn.) Rendle in Journ. Linn. Soc. Bot. 36: 376. (1904).

Taiwan: Wanshoushan, 1009.

Cynanchum alternilobum (M. G. Gilbert & P. T. Li) Liede & Khanum (APOCYNACEAE) New species

Sichuania alterniloba M. G. Gilbert & P. T. Li in Novon 5: 12. (1995).

Sichuan: South Wushan, 7262 (type of *Sichuania alterniloba* Gilbert & Li).

Endemic to China and so far only known from Sichuan province. One of many new plants that continue to be found in Henry's vast herbarium collections.

Cynanchum annularium (Roxb.) Liede & Khanum (APOCYNACEAE)

Holostemma rheedei Wallich Anonymous in Kew List of Determinations (volume 34).

Yunnan: Mengzi, 10285.

Cynanchum boudieri H. Lév. & Vaniot (APOCYNACEAE)

Cynanchum amphiboleum C. K. Schneider in Sargent Pl. Wilson. 3: 346. (1916).

Hubei: South Badong, 6030.

02

Cynanchum brevicoronatum Gilbert & Li (APOCYNACEAE) New species
    in Novon 5: 5. (1995).
    Hubei: Donghu, 6418 (type).
    A twining herb, distributed through the provinces of Hubei and Sichuan (where it was also collected by Farges).

Cynanchum gracillimum Wall. ex Wight (APOCYNACEAE)
    Adelostemma gracillimum (Wall. ex Wight) Hook. f. Tsiang in Sunyatsenia 2: 184. (1934).
    Yunnan: Mengzi, 9335, 9735. Pu'er at 1,300 metres 13495.
    Native to southern China and Myanmar.

Cynanchum hemsleyana (Oliv.) Liede & Khanum (APOCYNACEAE) New species
    Holostemma sinense Hemsley in Journ. Linn. Soc. Bot. 103. (1889). *Metaplexis hemsleyana* Oliver in Hooker's Icon. Pl. t. 1970. (1891), Schlecter & Diels in Engler's Bot. Jahrb. 29: 541. (1900), Chung in Mem. Sc. Soc. China 1: 222. (1924), Schneider in Illus. Handb. Laubholzk. 2: 855. (1912).
    Hubei: Yichang, 2755 (co-type of *Holostemma sinense* Hemsley), 3992 (type of *Holostemma sinense* Hemsley). Fang Xian, 6625.
    A climber to 2.5 metres tall, Wilson noted this species was abundant around Yichang.

Cynanchum insulanum (Hance) Hemsl. (APOCYNACEAE)
    Anonymous in Kew List of Determinations (volume 24).
    Hainan: Environs of Haikou, 8058.

Cynanchum officinale (Hemsl.) Tsiang & H. D. Zhang (APOCYNACEAE) New species
    Pentatropis officinalis Hemsley in Journ. Linn. Soc. Bot. 26: 110. (1889), Henry in Notes Econ. Bot. China 48. (1893).
    Hubei: Badong, cultivated, 4814 (type of *Pentatropis officinalis* Hemsley).
    This handsome herbaceous vine was collected by members of the Glasnevin Central China Expedition (GCCE586) near Henji town in Xingshan in September 2004. The root, according to Henry was employed as a drug, he knew this species as the *chu-sha-lien*.

Cynanchum ovalifolium Wight (APOCYNACEAE)
    Cynanchum formosanum (Maxim.) Hemsl. Henry in Trans. Asiat. Soc. Jap. 24 suppl: 60. (1896) (List Pl. Formosa), Li in Woody Flora of Taiwan 799. fig. 322. (1963).
    Taiwan: Kaohsiung, without number.
    Distributed from Taiwan south to northern Queensland.

Cynanchum spp. (APOCYNACEAE)
    Anonymous in Kew List of Determinations (volume 24 & 34). *Metaplexis* sp. Anonymous in Kew List of Determinations (volume 24).
    Hubei: Badong, 1047 (December 1885), 1812. Changleping, 6293. Donghu, 6414. Fang Xian, 6709.

Yunnan: Simao, 12524.

Cynanchum wilfordii (Maxim.) Hook. f. (APOCYNACEAE)
    Anonymous in Kew List of Determinations (volume 24).
    Hubei: Badong, 6118.

Cynodon dactylon (L.) Pers. (POACEAE)
    Henry in Journ. China Br. Roy. Asiat. Soc. 22: 250. (1887) (Chinese Names of Plants); in Trans. Asiat. Soc. Jap. 24 suppl: 109. (1896) (List Pl. Formosa), Rendle in Journ. Linn. Soc. Bot. 36: 402. (1904).
    Hubei: Yichang, 478, 1365. **Taiwan:** Kaohsiung, 1038.
    According to Henry this species was known colloquially in the Three Gorges region as *kou ya ken*. Abundant in hot valleys and the lower slopes of mountains in the Three Gorges region, it was collected by the Glasnevin Central China Expedition (GCCE 696) near Xiakou (Daxiakou) on the Xiang Xi River (a tributary of the Yangtze) in October 2004.

Cynoglossum amabile Stapf & J. R. Drumm. (BORAGINACEAE) New species
    in Kew Bull. Misc. Inform.202. (1906)
    Yunnan: Mengzi, 9365, 13759.
    Described from collections made by Soulie and Pratt near Kangding, Sichuan and by Hancock and Henry in southern Yunnan.

Cynoglossum lanceolatum Forssk. (BORAGINACEAE)
    Cynoglossum micranthum Desf. Henry in Trans. Asiat. Soc. Jap. 24 suppl: 63. (1896) (List Pl. Formosa).
    Taiwan: Tamsui, 1744.

Cynoglossum sp. (BORAGINACEAE)
    Anonymous in Kew List of Determinations (volume 24).
    Hubei: Badong, 538, 1878. Yichang, 1301.

Cynoglossum zeylanicum (Vahl.) Thunb. (BORAGINACEAE)
    Brand in Engler, Pflanzenr. iv. 252: 134. (1921).
    Cynoglossum micranthum Henry non Desf. in Journ. China Br. Roy. Asiat. Soc. 22: 277. (1887) (Chinese Names of Plants), Forbes & Hemsley in Journ. Linn. Soc. Bot. 26: 150. (1890). *Cynoglossum furcatum* Wallich Forbes & Hemsley in Journ. Linn. Soc. Bot. 26: 149. (1890), Henry in Trans. Asiat. Soc. Jap. 24 suppl: 63. (1896) (List Pl. Formosa).
    Hubei: Badong, 863 (September 1885), 4720, 4981. Liantuo, 1896, 2081, Jianshi, 5919. **Taiwan:** Wanshoushan, 1015.
    Henry stated that in the Three Gorges region this species was known colloquially as the *ts'ao tzu*.

Cyperus brevifolius (Rottb.) Hassk. (CYPERACEAE)
    Kukenthal in Engler, Pflanzenr. 20: 600. (1936).
    Kyllinga brevifolia Rottb. Henry in Trans. Asiat. Soc. Jap. 24 suppl: 104. (1896) (List Pl. Formosa), C. B. Clarke in Journ. Linn. Soc. Bot. 36: 223. (1903). *Kyllinga brevifolia* Rottb. var. *oligostachya* (Boeck.) C. B. Clarke

Henry in Trans. Asiat. Soc. Jap. 24 suppl: 104. (1896) (List Pl. Formosa), C. B. Clarke in in Journ. Linn. Soc. Bot. 36: 224. (1903). *Kyllinga monocephala* L.f. Henry in Trans. Asiat. Soc. Jap. 24 suppl: 104. (1896) (List Pl. Formosa), C. B. Clarke in Journ. Linn. Soc. Bot. 36: 224. (1903).

Hubei: Yichang, 2908, 3985. South Badong, 7375. Fang Xian, 6627. Antelope Glen, 3644 (in part). San You Dong Glen, 4268. Sichuan: South Wushan, 7456. **Hainan:** Without locality, 30. **Taiwan:** Kaohsiung, 1102, 1816. Wanshoushan, 1012. Kaohsiung, 1012a., 1012b. Oluanpi, A. 352, 1329. **Yunnan:** Mengzi at 1,650 metres, 10728.

A perennial to 30 cm tall, widely distributed as a weed in tropical and subtropical areas of the world.

Cyperus compressus L. (CYPERACEAE)

C. B. Clarke in Journ. Linn. Soc. Bot. 36: 210. (1903), Kukenthal in Engler, Pflanzenr. 20: 156. (1936).

Bulbostylis barbata Henry non (Rottb.) C. B. Clarke in Trans. Asiat. Soc. Jap. 24 suppl: 105. (1896) (List Pl. Formosa).

Hainan: Environs of Haikou, 8242. **Taiwan:** Kaohsiung, 1868.

A tufted annual, found over tropical, subtropical and warm regions of the world.

Cyperus cuspidatus Kunth. (CYPERACEAE)

Kukenthal in Engler, Pflanzenr. 20: 261. (1936).

Cyperus uncinatus Poir. C. B. Clarke in Journ. Linn. Soc. Bot. 36: 219. (1903).

Yunnan: Simao at 1,600 metres, 13390.

A cosmotropic annual to about 14 cm tall, often a weed of rice-fields.

Cyperus cyperinus (Retz.) Valck.Sur. (CYPERACEAE)

Mariscus cyperinus (Retz.) Vahl. Henry in Trans. Asiat. Soc. Jap. 24 suppl: 104. (1896) (List Pl. Formosa), C. B. Clarke in Journ. Linn. Soc. Bot. 36: 220. (1903).

Taiwan: Wanshoushan, 1118, 1700. **Yunnan:** At Hsienkei on the Red River near the Vietnamese border, 9563.

A perennial to 70 cm tall, a widespread species with a range from India to Malesia and northern Australia.

Cyperus cyperoides L.) Kuntze (CYPERACEAE)

Kukenthal in Engler, Pflanzenr. 20: 514. (1936).

Mariscus philippensis Steud. C. B. Clarke in Journ. Linn. Soc. Bot. 36: 221. (1903). *Mariscus sieberianus* Nees ex Steud. Henry in Trans. Asiat. Soc. Jap. 24 suppl: 104. (1896) (List Pl. Formosa) C. B. Clarke in Journ. Linn. Soc. Bot. 36: 221. (1903).

Hubei: Badong, 1330. **Hainan:** Without locality, without number. **Taiwan:** Tamsui, 1456. **Yunnan:** Mengzi, 13726.

Cyperus diaphanus Schrad. (CYPERACEAE)

Pycreus latespicatus (Boek.) C. B. Clarke C. B. Clarke in Journ. Linn. Soc. Bot. 36: 204. (1903).

Yunnan: Milê, 9119 (in part).

Cyperus dichrostachyus Hochst. Ex A. Rich. (CYPERACEAE)

C. B. Clarke in Journ. Linn. Soc. Bot. 36: 210. (1903).

Yunnan: South of the Red River from Manmei at 1,900 metres, 9473.

Cyperus difformis L. (CYPERACEAE)

Henry in Trans. Asiat. Soc. Jap. 24 suppl: 102. (1896) (List Pl. Formosa), C. B. Clarke in Journ. Linn. Soc. Bot. 36: 210. (1903), Kukenthal in Engler, Pflanzenr. 20: 237. (1936).

Hubei: Yichang, 2730, 4159. In dry rice fields near Yichang, 4294. **Taiwan:** Kaohsiung, 730. **Yunnan:** Near Mengzi at 1,500 metres, 11318.

An annual species found over temperate and subtropical regions of the world. Often found in rice fields.

Cyperus digitatus Roxb. (CYPERACEAE)

Henry in Trans. Asiat. Soc. Jap. 24 suppl: 102. (1896) (List Pl. Formosa), C. B. Clarke in Journ. Linn. Soc. Bot. 36: 211. (1903), Kukenthal in Engler, Pflanzenr. 20: 55. (1936).

Guangxi: Lungchow. 685 (in part). **Taiwan:** Kaohsiung, 1862. **Yunnan:** Mengzi at 1,600 metres, 11320.

A tall perennial to about 120 cm, found in wet places in tropical and subtropical regions of both hemispheres.

Cyperus distans L. f. (CYPERACEAE)

Henry in Trans. Asiat. Soc. Jap. 24 suppl: 102. (1896) (List Pl. Formosa). C. B. Clarke in Journ. Linn. Soc. Bot. 36: 102. (1903).

Cyperus distans L. f. var. *pseudonutans* Kukenth. Kukenthal in Engler, Pflanzenr. 20: 140. (1936).

Taiwan: Kaohsiung, 790 (in part), 1011, 1041, 1129 (in part), 1880.

A perennial to 80 cm with a small corm-like rhizome. A pantropic species and frequently seen in the rice fields of Asia.

Cyperus exaltatus Retz. (CYPERACEAE)

C. B. Clarke in Journ. Linn. Soc. Bot. 36: 212. (1903).

Hainan: Environs of Haikou, 36, 8369 (used to make mats).

Cyperus flavidus Retz. (CYPERACEAE)

Pycreus globosus Reichb. Henry in Trans. Asiat. Soc. Jap. 24 suppl: 102. (1896) (List Pl. Formosa), C. B. Clarke in Journ. Linn. Soc. Bot. 36: 203. (1903), Kukenthal in Engler, Pflanzenr. 20: 352. (1936). *Pycreus globosus* (All.) Reichb. var. *nilagiricus* (Hochst.) C.B. Clarke C. B. Clarke in Journ. Linn. Soc. Bot. 36: 204. (1903), Kukenthal in

Engler, Pflanzenr. 20: 355. (1936). *Cyperus globosus* All. var. *nilagiricus* (Hochst.) C. B. Clarke Kukenthal in Engler, Pflanzenr. 20: 355. (1936). *Pycreus globosus* (All.) Reichb. var. *stricta* (Roxb.) C. B. Clarke C. B. Clarke in Journ. Linn. Soc. Bot. 36: 205. (1903). *Cyperus globosus* All. var. *strictus* (Roxb.) C. B. Clarke. Kukenthal in Engler, Pflanzenr. 20: 356. (1936).

Hubei: Yichang, 83, 83a. Badong, 2574. Jianshi, 7520. **Taiwan:** Kaohsiung, 760. **Yunnan:** Mengzi at 1,500 metres, 9119 (in part). South of the Red River from Manmei at 1,900 metres, 9604, 9640.

Cyperus haspan L. (CYPERACEAE)

C. B. Clarke in Journ. Linn. Soc. Bot. 36: 213. (1903).

Hubei: Yichang, 2250. **Yunnan:** Simao, 12491.

Widely dispersed through temperate, subtropical and tropical regions of the world and often a weed of rice fields.

Cyperus imbricatus Retz. (CYPERACEAE)

Kukenthal in Engler, Pflanzenr. 20: 69. (1936).

Cyperus radiatus Vahl. Henry in Trans. Asiat. Soc. Jap. 24 suppl: 102. (1896) (List Pl. Formosa), C. B. Clarke in Journ. Linn. Soc. Bot. 36: 216. (1903).

Taiwan: Kaohsiung, 1819.

A perennial to 1 metre tall, pantropic in distribution and favouring wet places.

Cyperus iria L. (CYPERACEAE)

Henry in Trans. Asiat. Soc. Jap. 24 suppl: 102. (1896) (List Pl. Formosa), C. B. Clarke in Journ. Linn. Soc. Bot. 213. (1903).

Cyperus iria L. var. *paniciformis* (Franch.) C. B. Clarke. C. B. Clarke in Journ. Linn. Soc. Bot. 36: 214. (1903).

Hubei: Yichang, 302, 680, 2161, 2770. Badong, 2495. Jianshi, 7415. **Taiwan:** Kaohsiung, 731.

A fiberous rooted annual, widely distributed in temperate, subtropical and tropical regions across the globe, though possibly introduced in the American tropics. A weed of rice fields.

Cyperus javanicus Houtt. (CYPERACEAE)

Mariscus albescens Gaud. Henry in Trans. Asiat. Soc. Jap. 24 suppl: 104. (1896) (List Pl. Formosa), C. B. Clarke in Journ. Linn. Soc. Bot. 36: 220. (1903).

Taiwan: Kaohsiung, 763, 783, 1007 (in part), 1032, 1831.

Cyperus leptocarpus (F. Muell.) Bauters (CYPERACEAE)

Lipocarpha microcephala (R. Br.) Kunth C. B. Clarke in Journ. Linn. Soc. Bot. 36: 288. (1903).

Hubei: Liantuo, 4542.

Cyperus malaccensis Lam. subsp. monophyllus (Vahl.) T. Koyama (CYPERACEAE)

Cyperus tegetiformis Henry non Roxb. ex Arn. in Trans. Asiat. Soc. Jap. 24 suppl: 102. (1896) (List Pl. Formosa). C. B. Clarke in Journ. Linn. Soc. Bot. 36: 218. (1903). *Cyperus malaccensis* Lam. var. *brevifolius* Boeck. Kukenthal in Engler, Pflanzenr. 20: 86. (1936).

Taiwan: Kaohsiung, 790 (in part), 1129 (in part).

A tufted perennial to 80 cm tall, native to southern China and the Japan (south) where it favours wet places. Henry stated it was known as *kiam-ts'ao*, i.e. 'salt grass.' The dried culms are used as string for tying small packages. According to Hosie this species was extensively cultivated in Chinese Taiwan for mat making. It is still often cultivated in paddy fields for this purpose.

Cyperus melanospermus (Nees.) Valck.Sur. (CYPERACEAE)

Kukenthal in Engler, Pflanzenr. 20: 583. (1936).

Kyllinga melanosperma Nees Clarke in Journ. Linn. Soc. Bot. 36: 224. (1903).

Yunnan: Mengzi at 1,500 metres, 10164.

Cyperus niveus Retz. (CYPERACEAE)

C. B. Clarke in Journ. Linn. Soc. Bot. 36: 215. (1903), Kukenthal in Engler, Pflanzenr. 20: 288. (1936).

Yunnan: Tien Fang at 300 metres, 9511.

Cyperus nutans Vahl. (CYPERACEAE)

C. B. Clarke in Journ. Linn. Soc. Bot. 36: 215. (1903), Kukenthal in Engler, Pflanzenr. 20: 144. (1936).

Yunnan: Mengzi at 1,500 metres, 9737.

Cyperus nutans Vahl. var. eleusinoides (Kunth.) Haines (CYPERACEAE)

Cyperus eleusinoides Kunth. Henry in Trans. Asiat. Soc. Jap. 24 suppl: 102. (1896) (List Pl. Formosa), C. B. Clarke in Journ. Linn. Soc. Bot. 36: 212. (1903), Kukenthal in Engler, Pflanzenr. 20: 144. (1936).

Taiwan: Kaohsiung, 768, 1078. **Yunnan:** Mengzi at 1,650 metres, 10012.

Cyperus odoratus L. (CYPERACEAE)

Mariscus pohlianus Nees Henry in Trans. Asiat. Soc. Jap. 24 suppl: 104. (1896) (List Pl. Formosa). *Mariscus ferax* C. B. Clarke Henry in Trans. Asiat. Soc. Jap. 24 suppl: 104. (1896) (List Pl. Formosa). *Torulinium confertum* Ham. C. B. Clarke in Journ. Linn. Soc. Bot. 36: 222. (1903).

Taiwan: Kaohsiung, 716, 729, 1776.

Cyperus orthostachyus Franch. & Sav. (CYPERACEAE)

Cyperus truncatus Turcz. var. *orthostachya* (Franch. & Sav.) C. B. Clarke in Journ. Linn. Soc. Bot. 36: 218.(1903), Kukenthal in Engler, Pflanzenr. 20: 155. (1936).

Hubei: Yichang, without number.

Cyperus pilosus Vahl. (CYPERACEAE)

C. B. Clarke in Journ. Linn. Soc. Bot. 36: 215. (1903), Kukenthal in Engler, Pflanzenr. 20: 92. (1936).

Hubei: Yichang, 2358, 5019 (in part). **Yunnan:** Simao

at 1,300 metres, 13140.

Widespread across India, China and Japan often found growing by rivers or in ricefields.

Cyperus polystachyos Rottb. (CYPERACEAE)

Pycreus polystachyos (Rottb.) P. Beauv. Henry in Trans. Asiat. Soc. Jap. 24 suppl: 102. (1896) (List Pl. Formosa), C. B. Clarke in Journ. Linn. Soc. Bot. 36: 205. (1903).

Taiwan: Kaohsiung, 1047, 1060.

An annual or short lived perennial, pantropic in distribution.

Cyperus rotundus L. (CYPERACEAE)

Henry in Journ. China Br. Roy. Asiat. Soc. 22: 246. (1887) (Chinese Names of Plants); in Notes Econ. Bot. China 6. (1893); in Trans. Asiat. Soc. Jap. 24 suppl: 102. (1896) (List Pl. Formosa), C. B. Clarke in Journ. Linn. Soc. Bot. 36: 216. (1903), Kukenthal in Engler, Pflanzenr. 20: 107. (1936).

Hubei: Yichang, 214. **Hainan:** Environs of Haikou, 8240. **Taiwan:** Kaohsiung, 1014, 1035, 1165, 1774, 1958.

A slender, stoloniferous perennial to 40 cm tall. Distributed in all tropical, subtropical and temperate regions of the world, the most abundant of the sedges and a major pest in rice-fields. According to Henry this species was known colloquially at Yichang as *hui t'ou ch'ing* and was the source of the drug *xiang fu* used to regulate menstruation and to alleviate pain.

Cyperus sanguinolentus Vahl. (CYPERACEAE)

Pycreus sanguinolentus (Vahl.) Nees Henry in Trans. Asiat. Soc. Jap. 24 suppl: 102. (1896) (List Pl. Formosa), C. B. Clarke in Journ. Linn. Soc. Bot. 36: 206. (1903), Kukenthal in Engler, Pflanzenr. 20: 385. (1936). *Pycreus sanguinolentus* (Vahl.) Nees ex C. B. Clarke var. *korshinskii* (Meinsh.) Kukenth Kekenthal in Engler, Pflanzenr. 20: 387. (1936).

Hubei: Yichang, 2907. Badong, 5019 (in part). **Hainan:** Environs of Haikou, 7949. **Taiwan:** Kaohsiung, 1120. **Yunnan:** Mengzi, 13727.

Widely distributed through temperate and subtropical regions of the Old World.

Cyperus serotinus Rottb. (CYPERACEAE)

Juncellus inundatus Henry non C. B. Clarke in Trans. Asiat. Soc. Jap. 24 suppl: 102. (1896). *Juncellus serotinus* (Roxb.) C. B. Clarke C. B. Clarke in Journ. Linn. Soc. Bot. 36: 208. (1903), Kukenthal in Engler, Pflanzenr. 20: 316. (1936).

Hubei: Yichang, 2249, 2371. **Taiwan:** Kaohsiung, 702, 1842.

Widespread from the Mediterranean region through India and Japan.

Cyperus serotinus Retz. var. *inundatus* (Roxb.) Kukenth (CYPERACEAE)

Kukenthal in Engler, Pflanzenr. 20: 318. (1938).

Juncellus inundatus (Roxb.) C. B. Clarke Henry in Trans. Asiat. Soc. Jap. 24 suppl: 102. (1896) (List Pl. Formosa), C. B. Clarke in Journ. Linn. Soc. Bot. 36: 207. (1903).

Taiwan: Kaohsiung, 702a., 1089, 1117, 1130, 1163.

Cyperus sp. (CYPERACEAE)

Anonymous in Kew List of Determinations (volume 24).

Hubei: Yichang, 4254.

Cyperus stoloniferus Retz. (CYPERACEAE)

Henry in Trans. Asiat. Soc. Jap. 24 suppl: 102. (1896) (List Pl. Formosa), C. B. Clarke in Journ. Linn. Soc. Bot. 36: 216. (1903), Kukenthal in Engler, Pflanzenr. 20: 106. (1936).

Taiwan: Kaohsiung Spit, 1957.

A widespread species, found over much of tropical Africa through India and Indochina, China to Malesia and northern Australia.

Cyperus tuberosus Rottb. (CYPERACEAE)

C. B. Clarke in Journ. Linn. Soc. Bot. 36: 219. (1903).

Cyperus tenuiflorus Henry non Rottb. in Trans. Asiat. Soc. Jap. 24 suppl: 102. (1896) (List Pl. Formosa).

Taiwan: Kaohsiung, 778, 1037.

A stoloniferous perennial to 30 cm tall. Widespread from the Mediterranean region through to the Indian subcontinent to China and south to northern Australia.

Cyperus zollingeri Steud. (CYPERACEAE)

C. B. Clarke in Journ. Linn. Soc. Bot. 36: 219. (1903), Kukenthal in Engler, Pflanzenr. 20: 133. (1936).

Yunnan: Simao at 1,600 metres, 12186.

Cyphostemma oxyphyllum (A. Rich.) Vollesen (VITACEAE)

Vitis oxyphylla A. Rich. Anonymous in Kew List of Determinations (volume 34).

Yunnan: Simao, 11736.

Cyphotheca montana Diels (MELASTOMATACEAE) New genus

in Bot. Jahrb. Syst. 65: 103. (1932), Li in Journ. Arnold Arb. 25: 11. (1944).

Yunnan: In forests on the mountains of Fengchunlin, along the Red River at 2,130 metres, 10655 (isosyntype).

Endemic to Yunnan.

Cypripedium fargesii Franch. (ORCHIDACEAE) New species

S. C. Chen in Acta Phytotax. Sin. 23: 372. (1985), Cribb in The Genus Cypripedium. 280. (1997).

Cypripedium ebracteatum Rolfe in Kew Bull. Misc.

Inf. 204. (1896), in part, Bretschneider in Hist. Eur. Bot. Disc. China 2: 792. (1898), Rolfe in Journ. Linn. Soc. Bot. 36: 64. (1903).

Hubei: Yichang, 1404. Badong 1404a., (type of *Cypripedium ebracteatum* Rolfe).

Farges's slipper orchid is distributed through the provinces of Gansu, Hubei and Sichuan. This species was first collected by Henry in 1886 and was described as a new species from material gathered by the French missionary Farges in the early 1890s in Sichuan.

Cypripedium flavum P. F. Hunt & Summerh. (ORCHIDACEAE)

Cypripedium luteum Franchet (non *C. luteum* Raf.) Rolfe in Journ. Linn. Soc. Bot. 36: 65. (1903).

Hubei: Fang Xian, 5391e.

This handsome slipper orchid was discovered by Père Armand David near Baoxing in western Sichuan in June 1869 and was described by Franchet as *Cypripedium luteum* in 1887. However this specific epithet had already been used by C. S. Rafinesque for a North American species in 1828, despite the American species now being concidered synonymous with *Cypripedium parviflorum*, Franchet's *C. luteum* was renamed *C. flavum* by Hunt and Summerhayes in 1966. *Cypripedium flavum* was introduced to cultivation in April 1911 by E. H. Wilson through the Arnold Arboretum. Reginald Farrer referred to this beautiful species as 'Proud Margaret' and described it as having 'comely round flowers of a clear yellow, with a waxen sulphur lip. The segments are sometimes mottled with a few fleshy stains, the lip is freckled within, and the staminode in some forms, but not in all, is, or goes, of a rich chocolate which gives Proud Margaret her looks of well-fed intelligence'.

Cypripedium franchetii Rolfe (ORCHIDACEAE)

Cribb in The Genus Cypripedium. 279. (1997).

Cypripedium luteum Rolfe non Franchet in Journ. Linn. Soc. Bot. 36: 65. (1903), Pfitzer in Engler, Pflanzenr. 50: 32. (1903).

Hubei: South Badong, 5391. Fang Xian, 6740.

Franchet's slipper orchid, native to the provinces of Gansu, Henan, Hubei Shaanxi and Sichuan and commemorates the famous French botanist A. R. Franchet who was responsible for naming the thousands of herbarium specimens collected in China by the French missionaries, particularly those of Delavay and David. It was also collected in Sichuan by Père Paul Farges and the Irish missionary and plant collector Fr. Hugh Scallan (Pater Hugo).

Cypripedium henryi Rolfe (ORCHIDACEAE) New species

in Kew Bull. Misc. Inf. 211. (1892), Rolfe in Orch.

Rev. 4: 330. (1896), Bretschneider in Hist. Eur. Bot. Disc. China 2: 792. (1898), Morley in Glasra 3: 79. (1979), O'Brien in Augustine Henry, An Irish Plant Collector in China 17. (2002).

Cypripedium chinense Franchet in Journ. de Bot. (Morot.) 8: 230. (1894).

Hubei: South Badong, 5391a. Jianshi, 5391d. On the borders of Fang and Xinghan, 5391j. **Sichuan:** South Wushan, 5391c., (isosyntype of *Cypripedium chinense,* Franchet), 5391d., (isosyntype of *Cypripedium chinense* Franchet), 5391b., 5391c.

Henry's slipper orchid, native to north-west Yunnan, Guizhou, Sichuan, Gansu, Hubei and southern Shaanxi. A handsome lime-green flowered species and one of the easiest of the slipper orchids to cultivate. *Cypripedium henryi* was exhibited from Dublin at the Chelsea Flower Show in May 2002 in an exhibit in Augustine Henry's honour.

Cypripedium japonicum Thunb. (ORCHIDACEAE)

Henry in Notes Econ. Bot. China 59. (1893), Pfitzer in Engler, Pflanzenr. 50: 42. (1903), Cribb in The Genus Cypripedium. 280. (1997).

Hubei: Badong, 1404, 3777.

The Japanese slipper orchid was described by Carl Peter Thunberg (a student of Linnaeus) in 1784 from one of his own collections made during a visit to Japan. Native to central and eastern China where it groes in damp forests between 620 and 2,000 metres above sea level. Henry called this species the *lao-hu-ch'i.*

Cypripedium plectrochilum Franch. (ORCHIDACEAE)

Cribb in The Genus Cypripedium. 280. (1997).

Cypripedium arietinum Rolfe non R. Br. in Journ. Linn. Soc. Bot. 36: 63. (1903).

Sichuan: South Wushan, 5474.

Cypripedium plectrochilum was discovered by Père Delavay in south-west Sichuan and is distributed in northern Myanmar, south-east Tibet (Xizang), north-west Yunnan, Sichuan and the mountains of north-west Hubei. In these regions it has been collected by such collectors as Forrest, Handel-Mazzetti, Henry, Delavay, Kingdon Ward, Maire, Montbeig, Pratt, Rock, Schneider, Soulie and Wilson.

Cypripedium sp. (ORCHIDACEAE)

Anonymous in Kew List of Determinations (volume 24).

Hubei: Fang Xian, 6865.

Cyrtococcum patens (L.) A. Camus (POACEAE)

Panicum patens L. Rendle in Journ. Linn. Soc. Bot. 36: 332. (1904).

Taiwan: Wanchin, 1545.

Cyrtomium aequibasis (C. Chr.) Ching (POLYPODIACEAE) New species

in Bull. Chin. Bot. Soc. 2: 99. (1936).

Cyrtomium caryotideum (Wall. ex Hook. ex Grev.) C. Presl. var. *aequibasis* C. Chr. in Amer. Fern. J. 20(2): 51. (1930).

Yunnan: Mengzi, 9123b.

Cyrtomium caryotideum (Wall. ex Hook. & Grev.) C. Presl. (POLYPODIACEAE)

Cyrtomium falcatum (L. f.) C. Presl. subvar. *hastosum* Christ in Bull. Soc. Bot. France 5: 32. (1905). *Aspidium falcatum* Baker non (L. f.) Sw. in Journ. Bot. 26: 228. (1888).

Hubei: Yichang, 2177.

Cyrtomium falcatum (L. f.) C. Presl. (POLYPODIACEAE)

Aspidium falcatum (L. f.) Sw. Anonymous in Kew List of Determinations (volume 34).

Yunnan: Mengzi, 11830

Cyrtomium fortunei (Diels) J. Sm. (POLYPODIACEAE)

Aspidium falcatum Baker non (L. f.) Sw. in Journ. Bot. 26: 228. (1888). *Cyrtomium falcatum* (L. f.) Presl. var. *polypterum* (Diels) C. Chr. Christ in Bot. Jahrb. 29: 195. (1900), Christ in Bull. Soc. Bot. France 5: 33. (1905).

Hubei: Badong, 3686.

Cyrtomium lonchitoides (H. Christ) H. Christ (POLYPODIACEAE) New species

Aspidium lonchitoides H. Christ in Bull. Herb. Boissier 7: 16. (1899).

Yunnan: On the mountains to the north of Mengzi at 1,900 metres, 11829 (isotype of *Aspidium lonchitoides* H. Christ, collected 24th October 1897).

Distributed in China, Vietnam and Japan.

Cyrtomium macrophyllum (Makino) Tagawa (POLYPODIACEAE)

Aspidium falcatum Baker non (L. f.) Sw. in Journ. Bot. 26: 228. (1888). *Polystichum falcatum* (L. f.) C. Presl. f. *macropterum* Diels in Bot. Jahrb. 29: 194. (1900).

Hubei: Yichang, 836, 993. Badong, 1792, 2423, 2572, 2578 (in part), 3687, 5134, 5148, 5149, 5173. Liantuo, 1978, 1942, 1944, 2149, 3035.

Cyrtomium yamamotoi Tagawa (POLYPODIACEAE)

Aspidium falcatum Baker non Sw. in Journ. Bot. 26: 228. (1888). *Cyrtomium falcatum* (L. f.) Presl. var. *intermedium* Diels in Bot. Jahrb. 29: 195. (1900) Christ in Bull. Soc. Bot. France 5: 33. (1905).

Hubei: Yichang, 438. Badong, 5135.

Dacryopinax spathularia (Schwein.) G. W. Martin (DACRYMYCETACEAE)

Guepinia spathularia Anonymous in Kew List of Determinations (volume 24).

Hubei: Fang Xian, 6668.

A handsome fungus, cosmopolitan in distribution and generally found on dead wood and dead branches of trees.

Dactylicapnos scandens (D. Don) Hutch. (FUMARIACEAE)

Dicentra thalictrifolia (Wallich) Hook. f. & Thoms. Anonymous in Kew List of Determinations (volume 34).

Yunnan: Mengzi, 9191.

Distributed in Nepal, India, Bhutan, Myanmar and western China. A handsome climbing perennial, cultivated in the order beds at the National Botanic Gardens, Glasnevin (previously *Dicentra scandens*).

Dactylicapnos torulosa (Hook. f. & Thoms.) Hutch. (FUMARIACEAE)

Dicentra torulosa Hook. f. & Thoms. var. *yunnanensis* Fedde in Repert. Spec. Nov. Regni. Veg. 17: 198. (1921).

Yunnan: Simao, 13361 (isotype of *Dicentra torulosa* Hook. f. & Thoms. var. *yunnanensis* Fedde).

Native to India and western China.

Dactylis glomerata L. (POACEAE)

Rendle in Journ. Linn. Soc. Bot. 36: 421. (1904).

Hubei: Badong, 4673. Fang Xian, 6530. Xingshan, 6530a. **Yunnan:** On grassy mountains to the south of Mengzi at 1,650 metres, 10740.

Native to Europe, North Africa and temperate Asia, now widely naturalised in North America and southern Africa.

Dactyloctenium aegyptiacum (L.) Willd. (POACEAE)

Rendle in Journ. Linn. Soc. Bot. 36: 406. (1904).

Eleusine aegyptiaca Pers. Henry in Trans. Asiat. Soc. Jap. 24 suppl: 109. (1896) (List Pl. Formosa).

Taiwan: Oluanpi, 258. Wanshoushan, 755, 1087.

Dalbergia assamica Benth. (FABACEAE)

Dalbergia szemaoensis Prain in Ann. Roy. Bot. Gard. (Calcutta) 10: 91. (1904).

Yunnan: Simao at 1,600 metres, 11895 (type of *Dalbergia szemaoensis* Prain).

Dalbergia dyeriana Prain (FABACEAE) New species in Journ. Asiatic. Soc. Bengal 70: 45: (1901).

Dalbergia assamica Zheng non Benth. in Journ. Wuhan Bot. Res. 2.(1): 98. (1984).

Hubei: Antelope Glen, 3437. Yichang, 4132. Liantuo, 4561. **Yunnan:** Mengzi, 10503 (type).

Endemic to China.

Dalbergia foliosa T. S. Ralph (FABACEAE)

Dalbergia foliacea Wall. ex Benth. Anonymous in Kew List of Determinations (volume 34).

Yunnan: Simao, 11887, 11952, 12316, 12568.

Dalbergia henryana Prain (FABACEAE) New species in J. Asiat. Soc. Bengal 70: 46. (1901).

Yunnan: In Pingbian on the Daweishan Range at 1,640 metres, 11248 (isotype).

Endemic to China (Guizhou and Yunnan).

02

Dalbergia hupeana Hance (FABACEAE)

Forbes & Hemsley in Journ. Linn. Soc. Bot. 23: 198. (1887), Henry in Journ. China Br. Roy. Asiat. Soc. 22: 272. (1887) (Chinese Names of Plants); in Notes Econ. Bot. China 55. (1893), Hemsley in Hooker's Icon. Pl. t. 1968. (1891), Zheng in Journ. Wuhan Bot. Res. 2.(1): 98. (1984).

Hubei: Yichang, 3112, 3492, 3670, 4128. Without locality, 3112a., 3112b. Antelope Glen, 3492. Liantuo, 4558. Badong, 4923.

The *paitan*, in Henry's notes it is stated that it is a tree to about 13 metres tall, very common around Yichang; and the wood is used for the rammers of oil-presses. Its wood is white, heavy and very tough and durable, Henry stated the tree was known as the *tian shu* or *t'an* and the wood as *tian mu*. On the Chengdu plain in Sichuan Wilson stated it was almost exclusively used in building wheel-barrows. According to Henry this tree was first found near Ningbo in eastern China by the Kew collector Richard Oldham and was described from material later collected near Yichang by the Irish diplomat, Thomas Watters.

Dalbergia kingiana Prain (FABACEAE) New species

in J. Asiat. Soc. Bengal, pt. 2, Nat. Hist. 67: 289. (1901).

Yunnan: Simao, 12848 (type, collected February 26th 1898).

Distributed in Yunnan, Indochina and Myanmar.

Dalbergia lanceolaria L. f. subsp. ***paniculata*** (Roxb.) Thoth. (FABACEAE)

Dalbergia paniculata Roxb. Anonymous in Kew List of Determinations (volume 34).

Yunnan: Simao, 11667.

Dalbergia millettii (Benth.) Benth. (FABACEAE)

Forbes & Hemsley in Journ. Linn. Soc. Bot. 23: 198 (1887), Prain in Journ. Asiatic. Soc. Bengal. 70: 57. (1901) Zheng in Journ. Wuhan Bot. Res. 2.(1): 98. (1984).

Yunnan: Mengzi at 1,650 metres, 9975.

Bhutan to China and Myanmar.

Dalbergia pinnata (Lour.) Prain (FABACEAE)

Dalbergia tamarindifolia Roxb. Anonymous in Kew List of Determinations (volume 34).

Yunnan: Simao, 12147.

Dalbergia polyadelpha Prain (FABACEAE) New species

in Ann. Roy. Bot. Gard. (Calcutta) 10(1): 84. pl. 65. (1904).

Yunnan: Simao, 11688 (type, collected 6th March 1897).

Distributed in western China.

Dalbergia spp. (FABACEAE)

Zheng in Journ. Wuhan Bot. Res. 2.(1): 98. (1984). Anonymous in Kew List of Determinations (volume 34).

Hubei: South Badong, 6138. Changyang, 6138a.
Yunnan: Simao, 11740, 11898, 12454, 12502, 12988, 13287, 13356, 13428, 13429.

Dalbergia stenophylla Prain (FABACEAE) New species

in Journ. Asiatic Soc. Bengal 70: 56. (1901).

Dalbergia milletti Forbes & Hemsley non Benth. in Journ. Linn. Soc. Bot. 23: 198. (1887), in part, Zheng in Journ. Wuhan Bot. Res. 2.(1): 98. (1984).

Hubei: Yichang, 1355, 3095, 4135. On Moji Shan, 3562. Liantuo, 1950, 3852. South Badong, 6188. Changleping, 6286 (March 1888).

Dalbergia stipulacea Roxb. (FABACEAE)

Anonymous in Kew List of Determinations (volume 34).

Yunnan: Simao, 12117.

Distributed in Myanmar, China (Yunnan), Vietnam and Thailand.

Dalbergia yunnanensis Franch. (FABACEAE)

Prain in Journ. Asiatic Soc. Bengal. 70: (1901).

Yunnan: Mengzi, 10205.

Distributed from Bangladesh to western China and Myanmar, discovered in Yunnan by Delavay.

Dalechampia bidentata Blume (EUPHORBIACEAE)

Dalechampia bidentata Blume var. *yunnanensis* Pax & K. Hoffmann in Pflanzenr. (Engler) 4. Fam. 147(12): 32. (1919).

Yunnan: Simao at 1,500 metres, 12354, (isotype of *Dalechampia bidentata* Blume var. *yunnanensis* Pax & K. Hoffmann).

Damnacanthus angustifolius Hayata (RUBIACEAE) New species

Prismatomeris brevipes Hutchinson in Sargent Pl. Wilson. 3: 413. (1916). *Prismatomeris linearis* Hutchinson in Sargent Pl. Wilson. 3: 414. (1916).

Yunnan: In Pingbian on the Daweishan Range at 1,950 metres, 9040 (isotype of *Prismatomeris linearis* Hutchinson), 9040a., 9040e., (type of *Prismatomeris brevipes* Hutchinson).

Damnacanthus henryi (H. Léveille) H. S. Lo (RUBIACEAE) New species

Canthium henryi H. Léveille in Repert. Spec. Nov. regni. Veg. 13: 178. (1914). *Prismatomeris henryi* (H. Leveille) Rehder in Journ. Arnold Arbor. 16: 329. (1935).

Yunnan: In Pingbian on the Daweishan Range at 1,950 metres, 9045, 9040f., (isotype of *Canthium henryi* H. Léveille). Simao, 11637.

Damnacanthus indicus Gaertn. f. (RUBIACEAE)

Journ. Linn. Soc. Bot. 23. 386. (1888), Zheng in Journ. Wuhan Bot. Res. 2.(1): 188. (1984).

Hubei: Liantuo, 3871.

In current Chinese herbal medicine the dried stems and leaves of this species are used as a diuretic and to relieve the pain of traumatic injury.

Damrongia clarkeana (Hemsley) C. Puglisi (GESNERIACEAE) New species

Boea clarkeana Hemsley in Journ. Linn. Soc. Bot. 26: 232. (1890), Bretschneider in Hist. Eur. Bot. Disc. China 2: 789. (1898).

Hubei: Liantuo, 4544. South Donghu, 7584.

Central China to Vietnam.

Daphne acutiloba Rehder (THYMELAEACEAE) New species

in Sargent Pl. Wilson. 2: 539. (1916), O' Brien in Augustine Henry, An Irish Plant Collector in China 17. (2002).

Yunnan: Near Mengzi on wooded cliffs at 2,100 metres, 11321.

An evergreen shrub to 2 metres tall, white, unscented flowers carried in July. Introduced by Wilson in 1907 while collecting for the Arnold Arboretum. In Ireland this species in cultivated in the John F. Kennedy Arboretum, New Ross, County Wexford. Endemic to China and distributed in Hubei, Sichuan and Yunnan.

Daphne genkwa Sieb. & Zucc. (THYMELAEACEAE)

Henry in Journ. China Br. Roy. Asiat. Soc. 22: 258. (1887) (Chinese Names of Plants), Forbes & Hemsley in Journ. Linn. Soc. Bot. 26: 395. (1891), Rehder in Sargent Pl. Wilson. 2: 538. (1916), Zheng in Journ. Wuhan Bot. Res. 2.(1): 155. (1984).

Hubei: Yichang, 134, 261, 801, 1004, 1005, 1541, 7893.

A deciduous, lilac-flowered shrub to about one metre tall. Originally described from Japan where it is cultivated (and known there according to Henry as *genkwa*. He also stated that in the Three Gorges region this species was known colloquially as *men t'ou huo* and it was also called the *yuan hua*. The flowers are used in medicine and act as a diuretic. The poisonous leaves and roots when dropped in a pond will kill all its inhabitants, hense another common name *yu tou* or 'fish poison'. Wilson stated it was abundant about Yichang, growing amoungst stones that were piled up by the sides of tiny fields in the gorges. Introduced to cultivation by Robert Fortune from China in 1843 and later from Japan where it has been cultivated for centuries. The flowers are similar to common lilac (*Syringa vulgaris*) in colour and form but it is short lived in cultivation. Cultivated in the rock garden at the National Botanic Gardens, Glasnevin.

Daphne kiusiana Miq. (THYMELAEACEAE)

Daphne odora Forbes & Hemsley non Thunb. in Journ. Linn. Soc. Bot. 26: 395. (1891). *Daphne odora* Thunb. var. *atrocaulis* Rehder in Sargent Pl. Wilson. 2: 545. (1916).

Hubei: South Badong, 5160, 5350 (in part), 5369, 7914. Liantuo, 6350. **Sichuan:** South Wushan, 5502, 5768, 7903. North Wushan, 7119 (type of *Daphne odora* Thunb. var. *atrocaulis* Rehder).

An evergreen shrub to 1.5 metres tall, producing masses of white intensly fragrant flowers from February to April. Distributed across China, Korea and Japan.

Daphne papyracea Wallich apud Steudel. (THYMELAEACEAE)

Rehder in Sargent Pl. Wilson. 2: 564. (1916).

Yunnan: In mountain woods at Fengchunling at 2,400 metres, 10118. Mountains to the south of Mengzi at 1,800 metres, 10118a. In forests to the east of Mengzi at 2,100 metres, 11363. Yuanjiang at 1,800 metres, 13293.

A shrub to 1.5 metres tall, carrying clusters of 3 to 10 white tubular flowers at the apex of branches during the months of June and July. This species is similar to *Daphne bholua* though the flowers are white and not fragrant. Distributed in India, the Himalaya and China.

Daphne papyracea Wall. ex G. Don var. *crassiuscula* Rehder (THYMELAEACEAE) New variety

in Sargent Pl. Wilson. 2: 546. (1916).

Yunnan: Mountains to the north of Mengzi at 1,200 metres, 10859 (type).

Endemic to China and distributed in the provinces of Guizhou, Sichuan and Yunnan.

Daphne sp. (THYMELAEACEAE)

Anonymous in Kew List of Determinations (volume 34).

Yunnan: Simao, 12690.

Daphne tangutica Maxim.) var. *wilsonii* (Rehder) H. F. Zhou (THYMELAEACEAE) New variety

Daphne odora Forbes & Hemsley non Thunb. in Journ. Linn. Soc. Bot. 26: 395. (1891). *Daphne wilsonii* Rehder in Sargent Pl. Wilson. 2: 540. (1916), Zheng in Journ. Wuhan Bot. Res. 2.(1): 155. (1984).

Hubei: Fang Xian, 6838 (syntype of *Daphne wilsonii* Rehder).

A shrub to about 1.5 metres tall bearing fragrant pink flowers in early summer, discovered in the mountains of Fang in northern Hubei by Henry on the latter half of his six month plant collecting expedition. The variety was later collected in the same area by Wilson and also at Yichang, Changyang and at Changleping in Wufeng. It was also collected by the Glasnevin Central China Expedition (GCCE 517) near Zhenzi Town in Xingshan in September 2004.

Daphniphyllum calycinum Benth. (DAPHNIPHYLLACEAE)

Forbes & Hemsley in Journ. Linn. Soc. Bot. 26: 429. (1894), Rosenthal in Engler, Pflanzenr. iv. 147a. 6. (1919).

Hainan: Environs of Haikou, without number.

Daphniphyllum longeracemosum Rosenthal (DAPHNIPHYLLACEAE) New species

in Pflanzenr. (Engler) 4. Fam. 147a. 14. (1916).

Yunnan: Mountains to the south of Mengzi at 1,650 metres, 9652 (isosyntype). Simao, 13374.

Native to western China and Vietnam.

Daphniphyllum macropodum Miq. (DAPHNIPHYLLACEAE)

Forbes & Hemsley in Journ. Linn. Soc. Bot. 26: 429. (1894), Rosenthal in Engler, Pflanzenr. iv. 147a. 9. (1919), Zheng in Journ. Wuhan Bot. Res. 2.(1): 114. (1984).

Hubei: Jianshi, 5673a. **Sichuan:** South Wushan, 5673. North Wushan, 7102.

A tree to 6 metres tall, found in moist woods in western Hubei and Sichuan.

Daphniphyllum majus Müll.Arg. (DAPHNIPHYLLACEAE)

in Engler, Pflanzenr. 4. Fam. 147a. 13. (1919), Wang in Acta Phytotax. Sin. 19: 79. (1981).

Daphniphyllum candelabrum Croizat & F. P. Metcalf in Lignan Sci J. 20: 110. (1941).

Yunnan: Simao, 12793 (type of *Daphniphyllum candelabrum* Croizat & F. P. Metcalf).

Daphniphyllum paxianum Rosenthal (DAPHNIPHYLLACEAE)

in Engler, Pflanzenr. 4. Fam. 147a: 13. (1919), Wang in Acta Phytotax. Sin. 19: 79. (1981).

Yunnan: Simao, 12064, 12657 (syntype), 12657a., (isosyntype).

Endemic to China.

Daphniphyllum pentandrum Hayata (DAPHNIPHYLLACEAE)

Daphniphyllum glaucescens Henry non Blume in Trans. Asiat. Soc. Jap. 24 suppl: 83. (1896) (List Pl. Formosa).

Taiwan: Oluanpi, 518, 668, 996, 1290. Tamsui, 1399.

A tree to 15 metres tall, native to China, Cambodia and Vietnam. Widely distributed in Chinese Taiwan.

Daphniphyllum sp. (DAPHNIPHYLLACEAE)

Anonymous in Kew List of Determinations (volume 34).

Yunnan: Simao, 12089.

Dasiphora fruticosa (L.) Rydb. (ROSACEAE)

Potentilla fruticosa L. Rehder in Sargent Pl. Wilson. 2: 301. (1915).

Sichuan: Henry's Chinese collector with Pratt, no locality stated but may be Leshan, Emei Shan, Wa Shan or Kangding, 8955.

Dasiphora manshurica (Maxim.) Juz. (ROSACEAE)

Potentilla fruticosa L. var. *manchurica* Maxim. Rehder

in Sargent Pl. Wilson 2: 303. (1915), Zheng in Journ. Wuhan Bot. Res. 2.(1): 74. (1984). *Potentilla glabra* Lodd. var. *mandshurica* (Maxim.) Handel-Mazzetti in Acta Hort. Gotob. 13: (1939).

Hubei: Xingshan, 6967.

A dwarf deciduous shrub, first described from a collection made near Vladivostok in 1873, though earlier collected in 1859 by the Kew collector, Charles Wilford further north.

Datura metel L. (SOLANACEAE)

Datura alba Rumph. ex Nees Henry in Trans. Asiat. Soc. Jap. 24 suppl: 66. (1896) (List Pl. Formosa).

Taiwan: Oluanpi, without number. Wanchin, without number, Kaohsiung without number.

Henry stated this was the *nao-yang-hua* of southern China, in Hubei however, this name was also applied to *Rhododendron molle* (Blume) G. Don.

Datura stramonium L. (SOLANACEAE)

Henry in Journ. China Br. Roy. Asiat. Soc. 22: 277. (1887) (Chinese Names of Plants), Forbes & Hemsley in Journ. Linn. Soc. Bot. 26. 176. (1890).

Hubei: Liantuo, 2113. Changyang, 7481.

The Jamestown thorn-apple, a vigorous annual to 1.5 metres tall bearing fragrant, white tubular flowers. Native to tropical North America and now a cosmopolitan weed in temperate and tropical regions, in the Three Gorges area Henry stated it was colloquially known as *ts'ui hsin hua.*

Daucus carota L. (APIACEAE)

Forbes & Hemsley in Journ. Linn. Soc. Bot. 23: 336. (1887), Henry in Journ. China Br. Roy. Asiat. Soc. 22: 244. (1887) (Chinese Names of Plants).

Hubei: Yichang, 179, 491, 1665. Liantuo, 1911, 2019. Badong, 1771 (July 1886), 1845, 5190.

The carrot which Henry stated was called *hung lo po,* cultivated throughout China and also wild there (it has a range from Ireland to northwestern and central China). At Yichang according to Henry it was common as a wild plant. Wild carrot, *Daucus carota* ssp. *carota* has a thin, white inedible root, the common cultivated carrot, *Daucus carota* ssp. *sativus* has a thick, fleshy tap root. As a cultivated plant the carrot reached China in the Yuan Dynasty (A.D. 1280—1368) from western Asia. The carrot fly, *Psila rosae* can be a serious pest of this crop in Europe but is not found in southeast Asia.

Davallia bullata Wall. ex Hook. (POLYPODIACEAE)

Anonymous in Kew List of Determinations (volume 34).

Yunnan: Simao, 12428.

Davallia denticulata (Burm. f.) Mett. & Kuhn. (POLYPODIACEAE)

Davallia elegans Sw. Henry in Trans. Asiat. Soc. Jap. 24 suppl: 110. (1896) (List Pl. Formosa).

Taiwan: Wanchin, 1493. **Yunnan:** Simao, 13141.

Native to southern China, Laos, Thailand and Vietnam.

Davallia perdurans H. Christ (POLYPODIACEAE) New species

in Bull. Herb. Boissier ser. 2, 6: 970. (1898).

Yunnan: Mengzi, 10086 (isosyntype).

Davallia platylepis Baker (POLYPODIACEAE) New species

in Kew Bull. Misc. Inf. 229. (1898).

Davallia henryana Baker in Kew Bull. Misc. Inf. 8. (1906).

Yunnan: Near Mengzi on rocks at 1,550 metres, 10082 (type of *Davallia platylepis* Baker). Simao, 13111. In forests on mountains to the east of Simao, 10082b., (type of *Davallia henryana* Baker).

North-east India to China (Guizhou and Yunnan).

Davallia pulchra (D. Don) M. Kato & Tsutumi (POLYPODIACEAE) New species

Davallia rigidula Baker in Kew Bull. Misc. Inf. 8. (1906).

Yunnan: Simao on rocks at 1,650 metres, 13069.

Davallia sinensis (H. Christ.) Ching (POLYPODIACEAE) New species

Davallia solida (Forst.) Sw. var. *sinensis* H. Christ in Bull. Herb. Boissier, ser. 2, 7: 18. (1899).

Yunnan: In Pingbian on the Daweishan Range in forests at 1,650 metres, 11822 [isotype of *Davallia solida* (Forst.) Sw. var. *sinensis* H. Christ]. Simao, 13309.

Davallia solida (G. Forst.) Sw. (POLYPODIACEAE)

Henry in Trans. Asiat. Soc. Jap. 24 suppl: 110. (1896) (List Pl. Formosa).

Taiwan: Oluanpi, 1240.

An epiphytic fern, native to Myanmar, Malaya, the Philippines, China (Taiwan) and Polynesia.

Davallia sp. (POLYPODIACEAE)

Anonymous in Kew List of Determinations (volume 34).

Yunnan: Simao, 12898, 13034.

Davallia trichomanoides Blume (POLYPODIACEAE))

Davallia dissecta J. Sm., T. Moore & Houlston Henry in Trans. Asiat. Soc. Jap. 24 suppl: 110. (1896) (List Pl. Formosa).

Taiwan: Tamsui, 1414, 1423.

Davallodes pulchra (D. Don) M. Kato & Tsutumi (POLYPODIACEAE)

Davallia rigidula Baker in Kew Bull. Misc. Inf. 8. (1906). *Davallia pulchra* D. Don Nooteboom in Acta Phytotax. Sin. 34: 172. (1996).

Yunnan: On rocks near Mengzi at 1,650 metres, 10333 [lectotype of *Davallia rigidula* Baker (K)].

Davallodes yunnanensis (Christ) M. Kato & Tsutsumi (POLYPODIACEAE) New species

Davallia yunnanensis Christ in Bull. Herb. Boissier 6: 970. (1898). *Humata yunnanensis* (Christ) Ching in Bull. Fan. Mem. Inst. Biol. ser. 2, 1: 296. (1949). *Araiostegia yunnanensis* (Christ) Tagawa & K. Iwats. ex Panigrahi & S. K. Basu in J. Econ. Taxon. Bot. 5(4): 850. (1984).

Yunnan: Mengzi 10333a., [type of *Humata yunnanensis* (Christ) Ching]. Simao, 13192, 13349.

Davidia involucrata Baillon (NYSSACEAE)

Oliver in Hooker's. Icon. Pl. t. 1961. (1891), Henry in Notes on Econ. Bot. China 61. (1893), Bretschneider in Hist. Eur. Bot. Disc. China 2: 784. (1898), Diels in Bot. Jahrb. 29: 505. (1900), Andre in Rev. Hort. 377. fig. 158. (1902), Henry in Flora & Sylva 1: 218. (1903), Masters in Gard. Chron. ser. 3, 33: 236. fig. 98. (1903); 39: 346. fig. 138. (1906), Hemsley in Journ. Linn. Soc. Bot. 35: 556. fig. 19. (1903); in Kew Bull. Misc. Inf. 301. (1907), Veitch in Journ. Roy. Hort. Soc. 28: 57. fig. 11, 12. (1903), Vilmorin & Bois in Frut. Vilmorin. 145. (1904), Mottet in Rev. Hort. 297. fig. 124, 125, 126. (1906); 321. fig. 105. (1907), M. de Vilmorin in Rev. Hort. Belge. 34: 230. (1908), Browne and McDonald in Ir. Nat. J. 19(6): 208. (1978), Nelson in Arnoldia 43: 21-38. (1983), Nelson in Pim, The Wood & the Trees 222. (1984), Lancaster in Travels in China 379. (1993), Nelson in A Heritage of Beauty 311. (2000).

Davidia vilmoriniana Dode in Rev. Hort. (Paris) 80: 406 (1908). *Davidia involucrata* Baill. var. *vilmoriniana* (Dode) Wagner in Das Pflanzenreich 41 (iv. 220a.): 17. (1910), Hemsley in Bot. Mag. cxxxviii. t. 8432. (1912), E. H. Wilson in Sargent Pl. Wilson 2: 256. (1916), Osborn in Gard. Chron. ser. 3, 86: 501. (1929), O' Brien in Augustine Henry, An Irish Plant Collector in China. 17. (2002).

Sichuan: South Wushan at the village of Ma-huang-p'o at 1,800 metres, 5577 (syntype of *Davidia vilmoriniana* Dode, May 17th 1888), 5577a., (fruits, September 5th 1888).

Henry's association with the handkerchief tree has been confused over the passage of time. The tree was discovered by the French Jesuit missionary and naturalist, Père Armand David near Baoxing (formerly Moupin) in 1869. Henry rediscovered the handkerchief tree (the glabrous-leaved type formerly known as *Davidia involucrata* var. *vilmoriniana*) in May 1888, while on the first part of his six month plant collecting expedition. This expedition brought him to South Wushan in Sichuan, very close to the Hubei-Sichuan

02

frontier. It was while riding up a river valley with his porters in the area, that he found a single 9 metre tall tree growing at the base of a cliff. As he was to later relate the scene was one of the strangest sights he ever witnessed in China, the *Davidia* was in full flower and waving its innumerable ghost handkerchiefs. Henry's specimens caused great excitement on their arrival at Kew, the fruits at that time were still unknown so Henry had his Chinese collectors return that autumn, the fruits were collected and promptly despatched to London where Daniel Oliver had them drawn and published.

Alas the fruits proved so curious that they were pickled and the variety was not introduced to cultivation until Père Farges sent seeds to Maurice de Vilmorin in France in 1897. From 37 seeds, one plant was raised in June 1899 and flowered for the first time at Les Barres in May 1906. Unaware of Vilmorin's success, the enterprising English nursery firm, Veitch of Chelsea, who had been very closely following Henry's work in China decided to send one of their own collectors to China. By 1900 David had died and the only person in China to know the whereabouts of the handkerchief tree was Henry. That single tree, found by Henry a dozen years before was to initiate the greatest era of plant introductions from western and central China. The man Harry Veitch sent needs no introduction, Wilson was dispatched to China with the following secret instructions from his employers;

The object of the journey is to collect a quantity of seeds of a plant, the name of which is known to us. This is the object, do not dissipate time, energy or money on anything else. In furtherance of this you will endeavour to visit Dr. Augustine Henry and obtain precise data as to the habit of this particular plant and information on the flora of Central China in general.

The rest of the story is so well known that there is no need to repeat it here. Suffice to say Wilson accomplished his mission despite the fact that Henry's tree had been recently felled and from that same collection some 13,000 seedlings were raised. The first seedling flowered at Veitch's Coomb Wood nursery in 1911, five years after Farges' plant.

In Sichuan, where Henry first found his tree, it is known as the *kung-t'ung* tree. In Hubei it is known as the *shan-peh-k'o* tree. The handkerchief tree is part of a remarkable relict flora that has survived in central China. One can only imagine Henry's reaction when he stumbled across the tree in the glens of South Wushan in May 1888. It is likely that the old tree on the east side of the pond at Glasnevin are from the original Wilson-Veitch introduction. This old tree presents a spectacular sight in the last week of May each year when it carries thousands of pure white handkerchief-like bracts. Wilson likened its innumerable white bracts to butterflies hovering in the branches of the tree.

George Forrest later collected the handkerchief tree (Forrest 16,446) in the Mekong-Salween divide in May 1918. Distributed through Guizhou, Hubei, Hunan, Sichuan and northern Yunnan. During the 1970s, Miss Moira Scannell, then Keeper of the National Irish Herbarium at the National Botanic Gardens, Glasnevin noticed that in autumn the fruits of *Davidia* (at Glasnevin) became infested with a microfungus. This later proved to be a new species and was formally named *Dothiorella davidiae* Ch. Zambettakis, F. Waliyar & M. Scannell in Bull. Soc. Mycologique de France 93(1): 21. (1977).

Debregeasia edulis (Sieb. & Zucc.) Wedd. (URTICACEAE)
Forbes & Hemsley in Journ. Linn. Soc. Bot. 26: 492. (1899).

Yunnan: Mengzi, 10140.

Debregeasia longifolia (Burm. f.) Wedd. (URTICACEAE)
Schneider in Sargent Pl. Wilson. 3: 313. (1916).

Debregeasia edulis Forbes & Hemsley non (Sieb. & Zucc.)Wedd. Forbes & Hemsley in Journ. Linn. Soc. Bot. 26: 492. (1899).

Yunnan: In forests near Simao at 1,300 metres, 12387. In mountains to the east of Simao at 2,400 metres, 12397a. In Pingbian, in forests on the Daweishan Range at 1,600 metres, 10702, 10702a.

Debregeasia orientalis C. J. Chen (URTICACEAE)
New species
in Novon 1: 56. (1991).

Debregeasia edulis Henry non (Sieb. & Zucc.) Wedd. in Journ. China Br. Roy. Asiat. Soc. 22: 258. (1887) (Chinese Names of Plants), Forbes & Hemsley in Journ. Linn. Soc. Bot. 26: 492. (1899), Li in Woody Flora of Taiwan 134. fig. 42. (1963). *Debregeasia longifolia* Schneider non (Burm. f.) Wedd in Sargent Pl. Wilson. 3: 313. (1917). *Debregeasia velutina* Wilmot-Dear non Gaudich. in Kew Bull. 43: 675. (1988).

Hubei: Yichang, 1201 (syntype). **Taiwan:** Kaohsiung, without number. **Yunnan:** In forests near Mengzi at 1,800 metres, 10536 (syntype).

An evergreen shrub to 5 metres tall, distributed in India, the Himalaya (Bhutan and Nepal) and much of China. *Debregeasia* commemorates Prosper Justin de Bregeas, second in command of the ship *La Bonite* during 1836—1837 when Gaudichaud sailed around the world and during which travels the genus was discovered.

Decaisnea fargesii Franch. (LARDIZABALACEAE)
New species
Raffill in Gard. Chron. ser. 3, 37: 326. (1904), Rehder

& E. H. Wilson in Sargent Pl. Wilson. 1: 344. (1913), Zheng in Journ. Wuhan Bot. Res. 2.(1): 41. (1984).

Decaisnea insignis Henry non (Griff.) Hook. f. & Thoms. in Notes Econ. Bot. China 63. (1893).

Hubei: South Badong, 3784. Changyang, 5405a. Jianshi, 5405b. **Sichuan:** South Wushan, 5405.

A shrub to 7 metres tall in its native habitat, discovered by Henry in about 1886 and based on material later collected (during the early 1890s) by Père Paul Farges. The genus commemorates the Belgian-born Joseph Decaisne who lworked firstly as a gardener and later as a systematic botanist at the Natural History Museum in Paris. This large shrub is well known in cultivation for the sake of its handsome pinnate leaves and remarkable bunches of long dull-blue broadbean like pods. A common shrub in Sichuan and Hubei. Farges sent seeds to Messrs. Vilmorin in 1895 who first flowered and fruited it. A plant was received at Kew from Vilmorin in 1897 and this first flowered in the Himalayan House at Kew in 1901. Forrest also collected this species in north-west Yunnan in June 1917 (Forrest 13926). It was collected by members of the Glasnevin Central China Expedition (2002) in Sichuan on Emei Shan (GCCE 38 & GCCE 87) and in Hubei in Badong (GCCE 280). Wilson stated the fruits were eaten by the stumped-tailed macaques on Emei Shan. In Hubei the colloquial name for this small tree is *mao-erh-tzu* or *mao-erh-shih*.

Decalepis khasiana (Kurz) Ionta & Kimbale (APOCYNACEAE)

Pentanura khasiana Kurz Anonymous in Kew List of Determinations (volume 34).

Yunnan: Simao, 13097.

Decaneuropsis blanda (DC.) H. Rob. & Skvarla (ASTERACEAE)

Vernonia blanda DC. Anonymous in Kew List of Determinations (volume 34).

Yunnan: Mengzi, 10483.

Distributed in India, Myanmar, western China, Thailand, Laos and Vietnam.

Decaspermum gracilentum (Hance) Merr. & L. M. Perry (MYRTACEAE)

Li in Woody Flora of Taiwan 646. fig. 255. (1963).

Eugenia sp. Henry in Trans. Asiat. Soc. Jap. 24 suppl: 43. (1896) (List Pl. Formosa).

Taiwan: Oluanpi, 266.

An evergreen shrub to 3 metres tall, native to southeast China and Vietnam. In Chinese Taiwan this species is distributed in forests of the Hengchun Peninsula.

Decaspermum parviflorum (Lam.) A. J. Scott (MYRTACEAE)

Decaspermum fruticosum Merrill & Perry non J. R. &

G. Forster in Journ. Arnold Arb. 19: 201. (1938).

Yunnan: Simao, 11753, 11753a., 11753b., 11753c.

Distributed from Sikkim to the western Pacific.

Deeringia amaranthoides (Lam.) Merr. (AMARANTHACEAE)

Li in woody Flora of Taiwan 151. (1963).

Deeringia celosioides R. Br. Henry in Trans. Asiat. Soc. Jap. 24 suppl: 74. (1896) (List Pl. formosa). *Deeringia indica* Henry non Zoll. ex Moq. in Trans. Asiat. Soc. Jap. 24 suppl: 74. (1896) (List Pl. Formosa).

Taiwan: Kaohsiung, without number. Oluanpi, 233. Wanchin, 697. Wanshoushan, 1892.

A subdecumbent shrub with a wide geographical range across India to the Philippines to the Pacific Islands. Grows in thickets on the southern part of Chinese Taiwan.

Deeringia polysperma (Roxb.) Moq. (AMARANTHACEAE)

Li in Woody Flora of Taiwan 152. fig. 54. (1963).

Deeringia indica Zoll. Ex Moq. Henry in Trans. Asiat. Soc. Jap. 24 suppl: 74. (1896) (List Pl. Formosa).

Taiwan: Wanchin, 80, 191, 383. Oluanpi, 305, 639, 934.

A subdecumbent shrub ranging from the Malay peninsula to Moloccas, in Chinese Taiwan this species is distributed in the western and southern parts of the island. *Deeringia polysperma* was collected by the Glasnevin Taiwan Expedition (GTE 113) on Wanshoushan, in October 2004.

Deeringia sp. (AMARANTHACEAE)

Anonymous in Kew List of Determinations (volume 34).

Yunnan: Mengzi, 9093.

Delavaya toxocarpa Franch. (SAPINDACEAE)

Delavaya yunnanensis Franchet Radlkofer in Engler, Pflanzenr. iv. 165: 1476. (1934).

Yunnan: In woods near Mengzi between 1,200 and 1,650 metres, 9090, 9090a., 9090b., 9090c., 9090d.

Distributed in western China and Vietnam. Commemorates the French missionary and plant collector, Père Jean Marie Delavay.

Delonix regia (Bojer ex Hook.) Raf. (FABACEAE)

Poinciana regia Bojer ex Hook. Henry in Trans. Asiat. Soc. Jap. 24 suppl: 38. (1896) (List Pl. Formosa).

Hebei: Anping, cultivated 1898.

The peacock flower or flame tree, a deciduous tree with a widespreading crown to 10 metres tall bearing spectacular scarlet to orange flowers. Endemic to Madagascar (where it is endangered). Widely cultivated in the tropics.

Delphinium anthriscifolium Hance (RANUNCULACEAE)

Henry in Notes Econ. Bot. China 32. (1893), Munz in Journ. Arnold Arb. 48: 260. (1967); 49: 245. (1968).

Delphinium anthriscifolium Hance var. *calleryi*

(Franch.) Finet & Gagnep. Munz in Journ. Arnold Arb. 48: 260. (1967), Munz in Journ. Arnold Arb. 48: 260. (1967).

Hubei: Yichang, 1348. Liantuo, 1966. Jianshi, 5862. **Sichuan:** South Wushan, 7263. Henry's Chinese collector with Pratt, no locality stated but may be Leshan, Emei Shan, Wa Shan or Kangding, 8810.

A common annual in Hubei where it is the *huan-liang-tsao*. According to Bretschneider it was discovered by a Mr. Hay in May 1863 on Silver Island on the Yangtze.

Delphinium anthriscifolium Hance var. *savatieri* (Franch.) Munz (RANUNCULACEAE)

Munz in Journ. Arnold Arb. 48: 261. (1967), Munz in Journ. Arnold Arb. 49: 245. (1968).

Hubei: Yichang, 760, 1200.

Delphinium henryi A. R. Franchet (RANUNCULACEAE) New species

in Bull. Soc. Philom. Paris ser. 8, 5: 177. (1893), Bretschneider in Hist. Eur. Bot. Disc. China 2: 779. (1898), Munz in Journ. Arnold Arb. 48: 519. (1967), Morley in Glasra 3: 79. (1979).

Hubei: Fang Xian, 6952 (type). Without locality, 6952a.

Delphinium potaninii Huth var. *bonvalotii* (Franch.) W. T. Wang (RANUNCULACEAE)

Delphinium bonvalotii Franch. Bretschneider in Hist. Eur. Bot. Disc. China 2: 779. (1898). *Delphinium potaninii* Huth Munz in Journ. Arnold Arb. 49: 239. (1968).

Sichuan: Henry's Chinese collector with Pratt, no locality but most likely Kangding, 8792.

This species was also collected by Prince Henri d'Orleans while travelling enroute from near Lhasa to Kangding. At Kangding d'Orleans met Antwerp Pratt who took charge of his herbarium specimens and transported them to Shanghai.

Delphinium sp. (RANUNCULACEAE)

Anonymous in Kew List of Determinations (volume 24).

Hubei: Yichang, 412.

Delphinium trifoliolatum Finet & Gagnep. (RANUNCULACEAE) New species

Munz in Journ. Arnold Arb. 49: 119. & 245. (1968).

Hubei: Jianshi, 7448.

This species was based on a collection made by Wilson in July 1901 (Wilson 2496, published as Wilson 3496). Henry had collected it some 13 years previously.

Delphinium umbrosum Hand.-Mazz. (RANUNCULACEAE) New species

Munz in Journ. Arnold Arb. 49: 119. (1968).

Yunnan: Mengzi, 9668, 9668a.

Handel-Mazzetti based this species on his own 7809

from near Zhongdian in north-west Yunnan. Henry had found the plant decades previously. Distributed from western China to Nepal.

Delphinium yunnanense (Franch.) Franch. (RANUNCULACEAE)

Munz in Journ. Arnold Arb. 48: 543. (1967) Munz in Journ. Arnold Arb. 49: 245. (1968).

Yunnan: Mengzi, 9272, 9272a. Simao, 13440.

A handsome perennial to 90 cm tall, carrying loose racemes of soft blue-purple fragrant flowers in autumn. Distributed in Guizhou, Sichuan, and Yunnan. Discovered by Delavay on August 31st 1886, Forrest also later collected this species on the Doker La in the Mekong-Salween divide in August 1917 (Forrest 14653).

Dendrobium bellatulum Rolfe (ORCHIDACEAE) New species

in Journ. Linn. Soc. Bot. 36: 10. (1903), J. H. Veitch in Hortus Veitchii 126. (1906), Kranzlin in Engler, Pflanzenr. 50: 87. (1910).

Yunnan: In mountain forest in Pingbian, on the Daweishan Range at 1,600 metres, epiphytic on trees, 11109 (type).

In describing this species Rolfe states that this is a beautiful little plant with flowers like a miniature edition of *D. formosum* Roxb. ex Lindl. Introduced to cultivation by Wilson who sent living plants to Veitch's Coombe Wood nursery from Yunnan while visiting Henry in 1900. *Dendrobium bellatulum* is distributed from the central Himalaya to China, Thailand and Indochina.

Dendrobium chrysanthum Wall. ex Lindl. (ORCHIDACEAE)

Rolfe in Journ. Linn. Soc. Bot. 26: 10. (1910).

Dendrobium chrysanthum Wall. var. *microphthalama* Reichb. f Fr. Kranzlin in Engler, Pflanzenr. 50: 49. (1910).

Yunnan: On grassy banks and rocks near Mengzi at 1,650 metres, 11094, 11094a. Simao, 12966.

Dendrobium chrysotoxum Lindl. (ORCHIDACEAE)

Fr. Kranzlin in Engler, Pflanzenr. 50: 67. (1910).

Yunnan: Simao, 12212.

A spectacular epiphytic orchid bearing pendelous, many flowered racemes of golden yellow flowers in April to June. Distributed in India, Myanmar, Malaysia, Laos, Thailand, Vietnam and China (southern Yunnan).

Dendrobium compactum Rolfe ex Hemsl. (ORCHIDACEAE)

Fr. Kranzlin in Engler, Pflanzenr. 50: 84. (1910).

Yunnan: West of Simao at 1,500 metres, 11752a. Simao, 12752.

Native to the tropical and subtropical forests of Myanmar, China (Yunnan) and Thailand.

Dendrobium denneanum Kerr (ORCHIDACEAE)

Dendrobium clavatum Wall. ex Lind. Fr. Kranzlin in Engler, Pflanzenr. 50: 48. (1910).

Yunnan: Mengzi, 11836.

Dendrobium falconeri Hook. (ORCHIDACEAE)

Henry in Trans. Asiat. Soc. Jap. 24 suppl: 91. (1896) (List Pl. Formosa), Fr. Kranzlin in Engler, Pflanzenr. 50: 31. (1910).

Taiwan: Oluanpi, 1372.

A tufted epiphytic orchid, widespead across the Himalayan region, India, Myanmar, Thailand and southern China.

Dendrobium gibsonii Paxton (ORCHIDACEAE)

Rolfe in Journ. Linn. Soc. Bot. 36: 11. (1903), Fr. Kranzlin in Engler, Pflanzenr. 50: 44. (1910).

Yunnan: In Pingbian on the Daweishan Range, 10613. Mengzi, 11837. Simao, 12082, 12326, 13180.

A widespread species found over much of India, the Himalaya, Myanmar, western China and Thailand. Previous to Henry's collections this beautiful orchid was known only from northern India. An epiphytic species growing on trees or rocks in the forests of Guangxi and Yunnan at 1,400 to 2,400 metres above sea level. Flowers appear between April and June.

Dendrobium hainanense Rolfe (ORCHIDACEAE)
New species

in Kew Bull. Misc. Inf. 193. (1896), Bretschneider in Hist. Eur. Bot. Disc. China 2: 792. (1898), Kranzlin in Engler, Pflanzenr. 50: 221. (1910).

Hainan: Lingmen, without number.

Dendrobium henryi Schlechter (ORCHIDACEAE)
New species

in Fedde, Repert. Spec. Nov. Regni Veg. 17: 67. (1921).

Yunnan: Simao, on forest floor at 1,300 metres, 13179.

A beautiful yellow flowered, fragrant epiphytic orchid. Native to western China, Thailand and Vietnam.

Dendrobium lindleyi Steud. (ORCHIDACEAE)

Dendrobium aggregatum Roxb Fr. Kranzlin in Engler, Pflanzenr. 50: 62. (1910).

Yunnan: Simao, 12080.

Eastern Himalaya to China and Indochina.

Dendrobium loddigesii Rolfe (ORCHIDACEAE)

Rolfe in Journ. Linn. Soc. Bot. 36: 12. (1903), Fr. Kranzlin in Engler, Pflanzenr. 50: 29. (1910).

Hainan: 32 km (20 miles) west of Haikou, 8136. **Yunnan:** Simao, 12282.

A spectacular epiphytic species, native to Laos, Vietnam and south-west China and the eastern province of Guangdong where it grows on trees or on rocks in forests at 700 to 2,100 metres. Pale rose flowers appear in March and April and the stem is used in medicine.

Dendrobium moniliforme (L.) Sw. (ORCHIDACEAE)

Dendrobium zonatum Rolfe in Journ. Linn. Soc. Bot. 36: 13. (1903), Kranzlin in Engler, Pflanzenr. 50: 58. (1910).

Yunnan: In forests on Fengchunling at 2,250 metres, 10668 (type of *Dendrobium zonatum* Rolfe).

A widespread species, native to India, Myanmar, southern China and Indochina.

Dendrobium nobile Lindl. (ORCHIDACEAE)

Henry in Journ. China Br. Roy. Asiat. Soc. 22: 270. (1887) (Chinese Names of Plants), Rolfe in Journ. Linn. Soc. Bot. 36: 12. (1903), Fr. Kranzlin in Engler, Pflanzenr. 50: 35. (1910).

Sichuan: Chongqing, 2365, 3656.

According to Henry the dried stems of this orchid were sent from Chongqing (part of Sichuan province until 1997 and now a municipality) to Hankow (now a district of Wuhan) as a medicinal export. Henry stated that this species was known colloquially in the Three Gorges region as *shih hu* and the drug extracted is the *ya-t'ou*. It was also known in China as the *shi hu* and is widely distributed in south, western and central China where it grows on broadleaved trees and on rocks in forests at 500 to 1,700 metres. It is still an important plant in Chinese medicine and is also used there as an insecticide. In the *Index Florae Sinensis* Henry's 2365 is quoted by Forbes & Hemsley as being from Yichang, however in volume 24 of the Kew List of Determinations (housed in the library of the Royal Botanic Gardens, Kew) Henry's states this number is from Chongqing.

Dendrobium palpebrae Lindl. (ORCHIDACEAE)

Fr. Kranzlin in Engler, Pflanzenr. 50: 64. (1910).

Yunnan: Simao, 12213, 12213a., 12213b.

Dendrobium sp. (ORCHIDACEAE)

Henry in Trans. Asiat. Soc. Jap. 24 suppl: 91. (1896) (List Pl. Formosa).

Taiwan: Oluanpi, 1373.

Dendrobium spp. (ORCHIDACEAE)

Anonymous in Kew List of Determinations (vol. 34.).

Yunnan: Mengzi, 11850. Simao, 11804, 12083, 12733, 12977, 13084, 13612.

Dendrobium stenoglossum Schlechter (ORCHIDACEAE)
New species

in Fedde, Repert. Spec. Nov. Regni Veg. 17: 66. (1921).

Dendrobium strongylanthum Fr. Kranzlin non Reichb. f. in Engler, Pflanzenr. 50: 81. (1910).

Yunnan: Epiphytic on trees, on the mountains to the west of Simao at 1,800 metres, 12962.

Dendrobium trigonopus Reichb. f. (ORCHIDACEAE)

Fr. Kranzlin in Engler, Pflanzenr. 50: 63. (1910).

Yunnan: Simao, 12129.

A very beautiful yellow flowered epiphytic species, in southern Yunnan it is found growing on trees in forests at 1,500 metres. Also distributed in Myanmar, China, Thailand and Laos.

Dendrocalamus latiflorus Munro (POACEAE)

Henry in Trans. Asiat. Soc. Jap. 24 suppl: 109. (1896) (List Pl. Formosa), Rendle in Journ. Linn. Soc. Bot. 36: 448. (1904).

Sinocalamus latiflorus Munro McClure Li in Woody Flora of Taiwan. 914. (1963).

Taiwan: Wanchin, 14, 1760. Kaohsiung, 1803.

A large bamboo, culms to 25 metres tall and 20 cm. across. Native from Myanmar to southern China. Widely cultivated in Chinese Taiwan.

Dendrocnide meyeniana (Walp.) Chew (URTICACEAE)

Laportea pterostigma Wedd. Henry in Trans. Asiat. Soc. Jap. 24 suppl: 88. (1896) (List Pl. Formosa).

Taiwan: Wanshoushan, without number. Oluanpi, without number. Wanchin, without number.

A small tree, the leaves of which are covered with minute hairs which sting violently. According to Henry this handsome tree was known colloquially in Taiwan as *yao-jen-kon.* Native to the Philippines and China (Taiwan) where it grows throughout the island and on Lutao Island. *Dendrocnide meyeniana* was collected by the Glasnevin Taiwan Expedition (GTE 26) by the lighthouse at Oluanpi (South Cape of Henry and Wilson) in October 2004.

Dendrolobium triangulare (Retz.) Schindl. (FABACEAE)

Desmodium cephalotes (Roxb.) Wall. ex Wight & Arn. Henry in Trans. Asiat. Soc. Jap. 24 suppl: 33. (1896) (List Pl. Formosa). Anonymous in Kew List of Determinations (volume 34).

Hainan: Without locality, 8580. Environs of Kiungchow, 8681. **Taiwan:** Kaohsiung, without number. Wanchin, 847. Oluanpi, 954, 1219. **Yunnan:** In the Red River valley, 11189. Simao, 12107.

Dendrotrophe varians (Blume) Miq. (SANTALACEAE)

Forbes & Hemsley in Journ. Linn. Soc. Bot. 26: 409. (1894).

Hainan: Lingmen, 8527.

Dennstaedtia hirsuta (Sw.) Mett. Ex Miq. (DENNSTAEDTIACEAE)

Davallia hirsuta Sw. Baker in journ. Bot. 27: 176. (1889).

Hubei: South Badong, 6172.

Distributed from Siberia to China, Korea and Japan.

Dennstaedtia zeylanica (Sw.) Zink ex Fraser-Jenk. & Kandel (DENNSTAEDTIACEAE)

Dicksonia scabra Wall. ex Hook. Anonymous in Kew List of Determinations (volume 34).

Yunnan: Mengzi, 9920. Yuanjiang at 1,800 metres, 13296.

Widely distributed through the Himalaya, India, Sri Lanka, China (including Taiwan), Indochina, Korea, Japan, the Philippines, Borneo and Celebes.

Dentella repens (L.) J. R. Forst.& G. Forst. (RUBIACEAE)

Henry in Trans. Asiat. Soc. Jap. 24 suppl: 49. (1896) (List Pl. Formosa).

Taiwan: Kaohsiung, without number.

Deparia acrostichoides (Sw.) M. Kato (ASPLENIACEAE)

Asplenium thelypteroides Michx. Anonymous in Kew List of Determinations (volume 24).

Hubei: Badong, 4765, 5083, 5085, 5092, 5093, 6176. Changyang, 7641, 7642, 7649.

Distributed from Russia to China, Korea and Japan.

Deparia boryana (Willd.) M. Kato (ASPLENIACEAE)

Nephrodium boryanum (Willd.) Hook. Anonymous in Kew List of Determinations (volume 24 & 34).

Hubei: Liantuo, 2012. **Yunnan:** Simao, 13142.

Distributed from Nepal, India to Sri Lanka, Myanmar, China, Veitnam, Japan and the Philippines.

Deparia henryi (Baker) M. Kato (ASPLENIACEAE)
New species

Asplenium henryi Baker in Journ. Bot. 27: 176. (1889), Baker in Ann. Bot. 5: 306. (1890); in Summary of New Ferns 46. (1892), Bretschneider in Hist. Eur. Bot. Disc. China 2: 794. (1898), Christ in Bull. Soc. Bot. France 5: 50. (1905).

Hubei: Badong, 5033, 5058, 5087, 5124, 5140, 7646 (type of *Asplenium henryi* Baker).

Deparia heterophlebia (Mett. ex Baker) R. Sano (ASPLENIACEAE)

Diplazium hemionitideum Christ in Bull. Herb. Boiss. 7: 2. (1899).

Yunnan: In Pingbian on the Daweishan Range in forests at 1,650 metres, 11556 (syntype of *Diplazium hemionitideum* Christ).

Deparia japonica (Thunb.) M. Kato (ASPLENIACEAE)

Asplenium japonicum Thunb. Anonymous in Kew List of Determinations (volume 24).

Hubei: Yichang, 2055. Liantuo, 2147, 4428, 4605. Badong, 5107.

Deparia lancea (Wall. ex Hook. & Grev.) Fraser-Jenk. (ASPLENIACEAE)

Asplenium lanceum Thunb. Henry in Trans. Asiat. Soc. Jap. 24 suppl: 112. (1896) (List Pl. Formosa). *Diplazium lanceum* (Thunb.) Presl. Christ in Bull. Herb. Boiss. 7: 11. (1899), Christ in Bull. Soc. Bot. France 5: 51. (1905).

Hubei: Badong, 2380, 2858. **Taiwan:** Tamsui, 1377.

Yunnan: In a ravine 20 Milês north of Simao at 1,650 metres, 11561.

A widespread species, distributed through India, Sri Lanka, China and Japan.

Deparia stenoptera (Baker) Z. R. Wang (ASPLENIACEAE) New species

Polypodium stenopterum Baker in Journ. Bot. 26: 229. (1888), Baker in Ann. Bot. 5: 457. (1890) Bretschneider in Hist. Eur. Bot. Disc. China 2: 794. (1898).

Hubei: Badong, 3682 (type of *Polypodium stenopterum* Baker).

Derris fordii Oliv. (FABACEAE)

Anonymous in Kew List of Determinations (volume 24 & 34).

Hubei: Liantuo, 4448, 6389. **Yunnan:** Simao, 12897.

Endemic to China.

Derris henryi Thoth. (FABACEAE) New species

in J. Jap. Bot. 45(1): 6-8. f. 1. (1970).

Yunnan: Mengzi, 9398 (type).

Derris laxiflora Benth. (FABACEAE)

Henry in Trans. Asiat. Soc. Jap. 24 suppl: 37. (1896) (List Pl. Formosa), Li in Woody Flora of Taiwan 343. (1963).

Taiwan: Wanchin, 458, 565. Oluanpi, 1996.

A vigorous climber, endemic to Taiwan and scattered in low altitude forests, particularly on the southern part of the island.

Derris marginata (Roxb.) Benth. (FABACEAE)

Anonymous in Kew List of Determinations (volume 34).

Yunnan: Mengzi, 10893.

Distributed in India, Myanmar, southern China and Indochina.

Derris spp. (FABACEAE)

Anonymous in Kew List of Determinations (volume 24 & 34).

Hainan: Without locality, 8583. **Yunnan:** Mengzi, 9386, 11364.

Derris taiwaniana (Hayata) Z. Q. Song (FABACEAE)

Millettia pachycarpa Benth Dunn in Journ. Linn. Soc. Bot. 41: 168. (1912), Hemsley in Hooker's Icon. Pl. t. 2738. (1901).

Yunnan: Chuyuan, 10521. Simao, 13000, 13530.

Distributed in Nepal, India, Myanmar, China and Indochina.

Derris trifoliata Lour. (FABACEAE)

Li in Woody Flora of Taiwan 343. fig. 122. (1963).

Derris uliginosa (Willd.) Benth. in Trans. Asiat. Soc. Jap. 24 suppl: 38. (1896) (List Pl. Formosa).

Hainan: Without locality, 8228. **Taiwan:** Wanchin, 863. Kaohsiung, 1049.

A widely distributed climber in the Old World tropics. A strand plant in Taiwan.

Desmodium spp. (FABACEAE)

Anonymous in Kew List of Determinations (volume 24 & 34). Henry in Trans. Asiat. Soc. Jap. 24 suppl: 34. (1896) (List Pl. Formosa).

Hubei: Yichang, 47. **Hainan:** Ling-men, 8667. **Taiwan:** Wanchin, 1554. **Yunnan:** Mengzi, 9311, 10049, 11213. Simao, 11685, 11728, 125322, 13481, 13538.

Desmos chinensis Lour. (ANNONACEAE)

Unona discolor Vahl var. petalis angustioribus. Anonymous in Kew List of Determinations (volume 24).

Hainan: Hummocks near Haikou, 8094. Environs of Haikou, 8404. Ling-men (in the interior of the island), 8678. Without locality, 8094, 8404, 8526. Without locality, 8480.

Desmos sp. (ANNONACEAE)

Unona sp. Anonymous in Kew List of Determinations (volume 24).

Hainan: At Nodoa in the interior of the island, 8467.

Deutzia aspera Rehder (HYDRANGEACEAE) New species

in Sargent Pl. Wilson. 1: 149. (1912).

Yunnan: South of Red River from Manmei at 2,000 metres, 9475, (holotype).

Endemic to China, distributed in Yunnan and Tibet (Xizang).

Deutzia crassifolia Rehder (HYDRANGEACEAE) New species

in Sargent Pl. Wilson. 1: 148. (1913).

Deutzia purpurascens (Franch.) Rehder var. *pauciflora* Rehder in Sargent Pl. Wilson. 1: 19. (1913).

Yunnan: Mountains to the north of Mengzi at 1,800 metres 9475a., (type of *Deutzia purpurascens* (Franch.) Rehder var. *pauciflora* Rehder). Mountains to the north of Mengzi at 2,000 metres, 10978 (holotype of *Deutzia crassifolia* Rehder).

Endemic to China, distributed in Yunnan and Tibet (Xizang).

Deutzia crenata Sieb. & Zucc. (HYDRANGEACEAE)

Deutzia scabra Forbes & Hemsley non Thunb. in Journ. Linn. Soc. Bot. 23: 276. (1887).

Hubei: Yichang, 731, 791, 840, 1018. San You Dong Glen, 3509. Antelope Glen, 3581. Liantuo, 3191, 3908. Jianshi, 5823.

Native to Japan. Cultivated and naturalised in several parts of China.

Deutzia discolor Hemsley (HYDRANGEACEAE) New species

in Journ. Linn. Soc. Bot. 23: 275. (1887), Bretschneider in Hist. Eur. Bot. Disc. China 2: 783. (1898), Rehder in

Sargent Pl. Wilson. 1: 12. (1913), Nelson in A Heritage of Beauty 312. (2000), O' Brien in Augustine Henry, An Irish Plant Collector in China 19. (2002).

Hubei: Badong, 1745, 3766, 4056, 5426a. Changleping, 6291, 6335. **Sichuan:** South Wushan, 5426, 5718.

A very beautiful shrub to 1.5 metres tall bearing corymbs of pinkish-white flowers to 7 cm across in late May. Introduced to Irish gardens through the Royal (now National) Botanic Gardens, Glasnevin through Wilson 570 (Xingshan, Hubei) from seed sent by Professor Charles Sprague Sargent of the Arnold Arboretum. There also grows at Glasnevin a plant raised from a Farrer collection.

Deutzia henryi Rehder (HYDRANGEACEAE) New species

in Sargent Pl. Wilson. 1: 148. (1913), Schneider in Ill. Handb. Laubholzk. 2: 933. (1912).

Yunnan: In mountain forest near Simao at 1,600 metres, 10786. Simao at 1,640 metres, 11786 (type).

Endemic to Yunnan.

Deutzia pulchra S. Vidal (HYDRANGEACEAE)

Rehder in Sargent Pl. Wilson. 1: 18. (1911).

Deutzia near *D. pulchra* Vidal. Henry in Trans. Asiat. Soc. Jap. 24 suppl: 41. (1896) (List Pl. Formosa).

Taiwan: Wanchin, 38, 477.

A deciduous shrub to 2.5 metres tall bearing slender panicles of white flowers, sometimes flushed pink in May. Native to northern Luzon in the Philippines and to Taiwan in China. Henry's collection was a first record for Taiwan and it was introduced to cultivation by Wilson from Taiwan in 1918. It first flowered outside its native habitat in Ireland with the Marquess of Headford at Headford House in County Meath. According to H. G. Hillier the long sprays of flowers are like lily of the valley with as many as thirty-five blossoms in a single raceme.

Deutzia schneideriana Rehder (HYDRANGEACEAE) New species

in Sargent Pl. Wilson. 1: 7. (1911), Bean in Trees & Shrubs 3: 137. (1936), Zheng in Journ. Wuhan Bot. Res. 2.(1): 57. (1984), Nelson in A Heritage of Beauty 312. (2000), O' Brien in Augustine Henry, An Irish Plant Collector in China 19. (2002).

Deutzia staminea Forbes & Hemsley non R. Br. ex Wall. in Journ. Linn. Soc. Bot. 23: 277. (1887), Henry in Journ. China Br. Roy. Asiat. Soc. 22: 265. (1887) (Chinese Names of Plants).

Hubei: Liantuo, 1968, 2063. On Moji Shan, 3571.

A deciduous shrub to 2.5 metres tall, allied to *Deutzia scabra*. Wilson 2889 (from Changleping) forms the type, Henry had discovered the species twenty years earlier and

stated it was known colloquially at Badong as *p'ao tung ken*. In Ireland this handsome shrub is cultivated at the John F. Kennedy Arboretum and the National Botanic Gardens, Glasnevin.

Deutzia setchuenensis Franch. (HYDRANGEACEAE) New species

Rehder in Sargent Pl. Wilson. 1: 18. (1913), Zheng in Journ. Wuhan Bot. Res. 2.(1): 57. (1984).

Deutzia scabra var. *cymis paucifloris*. Hemsley in Journ. Linn. Soc. Bot. 23: 277. (1887). *Deutzia parviflora* Forbes & Hemsley non Bunge in Journ. Linn. Soc. Bot. 23: 276. (1887).

Hubei: Yichang, 1360, 3480. Xiling (Yichang) Gorge, 4139. Antelope Glen, 3585.

Described by Franchet from plants raised by Lemoine from seeds sent to him by the French missionary, Père Farges in 1895. Henry's 1360 date from 1886.

Deutzia sp. (HYDRANGEACEAE)

Anonymous in Kew List of Determinations (volume 24).

Hubei: North Badong, 7005. **Sichuan:** Henry's Chinese collector with Pratt, no locality stated but may be Leshan, Emei Shan, Wa Shan or Kangding, 8971.

Deutzia taiwanensis (Maxim.) Schneid. (HYDRANGEACEAE)

Deutzia scabra Henry non Thunb. in Trans. Asiat. Soc. Jap. 24 suppl: 41. (1896) (List Pl. Formosa).

Taiwan: Wanshoushan, without number.

A deciduous shrub to 2 metres tall bearing slender panicles of white flowers. Endemic to Taiwan where it was first collected by Richard Oldham in 1864. It is often confused with *Deutzia pulchra*.

Dianella ensifolia (L.) Redoute (ASPHODELACEAE)

Henry in Trans. Asiat. Soc. Jap. 24 suppl: 97. (1896) (List Pl. Formosa).

Dianella nemorosa Lam. Wright in Journ. Linn. Soc. Bot. 36: 119. (1903).

Hainan: Ling-men, 8663. **Taiwan:** Wanchin, 66, 198, 459. Wanshoushan, 792. Oluanpi, 911.

A widespread species distributed through the tropical and subtropical regions of Asia, Australia and the Pacific Islands.

Dianella sp. (ASPHODELACEAE)

Anonymous in Kew List of Determinations (volume 34).

Yunnan: Mengzi, 10288.

Dianthus chinensis L. (CARYOPHYLLACEAE)

Anonymous in Kew List of Determinations (volume 24).

Hubei: Jianshi, 7524.

A handsome annual or short-lived perennial, flowers

pink with a purple ring towards the base of petals. Introduced to cultivation through the Jardin des Plantes, Paris in 1705. Numerous cultivars were grown in China during the 19th century. Henry stated that both *Dianthus superbus* and *Dianthus chinensis* supplied the *qu mai,* a drug used to promote urination (the entire plant and flowers were used).

Dianthus superbus L. (CARYOPHYLLACEAE)

Anonymous in Kew List of Determinations (volume 24).

Hubei: Donghu, 6442. Fang Xian, 6761. Xingshan, 6957.

A perennial species with pink to lilac frilled petals. This beautiful little plant has a large natural range from France and Holland in the west to Siberia, China and Japan in the East. A multitude of cultivars selected from this species are available in the trade.

Dichanthium annulatum (Forssk.) Stapf (POACEAE)

Andropogon annulatus Forsk. var *genuinus* Hack. Rendle in Journ. Linn. Soc. Bot. 36: 369. (1904).

Hubei: Yichang, 200, 1090, 3528.

Widely distributed across northern Africa, India, Myanmar and China and cultivated elsewhere as a fodder crop.

Dichocarpum adiantifolium (Hook. f. & Thoms.) W. T. Wang & P. K. Hsiao (RANUNCULACEAE)

Isopyrum adiantifolium Hook.f. & Thoms. Anonymous in Kew List of Determinations (volume 24).

Hubei: South Badong, 5374. Changyang, 5252.

Dichocarpum fargesii (Franch.) W. T. Wang & P. K. Hsiao (RANUNCULACEAE) New species

Isopyrum fargesii Franchet in Journ. de Bot. 11: 231. (1897).

Sichuan: South Wushan, 5558, 5558a., (type of *Isopyrum fargesii* Franch.)

Dichondra repens J. R. Forst. & G. Forst. (CONVOLVULACEAE)

Anonymous in Kew List of Determinations (volume 34).

Yunnan: Mengzi, 11410.

Dichotomanthes tristaniicarpa Kurz. (ROSACEAE)

Rehder in Sargent Pl. Wilson. 2: 344. (1916), Bean in Trees & Shrubs 2: 52. (1989).

Yunnan: Mengzi at 1,500 to 1,700 metres, 9367, 9367a., 10255, 10255a.

An evergreen tree to 7 metres tall, discovered in May 1868 by John Anderson, a medical officer and naturaliast attached to the expedition of Major E. B. Sladen to south-west China. Sladen had been sent by the Goverment of India to explore trade routes to China. Along the way Anderson collected about 800 plants and gave these to the herbarium

at Calcutta Botanic Gardens, the duplicates were sent to Kew. *Dichotomanthes tristaniicarpa* was introduced to cultivation by George Forrest from Yunnan. It grows on the western end of the Chinese slope at the National Botanic Gardens, Glasnevin but is of botanical interest only.

Dichotomanthes tristaniicarpa Kurz var. *glabrata* Rehder & E. H. Wilson (ROSACEAE) New variety

in Sargent Pl. Wilson. 2: 344. (1915).

Yunnan: Simao at 1,300 to 1,500 metres, 11959, 11959a., (holotype).

Endemic to southern Yunnan and known only from the type locality at Simao.

Dichrocephala chrysanthemifolia (Blume) DC. (ASTERACEAE)

Anonymous in Kew List of Determinations (volume 34).

Yunnan: Mengzi, 10000.

Distributed in India, the Himalaya, China and Japan.

Dichrocephala integrifolia (L. f.) Kuntze (ASTERACEAE)

Dichrocephala latifolia (Pers.) DC. Henry in Trans. Asiat. Soc. Jap. 24 suppl: 52. (1896) (List Pl. Formosa).

Taiwan: Oluanpi, 340. Wanchin, 1541, 1714. **Yunnan:** Mengzi, 10968.

Dicliptera chinensis (L.) Juss. (ACANTHACEAE)

Henry in Trans. Asiat. Soc. Jap. 24 suppl: 69. (1896) (List Pl. Formosa).

Dicliptera roxburghiana Nees Anonymous in Kew List of Determinations (volume 34).

Taiwan: Wanchin, 1620. Wanshoushan, 1953. **Yunnan:** Mengzi, 11374.

Dicliptera japonica (Thunb.) Makino (ACANTHACEAE)

Dicliptera crinita (Thunb.) Nees Henry in Trans. Asiat. Soc. Jap. 24 suppl: 70. (1896) (List Pl. Formosa).

Taiwan: Oluanpi, 304. Wanchin, 1515. **Yunnan:** Simao, 12816.

Dicranella japonicum Mitt. (DICRANACEAE)

Dicranella japonicum Mitt. var. *yunnanense* Salmon in Journ. Linn. Soc. Bot. 34: 453. (1900).

Hubei: South Badong, 6165 (type of *Dicranella japonicum* Mitt. var. *yunnanense* Salmon, collected July 1888).

Dicranopteris linearis (Burm. f.) Underw. (GLEICHENIACEAE)

Baker in Journ. Bot. 26: 230. (1888)

Gleichenia linearis (Thunb.) Hook. Diels in Bot. Jahrb. 29: 205. (1900). *Polypodium lineare* Thunb. Baker in Journ. Bot. 26: 230. (1888), Christ in Bull. Soc. Bot. France 5: 14. (1905). *Gleichenia linearis* (Burm. f.) C. B. Clarke var. *longicaudata* H. Christ in Bull. Herb. Boissier ser. 1, 7: 19. (1899). *Polypodium lineare* Thunb. f. *steniste* Takeda in

02

Notes Roy. Bot. Gard. Edinburgh 8: 310. (1913—1915).

Hubei: Yichang, 787, 2163. Badong, 1723, 3683, 5027. Liantuo, 3049. Jianshi, 6246. Donghu, 6410. Changleping, 7817. **Sichuan:** South Wushan, 5654, 7532. North Wushan, 7028. **Yunnan:** Mengzi, 11528 (type of *Gleichenia linearis* (Burm. f.) C. B. Clarke var. *longicaudata* H. Christ).

A widespread species found in Africa, Asia, Australia and Indonesia.

Dicranopteris pedata (Houtt.) Nakaike (GLEICHENIACEAE)

Gleichenia dichotoma (Willd.) Hook. Henry in Trans. Asiat. Soc. Jap. 24 suppl: 109. (1896) (List Pl. Formosa).

Taiwan: Wanchin, without number. Oluanpi, 647.

Didissandra sp. (GESNERIACEAE)

Anonymous in Kew List of Determinations (volume 34).

Yunnan: Simao, 13024, 13400, 13569.

Didymocarpus glandulosus (W. W. Sm.) W. T. Wang (GESNERIACEAE) New species

Didymocarpus silvarum W. W. Sm. var. *glandulosum* W. W. Smith in Notes Roy. Bot. Gard. Edinburgh 5: 151. (1912).

Yunnan: South of the Red River from Manmei at 2,300 metres, 9745 (isotype of *Didymocarpus silvarum* W. W. Sm. var. *glandulosum* W. W. Smith).

Endemic to China and distributed in the western provinces.

Didymocarpus margaritae W. W. Smith (GESNERIACEAE) New species

in Notes Roy. Bot. Gard. Edinburgh 5: 151. (1912).

Yunnan: In the mountains to the north-west of Simao at 1,650 metres, 12380b., (type).

Endemic to Yunnan.

Didymocarpus mengtze W. W. Smith (GESNERIACEAE) New species

in Notes Roy. Bot. Gard. Edinburgh 5: 152. (1912).

Yunnan: In the mountains to north of Mengzi at 2,300 metres 10232 (isotype).

Endemic to Yunnan and known only from the type locality at Mengzi.

Didymocarpus purpureobracteatus W. W. Smith (GESNERIACEAE) New species

in Notes Roy. Bot. Gard. Edinburgh 5: 153. (1912).

Yunnan: In Pingbian, on the Daweishan Range at 1,650 metres, 9189 (type). South of the Red River from Manmei at 1,960 metres, 9189a. South of the Red River from Manpan at 1,950 metres, 9746. On Fengchunling at 2,300 metres, 9746a.

China (South-east Yunnan) and Indochina.

Didymocarpus silvarum W. W. Smith (GESNERIACEAE) New species

in Notes Roy. Bot. Gard. Edinburgh 5: 150. (1912).

Yunnan: In forests near Simao at 1,300 metres, 12463 (isotype).

Endemic to Yunnan and known only from the type locality at Simao.

Didymocarpus sp. (GESNERIACEAE)

Anonymous in Kew List of Determinations (volume 34).

Yunnan: Simao, 12448.

Didymoplexis pallens Griff. (ORCHIDACEAE)

Henry in Trans. Asiat. Soc. Jap. 24 suppl: 93. (1896) (List Pl. Formosa), Rolfe in Journ. Linn. Soc. Bot. 36: 47. (1903).

Taiwan: Kaohsiung, 1878.

A leafless, saprophytic terrestrial orchid to 20 cm tall with fleshy tuber-like rhizomes. A widespread species found over much of the Himalaya, Indochina, China, southern Japan, Malaysia, Indonesia and Australia.

Dienia ophrydis (J. Koenig) Siedenf. (ORCHIDACEAE)

Microstylis congesta (Lindl.) Reichb. f. Henry in Trans. Asiat. Soc. Jap. 24 suppl: 91. (1896) (List Pl. Formosa), Forbes & Hemsley in Journ. Linn. Soc. Bot. 5. (1903), (quoted as A. Henry 831.).

Taiwan: Wanchin, 835. Oluanpi, 1342. **Yunnan:** Simao, 12336.

A handsome terrestrial orchid to 40 cm tall, a widespread species native to India, Indochina, southern China, Japan (south), the Philippines and from Malaysia to Australia.

Digitaria bicornis (Lam.) Roem. & Schult. (POACEAE)

Digitaria barbata Willd. Rendle in Journ. Linn. Soc. Bot. 36: 322. (1904).

Hainan: Environs of Haikou, 7972. **Taiwan:** Kaohsiung Spit, 1073.

Digitaria ciliaris (Retz.) Koeler (POACEAE)

Digitaria sanguinalis (L.) Scop. var. *ciliaris* (Retz.) Parl. Rendle in Journ. Linn. Soc. Bot. 36: 325. (1904).

Taiwan: Oluanpi, 250. On a sea cliff at Kaohsiung, 727.

Digitaria cruciata (Nees) A. Camus (POACEAE)

Digitaria sanguinalis (L.) Scop. var. *cruciata* Hook. f. Rendle in Journ. Linn. Soc. Bot. 36: 326. (1904).

Hubei: Badong, 5043, 5044. Changyang, 7508. **Sichuan:** North Wushan, 7036.

Digitaria henryi Rendle (POACEAE) New species

in Journ. Linn. Soc. Bot. 36: 323. (1904).

Taiwan: In meadow on Wanshoushan, 1031 (type, collected 7th August 1894).

Native to Vietnam, southern China and Japan (south). A common littoral species in Chinese Taiwan. This species was collected by the Glasnevin Taiwan Expedition (GTE 110) in

Henry's type locality on Wanshoushan, in October 2004.

Digitaria longiflora (Retz.) Pers. (POACEAE)

Rendle in Journ. Linn. Soc. Bot. 36: 324. (1904).

Hubei: Badong, 2504 (in part). **Taiwan:** Kaohsiung, 1939.

Native to the tropics of the Old World. Common on dykes between rice-fields in Taiwan.

Digitaria radicosa J. (Presl.) Miq. (POACEAE)

Digitaria formosana Rendle in Journ. Linn. Soc. Bot. 36: 323. (1904).

Taiwan: On rocks on Wanshoushan, 1941 (type of of *Digitaria formosana* Rendle).

An annual species to 25 cm tall, native to China, south-east Asia and Polynesia.

Digitaria setigera Roth. (POACEAE)

Digitaria sanguinalis (L.) Scop. var. *extensa* (Hook.f.) Rendle Rendle in Journ. Linn. Soc. Bot. 36: 326. (1904).

Taiwan: Kaohsiung, 764. Wanshoushan, 1128.

Digitaria ternata (A. Rich.) Stapf (POACEAE)

Rendle in Journ. Linn. Soc. Bot. 36: 327. (1904).

Yunnan: On cliffs near Mengzi at 1,500 metres, 11180.

Dillenia indica L. (DILLENIACEAE)

Hoogland in Blumea 7: 108. (1952).

Yunnan: Simao, 12746.

The elephant apple, a beautiful tropical tree to about 17 metres tall and native from India to Java. One of the great aristocrats of the humid, tropical forests of Asia, *Dillenia indica* has much to be admired about it. The flaking bark is rusty red, the canopy is formed by large and striking deely veined leaves and the nodding, slightly fragrant *Magnolia*-like flowers (which appear in spring) are like enormous milky-white goblets encircling a coronet of golden stamens. The tree is mostly confined to river valleys and the large edible, apple-like fruits have evolved to float and are dispersed by river currents. The fruits taste like unripe apples and are used in some parts of Asia as a vegetable in curries, the dried leaves are used to polish ivory and the bark is used medicinally to treat the mouth infection, thrush.

Dillenia turbinata Finet. & Gagnep. (DILLENIACEAE)

Hoogland in Blumea 7: 103. (1952).

Hainan: Ling-men, 8622.

An evergreen tropical tree to 30 metres tall, native in China (Guangxi and Hainan) and to Indochina. In Hainan Henry stated the tree was known as *pai*. The fruits of this species are edible.

Dimetia capitellata (Wallich ex G. Don) Neupane & N. Wikstr. (RUBIACEAE)

Hedyotis capitellata Wallich ex G. Don Hutchinson in Sargent Pl. Wilson. 3: 408. (1917).

Yunnan: South of the Red River from Manmei, 9807.

In woods near Simao at 1,550 metres, 12264.

Distributed in India, Myanmar, China, Korea and Japan.

Dimetia hedyotidea (DC.) T. C. Hsu (RUBIACEAE)

Hedyotis macrostemon Hook. & Arn. Hutchinson in Sargent Pl. Wilson. 3: 409. (1917).

Hainan: 8 km (5 miles) south of Kuingchow, 8184. Without locality, 8486. Ling-men (in the interior of the island), 8627 Environs of Kiungchow, 8711. **Yunnan:** In Pingbian in forests on the Daweishan Range at 1,640 metres, 9555a.

Native to China and Vietnam.

Dimetia scandens (Roxb.) R. J. Wang (RUBIACEAE)

Hedyotis scandens Roxb. Hutchinson in Sargent Pl. Wilson. 3: 409. (1917).

Yunnan: In woods near Simao at 1,520 metres, 11700. In mountains to the south of Simao at 1,480 metres, 11700a. In mountains near Simao at 1,950 metres, 11700b. Simao at 1,640 metres, 13484.

Distributed in India, China (Yunnan) and Vietnam.

Dimocarpus longan Lour. (SAPINDACEAE)

Rehder & E. H. Wilson inSargent Pl. Wilson. 2: 193. (1916).

Nephelium longana (Lamm.) Cambess. Henry in Trans. Asiat. Soc. Jap. 24 suppl: 29. (1896) (List Pl. Formosa). *Euphoria longana* Lamark. Rehder & E. H. Wilson in Sargent Pl. Wilson. 2: 193. (1916).

Sichuan: Henry's Chinese collector with Pratt, no locality stated but may be Leshan, Emei Shan, Wa Shan or Kangding, without number. **Hainan:** Lung-yen, 7960. **Taiwan:** Kaohsiung, without number.

The longan, a tropical tree to 40 metres tall, sometimes butressed and cultivated as an orchard fruit. In China this tree is named *longyan* due to the shiny seeds which are said to look like a dragon's eye. It is also the *li-chi-nu* or 'slave of lychee' since it tolerates more cold than the lychee. This tree has a large geographical range from India, across Myanmar, to southern China and Malesia. In parts of southeast Asia the dried fruits are used to treat insomnia, amnesia and mental health, the seeds are also used as shampoo. This species was collected by the Glasnevin Taiwan Expedition (GTE 56) by the lighthouse at Oluanpi (South Cape of Henry and Wilson) in October 2004.

Dinebra chinensis (L.) P. M. Peterson & N. Snow (POACEAE)

Leptochloa chinensis (L.) Nees Henry in Trans. Asiat. Soc. Jap. 24 suppl: 109. (1896) (List Pl. Formosa).

Taiwan: Kaohsiung, without number.

Native to much of south-east Asia, a weed in paddy fields in Chinese Taiwan.

02

Dinetus dinetoides (Schneider) Staples
(CONVOLVULACEAE) New species

in Novon 3: 199. (1999).

Porana dinetoides Schneider in Sargent Pl. Wilson. 3: 360. (1916).

Yunnan: Near Mengzi at 1,700 metres, 9340 (holotype of *Porana dinetoides* Schneider).

Dinetus duclouxii (Gagnepain & Courchet) Staples
(CONVOLVULACEAE) New species

Staples in Novon 3: 199. (1993).

Porana triserialis Schneider in Sargent Pl. Wilson. 3: 357. (1916). *Porana triserialis* Schneider var. *lasia* Schneider in Sargent Pl. Wilson. 3: 362. (1917).

Sichuan: Henry's Chinese collector with Pratt, 8866 (syntype of *Porana triserialis* Schneider). **Yunnan:** In woods near Mengzi at 1,500 to 1,600 metres, 9229 (syntype of *Porana triserialis* Schneider), 9229b., 9229c. (in part. holotype of *Porana triserialis* Schneider var. *lasia* Schneider).

A climbing shrub with glabrous twining stems producing axillary racemose panicles of white (or red, purple, blue or yellow) funnel-shaped flowers during June and July. Distributed in Hubei, Sichuan and Yunnan. Discovered by Henry's Chinese collector while travelling with Pratt in 1889, it was described as a new species from material later collected in Milê by Ducloux on the 5th September 1907.

Dinetus racemosus (Roxb.) Sweet (CONVOLVULACEAE)

Porana racemosa Roxb. Forbes & Hemsley in Journ. Linn. Soc. Bot. 26: 166. (1890), Schneider in Sargent Pl. Wilson. 3: 361. (1916).

Hubei: Yichang (Xiling) Gorge, 2905. Badong, 2487. Liantuo, 2595, 3062, 4466, 4467. Yichang, 4357. North Badong, 7004. **Yunnan:** Milê, 9954.

A very beautiful deciduous climbing shrub to 10 metres tall, scaling the surrounding vegetation by means of twining stems. Between June to September this species carries masses of blue funnel-shaped flowers in axillary racemes. Native to India, the Himalaya, Myanmar, China, Thailand and Vietnam.

Dioscorea aspersa Prain & Burkill (DIOSCOREACEAE)

in J. Asiat. Soc. Bengal 4: 447. (1908).

Dioscorea pulverea Prain & Burkill in Journ. Asiat. Soc. Bengal x. 31. (1914).

Yunnan: Mengzi at 1,600 metres, 10287 (in part). Mengzi at 1,500 metres, 9288 (type of *Dioscorea pulverea* Prain & Burkill).

Discovered by William Hancock on the Great Black Mountain range (to the north of Mengzi) in 1894. Endemic to China and distributed in the provinces of Guizhou and Yunnan.

Dioscorea benthamii Prain & Burkill (DIOSCOREACEAE)

Dioscorea oppositifolia Henry non L. in Trans. Asiat. Soc. Jap. 24 suppl: 95. (1896) (List Pl. Formosa), Wright in Journ. Linn. Soc. Bot. 36: 92. (1904).

Taiwan: Wanchin, 1673.

Native to Fujian, Guangdong, Guangxi, Hong Kong and Taiwan.

Dioscorea bulbifera L. (DIOSCOREACEAE)

Knuth in Engler, Pflanzenr. 43: 88. (1924).

Yunnan: Simao, 12266a.

Tropical and Subtropical Old World.

Dioscorea cirrhosa Lour. (DIOSCOREACEAE)

Dioscorea rhipogonioides Oliv. Henry in Trans. Asiat. Soc. Jap. 24 suppl: 96. (1896) (List Pl. Formosa), Wright in Journ. Linn. Soc. Bot. 36: 93. (1903).

Taiwan: Oluanpi, 105, 589, 970, 1297. Wanchin, 1661.

According to Henry the tubers of this species were the *shu-lang*, 'dye root' of the English, and 'faux gambier' of the French. It gave a durable reddish brown colour to fishing nets, ropes, sails and cloths. In Vietnam this species is known as *cunao* and from there in 1893 1,737 tonnes were exported to Hong Kong, giving some indication of the products value.

Dioscorea collettii Hook. f. (DIOSCORIACEAE)

Dioscorea nigrescens R. Knuth in Engler, Pflanzenr. 87 (iv. 43): 253. (1924).

Yunnan: Simao, 12338c., (type of *Dioscorea nigrescens* R. Knuth).

Distributed in India, Myanmar, China and Indochina.

Dioscorea delavayi Franchet (DIOSCOREACEAE)

Knuth in Engler, Pflanzenr. 43: 143. (1924).

Dioscorea sp. undeterminable. Wright in Journ. Linn. Soc. Bot. 36: 94. (1903). *Dioscorea engleriana* Knuth in Engler, Pflanzenr. 43: 140. (1924). *Dioscorea kamponensis* Kunth. var. *henryi* Prain & Burkill in Journ. Asiat. Soc. Bengal 10: 22. (1914). *Dioscorea burkillii* (Prain & Burkill) Knuth. Knuth in Engler, Pflanzenr. 43: 143. (1924). *Dioscorea henryi* (Prain & Burkill) C. T. Ting Ting & Chang in Acta Phytotax. Sin. 20: 208. (1982).

Hubei: Donghu, 6419. **Yunnan:** Mengzi at 1,900 metres, 9495 (type of *Dioscorea kamponensis* Kunth. var. *henryi* Prain & Burkill). Mengzi 9495b., (type of *Dioscorea engleriana* Knuth), 9715 (syntype of *Dioscorea kamponensis* Kunth. var. *henryi* Prain & Burkill).

Dioscorea esculenta (Lour.) Burkill (DIOSCOREACEAE)

Dioscorea sativa Henry non L. in Trans. Asiat. Soc. Jap. 24 suppl: 95. (1896) (List Pl. Formosa), Wright in Journ. Linn. Soc. Bot. 36: 93. (1903). *Dioscorea* sp. Henry in Trans. Asiat. Soc. Jap. 24 suppl: 96. (1896) (List Pl.

Formosa). *Dioscorea spinosa* Roxb. ex Hook. f. Forbes & Hemsley in Journ. Linn. Soc. Bot. 93. (1904).

Hainan: Without locality, 8406. **Taiwan:** Wanchin, 854, 856. Oluanpi, 1210. Kaohsiung 1871, 1871a., 1871b.

The sweet Chinese yam, a crop of great antiquity. Henry seemed to be of the opinion that his collections were from wild plants but Herklots in *Vegetables in South-East Asia* 425. (1972) states it is not known in the wild but persists as a relic of cultivation throughout the moister parts of tropical Asia from Mumbai (Bombay) in the west to Chinese Taiwan in the east, south to the Pacific Islands and Tahiti. Henry stated that the cultivated yam in China was known as *shu,* and this term had been extended to other tubers, introductions from America, as the sweet potato, known as *fan-shu, hung-shu,* and the common potato (*Solanum tuberosum*) *shu-ts'ai.*

Dioscorea glabra Roxb. (DIOSCOREACEAE)

Dioscorea glabra Roxb. var. *longifolia* Prain & Burkill in J. & Proc. Asiat. Soc. Bengal 10: 38. (1914), Knuth in Engler, Pflanzenr. 43: 277. (1924).

Yunnan: Simao, 13540 (syntype of *Dioscorea glabra* Roxb. var. *longifolia* Prain & Burkill).

Native to India, Bhutan, Myanmar, western China, Cambodia and Laos.

Dioscorea hamiltonii Hook. f. (DIOSCOREACEAE)

Dioscorea persimilis Prain & Burkill in Journ. & Proc. Asiat. Soc. Bengal 4: 454. (1908), Knuth in Engler, Pflanzenr. 43: 267. (1924). *Dioscorea glabra* Wright non Roxb. in Journ. Linn. Soc. Bot. 36: 91. (1903).

Hainan: Environs of Haikou, 8407. Environs of Kiungchow, 8690 (type of *Dioscorea persimilis* Prain & Burkill).

Dioscorea hemsleyi Prain & Burkill (DIOSCOREACEAE) New species

in J. Asiat. Soc. Bengal 4: 451. (1908), Knuth in Engler, Pflanzenr. 43: 319. (1924).

Yunnan: Mengzi, 10287 (in part), 10287b., (type).

Distributed in Myanmar, western China and Indochina.

Dioscorea hispida Dennst. (DIOSCOREACEAE)

Dioscorea daemona Henry non Roxb. in Trans. Asiat. Soc. Jap. 24 suppl: 96. (1896) (List Pl. Formosa). *Dioscorea daemona* Roxb. var. *reticulata* Hook. f. Wright in Journ. Linn. Soc. Bot. 36: 91. (1903). *Dioscorea triphylla* L. var. *reticulata* (Hook. f.) Prain & Burkill Knuth in Engler, Pflanzenr. 43: 132. (1924).

Hainan: Hummocks near Haikou, 8106, 8288. Environs of Kiungchow, 8725. **Taiwan:** Kaohsiung, 1875.

A twining, tuberous perennial vine. Native to India and China to Malaysia and cultivated for its edible tubers.

Dioscorea japonica Thunb. (DIOSCOREACEAE)

Anonymous in Kew List of Determinations (volume 24).

Hubei: Badong, 1857.

Dioscorea kamoonensis Kunth (DIOSCORIACEAE)

Dioscorea sp. undeterminable. Wright in Journ. Linn. Soc. Bot. 36: 94. (1903). *Dioscorea kamoonensis* Knuth var. *fargesii* (Franch.) Prain & Burkill in Journ. Asiatic. Soc. Bengal 4: 451. (1908). *Dioscorea fargesii* Franch. Knuth in Engler, Pflanzenr. 43: 142. (1924). *Dioscorea subfusca* R. Knuth in Engler, Pflanzenr. 43: 143. (1924) *Dioscorea mengtzeana* R. Knuth in Engler, Pflanzenr. 87 (iv. 43): 142. (1924).

Hubei: Liantuo, 2666, 4486 (type of *Dioscorea subfusca* R. Knuth). Changyang, 7501. **Sichuan:** North Wushan, 7103 [isolectotype of *Dioscorea kamoonensis* Knuth var. *fargesii* (Franch.) Prain & Burkill]. **Yunnan:** Mengzi at 1,900 metres, 11301 (type of *Dioscorea mengtzeana* R. Knuth).

Dioscorea melanophyma Prain & Burkill (DIOSCOREACEAE) New species

in J. Proc. Asiat. Soc. Bengal 4: 452. (1908), Knuth in Engler, Pflanzenr. 43: 139. (1924).

Yunnan: Mengzi at 1,800 metres, 9495d., 10253 (isosyntype).

Native from Nepal to the western provinces of China.

Dioscorea nipponica Makino (DIOSCOREACEAE)

Prain & Burkill in Journ. Asiatic. Soc. Bengal. 10: 14. (1914).

Dioscorea quinqueloba Henry non Thunberg in Journ. China Br. Roy. Asiat. Soc. 22: 269. (1887) (Chinese Names of Plants), Wright in Journ. Linn. Soc. Bot. 36: 92. (1903). *Dioscorea nipponica* var. *vera* Prain & Burkill Knuth in Engler, Pflanzenr. 43: 314. (1924). *Dioscorea giraldii* Knuth. R. Knuth in Engler, Pflanzenr. 43: 315. (1924). *Dioscorea acerifolia* Rehder & E. H. Wilson non Uline apud Diels in Sargent Pl. Wilson. 3: 14. (1916).

Hubei: Yichang, 150, 183. Badong, 4769. Liantuo, 5870a. Jianshi, 5870, 5870b. South Badong, 7358.

According to Henry's notes this was the *shan yo* whose bitter roots were eaten by peasants in summer.

Dioscorea nitens Prain & Burkill (DIOSCORIACEAE) New species

in J. & Proc. Asiat. Soc. Bengal 10: 18. (1914), Knuth in Engler, Pflanzenr. 43: 319. (1924).

Yunnan: In forests near Simao at 1,500 metres, 12338 (isosyntype).

Distributed from China (Yunnan) to northern Thailand.

Dioscorea panthaica Prain & Burkill (DIOSCORIACEAE) New species

in Journ. Asiat. Soc. Bengal part 2. suppl. 73: 6. (1904),

Prain & Burkill in Fedde, Repert. Nov. Spec. Regni Veg. 1: 63. (1905), Knuth in Engler, Pflanzenr. 34: 177. (1924).

Yunnan: in the mountains to the north of Mengzi at 2,300 metres, 11065 (type).

Distributed from western China to Thailand.

Dioscorea polystachya Turcz. (DIOSCOREACEAE)
Wright in Journ. Linn. Soc. Bot. 36: 92. (1903).

Dioscorea doryphora Hance Henry in Trans. Asiat. Soc. Jap. 24 suppl: 96. (1896) (List Pl. Formosa), Wright in Journ. Linn. Soc. Bot. 36: 91. (1903), Knuth in Engler, Pflanzenr. 43: 261. (1924).

Hubei: On the floor of a cave on way to Moji Shan, (June 1886), 1538. Yichang, 2023. **Taiwan:** Wanchin, 878, 1672. Kaohsiung, 1922.

Native to China (including Taiwan) and Japan (south).

Dioscorea sp. (DIOSCORIACEAE)
Anonymous in Kew List of Determinations (volume 34).

Yunnan: Mengzi, 10935.

Dioscorea tokoro Makino ex Miyabe (DIOSCOREACEAE)
Dioscorea sativa Wright non L. in Journ. Linn. Soc. Bot. 36: 93. (1903). *Dioscorea enneaneura* (Uline ex Diels) Prain & Burkill in J. Asiat. Soc. Bengal. Pt. 2. Nat. Hist. 73, Suppl: 11. (1904), Prain & Burkill in Fedde, Repert. Nov. Spec. Regni Veg. 1: 64. (1905), Knuth in Engler, Pflanzenr. 34: 178. (1924).

Hubei: Liantuo, 2068. Antelope Glen, 3641, 4193. Yichang 3641a., [isotype of *Dioscorea enneaneura* (Uline ex Diels) Prain & Burkill].

Dioscorea yunnanensis Burkill (DIOSCORIACEAE) New species
in Journ. Asiat. Soc. Bengal 73: 186. (1904), Prain & Burkill in Fedde, Repert. Nov. Spec. Regni Veg. 1: 59. (1905).

Yunnan: In woods near Mengzi at 1,650 metres, 9288a., (type, collected 30th June 1896).

Endemic to China and distributed in the provinces of Guizhou and Yunnan.

Dioscorea zingiberensis C. H. Wright (DIOSCOREACEAE) New species
in Journ. Linn. Soc. Bot. 36: 93. (1903), Pampanini in Nuov. Giorn. Bot. Ital. n. ser. xvii. 243. (1910); xviii. 110. (1911), Prain & Burkill in Journ. Asiatic. Soc. Bengal. 10: 17. (1914), Rehder in Sargent Pl. Wilson. 3: 14. (1916), Knuth in Engler, Pflanzenr. 43: 177. (1924).

Dioscorea henryi Uline ex Diels in Bot. Jahrb. 29: 261. (1900).

Hubei: Yichang, 407. On Tsui Fu Shan near Yichang (May 1888), 1621, (type) San You Dong Glen, (July 1886) 1520.

According to Henry the root of this creeper was eaten by the mountain people of Hubei in winter. At Yichang he stated it was locally known as *hjang chiang* i.e. yellow ginger. This handsome species is cultivated in the order beds at the National Botanic Gardens, Glasnevin from collections made by the Glasnevin Central China Expedition (GCCE 759) in Changyang in October 2004.

Diospyros armata Hemsley (EBENACEAE) New species
in Journ. Linn. Soc. Bot. 26: 69. (1899), Bretschneider in Hist. Eur. Bot. Disc. China 2: 787. (1898), Bean in Kew Bull. Misc. Inf. 165. (1913), Rehder & E. H. Wilson in Sargent Pl. Wilson. 2: 591. (1916), Bean in Trees & Shrubs 1: 493. (1921), Zheng in Journ. Wuhan Bot. Res. 2.(1): 173. (1984), Nelson in Pim, The Wood & the Trees 220. (1984), Bean in Trees & Shrubs 2: 55. (1992), Nelson in A Heritage of Beauty 312. (2000).

Hubei: Liantuo, 4393. South Donghu, 7717 (type).

A tree to 13 metres tall with girths of up to 1.5 metres. Found at low altitudes near Yichang and Liantuo, the tree bears fragrant creamy-yellow flowers in May. Henry sent seeds to Kew on February 8th 1888 but no record of their germination was kept. Introduced to cultivation by Wilson in 1904 through Veitch's Coombe Wood nursery. There is an old Wilson plant at Glasnevin.

Diospyros balfouriana Diels (EBENACEAE)
in Notes Roy. Bot. Gard. Edinburgh. 5: 209. (1912).

Diospyros lotus Rehder & Wilson non L. in Sargent Pl. Wilson. 2: 587. (1916).

Yunnan: Mengzi at 1,600 metres, 9898d., (type).

Endemic to Yunnan.

Diospyros blancoi A. DC. (EBENACEAE)
Diospyros utilis Hemsley in Ann. of Bot. 9: 154. (1895), Henry in Trans. Asiat. Soc. Jap. 24 suppl: 58. (1896) (List Pl. Formosa), Bretschneider in Hist. Eur. Bot. Disc. China 2: 787. (1898). *Diospyros discolor* Willd. Li in Woody Flora of Taiwan 731. fig. 298. (1963), Lu and Yang in Bull. Taiwan For. Res. Inst. New Series 3(2): 133. (1988).

Taiwan: Wanchin, 523, 815 (type of *Diospyros utilis* Hemsley), 1677. Oluanpi, 677.

Henry described this species as a large tree furnishing a good wood that was used for making axels and an edible fruit called *mao-shih*, that is, hairy persimmon. *The Gazetteer* also gave this name and *fan-shi* or "savage persimmon" obviously an unfortunate reference to the indigenous inhabitants of Taiwan who ate the fruits. Native from Borneo to Taiwan where it grows on the eastern and southern parts of the island. This species was collected by the Glasnevin Taiwan Expedition (GTE 34) by the lighthouse at Oluanpi (South Cape of Henry and Wilson) in October 2004. A beautiful tree with handsome, large lanceolate leaves, silver-grey beneath.

Diospyros dumetorum W. W. Smith (EBENACEAE)

Diospyros mollifolia Rehder & E. H. Wilson in Sargent Pl. Wilson. 2: 591. (1916).

Yunnan: Yuanjiang at 1150 metres, 13303 (syntype of *Diospyros mollifolia* Rehder & E. H. Wilson).

Distributed from western China to Thailand.

Diospyros eriantha Champ. ex Benth. (EBENACEAE)

Henry in Trans. Asiat. Soc. Jap. 24 suppl: 58. (1896) (List Pl. Formosa), Li in Woody Flora of Taiwan 730. (1963).

Taiwan: Wanchin, 26, 132, 566, 827, 1527. Oluanpi, 623, 957. Kaohsiung, 1877.

A tree to 13 metres tall, distributed from southern China to Malaysia. According to Henry's Chinese collectors, this was a good timber tree growing to one foot in diameter. On Wanshoushan, at Kaohsiung however it only forms shrubs. The genus *Diospyros* consists of about 500 species, mostly found in the tropics of the Old World. The fruits are fleshy, edible and float on water and so are dispersed by birds, animals and water. This species was collected by the Glasnevin Taiwan Expedition (GTE 112) on Wanshoushan, in October 2004.

Diospyros kaki L. f. (EBENACEAE)

Forbes & Hemsley in Journ. Linn. Soc. Bot. 26: 69. (1899).

Diospyros kaki L. f. var. *silvestris* Makino Rehder & E. H. Wilson in Sargent Pl. Wilson. 2: 590. (1916), Zheng in Journ. Wuhan Bot. Res. 2.(1): 173. (1984). *Diospyros kaki* Henry non L. f. in Journ. China Br. Roy. Asiat. Soc. 22: 270. (1887) (Chinese Names of Plants), Forbes & Hemsley in Journ. Linn. Soc. Bot. 26: 69. (1889).

Hubei: San You Dong Glen near Yichang (July 1886), 1502. In the Lung Wang Tung Glen, 1560. Antelope Glen, 3441. Yichang 3485, 3567. Liantuo, 1962, 3861. **Yunnan:** Simao in forests and woods at 1,500 to 1,600 metres, 9341. Mountains to the north of Mengzi at 1,600 metres, 9898c. In woods and forests near Simao at 1,500 to 1,600 metres, 11618, 11818b., 11618c., 11618d.

The Chinese persimmon, a tree to 18 metres tall and widely cultivated in China. According to Wilson, cultivars of this fruit tree are as numerous as those of the apple or pear in the West. These cultivars differ in size, shape, season of ripening and in quantity or absence of seeds. In parts of China, especially around Beijing, the fruits are sometimes bletted before consumption and the tree is extremly handsome carrying a heavy crop of fruits. Henry collected from both cultivated (A. Henry 1560, 3567, 3861) and wild trees. Henry stated that the wild persimmon was the *yu-shih-tzu,* the oil or varnish persimmon. The fruits were cut

into halves and put into water to decompose; the oil then obtained was used for waterproofing Chinese umbrellas, water-proofs, rain-hats, etc. The varnish when extracted was clear and colourless but by adding a few leaves of the *la shu* (*Ligustrum lucidum*) a warm brown tint was obtained. Wilson called it the *yu shih tzu* (literally; oil persimmon) and stated it was abundant in the mountains of central China where it formed trees over 60 metres tall. In the milder parts of the British Isles and Ireland the Chinese Persimmon will fruit against a wall.

Diospyros lotus L. (EBENACEAE)

Henry in Journ. China Br. Roy. Asiat. Soc. 22: 270. (1887) (Chinese Names of Plants), Forbes & Hemsley in Journ. Linn. Soc. Bot. 26: 70. (1889), Henry in Notes Econ. Bot. China 47. (1993), Diels in Bot. Jahrb. 29: 527. (1900), Rehder & E. H. Wilson in Sargent Pl. Wilson 2: 587. (1916), Zheng in Journ. Wuhan Bot. Res. 2(1): 173. (1984).

Hubei: Liantuo, 1914, 3014, 3899. Changyang, 6235. Badong, 2871. Jianshi 5280. Without locality, 5820a. (March 1888). **Sichuan:** North Wushan, 7044. **Yunnan:** Milê district, 9898. Mengzi 9898a.

A large tree to 26 metres tall with girths up to 4 metres. In central China this tree was known as the *kou-shih tzu,* though at Yichang it was called *suan-tsao* according to Henry. The fruits of this species change to yellow before finally becoming a bloomy purplish black as they ripen. Cultivated in Europe since the 17th century, there are trees at the National Botanic Gardens, Glasnevin. *Diospyros lotus* was collected by the Glasnevin Central China Expedition (GCCE 817) on Tianthu Shan in Changyang in October 2004.

Diospyros spp. (EBENACEAE)

Anonymous in Kew List of Determinations (volume 24 & 34).

Hainan: Environs of Kiungchow, 8700. **Yunnan:** Mengzi, 9087. Simao, 13047.

Diospyros strigosa Hemsley (EBENACEAE) New species

in Kew Bull. Misc. Inf. 193. (1910).

Hainan: Environs of Kiungchow, 8741 (type).

Endemic to China and distributed in Guangdong and Hainan.

Diospyros yunnanensis Rehder & E. H. Wilson (EBENACEAE) New species

in Sargent Pl. Wilson. 2: 592. (1916).

Yunnan: In mountains to the south-west of Mengzi at 1,600 metres, 10938 (syntype). In forests near Simao at 1,300 to 1,600 metres, 12948, 12948a. In forests near Simao at 1,300 metres 12984 (syntype), 12984a., (syntype). In forests near Simao at 1,300 to 1,600 metres, 13288 (syntype).

A tree to 7 metres tall, endemic to southern Yunnan.

Dipelta floribunda Maxim. (CAPRIFOLIACEAE)

Zheng in Journ. Wuhan Bot. Res. 2.(1): 191. (1984).

Hubei: Xingshan at 1,400 metres, 6558. Fang Xian, 6699.

A deciduous shrub to 4 metres tall bearing in summer an abundance of fragrant pink bell-shaped flowers, yellow in the throat. Described as a new genus from a collection by the Russian explorer Pavel Jakovlevich Piasetski about 1874 in Shaanxi and introduced to cultivation by Wilson through Messrs Veitch by means of living plants in 1902 and by seeds in 1904. The first flowers opened in Veitch's Coombe Wood nursery in 1907. The allied *Dipelta yunnanensis* which was discovered by Delavay in 1886 flowered for the first time in cultivation at Glasnevin in May 1919 from a Forrest plant.

Dipentodon sinicus Dunn (DIPENTODONTACEAE) New species

in Kew Bull. Misc. Inf. 311. (1911).

Yunnan: Mengzi at 1,950 metres, 10741 (isotype).

Indian (Assam) to western China.

Diplachne fusca (L.) P. Beauv. ex Roem. & Schult. (POACEAE)

Rendle in Journ. Linn. Soc. Bot. 36: 411. (1904).

Taiwan: Kaohsiung Plain, in rice fields, 1822, 1859. Kaohsiung Spit 1822a., 1846.

Native to tropical South Africa and Egypt, to southeast Asia and Australia.

Diplaziopsis javanica (Blume) C. Chr. (ATHYRIACEAE)

Allantodia brunoniana Wallich Anonymous in Kew List of Determinations (volume 34).

Yunnan: Mengzi, 10355.

A handsome fern to about 50 cm tall, distributed across the Himalaya, India, Sri Lanka, Indochina, China, the Philippines and Malaysia to the Pacific Islands.

Diplazium bantamense Blume (ASPLENIACEAE)

Asplenium bantamense (Blume) Baker Henry in Trans. Asiat. Soc. Jap. 24 suppl: 112. (1896) (List Pl. Formosa).

Taiwan: Wanchin, 1635.

Diplazium esculentum (Retz.) Sw. (ASPLENIACEAE)

Asplenium esculentum (Retz.) C. Presl. Henry in Trans. Asiat. Soc. Jap. 24 suppl: 112. (1896) (List Pl. Formosa).

Taiwan: Tamsui, 1397. Kaohsiung Plain, 2040.

Diplazium hirtipes Christ (ASPLENIACEAE) New species

in Bull. Herb. Boiss. 7: 11. (1899)

Asplenium parallelosorum Baker in Kew Bull. Misc. Inf. 9. (1906).

Yunnan: In mountain forest to the east of Mengzi at 1,950 metres, 10103 (type of *Diplazium hirtipes* Christ, syntype of *Asplenium parallelosorum* Baker).

Diplazium latifolium T. Moore (ASPLENIACEAE)

Diels in Bot. Jahrb. 29: 197. (1900) Christ in Bull. Soc. Bot. France 5: 51. (1905).

Diplazium calogramma Christ in Notul. Syst. (Paris) 1: 45. (1906).

Hubei: Badong, 5211. **Yunnan:** In mountain forest on Fengchunling at 2,200 metres, without number.

Distributed in India, China, Korea, Japan and Vietnam. In his description of *Diplazium calogramma* Christ cited A. Henry 1125, rather than 5211.

Diplazium leptophyllum (Baker) Christ. (ASPLENIACEAE) New species

Asplenium leptophyllum Baker in Kew Bull. Misc. Inf. 10. (1906).

Yunnan: In forests near Simao, 13106 (isotype of *Asplenium leptophyllum* Baker, collected 3rd September 1898).

Diplazium maximum (D. Don) C. Chr. (ASPLENIACEAE)

Gymnogramma gigantea Baker in Journ. Bot. 27: 177. (1889), Baker in Ann. Bot. 5: 483. (1890).

Hubei: Xingshan, 6517 (type of *Gymnogramma gigantea* Baker).

Diplazium polypodioides Mett. var. ***sinensis*** Christ (ASPLENIACEAE) New variety

Christ in Bull. Herb. Boiss. 7: 12. (1899).

Yunnan: On Fengchunling in mountain forest at 2,200 metres, 11526.

Diplazium pullingeri (Baker) J. Sm. (ASPLENIACEAE)

Asplenium lepidorachis Baker Anonymous in Kew List of Determinations (volume 34).

Yunnan: Simao, 13567.

Diplazium spinulosum Blume (ASPLENIACEAE)

Asplenium spinulosum (Blume) Mett. Anonymous in Kew List of Determinations (volume 24).

Hubei: Yichang, 5088.

Diplazium squamigerum (Mett.) Matsum. (ASPLENIACEAE)

Diels in Bot. Jahrb. 29: 197. (1900).

Asplenium squamigerum Mett. Baker in Journ. Bot. 25: 171. (1887).

Hubei: Badong, 1790.

Distributed through northern India, China, Korea and Japan. This Badong collection proved to be a new record for China.

Diplazium sylvaticum (Bory) Sw. (ASPLENIACEAE)

Asplenium sylvaticum (Bory) C. Presl. Henry in Trans. Asiat. Soc. Jap. 24 suppl: 112. (1896) (List Pl. Formosa).

Taiwan: Wanchin, 41, 1511.

02

Diplazium wichurae (Mett.) Diels (ASPLENIACEAE)
Diels in Bot. Jahrb. 29: 197. (1900).

Asplenium wichurae Mett Baker in Journ. Bot. 27: 176. (1889).

Hubei: Antelope Glen near Yichang, 7882. Yichang, 7896.

Distributed through China, Korea and Japan.

Diploclisia affinis (Oliv.) Diels (MENISPERMACEAE)
New species

in Engler, Pflanzenr. iv.-94: 227. (1910), Zheng in Journ. Wuhan Bot. Res. 2.(1): 44. (1984).

Cocculus affinis Oliver in Hooker's Icon. Pl. t. 1760. (1888) Diels in Bot. Jahrb. xxix. 345. (1900).

Hubei: Liantuo, 1887, 3818.

According to Henry this species was known colloquially around Yichang as *ch'ien tan t'eng*.

Diploclisia glaucescens (Blume) Diels (MENISPERMACEAE)
Diels in Engler, Pflanzenr. iv. 94: 225. (1910).

Yunnan: Simao at 1,200 metres, 12238a.

Diplocyclos palmatus (L.) C. Jeffrey (CUCURBITACEAE)
Bryonia laciniosa Anonymous non L. in Kew List of Determinations (volume 24).

Hainan: Without locality, 8523. Environs of Kiungchow, 8776.

A slender vine, native to Africa, Indochina, Malesia, the Philippines and Australia.

Diploprora championii (Lindl. ex Benth.) Hook. f. (ORCHIDACEAE)
Anonymous in Kew List of Determinations (volume 34).

Henry in Trans. Asiat. Soc. Jap. 24 suppl: 93. (1896) (List Pl. Formosa), Rolfe in Journ. Linn. Soc. Bot. 36: 34. (1903).

Taiwan: Wanchin, 898, 1606; **Yunnan:** Simao, 13610.

A handsome epiphytic orchid found over much of the Himalaya, India (Deccan), Sri Lanka, Myanmar, southern China (Taiwan & Yunnan), Thailand and Vietnam.

Diplospora dubia (Lindl.) Masam. (RUBIACEAE)
Diplospora viridiflora DC. Henry in Trans. Asiat. Soc. Jap. 24 suppl: 50. (1896) (List Pl. Formosa). *Tricalysia dubia* (Lindl.) Ohwi Li in Woody Flora of Taiwan 875. fig. 355. (1963).

Taiwan: Wanshoushan, 317, 722, 1147. Wanchin, 467. Oluanpi, 634, 633, 1602.

An evergreen shrub or a small tree, native from southern China to Vietnam and Japan (south). Common in broad-leaved forest in Chinese Taiwan.

Diplospora fruticosa Hemsley (RUBIACEAE) New species

in Journ. Linn. Soc. Bot. 23: 383. (1888), Bretschneider

in Hist. Eur. Bot. Disc. China 2: 785. (1898), Zheng in Journ. Wuhan Bot. Res. 2.(1): 190. (1984).

Hubei: Antelope Glen, 3508. Yichang, 3508a. South Donghu, 7491. Changyang, 7680.

Diplospora mollissima Hutchinson (RUBIACEAE)
New species

in Sargent Pl. Wilson. 3: 401. (1916).

Yunnan: In forests to east of Simao at 1,460 metres, 12246. Forests to south-east of Simao at 1,320 metres, 12246a. In Xishuangbanna Dai autonomous prefecture in forests to the south of Yulo at 1,320 metres, 12928 (syntype).

Endemic to Yunnan.

Dipsacus asper Wall. ex DC. (DIPSACACEAE) New to China

Henry in Journ. China Br. Roy. Asiat. Soc. 22: 248. (1887) (Chinese Names of Plants), Forbes & Hemsley in Journ. Linn. Soc. Bot. 23: 399. (1888), Hemsley in Hooker's Ic. Pl. t. 1984. (1891), Henry in Notes Econ. Bot. China 19. (1893).

Hubei: Yichang, 160, 2941. On Pingshanba near Yichang, 2267 (August 1886). Liantuo, 4537. Badong, 4792.

This species was previously only known from the Khasi hills in eastern India.. Henry stated it was known colloquially in the Three Gorges region as *hsu tuan* (the *xu duan* or 'restore what is broken') and the root is used as a drug used as a tonic. *Dipsacus asper* is still a very common plant in the mountains of the Three Gorges region. It was collected by the Glasnevin Central China Expedition in Badong in October 2002 and flowered at Glasnevin and Airfield House, Dublin, in the summer of 2003. It was collected again by the 2004 Glasnevin Expedition to the Three Gorges region on Hann Shao Shan in Xingshan (GCCE 480), near Zhenzi town, also in Xingshan (GCCE 546) in September of that year. It is a very handsome and vigorous plant and the creamy-white flowers are sweetly scented. Alice Henry (Henry's second wife) grew *Dipsacus asper* in her garden at Ranelagh (Dublin) during the 1920s, as a Henry plant

Dipsacus sp. (DIPSACACEAE)
Anonymous in Kew List of Determinations (volume 24).
Hubei: Fang Xian, 6915.

Dipteris conjugata Reinw. (DIPTERIDACEAE)
Polypodium dipteris Blume Henry in Trans. Asiat. Soc. Jap. 24 suppl: 115. (1896) (List Pl. Formosa). *Dipteris horsfieldii* (R. Br. ex Hook.) Bedd. Christ in Bull. Herb. Boiss. 6: 880. (1898).

Taiwan: Tamsui, 1444. **Yunnan:** In Pingbian in mountain forest on the Daweishan Range at 1,650 metres, 9041.

Dipteronia dyeriana A. Henry (SAPINDACEAE) New species

in Gard. Chron. ser. 3, 33: 22. (1903).

Yunnan: In forests to the east of Mengzi at 2,300 metres, 11352 (type).

A small tree to 4 metres tall. Henry named this species in honour of his friend and colleague, William Thistleton Dyer, Director of the Royal Botanic Gardens, Kew as a means of acknowledging many years of help and encouragement while collecting in China. Endangered in its native habitat (Yunnan), Henry only ever found it once.

Dipteronia sinensis Oliver (SAPINDACEAE) New genus

in Hooker's. Icon. Pl. 19: t. 1898. (1889), Bretschneider in Hist. Eur. Bot. Disc. China 2: 781. (1898), Henry in Flora & Sylva 1: 217. (1903); in Gard. Chron. ser. 3, 33: 21. (1903), E. H. Wilson in Journ. Roy. Hort. Soc. 24: 657. (1905), Schneider in Illus. Handb. Laubholzk. 2: 192. (1912), Besant in Gard. Chron. ser. 3, 98: 335. (1935), Morley in Glasra 3: 79. (1979), Zheng in Journ. Wuhan Bot. Res. 2.(1): 134. (1984), Nelson in A Heritage of Beauty 312. (2000), O' Brien in Augustine Henry, An Irish Plant Collector in China 19. (2002).

Hubei: Jianshi, 5696a., 5696b., South Badong, 5696c. Without locality, 6505a., (syntype). Xingshan, 6505 (type). **Sichuan:** South Wushan, 5696 (syntype), 7259.

One of an astonishing number of new endemic genera found in Henry's collections. He described it as a very common tree, that statement is no longer true today. Its closest ally is *Acer* from which it differs in having pinnate leaves and fruits which are winged all around the margin. According to Ying and Zhang (see Acta Phytotax. Sin. 22: 268. 1984) the three centres of endemic plants in China are 1. eastern Sichuan-western Hubei, 2. south-east Yunnan, 3. western Sichuan-north-west Yunnan. *Dipteronia sinensis* was introduced to cultivation by Wilson in 1900 when collecting for Messrs Veitch.

Dischidia chinensis Champ. ex Benth. (APOCYNACEAE)

Anonymous in Kew List of Determinations (volume 24).

Hainan: Environs of Haikou, 8427.

Dischidia formosana Maxim. (APOCYNACEAE)

Henry in Trans. Asiat. Soc. Jap. 24 suppl: 61. (1896) (List Pl. Formosa).

Taiwan: Oluanpi, 1977.

Discovered in Taiwan by the Kew collector, Richard Oldham.

Discocleidion rufescens (Franch.) Pax & Hoffm. (EUPHORBIACEAE)

Zheng in Journ. Wuhan Bot. Res. 2.(1): 111. (1984).

Alchornea rufescens Franchet. Henry in Journ. China Br. Roy. Asiatic. Soc. 22: 234. (1887) (Chinese Names of Plants), Forbes & Hemsley in Journ. Linn. Soc. Bot. 26: 438. (1894).

Hubei: Yichang, 20, 293, 1581a. In the San You Dong Glen (May 1886), 1581. Liantuo, 2090, 2694. Antelope Glen, 3443. On Moji Shan, 3569. South Badong, 5336. Without locality, 5336a.

A bush to about 2 metres tall, common at low altitudes in Hubei and known colloquially at Yichang as *ch'ing cha pao yeh* according to Henry. It was discovered by Père Armand David in southern Shaanxi in 1873.

Disporopsis fuscopicta Hance (ASPARAGACEAE)

Wright in Journ. Linn. Soc. Bot. 36: 110. (1903).

Hubei: South Badong, 6168. Fang Xian, 6705.

Disporum cantoniense (Lour.) Merr. (COLCHICACEAE)

Disporum pullum Salisb. Henry in Notes Econ. Bot. China 59. (1893), Wright in Journ. Linn. Soc. Bot. 36: 142. (1903).

Hubei: Yichang, 577, 1179, 2170. Kan Chi Glen near Yichang, 2960. In a cave on the way to Moji Shan (October 1886) 2960a. Antelope Glen, 3321. Badong, 919 (November 1885), 2830, 3782, 2783, 5459. Without locality, 5459a. **Sichuan:** South Wushan, 5545.

A handsome perennial to 80 cm tall. Distributed through the Himalaya, China, Indochina and Indonesia. A handsome medicinal plant which Henry stated was known as the *chi-kung-ch'i*. This species was collected by the Glasnevin Central China Expedition (GCCE 814) on Tianthu Shan in Hubei in October 2004. In this region in grew in dense broad-leaved forest with *Arisaema, Cardiocrinum* and *Paris.*

Disporum sessile (Thunb.) D. Don (COLCHICACEAE)

Henry in Journ. China Br. Roy. Asiat. Soc. 22: 272. (1887) (Chinese Names of Plants), Wright in Journ. Linn. Soc. Bot. 143. (1903).

Hubei: Badong, 247 (May 1885), 1474. Liantuo, 7635.

According to Henry this species was known colloquially in the Three Gorges region as *t'an chu hua.*

Disporum smilacinum A. Gray (COLCHICACEAE)

Anonymous in Kew List of Determinations (volume 24).

Hubei: Fang Xian, 6731, 6827.

Disporum spp. (COLCHICACEAE)

Anonymous in Kew List of Determinations (volume 34).

Yunnan: Mengzi, 9286, 10362, 10880, 11839. Simao, 12074, 12306.

Distephanus henryi (Dunn) H. Rob. (ASTERACEAE) New species

Vernonia henryi Dunn in Journ. Linn. Soc. Bot. 35:

500. (1903).

Yunnan: Near Linan at 1,650 metres, 13343 (type).

Henry found this species once only when on a journey between Simao and Mengzi. This was probably in September 1899 when he was transferred back to Mengzi as head of that customs office. He was also accompanied on this trip by 23 year old Ernest Wilson who had just arrived in China to collect for Messrs. Veitch. Endemic to Yunnan.

Distyliopsis laurifolia (Hemsley) Endress (HAMAMELIDACEAE) New species

Sycopsis laurifolia Hemsley in Hooker's Icon. Pl. 29: tt. 2836. (1907), Fedde in Repert. Nov. Spec. Regni Veg. 5: 342. (1908), Rehder & E. H. Wilson in Sargent Pl. Wilson. 1: 431. (1913), Walker in Journ. Arnold Arb. 25: 337. (1944).

Yunnan: Mengzi at 2,300 metres, 11365 (isotype of *Sycopsis laurifolia* Hemsley).

Endemic to China and distributed in the provinces of Guizhou and Yunnan. The genus *Distyliopsis* contains about four species of trees distributed across southern China, Vietnam, Malaysia, Indonesia and the Philippines.

Distylium buxifolium (Hance) Merr. (HAMAMELIDACEAE)

Walker in Journ. Arnold Arb. 25: 326. (1944), Zheng in Journ. Wuhan Bot. Res. 2.(1): 65. (1984).

Distylium racemosum Sieb. & Zucc. var. *chinense* Forbes & Hemsley non Franch. in Journ. Linn. Soc. Bot. 23: 289. (1887), Henry in Journ. China Br. Roy. Asiat. Soc. 22: 271. (1887) (Chinese Names of Plants.) *Distylium chinense* Rehder & E. H. Wilson non (Franch.) Diels in Sargent Pl. Wilson. 1: 423. (1913), in part.

Hubei: Yichang, 3314, 4280, 7805 (October 1887). Liantuo, 3826. Without locality, 3314a., 3314b.

A common low-growing shrub on the banks of the Yangtze. According to Henry this species was known as *shih-t'ou-k'o tzu*.

Distylium chinense (Franch. ex Hemsl.) Diels (HAMAMELIDACEAE)

Rehder & E. H. Wilson in Sargent Pl. Wilson. 1: 423. (1913).

Distylium racemosum Sieb. & Zucc. var. *chinense* Forbes & Hemsley non Franch. in Journ. Linn. Soc. Bot. 23: 289. (1887), Henry in Journ. China Br. Roy. Asiat. Soc. 22: 271. (1887) (Chinese Names of Plants).

Hubei: Yichang, 1300.

A common low-growing shrub (or sometimes a small tree) on the banks of the Yangtze and its associated tributaries in the Three Gorges region. It is particularly abundant on the banks of the Daning River, a tributary of the Yangtze that flows through the Three Lesser Gorges. According to Henry this species was known as *shih-tou-koutzu*. Discovered by Père Delavay in 1882.

Distylium racemosum Sieb & Zucc. (HAMAMELIDACEAE)

Henry in Trans. Asiat. Soc. Jap. 24 suppl: 42. (1896) (List Pl. Formosa), Li in Woody Flora of Taiwan 265. (1963).

Taiwan: Oluanpi, 980.

A large tree to 25 metres tall, native to Korea, Japan to Japan (south). Found only on the southern part of Chinese Taiwan where it is rare.

Docynia delavayi (Franch.) C. K. Schneid. (ROSACEAE)

Rehder in Sargent Pl. Wilson. 2: 296. (1915).

Pyrus delavayi Franch. Henry in Flora & Sylva 1: 218. (1903).

Yunnan: On mountains to the west of Mengzi at 1,800 to 2,000 metres, 10036, 10036a. In mountain forest to the east of Simao at 1,600 metres, 11603.

An evergreen tree to 10 metres tall, carrying clusters of white flowers on ends of branches in March and April. This handsome tree was discovered in Yunnan by Delavay and is also native to the provinces of Sichuan and Guizhou. Wilson in *A Naturalist in Western China* (2: 28. 1913) stated this tree was abundant in Yunnan and the fresh fruits known as *tao yi* were used in ripening persimmons. The fruits of each were arranged in alternate layers in large jars and covered with rise husks, in ten hours the persimmons were bletted and ready for eating. To the Hani (an ethnic group) in southern Yunnan it is the *qkaqpyuq* and the bark, leaves and kernels are used to treat diarrhoea, gasteroenteritis and externally for cuts, sprains and fractures. Introduced to France in 1890, this species first flowered in the UK at the RHS Garden, Wisley in 1938.

Dodonaea viscosa Jacq. (SAPINDACEAE)

Henry in Trans. Asiat. Soc. Jap. 24 suppl: 29. (1896) (List Pl. Formosa), Li in Woody Flora of Taiwan 494. fig. 189. (1963).

Taiwan: Kaohsiung, without number.

An pantropic evergreen shrub to 6 metres tall, distributed in Taiwan from seashore to high altitude. A pioneer species, its fruits have been used by some groups to stun and catch fish and the leaves and roots are used medicinally. Hardy in the milder counties of Ireland where the cultivar *Dodonea viscosa* 'Purpurea' is often grown.

Dolichos junghuhnianus Benth. (FABACEAE)

Dolichos henryi Harms in Fedde, Repert. Spec. Nov. Regni. Veg. 17: 137. (1921).

Yunnan: In the Red River valley near Manpan at 320 metres, 10132 (type of *Dolichos henryi* Harms).

Dolichos purpureus (L.) Sweet (FABACEAE)

Dolichos lablab L. Forbes & Hemsley in Journ. Linn.

Soc. Bot. 23: 194. (1887), Henry in Journ. China Br. Roy. Asiat. Soc. 22: 274. (1887) (Chinese Names of Plants); in Notes Econ. Bot. China 12. (1893).

Hubei: Yichang, 650. Changyang, 7737, 7769. **Yunnan:** Simao, 12755.

According to Henry this is the *bian dou* or *bai gian dou*, better known in the west as the hyacinth bean, a stout, twining perennial. Native to the tropics of the Old World (though it may have originated in India) and very extensively cultivated since ancient times. In the Three Gorges region this species was also known colloquially as the *pien tou*. *Dolichos lablab* is grown in some regions of Asia to prevent soil erosion and said to be a good fodder plant for cattle, for hay it is said to be equal to alfalfa. In Chinese herbal medicine it is used in cases of chronic diarrhea and vomiting.

Dolichos trilobus L. (FABACEAE)

Dolichos falcatus J. G. Klein ex Willd. Anonymous in Kew List of Determinations (volume 34).

Yunnan: Mengzi, 9310. Simao, 12507.

The Hani (an ethnic group) of Xishuangbanna in southern Yunnan call this species *xaqbeevjavhhaq* and use its swollen roots to treat sprains and fractures.

Dorcoceras hygrometrica Bunge (GESNERIACEAE)

Boea hygrometrica (Bunge) R. Br. Forbes & Hemsley in Journ. Linn. Soc. Bot. 26: 234. (1890).

Hubei: Yichang, 188, 1566, 3959.

Introduced to cultivation from Beijing in 1876 by S. Woolton Bushell of London University. Bushell was appointed physician to the British Legation in 1868. According to Bretschneider he was a distinguished sinologist, was well versed in Chinese history and numismatics. In 1872 he sent a collection of about 200 dried plants from the mountains in the Beijing region to Kew, these apparently were the first plants from northern China to reach the Kew herbarium. Bushell's other claim to fame was his discovery of the ruins of the palace of Kublai Khan in south-east Mongolia in 1872.

Draba ladyginii Pohle (BRASSICACEAE)

Schulz in Engler, Pflanzenr. iv. 105: 301. (1927).

Hubei: Xingshan, 6969.

Dracaena angustifolia (Medik.) Roxb. (ASPARAGACEAE)

Henry in Trans. Asiat. Soc. Jap. 24 suppl: 97. (1896) (List Pl. Formosa).

Taiwan: Oluanpi, 699, 1304.

A shrub or small tree to 5 metres tall. Native to India, China (Taiwan), the Philippines, Java, Malaysia and Australia. In Taiwan this species is restricted to the southern tip of the island and Lanyu Island.

Dracocephalum sp. (LAMIACEAE)

Anonymous in Kew List of Determinations (volume 34).

Yunnan: Mengzi, 10528.

Dregea sinensis Hemsley (APOCYNACEAE) New species

Forbes & Hemsley in Journ. Linn. Soc. Bot. 26: 115. (1889), Bretschneider in Hist. Eur. Bot. Disc. China 2: 788. (1898), K. Schumann in Engler & Schlecter in Bot. Jahrb. 29: 544. (1900), Schneider in Sargent Pl. Wilson. 3: 352. (1917), Lancaster in Travels in China 408. (1989), Nelson in A Heritage of Beauty 312. (2000), O' Brien in Augustine Henry, An Irish Plant Collector in China 19. (2002).

Wattakaka sinensis (Hemsl.) Stapf. Bean in Trees & Shrubs 4: 742. (1989), Morley in Glasra 3: 80. (1979).

Hubei: Badong, 1767 (type, collected July 1886). Yichang, 3005, 3125 (co-type), 4164. Liantuo, 3868, 6387. Antelope Glen, 3125 (co-type). **Yunnan:** Mengzi, 11061.

A slender, scandent shrub to 2.5 metres tall. Introduced to cultivation by Wilson in 1907. In Ireland this species is rare in cultivation and is grown at Castlewellan, Mount Congreve and Glasnevin.

Drosera indica L. (DROSERACEAE)

Diels in Engler, Pflanzenr. iv. 112: 77. (1906).

Hainan: On hummocks near Haikou, 8122.

An annual insectiverous herb, common in tropical Africa, Asia and Australia.

Drosera lunata Buch.-Ham. ex DC. (DROSERACEAE)

Drosera peltata J. E. Smith var. *lunata* (Buch.-Ham. ex DC.) C. B. Clarke Anonymous in Kew List of Determinations (volume 34).

Yunnan: Mengzi, 10017.

A terrestrial insectiverous perennial herb, widespread in the eastern hemisphere where it grows from India to Japan and southwards from there to Tasmania.

Drymaria sp. (CARYOPHYLLACEAE)

Anonymous in Kew List of Determinations (volume 24).

Hubei: Badong, 3762.

Drynaria coronans (Wall. ex Mett.) J. Sm. Ex T. Moore (POLYPODIACEAE)

Drynaria conjugata (Baker) Bedd. Christ in Bull. Herb. Boiss. 6: 880. (1898).

Yunnan: In Pingbian in mountain forest on the Daweishan Range at 1,650 metres, 10394.

Drynaria delavayi Christ (POLYPODIACEAE)

Drynaria rivalis (Mett. ex Baker) Christ var. *yunnanensis* Christ in Bull. Herb. Boiss. 7: 6. (1899).

Yunnan: In mountain forest to the north of Mengzi at 2,800 metres, 11512 (type of *Drynaria rivalis* (Mett. ex

02

Baker) Christ var. *yunnanensis* Christ).

Drynaria meyeniana (Schott.) Christenh. (POLYPODIACEAE)

Polypodium meyenianum (Schott.) Hook. Henry in Trans. Asiat. Soc. Jap. 24 suppl: 115. (1896) (List Pl. Formosa).

Taiwan: Oluanpi, 586.

An epiphytic fern, native to China (Taiwan) and the Philippines.

Drynaria propinqua (Wall. ex Mett.) J. Sm. ex Bedd. (POLYPODIACEAE)

Christ in Bull. Herb. Boiss. 6: 880. (1898).

Polypodium propinquum Wall. ex Mett. Anonymous in Kew List of Determinations (volume 34).

Yunnan: On exposed rocks near Mengzi at 1,650 metres, 9993. Simao, 12625, 12737, 13092.

Drynaria quercifolia (L.) J. Sm. (POLYPODIACEAE)

Polypodium conjugatum Poir. Henry in Trans. Asiat. Soc. Jap. 24 suppl: 115. (1896) (List Pl. Formosa).

Taiwan: Wanchin, 1482.

Drynaria roosii Nakaike (POLYPODIACEAE)

Drynaria fortunei (Kunze ex Mett.) J. Sm. Christ in Bull. Herb. Boiss. 6: 880. (1898), Diels in Bot. Jahrb. 29: 207. (1900), Baker in Journ. Bot. 26: 230. (1888), Takeda in Notes Roy. Bot. Gard. Edinburgh viii. 285. (1913—1915). *Polypodium fortunei* Kunze ex Mett. A. Henry in Journ. China Br. Roy. Asiat. Soc. 22: 233. (1887) (Chinese Names of Plants), Takeda in Notes Roy. Bot. Gard. Edinburgh viii. 286. (1913—1915). *Polypodium simplex* Christ non Sw. in Bull. Soc. Bot. France 5: 16. (1905). *Polypodium normale* Christ non D. Don in Bull. Soc. Bot. France 5: 16. (1905).

Hubei: Yichang, 393, 748, 2230. Antelope Glen near Yichang, 2230. Liantuo, 1994. Badong, 3704. Changleping, 7842. **Sichuan:** South Wushan, 7247. **Yunnan:** On rocks and on trees in shade near Mengzi at 1,500 metres, 10177. On the mountains near Mengzi at 1,640 metres, 13363, 13634.

An epiphytic fern, found over much of China, Thailand, Vietnam and Laos where it grows on rocks and tree trunks. According to Henry this fern was known colloquially as *ai chiang* and the rhizome was used as a drug for making childrens hair strong and black.

Dryopteris chrysocoma (Christ) C. Chr. (POLYPODIACEAE) New species

in Ind. Fil. 257. (1905).

Aspidium felix-mas (L.) Sw. var. *chrysocoma* Christ in Bull. Herb. Boissier 6: 966. (1898).

Yunnan: Lunan, 9957.

Dryopteris cochleata (D. Don) C. Chr. (POLYPODIACEAE)

Nephrodium cochleatum D. Don Anonymous in Kew List of Determinations (volume 34).

Yunnan: Mengzi, 9102.

Native to Bangladesh, the Himalaya, Myanmar, western China, Thailand and the Philippines.

Dryopteris dickinsii (Franch. & Savat.) C. Chr. (POLYPODIACEAE) New to China

Aspidium dickinsii (Bak.) Franch. & Savat. Christ in Bull. Soc. Bot. France 5: 40. (1905). *Nephrodium dickinsii* (Franch. & Sav.) Bak. Baker in Journ. Bot. 25: 171. (1887), Baker in Journ. Bot. 26: 288. (1888).

Hubei: Badong, 1714, 3677.

Distributed in India, China and Japan.

Dryopteris dilatata (Hoffm.) A. Gray (POLYPODIACEAE)

Polypodium dilatatum Hoffm. Anonymous in Kew List of Determinations (volume 34).

Yunnan: Mengzi, 9554.

Dryopteris enneaphylla (Baker) C. Chr. (POLYPODIACEAE) New species

Nephrodium enneaphyllum Baker in Journ. Bot. 25: 170. (1887), Baker in Journ. Bot. 27: (1889); in Ann. Bot. 5: 316. (1890), Bretschneider in Hist. Eur. Bot. Disc. China 2: 794. (1898).

Hubei: Antelope Glen, 3217. Yichang, 7881.

Distributed in China.

Dryopteris erythrosa (D. C. Eaton) Kuntze (POLYPODIACEAE)

Aspidium erythrosorum D. C. Eaton Christ in Bull. Herb. Boiss. 7: 17. (1899). *Nephrodium erythrosorum* (D. C. Eaton) Hook. Anonymous in Kew List of Determinations (volume 34).

Yunnan: In mountain forest near Mengzi, 11533. Simao, 12736.

Native to China, Korea and Japan.

Dryopteris fructuosa (H. Christ) C. Chr. (POLYPODIACEAE) New species

Aspidium varium Sw. var. *fructuosum* H. Christ in Bull. Herb. Boissier 6: 967. (1898). *Aspidium fructuosum* (Christ) Christ in Bull. Soc. Bot. France 5: 38. (1905).

Yunnan: Manmei at 1,950 metres, 10095 (isotype of *Aspidium varium* Sw. var. *fructuosum* H. Christ).

Native to India, Bhutan, Nepal, Myanmar and China.

Dryopteris gymnophylla (Baker) C. Chr. (POLYPODIACEAE) New species

Nephrodium gymnophyllum Baker in Journ. Bot. 25: 170. (1887), Baker in Ann. Bot. 5: 323. (1890).

Hubei: Liantuo, 2643, (in part).

Dryopteris hirtipes (Blume) Kuntze (POLYPODIACEAE)

Aspidium hirtipes Blume Christ in Bull. Soc. Bot. France 5: 40. (1905). *Nephrodium hirtipes* (Blume) Hook. Anonymous in Kew List of Determinations (volume 34).

Hubei: Badong, 4764, 5090, 5172. Changyang, 7648.

Yunnan: Mengzi, 10266. Simao, 13061.

Distributed in Nepal, India, Sri Lanka, Myanmar, China, Thailand, Vietnam, Japan and the Philippines.

Dryopteris lacera (Thunb.) Kuntze (POLYPODIACEAE)

Aspidium lacerum (Thunb.) Sw. Christ in Bull. Soc. Bot. France 5: 39. (1905). *Nephrodium lacerum* (Thunb.) Baker Baker in Journ. Bot. 26: 228. (1887).

Hubei: Badong, 1724, 2571, 3680, 4775, 5061. Liantuo, 2118, 3043. Jianshi, 5966. **Hunan:** Shimen, 7542.

Distributed in China, Korea and Japan.

Dryopteris microlepis (Baker) C. Chr. (POLYPODIACEAE) New species

Nephrodium microlepis Baker in Kew Bull. Misc. Inf. 10. (1906).

Yunnan: In forest south of Simao at 1,650 metres, 13154 (type of *Nephrodium microlepis* Baker).

Dryopteris ochtodes (Kunze) C. Chr. (POLYPODIACEAE)

in Notul. Syst. (Paris) 1: 41. (1909).

Yunnan: South of the Red River from Manmei at 1,900 metres, 10094a.

Dryopteris odontoloma (T. Moore ex Bedd.) C. Chr. (POLYPODIACEAE)

Nephrodium filix-mas Henry non (L.) Rich. in Journ. China Br. Roy. Asiat. Soc. 22: 242. (1887) (Chinese Names of Plants). *Nephrodium filix-mas* (L.) Rich. var. *odontoloma* (T. Moore ex Bedd.) Bak. Baker in Ann. Bot. 5: 323. (1890).

Hubei: Fang Xian, 6951.

Dryopteris rosthornii (Diels) C. C. Chr. (POLYPODIACEAE)

Aspidium polylepis Christ non Franch. & Savat. in Bull. Soc. Bot. France 5: 38. (1905). *Dryopteris centrochinensis* R. C. Ching in Bull. Fan. Mem. Inst. Biol. 8: 423. (1938).

Hubei: Yichang, 1448 (type *Dryopteris centrochinensis* R. C. Ching).

Dryopteris scottii (Bedd.) Ching (POLYPODIACEAE)

Phegopteris grossa Christ in Bull. Herb. Boiss. 7: 13. (1899).

Yunnan: In ravines on the mountains to the south-west of Mengzi, 11558 (type of *Phegopteris grossa* Christ).

Distributed in India, Bhutan, Myanmar and China.

Dryopteris sikkimensis (Bedd.) Kunze (POLYPODIACEAE)

Aspidium sikkimense (Bedd.) Baker Anonymous in Kew List of Determinations (volume 34).

Yunnan: Mengzi, 9919.

Distributed in India (Sikkim,) China (Tibet (Xizang) and Yunnan).

Dryopteris sparsa (D. Don) Kuntze (POLYPODIACEAE)

Nephrodium sparsum D. Don Anonymous in Kew List of Determinations (volume 34).

Yunnan: Simao, 13053.

Distributed through the Himalaya, India, Myanmar, China, Japan, the Philippines, the Malay Peninsula, and New Guinea.

Dryopteris sphaeropteroides (Baker) C. Chr. (POLYPODIACEAE)

Phegopteris sphaeropteroides (Baker) H. Christ Christ in Bull. Herb. Boiss. 7: 14. (1899).

Yunnan: In mountain forest near Mengzi at 1,650 metres, 11539.

Endemic to China (Sichuan and Yunnan).

Dryopteris splendens (Desv.) Kuntze (POLYPODIACEAE)

Nephrodium splendens Desv. Baker in Journ. Bot. 25: 171. (1887) Baker in Ann. Bot. 5: 324. (1890).

Hubei: Badong, 2833.

Distributed from Nepal to China.

Dryopteris spp. Rich. (POLYPODIACEAE)

Nephrodium felix-mas Henry non (L.) Rich. in Journ. China Br. Roy. Asiat. Soc. 22: 242. (1887) (Chinese Names of Plants). Anonymous in Kew List of Determinations (volume 34).

Hubei: Badong, 1719, 1726, 2511, 5164. Liantuo, 1886, 2654, 3195, 3207 (in part), 4636. Jianshi, 5961. Fang Xian, 6950. Changyang, 7644. Changleping, 7811. **Yunnan:** Simao, 13108.

Dryopteris stenolepis (Baker) H. Christ (POLYPODIACEAE) New species

in Index Filic. 5: 294. (1905).

Polypodium stenolepis Baker in Kew Bull. Misc. Inf. 231. (1898). *Aspidium stenolepis* (Baker) H. Christ in Bull. Herb. Boissier 7: 22. (1899).

Yunnan: Mountains to the north of Mengzi at 2,700 metres, 9038 (type of *Polypodium stenolepis* Baker). Mengzi, 9038a., [isotype of *Aspidium stenolepis* (Baker) H. Christ].

Distributed in India, Bhutan and western China.

Dryopteris varia (L.) Kuntze (POLYPODIACEAE)

Polypodium adnascens Sw. var. *varians* Henry in Trans. Asiat. Soc. Jap. 24 suppl: 114. (1896) (List Pl. Formosa). *Aspidium varium* (L.) Sw. Christ in Bull. Soc. Bot. France 5: 43. (1905).

Hubei: Yichang, 2054, 2971, 2972. Liantuo, 2144, 2643, 3029, 3034. Badong, 3681, 5138. Changleping. 7810, 7820. **Taiwan:** Oluanpi, 594, 1247. Tamsui, 1432. Wanchin, 1540. Tamsui, 1425a.

Distributed in India, China, Korea, Japan and the Philippines.

Duhaldea cappa (Buch.-Ham. ex D. Don) Pruski & Anderb. (ASTERACEAE)

Inula cappa (Buch.-Ham. ex D. Don) DC. Anonymous in Kew List of Determinations (volume 34).

02

Yunnan: Mengzi, 10278.

A much branched perennial to 1 metre tall, the Hani (an ethnic group) of Xishuangbanna in southern Yunnan this species is called *haqzeilkeeltaov* and use it to treat flu, sprains, rheumatism and joint pain. Native to India, Myanmar, China, Thailand and Vietnam.

Duhaldea eupatorioides (DC.) Anderb. (ASTERACEAE)

Inula eupatorioides DC. Anonymous in Kew List of Determinations (volume 34).

Yunnan: Mengzi, 10482, 11324.

Distributed in India, Bhutan, Myanmar, China [Tibet (Xizang), Sichuan and Yunnan], Laos and Vietnam.

Duhaldea nervosa Wall. ex DC.) Anderb. (ASTERACEAE)

Inula nervosa (Wall. ex DC.) Anonymous in Kew List of Determinations (volume 34).

Yunnan: Simao, 11683, 12478.

Duhaldea pterocaula (Franch.) Anderb. (ASTERACEAE)

Inula pterocaula Franchet Anonymous in Kew List of Determinations (volume 34).

Yunnan: Mengzi, 9699.

Endemic to China and distributed in Sichuan and Yunnan

Duhaldea rubricaulis (Wall. ex DC.) Benth. & Hook. f. (ASTERACEAE)

Inula rubricaulis (Wall. ex DC.) Anderb. Anonymous in Kew List of Determinations (volume 34).

Yunnan: Yuanjiang, 13332.

Native to India, Bhutan, Nepal, Burma, Yunnan, Thailand and Vietnam.

Dumasia cordifolia Benth. ex Baker (FABACEAE)

Hemsley in Hooker's Icon. Pl. t. 2627. (1896).

Yunnan: In Pingbian in mountain forest on the Daweishan Range at 1,650 metres, 10326.

Distributed in India and western China.

Dumasia henryi (Hemsley) R. Sha & M. G. Gilbert (FABACEAE)

Fl. China 10: 244. (2010).

Rhynchosia henryi Hemsley in Journ. Linn. Soc. Bot. 23: 196. (1887), Bretschneider in Hist. Eur. Bot. Disc. China 2: 782. (1898).

Hubei: Liantuo, 3068, 4650.

Dumasia hirsuta Craib (FABACEAE) New species

in Sargent Pl. Wilson. 2: 116. (1916).

Hubei: Badong, in thickets at 1,000~1,300 metres, 6115 (type).

Dumasia sp. (FABACEAE)

Anonymous in Kew List of Determinations (volume 24).

Hubei: San You Dong Glen, 3319.

Dumasia villosa DC. (FABACEAE)

Forbes & Hemsley in Journ. Linn. Soc. Bot. 23: 489. (1887).

Hubei: Liantuo, 3068, 4650. **Yunnan:** Between Fengchunling and Yuanjiang, 9238. Simao, 12453, 12616, 12872, 13550.

Dunbaria circinalis (Benth.) Baker (FABACEAE)

Anonymous in Kew List of Determinations (volume 34).

Yunnan: Simao, 12384.

Dunbaria punctata (Wight & Arn.) Benth. (FABACEAE)

Dunbaria conspersa Benth. Anonymous in Kew List of Determinations (volume 24).

Hainan: Environs of Haikou, 8029. Environs of Kiungchow, 8692.

Dunbaria sp. (FABACEAE)

Anonymous in Kew List of Determinations (volume 34).

Yunnan: Simao, 12377.

Dunbaria trichodon (Dunn) Maesen (FABACEAE) New species

Atylosia trichodon Dunn in Journ. Linn. Soc. Bot. 35: 491. (1903).

Yunnan: Mountains to the west of Simao at 1,650 metres, 12474 (isotype of *Atylosia trichodon* Dunn, collected October 7th 1898).

Dunbaria villosa (Thunb.) Makino (FABACEAE)

Dunbaria subrhombea (Miq.) Hemsley Forbes & Hemsley Journ. Linn. Soc. Bot. 23: 195. (1887).

Hubei: Yichang, 2317, 2375.

Duranta erecta L. (VERBENACEAE)

Duranta plumieri Jacq. Henry in Trans. Asiat. Soc. Jap. 24 suppl: 70. (1896) (List Pl. Formosa).

Taiwan: Tamsui, cultivated, 1405.

The golden dewdrop, a shrub or small tree to 6 metres tall, native to tropical America.

Dysolobium pilosum (J. G. Klein ex Willd.) Marechal (FABACEAE)

Vigna pilosa (J. G. Klein ex Willd.) Baker Henry in Trans. Asiat. Soc. Jap. 24 suppl: 36. (1896) (List Pl. Formosa).

Taiwan: Kaohsiung, without number.

An annual twining herb, found in grassland on the southern tip of Taiwan.

Dysoxylum pallens Hiern (MELIACEAE)

Dysoxylum spicatum H. L. Li in Journ. Arnold Arbor. 25: 303. (1944).

Yunnan: Mountains to the south of Simao at 1,650 metres, 11748 (type of *Dysoxylum spicatum* H. L. Li). In forests near Simao at 1,300 metres, 11748a.

Distributed in India, Sri Lanka, China (Hainan and Yunnan) and Vietnam.

Echenbachia blinii (H. Lév.) Brouillet (ASTERACEAE)

Conyza pinnatifida Dunn in Journ. Linn. Soc. Bot. 35: 502. (1903), nom illeg.

Yunnan: Common on the grass mountains near Mengzi at 1,500 metres, 9982. Common on the grass mountains near Mengzi at 1,950 metres, 12176a. Simao, 12176.

Echenbachia japonica (Thunb.) J. Kost. (ASTERACEAE)

Conyza japonica (Thunb.) Less. ex DC Anonymous in Kew List of Determinations (volume 34).

Yunnan: Mengzi, 10187, 10269, 11139.

Echenbachia leucantha (D. Don) Brouillet (ASTERACEAE)

Conyza viscidula Wall. ex DC. Henry in Trans. Asiat. Soc. Jap. 24 suppl: 52. (1896) (List Pl. Formosa).

Taiwan: Kaohsiung, 1713. **Yunnan:** Mengzi, 11054. Simao, 12823.

Echenbachia stricta (Willd.) Raizada (ASTERACEAE)

Conyza stricta Willd. Anonymous in Kew List of Determinations (volume 34).

Yunnan: Mengzi, 9304.

Echinochloa colonum (L.) Link (POACEAE)

Panicum colonum L. Rendle in Journ. Linn. Soc. Bot. 36: 328. (1904).

Hainan: Environs of Haikou, 8239. **Hubei:** Yichang, 2769, 4286. Badong, 7313.

Shama millet, cultivated as a food source in temperate parts of the world. An annual to 80 cm tall, native to tropical Africa and Asia.

Echinochloa crus-galli (L.) P. Beauv. (POACEAE)

Panicum colonum Henry non L. in Trans. Asiat. Soc. Jap. 24 suppl: 106. (1896) (List Pl. Formosa). *Panicum crus-galii* L. Henry in Journ. China Br. Roy. Asiat. Soc. 22: 263. (1887) (Chinese Names of Plants); in Trans. Asiat. Soc. Jap. 24 suppl: 106. (1896) (List Pl. Formosa), Rendle in Journ. Linn. Soc. Bot. 36: 328. (1904).

Hubei: Yichang, 746, 2341, 4336. Changyang, 7743. **Taiwan:** Oluanpi, 276. Wanchin, 841. Kaohsiung, 1046, 1796, 1960, 2003.

Japanese barnyard millet, a weed in temperate zones of both hemispheres. Henry stated that iin the Three Gorges region it was known colloquially as *pai tzu ts'ao*.

Eclipta prostrata (L.) L. (ASTERACEAE)

Eclipta alba (L.) Hassk. Henry in Journ. China Br. Roy. Asiat. Soc. 22: 263. (1887) (Chinese Names of Plants); in Trans. Asiat. Soc. Jap. 24 suppl: 54. (1896) (List Pl. Formosa)

Hubei: Yichang, 170, 1670. **Hainan:** Environs of Haikou, 8332. Environs of Kiungchow, 8688. **Taiwan:** Kaohsiung, without number. Oluanpi, 367, 1278. Wanchin,

862, 973. **Yunnan:** Mengzi, 9860.

Edgeworthia chrysantha Lind. (THYMELAEACEAE)

Forbes & Hemsley in Journ. Linn. Soc. Bot. 26: 401. (1894), Rehder in Sargent Pl. Wilson.2: 551. (1916).

Edgeworthia gardneri Forbes & Hemsley non (Wall.) Meisn. in Journ. Linn. Soc. Bot. 26: 396. (1891), Henry in Notes Econ. Bot. China 22. (1893).

Hubei: Yichang, 3976. **Sichuan:** South Wushan, 7246.

A deciduous shrub to 2 metres tall, though rarely reaching those dimensions in cultivation. As with most other members of the Thymelaeceae this is a highly ornamental plant and should be grown wherever possible. A native of China, it was introduced to cultivation in 1845 and has been cultivated in China for centuries for ther manufacture of a high-class paper used for printing currency on. In China it was known as *chi hsiang* or "knot scent" since the stems are so pliable that they can be tied into tight knots without interfering with the development of the plant. It is an extremly beautiful sight when in full flower and the genus commemorates Irish siblings, the novelist Maria Edgeworth, and her youngest brother, the botanist, M. Pakenham Pakenham. Henry's specimens were cultivated and known at Yichang as *meng-hwa*.

Ehretia acuminata R. Br. (BORAGINACEAE)

Forbes & Hemsley in Journ. Linn. Soc. Bot. 143. (1890), Henry in Trans. Asiat. Soc. Jap. 24 suppl: 62. (1896) (List Pl. Formosa), E. H. Wilson in Sargent Pl. Wilson. 3: 363. (1917).

Ehretia ovalifolia Hassk. Henry in Journ. China Br. Roy. Asiat. Soc. 22: 277. (1887) (Chinese Names of Plants). *Ehretia taiwaniana* Nakai in Journ. Arnold Arb. 5: 38. (1924). *Ehretia acuminata* R. Br. var. *obovata* (Lind.) Johns. Johnson in Journ. Arnold. Arb. 32: 21. (1951). *Ehretia thyrsiflora* (Sieb. & Zucc.) Nakai Nakai Li in Woody Flora of Taiwan 814. (1963) Zheng in Journ. Wuhan Bot. Res. 2.(1): 184. (1984).

Hubei: Liantuo, 1941, 3907, 4556, 6358. Donghu, 6402. **Hainan:** Environs of Kiungchow, 8274. **Taiwan:** Oluanpi, 1287, 922 (syntype of *Ehretia taiwaniana* Nakai), 952 (syntype of *Ehretia taiwaniana* Nakai). Wanchin, 443 (syntype of *Ehretia taiwaniana* Nakai), 506 (syntype of *Ehretia taiwaniana* Nakai). Kaohsiung, 1135, 1136, 1778 (syntype of *Ehretia taiwaniana* Nakai), 1874 (syntype of *Ehretia taiwaniana* Nakai). **Yunnan:** Simao at 1,500 metres, 10454.

A small deciduous tree to 10 metres tall, native to Japan, China, India, Indonesia and Australia, rare in cultivation. According to Henry this was known as the *ta ts'u kang* tree in Hubei. The wood was light and used in the manufacture of furniture. There is a large specimen at Birr

02

Castle, County Offaly.

Ehretia asperula Zollinger & Moritzi (BORAGINACEAE)

Johnston in Journ. Arnold Arb. 32: 106. (1951).

Hainan: Environs of Haikou, 8027, 8186. On the sandy sea shore at Haikou, 8108. Environs of Kiungchow, 8217, 8699. Without locality, 8575.

Ehretia corylifolia C. H. Wright (BORAGINACEAE)

Johnston in Journ. Arnold Arb. 32: 104. (1951).

Yunnan: Mengzi, 10548, 10548a., 10548b.

Discovered by the Irish plant collector and Imperial Maritime Customs officer, William Hancock near Mengzi in 1894. Endemic to Yunnan.

Ehretia dicksonii Hance (BORAGINACEAE)

Johnston in Journ. Arnold Arb. 32: 99. (1951), Li in Woody Flora of Taiwan 812. fig. 328. (1963).

Ehretia macrophylla Forbes & Hemsley non Wallich in Journ. Linn. Soc. Bot. 26: 145. (1890), Henry in Trans. Asiat. Soc. Jap. 24 suppl: 62. (1896) (List Pl. Formosa), E. H. Wilson in Sargent Pl. Wilson. 3: 364. (1917). *Ehretia macrophylla* Wall. var. *tomentosa* Gagnep. & Lour. in Fl. Indo-Chine 4: 212. (1914).

Hubei: Yichang, 3522. Without locality, 3522a. **Sichuan:** South Wushan, 7526. **Taiwan:** Kaohsiung, without number. Oluanpi, 323. Wanchin, 190, 456. Kaohsiung, 313. **Hainan:** Environs of Kiungchow, 8294 (type of *Ehretia macrophylla* Wall. var. *tomentosa* Gagnep. & Lour.).

A common tree in western Hubei and Sichuan where it grows to about 15 metres tall with a girth of 2 metres on good specimens. The white flowers are powerfully fragrant. According to Henry and Wilson this tree was known colloquially in Hubei as *tzu kang* and its light, strong wood was said to make the best carrying poles used by the inhabitants of central and western China. Introduced to cultivation by Wilson.

Ehretia dunniana Leveille (BORAGINACEAE) New species

Johnston in Journ. Arnold Arb. 32: 104. (1951).

Yunnan: Mountains to the south of Simao, 11797. Mountains to the east of Simao, 11797a. Simao, 11797b., 13021.

Leveille based this species on material gathered by the French missionary Père Cavalerie. Cavalerie was based in Guizhou province (adjacent to Yunnan) between 1912—1921, Henry discovered the tree in 1897. Endemic to China and distributed in Guizhou and Yunnan.

Ehretia longiflora Champ. ex Benth. (BORAGINACEAE)

Henry in Trans. Asiat. Soc. Jap. 24 suppl: 62. (1896) (List Pl. Formosa), Johnston in Journ. Arnold Arb. 32: 105. (1951).

Taiwan: Wanchin, 432, 509, 563.

A small tree, native to Indochina and China (Hainan, Hong Kong and Taiwan).

Ehretia macrophylla Wall. var. glabrescens (Nakai) Y. L. Liu (BORAGINACEAE) New variety

Ehretia sp. Henry in Journ. China Br. Roy. Asiat. Soc. 22: 277. (1887) (Chinese Names of Plants). *Ehretia macrophylla* Forbes & Hemsley non Wallich in Journ. Linn. Soc. Bot. 26: 145. (1890), in part, Henry in Trans. Asiat. Soc. Jap. 24 suppl: 62. (1896) (List Pl. Formosa), in part, E. H. Wilson in Sargent Pl. Wilson. 3: 364. (1917), in part. *Ehretia dicksonii* Hance var. *glabrescens* Nakai in J. Arnold Arbor. 5: 38. (1924) Johnston in Journ. Arnold Arb. 32: 100. (1951).

Hubei: Liantuo, 3866. Yichang, 7177. **Yunnan:** Mengzi at 1,950 metres, 10515 (syntype of *Ehretia dicksonii* Hance var. *glabrescens* Nakai).

As pointed out by Johnson (1951) Nakai based this well marked geographical variety on three collections without designating a type; 1. Wilson 3554 from Yichang, 2. Wilson 3554b. from the Min valley, Sichuan and 3. A. Henry 10515 from Mengzi, Yunnan. Introduced to cultivation by Vilmorin from seeds collected by Farges in Sichuan.

Ehretia microphylla Lam. (BORAGINACEAE)

Li in Woody Flora of Taiwan 814. (1963)

Ehretia buxifolia Roxb. Henry in Trans. Asiat. Soc. Jap. 24 suppl: 62. (1896) (List Pl. Formosa), Henry in Annotation in the *Index Florae Sinensis* for Hainan collection (copy at Glasnevin). *Carmona microphylla* (Lam.) G. Don. Johnston in Journ. Arnold Arb. 32: 17. (1951).

Hainan: Environs of Haikou, 7952, 8041. Without locality, 8358. **Taiwan:** Kaohsiung, without number.

A small evergreen shrub to 2 metres tall, widely distributed from India to the Philippines, New Guinea and the Solomon Islands.

Ehretia resinosa Hance (BORAGINACEAE)

Li in Woody Flora of Taiwan 814. (1963), Johnston in Journ. Arnold Arb. 32: 103. (1951).

Ehretia formosana Hemsley Henry in Trans. Asiat. Soc. Jap. 24 suppl: 62. (1896) (List Pl. Formosa).

Taiwan: Wanchin, without number. Oluanpi, 1277.

A shrub or small tree, native to the Philippines and China (Taiwan) where it is confined to the southern part of the island. *Ehretia resinosa* was discovered by Robert Swinhoe near Kaohsiung (then Takow) in 1865 and it was also later collected on Wanshoushan, by E. H. Wilson. Nine decades later it was collected by the Glasnevin Taiwan Expedition (GTE 15) in the tropical rainforests that surround the lighthouse at Oluanpi (South Cape of Henry and Wilson) in October 2004.

02

Ehretia tsangii Johnston (BORAGINACEAE) New species

in Journ. Arnold Arb. 32: 104. (1951).

Yunnan: Simao, 13379.

Endemic to western China.

Elaeagnus conferta Roxb. (ELAEAGNACEAE)

Rehder in Sargent Pl. Wilson. 2: 417. (1915).

Yunnan: In forests to the south of Simao at 1,500 metres, 12680.

The Hani (an ethnic group) of Xishuangbanna in southern Yunnan call this species *pavqpeellovniov* and use its roots to treat asthma. This evergreen climbing shrub had not been reported from China before Henry's Simao collection. Distributed in Nepal, India, western China, Laos and Thailand.

Elaeagnus difficilis Servettaz (ELAEAGNACEAE) New species

in Bull. Herb. Boissier 8: 386. (1908), Zheng in Journ. Wuhan Bot. Res.2. (1): 156. (1984).

Elaeagnus glabra Forbes & Hemsley non Thunb. in Journ. Linn. Soc. Bot. 26: 402. (1894). *Elaeagnus cuprea* Rehder in Sargent Pl. Wilson 2: 414. (1915).

Hubei: Badong, 1451 (isotype), 3697, 5154. South Badong, 5473.

A handsome shrub to about 3 metres tall, this species was collected by the Glasnevin Central China Expedition (GCCE 496) near Zhenzi town in Xingshan in October 2004.

Elaeagnus glabra Thunb. (ELAEAGNACEAE)

Henry in Journ. China Br. Roy. Asiat. Soc. 22: 267. (1887) (Chinese Names of Plants), Forbes & Hemsley in Journ. Linn. Soc. Bot. 26: 402. (1894), Zheng in Journ. Wuhan Bot. Res. 2.(1): 156. (1984).

Hubei: Badong, 3169. Liantuo, 3193 (February 1887).

A large shrub allied to *Elaeagnus pungens* though differing from that species in the absence of thorns and its thinner, longer pointed leaves and its more rambling, almost climbing nature. As a bush it can reach 7 metres but when growing through trees it can grow to 14 metres or more tall in warm climates. White fragrant flowers appear in October and November. According to Henry this species was known colloquially in the Three Gorges region as *san yueh huang tzu*.

Elaeagnus cf. *glabra* Thunb. (ELAEAGNACEAE)

Anonymous in Kew List of Determinations (volume 24).

Hainan: Environs of Haikou, 8361.

Elaeagnus henryi O. Warburg (ELAEAGNACEAE) New species

in Bot. Jarhb. xxix. 483. (1900), Servettat in Bot.

Centralb. Beih. xxv. 77. fig. 4. (1909), Schneider in Illus. Handb. Laubholzk. 2: 414. fig. 283. (1909), Rehder in Sargent Pl. Wilson 2: 414. (1915), Zheng in Journ. Wuhan Bot. Res. 2.(1): 156. (1984).

Elaeagnus latifolia Forbes & Hemsley non L. in Journ. Linn. Soc. Bot. 26: 403. (1894).

Hubei: Yichang, 3307, 3307a., (type, October 1887).

Yunnan: Mengzi, 10425, 11392, 11449.

Elaeagnus lanceolata O. Warburg (ELAEAGNACEAE) New species

in Bot. Jahrb. Syst. 29: 483. (1900), Rehder in Sargent Pl. Wilson 2: 413. (1915), Zheng in Journ. Wuhan Bot. Res. 2.(1): 156. (1984).

Hubei: Badong, 5157 (type), 5283. Jianshi, 7424. Fang Xian, 6652. Without locality, 5483a.

Elaeagnus loureiroi Champion (ELAEAGNACEAE)

Rehder in Sargent Pl. Wilson. 2: 416. (1915).

Yunnan: Mengzi at 1,500 metres, 9857, 9858, 11357. Tonguan at 1,300 metres, 11574. Simao at 1,500 metres, 12787, 12787a., 12787b.

Distributed in China and Vietnam.

Elaeagnus macrantha Rehder (ELAEAGNACEAE) New species

in Sargent Pl. Wilson. 2: 416. (1915).

Yunnan: 50 miles (80 kms) east of Simao at 1,500 metres, 12818 (type).

An erect spiny shrub to 1.25 metres with white flowers. Endemic to southern Yunnan.

Elaeagnus magna (Ser.) Rehder (ELAEAGNACEAE) New species

in Sargent Pl. Wilson. 2: 411. (1915), Zheng in Journ. Wuhan Bot. Res. 2.(1): 156. (1984).

Elaeagnus sp. nov. Forbes & Hemsley in Journ. Linn. Soc. Bot. 26: 405. (1894). *Elaeagnus umbellata* Forbes & Hemsley non Thunb. in Journ. Linn. Soc. Bot. 26: 404. (1894). *Elaeagnus umbellata* Thunb. ssp. *magna* Servett in Bull. Herb. Boissier. viii. 383. (1908).

Hubei: In the San You Dong Glen (June 1886), 1637 (type). Yichang, 1637a., 7775.

Elaeagnus multiflora Thunb. (ELAEAGNACEAE)

Forbes & Hemsley in Journ. Linn. Soc. Bot. 26: 404. (1894).

Hubei: Badong, 1424.

A semi-evergreen shrub to 3 metres tall, flowering in April and May, the flowers are very fragrant and the fruits which ripen in June are edible and an ornamental feature.

Elaeagnus multiflora Thunb. f. *angustata* Rehder (ELAEAGNACEAE) New form

in Sarg. Pl. Wilson. 2: 413. (1915), Zheng in Journ. Wuhan Bot. Res.2. (1): 157. (1984).

Elaeagnus multiflora Forbes & Hemsley non Thunb. in Journ. Linn. Soc. Bot. 26: 404. (1894).

Hubei: Jianshi, 5484. Without locality, 5484b.

A form with long, narrow leaves.

Elaeagnus oldhamii Miq. (ELAEAGNACEAE)

Henry in Trans. Asiat. Soc. Jap. 24 suppl: 51. (1896) (List Pl. Formosa).

Taiwan: Kagee, 364.

A small tree, native to eastern China and to south China (Taiwan) where it was discovered by the Kew collector, Richard Oldham in 1864.

Elaeagnus pungens Thunb. (ELAEAGNACEAE)

Forbes & Hemsley in Journ. Linn. Soc. Bot. 26: 404. (1894), Zheng in Journ. Wuhan Bot. Res.2. (1): 157. (1984).

Hubei: Liantuo, 3065.

Elaeagnus sarmentosa Rehder (ELAEAGNACEAE)

New species

in Sargent Pl. Wilson. 2: 417. (1915).

Yunnan: In forests near Mengzi at 1,900 metres, 11439 (type).

Endemic to China and distributed in the provinces of Guangxi and Yunnan.

Elaeagnus sp. (ELAEAGNACEAE)

Rehder in Sargent Pl. Wilson. 2: 417. (1916).

Yunnan: Mengzi, 10436, 11309, 11332. Simao, 12684.

Elaeagnus umbellata Thunb. (ELAEAGNACEAE)

Forbes & Hemsley in Journ. Linn. Soc. Bot. 26: 404. (1894).

Hubei: Yichang, 1228.

A large wide-spreading deciduous shrub, native to Afghanistan, the Himalaya, China and Japan.

Elaeagnus viridis Servettaz (ELAEAGNACEAE) New species

in Bull. Herb. Boissier. ser. 2. 8: 388. (1908); in Bot. Centralb. Beih. xxv. 2, 88. fig. 28-30. (1909), Rehder in Sargent Pl. Wilson 2: 414. (1916), Zheng in Journ. Wuhan. Bot. Res. 2.(1): 157. (1984).

Elaeagnus glabra Forbes & Hemsley non Thunb. in Journ. Linn. Soc. Bot. 26: 402. (1894). *Elaeagnus pungens* Forbes & Hemsley non Thunb. in Journ. Linn. Soc. Bot. 26: 404. (1894).

Hubei: Yichang, on the banks of the Yangtze at 30 to 900 m., 1105, 2953.

Elaeocarpus decipiens F. B. Forbes & Hemsl. (ELAEOCARPACEAE)

Anonymous in Kew List of Determinations (volume 24).

Hainan: At Nodoa in the interior of the island, 8442.

Elaeocarpus glabripetalus Merr. var. *alatus* (Kunth) Hung T. Chang (ELAEOCARPACEAE) New variety

Elaeocarpus alatus Kunth in Repert. Spec. Nov. Regni Veg. 50. 81. (1941).

Hubei: South Donghu, 7703 (isotype of *Elaeocarpus alatus* R. Knuth).

See also comment on *Elaeocarpus omeiensis* Rehder & E. H. Wilson in Sargent Pl. Wilson 2: 360. (1915).

Elaeocarpus hainanensis Oliver (ELAEOCARPACEAE)

New species

in Hooker's Icon. Pl. t. 2462. (1896), Bretschneider in Hist. Eur. Bot. Disc. China 2: 780. (1898), Morley in Glasra 3: 79. (1979).

Hainan: Without locality, 8572.

Elaeocarpus japonicus Sieb. & Zucc. (ELAEOCARPACEAE)

Elaeocarpus yunnanensis Brandis ex Tutcher Anonymous in Kew List of Determinations (volume 34).

Yunnan: Mengzi, 10476.

Distributed in China, Vietnam and Japan.

Elaeocarpus prunifolius (Müll.Berol.) Wall. ex Mast. (ELAEOCARPACEAE)

Anonymous in Kew List of Determinations (volume 34).

Yunnan: Simao, 11663.

Elaeocarpus robustus Roxb. (ELAEOCARPACEAE)

Anonymous in Kew List of Determinations (volume 34).

Yunnan: Simao, 12038, 12135.

Elaeocarpus sp. (ELAEOCARPACEAE)

Anonymous in Kew List of Determinations (volume 34).

Yunnan: Simao, 12030, 11623, 12981, 13643, 13656.

Elaeocarpus sylvestris (Lour.) Poir. (ELAEOCARPACEAE)

Elaeocarpus sp. Henry in Trans. Asiat. Soc. Jap. 24 supp: 24. (1896) (List Pl. Formosa). *Elaeocarpus lanceifolius* Henry non Roxb. in Trans. Asiat. Soc. Jap. 24 suppl: 24. (1896) (List Pl. Formosa). *Elaeocarpus decipiens* Henry non F. B. Forbes & Hemsley in Trans. Asiat. Soc. Jap. 24 suppl: 24. (1896) (List Pl. Formosa), Li in Woody Flora of Taiwan 534. fig. 207. (1963).

Taiwan: Wanchin, 508, 823, 1483, 1563. Oluanpi, 945, 1990.

An evergreen tree to 15 metres tall. Native to southern China and Japan (south). The wood of this tree was known locally as *shih-nan* according to Henry. Found in forests at low altitudes in Chinese Taiwan. This species grows on the Chinese slope at the National Botanic Gardens, Glasnevin..

Elaeocarpus varunua Buch.-Ham. ex Mast. (ELAEOCARPACEAE)

Anonymous in Kew List of Determinations (volume 34).

Yunnan: Mengzi, 11468.

02

Elaphoglossum latifolium (Sw.) J. Sm. (POLYPODIACEAE)

Acrostichum latifolium Sw. Henry in Trans. Asiat. Soc. Jap. 24 suppl: 116. (1896) (List Pl. Formosa).

Taiwan: Oluanpi, 1360 (in part).

This taxon is a New World species and the identification is most likely incorrect.

Elaphoglossum marginatum T. Moore (POLYPODIACEAE)

Elaphoglossum fuscopunctatum Christ in Bull. Herb. Boiss. 6: 867. (1898).

Yunnan: On trees on the mountains to the south of Red River at 2,200 metres, 9158 (isotype of *Elaphoglossum fuscopunctatum* Christ).

Elaphoglossum petiolatum (Sw.) Urb. (POLYPODIACEAE)

Acrostichum viscosum Sw. Anonymous in Kew List of Determinations (volume 34).

Yunnan: Simao, 13110.

A New World species, this taxon may belong elsewhere.

Elaphoglossum stelligerum (Wall. ex Baker) T. Moore ex Salmon (POLYPODIACEAE)

Acrostichum yunnanense Baker in Kew Bull. Misc. Inf. 233. (1898). *Elaphoglossum viscosum* Christ in Bull. Herb. Boiss. 6: 867. (1898), non. (Sw.) Schott.

Yunnan: On grassy banks near Mengzi at 1,650 metres, 10310 (type of *Acrostichum yunnanense* Baker).

Elatostema ficoides (Wall.) Wedd. (URTICACEAE)

Forbes & Hemsley in Journ. Linn. Soc. Bot. 26: 482. (1889).

Elatostemma sp. Henry in Trans. Asiat. Soc. Jap. 24 suppl: 89. (1896) (List Pl. Formosa).

Hubei: Antelope Glen, 4169. **Taiwan:** Wanchin, 1617.

Elatostema ichangense H. Schroeter (URTICACEAE) New species

in Repert. Spec. Nov. Regni Veg. 47: 220. (1939).

Elatostema lineolatum Wright. var. *majus* Forbes & Hemsley non Thw. in Journ. Linn. Soc. Bot. 482. (1899).

Hubei: On the side of the Antelope Glen, 2303 (isosyntype). Yichang, 2731 (isosyntype).

This species was collected by the Sino-American Expedition in Shennongjia Forest District in 1980.

Elatostema lineolatum Wright. var. *majus* Wedd. (URTICACEAE)

Forbes & Hemsley in Journ. Linn. Soc. Bot. 26: 482. (1899).

Elatostemma sp. Henry in Trans. Asiat. Soc. Jap. 24 suppl: 89. (1896) (List Pl. Formosa).

Taiwan: Wanchin, 53, 53a.

A subshrub, native to Nepal, Bhutan, India and China. Growing on moist forest floors and in ravines at low to medium altitudes in Chinese Taiwan.

Elatostema papillosum Wedd. (URTICACEAE)

Anonymous in Kew List of Determinations (volume 34).

Yunnan: Mengzi, 11405.

Elatostema platyphyllum Wedd. (URTICACEAE)

Forbes & Hemsley in Journ. Linn. Soc. Bot. 26: 482. (1899).

Elatostemma sp. Henry in Trans. Asiat. Soc. Jap. 24 suppl: 89. (1896) (List Pl. Formosa).

Taiwan: Wanchin, 154.

Elatostema sessile J. R. Forst. & G. Forst. (URTICACEAE)

Forbes & Hemsley in Journ. Linn. Soc. Bot. 26: 483. (1889).

Hubei: Badong, 1791. Yichang, 2235, 2928, 3266a. Liantuo, 2593, 3266. **Yunnan:** Simao, 13029, 13113.

Native to India, Bhutan and western China.

Elatostema spp. (URTICACEAE)

Henry in Trans. Asiat. Soc. Jap. 24 suppl: 53. (1896) (List Pl. Formosa). Anonymous in Kew List of Determinations (volume 34).

Taiwan: Oluanpi, 202. **Yunnan:** Mengzi, 10415, 13676. Simao, 11964, 12001, 12335, 12573.

Elatostema trichocarpum Handel-Mazzetti (URTICACEAE) New species

in Symb. Sin. 7: 148. (1929).

Hubei: Jianshi, 5984 (isosyntype).

Elatostema umbellatum (Sieb. & Zucc.) Blume (URTICACEAE)

Forbes & Hemsley in Journ. Linn. Soc. Bot. 26: 483. (1899).

Hubei: Badong, 4766. Liantuo, 6388. **Yunnan:** Mengzi, 10501, 11229.

Eleocharis acicularis (L.) Roem. & Schult. (CYPERACEAE)

Henry in Journ. China Br. Roy. Asiat. Soc. 22: 261. (1887) (Chinese Names of Plants), C. B. Clarke in Journ. Linn. Soc. Bot. 36: 225. (1903) Henry in Trans. Asiat. Soc. Jap. 24 suppl: 104. (1896) (List Pl. Formosa).

Hubei: Yichang, 2368. South Badong, 7367. **Taiwan:** Kaohsiung, 1839.

A widespread species in subtropical regions of North America and Eurasia. Often a weed in unflooded paddy fields. According to Henry this species was known colloquially in the Three Gorges region as *niu mao chan*.

Eleocharis acutangula (Roxb.) Schult. (CYPERACEAE)

Eleocharis fistulosa (Poir.) Link C. B. Clarke in Journ. Linn. Soc. Bot. 36: 227. (1903).

Hubei: Yichang, 4102, 4102a.

Eleocharis capitata (L.) R. Br. (CYPERACEAE)

Forbes & Hemsley in Journ. Linn. Soc. Bot. 227.

(1904).

Heleocharis capitata (L.) R. Br. Henry in Trans. Asiat. Soc. Jap. 24 suppl: 104. (1896) (List Pl. Formosa).

Taiwan: Kaohsiung, 704, 1042, 1961.

Eleocharis congesta D. Don var. *japonica* (Miq.) T. Koyama (CYPERACEAE)

Eleocharis acicularis R. Br. var. *japonica* Miq. Anonymous in Kew List of Determinations (volume 24).

Hubei: Yichang, 405.

Eleocharis dulcis (Burm. f.) Trin. Ex Hensch. (CYPERACEAE)

Heleocharis plantaginea Boek. Henry in Trans. Asiat. Soc. Jap. 24 suppl: 104. (1896) (List Pl. Formosa). *Eleocharis plantaginea* (Retz.) Roem. & Schult. Clarke in Journ. Linn. Soc. Bot. 36: 228. (1903).

Hubei: Yichang, 4247. **Taiwan:** Kaohsiung, 734.

Eleocharis geniculata (L.) Roem. & Schult. (CYPERACEAE)

Heleocharis capitata (L.) R. Br. Henry in Trans. Asiat. Soc. Jap. 24 suppl: 104. (1896) (List Pl. Formosa).

Taiwan: Kaohsiung, 1797.

Eleocharis palustris (L.) Roem & Schult. (CYPERACEAE)

Clarke in Journ. Linn. Soc. Bot. 36: 227. (1903).

Yunnan: Mengzi at 1,400 metres, 10862.

Eleocharis pellucida J. Presl & C. Presl (CYPERACEAE)

Eleocharis afflata Steud. Clarke in Journ. Linn. Soc. Bot. 36: 226. (1903).

Hubei: Antelope Glen, 3644 (in part). In dry rice fields near Yichang, 4295, 7786.

Eleocharis sp. (CYPERACEAE)

Heleocharis sp. Henry in Trans. Asiat. Soc. Jap. 24 suppl: 104. (1896) (List Pl. Formosa).

Taiwan: Kaohsiung, 1889.

Eleocharis wichurae Bocklr. (CYPERACEAE)

Eleocharis tetraquetra Clarke non Nees in Journ. Linn. Soc. Bot. 36: 228. (1903).

Hubei: Yichang, 2298, 4232.

Elephantopus scaber L. (ASTERACEAE)

Henry in Trans. Asiat. Soc. Jap. 24 suppl: 51. (1896) (List Pl. Formosa).

Hainan: Environs of Kiungchow, 8746. **Taiwan:** Wanchin, 1644. **Yunnan:** Mengzi, 10371.

Widespread in China.

Eleusine coracana (L.) Gaertn. (POACEAE)

Rendle in Journ. Linn. Soc. Bot. 36: 405. (1904).

Hainan: Environs of Kiungchow, 8283. **Yunnan:** In mountain forest to the south of Mengzi on the side of a streamlet at 1,600 metres, 10962. Simao at 1,400 metres, 12431.

This species is widely cultivated in the tropics of the Old World.

Eleusine indica (L.) Gaertn. (POACEAE)

Rendle in Journ. Linn. Soc. Bot. 36: 406. (1904).

Taiwan: Wanshoushan, 1013. Kaohsiung, 1124.

Goosegrass, a pantropic species, in current Chinese herbal medicine this species is used to treat cases of rheumatic arthritis.

Eleutharrhena macrocarpa (Diels) Forman (MENISPERMACEAE) New genus

in Kew Bull. 30: 99. (1975).

Pycnarrhena macrocarpa Diels in Engler. Pflanzenr. iv. 94: 52. (1910), Forman in Kew Bull. 26: 407. (1972).

Yunnan: Simao, in a forested ravine at 1,200 metres, 12810 (type of *Pycnarrhena macrocarpa* Diels).

This genus is based on material gathered near Simao by Henry, though there is a sterile specimen at Kew collected in the Khasi Hills, NE India, by Joseph Hooker and Thomas Thomson.

Eleutherococcus giraldii (Harms.) Nakai (ARALIACEAE) New species

Acanthopanax giraldii Harms. Harms & Rehder in Sargent Pl. Wilson 2: 560. (1916), Zheng in Journ. Wuhan Bot. Res. 2(1): 159. (1984), Bean in Trees & Shrubs 1: 179. (1989).

Hubei: Fang Xian, 6891.

A deciduous shrub to 2.5 metres tall and native to central and northern China. Discovered by Henry in 1888 it commemorates the Italian missionary Fr. Guiseppe Giraldi who collected plants in Shaanxi province between 1890 and 1895 and sent them to Florence.

Eleutherococcus henryi Oliver (ARALIACEAE) New species

in Hooker's. Icon. Pl. 18(1): tt. 1711. (1887,) Henry in Journ. China Br. Roy. Asiat. Soc. 22: 279. (1887) (Chinese Names of Plants); in Notes Econ. Bot. China 9. (1903), Bretschneider in Hist. Eur. Bot. Disc. China 2: 784. (1898), Forbes & Hemsley in Journ. Linn. Soc. Bot. 23: 341. (1887), Anon. in Gard. Chron. ser. 3, 38: 402. fig. 151. (1905), J. H. Veitch in Hortus Veitchii 365. (1906), Nelson in A Heritage of Beauty 312. (2000), O' Brien in Augustine Henry, An Irish Plant Collector in China 19. (2002).

Acanthopanax henryi (Oliv.) Harms in Engler & Prantl. Nat. Pflanzenfam. 3: Abt. 8, 49. (1897); in Bot. Jahrb. 29: 488. (1900), Anon. in Gard. Chron. ser. 3, 38: 402. fig. 151. (1905), Schneider in Illus. Handb. Laubholzk. 2: 424. fig. 289-290. (1909), Stapf in Bot. Mag. cxxxvi. t. 8316. (1910), Pampanini in Nuov. Giorn. Ital. n. ser. 8: 130. (1911), Harms & Rehder in Sargent Pl. Wilson 2: 557. (1916), Zheng in Journ. Wuhan Bot. Res. 2.(1): 160. (1984).

Hubei: Badong, 1711, 2573 (type), 4832. Jianshi, 7907. South Badong, 7908. In glens near Yichang, 7609.

According to Henry this shrub grew on the cliffs at Badong. The bark of the root is an important drug and in Henry's time was used for making medicated wine. The drug extracted from the bark is said to strengthen the sinews and bones, it is used in cases of difficult urination and as an antiinflammatory. Large quantities were exported from Sichuan and Henry stated the plant was colloquially known in the Three Gorges region as *wu chia p'i*. Introduced to cultivation by Wilson from Hubei through Veitch's Coombe Wood nursery in 1900, it first flowered there in August 1905. The Veitch catalogue *New Hardy Plants from Western China* (Autumn 1912) had plants available for 3 shillings and six pence each. Cultivated in Ireland at the John F. Kennedy Arboretum, Malahide Castle and the National Botanic Gardens, Glasnevin.

Eleutherococcus leucorrhizus Oliver (ARALIACEAE) New species

in Hooker's Icon. Pl. xviii. tt. 1711. (1887), Henry in Journ. China Br. Roy. Asiat. Soc. 22: 279. (1887) (Chinese Names of Plants); in Notes Econ. Bot. China 9. (1893), Bretschneider in Hist. Eur. Bot. Disc. China 2: 784. (1898), Forbes & Hemsley in Journ. Linn. Soc. Bot. 23: 342. (1887), Anon. in Gard. Chron. ser. 3, 38: 404. fig. 152. (1905), J. H. Veitch in Hortus Veitchii 365. (1906), Bean in Trees & Shrubs 1: 180. (1989), Nelson in A Heritage of Beauty 313. (2000), O' Brien in Augustine Henry, An Irish Plant Collector in China 19. (2002).

Acanthopanax leucorrhizus (Oliv.) Harms in Engler & Prantl, Nat. Pflanzenfam. 3: Abt. 7, 49. (1897); in Bot. Jahrb. 29: 488. (1900), Bean in Trees & Shrubs. 1: 130. (1914), Zheng in Journ. Wuhan Bot. Res. 2.(1): 160. (1984).

Hubei: Badong, 116, 2580 (type). Glens near Yichang, 7909.

A deciduous shrub to 2.5 metres tall, Henry stated this was the *wu chia p'i* and the root-bark was a drug. Introduced to cultivation by Wilson in 1901 while collecting for Messrs Veitch. Cultivated at the National Botanic Gardens, Glasnevin.

Eleutherococcus leucorrhizus Oliver var. *scaberulus* (Harms & Rehd.) Nakai (ARALIACEAE) New variety

Acanthopanax leucorrhizus (Oliv.) Harms. var. *scaberulus* Harms & Rehder in Sargent Pl. Wilson 2: 558. (1916), Zheng in Journ. Wuhan Bot. Res. 2.(1): 160. (1916).

Hubei: Jianshi, 5950. Fang Xian, 5950 c. South Badong, 5950d., 5950e. Badong, 5950f. **Sichuan:** South Wushan, 5950a.

Eleutherococcus leucorrhizus Oliver var. *setchuenensis* (Harms) C. B. Shang & J. Y. Huang (ARALIACEAE) New variety

Acanthopanax setchuenensis Harms in Bot. Jahrb. 29:

488. (1900), Harms & Rehder in Sargent Pl. Wilson 2: 559. (1916), Zheng in Journ. Wuhan Bot. Res. 2.(1): 160. (1984).

Hubei: Xingshan, 5950b., 6521, 6630. Without locality, 6521a.

The type specimen was collected by A. von Rosthorn in Sichuan in August 1891, Henry had discovered three years previously.

Eleutherococcus nodiflorus (Dunn) S. Y. Hu (ARALIACEAE)

Acanthopanax spinosus Forbes & Hemsley non (L.) Miq. in Journ. Linn. Soc. Bot. 23: 341. (1887), Harms & Rehder in Sargent Pl. Wilson 2: 562. (1916). *Acanthopanax villosulus* Harms in Sargent Pl. Wilson 2: 562. (1916). *Acanthopanax gracilistylus* W. W. Sm. Zheng in Journ. Wuhan Bot. Res. 2.(1): 160. (1984).

Hubei: Yichang, 1224, 3406, 3406a. **Sichuan:** South Wushan, 5890 (syntype of *Acanthopanax villosulus* Harms).

Eleutherococcus rehderianus (Harms.) Nakai (ARALIACEAE) New species

Acanthopanax rehderianus Harms in Sargent Pl. Wilson 2: 561. (1916).

Hubei: Jianshi, 5930.

Eleutherococcus simonii Simon-Louis ex Mouill.) (ARALIACEAE)

Acanthopanax simonii (Simon-Louis ex Mouill.) C. K. Schneid. Harms & Rehder in Sargent Pl. Wilson 2: 559. (1916).

Hubei: Xingshan, 6503. 6503a.

A deciduous shrub to 3 metres tall. According to Bean, this species was first noticed in Europe in the nursery of Simon-Louis, near Metz having been raised from seed sent from China by one of the many French missionaries living in China. It was reintroduced by Wilson through Messrs Veitch in 1901.

Eleutherococcus sp. No. 1. (ARALIACEAE)

Acanthopanax sp. Hemsley in Anonymous in Kew List of Determinations (volume 24).

Hubei: Xingshan, 6518.

Eleutherococcus sp. No. 2. (ARALIACEAE)

Acanthopanax aculeatum (Aiton) Seem. forma Anonymous in Kew List of Determinations (volume 34).

Yunnan: Mengzi, 10158. Simao, 13460.

Eleutherococcus sp. No. 3. (ARALIACEAE)

Acanthopanax spinosus (L. f.) Miq. Harms & Rehder in Sargent Pl. Wilson. 2: 562. (1916).

Yunnan: Mengzi, 10639.

Eleutherococcus sp. No. 4. (ARALIACEAE)

Plectronia sp. Anonymous in Kew List of Determinations (volume 34).

Yunnan: Mengzi, 11154, 11259.

Eleutherococcus trifoliatus (L.) S. Y. Hu (ARALIACEAE)

Acanthopanax aculeatum (Aiton) Seem. Forbes & Hemsley in Journ. Linn. Soc. Bot. 23: 339. (1887), Henry in Journ. China Br. Roy. Asiat. Soc. 22: 267. (1887) (Chinese Names of Plants); in Trans. Asiat. Soc. Jap. 24 suppl: 47. (1896) (List Pl. Formosa). *Acanthopanax trifoliatus* (L.) Voss. Harms & Rehder in Sargent Pl. Wilson.2.(1): 563. (1916), Li in Woody Flora of Taiwan 663. fig. 270. (1963), Zheng in Journ. Wuhan Bot. Res. 2(1): 160. (1984).

Hubei: Yichang, 186, 1092, 2253, 2253a., 2702. Badong, 1011, 2554. Liantuo, 2639. Without locality, 2702a. In the Kan Chi Glen near Yichang (October 1886), 2914. **Taiwan:** Tamsui, 1460. Wanchin, 1516. Oluanpi, 1707, 335. **Yunnan:** Mengzi. 11158a. Simao, 12561, 12770.

A scandent shrub or climber of a very wide distribution from the Himalaya to China, Japan and the Philippines. According to Henry this species was known colloquially in the Three Gorges district as *san chia p'i*. To the Hani (an ethnic group) of Xishuangbanna in southern Yunnan this species is called *juldovq* and the roots are used to strenghten the spleen and in cases of kidney disease. *Eleutherococcus trifoliatus* was collected by the Glasnevin Central China Expedition (GCCE 1) on Emei Shan in Sichuan in September 2002, seedlings from this collection grow on the western end of the Chinese slope at Glasnevin. It was collected again by the 2004 Glasnevin Expedition (GCCE 704) in October of that year by the mouth of a cave in the San You Dong glen above Yichang. Henry, Pratt and Wilson knew this glen as the San Yu Tung (glen of the three pilgrim's) and frequently collected there and wrote of the cave and the nearby extensive colonies of *Primula rupestris* (q.v.).

Ellipanthus glabrifolius Merrill (CONNARACEAE)

Schellenberg in Engler, Pflanzenr. iv. 127: 183. (1938).

Hainan: Without locality, 20, 8566.

Elsholtzia blanda (Benth.) Benth. (LAMIACEAE)

Dunn in Notes Roy. Bot. Gard. Edinburgh 6: 151. (1915).

Yunnan: Mengzi at 1,750 metres, 9250, 9250a., 10070.

The Hani (an ethnic group) of Xishuangbanna in southern Yunnan use the leaves and flowers of this species to treat cases of laryngitis, flu and gasteroenteritis. Native to india, the Himalaya, western China and Indochina.

Elsholtzia bodinieri Vaniot (LAMIACEAE)

Dunn in Notes Roy. Bot. Gard. Edinburgh 6: 153. (1915).

Yunnan: Mengzi at 1,950 metres, 9368, 9368a. Yuanjiang at 1,650 metres, 9368b.

Endemic to China and distributed in the provinces of Guizhou and Yunnan.

Elsholtzia ciliata (Thunb.) Hyl. (LAMIACEAE)

Elsholtzia cristata Willd. Henry in Journ. China Br. Roy. Asiat. Soc. 22: 266. (1887) (Chinese Names of Plants), Forbes & Hemsley in Journ. Linn. Soc. Bot. 26: 277. (1890), Dunn in Notes Roy. Bot. Gard. Edinburgh 6: 151. (1915).

Hubei: Yichang, 135, 159, 955. Badong, 1028 (December 1885), 2852, 4882, 4953. Liantuo, 2607, 2674, 3069.

Henry's Badong collector called this species *po-ho*.

Elsholtzia communis (Collet & Hemsley) Dunn (LAMIACEAE)

in Notes Roy. Bot. Gard. Edinburgh 6: 150. (1915).

Yunnan: Mengzi at 1,500 metres, 9726. Milê, 9892.

Native to Yunnan, Myanmar and Thailand.

Elsholtzia flava (Benth.) Benth. (LAMIACEAE)

Dunn in Notes Roy. Bot. Gard. Edinburgh 6: 149. (1915).

Yunnan: Mountain forests to the north of Mengzi at 2,650 metres, 10239.

Distributed in Nepal, India and China.

Elsholtzia fruticosa (D. Don) Rehder (LAMIACEAE)

Rehder in Sargent Pl. Wilson. 3: 381. (1917).

Elsholtzia polystachya Benth. Dunn in Notes Roy. Bot. Gard. Edinburgh 6: 150. (1915).

Hubei: Fang Xian, 6755. **Yunnan:** Mengzi, 9696. On grassy mountains near Mengzi at 1,500 metres, 11305.

Native to India, Bhutan, Nepal and China.

Elsholtzia heterophylla Diels (LAMIACEAE)

Dunn in Notes Roy. Bot. Gard. Edinburgh 6: 151. (1915).

Yunnan: Mengzi at 1,750 to 2,300 metres, 9550, 9550a., 9550b., 10305. Milê, 9950b.

Distributed in Myanmar and China (Yunnan).

Elsholtzia kachinensis Prain (LAMIACEAE)

Dunn in Notes Roy. Bot. Gard. Edinburgh vi. 152. (1915).

Yunnan: Mountains to the east of Simao at 1,650 metres, 12670.

The Hani (an ethnic group) of Xishuangbanna in southern Yunnan call this species *laolqaol* and use it as a diuretic. Distributed in India, Myanmar, western and central China and northern Indochina.

Elsholtzia ochroleuca Dunn (LAMIACEAE) New species

in Notes Roy. Bot. Gard. Edinburgh 8: 161. (1913).

Yunnan: Mengzi at 1,600 metres, 9136 (isotype).

Endemic to China and distributed in Sichuan and Yunnan.

Elsholtzia pilosa (Benth.) Benth. (LAMIACEAE)

Dunn in Notes Roy. Bot. Gard. Edinburgh vi. 150.

(1915).

Yunnan: Mengzi at 1,950 metres, 9711, 9233.

A widespread species found over much of India, Nepal, Myanmar, western China and Vietnam.

Elsholtzia rugulosa Hemsley (LAMIACEAE)
Dunn in Notes Roy. Bot. Gard. Edinburgh 6: 149. (1915).

Yunnan: Simao at 1,550 metres, 9912.

Discovered by Frederick Bourne of the British Consular Service during the 1880s. Endemic to China.

Elsholtzia sp. (LAMIACEAE)
Anonymous in Kew List of Determinations (volume 34).

Yunnan: Mengzi, 13802.

Elsholtzia stachyodes (Link.) Raizada & H. O. Saxena (LAMIACEAE)

Elsholtzia incisa (Benth.) Benth. Forbes & Hemsley in Journ. Linn. Soc. Bot. 26: 277. (1890), Dunn in Notes Roy. Bot. Gard. Edinburgh 6: 150. (1915).

Hubei: San You Dong Glen, 2889. Liantuo, 2889a. Yichang, 3272. **Yunnan:** Mengzi, 11290.

A common weed of cultivated ground in western Hubei, this species is cultivated in the order beds at the National Botanic Gardens, Glasnevin from collections made by the Glasnevin Central China Expedition (GCCE 523) near Zhenzi town in Xingshan in September 2004. Distributed in Nepal, India, Myanmar and China.

Elymus caninus (L.) L. (POACEAE)
Agropyron caninum (L.) P. Beauv. Rendle in Journ. Linn. Soc. Bot. 36: 431. (1904).

Hubei: Yichang, 197, 402. Badong, 4726. Jianshi, 5873.

Elymus ciliaris (Trin.) Tzvelev (POACEAE)
Agropyron ciliare (Trin.) Franch. Rendle in Journ. Linn. Soc. Bot. 36: 432. (1904).

Hubei: Yichang, 201, 3486.

Elytranthe albida (Blume) Blume (LORANTHACEAE)
Danser in Blumea 2: 43. (1936).

Elytranthe henryi Lecompte in Sargent Pl. Wilson. 3: 317. (1917).

Yunnan: On trees near Simao at 1,600 metres, 11604 (type of *Elytranthe henryi* Lecompte).

Embelia henryi E. Walker (PRIMULACEAE) New species
in J. Wash. Acad. Sci. 27: 200. (1937).

Yunnan: Mengzi at 1,650 metres, 10913 (holotype).

Native to western China and Vietnam. A handsome climber.

Embelia laeta (L.) Mez (PRIMULACEAE)
Mez in Engler, Pflanzenr. iv. 236: 326. (1902).

Hainan: Environs of Haikou, 8008.

Embelia parviflora Wall. ex A. DC. (PRIMULACEAE)
Embelia myrtifolia Hemsley & Mez in Notizbl. Bot. Gart. Berlin, 107. (1901), Mez in Engler, Pflanzenr. iv. 236: 323. (1902).

Yunnan: Mengzi, 9384. Simao at 1,300 metres, 12826 (type of *Embelia myrtifolia* Hemsley & Mez).

The Hani (an ethnic group) of Xishuangbanna in southern Yunnan call this species *ho'qavqnaqpavq*, the roots and stem are used to treat menstrual disorders. Distributed in India, Myanmar, China, Thailand and Vietnam.

Embelia polypodioides Hemsley & Mez (PRIMULACEAE) New species
in Notizbl. Bot. Gart. Berlin-Dahlem. 108. (1901), Mez in Engler, Pflanzenr. iv. 236: 325. (1902).

Yunnan: South of Red River at Kuan Yen Shan, 10060 (isotype). Fengchunling at 2,400 metres, 10060a. In Pingbian, in forests on the Daweishan Range at 1,900 metres, 10060b.

Native to western China and Vietnam.

Embelia procumbens Hemsley (PRIMULACEAE) New species
in Hooker's Icon Pl. 28: t. 2724. (1901), Mez in Engler, Pflanzenr. iv. 236: 326. (1902).

Embelia saxatilis Hemsley in Hooker's Icon. Pl. 28: t. 2724. (1901), Mez in Engler, Pflanzenr. iv. 236: 325. (1902).

Yunnan: Mengzi, growing in mountain forests at 2,600 metres, creeping on the ground, 11160 (type). Mengzi, creeping on wooded cliffs at 2,600 metres, 9793, (isotype of *Embelia saxatilis* Hemsley).

Endemic to China and distributed in the provinces of Sichuan and Yunnan.

Embelia ribes Burm. f. (PRIMULACEAE)
Mez in Engler, Pflanzenr. iv. 236: 303. (1902).

Yunnan: Mengzi, 10690. Simao, 11680.

The Hani (an ethnic group) of Xishuangbanna in southern Yunnan call this species *laolqyulssaq* and the roots and young stem are used to treat cases of acute gastroenteritis and externally as an application to wounds to stop bleeding. A widespread species in Asia.

Embelia scandens (Lour.) Mez (PRIMULACEAE)
Mez in Engler, Pflanzenr. iv. 236: 317. (1902).

Hainan: Environs of Haikou, 8066.

Embelia subcoriacea (C. B. Clarke) Mez (PRIMULACEAE)
Mez in Engler, Pflanzenr. iv. 236: 408. (1902).

Yunnan: Mengzi, 9885.

Embelia vestita Roxb. (PRIMULACEAE)
Pipoly & Chen in Novon 5: 359. (1995).

Embelia prunifolia Mez in Pflanzenr. (Engler) Myrsin. 4. Fam. 236: 316. (1902). *Embelia oblongifolia* Hemsl. Mez

02

in Engler, Pflanzenr. iv. 236: 316. (2002).

Yunnan: Mengzi at 1,690 metres, 9380. Near the Red River to the south of Mengzi, 9380a., 9380d. Mengzi, 9380c., 11327a., (type of *Embelia prunifolia* Mez), 10260c. Mengzi, 11394. Simao, 12776a., 12776b.

Native to Nepal, India, Myanmar, Vietnam and China (Fujian, Guangdong, Guangxi, Guizhou, Hainan, Hunan, Sichuan, Taiwan, Tibet (Xizang), Yunnan and Zhejiang).

Emilia javanica (Burm. f.) C. B. Rob. (ASTERACEAE)

Emilia sonchifolia Henry non (L.) DC. in Trans. Asiat. Soc. Jap. 24 suppl: 55. (1896) (List Pl. Formosa).

Taiwan: Wanshoushan, 799, 1200.

Emilia sonchifolia (L.) DC. (ASTERACEAE)

Forbes & Hemsley in Journ. Linn. Soc. Bot. 26: 449. (1888).

Hubei: Yichang, 2135. Yichang (Xiling) Gorge, 2903. South Donghu, 7595. **Hainan:** Environs of Haikou, 8077.

Emilia sp. (ASTERACEAE)

Anonymous in Kew List of Determinations (volume 34).

Yunnan: Mengzi, 13790.

Emmenopterys henryi Oliver (RUBIACEAE) New genus

in Hooker's Icon. Pl. 19: t. 1823. (1889), Henry in Notes Econ. Bot. China 44. (1893); in Flora & Sylva 1: 218. (1903), Bretschneider in Hist. Eur. Bot. Disc. China 2: 785. (1898), E. Pritzel in Bot. Jahrb. 29: 580. (1901), Henry in Flora & Sylva 217-218. (1903), Pampanini in Nuov. Giorn. Bot. Ital. n. ser. xvii. 718. (1910), Schneider in Illus. Handb. Laubholzk. 2: 1055. (1912), Osborn in Gard. Chron. ser. 3, 86: 267. (1929), Bean in Trees & Shrubs 3: 143. (1936), Barmes in Journ. Roy. Hort. Soc. xcvi. 496. (1971), Morley in Glasra 3: 79. (1979), Zheng in Journ. Wuhan Bot. Res. 2.(1): 188. (1984), Nelson in Pim, The Wood & the Trees 231. (1984); in A Heritage of Beauty 313. (2000), O' Brien in Augustine Henry, An Irish Plant Collector in China 19. (2002), O' Brien in Ir. Garden 13(5): 55. (2004).

Hubei: Badong, 4857 (type), 4999, 5196. Changyang, 7480.

One of Henry's most remarkable finds. Wilson described *Emmenopterys henryi* as one of the most strikingly beautiful trees of the Chinese forests. He stated it was known colloquially in Hubei and Sichuan as *hsiang kou shu* and that it was common in moist forests. Wilson collected material near Yichang from trees 30 metres tall with girths of up to 4 metres. It was not until November 1907 that he secured ripe seeds and it was from W. 622 that plants were raised in Veitch's Coombe Wood nursery. The tree at Kew was obtained from this source in 1913 and by 1929 it was 6 metres tall. In 1919 Kew donated a plant to the National Botanic Gardens, Glasnevin and it is highly likely this was propagated from the Kew tree and so is derived from Wilson 622 from Yichang. Both the Kew and Glasnevin trees are amoungst the finest in cultivation, the Glasnevin tree flowered once, on a single branch high up in the canopy, during the 1990s.

Alas, in the passing of a century this beautiful tree is now endangered and finds refuge in national parks like Emei Shan in Sichuan and in that other botanical wonderland, Shennongjia in Hubei. In all of Badong only two trees remain. One of these trees was visited by members of the Glasnevin Central China Expedition (2002). The tree is estimated to be 250 years old and is 15 metres tall with a girth of 2.8 metres. From the same tree both herbarium specimens and seeds (GCCE204) were collected. The tree at Badong had obviously carried an abundant supply of blooms and held thousands of persistent pink-tinged bracts. Despite the lack of blossoms on trees in Britain and Ireland, *Emmenopterys* is a very conspicuous and handsome tree in spring when its new bronze foliage emerges.

Engelhardia roxburghiana Lindl. (JUGLANDACEAE)

Engelhardia wallichiana Lindl. ex C. DC. Skan in in Journ. Linn. Soc. Bot. 26: 495. (1899). *Engelhardtia wallichiana* Lindl. var. *chrysolepis* (Hance) C. DC. Forbes & Hemsley in Journ. Linn. Soc. Bot. 495. (1899), Rehder & E. H. Wilson Sargent Pl. Wilson. 3: 186. (1917).

Hainan: At Nodoa in the interior of the island, 8444.

Yunnan: Mengzi, 11149. Simao at 1,300 metres, 13049.

Engelhardia sp. (JUGLANDACEAE)

Anonymous in Kew List of Determinations (volume 34).

Yunnan: Simao, 11774, 12699, 12797.

Engelhardia spicata Lechen ex Blume (JUGLANDACEAE)

Engelhardia aceriifolia (Reinw.) Blume Skan in Journ. Linn. Soc. Bot. 26: 495. (1899).

Yunnan: Mengzi, 11269.

Engelhardia spicata Lechen ex Blume var. *integra* (Kurz) W. E. Manning ex Steenis (JUGLANDACEAE)

Engelhardia colebrookeana Lindl. Skan in Journ. Linn. Soc. Bot. 26: 495. (1899), Walsh in Aug. Henry Forst. Herb. 55. (1957).

Yunnan: Mengzi, 10502, 10502a., 11423.

Enhydra fluctuans Lour. (ASTERACEAE)

Anonymous in Kew List of Determinations (volume 24).

Hainan: Environs of Haikou, 7954.

Enkianthus chinensis Franch. (ERICACEAE)

Rehder & E. H. Wilson in Sargent Pl. Wilson. 1: 551. (1913), Zheng in Journ. Wuhan Bot. Res. 2.(1): 166. (1984).

Hubei: Changleping, 6277. Fang Xian, 6612. **Sichuan:** Henry's Chinese collector with Pratt, no locality stated but may be Leshan, Emei Shan, Wa Shan or Kangding, 8861.

A very beautiful shrub to 4 metres tall bearing pendelous racemes of bell-shaped pink striped flowers in early summer. Discovered in Yunnan by Delavay and introduced to cultivation in 1901 by Wilson. As with other members of the genus the autumn colour is spectacular.

Enkianthus serrulatus (E. H. Wilson) C. K. Schneider (ERICACEAE) New species

in Illus. Handb. Laubholzk. 2: 519. (1911), Zheng in Journ. Wuhan Bot. Res. 2.(1): 167. (1984).

Enkianthus quinqueflorus Lour. var. *serrulatus* E. H. Wilson in Gard. Chron. ser. 3, 41: 344. (1907), Fedde in Repert. Nov. Spec. Regni Veg. 6: 235. (1909), Rehder & E. H. Wilson in Sargent Pl. Wilson. 1: 550. (1913).

Sichuan: South Wushan, 5475. **Yunnan:** In mountains forest to the south of Mengzi at 1,650 metres, 11009.

Discovered by Henry in May of 1888 and introduced to cultivation by Wilson in 1900. Henry described his 5475 as a shrub 1 to 2.5 metres tall, his 11,009 from the mountains to the south of Mengzi at 1,650 metres as a bush to 2.5 metres tall.

Ensete sp. (MUSACEAE)

Musa sp. Wright in Journ. Linn. Soc. Bot. 36: 75. (1903).

Yunnan: Simao, 12325.

This plant was cultivated and the enlarged base of the stem was eaten as a vegetable. According to Forbes and Hemsley it may be the widely cultivated *Ensete superbum* (Roxb.) E. E. Cheesm. (syn. *Musa superba* Roxb.).

Entada phaseoloides (L.) Merr. (FABACEAE)

Entada scandens (L.) Benth. Henry in Trans. Asiat. Soc. Jap. 24 suppl: 38. (1896) (List Pl. Formosa).

Taiwan: Oluanpi, 951. **Yunnan:** Mengzi, 11156. Simao, 13008.

A large woody climber, widely distributed in the tropics. According to Henry the enormous pods of this remarkable climber (up to 120 cm long) were known to the indigenous people of Taiwan as *ku-la-li*. He also states that Emil Bretschneider found the seeds for sale in a drug shop in Beijing under the name *mu-yao-tze* and that the seeds could be found on the shores of southern Taiwan, just as he said they are occasionally found on the west coast of Ireland, having drifted from the West Indies.

Enteropogon dolichostachyus (Lag.) Keng. (POACEAE)

Chloris incompleta Roth. Rendle in Journ. Linn. Soc. Bot. 36: 404. (1904).

Taiwan: Kaohsiung, 1151.

A widespread species, native to a large geographical range from Afghanistan, throughout India to southern China and south-east Asia.

Enteropogon unispiceus (F. Muell.) Clayton (POACEAE)

in Journ. Linn. Soc. Bot. 36: 403. (1904).

Taiwan: Kaohsiung, 1119 (isotype). Wanshoushan, 1940 (type of *Enteropogon gracilior* Rendle).

Endemic to Taiwan and found only in the southern part of the island.

Eomecon chionantha Hance (PAPAVERACEAE)

Henry in Notes Econ. Bot. China 49. (1893), Fedde in Engler, Pflanzenr. iv. 104: 207. (1909).

Hubei: South Badong, 3722, 5404.

A rhizomatous perennial herb to 40 cm tall, bearing handsome ovate-heart-shaped leaves and bearing cymes of small white flowers in May and June. Discovered by the Rev'd. B. C. Henry of the American Presbyterian Board of Missions in Guangdong in April 1883. It was introduced to cultivation by Charles Ford, Superintendent of Kong Kong Botanic Gardens who sent living plants to Kew where they first flowered in 1886. Henry stated this was the *hsueh-shui-ts'ao*, the rhizomes are still used in Chinese herbal medicine.

Epigynum auritum (Schneid) Tsiang & P. T. Li (APOCYNACEAE) New species

Tsiang in Sunyatsenia 2: 145. (1934).

Trachelospermum auritum Schneider in Sargent Pl. Wilson. 3: 341. (1916).

Yunnan: Mountains to the south of Simao at 1,300 metres, 12136 (type of *Trachelospermum auritum* Schneider), 12852 (syntype of *Trachelospermum auritum* Schneider).

Distributed in China (Yunnan) and Thailand.

Epilobium angustifolium L. (ONAGRACEAE)

Anonymous in Kew List of Determinations (volume 24).

Hubei: Jianshi, 6031. Without locality 6031a. **Sichuan:** Wa Shan, Henry's collector with Pratt, 8945.

Henry's 8945 from Wa Shan is mixed with *Epilobium* aff. *japonicum* (Miq.) Hausskn.

Epilobium hirsutum L. (ONAGRACEAE)

Forbes & Hemsley in Journ. Linn. Soc. Bot. 23: 307. (1887), Diels in Bot. Jahrb. 29: 485. (1900).

Hubei: Liantuo, 2134, 2257, 2650. Yichang, 2753. Badong, 4843 (in part).

According to Henry this species was known colloquially in the Three Gorges region as *ta p'o chen*.

Epilobium pyrricholophum Franch. & Sav. (ONAGRACEAE)

Epilobium japonicum (Miq.) Haussk. Forbes & Hemsley in Journ. Linn. Soc. Bot. 23: 307. (1887).

Hubei: Badong, 6053, 7320. Fang Xian, 6722.

02

Epilobium aff. ***roseum*** (Schreb.) Schreb. (ONAGRACEAE)

Forbes & Hemsley in Journ. Linn. Soc. Bot. 23: 308. (1887).

Hubei: Badong, 2445.

Epilobium royleanum Hausskn. (ONAGRACEAE)

Epilobium himalayense Hausskn. Anonymous in Kew List of Determinations (volume 34).

Yunnan: South of the Red River from Manmei at 1,900 metres, 9642. Mengzi, 10638.

Native to India, Afghanistan, Nepal and central and western China.

Epilobium sp. nova (ONAGRACEAE)

Erectum, strictum, multiramosum, glabrum, ramis virgatis gracillimis, foliis caulinis linearibus remote calloso-serratis 2-3 pollicaribus, capsulis fere filiformibus, seminibus minutis oblongis laevibus longissime comosis.

Forbes & Hemsley in Journ. Linn. Soc. Bot. 23: 308. (1887).

Hubei: Badong, 1043 (December 1885).

Stated by Forbes and Hemsley to be a distinct new species; but the specimen consists only of the upper part of the plant in fruit.

Epilobium spp. (ONAGRACEAE)

Anonymous in Kew List of Determinations (volume 24 & 34.).

Hubei: Badong, 279 (July 1885), 1037 (December 1885), 1747, 1803, 1806, 1868, 4012, 4846, 5082, 6141. Donghu, 6409, 6429, 6453. Xingshan, 6941. Jianshi, 7521. **Yunnan:** Simao, 12627, 13457. Mengzi, 13723.

Epimedium lishihchenii Stearn (BERBERIDACEAE)

Epimedium macranthum Henry non Morren & Decne. in Journ. China Br. Roy. Asiat. Soc. 22: 243. (1887) (Chinese Names of Plants). *Epimedium membranaceum* K. Meyer ssp. *orientale* Stearn in Journ. Linn. Soc. Bot. LI. 497. (1937—1938).

Sichuan: Yichang 1251 (syntype of *Epimedium membranaceum* K. Meyer ssp. *orientale* Stearn). Badong, 1434 (syntype of *Epimedium membranaceum* K. Meyer ssp. *orientale* Stearn). South Wushan, 5526 (syntype of *Epimedium membranaceum* K. Meyer ssp. *orientale* Stearn). Without locality, 5526a.

Epimedium root is an important ingredient of Chinese herbal medicine. A number of species are grown for this purpose, including *Epimedium saggittatum* (Sieb. & Zucc.) Maxim. Professor H. H. Hu of Beijing informed Stearn, the monographer of *Epimedium* that the same plant was mentioned in Chinese herbals as an aphrodisiac for sheep. It has been used in China since ancient times and is mentioned in the Emperor Shen-nung's herbal *Pen ts'ao king*. Henry stated the plant was called *ying yang huo* and the drug extracted was said to lower blood pressure and to relieve hypertension.

Epipactis helleborine (L.) Crantz. (ORCHIDACEAE)

Anonymous in Kew List of Determinations (volume 24). *Epipactis consimilis* D. Don Rolfe in Journ. Linn. Soc. Bot. 36: 48. (1903).

Hubei: Liantuo, 2117. **Yunnan:** By a riverside near Lunan, 11108.

Epipactis mairei Schltr. (ORCHIDACEAE) New species

Chen & Luo in Acta Phytotax. Sin. 40: 141. (2002).

Epipactis gigantea Rolfe non Dougl. ex Hook. in Journ. Linn. Soc. Bot. 36: 49. (1903).

Hubei: Badong, 705 (July 1885), 1880. Liantuo, 3953. Xingshan, 4518. **Sichuan:** South Wushan, 5633.

A widespread orchid first found by Henry in July 1885 and later collected by Maire near Kunming in Yunnan. This species is also native to Gansu where it was found by Farrer and Purdom, in Sichuan where it was collected near Kangding by Pratt, in Bhutan where it was found by Ludlow and Sherriff and in Myanmar where it was collected by Kingdon Ward.

Epipremnum pinnatum (L.) Engler (ARACEAE)

Engler & Krause in Engler, Pflanzenr. iv. 23B: 60. (1908).

Epipremnum mirabile Schott. Henry in Trans. Asiat. Soc. Jap. 24 suppl: 101. (1896) (List Pl. Formosa).

Taiwan: Kaohsiung, without number. Wanchin, 839.

A large climber on trees, remarkable for its loop-holed and indented leaves. Native to China, Indochina, the Philippines, Polynesia, Malaysia, Papua New Guinea and Australia. This species was collected by the Glasnevin Taiwan Expedition (GTE 7) in Banana Bay Coastal Forest near Kenting in October 2004.

Epipremnum spp. (ARACEAE)

Anonymous in Kew List of Determinations (volume 34).

Yunnan: Mengzi, 11216. Simao, 12728.

Epiprinus siletianus (Baill.) Croizat (EUPHORBIACEAE)

Croizat in Journ. Arnold Arb. 23: 53. (1942).

Homonoia symphylliifolia Merrill non (Kurz.) Hook. f. in Lingn. Journ. Sci. 19: 188. (1940).

Yunnan: Simao, 13161.

Distributed in India, Myanmar, China (Hainan and Yunnan), Thailand and Vietnam.

Epithema taiwanensis S. S. Ying (GESNERIACEAE)

Epithema sp. Henry in Trans. Asiat. Soc. Jap. 24 suppl: 68. (1896) (List Pl. Formosa). *Epithema brunonis* (Wallich) Decaisne var. *fasciculata* C. B. Clarke Anonymous in Kew list of Determinations vol. 24.

Taiwan: Wanshoushan, 1916, on southern part of semishaded and moist coral rock at 200 metres.

Equisetum arvense L. (EQUISETACEAE)

Henry in Journ. China Br. Roy. Asiat. Soc. 22: 269. (1887) (Chinese Names of Plants) Diels in Bot. Jahrb. 29: 209. (1900).

Hubei: Yichang, 182.

The *ti sung* or field horsetail, a relic of the Carboniferous age. A perennial to 0.5 metres tall and native to the north temperate zone (north America, Europe and Asia). In herbal medicine this species is used as a diuretic and to treat inflammation of the lower urinary tract. This species was collected by the Sino-American expedition in Shennongjia in 1980.

Equisetum ramosissimum Desf. (EQUISETACEAE)

Henry in Journ. China Br. Roy. Asiat. Soc. 22: 238. (1887) (Chinese Names of Plants), Diels in Bot. Jahrb. 29: 209. (1900).

Hubei: In a glen near Pingshanba (near Yichang, May 1888), 1588. Yichang, 2297, 4324.

A widespread species, distributed across Europe, east and southern Africa and much of Asia. According to Henry this horsetail was known colloquially as *chieh ts'ao* in the regions of Yichang, Badong and Changyang.

Equisetum ramosissimum Desf. var. *haugelii* (Milde.) Christenh. & Husby (EQUISETACEAE)

Equisetum debile Roxb. ex Vaucher Henry in Trans. Asiat. Soc. Jap. 24 suppl: 117. (1896) (List Pl. Formosa).

Taiwan: Kaohsiung, without number.

Distributed from India to southern Asia, the Philippines, Indonesia and New Caladonia.

Eragrostis atrovirens Trin. ex Steud. (POACEAE)

Eragrostis elegantula Steud. Rendle in Journ. Linn. Soc. Bot. 36: 412. (1904).

Hainan: Environs of Haikou, 8072. **Taiwan:** Kaohsiung, on the east part of the plain on a path in rice fields, 1820.

Eragrostis cilianensis (All.) Vignolo-Lutati ex Jance (POACEAE)

Eragrostis major Host. Rendle in Journ. Linn. Soc. Bot. 36: 416. (1904).

Hubei: Yichang, 172. **Taiwan:** Wanshoushan, 1207.

Widely distributed through the warmer parts of the Old World, often used as a fodder crop.

Eragrostis cylindrica (Roxb.) Arn. (POACEAE)

Eragrostis geniculata Nees & Mey. Rendle in Journ. Linn. Soc. Bot. 36: 414. (1904).

Hainan: Environs of Haikou, 7970.

Native to Vietnam and southern China.

Eragrostis elongata (Willd.) J. Jacq. (POACEAE)

Rendle in Journ. Linn. Soc. Bot. 36: 413. (1904).

Taiwan: Kaohsiung, 1019. Wanshoushan, 1872.

Eragrostis ferruginea (Thunb.) P. Beauv. (POACEAE)

Rendle in Journ. Linn. Soc. Bot. 36: 413. (1904).

Hubei: Yichang, 2743, 4288. **Yunnan:** On grassy mountains to the north of Mengzi at 1,600 metres, without number.

A widespread species, distributed across India (Sikkim), China and Japan.

Eragrostis japonica (Thunb.) Trin. (POACEAE)

Eragrostis tenuissima Sclr. Henry in Journ. China Br. Roy. Asiat. Soc. 22: 257. (1887) (Chinese Names of Plants). *Eragrostis interrupta* Beauv. var. *tenuissima* (Sclr.) Stapf. Rendle in Journ. Linn. Soc. Bot. 36: 415. (1904).

Hubei: Yichang, 137, 2220. Changyang, 7738. **Taiwan:** Tamsui, 1450. Kaohsiung in fields, 2018.

Native to Africa, India, China, Korea and Japan and extending southwards to Australia. According to Henry this species was known colloquially in the Three Gorges region as *luan ts'ao*.

Eragrostis pilosa (L.) Beauv. (POACEAE)

Rendle in Journ. Linn. Soc. Bot. 36: 417. (1904).

Hubei: Badong, 5001.

Widely distributed through tropical and warm regions of the Old World.

Eragrostis spp. (POACEAE)

Anonymous in Kew List of Determinations (volume 34).

Yunnan: Simao, 13188. Mengzi, 13731.

Eranthemum pulchellum Andrews (ACANTHACEAE)

Daedalacanthus nervosus (Vahl.) T. Anderss. Anonymous in Kew List of Determinations (volume 34).

Yunnan: Mengzi, 10189. Simao, 13011.

Eranthemum spp. (ACANTHACEAE)

Anonymous in Kew List of Determinations (volume 34).

Yunnan: Simao, 11940. Mengzi, 13766.

Eremochloa ophiuroides (Munro) Hack. (POACEAE)

Rendle in Journ. Linn. Soc. Bot. 36: 363. (1904).

Hubei: Yichang, 4148.

Distributed through China and Vietnam.

Eremochloa zeylanica (Hack. ex Trimen) Hack. (POACEAE)

Rendle in Journ. Linn. Soc. Bot. 36: 364. (1904).

Hubei: Hills behind Yichang, 569, 3502.

Eria confusa Hook. f. (ORCHIDACEAE)

Anonymous in Kew List of Determinations (volume 34).

Yunnan: Simao, 11743.

Distributed in India, the Himalaya, China (Taiwan and Yunnan) and Indochina.

Eria sp. No. 1. (ORCHIDACEAE)

Henry in Trans. Asiat. Soc. Jap. 24 suppl: 91. (1896) (List Pl. Formosa).

Taiwan: Oluanpi, 1374.

Eria sp. No. 2. (ORCHIDACEAE)

Anonymous in Kew List of Determinations (vol. 34.).

Yunnan: Simao, 12419, 13487, 13583.

Ericaceae (ERICACEAE)

Leucothoe sp. Forbes & Hemsley in Journ. Linn. Soc. Bot. 26: 16. (1889).

Hubei: Badong, 328 (June 1885).

Hemsley stated the material before him was insufficient for description.

Erigeron acris L. (ASTERACEAE)

Forbes & Hemsley in Journ. Linn. Soc. Bot. 23: 417. (1888).

Hubei: Badong, 281 (July 1885), 1755, 1810, 1859, 4019, 6047.

Erigeron alpinus L. (ASTERACEAE)

Anonymous in Kew List of Determinations (volume 34).

Yunnan: Mengzi, 9596, 9658, 10604.

Erigeron banariensis L. (ASTERACEAE)

Erigeron linifolius Willd. Henry in Trans. Asiat. Soc. Jap. 24 suppl: 52. (1896) (List Pl. Formosa).

Hainan: Environs of Kiungchow, 8743. **Taiwan:** Oluanpi, 289, 985. Wanchin, 1588.

Erigeron canadensis L. (ASTERACEAE)

Forbes & Hemsley in Journ. Linn. Soc. Bot. 23: 418. (1888).

Hubei: Yichang, 213, 372, 639, 949, 1877. On Tsui Fu Shan near Yichang (May 1888), 1614. Badong, 360 (June 1885). **Yunnan:** Mengzi, 9819

Canadian fleabane, a native of America and Canada and a colonist of the Old World. In China the dried aerial part of the plant is used as a diuretic. The Hani (an ethnic group) of southern Yunnan call this species *gaqaqalmovq* and the tap roots are used in treating cases of pneumonia and jaundice.

Erigeron sp. (ASTERACEAE)

Anonymous in Kew List of Determinations (volume 24).

Hubei: South Badong, 6100.

Eriobotrya deflexa (Hemsl.) Nakai (ROSACEAE) New species

in Bot. Mag. (Tokyo) 30: 18. (1918), Li in Woody Flora of Taiwan 272. fig. 101. (1963), O' Brien in Augustine Henry, An Irish Plant Collect in China 19. (2002).

Photinia deflexa Hemsley in Ann. of Bot.9: 153. (1895),

Henry in Trans. Asiat. Soc. Jap. 24 suppl: 41.(1896) (List Pl. Formosa), Forbes & Hemsley in Journ. Linn. Soc. Bot. 153. (1892), Bretschneider in Hist. Eur. Bot. Disc. China 2: 783. (1898), Matsum. & Hayata in Journ. Coll. Sci. Univ. Tokyo 22: 129. (1906), Hayata in Ic. Pl. Formos. 1: 246. (1911).

Taiwan: Oluanpi, 282 (isotype), 631, 1333. Wanchin, 498. Kaohsiung, 1026.

A medium sized evergreen tree. Found throughout the broadleaved forests of Taiwan (where it is endemic), from low altitude to 1,800 metres. Henry called this the *k'o* tree. This species was first planted out-of-doors in Europe at Dunloe Castle, Co. Kerry by Roy Lancaster. It also grows happily at the National Botanic Gardens, Kilmacurragh.

Eriobotrya henryi Nakai (ROSACEAE) New species

in Journ. Arnold Arbor. 5: 70. (1924), Vidal in Adansonia 5: 562. pl. 3. 1-2. (1965).

Yunnan: Mountains to the east of Simao at 2,300 metres, 13018 (lectotype, collected 12th June 1898). In the mountains to the east of Simao at 1,650 metres, 11644 (syntype). Simao at 1,540 metres, 11644a., (syntype).

Named by Henry as the *zhai ye pi ba*, a shrub or small tree to 7 metres tall. Distributed in China (Guangxi, Guizhou and Yunnan) and in Myanmar.

Eriobotrya japonica (Thunb.) Lindl. (ROSACEAE)

Forbes & Hemsley in Journ. Linn. Soc. Bot. 23: 261. (1887), Henry in Journ. China Br. Roy. Asiat. Soc. 22: 266. (1887) (Chinese Names of Plants), Zheng in Journ. Wuhan Bot. Res. 2.(1): 69. (1984).

Hubei: Yichang, 3269. South Badong, 5343. On cliffs in the Antelope Glen, 5343a.

An evergreen tree to 7 metres tall, native to China and Japan and introduced to cultivation by Sir Joseph Banks in 1787. The loquat was common in both in Hubei and Sichuan according to Wilson and at Yichang it grew wild on the cliffs of the Three Gorges and according to Henry was known colloquially as the *p'i pa shu* (*pi pa ye*). It is still common around Yichang and in the mountains near Badong, it was collected by the Glasnevin Central China Expedition (GCCE 671) in the Red Dragon Reserve in Xingshan. The leaves are used in Chinese herbal medicine to expel phlegm, to treat nausea, vomiting and hiccups. In the south of Spain and other regions of southern Europe it is a valuable fruit crop, in the British Isles and Ireland it is grown more for its handsome foliage as it fruits only in favoured gardens like Bodnant in Wales and Malahide Castle in County Dublin.

Eriobotrya prinoides Rehder & E. H. Wilson (ROSACEAE) New species

in Sargent Pl. Wilson. 1: 194. (1912), Hand.-Mazz. in Symb. Sin. 7: 477. (1933), Vidal in Adansonia 5: 573. pl. 4. 4-5. (1965); in Fl. Camb., Laos & Viet. 6: 78. pl. 13: 4-5.

(1968).

Yunnan: By a village on the way to Mengzi at 1,600 metres, 9878 (type, collected 29th November 1896).

A handsome tree with fragrant white flowers. Native to Laos and China (Sichuan and Yunnan).

Eriocaulon brownianum Mart. (ERIOCAULACEAE)
Ma in Acta Phytotax. Sin. 29: 298. (1991).

Eriocaulon yunnanense Moldenke in Phytologia 2(7): 222. (1947).

Yunnan: Simao, 12362 (holotype of *Eriocaulon yunnanense* Moldenke).

Native to India, Sri Lanka, western China, Thailand and Vietnam.

Eriocaulon buergerianum Korn. (ERIOCAULACEAE)
Wright in Journ. Linn. Soc. Bot. 36: 198. (1903).

Hubei: Yichang, 37, 2767. Liantuo, 4424.

A stemless annual, distributed through China, Korea and Japan (south). Often found in paddy fields and other such wet places. According to Henry, in the Three Gorges region the genus *Eriocaulon* was known colloquially as *gu jing cao (ku ching ts'ao)*. When used with *long dan cao (Gentiana scabra)* in Chinese herbal medicine this species is used to treat cases of toothache and headache.

Eriocaulon cinereum R. Br. (ERIOCAULACEAE)
Wright in Journ. Linn. Soc. Bot. 36: 199. (1903).

Hubei: Yichang, 7786 (in part).

A stemless annual, widely distributed across Africa, Nepal, India, Sri Lanka, China, Korea, Japan, Vietnam, Indochina, Malaysia, the Philippines and Australia.

Eriocaulon henryanum Ruhland (ERIOCAULACEAE)
New species
in Pflanzenr. (Engler) Eriocaul. 4.(30): 86. (1903), Ma in Acta Phytotax. Sin. 29: 297. (1991).

Yunnan: Mengzi at 1, 650 metres, 9443 (isotype).
Distributed in Yunnan, Thailand and Vietnam.

Eriocaulon nepalense J. D. Prescott ex Bong. var. *luzulifolium* (Mart.) Praj. & J. Parn. (ERIOCAULACEAE)
Wright in Journ. Linn. Soc. Bot. 36: 199. (1903).

Eriocaulon wallichianum Henry non Mart. in Trans. Asiat. Soc. Jap. 24 suppl: 102. (1896) (List Pl. Formosa).

Taiwan: Kaohsiung Plain, 1801.

Eriocaulon sexangulare L. (ERIOCAULACEAE)
Eriocaulon wallichianum Mart. Wright in Journ. Linn. Soc. Bot. 36: 210. (1903).

Hainan: Environs of Haikou, 8073.

Eriocaulon sp. (ERIOCAULACEAE)
Anonymous in Kew List of Determinations (volume 34).

Yunnan: Mengzi, 13744.

Eriochloa villosa (Thunb.) Kunth (POACEAE)
Henry in Journ. China Br. Roy. Asiat. Soc. 22: 256. (1887) (Chinese Names of Plants), Rendle in Journ. Linn. Soc. Bot. 36: 321. (1904).

Hubei: Yichang, 453, 677. Badong, 4787, 5002.

Native to China, Vietnam and Japan. A common grass in the Three Gorges region where Henry stated it was known colloquially as *lo lo erh ts'ao*.

Eriolaena candollei Wall. (MALVACEAE)
Eriolaena glabrescens Hu in Journ. Arnold Arbor. 5: 231. (1924).

Yunnan: Mountains to the south of Simao at 1,300 metres, 12343 (type of *Eriolaena glabrescens* Hu).

Eriolaena spectabilis (DC.) Planch. ex Mast. (MALVACEAE)
Eriolaena szemaoensis Hu in Journ. Arnold. Arb. 5: 230. (1924).

Yunnan: Manpan, 11053. Simao, 12506 (syntypes of *Erioaena szemaoensis* Hu), 12506a., (syntypes of *Erioaena szemaoensis* Hu). Mountains to the west of Simao at 1,650 metres, 11873 (type of *Eriolaena szemaoensis* Hu).

Distributed in India and western China.

Erioscirpus comosus (Wall.) Palla (CYPERACEAE)
Eriophorum comosum (Wall.) Nees A. Henry in Journ. China Br. Roy. Asiat. Soc. 22: 234. (1887) (Chinese Names of Plants), C. B. Clarke in Journ. Linn. Soc. Bot. 36: 256. (1903).

Hubei: Yichang, 575, 4000. Liantuo, 2623. Changyang, 7734.

According to Henry this species was common on the cliffs of the gorges where it hung down in thufts, whence the colloquial name, *ai hu tzu ts'ao*.

Eriosema chinense Vogel (FABACEAE)
Anonymous in Kew List of Determinations (volume 24).

Hainan: Ling-men, 8674.

An erect herb, a widespread species distributed across India, Bangladesh, Myanmar, Vietnam, China, the Philippines and Australia.

Eriosema virgatum Benth. (FABACEAE)
Anonymous in Kew List of Determinations (volume 34).

Yunnan: Mengzi, 10765.

Eriosolena involucrata (Wall.) Van Tieghem (THYMELAEACEAE)
Rehder in Sargent Pl. Wilson. 2: 550. (1916).

Yunnan: Simao at 1,500 to 1,600 metres, 11564, 11564b., 11564c.

Erycibe glaucescens Wallich ex Choisy (CONVOLVULACEAE)
Erycibe laevigata Wallich ex Choisy Anonymous in

Kew List of Determinations (volume 34).

Yunnan: Simao, 11863.

Erycibe henryi Prain (CONVOLVULACEAE) New species

in Journ. Asiat. Soc. Bengal, Pt. 2, Nat. Hist. 73: 15. (1904), Dunn in Journ. Linn. Soc. Bot. 39: 446. (1911), Sasaki in Cat. Govt. Herb. Formosa 424. (1930), Yamamoto in J. Trop. Agric. 6: 552. (1934), How in Sunyatsenia 6: 225. (1946), Li in Woody Flora of Taiwan 809. fig. 326. (1963).

Erycibe sp. nova. Henry in Trans. Asiat. Soc. Jap. 24 suppl: 63. (1896) (List Pl. Formosa).

Taiwan: Wanshoushan, 1859 (type), 1884 (isotype). Oluanpi, 2072.

A scandent shrub, native from southern China to Japan (south). In Chinese Taiwan this species grows at low altitude throughout the island.

Erycibe sp. (CONVOLVULACEAE)

Anonymous in Kew List of Determinations (volume 34).

Yunnan: Simao, 12514.

Erycibe subspicata Wall. ex G. Don (CONVOLVULACEAE)

Anonymous in Kew List of Determinations (volume 34).

Yunnan: Simao, 12739, 13430.

Distributed in India, Myanmar, western China and Indochina.

Eryngium foetidum L. (APIACEAE)

Hemsley in Journ. Linn. Soc. Bot. 34: 475. (1900), Wolff in Engler, Pflanzenr. iv. 228: 203. (1913).

Yunnan: In forests south-east of Simao at 1,100 to 1,250 metres, 12245, 12245a.

Introduced from tropical America (Mexico to Brazil) where it was once used in medicine and for flavouring soups. Widely naturalised in the tropics and subtropics.

Erysimum macilentum Bunge (BRASSICACEAE)

Anonymous in Kew List of Determinations (volume 24 & 34).

Hubei: Yichang, 335, 852, 1565, 3380. Liantuo, 3890. Yunnan: Lunan, 10580.

Endemic to China.

Erythranthe nepalensis (Benth.) G. L. Nesom (PHRYMACEAE)

Mimulus nepalensis Benth. Forbes & Hemsley in Journ. Linn. Soc. Bot. 26: 181. (1890), Grant in Ann. Missouri Bot. Gard. 11: 206. (1924).

Hubei: Badong, 4682, 7445. Yichang, 5897. Henry's Chinese collector with Pratt, no locality stated but may be any of the following, Leshan, Emei Shan, Wa Shan or Kangding, 8878. Yunnan: Mengzi, 10450.

A more or less glabrous perennial, found in wet places

and native to the Himalaya, India, China, Indochina and Japan.

Erythrina sp. (FABACEAE)

Micropteryx sp. Anonymous in Kew List of Determinations (volume 24).

Hubei: North Badong, 7399.

Erythrina variegata L. (FABACEAE)

Erythrina indica Lam. Henry in Notes Econ. Bot. China 61. (1893); in Trans. Asiat. Soc. Jap. 24 suppl: 35. (1896) (List Pl. Formosa).

Taiwan: Kaohsiung, without number. Yunnan: South of Red River, 10061.

Of the 112 species of *Erythrina* worldwide this is by far the commonest and has a large geographical range from east Africa to the Indian Ocean Islands and west to Polynesia. It is also widely planted in subtropical and tropical regions of the world and carries masses of fiery orange-red flowers in winter and early spring before the emergence of leaves. Known as the *tz'u-t'ung'* according to Henry.

Erythrodes chinensis (Rolfe) Schltr. (ORCHIDACEAE)

Erythrodes henryi Schltr. in K. M. Schumann & C. A. G. Lauterbach, Fl. Schutzgeb. Südsee, Nachtr: 87. (1905). *Physurus* sp. Anonymous in Kew List of Determinations (volume 34).

Yunnan: Mengzi, 11844.

Erythropalum scandens Blume (OLACACEAE)

Anonymous in Kew List of Determinations (volume 34).

Yunnan: Simao, 12167, 12748.

Widespread in Asia from India to the Philippines.

Erythroxylum parishii (Hook. f.) ined. (ERYTHROXYLACEAE)

Erythroxylum kunthianum (Wall.) Kurz. Schulz in Engler, Pflanzenr.134: 160. (1907).

Yunnan: South of the Red River on the Fengchunling mountain range at 1,300 to 2,100 metres, 10695, 12048, 12849.

Eualiopsis binata (Retz.) C. E. Hubb. (POACEAE)

Ischaemum angustifolium (Trin.) Hack. ex Oliv. Henry in Trans. Asiat. Soc. Jap. 24 suppl: 108. (1896) (List Pl. Formosa), Rendle in Journ. Linn. Soc. Bot. 36: 364. (1904).

Taiwan: Wanshoushan, 1056.

Euchiton japonicum (Thunb.) Holub (ASTERACEAE)

Gnaphalium japonicum Thunb. Forbes & Hemsley in Journ. Linn. Soc. Bot. 23: 427. (1888).

Hubei: Badong, 1769. Hills behind Yichang, 3465.

Euchresta formosana (Hayata) Ohwi (FABACEAE)

Euchresta horsfieldii Henry non (Lesch.) Benn. in Trans. Asiat. Soc. Jap. 24 suppl: 38. (1896) (List Pl. Formosa).

02

Taiwan: Oluanpi, 1228.

An erect shrub, distributed from China (Taiwan) to the Philippines. In Taiwan it is mostly found on the southern part of the island where Li describes it as being rather scarce.

Eucommia ulmoides Oliver (EUCOMMIACEAE)
New family, genus and species

in Hooker's. Icon. Pl. 20: t. 1950. (1890), Henry in Notes Econ. Bot. China 16. (1893), Oliver in Hooker's Icon. Pl. 24: t. 2361. (1895), Weis in Trans. Linn. Soc. Bot. ser. 2, 3: 243. t. 57. (1892), Bretschneider in Hist. Eur. Bot. Disc. China 2 : 779. (1898), Diels in Bot. Jahrb. 29: 346. (1900); in Kew Bull. Misc. Inf. 14: 89. (1901), Watson in Gard. Chron. ser. 3, 33: 104. (1903), Henry in Flora & Sylva 1: 217. (1903); in Kew. Bull. Misc. Inf. 17: 4. (1904), E. H. Wilson in Journ. Roy. Hort. Soc. 24: 657. (1905), Schneider in Illus. Handb.Laubholzk. I: 424. fig. 270. (1904), Finet & Gagnepain in Bull. Soc. Bot. France. LII. Mem. iv. 23. (1905); Contrib. Fl. As. Or. 2: 23. (1907), Mottet in Rev. Hort. 226. fig. 89. (1909), Pampanini in Nouv. Giorn. Bot. Ital. n. ser. xviii. 114. (1911), Rehder in Moller's Deutsch. Gartn.-Zeit. xxvii. 11. (1912), Parkins in Kew Bull. Misc. Inf. 177. (1921), Hsen & Woon in Icon. Plant. Sin. 1: 26. (1927), Engler in Prantl. Nat. Pflanzenf. 18a, 69. (1930), Varossieau in Blumea 5: 81. (1942), Morley in Glasra 3: 79. (1979), Zheng in Journ. Wuhan Bot. Res. 2.(1): 66. (1984), Nelson in Pim, The Wood & the Trees 231. (1984), Li & Hsu in Acta Phytotax. Sin. 24: 160. (1986), Bean in Trees & Shrubs 2: 138. (1989), Nelson in A Heritage of Beauty 313. (2000), O' Brien in Augustine Henry, An Irish Plant Collector in China 19. (2002).

Ulmus sp. nova. Henry in Journ. China Br. Roy. Asiat. Soc. 22: 274. (1887) (Chinese Names of Plants).

Hubei: Changyang, 3182. Badong, 4683 (isosyntype), 7936.

The *du zhong* or *tu chung*, a tree to 21 metres tall in its native habitat (15 metres tall after a century in cultivation at Glasvevin), discovered by Henry in 1887 when he sent back specimens bearing male flowers to the herbarium at the Royal Botanic Gardens, Kew. Henry stated he never saw truly wild trees but had been informed that wild trees did exist near Fang Xian, a region to the north of Yichang. He went on to state that this dioecious tree was the source of the very valuable Chinese drug *tu-chung* (mentioned by the Emperor Shennong in his herbal). The sole species in the genus and was one of a number of new famlies found by Henry. Henry's original material lacked female flowers and in his description of the genus Daniel Oliver did not place it in any definate family, preferring to suggest a relation to both the Ulmaceae and the Euphorbaceae. On receiving

further material sent to Paris by Farges (from trees cultivated in Sichuan) in 1894 Oliver gave the first description of both male and female flowers. Engler finally placed the genus in the Eucommiaceae in 1919.

Both the bark and leaves of *Eucommia* contain large amounts of gutta percha, an elastic like substance, that when snapped and drawn across exhibit a silvery sheen of innumerable threads of gum. When heated they melt and burn with the characteristic smell of rubber. Farges introduced *Eucommia ulmoides* in 1890 and seedlings were raised at the Jardin Colonial, in the garden of the Faculty of Medicine and by the nursery of Mons. Maurice Vilmorin. In November 1897, Vilmorin presented a plant to the Royal Botanic Gardens, Kew the first to be cultivated in England. By 1901, the Jardin Colonial was growing experimental plots of *Eucommia* in Vietnam and North Africa.

Of the tree received at Kew from Maurice de Vilmorin in 1897, W. J. Bean stated Kew Bull. (1921) that since it had never been trained, the plant had retained a bush like habit, three cuttings taken from the original plant had been pruned and trained and the largest of these was 23 feet (7 metres) tall with a girth of 22 inches (55 cm.) in 1921. At the National Botanic Gardens, Glasnevin two plants were raised from Wilson's 383 from a collection made from a cultivated tree in Badong in April 1908. From information supplied by Sir Frederick Moore, then Keeper of Glasnevin, one of these trees was 15 feet (4.5 metres) tall by 1921 and the other tree was about 8 feet tall (2.5 metres) and bushy. The first tree mentioned still exists in the old Chinese shrubbery and is now about 15 metres tall and is still healthy and vigorous. Two trees raised at the Royal Botanic Garden, Edinburgh from Wilson's 383 obviously received no training as they were of a bushy habit. Both were planted in 1911 and by 1921 measured 8 feet (2.5 metres) and 5 feet (1.75 metres). It would appear that as a young tree *Eucommia* has a spreading, bushy habit and needs judicious pruning and training to produce a good tree. This has certainly been the case with a young tree on the new Chinese slope at the National Botanic Gardens, Glasnevin. Seeds were taken from A. Henry 4381 at the Kew herbarium on May 7th 1888 but no record of how they fared was ever kept. Wilson stated it was taken as a tonic (it has been used in China for over 2,000 years) diuretic, aphrodisiac and a cure for colds. It is also used as tonic for the liver, to stregnthen sinews and bones, to prevent miscarriage and to prevent significant back pain in pregnant women.

According to Henry an enormous trade of *Eucommia* bark existed in late 19th century China, from the central provinces of Hubei, Sichuan and Shaanxi, *tu chung* was brought to Hankow (now Wuhan) and from this port alone,

100 tonnes were exported annually by steamer to other treaty ports. The product commanded a high price and the Hankow exports reaped £18,000 (sterling) in 1888 for the Customs Service. The tree is still cultivated on a large scale in central China. In Badong County members of the Glasnevin Central China Expedition saw vast piles of harvested bark in several areas ready to be shipped down the Yangtze. It is commercially grown in the *Metasequoia* valley alongside another Henry discovery, *Coptis chinensis* Franch. and is sold by vendors on Emei Shan.

Eulalia aurea (Bory) Kunth. (POACEAE)

Erianthus fulvus (R. Br.) Kunth. Rendle in Journ. Linn. Soc. Bot. 36: 350. (1904).

Yunnan: On grassy mountains near Mengzi at 1,500 to 1,900 metres, 9818, 9818a., 10325.

Eulalia phaeotrix (Hack.) Kuntze (POACEAE)

Pollinia phaeothrix Hack. Rendle in Journ. Linn. Soc. Bot. 36: 356. (1904).

Hubei: Yichang, 289.

Eulalia villosa (Spreng.) Nees (POACEAE)

Pollinia quadrinervis Hack. var. *latifolia* Rendle in Journ. Linn. Soc. Bot. 36: 357. (1904).

Yunnan: On grassy mountains near Mengzi at 1,920 metres, 10993.

Eulaliopsis binata (Retz.) C. E. Hubb. (POACEAE)

Ischaemum angustifolium Hack ex Oliv. Rendle in Journ. Linn. Soc. Bot. 36: 364. (1904).

Hubei: Yichang, 568. **Yunnan:** Mengzi at 1,600 metres, 10390. Mengzi, 10391.

Native to Afghanistan, India, China, Korea and the Philipopines.

Eulophia dabia (D. Don) Hochr. (ORCHIDACEAE)

Chen & Luo in acta Phytotax. Sin. 40: 149. (2002). *Eulophia faberi* Rolfe in Journ. Linn. Soc. Bot. 28. (1903).

Hubei: Yichang, 494 (lectotype of *Eulophia faberi* Rolfe), 3589. **Yunnan:** In the valley of the Red River near Manpan, on grassy hills at 2,000 metres, 10612.

Eulophia nuda Lindl. (ORCHIDACEAE)

Rolfe in Journ. Linn. Soc. Bot. 36: 29. (1903).

Yunnan: On grassy mountains near Mengzi at 1,600 metres, 11141.

A widespread species in Asia.

Eulophia picta (R. Br.) Ormerod (ORCHIDACEAE)

Geodorum formosanum Rolfe in Ann.of Bot. 9: 157. (1895), Henry in Trans. Asiat. Soc. Jap. 24 suppl: 92. (1896) (List Pl. Formosa), Bretschneider in Hist. Eur. Bot. Disc. China 2: 792. (1898), Rolfe in Journ. Linn. Soc. Bot. 36: 32. (1903).

Taiwan: Kaohsiung, 1137 (isotype of *Geodorum formosanum* Rolfe). Oluanpi, 1375.

A handsome terrestrial orchid to 30 cm tall, native to Indochina, southern China, Japan (south), the Philippines, through Malaysia to Papua New Guinea and Australia. In Chinese Taiwan this species inhabits semideciduous forests at low altitudes in the southern parts of the island and on Lanyu Island.

Eulophia recurva (Roxb.) M. W. Chase, Kumar & Schuit. (ORCHIDACEAE)

Geodorum dilatatum R. Br. Rolfe in Journ. Linn. Soc. Bot. 36: 32. (1903).

Yunnan: In shady woods at Hsinkai by the Red River, 11103.

Eulophia sp. (ORCHIDACEAE)

Anonymous in Kew List of Determinations (volume 34).

Yunnan: Simao, 12084, 13050.

Eulophia zollingeri (Rchbf.) J. J. Sm. (ORCHIDACEAE)

Cyrtopera formosana Rolfe in Kew Bull. Misc. Inf. 198. (1896), Bretschneider in Hist. Eur. Bot. Disc. China 2: 792. (1898). *Eulophia formosana* (Rolfe) Rolfe in Journ. Linn. Soc. Bot. 36: 28. (1903).

Taiwan: Oluanpi, 1974 (type of *Cyrtopera formosana* Rolfe).

Eumachia straminae (Hutch.) Barrabé, C. M. Taylor & Razafim (RUBIACEAE) New species

Psychotria straminae Hutchinson in Sargent Pl. Wilson. 3: 416. (1917).

Yunnan: In Pingbian in forests on the Daweishan Range at 1,640 metres, 11138 (type of *Psychotria straminae* Hutchinson), 11428, 13461.

Euonymus acanthocarpus Franch. (CELASTRACEAE)

Zheng in Journ. Wuhan Bot. Res. 2.(1): 123. (1984).

Euonymus theifolius Wall. var. *scandens* Loesener in Bot. Jahrb. 30: 455. (1902).

Hubei: Liantuo, 2991, 6360. **Yunnan:** Mengzi at 2,000 to 2,300 metres, 10544.

Euonymus actinocarpus Loesener (CELASTRACEAE) New species

in Bot. Jahrb. Syst. 30: 459. (1902), Zheng in Journ. Wuhan Bot. Res. 2.(1): 123. (1984).

Hubei: Liantuo, 4399 (type).

Euonymus aculeatus Hemsley (CELASTRACEAE) New species

in Kew. Bull. Misc. Inform. 209. (1893), Bretschneider in Hist. Eur. Bot. Disc. China 2: 780. (1898), Loesener & Rehder in Sargent Pl. Wilson. 1: 490. (1913).

Sichuan: South Wushan, 5335a., (type), 6143.

Euonymus alatus (Thunb.) Sieb. (CELASTRACEAE)

Henry in Journ. China Br. Roy. Asiat. Soc. 22: 262.

(1887) (Chinese Names of Plants), Loesener & Rehder in Sargent Pl. Wilson.1: 493. (1913), Zheng in Journ. Wuhan Bot. Res. 2.(1): 123. (1984).

Euonymus subtriflorus Blume Anonymous in Kew List of Determinations (volume 24).

Hubei: Badong, 238 (June 1885). Yichang, 1006, 1071, 1172, 2096, 3087, 3394. North Badong, 7168. Fang Xian, 6708. Jianshi, 7412. Without locality, 3087a. **Sichuan:** North Wushan, 7054. Henry's Chinese collector with Pratt, no locality stated but may be Leshan, Emei Shan, Wa Shan or Kangding, 8797.

A deciduous wide spreading shrub to 2 metres tall and well known in cultivation on account of its elegant habit and its rich scarlet autumn colour. Native to China and Japan. In China the woody "wings" on the stems (which are a feature of his species) are used as a pain killer. According to Henry this species was known colloquially in the Three Gorges region as *pa shu.* Seeds from Wilson's 354a. from Yichang reached Glasnevin in 1908. *Euonymus alatus* was collected by the Glasnevin Central China Expedition (GCCE 494) near Zhenzi town in Xingshan in September 2004.

Euonymus bockii Loes. (CELASTRACEAE)

Euonymus orgyalis W. W. Smith in Notes Roy. Bot. Gard. Edinburgh. 13: 161. (1921), Hu in Journ. Arnold Arb. 35: 197. (1954).

Yunnan: South of the Red River, in forests on Fengchunling at 2,300 metres, 10661 (type of *Euonymus orgyalis* W. W. Smith). In Pingbian, in forests on the Daweishan Range at 1,650 metres, 11403 (isotype of *Euonymus orgyalis* W. W. Smith). Mengzi, 11404.

Euonymus clivicola W. W. Sm. (CELASTRACEAE)

Euonymus elegantissimus Loesener & Rehder in Sargent Pl. Wilson. 1: 496. (1913), Zheng in Journ. Wuhan Bot. Res. 2.(1): 124. (1984).

Hubei: Yichang, 6584.

Euonymus cornutus Hemsley (CELASTRACEAE) New species

in Kew Bull. Misc. Inf. 209. (1893), Bretschneider in Hist. Eur. Bot. Disc. China 2: 780. (1898), Loesener in Bot. Jahrb. xxix. 441. (1900); xxx. 458. (1902), Zheng in Journ. Wuhan Bot. Res. 2.(1): 124. (1984), O' Brien in Augustine Henry, An Irish Plant Collector in China 19. (2002).

Hubei: South Badong, 5442. Jianshi, 5442a., 5954 (type). Without locality, 5954a.

A very handsome shrub and much valued in the garden on account of its elegant fruits. Introduced to cultivation by Wilson in 1908, it was also later reintroduced by Forrest. In Ireland *Euonymus cornutus* is cultivated at the John F. Kennedy Arboretum and the National Botanic Gardens, Glasnevin. This species was collected by the Glasnevin Central China Expedition (GCCE 19 & GCCE 58) on Emei Shan (Mount Omei) in September 2002.

Euonymus cornutus Hemsl. var. *quinquecornutus* (Comber) Blakel (CELASTRACEAE) New variety

Euonymus cornutus Hemsley in Kew Bull. Misc. Inf. 209. (1893), in part. *Euonymus quinquecornuta* Comber in Notes Roy. Bot. Gard. Edinburgh18: 243. (1934).

Hubei: Without locality, 6815a.

The fruits are five-horned as opposed to four-winged in the type. Later collected by Forrest in Yunnan and reintroduced by Farrer from Shaanxi.

Euonymus dielsianus Loesener (CELASTRACEAE) New species

in Bot. Jahrb. Syst. 29: 440. (1900), Schneider in Illus. Handb. Laubholzk. 2: 175. (1912), Zheng in Journ. Wuhan Bot. Res. 2.(1): 124. (1984).

Euonymus dielsianus Loesener var. *latifolius* Loesener in Bot. Jahrb. 30: 455. (1902).

Hubei: Xiling (Yichang) Gorge, 3315 (isosyntype). Without locality, 3315a. Yichang, 3962 (October 1887). South Badong, 7300. **Yunnan:** Mengzi, 10733. Mengzi at 2,180 to 2,200 metres, 10810.

Euonymus echinatus Wall. (CELASTRACEAE)

Euonymus subsessilis Sprague in Kew Bull. Misc. Inf. 32: 34. (1908), Loesener & Rehder in Sargent Pl. Wilson. 1: 489. (1913), Zheng in Journ. Wuhan Bot. Res. 2.(1): 126. (1984).

Hubei: Antelope Glen, 3116. Yichang, 3511 (isosyntype), 3511a., 3511b., (isosyntype of *Euonymus subsessilis* Sprague, October 1887).

Euonymus fortunei (Turcz.) Hand.- Mazz. (CELASTRACEAE)

Zheng in Journ. Wuhan Bot. Res. 2.(1): 124. (1984).

Hubei: South Badong, 6110.

This species is seen at its best when allowed to scale other trees. It thrives particularly well at Wuhan Botanical Gardens in Hubei where it has climbed several mature trees of *Taxodium ascendens* and makes a handsome sight when carrying a heavy crop of fruits. Seedlings raised from the Wuhan plants grow at the National Botanic Gardens, Kilmacurragh from collections made by visiting members of the 2004 Glasnevin Central China Expedition (GCCE 881).

Euonymus giraldii Loesener (CELASTRACEAE)

Euonymus giraldii Loesener var. *ciliatus* Loesener in Bot. Jahrb. 29: 443. (1900).

Hubei: Fang at 1,900 to 3,000 metres, 6815.

Euonymus grandiflorus Wallich ex Roxb. (CELASTRACEAE)

Loesener in Bot. Jahrb. 30: 452. (1902).

Yunnan: Mengzi at 2,000 metres, 9706.

A semi-evergreen tree to 5 metres tall, native to the Himalaya and western China. The plant normally seen in cultivation is *Euonymus grandiflorus* Wallich. ex Roxb. f. *salicifolius* Stapf & F. Ballard, which has longer narrower leaves. There is a fine tree of this form raised from Cooper's collection in Bhutan in the *Euonymus* class at the National Botanic Gardens, Glasnevin.

Euonymus hamiltonianus Wallich (CELASTRACEAE)

Loesener in Bot. Jahrb. 30: 461. (1902), Zheng in Journ. Wuhan Bot. Res. 2.(1): 125. (1984).

Euonymus sieboldianus Blume Henry in Journ. China Br. Roy. Asiat. Soc. 22: 257. 1887) (Chinese Names of Plants). *Euonymus lanceifolius* Loesener in Bot. Jahrb. Syst. 30: 462. (1898), Loesener and Rehder in Sargent Pl. Wilson. 1: 491. (1913). *Euonymus yedoensis* Koehne var. *koehneana* Loesener in Sargent Pl. Wilson. 1: 491. (1913).

Hubei: Badong, 1736, 2517, 2864, 2865, 4057. Changyang, 5264. Fang Xian, 6648. **Sichuan:** Henry's Chinese collector with Pratt, no locality stated but may be Leshan, Emei Shan, Wa Shan or Kangding, 8795. **Yunnan:** Mengzi at 2,600 metres, 11165, (type of *Euonymus lanceifolia* Loesener). Simao, 13411.

The tallest known specimen in cultivation grows at the Royal Horticultural Society's garden at Wisley. When measured in 2000 it was 7 metres tall with a trunk diameter of 50 cm. According to Henry this species was known colloquially in the Three Gorges region as *lu yueh ling*.

Euonymus hemsleyanus Loesener (CELASTRACEAE) New species

in Bot. Jahrb. Syst. 30: 460. (1902).

Yunnan: Mengzi at 1,500 metres, 9120 (isosyntype).

Euonymus hupehensis (Loes.) Loesener (CELASTRACEAE) New species

in Bot. Jahrb. Syst. 30. 454. (1902), Zheng in Journ. Wuhan Bot. Res. 2.(1): 125. (1984).

Euonymus hupehensis Loesener var. *maculatus* Loesener in Bot. Jahrb. Syst. 30: 454. (1902).

Hubei: Changyang, 7764 (isotype of *Euonymus hupehensis* Loesener var. *maculatus* Loesener). **Yunnan:** Simao at 1,650 metres, 12446 (syntype of *Euonymus hupehensis* Loesener var. *maculatus* Loesener).

The tallest known specimen of this specimen in cultivation is a tree at the Royal Botanic Garden, Edinburgh. Raised from Forrest 8496, it was 9 metres tall with a trunk diameter of 35 cm when measured in 1985.

Euonymus japonicus Thunb. (CELASTRACEAE)

Loesener in Bot. Jahrb. 30: 453. (1902).

Hubei: Liantuo, 4479. **Hainan:** On Liang Shan, 16 km (10 miles) east of Haikou, 8198, 8689.

A small shrub or a vigorous climber. Native to China, Japan, the Philippines, Java, Sumatra and Malaysia.

Euonymus kiautschovicus Loes. (CELASTRACEAE)

Euonymus kiautschovica Loes. var. *patens* (Rehder) Loesener in Sargent Pl. Wilson. 1: 486. (1913).

Hubei: Badong, 6690.

An evergreen spreading shrub to 4 metres tall, introduced to cultivation in 1860 by Dr. G. R. Hall who sent seeds to the USA.

Euonymus laxiflorus Champ. ex Benth. (CELASTRACEAE)

Euonymus forbesianus Loesener in Bot. Jahrb. Syst. 30: 475. (1902).

Yunnan: South of the Red River from Manmei at 2,000 metres, 9478. Near Yuanjiang on the slopes of Fengchunling, 10841 (isotype of *Euonymus forbesianus* Loesener).

Euonymus mengtseanus (Loesener) Sprague (CELASTRACEAE) New species

in Kew Bull. Misc. Inf. 35. (1908).

Euonymus theifolia Wallich var. *mengziana* Loesener in Bot. Jahrb. Syst. 30: 455. (1902).

Yunnan: Mengzi, 10684 (type of *Euonymus theifolia* Wallich var. *mengziana* Loesener, isotype of *Euonymus mengtseanus* (Loes) Sprague).

Euonymus microcarpus (Oliver ex Loesener) Sprague (CELASTRACEAE) New species

in Kew Bull. Misc. Inf. 35. (1908), Loesener & Rehder in Sargent Pl. Wilson. 1: 487. (1913), Zheng in Journ. Wuhan. Bot. Res. 2(1): 125. (1984).

Euonymus chinensis Lindley var. *macrocarpa* Oliver ex Loesener in Bot. Jahrb. Syst. 30: 456. (1903).

Hubei: In glens near Yichang, 1397 (isosyntype of *Euonymus chinensis* Lindl. var. *microcarpa* Oliver ex Loesener). At side of rice fields in the Monastery Valley, 1650 (isosyntype). Liantuo, 3073. On the summit of Moji Shan, at 700 metres alt., 3580 (isosyntype), 3099 (isosyntype).

Henry stated this species was plentiful on the summit of Moji Shan.

Euonymus myrianthus Hemsley (CELASTRACEAE) New species

in Kew. Bull. Misc. Inform. 210. (1893), Bretschneider in Hist. Eur. Bot. Disc. China 2: 781. (1898), Loesener & Rehder in Sargent Pl. Wilson. 1: 487. (1913), Zheng in Journ. Wuhan Bot. Res. 2.(1): 125. (1984), Bean in Trees & Shrubs 2: 157. (1992), Nelson in A Heritage of Beauty 313. (2000), O' Brien in Augustine Henry, An Irish Plant Collector in China 21. (2002).

Hubei: South Badong, 5335. Jianshi, 5945 (type). Badong, 6126. Changleping, 7823. **Sichuan:** South Wushan, 5540. North Wushan, 7016.

A very handsome evergreen shrub to 4 metres

tall, carrying masses of orange scarlet fruits in autumn. Introduced to cultivation by Wilson when collecting for the Arnold Arboretum in 1908. Cultivated in several major Irish gardens, it is particularly good at Mount Usher in County Wicklow.

Euonymus nitidus Benth. (CELASTRACEAE)

Euonymus flavescens Loesener in Bot. Jahrb. Syst. 29: 437. (1900). *Euonymus oblongifolius* Loesener & Rehder Zheng in Journ. Wuhan Bot. Res. 2.(1): 125. (1984).

Hubei: Antelope Glen, 3337 (type of *Euonymus flavescens* Loesener, October 1887). Without locality, 3337a.

Euonymus salicifolius Loesener. (CELASTRACEAE)
New species

in Bot. Jahrb. 30: 458. (1902).

Yunnan: In mountains near Simao at 1,650 metres, 11718b., (type). Mountains to the east of Simao at 1,650 metres, 11718.

Euonymus sanguineus Loesener (CELASTRACEAE)
New species

Zheng in Journ. Wuhan Bot. Res. 2.(1): 125. (1894). *Euonymus sanguineus* Loes. var. *orthoneura* Loesener in Bot. Jahrb. 29: 442. (1900). *Euonymus sanguinea* Loes. var. *camptoneura* Loesener in Bot. Jahrb. 29: 442. (1900), Loesener in Sargent Pl. Wilson. 1: 494. (1913).

Hubei: South Badong, 5445, 6183. On the mountains above Jianshi at 1,650 metres, 6039. Xingshan at 1,400 metres, 6507, 6556. **Sichuan:** South Wushan, 5562, 7254.

A deciduous shrub to 7 metres tall, a handsome feature in autumn are the large red, yellow coated fruits and beautiful scarlet autumn colour. Introduced to cultivation by Wilson in 1908.

Euonymus spp. (CELASTRACEAE)

Anonymous in Kew List of Determinations (volume 24 & 34).

Hubei: Liantuo, 2621, 3929, 4398, 4468. Badong, 2805, 3706, 3742, 2822, 4671, 5121, 5122. Yichang, 3221. Jianshi, 5848, 5852 Changleping, 6321. Jianshi, 5978. **Sichuan:** South Wushan, 5737, 7257, 7268. North Wushan, 7082. Henry's Chinese collector with Pratt, no locality stated but may be Leshan, Emei Shan, Wa Shan or Kangding, 8788. **Yunnan:** Mengzi, 9106, 9414, 10268, 10304, 10367, 10435, 11011, 11360, 13631. South of the Red River near Manmei at 1,433 metres, 9631. Simao, 11861, 12023, 12648, 13043, 13376, 13437.

Euonymus tashiroi Maxim. (CELASTRACEAE)

Lu & Yang in Flora of Taiwan 3: 651. (1993). *Euonymus chinensis* Henry non Lindl. in Trans. Asiat. Soc. Jap. 24 suppl: 27. (1886) (List Pl. Formosa). *Euonymus acutorhombifolius* Hayata Li in Woody Flora of Taiwan 472. fig. 177. (1903).

Taiwan: Wanchin, 893. Oluanpi, 2051.

Native to Japan (south) and China (Taiwan).

Euonymus theifolius Wall. (CELASTRACEAE)

Anonymous in Kew List of Determinations (volume 34).

Yunnan: Mengzi, 11336.

Euonymus vagans Wall. (CELASTRACEAE)

Euonymus hupehensis Loesener var. *brevipedunculatus* Loesener in Bot. Jahrb. Syst. 30: 454. (1902).

Yunnan: Mengzi at 1,950 metres, 10514 (syntype *Euonymus hupehensis* Loesener var. *brevipedunculatus* Loesener), 10514a., (isosyntype *Euonymus hupehensis* Loesener var. *brevipedunculatus* Loesener). Simao at 1,610 metres, 13017 (syntype *Euonymus hupehensis* Loesener var. *brevipedunculatus* Loesener).

Euonymus venosus Hemsley (CELASTRACEAE)
New species

in Kew. Bull. Misc. Inform. 210. (1893), Bretschneider in Hist. Eur. Bot. Disc. China 2: 781. (1898), Loesener in Bot. Jahrb. 29: 441. (1900)); xxx. 458. (1902), Loesener & Rehder in Sargent Pl. Wilson. 1: 488. (1913).

Hubei: South Badong, 7284 (type). **Sichuan:** South Wushan, 5778. North Wushan, 7019.

Eupatorium chinense L. (ASTERACEAE)

Anonymous in Kew List of Determinations (volume 24).

Hubei: Badong, 1773 (July 1886). Liantuo, 2687.

Eupatorium fortunei Turcz. (ASTERACEAE)

Eupatorium stoechadosmum Hance Forbes & Hemsley in Journ. Linn. Soc. Bot. 23: 405. (1888)

Hubei: Badong, 2385.

Henry Fletcher Hance stated this species was cultivated in 19th century China on account of the fragrance of its flowers, which is the same as the odour of lavender.

Eupatorium japonicum Thunb. (ASTERACEAE)

Forbes & Hemsley in Journ. Linn. Soc. Bot. 23: 403. (1888), Henry in Trans. Asiat. Soc. Jap. 24 suppl: 51. (1896) (List Pl. Formosa). *Eupatorium wallichii* DC. Henry in Journ. China Br. Roy. Asiat. Soc. 22: 249. (1887) (Chinese Names of Plants).

Hubei: Yichang, 40, 2058, 2714, 2948, 3249. Badong, 362 (June 1885), 874, 917 (November 1885), 4685, 4848.. Liantuo, 4572. Fang Xian, 6776. **Taiwan:** Wanchin, 11, 36, 39, 182. Oluanpi, 281, 597. **Yunnan:** Mengzi, 9328, 10028. Simao, 12477, 12584.

A widespread species found over much of China, Korea and Japan.

Eupatorium lindleyanum DC. (ASTERACEAE)

Forbes & Hemsley in Journ. Linn. Soc. Bot. 23: 404. (1888), Henry in Trans. Asiat. Soc. Jap. 24 suppl: 52. (1896)

(List Pl. Formosa).

Hubei: Yichang, 58, 141, 398, 982. Badong, 518 (September 1885), 4967. Antelope Glen, 3259. **Hainan:** Ling-men, 8633. **Taiwan:** Kaohsiung, without number.

Eupatorium spp. (ASTERACEAE)

Anonymous in Kew List of Determinations (volume 24 & 34). Henry in Trans. Asiat. Soc. Jap. 24 suppl: 54. (1896) (List Pl. Formosa).

Hubei: Changyang, 7500. **Hainan:** Without locality, 8606. **Taiwan:** Oluanpi, 220.

Eupatorium tashiroi Hayata (ASTERACEAE)

Eupatorium reevesii Henry non Wall. ex DC. in Trans. Asiat. Soc. Jap. 24 suppl: 52. (1896) (List Pl. Formosa).

Taiwan: Oluanpi, 633.

Euphorbia atoto G. Forst. (EUPHORBIACEAE)

Henry in Trans. Asiat. Soc. Jap. 24 suppl: 81. (1896) (List Pl. Formosa).

Taiwan: Oluanpi, 248. Kaohsiung, 2001.

Native to India, southern China, Japan, the Philippines, Malaysia, Java and Polynesia. Occurs locally on uplifted coral reefs in coastal areas of southern and eastern Taiwan of China.

Euphorbia esula L. (EUPHORBIACEAE)

Forbes & Hemsley in Journ. Linn. Soc. Bot. 26: 412. (1894).

Hubei: Yichang, 1385.

Euphorbia helioscopia L. (EUPHORBIACEAE)

Forbes & Hemsley in Journ. Linn. Soc. Bot. 26: 413. (1894).

Hubei: Antelope Glen, 3428. Yichang 3428a.

Henry called this the *ze qi*, literally; 'marsh laquer', this species is used in Chinese herbal medicine as a diuretic, to stop coughing and to kill internal parasites.

Euphorbia humifusa Willd. (EUPHORBIACEAE)

Henry in Journ. China Br. Roy. Asiat. Soc. 22: 249. (1887) (Chinese Names of Plants), Forbes & Hemsley in Journ. Linn. Soc. Bot. 26: 414. (1894), Henry in Trans. Asiat. Soc. Jap. 24 suppl: 81. (1896) (List Pl. Formosa).

Hubei: Yichang, 3130. **Taiwan:** Kaohsiung, 1017.

According to Henry this was the *i tzu ts'ao*, a small plant mostly found on the dry, gravelly beds of streams. He also stated it was used locally at Yichang in treating eye diseases.

Euphorbia hypericifolia L. (EUPHORBIACEAE)

Henry in Trans. Asiat. Soc. Jap. 24 suppl: 81. (1896) (List Pl. Formosa).

Taiwan: Oluanpi, 212.

Euphorbia milii Des. Moul. (EUPHORBIACEAE)

Euphorbia bojeri Hook. Henry in Trans. Asiat. Soc. Jap. 24 suppl: 81. (1896) (List Pl. Formosa).

Taiwan: Wanchin, cultivated 426.

The crown of thorns, a native of Madagascar.

Euphorbia parviflora L. (EUPHORBIACEAE)

Euphorbia pilulifera L. Anonymous in Kew List of Determinations (volume 24). Henry in Trans. Asiat. Soc. Jap. 24 suppl: 81. (1896) (List Pl. Formosa), Forbes & Hemsley in Journ. Linn. Soc. Bot. 26: 416. (1894).

Hainan: Environs of Haikou, 8055, 8422. Without locality, 8585. Ling-men (in the interior of the island), 8652. **Taiwan:** Wanchin, 178. Oluanpi, 268. Kaohsiung, 800.

According to Henry, this species was known as *ju-a-ts'ao* and the milky juice of the plant was said to be applied to the eye in cases of corneal opacity and in other cases to increase the secretion of milk, when taken internally with pork. Native from India to China and Indochina.

Euphorbia pekinensis Rupr. (EUPHORBIACEAE)

Forbes & Hemsley in Journ. Linn. Soc. Bot. 26: 415. (1894).

Hubei: Yichang, 436, 476, 837, 2934, 3648, 6190.

Henry called this the *da ji* or *jing da ji*, (literally, big lance from the capital). Peking spurge.

Euphorbia pilosa L. (EUPHORBIACEAE)

Forbes & Hemsley in Journ. Linn. Soc. Bot. 26: 416. (1894).

Hubei: Yichang, 696, 818, 2588a. Badong, 2448, 4011, 4746. **Sichuan:** South Wushan, 5556. Without locality, 5556a., 5556b., 5556c.

Euphorbia sieboldiana C. Morren & Decne (EUPHORBIACEAE)

in Journ. Linn. Soc. Bot. 26: 412. (1894).

Euphorbia erythraea Hemsley in Journ. Linn. Soc. Bot. 26: 412. (1894). *Euphorbia hippocrepica* Hemsley in Journ. Linn. Soc. Bot. 26: 414. (1894), Bretschneider in Hist. Eur. Bot. Disc. China 2: 791. (1898). *Euphorbia henryi* Hemsley in Journ. Linn. Soc. Bot. 26: 413. (1894), Bretschneider in Hist. Eur. Bot. Disc. China 2: 791. (1898), Morley in Glasra 3: 79. (1979).

Hubei: Yichang, 1140, 1261(syntype of *Euphorbia erythraea* Hemsley), 1275, 3429a. Xiling (Yichang) Gorge, 3429, 3430. Antelope Glen, 3432. Changyang, 7853 (syntype of *Euphorbia hippocrepica* Hemsley), 7902. Yichang, 7867.

Euphorbia spp. (EUPHORBIACEAE)

Anonymous in Kew List of Determinations (volume 24 & 34).

Hubei: Yichang, 435, 972, 1259. **Yunnan:** Mengzi, 9121, 9237, 9331, 10115. Simao, 13416.

Euphorbia stricta L. (EUPHORBIACEAE)

Euphorbia serrulata Thuill. Forbes & Hemsley in Journ. Linn. Soc. Bot. 26; 417. (1894), Henry in Trans. Asiat. Soc. Jap. 24 suppl: 81. (1896) (List Pl. Formosa).

Hainan: Environs of Kiungchow, 8312. Ling-men (in the interior of the island), 8632. **Taiwan:** Kaohsiung, without number. Wanchin, 888.

Euphorbia thymifolia L. (EUPHORBIACEAE)

Henry in Trans. Asiat. Soc. Jap. 24 suppl: 81. (1896) (List Pl. Formosa).

Taiwan: Wanchin, 871. Oluanpi, 1319.

Pantropical in distribution, a common seashore plant in Taiwan.

Euphorbia tirucalli L. (EUPHORBIACEAE)

Henry in Trans. Asiat. Soc. Jap. 24 suppl: 82. (1896) (List Pl. Formosa).

Taiwan: Wanshoushan, without number.

A succulent spineless tree to 10 metres tall, native to tropical East and South Africa and naturalised in coastal areas of southern Taiwan of China. According to Henry this species was a very common shrub or small tree in 19th century Taiwan (it still is in parts today) and was known as *tan-nien* and at Xiamen (Amoy) as *t'ieh-shu*. According to Rehder and E. H. Wilson (in *A Monograph of Azaleas* 4. (1921)) there was a strong trade during the Han Dynasty (B.C. 206-A.D. 25) between China and India and Zanzibar. From this trade many alien plants were introduced to China including the cactus-like tree euphorbias. Many of these were used as hedge plants and were commonly naturalised around Hong Kong and Kaohsiung in the late 19th century.

Euphorbiaceae (EUPHORBIACEAE)

Anonymous in Kew List of Determinations (volume 24 & 34).

Hubei: Yichang, 682. **Yunnan:** Mengzi, 10561, 10867. Simao, 11903, 12156, 12160.

Euploca strigosa (Willd.) Diane & Hilger (BORAGINACEAE)

Heliotropium strigosum Willd. Henry in Trans. Asiat. Soc. Jap. 24 suppl: 63. (1896) (List Pl. Formosa).

Hainan: Environs of Haikou, 8158. 8433. **Taiwan:** Kaohsiung, 1105, 1790.

Euptelea pleiosperma Hook. f. & Thoms. (EUPTELEACEAE)

Rehder & E. H. Wilson in Sargent Pl. Wilson. 1: 313. (1913), in part. Zheng in Journ. Wuhan Bot. Res. 2.(1): 36. (1984).

Euptelea franchetii Van Teigh. Rehder & E. H. Wilson in Sargent Pl. Wilson 2: 314. (1916).

Hubei: Donghu, 6455 (March 1888). Fang Xian, 6918. South Badong, 7337. **Sichuan:** South Wushan, 7232. **Yunnan:** Mountains to the north of Mengzi, in woods at 2,300 metres, 10746.

A deciduous tree to 10 metres tall, native to the eastern Himalaya, Myanmar and central and western China. Discovered by William Griffith in the Mishmi Hills, NE India and described in 1864. *Euptelea pleiosperma* was introduced to cultivation by Père Farges who sent seeds to France where tree first flowered at Les Barres in 1900. According Wilson this was one of the most common of all small trees throughout western Hubei and eastern Sichuan at the turn of the 20th century. It is found mainly along streams and on the margins of moist woods and thickets. Forrest also collected this species in north-west Yunnan in June 1917 (Forrest 13917). Plants currently cultivated at the National Botanic Gardens, Glasnevin were raised from collections made by the Glasnevin Cental China Expedition (GCCE 91) on Emei Shan (Mount Omei) in September 2002.

Eurya acuminata DC. (PENTAPHYLACACEAE)

Rehder & E. H. Wilson in Sargent Pl. Wilson. 2: 400. (1915), Kobuski in Ann. Missouri Bot. Gard. 25: 321. (1938).

Yunnan: Mengzi, in woods at 1,600 to 2,300 metres, 9039d., 11171 In forests near Mengzi at 1,600 to 2,000 metres, 11414a. Mengzi, 10372. Yuanjiang, 13312.

A widespread species found over much of India, Sri Lanka, Nepal, Myanmar, southern China, Java, Sumatra, the Malay Peninsula and the Philippines.

Eurya chinensis R. Br. (PENTAPHYLACACEAE)

Rehder & E. H. Wilson in Sargent Pl. Wilson. 2: 400. (1916), Kobuski in Ann. Missouri Bot. Gard. 25: 319. (1938), Li in Woody Flora of Taiwan 589. (1963).

Eurya japonica Henry non Thunb. in Trans. Asiat. Soc. Jap. 24 suppl: 19. (1896) (List Pl. Formosa). *Eurya acuminata* Rehder & E. H. Wilson in part. non De Candolle in Sargent Pl. Wilson. 2: 400. (1915).

Taiwan: Wanchin, 196, 535. Oluanpi, 375.

Native to Sri Lanka and southern China. Common at low altitudes throughout Taiwan.

Eurya emarginata (Thunb.) Makino (PENTAPHYLACACEAE)

Li in Woody Flora of Taiwan 90. (1963).

Eurya japonica Henry non Thunb. in Trans. Asiat. Soc. Jap. 24 suppl: 19. (1896) (List Pl. Formosa).

Taiwan: Oluanpi, 1004.

A small tree or shrub, native to China, Korea to Japan. Common in Chinese Taiwan where it is scattered throughout the island at low altitude.

Eurya groffii Merrill (PENTAPHYLACACEAE)

Eurya acuminata De Candolle var. *multiflora* Rehder & E. H. Wilson non (DC.) Blume in Sargent Pl. Wilson. 2: 401.(1915). *Eurya acuminata* DC. var. *groffii* (Merrill) Kobuski in Ann. Missouri Bot. Gard. 25: 325. (1938).

Yunnan: Mengzi, 9021. In mountains near Simao at 1,600 metres, 10914b.

Endemic to China.

Eurya henryi Hemsley (PENTAPHYLACACEAE) New species

in Hooker's Icon. Pl. tt. 2761. (1903), Melchior in Engler & Prantl, Nat. Pflanzenfam. ed. 2, 21: 148. (1925).

Eurya distichophylla Hemsl. var. *henryi* (Hemsley) Kobuski in Ann. Missouri Bot. Gard. 25: 330. (1938).

Yunnan: Mountain to the east of Mengzi at 2,300 metres, 11342, (type).

Eurya nidita Korthals (PENTAPHYLACACEAE)

Kobuski in Ann. Missouri Bot. Gard. 25: 310. (1938), Zheng in Journ. Wuhan Bot. Res. 2.(1): 151. (1984).

Eurya japonica Henry non Thunb. in Journ. China Br. Roy. Asiat. Soc. 22: 234. (1887) (Chinese Names of Plants); in Trans. Asiat. Soc. Jap. 24 suppl: 19. (1896) (List Pl. Formosa). *Eurya japonica* Thunb. var. *nitida* Dyer Rehder & E. H. Wilson Sargent Pl. Wilson 2: 398. (1915). *Eurya japonica* Thunb. var. *aurescens* Rehder & E. H. Wilson in Sargent Pl. Wilson 2: 399. (1915). *Eurya nitida* Korthals var. *aurescens* (Rehder & E. H. Wilson) Kobuski in Ann. Missouri Bot. Gard. 25: 314. (1938), Zheng in Journ. Wuhan Bot. Res. 2.(1): 152. (1984).

Hubei: Liantuo, 1907, 3187 (February 1887). Yichang, 2344, 7946. Badong, 3673 (October 1887), 5167, 5162 (May 1888), 5167, 5170. Fang Xian, 6693. Changleping, 7830. **Sichuan:** South Wushan, 5616, 7099. **Hainan:** Environs of Kiungchow, 8721. **Taiwan:** Wanchin, 20, 21, 122. Oluanpi, 1987. **Yunnan:** Mengzi, in woods at 1,600 to 2,000 metres, 9039, 9039a.

A widespread shrub or small tree with a range from India to Indochina, China, the Philippines and south to Java and Borneo. Known as *yeh ch'a* at Yichang according to Henry.

Eurya obliquifolia Hemsley (PENTAPHYLACACEAE) New species

in Hooker's Icon. Pl. 28: tt. 2761. (1903), Kobuski in Ann. Missouri Bot. Gard. 25: 315. (1938).

Yunnan: In Pingbian, in forests on the Daweishan Range at 1,650 metres, 10914 (syntype), 10914a., (syntype).

Endemic to Yunnan.

Eurya sp. (PENTAPHYLACACEAE)

Eurya japonica Thunb. Anonymous in Kew List of Determinations (volume 24).

Hubei: Badong, 1710, 3728, 4988, 5183. Liantuo, 2646, 4414. Yichang, 2970. Changleping, 6298, 7824, 7826.

Either *Eurya nitida* Korthals or *Eurya nitida* Korthals var. *aurescens* (Rehd. & Wils.) Kobuski

Eurysolen gracilis Prain (LAMIACEAE)

Dunn in Notes Roy. Bot. Gard. Edinburgh 6: 190. (1915).

Yunnan: Simao, 12913.

Distributed in India, Myanmar and China (Yunnan).

Euscaphis japonica (Thunb.) Kanitz (STAPHYLEACEAE)

Rehder & E. H. Wilson in Sargent Pl. Wilson. 2: 187. (1916), Zheng in Journ. Wuhan Bot. Res. 2.(1): 126. (1984).

Euscaphis staphyleoides Sieb. & Zucc. Forbes & Hemsley in Journ. Linn. Soc. Bot. 23: 143. (1886), Henry in Journ. China Br. Roy. Asiat. Soc. 22: 246. (1887) (Chinese Names of Plants).

Hubei: Yichang, 148, 687, 844. Badong, 1425, 1732, 2578, 7394. South Badong, 5767c. Liantuo, 1912, 2610, 3001, 3860, 3916. **Sichuan:** South Wushan, 5767, 5767b.

A common deciduous shrub to 3 metres tall in Hubei and Sichuan where Henry stated it was colloquially known as *hung-liang*. The flowers are greeny-white and the black seeds were used as a drug according to Henry. Native to China, Korea and Japan, it is best grown in the milder counties of the British Isles and Ireland (USDA 9).

Eustigma lenticellatum C. Y. Wu (HAMAMELIDACEAE) New species

in Flora Yunnanica 1: 130. (1977).

Sycopsis griffithiana Oliv. Rehder & E. H. Wilson in Sargent Pl. Wilson. 1: 431. (1913), Walker in Journ. Arnold Arb. 25: 336. (1944).

Yunnan: Mengzi, 11464 (type).

Distributed from Inidia (Assam) to China (south-east Yunnan).

Eutrema violifolium (H. Lév.) Al-Shehbaz & Warwick (BRASSICACEAE) New species

Neomartinella violifolia (H. Leveille) Pilger Al-Shehbaz in Novon 10: 338. (2000).

Hubei: Badong, 5439.

Discovered by Henry in 1888 and described from material later collected in Guizhou in March 1898 by Martin and Bodinier.

Eutrema yunnanense Franch. (BRASSICACEAE)

Schulz in Engler, Pflanzenr. iv. 105: 37. (1924).

Sichuan: South Wushan, 5710.

Discovered by Père Jean Marie Delavay in May 1887.

Evolvulus alsinoides (L.) L. (CONVOLVULACEAE)

Forbes & Hemsley in Journ. Linn. Soc. Bot. 26: 166. (1890), Henry in Trans. Asiat. Soc. Jap. 24 suppl: 65. (1896) (List Pl. Formosa).

Hubei: Yichang, 25. On Tsui Fu Shan near Yichang (May 1888), 1609. **Hainan:** Environs of Haikou, 8048, 8355. **Taiwan:** Oluanpi, without number. Kaohsiung, 1091.

Generally spread in tropical and subtropical countries.

Exacum tetragonum Roxb. (GENTIANACEAE)

Anonymous in Kew List of Determinations (volume 34).

Yunnan: Mengzi, 11177. Simao, 12441.

A widespread species, native to Nepal, Sikkim, Myanmar, China (meadows of Guangdong, Guangxi, Guizhou, Jiangxi and Yunnan), Laos, Vietnam, Malaysia, the Philippines, New Guinea and Australia.

Exallage auricularia (L.) Bremek. (RUBIACEAE)

Hedyotis auricularia L. Anonymous in Kew List of Determinations (volume 24 & 34).

Hainan: 32 km (20 miles) west of Haikou, 8137. Environs of Haikou, 8321, 8430. Without locality, 8506. **Yunnan:** In Pingbian, in forests on the Daweishan Range at 1,650 metres, 11044.

A widespread species found across most of Asia.

Exallage vestita (Willd.) Nandikar & K. C. Kishor (RUBIACEAE)

Hedyotis vestita R. Br. ex G. Don Anonymous in Kew List of Determinations (volume 34).

Yunnan: Simao, 12193.

Exbucklandia populnea (R. Br. ex Griff.) R. W. Brown (HAMAMELIDACEAE)

Bucklandia populnea R. Br. Anonymous in Kew List of Determinations (volume 34).

Yunnan: Mengzi, 11068.

Distributed in Nepal, India, Bhutan, western China, Thailand and Vietnam.

Excoecaria agallocha L. (EUPHORBIACEAE)

Henry in Trans. Asiat. Soc. Jap. 24 suppl: 85. (1896) (List Pl. Formosa).

Taiwan: Kaohsiung, without number.

A small tree, native to the tropical seashores of Asia, Australia and western Polynesia. Henry collected this specimen on edge of the Kaohsiung Lagoon.

Excoecaria cochinchinensis Lour. (EUPHORBIACEAE)

Excoecaria orientalis Pax & K. Hoffm. in Engler, Pflanzenr. 52: 160. (1912).

Taiwan: Wanshoushan, 1857 (isosyntype of *Excoecaria orientalis* Pax & K. Hoffm).

Excoecaria formosana (Hayata) Hayata (EUPHORBIACEAE) New species

in Icon. Pl. Form. 3: 173. (1913) Li in Woody Flora of Taiwan 425. (1963).

Excoecaria sp. Henry in Trans. Asiat. Soc. Jap. 24 suppl: 85. (1896). *Excoecaria crenulata* Wight. var. *formosana* Hayata in Journ. Coll. Sci. Univ. Tokyo. 30 (1): 271. (1911) (Mat. Fl. Formosa).

Taiwan: Wanchin, 94, 1684. Oluanpi, 263, 1307.

A low growing shrub, native to Japan and China (Taiwan) where it grows in forests on the southern part of the island.

Fabaceae (FABACEAE)

Leguminosae Anonymous in Kew List of Determinations (volume 34).

Yunnan: Mengzi, 13821.

Faberia silhetensis (DC. ex Froel.) Sennikov (ASTERACEAE)

Crepis silhetensis Hook. f. Anonymous in Kew List of Determinations (volume 34).

Yunnan: Mengzi, 10271.

The genus commemorates the Rev. Ernst Faber, a German missionary and colleague of Augustine Henry.

Fagopyrum cymosum (Trevir.) Meissn. (POLYGONACEAE)

Henry in Journ. China Br. Roy. Asiat. Soc. 22: 238. (1887) (Chinese Names of Plants).

Polygonum cymosum Trevir. Forbes & Hemsley in Journ. Linn. Soc. Bot. 26: 337. (1891), Henry in Notes Econ. Bot. China 23. (1893).

Hubei: Yichang, 2716. South Badong, 7329. Changyang, 7688.

According to Henry, the colloquial name at Badong for this buckwheat was *ch'iao tang kui*, around Yichang it occured wild in the glens and was cultivated as a medicinal plant in the mountainous areas of Changyang and Badong.

Fagopyrum esculentum Moench. (POLYGONACEAE)

Henry in Journ. China Br. Roy. Asiat. Soc. 22: 238. (1887) (Chinese Names of Plants).

Hubei: Yichang, 463, 464, 1369. Badong, 1701, 2809, 6122.

Buckwheat, a spring annual to 70 cm tall, bearing pink or white flowers above triangular leaves. Native to central and northern Asia, it is cultivated in many parts of the world as a cereal crop. According to Henry this species was known as *t'ien ch'iao me* and cultivated at Yichang and Badong. Henry's 6122 from Badong is a selected variety for autumn harvest and as called *fu ch'iao*. An annual cereal crop thought to be derived from the perennial *Fagopyrum cymosum* (Trevir.) Meissn, also grown as a food crop. Buckwheat has been grown in China for at least 1,500 years now and reached Europe in the 15th century.

Fagopyrum gracilipes (Hemsl.) Dammer (POLYGONACEAE) New species

in Bot. Jahrb. Syst. 29: 315. (1901).

Polygonum gracilipes Hemsley in Journ. Linn. Soc. Bot. 26: 340. (1891.)

Hubei: Badong, 1807, 4742, 4789 (syntype of *Polygonum gracilipes* Hemsley), 5057.

Fagopyrum spp.(POLYGONACEAE)

Anonymous in Kew List of Determinations (volume 24 & 34).

Hubei: Badong, 947 (November 1885). **Yunnan:** Mengzi, 9278.

Fagopyrum tataricum (L.) Gaertn. (POLYGONACEAE)
Henry in Journ. China Br. Roy. Asiat. Soc. 22: 238. (1887) (Chinese Names of Plants).

Hubei: Yichang, 1351.

According to Henry this buckwheat was cultivated and known colloquially at Yichang as *k'u ch'iao me*. It thrives in poor soil, needs a short growing season and does well in the upland cultivated regions of Tibet (Xizang) and other parts of the Himalaya.

Fagus engleriana Seemen (FAGACEAE) New species
in Bot. Jahrb. 29: 285. fig. a-d. (1900), Henry in Elwes & Henry, Trees Gt. Brit. & Irel. 1: 5. (1906), Rehder & E. H. Wilson in Sargent Pl. Wilson. 3: 190. (1916), Bean in Trees & Shrubs 3: 160. (1936), Walsh in Aug. Henry Forst. Herb. 63. (1957), Y. T. Chang & C. C. Huang Acta Phytotax. Sin. 26: 114. (1988), Li & Zhang in Journ. Wuhan Bot. Res. 16: 243. (1998), Nelson in A Heritage of Beauty 313. (2000), O' Brien in Augustine Henry, An Irish Plant Collector in China 21. (2002).

Fagus sylvatica L. var. *longipes* Oliver in Hooker's. Icon. Pl. t. 1936. (1890), in textu., Bretschneider in Hist. Eur. Bot. Disc. China 2: 791. (1898). *Fagus sylvatica* L. var. *bracteolis* involucri exterioribus anguste spatulatim dilatatis, Oliver in Hooker's Icon. Pl. 20: tt. 1936. (1890), Skan in Journ. Linn. Soc. Bot. 26: 525. (1899).

Hubei: Fang Xian, 6797 (type)

A tree to about 23 metres tall, with a girth of up to 2 metres. Discovered by Henry in the mountainous terrain of Fang in northern Hubei in the autumn of 1888. Introduced to cultivation by Wilson through the Arnold Arboretum in 1907. There is a fine young tree in the arboretum at Glasnevin and there are good specimens at Birr Castle in County Offaly and at Mount Usher in County Wicklow. The tallest known specimen in Britain and Ireland grows at Westonbirt in Gloucestershire, when measured in 2002 it was 18 metres tall with a trunk diameter of 59 cm.

Fagus lucida Rehder & E. H. Wilson (FAGACEAE) New species
in Sargent Pl. Wilson. 3: 191. (1916), Walsh in Aug. Henry Forst. Herb. Cat. 63. (1957), Zheng in Journ. Wuhan Bot. Res. 2.(1): 24. (1984), Bean in Trees & Shrubs 2: 176. (1992), O' Brien in Augustine Henry, An Irish Plant Collector in China 21. (2002).

Hubei: Fang Xian at 2,250 metres, 6793.

A tree to 14 metres tall, discovered by Henry in 1887 and introduced to cultivation through the Arnold Arboretum by Wilson by means of living plants in 1911. In Ireland this tree is cultivated at Birr Castle and the John F. Kennedy Arboretum. The Birr Castle tree is the largest in Britain and Ireland and was 16 metres tall with a trunk diameter of 35

cm when last measured in 1990.

Fagus sinensis Oliv. (FAGACEAE) New species
in Hook. Icon. Pl. t. 1936. (1891), Henry in Elwes & Henry, Trees Gt. Brit. & Irel. 1: 5. (1906).

Fagus sylvatica L. var. *longipes* Oliver in Hooker's. Icon. Pl. 20: t. 1936. (1890), in textu., Henry in Notes Econ. Bot. China 44. (1893), Bretschneider in Hist. Eur. Bot. Disc. China 2: 791. (1898), Franchet in Journ. de Bot. 201. (1899), Skan in Journ. Linn. Soc. Bot. 26: 525. (1899). *Fagus longipes* (Oliv.) Leveille. Fl. Kouy-Tcheou. 126. (1914). *Fagus longipetiolata* Seemen in Bot. Jahrb. Syst. 23: Beibl. 57: 56. (1887), Rehder & E. H. Wilson in Sargent Pl. Wilson. 3: 190. (1917), Zheng in Journ. Wuhan Bot. Res. 2.(1): 24. (1984), Bean in Trees & Shrubs 2: 176. (1992), Li & Zhang in Journ. Wuhan Bot. Res. 16: 244. (1998).

Hubei: South Badong, 5334 (co-type of *Fagus longipetiolata* Seemen), 5334a., (type of *Fagus longipetiolata* Seemen). South Badong, 7444. **Yunnan:** On the Daweishan Range in Pingbian at 2,600 metres, 9027.

Wilson, who later collected this tree at Badong where Henry had discovered the species, stated it was the common beech of central and western China but also mentioned it was no where abundant. In western Sichuan near Baoxing he collected material from a tree 16 metres tall with a girth of 2 metres. The Austrian collector A. von Rosthorn also found it south of Chongqing in Sichuan in 1891. Introduced to cultivation in by Wilson in 1911. Native to China and Vietnam.

Farfugium japonicum (L.) Kitam. (ASTERACEAE)
Senecio kaempferi DC. Henry in Trans. Asiat. Soc. Jap. 24 suppl: 55. (1896) (List Pl. Formosa).

Taiwan: Oluanpi, 660, 1964.

Fargesia murielae (Gamble) T. P. Yi (POACEAE) New species
Arundinaria sparsiflora Rendle in Journ. Linn. Soc. Bot. 36: 436. (1904).

Hubei: In conifer woods in Shennongjia at 2,600 to 3,100 metres, 6938 (type of *Arundinaria sparsiflora* Rendle).

Discovered by Henry in the autumn of 1888 and named from material cultivated at the Royal Botanic Gardens, Kew that had been raised from seeds collected by Wilson in the same area in the early 20th century for the Arnold Arboretum. On Wilson's request it was named for his daughter, Muriel.

Fargesia nitida (Mitford ex Bean) Keng f. ex T. P. Yi (POACEAE) New species
Arundinaria nitida Mitford ex Bean in Gard. Chron. ser. 3, xviii. 186. t. 33. (1895), Stapf in Kew Bull. Misc. Inf. 20. (1896), Bretschneider in Hist. Eur. Bot. Disc. China 2:

793. (1898), Pilger in Engl. Bot. Jahrb. xxix. 226. (1900).

Hubei: On cliffs in Shennongjia between 1,900 and 3,100 metres, 6832, in part.

Henry sent good flowering specimens to Kew from his 1888 expedition on which the species was based. *Fargesia nitida* was raised at Kew from seeds received from St. Petersburgh, said to have been collected the following year (1889) in northern Sichuan by Potanin. Given the altitudes quoted by Henry, his collections are clearly from the upper slopes of present day Shennongjia Forerst District. This region was visited by the Glasnevin Central China Expedition in September 2004 where a decade ago all the colonies of this species had flowered and died leaving extensive thickets of dead canes.

When collecting in a tract of coniferous forest in the Fang region to the north of Yichang in the autumn of 1888 (this may also have been Shennongjia) Henry stated he travelled through an enormous area of dead bamboo canes. He was informed that three or four years before his visit, that the bamboo had flowered all over the country and produced seeds which had been gathered by the local mountain people and was utilized as food, comparable to rice in quality.

Fatoua pilosa Gaud. (MORACACEAE)
Forbes & Hemsley in Journ. Linn. Soc. Bot. 26: 454. (1894), Henry in Trans. Asiat. Soc. Jap. 24 suppl: 86. (1896) (List Pl. Formosa).

Hainan: Environs of Kiungchow, 8752. **Taiwan:** Kaohsiung, 769.

A perennial herb, native to China (Taiwan), the Philippines, Moluccas and New Guinea.

Fatoua villosa (Thunb.) Nakai (MORACEAE)
Fatoua pilosa Forbes & Hemsley non Gaudich in Journ. Linn. Soc. Bot. 26: 454. (1894).

Hubei: Yichang, 2779, 2950, 4309.

Favolus grammocaphalus (Berk.) Imazeki (POLYPORACEAE)
Polyporus grammocaphalus Berk. Anonymous in Kew List of Determinations (volume 24).

Hubei: Fang Xian, 6698. (Fungus).

Favolus sp. (POLYPORACEAE)
Favolus brasiliensis (Fr.) Fr. Anonymous in Kew List of Determinations (volume 24).

Hubei: Changyang, on trees, 7932. (Fungus).

Festuca modesta Steud. (POACEAE)
Rendle in Journ. Linn. Soc. Bot. 36: 429. (1904).

Hubei: South Badong, 5345.

Festuca ovina L. (POACEAE)
Rendle in Journ. Linn. Soc. Bot. 36: 429. (1904).

Hubei: Xingshan at 3,200 metres, 6966.

Widely distributed through temperate regions of the world and in alpine zones of tropical areas.

Festuca parvigluma Steud. (POACEAE)
Rendle in Journ. Linn. Soc. Bot. 36: 429. (1904).

Hubei: Jianshi, 5792.

Native to China, Korea and Japan.

Ficus altissima Blume (MORACEAE)
Forbes & Hemsley in Journ. Linn. Soc. Bot. 26: 457. (1899).

Hainan: Environs of Haikou, 8190.

Ficus ampelos Burm. f. (MORACEAE)
Ficus kingiana Hemsley in Hooker's Icon. Pl. 26: t. 2535. (1899), Henry in Trans. Asiat. Soc. Jap. 24 suppl: 88. (1896) (List Pl. Formosa), Forbes & Hemsley in Journ. Linn. Soc. Bot. 463. (1899), Li in Woody Flora of Taiwan 126. (1963). *Ficus hayatae* Sata Li in Woody Flora of Taiwan 127. (1963). *Ficus gibbosa* Henry non Blume in Trans. Asiat. Soc. Jap. 24 suppl: 88. (1896) (List Pl. Formosa), Forbes & Hemsley in Journ. Linn. Soc. Bot. 26: 460. (1899).

Taiwan: Wanchin, 77, 166. Oluanpi, 684, 925, 1337. Kaohsiung, 708 (type of *Ficus kingiana* Hemsley).

A small evergreen tree, native to China (Taiwan), the Philippines, Indonesia, western New Guinea, Sulawesi (Celebes) and Moluccas.

Ficus benjamina L. (MORACEAE)
Forbes & Hemsley in Journ. Linn. Soc. Bot. 26: 457. (1899).

Hainan: Environs of Haikou, 7994.

A large evergreen tree with aerial roots decending from large branches, reaching the ground and ultimately rooting to form new trunks and giving a banyan-like appearance. A widespread species across India to China and south to northern Australia. Henry collected his 7994 from a 9 metre tall tree and called this species the 'small-leaved *jung* tree'.

Ficus erecta Thunb. (MORACEAE)
Forbes & Hemsley in Journ. Linn. Soc. Bot. 459. (1899).

Hainan: Environs of Haikou, 8418.

A small deciduous tree or large shrub to 4 metres tall, native to the eastern Himalaya, southern China, Korea and Japan.

Ficus erecta Thunb. var. **beecheyana** (Hook. & Arn.) King (MORACEAE)
Ficus beecheyana Hook. & Arn. Henry in Trans. Asiat. Soc. Jap. 24 suppl: 87. (1896) (List Pl. Formosa), Forbes & Hemsley in Journ. Linn. Soc. Bot. 26: 457. (1899), Li in Woody Flora of Taiwan 124. (1963).

Taiwan: Wanchin, 130. Kaohsiung, 158, 245, 314, 1184, 1794, 2011. Oluanpi, 964. Without locality, 314a., 314b., 314c.

A small deciduous tree, native to the eastern Himalaya,

China (including Hong Kong and Taiwan) and Vietnam. Very common in Taiwan at low altitude.

Ficus fistulosa Reinw. ex Blume (MORACEAE)

Ficus obscura Henry non Blume in Trans. Asiat. Soc. Jap. 24 suppl: 88. (1896) (List Pl. Formosa), Forbes & Hemsley in Journ. Linn. Soc. Bot. 26: 464. (1899). *Ficus harlandii* Benth. Li in Woody Flora of Taiwan 121. (1963).

Taiwan: Wanchin, 124.

A small evergreen tree with a range from India through southern China to Indonesia and the Philippines. Common in Chinese Taiwan in broadleaved forests at low altitude.

Ficus formosana Maxim. (MORACEAE)

Henry in Trans. Asiat. Soc. Jap. 24 suppl: 87. (1896) (List Pl. Formosa), Li in Woody Flora of Taiwan 121. fig. 37. (1963).

Taiwan: Wanchin, 24.

A small evergreen shrub, native to southern China (including Hainan, Hong Kong and Taiwan) and Vietnam. A common undershrub in forests at low altitude in Taiwan where it was discovered by Richard Oldham in 1864.

Ficus formosana Maxim. f. *shimadai* (Hayata) Sata. (MORACEAE) New form

Li in Woody Flora of Taiwan 122. (1963).

Ficus formosana Maxim. var. *foliis angustis*. Henry in Trans. Asiat. Soc. Jap. 24 suppl: 87. (1896) (List Pl. Formosa).

Taiwan: Wanchin, 25, 497.

Leaves narrowly lanceolate to linear with lateral veins of 11 to 15 pairs. Native to China (including Hong Kong and Taiwan) and Vietnam. An undershrub in broadleaved forest near riversides in Taiwan. First collected there by Henry.

Ficus glaberrima Blume (MORACEAE)

Merrill & Chen in Sunyatsenia 2: 228. (1935).

Yunnan: Mengzi, 10845.

Native to India, Bhutan, Nepal, Myanmar, China and Vietnam.

Ficus henryi Warburg ex Diels (MORACEAE) New species

in Bot. Jahrb. Syst. 29: 299. (1900), Zheng in Journ. Wuhan Bot. Res. 2.(1): 32. (1984).

Ficus clavata Forbes & Hemsley non Wall. ex Miq. in Journ. Linn. Soc. Bot. 26: 458. (1899), Zheng in Journ. Wuhan Bot. Res. 2.(1): 32. (1984). *Ficus* sp. Zheng in Journ. Wuhan. Bot. Res. 2.(1): 33. (1984).

Hubei: Antelope Glen, 3347, 3497, 4261 (syntype). South Donghu, 7490. Changleping, 7843.

Ficus heteromorpha Hemsley (MORACEAE) New species

in Hooker's. Icon. Pl. 26: tt. 2533-2534. (1899), Forbes & Hemsley in Journ. Linn. Soc. Bot. 26: 461. (1899), Kew Bull. Misc. Inf. 210. (1899), Warburg in Bot. Jahrb. xxix. 300. (1900), Zheng in Journ. Wuhan Bot. Res. 2.(1): 32. (1984).

Hubei: Liantuo, 1933, 3924 (isosyntype), 6362. Antelope Glen, 3439. Yichang, 3965 (isosyntype, October 1887). Xingshan, 6550. Without locality, 6550a.

Ficus heterophylla L. f. (MORACEAE)

Forbes & Hemsley in Journ. Linn. Soc. Bot. 26: 461. (1899).

Hainan: Environs of Kiungchow, 8279.

Ficus macropodocarpa H. Lév. & Vaniot (MORACEAE) New species

Ficus stenophylla Hemsley in Hooker's. Icon. Pl. 26: t. 2536. (1897), Forbes & Hemsley in Journ. Linn. Soc. Bot. 26: 467. (1889), Bretschneider in Hist. Eur. Bot. Disc. China 2: 791. (1898), Zheng in Journ. Wuhan Bot. Res. 2.(1): 33. (1984).

Hubei: Yichang, 2963. Xiling (Yichang) Gorge, 4350 (isosyntype of *Ficus stenophylla* Hemsley). Liantuo, 4350a. **Hainan:** Environs of Kiungchow, 8716.

Distributed aqcross China and Indochina.

Ficus nervosa Heyne (MORACEAE)

Henry in Trans. Asiat. Soc. Jap. 24 suppl: 87. (1896) (List Pl. Formosa), Forbes & Hemsley in Journ. Linn. Soc. Bot. 464. (1899), Li in Woody Flora of Taiwan 118. (1963).

Taiwan: Wanchin, 116, 416, 1679. Oluanpi, 1999, 2000.

A medium-sized evergreen tree, native from India to Malaysia, the Philippines to southern China. Common in Taiwan at low altitude.

Ficus pumila L. (MORACEAE)

Henry in Trans. Asiat. Soc. Jap. 24 suppl: 88. (1896) (List Pl. Formosa), Forbes & Hemsley in Journ. Linn. Soc. Bot. 26: 465. (1899), Li in Woody Flora of Taiwan 124. (1963), Nelson in Pim, The Wood & the Trees 220. (1984).

Hubei: Liantuo, 3806, 4499. **Hainan:** Without locality, 8265. **Taiwan:** Wanshoushan, 1159. Oluanpi, 1370.

An evergreen scandent shrub, climbing trees and rock faces by means of aerial roots. Native to Indochina, China (including Taiwan) and Japan and introduced to cultivation in 1759. According to Henry, the fruits of this shrub, when cut in half and dried were known as *ok-gue*, and from these an excellent jelly was made. He sent seeds to Kew in February 1893. *Ficus pumila* has survived out-of-doors at Glasnevin for decades and on the arch to the alpine yard it carries ripe fruits after a hot summer. According to Wilson's field notes this species was a common epiphyte at Yichang. In Chinese herbal medicine the dried fruits are usecd as a tonic.

Ficus retusa L. (MORACEAE)

Henry in Trans. Asiat. Soc. Jap. 24 suppl: 87. (1896)

(List Pl. Formosa), Forbes & Hemsley in Journ. Linn. Soc. Bot. 26: 466. (1899).

Hainan: Without locality, 51. **Taiwan:** Kaohsiung, without number. Oluanpi, 201.

Widely distributed across India to Malaysia to Australia. Common at low altitude in Chinese Taiwan. This banyan, known as the *jung* according to Henry, is very common at low altitude in Taiwan and widely planted as an avenue tree according to Li.

Ficus ruficaulis Merr. (MORACEAE) New variety

Ficus roxburghii Henry non Wall. ex Steud. in Trans. Asiat. Soc. Jap. 24 suppl: 88. (1896) (List Pl. Formosa), Forbes & Hemsley in Journ. Linn. Soc. Bot. 26: 467. (1899). *Ficus antaoensis* Hayata Li in Woody Flora of Taiwan 118. (1963).

Taiwan: Oluanpi, 1310.

A medium-sized deciduous tree, distributed from the Hengchun Peninsula on the southern end of Taiwan to Sulawesi.

Ficus sarmentosa Buch.-Ham. ex Sm. (MORACEAE)

Ficus foveolata (Miq.) Miq. Henry in Trans. Asiat. Soc. Jap. 24 suppl: 87. (1896) (List Pl. Formosa), Forbes & Hemsley in Journ. Linn. Soc. Bot. 26: 460. (1899), Li in Woody Flora of Taiwan 123. (1963).

Taiwan: Oluanpi, 942, 1972.

A large climbing shrub, native to India, China and Japan.

Ficus sarmentosa Buch.-Ham. ex J. E. Sm. var. *henryi* (King) Corner (MORACEAE) New variety

Ficus foveolata Henry non Wallich in Notes Econ. Bot. 50. (1893). *Ficus foveolata* Wall. var. *henryi* King in Hooker's Icon. Pl. t. 1824. (1889), Bretschneider in Hist. Eur. Bot. Disc. China 2: 791. (1898), Schneider in Sargent Pl. Wilson. 3: 310. (1917) (1889).

Hubei: Liantuo, 3008, 4596. Antelope Glen, 3122, 3552. Yichang, 3302, 3302a., 4737.

This is the *ai-p'a-t'eng*, according to Henry it creeped over rocks in the Antelope Glen near Yichang.

Ficus sarmentosa Buch.-Ham. ex. J. E. Sm. var. *impressa* (Champ.) Corner (MORACEAE)

Ficus impressa Champ. Forbes & Hemsley in Journ. Linn. Soc. Bot. 26: 463. (1899), Zheng in Journ. Wuhan Bot. Res. 2.(1): 33. (1984).

Hubei: Yichang Gorge, hanging down walls of cliffs, 1095. Liantuo, 2691, 3012. Badong, 3698. Xingshan, 6506.

Ficus septica Burm. f. (MORACEAE)

Li in Woody Flora of Taiwan 118. (1963).

Ficus fistulosa Henry non Reinw. ex Blume in Trans. Asiat. Soc. Jap. 24 suppl: 88. (1896) (List Pl. Formosa), Forbes & Hemsley in Journ. Linn. Soc. Bot. 459. (1899).

Ficus harlandii Forbes & Hemsley non Bentham in Journ. Linn. Soc. Bot. 461. (1899). *Ficus leucantatoma* Forbes & Hemsley non Poir. in J. Linn. Soc. Bot. 26: 464. (1899).

Taiwan: Wanchin, 16, 499, 1759. Tamsui, 1731.

A large evergreen tree, native to tropical and subtropical south-east Asia. This species was collected by the Glasnevin Taiwan Expedition (GTE 105) on Wanshoushan, in October 2004.

Ficus simplicissima Lour. (MORACEAE)

Ficus hirta Vahl. Forbes & Hemsley in Journ. Linn. Soc. Bot. 26: 462. (1899).

Hainan: 32 km (20 miles) west of Haikou, 8130. Lingmen (in the interior of the island), 8610, 8642.

Ficus spp. (MORACEAE)

Anonymous in Kew List of Determinations (volume 24 & 36).

Hubei: Badong, 5158. Changyang, 7485. **Yunnan:** Mengzi, 9004, 9008, 9171, 9407, 9575, 9855, 9876, 9908, 9965, 10611, 10648, 10706, 10717, 10826, 11023, 11029, 11033, 11270, 13308, 11462, 11476, 11495, 11507, 11508, 11823. Milê district, 9946. By the Red River Valley near Manhao, 9531, 9543. Simao at 1,400 metres, 11567. Simao, 11599, 11606, 11642, 11657, 11761, 11777, 11796, 11996, 12077, 12090, 12109, 12149, 12180, 12182, 12320, 12427, 12438, 12473, 12602, 12603, 12624, 12762, 12867, 13046. Yuanjiang, 13315, 13316.

Ficus subpisocarpa Gagnep. (MORACEAE)

Ficus wightiana Henry non (Miq.) Bentham in Trans. Asiat. Soc. Jap. 24 suppl: 87. (1896), Forbes & Hemsley in Journ. Linn. Soc. Bot. 26: 469. (1899). *Ficus superba* Li non (Miq.) Miq. in Woody Flora of Taiwan 121. (1963).

Taiwan: Kaohsiung Plain, 1879, 2030.

A large deciduous tree native to southern China, Indochina, Indonesia and Japan. Very common in Chinese Taiwan at low altitude. Henry gave the local name as the *chio-jung*. This species was collected by the Glasnevin Taiwan Expedition (GTE 32) by the lighthouse at Oluanpi (South Cape of Henry and Wilson) in October 2004.

Ficus tikoua Bureau (MORACEAE)

Forbes & Hemsley in Journ. Linn. Soc. Bot. 26: 468. (1899), Zheng in Journ. Wuhan Bot. Res. 2.(1): 33. (1984).

Hubei: Yichang, 3524, 4161.

A ground hugging shrub to 8 cm tall,. at Yichang Henry stated it was called *ti fang ken*, while in Yunnan according to Delavay it was called *ti koua*. The Hani (an ethnic group) of Xishuangbanna in southern Yunnan call this species *siqguqni'qan* and use it to treat cases of upper respiratory tract infections, bladder infections and indigestion. This species was collected by the Glasnevin Central China Expedition (GCCE 598) near Henji town in Xingshan in

02

September 2004.

Ficus tinctoria G. Forst. (MORACEAE)

Li in Woody Flora of Taiwan 126. (1963).

Ficus gibbosa Henry non Blume in Trans. Asiat. Soc. Jap. 24 suppl: 88. (1896). *Ficus swinhoei* King Forbes & Hemsley in Journ. Linn. Soc. Bot. 26 468. (1899).

Hainan: Without locality, 8400. Ling-men (in the interior of the island), 8783. **Taiwan:** Kaohsiung, 316.

A scandent shrub with a native range from Hainan, Taiwan, the Philippines to the Pacific Islands in the exteme south. In Taiwan this species is confined to coastal areas of the Hengchun Peninsula where it is rare. *Ficus tinctoria* was collected by the Glasnevin Taiwan Expedition (GTE 10) on Shadao Shell beach near Oluanpi (South Cape of Henry and Wilson) in October 2004.

Ficus vasculosa Wall. ex Miq. (MORACEAE)

Forbes & Hemsley in Journ. Linn. Soc. Bot. 26: 468. (1899).

Hainan: Environs of Haikou, without number.

Ficus virens Ait. var. *sublanceolata* (Miq.) Corner (MORACEAE)

Zheng in Journ. Wuhan Bot. Res. 2.(1): 33. (1984).

Ficus infectoria Forbes & Hemsley non Roxb. in Journ. Linn. Soc. Bot. 26: 463. (1899), Henry in Notes Econ. Bot. China 44. (1893).

Hubei: Badong, at Wan Lui on the Yangtze, 7173.

According to Henry this was the *huang-ko*, a tree commonly planted around temples and shrines and called the Chinese banyan by foreigners. It forms massive trees to 26 metres tall and a combination of all its trunks can measure over 16 metres in circumference.

Ficus virgata Reinw. ex Blume (MORACEAE)

Li in Woody Flora of Taiwan 119. (1963).

Ficus gibbosa Henry non Blume in Trans. Asiat. Soc. Jap. 24 suppl: 88. (1896) (Listy Pl. Formosa). *Ficus vasculosa* Li non Wall. ex Miq. in Woody Flora of Taiwan 119. (1963).

Taiwan: Oluanpi, 902.

A medium-sized evergreen tree with a range from southern China to Australia. In Chinese Taiwan this species is found in low-altitude forests along the seashore.

Fimbristylis complanata (Retz.) Link (CYPERACEAE)

Henry in Trans. Asiat. Soc. Jap. 24 suppl: 105. (1896) (List Pl. Formosa), C. B. Clarke in Journ. Linn. Soc. Bot. 36: 231. (1903).

Hubei: Yichang, 4221. Xiling (Yichang) Gorge, 4311. **Taiwan:** Kaohsiung, 772, 785, 787, 1772, 1817, 1956. **Yunnan:** Simao, 11928.

Perennial to 70 cm tall, a pantropical species. Common in wet grass land and rice fields.

Fimbristylis cymosa R. Br. (CYPERACEAE)

Fimbristylis formosensis C. B. Clarke Henry in Trans. Asiat. Soc. Jap. 24 suppl: 105. (1896) (List Pl. Formosa), Forbes & Hemsley in Journ. Linn. Soc. Bot. 36: 244. (1904—1906). *Fimbristylis spathacea* Roth. Henry in Trans. Asiat. Soc. Jap. 24 suppl: 105. (1896) (List Pl. Formosa), C. B. Clarke in Journ. Linn. Soc. Bot. 36: 244. (1903).

Taiwan: Kaohsiung Spit, 1824, 1860. Kaohsiung, amoungst sea-shore rocks, 1071, 2013.

Fimbristylis dichotoma (L.) Vahl (CYPERACEAE)

Fimbristylis diphylla Vahl. Henry in Trans. Asiat. Soc. Jap. 24 suppl: 104. (1896) (List Pl. Formosa), C. B. Clarke in Journ. Linn. Soc. Bot. 36: 233. (1903).

Hubei: Yichang, 2248, 2296, 4248, 4251. **Taiwan:** Kaohsiung, 786a., 788, 789, 1040, 1101, 1148.

Widely distributed in the warmer regions of the world.

Fimbristylis ferruginea (L.) Vahl. (CYPERACEAE)

Henry in Trans. Asiat. Soc. Jap. 24 suppl: 105. (1896) (List Pl. Formosa), C. B. Clarke in Journ. Linn. Soc. Bot. 36: 236. (1903).

Taiwan: Kaohsiung, 712, 733, 773, 1066, 1835, 1848.

Perennial to 80 cm tall, pantropical in distribution. In Taiwan this species is mostly found near brackish waters at seashores.

Fimbristylis henryi C. B. Clarke (CYPERACEAE)
New species

in Journ. Linn. Soc. Bot. 36: 237. (1903).

Hubei: Yichang, 2774 (type).

The species was also collected by the Irish missionary, Fr. Hugh Scallan (Pater Hugo) in north-central China. C. B. Clarke in Journ. Linn. Soc. Bot. 36: 245. (1903) also quotes *Fimbristylis stauntonii* Debeaux & Franch. for this number. This collection may be in part, I have not checked this at the Kew herbarium.

Fimbristylis polytrichoides (Retz.) R. Br. (CYPERACEAE)

Henry in Trans. Asiat. Soc. Jap. 24 suppl: 105. (1896) (List Pl. Formosa), C. B. Clarke in Journ. Linn. Soc. Bot 36: 241. (1903).

Taiwan: Kaohsiung, 1100.

A perennial to 30 cm tall, native to the tropics of the Old World.

Fimbristylis quinquangularis (Vahl.) Kunth (CYPERACEAE)

Fimbristylis miliacea (L.) Vahl. Henry in Trans. Asiat. Soc. Jap. 24 suppl: 105. (1896) (List Pl. Formosa), C. B. Clarke in Journ. Linn. Soc. Bot. 36: 240. (1903).

Hubei: Yichang, 4147. In dry rice fields near Yichang, 4297, 7785. **Taiwan:** Oluanpi, 249. Kaohsiung 786, 1113, 1845, 1949.

A perennial (though sometimes growing as an annual) to 70 cm tall, native to the tropics of the Old World. Often found in rice-fields, abundant in China.

Fimbristylis rigidula Nees. (CYPERACEAE)

C. B. Clarke in Journ. Linn. Soc. Bot. 36: 242. (1903).

Hubei: Yichang, 545, 1234, (in part). **Yunnan:** Mengzi, 13728.

Native to India, China and the Philippines.

Fimbristylis schoenoides (Retz.) Vahl. (CYPERACEAE)

Henry in Trans. Asiat. Soc. Jap. 24 suppl: 105. (1896) (List Pl. Formosa), C. B. Clarke in Journ. Linn. Soc. Bot. 36: 243. (1903).

Taiwan: Kaohsiung, 1912.

A perennial to 40 cm tall, native to India, Malesia, Indochina, China, Japan and northern Australia.

Fimbristylis sericea (Poir.) R. Br. (CYPERACEAE)

Henry in Trans. Asiat. Soc. Jap. 24 suppl: 105. (1896) (List Pl. Formosa), C. B. Clarke in Journ. Linn. Soc. Bot. 36: 243. (1903).

Hainan: Environs of Haikou, 8160. On Liang Shan, 16 km (10 miles) east of Haikou, 8195. **Taiwan:** Kaohsiung, 1007 (in part), 1008.

A perennial to 30 cm tall with a creeping decumbent rhizome. Native to India, Malesia, Indochina, China, Japan and northern Australia.

Fimbristylis spp. (CYPERACEAE)

Anonymous in Kew List of Determinations (volume 24 & 34).

Hubei: Liantuo, 4541. **Yunnan:** Simao, 12229.

Fimbristylis thomsonii Boeck (CYPERACEAE)

C. B. Clarke in Journ. Linn. Soc. Bot. 36: 246. (1903).

Yunnan: Mengzi at 1,600 metres, 13729.

Widespread across India, Myanmar, southern China, Laos and Vietnam.

Firmiana simplex (L.) Wight (MALVACEAE)

Rehder & E. H. Wilson in Sargent Pl. Wilson. 2: 237. (1916), Li in Woody Flora of Taiwan 556. fig. 216. (1963).

Sterculia platanifolia L. f. Henry in Notes Econ. Bot. China 61. (1893); in Trans. Asiat. Soc. Jap. 24 suppl: 22. (1896) (List Pl. Formosa).

Taiwan: Oluanpi, 963. Tamsui, 1387.

Henry called this the *wu-t'ung* or parasol tree. A large deciduous tree, native to Vietnam, Japan, southern China including Taiwan where it grows at low altitude throughout the island. He also stated that the indigenous race who inhabited the mountainous parts of Taiwan at that time made a kind of cloth from the inner bark of young trees. In Hubei, where he knew it as *t'ung-ma*, Henry stated the fibre was made into shoes, sandals and ropes and the seeds were used in medicine. *Firmiana simplex* was introduced to cultivation by Robert Fortune. In China the parasol tree was commonly planted in the gardens of poets of poets and scholars, who made it their emblem. This tree needs a good deal of summer heat to ripen its wood in northern Europe and the only large tree in Britain or Ireland grows at the Fota Arboretum in Co. Cork where it was planted in 1909. When last measured in 1984 it was 8 metres tall with a girth of 26cm. There are young trees at the National Botanic Gardens, Glasnevin and its annex at Kilmacurragh, and near the Great Palm House in the Royal Botanic Gardens, Kew. *Firmiana major* has even larger leaves and grows behind the Purple Clouds Temple on Wudang Shan in north-west Hubei.

Fissidens dubius P. Beauv. (FISSIDENTACEAE)

Fissidens decipiens De Not. Salmon in Journ. Linn. Soc. Bot. 34: 454. (1900).

Sichuan: North Wushan (on a stone), 7058 (September 1888). (Moss).

Fissistigma glaucescens (Hance) Merr. (ANNONACEAE)

Melodorum glaucescens Hance Anonymous in Kew List of Determinations (volume 24).

Hainan: Ling-men, 8675.

A climbing shrub to 3 metres tall, native to the tropics of southern China and Vietnam.

Fissistigma oldhamii (Hemsl.) Merr. (ANNONACEAE)

Melodorum aff. *verrucosum* Hance (not Hook. f. & Thoms.) Anonymous in Kew List of Determinations (volume 24).

Hainan: Without locality, 8555.

Fissistigma polyanthum (Hook. f. & Thoms.) Merr. (ANNONACEAE)

Melodorum polyanthum Hook. f. & Thoms. Anonymous in Kew List of Determinations (volume 34).

Yunnan: Simao, 11646.

The Hani (an ethnic group) of Xishuangbanna in southern Yunnan call this species *miqbevsaqdu* and use its roots and stem to treat rheumatism, joint pains, sprains and external injuries. Distributed in Myanmar, China and Vietnam.

Fissistigma wallichii (Hook. f. & Thoms.) Merr. (ANNONACEAE)

Melodorum wallichii Hook. f. & Thoms. Anonymous in Kew List of Determinations (volume 34).

Yunnan: Mengzi, 10592.

Native to India and China (Guangxi, Guizhou and Yunnan).

Flacourtia mollis Hook. f. & Thoms. (SALICACEAE)

Flacourtia montana Anon. non J. Graham Anonymous in Kew List of Determinations (volume 34).

Yunnan: Simao, 11948.

02

Distributed in China (Yunnan) and Myanmar.

Flacourtia sp. (SALICACEAE)

Anonymous in Kew List of Determinations (volume 34).

Yunnan: Simao, 12937.

Flagellaria indica L. (FLAGELLARIACEAE)

Henry in Trans. Asiat. Soc. Jap. 24 suppl: 99. (1896) (List Pl. Formosa), Brown in Journ. Linn. Soc. Bot. 36: 160. (1903), Li in Woody Flora of Taiwan 920. fig. 369. (1963).

Hainan: Ling-men, 8657. **Taiwan:** Oluanpi, 1355.

A tall woody vine to 15 metres, climbing with the aid of spiral leaf-tendrils. Widely distributed across tropical Africa, Asia and Australia. Abundant in dense coastal forest in Chinese Taiwan. *Flagellaria indica* was collected by the Glasnevin Taiwan Expedition (GTE 22) by the lighthouse at Oluanpi (South Cape of Henry and Wilson) in October 2004.

Flavoparmelia caperata (L.) Hale (PARMELIACEAE)

Parmelia caperata (L.) Ach. Anonymous in Kew List of Determinations (volume 24).

Hubei: Badong, 5097. (Lichen).

Flemingia macrophylla (Willd.) Kuntze ex Merr. (FABACEAE)

Flemingia congesta Roxb. ex W. T. Aiton Forbes & Hemsley in Journ. Linn. Soc. Bot. 23: 197. (1887), Henry in Trans. Asiat. Soc. Jap. 24 suppl: 37. (1896) (List Pl. Formosa).

Hubei: Yichang, 322, 1640. **Taiwan:** Wanchin, 13, 167. **Yunnan:** Mengzi, 10413, 10991. Chuyuen, 10441. Simao, 11673, 12437, 12697.

Flemingia strobilifera (L.) W. T. Aiton (FABACEAE)

Henry in Trans. Asiat. Soc. Jap. 24 suppl: 37. (1896) (List Pl. Formosa).

Maughania strobilifera (L.) J. St. Hilaire ex Jacks Li in Woody Flora of Taiwan 352. fig. 125. (1963).

Sichuan: Emei Shan, Henry's Chinese collector with Pratt, 8983. **Taiwan:** Kaohsiung, without number. **Yunnan:** By the Red River, 10396.

An erect shrub to 1.5 metres tall, native to India, China and Malaysia. In Chinese Taiwan this species is restricted to the southern part of the island.

Floscopa scandens Lour. (COMMELINACEAE)

Floscopa paniculata Henry non Benth. (1861) in Trans. Asiat. Soc. Jap. 24 suppl: 99. (1896) (List Pl. Formosa).

Taiwan: Wanchin, 1663.

Native to the Himalaya, India, Indochina, Malaysia and southern China. In Taiwan this perennial inhabits swamps and marshes in broadleaved tropical forests at low altitudes.

Floscopa spp. (COMMELINACEAE)

Anonymous in Kew List of Determinations (volume 34).

Yunnan: Mengzi, 9714, 10729. Simao, 12392.

Flueggea sp. (PHYLLANTHACEAE)

Securinega sp. Anonymous in Kew List of Determinations (volume 34).

Yunnan: Simao, 12357.

Flueggea suffruticosa (Pallas) Baill. (PHYLLANTHACEAE)

Securinega fluggeoides (Müell.-Arg) Müell.-Arg. Forbes & Hemsley in Journ. Linn. Soc. Bot. 26; 426. (1894), Henry in Trans. Asiat. Soc. Jap. 24 suppl: 82. (1896) (List Pl. Formosa). *Securinega suffruticosa* (Pallas) Rehd. Li in Woody Flora of Taiwan 439. fig. 169. (1963).

Hubei: Yichang, 59, 1686. South Donghu, 7709. On Tsui Fu Shan near Yichang (May 1888), 1604, 1620. Antelope Glen, 3557. Xingshan, 4530. Without locality, 3557a. **Taiwan:** Wanshoushan, 1115.

Native to Siberia, Japan, Korea, Japan (south) and China. In Chinese Taiwan Li describes it as being rather scarce. Introduced to cultivation from Mancuria in 1783. Grows in the chain-tent shrubbery at the National Botanic Gardens, Glasnevin where it flowes in late May into June.

Flueggea virosa (Roxb. ex Willd.) Royle. (PHYLLANTHACEAE)

Flueggea microcarpa Blume Forbes & Hemsley in Journ. Linn. Soc. Bot. 26: 427. (1894), Henry in Trans. Asiat. Soc. Jap. 24 suppl: 83. (1896) (List Pl. Formosa). *Securinega virosa* (Roxb. ex Willd.) Royle Li in Woody Flora of Taiwan 439. (1963). *Flueggea obovata* (Willd.) Wall. ex Fern.-Vill. Anonymous in Kew List of Determinations (volume 24).

Hubei: Yichang, 622, 838. **Hainan:** Environs of Haikou, 8004, 8262. **Taiwan:** Oluanpi, 337, 920, 998. Kaohsiung, 676, 1065, 1072.

A shrub of very wide distribution from tropical Africa, Indo-Malaysia, China, the Philippines to Australia.

Foeniculum vulgare Mill. (APIACEAE)

Henry in Trans. Asiat. Soc. Jap. 24 suppl: 47. (1896) (List Pl. Formosa).

Hubei: Yichang, 512, 2040. **Taiwan:** Oluanpi, without number.

The *xiao hui xiang* or common fennel, a native of the Mediterranean region and cultivated worldwide. In Chinese herbal medicine the seeds are used to alleviate pain.

Fordia leptobotrya (Dunn) Schot (FABACEAE) New species

Millettia leptobotrya Dunn in Journ. Linn. Soc. Bot. 41: 189. (1912).

Yunnan: Simao, 12792, (type of *Millettia leptobotrya* Dunn), 12792a., (isotype of *Millettia leptobotrya* Dunn).

Fordiophyton strictum Diels (MELASTOMATACEAE)
New species

in Bot. Jahrb. Syst. 65: 113. (1932) Li in Journ. Arnold Arb. 25: 28. (1944).

Yunnan: Mengzi, 9037 (isotype), 9037a.

Distributed in China (Guangxi and Yunnan) and Vietnam.

Forsythia suspensa (Thunb.) Vahl. (OLEACEAE)

Forbes & Hemsley in Journ. Linn. Soc. Bot. 26: 82. (1889), Zheng in Journ. Wuhan Bot. Res. 2.(1): 176. (1984).

Forsythia suspensa (Tunb.) Vahl. var. *pubescens* (Rehder) Lingelsh. Lingelsheim in Engler, Pflanzenr. 72. (iv. 243): 112. (1920).

Hubei: Liantuo, 3939, 4580. Yichang at 1,400 metres, 6534. Fang Xian, 6534a.

The *lian qiao,* a rambling deciduous shrub to 2.5 metres tall but capable of reaching 9 to 10 metres if trained against a wall. The golden-yellow flowers are produced on the previous years growth in March and April. The fruits are used in various forms of medicine, when used with the fruits of *Arctium lappa* the resulting product is used to treat cases of tonsilitis.

Forsythia viridissima Lindl. (OLEACEAE)

Forbes & Hemsley in Journ. Linn. Soc. Bot. 26: 82. (1889), Zheng in Journ. Wuhan Bot. Res. 2.(1): 177. (1984).

Hubei: Yichang, 3395, 3400.

A deciduous or semi-evergreen shrub to 2.5 metres tall, flowers bright yellow and appearing in March and early April. Distributed through the Yangtze River basin and north-east China. *Forsythia viridissima* was introduced to cultivation by Robert Fortune who purchased a plant in 1845 which he sent to the garden of the Horticultural Society (now the Royal Horticultural Society) where it first flowered in 1847. Alongside *Forsythia suspensa* (q.v.) it is the parent of *Forsythia* x *intermedia,* one of the finest spring flowered shrubs.

Fragaria nilgerrensis Schlectendal ex J. Gay (ROSACEAE)

Anonymous in Kew List of Determinations (volume 34).

Yunnan: Mengzi, 10195.

A robust perennial herb to 25 cm tall, widely distributed across Nepal, eastern India (including Sikkim), China and northern Vietnam.

Fragaria orientalis Lozinsk (ROSACEAE)

Fragaria collina Henry non Maxim. in Flora & Sylva 1: 218. (1903). *Fragaria elatior* Forbes & Hemsley non (Thuill.) Ehrh. in Journ. Linn. Soc. Bot. 23: 239. (1887), Henry in Journ. China Br. Roy. Asiat. Soc. 22: 265. (1887) (Chinese Names of Plants); in Journ. China Br. Roy. Asiat.

Soc. 22: 265. (1887) (Chinese Names of Plants); in Notes Econ. Bot. China 46. (1893).

Hubei: Badong 229 (June 1885), 754, 1749, 4020, 5304. Fang at 3,000 metres, 6853.

According to Henry the oriental strawberry was known colloquially in the Three Gorges region as *ti p'ao tzu*. He also stated it grew at altitudes from 950 to 1,650 metres in the mountains of Hubei.

Frangula crenata (Sieb. & Zucc.) Miq. (RHAMNACEAE)

Rhamnus crenata Sieb. & Zucc. Forbes & Hemsley in Journ. Linn. Soc. Bot. 23: 128. (1886), Henry in Journ. China Br. Roy. Asiat. Soc. 22: 267. (1887) (Chinese Names of Plants), Schneider in Illus. Handb. Laubholzk. 2: 269. (1912); in Sargent Pl. Wilson 2: 232. (1914), Zheng in Journ. Wuhan Bot. Res. 2.(1): 138. (1984).

Hubei: Yichang, 60, 590, 698, 912, 1126, 1394, 2325. Badong, 120, 306 (June 1885), 1824, 2801, 6058. Liantuo, 2655. Changleping, 6263. Jianshi, 7273, 7405.

A widespread and variable species forming shrubs from 3 to 5 metres tall, distributed across Japan, Korea, China and Vietnam. Henry stated that in the Three Gorges region this species was known colloquially as *san huang* and according to Henry its roots were used as a drug.

Frangula henryi (C. K. Schneider) Grubov (RHAMNACEAE) New species

Rhamnus henryi C. K. Schneider in Sargent Pl. Wilson. 2: 244. (1914).

Yunnan: On the Daweishan Range, in Pingbian at 2,800 metres, 9185 (holotype of *Rhamnus henryi* C. K. Schneider).

Endemic to China and distributed in the western provinces of Guangxi, Sichuan, Tibet (Xizang) and Yunnan.

Fraxinus baroniana Diels (OLEACEAE)

Zheng in Journ. Wuhan Bot. Res. 2.(1): 177. (1984).

Hubei: South Badong, 6059.

Fraxinus chinensis Roxb. (OLEACEAE)

Forbes & Hemsley in Journ. Linn. Soc. Bot. 26: 85. (1889), Lingelsheim in Engler, Pflanzenr. 72. (iv. 243): 28. (1920), Zheng in Journ. Wuhan Bot. Res. 2.(1): 177. (1984).

Fraxinus sp. Henry in Journ. China Br. Roy. Asiat. Soc. 22: 252. (1887) (Chinese Names of Plants). *Fraxinus chinensis* Roxb. var. *rotundata* Lingelsheim in Engler, Pflanzenr. 72. (iv. 243): 29. (1920). *Fraxinus velutina* Lingelsheim (non Torrey) in Engler's Bot. Jahrb. xl. 216. (1907), O' Brien in Augustine Henry, An Irish Plant Collector in China 21. (2002). *Fraxinus yunnanensis* Lingelsheim in Engler, Pflanzenr. 72. (iv. 243): 31. (1920).

Hubei: Liantuo, 3815, 6341, 6380. Jianshi, 5869. Fang Xian, 5869a. **Sichuan:** Henry's Chinese collector with Pratt, locality stated, Leshan, Emei Shan, Wa Shan or Kangding,

8863 (type of *Fraxinus chinensis* Roxb. var. *rotundata* Lingelsheim). **Yunnan:** Mengzi, 9936. Simao, 11869. Mountains to the west of Simao at 1,650 metres, 11897 (type of *Fraxinus velutina* Lingelsheim (non Torrey)).

According to Henry this was the white wax tree or *peh la shu* on which the wax-insect was once farmed. Henry's specimens all appear to be from the wild. However in parts of Sichuan, especially around Emei Shan, Leshan (formerly Kiating) and the Min valley this tree was specially cultivated as a host to the scale insect, *Coccus pela* for the production of insect white-wax or *peh la*. This industry was once enormous in China, being second only to sericulture. The two host plants for the scale insect were *Ligustrum lucidum* (on which the insect was raised) and *Fraxinus chinensis* (on which it deposited the wax). Wilson was of the opinion that the insect's natural host was the *Ligustrum*. In central China it was called the *la shu* (wax tree) The insects were bred over 200 miles from the growing areas in Sichuan and were removed at 'gall' stage from the *Ligustrum* as a minute egg.

In the month of May hundreds of porters were emloyed to carry the larvae with all possible speed to the growing areas. This was done by night with the aid of lanterns, lest the larvae hatch in the heat of the day. The porters travelled 40 miles a day and by aid of relay, the 200-mile journey by foot over the mountains was carried out over a 6-day period. Trees of *Fraxinus chinensis* were kept heavily pruned (there is an old Wilson tree at Glasnevin that is heavily pruned each year and grows back vigorously in spring) to about 2 metres tall. From these developed lateral shoots and it was on these the insect was placed and on hatching immediately began to lay wax. The insect's natural predator is the ladybird, termed by the Chinese the *la gho* (wax-dog). At the end of August the branches were cut off and the wax on the underside was scraped off and thrown into boiling water, the wax then dissolved and floated to the surface. The wax was used in candle making and as it was thought to have medicinal properties it was used as a coating for pills. Since the product had to be shipped down the Yangtze though the dangerous rapids of the Three Gorges it was for the wealthy only.

Fraxinus depauperata (A. von Lingelsheim) Z. Wei (OLEACEAE) New species

in Fl. Reipubl. Pop. Sin. 61: 19. pl. 4, f. 9-10. (1992).

Fraxinus paxiana A. von Lingelsheim var. *depauperata* A. von Lingelsheim in Engler, Pflanzenr. 72. (iv. 243.): 22. (1920).

Hubei: South Badong, 6057 (type of *Fraxinus paxiana* A. von Lingelsheim var. de*paxiana* A. von Lingelsheim)

Fraxinus ferruginea Lingelsheim (OLEACEAE) New species

in Bot. Jahrb. Syst. 40: 212. (1907).

Yunnan: Mountains to the west of Simao at 1,950 metres, 11864 (isosyntype). Mountains to the west of Simao at 1,650 metres, 11864a., (isosyntype).

Distributed in the provinces of Guizhou, Yunnan, Tibet (Xizang) and also extending into Myanmar .

Fraxinus floribunda Wall. (OLEACEAE)

Lingelsheim in Engler, Pflanzenr. 72. (iv. 243.): 20. (1920).

Yunnan: Simao, 11897, 12004.

The Himalayan manna ash is a deciduous medium-sized tree of medium height and is a close ally of the European, *Fraxinus ornus*. During the months of March and April this species carries terminal spikes of small, white scented flowers in abundance. In the Himalaya the timber of this tree is used for oars, ploughs and carrying poles. Distributed in Afghanistan, Nepal, India, Bhutan, Myanmar, China, Thailand, Vietnam, Laos and Japan.

Fraxinus griffithii C. B. Clarke (OLEACEAE)

Lingelsheim in Engler, Pflanzenr. 72 (iv. 243): 15. (1920).

Fraxinus bracteata Hemsley in Journ. Linn. Soc. Bot. 26: 84. (1889), Bretschneider in Hist. Eur. Bot. Disc. China 2: 787. (1898), Diels in Bot. Jahrb. 29: 530. (1900), J. H. Veitch in Hortus Veitchii 366. (1906), Schneider in Illus. Handb.Laubholzk. 2: 818. (1912), Zheng in Journ. Wuhan Bot. Res. 2.(1): 177. (1984), Nelson in A Heritage of Beauty 313. (2000). *Fraxinus* sp. Henry in Trans. Asiat. Soc. Jap. 24 suppl: 59. (1896) (List Pl. Formosa). *Fraxinus formosana* Hayata Li in Woody Flora of Taiwan 756. (1963).

Hubei: Half a mile from the Monastery Valley near Yichang (July 1886), 1651 (type of *Fraxinus bracteata* Hemsley). Liantuo, 3937, 6302. Donghu, 6395. Changleping, 6281. Changyang, 7770. **Taiwan:** Oluanpi, 657, 926. Kaohsiung, 1863.

Belonging to the Ornus group, this species was described from a collection made by William Griffith in the Mishmi hills of Assam, it is also found in Myanmar and in south, central and western China. It was introduced to cultivation by Wilson in 1900 from Hubei but did not last long in cultivation.

Fraxinus insularis Hemsl. (OLEACEAE)

Fraxinus retusa Champ. Henry in Trans. Asiat. Soc. Jap. 24 suppl: 59. (1896) (List Pl. Formosa). *Fraxinus retusa* Champ. var. *henryana* Oliver in Hooker's Icon. Pl. x. t. 1930. (1890), Lingelsheim in Bot. Jahrb. xl. 213. (1907), Schneider in Illus. Handb.Laubholzk. 2: 818. (1912), Lingelsheim in Engler, Pflanzenr. 72. (iv. 243): 22. (1920),

O' Brien in Augustine Henry, An Irish Plant Collector in China 21. (2002).

Hubei: Changyang 5747a. **Sichuan:** South Wushan, 5614, 5747, 5891, 5493 (type of *Fraxinus retusa* Champ. var. *henryana* Oliver). **Taiwan:** Oluanpi, 597.

A large deciduous tree, native from central and eastern China to Japan (south). In Chinese Taiwan *Fraxinus insularis* is found in forests at low to medium altitude.

Fraxinus malacophylla Hemsley (OLEACEAE) New species

in Hooker's Icon. Pl. 26(4): t. 2598. (1899), Lingelsheim in Engler, Pflanzenr. 72. (iv. 243.): 15. (1920).

Yunnan: Mengzi at 1,500 metres, 9970 (type).

Native to China (Guangxi and Yunnan) and Thailand.

Fraxinus paxiana A. von Lingelsheim (OLEACEAE) New species

in Bot. Jahrb. Syst. 40: 213. (1907), Lingelsheim in Engler, Pflanzenr. 72. (iv. 243): 22. (1920), Schneider in Illus. Handb.Laubholzk. 2: 816. (1911), Zheng in Journ. Wuhan Bot. Res. 2.(1): 178. (1984).

Hubei: Fang Xian, 6803 (type).

The *qin pi,* a tree to 18 metres tall with a girth of about one metre. Discovered by Henry in 1888. It was introduced to cultivation by Wilson in 1901 and has ripened viable seed at Headfort (the former garden of the Marquis of Headford, who was a sponsor and personal friend of George Forrest) in County Meath. The bark of this tree, when used with the roots of the *bai tou wang* (*Pulsatilla chinensis*) and the bark of *Phellodendron,* is used to treat dysentry-like disorders.

Fraxinus platypoda Oliver (OLEACEAE) New species

in Hooker's. Icon. Pl. 20: t. 1929 (1890), Bretschneider in Hist. Eur. Bot. Disc. China 2: 787. (1898), Diels in Bot. Jahrb. 29: 531. (1900), Schneider in Illus. Handb.Laubholzk. 2: 822. (1912), Lingelsheim in Engler, Pflanzenr. 72. (iv. 243): 39. (1920), Bean in Trees & Shrubs 3: 166. (1936), Nelson in A Heritage of Beauty 313. (2000), O' Brien in Augustine Henry, An Irish Plant Collector in China 21. (2002).

Hubei: Fang Xian, 6800 (isotype).

A tree to 20 metres tall with girths of up to 2 metres. Rare in both Hubei and Sichuan, Wilson, like Henry, only made one collection for the Arnold Arboretum near Kangding in Sichuan in June 1908 when he also introduced this species to cultivation. In Ireland *Fraxinus platypoda* is cultivated at Abbeyleix, County Laois and the John F. Kennedy Arboretum. The Abbeyleix tree is the tallest in Britain and Ireland and was 17 metres tall withy a trunk diameter of 35 cm when measured in 2000.

Fraxinus sp. (OLEACEAE)

Anonymous in Kew List of Determinations (volume 24).

Hubei: Liantuo, 1964. Without locality, 1964a. **Sichuan:** South Wushan, 5603. North Wushan, 7092.

Freycinetia formosana Hemsl. (PANDANACEAE)

Henry in annotation in Henry in Trans. Asiat. Soc. Jap. 24 suppl: (1896) (List Pl. Formosa), copy at Glasnevin).

Taiwan: Oluanpi, 1143a.

A climbing shrub to 10 metres with aerial roots, native to Japan (south) and China (Taiwan) where it is found in the northern and southern extremes of the island along the coast. Discovered in Taiwan in 1864 by the Kew collector, Richard Oldham. This species was also later collected by Ernest Wilson near Taipei.

Frullania monciliata (Reinw., Blume & Nees) Mont. (FRULLANIACEAE)

Anonymous in Kew List of Determinations (volume 24).

Hubei: Fang Xian, 7928a. (Liverwort).

Fuirena cuspidata (Roth.) Kunth. (CYPERACEAE)

Fuirena wallichiana Kunth. C. B. Clarke in Journ. Linn. Soc. Bot. 36: 257. (1903).

Yunnan: Mengzi at 1,950 metres, 9327.

Fuirena umbellata Rottb. (CYPERACEAE)

Fuirena glomerata Lam. Henry in Trans. Asiat. Soc. Jap. 24 suppl: 105. (1896) (List Pl. Formosa), C. B. Clarke in Journ. Linn. Soc. Bot. 36: 256. (1903).

Taiwan: Kaohsiung, 1084.

Gahnia javanica Moritzi (CYPERACEAE)

C. B. Clarke in Journ. Linn. Soc. Bot. 36: 262. (1903).

Yunnan: Mengzi at 2,300 metres, 9168.

Galactia sp. (FABACEAE)

Henry in Trans. Asiat. Soc. Jap. 24 suppl: 35. (1896) (List Pl. Formosa).

Taiwan: Wanchin, 1625.

Galactia striata (Jacq.) Urb. var. *villosa* (Wight & Arn.) Verdc. (FABACEAE)

Galactia tenuiflora (Klein ex Willd.) Wight & Arn. Henry in Trans. Asiat. Soc. Jap. 24 suppl: 35. (1896) (List Pl. Formosa).

Taiwan: Wanchin, 890. Kaohsiung 1079.

A slender twining herb, native over a wide range from India to Malaysia and southern China to Australia.

Galium aparine L. (RUBIACEAE)

Henry in Journ. China Br. Roy. Asiat. Soc. 22: 241. (1887) (Chinese Names of Plants), Forbes & Hemsley in Journ. Linn. Soc. Bot. 23: 393. (1888).

Hubei: Yichang, 1266, 1787, 3371, 3543, 3555, 3662. In a ravine on Moji Shan, 3546.

Known colloquially in the Three Gorges area as *chu la ts'ao* according to Henry. In Chinese herbal medicine this plant in its dried state state is used as a diuretic.

Galium asprellum Michx. f. forma (RUBIACEAE)

Anonymous in Kew List of Determinations (volume 24).

Sichuan: South Wushan, 5662.

Galium aff. *brachypodum* Maxim. (RUBIACEAE)

Anonymous in Kew List of Determinations (volume 24).

Hubei: Yichang, 479.

Galium bungei Steud. (RUBIACEAE)

Galium gracile Bunge Forbes & Hemsley in Journ. Linn. Soc. Bot. 23: 394. (1888).

Hubei: Yichang, 486, 1341, 3542. San You Dong Glen, 3558. In a ravine on Moji Shan, 3547. Antelope Glen, 3550. Badong, 2444, 4048.

Galium hupehense Pamp. (RUBIACEAE)

Galium boreale L. var. *molle* Hemsley in Journ. Linn. Soc. Bot. 23: 394. (1888).

Hubei: Liantuo, 2036, 4400, 6361. 6361a. Xingshan, 4532.

Galium spp. (RUBIACEAE)

Anonymous in Kew List of Determinations (volume 24 & 34.).

Hubei: Yichang, 4116. Badong, 4741. **Yunnan:** Mengzi, 9824, 11001, 11019, 11152..

Galium trichornutum Dandy (RUBIACEAE)

Galium trichorne Stokes Forbes & Hemsley in Journ. Linn. Soc. Bot. 23: 395. (1888).

Hubei: Yichang, 780, 3545.

Galium triflorum Michx. (RUTACEAE)

Anonymous in Kew List of Determinations (volume 34).

Yunnan: Mengzi, 10658.

Garcinia cowa Roxb. ex Choisy (CLUSIACEAE)

Anonymous in Kew List of Determinations (volume 34).

Yunnan: Simao, 11943.

Garcinia multiflora Champ. ex Benth. (CLUSIACEAE)

Henry in Trans. Asiat. Soc. Jap. 24 suppl: 19. (1896) (List Pl. Formosa), Robson in Flora of China 2: 694. (1996).

Taiwan: Wanchin, 411, 512, 1678. Oluanpi, 1338, 1604, 2052.

A small tree, native to northern Indochina, southern China, including Taiwan where it grows on the southern part of the island.

Garcinia oblongifolia Champ. ex Benth. (CLUSIACEAE)

Anonymous in Kew List of Determinations (volume 24).

Hainan: Environs of Kiungchow, 8705.

Gardenia jasminoides Ellis (RUBIACEAE)

Li in Woody Flora of Taiwan 852. fig. 343. (1963).

Gardenia florida L. Henry in Journ. China Br. Roy. Asiat. Soc. 22: 239. (1887) (Chinese Names of Plants), Forbes & Hemsley in Journ. Linn. Soc. Bot. 23: 382. (1888), Henry in Trans. Asiat. Soc. Jap. 24 suppl: 50. (1896) (List Pl. Formosa).

Hubei: Yichang, 2039. On the mountains above Jianshi at 1,650 metres, 6037. **Hainan:** Environs of Haikou, 8052, 8343. **Taiwan:** Wanchin, 568, 1514. Wanshoushan, 742. Oluanpi, 1227.

According to Henry this handsome evergreen shrub was known colloquially around Yichang as *chih-tzu hua*. Its fruits, according to Wilson were known as *chih tzu hwa (zhi zi)* and were used for dyeing certain woods yellow and also as a yellow pigment in paint.

Gardenia sp. (RUBIACEAE)

Anonymous in Kew List of Determinations (volume 34).

Yunnan: Simao, 12909. Mengzi, 13699.

Gardneria multiflora Makino (LOGANIACEAE)

Rehder & E. H. Wilson in Sargent Pl. Wilson. 1: 563. (1913), Zheng in Journ. Wuhan Bot. Res. 2.(1): 183. (1984).

Gardneria nutans Forbes & Hemsley non Sieb. & Zucc. in Journ. Linn. Soc. Bot. 26: 121. (1890). *Gardneria chinensis* Nakai Zheng in Journ. Wuhan Bot. Res. 2.(1): 183. (1984).

Hubei: Antelope Glen, 3350. Jianshi, 6016. **Yunnan:** Mountains to the south-west of Mengzi at 1,600 metres, 9581a.

A common scandent shrub at low altitudes in western Hubei and Sichuan. Distributed in China and Japan.

Gastrochilus yunnanensis Schlecter (ORCHIDACEAE)

New species in Repert. Spec. Nov. Regni Veg. Beih. 4: 76. (1919)

Yunnan: Simao, mountains to the east at 1,150 metres, 12758.

Native to Bangladesh, Yunnan, Thailand and Vietnam.

Gastrodia elata Blume (ORCHIDACEAE)

Rolfe in Journ. Linn. Soc. Bot. 36: 48. (1903).

Gastrodia near *Gastrodia orobanchoides* (Falc.) Hook. f. Henry in Journ. China Br. Roy. Asiat. Soc. 22: 273. (1887) (Chinese Names of Plants), *Gastrodia mairei* Schlecter in Repert. Spec. Nov. Regni Veg. 12: 105. (1913).

Hubei: Badong, 868 (September 1885). South Badong, 4659, 6078. Donghu, 6078a. **Yunnan:** Simao, 11813 (syntype of *Gastrodia mairei* Schlecter).

According to Henry the roots of this terrestrial orchid when roasted were eaten like a potato in the Three Gorges region. He stated it was known as *tien-ma* i.e. 'heavenly hemp', this species is distributed across north, central and south-west China and in Jilin in the north-east, in forest and bamboo groves between 1,600 to 2,700 metres above sea level. Flowers appearing in June. Still an important item in Chinese medicine, it is used in cases of headache, dizziness, childhood convulsions, epilepsy and limb cramps and spasms. Chinese botanists have developed a sucessful method of commercially growing this crop which is dependant on the fungus, *Armillaria mellea* for both nutrition and moisture. By gathering soil containing *Armillaria mellea* near wild populations of this saprophytic orchid it was possible to develop a growing medium in which to cultivate immature tubers. These tubers were used in enormous quantities each year in China and so this new development has hopefully saved remaining wild plants which are under state protection.

Gaultheria discolor Nutt. Ex Hook. (ERICACEAE)

Gaultheria longibracteolata R. C. Fang in Novon 9: 166. fig. 4. (1999).

Yunnan: In the mountains to the north of Mengzi at 2,650 metres, 9460a. On the upper slopes of Fengchunling (south of the Red River) at 2,400 metres, 9460b. Near Kunming on Kuan Yin Shan at 2,000 metres, 9761.

Distributed from Bhutan to Yunnan, northern Indochina and Thailand.

Gaultheria pingbienensis (C. Y. Wu) Yi R. Li, Lu Lu & P. W. Fritsch (ERICACEAE) New species

Gaultheria leucocarpa Blume var. *pingbienensis* C. Y. Wu In the Harvard University Herbarium (A.).

Yunnan: Mengzi at 1,600 metres, 9914, 9914c. Mengzi at 1,650 metres, 9914a. Simao, 12388.

Endemic to Yunnan, where it was first collected by Henry.

Gaultheria semi-infera (C. B. Clarke) Airy Shaw (ERICACEAE)

In the Harvard University Herbarium (A.).

Yunnan: In the mountains to the north of Mengzi at 2,600 metres, 9460.

A handsome evergreen shrub to 2 metres tall, native to the Sikkim Himalaya, Nepal, Bhutan, Myanmar and western China. Allied to *Gaultheria tetramera* W. W. Sm. though taller and bearing elliptic leaves. Introduced to cultivation by Forrest.

Gaultheria sp. (ERICACEAE)

In the Harvard University Herbarium (A.).

Yunnan: Without locality, 9460c.

Gelidium amansii (J. V. Lamouroux) J. V. Lamouroux (GELIDIACEAE)

Henry in Trans. Asiat. Soc. Jap. 24 suppl: 118. (1896) (List Pl. Formosa).

Taiwan: Tamsui, 1763.

A seaweed exported from Tamsui in the late nineteenth century. Agar is extracted predominantly from the red algae *Gelidium*. Different species yield agars with varying properties, the same properties make it an unsurpassed microbiological and tissue culture medium. It is also used in bakery products to keep them moist and to clarify wines, juices and vinegars.

Gelidium pristoides (Turner) Kützing (GELIDIACEAE)

Suhria pristoides (Turner) J. Agardh Henry in Trans. Asiat. Soc. Jap. 24 suppl: 118. (1896) (List Pl. Formosa).

Taiwan: Tamsui, 1762.

Seaweed, exported from Tamsui

Gelsemium elegans (Gardn. & Champ.) Benth. (GELSEMIACEAE)

Anonymous in Kew List of Determinations (volume 34).

Yunnan: Mengzi, 10452.

Distributed in India, Myanmar, China, Thailand, Laos and Vietnam. The Hani (an ethnic group) of Xishuangbanna in southern Yunnan call this species *xiaq* and use the leaves and stems to treat leprosy.

Genianthus micranthus (Roxb.) I. M. Turner (APOCYNACEAE)

Genianthus laurifolius (Roxb.) Hook. f. Anonymous in Kew List of Determinations (volume 34).

Yunnan: Simao, 13009.

Gentiana chinensis Kusnezow (GENTIANACEAE) New species

in Bull. Acad. Petrop. xiii. 338. (1892).

Sichuan: Henry's Chinese collector with Pratt, no locality stated but may be Leshan, Emei Shan, Wa Shan or Kangding, 8867.

Gentiana crassa Kurz. subsp. *rigescens* (Franch. ex Hemsl.) Halda (GENTIANACEAE)

Gentiana rigescens Franch. ex Hemsl. Anonymous in Kew List of Determinations (volume 34).

Yunnan: Mengzi, 9230.

A glabrous perennial herb to 40 cm tall bearing crowded cymes of large pink tubular flowers in autumn. Discovered on the Cangshan Range above Dali by Delavay. Distributed in China (Guizhou, Guangxi, Hunan, Sichuan and Yunnan), and Myanmar.

Gentiana decemfida Buch.-Ham. ex D. Don (GENTIANACEAE)

Anonymous in Kew List of Determinations (volume 34).

Yunnan: Lunan, 10582. Simao, 11616.

Gentiana rubicunda Franch. (GENTIANACEAE)

Forbes & Hemsley in Journ. Linn. Soc. Bot. 26: 134.

(1890).

Hubei: Changyang, 5234, 7792. Badong, 5629. Fang Xian, 6872. Without locality, 6872a. **Sichuan:** South Wushan, 7271.

An elegant, annual species carrying in autumn a profusion of showy, funnel-shaped pink flowers. Discovered by Père Armand David near Baoxing in western Sichuan in May 1869, Père Delavay found it again in north-west Yunnan in 1882. *Gentiana rubicunda* is still a relatively common plant in the Three Gorges region, it was collected by the Glasnevin Central China Expedition (GCCE 771) on Tianthu Shan in Changyang in October 2004.

Gentiana rubicunda Franch. subsp. *samolifolia* (Franch.) Halda (GENTIANACEAE) New subspecies

Gentiana samolifolia Franchet in Bull. Soc. Bot. France xliii. 485. (1896), Bretschneider in Hist. Eur. Bot. Disc. China 2: 788. (1898). *Gentiana delicata* Forbes & Hemsley non Hance in Journ. Linn. Soc. Bot. 26: 127. (1890).

Hubei: Badong, 304, 462 (June 1885), 717, 5456. Without locality, 5456a. Changyang, 5456b. **Sichuan:** Henry's Chinese collector with Pratt, no locality stated but may be Leshan, Emei Shan, Wa Shan or Kangding, 8858.

Later collected in Sichuan by Farges.

Gentiana scabra Bunge (GENTIANACEAE)

Forbes & Hemsley in Journ. Linn. Soc. Bot. 26: 134. (1890), Henry in Notes Econ. Bot. China 6. (1893).

Hubei: Changyang, 7692.

This species furnished the drug known as *lung-tan-ts'ao* according to Henry.

Gentiana spp. (GENTIANACEAE)

Anonymous in Kew List of Determinations (volume 24 & 34).

Hubei: Yichang, 765, 3286 (in part). Badong, 233 (June 1885), 1438, 1453, 3718, 3778. **Sichuan:** North Wushan, 7123. **Yunnan:** Mengzi, 10505, 10903. Simao, 12098.

Gentiana vandellioides Hemsley (GENTIANACEAE) New species

in Journ. Linn. Soc. Bot. 26: 137. (1890), Bretschneider in Hist. Eur. Bot. Disc. China. 2: 788. (1898).

Hubei: At Fang Xian, in clefts of rocks at 2,300 metres, 6738 (isotype), 6871 (type).

Gentiana yokusai Burkill (GENTIANACEAE) New species

in Journ. Asiatic Soc. Bengal. 11: 316. (1906).

Gentiana squarrosa Forbes & Hemsley non Ledeb. in Journ. Linn. Soc. Bot. 26: 135. (1890)

Hubei: Yichang, 506. South Badong, 7377. **Sichuan:** Henry's Chinese collector with Pratt, no locality stated but may be Leshan, Emei Shan, Wa Shan or Kangding, 8854.

Gentianopsis detonosa (Rottb.) Ma (GENTIANACEAE)

Gentiana detonosa Rottb. Forbes & Hemsley in Journ. Linn. Soc. Bot. 26: 127. (1890), Burkill in Journ. Asiatic Soc. Bengal 11: 319. (1906).

Hubei: Xingshan, 6522. Baokang, 6522a.

Geophila herbacea (Jacq.) K. Schum. (RUBIACEAE)

Geophila reniformis Henry non D. Don in Trans. Asiat. Soc. Jap. 24 suppl: 50. (1896) (List Pl. Formosa).

Taiwan: Wanchin, 1531. Wanshoushan, 1858.

Geranium delavayi Franch. (GERANIACEAE)

Knuth in Engler, Pflanzenr. 53: 133. (1912).

Yunnan: Mengzi, 10228a.

Discovered by Delavay in August 1885. Endemic to China and distributed in Sichuan and Yunnan.

Geranium franchetii R. Knuth (GERANIACEAE) New species

in Engler, Pflanzenr. 53: 177. (1920).

Hubei: South Badong, 6154.

Geranium krameri Franch. & Sav. (GERANIACEAE)

Knuth in Engler, Pflanzenr. 53: 192. (1912).

Hubei: Yichang, 1687, 2232, 2777. Liantuo, 2645.

Distributed from Russia to China, Korea and Japan.

Geranium aff. *nodosum* L. (GERANIACEAE)

Anonymous in Kew List of Determinations (volume 24).

Hubei: South Badong, 7343.

Geranium ocellatum Jaquem. ex Cambess. (GERANIACEAE)

Anonymous in Kew List of Determinations (volume 34).

Yunnan: Chuyuan, 10606. Simao, 12882.

Native to Nepal and China (Guizhou, Sichuan and Yunnan).

Geranium platyanthum Duthie (GERANIACEAE)

Geranium eriostemon Fisch. ex DC. Diels in Bot. Jahrb. 29: 419. (1900).

Hubei: Yichang, 688.

Wilson sent seeds of this species to Veitch's Coombe Wood nursery in 1901.

Geranium robertianum L. (GERANIACEAE)

Knuth in Engler, Pflanzenr. 53: 64. (1912), Diels in Bot. Jahrb. 29: 419. (1900).

Sichuan: South Wushan, 5714.

Herb Robert, widespread across Europe, the Caucasus, the Himalaya and China.

Geranium rosthornii R. Knuth (GERANIACEAE) New Species

Geranium wlassovianum Knuth non Fisch. ex Link. in Engler, Pflanzenr. 53: 178. (1920), *Geranium henryi* R. Knuth in Repert. Spec. Nov. Regni. Veg. 19: 228. (1923).

Hubei: Fang Xian, 6752 (type of *Geranium henryi* R. Knuth).

First Found by Henry at Fang in Hubei, later collected by von Rosthorn for whom it is named. This species was collected in the Shennongjia Forest District by the Sino-American Expedition in 1980.

Geranium sibiricum L. (GERANIACEAE)

Knuth in Engler, Pflanzenr. 53: 596. (1912), Diels in Bot. Jahrb. 29: 419. (1900).

Hubei: Badong, 1772, 1774, 1745, 2406, 3765, 4018, 5072. Liantuo, 2633. Changyang, 5262. Jianshi, 5989.

Geranium sibiricum was based on a collection made in China by the French Jesuit missionary, Père d'Incarville (collected in China between 1740 to 1756) for whom the genus *Incarvillea* was named.

Geranium cf. *sibiricum* L. (GERANIACEAE)

Anonymous in Kew List of Determinations (volume 34).

Yunnan: Mengzi, 9270.

Geranium sp. (GERANIACEAE)

Anonymous in Kew List of Determinations (volume 24).

Hubei: South Badong, 5341. Liantuo, 6318.

Geranium thunbergii Sieb. & Zucc. (GERANIACEAE)

Geranium nepalense Diels non Sweet in Bot. Jahrb. 29: 419. (1900).

Hubei: Lung Tung Ch'i Glen (near Yichang), 4197.

A widespread species, native to both China and Japan. Collected by the Glasnevin Central China Expedition (GCCE 529) near Zhenzi town in Xingshan.

Gerbera piloselloides (L.) Cass. (ASTERACEAE)

Forbes & Hemsley in Journ. Linn. Soc. Bot. 23: 473. (1888).

Hubei: Yichang, 461, 3532. **Yunnan:** Mengzi, 9740.

Distributed in Nepal, India, Bhutan, Myanmar, China, Thailand, Laos and Vietnam.

Gerbera spp. (ASTERACEAE)

Anonymous in Kew List of Determinations (volume 34).

Yunnan: Mengzi, 10311. South of the Red River, 12836. Simao, 12099.

Gesneriaceae (GESNERIACEAE)

Cyrtandra sp. Anonymous in Kew List of Determinations (volume 24).

Hubei: Badong, 934 (November 1885), 1033 (December 1885).

Geum aleppicum Jacq. (ROSACEAE)

Geum strictum Forbes & Hemsley non Ait. in Journ. Linn. Soc. Bot. 29: 239. (1887), Diels in Bot. Jahrb. 29: 404. (1900).

Hubei: Yichang, 282, 4842. Badong, 282 (July 1885), 3693, 4678, 5114. Jianshi, 6024. Fang Xian, 6686. **Yunnan:** Mengzi, 10708.

Geum sp. (ROSACEAE)

Anonymous in Kew List of Determinations (volume 24).

Hubei: Changyang, 7754.

Girardinia diversifolia (Link.) Friis (URTICACEAE)

Forbes & Hemsley in Journ. Linn. Soc. Bot. 26: 475. (1899).

Girardinia aff. *heterophylla* Decne. Henry in Journ. China Br. Roy. Asiat. Soc. 22: 270. (1887) (Chinese Names of Plants).

Hubei: Badong, 926 (November 1885). **Yunnan:** Mengzi, 9065.

The Himalayan or Nilghiri nettle. According to Henry at Badong this plant was known as the *she ma ts'ao*.

Gironniera spp. (CANNABACEAE)

Anonymous in Kew List of Determinations (volume 24 & 34).

Hubei: Yichang, 1299. **Yunnan:** Simao, 12417.

Gironniera subaequalis Planch. (CANNABACEAE)

Forbes & Hemsley in Journ. Linn. Soc. Bot. 26: 452. (1894).

Hainan: At Nodoa in the interior of the island, 8459. Environs of Kiungchow, 8722.

Glabrella longipes (Hemsl. ex Oliv.) Mich.Möller & W. H. Chen (GESNERIACEAE)

Didissandra longipes Hemsley ex Oliver Anonymous in Kew List of Determinations (volume 34).

Yunnan: Mengzi, 9801.

Endemic to western China.

Glebionis coronaria (L.) Cass. ex Spach. (ASTERACEAE)

Chrysanthemum coronarium L. Henry in Journ. China Br. Roy. Asiat. Soc. 22: 243. (1887) (Chinese Names of Plants), Forbes & Hemsley in Journ. Linn. Soc. Bot. 23: 437. (1888).

Hubei: Yichang, 275.

Chop suey greens, an annual herb to 1 metre tall, native of the Mediterranean region and cultivated in the Three Gorges region according to Henry where it was known as *t'ung hao*. It is commonly grown as a vegetable in China (and Japan) and the young shoots are stir-fried. It will grow throughout the winter and can be harvested in 30 days from sowing.

Glebionis segetum (L.) Fourr. (ASTERACEAE)

Chrysanthemum segetum L. Henry in Trans. Asiat. Soc. Jap. 24 suppl: 54. (1896) (List Pl. Formosa).

Taiwan: Kaohsiung, without number.

Cultivated as a vegetable and stated by Henry to be

known as *t'ung-hao*.

Glechoma longituba (Nakai) Kupr. (LAMIACEAE)

Nepeta glechoma Forbes & Hemsley non Benth. in Journ. Linn. Soc. Bot. 26: 290. (1890), Dunn in Notes Roy. Bot. Gard. Edinburgh 6: 167. (1915).

Hubei: Yichang, 655, 3397, 3724, 3840. Changyang, 5259, 7798.

The dried stems and leaves of this ground ivy are currently used in Chinese herbal medicine to treat cases of acute urinary infection and jaundice.

Gleditsia japonica Loddiges ex W. Baxter var. *delavayi* (Franch.) L. C. Li (FABACEAE)

Gleditsia delavayi Franchet Henry in Elwes and Henry, The Trees Gt. Brit. and Irel. 6: 1513. (1912).

Yunnan: Mengzi, 9440.

Gleditsia japonica var. *delavayi* was discovered by its namesake, the French missionary, Père Jean Marie Delavay in Yunnan and was introduced to cultivation by Ernest Wilson who collected seeds in southern Yunnan in the autumn of 1899 while visiting Augustine Henry. These were raised in Veitch's Coomb wood nursery in 1900, though Henry states all plants there died in the winter of 1905—1906. Distributed from north-east India to China (Guizhou and Yunnan). Cultivated at the National Botanic Gardens, Glasnevin.

Gleditsia rolfei Vidal. (FABACEAE)

Gleditsia sp. Henry in Trans. Asiat. Soc. Jap. 24 suppl: 38. (1896) (List Pl. Formosa).

Taiwan: Oluanpi, 1345, 2066.

A medium sized deciduous tree, endemic to Taiwan and known only from the Hengchun Peninsula, where it grows at low altitude along streams. Wilson collected this species in the same locality.

Gleditsia sinensis Lam. (FABACEAE)

Zheng in Journ. Wuhan Bot. Res. 2.(1): 30. (1984)
Gleditsia sp. Henry in Journ. China Br. Roy. Asiat. Soc. 22: 276. (1887) Chinese Names of Plants). *Gleditsia officinalis* Hemsley in Kew Bull. Misc. Inf. 82. (1892), Henry in Notes Econ. Bot. China 17. (1893), Bretschneider in Hist. Eur. Bot. Disc. China 2: 782. (1898), Schneider in Illus. Handb. Laubholzk. 2: 9. (1912). *Gleditsia macrantha* Desf. Henry in Elwes and Henry, The Trees Gt. Brit. & Irel. 6: 1414. (1912).

Hubei: Yichang, 3989. Changyang, 7771. **Sichuan:** South Wushan, 5619 (isolectotype of *Gleditsia officinalis* Hemsley), 7230 (in part), (isosyntype of *Gleditsia officinalis* Hemsley). **Hainan:** Environs of Kiungchow, 8290, 8292.

The *zao jiao* or soap thorn according to Henry, a deciduous tree to 15 metres tall in its native habitat and described in 1786 by Lamark from a tree growing at

Versailles in France. It was reintroduced to cultivation by Wilson in 1908 from Sichuan. According to Henry the pods of this soap-tree (the *tsao k'o shu*) were used in medicine (hense Hemsley's *Gleditsia officinalis*) and were exported from Sichuan under the name of *ya-tsao*. Wilson stated it was a common tree in the Yangtze valley where it made a tree to 30 metres tall and the fruits being rich in saponin were used instead of soap in laundry and tanning work. It is still an important medicinal plant in China, the fruits are currently used to expel roundworm and to treat cases of constipation. The dried spines are also used as a medicine administered externally to patients suffering from scabies and leprosy.

Glinus hirtus (Thunb.) Sennikov. & Sukhor. (MULLUGINACEAE)

Mollugo hirta Thunb. Henry in Trans. Asiat. Soc. Jap. 24 suppl: 46. (1896) (List Pl. Formosa).

Taiwan: Kaohsiung, cultivated, without number.

Glinus oppositifolius (L.) Aug. DC. (MULLUGINACEAE)

Mollugo spergula L. Henry in Trans. Asiat. Soc. Jap. 24 suppl: 46. (1896) (List Pl. Formosa), Forbes & Hemsley in Journ. Linn. Soc. Bot. 324. (1904).

Hainan: On Liang Shan, 16 km (10 miles) east of Haikou, 8194. **Taiwan:** Wanchin, 442.

Globba racemosa Sm. (ZINGIBERACEAE)

Diels in Bot. Jahrb. 29: 263. (1900)
Globba strigulosa Schumann in Engler, Pflanzenr. 20: 137. (1904).

Hubei: Donghu, 6400. Changyang, 7679. **Yunnan:** Mountains near Simao at 1,600 metres, 12550 (type of *Globba strigulosa* Schumann, collected 7th October 1898).

Distributed in Nepal, India, Bhutan, Myanmar, China and Thailand.

Globba schomburgkii Hook. f. (ZINGIBERACEAE)

Globba chinensis Schumann in Engler, Pflanzenr. 20: 153. (1904).

Yunnan: Mountains near Simao at 1,300 metres, 12252 (type of *Globba chinensis* Schumann).

Distributed in tropical and subtropical forests in Myanmar, Yunnan, Thailand and Vietnam.

Glochidion lanceolarium (Roxb.) Voigt. (PHYLLANTHACEAE)

Glochidion macrophyllum Benth. Forbes & Hemsley in Journ. Linn. Soc. Bot. 26: 425. (1894).

Hainan: Without locality, 8557. Environs of Kiungchow, 8760.

Glochidion moluccanum Blume (PHYLLANTHACEAE)

Henry in Trans. Asiat. Soc. Jap. 24 suppl: 82. (1896) (List Pl. Formosa).

Taiwan: Kaohsiung Plain, 1929.

02

Glochidion puberum (L.) Hutchinson (PHYLLANTHACEAE)

Hutchinson in Sargent Pl. Wilson. 2: 518. (1916), Li in Woody Flora of Taiwan 429. (1963), Zheng in Journ. Wuhan Bot. Res. 2.(1): 111. (1984).

Glochidion obscurum Forbes & Hemsley non (Roxb. ex Willd.) Blume in Journ. Linn. Soc. Bot. 26: 425. (1894). *Glochidion fortunei* Henry non Hance in Trans. Asiat. Soc. Jap. 24 suppl: 82. (1896).

Hubei: Yichang, 242, 243, 1020. San You Dong Glen (July 1886), 1532. Liantuo, 1927, 3050. **Taiwan:** Kaohsiung, 1793.

A shrub or small tree, native to Japan and southern China (including Taiwan). A common small shrub found`at low altitudes in Hubei.

Glochidion rubrum Blume (PHYLLANTHACEAE)

Li in woody Flora of Taiwan 426. (1963).

Glochidion fortunei Hance Henry in Trans. Asiat. Soc. Jap. 24 suppl: 82. (1896) (List Pl. Formosa).

Taiwan: Oluanpi, 369, 1216. Kaohsiung, 1053.

A small tree native from Malaysia to China (Taiwan) where it is common at low altitude.

Glochidion spp. (PHYLLANTHACEAE)

Henry in Trans. Asiat. Soc. Jap. 24 suppl: 82. (1896) (List Pl. Formosa). Anonymous in Kew List of Determinations (volume 34).

Taiwan: Kaohsiung Plain, 2042. **Yunnan:** Mengzi, 9261, 9485, 10755, 10944, 11176. Simao, 11711, 11717, 11929, 12936, 13146, 13384, 13443.

Glochidion zeylanicum (Gaertn.) A. Juss (PHYLLANTHACEAE)

Glochidion honkongense Muell-Arg. Forbes & Hemsley in Journ. Linn. Soc. Bot. 26: 424. (1894), Henry in Trans. Asiat. Soc. Jap. 24 suppl: 82. (1896) (List Pl. Formosa), Li in Woody Flora of Taiwan 428. (1963).

Glochidion sp. Henry in Trans. Asiat. Soc. Jap. 24 suppl: 82. (1896) (List Pl. Formosa).

Hainan: Without locality, 8340. **Taiwan:** Kaohsiung, 707. Wanchin, 829.

A small or medium sized tree, native to India, China and Japan. southern China. In Chinese Taiwan it was known as *ta-hung-hsin* according to Henry.

Glochidion zeylanicum (Gaertn.) A. Juss. var. *tomentosum* (Dalzell) Trimen (PHYLLANTHACEAE)

Glochidion sp. Henry in Trans. Asiat. Soc. Jap. 24 suppl: 82. (1896) (List Pl. Formosa). *Glochidion arnottianum* Muell.-Arg. Henry in Trans. Asiat. Soc. Jap. 24 suppl: 82. (1896) (List Pl. Formosa), Forbes & Hemsley in Journ. Linn. Soc. Bot. 26: 424. (1894). *Glochidion moluccanum* Henry non Blume Henry in Trans. Asiat. Soc. Jap. 24 suppl: 82. (1896) (List Pl. Formosa). *Glochidion dasyphyllum* K. Koch. Li in Woody Flora of Taiwan 427. (1963).

Hainan: Ling-men, 8612. **Taiwan:** Wanchin, 117, 415. Kaohsiung, 713. Kaohsiung Plain, 2038. Oluanpi, 1335.

A small tree native India, China and Japan. In Chinese Taiwan it was known as *ch'ih-hsieh* according to Henry.

Glossogyne bidens (Retz.) Cass. ex Less. (ASTERACEAE)

Glossogyne tenuifolia (Labill.) Veldcamp Henry in Trans. Asiat. Soc. Jap. 24 suppl: 54. (1896) (List Pl. Formosa).

Hainan: Environs of Haikou, 8244. **Taiwan:** Kaohsiung, without number. Wanchin, 865.

Glycine max (L.) Merr. (FABACEAE)

Henry in Trans. Asiat. Soc. Jap. 24 suppl: 34. (1896) (List Pl. Formosa).

Glycine hispida (Moench.) Maxim. Henry in Journ. China Br. Roy. Asiat. Soc. 22: 273. (1887) (Chinese Names of Plants); in Notes Econ. Bot. China 13. (1893), Forbes & Hemsley in Journ. Linn. Soc. Bot. 23: 188. (1887).

Hubei: Liantuo, 2210. **Taiwan:** Kaohsiung, without number. Oluanpi, 264.

The *ta-tou* , the soy or soja bean, widely cultivated throughout China, Japan and other parts of Asia, in the Three Gorges region Henry stated it was known as *huang tou*. An ancient Chinese crop, it is thought to have been first grown in the eastern part of northern China in the 11th century B.C. In China, soybeans are referred to as 'poor man's meat' or 'cow-without-bones' indicating the important role this legume plays in that vast country. There are over 2,500 cultivars known today. It is a long day plant which has prevented its spread to lower altitudes in the past, though in recent decades short day plants have been developed. The cultivated soy bean was selected in China about 3,000 years ago from *Glycine max* subsp. *soja* (q.v.) to produce larger seeds, increased oil content, but a reduced protein level, non-shattering seed pods and over millenia it has lost its climbing habit. The germinated seedlings are 'bean sprouts' and are a popular stir fry ingredient. The soybean was introduced to the West from China by the French missionaries. It is considered by experts to be one of the plants that will in the future feed the planet's growing population.

Glycine max (L.) Merr. subsp. *soja* (Sieb. & Zucc.) H. Ohashi (FABACEAE)

Glycine soja Sieb. & Zucc. Forbes & Hemsley in Journ. Linn. Soc. Bot. 23: 188. (1887).

Hubei: Yichang, 638, 2279, 2279a. On Tsui Fu Shan near Yichang (May 1888), 1611. **Yunnan:** Mengzi, 9487.

The wild ancestor of the soy bean, a climbing annual.

Glycine tabacina (Labill.) Benth. (FABACEAE)

Henry in Trans. Asiat. Soc. Jap. 24 suppl: 34. (1896) (List Pl. Formosa).

Taiwan: Kaohsiung, 1180.

Native over a wide range from southern China to Australia.

Glycine tomentella Hayata (FABACEAE)

Glycine tomentosa (Benth.) Benth. Henry in Trans. Asiat. Soc. Jap. 24 suppl: 34. (1896) (List Pl. Formosa).

Taiwan: Kaohsiung, without number.

Glycosmis parviflora (Sims.) Little (RUTACEAE)

Glycosmis pentaphylla Henry non (Retz.) DC. in Trans. Asiat. Soc. Jap. 24 suppl: 25. (1896) (List Pl. Formosa), *Glycosmis cochinchinensis* Li non (Lour.) Pierre ex Engler in Woody Flora of Taiwan 375. fig. 135. (1963).

Hainan: Environs of Haikou, 8068, 8355. **Taiwan:** Wanchin, 833, 1487, 1587, 1614. Oluanpi, 1226, 1294.

A variable shrub or small tree, widely distributed from India to Malaysia and southern China to the Philippines. The *chui ping* tree, according to Henry the leaves of this species were used to make wine ferment in Hainan. A common inhabitant of tropical forests at low altitude in Chinese Taiwan.

Glycosmis pentaphylla (Retz.) DC. (RUTACEAE)

Anonymous in Kew List of Determinations (volume 34).

Yunnan: Simao, 12905, 13268.

Gnaphalium sp. (ASTERACEAE)

Anonymous in Kew List of Determinations (volume 24).

Hubei: Xingshan, 6945.

Gnetum edule (Willd.) Blume (GNETACEAE)

Gnetum scandens Roxb. Skan in Journ. Linn. Soc. Bot. 26: 539. (1902).

Yunnan: Mengzi, 10952.

Gnetum sp. (GNETACEAE)

Anonymous in Kew List of Determinations (volume 34).

Yunnan: Simao, 11723.

Gomphandra tetrandra (Wall. in Roxb.) Sleum. (STEMONURACEAE)

Sleumer in Blumea 17: 204. (1962).

Yunnan: Mengzi, 10492, 10738, 11273. Simao, 13007.

Gomphogyne cissiformis Griff. (CUCURBITACEAE)

Cogniaux in Engler, Pflanzenr. iv. 275. i: 38. (1916).

Yunnan: Mengzi, 10242.

Native to India, Bhutan, Yunnan, Vietnam and the Philippines.

Gomphostemma javanicum (Blume) Benth. (LAMIACEAE)

Gomphostemma lucidum Wall. ex Benth. Dunn in Notes Roy. Bot. Gard. Edinburgh 6: 190. (1915).

Yunnan: In Pingbian, in forests on the Daweishan Range at 1,650 metres, 11046. Simao, 12011, 12011a., 12307. In forests near Simao at 1,450 metres, 12609.

Distributed from India to China, Thailand and Malesia.

Gomphostemma microdon Dunn (LAMIACEAE)

New species

in Notes Roy. Bot. Gard. Edinburgh 8: 170.(1913).

Yunnan: In forests to the west of Simao at 1,450 metres, 12501 (isotype, collected 7th October 1898).

Native to tropical rainforest in Xishuangbanna in southern Yunnan of China and Laos.

Gomphostemma parviflorum Wall. ex Benth. (LAMIACEAE)

Dunn in Notes Roy. Bot. Gard. Edinburgh 6: 189. (1915).

Yunnan: In mountain-forests to the east of Simao at 1,300 metres, 12253.

Gomphostemma pedunculatum Benth. ex Hook. f. (LAMIACEAE)

Dunn in Notes Roy. Bot. Gard. Edinburgh 6: 189. (1915).

Yunnan: Simao, 12518, 12518a. Yuanjiang, 13594.

Distributed in India, Myanmar, Yunnan, Indochina and Thailand.

Gomphostemma sp. (LAMIACEAE)

Anonymous in Kew List of Determinations (volume 34).

Yunnan: Simao, 12314.

Gomphrena globosa L. (AMARANTHACEAE)

Henry in Trans. Asiat. Soc. Jap. 24 suppl: 75. (1896) (List Pl. Formosa).

Hubei: Liantuo, 4441. **Taiwan:** Oluanpi, cultivated, without number.

A tropical summer-flowered annual to 1 metre tall, native to India. Cultivated throughout the tropics and sometimes grown as a bedding plant in Europe.

Gomphrena sp. (AMARANTHACEAE)

Anonymous in Kew List of Determinations (volume 34).

Yunnan: Simao, 12713.

Gonatanthus pumilus (D. Don) Engl. & Krause (ARACEAE)

Engler & Krause in Engler, Pflanzenr. iv. 23Da: 19. (1912).

Yunnan: Simao at 1,600 metres, 12459a.

Gongronema napalense (Wall.) Decne. (APOCYNACEAE)

Anonymous in Kew List of Determinations (volume 34).

Yunnan: Simao, 12267, 13419.

Gongronemopsis tenacissima (Roxb.) S. Reuss, Liede & Meve (APOCYNACEAE)

Marsdenia tenacissima (Roxb.) Moon Anonymous in Kew List of Determinations (volume 34).

Yunnan: Simao, 11876.

Widespread in Asia.

Goniophlebium amoenum (Wall. ex Mett.) Bedd. (POLYPODIACEAE)

Polypodium amoenum Wall. ex Mett. Baker in Ann. Bot. 5: 470. (1890), Henry in Trans. Asiat. Soc. Jap. 24 suppl: 114. (1896) (List Pl. Formosa), Christ in Bull. Herb. Boiss. 6: 870. (1898), Diels in Bot. Jahrb. 29: 202. (1900), Christ in Bull. Soc. Bot. France 5: 13. (1905). *Polypodium valdealatum* H. Christ in Bull. Herb. Boissier 7: 4. (1899).

Hubei: Badong, 1740. **Sichuan:** South Wushan, 5754. **Taiwan:** Tamsui, 1433. **Yunnan:** In mountain woods to the south of Mengzi at 1,600 metres, 9785. In mountain forest to the north of Mengzi at 2,600 metres, 11513 (isotype of *Polypodium valdealatum* H. Christ). Simao, 13073.

Distributed through the Himalaya, China, Indochina and northern Thailand. Henry's Badong collection proved to be a new record for China.

Goniophlebium fieldingianum (Kunze ex Mett.) T. Moore (POLYPODIACEAE)

Polypodium taliense Christ Christ in Bull. Soc. Bot. France 5: 13. (1905).

Yunnan: Mengzi, 10168.

Distributed in Nepal, Bhutan, Myanmar, China and Thailand.

Goniophlebium manmeiense (H. Christ) Rodl-Linder (POLYPODIACEAE) New species
in Blumea 34(2): 394. (1990)

Polypodium manmeiense Christ in Bull. Herb. Boissier 6: 870. (1898), H. Christ in Bull. Soc. Bot. France 5: 13. (1905). *Metapolypodium manmeiense* (Christ) Ching in Acta Phytotax. Sin. 16(4): 29. (1978). *Polypodium pseudodimidiatum* Christ in Bull. Soc. Bot. France 5: 14. (1905). *Polypodium scalare* Christ in Bull. Soc. Bot. France 5: 14. (1905).

Yunnan: South of the Red River from Manmei at 1,950 metres, 10081 (isotype of *Polypodium manmeiense* Christ). Simao, 13036 (in part, type of *Polypodium pseudodimidiatum* Christ, type of *Polypodium scalare* Christ).

Distributed in Myanmar, China (Sichuan and Yunnan), Thailand and Laos.

Goniophlebium nipponicum (Mett.) Bedd. ex Rödl-Linder (POLYPODIACEAE)

Polypodium niponicum Mett. Christ in Bull. Herb. Boiss. 6: 870. (1890), Diels in Bot. Jahrb. 29: 202. (1900).

Polypodium niponicum Mett. var. *laevipes* Christ in Bull. Soc. Bot. France 5: 13. (1905).

Hubei: Liantuo, 2660, 3196. **Yunnan:** Mengzi, 10179.

Goniophlebium subauriculatum (Blume) C. Presl. (POLYPODIACEAE)

Polypodium subauriculatum Blume Anonymous in Kew List of Determinations (volume 34).

Yunnan: Simao, 13063, 13145.

Goniophlebium yunnanense (Franch. Bedd. (POLYPODIACEAE)

Polypodium yunnanense Franchet Christ in Bull. Herb. Boiss. 6: 870. (1898).

Yunnan: South of the Red River from Manmei at 2,250 metres, 9748.

Gonocarpus micranthus Thunb. (HALORAGACEAE)

Haloragis micrantha (Thunb.) R. Br. Anonymous in Kew List of Determinations (volume 24).

Hubei: South Badong, 6151, 7374.

Distributed through China, Japan (south), Malaysia, New Zealand and Australia.

Gonostegia hirta (Blume) Miq. (URTICACEAE)

Pouzolzia hispida Benn. Henry in Journ. China Br. Roy. Asiat. Soc. 22: 261. (1887) (Chinese Names of Plants), Forbes & Hemsley in Journ. Linn. Soc. Bot. 26: 489. (1899).

Hubei: Yichang, 467, 1646, 1682, 3637. At side of rice fields in the Monastery Valley, 1646. Liantuo, 1905. **Yunnan:** Mengzi, 10727.

A widespread species in China. According to Henry this species was known colloquially in the Three Gorges region as *no mi tuan erh*.

Gonostegia pentandra (Roxb.) Miq. (URTICACEAE)

Pouzolzia hypericifolia Blume Forbes & Hemsley in Journ. Linn. Soc. Bot. 26: 490. (1899).

Hainan: Without locality, 8119.

Goodyera fumata Thwaites (ORCHIDACEAE)

Goodyera formosana Rolfe in Ann. of Bot. 9: 159. (1895), Bretschneider in Hist. Eur. Bot. Disc. China 2: 792. (1898), Rolfe in Journ. Linn. Soc. Bot. 45. (1903). *Epipactis formosana* (Rolfe) Eaton in Proc. Biol. Soc. Washington 21: 64. (1908), Eaton Eaton in Repert. Nov. Spec. Regni Veg. 6: 41. (1908).

Taiwan: Wanchin, mountains, 409, (isotype of *Goodyera formosana* Rolfe). **Yunnan:** Simao, 13005.

A terrestrial orchid of widespread distribution. Native to the Himalaya, Sri Lanka, Myanmar (Burma), Thailand, southern China and Japan (south). In forests in the centre and southern part of Chinese Taiwan.

Goodyera henryi Rolfe (ORCHIDACEAE) New species
in Kew Bull. Misc. Inf. 201. (1896), Bretschneider in Hist. Eur. Bot. Disc. China 2: 792. (1898), Rolfe in Journ.

02

Linn. Soc. Bot. 36: 45. (1903), Morley in Glasra 3: 79. (1979).

Epipactis henryi (Rolfe) Eaton in Proc. Biol. Soc. Washington 21: 64. (1908); in Fedde, Repert. Nov. Spec. Regni Veg. 6: 42. (1908).

Hubei: Fang Xian, 6878 (isotype).

Goodyera procera (Ker-Gawl.) Hook. (ORCHIDACEAE)
Henry in Trans. Asiat. Soc. Jap. 24 suppl: 93. (1896) (List Pl. Formosa), Rolfe in Journ. Linn. Soc. Bot. 36: 45. (1903).

Taiwan: Wanchin, 165, 1826. **Yunnan:** Simao, 12079.

A terrestrial orchid to 30 cm tall, widely distributed across subtropical regions of the Himalaya, India, southwest China, Japan, Indochina and Malaya.

Goodyera repens (L.) R. Br. (ORCHIDACEAE)
Rolfe in Journ. Linn. Soc. Bot. 36: 45. (1903).

Hubei: Xingshan, 6513.

A widespread species in circumpolar temperate regions and in subtropical mountains.

Goodyera schlechtendaliana Reichb. f. (ORCHIDACEAE)
Rolfe in Journ. Linn. Soc. Bot. 36: 46. (1903).

Hubei: Liantuo, 2605. South Badong, 6133. South Donghu, 7580.

A handsome terrestrial species, widely distributed through most of the Chinese provinces to the south of the Yangtze River and Taiwan where it grows in forests between 1,100 to 2,500 metres above sea level. The leaves are beautifully marbled and blossoms appear in September and October. The entire plant is used in Chinese herbal medicine. Also found in Korea and Japan to Sumatera.

Goodyera similis Blume (ORCHIDACEAE)
Goodyera velutina Maxim. ex Regel. Rolfe in Journ. Linn. Soc. Bot. 36: 46. (1903).

Hubei: Fang Xian, 6638. Changyang, 7682.

A small terrestrial orchid, native to China, Korea and Japan.

Goodyera spp. (ORCHIDACEAE)
Anonymous in Kew List of Determinations (volume 34).

Yunnan: Mengzi, 11843. Simao, 13597.

Gossypium herbaceum L. (MALVACEAE)
Henry in Trans. Asiat. Soc. Jap. 24 suppl: 21. (1896) (List Pl. Formosa).

Hainan: Environs of Kiungchow, 8305, 8408. **Taiwan:** Kaohsiung (cultivated in a garden), 1889. **Yunnan:** Mengzi, 11024.

The *mei wha* or Levant cotton, a branching annual to 1 metre tall. Cotton cultivation was introduced to China in the 11th century though the product met with enormous opposition from the producers of silk, China-grass and other native textiles.

Gouania obtusifolia Vent. ex Brongn. (RHAMNACEAE)
Gouania javanica Miquel Schneider in Sargent Pl. Wilson. 2: 253. (1914).

Yunnan: On Fengchunling at 2,800 metres, 11188.

A tree to 2.5 metres tall with yellowish flowers. Distributed in China, Thailand, Laos, Cambodia, Vietnam and the Philippines.

Grammitis doniana (Spreng.) Christenh. (POLYPODIACEAE)
Polypodium sinicum Christ in Bull. Herb. Boissier 7: 3. (1899). *Polypodium convolutum* Baker in Kew Bull. Misc. Inf. 12. (1906).

Yunnan: On trees at Fengchunling at 2,650 metres, 10186 (type of *Polypodium sinicum* Christ).

Grangea maderaspatana (L.) Poir. (ASTERACEAE)
Henry in Trans. Asiat. Soc. Jap. 24 suppl: 52. (1896) (List Pl. Formosa).

Hainan: Environs of Haikou, 8042. **Taiwan:** Kaohsiung, without number.

Grewia arbutifolia Ventenat ex Jussieu (MALVACEAE)
Anonymous in Kew List of Determinations (volume 34).

Yunnan: Milê, 9887.

Grewia biloba G. Don (MALVACEAE)
Li in Woody Flora of Taiwan 539. (1896) (List Pl. Formosa), Zheng in Journ. Wuhan Bot. Res. 2.(1): 146. (1984).

Grewia parviflora Henry non Bunge in Trans. Asiat. Soc. Jap. 24 suppl: 23. (1896) (List Pl. Formosa). *Grewia parviflora* Bunge var. *glabrescens* (Benth.) Rehder & E. H. Wilson in Sargent Pl. Wilson 2: 371. (1915). *Grewia biloba* G. Don var. *glabrescens* (Benth.) Rehd. & Wils. Zheng in Journ. Wuhan Bot. Res. 2.(1): 146. (1984).

Hubei: Moji Shan (June 1886), 1555. Changyang, 7601. Badong, 2495. Liantuo, 1929, 2695. **Taiwan:** Wanchin, 406, 820, 876. Oluanpi, 1206. Kaohsiung Plain, 1845

Native to tropical Asia, in Taiwan this shrub is found in forests on the southern and eastern parts of the island.

Grewia biloba G. Don var. *parviflora* (Bunge) Hand.-Mazz. (MALVACEAE)
Grewia parviflora Bunge Henry in Journ. China Br. Roy. Asiat. Soc. 22: 266. (1887) (Chinese Names of Plants).

Hubei: Yichang, 312, 2100, 2281, 3629. On Tsui Fu Shan near Yichang (May 1888), 1600. In the Hsiao Pingshanba Glen (October 1886), 2916. Liantuo, 2608, 4616. Xingshan, 4515.

According to Henry this species was known colloquially in the Three Gorges area as *pien can k'o tzu*.

Grewia cuspidatoserrata Burret (MALVACEAE) New species

in Notizbl. Königl. Bot. Gart. Berlin 9: 718. (1926).

Yunnan: Mengzi, 10162, 10162a., (isotype).

Distributed in Guangxi and Yunnan.

Grewia eriocarpa Juss. (MALVACEAE)

Li in Woody Flora of Taiwan 541. (1963).

Grewia tiliifolia Henry nonVahl. in Trans. Asiat. Soc. Jap. 24 suppl: 23. (1896) (List Pl. Formosa).

Taiwan: Wanchin, 496, 557.

A tree to 8 metres tall, native from Thailand to southern China (including Hainan and Taiwan), in Taiwan it grows at low altitude in forests on the southern part of the island.

Grewia henryi Burret (MALVACEAE) New species

in Notizbl. Königl. Bot. Gart. Berlin 9: 674. (1926).

Yunnan: Mengzi, 10981. Mengzi at 1,800 metres, 10981a. Simao, 12520. Simao at 1,700 metres, 12520c., (isotype).

Endemic to China.

Grewia laevigata Vahl. (MALVACEAE)

Anonymous in Kew List of Determinations (volume 34).

Yunnan: At Hsienkei on the Red River near the Vietnamese border, 9560. Simao, 12358

Grewia piscatorum Hance (MALVACEAE)

Henry in Trans. Asiat. Soc. Jap. 24 suppl: 23. (1896) (List Pl. Formosa), Li in Woody Flora of Taiwan 541. (1963).

Taiwan: Kaohsiung, 705, 1086.

An evergreen tree or shrub to 3 metres tall, native to southern China including Taiwan, where it is found by the coast.

Grewia spp. (MALVACEAE)

Anonymous in Kew List of Determinations (volume 24 & 34).

Hainan: Without locality, 8588. **Yunnan:** Simao, 13489.

Grewia tiliifolia Vahl. (MALVACEAE)

Anonymous in Kew List of Determinations (volume 34).

Yunnan: Mengzi, 10652. Manpan, 10676. Simao, 12185, 12925.

Distributed in India, Myanmar and China (Yunnan).

Griffitharia hemsleyi (C. K. Schneid.) Rushforth (ROSACEAE) New species

Pyrus aria Forbes & Hemsley non (L.) Ehrh. in Journ. Linn. Soc. Bot. 23: 254. (1887). *Sorbus xanthoneura* Rehder in Sargent Pl. Wilson. 2: 272. (1915). *Pyrus xanthoneura* (Rehder) Cardot in Notul. Syst. (Paris) 3: 349. (1918); in Bull. Mus. Hist. Nat. (Paris) 24: 79. (1918). *Micromeles hemsleyi* Schneider in Illus. Handb. Laubholzk. 1: 704.

f. 388a. (1906); in Fedde, Repert. Nov. Spec. Regni Veg. 3: 152. (1907). *Sorbus hemsleyi* (C. K. Schneid.) Rehder in Sargent Pl. Wilson. 2: 276. (1915), Sponberg in Journ. Arnold Arb. 67: 261. (1986). O' Brien in Augustine Henry, An Irish Plant Collector in China 36. (2002).

Hubei: Fang Xian, 6831, 6830 (type of *Sorbus xanthoneura* Rehder). Changleping, 6269. Without locality, 6830a., [type of *Sorbus hemsley* (Schneid.) Rehder].

A beautiful tree to about 8 metres tall carrying handsome aria-type leaves which are brilliant silver on the undersides. Cultivated at the National Botanic Gardens, Glasnevin.

Griffitharia schwerinii (C. K. Schneid.) Rushforth (ROSACEAE) New species

Pyrus aria Forbes & Hemsley non (L.) Ehrh. in Journ. Linn. Soc. Bot. 23: 254. (1887), *Micromeles schwerinii* Schneider in Illus. Handb. Laubholzk. 1: 702. fig. 388. (1906); in Fedde. Rep. Spec. Regni. Veg. Nov. 3: 15. (1906). *Sorbus henryi* Rehder in Sargent Pl. Wilson. 2: 274. (1915).

Sichuan: Henry's Chinese collector with Pratt, no locality stated but may be any of the following, Leshan, Emei Shan, Wa Shan or Kangding, 8957 (type of *Sorbus henryi* Rehder).

Griffitharia sp. (ROSACEAE)

Pyrus aff. *vestita* Wall. Forbes & Hemsley in Journ. Linn. Soc. Bot. 23: 259. (1887).

Hubei: Badong, 862.

Grona griffithiana (Benth.) H. Ohashi & K. Ohashi (FABACEAE)

Desmodium griffithianum Benth. Anonymous in Kew List of Determinations (volume 34).

Yunnan: Simao, 13349.

Distributed in India, Myanmar, western China and Indochina.

Grona heterocarpos (L.) H. Ohashi & K. Ohashi (FABACEAE)

Desmodium buergeri Miq. Anonymous in Kew List of Determinations (volume 24). *Desmodium heterocarpon* (L.) DC. Zheng in Journ. Wuhan Bot. Res. 2.(1): 99. (1984).

Hubei: Yichang, 77. Badong, 4810, 4815, 4871. Liantuo, 4627. **Yunnan:** Simao, grassy mountains at 1676 m., 12349C

Grona heterocarpos (L.) H. Ohashi & K. Ohashi var. ***strigosa*** (Meeuwen) H. Ohashi & K. Ohashi (FABACEAE)

Desmodium polycarpon (Poir.) DC. Henry in Trans. Asiat. Soc. Jap. 24 suppl: 33. (1896) (List Pl. Formosa).

Hainan:. Environs of Haikou, 8393. Without locality 8474. Ling-men (in the interior of the island), 8614. **Taiwan:** Kaohsiung, without number. Wanchin, 853. **Yunnan:** Mengzi, 9234, 9336. Manpan, 11300.

02

Grona heterophylla (Willd.) H. Ohashi & K. Ohashi (FABACEAE)

Desmodium heterophyllum (Willd.) DC. Henry in Trans. Asiat. Soc. Jap. 24 suppl: 33. (1896) (List Pl. Formosa). *Desmodium triflorum* (L.) DC. Henry in Trans. Asiat. Soc. Jap. 24 suppl: 33. (1896) (List Pl. Formosa).

Taiwan: Wanchin, 881. Kaohsiung, 1894.

A trailing herb, native to Asia and Australia. A weed of paddy fields in the tropics of Chinese Taiwan.

Grona reticulata (Champ. ex Benth.) H. Ohashi & K. Ohashi (FABACEAE)

Desmodium reticulatum Champ. ex Benth. Anonymous in Kew List of Determinations (volume 24).

Hainan: On Liang Shan, 16 km (10 miles) east of Haikou, 8192.

Distributed in Myanmar (Burma), China, Thailand and Vietnam.

Grona retroflexa (L.) H. Ohashi & K. Ohashi (FABACEAE)

Desmodium retroflexum (L.) DC Anonymous in Kew List of Determinations (volume 24).

Hainan: Ling-men, 8615.

Gueldenstaedtia henryi E. Ulbrich (FABACEAE)
New species

in Bot. Jahrb. Syst. 36. Biebl. 82: 59. (1905).

Gueldenstaedtia multiflora Forbes & Hemsley non Bunge in Journ. Linn. Soc. Bot. 23: 164. (1887).

Hubei: Yichang, 1238 (syntype).

Gueldenstaedtia verna (Georgi) Boriss. (FABACEAE)

Gueldenstaedtia multiflora Bunge Forbes & Hemsley in Journ. Linn. Soc. Bot. 23: 164. (1887), Nelson in Pim, The Wood & the Trees 219. (1984); in A Heritage of Beauty 313. (2000).

Hubei: On the banks of the Yangtze near Yichang, 1308. Antelope Glen, 3384. **Sichuan:** Emei Shan, Henry's Chinese collector with Pratt, 8982.

Henry sent seeds of this species to Kew where they arrived on October 29th 1886.

Guilandina bonduc L. (FABACEAE)

Caesalpinia bonducella (L.) Fleming Anonymous in Kew List of Determinations (volume 24).

Hainan: Environs of Haikou, 8379.

Guilandina minax (Hance) G. P. Lewis (FABACEAE)

Caesalpinia morsei Dunn in Journ. Linn. Soc. Bot. 35: 192. (1903).

Yunnan: Mountains to the north-west of Simao at 1,650 metres, 10739 (isotype of *Caesalpinia morsei* Dunn, collected 22nd July 1898).

Eastern Himalaya to southern China and Indochina.

Gymnanthemum extensum (DC.) Steetz. (ASTERACEAE)
Vernonia cylindriceps C. B. Clarke Anonymous in Kew List of Determinations (volume 34).

Yunnan: Mengzi, 9129. Simao, 12856.

Gymnema spp. (ORCHIDACEAE)
Anonymous in Kew List of Determinations (volume 24 & 34).

Hubei: Changyang, 7808. **Yunnan:** Mengzi, 13800.

Gymnema sylvestre (Retz.) R. Br. ex Sm. (APOCYNACEAE)
Gymnema affine Henry non Decne in Trans. Asiat. Soc. Jap. 24 suppl: 61. (1896). *Gymnema alterniflorum* (Lour.) Merr. Li in Woody Flora of Taiwan 799. fig. 323. (1963).

Hainan: Environs of Haikou, 8049. At Nodoa in the interior of the island, 8462. Environs of Kiungchow, 8697. **Taiwan:** Kaohsiung, without number. Oluanpi, 1276. Tamsui, 1407.

A vigorous climber, native to southern and southeastern China, in Taiwan this species is found in low altitude forest. In the list of medicines prepared by the Imperial Maritime Customs Service this plant was named *wu-hsueh-teng* and was exported fron China.

Gymnocarpium oyamense (Baker) Ching (DRYOPTERIDACEAE)
Polypodium gymnogrammoides Baker Anonymous in Kew List of Determinations (volume 24).

Hubei: Donghu, 6440.

Distributed from Nepal, to China, Japan and the Philippines.

Gymnocladus chinensis Baill. (FABACEAE)
Forbes & Hemsley in Journ. Linn. Soc. Bot. 23: 207. (1887), Henry in Journ. China Br. Roy. Asiat. Soc. 22: 277. (1887) (Chinese Names of Plants), Nicholson in Garden and Forest 139. (1889), Henry in Notes Econ. Bot. China 18. (1893), Bretschneider in Hist. Eur. Bot. Disc. China 2: 782. (1898), Henry in The Trees of Gt. Brit. & Irel. 2: 427. (1907), Morley in Glasra 3: 79. (1979), Zheng in Journ. Wuhan Bot. Res. 2.(1): 100. (1984), Nelson in Pim, The Wood & the Trees 221. (1984); in A Heritage of Beauty 313. (2000).

Hubei: Yichang, 775. Liantuo, 1990, 2586, 3873, 5576a. **Sichuan:** South Wushan, 5576.

According to Hemsley, Henry's specimens were the finest to reach the Kew herbarium. Henry stated the tree was common at the first rapids at Nanto (now Liantuo) in the Xiling Gorge. This is the soap tree, which Henry stated was the *ju tsao chio* or *fei-tsao*, or fat black bean of the Chinese, once used by women in washing their hair. This tree was introduced through the Royal Botanic Gardens, Kew from seed Henry had sent from Yichang in 1888.

Gymnosporia diversifolia Maxim. (CELASTRACEAE)
Li in Woody Flora of Taiwan 473. fig. 178. (1963).

Celastrus diversifolius (Maxim.) Hemsl. Henry

in Trans. Asiat. Soc. Jap. 24 suppl: 27. (1896) (List Pl. Formosa).

Hainan: Environs of Kiungchow, 8685. **Taiwan:** Kaohsiung, without number. Oluanpi, 239, 308, 347. Wanchin, 1081.

A spiny shrub, native to Thailand, Indochina, southern China (including Taiwan and Hainan), the Philippines, Japan (south). In Taiwan it grows mainly along the seashore in the southern and eastern parts of the island.

Gymnosporia royleana Wall. ex M. A. Lawson (CELASTRACEAE)

Anonymous in Kew List of Determinations (volume 34).

Yunnan: Mengzi, 9391.

Gymnosporia variabilis (Hemsl.) Loesener (CELASTRACEAE)

in Bot. Jahrb. xxix. 446. (1900), Rehder & E. H. Wilson in Sarg. Pl. Wils. 2: 359. (1916), Zheng in Journ. Wuhan Bot. Res. 2.(1): 126. (1984).

Celastrus variabilis Hemsley in Journ. Linn. Soc. Bot. 23: 124. (1886), Henry in Journ. China Br. Roy. Asiat. Soc. 22: 235. (1887) (Chinese Names of Plants), Nelson in A Heritage of Beauty 311. (2000).

Hubei: Yichang, 130, 449, 1022, 1067, 3016. Liantuo, (October 1886) 3016, 3056, 4421. Without locality, 349a., 449a.

First found in the Yichang (Xiling) Gorge by Charles Maries in 1879 and introduced by him through James Veitch & Sons, Chelsea. Hemsley based this species on material collected by Henry and Maries. According to Augustine Henry it was known colloquially as *tz'u ch'a*. Seeds were taken off one of Henry's herbarium specimens at Kew on October 29th 1887 and sent for propagation.

Gymnotheca chinensis Decne. (SAURURACEAE)

Hooker's. Icon. Pl. t. 1873. (1889), Forbes & Hemsley in Journ. Linn. Soc. Bot. 26: 364. (1891).

Hubei: Lung Tung Ch'i Glen (near Yichang), 4180.

Gynochthodes umbellata (L.) Razafirm & B. Bremer (RUBIACEAE)

Morinda umbellata L. Henry in Trans. Asiat. Soc. Jap. 24 suppl: 50. (1896) (List Pl. Formosa), Hutchinson in Sargent Pl. Wilson. 3: 413. (1917), Li in Woody Flora of Taiwan 860. (1963).

Hainan: Near the Kiungchow Pagoda, 7975. Without locality, 8578. **Taiwan:** Kaohsiung, without number. Oluanpi, 936. **Yunnan:** In Pingbian in forests on the Daweishan Range at 1,640 metres, 11246.

A climbing shrub, native to the Philippines and southern China. Common on the southern part of Chinese Taiwan.

Gynostemma cardiospermum Cogniaux ex Oliver (CUCURBITACEAE) New species

in Hooker's Icon. Pl. 23: t. 2225. (1894), Bretschneider in Hist. Eur. Bot. Disc. China 2: 784. (1898).

Trirostellum cardiospermum (Cogn. ex Oliv.) Z. P. Wang & Q. Z. Xie in Acta Phytotax. Sin. 19: 463. (1981).

Hubei: Fang Xian, 6701 (isosyntype), 6779 (isosyntype), 7613 (type).

Gynostemma pentaphyllum (Thunb.) Makino (CUCURBITACEAE)

Gynostemma pedatum Blume Forbes & Hemsley in Journ. Linn. Soc. Bot. 23: 320. (1887).

Hubei: Liantuo, 2618, 4623. Yichang, 3303. Antelope Glen, 3458, 4346. Fang Xian, 6728. Jianshi, 7626. **Yunnan:** Mengzi, 9006.

Gynostemma sp. (CUCURBITACEAE)

Anonymous in Kew List of Determinations (volume 24).

Hubei: Antelope Glen, 4260. Without locality, 5377c. Donghu, 6428, 6436.

Gynura bicolor (Roxb. ex Willd.) DC. (ASTERACEAE)

Henry in Trans. Asiat. Soc. Jap. 24 suppl: 55. (1896) (List Pl. Formosa).

Taiwan: Wanchin, 1715. **Yunnan:** In Pingbian on the Daweishan Range at 1,650 metres, 9381b.

A handsome red-flowed climber, widespread in Asia.

Gynura divaricata (L.) DC. (ASTERACEAE)

Gynura ovalis DC. Henry in Trans. Asiat. Soc. Jap. 24 suppl: 46. (1896) (List Pl. Formosa).

Taiwan: Oluanpi, 325, 2048. Wanchin, 484, 485.

Gynura japonica (Thunb.) Juel (ASTERACEAE)

Gynura pinnatifida DC. Henry in Journ. China Br. Roy. Asiat. Soc. 22: 267. (1887) (Chinese Names of Plants), Forbes & Hemsley in Journ. Linn. Soc. Bot. 23: 448. (1888), Henry in Notes Econ. Bot. China 22. (1893).

Hubei: Yichang, 2959.

According to Henry this species was known colloquially as *chien chung hsiao* in the regions of Badong, Changyang and Yichang. It Yichang where it was cultivated in pots it was also called *san ch'i,* a name also gave to a drug extracted from a species of *Panax* from Yunnan.

Gynura pseudochina (L.) DC. (ASTERACEAE)

Anonymous in Kew List of Determinations (volume 34).

Yunnan: Mengzi, 10516, 10718.

Distributed in India, Sri Lanka, Myanmar, China and Thailand.

Gynura sp. (ASTERACEAE)

Anonymous in Kew List of Determinations (volume 34).

Yunnan: Simao, 12297.

Gypsophila sp. (CARYOPHYLLACEAE)

Saponaria vaccaria Henry non L. in Journ. China Br. Roy. Asiat. Soc. 22: 275. (1887) (Chinese Names of Plants).

Hubei: Yichang, 790, 2218. Badong, 1766.

According to Henry this species was known colloquially in the Three Gorges region as *t'ui ching ts'ao.*

Habenaria austrosinensis T. Tang & F. T. Wang (ORCHIDACEAE) New species

in Bull. Fan Mem. Inst. Biol. 7: 134. (1936).

Yunnan: Simao at 1,219 metres, 12467 (type).

Native to China (Yunnan) and Thailand.

Habenaria ciliolaris Kranzl. (ORCHIDACEAE)

Rolfe in Journ. Linn. Soc. Bot. 36: 58. (1903).

Hubei: South Donghu, 7487.

A handsome terrestrial orchid, widely distributed in China.

Habenaria davidii Franch. (ORCHIDACEAE)

Habenaria ensifolia Rolfe non Lindl. in Journ. Linn. Soc. Bot. 36: 58. (1903), *Habenaria leucopecten* Schlecter in Repert. Spec. Nov. Regni Veg. Beih. 4: 49. (1919).

Yunnan: In rocky and grassy mountains in Mengzi at 1,600 to 2,100 metres, 11119 (type of *Habenaria leucopecten* Schlecter).

Habenaria delavayi Finet (ORCHIDACEAE)

Habenaria yunnanensis Rolfe in J. Linn. Soc. Bot. 36: 61. (1903).

Yunnan: On rocky mountains in Mengzi at 1,900 metres, 11121 (type of *Habenaria yunnanensis* Rolfe).

Habenaria dentata (Sw.) Schltr. (ORCHIDACEAE)

Habenaria miersiana Champ. ex Benth. Henry in Trans. Asiat. Soc. Jap. 24 suppl: 93. (1896) (List Pl. Formosa), Rolfe in Journ. Linn. Soc. Bot.36: 60. (1903).

Hubei: Liantuo, 4576, 4611. **Taiwan:** Oluanpi, 1212. Wanchin, 1607. **Yunnan:** On grassy mountains near Mengzi at 1,600 to 1,900 metres, 11098, 11123, 11128. Simao, 12535.

A terrestrial orchid to 80 cm tall, distributed in the Himalaya, India, Myanmar, China, Indochina, Thailand and southern Japan.

Habenaria ensifolia Lindl. (ORCHIDACEAE)

Rolfe in Journ. Linn. Soc. Bot. 36: 58. (1903).

Hubei: Donghu at 1,300 metres, 6446. South Badong, 7384.

Habenaria fordii Rolfe (ORCHIDACEAE)

Rolfe in Journ. Linn. Soc. Bot. 36: 58. (1903).

Yunnan: In woods near Mengzi at 1,300 metres, 11101.

Endemic to China and distributed in the provinces of Guangdong, Guangxi and Yunnan. The specific epithet commemorates Charles Ford of Hong Kong Botanic Gardens who discovered this handsome orchid in Guangdong.

Habenaria polytricha Rolfe (ORCHIDACEAE) New species

in Hooker's Icon. Pl. t. 2496. (1896), Henry in Trans. Asiat. Soc. Jap. 24 suppl: 93. (1896) (List Pl. Formosa), Bretschneider in Hist. Eur. Bot. Disc. China 2: 792. (1898), Rolfe in Journ. Linn. Soc. Bot. 36: 60. (1903).

Taiwan: Oluanpi, 1246 (type).

A terrestrial orchid to 40 cm tall, native to southern China, including Taiwan where it is found in broadleaved forests below 1,500 metres.

Habenaria cf. *reniformis* (D. Don) Hook. f. (ORCHIDACEAE)

Anonymous in Kew List of Determinations (volume 34).

Yunnan: Simao, 12241.

Habenaria rostellifera Rchb. f. (ORCHIDACEAE)

Habenaria hancockii Rolfe Rolfe in Journ. Linn. Soc. Bot. 36: 59. (1903).

Yunnan: On grassy mountains near Mengzi at 1,500 to 1,900 metres, 11122.

Also previously collected by the Irish plant collector, William Hancock, who, like Augustine Henry worked in the Chinese Imperial Maritime Customs Service. Hancock was based at Mengzi in the early 1890s before Henry's arrival there (they had also met on one occasion in Shanghai).

Habenaria spp. (ORCHIDACEAE)

Anonymous in Kew List of Determinations (volume 24 & 34).

Hubei: Fang Xian, 6923. South Donghu, 7708. South Badong, 7349. **Yunnan:** Simao, 12373, 12634, 13133, 13591.

Hackelochloa granularis (L.) Kuntze (POACEAE)

Manisuris granularis (L.) L.f. Henry in Trans. Asiat. Soc. Jap. 24 suppl: 108. (1896) (List Pl. Formosa), Rendle in Journ. Linn. Soc. Bot. 36: 363. (1904).

Hainan: On hummocks near Haikou, 8120. **Taiwan:** Tamsui, 1478. Wanshoushan, 1865.

An annual species carrying culms to 30 cm tall, found in most tropical regions of the world.

Haldina cordifolia (Roxb.) Ridsdale (RUBIACEAE)

Adina cordifolia (Roxb.) Brandis Hutchinson in Sargent Pl. Wilson. 3: 390. (1917).

Yunnan: Near the Red River at Hsu Kai at 330 metres, 9570.

A large deciduous tree with a spreading crown, native to the Himalaya, Sri Lanka, Myanmar, Thailand, Vietnam and China (Yunnan).

Halenia elliptica D. Don (GENTIANACEAE)

Anonymous in Kew List of Determinations (volume 24 & 34).

Hubei: Badong, 2398, 2456. **Yunnan:** Mengzi, 9269.

Distributed in Nepal, India, Bhutan, Krygyzstan, Myanmar and China.

Halenia grandiflora (Hemsl.) M. Y. Chen (GENTIANACEAE) New species

Pl. Diversity 44: 297. (2021).

Halenia sp. nova Henry in Journ. China Br. Roy. Asiat. Soc. 22: 253. (1887) (Chinese Names of Plants). *Halenia elliptica* D. Don var. *grandiflora* Hemsley in Journ. Linn. Soc. Bot. 26: 141. (1890), Henry in Notes Econ. Bot. China 60. (1893).

Hubei: Badong, 517 (September 1885), 540, 4824, 4982 (type of *Halenia elliptica* D. Don var. *grandiflora* Hemsley), 5198, 5212, 6112. South Badong, 6151. Fang Xian, 6531. **Sichuan:** Henry's Chinese collector with Pratt, no locality stated but may be Leshan, Emei Shan, Wa Shan or Kangding, 8992.

A beautiful upright annual carrying long-spurred blue-voilet flowers with white centres in autumn. Henry Badong collector called this species the *lao hu hua*, it still a relatively common plant in Badong where it was collected by the Glasnevin Central China Expedition (GCCE 297). It was also collected by the 2004 Glasnevin Expedition to the Three Gorges area near Xingshan (GCCE 521 & GCCE 572) where it is particularly abundant.

Hamamelis mollis Oliver (HAMAMELIDACEAE) New species

in Hooker's. Icon. Pl. 18: t. 1742. (1888), Forbes & Hemsley in Journ. Linn. Soc. Bot. 23: 290. (1887), Bean in Gard. Chron. ser. 3, 24: 363. (1898), Henry in Flora & Sylva 1: 218. (1903), Rehder & E. H. Wilson in Sargent Pl. Wilson. 1: 431. (1913), Zheng in Journ. Wuhan Bot. Res. 2(1): 65. (1984), Nelson in Pim, The Wood & the Trees 231. (1984); in A Heritage of Beauty 313. (2000,) O' Brien in Augustine Henry, An Irish Plant Collector in China 21. (2002).

Hubei: Badong, 3791 (type), 3793, 3793a., 4918. Donghu, 6412.

The Chinese witch-hazel, one of the finest winter-flowering shrubs. This common shrub was described from material gathered by Henry in 1887 but had been previously discovered and introduced by Charles Maries in 1879 while collecting for Messrs Veitch. It is incredible to think that such a valuable shrub grew for 20 years unrecognised in their nursery and was merely concidered a superior form of *Hamamelis japonica* Sieb. & Zucc. The Veitch plant was first recognised as being distinct by George Nicholson,

the Curator of Kew and Messrs Veitch did not begin to propagate it until 1898 (according to W. J. Bean, Maries' original plant was a bush of about 75 cm tall in 1898). Wilson noted it was particularly abundant in Changyang. He reintroduced it by means of both living plants and seeds in 1907.

Hancenia forbesii (H. Boissieu) Pimenov & Kljuykov (APIACEAE) New species

Notopterygium forbesii H. Boissieu in Bull. Herb. Boiss. 2: 840. (1903), Pu et al. in Acta Phytotax. Sin. 38: 432. (2000).

Hubei: Fang Xian, 6629 (isotype of *Notopterygium forbesii* H. Boissieu).

Hanceola sinensis (Hemsl.) Kudo (LAMIACEAE)

Hancea sinensis Hemsl. Dunn in Notes Roy. Bot. Gard. Edinburgh 6: 153. (1915).

Yunnan: In Pingbian, on the Daweishan Range, 9196. Endemic to China.

Hancockia uniflora Rolfe (ORCHIDACEAE) New genus

in Journ. Linn. Soc. Bot. 36: 20. (1903).

Yunnan: In forests on the Daweishan Range in Pingbian at 1,600 metres, 11112 (type).

This genus was discovered by Henry 64 kilometres south-west of Mengtzi on the Daweishan Range near the Vietnamese border. Rolfe named the genus in honour of William Hancock, also from Ireland. Hancock, like Henry sent specimens from Taiwan and Yunnan to Kew. Distributed in Yunnan, Vietnam and Japan.

Haplanthus laxiflorus (Blume) Gnanasek, G. V. S. Murthy & Y. F. Deng (ACANTHACEAE)

Andrographis tenuiflora T. Anderson Anonymous in Kew List of Determinations (volume 34).

Yunnan: Yuanjiang, 13234.

Haplopteris elongata (L.) E. H. Crane (PTERIDACEAE)

Vittaria elongata (L.) Sw. Henry in Trans. Asiat. Soc. Jap. 24 suppl: 116. (1896) (List Pl. Formosa), Christ in Bull. Herb. Boiss. 6: 868. (1898).

Taiwan: Oluanpi, 654. **Yunnan:** South of the Red River from Manmei at 1,950 metres, 9195.

Haplopteris plurisulcata (Ching) X. C. Zhang (PTERIDACEAE) New species

Vittaria plurisulcata Ching in Sinensia 1(12): 186. pl. 4. f 1-3. (1931).

Yunnan: Mengzi, on trees, 9195c., (type of *Vittaria plurisulcata* Ching, collected 27th November 1896).

Harashuteria hirsuta (Baker) K. Ohashi & H. Ohashi (FABACEAE)

Shuteria hirsuta Baker Anonymous in Kew List of Determinations (volume 34).

Yunnan: Mengzi, 9312. Simao, 12480, 12585.

Distributed across Nepal, India, Bhutan, Myanmar, Yunnan, Thailand, Laos and Vietnam.

Harpullia cupanioides Roxb. (SAPINDACEAE)

Anonymous in Kew List of Determinations (volume 34).

Yunnan: Simao, 12140, 12144.

A widespread species, native to India, Bangladesh, China (Guangdong, Hainan and Yunnan) and the Philippines.

Harrisonia perforata (Blanco) Merr. (SIMAROUBACEAE)

Harrisonia bennettii Anonymous non A. W. Benn. in Kew List of Determinations (volume 24).

Hainan: Without locality, 8469, 8521.

Haymondia wallichii (DC.) A. N. Egan & B. Pan (FABACEAE)

Pueraria wallichii DC. Anonymous in Kew List of Determinations (volume 34).

Yunnan: Yuanjiang, 13233.

Hedera sinensis (Tobl.) Hand.-Mazz. (ARALIACEAE)

Hedera helix Henry non L. in Journ. China Br. Roy. Asiat. Soc. 22: 263. (1887) (Chinese Names of Plants), Forbes & Hemsley in Journ. Linn. Soc. Bot. 23: 434. (1887). *Hedera himalaica* Harms & Rehder non Tobler in Sargent Pl. Wilson 2: 555. (1916). *Hedera nepalensis* K. Koch var. *sinensis* (Tobl.) Rehd. Zheng in Journ. Wuhan Bot. Res. 2.(1): 161. (1984).

Hubei: Badong, 704 (July 1885), 3700, 5120. Yichang, 1258, 3261 (February 1887), 3326. Liantuo, 2617, 2984, 3836.. **Yunnan:** Mengzi at 1,500 to 2,100 metres, 9856, 9856a. Simao, 13304.

According to Henry this ivy was common on cliffs at Badong and Yichang and was called *p'a shan lu*. It is still abundant in the Three Gorges tegion and was collected by the Glasnevin Central China Expedition (GCCE 371) in Badong in October 2002 and again in the Red Dragon Rerseve in Xingshan (GCCE 675) in October 2004. Native to China and Vietnam.

Hedychium coronarium J. Koenig (ZINGIBERACEAE)

Anonymous in Kew List of Determinations (volume 24).

Sichuan: Henry's Chinese collector with Pratt, no locality stated, may be either Emei Shan or Kangding, 8831, 8893.

A perennial rhizomatous herb to 2 metres tall, bearing terminal spikes of white fragrant flowers in September. Widely distributed across India, China, Laos, Vietnam and Malaysia.

Hedychium gomezianum Wall. (ZINGIBERACEAE)

Schumann in Engler, Pflanzenr. iv. 46: 55. (1904).

Yunnan: On grassy mountains near Mengzi at 1,600 metres, 10034a. Simao at 1,900 metres, 13058a.

Hedychium spp. (ZINGIBERACEAE)

Anonymous in Kew List of Determinations (volume 34).

Yunnan: Mengzi, 10830, 11078, 13806. Simao, 12327, 12390.

Hedyotis spp. (RUBIACEAE)

Anonymous in Kew List of Determinations (volume 24 & 34).

Hubei: At Shih pan shan (off the Xiling (Yichang) Gorge), 4186. Yichang, 4345. Badong, 4821. Jianshi, 6042. Changyang, 6201. South Badong, 7731. **Yunnan:** Mengzi, 10134. In the Red River valley near Manhao, 9539. At Fengchunling near the Red River at 2,200 metres, 11207. Simao, 11898, 12221, 12420, 13198, 13483.

Hedyotis uncinella Hook.& Arn. (RUBIACEAE)

Henry in Trans. Asiat. Soc. Jap. 24 suppl: 49. (1896) (List Pl. Formosa).

Hainan: Ling-men, 8619. **Taiwan:** Wanchin, 111. Oluanpi, 993. **Yunnan:** Mengzi, 9016, 10279. Simao, 12240.

Distributed across India and China.

Helianthus annuus L. (ASTERACEAE)

Anonymous in Kew List of Determinations (volume 24).

Hubei: Changyang, 7509.

The sun flower or *heung yat kw'ai* (literally; facing the sun), a native of Canada, the USA and Mexico and cultivated both as an ornamental and for its seeds which are roasted and eaten in China.

Helichrysum indicum (L.) Grierson (ASTERACEAE)

Gnaphalium indicum L. Henry in Trans. Asiat. Soc. Jap. 24 suppl: 53. (1896) (List Pl. Formosa).

Taiwan: Oluanpi, 288, 638.

Helicia cochinchinensis Lour. (PROTEACEAE)

Henry in Trans. Asiat. Soc. Jap. 24 suppl: 80. (1896) (List Pl. Formosa), Sleumer in Blumea 8: 75. (1955), Li in Woody Flora of Taiwan 141. (1963).

Helicia annularis W. W. Smith in Notes Roy. Bot. Gard. Edinburgh 9: 178. (1916).

Taiwan: Wanchin, 819. **Yunnan:** Mengzi at 1,200 metres, 10756 (type of *Helicia annularis* W. W. Smith). In woods near Simao at 1,500 metres, 12315.

A small evergreen tree, native to Vietnam, Cambodia, Thailand, southern China and Japan. Found in broad-leaved forest at low altitude in Chinese Taiwan.

Helicia formosana Hemsley (PROTEACEAE) New species

in Ann. of Bot. 9: 156. (1895), Forbes & Hemsley in

Journ. Linn. Soc. Bot. 26: 156. (1892), Henry in Trans. Asiat. Soc. Jap. 24 suppl: 80. (1896) (List Pl. Formosa), Sleumer in Blumea 8: 52. (1955), Li in Woody Flora of Taiwan 141. fig. 49. (1963).

Taiwan: Wanchin, 511, 805. Oluanpi, 1225.

A small evergreen tree known as the *shan-p'i-p'a* and found according to Henry, in mountains near watercourses. Native to Vietnam and China including Taiwan. A common tree in broadleaved forest throughout Taiwan at low to medium altitudes.

Helicia grandis Hemsley (PROTEACEAE) New species

in Hooker's Icon. Pl. 27: t. 2631. (1900) Hemsley in Journ. Linn. Soc. Bot. 34: 177. (1900) Sleumer in Blumea 8: 41. (1955)

Yunnan: In Pingbian, in forests on the Daweishan Range at 1,650 metres, 10704 (isotype).

An evergreen tree to 10 metres tall, distributed in China (southern Yunnan) and Vietnam. A handsome species bearing terminal or axillary racemes of rusty coloured flowers.

Helicia nilagirica Bedd. (PROTEACEAE)

Sleumer in Blumea 8: 669. (1955).

Helicia stricta Diels in Fedde, Repert. Spec. Nov. Regni Veg. 13: 527. (1915).

Yunnan: Simao at 1,270 to 1,525 metres, 11935 (type of *Helicia stricta* Diels), 12858.

Distributed in Nepal, Bhutan, Myanmar, Yunnan, Thailand, Cambodia, Laos and Vietnam.

Helicia pyrrhobotrya Kurz (PROTEACEAE)

Sleumer in Blumea 8: 65. (1955).

Yunnan: In forests to the south-east of Simao at 1,200 to 1,800 metres, 12808.

Native to China (Guangxi and Yunnan) and Myanmar. This collection proved to be a new record for China.

Helicia silvicola W. W. Smith (PROTEACEAE) New species

in Notes Roy. Bot. Gard. Edinburgh 9: 181. (1916), Sleumer in Blumea 8: 64. (1955).

Yunnan: In forests near Simao at 2,100 metres, 13075 (isotype).

Endemic to southern Yunnan.

Helicia spp. (PROTEACEAE)

Anonymous in Kew List of Determinations (volume 34).

Yunnan: Mengzi, 11383. Simao, 13116, 13596.

Helicia vestita W. W. Smith (PROTEACEAE) New species

in Notes Roy. Bot. Gard. Edinburgh 9: 181. (1916), Sleumer in Blumea 8: 31. (1955).

Yunnan: In the mountains to the south of Simao at 1,370 metres, 12159 (isotype).

Distributed in China (Yunnan, Tibet (Xizang)) and Thailand.

Heliciopsis henryi (Diels) W. T. Wang (PROTEACEAE) New species

in Acta Phytotax. Sin. 5: 307. (1956).

Helicia henryi Diels in Repert. Spec. Nov. Regni. Veg. 13: 528. (1915). *Helicia pallidiflora* W. W. Smith in Notes Roy. Bot. Gard. Edinburgh 9: 179. (1916). *Heliciopsis terminalis* Sleumer non (Kurz) Sleumer Sleumer in Blumea 8: 80. (1916).

Yunnan: In woods south of Simao at 1,525 metres, 11910 (type of *Helicia henryi* Diels). In forests near Simao at 1,650 metres, 11910a., (isotype of *Helicia henryi* Diels, isotype of *Helicia pallidiflora* W. W. Smith).

Endemic to Yunnan.

Helicteres angustifolia L. (MALVACEAE)

Henry in Trans. Asiat. Soc. Jap. 24 suppl: 22. (1896) (List Pl. Formosa), Li in Woody Flora of Taiwan 559. fig. 217. (1963).

Hainan: Hummocks near Haikou, 8093. Environs of Haikou, 8349. **Taiwan:** Kaohsiung Plain, 6. Wanchin, 538, 1768.

A shrubby paleotropic weed, common in Taiwan.

Helicteres elongata Wall. ex Mast. (MALVACEAE)

Anonymous in Kew List of Determinations (volume 34).

Yunnan: Manpan, 11185. Simao, 12218.

Native to India, Myanmar, China (Guangxi and Yunnan) and Thailand.

Helicteres hirsuta Lour. (MALVACEAE)

Helicteres spicata Colebr. ex G. Don Anonymous in Kew List of Determinations (volume 24). *Helicteres spicata* Colebr. ex G. Don var. *hainanensis* Hance Anonymous in Kew List of Determinations (volume 24).

Hainan: Environs of Haikou, 8338. 8 km (5 miles) south of Kuingchow, 8172, 8247. Without locality, 8483.

Also earlier collected in Hainan by the Rev'd B. C. Henry, an American missionary who made extensive collections in Guangdong and visited the island of Hainan in October and November 1882. In the central mountains of the island he visited the Li, an independent group of native peoples.

Helicteres isora L. (MALVACEAE)

Anonymous in Kew List of Determinations (volume 34).

Yunnan: Yuanjiang, 13256.

Distributed in India, Sri Lanka, China (Yunnan and Hainan), Thailand and Vietnam.

Heliotropium arboreum (Blanco) Mabb. (BORAGINACEAE)

Tournefortia argentea L.f. Henry in Trans. Asiat. Soc.

02

Jap. 24 suppl: 62. (1896) (List Pl. Formosa). *Messerschmidia argentea* (L. f.) Johnston Johnston in Journ. Arnold Arb. 32: 121. (1951), Li in Woody Flora of Taiwan 815. (1963).

Taiwan: Oluanpi, 322.

A small tree, native from tropical Asia to Madagascar, Malaya, tropical Australia and Polynesia. A sea shore tree in Chinese Taiwan. The fruits of this species are well adapted for widespread dispersal by ocean currents. *Heliotropium arboreum* was collected by the Glasnevin Taiwan Expedition (GTE 11) at Shadao Shell beach near Oluanpi (South Cape of Henry and Wilson) in October 2004. A beautiful species, its brilliant grey foliage makes a spectacular contrast against the azure blue Taiwean Straits in southern Taiwan.

Heliotropium formosanum I. M. Johnston (BORAGINACEAE) New species

in Journ. Arnold Arb. 32: 114. (1951).

Heliotropium strigosum Henry non Willd. in Trans. Asiat. Soc. Jap. 24 suppl: 63.(1896) (List Pl. Formosa).

Taiwan: Oluanpi, 956.

According to Henry this species was known coloquially as *kou-i-tan*.

Heliotropium indicum L. (BORAGINACEAE)

Henry in Trans. Asiat. Soc. Jap. 24 suppl: 63. (1896) (List Pl. Formosa), Johnston in Journ. Arnold Arb. 32: 111. (1951).

Hainan: Environs of Haikou, 8414. **Taiwan:** Kaohsiung, without number. Oluanpi, 309.

A coarse annual weed, distributed in Asia throughout Indochina and southern China (including Hainan and Taiwan). Now widely naturalised in warmer parts of the world but thought to have originated in America.

Heliotropium montanum (Lour.) Rueangs. & Chantar. (BORAGINACEAE)

Tournefortia montana Loureiro Johnson in Journ. Arnold Arb. 32: 116. (1951).

Yunnan: In Yulo forests (Xishuangbanna) at 1,300 metres, 12862.

A large scandent shrub with yellow flowers, distributed in Guangdong, Yunnan, the South China Sea islands and Vietnam.

Heliotropium sarmentosum (Lam.) Craven (BORAGINACEAE)

Tournefortia sarmentosa Lam. Henry in Trans. Asiat. Soc. Jap. 24 suppl: 62. (1896) (List Pl. Formosa).

Taiwan: Kaohsiung, without number. Oluanpi, without number.

A scandent vine of indefinate length, widely dispersed from Mauritius to Indochina, Malaya, the Philippines and tropical Australia. A sea-shore tree in Chinese Taiwan where it was known coloquially in the late 19th century as *ching-pung-ting* according to Henry.

Heliotropium sp. (BORAGINACEAE)

Tournefortia sp. Anonymous in Kew List of Determinations (volume 34).

Yunnan: Simao, 11733.

Helixanthera parasitica Lour. (LORANTHACEAE)

Loranthus adpressus (Van Teigh.) Lecompte Lecompte in Sargent Pl. Wilson. 3: 317. (1917).

Yunnan: Simao, 11664.

A widespread species in Asia.

Hellenia lacera (Gagnep.) Govaerts (ZINGIBERACEAE)

Costus lacerus Gagnep. Maas in Blumea 25: 546. fig. 1. (1979).

Yunnan: Mengzi, 11265. Simao, 12296.

A handsome rhizomatous perennial to 2.5 metres tall, bearing terminal spikes of white tubular flowers blotched yellow in the centres during the months of May to July. Distributed in Nepal, India, Bhutan, Myanmar and western China.

Hellenia speciosa (J. Koenig) S. R. Dutta (ZINGIBERACEAE)

Costus speciosus (J. Koenig) Smith Henry in Trans. Asiat. Soc. Jap. 24 suppl: 95. (1896) (List Pl. Formosa), Wright in Journ. Linn. Soc. Bot. 36: 71. (1903).

Taiwan: Kaohsiung, 1204.

A perennial to 1.5 metres tall, native to Australia, Indo-Malaysia, Java, the Philippines and China (including Hainan and Taiwan).

Helwingia chinensis Batalin (HELWINGIACEAE) New species

in Acta. Hort. Petrop. 13: 97. (1893), Harms in Bot. Jahrb. xxix. 505. (1900), Wagnerin in Fedde. Rep. Spec. Regni. Spec. Nov. iv. 337. (1907); in Engler. Pflanzenr. iv.-229. 37. fig. 8. (1910), Rehder in Sargent Pl. Wilson 2: 571. (1916).

Helwingia rusciflora Forbes & Hemsley non Willd. in Journ. Linn. Soc. Bot. 23: 341. (1887), Henry in Journ. China Br. Roy. Asiat. Soc. 22: 275. (1887) (Chinese Names of Plants). *Helwingia chinensis* Batal. var. *macrocarpa* Pamp. Rehder in Sargent Pl. Wilson. 2: 571. (1916), Zheng in Journ. Wuhan Bot. Res. 2(1): 165. (1984).

Hubei: Badong, 2849, 5127, 5282d., 6719b., (isosyntype). South Badong, 5282. Fang Xian, 6719 (type). Jianshi, 5795. **Sichuan:** South Wushan, 6719a.

According to Henry this species was known colloquially in the Three Gorges region as *t'ung tiao hua*.

Helwingia chinensis Batalin var. *crenata* (Lingelsh.) Fang (HELWINGIACEAE) New variety

in Acta Phytotax. Sin. 1: 171. (1951).

Helwingia ruscifolia Forbes & Hemsley non Willd.

in Journ. Linn. Soc. Bot. 23: 341. (1887), Henry in Journ. China Br. Roy. Asiat. Soc. 22: 275. (1887) (Chinese Names of Plants). *Helwingia himalaica* Wagnerin non Hook. f. & Thoms. ex C. B. Clarke in Engler, Pflanzenr. iv. 229: 37. (1910), Rehder in Sargent Pl. Wilson. 2: 571. (1916). *Helwingia chinensis* Batal. var. *macrocarpa* Pamp. Rehder in Sargent Pl. Wilson 2: 571. (1916), Zheng in Journ. Wuhan Bot. Res. 2(1): 165. (1984).

Hubei: Liantuo, 4504, 5282a. Sichuan: South Wushan, 5282b. **Yunnan:** Simao at 1,300 to 1,600 metres, 11992b.

This variety was based on a collection made by Limpricht in Sichuan in April 1904, it had however been found by Henry in the spring of 1888. Endemic to China.

Helwingia himalaica Hook. f. & Thoms ex C. B. Clarke (HELWINGIACEAE)

Wagnerin in Engler, Pflanzenr. iv. 229: 37. (1910), Rehder in Sargent Pl. Wilson. 2: 571. (1916).

Helwingia himalaica Hook. f. & Thoms ex Clarke var. *parviflora* H. L. Li in Journ. Arn. Arb. 25: 310. (1944).

Yunnan: Near Mengzi in woods at 1,600 metres, 9032, 9032a., 9032b., 9032c. Simao at 1,300 to 1,600 metres 11992 (syntype of *Helwingia himalaica* Hook. f. & Thoms ex Clarke var. *parviflora* H. L. Li).

Distributed in Nepal, India, Bhutan, Myanmar, China and Vietnam. Joseph Rock also made collections of this species from the mountains above the village of Yongxi (Londre) on the Mekong-Salween watershed.

Helwingia japonica (Thunb.) F. G. Deitr. (HELWINGIACEAE)

Rehder in Sargent Pl. Wilson. 2: 570. (1916), Zheng in Journ. Wuhan Bot. Res. 2.(1): 166. (1984).

Helwingia rusciflora Forbes & Hemsley non Willd. in Journ. Linn. Soc. Bot. 341. (1887).

Hubei: Badong, 3775, 5282e. Zigui, 5282c.

A deciduous dioecious shrub to 1 metre tall, small, greenish-white flowers are produced on the mid-ribs of the centres of leaves followed by black fruits. Native to Japan and China, introduced to cultivation by Siebold in 1830. This species was collected by the Glasnervin Central China Expedition (GCCE 541) near Zhenzi town in Xingshan in September 2004.

Helwingia japonica (Thunb.) F. G. Deitr. var. *hypoleuca* Hemsley ex Rehder (HELWINGIACEAE)

New variety in Sargent. Pl. Wilson. 2: 570. (1916), Zheng in Journ. Wuhan Bot. Res. 2.(1): 165. (1984).

Hubei: Badong, 1706 (type). Without locality, 5852c.

Hemarthria compressa (L. f.) R. Br. (POACEAE)

Rottboellia compressa L. f. var. *genuina* Hack. Rendle in Journ. Linn. Soc. Bot. 36: 361. (1904).

Taiwan: Kaohsiung, 1896. **Yunnan:** Mengzi at 1,500

metres, 10169.

Distributed across India, Sri Lanka, Myanmar, Malaysia, China and Japan.

Hemerocallis fulva (L.) L. (HEMEROCALLIDACEAE)

Henry in Trans. Asiat. Soc. Jap. 24 suppl: 97. (1896) (List Pl. Formosa), Wright in Journ. Linn. Soc. Bot. 36: 115. (1903).

Hubei: South Badong, 6142. **Taiwan:** Wanchin, 818.

The day lily, this species was introduced to Europe at a very early date, probably along the northern silk route through central Asia. Henry stated that in China the day lily was once known as *Ii nan ts'ao* (boy favouring herb) because it was beleived to favour the birth of sons when worn in women's girdles. It is still used in parts of China to ease the pain of child birth, *Hemerocallis* contains asparagin, a pain relieving agent. The dried flower buds of *Hemerocallis fulva* are used as a vegetable and as a relish with meat dishes.

Hemerocallis lilioasphodelus L. (HEMEROCALLIDACEAE)

Hemerocallis flava (L.) L. Henry in Journ. China Br. Roy. Asiat. Soc. 22: 245. (1887) (Chinese Names of Plants), Wright in Journ. Linn. Soc. Bot. 36: 115. (1903).

Hubei: Yichang, 195. Badong 1851, 6063.

According to Henry this species was known colloquially in the Three Gorges region as *huang hua ts'ai*. Wilson stated the flowers of this species were eaten in Central China.

Hemerocallis minor Miller (HEMEROCALLIDACEAE)

Henry in Journ. China Br. Roy. Asiat. Soc. 22: 257. (1887), Wright in Journ. Linn. Soc. Bot. 36: 116. (1903).

Hubei: On Tsui Fu Shan near Yichang (May 1888), 1618. South Badong, 6182.

Native to eastern Siberia, China and Japan. According to Henry this species was known colloquially around Yichang as *lu ts'ung*.

Hemerocallis sp. (HEMEROCALLIDACEAE)

Anonymous in Kew List of Determinations (volume 34).

Yunnan: Mengzi, 9497, 10468.

Hemiboea cavaleriei H. Lév. (GESNERIACEAE) New species

Li in Acta Phytotax. Sin. 25: 221. (1987).

Yunnan: On the Daweishan Range, in Pingbian at 2,800 metres, 9184.

Leveille based this species on material collected by Cavalerie in September 1902, Henry had discovered it some six years previous to this. Distributed in China and Vietnam.

Hemiboea sp. (GESNERIACEAE)

Anonymous in Kew List of Determinations (volume 34).

Yunnan: Mengzi, 11431.

Hemiboea subcapitata Clarke (GESNERIACEAE)
New genus

in Hooker's Icon. Pl. 18: tt. 1798. (1888), Forbes & Hemsley in Journ. Linn. Soc. Bot. 26: 232. (1890), Henry in Notes Econ. Bot. China 49. (1893), Bretschneider in Hist. Eur. Bot. Disc. China 2: 789. (1898), Li in Acta Phytotax. Sin. 25: 223. (1987).

Hemiboea henryi C. B. Clarke in Hooker's Icon. Pl. 18: t. 1798. (1888), Forbes & Hemsley in Journ. Linn. Soc. Bot. 26: 232. (1890), Henry in Notes Econ. Bot. China 49. (1893), Bretschneider in Hist. Eur. Bot. Disc. China 2: 789. (1898), J. H. Veitch in Hortus Veitchii 422. (1906), Li in Acta Phytotax. Sin. 25: 223. (1987), Morley in Glasra 3: 79. (1979).

Hubei: Liantuo, 2107. Yichang, 4215 (type of _Hemiboea henryi_ C. B. Clarke), 4215a., 4331. Badong, 4894, 5175. Xingshan, 6493. Antelope Glen, 6493a. **Sichuan:** South Wushan, 7251.

Henry stated thios was the _hsiang-lung-ts'ao_, a hardy, herbaceous gesnerad to 30 cm. with white waxy tubular flowers, yellow in the throat. According to his notes a decoction of the roots of this species, mixed with spirits was used by local village doctors to treat cases of snakebite. Introduced to cultivation by Wilson through Veitch's Coombe Wood nursery.

Hemionitis achariorum Christenh. (PTERIDACEAE)
Pteris palmata Willd. Anonymous in Kew List of Determinations (volume 34).

Yunnan: Mengzi, 9491.

Central and South America, this number belongs to another taxon.

Hemionitis arifolia (Burm. f.) T. Moore (PTERIDACEAE)
Hemionitis cordata Henry non Roxb. in Trans. Asiat. Soc. Jap. 24 suppl: 116. (1896) (List Pl. Formosa).

Taiwan: Wanchin, 1601.

Native to India, Sri Lanka, Myanmar, Indochina, the Philippines and Malaysia.

Hemionitis bipinnata (Christ) Mickel (PTERIDACEAE)
New species

Gymnogramma bipinnata Christ in Notul. Syst. (Paris) 1: 55. (1909), _Gynogramme vestita_ Christ non Hook. Christ in Bull. Herb. Boiss. 6: 868. (1898).

Yunnan: In mountains near Mengzi at 2,100 metres, 9001, (in part).

Hemionitis chusana (Hook.) Christenh. (PTERIDACEAE)
Cheilanthes mysurensis Henry non Wall. ex Hook. in Trans. Asiat. Soc. Jap. 24 suppl: 111. (1896) (List Pl. Formosa).

Taiwan: Wanshoushan, 1063.

A widespread fern found over much of China, Indochina, Korea, Japan and the Philippines.

Hemionitis delavayi (Baker) Christenh. (PTERIDACEAE)
Gymnogramma delavayi Baker Christ in Bull. Herb. Boiss. 6: 868. (1898).

Yunnan: On mountains to the north of Mengzi at 1,650 metres, 9001 (in part).

Hemionitis dubia (C. Hope) Christenh. (PTERIDACEAE)
Wu in Acta Phytotax. Sin. 19: 69. (1981).

Cheilanthes subrufa Baker in Kew Bull. Misc. Inf. 8. (1906). _Aleuritopteris subrufa_ (Bak.) Ching in Hongk. Nat. 10: 202. (1941), Ching in Cormophy. Sinic. 1: 163. t. 325. (1972).

Yunnan: Mengzi, 9080 (type of _Cheilanthes subrufa_ Baker). Mountains to the north of Mengzi at 1,650 metres, 11831.

Distributed in Nepal, India and China.

Hemionitis farinosa (Forssk.) Christenh. (PTERIDACEAE)
Cheilanthes farinosa (Forssk.) Kaulf. Anonymous in Kew List of Determinations (volume 34).

Yunnan: Simao, 12513. Yuanjiang, 13223.

Hemionitis fragilis (Hook.) Christenh. (PTERIDACEAE)
Cheilanthes fragilis Hook. Anonymous in Kew List of Determinations (volume 34).

Yunnan: Simao, 13391.

Hemionitis hopeana (C. Chr. & C. H. Wright) Christenh. (PTERIDACEAE)
Pellaea squamosa Hope & C. H. Wight in Journ. Linn. Soc. Bot. 35: 518. (1903).

Yunnan: Yuanjiang at 800 metres, 13209 (type of _Pellaea squamosa_ Hope & C. H. Wight).

Hemionitis michelii (Christ) Christenh. (PTERIDACEAE)
Cheilanthes argentea (S. G. Gmel.) Kunze Anonymous in Kew List of Determinations (volume 24 & 34).

Hubei: Yichang, 4104. **Yunnan:** Mengzi, 9976.

A widespread species, distributed from Siberia to India, China, Indochina, Korea and Japan.

Hemionitis nitidula (Hook.) Christenh. (PTERIDACEAE)
Pellaea henryi Christ in Bull. Herb. Boiss. 7: 7. (1899). _Pellaea nitidula_ (Hook.) Baker Diels in Bot. Jahrb. 29: 199. (1900). _Mildella henryi_ (Christ) Hall & Lell. in Am. Fern. Journ. 57: 129. (1967).

Hubei: Liantuo, 2647. **Yunnan:** On the Mengzi hills at 1,650 metres, 11832.

Native to Bhutan, Nepal, China, Vietnam and Japan.

Hemionitis opposita (Kaulf.) Christenh. (PTERIDACEAE)
Cheilanthes mysurensis Wall. ex Hook. Diels in Bot. Jahrb. 29: 199. (1900).

Hubei: Yichang, 2168. Liantuo, 2211, 3209. Changleping, 7841. Badong, 7915. **Yunnan:** Yuanjiang,

13235.

Hemionitis patula (Baker) Christenh. (PTERIDACEAE) New species

Cheilanthes patula Baker in Journ. Bot. 26: 225. (1888), Baker in Ann. Bot. 5: 212. (1890), Bretschneider in Hist. Eur. Bot. Disc. China 2: 794. (1898), Diels in Bot. Jahrb. 29: 199. (1900).

Hubei: Yichang, 3998.

Hemionitis sp. (PTERIDACEAE)

Cheilanthes farinosa Anonymous non (Forsk.) Kaulf. in Kew List of Determinations (volume 24).

Hubei: Jianshi, 5878.

Hemionitis subcordata (D. C. Eaton ex Davenp.) Mickel (PTERIDACEAE) New species

Gymnogramma javanica Christ in Bull. Herb. Boiss. 6: 868.(1898) non Blume 1828. *Gymnogramma subcordata* D. C. Eaton & Davenp. in Cont. US. Natl. Herb. 5: 138. t. 16. (1897). *Coniogramme subcordata* (D. C. Eaton ex Davenp.) Maxon in Cont. US. Natl. Herb. 17: 174. (1913).

Yunnan: Mountains to the south of Mengzi at 1,900 metres, 9257a., (isotype of *Coniogramme subcordata* (D. C. Eaton ex Davenp.) Maxon.

Hemionitis trichophylla (Baker) Christenh. (PTERIDACEAE)

Cheilanthes undulata C. Hope & C. N. Wright in Gard. Chron. ser. 3, 34: 397. (1903).

Yunnan: Yuanjiang, 13220 (type of *Cheilanthes undulata* C. Hope & C. N. Wright).

Distributed from Sikkim to western China.

Hemionitis vestita (Hook.) J. Sm. (PTERIDACEAE)

Gymnogramma vestita Hook. Anonymous in Kew List of Determinations (volume 34).

Yunnan: Simao, 13190.

Hemipilia cucullata (L.) Y. Tang, H. Peng & T. Yukawa (ORCHIDACEAE)

Gymnadenia cucullata (L.) Rich. Rolfe in Journ. Linn. Soc. Bot. 36: 52. (1903).

Hubei: Fang at 2,600 to 2,900 metres (Shennongjia), 6873, 6873a., 6890. Without locality, 6890b.

Collected by the Sino-American Expedition in the Shennongjia Forest District in 1980.

Hemipilia gracilis (Blume) Y. Tang, H. Peng & T. Yukawa (ORCHIDACEAE)

Gymnadenia gracilis (Blume) Miq. Rolfe in Journ. Linn. Soc. Bot. 36: 53. (1903). *Amitostigma gracile* (Blume) Schlecter Tang & Lang in Acta Phytotax. Sin. 20: 79. (1982).

Hubei: Badong, 4460. Donghu at 1,300 metres, 6435.

A small terrestrial orchid to between 5 and 15 cm tall, native to China, Korea and Japan.

Hemipilia henryi Reichb. f. ex Bur. & Franchet (ORCHIDACEAE) New species

in Journ. de Bot. 5: 152. (1891), Rolfe in Kew Bull. Misc. Inf. 203. (1896), Bretschneider in Hist. Eur. Bot. Disc. China 2: 792. (1898), Rolfe in Journ. Linn. Soc. Bot. 36: 62. (1903).

Hubei: Yichang, 1534. Liantuo, 6347. Xingshan, 6347a., (isotype). Fang Xian, 6347b., (isosyntype). **Yunnan:** Simao, 12472.

A handsome terrestrial orchid to 30 cm tall, bearing racemes of pink to purple-red flowers above a single ovate-cordate leaf during July and August. Distributed in the mountains of Hubei, Sichuan and Yunnan.

Hemsleya chinensis Cogniaux (CUCURBITACEAE) New genus and species

in Hooker's. Icon. Pl. 9: t. 1822. (1889), Henry in Notes Econ. Bot. China 52. (1893), Bretschneider in Hist. Eur. Bot. Disc. China 2: 784. (1898), Cogniaux in Engler, Pflanzenr. iv. 275. i: 23. (1916), Morley in Glasra 3: 79. (1979), Chang & Shen in Acta Phytotax. Sin. 21: 185. (1983).

Gynostemma species nova, vel genus novum, Hemsley in Journ. Linn. Soc. Bot. 321. (1887)

Hubei: Badong, 2436 (isotype), 4771. Jianshi, 5928, 5928a. South Badong, 5928b., 5928c., 5928d. **Sichuan:** North Wushan, 7015.

In his *Notes on Economic Botany of China,* Henry states that this is the *ta-wu-yueh-wu,* a large twiner common in the mountains of Hubei and Sichuan. According to Chang and Shen (1983), the tuberous roots of *Hemsleya* are used as a traditional medicine called *jin-gui-lian* and in recent years has been chiefly against flatulence, diarrhea, dysentry, cardialgia and bronchitis with good effects. The genus is distributed in central, south and south-west China with a concentration in Sichuan.

Hemsleya graciliflora (Harms) Cogniaux (CUCURBITACEAE) New species

in Engler, Pflanzenr. iv. 275. i: 24. (1916), C. Y. Wu in Acta Phytotax. Sin. 23: 124. (1985).

Hemsleya chinensis Cogniaux in Hooker's Icon. Pl. 9: t. 1822. (1889), in part. *Alsomitra graciliflora* Harms in Bot. Jahrb. 29: 602. (1901).

Hubei: Yichang, 4303. Liantuo, 4452.

The 'Kew Inwards' book for 1888—1892 records this as 'a very interesting cucurbit,' seeds were taken from A. Henry 4452 for propagation on June 1st 1888.

Henckelia anachoreta (Hance) D. J. Middleton & Mich.Möller (GESNERIACEAE)

Chirita anachoreta Hance Henry in Trans. Asiat. Soc. Jap. 24 suppl: 68. (1896) (List Pl. Formosa). *Chirita dimidiata* R. Br. Anonymous in Kew List of Determinations (volume 34).

Taiwan: Wanchin, 1696 (in part). **Yunnan:** Simao, 13438.

Henckelia pumila (D. Don) A. Dietr. (GESNERIACEAE)

Chirita pumila D. Don Anonymous in Kew List of Determinations (volume 34).

Yunnan: In Pingbian, in forests on the Daweishan Range at 1,650 metres, 11040. Simao, 12281.

Distributed in the Himalaya, India, Myanmar, western China and Vietnam.

Henckelia pycnantha (W. T. Wang) D. J. Middleton & Mich.Möller (GESNERIACEAE) New species

Chirita pycnantha W. T. Wang in Novon 7: 425. (1997).

Yunnan: Yiwu (on the Laos border), 13571.

Endemic to Yunnan (Simao).

Henckelia spp. (GESNERIACEAE)

Chirita sp. Anonymous in Kew List of Determinations (volume 34).

Yunnan: Mengzi, 11275, 11421. Simao, 11789, 12037, 12443, 13521. Yuanjiang, 13252.

Henckelia urticifolia (Buch.-Ham. ex D. Don) A. Dietr. (GESNERIACEAE)

Chirita urticifolia Buch.-Ham. ex D. Don Anonymous in Kew List of Determinations (volume 34).

Yunnan: Mengzi, 9161.

Distributed in Nepal, India, Bhutan, Myanmar and western China.

Henslowia sp. (SANTALACEAE)

Anonymous in Kew List of Determinations (volume 34).

Yunnan: Simao, 11954.

Hepatica henryi (Oliv.) Steward (RANUNCULACEAE) New species

in Rhodora 29: 53. (1927), O' Brien in Augustine Henry, An Irish Plant Collector in China 21. (2002).

Anemone henryi Oliver in Hooker's. Icon. Pl. t. 1570. (1887), Morley in Glasra 3: 78. (1979).

Hubei: Badong, 1464, 5418.

Heptapleurum arboricola Hayata (ARALIACEAE) New species

Heptapleurum octophyllum Henry non (Lour.) Benth in Trans. Asiat. Soc. Jap. 24 suppl: 48. (1896) (List Pl. Formosa). *Schefflera arboricola* (Hayata) Merr. Li in Woody Flora of Taiwan 670. (1963).

Taiwan: Oluanpi, 297, 1358.

A shrub to 4 metres tall, native to Hainan and Taiwan, scattered throughout Taiwan and first collected there by Henry. This species was collected by the Glasnevin Taiwan Expedition (GTE 86) in Kenting National Park in October 2004.

Heptapleurum calyptratum (Hook. f. & Thoms.) Y. F. Deng (ARALIACEAE)

Tupidanthus calyptratus Hook. f. & Thoms. Hemsley in Hookers Icon Pl. t. 2672. (1901), Hemsley in Journ. Linn. Soc. Bot. 34: 470. (1900).

Yunnan: Simao at 1,150 metres, 12298, 12298a.

Native to India, Bangladesh, Myanmar, China (Yunnan), Laos, Cambodia and Vietnam.

Heptapleurum chinense (Dunn) Y. F. Deng (ARALIACEAE) New species

Oreopanax chinensis Dunn in Journ. Linn. Soc. Bot. 35: 500. (1903).

Yunnan: In forests near Simao at 1,450 metres, 12939 (type of *Oreopanax chinensis* Dunn).

A tree to 9 metres tall, endemic to China (Jiangxi and southwest Yunnan).

Heptapleurum delavayi Franch. (ARALIACEAE)

Schefflera delavayi (Franch.) Harms Harms & Rehder in Sargent Pl. Wilson. 2: 555. (1916).

Yunnan: Mengzi, 9214.

Distributed in China and northwest Vietnam.

Heptapleurum ellipticum (Blume) Seem. (ARALIACEAE)

Heptapleurum venulosum Anonymous non (Wight & Arnott) Seem. in Kew List of Determinations (volume 34).

Yunnan: Mengzi, 9403, 10541. Simao, 13044.

Distributed in India, China, Vietnam, Cambodia and Laos.

Heptapleurum griffithii (Seem.) ined. (ARALIACEAE)

Heptapleurum elatum (Buch.-Ham.) C. B. Clarke Anonymous in Kew List of Determinations (volume 34).

Yunnan: In Pingbian in forests on the Daweishan Range at 1,640 metres, 13474.

Native to Himalaya to China (Yunnan) and Indochina.

Heptapleurum heptaphyllum (L.) Y. F. Deng (ARALIACEAE)

Heptapleurum octophyllum (Lour.) Benth ex Hance Henry in Trans. Asiat. Soc. Jap. 24 suppl: 48. (1896) (List Pl. Formosa). *Agalma lutchuense* Nakai in Journ. Arnold. Arb. 5: 20. (1924). *Schefflera octophylla* (Lour.) Harms Li in Woody Flora of Taiwan 671. fig. 279. (1963).

Taiwan: Wanchin, 17, 57 (syntype of *Agalma lutchuense* Nakai), 1718. Tamsui, 1735 (syntype of *Agalma lutchuense* Nakai), Tamsui, 1735a., (syntype of *Agalma lutchuense* Nakai), 1780 (syntype of *Agalma lutchuense* Nakai). **Yunnan:** Simao, 12801.

A small tree or shrub with solid pith, native to southern China, Indochina, Japan (south) and the Philippines. Common in Chinese Taiwan at low altitude. The wood of this tree was onced used in the manufacture of clog-soles on mainland China. *Heptapleurum heptaphyllum* was collected by the Glasnevin Taiwan Expedition (GTE 82) in Kenting National Park in October 2004.

Heptapleurum hoi Dunn (ARALIACEAE) New species
in Journ .Linn. Soc. Bot. 35: 498. (1903).

Schefflera hoi (Dunn) R. Viguier in Ann. Sci. Nat. Bot. ser. 9, 9: 333. (1909).

Yunnan: South of Red River from Manmei, at 1,900 metres, 9723 (isotype). In Pingbian in forests on the Daweishan Range at 1,640 metres, 13462.

Heptapleurum hoi commemorates Augustine Henry's best Chinese collector, Ho, who collected for Henry while stationed at both Mengzi and Simao. Distributed from north-east India to China [Yunnan and Tibet (Xizang)] and Vietnam.

Heptapleurum hypoleucoides (Harms) Lowry & G. M. Plunkett (ARALIACEAE) New species

Schefflera hypoleucoides Harms in Repert. Spec. Nov. Regni Veg. 16: 246. (1919).

Yunnan: In forests to the south-west of Menszi at 1,300 metres, 11435 (isosyntype). Near Yuanjiang at 1,950 metres, 13301 (isosyntype of *Schefflera hypoleucoides* Harms).

Native to China (Guangxi and Yunnan) and Vietnam.

Heptapleurum hypoleucum Kurz. (ARALIACEAE)
Anonymous in Kew List of Determinations (volume 34).

Yunnan: Mengzi, 9654, 11840.

Distributed in India, Myanmar, Vietnam and southern Yunnan.

Heptapleurum macrophyllum Dunn (ARALIACEAE) New species
in Journ. Linn. Soc. Bot. 35: 499. (1903).

Yunnan: In forests near Simao at 1,650 metres, 13409 (isotype).

Distributed in China (Yunnan) and Vietnam.

Heptapleurum rhododendrifolium (Griff.) G. M. Plunkett & Lowry (ARALIACEAE)

Heptapleurum impressum C. B. Clarke Anonymous in Kew List of Determinations (volume 34).

Yunnan: Mengzi, 9455.

Native to India, Bhutan, Nepal and China [Tibet (Xizang) and Yunnan].

Heptapleurum sp. (ARALIACEAE)
Anonymous in Kew List of Determinations (volume 24).

Hainan: Environs of Kiungchow, 8747, 8754.

Heracleum aff. ***barbatum*** Ledeb. (APIACEAE)
Anonymous in Kew List of Determinations (volume 24).

Hubei: Badong, 113, 2528, 2529.

Heracleum hemsleyanum Diels (APIACEAE) New species
in Bot. Jahrb. 29: 503. (1900).

Hubei: Badong, 4873. Fang Xian, 6768. Xingshan, 6469. Without locality, 6469a., 6469b. 6768a.

Heracleum henryi Wolff (APIACEAE) New species
in Repert. Spec. Nov. Regni Veg. 33: 76. (1934).

Yunnan: Simao at 1,650 metres, 12486b.

Endemic to Yunnan.

Heracleum moellendorffii Hance (APIACEAE)
Heracleum lanatum Forbes & Hemsley non Michx. in Journ. Linn. Soc. Bot. 23: 336. (1887).

Hubei: Badong, 522 (September 1885), 4911, 4960.

Heritiera littoralis Dryand. (MALVACEAE)
Henry in Trans. Asiat. Soc. Jap. 24 suppl: 22. (1896) (List Pl. Formosa).

Taiwan: Oluanpi, 605.

A medium-sized tree, native to the tropics of the Old World and common along the coastal northern and southern regions of Taiwan. This species is adapted to life on the seashore and is often found associated with mangrove swamps and the mouths of rivers. The wallnut-like fruits are keeled to aid flotation and it are distributed on ocean currents. In October 2004 the Glasnevin Taiwan Expedition made collections of this species in Kenting National Park (GTE 87) and by the lighthouse at Oluanpi (South Cape of Henry and Wilson) in October 2004 (GTE 20). In both locations this species grows on fossalized coral that was raised off the ocean floor about 100,000 years ago. The shallow roots of this tree form butresses in old age.

Herminium coiloglossum Schlechter (ORCHIDACEAE) New species
in Repert. Spec. Nov. Regni Veg. 3: 15. (1906).

Yunnan: Simao at 1,700 metres, 13556 (isotype).

Herminium lanceum (Thunb. ex Sw.) Vuijk. (ORCHIDACEAE)
Herminium angustifolium Benth. ex. Hook. f. Rolfe in Journ. Linn. Soc. Bot. 36: 50. (1903).

Hubei: Badong, 856 (September 1885). Yichang, 7383. Liantuo, 2042, 3912. Donghu, 6421, 7730. Fang Xian, 6421a. South Badong, 7283. Changyang, 6202. **Sichuan:** North Wushan, 6421b. **Yunnan:** Mengzi, 9684.

A slender terrestrial orchid to 30 cm tall, widespread across the Himalaya, India, China, Korea, Japan, Thailand and Malaysia. Common in grassland and in thickets at 1,600 to 3,600 metres above sea level. The entire plant is used in Chinese herbal medicine.

Herminium sp. (ORCHIDACEAE)
Anonymous in Kew List of Determinations (volume

34).

Yunnan: Mengzi, 10145.

Hetaeria exigua (Rolfe) Schlecter (ORCHIDACEAE)
New species

Spiranthes exigua Rolfe in Kew Bull. Misc. Inf. 200. (1896), Bretschneider in Hist. Eur. Bot. Disc. China 2: 792. (1898), Rolfe in Journ. Linn. Soc. Bot. 36: 41. (1903).

Hubei: Fang Xian, 6585.

Heterodermia speciosa (Wulf.) Trev. (PHYSCIACEAE)
Wei in An Enumeration of Lichens in China 112. (1991).

Anaptychia speciosa (Wulf.) Mass. f. *sorediosa* Muell.-Arg. Muell.-Arg in Bull. Herb. Boissier 1: 236. (1893).

Hubei: Badong, 5098. Fang Xian, 6718 (in part). (Lichen).

Heterolamium debile (Hemsl.) C. Y. Wu (LAMIACEAE)
New species

Orhosiphon debilis Hemsley in Journ. Linn. Soc. Bot. 26: 267. (1890), Dunn in Kew Bull. Misc. Inf. 201. (1917).

Hubei: Jianshi, 5770a. **Sichuan:** South Wushan, 5770.

Heterolamium debile (Hemsl.) C. Y. Wu. var. *cardiophyllum* (Hemsl.) C. Y. Wu. (LAMIACEAE)
New variety. *Plectranthus cardiaphyllus* Hemsley in Journ. Linn. Soc. Bot. 26: 269. (1890), Bretschneider in Hist. Eur. Bot. Disc. China 2: 789. (1898).

Hubei: South Badong, 6050. Xingshan, 6993.

Heteropanax chinensis (Dunn) Li (ARALIACEAE)
New species

Heteropanax fragrans (Roxb. ex DC.) Seeman var. *chinensis* Dunn in Journ. Linn. Soc. Bot. 38: 360. (1908).

Yunnan: Simao at 1,150 metres, 12865 (isosyntype of *Heteropanax fragrans* (Roxb. ex DC.) Seeman. var. *chinensis* Dunn).

Native to China (Guangxi and Yunnan) and Vietnam.

Heteropogon contortus (L.) P. Beauv. ex Roem. & Shult. (POACEAE)
Henry in Journ. China Br. Roy. Asiat. Soc. 22: 246. (1887) (Chinese Names of Plants), Rendle in Journ. Linn. Soc. Bot. 36: 366. (1904).

Heteropogon hirtus Pers. Henry in Trans. Asiat. Soc. Jap. 24 suppl: 108. (1896) (List Pl. Formosa).

Hubei: Yichang, 442, 4143. **Taiwan:** Wanshoushan, without number.

The *huang ts'ao* or spear-grass. According to Henry this species was used for thatch at Yichang. Widely distributed in the tropics and ascending to 2,000 metres in the Himalaya.

Heterostemma sp. (APOCYNACEAE)
Anonymous in Kew List of Determinations (volume 34).

Yunnan: Mengzi, 13788.

Hewittia malabarica (L.) Suresh. (CONVOLVULACEAE)
Hewittia bicolor (Vahl.) Wight & Arnott Henry in Trans. Asiat. Soc. Jap. 24 suppl: 65. (1896) (List Pl. Formosa).

Hainan: Environs of Haikou, 8034, 8376. Environs of Kiungchow, 8316 (in part). **Taiwan:** Oluanpi, 237, 1757. **Yunnan:** Mengzi, 9352.

Widespread in Asia.

Heynea trijuga Roxb. ex Sims. (MELIACEAE)
Heynea trijuga Roxb. ex Sims. Anonymous in Kew List of Determinations (volume 34).

Yunnan: Simao, 11871, 11958.

Distributed from India to China and Vietnam to Malesia.

Hibiscus altissimus Hornby (MALVACEAE)
Hibiscus furcatus Willd. Anonymous in Kew List of Determinations (volume 34).

Yunnan: Simao, 13566.

Hibiscus mutabilis L. (MALVACEAE)
Henry in Journ. China Br. Roy. Asiat. Soc. 22: 243. (1887) (Chinese Names of Plants); in Trans. Asiat. Soc. Jap. 24 suppl: 21. (1896) (List Pl. Formosa), Zheng in Journ. Wuhan Bot. Res. 2.(1): 148. (1984).

Hubei: Yichang, 24. Near Yichang in the Monastery Valley, 1491. On the banks of a river in the San You Dong Glen, 1691. Liantuo, 2640. **Taiwan:** Kaohsiung, without number.

A small tree to 5 metres tall, native to southern China and extensively cultivated and naturalised in Taiwan. The flowers changing from white to red as the day advances. Henry stated that this species was known colloquially at Yichang as the *fu yung hua*. A beautiful plant, this species is currently used in Chinese herbal medicine to arrest bleeding and externally in cases of skin disease.

Hibiscus rosa-sinensis L. (MALVACEAE)
Henry in Trans. Asiat. Soc. Jap. 24 suppl: 21. (1896) (List Pl. Formosa).

Taiwan: Kaohsiung Spit, without number (cultivated). **Yunnan:** South of the Red River from Manmei at 2,300 metres, 9470. Milê district, 10575.

Hibiscus sp. nov. (MALVACEAE)
Anonymous in Kew List of Determinations (volume 34).

Yunnan: Simao, 12689.

Henry's note; grown at Kew from my seeds.

Hibiscus spp. (MALVACEAE)
Anonymous in Kew List of Determinations (volume 24 & 34).

Hubei: Badong, 374 (June 1885). **Yunnan:** Mengzi,

9132. Simao, 13263.

Hibiscus surattensis L. (MALVACEAE)

Henry in Trans. Asiat. Soc. Jap. 24 suppl: 21. (1896) (List Pl. Formosa).

Hainan: Environs of Haikou, 8075. Environs of Kiungchow, 8284. **Taiwan:** Wanchin, 168, 1584.

A sprawling herb, native to Hainan, Taiwan, the Philippines and Java.

Hibiscus syriacus L. (MALVACEAE)

Henry in Journ. China Br. Roy. Asiat. Soc. 22: 234. (1887) (Chinese Names of Plants); in Trans. Asiat. Soc. Jap. 24 suppl: 21. (1896) (List Pl. Formosa), Rehder & E. H. Wilson in Sargent Pl. Wilson 2: 374. (1916), Li in Woody Flora of Taiwan 545. (1963), Zheng in Journ. Wuhan Bot. Res. 2.(1): 234. (1984).

Hubei: Yichang, 297, 632, 2126, 2717. Liantuo, 2126a., 2627, without number. **Hainan:** Environs of Haikou, 8243. **Taiwan:** Wanshoushan, 1092.

A deciduous shrub to 2.5 metres tall, the flowers of this species are extremly variable in colour and may be white, red, violet or purple. Linnaeus believed it was native to Syria, hense the specific epithet but it is known to be native to India and China. According to Henry this well-known shrub had two colloquial names at Yichang, *ch'a hua* and *ch'a chin t'iao*. He stated both white and reddish flowered forms occured in the region. The Hani (an ethnic group) of Xishuangbanna in southern Yunnan call this species *laolcil* and the flowers and roots are used to cure dysentry and diarrhoea. Henry's Taiwan collection is the wild white form, known on mainland China as *mu-chin*.

Hibiscus tiliaceus L. (MALVACEAE)

Henry in Trans. Asiat. Soc. Jap. 24 suppl: 21. (1896) (List Pl. Formosa).

Hainan: Environs of Haikou, 8375, 8416. Lingmen, 8784. **Taiwan:** Kaohsiung, without number. Oluanpi, without number.

The *hai-ma*, a pantropic tree of rapid growth reaching about 10 metres tall. The attractive flowers are held in terminal panicles and are carried all year round. Individual flowers emerge light yellow with a maroon eye and gradually change to deep orange before falling from the tree the following evening. Polynesians use the lightweight wood of this tree for the outriggers of their canoes, the flowers as a laxative and the bark is used to manufacture a fine fibre for dance skirts used in religious festivals. A good tree in littoral parts of the tropics and easily propagated. This species was collected by the Glasnevin Taiwan Expedition (GTE 39) by the lighthouse at Oluanpi (South Cape of Henry and Wilson) in October 2004.

Hibiscus trionum L. (MALVACEAE)

Anonymous in Kew List of Determinations (volume 34).

Yunnan: Mengzi, 9996.

A short lived perennial herb, distributed in the tropics of the Old World and widely naturalised in China.

Hibiscus vitifolius L. var. fl. purpureis. (MALVACEAE)

Anonymous in Kew List of Determinations (volume 24).

Hainan: Environs of Haikou, 8025.

Hibiscus yunnanensis S. Y. Yu (MALVACEAE) New species

in Fl. China. Fam. 153, Malvaceae 55-56. (1955).

Yunnan: Yuanjiang, 13218 (holotype).

Endemic to China and known only from the Yuanjiang region of southern Yunnan.

Hieracium umbellatum L. (ASTERACEAE)

Forbes & Hemsley in Journ. Linn. Soc. Bot. 23: 477. (1888).

Hieracium umbellatum L. var. *subvirgatum* Zahn. Zahn in Engler, Pflanzenr. 79: 911. (1922).

Hubei: Badong, 936 (November 1885), 1035 (December 1885), 2476, 4827, 5192, 5056. South Badong, 7322. Fang Xian, 6532. **Sichuan:** South Wushan, 7454.

Hiptage benghalensis (L.) Kurz. (MALPIGHIACEAE)

Li in Woody Flora of Taiwan 401. fig. 148. (1963).

Hiptage madablota Gaert. Henry in Trans. Asiat. Soc. Jap. 24 suppl: 24. (1896) (List Pl. Formosa). *Hiptage obtusifolia* (Roxb.) DC. Niedenzu in Engler, Pflanzenr. iv.141: 79. (1928).

Taiwan: Oluanpi, 216, 630. **Yunnan:** Simao, 11776a.

A rampant vine to 30 metres tall, sometimes a shrub. Widely distributed from India and Malaysia to southern China and often cultivated for its fragrant flowers.

Hiptage minor Dunn (MALPIGHIACEAE) New species

in Journ. Linn. Soc. Bot. 35: 487. (1903), Niedenzu in Engler, Pflanzenr. iv. 141: 83. (1928).

Hiptage henryana Niedenzu in Arbeiten Bot. Inst. Konigl. Lyceums. Hosianum Braunsberg 6: 45. (1915).

Yunnan: At the base of the Mengzi mountains, covering rocks at 1,700 metres, 10792, 10792a., (type of *Hiptage minor* Dunn). 10792b., (type of *Hiptage henryana* Niedenzu).

Endemic to China and distributed in the provinces of Guizhou and Yunnan. This species had been previously collected near Mengzi by the Irish customs official, William Hancock.

Histiopteris incisa (Thunb.) J. Sm. (DENNSTAEDTIACEAE)

Pteris incisa Thunb. Henry in Trans. Asiat. Soc. Jap. 24 suppl: 111. (1896) (List Pl. Formosa).

02

Taiwan: Tamsui, 1438.

Distributed through the tropics and subtropics of the world.

Holcoglossum kimballianum (Rchb. f.) Garay. (ORCHIDACEAE)

Vanda kimballiana Rchb. f. Anonymous in Kew List of Determinations (volume 34).

Yunnan: Simao, 12751, 13545.

Epiphytic on trees throughout the tropical and subtropical forests of Myanmar, Thailand and China (Yunnan).

Homalium ceylanicum (Gardner) Benth. (SALICACEAE)

Homalium hainanense Gagnepain in Notul. Syst. (Paris) 3: 248. (1916).

Hainan: Environs of Kiungchow, 8765 (isotype of *Homalium hainanense* Gagnepain).

Distributed in China and Vietnam.

Homalium cochinchinensis (Lour.) Druce (SALICACEAE)

Li in Woody Flora of Taiwan 605. (1963).

Homalium fagifolium Benth. Henry in Trans. Asiat. Soc. Jap. 24 suppl: 45. (1896) (List Pl. Formosa).

Taiwan: Wanchin, 502.

A small tree native to Indochina and southern China. In Chinese Taiwan this species is distributed in broad-leaved forest on slopes and along streams at low altritude in the south-central part of the island.

Homalium sp. nova (SALICACEAE)

Anonymous in Kew List of Determinations (volume 24).

Hainan: 32 km (20 miles) west of Haikou, 8150.

Homocodon brevipes (Hemsl.) D. Y. Hong (CAMPANULACEAE) New species

Wahlenbergia brevipes Hemsley in Hooker's Ic. Pl. 24: t. 2768. (1903).

Yunnan: Forests south-west of Mengzi at 1,500 metres, 10941 (type of *Wahlenbergia brevipes* Hemsley).

Endemic to China and distributed in Guizhou, Sichuan and Yunnan.

Homomallium plagiangia (C. Mull.) Broth. (HYPNACEAE)

Pylaisia plagiangia C. Mull. Salmon in Journ. Linn. Soc. Bot. 34: 469. (1900).

Hubei: On dry rocks in south Badong, 5481. (Moss).

Homonoia riparia Lour. (EUPHORBIACEAE)

Forbes & Hemsley in Journ. Linn. Soc. Bot. 26: 443. (1894), Henry in Trans. Asiat. Soc. Jap. 24 suppl: 84. (1896) (List Pl. Formosa), Pax in Engler, Pflanzenr. iv. 147. ix-xi: 114. (1919), Li in Woody Flora of Taiwan 430. fig. 152. (1963).

Hainan: Without locality, 8584. **Taiwan:** Wanchin, 180. Oluanpi, 261. **Yunnan:** Mengzi, 10665, 10665a.

A widely distributed shrub, native to Indochina, southern China and Java. Abundant on banks and dry stream beds in Chinese Taiwan where it inhabits the southern and eastern parts of the island. Known as *shui-lui* according to Henry. The Hani (an ethnic group) of Xishuangbanna in southern Yunnan call this species *eelnmq* and the roots and leaves are used to treat chronic hepatitis and rheumatism.

Hordeum aegiceras Nees ex Royle (POACEAE)

Hordeum vulgare L. var. *aegiceras* (Nees ex Royle) Aitch. Anonymous in Kew List of Determinations (volume 24).

Hubei: Changyang, 5275. (Ricewheat or grim barley).

Hosiea sinensis (Oliv.) Hemsley & E. H. Wilson (ICACINACEAE) New species

in Kew Bull. Misc. Inf. 154. (1906), Rehder & E. H. Wilson in Sargent Pl. Wilson 2: 190. (1916).

Natsiatum sinense Oliver in Hooker's. Ic. Pl. 19: t. 1900. (1889).

Hubei: Jianshi, 5598b. South Badong, 5598d., 7342. **Sichuan:** South Wushan, 5598 (type), 5598a., 5598c.

Hosta plantaginea (Lam.) Aschers. (HOSTACEAE)

Diels in Bot. Jahrb. 29: 241. (1900).

Funkia subcordata Spreng. Wright in Journ. Linn. Soc. Bot. 36: 117. (1903).

Hubei: Badong, 526 (September 1885), 2480. Near Moji Shan, on top of cliffs in inaccessible places, 2356. Without locality, 2480a. **Sichuan:** Kangding, Henry's Chinese collector with Pratt, 8892.

Hosta ventricosa (Salisb.) Stearn (HOSTACEAE)

Funkia ovata Spreng. Wright in Journ. Linn. Soc. Bot. 36: 116. (1903). *Funkia ovata* Spreng. var. *minor* Baker Wright in Journ. Linn. Soc. Bot. 36: 117. (1903). *Hosta caerulea* (Andr.) Tratt. Diels in Bot. Jahrb. 29: 241. (1900).

Hubei: Badong, 287 (July 1885), 2577, 4657. Liantuo, 6209. Changleping, 6209. Without locality, 6209a. **Sichuan:** Kangding, Henry's Chinese collector with Pratt, 8898.

A common perennial in the Three Gorges region. *Hosta ventricosa* was collected by the Glasnevin Central China Expedition (GCCE 313) in October 2002 between Yesanguan and Badong and again by the 2004 Glasnevin Expedition near Dragon Gate River Forest Park in Xingshan in October 2004. Henry sent seeds (collected in Badong) of this species to the Royal Botanic Gardens, Kew where they arrived on March 17th 1887.

Houttuynia cordata Thunb. (SAURURACEAE)

Forbes & Hemsley in Journ. Linn. Soc. Bot. 26: 364. (1891).

Hubei: Yichang, 834. Badong, 1805. Liantuo, 1981. Jianshi, 5828. **Yunnan:** Mengzi, 10307a.

Fish wort, which Henry stated was known as the *yu xing cao* (literally; fish smelling), a creeping perennial belonging to the primitive family, Saururaceae. Widely distributed through the woodlands of Asia, from the Himalaya into China to Java. The leaves and young shoots are eaten raw or cooked like spinach in Sichuan. It is used in traditional Chinese herbal medicine and has exhibited anti-viral effects against influenza. The Hani (an ethnic group) of Xishuangbanna in southern Yunnan call this species *palhao* and use it to treat cases of whooping cough and mumps. *Houttuynia cordata* was collected by the Glasnevin Central China Expedition (GCCE 245) on the hills above Badong in September 2002.

Hovenia dulcis Thunb. (RHAMNACEAE)

Henry in Journ. China Br. Roy. Asiat. Soc. 22: 251. (1887) (Chinese Names of Plants), Schneider in Sargent Pl. Wilson. 2: 252. (1916), Zheng in Journ. Wuhan Bot. Res. 2.(1): 138. (1984).

Hubei: Yichang, 2954. **Yunnan:** South of the Red River from Manmei at 2,000 metres, 9465. Mountains to the east of Simao at 1,600 to 2,000 metres, 12034, 12034a.

The Japanese raisin tree, a tree to about 25 metres tall, though Wilson records having seen trees occasionally reach 35 metres tall. In cultivation this species generally reaches about 11 metres tall. Native to China, Japan and Korea, it is naturalised over much of the Himalaya. Henry stated that the tree was colloquially known as *kuai-tsao* and Wilson stated that the thickened pedicels of the fruits were used medicinally by the Chinese to offset the effects of over-indulgence with wine. They are also eaten in Japan. At Singleton Park, Glamorgan, UK it was 11 metres tall with a trunk diameter of 42 cm when measured in 2000.

Hoya carnosa (L. f.) R. Brown (APOCYNACEAE)

Henry in Trans. Asiat. Soc. Jap. 24 suppl: 61. (1896) (List Pl. Formosa), Li in Woody Flora of Taiwan 802. fig. 324. (1963).

Taiwan: Wanshoushan, 723.

A twining shrub with succulent stems. Widely distributed from India to southern China and Japan. In Chinese Taiwan this species grows throughout the island at low altitude over rocks and through trees.

Hoya manipurensis Deb (APOCYNACACEAE)

Hoya lantsangensis Tsiang & Li in Acta Phytotax. Sin. 12: 126. (1974). *Antiostelma lantsangense* (Tsiang & Li) Li in Novon 2: 218. (1992).

Yunnan: Mo-Jiang at 1,000 metres (near Tonguan), 13689 (type of *Hoya lantsangensis* Tsiang & Li). Simao at 1,640 metres, 11638.

Hoya mengtzeensis Tsiang & P. T. Li (APOCYNACEAE) New species

in Acta Phytotax. Sin. 12: 120. (1974).

Yunnan: Near Mengzi, covering rocks at 1,750 metres,

11368 (holotype, collected on 23rd May 1898).

An epilithic or epiphytic white-flowered shrub to 1.5 metres tall, endemic to China where it grows on cliffs and rock-faces in Guangxi and Yunnan.

Hoya nervosa Tsiang & P. T. Li (APOCYNACEAE) New species

in Acta Phytotax. Sin. 12: 122. pl. 28. (1974).

Yunnan: Simao, 12994 (holotype).

Endemic to China and distributed in the provinces of Guangxi and Yunnan.

Hoya pandurata Tsiang (APOCYNACEAE) New species

in Sunyatsenia 4: 125. (1939).

Yunnan: In forests to the east of Simao at 1,300 metres, 12258 (type).

An epiphytic subshrub to 1.5 metres tall, China (southern Yunnan) and Indochina.

Hoya spp. (APOCYNACEAE)

Anonymous in Kew List of Determinations (volume 24 & 34).

Hainan: Environs of Haikou, 8000. Environs of Kiungchow, 8234. **Yunnan:** Simao, 12429, 12946, 13065, 13173. Mengzi, 13780, 13781, 13782.

Huangtcia renifolia (L.) H. Ohashi & K. Ohashi (FABACEAE)

Desmodium reniforme (L.) DC. Henry in Trans. Asiat. Soc. Jap. 24 suppl: 33. (1896) (List Pl. Formosa).

Taiwan: Oluanpi, 1243. **Yunnan:** Simao, 12493, 13196.

Distributed in India, Myanmar, China and Indochina.

Huberantha cerasoides (Roxb.) Chaowasku (ANNONACEAE)

Polyalthia cerasoides (Roxb.) Benth. & Hook. f. ex Bedd. Merrill & Chun in Sunyatsenia 2: 231. (1935).

Hainan: Environs of Kiungchow, 8220. Without locality, 8518.

Humulus scandens (Lour.) Merr. (CANNABACEAE)

Humulus japonicus Sieb. & Zucc. Forbes & Hemsley in Journ. Linn. Soc. Bot. 26: 453. (1894).

Hubei: Yichang, 2224. **Yunnan:** Mengzi, 9697.

A rapidly growing, scandent, dioecious annual, native to China and Japan. In Chinese herbal medicine the aerial parts of this hop are used to treat cold and fever. *Humulus scandens* is cultivated at the National Botanic Gardens, Glasnevin from collections made by the Glasnevin Central China Expedition near Yichang (GCCE 454) in September 2004 and at Xiakou (Daxiakou) on the Xiang Xi River (GCCE 698) in October of that year.

Huodendron biaristatum (W. W. Sm.) Rehder (STYRACACEAE) New genus

in Journ. Arnold Arb. 16: 344. (1935).

02

Styrax biaristatum W. W. Smith in Notes Roy. Bot. Gard. Edinburgh 12: 233. (1920).

Yunnan: In Pingbian on the Daweishan Range, 10764. South of the Red River, 13662 (isotype of *Styrax biaristatum* W. W. Smith). Mengzi, 13662a., (syntype of *Styrax biaristatum* W. W. Smith).

Henry had first discovered this genus on the Daweishan Range in the extreme sourh-east of Yunnan near the Vietnamese border 25 years previous to George Forrest's collections (the type species of the genus is *Huodendron tibeticum* (Anthony) Rehder, a tree collected by Forrest in 1922). This species is native to Yunnan, Myanmar, northern Vietnam and Thailand and commemorates the famous Chinese botanist, Professor H. H. Hu. Cultivated at the National Botanic Gardens, Kilmacurragh.

Huperzia carinatum (Desv.) Trevis. (LYCOPODIACEAE)

Lycopodium carinatum Desv. Henry in Trans. Asiat. Soc. Jap. 24 suppl: 117. (1896) (List Pl. Formosa).

Taiwan: Oluanpi, 656.

A widespread species distributed in tropical Asia, Polynesia, Japan (south) and China (Taiwan and Hainan). Epiphytic in forests.

Huperzia henryi (Baker) Holub (LYCOPODIACEAE)
New species

in Folia Geobot. Phytotax. 20(1): 73. (1985).

Lycopodium henryi Baker in Kew Bull. Misc. Inf. 15. (1906). *Phlegmariurus henryi* (Baker) Ching in Acta Bot. Yunnan 4(2): 125. (1982).

Yunnan: On trees in forests to the south of Simao at 1,950 metres, 11551.

Distributed in China (Guangxi and Yunnan) and Vietnam.

Huperzia lucidula (Michx.) Trevis. (LYCOPODIACEAE)

Lycopodium lucidulum Michx. Diels in Bot. Jahrb. 29: 210. (1900).

Hubei: South Badong, 6174. Changleping, 6207. Xingshan, 6972. **Sichuan:** North Wuhan, 7090. **Yunnan:** Mengzi, 13725.

Huperzia salvinioides (Hert.) Holub (LYCOPODIACEAE)

Lycopodium filiforme Henry non Wall. ex Roxb. in Trans. Asiat. Soc. Jap. 24 suppl: 117.(1896) (List Pl. Formosa). *Urostachys formosanus* Hert. ex Nessel in Lycopodiac. 229. (1939).

Taiwan: Oluanpi, 595 (isotype of *Urostachys formosanus* Hert. ex Nessel).

An epiphytic species with stems to 30 cm long. Distributed through China (Taiwan), Japan (south) and the Philippines.

Huperzia serrata (Thunb.) Trevis (LYCOPODIACEAE)

Lycopodium serratum Thunb. Anonymous in Kew List

of Determinations (volume 34).

Yunnan: Mengzi, 10190. In Pingbian in forests on the Daweishan Range at 1,640 metres, 13471.

Native to China, Japan and Korea.

Huperzia squarrosa (G. Forst.) Trevis. (LYCOPODIACEAE)

Lycopodium squarrosum G. Forst. Henry in Trans. Asiat. Soc. Jap. 24 suppl: 117. (1896) (List. Pl. Formosa).

Taiwan: Oluanpi, 1424.

A widespread species, distributed through Africa, tropical Asia and Polynesia. Epiphytic in forests.

Hydrangea ampla (Chun) Y. De Smet & Granados (HYDRANGEACEAE)

Schizophragma integrifolium (Franch.) Oliver in Hooker's Icon. Pl. xx. t. 1934. (1890), Bretschneider in Hist. Eur. Bot. Disc. China 2: 783. (1898), Morley in Glasra 3: 80. (1979), Nelson in A Heritage of Beauty 319. (2000), O' Brien in Augustine Henry, An Irish Plant Collector in China 35. (2002).

Hubei: Badong, 5509. Jianshi, 6012. Fang Xian, 6687. **Sichuan:** Henry's Chinese collector with Pratt, no locality stated but may be any of the following, Leshan, Emei Shan, Wa Shan or Kangding, 8951.

A robust, deciduous shrub to 13 metres tall, climbing by means of aerial roots. Large flat cymes of white flowers similar to the lacecap hydrangeas appear in summer. Its native range extends from north-east India to China and Indochina.

Hydrangea anomala D. Don (HYDRANGEACEAE)

Rehder in Sargent Pl. Wilson. 1: 34. (1913), McClintock in Proc. Calif. Acad. Sci. 19: 197. (1957), Zheng in Journ. Wuhan Bot. Res.2. (1): 58. (1984).

Hubei: Badong, 4778. Xingshan, 6511. **Sichuan:** South Wushan, 5557, 5637, 5658.

A deciduous scandent shrub to 15 metres tall, climbing through trees by means of aerial roots. Native to the Hinalaya and China, introduced to cultivation in 1839.

Hydrangea aspera D. Don (HYDRANGACEAE)

Forbes & Hemsley in Journ. Linn. Soc. Bot. 23: 272. (1887), Rehder in Sargent Pl. Wilson. 1: 40. (1913).

Hubei: Badong, 1825. Liantuo, 2064, 4396. Yichang, 2167. Yichang (Xiling) Gorge, 2345. **Yunnan:** Mengzi at 1,800 metres, 9802.

A deciduous shrub to 3 metres tall, described in 1825 from a Nepalese collection. *Hydrangea aspera* is a widespread species ranging from the Himalaya to China (including Taiwan) and south to Java and Sumatra.

Hydrangea bretschneideri Dippel (HYDRANGEACEAE)

Hydrangea xanthoneura Diels var. *glabrescens* Rehder in Sargent Pl. Wilson. 1: 27. (1913). *Hydrangea bretschneideri* Dippel var. *glabrescens* (Rehd.) Rehder in

Sargent Pl. Wilson. 1: 579. (1913). *Hydrangea heteromalla* Rehder non D. Don in Sargent Pl. Wilson. 1: 579. (1913).

Yunnan: Mengzi at 2,700 metres, 10235.

Hydrangea caerulea (Stapf) Y. De Smet & Granados (HYDRANGEACEAE) New species

Deinanthe caerulea Stapf in Bot. Mag. 137: t. 8373. (1911), Besant in Gard. Chron. ser. 3, 98: 335. (1935), Morley in Glasra 3: 79. (1979), Nelson in Pim, The Wood & the Trees 230. (1984), Lancaster in The Garden 121: 6. (1996), Nelson in A Heritage of Beauty 312. (2000), O' Brien in Augustine Henry, An Irish Plant Collector in China 19. (2002).

Deinanthe bifida Diels non Maxim. in Bot. Jahrb. 29: 372. (1900), J. H. Veitch in Journ. Roy. Hort. Soc. 28: 62. non Maxim. (1903—1904).

Hubei: Liantuo, 6357 (type of *Deinanthe caerulea* Stapf), 6434.

Introduced to cultivation by Wilson.

Hydrangea febrifuga (Lour.) Y. De Smet & Grenados (HYDRANGEACEAE)

Dichroa henryi H. Leveille in Sert. Yunnanense 1. (1916). *Dichroa febrifuga* Lour Zheng in Journ. Wuhan Bot. Res. 2.(1): 58. (1984).

Hubei: South Donghu, 7706. **Yunnan:** Simao at 1,650 metres, 11050, (isotype of *Dichroa henryi* H. Leveille).

Hydrangea glaucescens (Rehder) Y. De Smet & Granados (HYDRANGEACEAE) New species

Schizophragma integrifolium (Franch.) Oliv. var. *minus* Rehder in Sargent Pl. Wilson. 1: 43. (1911), Zheng in Journ. Wuhan Bot. Res. 2.(1): 63. (1984).

Hubei: Jianshi, 5965 (type of *Schizophragma integrifolium* (Franch.) Oliv. var. *minus* Rehder).

First collected at Jianshi by Henry and later collected on Emei Shan in Sichuan by Wilson in July 1904.

Hydrangea hypoglauca Rehder (HYDRANGEACEAE) New species

in Sargent Pl. Wilson. 1: 26. (1913).

Hydrangea heteromalla McClintock non D. Don in Proc. Calif. Acad. Sci. 29: 214. (1957).

Hubei: South Badong, 6056 (type, March 1888).

Hydrangea longipes Franch. (HYDRANGACEAE)

Rehder in Sargent Pl. Wilson. 1: 33. (1913), Bean in Trees & Shrubs 1: 627. (1921), Zheng in Journ. Wuhan Bot. Res. 2.(1): 58. (1984), Nelson in A Heritage of Beauty 313. (2000), O' Brien in Augustine Henry, An Irish Plant Collector in China 21. (2002).

Hydrangea longipes Hemsley non Franchet in Journ. Linn. Soc. Bot. 23: 274. (1887), Henry in Journ. China Br. Roy. Asiat. Soc. 22: 260. (1887) (Chinese Names of Plants), Bretschneider in Hist. Eur. Bot. Disc. China 2: 783. (1898),

J. H. Veitch in Hortus Veitchii 368. (1908), Nelson in Pim, The Wood & the Trees 231. (1984). *Hydrangea aspera* McClintock non D. Don in Proc. Calif. Acad. Sci. 29: 190. (1957).

Hubei: Liantuo, 1979, 2008, 3042. North Badong, 6988. Jianshi, 5839, 5839a. South Badong, 5839b.

Hydrangea longipes was first described by Franchet in 1885 from material gathered in Baoxing (Mupin) by Père Armand David in April 1869. In 1887 Hemsley by coincidence described it again as a new species from Augustine Henry's collections using the same specific epithet. Henry stated that in the Badong area of the Three Gorges region it was known colloquially as *nui sahe t'iao*. This ratherr rare shrub was introduced to cultivation by E. H. Wilson through Messrs Veitch in 1901, it is cultivated in Ireland at the Fota Arboretum and at the National Botanic Gardens, Glasnevin. *Hydrangea longipes* was collected by the Glasnevin Central China Expedition (GCCE 456, GCCE 560, & GCCE 562) in the autumn of 2004 in Xingshan where it is a common shrub or small tree.

Hydrangea longipes Franch. var. *fulvescens* (Rehder) W. T. Wang ex C. F. Wei (HYDRANGACEAE) New variety

Hydrangea fulvescens Rehder in Sargent Pl. Wilson. 1: 39. (1913).

Hubei: Jianshi, 5949 (syntype of *Hydrangea fulvescens* Rehder).

Hydrangea macrophylla (Thunb.) Seringe (HYDRANGEACEAE)

E. H. Wilson in Journ. Arnold Arb. 4: 234. (1923), Zheng in Journ. Wuhan Bot. Res. 2.(1): 59. (1984).

Hydrangae opuloides K. Koch. var. *hortensia* Dippel Rehder in Sargent Pl. Wilson. 1: 37. (1913).

Hubei: South Badong, 7385. **Hunan:** Shimen, 7564.

Henry's 7385 is the cultivated form with all flowers sterile. This familiar plant has been cultivated for centuries in China. Introduced to Europe by Sir Joseph Banks through the Royal Botanic Gardens, Kew in 1789 from China.

Hydrangea obtusifolia (Hu) Y. De Smet & Granados (HYDRANGEACEAE)

Decumaria sinense Oliver in Hooker's. Icon. Pl. t. 1741. (1888), Bretschneider in Hist. Eur. Bot. Disc. China 2: 783. (1898), Rehder in Sargent Pl. Wilson. 1: 152. (1913), Bean in Trees & Shrubs 3: 133. (1936), Morley in Glasra 3: 79. (1979), Nelson in Pim, The Wood & the Trees 230. (1984); in A Heritage of Beauty 312. (2000).

Hubei: Xiling Gorge above Yichang, 3434. Changyang, 5219. Badong, 5219a. **Sichuan:** South Wushan, 5219b.

A beautiful evergreen climber to 4 metres tall, discovered in the Xiling (Yichang) Gorge by Henry in 1886.. Henry described this plant as a beautiful creeper, hanging from rocks in the Yichang (Xiling) Gorge, with white scented flowers. It was also later collected by Wilson.

Hydrangea pottingeri Prain (SAXIFRAGACEAE)

Hydrangea chinensis Maxim. Henry in Trans. Asiat. Soc. Jap. 24 suppl: 41. (1896) (List Pl. Formosa), Rehder in Sargent Pl. Wilson. 1: 37. (1913), Li in Woody Flora of Taiwan 245. fig. 90. (1963).

Hydrangea scandens (L. f.) Seringe ssp. *chinensis* (Maxim.) McClintock in Proc. Calif. Acad. Sci. 29: 206. (1957).

Taiwan: Wanchin, 98, 492, 1639, 1716. Oluanpi, 379, 590, 1321, 2054. Tamsui, 1451.

A small shrub, native from north-east India to China (Taiwan). A common species throughout Taiwan.

Hydrangea scandens (Linn. f.) Seringe (SAXIFRAGACEAE)

McClintock in Proc. Calif. Acad. Sci. 29: 205. (1957).

Hydrangae virens Henry non Sieb. in Trans. Asiat. Soc. Jap. 24 suppl: 41. (1896) (List Pl. Formosa).

Taiwan: Wanchin, 560.

Also native to Japan. Rare in China (Taiwan).

Hydrangea spp. (HYDRANGEACEAE)

Anonymous in Kew List of Determinations (volume 24 & 34).

Hubei: Badong, 370 (June 1885), 3721, 4050. **Hunan:** Shimen, 7545. **Yunnan:** In Pingbian on the Daweishan Range at 1,950 metres, 9048. Mengzi, 9760.

Hydrangea strigosa Rehder (HYDRANGEACEAE)
New species

in Sargent Pl. Wilson. 1: 31. (1911), Zheng in Journ. Wuhan Bot. Res. 2.(1.): 58. (1984). *Hydrangea strigosa* Rehd. var. *sinica* (Diels.) Rehder in Sargent Pl. Wilson. 1: 32. (1913). *Hydrangea strigosa* Rehder. var. *macrophylla* (Hemsl.) Rehder in Sargent Pl. Wilson. 1: 32. (1913). *Hydrangea villosa* Rehder. var. *strigosior* (Diels.) Rehder in Sargent Pl. Wilson. 1: 39. (1913).

Hydrangea aspera Henry non D. Don in Journ. China Br. Roy. Asiat. Soc. 22: 255. (1887) (Chinese Names of Plants). *Hydrangea aspera* D. Don var. *angustifolia* Hemsley in Journ. Linn. Soc. Bot. 23: 273. (1887) (type from Yichang. A. Henry without number) *Hydrangea aspera* D. Don var. *macrophylla* Hemsley in Journ. Linn. Soc. Bot 23: 273. (1887) Zheng in Journ. Wuhan Bot. Res. 2(1): 59. (1984) Bean in Trees & Shrubs 2: 388. (1992) (type from Liantuo without number). *Hydrangea longipes* Hemsley var. *lanceolata* Hemsley in Journ. Linn. Soc. Bot. 23: 274. (1887). *Hydrangea aspera* D. Don subsp. *strigosa* (Rehd.) McClintock in Proc. Calif. Acad. Sci. 29: 193. (1957), Nelson in A Heritage of Beauty 313. (2000), O' Brien in Augustine Henry, An Irish Plant Collector in China 21. (2002).

Hubei: Liantuo, 185a., 2083, 2206, 2473. Yichang, 185, 1083. Badong, 1786, (type of *Hydrangea longipes* Hemsley var. *lanceolata* Hemsley), 4844. Xingshan, 6477.

According to Henry this species was known colloquially in the Three Gorges region as *lia liao p'i*. This species was introduced to Irish gardens by the National Botanic gardens, Glasnevin, through Wilson 773 (Xingshan) and Wilson 757 (from the same locality). Both were collected at low-altitude and so flowered late in cultivation. John Besant in the *Gardeners' Chronicle* (October 19th 1935) mentions a plant raised at Glasnevin from Wilson's 757 that flowered in October. In the autumn of 2002 the Glasnevin Central China Expedition collected seeds of *Hydrangea aspera* subsp. *strigosa* at low altitude near the Fuhu temple on Emei Shan in Sichuan (GCCE 43) and at relatively high altitude at Yesanguan in Badong (GCCE 341). It was again collected by the 2004 Glasnevin Expedition to the Three Gorges region near the San You Dong Glen above Yichang (GCCE 714) and on Tianthu Shan in Changyang (GCCE 809) in October of that year. The plants raised from the Badong collection flower prolifically at Glasnevin.

Hydrangea viburnoides (Hook. f. & Thoms.) Y. De Smet & Granados (HYDRANGEACEAE)

Pileostegia viburnoides Hook. f. & Thoms. Forbes & Hemsley in Journ. Linn. Soc. Bot. 23: 274. (1887), Zheng in Journ. Wuhan Bot. Res. 2.(1): 61. (1984).

Hubei: Antelope Glen, 3220. Yichang, 4130. Liantuo, 6340.

An evergreen shrub to 7 metres tall, in its native habitat this species scales trees, cliffs and rock-faces by means of aerial roots. During September and October dense terminal panicles of white flowers are produced. Distributed in the Khasi hills in northeast India, China (including Taiwan) and Japan. *Hydrangea viburnoides* was introduced to cultivation by Wilson through the Arnold Arboretum in 1908. A handsome climber for a warm wall.

Hydrangea yunnanensis Rehder (HYDRANGEACEAE)
New species

in Sargent Pl. Wilson. 1: 37. (1911).

Hydrangea scandens (L. f.) Seringe ssp. *chinensis* McClintock non (Maxim.) McClint. in Proc. Calif. Acad. Sci. 29: 206. (1957).

Yunnan: In mountain forest to the north of Mengzi at 2,100 metres, 10226, 10226b. In mountain forests to the north of Mengzi at 2,100 metres, 10236, (syntype of *Hydrangae yunnanensis* Rehder). Mountain forests to the north of Mengzi at 2,300 metres, 10236b., (syntype of *Hydrangae yunnanensis* Rehder).

Distributed in China (Yunnan) and northern Vietnam.

Hydrilla verticillata (L. f.) Royle (HYDROCHARTACEAE)
Wright in Journ. Linn. Soc. Bot. 36: 1. (1903).
Hubei: Yichang, 4368 (in part).

Hydrocera triflora (L.) Wight & Arn. (BALSAMINACEAE)
Anonymous in Kew List of Determinations (volume 24).
Hainan: Environs of Kiungchow, 8744.

Hydrocharis dubia (Bl.) Backer (HYDROCHARTACEAE)
Hydrocharis morsus-ranae Wright non L. in Journ. Linn. Soc. Bot. 36: 2. (1903).
Hubei: Yichang, 127, 2251.

Hydrocotyle dielsiana Wolff (APIACEAE) New species
in Repert Spec. Nov. Regni Veg. 27: 112. (1930).
Hubei: Donghu, 6405b., (syntype).

Hydrocotyle hookeri (C. B. Clarke) Craib subsp. **chinensis** (Dunn ex Shan & S. L. Liou) M. F. Watson & M. L. Sheh (APIACEAE)
Hydrocotyle chinensis (Dunn) Craib non L. Craib in Kew Bull. Misc. Inf. 7-60. (1911), Shan & Lu in Acta Phytotax. Sin. 9: 119-134. (1964). *Hydrocotyle shanii* Boufford in Acta Phytotax. Sin. 28: 331. (1990), superflous name. *Hydrocotyle burmanica* Kurz ssp. *craibii* (H. Eichler) C. Y. Wu & Pu in Novon 8: 70. (1998).
Hubei: Antelope Glen, 4195. Changyang, 6230, 7598. **Yunnan:** In woods near Mengzi at 2,800 metres, 10224 (type of *Hydrocotyle burmanica* Kurz ssp. *craibii* (H. Eichler) C. Y. Wu & Pu). Mengzi, 11285.
Endemic to China and distributed in Hunan, Sichuan and Yunnan.

Hydrocotyle javanica Thunb. (APIACEAE)
Henry in Trans. Asiat. Soc. Jap. 24 suppl: 47. (1896) (List Pl. Formosa).
Taiwan: Wanchin, 1542.

Hydrocotyle sibthorpioides Lam. (APIACEAE)
Hydrocotyle rotundifolia Roxb. ex DC. Forbes & Hemsley in Journ. Linn. Soc. Bot. 23: 325. (1887), Henry in Trans. Asiat. Soc. Jap. 24 suppl: 47. (1896) (List Pl. Formosa).
Hubei: Liantuo, 1930, 6230a. **Hainan:** Near the Kiungchow Pagoda, 7981. **Taiwan:** Kaohsiung Plain, 1795, 2028.
A prostrate perennial herb, distributed in tropical and warm-temperate Asia. In Chinese herbal medicine this species of pennyworth is used to treat cases of cold and tonsilitis.

Hydrocotyle sp. (APIACEAE)
Hydrocotyle burmanica Kurz. Anonymous in Kew List of Determinations (volume 24).
Hubei: Badong, 4725, 6241b. Liantuo, 6241. Donghu,

6405. Jianshi, 6241. Without locality, 6241a. **Sichuan:** North Wushan, 6405a.

Hydrolea zeylanica (L.) Vahl. (HYDROPHYLLACEAE)
Henry in Trans. Asiat. Soc. Jap. 24 suppl: 62. (1896) (List Pl. Formosa).
Taiwan: Wanchin, 1649.

Hygrophila ringens (L.) R. Br. ex Spreng (ACANTHACEAE)
Hygrophila salicifolia (Vahl.) Nees Forbes & Hemsley in Journ. Linn. Soc. Bot. 26: 237. (1890), Henry in Trans. Asiat. Soc. Jap. 24 suppl: 69. (1896) (List Pl. Formosa).
Hubei: Yichang, 65, 2225, 2768, 2768a. **Taiwan:** Kaohsiung Plain, without number.

Hylodesmum duclouxii (Pamp.) Y. F. Deng (FABACEAE)
Desmodium henryi Schindler in Repert. Spec. Nov. Regni Veg. 22: 260. (1926).
Yunnan: Mengzi at 2,250 metres, 9700 (isolectotype of *Desmodium henryi* Schindler). In woods on Fengchunling at 2,100 metres, 9700b.

Hylodesmum leptopus (A. Gray ex Benth.) H. Ohashi & R. R. Mill (FABACEAE)
Desmodium gardneri Benth. Forbes & Hemsley in Journ. Linn. Soc. Bot. 23: 172. (1887), Henry in Trans. Asiat. Soc. Jap. 24 suppl: 33. (1896) (List Pl. Formosa). *Desmodiun laxum* DC. subsp. *leptopus* (A. Gray ex Benth.) H. Ohashi Zheng in Journ. Wuhan Bot. Res. 2.(1): 99. (1984).
Hubei: Yichang, 2121, 2378. Xingshan, 6449. Antelope Glen, 4262. Badong, 4801. Fang Xian, 6449a. South Donghu, 7672. **Taiwan:** Oluanpi, 987. Wanchin, 1664.

Hylodesmum oldhamii (Oliv.) H. Ohashi & R. R. Mill. (FABACEAE)
Desmodium oldhamii Oliv Anonymous in Kew List of Determinations (volume 24).
Hubei: Fang Xian, 6703.

Hylodesmum podocarpum (DC.) H. Ohashi & R. R. Mill (FABACEAE)
Desmodium podocarpum DC. Forbes & Hemsley in Journ. Linn. Soc. Bot. 23: 174. (1887), Zheng in Journ. Wuhan Bot. Res. 2.(1): 99. (1984), Anonymous in Kew List of Determinations (volume 34).
Hubei: Yichang, 2, 356, 2165, 2165a., 2165b., 2207, 2392, 4768, 4810, 4871a., 2316, 2316a., 2316b., 2376. Badong, 4810, 4815, 4871. **Yunnan:** In Pingbian on the Daweishan Range at 1,950 metres, 9049. Mengzi, 9066, 9309.
A widespread shrub, distributed in the Himalaya, India, Myanmar, China, Korea and Japan.

02

Hylodesmum podocarpum (DC.) H. Ohashi & K. Ohashi var. ***szechuenense*** (Craib.) H. Ohashi & R. R. Mill (FABACEAE) New variety

Lespedeza trichocarpa Forbes & Hemsley non (Stephan ex Willd.) Pers. in Journ. Linn. Soc. Bot. 183. (1887). *Desmodium podocarpum* DC. var. *szechuenense* Craib in in Sargent Pl. Wilson 2: 104. (1916). *Desmodium szechuense* (Craib) Schindl. Zheng in Journ. Wuhan Bot. Res. 2.(1): 99. (1984).

Hubei: Yichang, 167 (syntype of *Desmodium podocarpum* DC. var. *szechuenense* Craib.). Yichang (Xiling) Gorge, 4136. Without locality, 167a.

Hylomecon japonica (Thunb.) Prantl & Kundig (PAPAVERACEAE)

Fedde in Engler, Pflanzenr. 104: 209. (1909).

Hubei: Badong, 1407, 5420. Jianshi, 5420a. Changleping, 5420b.

Hylotelephium angustum (Maxim.) H. Ohba (CRASSULACEAE)

Sedum angustum Maxim. Forbes & Hemley in Journ. Linn. Soc. Bot. 26: 283. (1887).

Hubei: Changyang, 6198. Donghu, 6198a. Fang Xian, 6198b. Fang Xian, 6753. South Badong, 7406. Xingshan, 6753a.

Hylotelephium erythrostictum (Miq.) Ohba (CRASSULACEAE)

Sedum telephium L. Forbes & Hemsley in Journ. Linn. Soc. Bot. 23: 287. (1887).

Hubei: Badong, 2874.

Hylotelephium verticillatum (L.) H. Ohba (CRASSULACEAE)

Sedum verticillatum L. Anonymous in Kew List of Determinations (volume 24).

Hubei: Xiling (Yichang) Gorge, 3273.

Hymenachne amplexicaule (Rudge) Nees (POACEAE)

Panicum amplexicaule Rudge Rendle in Journ. Linn. Soc. Bot. 36: 327. (1904).

Taiwan: Kaohsiung, 1950.

Hymenasplenium cheilosorum (Kuntze ex Mett.) Tagawa (ASPLENIACEAE)

Asplenium heterocarpum Wall. Anonymous in Kew List of Determinations (volume 34).

Yunnan: Simao, 13522.

A widespread species found over much of India, Sri Lanka, Bhutan, Nepal, Myanmar, China, Vietnam, Japan, Thailand and the Philippines.

Hymenasplenium unilaterale (Lam.) Hayata (ASPLENIACEAE)

Asplenium unilaterale Lam. Henry in Trans. Asiat. Soc. Jap. 24 suppl: 112. (1896) (List Pl. Formosa). *Asplenium*

resectum Sw. Baker in Journ. Bot. 25: 171. (1887), Christ in Bull. Soc. Bot. France 5: 54. (1905).

Hubei: Yichang, 621, 2234, 4291. Liantuo, 4491. **Taiwan:** Wanchin, 96, 97. Wanshoushan, 793. **Yunnan:** Mengzi, 9068, 10339, 11516.

Distributed in Africa, India, Sri Lanka, Nepal, China, Korea, Myanmar, Thailand, Vietnam and the Philippines.

Hymenidium cristatum (H. Boissieu) Pimenov & Kljuykov (APIACEAE) New species

Pleurospermum cristatum H. Boissieu in Bull. Soc. Bot. France 50: 434. (1906).

Hubei: Xingshan, 6510.

Hymenochaete rubiginosa (Dicks.) Lev. (HYMENOCHAETACEAE)

Anonymous in Kew List of Determinations (volume 24).

Hubei: Fang Xian, 6675. (Fungus).

Hymenodictylon sp. (RUBIACEAE)

Anonymous in Kew List of Determinations (volume 34).

Yunnan: Simao, 12150.

Hymenophyllum acutum (C. Presl.) Ebihara & K. Iwats. (HYMENOPHYLLACEAE)

Trichomanes acutum C. Presl. Christ in Bull. Herb. Boiss. 7: 2. (1899).

Yunnan: In mountain forest to the east of Mengzi at 1,650 metres, 10836b.

Hymenophyllum barbatum (Bosch.) Baker (HYMENOPHYLLACEAE)

Hymenophyllum henryi Baker in J. Bot. 27: 176. (1889), Baker in Ann. Bot. 5: 194. (1890), Bretschneider in Hist. Eur. Bot. Disc. China 2: 793. (1898). *Hymenophyllum fastigiosum* H. Christ in Bull. Herb. Boissier 7: 3. (1899).

Hubei: South Badong, 5457 (isotype of *Hymenophyllum henryi* Baker). **Yunnan:** In Pingbian on the Daweishan Range on a tree in forest at 1,600 metres, 11859 (isotype of *Hymenophyllum fastigiosum* H. Christ).

A widespread species in Asia.

Hymenophyllum dilatatum Sw. var. ***amplum*** H. Christ (HYMENOPHYLLACEAE) New variety

in Bull. Herb. Boissier 7: 2. (1899)

Yunnan: Mengzi, 11545 (isotype).

Hymenophyllum javanicum Spreng. (HYMENOPHYLLACEAE)

Christ in Bull. Herb. Boiss. 6: 866. (1898).

Yunnan: South of the Red River from Manmei at 2,200 metres, on trees, 10098. South of the Red River from Manmei at 2,200 metres, on rocks, 10099.

A small epiphytic fern, distributed from Sri Lanka to China, Malaya, Indonesia, Australia and Fiji.

Hymenophyllum pallidum (Blume) Ebihara & K. Iwats (HYMENOPHYLLACEAE)

Trichomanes pallidum Blume Henry in Trans. Asiat. Soc. Jap. 24 suppl: 110. (1896) (List Pl. Formosa).

Taiwan: Tamsui, 1408.

Distributed through India, Sri Lanka, Thailand, Vietnam, China (Hainan and Taiwan), the Philippines, Malaya, Indonesia and Polynesia.

Hymenophyllum polyanthos (Sw.) Sw. (HYMENOPHYLLACEAE)

Christ in Bull. Herb. Boiss. 7: 2. (1899).

Yunnan: In mountain forest near Mengzi at 2,900 metres, on trees, 11860.

Distributed through tropical and subtropical regions of the world. Epiphytic on trees and rocks.

Hyoscyamus niger L. (SOLANACEAE)

Forbes & Hemsley in Journ. Linn. Soc. Bot. 26: 177. (1890), Henry in Notes Econ. Bot. China 50. (1893).

Hubei: Badong, 6746.

Henbane, an annual or short-lived perennial to 60 cm tall, native to Europe and Asia and naturalised in North America. Henry made his collection of this species, the *lang-t'ang-tzu* in a mountain garden in Badong.

Hypericum ascyron L. (HYPERICACEAE)

Anonymous in Kew List of Determinations (volume 24).

Hubei: Badong, 334 (June 1885), 2412, 6079. Yichang, 510, 729, 2061.

A perennial species to 1 metre tall, widespread across Russia, China, Korea, Japan, northeastern USA and Canada.

Hypericum attenuatum Fisch. ex Choisy (HYPERICACEAE)

Hypericum aff. *attenuatum* Chy. Anonymous in Kew List of Determinations (volume 24).

Hubei: Badong, 371 (June 1885). On the hills behind the San You Dong Glen (July 1886), 1506. Without locality, 6079a.

Hypericum augustinii N. Robson (HYPERICACEAE) New species

in Journ. Roy. Hort. Soc. 95: 495. (1970), Nelson in Pim, The Wood & the Trees 231. (1984); in A Heritage of Beauty 313. (2000), O' Brien in Augustine Henry, An Irish Plant collector in China 21. (2002).

Yunnan: Near Shiping on the Papien River at 1,200 metres, 13242.

A handsome shrub to 90 cm tall. Discovered by Henry in 1899, it may well have been introduced by him but as to how this species came to be introduced is a mystery. Robson who named this species stated that in 1970 there existed two cultivated specimens, one in Cornwall, the other in Ireland. (That garden may have been the Talbot Botanic Garden at Malahide Castle in Co Dublin. Lord Talbot supplied Robson with material for his paper on *Hypericum* and it was from the same source I obtained plants for the National Botanic Gardens, Glasnevin.) Henry sent various consignments of seeds and orchids to his friend Miss Evelyn Gleeson (for whom *Keteeleria evelyniana* was named) in Dublin and it might well have been raised by this lady. Endemic to China and distributed in the provinces of Guizhou and Yunnan.

Hypericum beanii N. Robson (HYPERICACEAE) New species

in Journ. Roy. Hort. Soc. 95: 495. (1970), Nelson in Pim, The Wood & the Trees 231. (1984), Lancaster in Travels in China 148. (1993), Nelson in A Heritage of Beauty 314. (2000), O' Brien in Augustine Henry, An Irish Plant Collector in China 21. (2002).

Hypericum patulum Thunb. var. *henryi* Veitch in Gard,. Chron. ser. 3, xxxvi. 229. (1904), without description, Bean in Gard. Chron. ser. 3, xxxviii. 179. (1905), Bean in Trees & Shrubs Brit. Isl. 1: 639. (1914), Rehder in Bailey, Standard Cycl. Hort. iii. 1631. (1915), Rehder in Sargent Pl. Wilson. 2: 403. (1916), Besant in Gard. Chron. ser. 3, 98: 334. (1935).

Yunnan: Mengzi at 1,600 metres, 9986.

Originally taken to be an improved form of *Hypericum patulum,* Henry introduced to this beautiful shrub to cultivation from Mengzi through the Royal Botanic Gardens, Kew in 1898. It was soon realised this new 'form' was hardier than *H. patulum* which it replaced in popularity and was widely distributed by Messrs R. Veitch of Exeter. The specific epithet commemorates William Jackson Bean, one time curator of the Royal Botanic Gardens, Kew and a personal friend of Henry. *Hypericum beanii* is cultivated in Ireland at Mount Usher and the National Botanic Gardens, Glasnevin. *Hypericum pseudohenryi* Robson is another plant mixed up in the original *H. patulum* var. *henryi* muddle. It has no associations with Augustine Henry, but was collected in Kangding in Sichuan by Wilson in November 1908 and later in Yunnan near Zhongdian by Rock in 1932.

Hypericum elodeoides Choisy (HYPERICACEAE)

Anonymous in Kew List of Determinations (volume 34).

Yunnan: Mengzi, 10273.

Distributed in the Himalaya, Myanmar and China.

Hypericum erectum Thunb. ex Murray (HYPERICACEAE)

Anonymous in Kew List of Determinations (volume 24).

Hubei: Badong, 2538.

A perennial species to 40 cm tall, distributed across Russia, central and eastern China, Korea and Japan.

Hypericum geminiflorum Hemsley (HYPERICACEAE)
New species

in Ann. of Bot. 9: 144. (1895), Henry in Trans. Asiat. Soc. Jap. 24 suppl: 19. (1896) (List Pl. Formosa), Bretschneider in Hist. Eur. Bot. Disc. China 2: 180. (1898), Robson in Flora of Taiwan 2: 702. (1996).

Hypericum trinervium Hemsley in Ann. of Bot. 9: 144. (1895), Bretschneider in Hist. Eur. Bot. Disc. China 2: 180. (1898).

Taiwan: Oluanpi, 906 (type of *Hypericum trinervium* Hemsley), 1342. Wanshoushan, 1155.

Henry sent seeds of this species (as *Hypericum trinervium* Hemsley) to Kew in February 1893.

Hypericum henryi Lev. & Van. (HYPERICACEAE)
New species

in Bull. Soc. France 54: 591. (1908), Robson in Jour. Roy. Hort. Soc. 95: 595. (1970), Lancaster in Travels in China 222. (1989), Nelson in A Heritage of Beauty 314. (2000), O' Brien in Augustine Henry, An Irish Plant Collector in China 21. (2002).

Guizhou: Mountains of Lou-tsong-koan, Père Emile Bodinier, without number.

A handsome shrub to 2 metres tall. Collected in 1897 by the French Jesuit, Père Emile Bodinier and named for Augustine Henry by Leveille and Vaniot. Native to Myanmar, western China, Thailand and Vietnam.

Hypericum japonicum Thunb. (HYPERICACEAE)

Henry in Trans. Asiat. Soc. Jap. 24 suppl: 19. (1896) (List Pl. Formosa), Diels in Bot. Jahrb. 29: 476. (1900).

Hypericum mutilum Henry non L. in Journ. China Br. Roy. Asiat. Soc. 22: 275. (1887) (Chinese Names of Plants).

Hubei: Yichang, 2052, 3658. Liantuo, 2219, 6344. Badong, 2395, 4817. South Badong, 7373. **Taiwan:** Kaohsiung, 1698, 1787. Wanchin, 1667.

An annual species varying in height from 2 to 40 cm tall. A widespread species with a range across Nepal, Sri Lanka, central and southern China, Korea, Japan, Australia and New Zealand. A charming little plant, this species gave a great splash of colour on the upper peaks of Shennongjia Forest District, Hubei during the Glasnevin Expedition visit there in September 2004. According to Henry this species was known colloquially in the Three Gorges region as *t'ui k'ou ts'ao*. *Hypericum japonicum* is currently employed in Chinese herbal medicine as a diuretic. The Hani (an ethnic group) of Xishuangbanna in southern Yunnan call this species *yalmovqxeel* and use it to treat hepatitis and conjunctivitis.

Hypericum* cf. *japonicum Thunberg (HYPERICACEAE)
Anonymous in Kew List of Determinations (volume 34).

Yunnan: Mengzi, 10018, 11124.

Hypericum longistylum Oliver (HYPERICACEAE)
New species

in Hooker's. Icon. Pl. xvi. t. 1534. (1888), Bretschneider in Hist. Eur. Bot. Disc. China 2: 180. (1898), Diels in Bot. Jahrb. 29: 476. (1900), Henry in Flora & Sylva 1: 218. (1903), Pampanini in Nuov. Giorn. Bot. Ital. n. ser. 17: 670. fig. 15. (1910), Rehder in Sargent Pl. Wilson 2: 404. (1916).

Hubei: Yichang, 217, 718, 582, 994, 1138, 1382. Liantuo, 1999.

Hypericum longistylum was introduced to cultivation by the 2002 Glasnevin Central Expedition (GCCE 169) from the mountains behind the city of Badong. For an illustration of this species see *Lancaster in Travels in China* 392. (1993).

Hypericum monogynum L. (HYPERICACEAE)

Hypericum chinense L. Rehder & E. H. Wilson in Sargent Pl. Wilson 2: 404. (1916).

Hubei: Yichang, common on banks of rivers, 249, 484, 1580 (May 1886), 1681, 3233, 3234. Without locality, 484b.

An evergreen shrub to 60 cm tall, native to China (including Taiwan), introduced to cultivation by the Earl of Northumberland in 1753 from seeds collected by Sir George Staunton and David Nelson who travelled on Lord Macartney's mission to the Emperor of China. Cultivated at the National Botanic Gardens, Glasnevin.

Hypericum patulum Thunb. (HYPERICACEAE)

Hypericum patulum Thunb. var. *parvifolium* Anonymous in Kew List of Determinations (volume 24).

Sichuan: Henry's Chinese collector with Pratt, no locality stated but may be Leshan, Emei Shan, Wa Shan or Kangding, 8787, 8813.

A semi-evergreen shrub to 1 metre tall, producing cymes of large golden-yellow flowers from May to July. Distributed in China (Anhui, Fujian, Guangxi, Guizhou, Hubei, Hunan, Jiangxi, Sichuan, Shaanxi and Zhejiang) and in northern Vietnam and Japan.

Hypericum perforatum L. (HYPERICACEAE)

Rehder & E. H. Wilson in Sargent Pl. Wilson. 3: 452. (1917).

Hubei: Badong, 283 (July 1885), 1846 (July 1886). Yichang, 346, 1569, 2162, 4157. Liantuo, 1939, 4621. **Yunnan:** Mengzi, 9688.

Perforate Saint John's worth, a subshrub to 0.5 metres tall, native to Europe and Asia and naturalised in North America and Australia. This species is currently used to treat mild depression and to treat wounds and burns. It was collected by the Sino-American Expedition in the Shennongjia Forest District in 1980.

02

Hypericum petiolatum L. (HYPERICACEAE)

Diels in Bot. Jahrb. 29: 476. (1900).

Hubei: Badong, 869 (September 1885), 1876, 4758. Jianshi, 7324. **Sichuan:** Kangding, Henry's Chinese collector with Pratt, 8804.

Hypericum petiolulatum Hook. f. & Thomson ex Dyer (HYPERICACEAE)

Anonymous in Kew List of Determinations (volume 34).

Yunnan: Mengzi, 10274.

Hypericum prattii Hemsley (HYPERICACEAE) New species

in Journ. Linn. Soc. Bot. 29: 303. (1892), Bretschneider in Hist. Eur. Bot. Disc. China 2: 780. (1898), Diels in Bot. Jahrb. xxix. 475. (1900), Rehder in Sargent Pl. Wilson 2: 404. (1916), O' Brien in Augustine Henry, An Irish plant Collector in China 21. (2002).

Sichuan: Kangding, Henry's Chinese collector with Pratt, 8808.

Also collected near the Min River in Sichuan by Faber.

Hypericum sampsonii Hance (HYPERICACEAE)

Henry in Journ. China Br. Roy. Asiat. Soc. 22: 275. (1887) (Chinese Names of Plants), Diels in Bot. Jahrb. 29: 476. (1900).

Hubei: Yichang, 585. Liantuo, 1940, 3888.

A perennial species to 70 cm tall, native to India (Assam), eastern Myanmar, China, northern Vietnam and Japan. According to Bretschneider this species was discovered by the Kew collector Richard Oldham in 1864 but was described by Hance from a collection made near Macao by Theo. Sampson in June 1865. According to Henry this species was known colloquially in the Three Gorges region as *t'ui ching ts'ao.*

Hypericum spp. (HYPERICACEAE)

Anonymous in Kew List of Determinations (volume 24 & 34).

Hubei: Yichang, 648. Badong, 5110, 5201, 7352. **Hainan:** Environs of Haikou, 8328. **Yunnan:** Mengzi, 11134.

Hypericum wilsonii N. Robson (HYPERICACEAE) New species

Hypericum patulum Thunb. var. *henryi* Veitch Rehder in Sargent Pl. Wilson 2: 403. (1916).

Hubei: South Badong, 6099.

Hypodematium crenatum (Forsk.) Kuhn. (HYPODEMATIACEAE)

Ching in Sunyatsenia 3: 11. (1935).

Hubei: Yichang, 433, 1089. Liantuo, 2212, 2641. **Yunnan:** Mengzi, 9492.

Distributed across Africa, Arabia, the Himalaya,

northern India, Myanmar, China, Japan, the Philippines and Malaysia.

Hypoestes cumingiana (Nees) Benth. & Hook. f. (ACANTHACEAE)

Henry in Trans. Asiat. Soc. Jap. 24 suppl: 70. (1896) (List Pl. Formosa).

Taiwan: Kaohsiung Plain, without number.

Hypoestes purpurea (L.) R. Br. (ACANTHACEAE)

Henry in Trans. Asiat. Soc. Jap. 24 suppl: 70. (1896) (List Pl. Formosa).

Hypoestes sp. nova Forbes & Hemsley in Journ. Linn. Soc. Bot. 26: 249. (1890).

Hainan: Environs of Haikou, 8419. **Taiwan:** Oluanpi, 290. Wanchin, 532.

Hypogymnia hypotrypa (Nyl.) Rassad. (PARMELIACEAE)

Parmelia hypotrypa Nyl. Muell.-Arg. in Bull. Herb. Boissier 1: 236. (1893).

Hubei: Xingshan, 6939, 6975. (Lichen).

Hypoxis aurea Lour. (HYPOXIDACEAE)

Wright in Journ. Linn. Soc. Bot. 36: 86. (1903).

Hubei: South Badong, 6242.

Hypserpa nitida Miers (MENISPERMACEAE)

Hypserpa laevifolia Diels in Engler, Pflanzenr. 46: (iv. 94) 210. (1910).

Hainan: Without locality, 8563 (type of *Hypserpa laevifolia* Diels, collected November 1889).

Hyptis capitata Jacq. (LAMIACEAE)

Henry in Trans. Asiat. Soc. Jap. 24 suppl: 72. (1896) (List Pl. Formosa).

Taiwan: Wanchin, without number. Oluanpi, 225.

An American native, naturalised in Chinese Taiwan.

Ichnocarpus frutescens (L.) W. T. Aiton (APOCYNACEAE)

Anonymous in Kew List of Determinations (volume 24).

Hainan: Environs of Kiungchow, 8231.

Ichnocarpus sp. (APOCYNACEAE)

Anonymous in Kew List of Determinations (volume 34).

Yunnan: Mengzi, 9580.

Ichtyoselmis macrantha (Oliver) Liden & Fukuhara (FUMARIACEAE) New species

Dicentra macrantha Oliver in Hooker's Icon. Pl. 20: t. 1937. (1890), Bretschneider in Hist. Eur. Bot. Disc. China 2: 779. (1898), Diels in Bot. Jahrb. 29: 354. (1900), Henry in Flora & Sylva 1: 217. (1903), Morley in Glasra 3: 79. (1979), Nelson in Pim, The Wood & the Trees 231. (1984); in A Heritage of Beauty 312. (2000), O' Brien in Augustine Henry, An Irish Plant Collector in China 19. (2002).

Hubei: Jianshi, 5846 (type of *Dicentra macrantha* Oliver).

Henry described finding this plant in a dark wood as a few individual specimens, growing with *Podophyllum versipelle* subsp. *boreale* and *Caulophyllum robustum*. Seeds were taken from his 5846 at Kew on April 6th 1889 and it was by this means the plant came into cultivation.

Idesia polycarpa Maxim. (SALICACEAE) New to China

Henry in Journ. China Br. Roy. Asiat. Soc. 22: 271. (1887) (Chinese Names of Plants); in Notes Econ. Bot. China 55. (1893); in Trans. Asiat. Soc. Jap. 24 suppl: 18. (1896) (List Pl. Formosa), Diels in Bot. Jahrb. 29: 478. (1900), Rehder & E. H. Wilson in Sargent Pl. Wilson. 1: 884. (1913), Li in Woody Flora of Taiwan 607. fig. 239. (1963), Zheng in Journ. Wuhan Bot. Res. 2.(1): 153. (1984).

Hubei: Badong, 1841, 2818. Jianshi, 5814. Without locality, 5814a., 5814b. **Taiwan:** Wanchin, 429.

A deciduous dioecious tree to 15 metres tall. Discovered in 1860 by the Russian botanist, Carl Maximowicz in Japan and soon after introduced to St Petersburg. It is extremly handsome in foliage and if both sexes are grown in close proximity heavy crops of berries are formed. Henry's Badong collector knew this tree as the *shui-tung-kua*. In Hubei its timber was used in the manufacture of ladles. At Stourhead in Wiltshire it has reached a height of 21 metres with a trunk of 37 cm. Li describes this tree as being scarce in Chinese Taiwan.

Ilex asprella (Hook. & Arn.) Champ. (AQUIFOLIACEAE)

Henry in Trans. Asiat. Soc. Jap. 24 suppl: 26. (1896) (List Pl. Formosa), Hu in Journ. Arnold Arb. 30: 269. (1949).

Taiwan: Oluanpi, 221, 254, 1334. Wanchin, 444, 572.

A small deciduous shrub to 3 metres tall. A widespread deciduous species in the coastal provinces of warm temperate and subtropical south-eastern China. Scattered at low to medium altitude throughout Chinese Taiwan and extending south to the Philippine Islands.

Ilex centrochinensis S. Y. Hu (AQUIFOLIACEAE) New species

in Journ. Arn. Arb. 30: 351. (1949), Zheng in Journ. Wuhan Bot. Res. 2.(1): 118. (1984).

Ilex aquifolium L. Henry in Journ. China Br. Roy. Asiat. Soc. 22: 253. (1887) (Chinese Names of Plants); in Notes Econ. Bot China 63. (1893). *Ilex aquifolium* L. var. *chinensis* Loesener ex Diels in Nova Acta Acad. Caes. Leop.-Carol. German. Nat. Cur. 78: 263. (1901). *Ilex dipyrena* Wall var. *leptacantha* Loesener in Bot. Jahrb. 29: 436. (1900).

Hubei: Yichang, on the crest of glens, 325, 1084 (syntype of *Ilex aquifolium* L. var. *chinensis* Loesener, type of *Ilex centrochinensis* S. Y. Hu). Liantuo, 2629. Glens off the Xiling (Yichang) Gorge, 3299 (syntype of *Ilex aquifolium* L. var. *chinensis* Loesener). Antelope Glen, 4239 (type of *Ilex dipyrena* Wall var. *leptacantha* Loesener).

An evergreen shrub to 3 metres tall. Endemic to the provinces of Hubei and Sichuan, the best known sites for this species are the Three Gorges region and the *Metasequoia* valley where it is found in forested areas and alongside streams and roadsides at 500 to 700 metres above sea level. Rare in cultivation.

Ilex chinensis Sims (AQUIFOLIACEAE)

Hu in Journ. Arnold Arb. 30: 299. (1949), Zheng in Journ. Wuhan Bot. Res. 2.(1): 118. (1984).

Ilex purpurea Hassk. var. *oldhamii* (Miq.) Loes. Loes in Sarg.Pl. Wils. 1: 76. (1913).

Hubei: In the Lung Wang Tung Glen (near Yichang, June 1886,) 1562. Yichang, 3103. Liantuo, 3214 (February 1887), 3911, 4440. Changleping, 6211.

The *tunching* (winter green) or *wan-sho-hong* (everlasting red) a large glabrous evergreen tree to 15 metres tall. This very handsome holly was collected by the Glasnevin Central China Expedition (GCCE 681) in the Red Dragon Reserve in Xingshan in October 2004.

Ilex cinerea Champ. (AQUIFOLIACEAE)

Hu in Journ. Arnold Arb. 31: 56. (1950).

China: Hong Kong, 84.

An evergreen shrub or small tree to 6 metres tall, endemic to Hong Kong. Allied to *Ilex formosana* Maxim.

Ilex cochinchinensis (Lour.) Loes. (AQUIFOLIACEAE)

Hu in Journ. Arnold Arb. 31: 239. (1950), Li in Woody Flora of Taiwan 455. (1963), Li in Flora of Taiwan 3: 624. (1993).

Ilex sp. Henry in Trans. Asiat. Soc. Jap. 24 suppl. 27. (1896) (List Pl. Formosa). *Ilex ardisioides* Loesener in Nova Acta Acad. Caes. Leop.-Carol. German. Nat. Cur. 78(1): 359. (1901), Hayata in Journ. Coll. Sci. Tokyo 30: 53. (1911), Yamamoto in Suppl. Ic. Pl. Form. 1: 30. fig. 10. (1925), Kanehira in Form. Trees 369. (1936), Hu & Tang in Bull. Fan. Mem. Inst. Biol. Bot. Ser. 9: 254. (1940).

Taiwan: Oluanpi,), 1002, (isotype of *Ilex mertensii* Maxim. var. *formosae* Loesener), 1251 (isosyntype of *Ilex mertensii* Maxim. var. *formosae* Loesener), 1311 (isotype of *Ilex ardisioides* Loesener 1991.

A small evergreen tree to 9 metres tall, native to northern Vietnam and China (Hainan and Taiwan). In Taiwan it is only known from Oluanpi.

Ilex corallina Franch. (AQUIFOLIACEAE)

Henry in Flora & Sylva 1: 127. (1903), Hu in Journ. Arnold Arb. 31: 64. (1950), Zheng in Journ. Wuhan Bot. Res. 2.(1): 118. (1984).

Ilex sp. Henry in Notes Econ. Bot. China 64. (1893). *Ilex corallina* Franch. var. *macrocarpa* S. Y. Hu in Journ.

Arnold. Arb. 31: 67. (1950).

Hubei: Xiling (Yichang) Gorge, 3344. Yichang, 3376. Changleping, 7847. Changyang, 7600. **Yunnan:** Milê, 10024. Mengzi, 10024a.

An evergreen glabrous tree to 10 metres tall, growing in mixed forest at altitudes of 2,100 to 3,100 metres. Endemic to China where it is found in the provinces of Yunnan, Sichuan, Guizhou and Hubei in mixed forest at 600 to 2,000 metres above sea level. Flowering occurs in May followed by red fruits in September. According to Henry the leaves of this species were used as a tea substitute by poorer natives. It was introduced to cultivation by Wilson in 1900 for Messrs Veitch. Forrest also sent seeds from Yunnan and collected material in the Mekong-Salween divide in November 1917 (Forrest 15047).

Ilex cornuta Lindl. ex Paxt. (AQUIFOLIACEAE)

Forbes & Hemsley in Journ. Linn. Soc. Bot. 23: 115. (1886), Henry in Journ. China Br. Roy. Asiat. Soc. 22: 253. (1887) (Chinese Names of Plants); in Notes Econ. Bot. China 63. (1893).

Ilex cornuta Lindl. ex Pax. var. *longipedicellata* Henry in Journ. China Br. Roy. Asiat. Soc. 22: 253. (1887) (Chinese Names of Plants). *Ilex cornuta* Lindl. ex Pax. var. *fortunei* (Lindl. ex. Paxt.) S. Y. Hu Zheng in Journ. Wuhan Bot. Res. 2.(1): 119. (1984).

Hubei: Yichang, 969, 3292 (October 1887), 3292a.

An evergreen, large shrub or small tree, native to China westwards from from Hubei to the lower Yangtze provinces, also found in Korea. Introduced to cultivation by Robert Fortune in 1846. Fortune's introduction (*Ilex cornuta* f. *fortunei*) was almost spineless. According to Henry this species occurred at low altitudes and was known colloquially as *lao shu tz'u.* The bark, leaves and fruits of this species are used as a tonic in Chinese herbal medicine and as a remedy for diseases of the kidney. The seed oil has been used in China in the manufacture of soap and a dye and gum are extracted from the bark. It was also once used as a host for scale insects during the production of insect wax in China. *Ilex cornuta* was collected by the Glasnevin Central China Expedition (GCCE 608) near Henji town in Xingshan in September 2004.

Ilex excelsa (Wall.) Hook. f. (AQUIFOLIACEAE)

Hu in Journ. Arnold Arb. 30: 306. (1949).

Yunnan: Mengzi, 13691.

An evergreen tree to 10 metres tall, native to Bhutan, Nepal, the Himalaya region of northern India, Vietnam and in southern China in the provinces of Guangxi and Yunnan.

Ilex fargesii Franch. (AQUIFOLIACEAE) New species

Hu in Journ. Arnold Arb. 31: 44. (1950), Zheng in Journ. Wuhan Bot. Res. 2.(1): 119. (1984).

Hubei: Fang Xian, 6760, 6899. **Sichuan:** North Wushan, 7147.

An evergreen tree to 7 metres tall, growing in forests between 900 to 3,350 metres above sea level. Native to Myanmar and China (Yunnan, Sichuan, Hubei and Xizang). Described by Franchet from material collected by Père Paul Farges in north-west Sichuan in the early 1890s, it had been found previously by Henry in the autumn of 1888. Seeds of this species from Wilson 231 from Xingshan were received at Glasnevin from the Arnold Arboretum in 1908. In cultivation *Ilex fargesii* has remained exceptionally rare, it had reached 14.5 metres tall at Abbotsbury Subtropical Gardens in Dorset by 1980.

Ilex formosana Maxim. (AQUIFOLIACEAE)

Henry in Trans. Asiat. Soc. Jap. 24 suppl: 27. (1896) (List Pl. Formosa), Hu in Journ. Arnold Arb. 31: 70. (1950), Li in Woody Flora of Taiwan 456. fig. 176. (1963).

Taiwan: Wanchin, 445, 830, 1582. Kaohsiung Plain, 1852.

A medium-sized evergreen tree to 12 metres tall, native to southern China, including Taiwan, where it grows in broad-leaved forest at low altitude.

Ilex hainanensis (Loes.) Merr. (AQUIFOLIACEAE) New species

in Lignan Sci. Journ. 13: 60. (1935), Tanaka & Odashima in Journ. Soc. Trop. Agr. 10: 372. (1938), Masamune in Fl. Kainant. (Hainan) 174. (1943).

Ilex rotunda Thunb. var. *hainanensis* Loesener in Nov. Act. Acad. Caes. Leop.-Carol. Nat. Cur. 78: 108. (Monog. Aquif. 1: 108) (1901).

Hainan: At Nodoa in the interior of the island, 8438.

A small evergreen tree to 5 metres tall, endemic to China and found on the island of Hainan and in the mountains of southeastern Guangxi, Guangdong, Guizhou and Yunnan.

Ilex intermedia Loesener (AQUIFOLIACEAE) New species

in Nova Acta Acad. Caes. Leop.-Carol. German. Nat. Cur. 78: 274. (1901), Pampanini in Nuov. Giorn. Bot. Ital. n. s. 17: 417. (1910), Hand.-Mazz. in Symb. Sin. 7: 655. (1933), Hu in Journ. Arnold Arb. 31: 78. (1950), Zheng in Journ. Wuhan Bot. Res. 2.(1): 119. (1984).

Sichuan: South Wushan, 5549 (isotype).

According to Hu this evergreen tree is endemic to the *Metasequoia* area (Lichuan), being confined to the Hubei-Sichuan border, it was discovered by Henry in South Wushan in the spring of 1888. South Wushan lies to the north of the *Metasequoia* valley. Both Henry and Wilson collected close to the region.

Ilex macrocarpa Oliver (AQUIFOLIACEAE) New species

in Hooker's Icon. Pl. 18: t. 1787. (1888), Henry in Notes Econ. Bot. China 63. (1893), Bretschneider in Hist. Eur. Bot. Disc. China 2: 780. (1898), Loes. ex Diels in Bot. Jahrb. 29: 436. (1900); in Nov. Act. Acad. Caes. Leop.-Carol. Nat. Cur. 78: 489. (1901), Pampanini in Nuov. Giorn. Bot. Ital. n. s. 17: 147. (1900), Henry in Notes Econ. Bot. China 63. (1893), Dunn & Tutcher in Kew Bull. Misc. Inf. 10: 60. (1912), Loesener in Sargent Pl. Wilson.1: 81. (1913), Pampanini in Nuov. Giorn. Bot. Ital. 17: 417. (1910), Rehder in Journ. Arnold. Arb. 8: 158. (1927); 14: 242. (1933), Chun in Sunyatsenia 1: 181. (1933), Handel-Mazzetti in Symb. Sin. 7: 659. (1933), Comber in Notes Roy. Bot. Gard. Edinburgh 18: 54. (1933), Tardieu-Blot in Fl. Gen. Indo-Chine Suppl. 1: 780. (1948), Hu in Journ. Arnold Arb. 30: 271. (1949), Morley in Glasra 3: 79. (1979), Zheng in Journ. Wuhan Bot. Res. 2.(1): 120. (1984), Nelson in A Heritage of Beauty 314. (2000), O' Brien in Augustine Henry, An Irish Plant Collector in China 22. (2002).

Ilex aff. *costata* Bl. Henry in Journ. China. Br. Roy. Asiat. Soc. 22: 257. (1887) (Chinese Names of Plants). *Ilex henryi* Loesener in Nova Acta. Acad. Caes Leop.-Carol. German. Nat. Cur. 78: 491. (1901), Schneider in Illus. Handb.Laubholzk. 2: 168. (1912). *Ilex dubia* (G. Don.) Trelease var. *hupehensis* Loesener in Nova Acta Acad. Caes. Leop.-Carol.German. Nat. Cur. 78: 488. (1901), in part.

Hubei: Liantuo, 1895, 3874, 4179 (May 1888), 2981 (isotype), 4633 (type). Changleping 6214 (type of *Ilex dubia* (G. Don) Trelease var. *hupehensis* Loesener). Yichang, 3451, 4179. Antelope Glen, 3445. South Badong, 7382. Jianshi, 7720. Without locality, 4179a. **Sichuan:** Emei Shan, Henry's Chinese collector with Pratt, 8977. **Yunnan:** Milê, 10308.

A deciduous tree to 17 metres tall, native central and south-west China and Vietnam and a variable species over this range. According to Henry this species was known colloquially in the Three Gorges region as *ku t'ou ch'ing*. Discovered by Henry in 1886, introduced to cultivation by Wilson in 1907 through the Arnold Arboretum.

Ilex macropoda Miquel (AQUIFOLIACEAE)

Hu in Journ. Arnold Arb. 30: 279. (1949), Zheng in Journ. Wuhan Bot. Res. 2.(1): 120. (1984).

Hubei: South Badong, 6107.

A deciduous tree to 13 metres tall. First described from Japan (where it is a common mountainous plant in areas of Hokkaido, Honshu, Shikoku and Kyushu), in China it is a widespread and common tree in the forests of the lower Yangtze provinces. This species is also recorded from Korea.

Ilex manneiensis S. Y. Hu (AQUIFOLIACEAE) New species

in J. Arnold Arbor. 30: 298. (1949).

Yunnan: South of Red River from Manmei at 1,950 metres. (Hu calls the town 'Mannei', hense the specific epithet), 9628, (holotype). Mengzi, 11014.

An evergreen tree to 9 metres tall, endemic to Yunnan.

Ilex metabaptista Loesener ex Diels (AQUIFOLIACEAE) New species

in Bot. Jahrb. 29: 435. (1900), Hu in Journ. Arnold Arb. 31: 227. (1950), Zheng in Journ. Wuhan Bot. Res. 2.(1): 120. (1984).

Hubei: Xiling (Yichang) Gorge, 3343 (isotype). Yichang, 3472 (isotype). Badong, 1764 (isotype, collected July 1886).

A pilose evergreen shrub or small tree to 4 metres tall growing at altitudes of 300 to 600 metres. Endemic to China where it is found in the provinces of Hubei, Hunan, Guizhou and Guangxi. Allied to *Ilex salicina* Hand.-Mazz., red fruits are carried in December.

Ilex micrococca Maxim. (AQUIFOLIACEAE)

Ilex polyneura (Hand.-Mazz.) Hu var. *glabra* Hu in Journ. Arnold Arb. 30: 265. (1944). *Ilex micrococca* Maxim. f. *pilosa* Hu in Journ. Arnold Arb. 30: 263. (1949). *Ilex polyneura* (Hand.-Mazz.) Hu var. *glabra* Hu in Journ. Arnold Arb. 30: 265. (1944).

Yunnan: Mengzi, 10329, 10329a., 10329c. Simao 11953, 11953a., 11953b., 11953c., 11953d., 11974a., (syntype of *Ilex micrococca* Maxim. f. *pilosa* Hu), 13702 (syntype of *Ilex micrococca* Maxim. f. *pilosa* Hu).

A deciduous tree to 20 metres tall, native to Japan, China and Vietnam.

Ilex pedunculosa Miq. (AQUIFOLIACEAE)

Hu in Journ. Arnold Arb. 30: 334. (1949), Zheng in Journ. Wuhan Bot. Res. 2.(1): 120. (1984).

Ilex pedunculosa Miq. f. *continentalis* Loesener ex Diels in Bot. Jahrb. 29: 435. (1900); Nova Acta Acad. Caes. Leop.-Carol. German. Nat. Cur. 78. (1901), Loesener in Sargent Pl. Wilson.1: 76. (1913), Rehder in Journ. Arnold. Arb. 7: 156. (1927); 14: 240. (1933).

Hubei: Badong, on high hills, 1702 (isotype of *Ilex pedunculosa* Miq. f. *continentalis* Loesener), 2808, 4913, 5168, Jianshi, 5910. Fang Xian, 6614 (isotype of *Ilex pedunculosa* Miq. f. *continentalis* Loesener).

An evergreen shrub or small tree to 10 metres tall. Also found in Korea, China (Taiwan) and Japan from where it was introduced by Charles Sargent in 1893 and later reintroduced by Wilson from Hubei as the f. *continentalis*. This species is commonly grown in gardens in both China and Japan. A relatively common plant in the Three Gorges district where it was collected by the Glasnevin Central China Expedition in the autumn of 2002.

Ilex cf. *pernyi* Franchet (AQUIFOLIACEAE)

Anonymous in Kew List of Determinations (volume 34).

Yunnan: Mengzi, 11169.

Ilex pernyi Franchet (AQUIFOLIACEAE)

Forbes & Hemsley in Journ. Linn. Soc. Bot. 23: 117. (1986), Henry in Journ. China Br. Roy. Asiat. Soc. 22: 258. (1887) (Chinese Names of Plants); in Notes Econ. Bot. China 63. (1893); in Flora & Sylva 1: 217. (1903), Zheng in Journ. Wuhan Bot. Res. 2.(1): 121. (1984), Nelson in Pim, the Wood & the Trees 220. (1984); in A Heritage of Beauty 314. (2000), O' Brien in Augustine Henry, An Irish Plant Collector in China 22. (2002).

Hubei: Badong, 900 (October 1885), 3158. Yichang, 5298 (March 1888). Jianshi, 5298a.

In its native habitat Perny's holly makes a small tree to 8 metres tall. Henry's Badong collector knew this species as *mao-erh-tz'u*. At Hankow (now part of the city of Wuhan on the eastern plains of Hubei) the same name was given to *Ilex cornuta* Lindl. ex Pax. Discovered in 1858 by the French missionary Paul Perny, this species is endemic to China and was introduced to cultivation by Wilson through Veitch's Coombe Wood nursery in 1900 and to the Arnold Arboretum in 1908. Seeds had been taken off Henry's 5298 at Kew on April 10th 1889 but no records of how they fared were kept. In the Veitch catalogue of Novelties for 1908, plants were available for ten shillings and six pence each, larger plants were available for the staggering sum of twenty one shilling each. *Ilex pernyi* is still a common shrub in the Three Gorges region, it was collected by the Glasnevin Central China Expedition (GCCE 314) between Badong and Yesanguan in October 2002 and again by the 2004 Glasnevin Expedition (GCCE 514) near Zhenzi town in Xingshan.

Ilex rotunda Thunb. (AQUIFOLIACEAE)

Henry in Trans. Asiat. Soc. Jap. 24 suppl: 27. (1896) (List Pl. Formosa), Hu in Journ. Arnold Arb. 30: 308. (1949), Li in Woody Flora of Taiwan 460. (1963).

Taiwan: Oluanpi, 929, 1973. **Yunnan:** Mengzi, 10433.

A large evergreen tree to 20 metres, native to Korea, Japan, southern China and Vietnam. Grows in broad-leaved forest throughout Taiwan, China.

Ilex sinica (Loes.) S. Y. Hu (AQUIFOLIACEAE) New species

in Journ. Arnold Arb. 31: 231. (1950).

Ilex malabaric Beddome, var. *sinica* Loesener in Nova Acta Acad. Caes. Leop.-Carol. German. Nat. Cur. 89: 281. (1908).

Yunnan: Mengzi, 10471. Near Simao at 1,300 metres 12595 (isotype of *Ilex malabarica* Beddome var. *sinica* Loesener). Simao, 12595a.

Found as a small evergreen tree to 8 metres, endemic to China where it is found in subtropical forests of south-west Guangxi and Yunnan provinces. Bears white flowers in May followed by red berries in November.

Ilex spp. (AQUIFOLIACEAE)

Anonymous in Kew List of Determinations (volume 24 & 34).

Hubei: Badong, 700, 709 (July 1885). **Yunnan:** Mengzi, 10906, 10930, 11366. Simao, 12020.

Ilex szechwanensis Loesener (AQUIFOLIACEAE) New species

in Nova Acta Acad. Caes. Leop.-Carol. German. Nat. Cur. 78. (1): 347. (1901), Hu in Journ. Arnold Arb. 30: 325. (1949), Zheng in Journ. Wuhan Bot. Res. 2.(1): 121. (1984).

Ilex szechwanensis Loes. var. *puberula* Loesener in Nov. Act. Acad. Caes. Leop.-Carol. Nat. Cur. 78: 347. (1901). *Ilex crenata* Thunb. var. *scoparium* W. W. Smith in Notes Roy. Bot. Gard. Edinburgh 9: 41.(1917).

Hubei: Jianshi 5808 (type of *Ilex szechwanensis* Loes. var. *puberula* Loesener). Fang Xian, 6912 (isotype). **Sichuan:** South Wushan, 5716. **Yunnan:** Mengzi, 11012. In the mountains to the north of Mengzi at 2,950 metres, 11303.

An evergreen shrub from 1 to 5 metres tall, sometimes a flat and creeping. Endemic to China.

Ilex tetramera (Rehder) C. J. Tseng (AQUIFOLIACEAE) New species

Ilex odorata Hamilton ex D. Don var. *tephrophylla* Loesener in Nova acta Acad. Ceas. Leop.-Carol. German. Nat. cur. 89: 286. (1908). *Ilex wardii* Hu non Merr. in Journ. Arnold Arb. 31: 257. (1950). *Symplocos tetramera* Rehder in Sargent Pl. Wilson. 3: 598. (1916). *Ilex tephrophylla* (Loes.) S. Y. Hu in Journ. Arnold Arb. 31: 68. (1950).

Yunnan: Near Simao at 1,300 metres, 12597 (type of *Ilex odorata* Hamilton ex D. Don var. *tephrophylla* Loesener). Simao, 12597a., (type of *Symplocos tetramera* Rehder). Pu'er at 1,600 metres, 13273 (syntype of *Symplocos tetramera* Rehder).

An evergreen tree to 10 metres tall. Yellow flowers are carried in March followed by red berries in October. Endemic to China where it is distributed in the western provinces. Hu in her monograph of *Ilex* Journ. Arnold Arb. 31: 257. (1950) includes A. Henry 12597 in *Ilex wardii* Merr.

Ilex triflora Blume (AQUIFOLIACEAE)

Hu in Journ. Arnold Arb. 30: 328. (1949).

Yunnan: Simao, 12018, 12018a., 12018b., 12018c., 12018d.

A large shrub or small tree with zigzag pubescent branches and black fruits. Widespread throughout India, Bangladesh, the Malay Peninsula, China, Vietnam, Indonesia, Sumatra, Java and Borneo.

Ilex umbellulata (Wall.) Loes. (AQUIFOLIACEAE)

Hu in Journ. Arnold Arb. 30: 315. (1949).

Yunnan: Simao, 11926.

A large evergreen tree to 18 metres tall, native to India(including Silhet and Assam), China (southwestern Yunnan) and Vietnam in mixed forest between 780 to 1,350 metres above sea level.

Ilex uraiensis Mori & Yamamoto (AQUIFOLIACEAE)

Ilex sp. Henry in Trans. Asiat. Soc. Jap. 24 suppl: 27. (1896) (List Pl. Formosa). *Ilex mertensii* Maxim. var. *formosae* Loes in Nova Acta Acad. Caes. Leop.-Carol. 78: 338. (1901) (Monogr. Aquifol. 1: 338.). *Ilex formosae* (Loes.) Li in Woody Flora of Taiwan 463. (1963).

Taiwan: Oluanpi, 938, 1002 (type of *Ilex mertensii* Maxim. var. *formosae* Loesener).

A small evergreen tree, native to Japan (south) and China (Taiwan) where it is found in broadleaved forest.

Ilex venulosa Hook. f. (AQUIFOLIACEAE)

Ilex umbellata (Wall.) Loes. var. *megalophylla* Loesener in Nova Acta Acad. Caes. Leop.-Carol. German. Nat. Cur. 89: 272. (1908). *Ilex umbellata* Hu non (Wall.) Loes. in Journ. Arnold Arb. 30: 315. (1949).

Yunnan: Near Simao at 1,650 metres, 13486 [type of *Ilex umbellata* (Wall.) Loes. var. *megalophylla* Loesener].

An evergreen glabrous shrub or small tree to 8 metres tall, first described by Hooker from India, also native to China (Yunnan) and Myanmar. Allied to *Ilex omeiensis* H. H. Hu & Tang, an interesting species endemic to Emei Shan in Sichuan province.

Ilex yunnanensis Franch. (AQUIFOLIACEAE)

Hu in Journ. Arnold Arb. 30: 338. (1949).

Ilex yunnanensis Franch. var. *gentilis* (Franch.) Loes. ex Diels Hu in Journ. Arnold Arb. 30: 340. (1949), Zheng in Journ. Wuhan Bot. Res. 2.(1): 121. (1984).

Hubei: Fang Xian, 6901. **Sichuan:** South Wushan, 7144.

A shrub or small tree to 12 metres tall. Common in the western provinces of China where it is the *shu cha tze* or 'water tea' i.e. tea grown by the water (the leaves of this species are used as a tea substitute on the Sino-Tibetan border). Introduced to cultivation by Wilson in 1911 through the Arnold Arboretum, this species is also native to Myanmar where it was collected by Kingdon Ward. Forrest also collected this species in the Mekong-Salween divide in July 1917 (Forrest 14074). *Ilex yunnanensis* is generally white flowered but plants from the snow range of north-west Yunnan and Hubei are said to be pink and red.

Illicium burmanicum E. H. Wilson (SCHISANDRACEAE) New species

Lin in Acta Phytotax. Sin. 38: 170. (2000).

Yunnan: To the south of Hsin-hsien near Pingbian at 1,950 metres, 10182 (holotype). In Pingbian on the Daweishan Range on a tree in forest at 1,600 metres, 10182a., 10182b. Mengzi, 9451.

Wilson described this species from material collected in Kachin, Myanmar by Rock. Henry had found this species decades earlier and described the trees from wich he collected his specimens as being 6 metres tall with greenish with yellow stamens. Distributed in Myanmar and China (Yunnan).

Illicium henryi Diels (SCHISANDRACEAE) New species

in Bot. Jahrb. Syst. 29: 323. (1900), Finet & Gagnepain in Bull. Soc. Bot. France. LII. Mem. iv. 28. (1905), Finet & Gagnepain in Contrib. Fl. As. Or. II. 28. (1907), Pampanini in Nuov. Giorn. Bot. Ital. n. ser. xvii. 274. (1910), Wilson in A Naturalist in Western China 1: 33. (1913), Rehder & E. H. Wilson in Sargent Pl. Wilson. 1: 417. (1913), Zheng in Journ. Wuhan Bot. Res. 2.(1): 47. (1984), Bean in Trees & Shrubs 2: 455. (1992), Nelson in A Heritage of Beauty 314. (2000), O' Brien in Augustine Henry, An Irish Plant Collector in China 22. (2002).

Illicium anisatum Henry non L. in Journ. China Br. Roy. Asiat. Soc. 22: 262. (1887) (Chinese Names of Plants). *Illicium cambodianum* Henry non Hance in Notes Econ. Bot. China 29. (1893).

Hubei: Antelope Glen, 4156. Yichang, 5547. Antelope Glen, 3388 (type). Liantuo, 3848.

According to both Henry's and Wilson's field notes this evergreen shrub was common on the cliffs of glens near Yichang where it was known locally as *pa-kou-wei-shu*. Henry stated it was also known as *pa chio hiu*. It makes a small tree in the wild though is generally seen as a shrub in cultivation. In its native habitat the flowers vary from pink to deep crimson. *Illicium henryi* is still certainly a common plant north of the Yangtze River in the Three Gorges region where the forest cover has suffered less than on the mountains to the south. It is abundant in certain parts of the Shennongjia Forest District and was collected by the Glasnevin Central China Expedition (GCCE 623) in the Dragon Gate River Forest Park in Xingshan in October 2004. This interesting relict species is cultivated in several Irish gardens.

Illicium micranthum Dunn (SCHISANDRACEAE) New species

in Hooker's Icon. Pl. 28: t. 2714. (1901), Lin in Acta Phytotax. Sin. 38: 174. (2000).

Yunnan: Mountains to the south of Simao at 1,650 metres, 12108a., (isosyntype). Mountains to the south of Simao at 1,100 metres, 12224 (lectotype). Mountains to the south of Simao at 1,150 to 1,650 metres, 12224a, (isosyntype), 12224c, (syntype). Mountains to the south of Simao at 1,450 metres, 12224b, (isosyntype).

Endemic to China.

Illicium sp. (SCHISANDRACEAE)

Anonymous in Kew List of Determinations (volume 24).

Illicium anisatum L. Henry in Journ. China Br. Roy. Asiat. Soc. 22: 262. (1887) (Chinese Names of Plants).

Hubei: Yichang, 106, 676. Badong, 4084. Liantuo, 4500.

Illicium tashiroi Maxim. (SCHISANDRACEAE)

Li in Woody Flora of Taiwan 157. (1963).

Illicium anisatum Henry non L. in Trans. Asiat. Soc. Jap. 24 suppl: 14. (1896) (List Pl. Formosa).

Taiwan: Oluanpi, 1316.

A small tree, native to Japan (south) and China (Taiwan).

Illigera cordata Dunn (HERNANDIACEAE) New species

in Journ. Linn. Soc. Bot. 38: 296. (1908).

Yunnan: In ravines near Mengzi at 1,500 metres, 9902 (isosyntype). Mengzi, 9902a., (lectotype), 10649.

Discovered by the Irish plant collector William Hancock near Mengzi in 1894, Dunn based this species on Henry's and Hancock's Mengzi colllections. Endemic to China and distributed in the provinces of Guangxi, Guizhou, Sichuan and Yunnan.

Illigera henryi W. W. Smith (HERNANDIACEAE) New species

in Notes Roy. Bot. Gard. Edinburgh 9: 42. (1917).

Yunnan: In Pingbian, in forests on the Daweishan Range at 1,650 metres, 11043 (isosyntype), 11402 (type).

A large climber, endemic to China and distributed in Guangxi and Yunnan.

Illigera luzonensis (Presl.) Merr. (HERNANDIACEAE)

Li in Woody Flora of Taiwan 223. (1963).

Illigera sp. Henry in Trans. Asiat. Soc. Jap. 24 suppl: 43. (1896) (List Pl. Formosa).

Taiwan: Kaohsiung, 703.

A scandent shrub found in primary forest on Japan (south), China (Taiwan) and the Philippines.

Illigera sp. (HERNANDIACEAE)

Anonymous in Kew List of Determinations (volume 24).

Hainan: Without locality, 8554.

Impatiens alborosea Tardieu (BALSAMINACEAE)

New species

in Not. Syst. ed. Humbert 11: 171. (1944).

Yunnan: Mengzi, 11234 (type).

Impatiens aquatilis Hook. f. (BALSAMINACEAE)

New species

in Nouv. Arch. Mus. iv. x. 247. (1908).

Yunnan: Near Mengzi at 1,700 metres, 9430. Growing as a subaquatic in mountains to the east of Simao at 1,650 metres, 12580.

Endemic to Yunnan.

Impatiens balsamina L. (BALSAMINACEAE)

Henry in Journ. China Br. Roy. Asiat. Soc. 22: 239. (1887) (Chinese Names of Plants).

Hubei: Yichang, 420.

An annual species from central and southern India bearing white, pink, violet, crimson or orange flowers. Widely naturalised in the tropics, according to Henry the flowers of this particular species were once valued for dyeing costly silk fabrics.

Impatiens blepharosepala E. Pritzel (BALSAMINACEAE)

New species

in Bot. Jahrb. Syst. 29: 455. (1900); in Nouv. Arch. Mus. iv. x. 265. (1908).

Hubei: Jianshi, 5847 (type).

A very beautiful fleshy purple-flowered annual balsam to 60 cm tall. This species was collected by the Sino-American Expedition in the Shennongjia Forest District in 1980. When travelling on the 2004 Glasnevin Central China Expedition to Shennongjia our group found this to be a common species, we also collected material at Changyang to the south of Yichang.

Impatiens chinensis L. (BALSAMINACEAE)

Anonymous in Kew List of Determinations (volume 34).

Yunnan: Simao, 12133.

Native to India, Myanmar, China, Thailand and Vietnam.

Impatiens clavicuspis Hook. f. (BALSAMINACEAE)

W. W. Smith in Notes Roy. Bot. Gard. Edinburgh viii. 337. (1913—1915).

Yunnan: Mengzi at 1,800 metres, 9762, collected 14th September 1896.

Distributed in China (Yunnan) and northern Myanmar.

Impatiens corchorifolia Franch. (BALSAMINACEAE)

New species

in Nouv. Arch. Mus. iv. x. 261. (1908).

Yunnan: In the mountains near Mengzi at 2,200 metres, 11304.

A very beautiful annual yellow-flowered balsam to 50

cm tall. Distributed in north-east India and China (Guizhou, Sichuan and Yunnan. The roots are used medicinally and for extracting tannin. Discovered by Delavay on the Cangshan Range above Dali, it was also later collected in Yunnan by Forrest.

Impatiens dicentra Franchet ex Hook. f. (BALSAMINACEAE) New species

in Nouv. Arch. Mus. iv. x. 268. (1908).

Hubei: Badong, 4975. Fang Xian, 6645.

Impatiens duclouxii Hook. f. (BALSAMINACEAE) New species

in Nouv. Arch. Mus. iv. x. 245. (1908).

Yunnan: In woods near Mengzi at 2,200 metres, 9267. In woods near Mengzi at 2,700 metres, 11080. On mountains near Simao at 1,650 metres, 12559 (type).

Based on collections made by the French missionaries Père Ducloux, Père Delavay and Augustine Henry in Yunnan. Distributed in north-east India, China (Yunnan) and Thailand.

Impatiens ernestii Hook. f. (BALSAMINACEAE) New species

in Nouv. Arch. Mus. iv. x. 256. (1908).

Hubei: Yichang, 863 in part.

This species, which commemorates Wilson was based on collections made in Yunnan by Delavay, in Hubei by Henry and on Emei Shan in Sichuan by Faber and Wilson.

Impatiens exiguiflora Hook. f. (BALSAMINACEAE) New species

in Hooker's. Icon. Pl. t. 2975. (1913).

Hubei: Badong, 4820.

Impatiens fissicornis Maximowicz (BALSAMINACEAE) in Acta Hort. Petrop. 11: 87. (1889).

Impatiens fissicornis Maxim. var. _henryi_ E. Pritzel in Bot. Jahrb. 29: 453. (1900).

Hubei: Liantuo, 4480.

Impatiens furcillata Hemsl. (BALSAMINACEAE) Pritzel in Bot. Jahrb. 29: 456. (1900).

Sichuan: North Wushan, 7056.

Impatiens henryi E. Pritzel (BALSAMINACEAE) New species

in Bot. Jahrb. 29: 455. (1900).

Hubei: 6710, 6769.

Endemic to Hubei.

Impatiens noli-tangere L. (BALSAMINACEAE) Pritzel in Bot. Jahrb. 29: 454. (1900), in Nouv. Arch. Mus. iv. x. 262. (1908).

Hubei: Badong, 1836, 1838, 4753, 5955a, 6589.

Impatiens potaninii Maxim. f. **_rubrobrunnea_** E. Pritzel (BALSAMINACEAE) New form

in Bot. Jahrb. 29: 455. (1900).

Impatiens sp. Henry in Journ. China Br. Roy. Asiat. Soc. 22: 254. (1887) (Chinese Names of Plants).

Sichuan: South Wushan, 5687.

According to Henry this species was known colloquially at Badong as _leng ts'ao hua_.

Impatiens pritzelii Hook. f. (BALSAMINACEAE) New species

in Nouv. Arch. Mus. iv. x. 243. (1908).

Impatiens tubulosa Pritzel non Hemsl. in Bot. Jahrb. 29: 456. (1900). _Impatiens pritzelii_ Hook. f. var. _hupehensis_ Hook. f. in Nouv. Arch. Mus. iv. x. 243. (1908).

Hubei: Yichang, 2730. South Donghu, 7492. Changyang, 7742. Without locality, 2730a. **Sichuan:** South Wushan, 7245.

Impatiens pterosepala Hook. f. (BALSAMINACEAE) New species

in Kew Bull. Misc. Inf. 274. (1890).

Hubei: Xingshan at 1,400 metres, 6551. Jianshi, 7419 (type).

Later collected in the same region by Wilson (W. 2692), it was also collected by the Sino-American Expedition in the Shennongjia Forest District in 1980.

Impatiens pulchra Hook. f. & Thomson (BALSAMINACEAE)

Impatiens mengtszeana Hook. f. in Nouv. Arch. Mus. Hist. Nat. ser. 4(10): 256. (1908). _Impatiens monticola_ Hook. f. in Nouv. Arch. Mus. iv. x. 257. (1908).

Hubei: Yichang, 863 in part. **Yunnan:** Mengzi, 9808. At Fengchunling near the Red River at 2,200 metres, 11208 (type of _Impatiens mengtszeana_ Hook. f.). In the mountains near Simao, 12536, 12536b., 12536c.

Impatiens siculifer Hook. f. (BALSAMINACEAE) New species

in Nouv. Arch. Mus. iv. x. 246. (1908).

Yunnan: In woods near Mengzi at 1,650 metres, 10030, 10030a., 10030b.

Endemic to China.

Impatiens siculifer Hook. f. var. **_porphyrea_** Hook. f. (BALSAMINACEAE) New variety

in Nouv. Arch. Mus. iv. x. 246. (1908).

Yunnan: In montane forest on Fengchunling at 2,200 metres, 11206.

Impatiens spp. (BALSAMINACEAE)

Anonymous in Kew List of Determinations (volume 24 & 34).

Hubei: Antelope Glen, 3254, 4172. In a ravine on Moji Shan, 4168. Badong, 109, 1837, 2433, 2443, 2453, 4738, 7365. Liantuo, 2594, 7919. Fang Xian, 6634, 6723, 6737, 6757, 6528, 6758. Donghu, 6408. Xingshan, 4514. Jianshi, 7435, 7436. Without locality, 3254a., 3254b. **Yunnan:** Mengzi, 9207, (collected September 23rd 1896), 9268,

9637, 9765, 10046, 11235, 11280, 11282. In montane forest on Fengchunling at 2,200 metres, 11205. Simao 12003, 12279, 12279a., (collected June 12th 1898), 13099. Yunxian, 13563, (collected October 28th 1899).

Impatiens stenantha Hook. f. (BALSAMINACEAE)

Anonymous in Kew List of Determinations (volume 34).

Yunnan: Mengzi, 10038. Simao, 12615.

Native to India, Bhutan, Nepal and China (Tibet (Xizang) and Yunnan).

Impatiens stenosepala E. Pritzel (BALSAMINACEAE) New species

in Bot. Jahrb. 29: 453. (1900).

Hubei: Yichang, 4172a., 3254b.

An annual species to 60 cm tall with red-brown stems and carrying from June to August, 1 to 3 purple-red flowers in the axils of leaves. Native to Guizhou, Hubei, Henan, Shaanxi and Sichuan where it grows in forests and on river banks between 1,0900 to 2,400 metres.

Impatiens sutchuenensis Franchet ex Hook. f. (BALSAMINACEAE) New species

in Nouv. Arch. Mus. iv. x. 262. (1908).

Hubei: Jianshi, 5955.

Distributed in Hubei, Shaanxi and Sichuan.

Imperata cylindrica (L.) Raeusch. (POACEAE)

Imperata arundinacea Cyrillo Henry in Journ. China Br. Roy. Asiat. Soc. 22: 258. (1887) (Chinese Names of Plants); in Trans. Asiat. Soc. Jap. 24 suppl: 107. (1896) (List Pl. Formosa). *Imperata arundinacea* Cyrill. var. *koenigii* (Retz.) Hack Rendle in Journ. Linn. Soc. Bot. 36: 346. (1904).

Hubei: Yichang, 1136. **Taiwan:** Oluanpi, 955. **Yunnan:** Mengzi, 9506.

Widespread in Asia. Henry stated that it was known as *mao* and used for thatch in 19th century Taiwan and was known colloquially in the Three Gorges region as *mao ts'ao*, it was also the *bai mao gen,* the rhizome is the source of a drug used to stop bleeding.

Indigofera atropurpurea Buch.-Ham. ex Hornem. (FABACEAE)

Craib in Notes Roy. Bot. Gard. Edinburgh 8: 71. (1913).

Yunnan: Mengzi at 1,380 to 1,500 metres, 9720, 9720a. Simao, 12516.

Indigofera bungeana Walp. (FABACEAE)

Forbes & Hemsley in Journ. Linn. Soc. Bot. 23: 156. (1886), Henry in Journ. China Br. Roy. Asiat. Soc. 22: 253. (1887) (Chinese Names of Plants), Zheng in Journ. Wuhan Bot. Res. 2.(1): 100. (1984).

Indigofera tinctoria Henry non L. in Journ. China Br. Roy. Asiat. Soc. 22: 253. (1887) (Chinese Names of

Plants), Zheng in Journ. Wuhan Bot. Res. 2.(1): 101. (1984).

Indigofera pseudotinctoria Matsum. Craib in Notes Roy. Bot. Gard. Edinburgh 8: 69. (1903), Bean in Trees & Shrubs 2: 457. (1992), O' Brien in Augustine Henry, An Irish Plant Collector in China 22. (2002).

Hubei: Yichang, 36, 909, 1530, 3237. In the San You Dong Glen near Yichang (July 1886), 1496. Liantuo, 1926. South Badong, 6062. **Yunnan:** Mengzi at 1,380 metres, 10387a.

According to Henry this species was known colloquially at Yichang as *yeh lan chih-tzu.* Introduced to cultivation by Henry from Mengzi in 1897 through the Royal Botanic Gardens, Kew. In Ireland this shrub is cultivated at Annes Grove, the John F. Kennedy Arboretum, Malahide Castle and Mount Congreve.

Indigofera cassioides Rottler ex DC. (FABACEAE)

Indigofera elliptica Roxb. Craib in Notes Roy. Bot. Gard. Edinburgh 8: 65. (1913).

Yunnan: South of the Red River from Manpan, 10980. In forests near Simao at 1,200 metres, 12781.

Native to Pakistan, India, China (Guangxi and Yunnan), Thailand and Vietnam.

Indigofera caudata Dunn (FABACEAE) New species

in Gard. Chron. ser, 3. 31: 210. (1902), Craib in Notes Roy. Bot. Gard. Edinburgh 8: 64. (1913).

Yunnan: In forests near Simao at 1,350 to 1,500 metres, 12166 (type), 12166a., 12166b.

A handsome shrub with copper-coloured young shoots and white flowers aranged in very long tail-like racemes. Introduced to cultivation by Henry in 1899 through the Royal Botanic Gardens, Kew. Native to China (Guangxi and Yunnan) and Laos.

Indigofera decora Lindl. (FABACEAE)

Henry in Trans. Asiat. Soc. Jap. 24 suppl: 31. (1896) (List Pl. Formosa).

Indigofera decora Lindl. var. *ichangensis* (Craib) Y. Y. Fang & C. Z. Zheng in Acta Phytotax. Sin. 27: 164. (1989). *Indigofera ichangensis* Craib in Notes Roy. Bot. Gard. Edinburgh 8: 55. (1913), Zheng in Journ. Wuhan Bot. Res. 2.(1): 100. (1984). *Indigofera ichangensis* Craib f. *leptantha* Craib in Notes Roy. Bot. Gard. Edinburgh 8: 67. (1913). *Indigofera ichangensis* Craib f. *rigida* Craib in Notes Roy. Bot. Gard. Edinburgh 8: 67. (1913), Zheng in Journ. Wuhan Bot. Res. 2.(1): 101. (1984).

Hubei: Yichang, 3512 (type of *Indigofera ichangensis* Craib), 3865. In the Lung wang tung valley, 3568. Antelope Glen, 4259. Changleping, 6280. Liantuo, 6343. **Taiwan:** Wanchin, 696, 1652, 1717.

A deciduous shrub to 60 cm, native to China and Japan and introduced to cultivation by Robert Fortune in 1845 from Shanghai.

Indigofera hancockii Craib (FABACEAE)

Craib in Notes Roy. Bot. Gard. Edinburgh 8: 71. (1913).

Yunnan: In mountains near Mengzi at 1,500 metres, 13719.

Endemic to China where it is distributed in the provinces of Sichuan and Yunnan. This species was discovered near Mengzi by the Irish plant collector, William Hancock.

Indigofera hendecaphylla Jacq. (FABACEAE)

Craib in Notes Roy. Bot. Gard. Edinburgh 8: 68. (1913).

Yunnan: Simao at 1,350 metres, 12308, 12495.

Tropical and subtropical Old World.

Indigofera henryi Craib (FABACEAE) New species

in Notes Roy. Bot. Gard. Edinburgh 8: 54. (1913), Fang & Zheng in Acta Phytotax. Sin. 27: 164. (1989).

Yunnan: On rocky mountains near Mengzi at 1,950 metres, 10050a., (holotype). In wooded glens near Mengzi at 2,120 metres, 10050b.

Endemic to China and distributed in the provinces of Guizhou, Sichuan and Yunnan.

Indigofera heterantha Wall. (FABACEAE)

Anonymous in Kew List of Determinations (volume 34).

Yunnan: Mengzi, 11502.

Indigofera hirsuta L. (FABACEAE)

Henry in Trans. Asiat. Soc. Jap. 24 suppl: 31. (1896) (List Pl. Formosa).

Hainan: Environs of Haikou, 8384. **Taiwan:** Kaohsiung, without number.

An annual or biennial species to 60 cm tall, pantropical in distribution.

Indigofera linifolia (L. f.) Retz. (FABACEAE)

Henry in Trans. Asiat. Soc. Jap. 24 suppl: 31. (1896) (List Pl. Formosa).

Taiwan: Kaohsiung, without number.

An annual species to 20 cm tall, native from India across China and south from the Philippines to Australia.

Indigofera mengtzeana Craib (FABACEAE) New species

in Notes Roy. Bot. Gard. Edinburgh 8: 57. (1913).

Yunnan: In woods near Mengzi at 1,830 metres, 10627 (type).

Endemic to China and distributed in the provinces of Sichuan and Yunnan.

Indigofera nigrescens Kurz ex King & Prain (FABACEAE)

Craib in Notes Roy. Bot. Gard. Edinburgh 8: 71. (1913).

Yunnan: In woods near Mengzi at 1,650 metres, 11212. On grassy mountains near Simao at 1,800 metres,

13716.

A widespread species found over much of India, Malaysia, South China and the Philippines.

Indigofera scabrida Dunn (FABACEAE) New species

in Journ. Linn. Soc. Bot. 35: 487. Craib in Notes Roy. Bot. Gard. Edinburgh 8: 64. (1913).

Yunnan: In exposed grassy parts of the Mengzi mountains at 1,500 to 1,900 metres, 9686 (isosyntype), 9686a., 9686b., (type).

A handsome little shrub to 90 cm with red flowers. Distributed in China (Sichuan and Yunnan) and Myanmar.

Indigofera spp. (FABACEAE)

Henry in Trans. Asiat. Soc. Jap. 24 suppl: 31. (1896) (List Pl. Formosa).

Hubei: Donghu, 6425. **Taiwan:** Oluanpi, 294, 1234. **Yunnan:** Mengzi, 9687, 10051.

Indigofera stachyodes Lindl. (FABACEAE)

Craib in Notes Roy. Bot. Gard. Edinburgh 8: 64. (1913).

Yunnan: On the hills near Simao at 1,500 metres, 12276.

Native to India, Bhutan, Nepal, Myanmar and western China.

Indigofera sticta Craib (FABACEAE) New species

in Notes Roy. Bot. Gard. Edinburgh 8: 61. (1913).

Yunnan: On the rocky mountains beyond the summit of Tank'ay near Mengzi at 1,800 metres, 13720 (type).

Indigofera suffruticosa Mill. (FABACEAE)

Indigofera anil L Henry in Trans. Asiat. Soc. Jap. 24 suppl: 31. (1896) (List Pl. Formosa).

Hainan: Environs of Haikou, 8424. **Taiwan:** Wanchin, 170. Kaohsiung Plain, 1799.

Cultivated for indigo used in dyeing, apparently Taiwanese seed were the best kind for this purpose and were an important export in the late 19th century.

Indigofera tinctoria L. (FABACEAE)

Henry in Trans. Asiat. Soc. Jap. 24 suppl: 31. (1896) (List Pl. Formosa).

Taiwan: Kaohsiung, without number. Wanchin, 1669.

An erect shrub to 60 cm tall, pantropic in its distribution and once widely cultivated in China for indigo.

Indigofera trifoliata L. (FABACEAE)

Forbes & Hemsley in Journ. Linn. Soc. Bot. 157. (1886), Craib in Notes Roy. Bot. Gard. Edinburgh 8: 63. (1913).

Hubei: Yichang, 683, 999, 4389, 4389a., 4389b. **Yunnan:** Mengzi, 9347, 9347a.

A small shrub to about 30 cm tall, native to India, China, Malaysia, the Philippines and Australia.

Inula japonica Thunb. (ASTERACEAE)

Anonymous in Kew List of Determinations (volume

24).

Hubei: Badong, 537. In glens near Yichang, 1641.

Inula lineariifolia Turcz. (ASTERACEAE)

Inula britannica L. var. *linariifolia* Regel. Forbes & Hemsley in Journ. Linn. Soc. Bot. 23: 429. (1888).

Hubei: Yichang, 63, 401, 2255, 2672, 2909. Badong, 520 (September 1885), 2478, 5015. Liantuo, 2663.

The *xuan fu hua* or 'revolved, upturned flower', the flowers are used as a drug to control vomiting and hiccups.

Distributed in Nepal, India, Bhutan, Myanmar, western China, Thailand and Vietnam.

Inula racemosa Hook.f. (ASTERACEAE)

Forbes & Hemsley in Journ. Linn. Soc. Bot. 23: 430. (1888), Hemsley in Hooker's. Icon. Pl. t. 1975. (1891) Henry in Notes Econ. Bot. China 19. (1893).

Hubei: Badong, (cultivated), 4744, 4928.

Henry stated that this species was cultivated for medicinal purposes and was known as the *k'uang-mu-hsiang*.

Inula sp. (ASTERACEAE)

Anonymous in Kew List of Determinations (volume 24).

Hubei: South Badong, 6095.

Iodes sp. (ICACINACEAE)

Anonymous in Kew List of Determinations (volume 24).

Sichuan: Henry's Chinese collector with Pratt, no locality stated but may be Leshan, Emei Shan, Wa Shan or Kangding, 8809.

Iodes vitiginea (Hance) Hemsley (ICACINACEAE)

Sleumer in Blumea 17: 223. (1969).

Hainan: Environs of Haikou, 8063, 8228, 8297. **Yunnan:** On the banks of the Red River at Manpan, 9608. **Vietnam:** Tongking, 9527.

A handsome woody climber supporting itself by means of axillary tendrils and producing cymes of yellow flowers in December and continueing to flower until June of the following year. Also distributed in the provinces of Guangxi, Guangdong and Guizhou, and in Laos.

Ipomoea aquatica Forssk. (CONVOLVULACEAE)

Forbes & Hemsley in Journ. Linn. Soc. Bot. 26: 157. (1890), Henry in Trans. Asiat. Soc. Jap. 24 suppl: 64. (1896) (List Pl. Formosa).

Ipomoea reptans Chy. Henry in Journ. China Br. Roy. Asiat. Soc. 22: 279. (1887) (Chinese Names of Plants).

Hubei: Badong, 368 (June 1885). **Taiwan:** Kaohsiung, cultivated and perhaps also wild, 1188, 1189.

Water spinach, cultivated in the Three Gorges area where Henry stated it was known as *weng ts'ai*. A semi-aquatic species whose young foliage is eaten like spinach. Found all over the tropics it is a serious weed in Florida.

Ipomoea batatas (L.) Lam. (CONVOLVULACEAE)

Henry in Trans. Asiat. Soc. Jap. 24 suppl: 63. (1896) (List Pl. Formosa).

Ipomoea fastigiata (Roxb.) Sweet Forbes & Hemsley in Journ. Linn. Soc. Bot. 26: 159. (1890).

Hubei: Changyang, cultivated, 7670. **Taiwan:** Kaohsiung, without number. **Yunnan:** Milê, 9290.

The sweet potato or *fan-shu* (literally; foreign tuber), a South American native first grown in China's Fujian province during the Wan-li reign (Ming dynasty), after that province was struck by a famine in 1593. From Fujian it gradually spread around the Chinese empire and reached Taiwan in the early 1600s, eventually making its way to Japan in 1698. In China it is very often planted as a follow-on crop from summer rice. Sweet potato is now one of the most important crops in the Old World tropics and China is the world's leading producer. While known as *fan-shu*, Henry points out that the *shu* proper, is *Dioscorea* or the yam, a plant known in China since ancient times. Henry also states that the sweet potato was commonly cultivated on the hills around Yichang (as was observed by members of the 2002 Glasnevin Central China Expedition) and in Yunnan at Mengzi. According to W. B. Hemsley's note in the *Index Florea Sinensis* this species flowers very rarely in China, though Henry did send flowering material. He stated it was cultivated in Hubei between 500 to 3000 feet. Native to South America and the West Indies.

The sweet potato, reached the Old World (Polynesia, c. 1,200 A.D.) before the discovery of America in 1492. Ancient pits thought to have stored sweet potatoes have been unearthed in New Zealand and the roots figure prominently in the mythology of New Zealand's Maori tribes. Columbus saw the first sweet potato on the 4th of November 1492 and later introduced it to Europe.

Ipomoea biflora (L.) Pers. (CONVOLVULACEAE)

Ipomoea hardwickii (Spreng.) Hemsl. Henry in Trans. Asiat. Soc. Jap. 24 suppl: 64. (1896) (List Pl. Formosa).

Taiwan: Kaohsiung, without number.

Ipomoea cairica (L.) Sweet (CONVOLVULACEAE)

Ipomoea palmata Forssk. Anonymous in Kew List of Determinations (volume 24).

Hainan: Environs of Haikou, 7953.

Distributed through tropical and subtropical Africa and Asia.

Ipomoea cynanchifolia Meisn. (CONVOLVULACEAE)

Anonymous in Kew List of Determinations (volume 34).

Yunnan: Mengzi, 9676.

Ipomoea digitata L. (CONVOLVULACEAE)

Henry in Trans. Asiat. Soc. Jap. 24 suppl: 64. (1896)

(List Pl. Formosa).

Taiwan: Kaohsiung, 1125.

Ipomoea indica (Burm.) Merr. (CONVOLVULACEAE)

Ipomoea congesta R. Br. Henry in Trans. Asiat. Soc. Jap. 24 suppl: 64. (1896) (List Pl. Formosa).

Taiwan: Wanchin, 101, 1538.

Ipomoea littoralis Blume (CONVOLVULACEAE)

Anonymous in Kew List of Determinations (volume 24).

Hainan: On the sandy sea shore at Haikou, 8126.

Ipomoea muricata (L.) Jacq. (CONVOLVULACEAE)

Forbes & Hemsley in Journ. Linn. Soc. Bot. 26: 161. (1890), Henry in Notes Econ. Bot. China 50. (1893); in Trans. Asiat. Soc. Jap. 24 suppl: 64. (1896) (List Pl. Formosa).

Hubei: Liantuo, 4612, cultivated and seeds eaten. **Taiwan:** Wanchin, 1600.

Cultivated in Hubei where Henry stated it was known as the *mu-tzu-ts'ai* and that it was the seeds that were eaten. Native to India.

Ipomoea nil (L.) Roth. (CONVOLVULACEAE)

Anonymous in Kew List of Determinations (volume 34).

Yunnan: Mengzi, 9582.

A pantropic annual climbing herb to 5 metres tall.

Ipomoea obscura (L.) Ker Gawl. (CONVOLVULACEAE)

Henry in Trans. Asiat. Soc. Jap. 24 suppl: 64. (1896) (List Pl. Formosa).

Hainan: Environs of Kiungchow, 8271. **Taiwan:** Kaohsiung, without number. **Yunnan:** Yuanjiang, 13255.

Widespread in Asia.

Ipomoea pes-caprae (L.) R. Br. (CONVOLVULACEAE)

Ipomoea biloba Forsst. Henry in Trans. Asiat. Soc. Jap. 24 suppl: 64. (1896) (List Pl. Formosa).

Hainan: Environs of Haikou, 8043. **Taiwan:** Oluanpi, without number.

A perennial species, common on tropical shores, a beautiful plant.

Ipomoea pes-tigridis L. (CONVOLVULACEAE)

Henry in Trans. Asiat. Soc. Jap. 24 suppl: 65. (1896) (List Pl. Formosa).

Hainan: Environs of Haikou, 8392. **Taiwan:** Kaohsiung, without number.

Ipomoea quamoclit L. (CONVOLVULACEAE)

Henry in Trans. Asiat. Soc. Jap. 24 suppl: 65. (1896) (List Pl. Formosa).

Taiwan: Wanchin, 864. Oluanpi, without number. **Yunnan:** Simao, 13206.

The star glory, an annual climber to 3 metres tall, native to tropical America and now naturalised in many parts of Asia.

Ipomoea sagittifolia Burm. f. (CONVOLVULACEAE)

Ipomoea sepiaria Koenig ex Roxb. Henry in Trans. Asiat. Soc. Jap. 24 suppl: 65. (1896) (List Pl. Formosa).

Hainan: Without locality, 8418. **Taiwan:** Kaohsiung Plain, 1954.

Ipomoea spp. (CONVOLVULACEAE)

Anonymous in Kew List of Determinations (volume 24 & 34).

Hainan: Without locality, 8574. **Taiwan:** Wanchin, 855, 1507, 1675. **Yunnan:** Mengzi, 10987. Simao, 12632, 13441.

Iris domestica (L.) Goldblatt & Mabb. (IRIDACEAE)

Belamcanda chinensis (L.) DC. Henry in Journ. China Br. Roy. Asiat. Soc. 22: 266. (1887) (Chinese Names of Plants); in Trans. Asiat. Soc. Jap. 24 suppl: 95. (1896) (List Pl. Formosa). *Belamcanda punctata* Moench. Wright in Journ. Linn. Soc. Bot. 36: 86. (1903).

Hubei: Yichang, 96. Liantuo, 2112a. Badong, 4905. Donghu, 6399. **Taiwan:** Oluanpi, 968. Wanchin, 1568. **Yunnan:** Simao, 13175.

The blackberry lily, which Henry stated was known colloquially in the Three Gorges region as *pien chu*, the rhizomes were called *she gan* and when used with *da qing ye* (*Isatis tinctoria*) the product was used in Chinese herbal medicine in cases of swelling of the throat. The Hani (an ethnic group) of Xishuangbanna in southern Yunnan call this species *qail'aqhhyu* and use the roots and stem to treat tuberculosis. Distributed in India, China, Vietnam, Korea and Japan.

Iris henryi Baker (IRIDACEAE) New species

in Handbook of Iridaceae 6. (1892), Wright in Journ. Linn. Soc. Bot. 36: 82. (1903), Dykes in Gen. Iris 49. (1913), Liu Yin in Chinese Mag. Bot. 3(2): 591. (1936), Morley in Glasra 3: 79. (1979).

Iris gracilipes Pampanini in Nuov. Giorn. Bot. Ital. 22: 269. (1915).

Hubei: On the banks of the Yangtze at Liantuo, 6372 (type).

A handsome plant with narrow strap-like leaves to 40 cm long, carrying blue or violet coloured flowers to 3 cm in diameter in May. Two flowers per stem are carried and the lower parts of the fall and standard petals are mottled yellow. Native to Hubei, Hunan and Sichuan, growing in grass in forests or on forest edges.

Iris japonica Thunb. (IRIDACEAE)

Wright in Journ. Linn. Soc. Bot. 36: 83. (1903).

Hubei: Yichang, 1161, 1254. Badong, 786 (May 1885). Liantuo, 1967.

Iris japonica is a vigorous and stoloniferous crested

Iris and is native to Central China and Japan where it is found in wet grassland or on forest verges. The most popular of the Evansia series in western cultivation, it is also a popular plant in cultivation in both China and Japan. It is a handsome species and thrives best in the warmest counties of Britain and Ireland, in colder regions it is best placed in a cool greenhouse where its fine bamboo-like foliage remains unblemished. This handsome Iris is abundant in damp forests and by streams and waterfalls in the Three Gorges region, especially in the mountains to the north of the Yangtze. It was collected by the Glasnevin Central China Expedition (GCCE 556) near Zhenzi town in Xingshan in October 2004.

Iris spp. (IRIDACEAE)

Anonymous in Kew List of Determinations (volume 34).

Yunnan: Mengzi, 9117, 10775, 10823. Simao, 11927.

Iris tectorum Maxim. (IRIDACEAE)

Wright in Journ. Linn. Soc. Bot. 36: 85. (1903).

Iris dichotoma Henry non Pall. in Journ. China Br. Roy. Asiat. Soc. 22: 268. (1887) (Chinese Names of Plants).

Hubei: Badong, 3781. Yichang, 3970. Changleping, 6330.

A beautiful plant, well known in cultivation and one of the finest of the crested Asian irises. Though long cultivated in Japan and known as the roof iris of Japan, since it often establishes itself on straw thacted roofs there (*tectorum* is the latin for roof). *Iris tectorum* was probably introduced to Japan from China. Native to central and south-western China it is also recorded from northern Myanmar. It was first introduced to cultivation by Philipp Franz von Siebold (1796—1866) who sent plants from Japan to St. Petersburgh, though the year is not recorded. In Chinese herbal medicine this species in currently used to treat cases of constipation.

Iris wattii Baker (IRIDACEAE)

Dykes in Gard. Chron. ser. 3, lvii. 95. (1915), Sealy in Gard. Chron. ser. 3, cii. 413. (1937), O' Brien in Augustine Henry, An irish Plant Collector in China 22. (1922).

Yunnan: Mengzi, 10599, 11821, 11821a.

Iris wattii is native to India, Bhutan, Myanmar and China [Tibet (Xizang) and Yunnan]. It was described by Baker from George Watt's 6337, a collection made in 1882 on the summit of Khongui Hill in the Manipur district of eastern India. It has been confused with *Iris confusa* to which it is similar, though *Iris wattii* is a taller plant with erect bamboo-like stems and is capable of reaching 2 metres tall when well grown. The flowers are lavander-blue with a white marking on the centre of the falls which are further spotted orange yellow and deep lavander. The flowers are also larger than *Iris confusa* (6-8 cm in diameter) and the falls are almost vertical unlike *Iris confusa* in which they are horizontal. *Iris wattii* was introduced to cultivation by Major Lawerence Johnson from the China-Myanmar border in 1931 though there is some speculation that this clone is a hybrid. It has been reintroduced in more recent times from Kunming Botanical Gardens. In the archives at the National Botanic Gardens is a hand-written list "Henry plants in our garden" compiled by Henry's wife Elsie, this species features on that list as being cultivated in their Dublin garden. It is currently growing on the Chinese slope at Glasnevin from material exhibited at the Chelsea Flower Show by the Irish Garden Plant Society and Bord Glas (the Irish horticultural development authority) in the silver gilt award winning display in Henry's honour. Augustine Henry's collections proved to be a new record for China.

Irpex lacteus (Fr.) Fr. (IRPICACEAE)

Irpex sinuosus Fr. Anonymous in Kew List of Determinations (volume 24).

Hubei: Changyang, on trees, 7933. (Fungus).

Isachne albens Trin. (POACEAE)

Rendle in Journ. Linn. Soc. Bot. 36: 321. (1904).

Yunnan: In mountain forest near Mengzi at 1,700 metres, 9379, 9379a., 9379b.

Widely distributed across Asia.

Isachne globosa (Thunb.) O. Kuntze (POACEAE)

Isachne australis R. Br. Henry in Trans. Asiat. Soc. Jap. 24 suppl: 106. (1896) (List Pl. Formosa), Rendle in Journ. Linn. Soc. Bot. 36: 321. (1904).

Hubei: Yichang, 224, 3520, 3649. **Taiwan:** Kaohsiung, 1076. Wanshoushan, 1920.

Widely distributed in Asia and Australia. Sometimes a troublesome weed of paddy fields.

Isatis tinctoria L. (BRASSICACEAE)

Rendle in Journ. Linn. Soc. Bot. 36: 364. (1904).

Yunnan: Chuyuan, 10588.

Ischaemum aristatum L. (POACEAE)

Ischaemum aristatum L. subsp. *imberbe* Hack. Rendle in Journ. Linn. Soc. Bot. 36: 364. (1904). *Spodiopogon obliquivalvis* Henry non Nees in Trans. Asiat. Soc. Jap. 24 suppl: 108. (1896) (List Pl. Formosa).

Hubei: Yichang, 2299, 4242. **Taiwan:** Kaohsiung, 1807, 2046.

Ischaemum rugosum Salisb. (POACEAE)

Rendle in Journ. Linn. Soc. Bot. 36: 366. (1904).

Taiwan: Kaohsiung, 2001.

Isodon adenanthus (Diels) Kudo (LAMIACEAE) New species

Plectranthus adenanthus Diels Dunn in Notes Roy. Bot. Gard. Edinburgh 6: 142. (1915).

Yunnan: Mengzi at 1,950 to 2,300 metres, 10067, 10067a., 13792

Diels based his original description on specimens collected by Forrest in August 1906 on the eastern flank of the Cangshan Range above Dali in north-west Yunnan, Henry had made his collections almost two decades previously. Endemic to China and distributed in the western provinces of Guizhou, Sichuan and Yunnan.

Isodon angustifolius (Dunn) Kudo (LAMIACEAE) New species

Plectranthus angustifolius Dunn in Notes Roy. Bot. Gard. Edinburgh 8: 154. (1915).

Yunnan: Mengzi at 1,650 metres, 10069 (isosyntype of *Plectranthus angustifolius* Dunn), 10069a.

Endemic to Yunnan.

Isodon coetsa (Buch.-Ham. ex D. Don) Kudô (LAMIACEAE)

Plectranthus coesta (Buch.-Ham. ex D. Don) Hara Dunn in Notes Roy. Bot. Gard. Edinburgh 6: 142. (1915).

Hubei: Liantuo, 2983. **Yunnan:** Mountains to the north of Mengzi at 2,600 metres, 10073, 10075. Mountain forests to the north of Mengzi at 2,100 metres, 10238. Mountains to the north of Mengzi at 1,950 metres, 11279. In forests to the south of Simao at 1,650 metres, 12698. Mountains to the north of Mengzi, 13774.

Widespread in Asia.

Isodon eriocalyx (Dunn) Kudo (LAMIACEAE) New species

Plectranthus eriocalyx Dunn in Notes Roy. Bot. Gard. Edinburgh 8: 155. (1913).

Yunnan: Mengzi at 1,450 metres, 9811 (syntype of *Plectranthus eriocalyx* Dunn, collected 11th October 1896).

Isodon excisus (Maxim.) Kudo (LAMIACEAE)

Plectranthus excisus Maxim. Forbes & Hemsley in Journ. Linn. Soc. Bot. 26: 270. (1890), Dunn in Notes Roy. Bot. Gard. Edinburgh vi. 141. (1911—1917).

Hubei: Fang Xian, 6695. Badong, 7357. **Sichuan:** South Wushan, 7049.

Distributed from Russia to China and Korea.

Isodon henryi (Hemsl.) Kudo (LAMIACEAE) New species

in Mem. Fac. Sci. Taihoku Univ. 2(2): 123. (1929).

Plectranthus (Isodon) henryi Hemsley in Journ. Linn. Soc. Bot. 26: 271. (1890), Bretschneider in Hist. Eur. Bot. Disc. China 2: 789. (1898), Dunn in Notes Roy. Bot. Gard. Edinburgh 6: 141. (1915), Morley in Glasra 3: 79. (1979).

Hubei: Yichang, 2727 (syntype of *Plectranthus henryi* Hemsley), 2763, 3086, 3094, (syntype of *Plectranthus henryi* Hemsley).

Distributed in central and western China.

Isodon japonicus (Burm. f.) H. Hara var. *glaucocalyx* (Maxim.) H. W. Li (LAMIACEAE)

Plectranthus glaucocalyx Maxim. Anonymous in Kew List of Determinations (volume 34).

Yunnan: Mengzi, 11419.

Isodon lophanthoides (Buch.-Ham. ex D. Don) Hara (LAMIACEAE)

Plectranthus striatus Benth. Forbes & Hemsley in Journ. Linn. Soc. Bot. 26: 274. (1890), Dunn in Notes Roy. Bot. Gard. Edinburgh 6: 140. (1915); in Kew Bull. Misc. Inf. 202. (1917)

Hubei: Lung Tung Chi Glen near Yichang, 2926. Changyang, 7689. **Yunnan:** In Pingbian on the Daweishan Range at 1,950 metres, 9042, 9043. In Pingbian, on the Daweishan Range at 1,650 metres, 11166. In Pingbian, on the Daweishan Range at 1,500 metres, 9073. Mengzi at 1,650 metres, 9796. Mountains to the north of Mengzi at 1,650 metres, 10071, 10071a. Mountains to the north of Mengzi at 1,950 metres, 11266. Milê, 10071a., 10072. Yuanjiang, 11576. Mountains to the west of Simao at 1,650 metres, 12475. Mountains to the west of Simao at 1,560 metres, 12581.

Native to India, Nepal, Myanmar, China, Thailand, Laos and Vietnam.

Isodon nervosus (Hemsl.) Kudo (LAMIACEAE) New species

Plectranthus (Isodon) nervosus Hemsley in J. Linn. Linn. Soc. Bot. 26: 272. (1890), Dunn in Notes Roy. Bot. Gard. Edinburgh 6: (1915).

Hubei: Yichang, 103, 992, 1055, 2725, 2260. Badong, 2821, 5105.

Endemic to China.

Isodon racemosus (Hemsl.) Hara (LAMIACEAE) New species

Plectranthus (Isodon) racemosus Hemsley in Journ. Linn. Soc. Bot. 26: 272. (1890), Bretschneider in Hist. Eur. Bot. Disc. China 2: 789. (1898), *Plectranthus excisus* Maxim. Dunn in Notes Roy. Bot. Gard. Edinburgh 6: 141. (1915).

Hubei: Yichang, 417, 418. Badong, 2463, 4855. **Sichuan:** South Wushan, 7233, 7253.

Distributed in China (Hubei, Sichuan and Taiwan) and Thailand.

Isodon rubescens (Hemsl.) Hara (LAMIACEAE) New species

Plectranthus (Isodon) rubescens Hemsley in Journ. Linn. Soc. Bot. 26: 272. (1890), Bretschneider in Hist. Eur. Bot. Disc. China 2: 789. (1898), Dunn in Notes Roy. Bot. Gard. Edinburgh 6: 141. (1915).

Hubei: Yichang, 974 (syntype of *Plectranthus*

rubescens Hemsley).

In current Chinese herbal medicine this species is is used to treat breast cancer. Endemic to China.

Isodon sculponeatus (Van.) Kudo (LAMIACEAE)

Plectranthus sculponiatus Vaniot Dunn in Notes Roy. Bot. Gard. Edinburgh 6: 140. (1915).

Yunnan: At the base of a mountain near Mengzi, 9235. Pu'er at 1,400 metres, 13492.

Native to India, Nepal and China.

Isodon sp. No. 1. (LAMIACEAE)

Plectranthus sp. Anonymous in Kew List of Determinations (volume 24).

Hubei: Badong, 422 (June 1885), 935 (November 1885), 2819.

Isodon sp. No. 2. (LAMIACEAE)

Plectranthus sp. Anonymous in Kew List of Determinations (volumes 34.).

Yunnan: Mengzi, 9024, 9122, 9345, 9739, 11286.

Isodon ternifolius (D. Don) Kudo (LAMIACEAE)

Plectranthus ternifolius D. Don Dunn in Notes Roy. Bot. Gard. Edinburgh 6: 138. (1915).

Yunnan: Mountains to the east of Simao at 1,650 metres, 9020, 9020a.

Distributed in Nepal, India, Bangladesh, Myanmar, western China and Vietnam.

Isodon walkeri (Arn.) H. Hara (LAMIACEAE)

Plectranthus stracheyi Benth Dunn in Notes Roy. Bot. Gard. Edinburgh 6: 139. (1915).

Yunnan: In forests to the south of Simao at 1,300 metres, 12721.

Native to Sri Lanka, Myanmar and western China.

Itea chinensis Hook. & Arn. (ITEACEAE)

Anonymous in Kew List of Determinations (volume 34).

Yunnan: Simao, 13109.

A widespread species occupying a range across India, Bhutan, Myanmar, China, Thailand, Laos and Vietnam. George Forrest later collected this species (Forrest 15,751) in Yunnan (in the Shelwi-Salween divide) in August 1917 and described it as a shrub to 2.5 metres tall inhabiting open situations by streams.

Itea ilicifolia Oliver (ITEACEAE) New species

in Hooker's Icon. Pl. 16: t. 1538. (1886), Henry in Journ. China Br. Roy. Asiat. Soc. 22: 253. (1887) (Chinese Names of Plants); in Notes Econ. Bot. China 63. (1893); in Flora & Sylva 1: 217. (1903), Bretschneider in Hist. Eur. Bot. Disc. China 2: 783. (1898), Forbes & Hemsley in Journ. Linn. Soc. Bot. 23: 278. (1887), E. H. Wilson in Journ. Roy. Hort. Soc. 24: 657. (1905), Besant in Gard. Chron. ser. 3, 335. (1935), Morley in Glasra 3: 79. (1979), Zheng in Journ.

Wuhan. Bot. Res. 2.(1): 60. (1984), Nelson in Pim, The Wood & the Trees 219. (1984), Bean in Trees & Shrubs 2; 460. (1992), Nelson in A Heritage of Beauty 314. (2000), O' Brien in Augustine Henry, An Irish Plant Collector in China 22. (2002).

Hubei: In glens near Yichang, 470, 1087 (type). On Tsui Fu Shan near Yichang (May 1888), 1623. Liantuo, 1936, 3919. Without locality, 1087a. Badong, 4865.

An evergreen shrub to 5 metres tall with holly like leaves, bearing pendelous racemes of fragrant greenish-white flowers in the month of August. Introduced to cultivation by Henry who sent seeds to Lord Kesteven who raised plants at Casewick House in Lincolnshire (where it first flowered in 1895). Lord Kesteven had visited Henry at Yichang in April 1886. A later introduction is credited to Wilson who collected seeds of this species for Messrs Veitch in 1900. (Henry sent seeds to Kew where they arrived on November 27th 1886, though no records of how they fared were kept). According to Henry (the holly-like) Yichang tassel bush was known colloquially in the Three Gorges area as *lao shu tz'u,* a name also applied to *Ilex cornuta* and *Ilex centrochinensis.* There is a magnificent plant of *Itea ilicifolia* in Henry's garden at Ranelagh in the south-side suburbs of Dublin planted by either Henry himself or his (second) wife Alice. Helen Dillon, who gardened next door to Henry's former residence, grew a plant propagated from this in her front garden. It is likely that the Ranelagh plants are descended from Wilson-Veitch plants or came directly from Lord Kesteven. *Itea ilicifolia* is still a common plant in the Three Gorges region, particularly in the mountains to the north of the Yangtze. It was collected by the Glasnevin Central China Expedition in Xingshan (GCCE 457) and in the Dragon Gate River Forest Park (GCCE 644), also in Xingshan in October 2004. In the Dragon Gate River Forest Park it scaled limestone cliffs (so characteristic of the landscape of western Hubei) in thousands. Seedlings from this collection were raised at Glasnevin in March 2005.

Itea macrophylla Wall. (ITEACEAE)

Anonymous in Kew List of Determinations (volume 34).

Yunnan: Mengzi, 9401. Simao, 11874.

Distributed in India, Bhutan, Myanmar, China (Guangxi, Hainan and Yunnan), Thailand, Vietnam and the Philippines.

Itea parviflora Hemsley (ITEACEAE) New species

in Ann. of Bot. 9: 154. (1895), Henry in Trans. Asiat. Soc. Jap. 24 suppl: 41. (1896) (List Pl. Formosa), Bretschneider in Hist. Eur. Bot. Disc. China 2: 783. (1898).

Itea parviflora Hemsl. var. *latifolia* Li in Journ. Wash. Acad. Sci. 42: 44. (1952), Li in Woody Flora of Taiwan 251.

(1963).

Taiwan: Oluanpi, 965, 1263 (type of *Itea parviflora* Hemsley), 1322. Wanchin, 145, 548, 550 (type of *Itea parviflora* Hemsl. var. *latifolia* Li), 1486.

A shrub to 5 metres tall, endemic to Taiwan and distributed throughout the island.

***Itea* sp. (ITEACEAE)**
Anonymous in Kew List of Determinations (volume 24).

Hubei: Liantuo, 3872.

***Itea yunnanensis* Franchet (ITEACEAE)**
Itea mengtzeana Engler i Pflanzenfam. 18(a): 184. (1930).

Yunnan: Mengzi, 9297.

An evergreen shrub or small tree growing from anything between 1 to 10 metres tall. During the months of April and May (in China) pendelous racemes of many flowers appear and are usually up to 20 cm long, flowering is later in cultivation usually in late July or August at Glasnevin. Endemic to China and distributed through the provinces of Guangxi, Guizhou, Sichuan, Tibet (Xizang) and Yunnan, this handsome shrub was discovered by Delavay in 1883 and was introduced to cultivation by Forrest in 1918. Similar to *Itea ilicifolia* Oliv. though as pointed out by W. J. Bean the holly-like leaves are not as strongly and spinily toothed and are narrower in proportion to their length and are longer stalked. This species is not quite as hardy as *Itea ilicifolia*.

***Itoa orientalis* Hemsley (SALICACEAE) New genus** in Hooker's Icon. Pl. t. 2688. (1901).

Yunnan: In Pingbian, in forests on the Daweishan Range at 1,650 metres, 9408 (isosyntype), 10703 (isosyntype).

Discovered by Henry in 1896. Hemsley named the genus in honour of Dr. Keisuke Ito, a pioneeer in the modern botany of Japan and his grandson Dr. Tokutaro Ito. Ito was a friend and pupil of P. F. von Siebold, another scholar of the Japanese flora. *Itoa orientalis* has formed a small tree in the very mild garden of the late Dr. David Robinson at Earlscliffe, at Howth in Co Dublin and has grown in a sheltered corner at the National Botanic Gardens, Glasnevin for a number of years now. Distributed in Vietnam and China (Guizhou, Guangxi, Sichuan and Yunnan).

***Ixeridium gracile* (DC.) Pak & Kawano (ASTERACEAE)**
Lactuca gracilis DC. Forbes & Hemsley in Journ. Linn. Soc. Bot. 23: 482. (1888).

Hubei: Badong, 1879. Liantuo, 3935.

***Ixeridium laevigatum* (Blume) Pak & Kawano (ASTERACEAE)**
Lactuca thunbergiana Maxim. var. *oldhamii* Maxim.

Henry in Trans. Asiat. Soc. Jap. 24 suppl: 56. (1896) (List Pl. Formosa).

Taiwan: Wanchin, 107, 527.

***Ixeris chinensis* (Thunb.) Nakai subsp. *versicolor* (Fisch. & Link.) Kitam. (ASTERACEAE)**
Lactuca versicolor (Fisch.) Schultz-Bip. ex Herder. Henry in Journ. China Br. Roy. Asiat. Soc. 22: 238. (1887) (Chinese Names of Plants), Forbes & Hemsley in Journ. Linn. Soc. Bot. 23: 485. (1888), Henry in Trans. Asiat. Soc. Jap. 24 suppl: 56. (1896) (List Pl. Formosa).

Hubei: Yichang, 344, 817, 1190, 1191, 1958 (in part). Badong, 1753, 4046 (in part), 4047, 5142. Without locality, 344a. **Taiwan:** Wanchin, 83. Oluanpi, 1399. Tamsui, 1745. Kaohsiung, 1808.

According to Henry this species was known colloquially at Yichang as *chien tao tzu ts'ao*.

***Ixeris gracilis* (DC.) Pak & Kawano (ASTERACEAE)**
Lactuca gracilis DC. Anonymous in Kew List of Determinations (volume 34).

Yunnan: Mengzi, 9731.

***Ixeris japonica* (Burm. f.) Nakai (ASTERACEAE)**
Lactuca debilis (Thunb.) Benth. ex Maxim. Henry in Trans. Asiat. Soc. Jap. 24 suppl: 56. (1896) (List Pl. Formosa).

Taiwan: Oluanpi, 279, 312.

***Ixeris polycephala* Cass. (ASTERACEAE)**
Lactuca polycephala (Cass.) Benth. Anonymous in Kew List of Determinations (volume 34).

Yunnan: Mengzi, 11063.

Distributed in Kazakastan, Nepal, India, Bangladesh, China and Japan.

***Ixeris repens* (L.) A. Gray (ASTERACEAE)**
Lactuca repens Henry non Maxim. in Trans. Asiat. Soc. Jap. 24 suppl: 56. (1896) (List Pl. Formosa).

Taiwan: Oluanpi, 353.

***Ixora chinensis* Lam. (RUBIACEAE)**
Anonymous in Kew List of Determinations (volume 24).

Hainan: Environs of Haikou, 8383. Environs of Kiungchow, 8763.

***Ixora finlaysoniana* Wall. ex G. Don (RUBIACEAE)**
Anonymous in Kew List of Determinations (volume 24).

Hainan: Hummocks near Haikou, 8096, 8234.

***Ixora henryi* H. Leveille (RUBIACEAE) New species** in Repert. Spec. Nov. Regni Veg. 13: 178. (1914).

Yunnan: In mountain forests to the south of Simao at 1,650 metres, 11637a., (type).

***Ixora philippinensis* Merr. (RUBIACEAE)**
Ixora chinensis Henry non Lam. in Trans. Asiat. Soc.

Jap. 24 suppl: 50. (1896) (List Pl. Formosa).

Taiwan: Wanchin, without number.

Henry gave *shan-tan* and *hsien-tan* as local names of this shrub.

Ixora spp. (RUBIACEAE)

Anonymous in Kew List of Determinations (volume 24 & 34).

Hainan: Environs of Kiungchow, 8232, 8254, Lingmen (in the interior of the island), 8644. Without locality, 8542, 8476. **Yunnan:** Mengzi, 10407.

Ixora yunnanensis Hutchinson (RUBIACEAE) New species

in Sargent Pl. Wilson. 3: 412. (1916).

Yunnan: Below Hsinkei on the Red River, 9584 (type). On the banks of the Red River at Manpan 10370.

Jacobaea analoga (DC.) Veldkamp (ASTERACEAE)

Senecio chrysanthemoides DC. Anonymous in Kew List of Determinations (volume 34). *Senecio laetus* Edgew. Jeffrey & Chen in Kew Bull. 39: 404. (1985). *Senecio argunensis* Forbes & Hemsley non Turcz. in Journ. Linn. Soc. Bot. 23: 450. (1888).

Hubei: Yichang, 562, 3651. **Yunnan:** Mengzi, 10008. Simao, 12997.

A perennial herb to 80 cm tall producing compound corymbs of yellow flowers in April to June. A widespread species, distributed from north-west Pakistan, Nepal, northern India, Bhutan and China. Henry's central Chinese collections were of great interest at the time of collectuing when the population at Yichang was 500 km disjunct from the then-known main centre of distribution. This species also grows on Mount Everest.

Jacquemontia paniculata (Burm. f.) Hallier. f. (CONVOLVULACEAE)

Ipomoea paniculata Burm. f. Anonymous in Kew List of Determinations (volume 24). *Convolvulus parviflorus* Vahl. Henry in Trans. Asiat. Soc. Jap. 24 suppl: 65. (1896) (List Pl. Formosa).

Hainan: 8 km (5 miles) south of Kuingchow, 8183. **Taiwan:** Kaohsiung, 1179, 1702.

Janochloa anditotalis (Retz.) Zuloaga & Delfini (POACEAE)

Panicum proliferum Lam. Rendle in Journ. Linn. Soc. Bot. 36: 332. (1904).

Taiwan: Kaohsiung, in ditches, 1948.

Japonicalia delphiniifolia (Sieb. & Zucc.) C. Ren & Q. E. Yang (ASTERACEAE)

Senecio zuccarinii Maxim. Anonymous in Kew List of Determinations (volume 24).

Hubei: Xingshan, 6487. Without locality, 6487a.

Jasminum coarctatum Roxb. (OLEACEAE)

Kobuski in Journ. Arnold Arb. 13: 173. (1932).

Yunnan: In the Red River valley near Manpan at 900 metres, 10887. Near Simao on the south road at 1,525 metres, 11653, 11653a.

Native to China (Guangxi, Guizhou and Yunnan) and Vietnam.

Jasminum duclouxii (H. Lév.) Rehder (OLEACEAE)

Jasminum dumicola W. W. Smith in Notes Roy. Bot. Gard. Edinburgh 12: 207. (1920), Kobuski in Journ. Arnold Arb. 13: 166. (1932).

Yunnan: In Pingbian on the Daweishan Range at 1,950 metres, 10634. South of the Red River on Fengchunling at 2,300 metres, 10634a., (isotype). Simao, 11708. In forests to the east of Simao at 1,450 metres, 11708a.

Distributed from Assam in north-east India to China.

Jasminum elongatum (P. J. Bergius) Willd. (OLEACEAE)

Jasminum undulatum Ker-Gawler Anonymous in Kew List of Determinations (volume 24 & 34).

Hainan: Environs of Haikou, 8162. Without locality, 8590. **Yunnan:** Mengzi, 13749.

Distributed in India, Myanmar, China and Vietnam to N. Australia.

Jasminum extensum Wall. ex G. Don (OLEACEAE)

Kobuski in Journ. Arnold Arb. 13: 165. (1932).

Jasminum taliense W. W. Smith in Notes Roy. Bot. Gard. Edinburgh 12: 210. (1920).

Yunnan: In forests to the south of Simao at 1,525 metres, 12661. In forests to the south of Simao at 1,650 metres, 12661a., (syntype of *Jasminum taliense* W. W. Smith).

Jasminum grandiflorum L. (OLEACEAE)

Jasminum officinale L. f. *grandiflorum* (L.) Kobuski Kobuski in Journ. Arnold Arb. 13: 161. (1932).

Yunnan: On cliffs near Pu'er at 1,525 metres, 13397.

Jasminum lanceolaria Roxb. (OLEACEAE)

Rehder in Sargent Pl. Wilson. 2: 612. (1916), Kobuski in Journ. Arnold Arb. 13: 156. (1932).

Jasminum lanceolaria Roxb. var. *puberulum* Hemsley in Journ. Linn. Soc. Bot. 26: 78. (1889), Rehder in Sargent Pl. Wilson 2: 612. (1916), Kobuski in Journ. Arnold Arb. 13: 158. (1932), Zheng in Journ. Wuhan Bot. Res. 2.(1): 179. (1984).

Hubei: Yichang, 2729. Liantuo, 3000, 4562. In a ravine on Moji Shan, 3669. Antelope Glen, 3669a. San You Dong Glen, 3669b. **Yunnan:** In forests near Simao at 1,600 metres, 11713.

A twining woody climber to 7 metres tall, bearing cymes of intensely fragrant white flowers from April to June. Distributed in India, Bhutan, Myanmar, China, Thailand and

Vietnam. Numerous cultivars have been selected in China.

Jasminum mesnyi Hance (OLEACEAE)

Kobuski in Journ. Arnold Arb. 13: 152. (1932), Lancaster in Travels in China 170. (1993).

Jasminum primulinum Hemsl. Henry in Flora & Sylva 2: 168. (1904), J. H. Veitch in Hortus Veitchii 401. (1906).

Yunnan: Mengzi at 1,370 to 1,525 metres, 9319. On mountains near Simao at 1,525 metres, 9319a.

An evergreen rambling shrub to 2.5 metres tall bearing semi-double yellow flowers in summer. Discovered by William Mesny in Guizhou in April 1880 and introduced to cultivation from Mengzi by Wilson by means of living plants in the autumn of 1899 while visiting Henry. The same plants first flowered at Veitch's Coombe Wood nursery in October 1901. In Veitch's *Novelties for 1908—1909* plants were selling for up to three shillings and six pence each. Henry stated it never set seed, the flowers were often double and the shrub was often seen growing in gardens or in hedges in the vacinity of villages. Having grown in the temperate pits at Glasnevin for many years this species was planted onto the western end of the Chinese slope where it grows quite well. Vietnam and China (Fujian, Guangdong, Guizhou, Hong Kong and Yunnan).

Jasminum microcalyx Hance (OLEACEAE)

Anonymous in Kew List of Determinations (volume 24).

Hainan: Environs of Haikou, 8415.

Discovered by the American missionary, the Rev'd B. C. Henry on Hainan in October 1882.

Jasminum nervosum Lour. (OLEACEAE)

Jasminum undulatum Ker. var. *elegans* Henry non Hemsl. in Trans. Asiat. Soc. Jap. 24 suppl: 59. (1896) (List Pl. Formosa). *Jasminum anastomosans* Wall. ex A. DC. Kobuski in Journ. Arnold Arb. 13: 176. (1932). *Jasminum hemsleyi* Yamamoto Li in Woody Flora of Taiwan 758. fig. 305. (1963).

Hainan: At Nodoa in the interior of the island, 8456. **Yunnan:** Simao at 1,370 metres, 11969, collected April 26th 1898.

Native to India, Nepal, Myanmar, China, Vietnam and Laos.

Jasminum nintooides Rehder (OLEACEAE) New species

in Sargent Pl. Wilson. 2: 615. (1916), Kobuski in Journ. Arnold Arb. 13: 171. (1932).

Yunnan: Mengzi at 1,600 metres, trailing and climbing over rocks, 9433 (type), 9433a., 9433b.

Endemic to Yunnan.

Jasminum polyanthum Franch. (OLEACEAE)

Kobusji in Journ. Arnold Arb. 13: 163. (1932), Green

in Kew Bull.52: 17. (1997).

Yunnan: Mengzi at 1,525 metres, 10314. Chuyuan, 10314a. In woods at Konyahua near Mengzi at 1,400 metres, 10314b. Simao at 1,525 metres, 11656.

An evergreen climber to 5 metres tall, flowers are rose coloured on the reverse of petals, white on the inside and powerfully and deliciously fragrant. Native to Yunnan where it was discovered by Delavay in 1883 and later found by Forrest on the eastern flank of the Cangshan Range above the city of Dali in north-west Yunnan. Distributed in Myanmar and China (Guizhou, Sichuan and Yunnan).

Jasminum sambac (L.) Ait. (OLEACEAE)

Henry in Trans. Asiat. Soc. Jap. 24 suppl: 59. (1896) (List Pl. Formosa).

Taiwan: Kaohsiung, cultivated, 1882.

A popular plant in Chinese gardens on account of its exquisite scent.

Jasminum sinense Hemsley (OLEACEAE) New species

in Journ. Linn. Soc. Bot. 26: 80. (1889), Bretschneider in Hist. Eur. Bot. Disc. China 2: 787. (1898), Diels in Bot. Jahrb. xxix. 533. (1900), Leveille in Fl. Kouy-Tcheou. 294. (1914; in Fedde, Rep. Spec. Nov. 13: 150. (1914), Rehder in Sargent Pl. Wilson 2: 612. (1916), Chung in Mem. Sci. Soc. China i. 217. (1924), Kobuski in Journ. Arnold Arb. 13: 159. (1932), Morley in Glasra 3: 79. (1979, (Zheng in Journ. Wuhan Bot. Res. 2(1): 179. (1984), Nelson in A Heritage of Beauty 314. (2000).

Hubei: Liantuo, 2106, 4464 (isotype). **Sichuan:** Henry's Chinese collector with Pratt, no locality stated but may be Leshan, Emei Shan, Wa Shan or Kangding, 8885. **Yunnan:** Mountains to the north of Mengzi at 1,525 to 1,830 metres, 9657, 9657a., 9657b. Simao at 1,830 metres, 13354.

Distributed in Vietnam and China (including Taiwan).

Jasminum spp. (OLEACEAE)

Anonymous in Kew List of Determinations (volume 24 & 34).

Hainan: At Nodoa in the interior of the island, 8439. Without locality, 8496. **Yunnan:** Mengzi, 9409.

Jasminum urophyllum Hemsl. (OLEACEAE)

in Journ. Linn. Soc. Bot. 26: 81. (1889).

Jasminum urophyllum Hemsl. var. *henryi* Rehder in Sargent Pl. Wilson 2: 613. (1916), Chun in Mem. Sci. Soc. China i. 217. (1924), Zheng in Journ. Wuhan Bot. Res. 2.(1): 179. (1984). *Jasminum urophyllum* Hemsl. var. *wilsonii* Rehder Kobuski in Journ. Arnold Arb. 13: 155. (1932).

Hubei: Jianshi, 5944 (type of *Jasminum urophyllum* Hemsley, type of *Jasminum urophyllum* Hemsl. var. *henryi* Rehder). Without locality, 5944a.

Jatropha curcas L. (EUPHORBIACEAE)

Henry in Notes Econ. Bot. China 61. (1893), Forbes & Hemsley in Journ. Linn. Soc. Bot. 26: 433. (1894), Henry in Trans. Asiat. Soc. Jap. 24 suppl: 83. (1896) (List Pl. Formosa).

Hainan: Environs of Haikou, 7995, 8266. **Taiwan:** Kaohsiung, 739. Lambay Isle, without number (collected August 15th 1893.

The *ma-feng-shu* or Barbados nut, a shrub or small tree to 6 metres tall, native to tropical America. According to Henry, the fruits of this species were pressed and the extracted oil (called *k'u-yu*) was used for lighting lamps.

Jatropha multifida L. (EUPHORBIACEAE)

Henry in Trans. Asiat. Soc. Jap. 24 suppl: 83. (1896) (List Pl. Formosa).

Taiwan: Oluanpi, 967.

The coral plant or physic tree, a shrub or small tree to 7 metres tall, native to tropical America and cultivated in Chinese Taiwan.

Jatropha sp. (EUPHORBIACEAE)

Anonymous in Kew List of Determinations (volume 34).

Yunnan: Mengzi, 10713.

Juglans mandshurica Maxim. (JUGLANDACEAE)

Forbes & Hemsley in in Journ. Linn. Soc. Bot. 26: 493. (1899).

Juglans regia Henry non L. in Notes Econ. Bot. China 48. (1893). *Juglans cathayensis* Dode in Bull. Soc. Dendr. France 47. (1909), Dode in Fedde, Repert. Spec. Nov. Regni Veg. 10: 298. (1912), Rehder in Sargent Pl. Wilson. 3: 185. (1917), Zheng in Journ. Wuhan Bot. Res. 2(1): 17. (1984). *Juglans draconis* Dode in Bull. Soc. Dendrol. France 49. (1909); in Fedde, Repert. Spec. Nov. Regni Veg. 10: 298. (1912).

Hubei: Liantuo, 3834, 5233a. Changyang, 5233. Without locality, 5233c. **Sichuan:** South Wushan, 5233b. **Yunnan:** Near Mengzi in woods at 1,800 to 2,000 metres, 10498, 10498a. Mengzi, 10498b., (type of *Juglans draconis* Dode).

The Chinese butternut or *shan-he,* a common bush or small tree in Hubei. Wilson collected seeds of *Juglans mandshurica* in 1903 for Veitch's Coombe Wood nursery where young trees of 2.5 metres tall later bore fruits. Veitch's catalogue *New Hardy plants from Western China* (Autumn 1912) had this species (as *J. cathayensis* Dode) available for ten shillings and six pence each. There are trees at Birr Castle in Co Offaly and at the National Botanic Gardens, Glasnevin. The tallest specimen in the British Isles and Ireland is a tree at Hergest Croft, Herefordshire. Planted in 1912, when measured in 1995 it was 17 metres tall with a trunk diameter of 59 cm.

Juglans regia L. (JUGLANDACEAE)

Skan in Journ. Linn. Soc. Bot. 26: 493. (1899), Rehder & E. H. Wilson in Sargent Pl. Wilson. 3: 184. (1917).

Juglans duclouxiana Dode in Bull. Soc. Dendrol. France 1: 81. (1906); in Fedde, Repert. Nov. Spec. Regni Veg. 6: 90. (1909).

Yunnan: Mengzi (cultivated) at 1,600 metres, 10507, 10507a., (syntype of *Juglans duclouxiana* Dode).

The common walnut is widely cultivated in many parts of China, it is a popular fruiting tree in south-east Tibet (Xizang) where it makes magnificent trees up to 25 metres tall, some with enormous girths.

Juncus alatus Franch. & Savat. (JUNCACEAE)

Brown in Journ. Linn. Soc. Bot. 36: 162. (1903), Buchenau in Engler, Pflanzenr. 36: 180. (1906).

Hubei: Badong, 1794. Jianshi, 5785.

Juncus allioides Franch. (JUNCACEAE)

Brown in Journ. Linn. Soc. Bot. 36: 162. (1903), Buchenau in Engler, Pflanzenr. 36: 229. (1906).

Hubei: Fang Xian, 6927. **Sichuan:** Kangding at 2,900 to 4,200 metres, Henry's Chinese collector with Pratt, 8916.

Juncus bufonius L. (JUNCACEAE)

Brown in Journ. Linn. Soc. Bot. 36: 162. (1903).

Sichuan: Henry's Chinese collector with Pratt, no locality stated but may be Leshan, Emei Shan, Wa Shan or Kangding, 8843.

Juncus effusus L. (JUNCACEAE)

Henry in Journ. China Br. Roy. Asiat. Soc. 22: 272. (1887) (Chinese Names of Plants), Brown in Journ. Linn. Soc. Bot. 36: 163. (1903).

Hubei: Badong, 1781, 4049. Liantuo, 1953, 3855.

According to Henry the rush occured both wild and cultivated in the Three Gorges district where it was known as *teng ts'ao.* It was cultivated for its pith which was used for lamp-wicks. He also went on to say that large quantities of this pith were exported from Sichuan and came down through the gorges on two or three boats lashed together. Before the importation of mineral oil, only lamps were in use. These were filled with vegetable oil and fitted with rush-wicks. Mats were also made from this species.

Juncus modestus Buchenau (JUNCACEAE)

Brown in Journ. Linn. Soc. Bot. 36: 164. (1903).

Hubei: Fang Xian, 6846, 6868.

Juncus modicus N. E. Brown (JUNCACEAE) New species

in Journ. Linn. Soc. Bot. 36: 165. (1903), Buchenau in Engler, Pflanzenr. 36: 231. (1906).

Hubei: Fang Xian, 6854 (isosyntype), 6868a.

Juncus potaninii Buchenau (JUNCACEAE)

Brown in Journ. Linn. Soc. Bot. 36: 165. (1903).

Juncus luzuliformis Franch. var. *potaninii* (Buchen.) Buchenau in Engler, Pflanzenr. 36: 228. (1906).

Hubei: Zigui, on wooded precipices at 1,900 metres, 6169.

Juncus prismatocarpus R. Br. subsp. ***leschenaultii*** (J. Gay ex Laharpe) Kirschner (JUNCACEAE)

Juncus prismatocarpus Brown non R. Br. in Journ. Linn. Soc. Bot. 36: 165. (1903).

Hubei: Badong, 2471, 2527, 4802. Yichang, 3504, 3518, 3984. Liantuo, 3863.

Distributed through India, China, Korea and Japan.

Juncus setchuenensis Buchen. (JUNCACEAE)

Juncus pauciflorus Brown non R. Br. Brown in Journ. Linn. Soc. Bot. 36: 165. (1903).

Hubei: Yichang, 222, 592, 3653.

Juncus sp. No. 1. (JUNCACEAE)

Anonymous in Kew List of Determinations (volume 24).

Sichuan: South Wushan, 5736.

Juncus sp. No. 2. (JUNCACEAE)

Anonymous in Kew List of Determinations (volume 34).

Yunnan: Simao, 11920.

Juniperus chinensis L. (CUPRESSACEAE)

Henry in Notes Econ. Bot. China 37. (1893), Masters in Journ. Linn. Soc. Bot. 26: 541. (1902), Henry in Elwes and Henry, The Trees Gt. Brit. & Irel. 6: 1430. (1912), Rehder & E. H. Wilson in Sargent Pl. Wilson 2: 60. (1916), Walsh in Aug. Henry Forst Herb. Cat. 16. (1957), Zheng in Journ. Wuhan Bot. Res. 2.(1): 12. (1984).

Hubei: Fang Xian, 6576.

A tree to 18 metres tall, native to Japan, Mongolia and China and introduced to Kew in 1804 by William Kerr. Kerr (who was a collector for the Royal Botanic Gardens, Kew) was sent to China as a young man in 1803 and later visited Java and the Philippines.

Juniperus formosana Hayata (CUPRESSACEAE)

Rehder & E. H. Wilson in Sargent Pl. Wilson. 2: 56. (1916), Walsh in A. Henry. Forst. Herb. 17. (1957), Zheng in Journ. Wuhan Bot. Res. 2.(1): 12. (1984).

Juniperus taxifolia Henry non Hook. & Arn. in Journ. China Br. Roy. Asiat. Soc. 22: 265. (1887) (Chinese Names of Plants); in Notes Econ. Bot. China 37. (1893), Masters in J. Linn. Soc Bot. 26: 543 (1902).

Hubei: Badong, 5 (collected September 1885, received at Kew on March 12th 1886), 2876 (October 1886). Yichang, 2876a. **Sichuan:** South Wushan, 5653.

A shrubby tree to 13 metres tall, widespread across China. Henry stated that the colloquial name for *Juniperus formosana* in Hubei ws *tz'u-peh-shu*. This species was collected by the Glasnevin Central China Expedition (GCCE 727) on Moji Shan near Yichang in October 2004. Moji Shan was known to Henry and later collectors as 'the Dome', the last western plant collector to visit this hill before the Glasnevin group was E. H. Wilson.

Juniperus squamata Buch. -Ham. ex Lamb. (CUPRESSACEAE)

Rehder & E. H. Wilson in Sargent Pl. Wilson. 2: 57. (1916), Zheng in Journ. Wuhan Bot. Res. 2.(1): 13. (1984).

Juniperus recurva Buch.-Ham. ex D. Don var. *squamata* (D. Don) Parl. Masters in Journ. Linn. Soc. Bot. 26: 543. (1902).

Hubei: Fang Xian, 6896, 6935. Xingshan, 6990, 6991.

A low growing shrub to 60 cm tall, native to the Himalaya and China and introduced from Nepal in 1836.

Jurinea chenopodiifolia (Klatt) N. Garcia, Herrando & Susanna (ASTERACEAE)

Saussurea decurrens Hemsley in Journ. Linn. Soc. Bot. 29: 310. (1892), Bretschneider in Hist. Eur. Bot. Disc. China 2: 786. (1898).

Hubei: Fang Xian, 6775 (type of *Saussurea decurrens* Hemsley).

Jurinea deltoidea (DC.) N. Garcia, Herrando & Susanna (ASTERACEAE)

Saussurea lamprocarpa Hemsley in Journ. Linn. Soc. Bot. 23: 465. (1888), Bretschneider in Hist. Eur. Bot. Disc. China 2: 786. (1898), J. H. Veitch in Hortus Veitchii 434. (1906), Nelson in A Heritage of Beauty 319. (2000).

Hubei: Badong, 2470 (syntype of *Saussurea lamprocarpa* Hemsley), 4972, 6181. Liantuo, 2675 (syntype of *Saussurea lamprocarpa* Hemsley). Without locality, 6181a., 6181b.

Introduced to cultivation by Wilson through Messrs Veitch. Henry had sent seeds to Kew on September 14th 1888 but no record of how they fared was kept.

Jurinea peguensis (C. B. Clarke) N. Garcia, Herrando & Susanna (ASTERACEAE)

Saussurea phyllocephala Collett & Hemsl. Anonymous in Kew List of Determinations (volume 34).

Yunnan: Mengzi, 9768 (collected 25th October 1896).

Justicia gendarussa Burm. f. (ACANTHACEAE)

Henry in Trans. Asiat. Soc. Jap. 24 suppl: 69. (1896) (List Pl. Formosa).

Taiwan: Wanchin, 155. Oluanpi, 247.

Justicia latiflora Hemsley (ACANTHACEAE) New species

in Journ. Linn. Soc. Bot. 26: 245. (1890).

Hubei: San You Dong Glen, 3412 (isosyntype).

Yichang, 3412a., 3412b.

Justicia patentiflora Hemsley (ACANTHACEAE)
New species

in Hooker's Icon. Pl. t. 28: 2792. (1905).

Yunnan: In the mountains to the south-east of Simao at 1,650 metres, 12773 (type).

Justicia quadrifaria (Wall. ex Nees) T. Anderson (ACANTHACEAE)

Forbes & Hemsley in Journ. Linn. Soc. Bot. 26: 246. (1890).

Hubei: Yichang, 2719, 4154. Liantuo, 4594.

Justicia sp. No. 1. (ACANTHACEAE)

Anonymous in Kew List of Determinations (volume 24).

Hubei: Yichang, 2752.

Justicia sp. No. 2. (ACANTHACEAE)

Anonymous in Kew List of Determinations (volume 34).

Yunnan: Simao, 12935.

Kadsura coccinea (Lemaire) A. C. Smith (SCHISANDRACEAE)

Saunders in Systematic Botany Monographs 54: 44. (1998).

Hong Kong: without number. **Yunnan:** Mengzi, 10734. Simao, 11810, 12049.

Native to China and Vietnam.

Kadsura japonica (L.) Dunal (SCHISANDRACEAE)

Henry in Trans. Asiat. Soc. Jap. 24 suppl: 16. (1896) (List Pl. Formosa), Li in Woody Flora of Taiwan 160. fig. 58. (1963), Saunders in Systematic Botany Monographs 54: 65. (1998).

Taiwan: Oluanpi, 1284. Wanchin, 1553, 1681.

A climbing evergreen shrub to 4 metres tall, climbing by means of twining stems. Solitary whitish-yellow flowers are carried in early summer. Native to southern Korea, Japan and China (Taiwan) where it grows in montane forest up to 2,000 metres. From *Schisandra*, *Kadsura* differs only in that its fruits are held in a cluster not strung along an extended floral axis.

Kadsura longipedunculata Finet & Gagnepain (SCHISANDRACEAE) New species

Saunders in Systematic Botany Monographs 54: 58. (1998).

Schisandra axillaris Diels non (Blume) Hook. f. & Thoms. in Bot. Jahrb. 29: 322. (1900). *Kadsura peltigera* Rehder & E. H. Wilson in Sargent Pl. Wilson. 1: 410. (1913).

Hubei: Donghu, 6433. Changyang, 7496. **Yunnan:** Simao, 12549. 12312a., (paratype of *Kadsura peltigera* Rehder & E. H. Wilson). In forests near Simao at 1,300 to 1,600 metres, 12312 (paratype of *Kadsura peltigera* Rehder & E. H. Wilson).

Discovered by Henry in 1888, this species was described from material later collected by Farges near Chengkou in Sichuan. Endemic to China where it is widely distributed.

Kadsura sp. No. 1. (SCHISANDRACEAE)

Anonymous in Kew List of Determinations (volume 24.).

Sichuan: Henry's Chinese collector with Pratt, no locality stated but may be any of the following, Leshan, Emei Shan, Wa Shan or Kangding, 8796.

Kadsura sp. No. 2. (SCHISANDRACEAE)

Anonymous in Kew List of Determinations (volume 34).

Yunnan: Simao, 11749, 13371.

Kaempferia rotunda L. (ZINGIBERACEAE)

Henry in Trans. Asiat. Soc. Jap. 24 suppl: 93. (1896) (List Pl. Formosa), Wright in Journ. Linn. Soc. Bot. 36: 69. (1903).

Taiwan: Wanchin, wild in the mountains, 851.

The resurrection lily, though Henry stated it was wild in the mountains near Wanchin, the origin of this species is obscure. It is widely cultivated in south-east Asia.

Kalanchoe ceratophylla Haw. (CRASSULACEAE)

Kalanchoe gracilis Hance Henry in Trans. Asiat. Soc. Jap. 24 suppl: 42. (1896) (List Pl. Formosa).

Taiwan: Kaohsiung, without number.

A fleshy perennial herb to 1 metre tall, native to southern China, southern Asia and Malaysia. According to Bretschneider this species was discovered in Chinese Taiwan by Robert Swinhoe in 1861.

Kalanchoe integra (Medik.) Kuntze. (CRASSULACEAE)

Kalanchoe spathulata (Poir.) DC Anonymous in Kew List of Determinations (volume 34).

Yunnan: Chuyuan, 10801.

A fleshy perennial to a metre tall, native to southern China, southern Asia and Malaysia.

Kalanchoe pinnata (Lam.) Pers (CRASSULACEAE)

Bryophyllum calycinum Salisb. Henry in Trans. Asiat. Soc. Jap. 24 suppl: 41. (1896) (List Pl. Formosa).

Taiwan: Wanchin, 359.

A succulent herb to 2 metres tall, widespread in the tropics.

Kalopanax septemlobus (Thunb.) Koidz. (ARALIACEAE)

Acanthopanax ricinifolium (Sieb. & Zucc.) Seem. Henry in Journ. China Br. Roy. Asiat. Soc. 22: 240. (1887) (Chinese Names of Plants), Forbes & Hemsley in Journ. Linn. Soc. Bot. 23: 340. (1887). *Kalopanax ricinifolius* (Sieb. & Zucc.) Miq. in Sargent Pl. Wilson 2: 564. (1916). *Kalopanax pictus* (Thunb.) Nakai Zheng in Journ. Wuhan Bot. Res. 2.(1): 161. (1984).

Hubei: Yichang, 2246, 2246a., 3101. Liantuo, 4573.

A large deciduous tree to 30 metres tall in its native habitat, to 14 metres in cultivation. Branches are armed with large yellow prickles and handsome palmate leaves up to 35 cm across. A widespread species ranging from Russia across China and into Korea. Introduced to cultivation by the Russian botanist, Carl Maximowicz in 1865. It is well represented in the gardens of Britian and Ireland. Henry stated that because of the resemblance of the foliage of this tree to that of *Catalpa ovata* it was known colloquially around Yichang as *tz'u ch'iu.* This species was collected by the Glasnevin Central China Expedition (GCCE 876) on Moji Shan, Yichang in October 2004.

Kerria japonica (L.) DC. (ROSACEAE)

Forbes & Hemsley in Journ. Linn. Soc. Bot. 23: 229. (1887), Henry in Journ. China Br. Roy. Asiat. Soc. 22: 276. (1887) (Chinese Names of Plants), Zheng in Journ. Wuhan Bot. Res. 2.(1): 70. (1984).

Hubei: Yichang, 854, 1380. Liantuo, 3921, 4554. Badong, 785 (May 1885), 1411, 4005. Yichang (Xiling) Gorge, 4138.

A monotypic genus allied to *Rhodotypos* and named to commemorate William Kerr who intoduced the double flowered form (*Kerria japonica* 'Pleniflora'). The type i.e. *Kerria japonica* was introduced to cultivation by Reeves in 1834. It has been cultivated for centuries in Japan but is native to China. The wild plant is common in the glens and higher mountains near Yichang and Henry stated it was known there as the *hsiao t'ung ts'ao.* Wilson stated that the double flowered form was also cultivated at Yichang. Forrest collected *Kerria japonica* in north-west Yunnan in June 1917. Young plants currently cultivated at Glasnevin were raised from a collection made by the Glasnevin Central China Expodition (GCCE 763) near Tianthu Shan in Changyang in October 2004.

Keteleeria davidiana (Bertr.) Beissn. (PINACEAE)

Masters in Journ. Linn. Soc. Bot. 26: 554. (1902), Henry in The Trees of Gt. Brit. & Irel. 6. 1475. (1912), Rehder & E. H. Wilson in Sargent Pl. Wilson. 2: 39. (1916), E. H. Wilson in Journ. Arnold Arb. 5: 53. (1926), Zheng in Journ. Wuhan Bot. Res. 2.(1): 8. (1984), Nelson in Pim, The Wood & the Trees 221. (1984), Bean in Trees & Shrubs 2: 507. (1992), Lancaster in Travels in China 324. (1993), Nelson in A Heritage of Beauty 314. (2000), O' Brien in Augustine Henry, An Irish Plant Collector in China 22. (2002).

Abies davidiana (Bertr.) Franch. Henry in Journ. China Br. Roy. Asiat. Soc. 22: 268. (1887) (Chinese Names of Plants); in Notes Econ. Bot. China 36. (1893), Nelson in Pim, The Wood & the Trees 220. (1984).

Hubei: Liantuo, 3276, 3878, 3878a. South Donghu, 7576. **Sichuan:** North Wushan, 7098. **Yunnan:** In mountains near Mengzi, 10774, 1774a, 11345, 11358. In mountains near Mengzi at 1,600 metres, 11355. In the mountains near Simao at 1,600 metres, 12734, 12855.

According to Henry's notes this was the *ma wei sung* and it formed woods on dry hills and valleys between altitudes of 600 and 1,200 metres in Hubei and 1,200 and 1,500 metres in Yunnan (lower than *Abies* or *Picea*). Wilson stated this fine tree grew as solitary individuals, shading tombs and wayside shrines and in other regions it formed extensive woods, with scrub oak as an undergrowth. Both Henry and Wilson recorded colossal specimens, in Hubei it grew to 40 metres tall with trunks to 2.5 metres in diameter, with large buttress like roots spreading from the bole (Henry described his 7108 as being 26 metres tall and 6 metres in circumference). The wood, being highly prized in construction led to its rapid decline in central and western China. *Keteleeria davidiana* was introduced to cultivation by Henry through the Royal Botanic Gardens, Kew in 1888 (one of his seedlings was 1.2 metres in the temperate house by 1912). Wilson later sent seeds to Veitch's Coombe Wood nursery in 1901. Rare in cultivation, in Ireland there are trees at the National Botanic Gardens, Glasnevin, Headfort and the John. F. Kennedy Arboretum, though it does not thrive in our dull, insular climate. Nelson (2000) states this species needs a daytime temperature of 25-35 degrees Celsius for four months of the year. At Wakehurst Place in Sussex a tree planted in 1918 measured 11 metres tall with a trunk diameter of 26 cm in 1997. Endemic to China.

Keteleeria evelyniana Masters (PINACEAE) New species

in Gard. Chron. ser. 3, 33: 194. fig. 82. (1903), Beissner in Mitt. Deutsch. Dendr. Ges. 12: 66. (1903), E. H. Wilson in Journ. Arnold Arb. 5: 54. (1926).

Pinus evelyniana (Mast.) A. Voss in Putlitz & Meyer, Landex. 4: 773. (1913).

Yunnan: Yuanjiang at 1,300 metres, 11815 (holotype).

A tree to 40 metres tall, distributed in China (Guizhou, Sichuan and Yunnan), Laos and Vietnam. At Henry's request this tree was named in compliment to Miss Evelyn Gleeson from Benown House, Athlone in County Westmeath, Ireland, and later of Dun Emer in Dundrum, Dublin. Evelyn Gleeson was a life-long friend and correspondent of Henry's and it is thanks to the many letters she preserved from Henry's years in China that we know so much about his activities there. Henry had the following to say about this tree in a letter to Dr. Masters:

'This beautiful new *Keteleeria,* so far as I know, occurs at one spot only on the side of the mountain which forms the southern border of the wide gorge of the Red River at Yuanjiang, in Yunnan, where there are five or six trees at about 4,000 feet above the level of the sea. Yuanjiang lies midway between Mengtse and Szemao, at about eight days march from either place. I first saw the tree in February, 1898, while on the march to Szemao from Mengtse, and obtained then two cones, all that remained on the trees. I expected to find the tree further south, but during the year and a half that I was stationed at Szemao, neither myself nor my collectors saw any trace of it or three shrubs were seen in a single locality and Mr. Wilson never met with it. It is possible that some species are very nearly on the point of extinction. However, with regard to the *Keteleeria,* the mountains about Yuanjiang were only partially explored by myself and my collectors, and from certain indications I believe they are very rich in new species quite distinct from those of the Szemao and Mengtse districts. This *Keteleeria* may be met with further east and west of the point where I saw it. I may add that *Archangiopteris,* a remarkable new genus of Marattiaceae (ferns), in spite of all our researches, was never seen except in one bit of forest. It is very likely a type that is dying out.'

For an illustration of this tree see Lancaster, *Travels in China* 151. (1993).

Kleinhovia hospita L. (MALVACEAE)

Henry in Trans. Asiat. Soc. Jap. 24 suppl: 22. (1896) (List Pl. Formosa), Li in Woody Flora of Taiwan 559. fig. 219. (1963).

Taiwan: Kaohsiung, without number. Wanchin, 12. Oluanpi, 1280.

A small to medium-sized tree, monotypic and native to the tropics of the Old World. Linnaeus gave this fine tropical tree the specific epithet *hospita* because of its many resident guests - epiphytic flowering angiosperms, ferns, bryophytes, lizards and snakes. The bark and leaves smell of bitter almonds (prussic acid) and are poisonous. In Malaya an insecticide used to kill head lice is made from the leaves. Found in secondary forest on the southern part of Taiwan, China.

Knoxia sumatrensis (Retz.) DC. (RUBIACEAE)

Knoxia corymbosa Willd. Forbes & Hemsley in Journ. Linn. Soc. Bot. 23: 384. (1888). *Hedyotis* sp. Henry in Trans. Asiat. Soc. Jap. 24 suppl: 49. (1896) (List Pl. Formosa).

Hubei: Yichang, 600, 7617. **Hainan:** Without locality, 8336. **Taiwan:** Wanchin, 889. Oluanpi, 1215, 1292, 2007. **Yunnan:** Mengzi, 9983. In the Red River Valley near Manhao, 9529. Simao, 13162.

Koeleria pyramidata (Lam.)P. Beauv. (POACEAE)

Koeleria cristata (L.)Pers. Rendle in Journ. Linn. Soc. Bot. 36: 410. (1904).

Hubei: Yichang, 3505.

Koelreuteria bipinnata Franch. (SAPINDACEAE)

Henry in Flora & Sylva 1: 218. (1903), Elwes & Henry in Trees Gr. Brit. & Ireland 7: 1931. (1913), Rehder & E. H. Wilson in Sargent Pl. Wilson 2: 193. (1916), Radlkofer in Engler, Pflanzenr. iv. 165: 1332. (1934), Meyer in Journ. Arnold Arb. 57: 149. (1976), Zheng in Journ. Wuhan Bot. Res. 2.(1): 134. (1984).

Hubei: South Donghu, 7591. **Yunnan:** Mountains to the north Mengzi at 2,300 metres, 9099. To the south of Mengzi near the village of Hsinkei at 1,800 metres, 9103.

A tree to 20 metres tall with girths of up to 60 cm. According to Wilson this tree is rare in Hubei and Sichuan but is common in Yunnan where it grows to a conciderable size. It was discovered by Delavay on July 26th 1885. *Koelreuteria bipinnata* was collected by the Glasnevin Central China Expedition (GCCE 373) near Badong in October 2002 and was collected again by the 2004 Glasnevin Expedition to the Three Gorges region (GCCE 876) near Dragon Gate River Forest Park in Xingshan in October of that year.

Koelreuteria elegans (Seem.) A. C. Smith. subsp. *formosana* (Hayata) F. G. Meyer (SAPINDACEAE) New subspecies

Clarke in Bean, Trees & Shrubs, suppl. 299. (1988).

Koelreuteria bipinnata Henry non Franch. in Trans. Asiat. Soc. Jap. 24 suppl: 29. (1896) (List Pl. Formosa). *Koelreuteria henryi* Dummer in Gard. Chron. ser 3, 52: 148. (1912), Radlkofer in Engler, Pflanzenr. iv. 165: 1332. (1934), Li in Woody Flora of Taiwan 497. fig. 191. (1963).

Taiwan: Wanchin, 1594 (type of *Koelreuteria henryi* Dummer).

A deciduous tree to 10 metres tall, this subspecies was based on a collection made by the Japanese botanist U. Mori (Mori 1736) in October 1906, Henry had discovered it some 12 years earlier. *Koelreuteria elegans* ssp. *formosana* exhibits various degrees of hardiness depending on its provinance. It is common in the tropical rainforests of the Hengchun Peninsula in southern Taiwan Province of China and was collected there by the Glasnevin Taiwan Expedition (GTE 69) in Kenting National Park in October 2004 and in the campus of the National Sun Yat Sen University in Kaohsiung (GTE 121). The original material introduced to Kew in 1976 from the Taiwan Forestry Institute in Taipei on the north of the island has proved hardy.

Kummerowia striata (Thunb.) Schindl. (FABACEAE)

Lespedeza striata (Thunb.) Hook. & Arn. Forbes & Hemsley in Journ. Linn. Soc. Bot. 23: 182. (1887), Henry in Journ. China Br. Roy. Asiat. Soc. 22: 263. (1887) (Chinese Names of Plants).

Hubei: Yichang, 1673, 2314. Without locality, 2314a.

An annual species native to east Asia where it is sometimes grown as a forage crop.

Kyllinga sesquiflora Torrey subsp. *cylindrica* (Nees) T. Koyama (CYPERACEAE)

Kyllinga cylindrica Nees Clarke in Journ. Linn. Soc. Bot. 36: 223. (1903).

Yunnan: Mengzi at, 1,500 metres, 10966.

A widespread subspecies, found over most of tropical Asia.

Lablab purpureus (L.) Sweet (FABACEAE)

Dolichos lablab L. Henry in Trans. Asiat. Soc. Jap. 24 suppl: 37. (1896) (List Pl. Formosa).

Taiwan: Kaohsiung, without number. **Yunnan:** Simao, 13496.

A handsome annual twining vine to 10 metres tall, native to Africa and cultivated and naturalised throughout the tropics. According to Henry it is the *pien-tou*.

Lactuca formosana Maxim. var. caulibus foliisque glabris glaucis capitulis minoribus. (ASTERACEAE)

Forbes & Hemsley in Journ. Linn. Soc. Bot. 23: 482. (1888).

Hubei: Yichang 4100. In a ravine on Moji Shan, 4276. Badong, 4847.

Lactuca indica L. (ASTERACEAE)

Lactuca amurensis Regel & Maxim. Henry in Journ. China Br. Roy. Asiat. Soc. 22: 250. (1887) (Chinese Names of Plants). *Lactuca squarrosa* (Thunb.) Miq. Henry in Journ. China Br. Roy. Asiat. Soc. 22: 250. (1887) (Chinese Names of Plants). *Lactuca brevirostris* Champ. ex Benth. Henry in Trans. Asiat. Soc. Jap. 24 suppl: 55. (1896) (List Pl. Formosa). *Lactuca brevirostris* Champ. ex Benth. var. foliis indivisis(or foliis laciniatus) Forbes & Hemsley in Journ. Linn. Soc. Bot. 23: 480. (1888).

Hubei: In the San You Dong Glen (June 1886), 1638. At side of rice fields in the Monastery Valley, 1645. Yichang, 2132, 4365. Liantuo, 3058, 4457, 4459. Fang Xian, 6678. **Hainan:** Environs of Haikou, 7955. **Taiwan:** Wanchin, 195. Oluanpi, 295. Kaohsiung Plain, 1122, 1800. Tamsui, 1453. **Yunnan:** Mengzi, 9645, 9646, 10321.

According to Henry this species was known colloquially in the Three Gorges region as *k'u ts'ai*.

Lactuca raddeana Maxim (ASTERACEAE)

Lactuca elata Hemsley (nom. illeg.) in Journ. Linn. Soc. Bot. 23: 481. (1888).

Hubei: Liantuo, 4620. South Badong, 7372. Fang Xian, 7572.

Lactuca sagittarioides C. B. Clarke (ASTERACEAE)

Anonymous in Kew List of Determinations (volume 34).

Yunnan: Simao, 12998.

Native to India, Bhutan, Nepal and China (Yunnan).

Lactuca sativa L. (ASTERACEAE)

Lactuca scariola Forbes & Hemsley non L. in Journ. Linn. Soc. Bot. 23: 483. (1888).

Hubei: Badong, 3181. Yichang 3995, 4333.

Cultivated lettuce, first cultivated by the Egyptians in about 4,500 B.C., according to Bretschneider lettuce was cultivated in China in the fifth century A.D. Celtuce was developed in China and was sent back to Vilmorin by one of the French missionaries.

Lactuca sibirica (L.) Benth. ex Maxim. (ASTERACEAE)

Anonymous in Kew List of Determinations (volume 24).

Hubei: Badong, 1839.

Lactuca sp. No. 1. (ASTERACEAE)

Anonymous in Kew List of Determinations (volume 24).

Hubei: Badong, 1881, 3168. Yichang, 2958. Fang Xian, 6697. **Sichuan:** South Wushan, 5571, 5648.

Lactuca sp. No. 2. (ASTERACEAE)

Anonymous in Kew List of Determinations (volume 34).

Hubei: Badong, 1881, 3168. Yichang, 2958. Fang Xian, 6697. **Sichuan:** South Wushan, 5571, 5648.

Laetiporus sulphureus (Bull.) Murril (FOMITOPSIDACEAE)

Polyporus sulphureus (Bull.) Fr. Anonymous in Kew List of Determinations (volume 24).

Sichuan: South Wushan, 7040. (Fungus).

Chicken of the woods fungus.

Lagerstroemia glabra (Koehne) Koehne var. *latifolia* Furt. & Mont. (LYTHRACEAE)

Zheng in Journ. Wuhan Bot. Res. 2.(1): 157. (1984).

Lagerstroemia subcostata Koehne var. *glabra* Koehne in Engler, Pflanzenr. 17. (iv. 216.): 260. (1903).

Hubei: Near Badong at Wan Lui on the Yangtze, 7169 (isoholotype of *Lagerstroemia glabra* (Koehne) Koehne var. *latifolia* Furt. & Mont., type of *Lagerstroemia subcostata* Koehne var. *glabra* Koehne).

Lagerstroemia indica L. (LYTHRACEAE)

Forbes & Hemsley in Journ. Linn. Soc. Bot. 23: 306. (1887), Henry in Journ. China Br. Roy. Asiat. Soc. 22: 278. (1887) (Chinese Names of Plants), Rehder & E. H. Wilson in Sargent Pl. Wilson 2: 418. (1916), Zheng in Journ. Wuhan Bot. Res. 2.(1): 157. (1984), Koehne in Engler, Pflanzenr. 216: 285. (1903).

Hubei: Yichang, 32. Liantuo, 2125. Badong, 2507. **Hainan:** Environs of Haikou, 8320.

The crepe myrtle, a deciduous tree to 9 metres tall bearing flowers varying from pink to deep red in July,

August and September, the bark on mature trees is an attractive smooth grey. Despite the specific epithet coined by Linnaeus this species is native to China and Korea. In Henry's time the crepe myrtle or *tzu ching* occured both wild and cultivated in the area around the Three Gorges. It was planted in gardens and around temples where selected forms with white and pink to carmine coloured flowers were common. The genus commemorates the famous Swedish naturalist Magnus von Lagerstroem (1696—1759) who was responsible for the acquisition of the type material and the introduction of this species from China. Lagerstroem's position as Director of the Swedish East India Company allowed him to acquire herbarium specimens from China which he sent with living plants to the Botanic Gardens at Upsala. The crepe myrtle was introduced to the Royal Botanic Gardens, Kew in 1759. At Glasnevin this species flowers only after a long, hot summer.

Lagerstroemia speciosa (L.) Pers. subsp. *intermedia* (Koehne) Deepu & Pandur. (LYTHRACEAE) New subspecies

Lagerstroemia intermedia Koehne in Engler, Pflanzenr. 17. (iv. 216.): 260. (1903).

Yunnan: Mountains to the south of Simao at 1,220 metres, without number (type of *Lagerstroemia intermedia* Koehne).

An evergreen tree to 8 metres tall, carrying terminal panicles of pink flowerrs to 17 cm long from May to June. China (Southern Yunnan) and Myanmar.

Lagerstroemia subcostata Koehne (LYTHRACEAE)

Henry in Trans. Asiat. Soc. Jap. 24 suppl: 44. (1896) (List Pl. Formosa), Li in Woody Flora of Taiwan 626. fig. 247. (1963).

Taiwan: Kaohsiung, without number. Wanchin, 43, 561, 1491, without number. Oluanpi, 958, 966, 1220. Tamsui, 1736.

A deciduous tree to 10 metres tall. Native from central China to Japan (south). A common forest tree in Chinese Taiwan at low to medium altitude. According to Henry the wood of this tree was used for house pillars as it did not rot in the ground. *Lagerstroemia subcostata* was first found in Taiwan by the Kew collector, Richard Oldham. This species was collected by the Glasnevin Taiwan Expedition (GTE 91) in Kenting National Park in October 2004.

Laggera alata (D. Don) Sch.-Bip. ex Oliv. (ASTERACEAE)

Forbes & Hemsley in Journ. Linn. Soc. Bot. 23: 422. (1888), Henry in Trans. Asiat. Soc. Jap. 24 suppl: 53. (1896) (List Pl. Formosa).

Hubei: Yichang, 2256, 2974, 3228. **Taiwan:** Wanchin, 1611, 1629. Tamsui. 1740. **Yunnan:** Chuyuen, 10442. Mengz,i 10320. Simao, 12778.

Widespread in Asia.

Laggera crispata (Vahl) Hepper & J. R. I. Wood (ASTERACEAE)

Laggera pterodonta (DC.) Sch. Bip. ex Oliv. Anonymous in Kew List of Determinations (volume 34).

Yunnan: Simao, 11641.

To the Hani (an ethnic group) of Xishuangbanna in southern Yunnan this species is called *hhaoqsaqlaqma* and is used to treat laryngitis, bronchitis, malaria and influenza. Distributed in India, China, Laos, Cambodia and Vietnam.

Lagopsis supina (Steph.) Ik.-Gal. ex Knorr. (LAMIACEAE)

Marrubium incisum Benth. Forbes & Hemsley in Journ. Linn. Soc. Bot. 26: 299. (1890), Dunn in Notes Roy. Bot. Gard. Edinburgh 6: 178. (1915).

Hubei: Yichang, 1233.

Lamiaceae undetermined. (LAMIACEAE)

Henry in Trans. Asiat. Soc. Jap. 24 suppl: 73. (1896) (List Pl. Formosa).

Taiwan: Wanchin, 82, 544, 1565, 1646, 1668. Oluanpi, 362. Kaohsiung, 1194. **Yunnan:** Mengzi, 9197.

Lamium album L. subsp. *barbatum* (Sieb. & Zucc.) Mennema (LAMIACEAE)

Lamium album Forbes & Hemsley non L. in Journ. Linn. Soc. Bot. 26: 302. (1890).

Hubei: Badong, 4672, 5403.

Lamium amplexicaule L. (LAMIACEAE)

Forbes & Hemsley in Journ. Linn. Soc. Bot. 26: 303. (1890), Dunn in Notes Roy. Bot. Gard. Edinburgh 6: 182. (1915).

Hubei: Yichang, 3372. **Yunnan:** Lunan, 10585.

Distributed in Eurasia, Ethiopia and Macaronesia.

Lamprocapnos spectabilis (L.) Fukuhara (FUMARIACEAE)

Dicentra spectabilis (L.) Lem. Diels in Bot. Jahrb. 29: 354. (1900).

Hubei: Badong, 3725.

Commonly called the bleeding heart this is one of the most beautiful of all spring-flowered perennials. In China it is the *ho pao moutan* or 'purse moutan' because of the similarity of its foliage to that of the peony rose. Discovered by the French Jesuit missionary Père Pierre d'Incarville (1706—1757) who collected in China (mostly around Beijing and also at Macao) between 1740 and 1756. d'Incarville sent his collections to the famous French botanist Bernard de Jussieu (1699—1776), Professor of Botany at the Royal Gardens, Paris. It was introduced to cultivation by Robert Fortune who brought living plants to England for the Horticultural Society (now the Royal Horticultural Society) in May 1846. Fortune stated it was cultivated with great pride by the Chinese

Mandarins who called it the *hong pak moutan* or 'red and white moutan'.

Lannea coromandelica (Houtt.) Merr. (ANACARDIACEAE)

Odina wodier Roxb. Anonymous in Kew List of Determinations (volume 24).

Hainan: Environs of Haikou, 8003.

Laportea bulbifera (Sieb. & Zucc.) Wedd. (URTICACEAE)

Laportea sinensis C. H. Wright in Journ. Linn. Soc. Bot. 26: 474. (1899), *Laportea oleracea* Wedd. Forbes & Hemsley in Journ. Linn. Soc. Bot. 26: 474. (1899).

Hubei: Badong, 4748, 4892. **Sichuan:** South Wushan, 7212 (type of *Laportea sinensis* C. H. Wright), 7364. **Yunnan:** On Fengchunling at 2,300 metres, 11198.

Laportea cuspidata (Wedd.) Friis (URTICACEAE)

Laportea grossedentata C. H. Wright in Journ. Linn. Soc. Bot. 26: 474. (1899).

Hubei: Badong, 4691, 6066 (syntype of *Laportea grossendentata* C. H. Wright), 6077.

Laportea interrupta (L.) Chew (URTICACEAE)

Fleurya interrupta (L.) Gaudich. Henry in Trans. Asiat. Soc. Jap. 24 suppl: 88. (1896) (List Pl. Formosa), Forbes & Hemsley in Journ. Linn. Soc. Bot. 26: 473. (1899).

Taiwan: Wanchin, 368, 386.

Native to tropical and subtropical Africa, Asia and the Pacific Islands.

Laportea sp. No. 1. (URTICACEAE)

Anonymous in Kew List of Determinations (volume 24).

Hubei: Yichang, 413. Fang Xian, 6767.

Laportea sp. No. 2. (URTICACEAE)

Anonymous in Kew List of Determinations (volume 34).

Yunnan: Simao, 12070.

Lapsanastrum apogonoides (Maxim.) Pak & K. Bremer (ASTERACEAE)

Lapsana apogonoides Maxim. Forbes & Hemsley in Journ. Linn. Soc. Bot. 23: 474. (1888), Henry in Notes Econ. Bot. China 53. (1893).

Hubei: Yichang, 1270. Xiling (Yichang) Gorge, 3330. Without locality, 3330a.

The *huang-hua-ts'ai,* a very small weed of tilled fields.

Lasianthus attenuatus Jack (RUBIACEAE)

Lasianthus wallichii (Wight & Arn.) Wight var. *hispidocostatus* H. Zhu in Acta Bot. Yunnan 20: 158. (1998), Zhu in Acta Phytotax. Sin. 39: 148. (2001).

Yunnan: In forests near Simao at 1,300 metres, 12789 [holotype of *Lasianthus wallichii* (Wight & Arn.) Wight var. *hispidocostatus* H. Zhu].

Tropical and subtropical Asia.

Lasianthus biermanni King ex Hook. f. subsp. **crassipedunculatus** C. Y. Wu & H. Zhu (RUBIACEAE)

New subspecies. in Acta Phytotax. Sin. 32: 75. (1994).

Lasianthus biermannii Hutchinson non King ex Hook. f. in Sargent Pl. Wilson. 3: 402.(1917).

Yunnan: In Pingbian in forests on the Daweishan Range at 1,640 metres, 11148.

Endemic to China and distributed in the provinces of Guizhou, Hainan and Yunnan.

Lasianthus chinensis (Champ.) Benth. (RUBIACEAE)

Henry in Trans. Asiat. Soc. Jap. 24 suppl: 50. (1896) (List Pl. Formosa), Li in Woody Flora of Taiwan 855. fig. 345. (1963).

Taiwan: Wanchin, 162. Oluanpi, 1209.

Native to southern China including Taiwan where it is a common woodland inhabitant.

Lasianthus fordii Hance (RUBIACEAE)

Zhu in Acta Phytotax. Sin. 39: 127. (2001).

Lasianthus japonicus Henry non Miq. in Trans. Asiat. Soc. Jap. 24 suppl: 50. (1896) (List Pl. Formosa), *Lasianthus wallichii* Henry non Wight in Trans. Asiat. Soc. Jap. 24 suppl: 51. (1896) (List. Pl. Formosa), *Lasianthus plagiophyllus* Li non Hance in Woody Flora of Taiwan 857. (1963).

Taiwan: Wanchin, 106, 112.

A widespread species, recently reported from Thailand and also found in Vietnam, Cambodia, Indonesia, China, Japan, the Philippines and New Guinea.

Lasianthus henryi Hutchinson (RUBIACEAE) New species

in Sargent Pl. Wilson. 3: 402. (1916), Zhu in Acta Phytotax. Sin. 32: 76. (1994).

Lasianthus inconspicuus Hook. f. var. *hirtus* Hutchinson in Sargent Pl. Wilson. 3: 402. (1917).

Yunnan: In the mountains to the south of Mengzi at 1,970 metres, 9775 (type of *Lasianthus inconspicuus* Hook. f. var. *hirtus* Hutchinson). In Pingbian in forests on the Daweishan Range at 1,650 metres, 11253 (type of *Lasianthus henryi* Hutchinson).

Endemic to China.

Lasianthus hookeri C. B. Clarke apud Hooker. f. (RUBIACEAE)

Hutchinson in Sargent Pl. Wilson. 3: 402. (1916), Zhu in Acta Phytotax. Sin. 39: 131. (2001).

Yunnan: Mountains to the east of Simao at 1,350 metres, 12031. In mountain forest near Simao at 1,450 metres, 12031a. Simao, 13570

Distributed through north-east India, Myanmar, China, Thailand and Vietnam.

02

Lasianthus japonicus Miq. subsp. ***longicaudus*** (Hook. f.) C. Y. Wu & H. Zhu (RUBIACEAE)

Lasianthus longicaudus Hook. f. Hutchinson in Sargent Pl. Wilson. 3: 402. (1916).

Yunnan: In Pingbian on the Daweishan Range at 1,950 metres, 9035, 10633.

Lasianthus lucidus Bl. var. ***inconspicuus*** (Hook. f.) Zhu (RUBIACEAE)

Zhu in Acta Phytotax. Sin. 39: 135. (2001).

Lasianthus inconspicuus Hook. f. Hutchinson in Sargent Pl. Wilson. 3: 402.(1916). *Lasianthus lucidus* Zhu non Blume in Acta Phytotax. Sin. 32: 63. (1994).

Yunnan: Mountains to the east of Simao at 1,300 metres, 12608. In forests near Simao at 1,450 metres, 12608a.

Lasianthus micranthus Hook. f. (RUBIACEAE)

Hutchinson in Sargent Pl. Wilson. 3: 402. (1916), Zhu in Acta Phytotax. Sin. 39: 137. (2001).

Yunnan: South of the Red River from Manmei, 13670.

Native to north-east India, northern Vietnam, China and recently recorded from Chiang Mai in Thailand.

Lasianthus schmidtii K. Schum. (RUBIACEAE)

Lasianthus kerrii Craib Hutchinson in Sargent Pl. Wilson. 3: 402. (1916), Zhu in Acta Phytotax. Sin. 32: 59. (1994); 39: 133. (2001).

Yunnan: In forests near Simao at 1,320 metres, 12571.

Lasianthus sp. No. 1. (RUBIACEAE)

Anonymous in Kew List of Determinations (volume 24.).

Hainan: Ling-men, 8609.

Lasianthus sp. No. 2. (RUBIACEAE)

Anonymous in Kew List of Determinations (volume 34).

Yunnan: Simao, 12606, 12788.

Lasianthus (***Pedunculatae***) sp. (RUBIACEAE)

Henry in Trans. Asiat. Soc. Jap. 24 suppl: 51. (1896) (List Pl. Formosa).

Taiwan: Wanchin, 1682.

Lathyrus palustris L. (FABACEAE)

Forbes & Hemsley in Journ. Linn. Soc. Bot. 23: 186. (1887).

Hubei: Yichang, 3478. **Sichuan:** Emei Shan, Henry's Chinese collector with Pratt, 8981.

Lathyrus pratensis L. (FABACEAE)

Forbes & Hemsley in Journ. Linn. Soc. Bot. 23: 186. (1887).

Sichuan: Emei Shan, Henry's Chinese collector with Pratt, 8986.

Lathyrus sp. (FABACEAE)

Anonymous in Kew List of Determinations (volume 24).

Hubei: Fang Xian, 6595.

Lauraceae (LAURACEAE)

Anonymous in Kew List of Determinations (volume 34).

Yunnan: Mengzi, 10328, 10330, 11396. Simao, 12191, 12813, 12893, 13102.

Lawsonia inermis L. (LYTHRACEAE)

Anonymous in Kew List of Determinations (volume 24).

Hainan: Environs of Haikou, 8031.

Henna, a small monotypic tree to 7 metres tall native to the Old World tropics. The flowers are distilled for their perfume. From the dried leaves of this tree ancient Egyptians made a dull orange-yellow to red or blackish dye called henna to dye their bodies or the swathing cloths of their mummies. Henna is still used as a hair conditioner.

Lecanthus obtusus (Royle) Hand.-Mazz. (URTICACEAE)

Lecanthus wightii Wedd. Forbes & Hemsley in Journ. Linn. Soc. Bot. 26: 480. (1899).

Yunnan: In Pingbian in woods on the Daweishan Range at 2,600 metres, 9031.

Lecanthus peduncularis (Royle) Wedd. (URTICACEAE)

Lecanthus wallichii Wedd. Forbes & Hemsley in Journ. Linn. Soc. Bot. 26: 480. (1899).

Yunnan: Mengzi, 9738, 9738a.

Lecanthus sp. (URTICACEAE)

Anonymous in Kew List of Determinations (volume 34).

Yunnan: Simao, 12594.

Leea asiatica (L.) Ridsdale (VITACEAE)

Leea crispa L. Anonymous in Kew List of Determinations (volume 34).

Yunnan: Simao, 12104.

Leea indica (Burm. f.) Merr. (VITACEAE)

Leea sambucina (L.) Willd. Anonymous in Kew List of Determinations (volume 34).

Yunnan: Simao, 11621, 12125.

A shrub to 4 metres tall, distributed through the tropics of Asia, Polynesia, New Guinea and Northern Australia.

Leea manillensis Walp. (VITACEAE)

Li in Woody Flora of Taiwan 524. fig. 203. (1963).

Leea sambucina Henry non (L.) Willd. in Trans. Asiat. Soc. Jap. 24 suppl: 28. (1896) (List Pl. Formosa).

Taiwan: Wanchin, without number. Kaohsiung, 79. Oluanpi, 632.

A shrub or small tree to 6 metres tall, native to the Philippines and China (Taiwan) where it grows at low altitude.

Leea sp. (VITACEAE)

Anonymous in Kew List of Determinations (volume 24).

Hainan: Ling-men, 8658.

May be *Leea indica* Burm. f. or *Leea longifoliola* Merr.

Leersia hexandra Sw. (POACEAE)

Henry in Trans. Asiat. Soc. Jap. 24 suppl: 107. (1896) (List Pl. Formosa), Rendle in Journ. Linn. Soc. Bot. 36: 345. (1904).

Taiwan: Kaohsiung, 1170, 2034.

Pantropical in distribution, in China it is found in Hainan and Taiwan and favours wet areas.

Leersia japonica (Makino ex Honda) Honda (POACEAE)

Leersia hexandra Rendle non Swartz. in Journ. Linn. Soc. Bot. 36: 345. (1904), in part.

Hubei: Yichang, 2783, 4249.

Leibnitzia anandria (L.) Nakai (ASTERACEAE)

Gerbera anandria (L.) Schultz-Bip. Forbes & Hemsley in Journ. Linn. Soc. Bot. 23: 472. (1888).

Hubei: Yichang, 803, 1250. Badong, 3717. Fang Xian, 6616, 6684. Changyang, 7691. Without locality, 6684a., 6684b. **Sichuan:** South Wushan, 7115.

Lemmaphyllum carnosum (J. Sm. ex Hook.) C. Presl. (POLYPODIACEAE)

Drymoglossum carnosum J. Sm. ex Hook. Anonymous in Kew List of Determinations (volume 34).

Yunnan: Mengzi, 9869.

Distributed from Nepal to western China, Thailand and Vietnam.

Lenzites sinensis Cooke (POLYPORACEAE) New species

in Grevillea 17: 75. (1889).

Hubei: Changyang, on trees, 7926, 7927. **Sichuan:** North Wushan, 7131 (type). (Fungus).

Leontopodium andersonii C. B. Clarke (ASTERACEAE)

Leontopodium subulatum (Franch.) Beauverd Anonymous in Kew List of Determinations (volume 34).

Yunnan: Mengzi, 10151.

Endemic to China and distributed in Guizhou, Sichuan and Yunnan.

Leontopodium dedekensii (Bureau & Franch) Beauverd (ASTERACEAE)

Anonymous in Kew List of Determinations (volume 34).

Yunnan: Mengzi, 9422, 9635.

Native to Burma and China.

Leontopodium japonicum Miq. (ASTERACEAE)

Forbes & Hemsley in Journ. Linn. Soc. Bot. 23: 424. (1888).

Hubei: Badong, 4830, 6186. Without locality, 6186a.

Japanese edelweiss, this species was collected by the Sino-American Expedition in the Shennongjia Forest District in 1980.

Leontopodium sp. (ASTERACEAE)

Anonymous in Kew List of Determinations (volume 34).

Yunnan: Mengzi, 9963, 9974.

Leonurus japonicus Houtt. (LAMIACEAE)

Leonurus sibiricus Forbes & Hemsley non L. in Journ. Linn. Soc. Bot. 26: 302. (1890), Henry in Trans. Asiat. Soc. Jap. 24 suppl: 73. (1896) (List Pl. Formosa), Dunn in Notes Roy. Bot. Gard. Edinburgh 6: 200. (1915).

Hubei: Yichang, 629. Near Yichang, on the hills behind the San You Dong Glen (July 1886), 1517. Liantuo, 4481. **Taiwan:** Oluanpi, without number. **Yunnan:** Mengzi, 10382.

A tropical weed, widely distributed in tropical Asia, including the Pacific, Australia and in the Neotropics where it was introduced from the Old World. Harley and Paton provide a short note on the correct nomenclature for this species with historical notes in Kew Bull. 56: 243. (2001)

Lepidagathis incurva Buch.-Ham. ex D. Don (ACANTHACEAE)

Lepidagathis hyalina Nees Henry in Trans. Asiat. Soc. Jap. 24 suppl: 69. (1896) (List Pl. Formosa).

Taiwan: Wanchin, 102. Tamsui, 1748, 1751. Oluanpi, 1754. **Yunnan:** Simao, 12941.

Lepidagathis stenophylla C. B. Clarke ex Hayata (ACANTHACEAE) New species

in J. Coll. Sci. Imp. Univ. Tokyo. 30(Art. 1): 214. (1911).

Lepidagathis hyalina Henry non Nees in Trans. Asiat. Soc. Jap. 24 suppl: 69.(1896) (List Pl. Formosa).

Taiwan: Oluanpi, 622, 622a.

Lepisanthes sp. (SAPINDACEAE)

Erioglossum sp. Anonymous in Kew List of Determinations (volume 34).

Yunnan: Simao, 11900.

Lepisorus angustus Ching (POLYPODIACEAE)

Polypodium lineare Thunb. f. *caudato-attenuatum* Takeda in Notes Roy. Bot. Gard. Edinburgh viii. 269. (1915).

Yunnan: In the mountains to the north-west of Simao at 1,650 metres, 10062a.

Endemic to China.

Lepisorus bicolor (Takeda) Ching (POLYPODIACEAE) New species

Polypodium excavatum Bory ex Willd. var. *bicolor* Takeda in Notes Roy. Bot. Gard. Edinburgh 8: 281. (1915).

Hubei: Badong, 1739, 2465. **Yunnan:** In mountain forest to the north of Mengzi at 2,800 metres, 10088.

Native to northern India and China (including Taiwan).

Lepisorus buergerianus (Miq.) C. F. Zhao, R. Wei & X. C. Zhang (POLYPODIACEAE)

Polypodium subhastatum Baker in J. Bot. 177. (1889); in Ann. Bot. 5: 478. (1890), Bretschneider in Hist. Eur. Bot. Disc. China 2: 794. (1898), Diels in Bot. Jahrb. 29: 205. (1900). *Polypodium subhastatum* Baker var. *hederaceum* (Christ) Mihi Takeda in Notes Roy. Bot. Gard. Edinburgh viii. 292. (1913—1915).

Hubei: South Badong, 5450, (type of *Polypodium subhastatum* Baker).

Lepisorus carnosus (Hook.) C. F. Zhao, R. Wei & X. C. Zhang (POLYPODIACEAE)

Polypodium drymoglossoides Baker in Journ. Bot. 25: 171. (1887), Baker in Ann. Bot. 5: 474. (1898), Bretschneider in Hist. Eur. Bot. Disc. China 2: 794. (1898), Diels in Bot. Jahrb. 29: 204. (1900), Christ in Bull. Soc. Bot. France 5: 16. (1905). *Drymoglossum carnosum* Hook. Henry in Trans. Asiat. Soc. Jap. 24 suppl: 116. (1896) (List Pl. Formosa).

Hubei: In grass in a glen near Pingshanba (near Yichang, May 1886), 1576. Jianshi, 5963. Liantuo, 2965, 4392. **Sichuan:** South Wushan, 7532a. **Taiwan:** Tamsui, 1406.

A handsome little fern often seen growing epilithically on the forests edge at mid-altitude in Hubei. The plant in a dried state is currently used in Chinese herbal medicine to treat patients suffering from pulmonary tuberculosis and to treat skin diseases.

Lepisorus chinensis (Mett. Ex Kuh.) ined. (POLYPODIACEAE)

Polypodium henryi Christ in Bull. Herb. Boiss. 6: 873. (1898). *Polypodium fortunei* Takeda non Kunze ex Mett. in Notes Roy. Bot. Gard. Edinburgh 8: 285. (1913—1915). *Microsorum henryi* (Christ) C. M. Kuo in Taiwania 30: 67. (1985).

Yunnan: In the mountains near Mengzi at 1,640 metres, 9780 (type of *Polypodium henryi* Christ, collected 22nd November 1896).

Lepisorus eilophyllus (Diels) Ching (POLYPODIACEAE) New species

Polypodium involutum Baker in Journ. Bot. 27: 177. (1889), Baker in Ann. Bot. 5: 473. (1890), Bretschneider in Hist. Eur. Bot. Disc. China 2: 794. (1898), Christ in Bull. Soc. Bot. France 5: 16. (1905). *Polypodium eilophyllum* Diels in Bot. Jahrb. Syst. 29: 204. (1900), Takeda in Notes Roy. Bot. Gard. Edinburgh. viii. 274. (1913—1915).

Hubei: Fang Xian, 6859 (isotype of *Polypodium eilophyllum* Diels, type of *Polypodium involutum* Baker). **Yunnan:** Epiphytic on trees in a wood near Mengzi at 1,950

metres, 9249b.

Distributed in India, China and Thailand.

Lepisorus ensatus (Thunb.) C. F. Zhao, R. Wei & X. C. Zhang (POLYPODIACEAE)

Polypodium oligolepis Baker in Kew Bull. Misc. Inf. 231. (1898). *Polypodium ensatum* Thunb. Christ in Bull. Herb. Boiss. 6: 874. (1898), Diels in Bot. Jahrb. 29: 203. (1900), Takeda in Notes Roy. Bot. Gard. Edinburgh viii. 288. (1913—1915).

Sichuan: South Wushan, 7133. **Hubei:** Changyang, 7678. **Yunnan:** Milê, 9896, (isotype of *Polypodium oligolepis* Baker).

Distributed through northern India, China, Japan and the Philippines.

Lepisorus excavatus (Bory ex Willd.) Ching (POLYPODIACEAE)

Polypodium excavatum Bory ex Willd. var. *concolor* Takeda in Notes Roy. Bot. Gard. Edinburgh 8: 281. (1915).

Yunnan: In mountain forest to the west of Mengzi at 1,950 metres, epiphytic on trees, 13070a.

Lepisorus henryi (G. Hieronymus) Li Wang (POLYPODIACEAE) New species

in J. Linn. Soc. 162: 35. (2010).

Hymenolepis henryi G. Hieronymus in Dansk. Bot. Ark. 6(3): 67. (1929). *Belvisia henryi* (G. Hieronymus) Raymond in Mem. Jard. Bot. Montreal 55: 32. (1962).

Yunnan: Mengzi, 11461, (type of *Hymenolepis henryi* G. Hieronymus).

Distributed in Bhutan, Myanmar, Yunnan, Thailand and Vietnam.

Lepisorus loriformis (Wall. ex Mett.) Ching (POLYPODIACEAE)

Polypodium xiphiopteris Baker in Kew Bull. 16. (1906). *Polypodium subimmersum* Baker in Kew Bull. Misc. Inf. 55. (1895). *Polypodium subimmersum* Baker f. *angustifrons* Mihi Takeda in Notes Roy. Bot. Gard. Edinburgh viii. 276. (1913—1915). *Polypodium mengtzeanum* Baker in Kew Bull. 14. (1906). *Polypodiastrum mengtzeense* (H. Christ) Ching in Acta Phytotax. Sin. 16(4): 28. (1978). *Polypodium subimmersum* Baker f. *mengtzeanum* (Bak.) Mihi Takeda in Notes Roy. Bot. Gard. Edinburgh viii. 276. (1913—1915). *Polypodium lineare* Thunb. var. *loriforme* (Wall.) Takeda Takeda in Notes Roy. Bot. Gard. Edinburgh. viii. 272. (1913—1915). *Polypodium asperum* Baker in Kew Bull. Misc. Inf. 231. (1898) non L. (1753). *Polypodium oblongisorum* C. Christensen in Ind. Filicum fasc. 9: 549. (1906).

Yunnan: Mengzi, 9194 (in part, isotype of *Polypodium oblongisorum* C. Christensen), 11827b., (type of *Polypodium mengtzeanum* Baker). On the mountains to the east of Mengzi at 1,950 metres, 11826 (type of *Polypodium xiphiopteris* Baker). Yuanjiang, 13330, 13425. In Pingbian on the Daweishan Range, in mountain forest at 1,650 metres, 9054 (syntype of *Polypodium mengtziensum* Baker, isotype of *Polypodium asperum* Baker). In Pingbian on the Daweishan Range, in mountain forest at 1,600 metres, on trees, 11025 (syntype of *Polypodium mengtziensum* Baker).

Himalaya, China, Myanmar.

Lepisorus macrosphaerus (Baker) Ching (POLYPODIACEAE)

Polypodium intramarginale Baker in Kew Bull. Misc. Inf. 13. (1906). *Polypodium asterolepis* Baker Takeda in Notes Roy. Bot. Gard. Edinburgh viii. 283. (1913—1915).

Hubei: Badong, 2556. **Yunnan:** Mengzi, 9203, 10042a., (in part), 13633. In woods, in a mountainous valley to the north of Mengzi at 1,950 metres, 10042 (type of *Polypodium intramarginale* Baker, collected 5th August 1897). Simao, 13363 (in part).

Native to the Himalaya, China and Indochina.

Lepisorus maculosus (H. Christ) C. F. Zhao, R. Wei & X. C. Zhang (POLYPODIACEAE) New species

Polypodium maculosum H. Crist in Bull. Herb. Boissier 6: 842. (1898). *Polypodium normale* Takeda non D. Don in Notes Roy. Bot. Gard. 8: 286. (1913—1915).

Yunnan: On the mountains to the south of Mengzi at 1,960 metres, on trees, 10090 (isotype of *Polypodium maculosum* H. Christ). On the mountains to the south of Simao at 1,300 metres, epilithic on rocks, 12947.

Distributed in China (Guangxi and Yunnan), Laos, Thailand and Vietnam to Sumatera.

Lepisorus nudus (Hook.) Ching (POLYPODIACEAE)

Polypodium nudum (Hook.) Kunze Takeda in Notes Roy. Bot. Gard. Edinburgh viii. 279. (1913—1915).

Yunnan: On the mountains to the north-west of Simao at 1,950 metres, epiphytic on trees, 13129.

Native to India, Nepal, Myanmar, China (Yunnan) and Thailand.

Lepisorus oligolepidus (Baker) Ching (POLYPODIACEAE) New species

Polypodium simplex Christ non Sw. in Bull. Herb. Boiss. vi. 875. (1895). *Polypodium oligolepidum* Baker in Notes Roy. Bot. Gard. Edinburgh 8: 276. (1913—1915).

Yunnan: On the mountains to the south-east of Mengzi, in woods at 1,950 metres, 10192.

Distributed in Himalaya, China, Indochina and Japan.

Lepisorus palmatopedatus (Baker) C. F. Zhao, R. Wei & X. C. Zhang (POLYPODIACEAE) New species

Polypodium palmatopedatum Baker in Kew Bull. Misc. Inf. 232. (1898). *Cheiropteris henryi* Christ in Bull. Herb. Boissier 6: 876. (1898). *Neocheiropteris henryi* (Christ)

Christ in Bull. Soc. Bot. France 5: 21. (1905).

Yunnan: Milê, 9289 (type of *Polypodium palmatopedatum* Baker, isotype of *Cheiropteris henryi* Christ).

Endemic to China and distributed in the western provinces of Guizhou, Sichuan and Yunnan.

Lepisorus rostratus (Bedd.) B. K. Noyar & S. Kaur (POLYPODIACEAE)

Polypodium subrostratum Christ Takeda in Notes Roy. Bot. Gard. Edinburgh viii. 311. (1913—1915).

Yunnan: Fengchunling, 13302.

Nepal to China (Taiwan) to northern Sumatra.

Lepisorus scolopendrium (Buch.-Ham. ex. D. Don) Mehra & Bir. (POLYPODIACEAE)

Polypodium excavatum Bory apud Willd. var. *concolor* Takeda in Notes Roy. Bot. Gard. Edinburgh viii. 281. (1913—1915).

Yunnan: In mountain forest to the north of Mengzi at 2,660 metres, epiphytic on trees, 10087.

Lepisorus sinensis (H. Christ) Ching (POLYPODIACEAE) New species

Neurodium sinense Christ in Bull. Herb. Boiss. 6: 880. (1898).

Yunnan: In Pingbian in mountain forest on the Daweishan Range at 1,650 metres, 10434. Simao, 13072.

Native to Bhutan, Myanmar, China (Yunnan) and Thailand.

Lepisorus subhemionitideus (H. Christ) C. F. Zhao, R. Wei & X. C. Zhang (POLYPODIACEAE) New species

Polypodium hymenodes Kunze var. *marginale* Takeda in Notes Roy. Bot. Gard. Edinburgh viii. 288. (1913—1915). *Polypodium subhemionitideum* H. Christ in Bull. Herb. Boiss. 7: 5. (1899). *Polypodium hymenodes* Kurze var. *sparisorum* Takeda in Notes Roy. Bot. Gard. Edinburgh viii. 287. (1913—1915).

Yunnan: Mengzi, 9265a., (syntype of *Polypodium hymenodes* Kunze var. *marginale* Takeda). On mountains to the east of Mengzi at 2,200 metres, 9265b., (isotype of *Polypodium subhemionitideum* H. Christ). Yuanjiang, 13340.

Distributed in Nepal, Myanmar, China and Vietnam.

Lepisorus sublinearis (Baker ex Takeda) Ching (POLYPODIACEAE) New species

Polypodium simplex Christ non Sw. in Bull. Herb. Boiss. 6: 875. non Sw. (1898). *Polypodium sublineare* Baker ex Takeda in Notes Roy. Bot. Gard. Edinburgh viii. 276. (1913—1915).

Yunnan: In mountain forest near Fengchunling at 2,950 metres, 9062a. On the mountains to the east of Mengzi at 1,950 metres, 11828. Yuanjiang at 1,950 metres, 13603 (syntype of *Polypodium sublineare* Baker ex Takeda).

Lepisorus superficialis (Blume) C. F. Zhao, R. Wei & X. C. Zhang (POLYPODIACEAE)

Polypodium superficiale Blume Henry in Trans. Asiat. Soc. Jap. 24 suppl: 115. (1896) (List Pl. Formosa), Christ in Bull. Herb. Boiss. 6: 874. (1898), Takeda in Notes Roy. Bot. Gard. Edinburgh 8: 293. (1913—1915). *Polypodium nigrocinctum* Christ in Bull. Herb. Boiss. 6: 874. (1898).

Taiwan: Wanchin, 1489. **Yunnan:** In the mountains to the south-west of Mengzi at 2,300 metres, 9264 (lectotype of *Polypodium nigrocinctum* Christ). On the mountains to the west of Simao at 1,650 metres, growing on rocks, 9264c. On a wooded cliff near Mengzi at 2,600 metres, 11454. In Pingbian, on the Daweishan Range at 1,650 metres, epiphytic on trees, 11454a.

An epiphytic or epilithic fern, native to Vietnam, China (including Taiwan) and Japan.

Lepisorus thunbergianus (Kaulf.) Ching (POLYPODIACEAE)

Polypodium lineare Thunb. Henry in Trans. Asiat. Soc. Jap. 24 suppl: 114. (1896) (List Pl. Formosa). *Polypodium lineare* Thunb. f. *thunbergianum* (Kaulf.) Takeda in Notes Roy. Bot. Gard. Edinburgh viii. 309. (1913—1915).

Taiwan: Wanchin, 1490. **Yunnan:** Milê district at 1,800 metres, 9893. Simao, 10062. Mengzi, 10087a.

A widespread fern, distributed from Pakistan to the south and east of China (including Taiwan), Japan and the Philippines.

Lepisorus triglossus (Baker) C. F. Zhao, R. Wei & X. C. Zhang (POLYPODIACEAE) New species

Polypodium triglossum Baker in Kew Bull. Misc. Inf. 232. (1898), Takeda in Notes Roy. Bot. Gard. Edinburgh viii. 295. (1913—1915). *Selliguea triphylla* H. Christ in Bull. Herb. Boissier 6: 878. (1898).

Yunnan: Milê, 9953 (type of *Polypodium triglossum* Baker, type of *Selliguea triphylla* H. Christ).

Endemic to Yunnan.

Lepistemon intermedius Hallier f. (CONVOLVULACEAE) New species

in Bot. Jahrb. Syst. 28: 32 (1901).

Ipomoea sp. Henry in Trans. Asiat. Soc. Jap. 24 suppl: 65. (1896) (List Pl. Formosa).

Taiwan: Wanchin, 1530 (isotype, collected 11th October 1893).

Leptoboea sp. (GESNERIACEAE)

Anonymous in Kew List of Determinations (volume 34).

Yunnan: Mengzi, 11498.

Leptochilus decurrens Blume (POLYPODACEAE)

Gymnopteris variabilis (Hook.) Bedd. Christ in Bull. Herb. Boiss. 6: 867. (1898).

Yunnan: Amidst rocks near Mengzi at 1,600 metres, 9246.

Leptochilus ellipticus (Thunb.) Noot. (POLYPODIACEAE)

Gymnogramma elliptica (Thunb.) Baker Baker in Journ. Bot. 26: 230. (1888), Henry in Trans. Asiat. Soc. Jap. 24 suppl: 116. (1896) (List Pl. Formosa), Christ in Bull. Soc. Bot. France 5: 21. (1905). *Polypodium ellipticum* Thunb. Diels in Bot. Jahrb. 29: 205. (1900). *Selliguea elliptica* (Thunb.) Christ Christ in Bull. Herb. Boiss. 6: 879. (1898). *Selliguea elliptica* Thunb. var. *flagellaris* Christ in Bull. Herb. Boiss. 7: 6. (1899).

Hubei: Antelope Glen, 3348. Xiling (Yichang) Gorge, 7885. **Taiwan:** Kaohsiung, without number. Wanchin, without number. **Yunnan:** Mengzi, 10769a. On the mountains to the south of Mengzi at 1,650 metres, 9259, 9259a. Simao, 13148.

Of widespread distribution in Asia.

Leptochilus henryi (Baker) X. C. Zhang (POLYPODIACEAE) New species

Colysis henryi (Baker) Ching in Bull. Fan. Mem. Inst. Biol. 4: 325. (1933). *Gymnogramma henryi* Baker in J. Bot. 25(6): 170. (1898). *Polypodium henryi* (Baker) Diels in Bot. Jahrb. 29: 204. (1900). *Selliguea henryi* (Baker) Christ in Bull. Herb. Boiss. 6: 879. (1898); in Bull. Soc. Bot. France 5: 21. (1905).

Hubei: Liantuo, 2114, (isotype of *Gymnogramma henryi* Baker). Antelope Glen, Yichang, 7880 (isotype of *Gymnogramma henryi* Baker). **Yunnan:** In Pingbian on the Daweishan Range in mountain forest at 1,650 metres, 10342.

Bangladesh, China and Vietnam.

Leptochilus pentaphyllus (Baker) Liang (POLYPODIACEAE) New species

Gymnogramma pentaphylla Baker in Kew Bull. Misc. Inf. 233.(1898).

Yunnan: Mengzi, 9033 (type, collected 16th December 1896). On mountains near Mengzi at 1,950 metres, 9033a., (type of *Gymnogramma pentaphylla* Baker).

A. Henry 9033a., also appears as *Polypodium mediosorum* Ching in Bull Fan. Mem. Inst. Biol. 2(2): 19 pl. 4. (1931)

Leptochilus pteropus subsp. **minor** (Bedd.) Fraser-Jenk. (POLYPODIACEAE)

Polypodium micropteris Baker in Kew Bull. 14: (1906), nom. illeg. *Polypodium pteropus* Takeda non Blume in Notes Roy. Bot. Gard. Edinburgh viii. 294. (1913—1915).

Yunnan: On the mountains to the east of Simao, in shade, at 1,950 metres, 12630 (type of *Polypodium micropteris* Baker).

Distributed across India, China, Vietnam, the

Philippines and Malaysia.

Leptochilus wrightii (Hook.) X. C. Zhang (POLYPODIACEAE)

Gymnogramma wrightii Hook. Henry in Trans. Asiat. Soc. Jap. 24 suppl: 116. (1896) (List Pl. Formosa).

Taiwan: Wanchin, 91. Tamsui, 1428. Wanchin, 1428a.

A widespread fern, often growing on moss covered rocks in streams. Native to Vietnam, southern China (including Taiwan) and Japan (south).

Leptodermis glomerata Hutchinson (RUBIACEAE) New species

in Sargent Pl. Wilson. 3: 406. (1917), Winkler in Fedde, Repert. Spec. Nov. Regni Veg. 18: 165. (1922).

Yunnan: In mountain forest near Milê, 9949 (type).

Leptodermis oblonga Bunge (RUBIACEAE)

Anonymous in Kew List of Determinations (volume 24).

Hubei: Liantuo, 6354. **Sichuan:** South Wushan, 7029.

Leptodermis tomentella Franchet ex H. Winkler (RUBIACEAE) New species

in Fedde, Repert. Spec. Nov. Regni Veg. 18: 159. (1922).

Leptodermis pilosa Hamilton non Diels in Sargent Pl. Wilson. 3: 405. (1917)

Yunnan: On grassy mountains near Mengzi at 1,640 metres, 10974 (isotype).

Leptodesmia microphylla (Thunb.) H. Ohashi & K. Ohashi (FABACEAE)

Zheng in Journ. Wuhan Bot. Res. 2.(1): 99. (1984).

Desmodium parvifolium DC. Forbes & Hemsley in Journ. Linn. Soc. Bot. 23: 174. (1887). *Desmodium parvifolium* DC. Anonymous in Kew List of Determinations (volume 34).

Hubei: Yichang, 321. Badong, 2501. By the monastery in the Monastery Valley near Yichang, 1658. **Yunnan:** Mengzi, 10143.

Leptogium menziesii Mont. f. *fuliginosum* Muell. Arg. (COLLEMATACEAE)

J. Muell.-Arg. in Bull. Herb. Boissier 1: 235. (1893), Wei in An Enumeration of Lichens in China 142. (1991).

Hubei: Donghu, 6441. Xingshan, 7633.

This lichen was also collected by the Glasnevin Central China Expedition (GCCE 485) on Hann Shao Shan in Xingshan in September 2004.

Leptopus chinensis (Bunge) Pojark. (PHYLLANTHACEAE)

Pax & Hoffmann in Engler, Pflanzenr. iv. 147. xv: 171. (1922).

Hubei: Changyang, 6221.

Leptopus cordifolius Decne. (PHYLLANTHACEAE)

Zheng in Journ. Wuhan Bot. Res. 2.(1): 110. (1984).

Andrachne cordifolia (Decne) Müell.Arg. Forbes & Hemsley in Journ. Linn. Soc. Bot.26: 420. (1894).

Hubei: Changyang, 5243.

Lepturus repens (G. Forst.) R. Br. (POACEAE)

Rendle in Journ. Linn. Soc. Bot. 36: 433. (1904).

Taiwan: Kaohsiung, 756, 1174, 1202.

Native to the seacoasts of Kenya, Sri Lanka, the Mascarenes to China (Taiwan) and Australia.

Lepturus sp. (POACEAE)

Anonymous in Kew List of Determinations (volume 34).

Yunnan: Mengzi, 13733.

Lespedeza buergeri Miq. (FABACEAE)

Zheng in Journ. Wuhan Bot. Res. 2.(1): 101. (1984).

Lespedeza bicolor Forbes & Hemsley non Turcz. in Journ. Linn. Soc. Bot. 23: 179. (1887).

Hubei: Yichang, 33, 400, 2050, 2273, 2880, 2933. Liantuo, 3047. Badong, 2880 (October 1886), 3167. Donghu, 6444.

This species is cultivated at the National Botanic Gardens, Glasnevin from collections made by the Glasnevin Central China Expedition in Xingshan (GCCE 497 & GCCE 571) in October 2004.

Lespedeza chinensis G. Don (FABACEAE)

Forbes & Hemsley in Journ. Linn. Soc. Bot. 23: 180. (1887).

Lespedeza canescens Rick. Zheng in Journ. Wuhan Bot. Res. 2.(1): 101. (1984).

Hubei: Yichang, 2746, 2277.

Lespedeza cuneata (Dum. Cours.) G. Don. (FABACEAE)

Zheng in Journ. Wuhan Bot. Res. 2.(1): 102. (1984)

Lespedeza juncea Pers. var. *sericea* Miq. Forbes & Hemsley in Journ. Linn. Soc. Bot. 23: 181. (1887).

Hubei: Yichang, 21, 251, 477, 884, 1955. On way to Moji Shan, 2227. Near Yichang, on the hills behind the San You Dong Glen (July 1886), 1509. Badong, 523 (September 1885), 2391, 2500, 2503, 5000. **Yunnan:** In the Red River valley near Manhao, 9528.

A perennial species native to east Asia and Australia. Sometimes cultivated as a forage crop.

Lespedeza daurica (Laxm.) Schindl. (FABACEAE)

Zheng in Journ. Wuhan Bot. Res. 2.(1): 101. (1984).

Lespedeza medicaginoides Bunge Forbes & Hemsley in Journ. Linn. Soc. Bot. 23: 182. (1887). *Lespedeza trichocarpa* (Stephan ex Willd.) Pers. Forbes & Hemsley in Journ. Linn. Soc. Bot. 23: 183. (1887).

Hubei: Yichang, 166, 400a., 2278, 2309.

Lespedeza inschanica (Maxim.) Schindl. (FABACEAE)

Zheng in Journ. Wuhan Bot. Res. 2.(1): 102. (1984).

Hubei: Yichang, 30.

Lespedeza juncea (L. f.) Pers. (FABACEAE)

Anonymous in Kew List of Determinations (volume 34).

Yunnan: Mengzi, 13764.

Native to Russia, Mongolia, China, Korea and Japan.

Lespedeza sp. No. 1. (FABACEAE)

Anonymous in Kew List of Determinations (volume 24).

Hubei: Yichang, 54, 739, 1326. Badong, 516, 528 (September 1885). Liantuo, 3944.

Lespedeza sp. No. 2. (FABACEAE)

Anonymous in Kew List of Determinations (volume 34).

Yunnan: Milê, 9889.

Lespedeza thunbergii (DC.) Nakai subsp. ***elliptica*** (Benth. ex Maxim.) H. Ohashi (FABACEAE)

Lespedeza elliptica Benth. ex Maxim. Forbes & Hemsley in Journ. Linn. Soc. Bot. 23: 180. (1887).

Hubei: Yichang, 2410. Badong, 5214.

This subspecies was collected by the Glasnevin Central China Expedition (GCCE 611 & GCCE 607) in Xingshan in October 2004.

Lespedeza tomentosa (Thunb.) Sieb. ex Maxim. (FABACEAE)

Schindler in Sargent Pl. Wilson. 2: 110. (1914), Zheng in Journ. Wuhan Bot. Res. 2.(1): 102. (1984).

Lespedeza villosa Pers. Forbes & Hemsley in Journ. Linn. Soc. Bot. 23: 183. (1887).

Hubei: Yichang, 2199. Liantuo (October 1886), 3076. **Yunnan:** Mengzi at 1,380 metres, 9815.

Native to Russia, India, China and Japan.

Lespedeza virgata (Thunb. ex Murray) DC. (FABACEAE)

Forbes & Hemsley in Journ. Linn. Soc. Bot. 23: 183. (1887), Zheng in Journ. Wuhan Bot. Res. 2.(1): 103. (1984).

Hubei: Yichang, 184, 618, 851, 1021, 2129, 2331. Near Yichang, on the hills behind the San You Dong Glen (July 1886), 1510. **Yunnan:** Simao, 11668.

An undershrub to about 60 cm tall, native to China, Korea and Japan.

Leucas chinensis (Retz.) Sm. (LAMIACEAE)

Leucas mollissima Forbes & Hemsley non Wall. in Journ. Linn. Soc. Bot. 26: 304. (1890), Henry in Trans. Asiat. Soc. Jap. 24 suppl: 73. (1896) (List Pl. Formosa), Dunn in Notes Roy. Bot. Gard. Edinburgh 6: 184. (1915).

Hainan: On hummocks near Haikou, 8114. **Taiwan:** Wanchin, 169. Oluanpi, 1235.

Leucas decemdentata (Willd.) Sm. (LAMIACEAE)

Leucas mollissima Wall. ex Benth. Dunn in Notes Roy. Bot. Gard. Edinburgh 6: 184. (1915).

Yunnan: Mengzi, 10029. Simao, 10029a. Widespread in Asia.

Leucas mollissima Wall. ex Benth. (LAMIACEAE)

Forbes & Hemsley in Journ. Linn. Soc. Bot. 26: 304. (1890), Dunn in Notes Roy. Bot. Gard. Edinburgh 6: 184. (1915).

Hubei: Yichang, 2349. Liantuo, 2599.

Leucas nubica Benth. (LAMIACEAE)

Leucas ciliata Hochst. ex Benth. Dunn in Notes Roy. Bot. Gard. Edinburgh 6: 184. (1915).

Yunnan: Mengzi at 1,650 metres, 10292, 10292a.

Native to India, Bhutan, Nepal, Myanmar, China, Laos and Vietnam.

Leucas zeylanica (L.) R. Br. (LAMIACEAE)

Forbes & Hemsley in Journ. Linn. Soc. Bot. 26: 304. (1890), Dunn in Notes Roy. Bot. Gard. Edinburgh 6: 184. (1915).

Hainan: Without locality, 8509.

Leucomeris decora Kurz (ASTERACEAE)

Anonymous in Kew List of Determinations (volume 34).

Yunnan: Simao, 11676, 12888.

Distributed in India (Assam), Myanmar, China (Yunnan), Thailand and Vietnam.

Leucosceptrum canum Sm. (LAMIACEAE)

Dunn in Notes Roy. Bot. Gard. Edinburgh 6: 193. (1915).

Yunnan: Mengzi at 1,950 metres, 9397, 9397a.

An evergreen shrub or a small tree to about 7 metres tall. A relitavely common tree in the Himalaya, Myanmar, western China, Laos and Vietnam.

Leucostegia immersa C. Presl. (DAVALLIACEAE)

Davallia immersa (C. Presl.) Wall. Anonymous in Kew List of Determinations (volume 34).

Yunnan: Mengzi, 10083.

Distributed across India, southern China, the Philippines and Indonesia.

Leycesteria formosa Wallich (CAPRIFOLIACEAE)

Rehder in Sargent Pl. Wilson. 1: 311. (1913).

Leycesteria formosa Wall. var. *brachysepala* Airy Shaw in Kew Bull. Misc. Inf. 169. (1932). *Leycesteria sinensis* Hemsley in Hooker's Icon. Pl. t. 2633. (1901), Hemsley in Journ. Linn. Soc. Bot. 34: 477. (1900); in Gard. Chron. ser. 3, 30: 85. (1901), Rehder in Sargent Pl. Wilson. 1: 312. (1913), in obs. Airy-Shaw in Kew Bull. Misc. Inf. 166. (1932), O' Brien in Augustine Henry, An Irish Plant Collector in China 22. (2002).

Yunnan: South of the Red River from Manmei at 1,800 metres, 9692 (isosyntype of *Leycesteria formosa* Wall. var. *brachysepala* Airy Shaw). Mountains to the north of Mengzi at 1,650 metres, 9692a., 9692b., (type of *Leycesteria formosa* Wall. var. *brachysepala* Airy Shaw). Mountains north of Mengtse at 2,300 metres, 9692c.

The pheasant berry or Himalayan honeysuckle. Naturalised in parts of Ireland.

Leycesteria gracilis (Kurz) Airy-Shaw (CAPRIFOLIACEAE)
Airy Shaw in Kew Bull. Misc. Inf. 174. (1932).

Leycesteria glaucophylla Rehder non (Hook. f. & Thoms.) Hook. f. in Sargent Pl. Wilson. 1: 312. (1913).

Yunnan: South of the Red River from Manmei at 1,800 metres, 9767.

Distributed in Bhutan, Myanmar, China [Tibet (Xizang) and Yunnan].

Leymus duthiei (Stapf ex Hook. f.) C. Yen, J. L. Yang & B. R. Baum (POACEAE)
Asperella duthiei Stapf ex Hook. f. in Hooker's Icon. Pl. t. 2525. (1897), Rendle in Journ. Linn. Soc. Bot. 36: 434. (1904).

Hubei: Jianshi, 5918.

New to China, found shortly before this collection by Duthie in India.

Ligularia dentata (A. Gray) H. Hara (ASTERACEAE)
Senecio clivorum Maxim. Forbes & Hemsley in Journ. Linn. Soc. Bot. 23: 451. (1888).

Hubei: Badong, 524 (September 1885), 2451, 5207. Xingshan, 6457. Fang Xian, 6457a.

A handsome plant with bold, striking foliage and handsome orange-yellow flowers. Introduced to cultivation by Wilson from Hubei through Messrs Veitch who first flowered it in their Coombe Wood nursery in July 1901. This species was collected by the Glasnevin Central China Expedition (GCCE 513) in Xingshan in September 2004.

Ligularia hookeri (C. B. Clarke) Handel-Mazzetti. (ASTERACEAE)
In the Harvard University Herbarium (GH.).

Senecio clivorum Forbes & Hemsley non Maxim. in Journ. Linn. Soc. Bot. 23: 451. (1888)

Sichuan: South Wushan, 7459.

Ligularia japonica (L. f.) Less. ex DC. (ASTERACEAE)
Senecio japonicus (L. f.) Schultz-Bip. Forbes & Hemsley in Journ. Linn. Soc. Bot. 23: 453. (1888).

Hubei: Yichang, on a moist bank in the Pingshanpa glen, off the Yichang (Xiling) Gorge, 3971, 4390.

Ligularia przewalskii (Maxim.) Diels (ASTERACEAE)
Senecio przewalskii Maxim. Anonymous in Kew List of Determinations (volume 24).

Sichuan: Kangding, Henry's Chinese collector with Pratt, 8918.

Described as a new species from a collection made in Mongolia by Przewalski, Bretschneider states that Père Armand David had found it several years previously.

Ligularia sibirica (L.) Cass. (ASTERACEAE)
Senecio ligularia Hook. f. Forbes & Hemsley in Journ. Linn. Soc. Bot. 23: 454. (1888), Henry in Notes Econ. Bot. China 29. (1893).

Hubei: Xingshan, 6459, 6470. Fang Xian, 6622. Without locality, 6470a., 6470b.

In the mountains of Hubei Henry stated this species was called *hu-lu-pao-yeh*.

Ligularia (Cacalia) sp. (ASTERACEAE)
Anonymous in Kew List of Determinations (volume 24).

Hubei: Badong, 6076. Fang Xian, 6826. **Sichuan:** South Wushan, 5689.

Ligularia veitchiana (Hemsl.) Greenm. (ASTERACEAE)
New species
in Stand. Cycl. Hort. 6: 3153. (1917).

Ligularia sp. Henry in Journ. China Br. Roy. Asiat. Soc. 22: 243. (1887) (Chinese Names of Plants). *Senecio ligularia* Hook. f. var. *speciosa* Hemsl. Hemsley in Journ. Linn. Soc. Bot. 23: 454. (1888). *Senecio veitchianus* Hemsley in Gard. Chron. ser. 3, 38: 212. (1905), Fedde in Repert. Spec. Nov. Regni Veg. 4: 173. (1907).

Hubei: Badong, 4889. Xingshan, 6460.

An herbaceous perennial to 2 metres tall, with bold foliage and erect spikes of yellow flowes in August and September. Henry stated that at Badong this species was known as *ch'ing ho yeh* and was common in marshes between 1,300 and 1,900 metres. Discovered by Henry in 1887 and later introduced to cultivation by Wilson through his employers, the nurserymen, Messrs Veitch, for whom this species was named. Distributed in western and central China.

Ligularia wilsoniana (Hemsl.) Greenm. (ASTERACEAE)
New species
Senecio ligularia Hook. f. var. *polycephalus* Hemsley in Journ. Linn. Soc. Bot. 23: 455. (1888). *Senecio wilsonianus* Hemsley in Gard. Chron. ser. 3, 38: 212. (1905), Anon. in Gard. Chron. ser. 3, 42: 201. (1907).

Hubei: Badong, 111, 2457.

Discovered in 1885 by Henry and named in compliment of Ernest Wilson who introduced it through Veitch's Coombe Wood nursery. Distributed in Hubei and Sichuan.

Ligusticum daucoides Franchet (APIACEAE)
Trachydium daucoides Franchet Diels in Bot. Jahrb. 29: 499. (1900).

Hubei: Fang Xian, 6934.

Ligustrum compactum Hook. f. & Thoms. apud Brandis (OLEACEAE)
Rehder in Sargent Pl. Wilson. 2: 604. (1916).

Yunnan: Mountains to the east of Mengzi at 2,000 metres, 9968.

A deciduous shrub to 4 metres tall, native to the northwest Himalaya and China (Yunnan). It was first introduced to cultivation from the Himalaya in 1874 and was later raised in Paris from seeds sent there in 1888 by Père Armand David.

Ligustrum confusum Decaisne (OLEACEAE)

Green in Kew Bull. 50: 382. (1995).

Yunnan: Mengzi, 9353. Simao, 12916.

An evergreen tree (deciduous in colder climates) to 13 metres tall, native to the Himalaya and central and western China and introduced to cultivation via the Royal Botanic Gardens, Kew from Calcutta in 1919.

Ligustrum expansum Rehder (OLEACEAE) New species

in Sargent Pl. Wilson. 2: 600. (1916), Zheng in Journ. Wuhan Bot. Res. 2.(1): 180. (1984).

Hubei: Jianshi, 5908 (holotype). Without locality, 5908a.

Ligustrum henryi Hemsley (OLEACEAE) New species

in Journ. Linn. Soc. Bot. 26: 90. (1889), Koehne in Festschr. Siebzig. Geburtst. Ascherson. 203. fig. 4a. (1904), J. H. Veitch in Hortus Veitchii 402. (1906), Schneider in Illus. Handb. Laubholzk. 2: 808. fig. 509.-510. (1911), Rehder in Sargent Pl. Wilson 2: 601. (1916), Morley in Glasra 3: 79. (1979), Zheng in Journ. Wuhan Bot. Res. 2.(1): 180. (1984), Bean in Trees & Shrubs 2: 570. (1992), Nelson in A Heritage of Beauty 314. (2000), O' Brien in Augustine Henry, An Irish Plant Collector in China 22. (2002).

Hubei: Antelope Glen, 3124, 3310, 3575 (type).

An evergreen shrub to 4 metres tall, of neat habit, bearing white fragrant flowers followed by black fruits. Introduced to cultivation by Wilson when collecting for Messrs Veitch in 1901. Veitch's catalogue *New Hardy Plants from Western China* (Autumn 1912) had this species available for two shillings and six pence. Henry and his wife Alice grew *Ligustrum henryi* in their garden at Ranelagh in the south Dublin suburbs. In Ireland it is currently cultivated at the John F. Kennedy Arboretum and at Rowallane. *Ligustrum henry* was collected by the Glasnevin Central China Expedition (GCCE 677) in the Red Dragon Reserve in Xingshan (where it is locally abundant) in October 2004).

Ligustrum leucanthum (S. Moore) P. S. Green (OLEACEAE) New species

Green in Kew Bull. 50: 384. (1995).

Ligustrum ibota Henry non Sieb. in Journ. China Br. Roy. Asiat. Soc. 22: 252. (1887) (Chinese Names of Plants), Forbes & Hemsley in Journ. Linn. Soc. Bot. 26: 91. (1889). *Ligustrum acutissimum* Koehne in Urb. & Graebn. Festschr.

Aschers. 201. (1904), Koehne in Fedde, Repert. Spec. Regni. Veg. Nov. 1: 8. fig. 1a. (1905), Schneider in Illus. Handb. Laubholzk. 2: 806.fig. 505. (1911), Rehder & E. H. Wilson in Sargent Pl. Wilson 2: 600. (1916), Zheng in Journ. Wuhan Bot. Res. 2(1): 179. (1984). *Ligustrum molliculum* Zheng non Hance in Journ. Wuhan Bot. Res. 2.(1): 181. (1984).

Hubei: Badong, 612 (July 1885), 1873, 2869, 3165. Fang Xian, 6583. Jianshi, 7447, 5881 (holotype of *Ligustrum acutissimum* Koehne, isotype of *Ligustrum leucanthum* (S. Moore) P. S. Green). North Badong, 7158.

Ligustrum lucidum W. T. Aiton (OLEACEAE)

Henry in Journ. China Br. Roy. Asiat. Soc. 22: 252. (1887) (Chinese Names of Plants), Forbes & Hemsley in Journ. Linn. Soc. Bot. 26: 92. (1889), Rehder in Sargent Pl. Wilson 2: 603. (1916), Zheng in Journ. Wuhan Bot. Res. 2.(1): 180. (1984).

Hubei: Yichang, 579, 970, 1073, 1396. Liantuo, 2603. Changleping, 6217. **Yunnan:** Simao, 12659.

According to Henry this was the commonest tree in Yichang where it formed small trees between 7 to 10 metres high and was known locally as *la shu*. It was also once used in the breeding of the white wax insect, *Ericerus pela* (formerly *Coccus pela,* belongs to the scale insect family Coccidae). Native to China from where it was introduced in 1794. In Chinese herbal medicine this species is greatly valued. The fruits are said to aid kidney function, strengthen muscles, relieve rheumatism and prevent hair loss. It is also claimed to prevent bone marrow loss in chemotherapy patients as well as being a potential source of treatment for HIV. An oil infusion from the flowers is also used to treat sun burn. *Ligustrum lucidum* was collected by the Glasnevin Central China Expedition on Moji Shan near Yichang (GCCE 729) and in Dragon Gate River Forest Park in Xingshan (GCCE 866) in Ovctober 2004.

Ligustrum pricei Hayata (OLEACEAE) New species

Li in Woody Flora of Taiwan 762. (1963), Bean in Trees & Shrubs 3: 213. (1936), Nelson in A Heritage of Beauty 314. (2000).

Ligustrum sp. Henry in Trans. Asiat. Soc. Jap. 24 suppl: 59. (1896) (List Pl. Formosa). *Ligustrum formosanum* Rehder in Sargent Pl. Wilson. 2: 608. (1916), O' Brien in Augustine Henry, An Irish Plant Collector in China 22. (2002). *Ligustrum pedunculare* Rehder in Sargent Pl. Wilson 2: 609. (1916), Zheng in Journ. Wuhan Bot. Res. 2.(1): 181. (1984).

Hubei: South Badong, 5338. **Sichuan:** North Wushan, 7117. **Taiwan:** Wanshoushan, 331 (type of *Ligustrum formosanum* Rehder).

A shrub from one to three metres tall, native from

Central China to Taiwan where it grows in broad-leaved forests below 1,800 metres. Discovered by Henry in 1893 and described from material gathered by Price in 1912 at Pu-li (formerly Horisha) in Nantou Co. Introduced to cultivation by Wilson.

Ligustrum quihoui Carr. (OLEACEAE)

Rehder in Sargent Pl. Wilson. 2: 607. (1916), Zheng in Journ. Wuhan Bot. Res. 2.(1): 181. (1984).

Ligustrum brachystachyum Decne. Forbes & Hemsley in Journ. Linn. Soc. Bot. 26: 389. (1889).

Hubei: In glens near Yichang, 215, 3973, 4372. Changyang, 7744. In the San You Dong Glen (June 1886), 1631. Liantuo, 1937, 6352. South Donghu, 7697. **Yunnan:** Mengzi, 10184.

A deciduous shrub to 3.5 metres tall, flowers white and fragrant and produced in late autumn. The best of the privets in terms of floral effect. In 1908 seeds of Wilson's 631 and 780 from Fang and his 778 from Badong arrived at Glasnevin from the Arnold Arboretum. This species commemorates M. Quihou, superintendent of the Jardin d'Acclimatation in Paris where it was first cultivated having been sent there by Gabriel Eugene Simon in 1862. In Chengdu while travelling on the Glasnevin Central China Expedition in 2002 we encountered a very dwarf form in the Chengdu Bamboo Garden that was used to make low hedges to 20 cm tall. This cultivar has not yet reached the west but has great potential and might be used in the same way as *Buxus sempervirens* 'Suffruticosa'.

Ligustrum robustum (Roxb.) Blume (OLEACEAE)

Anonymous in Kew List of Determinations (volume 24).

Sichuan: Henry's Chinese collector with Pratt, no locality stated but may be Leshan, Emei Shan, Wa Shan or Kangding, 8851.

Ligustrum sinense Lour. (OLEACEAE)

Henry in Journ. China Br. Roy. Asiat. Soc. 22: 252. (1887) (Chinese Names of Plants), Forbes & Hemsley in Journ. Linn. Soc. Bot. 26: 92. (1889), Rehder in Sargent Pl. Wilson 2: 605. (1916), Zheng in Journ. Wuhan. Bot. Res. 2.(1): 605. (1916).

Ligustrum sinense Lour. var. *stauntonii* (DC.) Rehd. Rehder in Sargent Pl. Wilson. 2: 606. (1916). *Ligustrum deciduum* Hemsley in Journ. Linn. Soc. Bot. 26: 90. (1889), Bretschneider in Hist. Eur. Bot. Disc. China 2: 787. (1898). *Ligustrum sinense* Lour. var. *nitidum* Rehder in Bailey, standard. Cycl. Hort. iv. 1700. (1915), Rehder in Sargent Pl. Wilson 2: 606. (1916), Zheng in Journ. Wuhan Bot. Res. 2.(1): 181. (1984).

Hubei: Yichang, 906, 2955, 3549, 3549b., 3619. Antelope Glen, 3320. South Badong, 7299. Xiling (Yichang) Gorge, 3619, 4171. Liantuo, 2683, 1922, 3072, 3190, 3904, 4403. Without locality, 2955a., 3549a. Badong, 1763. Lung wang tung valley near Yichang, 3561.

According to Henry this species was known colloquially in the Three Gorges region as *shan la shu*. He stated that his 906 was reported to be a tree on which the white wax-insect lived.

Ligustrum sinense Lour. var. *chayuense* (P. Y. Pai) Y. F. Deng & B. Q. Xu (OLEACEAE)

Ligustrum rugosulum W. W. Smith in Notes Roy. Bot. Gard. Edinburgh 9: 44. (1916).

Yunnan: In forests to the south-east of Simao at 1,300 metres, 12803. In the mountains to the south of Mengzi at 1,650 metres 12804b., (syntype of *Ligustrum rugosulum* W. W. Smith).

Distributed in China (Yunnan, Tibet (Xizang)) and Vietnam.

Ligustrum sp. No. 1. (OLEACEAE)

Anonymous in Kew List of Determinations (volume 24).

Hubei: Badong, 715.

Ligustrum sp. No. 2. (OLEACEAE)

Anonymous in Kew List of Determinations (volume 34).

Yunnan: Mengzi, 9659, 10183. Simao, 12918.

Ligustrum strongylophyllum Hemsley (OLEACEAE) New species

in Journ. Linn. Soc. Bot. 26: 93. (1889), Bretschneider in Hist. Eur. Bot. Disc. China 2: 787. (1898), Schneider in Illus. Handb. Laubholzk. 2: 799. (1912), Rehder in Sargent Pl. Wilson 2: 605. (1916), Zheng in Journ. Wuhan Bot. Res. 2.(1): 181. (1984), O' Brien in Augustine Henry, An Irish Plant Collector in China 25. (2002).

Hubei: Moji Shan (June 29th 1886) 1559 (isosyntype). Yichang, 2029, 3104 (type). Badong, 1286. Changleping, 6299.

A small handsome small tree in its native habitat, though normally seen as a tall shrub in cultivation. Described from Augustine Henry's collections, it had apparently been previously collected by Charles Maries for Veitch's nursery in the Xiling Gorge in 1879. In Ireland this species is cultivated at the National Botanic Gardens, Glasnevin and at the John F. Kennedy Arboretum.

Lilium amoenum E. H. Wilson ex Sealy (LILIACEAE) New species

in Tr. B. S. Edinburgh 28: 159. (1922), Sealy in Bot. Mag. t. 73. (1949), Woodcock & Stearn in Lilies of the World 140. (1950).

Lilium yunnanense Henry non Franchet in Journ. Roy. Hort. Soc. 25: 349. (1901—1902), Wright in Journ. Linn.

Soc. Bot. 36: 136. (1903). *Lilium sempervivoideum* E. H. Wilson non Lev. in Lilies of Eastern Asia 44. (1925).

Yunnan: Mengzi at 1,950 metres, 10743 (syntype).

A beautiful little lily, bearing, during the months of May and June, sweetly scented, pendant pink-purple bell-shaped flowers with fine purple blothes on the inside. Endemic to Yunnan. Leveille based this species on material gathered by Hancock (1895), Henry (1897) and Père E. E. Maire (June 1911). It had been first found by William Hancock on April 24th 1895 and the following year by Henry. *Lilium amoenum* was first flowered by Lord Aberconway from bulbs collected in the region of Lijiang by George Forrest.

Lilium bakerianum Collet & Hemsley (LILIACEAE)

Henry in Journ. Roy. Hort. Soc. 25: 349. (1901—1902), Wright in Journ. Linn. Soc. Bot. 36: 128. (1903) E. H. Wilson in Lilies of Eastern Asia 42. (1925).

Yunnan: Simao at 1,900 metres, 13026.

Previously only known from Myanmar, the discovery of this species in southern Yunnan proved the theory advanced by Augustine Henry that the flora of Simao was closely related to that of Myanmar.

Lilium bakerianum Collett & Hemsley var. *delavayi* (Franch.) E. H. Wilson (LILIACEAE)

E. H. Wilson in Lilies of Eastern Asia 43. (1925).

Lilium pseudotigrinum Henry non Carriere in Journ. Roy. Hort. Soc. 25: 349. (1901—1902), Wright in Journ. Linn. Soc. Bot. 36: 133. (1903).

Yunnan: On grassy mountains near Mengzi at 2930 metres, 10774b.

Discovered by Delavay in 1888 in north-west Yunnan. This handsome lily was introduced to cultivation by Forrest from the Cangshan Range above Dali in north-west Yunnan in 1911 and was also collected by Schneider in nearby Lijiang in 1914. Distributed in Myanmar and western China.

Lilium bakerianum Collett & Hemsley var. *rubrum* Stearn (LILIACEAE)

in Gard. Chron. ser. 3, cxxiv. 4. (1948).

Lilium lowii Henry non Baker in Journ. Roy. Hort. Soc. 25: 349. (1901—1902), Wright in Journ. Linn. Soc. Bot. 36: 132. (1903). *Lilium bakerianum* Forbes & Hemsley non Collett & Hemsley in Journ. Linn. Soc. Bot. 36: 132. (1903), E. H. Wilson in Lilies of Eastern Asia 43. (1925).

Yunnan: Mengzi at 1,600 to 2,200 metres, 10774, 10774a.

The eastern most variety of *Lilium bakerianum*. It was discovered by the Irish collector, William Hancock near Mengzi and is endemic to western China (Guizhou and Yunnan).

Lilium brownii F. E. Brown ex Miellez (LILIACEAE)

Henry in Journ. China Br. Roy. Asiat. Soc. 22: 265. (1887) (Chinese Names of Plants), Henry in Journ. Roy. Hort. Soc. 25: 346. (1901—1902), Wright in Journ. Linn. Soc. Bot. 36: 128. (1903).

Lilium brownii F. E. Br. ex Miell. var. *colchesteri* (Van Houtte) E. H. Wils. ex Elwes E. H. Wilson in Lilies of Eastern Asia 31. (1925)

Hubei: In glens near Yichang, 514, 4160. Liantuo, 2047, 2047a. **Yunnan:** Mengzi at 1,250 to 2,250 metres, 9453, 9453a., 9453b., 9453c.

The musk lily, a very beautiful species to 1.5 metres tall, producing 2 to 3 large fragrant white pendelous flowers in a terminal umbel in early summer. A widespread trumpet lily in China and known as *peh-ho* according to Wilson who also stated it was often cultivated for its bulbs which were esteemed as a table delicacy (several species of *Lilium* are eaten in China). He also stated that this lily was a feature of the glens leading off the Three Gorges and this is where Augustine collected some of his herbarium material. Endemic to China and distributed in Gansu, Henan, Hubei, Yunnan, Quinghai and Shaanxi.

Lilium callosum Sieb. & Zucc. (LILIACEAE)

Henry in Journ. Roy. Hort. Soc. 26: 346. (1901—1902), Wright in Journ. Linn. Soc. Bot. 36: 129. (1903), E. H. Wilson in Lilies of Eastern Asia 87. (1925).

Hubei: Yichang, 511, 2327.

Originally described from Nagasaki, this lily is also native to China and Korea. Wilson later collected it near Yichang.

Lilium davidii Duchartre var. *wilmottiae* (Wils.) Raffil. (LILIACEAE) New variety

O' Brien in Augustine Henry, An Irish Plant Collector in China 25. (2002).

Lilium willmottiae E. H. Wilson in The Lilies of Eastern Asia 67. (1925), Nelson in A Heritage of Beauty 315. (2000).

Hubei: without locality.

According to Wilson this graceful lily was discovered by Henry in western Hubei in 1888 and was later collected in north-eastern Sichuan by Farges in 1892. It was introduced to cultivation by Giraldi who sent bulbs to the Florence Botanic Gardens in 1895, these flowered in the summer of the same year. Farges sent seeds to Messrs. Vilmorin in 1896, some of these seedlings were sent to Kew where they flowered in 1899. In 1908—1909 Wilson sent bulbs to the Arnold Arboretum and to Miss Ellen Willmott. This consignment proved to be of a more vigorous stock and fared better in cultivation than the previous introductions. The specific epithet commemorates Miss Ellen Ann

Willmott of Great Warley, Essex (1858—1934), a friend of Henry and a sponsor of Wilson's later expeditions to China. This lily is cultivated in several Irish gardens.

Lilium fargesii Franch. (LILIACEAE) New species

Henry in Journ. Roy. Hort. Soc. 25: 438. (1901—1902), E. H. Wilson in Lilies of Eastern Asia 85. (1925).

Lilium tenuifolium Wright non Fisch. ex Hook. f. in Journ. Linn. Soc. Bot. 36: 135. (1903).

Hubei: Jianshi, 5917. Fang Xian, 6786.

A rare lily in western Hubei and eastern Sichuan. It was first found by Henry in the mountains to the north of Yichang in 1885 and was later collected by Père Paul Farges for whom it was named. This was the only lily that Wilson saw in China and did not introduce. Henry's Jianshi collection dates from May of 1888.

Lilium formosanum (Baker) Wallace (LILIACEAE)

Lilium longiflorum Henry non Thunb. in Trans. Asiat. Soc. Jap. 24 suppl: 97.(1896) (List Pl. Formosa); in Journ. Roy. Hort. Soc. 26: 348. (1901—1902). *Lilium philippinense* Baker var. *formosanum* (Bak.) E. H. Wilson apud Groves E. H. Wilson in Lilies of Eastern Asia 21.(1925).

Taiwan: Oluanpi, 927, 1569.

A beautiful lily to about 40 cm tall carrying large, delightfully fragrant white flowers, striped reddish-purple the length of petals on the reverse. Endemic to Chinese Taiwan where it is very common throughout the island. Wilson stated he never collected this species during his travels in Taiwan. He also stated it was common in the north of Taiwan but was much less frequent on the coral formations in the south from where Henry obtained his material. It was from Henry's Oluanpi collection that he based his description in *The Lilies of Eastern Asia*. Discovered in June 1858 by Charles Wilford in northern Taiwan, *Lilium formosanum* was introduced to cultivation by Messrs. Veitch in 1880 by their collector Charles Maries. It is a handsome lily, easily raised from seed and from seed it blooms within 6 months. It is most often seen in cool alpine house conditions.

Lilium henryi Baker (LILIACEAE) New species

in Gard. Chron. ser. 3, 4: 660.(1890) & ii. 380. fig. 75. (1890); in Gard. Chron. ser. 3, 8: 380. fig. 75. (1890); in Curtis's Bot. Mag. t. 7177. (1891), Watson in The Garden xl. 422. (1891), Bretschneider in Hist. Eur. Bot. Disc. China 2: 793. (1898), Watson in Garden lv. 233. (1899), Gertrude Jekyll in Lilies English Gard. 23. (1901); in Journ. Hort. ser. 3. xlii. 131. (1901), Henry in Journ. Roy. Hort. Soc. 25: 346. fig. 180. (1901—1902); in Garden lxii. 284. (1902), Mottet in Rev. Hort. 231. fig. 97. (1903), Wright in Journ. Linn. Soc. Bot. 36: 131. (1903), Grove in Lilies 56. (1911), E. H. Wilson in A Naturalist in Western China 1: 21. (1913), Tilton in Bailey, Stand. Cycl. Hort. 4: 1871. fig.

2163. (1916), Boynton in Addisonia 4: 65. t. 153. (1919), Dillistone in Garden lxxxvii. 102. (1923), Stout in Gard. Chron. ser. 3, lxxv. 256. figs. 107, 108. (1924), E. H. Wilson in Lilies of Eastern Asia 78. (1925), Stern in Journ. Roy. Hort. Soc. lvii. 292. (1932), Besant in Gard. Chron. ser. 3, 98: 334. (1935), Morley in Glasra 3: 79. (1979), Nelson in Pim, The Wood & the Trees 222. (1984); in A Heritage of Beauty 314. (2000), O' Brien in Augustine Henry, An Irish Plant Collector in China 25. (2002), Oakeley in The Garden 129(2): 103. (2004), O' Brien in Ir. Garden 13(5): 53. (2004).

Hubei: On Moji Shan, 4205. Yichang 4378, 4245 (type).

To my mind this lovely lily was Augustine Henry's most sensational discoveries. Henry found this plant in 1887 on the limestone cliffs of the Xiling (Yichang) Gorge and a few scattered plants on Moji Shan, a pyramidal-shaped hill known to 19th century European residents at Yichang as 'the Dome.' Moji Shan is located on the opposite side of the Yangtze from the former site of Customs building and foreign compound, it is a clearly visible landmark in Yichang even today, though the last buillding of the foreign compound was demolished as the Glasnevin expedition members stayed in Yichang in October 2002.

In the Xiling Gorge Henry stated it was common on the right bank of the Yangtze between the villages of Ping-Shan-pa and Shih-pi-shan and on the grassy slopes of limestone cliffs inland from Shih-pi-shan from which a path led up top a taoist monastery called *Yang-tai-kuan*. In March of 1889 before he departed Yichang for Hainan Henry collected bulbs from the gorges and gave them to Charles Ford of Hong Kong Botanic Gardens with instructions to forward them to Kew (Kew had requested bulbs of several lilies). At Kew, *Lilium henryi* flowered for the first time in cultivation in the summer of the same year.

According to Wilson, Henry's lily had been virtually eliminated from the type locality by the time of his arrival in 1900. In the wild it grows from 0.75 to 1.5 metres tall bearing 1 to 6 flowers, 2 to 4 flowers being the usual number. In cultivation it performs far better and is one of the beautiful of all Asiatic lilies, thriving in alkaline soils and hating a peat or acid soil. As Wilson was to point out:

It is peculiarly fitting that such a noble addition to our gardens should bear the honoured name of a pioneer who has done so much to acquaint a sceptical world of the rich floral wealth of interior China, - Professor Augustine Henry.

On his return to Ireland, Henry and his wife Alice grew this species in their garden at Ranelagh in the south Dublin suburbs.

Lilium lancifolium Thunb. (LILIACEAE)

Lilium tigrinum Ker-Gawl. Henry in Journ. China Br.

Roy. Asiat. Soc. 22: 265. (1887) (Chinese Names of Plants), Wright in Journ. Linn. Soc. Bot. 36: 135. (1903).

Hubei: Badong, 2397. Yichang, cultivated 4140.

According to Henry this species was cultivated at Yichang for its edible bulbs and was known colloquially as *chia peh ho*. The dried flowers were used by the Chinese for flavouring soups and as a remedy for pulmonary disease. *Lilium lancifolium* was collected by the Sino-American Expedition in the shennongjia Forest District in 1980.

Lilium leucanthum (Baker) Baker (LILIACEAE) New species

in Journ. Roy. Hort. Soc. 25: 337. (1901—1902), E. H. Wilson in Lilies of Eastern Asia 39. (1925).

Lilium longiflorum Thunb. var. *chloraster* Baker in Gard. Chron. 10: 66. (1891), *Lilium chloraster* (Bak.) E. H. Wilson in Journ. Roy. Hort. Soc. xlii. 36. (1916). *Lilium leucanthum* (Bak.) Baker var. *chloraster* (Bak.) E. H. Wilson in Lilies of Eastern Asia 41. (1925). *Lilium brownii* Baker non F. E. Brown ex Milliez. in Gard. Chron. ser. 3, 10: 225. (1895), Bretschneider in Hist. Eur. Bot. Disc. China 2: 793. (1898), Baker in Journ. Roy. Hort. Soc. 26: 338. (1901—1902), Watson in Garden xlvii. 97. (1895); in Gard. Chron. ser. 3, 32: 236. (1902), Wright in Journ. Linn. Soc. Bot. 36: 159. (1903), Nelson in A Heritage of Beauty 314. (2000). *Lilium brownii* F. E. Brown ex Miellez. var. *leucanthemum* Baker in Gard. Chron. ii. 180. (1894) ii. 180. (1894), Bretschneider in Hist. Eur. Bot. Disc. China 2: 793. (1898), Henry in Journ. Roy. Hort. Soc. 25: 346. (1901—1902), Wright in Journ. Linn. Soc. Bot. 36: 128. (1903), Nelson in Pim, The Wood & the Trees 221. (1984).

Lilium centifolium Stapf ex Elwes in Gard. Chron. ser. 3, 70: 101. (1921), Stern in Journ. Roy. Hort. Soc. lvii. 288. (1932), Farrer collections from Gansu, not A. Henry 4162.

Hubei: Yichang, without number. Antelope Glen, 4162.

This handsome lily is native to the mountains of western Hubei and Sichuan, reaching its western limits at the Chengdu plain (red basin). It was discovered in the glens off the Xiling (Yichang) Gorge by Henry and in 1888 he sent small bulbs of it to Kew on May 7th 1889. These first flowered there in 1891 (and were mistakenly identified as the Gansu native, *Lilium leucanthum* var. *centifolium*). It was crossed at Kew in July 1897 with *Lilium henryi*, the resulting seedlings first flowered in 1900 and were named *Lilium* x *kewense* Watson. The flowers were said to be creamy-buff changing to almost pure white with age. According to Wilson this hybrid had became extinct by 1925 though he merited it highly and believed that it should be raised anew. In 1900 and 1901 Wilson sent consignments of this lily to Messrs Veitch, these flowered in September 1902. Again he sent bulbs in 1908 and 1910 to the Arnold Arboretum.

Lilium primulinum Baker var. *ochraceum* (Franch.) Stearn (LILIACEAE)

Lilium nepalense Henry non D. Don in Journ. Roy. Hort. Soc. 25: 350. fig. 182. (1901—1902), Wright in Journ. Linn. Soc. Bot. 36: 133. (1903). *Lilium ochraceum* Franch. E. H. Wilson in Lilies of Eastern Asia 62. (1925).

Yunnan: Mengzi at 1,700 to 2, 300 metres, 9320.

A handsome lily from 40 to 200 cm tall, carrying pendant racemes (or in some cases, solitary) yellow-green flowers with reflexed petals in July and August. Northern Indochina and China (Yunnan, Guizhou and Sichuan) where the bulbs are used medicinally. Père Delavay first found this beautiful lily in north-west Yunnan. In 1894 it was gathered near Mengzi in Yunnan by William Hancock and two years later it was recollected in the same region by Augustine Henry. In the autumn of 1899 while visiting Henry, Ernest Wilson sent back about a hundred bulbs to Messrs. Veitch from the same region though it had been previously introduced from Myanmar by William Boxhall who sent bulbs to Messrs. Hugh Low & Co in 1888.

Lilium sulphureum Baker ex Hook. f. (LILIACEAE)

Woodcock and Stearn in Lilies of the World 4. (1950), O' Brien in Augustine Henry, An Irish Plant Collector in China 25. (2002).

Lilium leucanthum Hemsley non Baker in Curtis's Bot. Mag. 126: t. 7722. (1900), Henry in Journ. Roy. Hort. Soc. 26: 346. (1901—1902). *Lilium myriophyllum* Franch. E. H. Wilson in Lilies in Eastern Asia 33. (1901—1902), Nelson in A Heritage of Beauty 314. (2000).

Yunnan: Mengzi, without number.

Sent to Kew by Henry as seed in 1897. This beautiful lily has long been grown at Glasnevin, in the 1920s it grew in a south facing border in front of the Camellia house, it currently grows in a warm corner in the woodland garden. Discovered in north-west Yunnan by Delavay on July 29th 1888. According to Wilson it was introduced to cultivation by Henry who sent seeds to Kew in 1897 from Mengzi. From seedling raised, plants first flowered at Kew in 1899 and one of them was figured by Joseph Hooker in Bot. Mag. t. 7722. (as *Lilium leucanthum)*. Unfortunately the Yunnan seedlings were confused with plants raised from bulbs Henry had sent from central China (Yichang) in 1888 and were named *Lilium leucanthum* (Bak.) Baker. The mistake was not corrected until 1912 (see Watson, Gard. Chron. ser. 3, 51: 404. 1912). Henry stated it was a common plant around Mengzi between 1,300 and 1,650 metres in altitude. In November 1899, when Wilson was visiting Henry at Mengzi, Henry directed him to the place where it grew, Wilson dug up 50 or so bulbs and sent them back to Messrs. Veitch. Distributed in Myanmar and western China.

Limnophila aromatica (Lam.) Merr. (PLANTAGINACEAE)

Philcox in Kew Bull. 24: 145. (1970).

Limnophila gratissima Blume Henry in Trans. Asiat. Soc. Jap. 24 suppl: 67. (1896) (List Pl. Formosa).

Taiwan: Oluanpi, 1344. Wanchin, 1648. Kaohsiung, 2044.

A widespread aquatic perennial, distributed throughout south-east Asia.

Limnophila chinensis (Osbeck) Merr. (PLANTAGINACEAE)

Philcox in Kew Bull. 24: 151. (1970).

Yunnan: Simao, 12563, 12569.

A widespread subtropical aquatic annual, distributed in India, China, Thailand, Cambodia, Laos and Vietnam.

Limnophila connata (Buch.-Ham. ex D. Don) Handel-Mazzetti (PLANTAGINACEAE)

Philcox in Kew Bull. 24: 133. (1970).

Limnophila hypericifolia (Benth.) Benth. Forbes & Hemsley in Journ. Linn. Soc. Bot. 26: 186. (1890).

Hubei: Yichang, 409, 2354, 2354a. 2355. **Yunnan:** Mengzi, 10147, 10147a

A robust annual found in swampy grassland and verges of ponds and similar places, native of Nepal, India, Myanmar, Indochina, China, and Thailand.

Limnophila heterophylla (Roxb.) Benth. (PLANTAGINACEAE)

Philcox in Kew Bull. Misc. Inf. 24(1): 124. (1970).

Limnophila sp. Henry in Trans. Asiat. Soc. Jap. 24 suppl: 67. (1896) (List Pl. Formosa).

Taiwan: Kaohsiung Plain, 1849.

Limnophila indica (L.) Druce (PLANTAGINACEAE)

Henry in Trans. Asiat. Soc. Jap. 24 suppl: 67. (1896) (List Pl. Formosa), Philcox in Kew Bull. 24: 115. (1970).

Limnophila sessiliflora Henry non (Vahl.) Blume in Trans. Asiat. Soc. Jap. 24 suppl: 67. (1896) (List Pl. Formosa), in ref. to A. Henry 1726.

Taiwan: Wanchin, 1595, 1726.

A widespread aquatic perennial.

Limnophila repens (Benth.) Benth. (PLANTAGINACEAE)

Philcox in Kew Bull. 24: 154 & 165. (1970).

Hainan: Environs of Haikou, 8016.

Limnophila rugosa (Roth.) Merr. (PLANTAGINACEAE)

Philcox in Kew Bull. 24: 135. (1970).

Limnophila roxburghii Henry non G. Don. in Trans. Asiat. Soc. Jap. 24 suppl: 67. (1896) (List. Pl. Formosa).

Taiwan: Kaohsiung, 1116. Wanchin, 1552.

A widespread tropical aquatic annual, often a weed of rice-fields.

Limnophila sessiliflora (Vahl.) Blume (PLANTAGINACEAE)

Forbes & Hemsley in Journ. Linn. Soc. Bot. 26: 186. (1890), Philcox in Kew Bull. 24: 112. (1970).

Hubei: Yichang, 2339, 4302.

An aquatic perennial of widespread distribution in India, Nepal, eastern Pakistan, Myanmar, China, Vietnam, Korea, Japan, Malaya, Java and Indonesian Borneo.

Limnophila sp. (PLANTAGINACEAE)

Anonymous in Kew List of Determinations (volume 34).

Yunnan: Simao, 12567.

Limonium wrightii (Hance) Kuntze (PLUMBAGINACEAE)

Statice wrightii Hance Henry in Trans. Asiat. Soc. Jap. 24 suppl: 57. (1896) (List Pl. Formosa).

Taiwan: Oluanpi, 1326.

Lindenbergia muraria (Roxb. ex D. Don) Brühl (OROBANCHACEAE)

Lindenbergia urticifolia Lehm. Forbes & Hemsley in Journ. Linn. Soc. Bot. 26: 184. (1890).

Hubei: Yichang, 323, 390, 640, 902, 956. On Tsui Fu Shan near Yichang (May 1888), 1608. Badong, 848, 2452. Liantuo, 2628, 4582. **Yunnan:** Milê district, 9931. Simao, 13605.

Lindenbergia philippensis (Champ. & Schltdl.) Benth. (OROBANCHACEAE)

Forbes & Hemsley in Journ. Linn. Soc. Bot. 26: 184. (1890).

Hubei: Yichang, 1156. **Yunnan:** Mengzi, 9602.

Widespread in Asia.

Lindenbergia sp. (OROBANCHACEAE)

Anonymous in Kew List of Determinations (volume 34).

Yunnan: Simao, 12912.

Lindera akoensis Hayata (LAURACEAE) New species

Li in Woody Flora of Taiwan 211. (1963).

Litsea sp. Henry in Trans. Asiat. Soc. Jap. 24 suppl: 79. (1896) (List Pl. Formosa).

Taiwan: Oluanpi, 204, 940.

A small evergreen tree, endemic to Taiwan where it grows in broadleaved forest on the southern part of the Island. First collected there by Henry.

Lindera caudata (Nees) Hook f. (LAURACEAE)

Liou Ho in Laurac. Chine & Indochine 133. (1934).

Yunnan: Mengzi, 9172, 10401, 10401c., 11388.

Native to India, Myanmar, China (Guangxi and Yunnan), Thailand, Laos and Vietnam.

Lindera communis Hemsley (LAURACEAE) New species

in Journ. Linn. Soc. Bot. 26: 387. (1891), Bretschneider

in Hist. Eur. Bot. Disc. China 2: 791. (1898), Liou Ho in Laurac. Chine & Indochine 130. (1934), Zheng in Journ. Wuhan Bot. Res. 2.(1): 50. (1984).

Hubei: Yichang, 1146, 1204, 1207, 1262, 1296, 3413a., 3413b., (type). Antelope Glen, 3413. San You Dong Glen (July 1886), 1523. Liantuo, 2613, 2662, 3947, 4551, 4574. **Sichuan:** North Wushan, 7628. **Yunnan:** Mengzi, 10945. Simao, 9047c., 11681a.

According to Wilson, this and other evergreen shrubby species of *Lindera* abounded on the cliffs of the glens and gorges near Yichang. In the late 19th century branches and leaves of these shrubs were cut down and tied into bundles and thoroughly dried in the sun. These were then pounded into powder in a stamping mill driven by water power and the powder was then treated with a glutinous rice-water to make it adhesive and then made into incense sticks. These sticks, known by foreigners as 'joss-sticks' were used in enormous quantities at religious ceremonies and when burning emit a pleasant fragrance. Widely distributed in China.

Lindera erythrocarpa Makino (LAURACEAE)
Zheng in Journ. Wuhan Bot. Res.2. (1): 52. (1984).

Lindera umbellata Forbes & Hemsley non Thunb. in Journ. Linn. Soc. Bot. 26: 393. (1891).

Hubei: Badong, 2836. Donghu, 6430.

Lindera fragrans Oliver (LAURACEAE) New species
in Hooker's Icon. Pl. 18: t. 1788. (1888), Forbes & Hemsley in Journ. Linn. Soc. Bot. 26: 388. (1891), Henry in Notes Econ. Bot. China 20. (1893), Bretschneider in Hist. Eur. Bot. Disc. China 2: 791. (1898), Zheng in Journ. Wuhan. Bot. Res. 2.(1): 50. (1984), O' Brien in Augustine Henry, An Irish Plant Collector in China 25. (2002).

Hubei: Yichang, 1056 (syntype), 1298. Liantuo, 4553. Xiling (Yichang) Gorge, 3295 (syntype). Xingshan, 7616. Badong, 2832, 5297. Changyang, 5268 (syntype).

According to Henry the leaves of this species, the *hsiang-yeh-tzu*, were pounded by water-mills into powder and used with the branches and roots of arbor vitae (*Platycladus orientalis*) to make incense. It was abundant on the cliffs and glens of the Three Gorges during his time in Yichang. The plant currently growing in the central section of the Chinese slope at Glasnevin was raised from seeds collected by the Sino-American expedition to Shennongjia in 1980.

Lindera glauca (Sieb. & Zucc.) Blume (LAURACEAE)
Henry in Journ. China Br. Roy. Asiat. Soc. 22: 261. (1887) (Chinese Names of Plants), Forbes & Hemsley in Journ. Linn. Soc. Bot. 26: 388. (1891), Henry in Notes Econ. Bot. China 40. (1893), Liou Ho in Laurac. Chine & Indochine 129. (1934), Zheng in Journ. Wuhan Bot. Res.

2.(1): 51. (1984).

Hubei: Yichang, 347, 2284, 3239, 7774. In the Lung Wang Tung Glen near Yichang (June 1886), 1563. San You Dong Glen near Yichang (July 1886), 1501. Badong, 2817. Donghu, 6393. Changyang, 6200. **Sichuan:** South Wushan, 7217.

Henry stated this small tree was common around Yichang where it was known as the *niu chin t'iao*. The wood of this tree is tough and durable and was used to make carrying poles and tool handles.

Lindera latifolia Hook. f. (LAURACEAE)
Allen in Journ. Arnold Arb. 22: 7. (1941).

Yunnan: Yuanjiang, 13269.

Native to India, Bangladesh, China [Tibet (Xizang) and Yunnan] and Vietnam. Discovered by William Griffith in east Bengal.

Lindera longistaminata (H. Liu) Dao (LAURACEAE) New species

Litsea longistaminata (H. Liu) Kosterm. in Bull. Bot. Survey India 10: 286. (1968). *Litsea garrettii* Gamble var. *longistaminata* Liou in Laurac. Chine et Indo-Chine 196. (1934), Allen in Ann. Missouri Bot. Gard. 25: 388. (1938).

Yunnan: Simao at 1,650 metres, 12769 (isosyntype of *Litsea garrettii* Gamble var. *longistaminata* Liou), 12802 (isosyntype of *Litsea garrettii* Gamble var. *longistaminata* Liou).

Native to south-west China [Tibet (Xizang), Yunnan] and Vietnam.

Lindera megaphylla Hemsley (LAURACEAE) New species
in Journ. Linn. Soc. Bot. 26: 389. (1891), Henry in Notes Econ. Bot. China 43. (1893), Liou Ho in Laurac. Chine & Indochine 125. (1934), Morley in Glasra 3: 79. (1979), Zheng in Journ. Wuhan Bot. Res. 2.(1): 51. (1984), Nelson in A Heritage of Beauty 315. (2000), O' Brien in Augustine Henry, An Irish Plant Collector in China 25. (2002).

Lindera oldhamii Hemsl. Liou Ho in Laurac. Chine & Indochine 123. (1934). *Benzion touyunense* (Levl.) Rehd. f. *megaphyllum* (Hemsl.) Rehder in Journ. Arnold Arb. 11: 156. (1930).

Hubei: Badong, 1284. Yichang, 2195, 3010, 3345a., 1112 (in part). Xiling (Yichang) Gorge, 3345, 7618 (type). Antelope Glen, 3345a. Liantuo, 3010, 3345b. Changleping 6215, 6216, 7848. Without localiy, 7848a. Xingshan, 4508. Jianshi, 7525. **Yunnan:** Yuanjiang, 13275, 13275a.

A common dioecious tree in western Hubei, with a massive trunk and a broad oval crown of erect spreading branches. Henry mentioned a fine tree that once grew at the temple of Shih Pai Shan in the Yichang (Xiling) Gorge.

Introduced to cultivation by Wilson in 1900 through Veitch's Coombe Wood nursery. *Lindera megaphylla* is cultivated in Ireland at the John F. Kennedy Arboretum. The finest specimen in Britain and Ireland grows in the garden at Caerhays Castle, planted in 1914 by 1984 this tree was 16 metres tall with a trunk diameter of 29 cm.

Lindera meissneri Hook. f. (LAURACEAE)

Liou Ho in Laurac. Chine & Indochine 126. (1934).

Yunnan: Mengzi, 10961.

Lindera metcalfiana Allen var. *dictophylla* (C. K. Allen) H. P. Tsui (LAURACEAE) New variety

Lindera meissneri Liou Ho non Hook. f. in Laurac. Chine & Indochine 126. (1934). *Lindera dictyophylla* Allen in Journ. Arnold Arb. 22: 5. (1941).

Yunnan: Simao, 12822, 12822a., 12822b., 12822c., 13285.

The type specimen of *Lindera dictyophylla* was collected by C. W. Wang in Yunnan in February 1936 but this species had been previously found by Henry in 1897. Native to southern China and Vietnam.

Lindera neesiana (Wall. ex Nees) Kurz (LAURACEAE)

Lindera fruticosa Hemsley in Journ. Linn. Soc. Bot. 26: 388. (1891), Bretschneider in Hist. Eur. Bot. Disc. China 2: 791. (1898), Allen in Ann. Missouri Bot. Gard. 25: 399. (1938).

Hubei: Badong, 4750 (isosynype of *Lindera fruticosa* Hemsley). Fang Xian, 6571 (isosyntype of *Lindera fruticosa* Hemsley).

Lindera obtusiloba Blume (LAURACEAE)

Zheng in Journ. Wuhan Bot. Res. 2.(1): 50. (1984). *Lindera cercidifolia* Hemsley in Journ. Linn. Soc. Bot. 26: 387. (1891), Bretschneider in Hist. Eur. Bot. Disc. China 2: 791. (1898), Morley in Glasra 3: 79. (1979), O' Brien in Augustine Henry, An Irish Plant Collector in China 25. (2002).

Hubei: Badong, 2523, 3792 (type of *Lindera cercidifolia*), 4919.

According to Wilson, this deciduous shrub or small tree was very common in the woods of western Hubei where it was very conspicuous in early spring on account of the brilliant colour of the young leaves. Native to Japan, China and Korea and introduced to cultivation by Charles Maries through Veitch's Coombe Wood nursery in 1880 from Japan. Wilson reintroduced it in 1907 while collecting for the Arnold Arboretum. Cultivated in Ireland at Mount Congreve, County Waterford.

Lindera pulcherrima (Nees) Benth. ex Hook. f. var. *hemsleyana* (Diels.) H. P. Tsui (LAURACEAE) New variety

Lindera strychnifolia (Sieb. & Zucc.) Villar. var.

Hemsley in Journ. Linn. Soc. Bot. 26: 392. (1891). *Lindera strychnifolia* Henry non (Sieb. & Zucc.) Villar. in Notes Econ. Bot. China 20. (1893), Zheng in Journ. Wuhan Bot. Res. 2.(1): 52. (1984). *Lindera strychnifolia* (Sieb. & Zucc.) Villar. var. *hemsleyana* Diels in Bot. Jahrb. 29: 352. (1900), Gamble in Sargent Pl. Wilson 2: 82. (1916). *Lindera hemsleyana* (Diels.) Allen in Journ. Arnold Arb. 22: 25. (1922).

Hubei: Yichang, 2975, 3375 [syntype of *Lindera strychnifolia* (Sieb. & Zucc.) Villar. var. *hemsleyana* Diels]. Liantuo, 1894, 2985, 3805, 4552. Jianshi, 5497a. In a ravine on Moji Shan, 3375a. Without locality, 5497b.

A common shrub at Yichang according to Henry who called it the *san-t'iao-chin*. From its roots the *drug wu-yao* was extracted.

Lindera sericea (Sieb.& Zucc.) Blume (LAURACEAE)

Forbes & Hemsley in Journ. Linn. Soc. Bot. 391. (1891), Zheng in Journ. Wuhan Bot. Res.2. (1): 52. (1984).

Hubei: Badong, 2829. Donghu, 6452.

Lindera spp. (LAURACEAE)

Anonymous in Kew List of Determinations (volume 24 & 34). Henry in Trans. Asiat. Soc. Jap. 24 suppl: 80. (1896) (List Pl. Formosa).

Hubei: Yichang, 2101. Changleping, 7656. **Taiwan:** Oluanpi, 1328. **Yunnan:** Mengzi, 10439.

Lindera supracostata Lecompte (LAURACEAE)

Allen in Journ. Arnold Arb. 22: 24. (1922), Liou Ho in Laurac. Chine & Indochine 132. (1934).

Yunnan: Mengzi, 10873.

A handsome species was`first found by Delavay. Vietnam and China (Guizhou, Sichuan and Yunnan).

Lindera thomsonii Allen (LAURACEAE)

in Journ. Arnold Arb. 22: 22. (1941).

Lindera pulcherrima Liou Ho non (Nees) Benth. ex Hook. f. in Laurac. Chine & Indochine 134. (1934).

Yunnan: Mengzi, 9629.

This species was first found in India by Hooker and Thomson in the Khasi Hills, north-east India. Allen described it using material also collected by Henry in Yunnan, Poilane in Vietnam and Kingdon Ward in Myanmar. Distributed in India, Myanmar, western China and Vietnam.

Lindera tonkinensis Lecompte (LAURACEAE) New species

Allen in Journ. Arnold Arb. 22: 20. (1941)

Lindera strychnifolia Liou Ho non (Sieb. & Zucc.) Villar. in Laurac. Chine & Indochine 136. (1934).

Yunnan: Simao, 11686, 11686a., 11686b., 11686c.

Leveille based this species on material gathered by E. Poilane in Vietnam and Laos during the 1930s. Simao lies adjacent to the borders of these countries and as can be expected there are great similarities between the floras of the regions. Henry first collected this species about 1897. It was also later collected by Joseph Rock (probably in 1922) in southern Yunnan.

Lindera umbellata Thunb. var. *membranacea* (Maxim.) Momiy. (LAURACEAE)

Lindera membranacea Maxim. Forbes & Hemsley in Journ. Linn. Soc. Bot. 26: 389. (1891), Liou Ho in Laurac. Chine & Indochine 128. (1934).

Hubei: Changyang, 5248. Jianshi, 5885.

The *mu-chiangtzu*, the pungent fruits of this species were used as a spice in 19th century Hubei.

Lindernia anagallis (Burm. f.) Wannan, W. R. Barker & Y. S. Liang (LINDERNIACEAE)

Vandellia pedunculata Benth. Henry in Trans. Asiat. Soc. Jap. 24 suppl: 67. (1896) (List Pl. Formosa).

Hainan: Environs of Haikou, 8333, 8378. **Taiwan:** Wanchin, 1525. **Yunnan:** Simao, 12134.

Widely distributed in China.

Lindernia procumbens (Krock.) Philcox (LINDERNIACEAE)

Vandellia erecta Benth. Forbes & Hemsley in Journ. Linn. Soc. Bot. 26: 190. (1890).

Hubei: Yichang, 2367.

Lindernia ruellioides (Colm.) Spreng. (LINDERNIACEAE)

Bonnaya reptans (Roxb.) Spreng. Anonymous in Kew List of Determinations (volume 34).

Yunnan: Mengzi, 10766. Simao, 12231.

Lindernia sp. No. 1. (LINDERNIACEAE)

Vandellia sp. Anonymous in Kew List of Determinations (volume 24.).

Hubei: Yichang, 403, 404.

Lindernia sp. No. 2. (LINDERNIACEAE)

Vandellia sp. Anonymous in Kew List of Determinations (volume 34).

Yunnan: Mengzi, 13762.

Lindsaea cultrata (Willd.) Sw. (LINDSAEACEAE)

Anonymous in Kew List of Determinations (volume 34).

Yunnan: Mengzi, 9742.

A handsome little fern, distributed through southern China, Indochina, the Philippines, the Malay Peninsula, and Java.

Lindsaea ensifolia Sw. (LINDSAEACEAE)

Henry in Trans. Asiat. Soc. Jap. 24 suppl: 110. (1896) (List Pl. Formosa).

Taiwan: Wanchin, 1497, 1502.

Widely distributed across the tropics of the Old World.

Lindsaea orbiculata (Lam.) Mett. (LINDSAEACEAE)

Lindsaea flabellulata Dryand. Henry in Trans. Asiat. Soc. Jap. 24 suppl: 110. (1896) (List Pl. Formosa).

Taiwan: Tamsui, 1391.

Lindsaea repens (Bory) Thwaites (LINDSAEACEAE)

Davallia repens (Bory) Desv. Henry in Trans. Asiat. Soc. Jap. 24 suppl: 110. (1896) (List Pl. Formosa).

Taiwan: Oluanpi, 1362.

Linum perenne L. (LINACEAE)

Henry in Journ. China Br. Roy. Asiat. Soc. 22: 239. (1887) (Chinese Names of Plants).

Hubei: Badong, 2405.

Linum stelleroides Planch. (LINACEAE)

Anonymous in Kew List of Determinations (volume 24).

Hubei: Badong, 530 (September 1885), 2400, 2493. Liantuo, 2631, 4595. **Sichuan:** South Wushan, 7209.

Linum usitatissimum L. (LINACEAE)

Henry in Journ. China Br. Roy. Asiat. Soc. 22: 239. (1887) (Chinese Names of Plants), Henry in Kew Bull. Misc. Inf. 250. (1891); in Notes Econ. Bot. China 64. (1893).

Hubei: Badong, 288 (July 1885). Zigui, 6204.

Linseed or flax, an erect, annual herb to 1 metre tall. Thought to be native to Europe and eastern Asia, its geographical distribution has been so altered by humans that it is now difficult to determine from where it originally arose and there are no flax populations extant today that can be concidered truly wild. Linseed is thought to be the oldest domesticated oilseed crop and was domesticated by human selection by 6,000 B.C. Archeoligists have found irrigation systems near Mesopotamian cities that supported the cultivation seven thousand years ago. *Linum usitatissimum* is also the source of flax fibre, it was cultivated as an oil source for several thousand years before it before becoming a fibre provider. Henry called it the *shan chih* ma or *hu ma* and stated it was widely cultivated in the mountainous parts of Hubei and Sichuan where it was entirely cultivated for its seeds (rather than for the production of linen), which at the time were a common article in Chinese drug shops, and were used locally for their oil (linseed oil) which was used for cooking and lighting purposes.

Liparis bootanensis Griff. (ORCHIDACEAE)

Liparis plicata Franch. & Sav. Henry in Trans. Asiat. Soc. Jap. 24 suppl: 8. (1896) (List Pl. Formosa), Rolfe in Journ. Linn. Soc. Bot. 36: 8. (1903).

Taiwan: Wanchin, in mountains, 138, 1627.

An epiphytic species, widely distributed across the Himalaya to southern China, Indochina, the Philippines, south to Java and Sumatra.

Liparis campylostalix Rchb. f. (ORCHIDACEAE)

Liparis pauciflora Rolfe in Kew Bull. Misc. Inf. 193. (1896), Bretschneider in Hist. Eur. Bot. Disc. China 2: 792.

(1898), Rolfe in Journ. Linn. Soc. Bot. 36: 8. (1903).

Sichuan: South Wushan, 5675. **Hubei:** Without locality, 5675a.

Liparis distans C. B. Clarke (ORCHIDACEAE)

Liparis yunnanensis Rolfe in Journ. Linn. Soc. Bot. 36: 8. (1903). *Liparis oxyphylla* Schlecter Repert. Spec. Nov. Regni. Veg. Beih. 4: 63. (1919).

Yunnan: On mountains near Mengzi at 1, 650 metres, lithophyte, yellow flowers, amoung rocks, 10485 (type of *Liparis yunnanensis* Rolfe).

Liparis gigantea C. L. Tso (ORCHIDACEAE)

Liparis macrantha Rolfe in Ann. of Bot. 9: 156. (1895) (nom. illeg.), Bretschneider in Hist. Eur. Bot. Disc. China 2: 792. (1898), Rolfe in Journ. Linn. Soc. Bot. 36: 7. (1903).

Taiwan: Siko Hill, Tamsui, 1695 (type of *Liparis macrantha* Rolfe).

North-east India to China (Taiwan).

Liparis henryi Rolfe (ORCHIDACEAE) New species

in Kew Bull. Misc. Inf. 193. (1896), Henry in Trans. Asiat. Soc. Jap. 24 suppl: 91. (1896) (List Pl. Formosa), Bretschneider in Hist. Eur. Bot. Disc. China 2: 792. (1898), Rolfe in Journ. Linn. Soc. Bot. 36: 6. (1903).

Taiwan: Oluanpi, 2074.

A terrestrial orchid to 20 cm tall, endemic to Taiwan where it is confined to the tropical forests on the Hengchun Peninsula on the island's southern tip.

Liparis nervosa (Thunb.) Lindl. (ORCHIDACEAE)

Rolfe in Journ. Linn. Soc. Bot. 36: 7. (1903).

Taiwan: Oluanpi, 646.

A terrestrial orchid to 15 cm tall, native to China and Japan.

Liparis odorata (Willd.) Lindl. (ORCHIDACEAE)

Henry in Trans. Asiat. Soc. Jap. 24 suppl: 91. (1896) (List Pl. Formosa), Rolfe in Journ. Linn. Soc. Bot. 36: 8. (1903).

Taiwan: Wanchin, 896. **Yunnan:** Mengzi, 11105. Simao, 12300.

A widespread orchid in Asia.

Liparis spp. (ORCHIDACEAE)

Henry in Trans. Asiat. Soc. Jap. 24 suppl: 91. (1896) (List Pl. Formosa). Anonymous in Kew List of Determinations (volume 34).

Taiwan: Oluanpi, 581. **Yunnan:** Mengzi, 11852.

Liparis stricklandiana Rchb. f. (ORCHIDACEAE)

Liparis malleiformis W. W. Smith in Notes Roy. Bot. Gard. Edinburgh 13: 212. (1921).

Yunnan: Mengzi at 1,640 metres, 11820a.

Liparis viridiflora (Blume) Lindl. (ORCHIDACEAE)

Liparis longipes Lindl. Henry in Trans. Asiat. Soc.

Jap. 24 suppl: 91. (1896) (List Pl. Formosa), Rolfe in Journ. Linn. Soc. Bot. 36: 7. (1903). *Liparis pleistantha* Schlecter in Repert. Spec. Nov. Regni Veg. Beih. 4: 64. (1919).

Taiwan: Oluanpi, 1592. Wanchin, 1979. **Yunnan:** Simao, 12640a., (type of *Liparis pleistantha* Schlecter).

An epiphytic species to 20 cm tall, widespread in its distribution and ranging from the Himalaya, across India, Indochina, southern China, the Philippines to Malaya and Indonesia.

Lipoblepharis urticifolia (Blume) Orchard (ASTERACEAE)

Wedelia wallichii Less. Anonymous in Kew List of Determinations (volume 34).

Yunnan: Mengzi, 11022.

Distributed in India, China, Laos, Cambodia and Vietnam.

Liquidambar formosana Hance (HAMAMELIDACEAE)

Forbes & Hemsley in Journ. Linn. Soc. Bot. 23: 291. (1887), Henry in Journ.China Br. Roy. Asiat. Soc. 22: 242. (1887) (Chinese Names of Plants); in Notes Econ. Bot. China 14 & 56. (1893); in Trans. Asiat. Soc. Jap. 24 suppl: 42. (1896) (List Pl. Formosa), Henry in Elwes and Henry The Trees Gt. Brit. & Irel. 3: 506. (1908), Rehder & E. H. Wilson in Sargent Pl. Wilson. 1: 421. (1913), Zheng in Journ. Wuhan Bot. Res. 2.(1): 65. (1984).

Hubei: Yichang, 2611, 5218 (April 16th 1888), 5218a., (July 13th 1888). Liantuo, 2611. South Donghu, 7630. **Sichuan:** South Wushan, 5218b. (May 9th 1888), 7630. **Hainan:** Ling-men, 8638. **Taiwan:** Wanchin, 425.

A tree to 40 metres tall in its native habitat and first described from Chinese Taiwan from a collection made by the Kew collector Richard Oldham though Bretschnider states it was first found in China by the Russian botanist, Carl Maximowicz. Widespread on mainland China and extending into Korea and Indochina. *Liquidamber formosana* is the *feng-hsiang* and according to Henry the timber of this tree was used for making tea chests at Hankow (now part of Wuhan) the reason being that it imparted no flavour to the tea. At Shanghai Henry stated *Cunninghamia lanceolata* was used for the same purpose. The timber also seasoned quickly, planks sawn in spring could be used in autumn. The *feng-hsiang* is a common tree in warm-temperate areas of China and was particularly abundant in Hubei in the late 19th century. It was introduced to Kew in 1884 when seeds were sent from Hankow (now part of the city of Wushan, on the Yangtze to the east of Yichang). Most of the trees in cultivation today are said to derive from Wilson's Hubei collections made in 1907 for the Arnold Arboretum.

02

Liquidambar yunnanensis (Rehder & E. H. Wilson) Ickert-Bond & J. Wen (HAMAMELIDACEAE) New species

in Sargent Pl. Wilson. 1: 422. (1913).

Yunnan: In forests on the Daweishan Range in Pingbian at 2,000 metres, 10395, (type of *Altingia yunnanensis* Rehder & E. H. Wilson, collected 5th January 1896). In forests on the Daweishan range in Pingbian at 1,600 metres, 11082.

China (Yunnan) and Vietnam.

Liriodendron chinense (Hemsl.) Sargent (MAGNOLIACEAE) New species

in Trees & Shrubs 1: 103. fig. lii. (1985), Henry in Elwes & Henry, Trees Gt. Brit. & Irel. 1: 64. (1906), Morley in Glasra 3: 79. (1979), Sponberg & Bell in The New Plantsman 3(4): 213. (1996), O' Brien in Augustine Henry, An Irish Plant Collector in China 25. (2002).

Liriodendron tulipifera Henry non L. in Notes Econ. Bot. China 44. (1893). *Liriodendron tulipifera* L. var. *chinensis* Hemsley in in Gard. Chron. ser. 3, 6: 718. (1889), Bretschneider in Hist. Eur. Bot. Disc. China 2: 779. (1898).

Hubei: Jianshi, 5836, 5836b. Baokang, at 1,900 metres, 5836a.

A tree to 40 metres tall in its nativebabitat and already 23 metres tall at Glasnevin and Mount Usher in Ireland. Until the late 19th century *Liriodendron* was thought to be a monotypic genus with one taxon endemic to North America. In 1875 Dr. George Shearer collected a sterile specimen on the Lushan mountains. Shearer, a native of Liverpool, was sent in 1868 to take charge of the London Missionary Society's hospital at Hankow (now part of Wuhan) downriver from Yichang. In 1873 Shearer made a collection of over 600 species which he sent to Liverpool and eventually made their way to Kew in 1895. His sterile collection of *Liriodendron* was taken to be the American *Liriodendron tulipifera* by W. B. Hemsley and it was thought to have been a result of naturalisation from a cultivated tree. Again in 1878, Charles Maries, the Veitchian collector, sent a specimen from the same region, this collection was young and neither the leaves or flowers were fully developed.

In 1888 Henry sent back to Europe the first complete specimens of the Chinese tulip tree thus enabling botanists to distinguish it from the American tree, in 1903 Charles Sprague Sargent raised it from Hemsley's var. *chinense* to specific rank. At Baokang Henry found it forming a spreading shrub to almost 2 metres tall, at Jianshi in the spring of 1888 he collected good flowering specimens from a tree almost 10 metres tall with leaves up to 30 cm across. According to Wilson (who introduced it through Veitch's Coombe Wood nursery by means of living plants in 1901)

the colloquial name for this tree in Hubei was *wo-chang-chiu* or goose-foot, which referred to the shape of the leaves. It is still relatively common at high altitude in Badong in Hubei where it was observed by members of the Glasnevin Central China Expedition (2002) growing with *Populus lasiocarpa, Cornus kousa* ssp. *chinensis* and *Rosa multiflora* var. *cathayensis*. It was also collected by the 2004 Glasnevin Central China Expedition (GCCE 804) on Tianthu Shan in Changyang in October of that year. Despite having been separated from each other for tens of thousands of years both the American and Chinese species are inter-fertile and a hybrid has been produced.

Liriope graminifolia (L.) Baker (ASPARAGACEAE)

Henry in Journ. China Br. Roy. Asiat. Soc. 22: 253. (1887) (Chinese Names of Plants).

Hubei: Antelope Glen, 3223.

According to Henry this species was known colloquially in the Three Gorges region as *yeh lan ts'ao and hsiao me huang.*

Liriope sp. (ASPARAGACEAE)

Anonymous in Kew List of Determinations (volume 34).

Yunnan: Mengzi, 9799, 10472, 11845.

Liriope spicata (Thunb.) Lour. (ASPARAGACACEAE)

Henry in Journ. China Br. Roy. Asiat. Soc. 22: 253. (1887) (Chinese Names of Plants), Henry in Trans. Asiat. Soc. Jap. 24 suppl: 95. (1896) (List Pl. Formosa), Wright in Journ. Linn. Soc. Bot. 36: 79. (1903).

Hubei: Yichang, 353, 498, 3864. Badong, 911, 2550, 2806. Liantuo, 3864. Donghu, 6439. **Hainan:** Environs of Haikou, 8385. **Taiwan:** Wanchin, 146. Kaohsiung, 1098. Oluanpi, 1703, 1704.

Native to Vietnam, China, Korea and Japan.

Litchi chinensis Sonn. (SAPINDACEAE)

Nephelium litchi Steud. Anonymous in Kew List of Determinations (volume 34).

Yunnan: Yuanjiang, 13276.

A subtropical tree to about 20 metres tall, the lychee is an important fruit and is grown over much of southern China across to western Malaysia. It has been long cultivated in China and was discussed by the Chinese scholar Tsai Hsiang in 1079 A.D. in his early treatise on fruit cultivation. *Litchi chinensis* is extinct in the wild but has naturalised itself in parts of tropical and subtropical Asia.

Lithocarpus amygdalifolius (Skan) Rehder (FAGACEAE) New species

Li in woody Flora of Taiwan 89. (1963).

Quercus truncata Henry non King ex Hook. f. in Trans. Asiat. Soc. Jap. 24 suppl: 90. (1896) (List Pl. Formosa). *Quercus (Lithocarpus) amygdalifolia* Skan in Journ.

Linn. Soc. Bot. 26: 506. (1899). *Pasania amygdalifolia* (Skan) Schotty in Bot. Jahrb. xlvii. 660. (1912). *Synaedrys amygdalina* (Skan.) Koidzumi in Tokyo Bot. Mag. 188. (1916).

Taiwan: Oluanpi, 1254 (type of *Quercus amygdalifolia* Skan).

An evergreen tree with trunks often up to 2 metres in diameter. Native to China (including Taiwan) and Vietnam.

Lithocarpus brevicaudatus (Skan) Hayata (FAGACEAE) New species

in Gen. Ind. Fl. formos. 72. (1917), Rehder in Journ. Arnold Arb. 1: 123. (1920), Li & Zhang in Wuhan Bot. Res. 16: 244. (1998).

Quercus (*Pasania*) *brevicuadata* Skan in Journ. Linn. Soc. Bot. 26: 308. (1899). *Synaedrys brevicaudata* (Skan.) Koidzumi in Bot. Mag. Tokyo. 30: 194. (1916). *Pasania brevicaudata* (Skan) Schotty in Bot. Jahrb. 47: 666. (1912), Nakai in Journ. Jap. Bot. 15: 270. (1939), Li in Woody Flora of Taiwan 93. (1963).

Taiwan: Oluanpi, 1368 (type of *Quercus brevicuadata* Skan).

A tree to 20 metres tall in Taiwan where it is very common at altitudes between 500 to 2,000 metres. Also found on the mainland in Fujian and Guangdong.

Lithocarpus carolinae (Skan) Rehder (FAGACEAE) New species

in Journ. Arnold Arb. 1: 123. (1920).

Quercus carolinae Skan in Journ. Linn. Soc. Bot. 35: 518. (1903).

Yunnan: Tonguan at 1,950 metres, 13239 (isotype of *Quercus carolinae* Skan).

An 18 metre tall tree, endemic to Yunnan. This species commemorates Augustine Henry's first wife Caroline, who died in Denver, Colorado in 1894. Caroline Henry collected herbarium specimens for the Royal Botanic Gardens, Kew in Japan and Colorado, USA. The Colorado specimens were collected with the assistance of Augustine Henry's sister Mary, who had moved to the states to nurse Caroline through bad health. Caroline Henry is also commemorated by *Primula carolinehenryae* S. O'Brien.

Lithocarpus cleistocarpus (Seem.) Rehder & E. H. Wilson (FAGACEAE) New species

in Sargent Pl. Wilson. 2: 651. (1916), Zheng in Journ. Wuhan Bot. Res. 2.(1): 24. (1984), Li & Zhang in Wuhan Bot. Res. 16: 254. (1998).

Quercus cleistocarpa O. Seemen in Engl. Bot. Jahrb. xxiii. Beiblatt. 57: 52. (1897), Skan in Journ. Linn. Soc. Bot. 26: 510. (1899).

Hubei: Badong, 4924, Xingshan at 1,400 metres, 6567 (type of *Quercus cleistocarpa* O. Seemen), 6524, 6538.

Fang Xian, 6567a., 6715. Jianshi, 6002. Without locality, 6002a.

An evergreen tree to 15 metres tall, native to central China in the provinces of Hubei and Sichuan, Wilson, who introduced *Lithocarpus cleistocarpus* to cultivation through Messrs Veitch in 1901, described it as a much branched, wide-spreading flat-crowned tree. Two trees of this species are grown at Caerhays Castle in Cornwall and are the tallest in the British Isles and Ireland. The taller of these is 22 metres tall with a trunk diameter of 67 cm. The species was also later collected near Kangding (Tchenkeoutin) in Sichuan by Farges.

Lithocarpus dealbatus (Hook. f. & Thoms.) Rehder (FAGACEAE)

Quercus polystachya Skan non Wallich ex DC. in Journ. Linn. Soc. Bot. 26: 519. (1899). *Quercus dealbata* Hook. f. & Thoms. Skan in Journ. Linn. Soc. Bot. 26: 510. (1899). *Pasania viridis* Schotty in Bot. Jahrb. 47: 668. (1912). *Synaedrys viridis* (Schott.) Koidumi in Tokyo Bot. Mag. xxx. 198. (1916). *Lithocarpus viridis* (Schott.) Rehder & E. H. Wilson in Sargent Pl. Wilson. 3: 210. (1917).

Yunnan: In woods near Mengzi at 1,600 to 1,800 metres, 10520, 11434, 11434a. In mountains near Simao at 1,640 metres, 12329 (type of *Pasania viridis* Schotty). In mountains near Simao at 1,500 to 1,600 metres, 12329, 12329a., 12329b., 12329c., 12329d., 12329e. Mengzi at 1,550 to 1890 metres, 9636, 9636a., 9636b.

Rehder & E. H. Wilson disputed that A. Henry 9636 belongs to *Lithocarpus viridis* (q. v.) stating that A. Henry 9636 has leaves pubescent on the undersides and the male inflorescence is much shorter than *Lithocarpus viridis* and the flowers are different. Distributed in India, Bhutan, Myanmar, western China, Thailand, Laos and Vietnam.

Lithocarpus elaeagnifolius (Seem.) Chun (FAGACEAE)

Quercus elaeagnifolia Seemen Skan in Journ. Linn. Soc. Bot. 26: 512. (1899).

Hainan: Haikou, without number.

Lithocarpus elegans (Blume) Hatus. ex Soepadmo (FAGACEAE)

Lithocarpus spicatus (Blume) Rehder & E. H. Wilson in Sargent Pl. Wilson. 3: 207. (1916).

Yunnan: Yuanjiang at 1,150 metres, 13238.

Lithocarpus fordianus (Hemsley) Chun (FAGACEAE) New species

in Journ. Arn. Arb. 8: 21. (1927).

Quercus fordiana Hemsley in Hookers Icon. Pl. t. 2664. (1901), Forbes & Hemsley in Journ. Linn. Soc. Bot. 34: 478. (1900).

Yunnan: In the mountains to the east of Simao at 1,300 metres, 12054 (isosyntype of *Quercus fordiana* Hemsley).

In the mountains to the south-west of Simao at 1,650 metres, 12054a., (isosyntype of *Quercus fordiana* Hemsley). 12054c., (isosyntype of *Quercus fordiana* Hemsley).

Native to China (Yunnan and Guizhou) and Vietnam. The specific epithet commemorates Charles Ford of Hong Kong Botanic Gardens.

Lithocarpus formosana (Skan) Hayata (FAGACEAE) New species

Rehder in Journ. Arnold Arb. 1: 126. (1920).

Quercus (*Pasania*) *formosana* Skan in Journ. Linn. Soc. Bot. 26: 513. (1899). *Pasania formosana* (Skan) Schotty in Bot. Jahrb. 47: 670. (1912), Li in Bull. Torrey Bot. Club. 30: 320. (1953); in Woody Flora of Taiwan 93. (1963). *Synaedrys formosana* Koidzumi in Tokyo Bot. Mag. 30: 195. (1916).

Taiwan: Oluanpi, 1995, (type of *Quercus formosana* Skan).

A medium-sized evergreen tree, endemic to Taiwan where it inhabits the extreme southern part of the Hengchun Peninsula on the south of the island.

Lithocarpus henryi (Seem.) Rehder & E. H. Wilson (FAGACEAE) New species

in Sargent Pl. Wilson. 3: 209. (1917), Bean in Trees & Shrubs 3: 335. (1936), Walsh in Aug. Henry Forst. Herb. Cat. 64. (1957), Li & Zhang in Wuhan Bot. Res. 16: 245. (1998), Nelson in A Heritage of Beauty 315. (2000), O' Brien in Augustine Henry, An Irish Plant Collector in China 25. (2002).

Quercus henryi Seem. in Bot. Jahrb. Syst. 23. (Biebl. 57): 50. (1897), Zheng in Journ. Wuhan Bot. Res. 2.(1): 27. (1984). *Quercus spicata* Skan non Sm. in Journ. Linn. Soc. Bot. 26: 521. (1899). *Quercus* sp. Skan in Journ. Linn. Soc. Bot. 26: 522. (1899). *Pasania henryi* (Seem.) Schotty in Bot. Jahrb. xlvii. 665. (1912).

Hubei: Jianshi, 5805, 6023. Without locality, 5805a. **Sichuan:** North Wushan, 7030. Wushan, 7030a.

A magnificent evergreen tree growing to about 15 metres tall in the woods of Hubei and eastern Sichuan. Introduced to cultivation by Wilson in 1901 through Veitch's Coombe Wood nursery. This species is rare in cultivation though there are old trees in the gardens at Kilmacurragh in County Wicklow, a satellite garden of the National Botanic Gardens, Glasnevin. The finest specimens in Britain and Ireland grow at Caerhays Castle in Cornwall.

Lithocarpus konishii (Hayata) Hayata (FAGACEAE) New species

Pasania konishii (Hayata) Schottky Li in Woody Flora of Taiwan 93. (1963).

Castanopsis sp. Henry in Trans. Asiat. Soc. Jap. 24 suppl: 90. (1896) (List Pl. formosa).

Taiwan: Wanchin, 474, 1536.

A evergreen tree or shrub to 10 metres tall, distributed in Hainan and Taiwan where on the latter island it is common in the mountains of the central and southern regions, first collected there by Henry.

Lithocarpus lycoperdon (Skan) A. Camus (FAGACEAE) New species

Quercus lycoperdon Skan in Journ. Linn. Soc. Bot. 26: 518. (1899).

Yunnan: Mengzi at 2,300 metres, 9069 (isotype of *Quercus lycoperdon* Skan).

Distributed in western China, Laos and Vietnam.

Lithocarpus naiadarum (Hance) Chun (FAGACEAE)

Quercus naiadarum Hance Skan in Journ. Linn. Soc. Bot. 26: 519. (1899).

Hainan: 32 km (20 miles) west of Haikou, 8142. Environs of Kiungchow, 8739.

Discovered by the Rev'd B. C. Henry in the mountainous interior of Hainan in November 1882. Endemic to Hainan.

Lithocarpus obscurus C. C. Huang & Y. T. Chang (FAGACEAE)

Lithocarpus spicatus (Smith) Rehd. & Wils. var. *brevipetiolatus* (A. De Candolle) Rehder & E. H. in Sargent Pl. Wilson. 3: 208. (1917), Walsh in Aug. Henry Forst. Herb. 64. (1957).

Yunnan: Mountains to the east and south-west of Simao at 1,300 to 1,600 metres, 11614, 11614a., 11614b., 11614c., 11614d.

Lithocarpus polystachyus (Wall. ex A. DC.) Rehder (FAGACEAE)

Quercus polystachya Wall. ex A. DC. Forbes & Hemsley in Journ. Linn. Soc. Bot. 519. (1899).

Yunnan: Mengzi, 10434, 10434a.

Lithocarpus rotundatus (Blume) A. Camus (FAGACEAE)

Rehder & E. H. Wilson in Sargent Pl. Wilson. 3: 207. (1917), Chun in Journ. Arnold Arb. 9: 153. (1928).

Quercus cathayana Seemen in Repert. Spec. Nov. Regni. Veg. 3: 53. (1906). *Pasania cathayana* Schotty in Bot. Jahrb. xlvii. 666. (1912). *Lithocarpus cathayana* (Seemen) Rehder in Journ. Arnold Arb. i. 123. (1919).

Yunnan: Simao at 1,650 metres, 12330, 12330a., (type of *Quercus cathayana* Seemen). Simao at 1,500 metres, 12330b.

Native to India, Myanmar, China [Tibet (Xizang) and Yunnan], Thailand and Vietnam.

Lithocarpus sp. No. 1. (FAGACEAE)

Quercus neriifolia Seem. Anonymous in Kew List of Determinations (volume 34).

Yunnan: Mengzi, 10937.

Quercus nerifolia is synonymous with *Lithocarpus naiadarum*, a Hainan endemic. This taxon belongs to an allied species.

Lithocarpus sp. No. 2. (FAGACEAE)

Walsh in Aug. Henry Forst. Herb. 64. (1957).

Quercus glabra Walsh non Thunb. in Aug. Henry Forst. Herb. 111. (1957)

Yunnan: Mengzi, 13625.

Lithospermum erythrorrhizon Sieb. & Zucc. (BORAGINACEAE)

Henry in Journ. China Br. Roy. Asiat. Soc. 22: 277. (1887) (Chinese Names of Plants).

Lithospermum officinale Forbes & Hemsley non L. in Journ. Linn. Soc. Bot. 26: 154. (1890). *Lithospermum officinale* L. var. *erythrorhizon* (Sieb. & Zucc.) Maxim. Henry in Notes Econ. Bot. China 6. (1893).

Hubei: Yichang, 1145. In a glen near Pingshanba (near Yichang), 1585.

Known colloquially in the Three Gorges region as *tzu ts'ao* (*zi cao*). The red roots, according to Henry were used as both a dye and a drug.

Lithospermum zollingeri A. DC. (BORAGINACEAE)

Forbes & Hemsley in Journ. Linn. Soc. Bot. 26: 155. (1890).

Hubei: Yichang, 258, 1294.

According to Bretschneider, *Lithospermum zollingeri* was discovered in 1876 by Mr. J. P. Martin, the Postmaster of Shanghai on the Feng Wang Shan hills, 35 miles west of Shanghai.

Litsea acutivena Hayata (LAURACEAE) New species

Li in Woody Flora of Taiwan 215. (1963).

Lindera citriodora (Sieb. & Zucc.) Hemsl. Henry in Trans. Asiat. Soc. Jap. 24 suppl: 79. (1896) (List Pl. Formosa).

Taiwan: Wanchin, 114.

A medium-sized tree, native to southern China (including Hainan and Taiwan) and Indochina. This species was first collected by Henry who stated it was known as *shan-hu-chiao*, a name given to another species on mainland China.

Litsea chartacea (Wall. ex Nees) Hook. f. (LAURACEAE)

Allen in Ann. Missouri Bot. Gard. 25: 377. (1938).

Yunnan: Simao, 12013.

A widespread species first found in Nepal by Wallich and distributed from India to Vietnam and China.

Litsea cubeba (Lour.) Persoon (LAURACEAE)

Liou Ho in Laurac. Chine & Indochine 184. (1934), Allen in Ann. Missouri Bot. Gard. 25: 368. (1938).

Yunnan: Mengzi, 11326, 11395, 11395a. Simao, 12838.

A deciduous shrub or small tree to 7 metres tall with an aromatic odour of ginger. The Hani (an ethnic group) of Xishuangbanna in southern Yunnan call this species *siqbil* and use the leaves to treat cases of indigestion, vomiting and sunstroke. Native to India, southern China, Malaysia and Java.

Litsea elongata (Wall.) Benth. & Hook. f. (LAURACEAE)

Forbes & Hemsley in Journ. Linn. Soc. Bot. 26: 381. (1891), Liou Ho in Laurac. Chine & Indochine 195. (1934).

Hubei: South Badong, 6160. **Yunnan:** Mengzi, 9648.

Native to India, Nepal, China and Vietnam.

Litsea glutinosa (Lour.) C. B. Rob. (LAURACEAE)

Allen in Ann. Missouri Bot. Gard. 25: 384. (1938).

Litsea sebifera Pers. Liou Ho in Laurac. Chine & Indochine 196. (1934). *Litsea glutinosa* (Lour.) C. B. Rob. var. *brideliifolia* (Hay.) Merr. Allen in Ann. Missouri Bot. Gard. 25: 385. (1938).

Guangxi: without locality, 623. **Hainan:** Environs of Haikou, 8054, 8081, 8081a. Without locality, 8472. **Yunnan:** Simao, 12223, 12223a., 12223b., 12223c., 12223d., 12223e., 12223f.

A widespread species, distributed from tropical Asia to Malaysia.

Litsea hayatae Kanehira (LAURACEAE) New species

Li in Woody Flora of Taiwan 218. (1963).

Litsea sp. Henry in Trans. Asiat. Soc. Jap. 24 suppl: 79. (1896) (List Pl. Formosa).

Taiwan: Wanchin, 34, 73, 460, 1501.

A small evergreen tree, endemic to Taiwan where it grows in broad-leaved forest at low altitude. First collected by Henry.

Litsea honghoensis Liou Ho (LAURACEAE) New species

in Bull. Soc. Bot. France 80: 568. (1933), Allen in Ann. Missouri Bot. Gard. 25: 375. (1938).

Yunnan: Fengchunling at 2,300 metres, 10586a., (isotype).

Endemic to Yunnan.

Litsea hupehana Hemsley (LAURACEAE) New species

in Journ. Linn. Soc. Bot. 26: 382. (1891), Bretschneider in Hist. Eur. Bot. Disc. China 2: 790. (1898), Allen in Ann. Missouri Bot. Gard. 25: 391. (1938), Zheng in Journ. Wuhan Bot. Res. 2.(1): 53. (1984).

Hubei: Fang Xian, 6607 (isosyntype), 6660 (syntype).

Litsea lancifolia (Roxb.) Benth. & Hook. f. var. *pedicellata* Hook. f. (LAURACEAE)

Allen in Ann. Missouri Bot. Gard. 25: 396. (1938).

Lindera lancifolia Liou Ho non (Roxb.) Benth. &

Hook. f. in Laurac. Chine & Indochine 198. (1934).

Yunnan: Mengzi, 10759, 11143. Simao, 12235, 12301, 12301a., 12301b., 12301c.

Native to India, Myanmar and China (Yunnan).

Litsea martabanica (Kurz.) Hook. f. (LAURACEAE)

Litsea garrettii Gamble Allen in Ann. Missouri Bot. Gard. 25: 388. (1938), Liou Ho in Laurac. Chine & Indochine 196. (1934).

Yunnan: Simao, 11649, 11649b., 11649i.

China (Yunnan) and Indochina.

Litsea mollis Hemsley (LAURACEAE) New species

in Journ. Linn. Soc. Bot. 26: 383. (1891), Bretschneider in Hist. Eur. Bot. Disc. China 2: 790. (1898), Chun in Conmrt. Biol. Lab. Sci. Soc. China 1: 60. (1925), Liou in Laurac. Chine & Indoch. 186. (1932), Zheng in Journ. Wuhan Bot. Res. 2.(1): 53. (1984).

Litsea euosma W. W. Smith in Notes Roy. Bot. Gard. Edinburgh 13: 166. (1921), Liou Ho in Laurac. Chine & Indochine 187. (1934).

Hubei: Yichang, 2956. Liantuo, 2614, 4434. Badong, 1026 (December 1885), 3177 (syntype), 5035 (type). Yunnan: South of the Red River from Manmei at 1,950 metres, 9185 (syntype of Litsea euosma W. W. Smith), 10857, 10982.

Litsea monopetala (Roxb.) Pers. (LAURACEAE)

Litsea polyantha Juss. Forbes & Hemsley in Journ. Linn. Soc. Bot. 26: 384. (1891), Liou Ho in Laurac. Chine & Indochine 192. (1934).

Hainan: Environs of Kiungchow, 8219. Yunnan: Simao, 11794, 12005.

Native to southern Asia and the Pacific Islands.

Litsea pungens Hemsley (LAURACEAE) New species

in Journ. Linn. Soc. Bot. 26 384. (1891), Henry in Notes Econ. Bot. China 40. (1893), Bretschneider in Hist. Eur. Bot. Disc. China 2: 791. (1898), Chun in Cont. Biol. Lab. Sci. Soc. China 1: 59. (1925), Liou in Laurac. Chine & Indoch. 186. (1932), Allen in Ann. Missouri Bot. Gard. 25: 371. (1938), Zheng in Journ. Wuhan Bot. Res. 2.(1): 53. (1984).

Hubei: Yichang, 1302, 2617b., (type). Antelope Glen, 3617 (syntype). Badong, 230 (June 1885). Sichuan: South Wushan, 5579.

According to Henry the fruits of this small shrub were gathered in a green state, then salted and eaten. It was known at Yichang as the shan-hu-chiao and Henry also called it the mountain pepper.

Litsea sericea (Wall. ex Nees) Hook. f. (LAURACEAE)

Forbes & Hemsley in Journ. Linn. Soc. Bot. 26: 386. (1891), Zheng in Journ. Wuhan Bot. Res.2. (1): 54. (1984).

Hubei: Badong, 1452.

Litsea spp. (LAURACEAE)

Henry in Trans. Asiat. Soc. Jap. 24 suppl: 79. (1896) (List Pl. Formosa). Anonymous in Kew List of Determinations (volume 34).

Taiwan: Wanchin, 153, 1484. Yunnan: Mengzi, 10856, 11377. Simao, 12840.

Litsea szemaois (H. Liu) J. Li & H. W. Li LAURACEAE) New species

Litsea pierrei Lecompte var. szemaois H. Liu in Laurac. Chine & Indochine 174. (1934). Litsea baviensis Lecompte var. szemaois (H. Liu) C. K. Allen in Ann. Missouri Bot. Gard. 25: 377. (1938).

Yunnan: Simao at 1,450 metres, 12025 (isotype of Litsea pierrei Lecompte var. szemaois H. Liou).

Endemic to southern Yunnan and known only from the type locality at Simao.

Litsea variabilis Hemsley (LAURACEAE) New species

in Journ. Linn. Soc. Bot. 26: 386. (1891), Bretschneider in Hist. Eur. Bot. Disc. China 2: 791. (1898), Allen in Ann. Missouri Bot. Gard. 25: 393. (1938).

Hainan: Environs of Haikou, without number (syntype), 8431. Without locality, 8540 (syntype). Environs of Kiungchow, 8729 (syntype), 8761.

Native to Hainan, Guangdong, Guangxi, Yunnan and from there to Laos and Vietnam.

Livistona chinensis (Jacq.) R. Brown ex Mart. (ARACEAE)

Li in Woody Flora of Taiwan 918. (1963).

Trachycarpus excelsus Henry non Benth. & Hook. f. in Trans. Asiat. Soc. Jap. 24 suppl: 99. (1896) (List Pl. Formosa).

Taiwan: Wanchin, 821.

A tall palm, native to China, southern Japan and Micronesia. The dried seeds of this species are currently used in Chinese herbal medicine to relieve inflammation, as a painkiller and in treating cancer.

Lobaria isidiosa (Muell. Arg.) Vain. (LOBARIACEAE)

Stictina retigera Muell.-Arg. f. isidiosa Ejusd. Muell.-Arg. in Bull. Herb. Boissier 1: 236. (1893).

Hubei: Fang Xian, 6635. Changyang, on trees, 7928. (Lichen).

Lobaria meridionalis Vain. (LOBARIACEAE)

Sticta pulmonacea Ach. var. papillaris Del. Muell.-Arg. in Bull. Herb. Boissier 1: 236. (1893).

Hubei: Fang at 2,500 to 2,900 metres, 6799. (Lichen).

Lobelia chinensis Lour. (CAMPANULACEAE)

Wimmer in Engler, Pflanzenr. iv. 276b: 609. (1957).

Lobelia radicans Thunb Forbes & Hemsley in Journ. Linn. Soc. Bot. 26: 3. (1889).

Hubei: Yichang, 469, 1375. South Badong, 7301.

02

The *ban bian lian*, literally; 'the half-edged lily', used in Chinese herbal medicine to as a diuretic and in cases of snake and insect-bite.

Lobelia clavata F. E. Wimmer (CAMPANULACEAE) New species

in Repert. Spec. Nov. Regni Veg. 38: 78. (1935) Wimmer in Engler, Pflanzenr. iv. 276b: 653. (1957).

Yunnan: Simao, 12663 (type).

India (Assam) to China (Yunnan and Guizhou), Indochina and Myanmar.

Lobelia davidii Franch. (CAMPANULACEAE)

Lobelia pyramidalis Forbes & Hemsley non Wallich in Journ. Linn. Soc. Bot. 26. 3. (1889), Henry in Journ. China Br. Roy. Asiat. Soc. 22: 265. (1887) (Chinese Names of Plants), Nelson in Pim, The Wood & the Trees 219. (1984); in A Heritage of Beauty 315. (2000).

Hubei: Yichang, 145. Liantuo, 2995.

According to Henry this species was known colloquially at Liantuo (near Yichang) as *p'ao tung ken*. This species was discovered by Père Armand David in Jiujing in Jiangxi province.

Lobelia dopatrioides Kurz. var. **cantonensis** (E. Wimm.) W. J. de Wilde& Duyfjes (CAMPANULACEAE)

Lobelia alsinoides Lam. var. *cantonensis* E. Wimm. Wimmer in Engler, Pflanzenr. iv. 276b: 573. (1957)

Hainan: Environs of Kiungchow, 8302. Environs of Haikou, 8327.

Lobelia erectiuscula H. Hara (CAMPANULACEAE)

Lobelia erecta Hook. f. & Thoms. Wimmer in Engler, Pflanzenr. iv. 276b: 656. (1957).

Guizhou: without locality, 9271b.

Lobelia montana Reinw. ex Blume (CAMPANULACEAE)

Pratia montana (Reinw. ex Blume) Hassk. Anonymous in Kew List of Determinations (volume 34).

Yunnan: Mengzi, 9370.

Native to India, Nepal, China (Tibet (Xizang) and Yunnan) and Vietnam.

Lobelia nicotianifolia Roth (CAMPANULACEAE)

Lobelia colorata Wallich Wimmer in Engler, Pflanzenr. iv. 276b: 655. (1957).

Yunnan: Simao, 9271c.

Distributed in India, China (Guizhou and Yunnan) and Thailand.

Lobelia nummalaria Lam. (CAMPANULACEAE)

Pratia begonifolia (Wall.) Lindl. Henry in Trans. Asiat. Soc. Jap. 24 suppl: 56. (1896) (List Pl. Formosa), Forbes & Hemsley in Journ. Linn. Soc. Bot. 26: 2. (1889). *Pratia nummalaria* (Lam.) A. Br. & Aschers Wimmer in Engler, Pflanzenr. iv. 276b: 112. (1957).

Hubei: Liantuo, 3204, 4425. **Taiwan:** Wanchin, 1537.

Yunnan: Mengzi, 9842. Simao, 12010.

A prostrate perennial herb with creeping stems which root on contact with soil beneath. Distributed in Asia, the Hani (an ethnic group) of southern Yunnan call this species *dalgmq* and is used to treat cases of rheumatism, sprains and contusions.

Lobelia seguinii Levl. (CAMPANULACEAE)

Wimmer in Engler, Pflanzenr. iv. 276b: 648. (1957).

Lobelia pyramidalis Forbes & Hemsley non Wallich in Journ. Linn. Soc. Bot. 26. 3. (1889), Henry in Journ. China Br. Roy. Asiat. Soc. 22: 265. (1887) (Chinese Names of Plants), Nelson in Pim, The Wood & the Trees 219. (1984); in A Heritage of Beauty 315. (2000). *Lobelia seguinii* Levl. f. *longisepala* Wimmer in Akad. Anzeiger. Wein n. 14: 4. (1924), Wimmer in Engler, Pflanzenr. iv. 276b: 650. (1957).

Hubei: Yichang, 1085. In a cave at Moji Shan, 2239. **Yunnan:** Mengzi, 9671.

A handsome perennial herb to 1.5 metres tall. The Hani (an ethnic group) in southern Yunnan call this species *dovqyaol* and external applications of the plant are used to treat rheumatism, joint pain, sprains, bruises and infected wounds. Henry sent seeds of this species to Kew where they were received on April 5th 1887. *Lobelia seguinii* was collected by the Glasnevin Central China Expedition near Yichang (GCCE 453 & GCCE 477) in September 2004.

Lobelia sessilifolia Lamb. (CAMPANULACEAE)

Henry in Trans. Asiat. Soc. Jap. 24 suppl: 56. (1896) (List Pl. Formosa).

Taiwan: Wanchin, 1721.

Distributed in China (northeast and Taiwan), Korea and Japan.

Lobelia zeylanica L. (CAMPANULACEAE)

Lobelia affinis Wall. ex G. Don Anonymous in Kew List of Determinations (volume 34).

Yunnan: Mengzi, 11053.

Loeseneriella africana (Willd.) R. Wilczek. var. **obtusifolia** (Roxb.) N. Hallé (CELASTRACEAE)

Hippocratea obtusifolia Roxb. Anonymous in Kew List of Determinations (volume 34).

Yunnan: Pu'er, 13203. Yuanjiang, 13274.

Lomaria sp. (BLECHNACEAE)

Henry in Trans. Asiat. Soc. Jap. 24 suppl: 75. (1896) (List Pl. Formosa).

Taiwan: Oluanpi, 255.

Lomatogonium bellum (Hemsl.) H. Smith (GENTIANACEAE) New species

in Grana Palyn. 7(1): 145. (1967), T. N. Ho in Fl. Reip. Pop. 62: 335. (1988), Liu & Ho in Acta Phytotax. Sin. 30: 309. (1992).

Swertia bella Hemsley in Journ. Linn. Soc. Bot. 26:

138. (1890), Bretschneider in Hist. Eur. Bot. Disc. China 2: 788. (1898). *Pleurogyne rotata* (L.) G. Don var. *bella* (Hemsl.) Franchet in Bull. Soc. Bot. France 46: 310. (1899). *Pleurogyne carinthiaca* Batalin non Griseb. in Act. Hort. Petrop. xiii. 378. (1894).

Hubei: Fang at 2,750 to 2,900 metres (Shennongjia Forest District), 6919 (holotype of *Swertia bella* Hemsley).

A pretty subalpine species, also native to Tibet (Xizang) where it was collected by Younghusband and later again by Ludlow and Sherriff. This species was collected by the Sino-American Expedition in the Shennongjia Forest District in 1980.

Lomatogonium sp. (GENTIANACEAE)

Pleurogyne sp. Anonymous in Kew List of Determinations (volume 24).

Hubei: Badong, 2450, 2539.

Lonicera bournei Hemsl. (CAPRIFOLIACEAE)

Rehder in Ann. Rep. Missouri Bot. Gard. 14: 162. (1903).

Yunnan: Mengzi, 10570. Yuanjiang, 11573, 11573a.

Discovered by Frederick S. Bourne of the British Consular Service near Yuanjiang in January 1886. Native to China (Guangxi and Yunnan), Myanmar and Laos.

Lonicera calcarata Hemsley (CAPRIFOLIACEAE)
New species

in Hooker's Icon. Pl. 27: t. 2632. (1900), Hemsley in Journ. Linn. Soc. Bot. 34: 476. (1900), Schneider in Illus. Handb. Laubholzk. 2: 726. (1912).

Sichuan: Kangding at 2,900 to 4,500 metres, Henry's Chinese collector with Pratt, 8937. **Yunnan:** Mountains to the south-west of Mengzi at 1,650 metres, 10721 (isosyntype), 10721a., (type), 11721a. In a rocky ravine at Wulichun near Mengzi at 1,650 metres, 10721b., (isosyntype).

Endemic to western China.

Lonicera chrysantha Turcz. var. *koehneana* (Rehder) Q. E. Yang, Landrein, Borosova & Osborne (CAPRIFOLIACEAE)

New variety. *Lonicera koehneana* Rehder in Sargent, Trees & Shrubs. 1: 41. fig. xxi. (1902), Rehder in Ann. Rep. Missouri Bot. Gard. 14: 141. (1903), J. H. Veitch in Hortus Veitchii 369. (1906), O' Brien in Augustine Henry, An Irish Plant Collector in China 26. (2002).

Sichuan: South Wushan, 5613 (syntype of *Lonicera koehneana* Rehder), 5894 (type of *Lonicera koehneana* Rehder). **Hubei:** Badong, 6052.

Introduced to cultivation by Wilson through Messrs Veitch who first flowered it at their Coombe Wood nursery during the summer of 1905 (W. J. Bean in error states Wilson introduced in in 1908). In Ireland this subspecies

is represented by a single plant grown in the chain-tent shrubbery at the National Botanic Gardens, Glasnevin, this old shrub is surrounded by other Wilson-Veitch collections and so it may be from the original introduction.

Lonicera confusa DC. (CAPRIFOLIACEAE)

Rehder in Ann. Rep. Missouri Bot. Gard. 14: 156. (1903).

Hainan: 32 km (20 miles) west of Haikou, 8134. Lingmen (in the interior of the island), 8602.

A species sometimes confused with *Lonicera japonica* Thunb. It was also collected by Burbidge in Borneo.

Lonicera crassifolia Batalin (CAPRIFOLIACEAE)
New species

in Act. Hort. Petrop. 12: 172. (1892) Wolf in Gartenfl. 42: 332. (1893), Bretschneider in Hist. Eur. Bot. Disc. China 2: 785. (1898).

Sichuan: South Wushan, 5896. Henry's Chinese collector with Pratt, Wa Shan, 8927.

A beautiful, very prostrate species that has recently been introduced to cultivation.

Lonicera elisae Franchet (CAPRIFOLIACEAE)

Lonicera sp. nov; foliis deciduis non visis, floribus albis praecocibus glabris, corolla fere aequaliter 5-lobata. Forbes & Hemsley in Journ. Linn. Soc. Bot. 23: 368. (1888). *Lonicera infundibulum* Franchet in Journ. de Bot. 315. (1886), Bretschneider in Hist. Eur. Bot. Disc. China 2: 785. (1898).

Hubei: Badong, 3790.

Lonicera ferdinandi Franch. (CAPRIFOLIACEAE)

Forbes & Hemsley in Journ. Linn. Soc. Bot. 23: 361. (1888), Henry in Notes Econ. Bot. China 62. (1893), Rehder in Ann. Rep. Missouri Bot. Gard. 14: 78. (1903).

Hubei: Yichang, 1107. Liantuo, 3011, 3188.

A robust deciduous shrub producing pairs of yelllow flowers in the upper leaf axils in June followed by bright-red fruits. Native to China and Mongolia, it was discovered in June 1866 by Père Armand David and was introduced to cultivation in 1900 by E. H. Wilson.

Lonicera ferruginea Rehder (CAPRIFOLIACEAE)
New species

in Sargent, Trees & Shrubs. 1: 43. fig. xxii. (1902), Rehder in Ann. Rep. Missouri Bot. Gard. 14: 154. (1903).

Yunnan: In mountain forest to the west of Simao at 1,700 metres, 11921 (type). In forests near Simao at 1,700 metres, 11921a., (syntype). In forests to the south of Simao at 1,700 metres 11921b., (syntype). Simao, 13453.

Southern China and Indochina.

Lonicera fragrantissima Lindl. & Pax. var. *lancifolia* (Rehder) Q. E. Yang, Landrien, Borosova & Osborne (CAPRIFOLIACEAE)

Lonicera standishii Forbes & Hemsley non Carr. in

Journ. Linn. Soc. Bot. 23: 367. (1888), Zheng in Journ. Wuhan Bot. Res. 2(1): 194. (1984). *Lonicera standishii* Carr. f. *lancifolia* Rehder in Ann. Rep. Missouri Bot. Gard. 14: 81. (1903).

Hubei: Yichang, 1290, 3373, 5199. Liantuo, 3801, 6373. Changleping, 7818.

A semi-evergreen shrub to 2.5 metres tall, bearing creamy-white flowers produced in pairs in the leaf axils between November to March. Henry stated the edible fruits were called *k'u-t'ang-kuo*

Lonicera gynochlamydea Hemsley (CAPRIFOLIACEAE) New species

in Journ. Linn. Soc. Bot. 23: 362. (1888), Bretschneider in Hist. Eur. Bot. Disc. China 2: 785. (1898), Rehder in Ann. Rep. Missouri Bot. Gard. 14: 75. (1903), Schneider in Illus. Handb. Laubholzk. 2: 695. (1912), Zheng in Journ. Wuhan Bot. Res. 2.(1): 192. (1984), Bean in Trees & Shrubs 2: 605. (1992).

Caprifolium gynochlamydeum (Hemsl.) Kuntze in Rev. Gen. Pl. 1: 274. (1891).

Hubei: Badong, 3751, 5375, 5428 (isotype), 6320, 7341. Fang Xian, 6529. Changyang, 5241. On the mountains above Jianshi at 1,650 metres, 6036. Yichang, 7610.

An erect deciduous shrub bearing white, pink tinged flowers in pairs from leaf axils on purplish shoots in May. Introduced to cultivation by Wilson through Veitch's Coombe Wood nursery in 1901. While travelling through Badong (the type locality) with members of the Glasnevin Central China Expedition we found this species locally common and in places it grew on roadside cliffs.

Lonicera henryi Hemsley (CAPRIFOLIACEAE) New species

in Journ. Linn. Soc. Bot. 23: 359. (1888), Henry in Notes Econ. Bot. China 62. (1893), Bretschneider in Hist. Eur. Bot. Disc. China 2: 785. (1898), Rehder in Ann. Rep. Missouri Bot. Gard. 14: 148. (1903), Hemsley in Curtis's Bot. Mag. 137: t. 8375. (1911), Morley in Glasra 3: 79. (1979), Zheng in Journ. Wuhan Bot. Res. 2.(1): 192. (1984), Pim in The Wood & the Trees 35. (1984); in A Heritage of Beauty 315. (2000), O' Brien in Augustine Henry, An Irish Plant Collector in China 25. (2002).

Caprifolium henryi (Hemsl.) Kuntze in Rev. Gen. Pl. 1: 274. (1891).

Hubei: Badong, 1789, 2804, 2844, 4015, 5064. Jianshi, 5845.

An evergreen climber with oblong-lanceolate leaves, carrying in June purplish-red flowers. Wilson described this species as being abundant in the areas of Yichang and Badong and common in Changyang, Fang and Xingshan. Introduced to cultivation by Wilson through the Arnold Arboretum in 1908 and later by Forrest (Forrest 14955 was collected in the Mekong-Salween divide in September 1917). In the Veitch catalogue *New Hardy Plants from Western China* (Autumn 1912) this species was available for two shillings and six pence. Augustine Henry and his wife Alice grew this species in their garden at Ranelagh in the south Dublin suburbs. *Lonicera henryi* was collected by the Glasnevin Central China Expedition (GCCE 306) in Badong in October 2002 and by the 2004 Glasnevin Expedition to the Three Gorges region in Xingshan (GCCE 508). A popular climber in Irish gardens.

Lonicera hildebrandiana Collett & Hemsley (CAPRIFOLIACEAE) New to China

Rehder in Ann. Rep. Missouri Bot. Gard. 14: 163. (1903), Henry in Flora & Sylva 1: 217. (1903), Bean in Trees & Shrubs 2: 606. (1992), O' Brien in Augustine Henry, An Irish Plant Collector in China 25. (2002).

Yunnan: Mountains to the west of Simao at 1,600 metres, 11905. Mengzi, 10934.

A very beautiful and vigorous evergreen climbing shrub to 27 metres tall. Previous to Henry's collections this species was known only from the Shan hills in Myanmar where it had been discovered by Sir Henry Collett in 1888. Distributed in India, Bangladesh, Myanmar, China (Guangxi and Yunnan) and Thailand. In terms of leaf, flower and fruit it is the largest of the honeysuckles. In Ireland it is grown at Ilnacullin (Garinish Island), County Cork and at the Talbot Botanic Gardens, Malahide Castle in Dublin. It first flowered in cultivation at the National (then the Royal) Botanic Gardens, Glasnevin in August 1898.

Lonicera hypoglauca Miq. (CAPRIFOLIACEAE)

Lonicera affinis Hook. & Arn. var. *hypoglauca* (Miq.) Rehd. Rehder in Ann. Rep. Missouri Bot. Gard. 14: 158. (1903). *Lonicera affinis* Bean non Hook. & Arn. in Gard. Chron. ser. 3, 35: 372. (1904).

Yunnan: Mengzi, 10202, 10302, 10313. Simao, 11741, 12623.

Native to China and Japan. Introduced to cultivation by Henry through the Royal Botanic Gardens, Kew from seeds sent by him from Mengzi in 1897. An evergreen honeysuckle, it was trained on wires near the roof of the temperate house where it flowered in the summer of 1902 but apparently better showed its merits as a greenhouse climber in May 1904. Henry's plant was described as, 'a sheet of yellow flowers and possessed of a powerful fragrance which pervaded the whole house.'

Lonicera japonica Thunb. (CAPRIFOLIACEAE)

Henry in Journ. China Br. Roy. Asiat. Soc. 22: 239. (1887) (Chinese Names of Plants), Forbes & Hemsley in Journ. Linn. Soc. Bot. 23: 364. (1888), Henry in Notes

Econ. Bot. China 62. (1893), Rehder in Ann. Rep. Missouri Bot. Gard. 14: 159. (1903), Zheng in Journ. Wuhan Bot. Res. 2.(1): 193. (1984).

Hubei: Yichang, 90, 557, 1359. Jianshi, 5880. Badong, 5025, 5103.

A very vigorous climber to 10 metres tall bearing pairs of white, purple tinged fragrant flowers from the leaf axils in June. Native to China, Japan and Korea and introduced to cultivation from China during the early 19th century. The commonest of the many climbing honeysuckles in Hubei and known colloquially around Yichang as *chin-yin-hua*, according to Henry.

Lonicera ligustrina Wall. var. *pileata* (Oliver) Franch. (CAPRIFOLIACEAE) New variety

Lonicera pileata Oliver in Hooker's Icon. Pl. 16: t. 1585. (1887), Forbes & Hemsley in Journ. Linn. Soc. Bot. 23: 365. (1888), Bretschneider in Hist. Eur. Bot. Disc. China 2: 785. (1898), Rehder in Ann. Rep. Missouri Bot. Gard. 14: 76. (1903), Bean in Gard. Chron. ser. 3, 35: 243. (1904), Schneider in Illus. Handb. Laubholzk. 2: 695. (1912), Besant in Gard. Chron. ser. 3, 98: 335. (1935), Morley in Glasra 3: 79. (1979), Zheng in Journ. Wuhan Bot. Res. 2.(1): 194. (1984), Nelson in Pim, The Wood & the Trees 220. (1984), Bean in Trees and Shrubs 2: 617. (1992), Nelson in A Heritage of Beauty 315. (2000). *Lonicera pileata* Oliv. var. *linearis* Rehder in Sargent, Pl. Wilson. 1: 143 (1911).

Hubei: Xiling (Yichang) Gorge, 1236 (type of *Lonicera pileata* Oliver), 4358. Changyang, 7802. **Sichuan:** South Wushan, 7120. Henry's native collctor with Pratt, no locality stated but may be Leshan, Emei Shan, Wa Shan or Kangding, 8923, 8933. **Yunnan:** Simao at 1,600 metres, 11800 (type of *Lonicera pileata* Oliv. var. *linearis* Rehder

A low-growing evergreen spreading shrub, valuable in ground-cover plantings. The flowers, though insignificant are sweetly fragrant and are followed in autumn by handsome violet coloured fruits. Introduced to cultivation by Wilson when collecting for Messrs Veitch in 1900, this species first flowered at the Royal Botanic Gardens, Kew in April 1904. Wilson's 833 from Sichuan, originally ennummerated by Rehder (in Sargent Pl. Wilson. 1: 135. 1913) under this variety is the type of *Lonicera nitida* Wilson (see Gard. Chron. ser. 3. 102. (1911). From Wilson's 833 plants of *Lonicera nitida* were raised at Glasnevin. Bean states that the clone 'Ernest Wilson, once a popular hedge plant was probably raised from the same seed collection, the Glasnevin plant certainly matches it perfectly and has been labeled as such. Alice Henry grew *Lonicera ligustrina* var. *pileata* in her garden at Ranelagh in Dublin and Henry stated it was common in the Yichang Gorges. That statement still holds true, *Lonicera ligustrina* var. *pileata* was collected by

the Glasnevin Central China Expedition (GCCE682) in the Red Dragon Reserve in Xingshan in October 2004.

Lonicera maackii (Rupr.) Maxim. (CAPRIFOLIACEAE)

Forbes & Hemsley in Journ. Linn. Soc. Bot. 23: 364. (1888), Henry in Notes Econ. Bot. China 63. (1893), Rehder in Ann. Rep. Missouri Bot. Gard. 14: 141. (1903), Zheng in Journ. Wuhan Bot. Res. 2.(1): 193. (1984).

Lonicera maackii (Rupr.) Maxim. f. *podocarpa* Franch. ex Rehd. Zheng in Journ. Wuhan Bot. Res. 2(1): 193. (1984). *Lonicera quinquelocularis* Forbes & Hemsley non Hardw. in Journ. Linn. Soc. Bot. 23: 365. (1888), Zheng in Journ. Wuhan Bot. Res.2. (1): 194. (1984).

Hubei: Badong, 726, 3184, 4046. Liantuo, 2966, 3853, 4454, 4645. Changyang, 5228, 5266. Fang Xian, 6579. **Yunnan:** Mengzi, 10016a.

A deciduous shrub to 3.5 metres tall bearing fragrant, pure white flowers in pairs on the upper side of branches. Introduced to cultivation to St. Petersburg in 1880 from China and in 1900 by Wilson through Messrs Veitch. In Veitch's *Novelties for 1908—1909* plants were available for one shilling and six pence each. A handsome sight in autumn when laden with fruits. Native to Russia, China, Korea and Japan.

Lonicera microphylla Willd. ex Roem. & Schult. (CAPRIFOLIACEAE)

Zheng in Journ. Wuhan Bot. Res. 2.(1): 194. (1984).

Hubei: Changleping, 7657.

Lonicera mucronata Rehder (CAPRIFOLIACEAE) New species

in Ann. Rep. Missouri Bot. Gard. 14: 83. (1903), Rehder in Ann. Rep. Missouri Bot. Gard. 14: 47. (1907).

Sichuan: South Wushan, 5519 (type).

Lonicera nervosa Maxim. (CAPRIFOLIACEAE)

Rehder in Ann. Rep. Missouri Bot. Gard. 14: 121. (1903), Zheng in Journ. Wuhan Bot. Res.2. (1): 194. (1984).

Hubei: Fang Xian, 6862.

Discovered and introduced to cultivation by the Russian botanical explorer, Nicolai Mikailovich Przewalski who sent seeds to the Botanic Gardens at St. Petersburg from Gansu. Plants first flowered at St. Petersburgh in April 1886.

Lonicera similis Hemsley (CAPRIFOLIACEAE) New species

in Journ. Linn. Soc. Bot. 23: 366. (1888), Bretschneider in Hist. Eur. Bot. Disc. China 2: 785. (1898), Rehder in Ann. Rep. Missouri Bot. Gard. 14: 155. (1903), Zheng in Journ. Wuhan Bot. Res. 2.(1): 194. (1984).

Lonicera macrantha Forbes & Hemsley non (D. Don) Spreng. in Journ. Linn. Soc. Bot. 23: 365. (1888), Henry in Notes Econ. Bot. China 62. (1893). *Lonicera delavayi*

Franch. Rehder in Ann. Rep. Missouri Bot. Gard. 14: 155. (1903).

Hubei: Yichang, 1359 (in part), 3510 (type). Antelope Glen, 4155. **Sichuan:** South Wushan, 5543. **Yunnan:** Mengzi, 10797.

The *jin yin hua*, used in medicine. Distributed in Myanmar and China.

Lonicera sp. no. 1. (CAPRIFOLIACEAE)

Anonymous in Kew List of Determinations (volume 24).

Hubei: Yichang, 685, 1097, 1384. South Donghu, 7705. Liantuo, 3070. Changyang, 5250. **Sichuan:** South Wushan, 5561.

Lonicera sp. no. 2. (CAPRIFOLIACEAE)

Rehder in Ann. Rep. Missouri Bot. Gard. 14: 202. (1903).

Sichuan: South Wushan, 5707.

Lonicera subaequalis Rehder (CAPRIFOLIACEAE)
New species

in Ann. Rep. Missouri Bot. Gard. 14: 172. (1903).

Sichuan: Henry's Chinese collector with Pratt, no locality stated but may be Leshan, Emei Shan, Wa Shan or Kanding, 8936.

Lonicera tangutica Maxim. (CAPRIFOLIACEAE)

Lonicera sp. nov. Hemsley in Journ. Linn. Soc. Bot. 23: 368. (1888). *Lonicera longa* Rehder in Ann. Rep. Missouri Bot. Gard. 14: 61. (1903). *Lonicera saccata* Rehder in Sargent, Trees & Shrubs. 1: 39. pl. xx. (1905), Rehder in Rep. Missouri Bot. Gard. 14: 60. (1903).

Hubei: Badong, 4053. South Badong, 5306, 5311 (syntype of *Lonicera saccata* Rehder). Changyang, 5680b. Without locality, 5680a., (syntype of *Lonicera saccata* Rehder). Xingshan, 6960 (type of *Lonicera longa* Rehder). **Sichuan:** South Wushan, 5680 (type of *Lonicera saccata* Rehder). Henry's Chinese collector with Pratt, no locality stated but may be Leshan, Emei Shan, Wa Shan or Kanding, 8935.

Lonicera tangutica was discovered by Przewalski in Gansu in 1872, he introduced it through St. Petersburgh Botanic Garden where this species flowered in 1886.

Lonicera tragophylla Hemsley (CAPRIFOLIACEAE)
New species

in Journ. Linn. Soc. Bot. 23: 367. (1888), Henry in Notes Econ. Bot. China 62. (1893), Bretschneider in Hist. Eur. Bot. Disc. China 2: 785. (1898), Rehder in Sargent, Trees & Shrubs. 1: 91. fig. xlvi. (1903); in Ann. Rep. Missouri Bot. Gard. 14: 193. (1903), Schneider in Illus. Handb. Laubholzk. 2: 742. (1912), Morley in Glasra 3: 79. (1979), Zheng in Journ. Wuhan Bot. Res. 2.(1): 195. (1984), Nelson in Pim, The Wood & the Trees 232. (1984), Bean in Trees & Shrubs 2: 626. (1992), Nelson in A Heritage of Beauty 315. (2000), O' Brien in Augustine Henry, An Irish Plant Collector in China 26. (2002).

Hubei: Badong, on cliffs, scrambling through trees, 1707, 4010, 5898, 5898b., (type). Fang Xian, 5898a.

A deciduous climbing shrub to 4 metres tall carrying terminal clusters of up to twenty large golden yellow flowers held by an unusual disk-like bract in late June and July. Perhaps the finest of the hardy honeysuckles but not scented. It would be worth crossing with the native Irish *Lonicera peryiclymenum* L. to introduce this scent. Henry claimed this species was rare in Badong and it was known colloquially in that region as *ta chen yen hua*. *Lonicera tragophylla* was introduced to cultivation by Wilson through Veitch's Coombe Wood nursery in 1900. It first flowered there in July 1905 and in their *Novelties for 1908—1909* plants were selling for one shilling and six pence each. Distributed in Anhui, Gansu, Henan, Hubei, Sichuan, Shaanxi, Shanxi and Zhejiang.

Lonicera yunnanensis Franch. (CAPRIFOLIACEAE)

Lonicera yunnanensis Franch. var. *tenuis* Rehder in Ann. Rep. Missouri Bot. Gard. 14: 179. (1903), Bean in Trees & Shrubs 2: 627. (1992).

Yunnan: On rocky mountains near Mengzi at 2,000 metres, 10798 (type of *Lonicera yunnanensis* Franch. var. *tenuis* Rehder).

A low growing creeper bearing stalkless white flowers (later fading to yellow) in pairs in the axils of leaves. Introduced to cultivation by Wilson in 1901 through Veitch's Coombe Wood nursery.

Lophatherum gracile Brongn. (POACEAE)

Rendle in Journ. Linn. Soc. Bot. 36: 420. (1904).

Lophatherum lehmannii Nees Henry in Trans. Asiat. Soc. Jap. 24 suppl: 109. (1896) (List Pl. Formosa).

Hubei: Liantuo, 4635. **Taiwan:** Wanchin, 1647.

A tufted perennial to 30 cm tall, distributed from India to central and southern China, Japan, Malesia and northern Quensland. The dried stems and leaves of this grass are currently used in Chinese herbal medicine as a diuretic.

Loranthus delavayi Tieghem (LORANTHACEAE)

Lecompte in Sargent Pl. Wilson. 3: 316. (1917), Zheng in Journ. Wuhan Bot. Res. 2.(1): 35. (1984).

Loranthus odoratus Forbes & Hemsley non Wallich in Journ. Linn. Soc. Bot. 406. (1894). *Loranthus delavayi* Teighem var. *latifolius* Teighem Lecompte in Sargent Pl. Wilson. 3: 316. (1917). *Loranthus yadoriki* Henry non Sieb. ex Maxim. in Trans. Asiat. Soc. Jap. 24 suppl: 80. (1896) (List Pl. Formosa). *Hyphear delavayi* (Teigh.) Dans. Danser in Blumea 2: 45. (1936).

Hubei: Changyang on *Zanthoxylum* spp. and *Quercus*

variabilis Blume, 7849. **Taiwan:** Wanchin, 58. **Yunnan:** Mengzi at 2,000 metres, 9112. Simao at 2,700 metres, 12892.

A semi-parasitic shrub, native to Myanmar, Vietnam, central and southern China including Taiwan. In Taiwan this species grows on the branches of *Alnus japonica*, *Cyclobalanopsis morii*, *Cyclobalanopsis stenophylla* var. *stenophylloides*, *Pasania harlandii*, *Liquidamber formosana*, *Quercus variabilis*, *Carpinus kawakamii*, *Machilus thunbergii* and *Machilus acuminatissima*. According to Henry this species was parasitic on oaks (*Quercus* spp.) at Mengzi.

Loranthus spp. (LORANTHACEAE)
Anonymous in Kew List of Determinations (volume 34).

Yunnan: Mengzi, 9989, 10672, 10912, 11085, 13786. Simao, 13589.

Loropetalum chinense (R. Br.) Oliv. (HAMAMELIDACEAE)
Forbes & Hemsley in Journ. Linn. Soc. Bot. 23: 290. (1887), Henry in Journ. China Br. Roy. Asiat. Soc. 22: 235. (1887) (Chinese Names of Plants), Rehder & E. H. Wilson in Sargent Pl. Wilson. 1: 420. (1913), Zheng in Journ. Wuhan Bot. Res. 2(1): 65. (1984).

Hubei: Yichang, 254, 503, 998. In the San You Dong Glen (June 1886), 1634. Badong, 1634. **Yunnan:** In the vacinity of Simao at 1,300 to 1,500 metres, 12490, 12490a.

A wiry, evergreen shrub to 1.5 metres tall , producing spidery-white flowers in February and March like those of *Hamamelis*. Native from India (Assam) to China, Laos, Thailand and one region of Japan. This evergreen, spring flowered shrub is abundant in warmer parts of China and in Henry's and Wilson's time at Yichang it was abundant on the cliffs of the gorges and in the surrounding glens where it was known as *chi-mu* according to Henry. Introduced to cultivation by Charles Maries in 1880. It grows well at the Fota Arboretum in County Cork, the former residence of Lord Barrymore. This species was collected by the Glasnevin Central China Expedition (GCCE 495) in the hills above Yichang past the San You Dong Glen in October 2004. The genus is monotypic and numerous cultivars have been selected in China from the red-pink flowered *Loropetalum chinense* (R. Br.) Oliv. f. *rubrum* H. T. Chang.

Lotus corniculatus L. (FABACEAE)
Forbes & Hemsley in Journ. Linn. Soc. Bot. 23: 155. (1886) Diels in Bot. Jahrb. 29: 411. (1900).

Hubei: Yichang, 672, 719. Antelope Glen, 3383. Liantuo, 2001. Badong, 202, 1751, 1752, 4807. **Yunnan:** Mengzi, 9741.

Bird's foot trefoil, a major forage plant in the north temperate region.

Loxocalyx sp. (LAMIACEAE)
Anonymous in Kew List of Determinations (volume 34).

Yunnan: Mountain forests to the north of Mengzi at 2,100 metres, 10237.

Loxocalyx urticifolius Hemsley (LAMIACEAE) New genus
in Journ. Linn. Soc. Bot. 26: 309. fig. 5. (1890), Bretschneider in Hist. Eur. Bot. Disc. China 2: 790. (1898), Dunn in Notes Roy. Bot. Gard. Edinburgh 6: 183. (1915), Morley in Glasra 3: 79. (1979).

Hubei: Xingshan, 6482. Fang at 2,300 to 2,950 metres, 6795. **Sichuan:** South Wushan, 7266.

Loxogramme grammitoides (Bak.) C. Chr. (LOXOGRAMMACEAE) New species
Gymnogramma grammitoides Baker in Ann. Bot. 5: 486. (1890). *Selliguea grammitoides* (Baker) C. Chr. in Bull. Soc. Bot. France 5: 21. (1905). *Selliguea lanceolata* (Sw.) Fée Christ in Bull. Herb. Boiss. 6: 880. (1898).

Hubei: South Badong, 5451. Jianshi, 5451a. **Sichuan:** South Wushan, 7531 (type of *Gymnogramma grammitoides* Baker).

Loxogramme involuta (D. Don) C. Presl. (POLYPODIACEAE)
Selliguea involuta (D. Don) Kunze Christ in Bull. Herb. Boiss. 6: 879. (1898).

Yunnan: In the mountains to the east of Mengzi at 2,200 metres, 9059. On mountains to the south-west of Mengzi at 1,650 metres, 10343.

Loxogramme lanceolata (Sw.) C. Presl. (POLYPODIACEAE)
Polypodium loxogramme Mett. Diels in Bot. Jahrb. 29: 204. (1900). *Gymnogramma lanceolata* (Sw.) Hook. Anonymous in Kew List of Determinations (volume 34).

Hubei: Badong, 5131, 5452. Yichang, 7947. **Yunnan:** Mengzi, 9206, 11554. On the great mountain range at Fengchunling at 2,600 metres, on a tree, 10171.

Loxostigma griffithii (Wight) C. B. Clarke (GESNERIACEAE)
Anonymous in Kew List of Determinations (volume 34).

Yunnan: Mengzi, 11245.

Native to India, Bhutan, Nepal, Myanmar, western China and Vietnam.

Loxostigma spp. (GESNERIACEAE)
Anonymous in Kew List of Determinations (volume 34).

Yunnan: Mengzi, 10882. Simao, 12587, 13290.

Luculia gratissima (Wallich) Sweet (RUBIACEAE)
Hutchinson in Sargent Pl. Wilson. 3: 407. (1917).

Yunnan: In the mountains to the west of Simao at 1,500

metres, 12489. In the mountains to the west of Simao at 1,660 metres, 12489b.

An evergreen shrub to 4 metres tall carrying terminal corymbs of intensly fragrant, pink tubular flowers from May to October. Native to India, the Himalaya, China [Tibet (Xizang) and Yunnan] and Vietnam.

Luculia pinceana Hook (RUBIACEAE)

Luculia intermedia Hutchinson in Sargent Pl. Wilson. 3: 408. (1917).

Yunnan: In Pingbian in forests on the Daweishan range at 2,600 metres, 9023 (type of *Luculia intermedia* Hutchinson). Mengzi 9023a. Mountains to the southwest of Mengzi at 2,600 metres, 9023b. On the slopes of Fengchunling at 2,000 metres, 9023c.

An evergreen shrub to 2 metres tall, bearing cymes of white or pink tubular, intensely fragrant flowers in autumn. Native to India, the Himalaya, western China and Vietnam.

Ludwigia adscendens (L.) H. Hara (ONAGRACEAE)

Jussiaea repens L. Anonymous in Kew List of Determinations (volume 24).

Hainan: Near the Kiungchow Pagoda, 7977.

Ludwigia epilobioides Maxim. (ONOGRACEAE)

Ludwigia prostrata Henry non Roxb. in Trans. Asiat. Soc. Jap. 24 suppl: 45. (1896) (List Pl. Formosa).

Taiwan: Wanchin, 1730. Kaohsiung, 2029.

A well branched annual herb to 60 cm tall. Native to much of China, Vietnam, Korea and southern Japan. Often a weed of rice paddies.

Ludwigia octovalvis (Jacq.) P. H. Raven. (ONAGRACEAE)

Jussiaea suffruticosa L. Henry in Trans. Asiat. Soc. Jap. 24 suppl: 45. (1896) (List Pl. Formosa).

Hainan: Environs of Haikou, 8024. At Nodoa in the interior of the island, 8455 (in part). **Taiwan:** Oluanpi, 236, 986. Kaohsiung, 243. **Yunnan:** Mengzi, 11281.

A woody herb to 4 metres tall, widely distributed through the tropics and subtropics of the world. Widespread in China.

Ludwigia perennis L. (ONOGRACEAE)

Jussiaea caryophyllaea Lam. Anonymous in Kew List of Determinations (volume 24). *Ludwigia parviflora* Roxb. Henry in Trans. Asiat. Soc. Jap. 24 suppl: 45. (1896) (List Pl. Formosa).

Hainan: At Nodoa in the interior of the island, 8455 (in part). **Taiwan:** Wanchin, without number.

An annual herb to 1 metre tall, native to Africa, tropical and subtropical Asia to Malaysia and throughout tropical Australia and New Caledonia. Rare in Chinese Taiwan where it is found at low altitudes.

Ludwigia prostrata Roxb. (ONOGRACEAE)

Forbes & Hemsley in Journ. Linn. Soc. Bot. 23: 309.

(1887).

Hubei: Yichang, 399, 2797, 2798. Liantuo, 3037.

Luffa cylindrica (L.) M. Roem. (CUCURBITACEAE)

Anonymous in Kew List of Determinations (volume 24).

Hainan: Environs of Haikou, 8429.

According to Henry this was the *shui kwa* (watergourd) or the *si gua luo* i.e. 'net of string melon', perhaps better known as the smooth loofah or dish-cloth gourd. A vigorous annual found in tropical Asia and probably native to India. It is thought to have been introduced to China in the Tang dynasty, 600 A. D. The mature fruits may be over 0.75 metres in length. Japan is the main producer of the loofah and a single acre of land there can produce up to 24,000 fruits. The fiberous tissues are used in the USA for absorbing oil from water, as shock absorbers or in Asia for stuffing pillows and saddles or simply as a back scrub.

Luisia morsei Rolfe (ORCHIDACEAE)

Luisia teretifolia Rolfe non Gaud. in Journ. Linn. Soc. Bot. 36: 33. (1903).

Hainan: Environs of Kiungchow, 8686. **Yunnan:** Simao, 12423.

Distributed in southern and western China, Laos, Vietnam and Thailand.

Luisia aff. *teretifolia* Gaud. (ORCHIDACEAE)

Rolfe in Journ. Linn. Soc. Bot. 36: 34. (1903). *Luisia* sp. Henry in Trans. Asiat. Soc. Jap. 24 suppl: 92. (1896) (List Pl. Formosa).

Taiwan: Oluanpi, 695.

Impossible to determine without flowers. Three species are recorded in the current Flora of Taiwan 5: 959. (2000), these are *Luisia cordata* Fukuy., *Luisia megasepala* Hayata and *Luisia teres* (Thunb.) Blume.

Lumnitzera racemosa Willd. (COMBRETACEAE)

Henry in Trans. Asiat. Soc. Jap. 24 suppl: 43. (1896) (List Pl. Formosa), Li in Woody Flora of Taiwan 639. fig. 253. (1963).

Taiwan: Kaohsiung, without number.

A small tree to 10 metres tall, widely distributed across tropical Africa, Asia to the Pacific Islands and Australia. Henry described it as a shrub with white flowers, occuring in swampy ground (mangrove) alongside the Kaohsiung lagoon.

Luzula effusa Buchen. (JUNCACEAE)

Brown in Journ. Linn. Soc. Bot. 36: 161. (1903).

Hubei: Fang at 1,900 to 3,000 metres, 6809.

Distributed in the eastern Himalaya to China (including Taiwan) and New Guinea. A. Henry 6809 was collected in what is now the Shennongjia Forest District.

02

Luzula effusa Buchenau var. *chinensis* (N. E. Brown) K. F. Wu (JUNCACEAE) New variety

Luzula chinensis N. E. Brown in Journ. Linn. Soc. Bot. 36: 161. (1903), Buchenau in Engler, Pflanzenr. 36: 61. (1906).

Sichuan: Henry's Chinese collector with Pratt, no locality stated but may be Leshan, Emei Shan, Wa Shan, Emei Shan or Kangding, 8829 (isotype of *Luzula chinensis* N. E. Brown).

Luzula multiflora (Ehrh.) Lej. (JUNCACEAE)

Luzula campestris Brown non (L.) DC. in Journ. Linn. Soc. Bot. 36: 160. (1903).

Hubei: Badong, 1793, 3739.

Luzula pilosa (L.) Willd. (JUNCACEAE)

Luzula vernalis (Reichards) DC. Brown in Journ. Linn. Soc. Bot. 36: 162. (1903).

Hubei: South Badong, 5206, 5315, 5315a. Without locality, 5315b.

Luzula plumosa Wall. ex E. Mey. subsp. *reflexa* (Ebinger) Z. Kaplan (JUNCACEAE) New subspecies

Luzula vernalis Brown non (Reichard) DC. in Journ. Linn. Soc. Bot. 162. (1903). *Luzula plumosa* Wall. ex E. Mey. var. *reflexa* Ebinger in Mem. New York Bot. Gard. (10)5: 298. (1964).

Hubei: Changleping, 6316, (isotype of *Luzula plumosa* Wall. ex E. Mey. var. *reflexa* Ebinger).

Luzula sp. No. 1. (JUNCACEAE)

Anonymous in Kew List of Determinations (volume 24).

Hubei: Badong, 3733.

Luzula sp. No. 2. (JUNCACEAE)

Anonymous in Kew List of Determinations (volume 34).

Yunnan: Mengzi, 10947.

Lychnis sp. (CARYOPHYLLACEAE)

Anonymous in Kew List of Determinations (volume 24).

Hubei: Xingshan, 6958.

Lycianthes biflorum (Lour.) Bitter (SOLANACEAE)

Henry in Trans. Asiat. Soc. Jap. 24 suppl: 66. (1896) (List Pl. Formosa).

Solanum biflorum Lour. Henry in Trans. Asiat. Soc. Jap. 24 suppl: 66. (1896) (List Pl. Formosa).

Taiwan: Oluanpi, 307. Wanchin, 575. Wanshoushan, 758.

Lycianthes hupehensis (Bitter) Wu & Huang (SOLANACEAE) New species

Solanum biflorum Forbes & Hemsley non Lour. in Journ. Linn. Soc. Bot. 26: 169. (1890). *Lycianthes biflora* (Lour.) Bitter ssp. *hupehensis* Bitter in Abh. Naturwiss.

Vereine Bremen 24: 466. (1920).

Hubei: Yichang, 4304, (type of *Lycianthes biflora* (Lour.) Bitter ssp. *hupehensis* Bitter).

Lycianthes lysimachioides (Wall.) Bitter (SOLANACEAE)

Forbes & Hemsley in Journ. Linn. Soc. Bot. 26: 171. (1890).

Hubei: Liantuo, 3063. Xiling (Yichang) Gorge, 3268. Badong, 4757, 4762, 6080. Jianshi, 5912. Fang Xian, 6670. Without locality, 5912a. **Sichuan:** South Wushan, 7207. **Yunnan:** Mengzi, 10988. Simao, 12911.

Native to India, Nepal and China.

Lycianthes macrodon (Wall. ex Nees) Bitter (SOLANACEAE)

Solanum macrodon Wall. ex Nees Anonymous in Kew List of Determinations (volume 34).

Yunnan: Mengzi, 9218, 13652. Simao, 12009.

Distributed in Nepal, India, Bangladesh, Bhutan, China (Taiwan and Yunnan) and Thailand.

Lycianthes neesiana (Wall. ex Nees) D'Arcy & Zhi Y. Zhang (SOLANACEAE)

Solanum subtruncatum Wall. ex Dunal Anonymous in Kew List of Determinations (volume 34).

Yunnan: Simao, 12352.

Distributed in India, China and Thailand.

Lycium chinense Mill. (SOLANACEAE)

Henry in Journ. China Br. Roy. Asiat. Soc. 22: 250. (1887) (Chinese Names of Plants), Forbes & Hemsley in Journ. Linn. Soc. Bot. 26: 175. (1887), Henry in Trans. Asiat. Soc. Jap. 24 suppl: 66. (1896) (List Pl. Formosa), Schneider in Sargent Pl. Wilson. 3: 385. (1917), Zheng in Wuhan Bot. Res .2 (1): 186. (1984).

Hubei: Yichang, 23, 44, 3346. **Taiwan:** Kaohsiung Spit, (cultivated), without number.

The Chinese wolfberry or Chinese boxthorn, a deciduous shrub to 2 metres tall (higher against a wall) bearing pairs of small funnel-shaped flowers in May to July followed in autumn by small egg-shaped scarlet fruits. According to Hemsley this was commonly known in the late 19th century Lord Macartney's tea (when it was introduced to England it was thought to have been the true tea plant). The soft sweet leaves of this shrub are cooked with pork or used in soup with ducks eggs in China. In the Three Gorges region Henry stated it was colloquially known as *kou ch'i tzu* (in modern Chinese the *gou qi zi* or 'fruit of the matrimony vine'), it is used to treat impotence, abdominal pain and is used as a tonic.

Lycopersicum esculentum Mill. (SOLONACEAE)

Henry in Trans. Asiat. Soc. Jap. 24 suppl: 65. (1896) (List Pl. Formosa), Henry in Notes Econ. Bot. China 50.

(1893).

Taiwan: Kaohsiung, without number. Oluanpi, without number.

The *fan-ch'ieh* or tomato, a native of western South America, naturalised in Chinese Taiwan and probably introduced by the Dutch in the 16th century. It was not extensively cultivated in China until the end of the 19th century.

Lycopodiella cernua (L. f.) Pic. Serm. (LYCOPODIACEAE)

Lycopodium cernuum L. f. Henry in Trans. Asiat. Soc. Jap. 24 suppl: 117. (1896) (List Pl. Formosa), Diels in Bot. Jahrb. 29: 210. (1900).

Hubei: Yichang, 1163. **Hainan:** Environs of Haikou, 8411. Without locality, 8558. **Taiwan:** Oluanpi, 585. Tamsui, 1384, 1454. Wanchin, 1559. **Yunnan:** Mengzi, 10431.

Distributed in tropical and subtropical regions of the world.

Lycopodium annotinum L. (LYCOPODIACEAE)

Anonymous in Kew List of Determinations (volume 24).

Hubei: Fang Xian, 6836. North Badong, 6986.

Distributed through the cooler regions of the northern hemisphere.

Lycopodium clavatum L. (LYCOPODIACEAE)

Diels in Bot. Jahrb. 29: 210. (1900).

Hubei: Badong, 606 (July 1885), 2393, 5169. North Badong 6987. Liantuo, 6345.

Widely distributed in temperate areas and alpine mountainous regions of the tropics.

Lycopodium complanatum L. (LYCOPODIACEAE)

Anonymous in Kew List of Determinations (volume 24).

Sichuan: Henry's Chinese collector with Pratt, no locality stated but may be Leshan, Emei Shan, Wa Shan or Kangding, 8913.

Distributed in north temperate regions and alpine zones in the tropics.

Lycopodium obscurum L. (LYCOPODIACEAE)

Anonymous in Kew List of Determinations (volume 24).

Hubei: North Badong, 7165.

Distributed in temperate regions of the northern hemisphere.

Lycopus lucidus Turcz. ex Benth. (LAMIACEAE)

Forbes & Hemsley in Journ. Linn. Soc. Bot. 26: 282. (1890), Henry in Notes Econ. Bot. China 52. (1893), Dunn in Notes Roy. Bot. Gard. Edinburgh 6: 156. (1915).

Hubei: Liantuo, 4599. Badong, cultivated 5197. **Yunnan:** Simao, 12618, 12620.

The shiny bungleweed or *han-ou*, according to Henry it

was cultivated in the vacinity of Yichang and Badong for its roots which were eaten when salted. This specimen is from a wild population. The Hani (an ethnic group) of southern Yunnan call this species *biaqxang* and it is used by them as a diuretic and is used to reduce menstrual and post-delivery pelvic pain.

Lycoris aurea (L'Herit.) Herb. (AMARYLLIDACEAE)

Henry in Journ. China Br. Roy. Asiat. Soc. 22: 257. (1887) (Chinese Names of Plants), Wright in Journ. Linn. Soc. Bot. 36: 89. (1903), O' Brien in Ir. Garden 13(5): 54. (2004).

Hubei: Antelope Glen near Yichang, 2326. Yichang, 2326b.

A handsome *Nerine* like bulb producing umbels of 5 to 10 yellow flowers before the appearance of narrow strap-like foliage in autumn. Distributed in Myanmar, China and Japan. *Lycoris aurea* is locally common on steep inaccessible cliffs in the Three Gorges and their associated side gorges and tributaries. This species is particularly abundant in the Three Lesser Gorges on the Daning river (a tributary of the Yangtze) in north Wushan County, giving a splash of golden colour to the dark towering limestone cliffs. The same region was visited by the Glasnevin Central China Expedition in October 2002.

Lycoris radiata (L'Herit.) Herb. (AMARYLLIDACEAE)

Henry in Journ. China Br. Roy. Asiat. Soc. 22: 257. (1887) (Chinese Names of Plants), Wright in Journ. Linn. Soc. Bot. 36: 89. (1903), Nelson in Pim, The Wood & the Trees 219. (1984).

Hubei: Yichang, 2428, 1094, 326a., 2326c.

A *Nerine* like bulb producing umbels of bright red flowers during August and September. Henry stated that in the Three Gorges region the flowers of both *Lycoris aurea* and *Lycoris radiata* were known as *lung chao hua* or *lao wa suan*. Henry sent bulbs of this handsome plant to Kew where they arrived on November 27th 1886. Distributed in China and Japan. The bulbs are highly poisonous and are used medicinally.

Lycoris spp. (AMARYLLIDACEAE)

Anonymous in Kew List of Determinations (volume 34).

Yunnan: Mengzi, 9317. Simao, 13174.

Lygodium japonicum (Thunb.) Sw. (SHIZAEACEAE)

Henry in Trans. Asiat. Soc. Jap. 24 suppl: 116. (1896) (List Pl. Formosa), Christ in Bull. Herb. Boiss. 7: 19. (1899), Diels in Bot. Jahrb. 29: 208. (1900), Christ in Bull. Soc. Bot. France 5: 67. (1905).

Hubei: Yichang, 99. Liantuo, 4647. On the banks of the Yangtze near Yichang, 4384. Jianshi, 5796, 5796a., 5796b. **Hainan:** Environs of Haikou, 8079. **Taiwan:**

Wanchin, 867. **Yunnan:** Mengzi, 10916. In Pingbian on the Daweishan Range at 1,650 metres, 11529.

The *jin sha teng* or *hai jin sha cao* or gold sand vine, a common climbing fern at lower altitudes in woods and thickets in Hubei. Distributed in India, China, Korea, Japan, the Philippines and Australia. This species was collected by the Glasnevin Central China Expedition (GCCE 728) on Moji Shan (the Dome of Henry and Wilson) near Yichang in October 2004 and a few weeks later by the lighthouse at Oluanpi (South Cape of Henry and Wilson) in southern Taiwan (GTE 35).

Lygodium microphyllum (Cav.) R. Brown (SCHIZAEACEAE)

Lygodium scandens Henry non (L.) Sw. in Trans. Asiat. Soc. Jap. 24 suppl: 116. (1896) (List Pl. Formosa).

Taiwan: Tamsui, 1419.

A twining fern of widespread distribution throughout Africa, south-east Asia and Australia.

Lyonia ovalifolia (Wall.) Drude (ERICACEAE)

Pieris ovalifolia (Wall.) D. Don Forbes & Hemsley in Journ. Linn. Soc. Bot. 26: 17: (1889), Rehder & E. H. Wilson in Sargent Pl. Wilson. 1: 552. (1913).

Hubei: Liantuo, 4568. **Yunnan:** Mountains near Simao at 1,600 metres, 9091 (in part). On the mountains to the north of Mengzi at 1,600 metres, 9091b., 9091e. In Pingbian on the Daweishan Range at 1,650 metres, 9091f.

A semi-evergreen shrub or small tree (sometimes to as much as 12 metres tall) bearing axillary, one-sided racemes of white *Pieris*-like flowers in May and June. Native to Pakistan to the Himalaya (Nepal, Sikkim and Bhutan), Myanmar, China (Guangxi and Yunnan) south to Thailand, Vietnam, Laos and the Malay Peninsula. Introduced from the Himalaya in 1825 and reintroduced by George Forrest in 1930 from Yunnan. It is an extremly beautiful shrub and deserves to be more widely grown.

Lyonia ovalifolia (Wall.) Drude var. *elliptica* (Sieb. & Zucc.) Hand.-Mazz. (ERICACEAE)

Pieris ovalifolia D. Don var. *elliptica* (Sieb. & Zucc.) Rehd. & Wils. Rehder & E. H. Wilson in Sargent Pl. Wilson. 1: 552. (1913).

Yunnan: Mengzi, 9091d.

Judd states that A. Henry 9091 and 9091d. are puzzling, intermediate plants between *Lyonia ovalifolia* vars. *hebecarpa, ovalifolia* and/or *doyonensis* in leaf shape and venation characters.

Lyonia ovalifolia (Wall.) Drude var. *hebecarpa* Franch. ex Forbes & Hemsl.) Chun. (ERICACEAE)

Judd in Journ. Arnold Arb. 62: 157. (1981), Zheng in Journ. Wuhan Bot. Res. 2.(1): 167. (1984).

Pieris ovalifolia D. Don Forbes and Hemsley non

D. Don in Journ. Linn. Soc. Bot. 26: 17: (1889). *Pieris ovalifolia* D. Don var. *elliptica* Rehder & E. H. Wilson non (Sieb. & Zucc.) Hand.-Mazz. in Sargent Pl. Wilson. 1: 552. (1913). *Pieris henryi* Leveille in Bull. Soc. Bot. France 53: 204. (1906); in Fedde, Repert. Nov. Spec. Regni Veg. 3: 262. (1906).

Hubei: Jianshi, 5806, 5806a., 7432. Badong, 6128. **Yunnan:** Mengzi, 9681 (holotype of *Pieris henryi* Leveille).

The var. *hebecarpa* was discovered by Henri de Poli, a Frenchman in the service of the Messageries Maritimes and commissary on board one of their steamers running between Marseilles and Shanghai. In 1875 he travelled through Zhejiang and Guangxi and sent a collection of 325 dried plants to Paris. Endemic to China

Lyonia ovalifolia (Wall.) Drude var. *lanceolata* (Wall.) Handel.-Mazzetti (ERICACEAE)

Judd in Journ. Arnold Arb. 62: 164. (1981).

Pieris ovalifolia D. Don var. *lanceolata* (Wall.) Clarke Rehder in Sargent Pl. Wilson. 1: 552. (1913).

Yunnan: Vacinity of Mengzi at 1,600 to 1,800 metres, 9623, 10510, 10510a., 10510c., 11268. South of the Red River on Fengchunlin, 10510b.

Leaves lanceolate and narrower at the base than the type. Sepals narrower and longer. Widely distributed through the Himalaya and China (including Taiwan).

Lysidice rhodostegia Hance (FABACEAE)

Anonymous in Kew List of Determinations (volume 34).

Yunnan: In the Red River valley near Manhao, 9541. Mengzi, 13641.

A shrub or sometimes a tree of 20 metres tall. A very handsome species, bearing large panicles of small purple red flowers during the months of July and August. Distributed in western China and Vietnam, the roots, stems and leaves are used medicinally.

Lysimachia cf. *alternifolia* Wall. (PRIMULACEAE)

Henry in Trans. Asiat. Soc. Jap. 24 suppl: 57. (1896) (List Pl. Formosa).

Taiwan: Wanchin, 875

Lysimachia auriculata Hemsley (PRIMULACEAE)
New species

in Journ. Linn. Soc. Bot. 26: 47. (1889), Bretschneider in Hist. Eur. Bot. Disc. China 2: 786. (1898), Pax & Knuth in Engler, Pflanzenr. iv. 237: 295. (1905), Handel-Mazzetti in Notes Roy. Bot. Gard. Edinburgh 16: 107. (1928—1932).

Hubei: Yichang, 474, 572, 752. Badong 641 (July 1885). In a cave in the San You Dong Glen (October 1886), 2892. Liantuo, 3892 (type).

This species was later collected by the Irish missionary, Fr. Hugh Scallan (Pater Hugo) in Sichuan.

Lysimachia barystachys Bunge (PRIMULACEAE)

Forbes & Hemsley in Journ. Linn. Soc. Bot. 26: 47. (1889), Handel-Mazzetti in Notes Roy. Bot. Gard. Edinburgh 16: 109. (1928—1932).

Hubei: Near Yichang, on the hills behind the San You Dong Glen (July 1886), 1514.

A handsome perennial, closely allied to *Lysimachia clethroides*, though an altogether better plant with larger, more showy, long lasting flowers. Material presently cultivated at the National Botanic Gardens, Kilmacurragh was raised from collections made by the Glasnevin Central China Expedition (GCCE 357) at Yesanguan in Badong in October 2002 and from collections made by the 2004 Glasnevin Expedition to the Three Gorges region on Tianthu Shan in Changyang (GCCE 807).

Lysimachia brittenii R. Knuth (PRIMULACEAE) New species

in Pflanzenr. iv. 237. (1905), Handel-Mazzetti in Notes Roy. Bot. Gard. Edinburgh 16: 103. (1928—1932).

Lysimachia decurrens Forbes & Hemsley non G. Forst. in Journ. Linn. Soc. Bot. 26: 51. (1889).

Hubei: Antelope Glen, 4194. Liantuo, 4550.

Lysimachia candida Lindl. (PRIMULACEAE)

Forbes & Hemsley in Journ. Linn. Soc. Bot. 26: 48. (1889), Handel-Mazzetti in Notes Roy. Bot. Gard. Edinburgh 16: 119. (1928—1932).

Lysimachia candida Lindl. ssp. *eucandida* Knuth Pax & Knuth in Engler, Pflanzenr. iv. 237: 300. (1905), *Lysimachia obovata* Buch.-Ham. ex Wall. Handel-Mazzetti in Notes Roy. Bot. Gard. Edinburgh 16: 119. (1928—1932).

Hubei: Yichang, 465, 595, 766. **Hainan:** Environs of Haikou, without number. On Liang Shan, 16 km (10 miles) east of Haikou, 8193. **Yunnan:** Mengzi, 10601, 10885.

Native to Myanmar, China, Vietnam and Japan.

Lysimachia capillipes Hemsley (PRIMULACEAE) New species

in Journ. Linn. Soc. Bot. 26: 48. (1889), Bretschneider in Hist. Eur. Bot. Disc. China 2: 786. (1898), Pax & Knuth in Engler, Pflanzenr. iv. 237: 270. (1905).

Hubei: Yichang, 679. In a ravine on Moji Shan, 4176 (isosyntype). In grass in the San You Dong Glen, 1529.

Discovered by Henry in 1885 and collected by Faber about two years later (1887) near Chongqing.

Lysimachia christinae Hance (PRIMULACEAE)

Henry in Journ. China Br. Roy. Asiat. Soc. 22: 252. (1887) (Chinese Names of Plants), Forbes & Hemsley in Journ. Linn. Soc. Bot. 26: 49. (1889), Pax & Knuth in Engler, Pflanzenr. iv. 236: 259. (1905), Handel-Mazzetti in Notes Roy. Bot. Gard. Edinburgh 16: 100. (1928—1932).

Lysimachia christinae Hance var. *pubescens* Franch.

Pax & Knuth in Engler, Pflanzenr. iv. 236: 260. (1905).

Hubei: Yichang, 3138. Liantuo, 7659. Badong, 333 (June 1885). **Yunnan:** In the mountains to the north of Mengzi at 2,300 metres, 10212. Mengzi at 1,500 metres, 13742.

This species was discovered in 1872 near Ningbo in eastern China by Robert Swinhoe of the British Consular Service, it was named by Hance in compliment to Swinhoe's wife, Christin. According to Henry this species was known colloquially in the Three Gorges area as *kou lu huang*, it is currently used in Chinese herbal medicine as a diuretic. Endemic to China.

Lysimachia circaeoides Hemsley (PRIMULACEAE) New species

in Journ. Linn. Soc. Bot. 26: 49. (1889), Bretschneider in Hist. Eur. Bot. Disc. China 2: 787. (1898), Pax & Knuth in Engler, Pflanzenr. iv. 237: 299. (1905), Handel-Mazzetti in Notes Roy. Bot. Gard. Edinburgh 16: 113. (1928—1932).

Hubei: Yichang, 1991. San You Dong Glen, 3513. South Badong, 7287 (type).

Lysimachia clethroides Duby (PRIMULACEAE)

Henry in Journ. China Br. Roy. Asiat. Soc. 22: 250. (1887) (Chinese Names of Plants), Forbes & Henry in Journ. Linn. Soc. Bot. 26: 49. (1889), Pax & Knuth in Engler, Pflanzenr. iv. 237: 290. (1905), Handel-Mazzetti in Notes Roy. Bot. Gard. Edinburgh 16: 109. (1928—1932).

Hubei: In the San You Dong Glen (May 1886), 1582. Yichang, 3660. Badong, 241, 377 (June 1885), 1809, 3163, 5079. Liantuo, 1987. Without locality, 3163a. **Yunnan:** Mengzi, 10249.

Native to China, Laos, Korea and Japan, a common plant in the mountains above the gorges where it is known colloquially as either *kou wei pa* or *mao kou wei pa*. According to Henry this species was also known as *mao kou* (or the fox) at Yichang, the flower spike resembles a fox's tail.

Lysimachia congestiflora Hemsley (PRIMULACEAE)

in Journ. Linn. Soc. Bot. 26: 50. (1889), Pax & Knuth in Engler, Pflanzenr. iv. 237: 283. (1905), Handel-Mazzetti in Notes Roy. Bot. Gard. Edinburgh 16: 93. (1928—1932), C. M. Hu in Acta Phytotax. Sin. 23: 362. (1985).

Lysimachia christinae Hance var. *pubescens* Pax & Knuth non Franch. in Engler, Pflanzenr. iv. 236: 260. (1905). *Lysimachia gymnocephala* Handel-Mazzetti in Notes Roy. Bot. Gard. Edinburgh 16: 95. (1928—1932).

Hubei: Badong, 862 (type), 1822, (in part), 4727. **Sichuan:** Henry's Chinese collector with Pratt, no locality stated but may be Leshan, Emei Shan, Wa Shan or Kangding, 8855. **Yunnan:** Mengzi, 10936, 10742,(syntype of *Lysimachia gymnocephala* Handel-Mazzetti). Simao,

11924.

A creeping perennial herb to 15 cm tall producing clusters of yellow flowers in September and October. Discovered by Dr. John Anderson in May 1868 while travelling on the Sladen expedition to Yunnan from India. Native to India, the Himalaya, Myanmar, China, Thailand and Vietnam.

Lysimachia crispidens (Hance) Hemsley (PRIMULACEAE) in Journ. Linn. Soc. Bot. 26: 50. pl. 1. (1889), Irving in The Garden 63: 389. (1903), Pax & Knuth in Engler, Pflanzenr. iv. 237: 295. (1905), Handel-Mazzetti in Notes Roy. Bot. Gard. Edinburgh 16: 120. (1928—1932).

Hubei: Yichang, 125 (in part), 259, 276, 1257, 3368 (type). Liantuo, 3850.

A small, handsome pink-flowered species. Discovered by the Irish Diplomat and plant collector Thomas Watters in the Xiling (Yichang) Gorge in 1880, introduced to cultivation by E. H. Wilson through Veitch's Coombe Wood nursery, it was also grown at Kew at the turn of the 20th century in their alpine house.

Lysimachia decurrens G. Forst. (PRIMULACEAE) Henry in Trans. Asiat. Soc. Jap. 24 suppl: 57. (1896) (List Pl. Formosa), Handel-Mazzetti in Notes Roy. Bot. Gard. Edinburgh 16: 114. (1928—1932).

Lysimachia acroadenia Pax & Knuth non Maxim. in Engler, Pflanzenr. iv. 237: 300. (1905)

Hainan: On hummocks near Haikou, 8116. **Taiwan:** Wanchin, 188.

Distributed from the Himalaya to Macronesia.

Lysimachia deltoidea Wight var. **cinerascens** Franch. (PRIMULACEAE) Pax & Knuth in Engler, Pflanzenr. iv. 237: 263. (1905), Handel-Mazzetti in Notes Roy. Bot. Gard. Edinburgh 16: 88. (1928—1932).

Yunnan: Mengzi, 10970.

Native to Myanmar, western China, Thailand, Laos and Vietnam.

Lysimachia drymarifolia Franch. (PRIMULACEAE) *Lysimachia fargesii* Handel-Mazzetti non Franchet in Notes Roy. Bot. Gard. Edinburgh 16: 99. (1928—1932).

Hubei: Without locality, 5704e.

Lysimachia engleri R. Knuth (PRIMULACEAE) New species Pax & Knuth in Engler, Pflanzenr. iv. 237: 265. (1905), Handel-Mazzetti in Notes Roy. Bot. Gard. Edinburgh 16: 75. (1928—1932).

Yunnan: In forests to the east of Simao at 1,650 metres, 13025.

Endemic to China and distributed in the provinces of Sichuan and Yunnan.

Lysimachia fistulosa Handel-Mazzetti (PRIMULACEAE) New species in Notes Roy. Bot. Gard. Edinburgh 26: 84. (1928—1932).

Lysimachia klattiana Forbes & Hemsley non Hance in Journ. Linn. Soc. Bot. 23: 53. (1889)

Hubei: Changyang, 6229. Yichang, 1374 (in part).

Lysimachia foenum-graecum Hance (PRIMULACEAE) Handel-Mazzettii in Notes Roy. Bot. Gard. Edinburgh 16: 70. (1928—1932).

Yunnan: South of the Red River from Mengting, 10643.

Endemic to China, distributed in Guangdong, Guangxi, Hunan and Yunnan.

Lysimachia grammica Hance (PRIMULACEAE) Forbes & Hemsley in Journ. Linn. Soc. Bot. 26: 52. (1889).

Hubei: Badong, 4704.

Lysimachia hemsleyana Maximowicz ex Oliver (PRIMULACEAE) New species in Hooker's. Icon. Pl. 20: t 1980. (1891), Bretschneider in Hist. Eur. Bot. Disc. China 2: 787. (1898), Diels in Engler's Bot. Jahrb. 29: 523. (1901), Pax & Knuth in Engler, Pflanzenr. iv. 236: 259. (1905), Handel-Mazzetti in Notes Roy. Bot. Gard. Edinburgh 16: 97. (1928—1932).

Lysimachia christinae Forbes & Hemsley non Hance in Journ. Linn. Soc. Bot. 23: 49. (1889).

Hubei: Yichang, 489, 1381, 1822.

Lysimachia henryi Hemsley (PRIMULACEAE) New species in Journ. Linn. Soc. Bot. 26: 52. (1889), Bretschneider in Hist. Eur. Bot. Disc. China 2: 787. (1898), Irving in The Garden 64: 269. (1903), Hemsley in Curtis's Bot. Mag. 130: t. 7961. (1904), Pax & Knuth in Engler, Pflanzenr. iv. 237: 282. (1905), J. H. Veitch in Hortus Veitchii 423. (1906), Handel-Mazzetti in Notes Roy. Bot. Gard. Edinburgh 16: 85. (1928—1932), Morley in Glasra 3: 79. (1979), Nelson in Pim, The Wood & the Trees 232. (1984); in A Heritage of Beauty 315. (2000), O' Brien in Augustine Henry, An Irish Plant Collector in China 26. (2002).

Hubei: In glens near Yichang, 250 (in part), 670, 1374, 3579. On Tsui Fu Shan near Yichang (May 1888), 1624. Badong, 558 (type).

A very vigorous, handsome plant with a tufted trailing habit bearing large yellow flowers in the angles of crowded leaves, the whole plant forming a spreading carpet to about 10 cm tall and rooting at the nodes as it spreads. Discovered by Henry in 1885, it was introduced to cultivation by Wilson through Messrs Veith who flowered it in their Coombe Wood Nursery in 1902, it flowered at Kew the following summer.

At the National Botanic Gardens, Glasnevin it is grown in the damp shade of the woodland garden where it provides a colourful show each autumn. Easy to grow in shade or full sun given a humus-rich, moist growing medium. It certainly deserves to be more widely grown.

Lysimachia heterogenea Klatt (PRIMULACEAE)

Handel-Mazzetti in Notes Roy. Bot. Gard. Edinburgh 16: 107. (1928—1932).

Lysimachia paludicola Hemsley in Journ. Linn. Soc. Bot. 26: 54. (1889), Bretschneider in Hist. Eur. Bot. Disc. China 2: 787. (1898), Pax & Knuth in Engler, Pflanzenr. iv. 237: 294. (1905).

Hubei: Yichang, 3519 (type of *Lysimachia paludicola* Hemsley).

Lysimachia hypericoides Hemsley (PRIMULACEAE) New species

in Journ. Linn. Soc. Bot. 29: 314. (1892), Btetschneider in Hist. Eur. Bot. Disc. China 2: 787. (1898), Pax & Knuth in Engler, Pflanzenr. iv. 237: 265. (1905), Handel-Mazzetti in Notes Roy. Bot. Gard. Edinburgh 16: 78. (1928—1932).

Hubei: Jianshi, 5842 (type).

Lysimachia insignis Hemsley (PRIMULACEAE) New species

in Hooker's Icon. Pl. 27: t. 2634. (1901), Hemsley in Journ. Linn. Soc. Bot. 34: 477. (1900), Pax & Knuth in Engler, Pflanzenr. iv. 236: 308. (1905), Handel-Mazzetti in Notes Roy. Bot. Gard. Edinburgh 16: 69. (1928—1932), C. M. Hu in Acta Phytotax. Sin. 23: 357. (1985).

Yunnan: In Pingbian, in forests on the Daweishan Range at 1,640 metres, 10406. Mengzi at 1,600 metres, 10406a., (type).

This species was later collected by Handel-Mazzetti in southern Yunnan. It has recently been found in Vietnam.

Lysimachia klattiana Hance (PRIMULACEAE)

Pax & Knuth in Engler, Pflanzenr. iv. 237: 282. (1905), Handel-Mazzetti in Notes Roy. Bot. Gard. Edinburgh 16: 103. (1928—1932).

Hubei: Yichang, 250 (in part), 475 (in part), 1374 (in part).

Lysimachia lancifolia Craib (PRIMULACEAE) New species

Handel-Mazzetti in Notes Roy. Bot. Gard. Edinburgh 16: 72. (1928—1932).

Yunnan: South of the Red River from Manmei at 2,300 metres, 9471. In forests near Mengzi at 1,800 metres, 9471a.

This species was discovered by Henry in 1896 and was described from material later collected by Kerr in Thailand. According to Henry, the leaves of this species when dried have a strong odour and are were used for scenting hair oil. Distributed in China (Yunnan) and Thailand.

Lysimachia lobelioides Wall. (PRIMULACEAE)

Pax & Knuth in Engler, Pflanzenr. iv. 237: 296. (1905), Handel-Mazzetti in Notes Roy. Bot. Gard. Edinburgh 16: 113. (1928—1932).

Yunnan: Mengzi, 10002.

A widespread species, distributed in Nepal, India, Bhutan, Myanmar, western China and Thailand.

Lysimachia mauritiana Lam. (PRIMULACEAE)

Lysimachia lineariloba Hook. & Arn. Henry in Trans. Asiat. Soc. Jap. 24 suppl: 57. (1896) (List Pl. Formosa).

Taiwan: Oluanpi, 641, 1814.

Lysimachia paridiformis Franch. (PRIMULACEAE)

Forbes & Hemsley in Journ. Linn. Soc. Bot. 26: 55. (1889), Hemsley in Bot. Mag. 118: t. 7226. (1892), Bretschneider in Hist. Eur. Bot. Disc. China 2: 787. (1898), Hemsley in Hooker's. Icon. Pl. t. 1982. (1891), Pax in Engler, Pflanzenr. iv. 237: 280. (1905), Handel-Mazzetti in Notes Roy. Bot. Gard. Edinburgh 16: 86. (1928—1932), Nelson in Pim, The Wood & the Trees 222. (1984), Lancaster in Travels in China 371 (1993) Nelson in A Heritage of Beauty 315. (2000).

Hubei: Yichang, 3500, 3500a. Xiling (Yichang) Gorge, 4202. Changyang, 6228. Without locality, 6228a., 6228b.

This species was first collected by the French missionay Père Paul Perny in Guizhou and was figured in *Curtis's Botanical Magazine* from a living plant raised at Kew from seed sent there by Henry from Yichang in the spring of 1889, the seedling first flowered at Kew in 1891.

Lysimachia phyllocephala Handel-Mazzetti (PRIMULACEAE)

in Notes Roy. Bot. Gard. Edinburgh 16: 83. (1928—1932)

Lysimachia cephalantha (Franch.) R. Knuth Knuth in Engler, Pflanzenr. iv. 237: 284. (1905).

Yunnan: In ravines near Mengzi at 1,810 metres, 9428.

Handel-Mazzetti used Henry's material in describing the species, it had previously been collected by Delavay in Yunnan and was collected in May 1904 on Emei Shan in Sichuan by Wilson. Endemic to China.

Lysimachia pseudohenryi Pampanini (PRIMULACEAE) New species

Handel-Mazzetti in Notes Roy. Bot. Gard Edinburgh 16: 84. (1928—1932).

Lysimachia japonica Forbes & Hemsley non Thunb. in Journ. Linn. Soc. Bot. 23: 53. (1889), Pax & Knuth in Engler, Pflanzenr. iv. 236: 262. (1905), *Lysimachia henry* Hemsley in Journ. Linn. Soc. Bot. 26: 52. (1889), in part. *Lysimachia esquirolii* Handel-Mazzetti non Bonti. in Notes Roy. Bot. Gard. Edinburgh 16: 91. (1928—1932).

Hubei: Yichang, 482. Liantuo, 1983. **Sichuan:** North

Wushan, 7154.

Pampanini based the species on collections made by the Italian missionary Sylvestri who was based in Baokang in northern Hubei (Henry was the first westerner to visit the area) in about 1907. Henry had found the species over two decades earlier in 1885.

Lysimachia pseudotrichopoda Handel-Mazzetti (PRIMULACEAE) New species

in Notes Roy. Bot. Gard. Edinburgh 16: 71. (1928—1932).

Hubei: Badong, 5386, 7326. Jianshi, 5942.

This species was later collected by Père Paul Farges in north-east Sichuan on the 24th of May 1899.

Lysimachia rubiginosa Hemsley (PRIMULACEAE) New species

in Journ. Linn. Soc. Bot. 26: 56. (1889), Hemsley in Hooker's. Icon. Pl. t. 1981. (1891), Bretschneider in Hist. Eur. Bot. Disc. China 2: 787. (1898), Pax & Knuth in Engler, Pflanzenr. iv. 237: 284. (1905), Handel-Mazzetti in Notes Roy. Bot. Gard. Edinburgh 16: 101. (1928—1932).

Lysimachia involucrata Hemsley in Journ. Linn. Soc. Bot. 29: 315. (1892), Bretschneider in Hist. Bot. Disc. China 2: 787. (1898), Pax & Knuth in Engler, Pflanzenr. iv. 237: 284. (1905).

Hubei: Badong, by the sides of ditches, 1823, 2440, 4680, 4945, 6134. **Sichuan:** Wushan, 6244. Henry's Chinese collector with Pratt, no locality stated but may be Leshan, Emei Shan, Wa Shan or Kangding, 8884 (syntype of *Lysimachia involucrata* Hemsley). **Hunan:** Shimen, 7559.

Lysimachia sp. No. 1. (PRIMULACEAE)

Anonymous in Kew List of Determinations (volume 24).

Hubei: Badong, 701 (July 1885). Liantuo, 1995. **Sichuan:** South Wushan, 5730.

Lysimachia sp. No. 2. (PRIMULACEAE)

Anonymous in Kew List of Determinations (volume 34).

Yunnan: Simao, 12209.

Lysimachia stenosepala Hemsley (PRIMULACEAE) New species

in Journ. Linn. Soc. Bot. 23: 57. (1889), Irving in The Garden 64: 269. (1903), Pax & Knuth in Engler, Pflanzenr. iv. 237: 298. (1905), Handel-Mazzetti in Notes Roy. Bot. Gard. Edinburgh 16: 108. (1928—1932).

Lysimachia candida Lindl. ssp. *oppositifolia* Pax & Knuth non Knuth. in Engler, Pflanzenr. iv. 237: 301. (1905).

Hubei: Badong, 643 (July 1885), 1804, 1819. Xingshan, 4727. Liantuo, 4527. Jianshi, 5866. Changleping, 6265. Fang Xian, 6883. **Yunnan:** Mengzi, 13743.

A vigorous perennial bearing numerous white flowers in long racemes. Introduced to cultivation by Wilson through Veitch's Coombe Wood nursery. Endemic to China.

Lysionotus heterophyllus Franch. (GESNERIACEAE) New species

Lysionotus brachycarpus Rehder in Sargent Pl. Wilson. 3: 387. (1917).

Sichuan: Henry's Chinese collector with Pratt, no locality stated but may be any of the following, Leshan, Emei Shan, Wa Shan or Kangding, 8997.

Lysionotus pauciflorus Maxim. (GESNERIACEAE)

Forbes & Hemsley in Journ. Linn. Soc. Bot. 26: 225. (1890), Schneider in Sargent Pl. Wilson. 3: 387. (1917).

Lysionotus carnosus Hemsley in Gard. Chron. ser. 3, 28: 349. (1900), Nelson in Pim, The Wood & the Trees 222. (1984); in A Heritage of Beauty 315. (2000).

Hubei: Liantuo, 2111. Badong, 3691, 6139. **Yunnan:** Mengzi, 9224, 11848 (type of *Lysionotus carnosus* Hemsley).

An evergreen subshrub with prostrate stems to 60 cm long with adventitious roots and branching upwards at the nodes. During July and August axillary cymes of tubular white, lilac-tinged flowers are abundantly produced in the axils of the upper leaves. This handsome shrub was cultivated at Kew at the turn of the 20th century from seed received from Augustine Henry. Henry had also sent seeds to A. K. Bulley who raised plants in his garden at Neston in Cheshire. Distributed in China, Vietnam and Japan.

Lysionotus pauciflorus Maxim. var. ikedae (Hatus.) W. T. Wang (GESNERIACEAE)

Lysionotus pauciflorus Maxim. var. Henry in Trans. Asiat. Soc. Jap. 24 suppl: 68. (1896) (List Pl. Formosa).

Taiwan: Oluanpi, 1343.

Lysionotus serratus D. Don (GESNERIACEAE)

Anonymous in Kew List of Determinations (volume 34).

Yunnan: Mengzi, 11003.

Native to India, Bhutan, Nepal, Myanmar, western China, Thailand and Vietnam.

Lysionotus sp. (GESNERIACEAE)

Anonymous in Kew List of Determinations (volume 34).

Yunnan: Mengzi, 9009, 9187, 9712, 10117, 11466, 11470. On Fengchunling, in forests at 2,300 metres, 11201. In Pingbian in forests on the Daweishan Range at 1,640 metres, 13472.

Lythrum salicaria L. (LYTHRACEAE)

Forbes & Hemsley in Journ. Linn. Soc. Bot. 23: 304. (1887) Koehne in Engler, Pflanzenr. 216: 285. (1903).

Hubei: Yichang, 2311.

02

Macaranga andamanica Kurz (EUPHORBIACEAE)

Macaranga rosuliflora Croizat in Journ. Arnold Arb. 23: 51. (1942).

Yunnan: Simao, 12143.

Macaranga denticulata (Blume) Müell.-Arg. (EUPHORBIACEAE)

Pax in Engler, Pflanzenr. iv. 147. vii: 334. (1914).

Macaranga henricorum Hemsley in Journ. Linn. Soc. Bot. 26: 442. (1894).

Hainan: Without locality, 19 (syntype of *Macaranga henricorum* Hemsley) **Yunnan:** Simao, 12991, 12991a.

Hemsley's *Macaranga henricorum* was described from material gathered by the American missionary and plant collector, the Rev'd. B. C. Henry and Augustine Henry. Distributed in the Himalaya, India, China, Thailand, Laos and Vietnam. (The Henry collection number quoted by Hemsley (A. Henry 19) cannot be correct).

Macaranga henryi (Pax & K. Hoffm.) Rehder (EUPHORBIACEAE) New species

in Sunyatsenia 3: 240. (1936)

Mallotus henryi Pax & K. Hoffmann in Pflanzenr. (Engler) 4. Fam. 147(7): 177. (1914).

Yunnan: South of the Red River from Manmei, 13665 [type of *Mallotus henryi* Pax & K. Hoffmann, isotype of *Macaranga henryi* (Pax & K. Hoffm.) Rehder].

Distributed in western China and Vietnam.

Macaranga indica Wight (EUPHORBIACEAE)

Pax in Engler, Pflanzenr. iv. 147. vii: 349. (1914).

Yunnan: Mengzi, 10409a.

Native to India, Sri Lanka, China (Yunnan and Tibet (Xizang)) and Thailand.

Macaranga kurzii (O. Ktze.) Pax & Hoffm. (EUPHORBIACEAE)

Pax in Engler, Pflanzenr. iv. 147. vii: 360. (1914).

Yunnan: Mengzi, 10778, 13684. Simao, 12057a.

Distributed in Myanmar, western China, Thailand, Laos and Vietnam.

Macaranga sampsonii Hance (EUPHORBIACEAE)

Mallotus populifolius in Journ. Linn. Soc. Bot. 26: 441. (1894).

Hainan: Haikou and environs, without number.

Macaranga sp. (EUPHORBIACEAE)

Anonymous in Kew List of Determinations (volume 24).

Hainan: Without locality, 8543.

Macaranga tanarius (L.) Muell.-Arg. (EUPHORBIACEAE)

Henry in Trans. Asiat. Soc. Jap. 24 suppl: 84. (1896) (List Pl. Formosa), Pax in Engler, Pflanzenr. iv. 147. vii: 352. (1914), Li in Woody Flora of Taiwan 433. (1963).

Taiwan: Kaohsiung, without number. Oluanpi, 1968.

A tree of wide distribution from southern Asia to Australia. Common in Chinese Taiwan where it is often cultivated as a shade tree. Of rapid growth this is a pioneer species (it will rapidly colonise clearings in tropical rainforest) and can put on growths of up to 2 metres a year under favourable conditions. A very handsome foliage plant, this species was collected by the Glasnevin Taiwan Expedition (GTE 41) by the lighthouse at Oluanpi (South Cape of Henry and Wilson) in October 2004.

Machilus forrestii (W. W. Sm.) L. Li, J. Li, & W. H. Li (LAURACEAE) New species

Phoebe forrestii W. W. Smith in Notes Roy. Bot. Gard. Edinburgh 13: 176. (1921), Liou Ho in Laurac. Chine & Indoch. 68. (1934).

Yunnan: Mengzi, 10946.

Endemic to China and distributed in Tibet (Xizang) and Yunnan.

Machilus hemsleyi Nakai (LAURACEAE) New species in J. Jap. Bot. 16(3): 122. (1940).

Hubei: Badong, 6121 (isotype).

Machilus microcarpus Hemsley (LAURACEAE)

Hemsley in Journ. Linn. Soc. Bot. 26: 376. (1891).

Sichuan: South Wushan, 5615 (type).

Machilus nanmu (Oliv.) Hemsl. (LAURACEAE)

Phoebe nanmu (Oliv.) Gamble Zheng in Journ. Wuhan Bot. Res.2. (1): 56. (1984).

Hubei: Yichang, 1252.

In Hubei and Sichuan the name *nanmu shu* is applied to various species of *Actinodaphne*, *Machilus* and *Phoebe*. All three genera are abundant on the lower forested slopes of Emei Shan and give a good indication to what the flora of the Chengdu plain must have looked like before its vast tracts of low-altitude forest were cleared for agricultural purposes about 5,000 years ago. All make trees of a large size, with good, durable close-grained timber used during the 19th century for pillars and beams in temples. *Phoebe chinensis* is the most widespread of *nanmu* trees.

Machilus obovatifolius (Hayata) Kanehira & Sasaki (LAURACEAE) New species

Persea obovatifolia (Hayata) Kostermans Li in Woody Flora of Taiwan 226. (1963).

Taiwan: Oluanpi, 1709.

A small, evergreen tree, endemic to the Hengchun Peninsula in southern Taiwan where it was first collected by Henry.

Machilus spp. (LAURACEAE)

Anonymous in Kew List of Keterminations (vol. 24 & 34), Henry in Trans. Asiat. Soc. Jap. 24 suppl: 79. (1896) (List Pl. Formosa).

Hubei: South Badong, 5339. **Taiwan:** Oluanpi, 611.

Yunnan: Mengzi, 10617. Simao, 11669.

Machilus thunbergii Sieb. & Zucc. (LAURACEAE)

Henry in Trans. Asiat. Soc. Jap. 24 suppl: 79. (1896) (List Pl. Formosa).

Machilus sp. Henry in Trans. Asiat. Soc. Jap. 24 suppl: 79. (1896) (List Pl. Formosa). *Persea thunbergii* (Sieb. & Zucc.) Kostermans Li in Woody Flora of Taiwan 226. fig. 82. (1963).

Taiwan: Wanchin, 119, 131, 197, 432. Taipei, 131a. Oluanpi, 208, 663, 664, 671, 2055.

A large evergreen tree, native to China, southern Korea and Japan. Henry called this the 'Formosan laurel' and describes the Taiwan plants as enormous trees occcuring in the depts of the mountains. Henry stated it was known colloqually as *hsiu-lang, hsiu-lan-mu, shau-lam* or *shuai-a-nan*, stating it was also one of a number of trees known to the Chinese as *nan-mu*, called *lam-a* in native Taiwanese language.

Mackaya neesiana (Wall.) Das (ACANTHACEAE)

Asystasia chinensis S. Moore Forbes & Hemsley in Journ. Linn. Soc. Bot. 26: 243. (1890).

Hubei: Liantuo, 2067, 2205, 4502. On the side of the Antelope Glen, 2304. Yichang, 4124. Fang Xian, 6734. Without locality, 2304a. **Yunnan:** Mengzi, 9833, 11329. Simao, 13526.

Macleaya cordata (Willd.) R. Br. (PAPAVERACEAE)

Fedde in Engler, Pflanzenr. 104: 217. (1909).

Bocconia cordata Willd. Henry in Journ. China Br. Roy. Asiat. Soc. 22: 271. (1887) (Chinese Names of Plants).

Hubei: Badong, 3162, 5213. Liantuo, 1882, 2213. Without locality, 2213a.

The plume poppy, a handsome perennial to 3 metres tall, this species is a common inhabitant of the mountains north of the Yangtze in the Three Gorges region. Several collections were made by the Glasnevin Central China Expedition in the region of Xingshan (GCCE 462 & GCCE 583) and near Tianthu Shan in Changyang (GCCE 754). In volume 24 of the Augustine Henry 'Kew List' (Henry's own field notes on his central China collections, in the library of the Royal Botanic Gardens, Kew) Henry stated Henry 1882 from Liantuo was the *ta huang* i.e. rhubarb root and was used in the same region as a drug.

Maclura cochinchinensis (Lour.) Corner (MORACEAE)

Cudrania javanensis Trecul. Henry in Trans. Asiat. Soc. Jap. 24 suppl: 88. (1896) (List Pl. Formosa), Forbes & Hemsley in Journ. Linn. Soc. Bot. 26: 469. (1899), Schneider in Sargent Pl. Wilson. 3: 308. (1917). *Cudrania rectispina* Hance Henry in Trans. Asiat. Soc. Jap. 24 suppl: 88. (1896) (List Pl. formosa). *Cudrania obovata* Trecul Forbes & Hemsley in Journ. Linn. Soc. Bot. 26: 469. (1899),

Schneider in Sargent Pl. Wilson. 3: 308. (1917). *Cudrania cochinchinensis* (Lour.) Kudo & Masamune var. *gerontogea* (Sieb. & Zucc.) Kudo & Masamune Li in Woody Flora of Taiwan 113. (1963).

Hainan: Environs of Kiungchow, 8228. **Taiwan:** Wanchin, 135, 135b., 164. Tamsui, 135a. Without locality, 135c., 720a., 720b., 720c. Kaohsiung, 720. **Yunnan:** In Pingbian in forests on the Daweishan Range at 1,600 metres, 9987a. In Pingbian in a forested, rocky ravine on the Daweishan Range at 1,400 metres, 10821. In mountains forest Simao at 1,400 metres, 11937. Mountains to the south-west of Simao at 1,600 metres, 11937b. In forests near Simao at 1,400 metres, 12385, 12385b. Simao at 1,400 metres, 12985.

An evergreen climbing shrub, native to China (including Taiwan), Indochina, southern Japan to New Caledonia.

Maclura fruticosa (Roxb.) Corner (MORACEAE)

Cudrania fruticosa (Roxb.) Wight. ex Kurz. Schneider in Sargent Pl. Wilson. 3: 307. (1917).

Yunnan: In forests to the south-east of Simao at 1,600 metres, 13156.

Distributed in India, Bangladesh, Myanmar, Thailand and Vietnam.

Maclura pubescens (Trecul.) Z. K. Zhou & M. G. Gilbert (MORACEAE)

Cudrania pubescens Trecul. Schneider in Sargent Pl. Wilson. 3: 307. (1917).

Yunnan: Mountains to the west of Simao at 1,600 metres, 11906, 11906b. Mountains to the east of Simao at 1,400 metres, 11906a.

Maclura tricuspidata Carr. (MORACEAE)

Cudrania triloba Hance Henry in Journ. China Br. Roy. Asiat. Soc. 22: 235. (1887) (Chinese Names of Plants), Forbes & Hemsley in in Journ. Linn. Soc. Bot. 26: 470. (1899). *Cudrania tricuspidata* (Carr.) Bureau Schneider in Sargent Pl. Wilson. 3: 306. (1917), Zheng in Journ. Wuhan Bot. Res. 2.(1): 32. (1984).

Hubei: Yichang, 500, 500a., 1334, 2322, 2322a., 3140. Liantuo, 2102. **Yunnan:** Mengzi at 1,400 metres, 9987.

This is the 'silkworm thorn' the *che* or *tsa* tree, the leaves were used to feed silkworms when the supply of mulberry leaves failed. Usually a tree to about 8 metres tall, Wilson mentioned meeting with trees to 18 metres tall with girths of up to 2 metres. It was introduced to cultivation through the garden of the Paris Museum about 1862, Franchet believed the seeds were sent by Gabriel Eugene Simon. A very handsome foliage plant, though rare in cultivation. Native to China, Vietnam, Korea and Japan.

Macropanax dispermus (Blume) Kuntze (ARALIACEAE)

Macropanax oreophilus Miq. Anonymous in Kew List of Determinations (volume 34).

Yunnan: Simao, 13082, 13083.

Native to India, Bhutan, Nepal, Myanmar, Thailand and Vietnam.

Macropanax rosthornii (Harms.) C. Y. Wu ex G. Hoo (ARALIACEAE)

Macropanax sp. Anonymous in Kew List of Determinations (volume 24).

Hubei: Changyang, 7673.

Discovered by Henry in 1888 and based on material collected by von Rosthorn in Sichuan between 1891—1892.

Macropanax undulatus (Wall. ex G. Don) Seem. (ARALIACEAE)

Anonymous in Kew List of Determinations (volume 34).

Yunnan: Simao, 12402, 12644, 13448.

Macrosolen ampullaceus (Roxb.) Tiegh (LORANTHACEAE)

Loranthus ampullaceus Roxb. Forbes & Hemsley in Journ. Linn. Soc. Bot. 26: 405. (1894).

Hainan: On hummocks near Haikou, 8112.

Macrosolen bibracteolatus (Hance) Dancer (LORANTHACEAE)

Elytranthe bibracteolata (Hance) Lecomte var. *sinensis* Lecompte in Sargent Pl. Wilson. 3: 317. (1917).

Yunnan: Simao at 1,600 metres, 12631 [type of *Elytranthe bibracteolata* (Hance) Lecomte var. *sinensis* Lecompte].

Macrosolen cochinchinensis (Lour.) Teigh. (LORANTHACEAE)

Danser in Blumea 2: 42.(1936).

Elytranthe ampullacea Lecompte non (Roxb.) G. Don in Sargent Pl. Wilson. 3: 317. (1917). *Elytranthe ampullacea* (Roxb.) G. Don var. *tonkinensis* Lecompte in Notul. Syst. (Paris) 3: 99. (1916), Lecompte in Sargent Pl. Wilson. 3: 317. (1917).

Yunnan: Mengzi, 11473. On trees and shrubs near Simao at 2,000 metres, 11755a, (isotype of *Elytranthe ampullacea* (Roxb.) G. Don var. *tonkinensis* Lecomte). Simao, 11755a.

A semiparasitic shrub, native to India, Bhutan, Nepal, Myanmar, China, Vietnam and Cambodia.

Macrothelypteris setigera (Blume) Ching (ASPLENIACEAE)

Aspidium setigerum (Blume) Kuhn Christ in Bull. Soc. Bot. France 5: 34. (1905). *Nephrodium setigerum* (Blume) Baker Henry in Trans. Asiat. Soc. Jap. 24 suppl: 113. (1896) (List Pl. Formosa).

Hubei: Liantuo, 3061. **Taiwan:** Oluanpi, 680. Tamsui, 1385 (in part).

Macrothelypteris torresiana (Gaudich.) Ching (ASPLENIACEAE)

Hypolepis punctata (Thunb.) Mett. var. *henryi* Christ in Bull. Soc. Bot. France 5: 61. (1905). *Macrothelypteris torresiana* (Gaud.) Ching var. *calvata* (Bak.) Holttum Holttum in Blumea 17: 28. (1969).

Hubei: Yichang, 2154 [type of *Hypolepis punctata* (Thunb.) Mett. var. *henryi* Christ].

Maesa hupehensis Rehder (PRIMULACEAE) New species

Maesa doraena Forbes & Hemsley non Blume ex Sieb. & Zucc. in Journ. Linn. Soc. Bot. 26: 59. (1889). *Maesa japonica* (Thunb.) Moritzi var. *elongata* Mez in Engler, Pflanzenr. iv. 236: 51. (1936).

Sichuan: North Wushan, 7124. **Hubei:** Changyang, 7757.

Rehder based this species on material collected by Wilson near Changleping (a town to the south of Yichang) in May 1907, Henry discovered it in May of 1888.

Maesa indica (Roxb.) A. DC. (PRIMULACEAE)

Anonymous in Kew List of Determinations (volume 34).

Yunnan: Mengzi, 10416.

The Hani (an ethnic group) of Xishuangbanna in southern Yunnan call this evergreen shrub *geeqcml* and is used to treat cases of hepatitis, gastritis and hypertension. Distributed in India, China (Yunnan) and Vietnam.

Maesa japonica (Thunb.) Moritzi ex Zoll (PRIMULACEAE)

Mez in Engler, Pflanzenr. iv. 236: 50. (1936), Li in Woody Flora of Taiwan 718. (1963), Zheng in Journ. Wuhan Bot. Res. 2.(1): 172. (1984).

Maesa doraena Blume ex Sieb. & Zucc. Forbes & Hemsley in Journ. Linn. Soc. Bot. 26: 59. (1889), Henry in Trans. Asiat. Soc. Jap. 24 suppl: 57. (1896) (List Pl. Formosa).

Hubei: Antelope Glen, 3457. Liantuo, 3804, 4438, 4563, 6371. **Taiwan:** Wanchin, 126. **Yunnan:** Simao, 11625a.

A shrub to 3 metres tall, native to southwest and southern China, Vietnam and Japan. In forests from high to low altitude in Chinese Taiwan.

Maesa macilenta E. Walker (PRIMULACEAE) New species

in J. Wash. Acad. Sci. 21: 479. (1931).

Yunnan: In forests near Simao at 1,650 metres, 11704. Simao, 11704a., (holotype).

Endemic to Yunnan.

Maesa montana A. DC. (PRIMULACEAE)

Maesa castaneifolia Mez in Engl. Pflanzenr, Mrysin. 44. (1902), Rehder in Sargent Pl. Wilson. 2: 583. (1916). *Maesa henryi* Hu in Journ. Arnold Arb. 5: 232. (1924).

Yunnan: Mengzi at 1,500 metres, 9464, 9464a., 10153 (syntype of *Maesa castaneifolia* Mez). Milê district, 9464b. Simao at 1,500 metres, 11625 (type of *Maesa henryi* Hu).

Nepal to China and Indochina.

Maesa perlaria (Lour.) Merr. (PRIMULACEAE)

Maesa sinensis A. DC. Mez in Engler, Pflanzenr. iv. 236: 34. (1902).

Hainan: 32 km (20 miles) west of Haikou, 8152. Without locality, 8525. Ling-men (in the interior of the island), 8669.

Maesa permollis Kurz (PRIMULACEAE)

Mez in Engler, Pflanzenr. iv. 236: 51. (1902).

Yunnan: Simao, 11707, 11707d.

Native to Myanmar, Yunnan, Laos and Thailand.

Maesa ramentacea (Roxb.) A. DC. (PRIMULACEAE)

Mez in Engler, Pflanzenr. iv. 236: 27. (1902).

Yunnan: Mengzi, 10632, 13655.

Distributed in Nepal, India, Bangladesh, Myanmar, China, Laos, Cambodia, Vietnam and the Philippines.

Maesa sp. No. 1. (PRIMULACEAE)

Anonymous in Kew List of Determinations (volume 24).

Hubei: South Donghu, 7920.

Maesa sp. No. 2. (PRIMULACEAE)

Anonymous in Kew List of Determinations (volume 34).

Yunnan: Mengzi, 9649.

Maesa tenera Mez (PRIMULACEAE)

Li in Woody Flora of Taiwan 719. fig. 293. (1963).

Maesa sinensis Henry non A.DC. Henry in Trans. Asiat. Soc. Jap. 24 suppl: 57. (1896) (List Pl. Formosa).

Taiwan: Wanshoushan, 64. Oluanpi, 203, 1980.

A shrub to 3 metres tall, native from southern China, Indochina to Japan. Very common in Chinese Taiwan.

Magnolia baillonii Pierre (MAGNOLIACEAE)

Michelia baillonii (Pierre) Finet & Gagnep. Dandy in Notes Roy. Bot. Gard. Edinburgh 16: 129. (1928—1932).

Yunnan: Mengzi, 9545. Simao, 13421, 13513.

A large evergreen tree to 35 metres tall, bearing richly fragrant, small solitary yellowish-white flowers between April and July. This species is distributed in rmonsoon rainforest and subtropical evergreen broadleaved forest in southern Yunnan and is a highly ornamental tree worthy of trial in the milder parts of Europe and North America. Native to north-east India, Myanmar, Yunnan, Thailand, Vietnam and Cambodia, Henry's collections proved to be a new record for China. At Kunming Botanical Gardens in Yunnan province there is a very fine collection of Magnoliaceae, including young trees of this species.

Magnolia biondii Pamp. (MAGNOLIACEAE) New species

Dandy in Notes Roy. Bot. Gard. Edinburgh 16: 123. (1928—1932), Zheng in Journ. Wuhan Bot. Res. 2.(1): 45. (1984).

Hubei: Badong, 2522. Changyang, 7733.

Magnolia biondii was originally described by Pampaninii from a leafless flowering specimen collected in northern Hubei (probably at Baokang, an area first explored by Henry) by the Italian missionary, Giraldi (in Herb. Biondi) in about 1907. Henry first found this species two decades previously.

Magnolia champaca (L.) Baill. ex Pierre (MAGNOLIACEAE)

Michelia champaca L. Anonymous in Kew List of Determinations (volume 34).

Yunnan: Simao, 13147. Mengzi, 13717.

A spectacular flowering tree to about 30 metres tall and native to the subtropical forests of the Himalayan foothills, India, Myanmar, Vietnam and China where it has become rare. In Yunnan this fine tree is distributed in evergreen broad-leaved forest in subtropical regions in the south of the province. The intensely fragrant, showy yellow flowers are carried from spring to autumn and are held in special reverence by Hindus and are concidered sacred to the god, Vishnu. The tree yields excellent timber and for this reason it has become rare in the wild. *Magnolia × alba* (*M. champaca × M. montana*) from Java is cultivated in China, especially around Taoist temples. There are fine potted specimens at the Purple Clouds temple on Wudang Shan in northern Hubei in Central China.

Magnolia compressa Maxim (MAGNOLIACEAE)

Michelia compressa (Maxim.) Sargent Henry in Trans. Asiat. Soc. Jap. 24 suppl: 16. (1896) (List Pl. Formosa).

Michelia sp. nova. aff. fuscatae. Henry in Trans. Asiat. Soc. Jap. 24 suppl: 16. (1896) (List Pl. Formosa). *Michelia formosana* (Kanehira) Masam. & S. Suzuki Li in Woody Flora of Taiwan 187. fig. 68. (1963).

Taiwan: Oluanpi, 946, 1984. Tamsui, 1467. Wanchin, 1543.

A tall evergreen tree to 17 metres and up to a metre in diameter. Native to southern Japan and China (Yunnan and Taiwan). In Taiwan it is a very important timber tree and found throughout the island between altitudes of 200 to 2,800 metres. The higher altitude forms would be worth introducing and might possibly prove hardy in the milder counties of Britain and Ireland.

Magnolia delavayi Franchet (MAGNOLIACEAE)

Walsh in Aug. Henry Forst. Herb. 80. (1957).

Yunnan: Mengzi, 9450.

An evergreen tree to 12 metres tall (though taller in cultivation), carrying large alternate leathery leaves. The large, solitary, ivory-white fragrant flowers appear in summer and early autumn. Discovered by Père Delavay in May 1886, this tree is most commonly distributed in Yunnan province and found over most of the region except Xishuangbanna and Dehong prefectures in the south and the high mountains in the north-west. *Magnolia delavayi* also extends into the provinces of Guizhou and Sichuan and the bark is of medicinal use. It was introduced to cultivation from southern Yunnan by E. H. Wilson in the autumn of 1899 while visiting Augustine Henry (Henry persuaded Wilson to introduce a number of plants from the Mengzi area including *Magnolia delavayi* and *Jasminum mesnyi*) before travelling on to Yichang. This fine tree first flowered under glass at the Royal Botanic Gardens, Kew in 1908. *Magnolia delavayi* Franch. f. *rubra* K. M. Feng was recently discovered in Mouding in Central Yunnan and bears pinkish-white to red flowers and would certainly be worth introducing to the west. It has been suggested (Feng in *Rare and Precious Wild Flowers of China* 2: 2. 1999) that this form may be a hybrid between *Magnolia delavayi* Franch. and *Magnolia insignis* Wall.

Magnolia figo (Lour.) DC. (MAGNOLIACEAE)

Michelia fuscata (Andr.) Blume Henry in Trans. Asiat. Soc. Jap. 24 suppl: 16. (1896) (List Pl. Formosa).

Taiwan: Wanchin, 163.

An evergreen shrub to 5 metres tall, popular as a garden plant on account of its powerfully, sweetly scented yellowish green flowers. Introduced to cultivation in 1789. Henry stated that this collection was made from a cultivated plant and was known in Taiwan as *han-hsiao*.

Magnolia fordiana (Oliv.) Hu var. ***forrestii*** (W. W. Smith ex Dandy) V. S. Kumar (MAGNOLIACEAE) New variety

Manglietia forrestii W. W. Smith ex Dandy in Notes Roy. Bot. Gard. Edinburgh 16: 126. (1928—1932).

Yunnan: In forests near Simao at 1,500 metres, a tree of 8 metres tall, flowers white, 11988.

A handsome evergreen tree to 20 metres tall, carrying white, fleshy fragrant terminal flowers to 10 cm across in June. Endemic to China and distributed in Guangxi and Yunnan where it is found in evergreen broad-leaved forest between 600 and 2,400 metres. In his field notes for 1925 George Forrest describes his F. 26,694 (collected near Tengchong in Yunnan in June 1925) as a tree of 50 to 60 feet (15 to 18 metres) tall with fleshy, fragrant pure soft white flowers. He stated it was an excellent tree in every way, it was straight in growth with a heavy bole to 8 feet (2.4 metres) in circumference and was widely branched with heavy glossy, richly coloured and beautiful flowers.

Magnolia henryi Dunn (MAGNOLIACEAE) New species

in Journ. Linn. Soc. Bot. 35: 484. (1903), E. H. Wilson in Gard. Chron. ser. 3, 39: 234. (1906), Dandy in Notes Roy. Bot. Gard. Edinburgh 16: 125. (1928—1932).

Yunnan: 8 miles south of Simao in forests, in a ravine at 1,300 metres, 12782, 12782a., (type).

An evergreen tree to 7 metres tall with large coriaceous, oblong, acute leaves to 60 cm long and 20 cm wide. The fragrant flowers appear in May and are saucer shaped with thick fleshy petals. Extremly rare in the wild, Henry himself only saw one tree ever. The species extends into Xishuangbanna in the extreme south-west of Yunnan, Laos, Myanmar and northern Thailand where it was later found by the Irish plant collector, Kerr.

Magnolia kachirachirai (Kanehira & Yamamoto) Dandy (MAGNOLIACEAE) New species

Magnolia sp. Henry in Trans. Asiat. Soc. Jap. 24 suppl: 16. (1896) (List Pl. Formosa).

Taiwan: Oluanpi, 2060.

A large evergreen tree to 17 metres tall with trunk diameters to 1.2 metres. Endemic to Chinese Taiwan where it is found as a small population on the island's southernmost tip. This species was discovered by Henry in 1894 and described from material later collected by Japanese botanists.

Magnolia officinalis Rehder & E. H. Wilson (MAGNOLIACEAE) New species

in Sargent Pl. Wilson. 1: 391. (1913), Zheng in Journ. Wuhan Bot. Res. 2.(1): 45. (1984), Nelson in Pim, The Wood & the Trees 232. (1984), Bean in Trees & Shrubs 2: 659. (1992), Nelson in A Heritage of Beauty 315. (2000), O' Brien in Augustine Henry, An Irish Plant Collector in China 26. (2002).

Magnolia sp. nova. Henry in Journ. China Br. Roy. Asiat. Soc. 22: 244. (1887) (Chinese Names of Plants). *Magnolia hypoleuca* Henry non Sieb. & Zucc. in Notes Econ. Bot. China 27. (1893), Diels in Bot. Jahrb. 29: 322. (1900), Henry in Elwes and Henry, The Trees Gt. Brit. & Irel. 6: 1592. (1912).

Hubei: Badong, 549. South Badong, 5389 (syntype). Changyang, 5389b. **Sichuan:** South Wushan, 5389a.

The *hon po*, one of Henry's earliest finds (1885), *Magnolia officinalis* is widely cultivated in Hubei and Sichuan for its bark and flower buds which make a valuable Chinese medicine which in Henry's time was exported all over China. A deciduous tree to 15 metres tall, *Magnolia officinalis* carries foliage similar to the Japanese *Magnolia obovate* and bears large white, cupped flowers in spring. The removal of its bark has led to its disappearance from the forests of central China. According to Wilson, the bark when boiled yielded an extract which when taken internally acted as an aphrodisiac, a cure for coughs, colds and as a tonic and stimulant during convalescence. A similar extract from the flower buds was called *yu-p'o* and was taken by women to correct irregularities of menstruation. Henry sent seeds to Kew in 1887, these apparently never germinated and so it was later introduced by Wilson in 1900. *Magnolia officinalis* is a rare tree in the British Isles and Ireland, being usually represented in cultivation by the var. *biloba*, which was introduced by Sir Harold Hillier, who received seeds from Lushan Botanical Gardens in 1936. From this consignment five seedlings were raised. Both the type and its variety grow at Glasnevin and Kilmacurragh. *Magnolia officinalis* was collected by the Glasnevin Central China Expedition (GCCE 343) in a valley to the north of Yesanguan in Badong in October 2002 and again by the 2004 Glasnevin Expedition to Central China on Hann Shao Shan in Xingshan (GCCE 488), in both cases these collections were from cultivated plants.

Magnolia sp. No. 1. (MAGNOLIACEAE)

Anonymous in Kew List of Determinations (volume 24).

Hubei: Badong, 3711, 4886.

May be *M. heptapeta, M. diva, M. sprengeri* var. *elongata, M. cylindrica, M. quinqepeta* or *M. officinalis* var. *biloba*.

Magnolia sp. No. 2. (MAGNOLIACEAE)

Michelia sp. Anonymous in Kew List of Determinations (volume 34).

Yunnan: Mengzi, 9429, 10892, 11441. Yuanjiang, 13277.

Magnolia sp. nova (MAGNOLIACEAE)

Anonymous in Kew List of Determinations (volume 24).

Hubei: Badong, 548.

A leafless specimen, Henry stated the flowers of this species were used as a drug.

Magnolia sprengeri Pamp. (MAGNOLIACEAE) New species

Bean in Trees & Shrubs 2: 668. (1992), O' Brien in Augustine Henry, An Irish Plant Collector in China 26. (2002), O' Brien in Ir. Garden 13(5): 53. (2004).

Magnolia yulan Henry non Desf. in Journ. China Br. Roy. Asiat. Soc. 22: 260. (1887) (Chinese Names of Plants). *Magnolia conspicua* Henry non Salisb. in Notes Econ. Bot. China 28. (1893), Diels in Bot. Jahrb. 29: 321. (1900),

Henry in Elwes and Henry, The Trees Gt. Brit. & Irel. 6: 1596. (1912). *Magnolia denudata* Desr. var. *purpurascens* (Maxim.) Rehder & E. H. Wilson in Sargent Pl. Wilson. 1: 401. (1903).

Hubei: 1475, 6597. **Sichuan:** South Wushan, 5651.

A deciduous tree to 20 metres tall with grey peeling bark and carrying in spring before the appearance of leaves, large erect goblet-shaped flowers. According to Wilson this *Magnolia* was known colloquially as *yin-tuen-shu* and like other species in central China its bark provided a drug known as *mu-pi*. In Badong, according to Henry the flowers were also used as a drug and were known as *ying-ch'un-hua* or *ch'un-hua*. Introduced by Wilson from Changyang (south of Yichang, a region visited by the Glasnevin Central China Expedition in October 2004) in 1900. It was described by Pampanini in 1915 from a collection made by the Italian missionary, Silvestri a few years previously but was first collected by Henry in 1885. These collections represent a mix of *Magnolia sprengeri* Pamp. and *Magnolia sprengeri* Pamp. var. *diva* (Dandy) Stapf. Rehder and E. H. Wilson (q.v.) cite A. Henry 5651as *Magnolia denudata* Desr. var. *purpurascens* (Maxim.) Rehder & E. H. Wilson, which is synonymous with *Magnolia sprengeri* Pamp, var. *diva* (Dandy) Stapf.

Maianthemum henryi (Baker) La Frankie (ASPARAGACEAE) New species

in Taxon 35: 588. (1986).

Oligobotrya henryi Baker in Hooker's Icon. Pl. t. 1537. (1886), Henry in Notes Econ. Bot. China 59. (1893); in Flora & Sylva 1: 218. (1903), Bretschneider in Hist. Eur. Bot. Disc. China 2: 793. (1898), Diels in Bot. Jahrb. 29: 246. (1900), Wright in Journ. Linn. Soc. Bot. 36: 109. (1903), Morley in Glasra 3: 80. (1979), Nelson in a Heritage of Beauty 316. (2000). *Smilacina henryi* (Baker) Hara in Journ. Jap. Bot. 50: 226. (1975), O' Brien in Augustine Henry, An Irish Plant Collector in China 35. (2002).

Hubei: Badong, 914 (June 1885), 5583b, 5583c., 5583e. Jianshi, 5583a. Zigui, 5583d. Xingshan, 7535a. Changyang, 7535b., 7535c., 7535d., 7535e. **Hunan:** Shimen, 7535.

Originally the type of a new genus (*Oligobotrya*), *Maianthemum henryi* was also collected in Sichuan by Faber on Emei Shan and in Kangding by Pratt and Soulie. It is a medicinal plant of the higher mountains and is the *ping-p'an-ch'i*. Cultivated at the National Botanic Gardens, Glasnevin and the Royal Botanic Gardens, Edinburgh. This species was collected by the Sino-American Expedition in the Shennongjia Forest District in 1980.

Maianthemum japonicum (A. Gray) Baker (ASPARAGACEAE)

Tovaria japonica (A. Gray) Baker Wright in Journ.

Linn. Soc. Bot. 36: 110. (1903). *Smilacina japonica* A. Gray Henry in Notes Econ. Bot. China 59. (1893), Diels in Bot. Jahrb. 29: 246. (1900).

Sichuan. South Wushan, 5366, 5366b. **Hubei:** South Badong, 5366a., Jianshi, 5366c.

Maianthemum sp. (ASPARAGACEAE)

Smilacina sp. Anonymous in Kew List of Determinations (volume 34).

Yunnan: Mengzi, 10879, 13737.

Maianthemum tatsienense (Franch.) LaFrankie (ASPARAGACEAE)

Streptopus paniculatus Baker in Hooker's Icon. Pl. t. 1932. (1890), Henry in Notes Econ. Bot. China 59. (1893), Bretschneider in Hist. Eur. Bot. Disc. China 2: 793. (1898), Wright in Journ. Linn. Soc. Bot. 36: 110. (1903), Morley in Glasra 3: 80. (1979).

Sichuan: South Wushan, 5723. **Hubei:** Jianshi, 5723a., 5723b.

A medicinal plant known as the *suan-p'an-ch'i* according to Henry.

Maianthemum tubiferum (Batal.) LaFrankie (ASPARAGACEAE)

Smilacina tubifera Batal. Diels in Bot. Jahrb. 29: 246. (1900). *Tovaria tubifera* (Batal.) C. H. Wright Wright in Journ. Linn. Soc. Bot. 36: 111. (1903).

Hubei: Jianshi, 5791, 5858. Fang Xian, 6845. Changleping, 6252. Xingshan, 6845a. Without locality, 6252a. **Sichuan:** South Wushan, 7527.

Malaisia scandens (Lour.) Planch. (MORACEAE)

Li in Woody Flora of Taiwan 701. (1963).

Malaisia tortuosa Blanco Forbes & Hemsley in Journ. Linn. Soc. Bot. 26: 454. (1894), Henry in Trans. Asiat. Soc. Jap. 24 suppl: 86. (1896).

Hainan: Environs of Haikou, 8006, 8397. At Nodoa in the interior of the island, 8450. **Taiwan:** Oluanpi, 669. Wanshoushan, 701.

A scandent shrub, native to Indochina, China, the Philippines, Malaysia and Australia. This species was collected by the Glasnevin Taiwan Expedition (GTE 24) in *Pandanus* jungle by the lighthouse at Oluanpi (South Cape of Henry and Wilson) in October 2004.

Malaxis monophyllos (L.) Sw. (ORCHIDACEAE)

Microstylis monophyllos Rolfe non (L.) Lindl. in Journ. Linn. Soc. Bot. 36: 5. (1903).

Hubei: Fang Xian, 6869, 6875.

Mallotus apelta (Lour.) Müell.-Arg. (EUPHORBIACEAE)

Forbes & Hemsley in Journ. Linn. Soc. Bot. 26: 439. (1894), Croizat in Journ. Arnold Arb. 19: 142. (1938), Zheng in Journ. Wuhan Bot. Res. 2.(1): 112. (1984).

Rottlera chinensis A. Juss. Henry in Journ. China Br.

Roy. Asiat. Soc. 22: 250. (1887) (Chinese Names of Plants). *Mallotus japonicus* Henry non (L. f.) Müell.-Arg. in Notes Econ. Bot. China 61. (1893), Forbes & Hemsley in Journ. Linn. Soc. Bot. 26: 440. (1894), Pax in Engler, Pflanzenr. iv. 147. vii: 169. (1914). *Mallotus apelta* (Lour.) Muell.-Arg. var. *chinensis* (Geisel.) Pax & Hoffm. Pax in Engler, Pflanzenr. iv. 147. iv: 171. (1914).

Hubei: Badong, 1285, 4021. Liantuo, 1971, 4559. Yichang, 2122, 3615, 3616. At Shih Pan Shan [off the Xiling (Yichang) Gorge], 4211. **Hainan:** Environs of Haikou, 8057. **Vietnam:** Tonkin, 13640.

A shrub to about 3 metres tall, common in warm, rocky valleys and on cliffs in the Three Gorges region where it was known colloquially as *kou lan ma or yeh-t'ung* according to Henry. Native to China and Vietnam.

Mallotus barbatus (Wall.) Müell.-Arg. (EUPHORBIACEAE)

Mallotus esquirolii Levl. Pax in Engler, Pflanzenr. iv. 147. vii: 196. (1914).

Forbes & Hemsley in Journ. Linn. Soc. Bot. 26: 439. (1894), Pax in Engler, Pflanzenr. iv. 147. vii: 164. (1914), Croizat in Journ. Arnold Arb. 19: 134. (1938).

Hubei: Liantuo, 3917. **Yunnan:** Mengzi, 9525, 9525a., 9525b. Simao at 1,500 metres, 13023.

Distributed in India and China.

Mallotus japonicus (Thunb.) Muell.-Arg. (EUPHORBIACEAE)

Henry in Trans. Asiat. Soc. Jap. 24 suppl: 84. (1896) (List Pl. Formosa), Croizat in Journ. Arnold Arb. 19: 139. (1938), Li in Woody Flora of Taiwan 433. (1963).

Mallotus nepalensis Müell.-Arg.var. *ochraceoalbidus* (Müell.-Arg.) Pax & K. Hoffmann in Pflanzenr. (Engler) 63. (iv. 147. vii): 166. (1914), Croizat in Journ. Arnold Arb. 19: 136. (1938).

Taiwan: Wanchin, 143, 436, 504. Tamsui, 1734. **Yunnan:** Mengzi, 10925 (collected 5th June 1898).

A handsome small tree, native to southern China and Japan, in Chinese Taiwan at low altitude. This species grows out-of-doors near the Temperate House at the Royal Botanic Gardens, Kew.

Mallotus millietii Levl. (EUPHORBIACEAE)

Croizat in Journ. Arnold Arb. 19: 147. (1938)

Mallotus contubernalis Pax non Hance in Engler, Pflanzenr. iv. 147. vii: 180. (1914).

Yunnan: Mengzi, 10669. In mountain forest to the south of Mengzi at 1,950 metres, 10700, 10700a.

Endemic to China.

Mallotus nepalensis Müell.-Arg. (EUPHORBIACEAE)

Mallotus nepalensis Müell.-Arg. var. *floccosus* (Müell.-Arg.) Pax & Hoffmann in Pflanzenr. (Engler) 63: 166. (1914). *Mallotus tenuifolius* Croizat non Pax in Journ.

Arnold Arb. 19: 137. (1938).

Yunnan: Simao, 13060, 13060a.

Native to India, Nepal, Bhutan, Myanmar and China.

Mallotus paniculatus (Lam.) Müell.-Arg. (EUPHORBIACEAE)

Li in Woody Flora of Taiwan 434. (1963).

Mallotus cochinchinensis Lour. Forbes & Hemsley in Journ. Linn. Soc. Bot. 26: 439. (1894), Henry in Trans. Asiat. Soc. Jap. 24 suppl: 84.(1896) (List Pl. Formosa), Pax in Engler, Pflanzenr. iv. 147. vii: 166. (1914). *Mallotus albus* (Roxb. ex Jack) Müell.-Arg. Pax & Hoffmann in Pflanzenr. 63. (iv. 147. vii.): 168. (1914), Croizat in Journ. Arnold Arb. 19: 144. (1938).

Hainan: Environs of Kiungchow, 8718. **Taiwan:** Wanchin, 63, 834, 1598. Oluanpi, 962. **Yunnan:** Simao, 11991c., 11991d., 11991e., 12365.

This tree was known as *peh-yeh-tze* according to Henry and grows from southern Asia through to tropical Australia.

Mallotus philippensis (Lam.) Müell.-Arg. (EUPHORBIACEAE)

Forbes & Hemsley in Journ. Linn. Soc. Bot. 26: 440. (1894), Henry in Trans. Asiat. Soc. Jap. 24 suppl: 84. (1896) (List Pl. Formosa), Pax in Engler, Pflanzenr. iv. 147. vii: 184. (1914), Li in Woody Flora of Taiwan 433. (1963), Zheng in Journ. Wuhan Bot. Res. 2.(1): 112. (1984).

Hubei: In a glen near Pingshanba (near Yichang, May 1886) 1587. **Hainan:** Near the Kiungchow Pagoda, 7988, Hummocks near Haikou, 8092. Environs of Kiungchow, 8259, 8313 (in part). Without locality, 8479. **Taiwan:** Wanchin, 473, 420, 828. Wanshoushan, 1887. Oluanpi, 1967.

A large shrub or small tree to about 7 metres tall, native to the Himalaya, China, south-east Asia and northern Australia. In Nepal, a silk dye known as kamela is obtained from the crimson powder on its fruits, the same powder in used in Hindu religious festivals.

Mallotus repandus (Rottler) Muell.-Arg. (EUPHORBIACEAE)

Pax in Engler, Pflanzenr. iv. 147. vii: 180. (1914), Henry in Journ. China Br. Roy. Asiat. Soc. 22: 249. (1887) (Chinese Names of Plants), Forbes & Hemsley in Journ. Linn. Soc. Bot. 26: 441. (1894), Zheng in Journ. Wuhan Bot. Res. 2.(1): 113. (1984).

Mallotus contubernalis Hance Henry in Journ. China Br. Roy. Asiat. Soc. 22: 249. (1887) (Chinese Names of Plants), Forbes & Hemsley in Journ. Linn. Soc. Bot. 26: 441. (1894), Pax in Engler, Pflanzenr. iv. 147. vii: 180. (1914), Zheng in Journ. Wuhan Bot. Res. 2.(1): 113. (1984). *Mallotus chrysocarpus* Pamp. Croizat in Journ. Arnold Arb. 19: 147. (1938). *Mallotus repandus* Forbes & Hemsley

non Müell.-Arg. in Journ. Linn. Soc. Bot. 26: 441. (1894). *Mallotus repandus* (Willd.) Müell.-Arg. var. *megaphyllus* Croizat in Journ. Arnold Arb. 19: 146. (1938).

Hubei: Yichang, 678, 1354, 3590. Antelope Glen, 4151. In a ravine on Moji Shan, 3592. Near Yichang in the Monastery Glen, 1494. Badong, 1765. Liantuo, 3858. Jianshi, 5861. South Badong, 7292. **Sichuan:** South Wushan, 5542. **Hainan:** Environs of Haikou, 7991, 8038, 8319. 8 km (5 miles) south of Kuingchow, 8174. **Taiwan:** Kaohsiung, without number. Wanchin, 714. Oluanpi, 918. **Yunnan:** Mengzi, 13696 [syntype of *Mallotus repandus* (Willd.) Müell.-Arg. var. *megaphyllus* Croizat].

A common climbing shrub to about 3 metres tall, found in the warmer parts of Hubei and Sichuan. Southern Asia to northern Australia, in the Three Gorges area it was known colloquially as *kan hsiang t'eng* according to Henry.

Mallotus spp. (EUPHORBIACEAE)

Anonymous in Kew List of Determinations (volume 24 & 34).

Hubei: Fang Xian, 6628. **Yunnan:** In the Red River Valley near Manhao, 9537, 9544. Mengzi, 10714. In Pingbian, on the Daweishan Range at 2,300 metres, 11252. Simao, 11973, 11990, 12455, 13149.

Mallotus tenuifolius Pax (EUPHORBIACEAE) New species

Hutchinson in Sargent Pl. Wilson. 2: 525. (1916).

Mallotus japonicus Forbes & Hemsley non (L. f.) Müell.-Arg. in Journ. Linn. Soc. Bot. 26: 440. (1894), Pax in Engler, Pflanzenr. iv. 147. vii: 169. (1914).

Hubei: Changleping, 6260. Liantuo, 5742c. Jianshi, 5742a. **Sichuan:** South Wushan, 5742, 5742b.

A common roadside bush or small tree in Hubei to 5 metres tall and particularly abundant in warm, rocky valleys. Discovered by Augustine Henry about 1888, Pax based his description on a collection made by A. von Rosthorn in Sichuan between 1891 and 1892.

Mallotus tiliifolius (Blume) Müell.-Arg. (EUPHORBIACEAE)

Pax in Engler, Pflanzenr. iv. 147. vii: 148. (1914), Li in Woody Flora of Taiwan 435. (1963).

Mallotus playfairii Hemsl. Henry in Trans. Asiat. Soc. Jap. 24 suppl: 84. (1896) (List Pl. Formosa).

Taiwan: Kaohsiung, 766, 1766, 1766a.

A widely distributed shrub, native from the Philippines to Sumatra, New Guinea and northern Australia. Confined to the southern part of Chinese Taiwan.

Mallotus yunnanensis Pax & K. Hoffm. (EUPHORBIACEAE) New species

in Engler, Pflanzenr. iv. 147. vii: 188. (1914).

Yunnan: Mengzi at 1,500 metres, 10794 (syntype).

02

13629 (syntype).

Endemic to China and distributed in Guangxi, Guizhou and Yunnan.

Malus asiatica Nakai (ROSACEAE)

Pyrus malus Anonymous non L. in Kew List of Determinations (volume 24).

Hubei: South Badong, 5471.

Malus hupehensis (Pamp.) Rehder (ROSACEAE) New species

in Journ. Arnold Arb. 14: 207. (1933).

Pyrus baccata Henry non L. in Notes Econ. Bot. China 64. (1893) .*Pyrus spectabilis* Forbes & Hemsley non Ait. in Journ. Linn. Soc. Bot. 23: 258. (1887), Anon in Gard. Chron. ser. 3. 786. (1887), Henry in Notes Econ. Bot. China 64. (1893). *Malus theifera* Rehder in Sargent Pl. Wilson 2: 283. (1916).

Hubei: Yichang, 7398. Badong, 783 (May 1885). Liantuo, 3814, 4547, 7398. Xingshan, 6509.

A deciduous tree to 15 metres tall, flowers fragrant, white and tinged rose on emererging, followed in autumn by green fruits, tinged red on the exposed side. *Malus hupehensis* is the *hung-ch'a* or *lin-ch'in-ch'a* (or in Fang district the *hai-t'ang-ch'a*) according to Henry. Rehder probably named his *Malus theifera* from Henry's field notes which stated;

In the Patung district (Badong) there are two tea plants (substitutes for *Camellia sinensis*): one 'white tea' used for tea drunk by natives, and the other 'red tea' used for making black tea exported for making black tea exported to foreign countries. This specimen of *Pyrus spectabilis* (*Malus hupehensis*) is the 'red tea, and the shrub is larger than the 'white tea', the flower is redder (that of the 'white tea' being white, or perhaps a pale pink) and hangs down more.

Henry's 'red tea' was named *Malus theifera* f. *rosea* by Alfred Rehder. E. H. Wilson stated it was abundant in western Hubei and was occasionally cultivated. Seeds of Wilson's 1129 from Wa Shan in Sichuan reached Glasnevin in 1909 (he introduced it through Messrs Veitch in 1900). *Malus hupehensis* is allied to *Malus baccata,* it is a triploid and produces its seeds apomictically.

Malus kansuensis (Batal.) Schneid. var. **calva** (Rehder) T. C. Ku & Sponberg (ROSACEAE) New variety

Malus kansuensis (Batal.) Schneid. f. *calva* Rehder in Journ. Arnold Arb. 2: 50. (1922), Zheng in Journ.Wuhan Bot. Res. 2.(1): 71. (1984). *Pyrus kansuensis* Bretschneider non Batal. in Hist. Eur. Bot. Disc. China 2: 1028. (1898). *Malus kansuensis* Rehder non (Batal.) Schneid. in Sargent Pl. Wilson. 2: 286. (1916).

Hubei: Fang Xian, 6754. Xingshan, 6754a.

A deciduous tree to 7 metres tall, carrying clusters

of four to ten white flowers in terminal corymbs in May followed by scarlet egg-shaped fruits. Discovered by Henry in 1888 and described from material later gathered by Wilson. Native to Gansu, Hubei and Sichuan and introduced by Wilson in 1910, though W. J. Bean mentions he may have previously sent it to Messrs Veitch in 1900.

Malus toringo (Siebold) de Vriese (ROSACEAE)

Pyrus toringo (Sieb.) Miq. var. *baccata* Anonymous in Kew List of Determinations (volume 24).

Hubei: Changyang, 5244. **Sichuan:** South Wushan, 5244a.

Malus yunnanensis (Franch.) C. K. Schneid. var. **veitchii** (Osborn) Rehder (ROSACEAE)

New variety. in J. Arnold Arbor. 4: 115. (1923).

Malus yunnanensis Rehder non (Franch.) C. K. Schneid. Rehder in Sargent Pl. Wilson. 2: 287. (1916).

Sichuan: South Wushan, 5638 (syntype). **Hubei:** Changyang, 5638a.

The type, i.e. *Malus yunnanensis* was discovered by Père Delavay in Yunnan, the var. *veitchii* differs in having more lobed leaves and was discovered by Henry in the spring of 1888. Wilson's 1000a. from Mao-chou in Hubei reached Glasnevin in 1909, it grows in the old cemetry border where it has formed a handsome tree to about 10 metres and fruits heavily each autumn. This variety was also collected in Yunnan, near Lijiang by Forrest in May 1906. Bean concidered this the best of Wilson's introductions amoungst crabs, it bears a heavy crop of very ornamental red crabs and the leaves turn to scarlet and orange in the fall.

Malva parviflora L. (MALVACEAE)

Henry in Journ. China Br. Roy. Asiat. Soc. 22: 248. (1887) (Chinese Names of Plants).

Hubei: Yichang, 380.

Cultivated at Yichang as a vegetable where Henry stated it was known colloquially as *mao tung han ts'ai.*

Malva sylvestris L. (MALVACEAE)

Anonymous in Kew List of Determinations (volume 34).

Yunnan: Mengzi, 10820.

The *chin k'uei* or silk mallow, a native of Europe and temperate Asia. This species is cultivated in Chinese gardens for its ornamental flowers.

Malva verticillata L. (MALVACEAE)

Henry in Journ. China Br. Roy. Asiat. Soc. 22: 247. (1887) (Chinese Names of Plants).

Hubei: Badong, 425 (June 1885), 1337. North Badong, 7164. **Yunnan:** Mengzi, 10348.

The winter mallow or *tung k'uei,* according to Henry this plant was known colloquially at Yichang as *tung han ts'ai* and occurred both wild and cultivated there. This annual is still widely cultivated in parts of China today. All parts of the plant are edible while young and are eaten in soups or boiled like spinach. The seeds also supplies the drug *dong kui zi* which promotes urination.

Malvaceae, subfamily Tiloideae (MALVACEAE)

Tiliaceae Anonymous in Kew List of Determinations (volume 34).

Yunnan: Simao, 13445.

Malvastrum coromandelianum (L.) Gurcke (MALVACEAE)

Malvastrum tricuspidatum (R. Br.) A. Gray Henry in Trans. Asiat. Soc. Jap. 24 suppl: 20. (1896) (List Pl. Formosa).

Taiwan: Kaohsiung, without number.

A native of tropical and subtropical America, now naturalised throughout the tropics and common at Kaohsiung.

Mandragora caulescens C. B. Clarke (SOLANACEAE)

Anonymous in Kew List of Determinations (volume 24).

Hubei: Changyang, 7736.

Mangifera indica L. (ANACARDIACEAE)

Henry in Trans. Asiat. Soc. Jap. 24 suppl: 30. (1896) (List Pl. Formosa).

Taiwan: Kaohsiung, without number. Wanchin, 817. **Yunnan:** Simao, 11643.

The mango, a large evergreen tree native from India (Assam) and Myanmar to China (Yunnan) and now extensively naturalised over much of tropical Asia. The fruits of the wild naturalised type are small ands inedible according to Henry and were known in Chinese Taiwan as *shuain*. The Mango was introduced to Taiwan by the early Dutch settlers and it is now one of the most popular of all tropical fruits and is grown on a worldwide basis. In southeast Asia, in times of famine, the cotyledons of the seeds have been ground into a flour substitute and the leaves can be used as fodder for cattle. The fruit is revered as a sacred symbol in India and Nepal, various parts of it are used in religious ceremonies, especially for marriages, births and funerals. The trees are warshipped, being homes to the Gods and Godesses and in places it is concidered to be sinful to cut down a mango. In Nepal it is planted at meeting or resting places as its thick crown casts a dense cool shade. A tree to 30 metres tall, the mango belongs to the same family as poison ivy and the Chinese varnish tree [*Toxicodendron* (*Rhus*) *verniciflua*] and produces an irritating latex. Mangoes were introduced to Brazil by the Portuguese in the early 1700s and from there reached the West Indies in 1742. They are now an important food source for the poorer inhabitants of Jamaica and Haiti.

Manihot sp. (EUPHORBIACEAE)

Anonymous in Kew List of Determinations (volume 34).

Yunnan: Simao at 1,300 metres, 12754. Simao, 13305.

Probably the tapioca plant, *Manihot esculenta* Crantz., a native of Brazil and widely cultivated in the tropics.

Mappia pittosporoides Oliver (ICACINACEAE) New species

in Hooke's Icon. Pl. 18: t. 1762. (1888); in Journ. Bot. 27: 154. (1889), Diels in Bot. Jahrb. 29: 447. (1900), Rehder & E. H. Wilson in Sargent Pl. Wilson 2: 190. (1916), Hand.-Mazz. in Symb. Sin. 7: 667. (1933), Chun in Sunyatsenia 2: 72. (1934); 4: 229. (1940). *Neoleretia pittosporoides* (Oliv.) Baehni in Candollea 7: 178. (1936).

Nothapodytes pittosporoides (Oliv.) Sleumer in Notizbl. Berl.-Dahl. 15: 247. (1940), Howard in Journ. Arnold. Arb. 68: 23. (1942); 33: 273. (1952), Wu & Wang in Acta Phytotax. Sin. 6: 281. (1957), Zheng in Journ. Wuhan Bot. Res. 2.(1): 128. (1984).

Hubei: Yichang, 3537, 3537a., 4118, 3536 (syntype), 3990 (type).

A shrub to 3 metres tall with yellow, unpleasantly scented flowers and dark red fruits.

Marchantia nitida Lehm. & Lindenebg. (MARCHANTIACEAE)

Anonymous in Kew List of Determinations (volume 24).

Hubei: Yichang, 7862. (Liverwort).

Marchantia sp. (MARCHANTIACEAE)

Anonymous in Kew List of Determinations (volume 24).

Sichuan: North Wushan at 2,650 metres, 7149. (Liverwort).

Markhamia stipulata (Wall.) Seem. (BIGNONIACEAE)

Dolichandrone cauda-felina (Hance) Benth. & Hook. f. Anonymous in Kew List of Determinations (volume 34).

Yunnan: In the Red River valley to the south of Mengzi, 10121, 10121a.

A deciduous tree to 10 metres tall, the young branches are covered in a dense brown pubescence. During the winter months purple-red to pink bell-shaped flowers are produced in terminal racemes followed by large seed pods to 40 cm tall, similar to a cat's tail. This highly ornamental species is distributed in Bangladesh, Myanmar, China (Guangdong, Guangxi, Hainan and Yunnan) and Indochina.

Marrubium sp. (LAMIACEAE)

Anonymous in Kew List of Determinations (volume 34).

Yunnan: Mengzi, 10907.

02

Marsdenia brachyloba Gilbert & Li (APOCYNACEAE)
New species

in Novon 5: 11. (1995).

Yunnan: South of the Red River on Fungchunling, 11073 (paratype), 11196 (holotype).

Endemic to Yunnan.

Marsdenia griffithii Hook. f. (APOCYNACEAE)

Anonymous in Kew List of Determinations (volume 34).

Yunnan: Yuanjiang, 13236.

Distributed in India and China (Guizhou, Hunan and Yunnan).

Marsdenia sinensis Hemsley (APOCYNACEAE) New species

in Journ. Linn. Soc. Bot. 26: 113. (1889), Bretschneider in Hist. Eur. Bot. Disc. China 2: 788. (1898), Morley in Glasra 3: 79. (1979).

Hubei: On the banks of the Yangtze at Yichang, 4364. Liantuo, 7185.

Marsdenia sp. (APOCYNACEAE)

Anonymous in Kew List of Determinations (volume 34).

Yunnan: Mengzi, 11857. Simao, 11878.

Marsdenia tinctoria R. Br. (APOCYNACEAE)

Forbes & Hemsley in Journ. Linn. Soc. Bot. 26: 114. (1888), Henry in Trans. Asiat. Soc. Jap. 24 suppl: 61. (1896) (List Pl. Formosa).

Marsdenia tomentosa Li non Morr. & Decne. in Woody Flora of Taiwan 804. (1963).

Hubei: Antelope Glen, 3282. Antelope Glen, 4092, 4093. Yichang, 4093a. **Hainan:** Environs of Haikou, 8059, 8175. **Taiwan:** Kaohsiung, without number. Wanchin, 470.

A tall climber with fleshy yellow flowered climber, widely distributed from India to southern China, Japan and Malaysia. Distributed in Chinese Taiwan throughout the island.

Mastixia pentandra Blume subsp. ***chinensis*** (Merrill) K. M. Matthew (NYSSACEAE) New subspecies

Mastixia chinensis Merrill in Sunyatsenia 3: 256. (1937).

Yunnan: In forests to the south of Simao at 1,300 metres, 12414 (isotype of *Mastixia chinensis* Merrill).

Distributed in India, Myanmar, China (Yunnan), Thailand and Vietnam.

Matsumurella chinense (Benth.) Bendiksby (LAMIACEAE)

Lamium chinense Benth. Forbes & Hemsley in Journ. Linn. Soc. Bot. 26: 303. (1890), Dunn in Notes Roy. Bot. Gard. Edinburgh vi. 183. (1915).

Zhejiang: Pu-to Island, 31. **Hubei:** Yichang 1182.

Changyang 5220.

Mayodendron igneum (Kurz.) Kurz. (BIGNONIACEAE)

Anonymous in Kew List of Determinations (volume 34).

Yunnan: Simao, 11744.

An evergreen tree to 15 metres tall, bearing short racemose inflorescences of tubular orange-yellow flowers directly from the trunk and mature branches. Distributed in India, Myanmar, Vietnam and southern China. A handsome tree known in China as *huo shao hua shu* according to Henry.

Mazus gracilis Hemsley (MAZACEAE) New species

in Journ. Linn. Soc. Bot. 26: 181. (1890), Bretschneider in Hist. Eur. Bot. Disc. China 2: 788. (1898), Morley in Glasra 3: 79. (1979).

Hubei: Badong, 4063 (isotype).

Mazus henryi P. C. Tsoong (MAZACEAE) New species

in Kew Bull. Misc. Inf. 9: 444. (1944).

Mazus henryi P. C. Tsoong var. *elatior* P. C. Tsoong in Kew Bull. Misc. Inf. 9: 444. (1944).

Yunnan: In mountain forest to the east of Simao at 1,650 metres, 11620, 12949. Simao at 1,300 metres, 12906.

Mazus procumbens Hemsley (MAZACEAE) New species

in Journ. Linn. Soc. Bot. 26: 182. (1890).

Hubei: Yichang, 636, 4182. On Tsui Fu Shan near Yichang (May 1888), 1628.

Mazus pulchellus Hemsley (MAZACEAE) New species

in Journ. Linn. Soc. Bot. 26: 182. (1890).

Hubei: Yichang, 800, 1377, 1377a. Antelope Glen, 3418 (isosyntype).

Originally discovered by Charles Maries in the Xiling (Yichang) Gorge in 1879, Hemsley based the species on material gathered in the region by both Henry and Maries.

Mazus pumilis (Burm. f.) Steenis. (MAZACEAE)

Mazus rugosus Lour. Forbes & Hemsley in Journ. Linn. Soc. Bot. 26. 183. (1888), Henry in Trans. Asiat. Soc. Jap. 24 suppl: 67. (1896) (List Pl. Formosa).

Hubei: Yichang, 653, 1171, 1318. Changyang, 7749. **Sichuan:** South Wushan, 7242. **Yunnan:** Mengzi, 9275, 13761. **Taiwan:** Kaohsiung, 2032. Wanchin, 1724.

Widespread in Asia.

Mazus sp. (MAZACEAE)

Anonymous in Kew List of Determinations (volume 24).

Hubei: Changyang, 7756.

Mazus stachydifolius (Turcz.) Maxim. (MAZACEAE)

Forbes & Hemsley in Journ. Linn. Soc. Bot. 26: 183. (1890).

Hubei: Yichang, 686, 804, 831, 1231, 1324. Liantuo, 6368. **Sichuan:** South Wushan, 5757.

Meconopsis sp. (PAPAVERACEAE)

Anonymous in Kew List of Determinations (volume 24).

Sichuan: Henry's Chinese collector with Pratt, no locality stated but may be Emei Shan, Wa Shan or Kangding, 8793.

Material too young for determination.

Medicago lupulina L. (FABACEAE)

Henry in Journ. China Br. Roy. Asiat. Soc. 22: 263. (1887) (Chinese Names of Plants).

Hubei: Yichang, 490.

According to Henry this species was known colloquially in the Three Gorges region as *pan chui*.

Medicago sativa L. (FABACEAE)

Henry in Journ. China Br. Roy. Asiat. Soc. 22: 263. (1887) (Chinese Names of Plants)

Hubei: Yichang, 654, 1264.

Alfalfa or lucerne, a perennial pasture crop that may also be harvested for hay, in China it is the *kam fa ts'oi*. In suitable climates it may be harvested up to nine times a year. Cultivated since prehistoric times it is now widely grown all over the world with over 33 million hectares harvested in 1995. Primitive diploids occur in Iran (the cultivated plant is tetraploid) and it is believed that this was the area of domestication. The young shoot tips are cooked and eaten like spinach. Introduced to Hong Kong in 1945 by emmigrant Shanghai workers.

Medicago sp. (FABACEAE)

Anonymous in Kew List of Determinations (volume 24).

Hubei: Yichang, 820.

Medinilla formosana Hayata (MELASTOMATACEAE) New species

Li in Woody Flora of Taiwan 657. (1963).

Medinilla sp. Henry in Trans. Asiat. Soc. Jap. 24 suppl: 44. (1896) (List Pl. Formosa).

Taiwan: Oluanpi, 1346.

A scandent shrub, endemic to Taiwan where it is found on the southern part of the island.

Medinilla rubicunda (Jack) Blume (MELASTOMATACEAE)

Medinilla yunnanensis H. L. Li in J. Arnold Arbor. 35: 39. (1944), Li in Journ. Arnold Arb. 25: 310. (1944).

Yunnan: In the mountains to the west of Simao at 1,650 metres, 12075 (type of *Medinilla yunnanensis* H. L. Li), 12075a.

Medinilla sp. (MELASTOMATACEAE)

Anonymous in Kew List of Determinations (volume 34).

Yunnan: Mengzi, 10463.

Meehania faberi (Hemsley) C. Y. Wu (LAMIACEAE) New species

Dracocephalum faberi Hemsley in Journ. Linn. Soc. Bot. 26: 291. (1890).

Hubei: Badong, 4700. **Sichuan:** North Wushan, 7084 (isosyntype of *Dracocephalum faberi* Hemsley).

Hemsley based this species on material collected in 1887 by Henry at Badong and by the German missionary and botanist, the Rev'd Dr. Ernst Faber who gathered his specimens on Emei Shan (Mount Omei) in the same year. The Glasnevin Central China Expedition visited both regions in the autumn of 2002.

Meehania fargesii (Levl.) C. Y. Wu (LAMIACEAE) New species

Dracocephalum urticifolium Henry non Miq. in Journ. China Br. Roy. Asiat. Soc. 22: 243. (1887) (Chinese Names of Plants), Forbes & Hemsley in Journ. Linn. Soc. Bot. 26: 293. (1890), Dunn in Kew Bull. Misc. Inf. 202. (1917).

Hubei: Badong, 205, 237 (June 1885), 751. Changleping, 6322.

According to Henry this plant was known as *ho hsiang* at Yichang. Cultivated at the Royal Botanic Gardens, Kew.

Meehania fargesii (Levl.) C. Y. Wu var. *pedunculata* (Hemsl.) C. Y. Wu (LAMIACEAE) New variety

Dracocephalum urticifolium Miq. var. *pedunculatum* Hemsley in Journ. Linn. Soc. Bot. 26: 393. (1890). *Dracocephalum urticifolium* Miq. var. *typica* f. *racemosa* Dunn in Notes Roy. Bot. Gard. Edinburgh 6: 170. (1915).

Hubei: Badong, 1408, 5330. Jianshi, 5798. **Sichuan:** South Wushan, 5330a. North Wushan, 7041.

Meehania henryi (Hemsley) Y. Z. Sun ex C. Y. Wu (LAMIACEAE) New species

Dracocephalum henryi Hemsley in Journ. Linn. Soc. Bot. 26: 291. (1890), Bretschneider in Hist. Eur. Bot. Disc. China 2: 789. (1898), Morley in Glasra 3: 79. (1979).

Hubei: South Badong, 6109. Zigui 6109a., (isotype).

Meehania pinfaensis (H. Lev.) Sun & C. Y. Wu (LAMIACEAE)

Dracocephalum urticifolium Miq. var. *angustifolium* Dunn f. *normalis* Dunn in Notes Roy. Bot. Gard. Edinburgh 6: 170. (1915).

Yunnan: Mengzi, 13695.

Megacodon venosus (Hemsl.) Harry Sm. (GENTIANACEAE) New species

Gentiana (Megacodon) venosa Hemsley in Journ. Linn. Soc. Bot. 26: 137. (1890), Bretschneider in Hist. Eur. Bot. Disc. China 2: 788. (1898), Henry in Flora & Sylva 1: 217. (1903), Nelson in Pim, The Wood & the Trees 227. (1984); in A Heritage of Beauty 313. (2000).

02

Sichuan: North Wushan, 7134 (type of *Gentiana venosa* Hemsley).

A magnificent perennial to 1.75 metres tall with large white flowers to 7.5 cm. in diameter, spotted green.

Melampyrum roseum Maxim. (OROBANCHACEAE)

Forbes & Hemsley in Journ. Linn. Soc. Bot. 26: 220. (1890), Diels in Englers, Bot. Jahrb. 29: 569. (1901).

Hubei: Badong, 551 (September 1885), 2472, 4856, 6149. Xingshan, 4519.

Melampyrum roseum Maxim. var. *obtosifolium* (Bonati) D. Y. Hong (OROBANCHACEAE)

Melampyrum roseum Forbes & Hemsley non Maxim. in Journ. Linn. Soc. Bot. 26: 220. (1890), Diels in Englers, Bot. Jahrb. 29: 569. (1901). *Melampyrum laxum* Miq. var. *henryanum* Beauvard in Mem. Soc. Phys. Geneve 38: 543. (1916). *Melampyrum henryanum* (Beauvard) Soo in Journ. Bot. 65: 143. (1927).

Hubei: Xingshan, 4513 (type of *Melampyrum laxum* Miq. var. *henryanum* Beauvard).

In Soo's ennumeration, The Melampyrum Species of East Asia, he lists a Henry collection of *Melampyrum japonicum* (Fr. & Sav.) Soo made near Arima in Japan. This collection was not made by Augustine Henry, but by his first wife Caroline Henry (*d.* 1894). Caroline Henry collected specimens in both Japan and Colorado, USA. Her collections are housed at the Royal Botanic Gardens, Kew and she is commemorated in *Primula carolinehenryae* S. O'Brien (q.v.).

Melampyrum sp. (OROBANCHACEAE)

Anonymous in Kew List of Determinations (volume 24).

Hubei: Yichang, 2059. Liantuo, 3915. Badong, 2422.

Melanolepis multiglandulosa (Reinw.) Reich. f. & Zoll. (EUPHORBIACEAE)

Li in Woody Flora of Taiwan 435. (1963).

Mallotus moluccanus Henry non (L.) Müell.-Arg. in Trans. Asiat. Soc. Jap. 24 suppl: 84. (1896) (List Pl. Formosa).

Taiwan: Wanchin, 45, 151. Kaohsiung, 735, 1847. Oluanpi, 919.

A shrub or small tree, native to tropical Asia and common in Chinese Taiwan where it is also widely cultivated. This species was collected by the Glasnevin Taiwan Expedition (GTE 31) by the lighthouse at Oluanpi (South Cape of Henry and Wilson) in October 2004. A very handsome foliage plant.

Melanoseris cyanea (D. Don) Edgew. (ASTERACEAE)

Lactuca hastata DC. Anonymous in Kew List of Determinations (volume 34).

Yunnan: Mengzi, 9085, 11291. Milê district, 9943.

Melanoseris graciliflora (DC.) N. Kilian (ASTERACEAE)

Lactuca graciliflora DC. Forbes & Hemsley in Journ. Linn. Soc. Bot. 23: 482. (1888).

Hubei: Badong, 2544, 2396.

This species was collected by the Sino-American Expedition in the Shennongjia Forest District in 1980.

Melanoseris henryi (Dunn) N. Kilian (ASTERACEAE)
New species

Lactuca henryi Dunn in Journ. Linn. Soc. Bot. 35: 512. (1903).

Yunnan: Pu'er at 1,450 metres, 13494.

Melastoma malabathricum L. (MELASTOMATACEAE)

Melastoma polyanthum Blume Li in Journ. Arnold Arb. 5: 9. (1944). *Melastoma candidum* D. Don Henry in Trans. Asiat. Soc. Jap. 24 suppl: 43. (1896) (List Pl. Formosa), Li in Journ. Arnold Arb. 25: 8. (1944), Li in Woody Flora of Taiwan 659. fig. 265. (1963).

Sichuan: Henry's Chinese collector with Pratt, no locality stated but may be any of the following, Leshan, Emei Shan, Wa Shan or Kangding, 8939. **Hainan:** Environs of Haikou, 7999, 8161, 8402a. **Taiwan:** Kagee, without number. Wanchin, 179. Oluanpi, 1758. **Yunnan:** Simao, 11712, 11712a.

An evergreen shrub to 1.5 metres tall, flowers are borne in terminal corymbs and are pink or very rarely white. Distributed in China (southern Guizhou, Taiwan and Yunnan) and Indochina to Australia. A very common shrub at low altitude in Taiwan and known as *shan-shih-liu* according to Henry. The Hani (an ethnic group) of Xishuangbanna in southern Yunnan call this species *byuqbaivq* and is used to treat cases of indigestion, diarrhoea and hepatitis.

Melastoma malabathricum L. subsp. *normale* (D. Don) Karst. Mey. (MELASTOMATACEAE)

Melastoma normale D. Don Rehder & E. H. Wilson in Sargent Pl. Wilson. 2: 421. (1916), Li in Journ. Arnold Arb. 25: 8. (1944).

Sichuan: Emei Shan, Henry's Chinese collector with Pratt, 8976. **Yunnan:** In Pingbian in forests on the Daweishan Range at 1,640 metres, 10954. Mengzi at 1,600 metres, 10964.

A showy, evergreen shrub to 1 metre tall, bearing terminal corymbs of purple-red flowers in spring and summer. Widely distributed through Nepal, India, Myanmar, south-west China, the Philippines and Malaysia.

Melastoma sanguineum Sims. (MELASTOMATACEAE)

Anonymous in Kew List of Determinations (volume 24).

Hainan: 32 km (20 miles) west of Haikou, 8144.

Environs of Kiungchow, 8748.

Melastomataceae No. 1. (MELASTOMATACEAE)

Anonymous in Kew List of Determinations (volume 24).

Hubei: Liantuo, 4446.

Melastomataceae No. 2. (MELASTOMATACEAE)

Anonymous in Kew List of Determinations (volume 34).

Yunnan: Mengzi, 10456.

Melia azedarach L. (MELIACEAE)

Henry in Journ. China Br. Roy. Asiat. Soc. 22: 255. (1887) (Chinese Names of Plants); in Notes Econ. Bot. China 37. (1893); in Trans. Asiat. Soc. Jap. 24 suppl: 26. (1896) (List Pl. Formosa), Rehder & E. H. Wilson in Sargent Pl. Wilson 2: 157. (1916), Zheng in Journ. Wuhan Bot. Res. 2.(1): 109. (1984).

Melia toosendon Sieb. & Zucc. Henry in Notes Econ. Bot. China 37. (1893).

Hubei: Liantuo, 3882. In the Shih Pai Gorge off the Yichang (Xiling) Gorge, 7620. **Taiwan:** Oluanpi, 1281. **Yunnan:** Chu-yuan, 10864. In forests near Simao at 1,500 metres, 12889.

The Persian lilac or China-berry, a deciduous tree to 13 metres tall, indigenous to northern India and central and western China and now widely spread in warm-temperate and subtropical parts of the world. According to Wilson it was colloquially known as *lien shu* or *ch'uan-lien shu* in Hubei and Sichuan and the fruits were occasionally employed as a drug. Henry called it the *chuan lian zi* or Sichuan pagoda tree, stating the fruits were the source of a medicine used to expel parasites like roundworm and tapeworm.

In Nepal it is also much esteemed by rural villagers since most parts of the tree are used as medicine. To the Nepalese it is known as *neem* and from it preparations are made for the relief of snake bite and scorpion stings as well as tonics, poltices and antiseptics. The strong smelling oil from the fruits is used to alleviate rheumatism. The tree is also revered by Hindus, and its timber is used to make idols. The name bead tree originated from the habit of monks stringing them into rosaries.

This species contains a bitter narcotic which is poisonous. The fruits are the most poisonous part and cause a feeling of excitement and causes the heart to beat rapidly. This can be followed by paralysis, heart weakness, difficulty in breathing and in some cases, death due to asphyxia. In Asia and Central America an extract of the bark is used as a fish poison while in East Africa a preparation made from the leaves is used to repel locusts. In Ireland it is cultivated at Glanleam on Valentia Island.

Melica onoei Franch. & Savat. (POACAEA)

Rendle in Journ. Linn. Soc. Bot. 36: 419. (1894.)

Hubei: Fang Xian, 6641, 6716. **Sichuan:** North Wushan, 7034.

Distributed from Pakistan to China, Korea and Japan.

Melicope pahangensis T. G. Hartley (RUTACEAE)

Euodia simplicifolia Ridley Rehder & E. H. Wilson in Sargent Pl. Wilson. 2: 135. (1914).

Yunnan: In woods near Simao at 1,500 metres, 12092a.

Southern Yunnan to Peninsular Malaysia.

Melicope pteleifolia (Champ. ex Benth.) T. G. Hartley (RUTACEAE)

Euodia roxburghiana Henry non (Champ.) Benth. in Trans. Asiat. Soc. Jap. 24 suppl: 25. (1896) (List Pl. Formosa).

Taiwan: Wanchin, 121. Oluanpi, 1262, 2065. **Yunnan:** At Yulo, south of Simao (Xishuangbanna), 11658.

A widespread species, distributed across China, Myanmar, Indochina and Thailand.

Melicope triphylla (Lam.) Merr. (RUTACEAE)

Euodia triphylla (Lam.) DC. Anonymous in Kew List of Determinations (volumes 24 & 34).

Hainan: Environs of Kiungchow, 8719. **Yunnan:** Mengzi, 10351.

Melilotus indicus (L.) All. (FABACEAE)

Melilotus parviflorus Desf. Anonymous non Desv. in Kew List of Determinations (volume 24).

Hubei: Yichang, 311.

An upright perennial herb to 1.5 metres tall bearing yellow flowers in autumn. At the present time this species is cultivated in the order beds in front of the Great Palm House at Glasnevin from collections made by the Glasnevin Central China Expedition (GCCE 530) in Xingshan in September 2004.

Melilotus officinalis (L.) Lam. (FABACEAE)

Melilotus arvensis Wallr. Anonymous in Kew List of Determinations (volume 34).

Yunnan: Mengzi, 10386.

Melilotus sp. (FABACEAE)

Anonymous in Kew List of Determinations (volume 24).

Hubei: Yichang, 740.

Melilotus suaveolens Ledeb. (FABACEAE)

Anonymous in Kew List of Determinations (volume 24).

Hubei: Changyang, 7732.

Meliosma arnottiana (Wight) Walp. (SABIACEAE)

Meliosma wallichii Planchon ex Hook. f. Rehder & E. H. Wilson in Sargent Pl. Wilson. 2: 207. (1914). *Meliosma pinnata* (Roxb.) Walp. ssp. *arnottiana* (Walp.) Beus. Beusekom in Blumea 19: 499. (1971).

Yunnan: In woods near Mengzi at 1,800 metres, 10835. In mountain forest to the east of Simao, 12016, 12016a., 12016b. In woods near Mengzi at 1,800 metres, 13692.

Distributed in Nepal, India, Sri Lanka and China [Tibet (Xizang), Guangxi and Yunnan].

Meliosma cuneifolia Franch. (SABIACEAE)

Rehder & E. H. Wilson in Sargent Pl. Wilson. 2: 199. (1914), Zheng in Journ. Wuhan Bot. Res. 2.(1): 135. (1984).

Hubei: Jianshi, 5849, 5849a., (August 15th 1888). Fang Xian, 5849b. South Badong, 5849dc. **Hunan:** Shimen, 7550.

A small tree or large shrub, common in western Hubei and Sichuan at the margins of woods. Discovered by Père Armand David in Baoxing in western Sichuan in June 1869 and introduced to cultivation by E. H. Wilson in 1901 through Messrs Veitch's Coombe Wood Nursery where it first flowered in July 1909. It is a handsome shrub when carrying its deliciously fragrant panicles of yellowish-white flowers in May.

Meliosma flexuosa Pamp. (SABIACEAE) New species

Zheng in Journ. Wuhan Bot. Res. 2.(1): 135. (1984).

Meliosma pendens Rehder & E. H. Wilson in Sargent Pl. Wilson 2: 200. (1914).

Hubei: Jianshi, 6000. Changyang, 5849d., (paratype of *Meliosma pendens* Rehder & E. H. Wilson).

Native to eastern Central China where it occurs in dense shaded woods and open thickets and was sted by Wilson to be known as *chui chih pao hua shu*. This tree was first found in central China by Augustine Henry and was later collected by the Italian missionary C. Sylvestri (on whose material Pampanini based the type) who was based in Fang in northern Hubei about 1906.

Meliosma henryi Diels (SABIACEAE) New species

in Bot. Jahrb. Syst. 29: 452. (1900), Hemsley in Hooker's Icon. Pl. 9: t. 2832. (1907), Hand.-Mazz. in Symb. Sin. 7: 644. (1933), Cufod in Oest. Bot. Z.88: 262. (1939), How in Acta Phytotax. Sin. 3: 430. t. 56 f. 5-11. (1955), Zheng in Journ. Wuhan Bot. Res. 2.(1): 135. (1984).

Hubei: Jianshi, 5865 (type).

An evergreen tree to 18 metres tall, said to be rare this species grows in subtropical evergreen broad-leaved forest and is sometimes planted by temples. India (Assam) to central China.

Meliosma myriantha Sieb. & Zucc. (SABIACEAE)

Anonymous in Kew List of Determinations (volume 24).

Hubei: Badong, 2417, 4042, 7308.

Meliosma oldhamii Miq. (SABIACEAE)

Bean in Kew Bull. Misc. Inf. 166. (1900), Rehder & E. H. Wilson in Sargent Pl. Wilson 2: 206. (1914), Zheng in Wuhan Bot. Res. 2.(1): 135. (1984).

Meliosma wallichii Forbes & Hemsley non Planch. ex Hook. f. in Journ. Linn. Soc. Bot. 23: 146. (1886). *Meliosma sinensis* Nakai in Journ. Arnold Arb. 5: 80. (1924).

Hubei: Jianshi, 5929, 5863 (syntype of *Meliosma sinensis* Nakai).

A deciduous tree to 17 metres tall. The Kew collector, Richard Oldham first found this tree in Korea in 1863. Over 20 years later Henry found it in Hubei and Farges later found it in Sichuan. Introduced to cultivation by E. H. Wilson through Messrs Veitch in 1900.

Meliosma rigida Sieb. & Zucc. (SABIACEAE)

Meliosma glomerulata Rehder & E. H. Wilson in Sargent Pl. Wilson. 2: 203. (1914). *Meliosma simplicifolia* (Roxb.) Walp. ssp. *rigida* (Sieb. & Zucc.) Beus. Beusekom in Blumea 19: 473. (1971).

Yunnan: In forests near Simao at 1,600 metres, 11737 (type of *Meliosma glomerulata* Rehder & E. H. Wilson), 11737a., (paratype of *Meliosma glomerulata* Rehder & E. H. Wilson), 11737b., (paratype of *Meliosma glomerulata* Rehder & E. H. Wilson).

Native to China and Japan.

Meliosma simplicifolia (Roxb.) Walp (SABIACEAE)

Anonymous in Kew List of Determinations (volume 34).

Vietnam: Northern Vietnam (Tongkin), 9556.

Distributed in Nepal, Bhutan, India, Sri Lanka, Bangladesh, Myanmar, China [Tibet (Xizang) and Yunnan], Thailand, Laos and Vietnam.

Meliosma sp. (SABIACEAE)

Anonymous in Kew List of Determinations (volume 24).

Hubei: Badong, 921 (November 1885).

Meliosma squamulata Hance (SABIACEAE)

Henry in Trans. Asiat. Soc. Jap. 24 suppl: 29. (1896) (List Pl. Formosa), Li in Woody Flora of Taiwan 503. fig. 194. (1963).

Meliosma lepidota Blume ssp. *squamulata* (Hance) Bues. Buesekom in Blumea 19: 454. (1971).

Taiwan: Oluanpi, 1989.

An evergreen tree to 15 metres tall, native to southern China, including Taiwan where it grows in broad-leaved forest at low altitude.

Meliosma thorelii Lecompte (SABIACEAE)

Meliosma henryi Diels ssp. *thorelii* (Lecompte) Beus. Beusekom in Blumea 19: 445. fig. e1, 2. (1971).

Yunnan: Simao at 1,500 metres, 12647.

Native to China, Vietnam and Laos.

Meliosma velutina Rehder & E. H. Wilson (SABIACEAE)
in Sargent Pl. Wilson. 2: 202. (1914), Cufod in Oest.
Bot. Z.88: 267. (1939), Vidal in Not. Syst. 16: 307. (1960).
Meliosma simplicifolia (Roxb.) Walp. ssp. *fordii* Beusekom
non (Hemsl.) Beus. Beusekom in Blumea 19: 480. (1971).

Yunnan: In the mountains to the south of Simao at
1,500 metres, 12114 (holotype of *Meliosma velutina* Rehder
& E. H. Wilson).

Native to southern China, Laos and Vietnam.

Melissa axillaris (Benth.) Bakh. f. (LAMIACEAE)
Melissa parviflora Benth. Dunn in Notes Roy. Bot.
Gard. Edinburgh 6: 160. (1915).

Yunnan: Forests to the north of Mengzi at 2,650
metres, 10214, 10214a.

Widespread in Asia.

Melochia corchorifolia L. (MALVACEAE)
Henry in Trans. Asiat. Soc. Jap. 24 suppl: 22. (1896)
(List Pl. Formosa), Li in Woody Flora of Taiwan 562. fig.
220. (1963).

Hubei: Yichang, 2315. **Hainan:** Without locality, 8553
Taiwan: Kaohsiung Plain, 573. Wanchin, 2073

An undershrub to about 1 metre tall, found throughout
tropical Asia and a common weed in Taiwan.

Melodinus cochinchinensis (Lour.) Merr.
(APOCYNACEAE)
Melodinus henryi Craib in Kew Bull. Misc. Inf. 411.
(1911). *Melodinus monogynus* Roxb. ex Lindl. Anonymous
in Kew List of Determinations (volume 34).

Yunnan: In mountain forest near Simao at 1,200 to
1,500 metres, 11944, 12725. Mountains to the east of Simao
at 1,300 metres 11944a., (isolectotype of *Melodinus henryi*
Craib). Mengzi, 13794.

Native to Myanmar, China (Yunnan), Thailand and
Vietnam.

Melodinus sp. (APOCYNACEAE)
Anonymous in Kew List of Determinations (volume
34).

Yunnan: Simao, 12765.

Memecylon pauciflorum Blume (MELASTOMATACEAE)
Li in Journ. Arnold Arb. 25: 40. (1944).

Hainan: Environs of Haikou, 8349. At Nodoa in the
interior of the island, 8460. Without locality, 8577.

A widespread species found in India, Indochina,
Malaysia and northern Australia.

Mengzia foliosa (King & Pantl.) W. C. Huang, Z. J. Liu
& C. Hu (ORCHIDACEAE)
Arethusa sinensis Rolfe in Journ. Linn. Soc. Bot. 36:
46. (1903).

Yunnan: Mengzi, 11102 (isotype of *Arethusa sinensis*

Rolfe).

The genus commemorates Henry's base, Mengzi, in
southern Yunnan.

Menispermaceae (MENISPERMACEAE)
Anonymous in Kew List of Determinations (volume
34).

Yunnan: Mengzi, 11379.

Mentha canadensis L. (LAMIACEAE)
Mentha arvensis Henry non L. in Journ. China Br.
Roy. Asiat. Soc. 22: 266. (1887) (Chinese Names of Plants),
Forbes & Hemsley in Journ. Linn. Soc. Bot. 26: 281. (1890),
Dunn in Notes Roy. Bot. Gard. Edinburgh 6: 156. (1915).

Hubei: Yichang, 66, 72, 4204, 4453. Liantuo, 2637,
2735, 4453. Without locality, 2735a. **Sichuan:** South
Wushan, 7210. **Yunnan:** Mengzi at 1,450 metres, 10428,
10428a. In the mountains to the west of Simao, 10428b.

Chinese field mint, known colloquially in the Three
Gorges area as *yeh po ho* according to Henry. When mixed
with the root of *Platycodon grandiflorus* this species is
used in cases of painful swollen throats, used with the *jua
hua* or *Chrysanthemum × morifolium* in cases of headache.
To the Hani (an ethnic group) of southern Yunnan it is
anljilpavqdao and is used to treat rheumatism and influenza.
This species was collected by the Glasnevin Central China
Expedition (GCCE 520) in Xingshan in September 2004.

Mercurialis leiocarpa Sieb. & Zucc. (EUPHORBIACEAE)
Forbes & Hemsley in Journ. Linn. Soc. Bot. 26: 436.
(1894), Pax in Engler, Pflanzenr. iv. 147. vii: 280. (1914).

Hubei: Badong, 3714. Changyang (April 1888), 5229.
Jianshi, 5782. **Yunnan:** Mengzi, 10477.

Merremia hederacea (Burm. f.) Hallier. f.
(CONVOLVULACEAE)
Ipomoea chryseides Ker Gawl. Henry in Trans. Asiat.
Soc. Jap. 24 suppl: 64. (1896) (List Pl. Formosa).

Taiwan: Kaohsiung, without number.

Merremia hirta (L.) Merr. (CONVOLVULACEAE)
Ipomoea linifolia Blume Henry in Trans. Asiat. Soc.
Jap. 24 suppl: 65. (1896) (List Pl. Formosa).

Taiwan: Hills near Kaohsiung Lake, 1946.

Mesophaerum suaveolens (L.) Kuntze (LAMIACEAE)
Hyptis suaveolens (L.) Poit. Henry in Trans. Asiat. Soc.
Jap. 24 suppl: 72. (1896) (List Pl. Formosa), Dunn in Notes
Roy. Bot. Gard. Edinburgh 6: 145. (1915).

Hainan: Environs of Kiungchow, 8736. **Taiwan:**
Oluanpi, 231.

A Central and South American native naturalised in
Chinese Taiwan. According to Henry, this fragrant herb was
known as *shang-hsiang* or *p'ai-ku-hsiao* and the seeds were
used in making cakes.

Metagentiana eurycolpa (Marq.) T. N. Ho & S. W. Liu (GENTIANACEAE) New species

in Bot. Bull. Acad. Sin. 43: 83 - 91. (2002).

Gentiana eurycolpa Marquand in Kew Bull. Misc. Inf. 71. (1931).

Yunnan: Mengzi, 10023.

The genus *Metagentiana* was discovered by Père Delavay in north-west Yunnan on the 3rd of October 1882.

Metagentiana eurycolpa was first collected by Henry near Mengzi and was later collected by Maire at Lagu near Kunming, the capital of Yunnan province.

Metagentiana rhodantha (Franch.) T. N. Ho & S. W. Liu (GENTIANACEAE) New species

in Bot. Bull. Acad. Sin. 43: 83 - 91. (2002).

Gentiana rhodantha Franchet in Journ. Linn. Soc. Bot. 26: 133. (1890).

Hubei: Yichang, 964 (type), 3986 (in part). Liantuo, 2990, 4401, 4606. **Yunnan:** Mengzi, 9832. Simao, 12767.

A handsome perennial to 60 cm tall, bearing purple-red funnel-shaped flowers, striped deep purple from June to September. Franchet chose Henry's material on which to base his type of *Gentiana rhodantha*. *Metaqentiana rhodantha* had been previously collected by Delavay in Yunnan in June 1885 and by Maries in the Xiling (Yichang) Gorge in 1879. Seeds were taken from A. Henry 4401 in the Kew herbarium and sent for propagation on June 1st 1888. Endemic to China and distributed in the provinces of Gansu, Guangxi, Henan, Hubei, Shaanxi, Sichuan and Yunnan.

Metapanax davidii (Franch.) J. Wen & Frodin. (ARALIACEAE)

Harms & Rehder in Sargent Pl. Wilson. 2: 556. (1916).

Acanthopanax diversifolius Hemsley in Journ. Linn. Soc. Bot. 23: 340. (1887), Bretschneider in Hist. Eur. Bot. Disc. China 2: 784. (1898). *Nothopanax diversifolius* (Hemsl.) Harms in Engler & Prantl. Nat. Pflanzen Fam. 3: Abt. 8, 48. (1894).

Hubei: Yichang, 4337. Xingshan, 4511. Changyang, 7498. Without locality, 4337a. Liantuo, 2969 (type of *Acanthopanax diversifolium* Hemsley). Fang Xian, 6608. South Badong, 6608a. Jianshi, 6608b.

A small evergreen tree to 6 metres tall, carrying rounded panicles of greenish-yellow flowers in July and August. Native to western and central China and discovered by Père Armand David in 1869 near Baoxing (formerly Moupin) in Sichuan. This species is common in the forests to the north of the Yangtze in the Three Gorges region and was collected by the Glasnevin Central China Expedition (GCCE 625) in the Dragon Gate River Foprest Park in Xingshan in September 2004. Grows well at Mount Usher in Co Wicklow.

Metapanax delavayi (Franch.) J. Wen & Frodin. (ARALIACEAE)

Panax delavayi Franch. Anonymous in Kew List of Determinations (volume 34).

Yunnan: Mengzi, 9927.

China (distributed in Guizhou and Yunnan) and Vietnam.

Mezoneuron cucullatum (Roxb.) Wight & Arn. (FABACEAE)

Anonymous in Kew List of Determinations (volume 34).

Yunnan: Simao, 12215, 12835, 13528.

Mezoneuron sp. (FABACEAE)

Anonymous in Kew List of Determinations (volume 34).

Yunnan: Simao, 12876.

Microchloa indica (L. f.) P. Beauv. (POACEAE)

Microchloa setacea R. Br. Rendle in Journ. Linn. Soc. Bot. 36: 402. (1904).

Yunnan: On rocky mountains near Mengzi at 2,100 metres, 10998.

Microcos paniculata L. (MALVACEAE)

Grewia microcos L. Anonymous in Kew List of Determinations (volume 24).

Hainan: Hummocks near Haikou, 8089. Environs of Kiungchow, 8270. Environs of Haikou, 8345. 8503.

Microdesmis caseariifolia Planch. ex Hook. (PANDACEAE)

Forbes & Hemsley in Journ. Linn. Soc. Bot. 26: 433. (1894).

Hainan: Without locality, 8554.

Microglossa pyrifolia (Lamarck) Kuntze (ASTERACEAE)

Microglossa volubilis DC. Henry in Trans. Asiat. Soc. Jap. 24 suppl: 52. (1896) (List Pl. Formosa).

Hubei: Liantuo, 2103. **Hainan:** At Nodoa in the interior of the island, 8452. Without locality, 8587. **Taiwan:** Wanshoushan, without number. Oluanpi, without number. Wanchin, 118. **Yunnan:** In Pingbian on the Daweishan Range at 1,650 metres, 10870a. In forests to the south of Simao at 1,650 metres, 12775.

Microlepia marginata (Panz.) C. Chr. (DENNSTAEDTIACEAE)

Davallia marginalis (Thunb.) Baker Anonymous in Kew List of Determinations (volume 24).

Hubei: Liantuo, 2582, 4406, 4427.

Microlepia platyphylla (D. Don) J. Sm. (DENNSTAEDTIACEAE)

Davallia platyphylla D. Don Anonymous in Kew List of Determinations (volume 34).

Yunnan: Mengzi, 9848.

Distributed across Nepal, India, Sri Lanka, Burma, southern China and Indochina.

Microlepia rhomboidea (Hook.) C. Presl. ex Prantl (DENNSTAEDTIACEAE)

Davallia rhomboidea (Hook.) Wall. ex Kuntze Henry in Trans. Asiat. Soc. Jap. 24 suppl: 110. (1896) (List Pl. Formosa).

Taiwan: Tamsui, 1441a.

Microlepia strigosa (Thunb.) C. Presl (DENNSTAEDTIACEAE)

Davallia strigosa (Thunb.) Kunze Henry in Trans. Asiat. Soc. Jap. 24 suppl: 110. (1896) (List Pl. Formosa).

Hubei: Liantuo, 3208. **Taiwan:** Tamsui, 1441, 1492. **Yunnan:** Mengzi, 10093. Simao, 13100.

Microlepia tenera Christ (DENNSTAEDTIACEAE)
New species

in Notul. Syst. (Paris) 1: 53. (1909).

Yunnan: In forests to the south of Simao at 1,650 metres, 13155.

Endemic to China and distributed in Taiwan and Yunnan.

Micromeles cuspidata (Bertol.) C. K. Schneid. (ROSACEAE)

Sorbus polycarpa (Hook. f.) Rehder in Sargent Pl. Wilson. 2: 274. (1915).

Yunnan: Mengzi, 11296.

Micromeles zahlbruckneri (C. K. Schneid.) Mezhenskyj (ROSACEAE) New species

Sorbus zahlbruckneri C. K. Schneider in Bull. Herb. Boissier ser. 2. 6: 318. (1906), Schneider in Illus. Handb. Laubholzk. 1: 685. fig. 3790. (1906), Rehder in Sargent Pl. Wilson 2: 274. (1915), Bean in Trees & Shrubs 4: 428. (1989).

Sichuan: North Wushan, 7021.

Micromelum minutum (G. Forst.) Wight & Arn. (RUTACEAE)

Micromelum pubescens Blume Anonymous in Kew List of Determinations (volume 24 & 34).

Hainan: Environs of Haikou, 8035, 8112. Without locality, 8586. **Yunnan:** Simao, 11877, 12897.

Micromelum sp. (RUTACEAE)

Anonymous in Kew List of Determinations (volume 34).

Yunnan: Simao, 12190.

Microsorum insigne (Blume) Copel. (POLYPODIACEAE)

Selliguea anceps H. Christ in Bull. Herb. Boissier 6: 879. (1898).

Yunnan: In the mountains to the south of Mengzi at 1,650 metres, 10089 (isotype of *Selliguea anceps* H. Christ).

Microsorum punctatum (L.) Copel (POLYPODIACEAE)

Polypodium irioides Poir. Henry in Trans. Asiat. Soc. Jap. 24 suppl: 115. (1896) (List Pl. Formosa), Christ in Bull. Herb. Boiss. 6: 873. (1898).

Taiwan: Oluanpi, 1360 (in part). **Yunnan:** Near Manpan by the Red River in woods on rocks growing in clumps, 10899.

Native to India, Sri Lanka, China, Thailand and Vietnam.

Microsorum steerei (Harrington) Ching (POLYPODIACEAE)

Takeda in Notes Roy. Bot. Gard. Edinburgh 8: 293. (1913—1915).

Polypodium playfairii Baker Henry in Trans. Asiat. Soc. Jap. 24 suppl: 114. (1896) (List Pl. Formosa).

Taiwan: Wanshoushan, without number.

Native to Vietnam and southern China including Taiwan. Discovered on Wanshoushan, by Professor J. B. Steere in 1876. Steere was Professor of Zoology at Michigan University and collected in Equador, Peru and Brazil before making extensive zoological and botanical collections in China between 1870 and 1873.

Microstegium biaristatum (Steud.) Keng (POACEAE)

Andropogon formosanus Rendle in Journ. Linn. Soc. Bot. 36: 371. (1904).

Taiwan: Kaohsiung, on cliffs, 1167 (type of *Andropogon formosanus* Rendle).

Microstegium fasciculatum (L.) Henrard (POACEAE)

Pollinia monantha (Trin.) Nees ex Steud. Rendle in Journ. Linn. Soc. Bot. 36: 410. (1904). *Pollinia ciliata* Trin. Rendle in Journ. Linn. Soc. Bot. 36: 335. (1904). *Pollinia ciliata* Trin. var. *breviaristata* Rendle in Journ. Linn. Soc. Bot. 36: 355. (1904).

Hubei: Antelope Glen, 3133. **Taiwan:** Kaohsiung, 1140. **Yunnan:** Mengzi at 1,450 metres, 9210. Mengzi at 1,600 metres, 11412.

Distributed in India, Myanmar, China, Indochina and Thailand.

Microstylis sp. (ORCHIDACEAE)

Anonymous in Kew List of Determinations (volume 34).

Yunnan: Simao, 13131, 13543.

Microtoena patchouli (C. B. Clarke) C. Y. Wu & Hsuan ex Hsuan (LAMIACEAE)

Microtoena insuavis (Hance) Prain Dunn in Notes Roy. Bot. Gard. Edinburgh 6: 189. (1915).

Yunnan: Mengzi, 9850. Shiping, 11583. Simao, 12588.

Native to India, Myanmar and China (Yunnan).

Microtoena robusta Hemsley (LAMIACEAE) New species

in Journ. Linn. Soc. Bot. 26: 307. (1890), Bretschneider in Hist. Eur. Bot. Disc. China 2: 789. (1898), Dunn in Notes Roy. Bot. Gard. Edinburgh 6: 188. (1915).

Hubei: Xingshan, 6482a. Fang Xian, 7631.

Microtoena urticifolia Hemsley (LAMIACEAE) New species

in Journ. Linn. Soc. Bot. 26: 308. (1890), Bretschneider in Hist. Eur. Bot. Disc. China 2: 789. (1898), Dunn in Notes Roy. Bot. Gard. Edinburgh 6: 188. (1915).

Hubei: Badong, 2536, 4902, 7339.

Microtropis discolor (Wall.) Arn. (CELASTRACEAE)

Anonymous in Kew List of Determinations (volume 34).

Yunnan: Simao, 12686.

Distributed in India, Myanmar, China (Yunnan), Thailand and Vietnam.

Microtropis henryi Merrill & O. M. Freeman (CELASTRACEAE) New species

in Proc. Amer. Acad. Arts. 73: 289. (1940), Wu in Index Fl. Yunnan 1: 759. (1984), Cheng & Kao in Iconogr. Cormophyt. Sin. Suppl. 2: 247. fig. 8825. (1983), Fl. Reipubl. Popularis Sin. 45(3): 160. (1999).

Yunnan: In Pingbian on the Daweishan Range at 1,666 to 2,000 metres, 11443 (type).

Endemic to Yunnan.

Microtropis oligantha Merrill & F. L. Freeman (CELASTRACEAE) New species

in Proc. Amer. Acad. Arts 73: 288. (1940).

Yunnan: Near Yuangyang at Fengchunling at 2,700 metres, 10851 (holotype).

Endemic to Yunnan.

Microtropis petelotii Merrill & F. L. Freeman (CELASTRACEAE) New species

in Am. Acad. Arts & Sci. 73: 291. (1938), Shui and Chen in Bot. Bull. Acad. Sin. 43: 305 - 312. (2002), Ding in Blumea 13: 407. (1966).

Yunnan: In Pingbian on the Daweishan Range, in forests at 1,500 metres, 11417. In Pingbian on the Daweishan Range at 1,666 to 2,000 metres, 11491, 11491a.

Discovered in Yunnan by Henry in 1897, the species was named for Petelot who collected in northern Vietnam between 1931 to 1940. In China this species is distributed in Guangxi and Yunnan.

Mikania cordata (Burm. f.) B. L. Rob. (ASTERACEAE)

Mikania scandens Henry non (L.) Willd. in Trans. Asiat. Soc. Jap. 24 suppl: 52. (1896) (List Pl. Formosa).

Taiwan: Wanshoushan, without number. Wanchin, 1642.

Milium effusum L. (POACEAE)

Rendle in Journ. Linn. Soc. Bot. 36: 383. (1904).

Hubei: Badong, 4032, 5413. Changleping, 6331. **Sichuan:** South Wushan, 5661.

Widely distributed in Europe and Asia.

Millettia cubittii Dunn (FABACEAE) New species

in Journ. Linn. Soc. Bot. 41: 188. (1912).

Yunnan: In the Red River valley near Manpan at 950 metres, 10939 (type).

Distributed in Myanmar and China (Yunnan).

Millettia pachyloba Drake (FABACEAE)

Dunn in Journ. Linn. Soc. Bot. 41: 185. (1912).

Hainan: Without locality, 8489.

Millettia pulchra (Benth.) Kurz. (FABACEAE)

Fordia microphylla Dunn ex Z. Wei in Acta Phytotax. Sin. 27: 75. (1989).

Yunnan: Mengzi, 9439. Simao, 11758.

Distributed in India, Myanmar, China and Laos.

Millettia pulchra (Benth.) Kurz. var. ***chinensis*** Dunn (FABACEAE) New variety

in Journ. Linn. Soc. Bot. 41: 149. (1912).

Yunnan: In forests near Simao at 1,300 metres, 12322 (isotype). Mountains to the east of Simao at 1,500 metres, 13031 (syntype).

Endemic to China and distributed in Guangxi and Yunnan.

Millettia pulchra (Benth.) Kurz. var. ***microphylla*** Dunn (FABACEAE) New variety

in Journ. Linn. Soc. Bot. 41: 152. (1912), Hosokawa in Journ. Soc. Trop. Agr. 5: 57. (1933), Li in Woody Flora of Taiwan 352. (1963).

Taiwan: Oluanpi, 944 (type).

A tree or large shrub, endemic to Taiwan and confined to the Hengchun Peninsula in the extreme south of the island.

Millettia sp. No. 1. (FABACEAE)

Anonymous in Kew List of Determinations (volume 24).

Hubei: On the road to the Dome, (June 1886), 1554.

Millettia sp. No. 2. (FABACEAE)

Anonymous in Kew List of Determinations (volume 24).

Hainan: Near the Kiungchow Pagoda, 7974.

Millettia sp. No. 3. (FABACEAE)

Anonymous in Kew List of Determinations (volume 34).

Yunnan: Simao, 11896.

Millettia velutina Dunn (FABACEAE) New species

in Journ. Linn. Soc. Bot. 41: 149. (1912).

Yunnan: Mengzi at 1,550 to 1,650 metres, 9728.

Mengzi at 1,650 metres, 9728b., (type). Mengzi at 1,550 metres, 9728f.

Native to China (where it is distributed in the provinces of Guangdong, Guangxi, Guizhou, Hunan and Yunnan) and northern Indochina.

Mimosa pudica L. (FABACEAE)

Henry in Trans. Asiat. Soc. Jap. 24 suppl: 39. (1896) (List Pl. Formosa).

Hainan: Without locality, 8264. **Taiwan:** Oluanpi, cultivated, 222.

The sensitive plant, a small subscandent shrub to 1 metre tall, native to tropical America and now widely naturalised throughout the tropics. The Hani (an ethnic group) of Xishuangbanna in southern Yunnan call this species *tyuqdaivqtyuqxil* and use it to treat cases of influenza, malaria, fever in children, bronchitis, gastroenteritis, fatigue and bladder stones.

Mimosa sp. (FABACEAE)

Anonymous in Kew List of Determinations (volume 34).

Yunnan: In Pingbian on the Daweishan Range at 1,666 to 2,000 metres, 11492.

Mimulus sp. (PHRYMACEAE)

Anonymous in Kew List of Determinations (volume 24).

Hubei: Yichang, 1750.

Mirabilis jalapa L. (NYCTAGINACEAE)

Henry in Journ. China Br. Roy. Asiat. Soc. 22: 247. (1887) (Chinese Names of Plants); in Trans. Asiat. Soc. Jap. 24 suppl: 74. (1896) (List Pl. Formosa), Diels in Bot. Jahrb. 29: 317. (1900).

Hubei: Yichang, 617, 1679. Liantuo, 2018, 4415, 4651. **Taiwan:** Kaohsiung, without number.

The four-o-clock plant or marvel of Peru, a tuberous perennial to 1 metre tall. A colonist and widely naturalised in China, native to Peru. According to Henry this species was known colloquially at Yichang as *hsi tsao hua*. The Hani (an ethnic group) of Xishuangbanna in southern Yunnan use this species to treat cases of mastitis and insect bites.

Miscanthus floridulus (Labill.) Warb. ex K. Schum. & Lauterb. (POACEAE)

Miscanthus japonicus Anderss. Henry in Trans. Asiat. Soc. Jap. 24 suppl: 107. (1896) (List Pl. Formosa), Rendle in Journ. Linn. Soc. Bot. 36: 437. (1904).

Taiwan: Oluanpi, 1332. Kaohsiung, 1690, 1692, 1823. Wanshoushan, 1691.

Native from eastern Asia to Polynesia. Abundant on the eastern plains of Chinese Taiwan and often grown as a windbreak between fields. Henry stated that this tall bamboo-like grass was known colloquially in Taiwan as

kuan. *Miscanthus floridulus* was collected by the Glasnevin Taiwan Expedition (GTE 42 & GTE 63) by the lighthouse at Oluanpi (South Cape of Henry and Wilson) in October 2004.

Miscanthus sacchariflorus (Maxim.) Benth. & Hook. f. ex Franch. (POACEAE)

Rendle in Journ. Linn. Soc. Bot. 36: 347. (1904).

Hubei: Badong, 5123. Changyang, 7507.

Miscanthus sinensis Anderss. (POACEAE)

Rendle in Journ. Linn. Soc. Bot. 36: 348. (1904).

Hubei: Yichang, 93. Badong, 5065. **Taiwan:** Kaohsiung, 1173, 1175, 1185. Wanshoushan, 1689. Without locality, 1173a. **Yunnan:** Mengzi, 10014.

Miscanthus sinensis was collected by the Glasnevin Central China Expedition (GCCE 114 & GCCE 115) in the *Metasequoia* valley near Modaoqi village in September 2002. It was also collected by members of the 2004 Glasnevin Expedition to China in Xingshan (GCCE 503) and above the San You Dong Glen near Yichang (GCCE 725) in the autumn of that year.

Mischocarpus pentapetalus (Roxb.) Radlk. (SAPINDACEAE)

Radlkofer in Engler, Pflanzenr. iv. 165: 1293. (1934).

Yunnan: Simao 11775. Simao at 1,650 metres, 12978a. Indian (Assam) to China and Malesia.

Mitrasacme indica Wight (LOGANIACEAE)

Mitrasacme alsinoides R. Br. Henry in Trans. Asiat. Soc. Jap. 24 suppl: 61. (1896) (List Pl. Formosa).

Taiwan: Kaohsiung Plain, 1902.

Mitreola pedicellata Benth. (LOGANIACEAE) New to China

Forbes & Hemsley in Journ. Linn. Soc. Bot. 26: 117. (1889).

Hubei: Yichang, 1346. Lung Tung Ch'i Glen (near Yichang), 4201.

Mnesithea laevis (Retz.) Kunth var. *cochinchinensis* (Lour.) de Koning & Sosef (POACEAE)

Ophiuros monostachyus J. S. Presl. Rendle in Journ. Linn. Soc. Bot. 36: 362. (1904).

Taiwan: Kaohsiung, 1201.

Momordica charantia L. (CUCURBITACEAE)

Forbes & Hemsley in Journ. Linn. Soc. Bot. 23: 315. (1887).

Hubei: Changyang, 7762.

A vigorous climbing annual gourd to 4 metres tall, native to tropical Asia where it has been long cultivated for its unusual fruits which contain a substance similar to insulin. The young shoots are eaten in the Philippines. The fruits vary in length, the longer fruits being preferred by the Chinese, the shorter fruits in India.

Momordica cochinchinensis (Lour.) Spreng. (CUCURBITACEAE)

Henry in Trans. Asiat. Soc. Jap. 24 suppl: 45. (1896) (List Pl. Formosa).

Hainan: Environs of Kiungchow, 8236. **Taiwan:** Kaohsiung Plain, in rice fields, 1802. Oluanpi, without number, 1318. Wanchin, 1653.

The spiny bitter cucumber, a large climber cultivated for its fruits which are eaten while green and not fully grown. The root froths in water and can be used as a substitute for soap. Especially abundant in southern Taiwan.

Momordica sp. (CUCURBITACEAE)

Anonymous in Kew List of Determinations (volume 34).

Yunnan: Simao, 12727.

Monachosorum henryi H. Christ (DENNSTAEDTIACEAE)
New species

in Bull. Herb. Boissier 6: 869. (1898), H. Ito in Hara, Fl. East Himal. 463. (1966), Shieh in Quart. Journ. Chin. For. 6(4):95. (1973).

Yunnan: In Pingbian on the Daweishan Range, in mountain forest at 1,650 metres, 10457 (isotype). Mengzi 10457a., (collected 15th of November 1896).

A very handsome fern to about 50 cm tall. Native to the Himalaya, Vietnam, southern China (including Taiwan) and the Philippines.

Monochasma sheareri (S. Moore) Maxim. ex Franch. & Sav. (OROBANCHACEAE)

Forbes & Hemsley in Journ. Linn. Soc. Bot. 26: 202. (1890).

Hubei: Yichang, 3408.

Discovered by Dr. George Shearer of the London Missionary Society near Jiujing in Jiangxi province (formerly Kiukiang) on the Yangtze in 1873.

Monosis volkameriifolia (DC.) H. Rob. & Skvarla (ASTERACEAE)

Vernonia volkameriifolia DC. Anonymous in Kew List of Determinations (volume 34).

Yunnan: Mengzi, 10567. Simao, 11632.

Distributed in Nepal, Myanmar, western China, Thailand, Laos and Vietnam.

Monotropa hypopitys L. (ERICACEAE)

Hypopitys multiflora Scop. Anonymous in Kew List of Determinations (volume 24).

Hubei: Xingshan at 1,400 metres, 6564. Fang Xian, 6564a.

Monotropa uniflora L. (ERICACEAE)

Forbes & Hemsley in Journ. China Br. Roy. Asiat. Soc. 22: 234. (1887) (Chinese Names of Plants), Forbes & Hemsley in Journ. Linn. Soc. Bot. 26: 34. (1889).

Hubei: Badong, 894, 6159. **Sichuan:** South Wushan, 7464. **Hunan:** Shimen, 7565. **Yunnan:** Milê District, 9939.

Henry's Badong collector knew this species as the *ai chiang ts'ao*. *Monotropa uniflora* was collected by the Glasnevin Central China Expedition (GCCE 540) in Xingshan in September 2004. A widespread holoparasite in Asia.

Morinda citrifolia L. (RUBIACEAE)

Li in Woody Flora of Taiwan 860. (1963).

Sarcocephalus sp. Henry in Trans. Asiat. Soc. Jap. 24 suppl: 49. (1896) (List Pl. Formosa).

Hainan: 8 km (5 miles) south of Kuingchow, 8171. **Taiwan:** Oluanpi, 328, 350.

A small strand tree, native from India and Malaysia to tropical Australia and eastern Polynesia. This species is much used in modern homoeopathy.

Morus alba L. (MORACEAE)

Henry in Journ. China Br. Roy. Asiat. Soc. 22: 267. (1887) (Chinese Names of Plants), Forbes & Hemsley in Journ. Linn. Soc. Bot. 26: 456. (1894), Henry in Elwes & Henry. Trees of Great Brit. & Irel. vi. 1609. (1912), Schneider in Sargent Pl. Wilson. 3: 294. (1916), Zheng in Journ. Wuhan. Bot. Res. 2.(1). 34. (1984).

Hubei: Yichang, 1335, 1339, 1392, 3498, 3756. Liantuo, 3843. Jianshi, 5926. **Sichuan:** South Wushan, 5564, 5749. **Hainan:** Environs of Kiungchow, 8282.

The white mulberry, a deciduous tree to 15 metres tall, a native of central China and cultivated since ancient times in southern Europe. During Henry's time in central China this mulberry was extensively cultivated around Yichang for its leaves which were fed to silkworms, *Bombyx mori*. Sericulture at this time was one of Sichuan's most important industries (as it is today). The Chinese call this tree *sang shu* and kept cultivated trees pollarded to ease the gathering of leaves. In the first three weeks of their lives infant silkworms were fed with finely chopped leaves of *Maclura tricuspidata* (q.v.) which improved the quality of the silk. With the banning of opium growing in China at the beginning of the 20th century sericulture took on an even more important role and that country continues to be the world's leading producer of silk. The finest silk is said to be produced from silkworms fed on the leaves of the white mulberry.

Morus australis Poir. (MORACEAE)

Li in Woody Flora of Taiwan 128. fig. 39. (1963).

Morus alba Forbes & Hemsley non L. in Journ. Linn. Soc. Bot. 23: 456. (1894), in part. Henry in Trans. Asiat. Soc. Jap. 24 suppl: 86. (1896) (List Pl. Formosa). *Morus acidosa* Griffith Schneider in Sargent Pl. Wilson. 3: 297. (1916).

Hubei: South Badong, 5453, 6094. Changleping, 6248,

6249. Without locality, 6249a. **Sichuan:** South Wushan, 4745, 5668, 5669, 5670, 5740, 5741. **Taiwan:** Wanchin, 100, 134. Oluanpi, 269, 1336. Kaohsiung, 744, 1784. **Yunnan:** In mountain forest near Mengzi at 1,900 metres, 10535. In the mountains north of Mengzi at 1,900 metres, 10535b. In Pingbian on the Daweishan Range at 1,600 metres, 12980.

A small bushy tree to 8 metres tall, native to India, Bhutan, Nepal, Myanmar, Indochina, China (including Taiwan), Korea and Japan. Wilson sent seeds in 1907 though it was already in cultivation previous to this date. Widespread and common in Taiwan. According to Henry the indigenous inhabitants of the mountainous area of Wanchin, made a cloth from the root-bark of young trees. Specimens of the root, fibre, cloth and game-bag were sent by Henry to the Kew museum. It is not used to feed silkworms.

Morus cathayana Hemsley (MORACEAE) New species

in Journ. Linn. Soc. Bot. 26: 456. (1894), Bretschneider in Hist. Eur. Bot. Disc. China 2: 791. (1898,) E. Pritzel in Bot. Jahrb. xxix. 298. (1900), Henry in Elwes & Henry. Trees of Great Brit. & Irel. 6: 1606. (1912), Bean in Trees & Shrubs. 2: 85. (1914), Schneider in Sargent Pl. Wilson. 3: 292. (1916), Morley in Glasra 3: 79. (1979), Zheng in Journ. Wuhan Bot. Res. 2.(1): 36 (1984), Nelson in A Heritage of Beauty 315. (2000), O' Brien in Augustine Henry, An Irish Plant Collector in China 26. (2002).

Morus sp. Henry in Journ. China Br. Roy. Asiat. Soc. 22: 267. (1887) (Chinese Names of Plants).

Hubei: South Badong, 1409. Jianshi, 5860. Liantuo, 6378. **Sichuan:** South Wushan, 5487, 6378.

A tree to 6 metres tall, though accasionally seen to 15 metres tall with trunk girths up to 1.5 metres. The fruits when ripe may be white, red or black. It was known colloquially in Hubei as *hu lu sang* according to Henry who stated the leaves of this species were not fed to silk-worms. Introduced to cultivation by Wilson in 1907. In Ireland it is grown at the John F. Kennedy Arboretum.

Morus macroura Miq. (MORACEAE)

Morus laevigata Wallich ex Brandis Schneider in Sargent Pl. Wilson. 3: 301. (1916).

Yunnan: Mountains to the south of Simao at 1,200 metres, 11975. In the mountains to the east of Simao at 1,500 metres, 12019a.

Distributed in Myanmar, China [Tibet (Xizang) and Yunnan], Thailand, Laos, Cambodia and Vietnam.

Mosla chinensis Maxim. (LAMIACEAE)

Forbes & Hemsley in Journ. Linn. Soc. Bot. 26: 280. (1890), Dunn in Notes Roy. Bot. Gard. Edinburgh 6: 155.

(1915).

Mosla fordii Maxim. Forbes & Hemsley in Journ. Linn. Soc. Bot. 26: 280. (1890).

Hubei: Yichang, 2240, 4325.

Mosla dianthera (Buch.-Ham. ex Roxb.) Maxim. (LAMIACEAE)

Henry in Journ. China Br. Roy. Asiat. Soc. 22: 249. (1887) (Chinese Names of Plants), Forbes & Hemsley in Journ. Linn. Soc. Bot. 26: 280. (1890), Dunn in Notes Roy. Bot. Gard. Edinburgh 6: 155. (1915).

Calamintha clinopodium Benth. var. *nepalensis* (Benth.) Dunn in Notes Roy. Bot. Gard. Edinburgh 6: 160. (1917). *Mosla lanceolata* Forbes & Hemsley non (Benth.) Maxim. in Journ. Linn. Soc. Bot. 26: 281. (1890), Dunn in Notes Roy. Bot. Gard. Edinburgh 6: 153. (1915). *Mosla punctata* Forbes & Hemsley non Maxim. in Journ. Linn. Soc. Bot. 26: 281. (1890).

Hubei: Yichang, 61, 865, 2274, 2319, 2790 (in part). Liantuo, 2609. Badong, 2489, 2560, 2742, 3661, 4812, 4813, 4841, 4895, 4896, 4897, 5017.

According to Henry this species was known colloquially at Yichang as *kan han ts'ao.*

Mucuna bracteata DC. ex Kurz (FABACEAE)

Wilmot-Dear in Kew Bull. 39: 59. (1984).

Yunnan: Mengzi, 10417. Simao, 12749.

An herbaceous or semi-woody climber with a wide range across the eastern Himalaya, China, Thailand, Laos and Vietnam.

Mucuna macrocarpa Wall. (FABACEAE)

Wilmot-Dear in Kew Bull. 39: 36. (1984), Wilmot-Dear in Kew Bull. 47: 205. (1992).

Mucuna collettii Lace in Kew Bull. Misc. Inf. 9: 398. (1915).

Yunnan: Mengzi, 9255. Simao, 11631, 11631a. 11702 (syntype of *Mucuna collettii* Lace).

A rampant climber to 70 metres long, distributed through the Himalaya to China, Myanmar, Thailand and Vietnam. The Hani (an ethnic group) of Xishuangbanna in southern Yunnan call this species *xavxil'albev* and use its stems to treat the effects of polio, rheumatism, menstrual disorders and joint pain.

Mucuna pruriens (L.) DC. (FABACEAE)

Wilmot-Dear in Kew Bull. 39: 61. (1984).

Yunnan: In mountain forest near Yuanjiang, 10963, 13499.

A semi-woody climber to 4 metres long, widely distributed through tropical Africa, Madagascar, Asia and tropical America, in China Henry stated it was known as the *zhe maoli dou* or dragon's teeth bean.

Mucuna pruriens (L.) DC. var. ***utilis*** (Wall. ex Wight) Bak. ex Burck. (FABACEAE)

Wilmot-Dear in Kew Bull. 39: 61. (1984).

Taiwan: Wanchin, an escape from cultivation, 1719.

This is the cultivated form, widely grown in the tropics as a fodder crop and for green manure, it is now naturalised in parts of China. *Mucuna pruriens* is an annual climbing legume from India and the specific epithet comes from the Latin prurire, to itch, a reference to the irration caused by tiny barbs on the pods that attach to the skin of the passer-by or harvester. The seeds are used as medicine in many parts of the world and have been regarded as a powerful aphrodisiac for thousands of years. Modern studies have shown a theraputic role for it in the treatment of Parkinson's disease and diabetes. It is also used in cases of snake bite, some venoms act by causing the blood to clot, the effect of this plant is to inhibit clotting.

Mucuna sempervirens Hemsley (FABACEAE) New species

in Journ. Linn. Soc. Bot. 23: 190. (1887), Henry in Journ. China Br. Roy. Asiat. Soc. 22: 259. (1887) (Chinese Names of Plants); I. C. 261. (1887), Hemsley in Curtis' Bot. Mag. t. 7978 (1904), Bretschneider in Hist. Eur. Bot. Disc. China 2: 697 & 781 (1898), Hemsley in Gard. Chron. ser. 3, 349. (1900); Hemsley in Gard. Chron. ser. 3, 35: 282. (1904), H. Levl., in Bull. Soc. Bot. France 55: 408. (1908) & Fl. Kouy. Tcheou: 240. (1914—1915), Rehder in Journ. Arnold. Arbor. 13: 331. (1932); Icon. Cormophyt. Sin. 2: 497. fig. 2724. (1972); Fl. Hupeh 2: 266. pl. 1109. (1979), Morley in Glasra 3: 54. (1979), Tateishi & Ohsashi in Bot. Mag. Tokyo 94: 102. fig. 7. (1981), Wilmot-Dear in Kew Bull. 39: 39. (1984), Nelson in Pim, The Wood & the Trees 222. (1984); in A Heritage of Beauty 315. (2000).

Hubei: Yichang, 1065 (holotype). In a glen off the Xiling (Yichang) Gorge, 3517. An immense climber on the banks of the Yangtze 4 miles above Yichang, 3517a.

A vigorous, evergreen woody climber to 25 metres tall, with stem diameters exceeding 40 cm. Henry introduced *Mucuna sempervirens* from Yichang in 1888 to the Royal Botanic Gardens, Kew where an original plant still grows in the Temperate House. This is a woody climber of gigantic proportions and was once very common on the cliffs of glens and gorges at low altitudes around Yichang. Two enormous vines grew at Tzuyang, on the riverbank above Yichang in Henry's time. It is very floriferous on old wood and its large, fleshy red to maroon-purple, foul-smelling flowers are carried in May on old wood or on the stems and are rich in nectar and attract a wide variety of insects. The flat seed-pods are bean-like and can be almost a metre long. Henry stated that around Yichang it was known as *mien-*

ma-teng or *niu-ma-t'eng.* There are good plants, one in particular of enormous girth, at Wuhan Botanical Gardens, where the species is grown up pines to demonstrate to the visiting public its destructive nature. Many of the trees that act as its support are in the process of being toppled under the sheer weight of this monstrous climber. Henry reported a single plant near Yichang covering an area of half an acre.

Mucuna sp. No. 1. (FABACEAE)

Henry in Trans. Asiat. Soc. Jap. 24 suppl: 35. (1896) (List Pl. Formosa).

Taiwan: Wanchin, 490.

'My native says, a large climber, *hsuch-t'eng,* the root is used as a drug.' A. Henry.

Mucuna sp. No. 2. (FABACEAE)

Anonymous in Kew List of Determinations (volume 34).

Yunnan: Mengzi, 13679.

Muhlenbergia hugelii Trin. (POACEAE)

Rendle in Journ. Linn. Soc. Bot. 36: 385. (1904). *Muhlenbergia viridissima* Nees ex Steud. Henry in Journ. China Br. Roy. Asiat. Soc. 22: 257. (1887) (Chinese Names of Plants).

Hubei: Badong, 2513. Lung tung ch'i (near Yichang), 4289. Fang Xian, 6720. Jianshi, 7410.

Distributed in China, Korea and Japan. According to Henry this species was known colloquially in the Three Gorges area as *luan-tzu-ts'ao.*

Munronia pinnata (Wall.) W. Theob. (MELIACEAE)

Munronia henryi Harms in Ber. Deutsch. Bot. Ges. 35: 77. (1917).

Yunnan: Simao at 1,650 metres, 12202 (isotype of *Munronia henryi* Harms).

Endemic to China, distributed in Guizhou and Yunnan.

Munronia unifoliolata Oliver (MELIACEAE) New species

in Hooker's. Icon. Pl. t. 1709. (1887), Bretschneider in Hist. Eur. Bot. Disc. China 2: 780. (1898).

Hubei: In the Lung Tung Ch'i Glen near Yichang (October 1886), 2901, 3963, 4198.

Murdannia divergens (C. B. Clarke) G. Brückn. (COMMELINACEAE)

Aneilema divergens C. B. Clarke Brown in Journ. Linn. Soc. Bot. 36: 151. (1903). *Aneilema sinicum* Henry non Lindl. in Trans. Asiat. Soc. Jap. 24 suppl: 98. (1896) (List Pl. Formosa) in part.

Taiwan: Wanchin, 899.

Murdannia edulis (Stokes) Faden (COMMELINACEAE)

Aneilema loureiroi Henry non Hance in Trans. Asiat. Soc. Jap. 24 suppl: 98. (1896) (List Pl. Formosa). *Aneilema formosanum* N. E. Brown in Journ. Linn. Soc. Bot. 36: 152.

(1903). *Aneilema scapiflorum* Wight var. *latifolium* N. E. Brown in Journ. Linn. Soc. Bot. 36: 154. (1903).

Hainan: Lingmen, without number, (syntype of *Aneilema scapiflorum* Wight var. *latifolium* N. E. Brown). **Taiwan:** Wanchin, 872 (isotype of *Aneilema formosanum* N. E. Brown).

A widespread species found over a large geographical range from the Himalaya to Papua New Guinea.

Murdannia keisak (Hassk.) Hand.-Mazz. (COMMELINACEAE)

Aneilema keisak Hassk. Brown in Journ. Linn. Soc. Bot. 36: 152. (1903).

Hubei: Yichang, 128, 2757.

An annual aquatic, distributed in the eastern United States, Italy, India, China, Vietnam, Laos, Cambodia, Korea and Japan. Usually in swamps, marshes and ricefields.

Murdannia loriformis (Hassk.) R. S. Rao & Kammathy (COMMELINACEAE)

Aneilema angustifolium N. E. Brown in Journ. Linn. Soc. Bot. 36: 151. (1903). *Aneilema sinicum* Henry non Ker. Gawl. in Trans. Asiat. Soc. Jap. 24 suppl: 98. (1896) (List Pl. Formosa), in part.

Taiwan: Kaohsiung Plain, 1897 (isosyntype of *Aneilema angustifolium*).

Murdannia nudiflora (L.) Brenan (COMMELINACEAE)

Aneilema nudiflorum (L.) R. Br. Henry in Journ. China Br. Roy. Asiat. Soc. 22: 248. (1887) (Chinese Names of Plants); in Trans. Asiat. Soc. Jap. 24 suppl: 98. (1896) (List Pl. Formosa), Brown in Journ. Linn. Soc. Bot. 36: 153. (1903).

Hubei: Yichang, 2252. **Hainan:** Environs of Haikou, 8334. **Taiwan:** Wanchin, 882.

According to Henry this species was known colloquially at Yichang as *hsia-tzu ts'ao*.

Murdannia simplex (Vahl.) Brenan (COMMELINACEAE)

Aneilema sinicum Ker. Gawl. Henry in Trans. Asiat. Soc. Jap. 24 suppl: 98. (1896) (List Pl. Formosa).

Taiwan: Oluanpi, 1351.

Murdannia sp. (COMMELINACEAE)

Aneilema sp. Anonymous in Kew List of Determinations (volume 34).

Yunnan: Mengzi, 10009, 10878, 11381, 13746. Simao, 12324, 12395, 13177, 13381, 13564.

Murraya paniculata (L.) Jack. (RUTACEAE)

Li in Woody Flora of Taiwan 377. fig. 136. (1963).

Murraya exotica L. Henry in Trans. Asiat. Soc. Jap. 24 suppl: 18. (1896) (List Pl. Formosa).

Taiwan: Oluanpi, without number. Kaohsiung, without number. Wanchin, 978. **Yunnan:** Mengzi, 9915. Pu'er, 13383.

A small tree with fragrant, white flowers in dense terminal or axillary corymbs. Native to tropical Asia and Australia and common at low altitudes in Chinese Taiwan where it is widely distributed throughout the island and is often cultivated as a hedge plant. Known as *shih-ling*, according to Henry it was used by indigenous tribes that inhabited the mountainous interior of Taiwan for making tobacco-pipes. This species was collected by the Glasnevin Taiwan Expedition (GTE 28) by the lighthouse at Oluanpi (South Cape of Henry and Wilson) in October 2004.

Mussaenda breviloba S. Moore (RUBIACEAE) New species

Hutchinson in Sargent Pl. Wilson. 3: 398. (1916).

Yunnan: Simao at 1,450 metres, 11931.

Discovered in Yunnan by Henry in 1897, this species was described from material later collected in Myanmar by Major Melville between November 1903 and January 1904.

Mussaenda divaricata Hutchinson (RUBIACEAE) New species

in Sargent Pl. Wilson. 3: 394. (1916).

Mussaenda divaricata Hutchinson var. *mollis* Hutchinson in Sargent Pl. Wilson. 3: 398. (1916).

Hubei: Antelope Glen, 3118, 4120. Liantuo, 6366. **Yunnan:** Mengzi, 9650a. Mengzi, 9884, 11391. **Vietnam:** In the province of Tonkin at Yenbay on the Red River, 9519.

An evergreen climbing shrub to 4 metres tall, producing from April to June cymes of small golden yellow flowers accompanied by large white leaf-like bracts. Distributed in China (Guizhou, Jiangxi, Sichuan and Yunnan) and Vietnam.

Mussaenda erosa Champion ex Bentham (RUBIACEAE)

Hutchinson in Sargent Pl. Wilson. 3: 399. (1916).

Yunnan: Near Manhao, south of the Red River at 1,340 metres, 10646. In the Red River valley near Manpan at 660 metres, 10646a. Mengzi 13694. Mengzi at 1,600 metres, 13648.

China, Vietnam.

Mussaenda hirsutula Miq. (RUBIACEAE)

Hutchinson in Sargent Pl. Wilson. 3: 399. (1916).

Hainan: Environs of Haikou, 8080. On hummocks near Haikou, 8124. Without locality, 8548.

Mussaenda laxiflora Hutchinson (RUBIACEAE) New species

in Sargent Pl. Wilson. 3: 399. (1916).

Yunnan: In Pingbian in forests on the Daweishan Range at 1,600 metres, 9650b., 11493 (type).

Endemic to Yunnan.

Mussaenda macrophylla Wall. (RUBIACEAE)

Hutchinson in Sargent Pl. Wilson. 3: 396. (1916).

Yunnan: In forests near Simao at 1,300 metres, 12265.

A climbing shrub to 2 metres tall, distributed in Nepal,

India, China, the Philippines and Malaysia.

Mussaenda parviflora Miquel. (RUBIACEAE)

Li in Woody Flora of Taiwan 862. (1963).

Mussaenda glabra Henry non Vahl. in Trans. Asiat. Soc. Jap. 24 suppl: 50. (1896) (List Pl. Formosa)

Taiwan: Wanchin, 567. Kaohsiung, 1476.

A scandent shrub, native to Japan (South) and China (Taiwan) where it common at low altitude.

Mussaenda pubescens Dryand. (RUBIACEAE)

Forbes & Hemsley in Journ. Linn. Soc. Bot. 23: 379. (1888), Hutchinson in Sargent Pl. Wilson. 3: 399. (1916), Zheng in Journ. Wuhan Bot. Res. 2.(1): 189. (1984).

Fujian: Xiamen (ex de la Touche), 1855. **Hainan:** Without locality, 8270. **Hubei:** Yichang, 2703. **Yunnan:** In forests near Simao at 1,500 metres, 12157.

A handsome and vigorous climber bearing small yellow flowers surrounded by spectacular white leaf-like bracts. The Hani (an ethnic group) of Xishuangbanna in southern Yunnan call this species *haqlaqmiaqba* and use it to treat cases of flu, bronchitis and laryngitis. Endemic to China. This species was collected by the Glasnevin Taiwan Expedition (GTE 102) by the lighthouse on Wanshoushan (Ape's Hill of Henry and Wilson) in October 2004.

Mussaenda sanderiana Ridl. (RUBIACEAE)

Mussaenda rehderiana Hutchinson in Sargent Pl. Wilson. 3: 397. (1916). *Mussaenda hossei* Craib Li in Journ. Arnold Arb. 25: 318. (1944).

Yunnan: Mountains to the south of Simao in forests at 1,600 metres, 11790, (type of *Mussaenda rehderiana* Hutchinson).

Native to southern China and Indochina.

Mussaenda sessilifolia Hutchchinson (RUBIACEAE) New species

in Sargent Pl. Wilson. 3: 397. (1916).

Yunnan: Forests to the south of Simao at 1,300 metres., 12774 (type).

Endemic to Yunnan.

Mussaenda shikokiana Makino (RUBIACEAE)

Mussaenda wilsonii Hutchinson in Sargent Pl. Wilson. 3: 393. (1916).

Hubei: Changyang, 6233.

Mussaenda sp. (RUBIACEAE)

Anonymous in Kew List of Determinations (volume 24).

Hubei: Liantuo, 3192.

Mycetia bracteata Hutchchinson (RUBIACEAE) New species

in Sargent Pl. Wilson. 3: 409. (1916).

Yunnan: In the mountains to the south of Simao at 1,300 metres, 11930a., (type).

Endemic to Yunnan and found only in the area of Simao.

Mycetia faberi (Hemsley) Razafim. & B. Bremer (RUBIACEAE)

Myrioneuron faberi Hemsley Hutchinson in Sargent Pl. Wilson. 3: 410. (1916).

Hubei: Lung Tung Ch'i Glen (near Yichang), 4200. **Yunnan:** In the mountains to the south of Mengzi, 9376. In Pingbian in forests on the Daweishan range at 1,640 metres, 9376a. In Pingbian in a ravine on the Daweishan Range at 1,320 metres, 9376b., 9376c. In mountains to the west of Simao at 1,460 metres, 9376e. Simao, 12168.

Southern China, Vietnam.

Mycetia glandulosa Craib (RUBIACEAE) New species

Hutchinson in Sargent Pl. Wilson. 3: 410. (1916)

Yunnan: In forests near Mengzi at 1,650 metres, 11930. In forests near Simao at 1,300 metres, 12700.

This species was discovered by Henry in 1896 and was based on material later collected by Kerr in Thailand.

Mycetia gracilis Craib (RUBIACEAE) New species

Hutchinson in Sargent Pl. Wilson. 3; 409. (1916).

Yunnan: In forests to the east of Simao at 1,420 metres, 12284. In forests near Simao at 1,400 metres, 12284a., 12284c., 12284d.

This species was discovered by Henry in 1897 and was based on material later collected in Thailand by Kerr.

Mycetia hirta Hutchinson (RUBIACEAE) New species

in Sargent Pl. Wilson. 3: 410. (1916).

Yunnan: In mountain forest to the east of Simao at 1,600 metres, 11633, 12299 (type), 12299a.

Endemic to China and distributed in Hainan and Yunnan.

Mycetia sp. (RUBIACEAE)

Adenosacme sp. & *Myrioneuron* sp. Anonymous in Kew List of Determinations (volume 34).

Yunnan: Mengzi, 11075, 13755.

Myoporum bontioides (Sieb. & Zucc.) A. Gray (SCROPHULARIACEAE)

Henry in Trans. Asiat. Soc. Jap. 24 suppl: 70. (1896) (List Pl. Formosa); in The Garden 61: 6. (1902).

Hebei: Anping, without number.

A small evergreen sea-shore shrub to 1.5 metres. Native from southern China to Japan, Li reports this species as being rather scarce in Chinese Taiwan.

Myosotis sylvatica Ehrh. ex Hoffm. (BORAGINACEAE)

Myosotis sp. Anonymous in Kew List of Determinations (volume 24).

Hubei: Yichang, 3294.

Myriactis nepalensis Less. (ASTERACEAE)

In the Harvard University Herbarium (GH.).

Sichuan: South Wushan, 7234. **Yunnan:** At Hsienkei on the Red River near the Vietnamese border, 9562. Mengzi, 9675.

Native to India, Nepal, China and Vietnam.

Myrica anenophora Hance (MYRICACEAE)

Myrica vidaliana Henry non Rolfe in Trans. Asiat. Soc. Jap. 24 suppl: 90. (1896) (List Pl. Formosa). *Myrica* aff. *adenophora* sp. nov. Skan in Journ. Linn. Soc. Bot. 496. (1899).

Taiwan: Oluanpi, 1259.

A small evergreen tree, native to China, including Taiwan where it grows in the southern and eastern parts of the island at low altitude.

Myrica esculenta Buch.-Ham. ex D. Don (MYRICACEAE)

E. H. Wilson in Sargent Pl. Wilson. 3: 189. (1916).

Myrica rubra Skan non (Lour.) Sieb. & Zucc. in Journ. Linn. Soc. Bot. 26: 496. (1899).

Yunnan: In the mountains of Mengzi, 9015, 9015a., 9015c., 9015d. In mountains near Simao at 1,600 metres, 9015e.

Distributed in India, Bhutan, Myanmar, China, Vietnam and Thailand.

Myrica aff. *rubra* (Lour.) Sieb. & Zucc. var. or sp. nov. (MYRICACEAE)

Skan in Journ. Linn. Soc. Bot. 26: 496. (1899).

Yunnan: Mengzi, 11057.

Myricaria laxiflora (Franch.) P. Y. Zhang & Y. J. Zhang (TAMARICACEAE)

Myricaria germanica Forbes & Hemsley non (L.) Desv. in Journ. Linn. Soc. Bot. 23: 317 (1887), Henry in Journ. China Br. Roy. Asiat. Soc. 22: 265. (1887) (Chinese Names of Plants), Diels in Bot. Jahrb. 29: 476. (1900). *Myricaria bracteata* Rehder non Royle in Sargent Pl. Wilson. 2: 407. (1916), Zheng in Journ. Wuhan Bot. Res. 2.(1): 153. (1984).

Hubei: Yichang, 22, 843. Liantuo, 3202.

A common shrub along the banks of the Yangtze, in the Three Gorges region Henry stated it was known colloquially as *shui peh chih*.

Myriophyllum spicatum L. (HALORAGACEAE)

Anonymous in Kew List of Determinations (volume 24).

Hubei: Yichang, 4362.

Myriopteron extensum (Wight) K. Schum. (APOCYNACEAE)

Tsiang in Sunyatsenia 2: 177. (1934).

Yunnan: Mengzi, 9524. Mountain to the north-west of Simao at 160 metres, 13114. Simao, 13541.

Distributed in India, Myanmar, western China, Thailand, Laos and Vietnam.

Myristica sp. (MYRISTICACEAE)

Anonymous in Kew List of Determinations (volume 34).

Yunnan: Mengzi, 10432, 11406. Simao, 11780, 12234, 13532.

Myrsine africana L. (PRIMULACEAE)

Mez in Engler, Pflanzenr. iv. 236: 340. (1902), Rehder in Sargent Pl. Wilson. 2: 581. (1916), Zheng in Journ. Wuhan Bot. Res. 2.(1): 172. (1984).

Hubei: Yichang, 337, 1114. Liantuo, 3066, 3808. Without locality, 1114a. **Yunnan:** In woods near Mengzi at 1,500 metres, 9972, 9972b.

An evergreen, densely branched shrub to over a metre tall. Native to the mountains of eastern and southern Africa, the Azores, the Himalaya and China. Introduced to cultivation from South Africa in the 17th century and from the Azores in 1778. According to W. J. Bean most of the plants now in cultivation are of Himalayan or Chinese origin. This species was collected by the Glasnevin Central China Expedition (GCCE 601) in Xingshan in September 2004. It is a handsome shrub of varying habit given its very large geographical range, there is a particularly fine form at Glanleam on Valentia Island in County Kerry, the former seat of the Knights of Kerry.

Myrsine faberi (Mez) Pipoly & Chen (PRIMULACEAE) New species

in Novon 5: 360. (1995).

Rapanea faberi Mez in Pflanzenr. (Engler) 4. Fam. 236: 358. (1902).

Yunnan: Mountains to the south of Mengzi at 1,950 metres, 9173a., (isolectotype of *Rapanea faberi* Mez). **Yunnan:** Simao, 13037.

China, Indochina.

Myrsine linearis (Lour.) Poir. (PRIMULACEAE)

Rapanea playfairii (Hemsl.) Mez Mez in Engler, Pflanzenreich. iv. 236: 361. (1902).

Hainan: On hummocks near Haikou, 8121. At Nodoa in the interior of the island, 8463. Without locality, 8524.

Myrsine seguinii H. Lév. (PRIMULACEAE) New variety

Rapanea yunnanensis Mez in Pflanzenr. (Engler) 4. Fam. 236: 359. (1902). *Rapanea neriifolia* (Sieb. & Zucc.) Mez var. *yunnanensis* (Mez.) Walker in Bull. Fan Mem. Inst. Biol. Bot. Ser. 9: 189. (1939), Walker in Journ. Arnold Arb. 23: 355. (1942).

Yunnan: In forests near Yuanjiang at 1,300 metres, 11570 (type of *Rapanea yunnanensis* Mez). Simao, 11570c., (syntype of *Rapanea yunnanensis* Mez).

Native to Myanmar, China, Vietnam and Japan.

Myrsine semiserrata Wall. (PRIMULACEAE)

Forbes & Hemsley in Journ. Linn. Soc. Bot. 26: 61. (1889), Mez in Engler, Pflanzenr. iv. 236: 339. (1902), Rehder in Sargent Pl. Wilson 2: 580. (1916), Zheng in Journ. Wuhan Bot. Res. 2.(1): 173. (1984).

Hubei: Antelope Glen, 3277. In Glens near Yichang, 3277a., 3277b., 3277c., 3277d. Changyang, 7685. **Yunnan:** South of the Red River from Manmei at 2,000 metres, 9173b. Without locality, 9173c. Mengzi at 1,600 to 1,800 metres, 9173c., 9173d., 9173e. Yuanjiang at 1,500 metres, 13270, 13311.

A small evergreen tree to about 3.5 metres tall, carrying in autumn berries that slowly ripen from pink to mauve, then finally blue. A handsome plant when seen in fruit but suited only to the mildest counties of Britain and Ireland. Native to India, Nepal, Myanmar and China.

Myrsine sequinii Levl. (PRIMULACEAE)

Li in Woody Flora of Taiwan 723. fig. 294. (1963).

Ilex sp. Henry in Trans. Asiat. Soc. Jap. 24 suppl: 27. (1896) (List Pl. Formosa).

Taiwan: Oluanpi, 1250.

A shrub or small tree from 2 to 12 metres tall. Native to eastern Asia from Indochina to Japan, a widespread species in Chinese Taiwan.

Nabalus tatarinowii (Maxim.) Nakai (ASTERACEAE)

Prenanthes tatarinowii Maxim. Forbes & Hemsley in Journ. Linn. Soc. Bot. 23: 486. (1888).

Hubei: Badong, 4995. Fang Xian, 6593. Xingshan, 6748. South Badong, 7624. **Sichuan:** South Wushan, 7196.

Nageia nagi (Thunb.) Kuntze (PODOCARPACEAE)

Li & Keng in Flora of Taiwan 1: 559. (1994).

Podocarpus argotaenia Henry non Hance Henry in Trans. Asiat. Soc. Jap. 24 suppl: 91. (1896) (List Pl. Formosa), Masters in Journ. Linn. Soc. Bot. 26: 547. (1902). *Podocarpus latifolia* Masters non Wall. in Journ. Linn. Soc. Bot. 26: 547. (1902). *Podocarpus macrophyllus* Masters non (Thunb.) Sweet in Journ. Linn. Soc. Bot. 26: 548. (1902). *Podocarpus formosensis* Dummer in Gard. Chron. ser. 3, 52: 295. (1904). *Nageia formosensis* (Dummer) C. N. Page in Notes Roy. Bot. Gard. Edinburgh 45: 382. (1989), Mill in Novon. 9(1): 77. (1999). *Podocarpus nagi* (Thunb.) Makino Walsh in Aug. Henry Forst. Herb. 111. (1957).

Taiwan: Wanchin, 403. Oluanpi, 1357 (isotype of *Podocarpus formosensis* Dummer). Tamsui, 1446 (type of *Podocarpus formosensis* Dummer).

A small to medium-sized tree, native to southern China (including Taiwan) and Japan, scattered throughout Taiwan in broadleaved forest.

Nandina domestica Thunb. (BERBERIDACEAE)

Henry in Journ. China Br. Roy. Asiat. Soc. 22: 260.

(1887) (Chinese Names of Plants), Schneider in Sargent Pl. Wilson. 1: 386. (1913), Zheng in Journ. Wuhan Bot. Res. 2.(1): 43. (1984).

Hubei: Yichang, 1295, 2705. Liantuo, 1935. Without locality, 2705a.

The Chinese name for this handsome shrub is *t'ien chu* or 'heavenly bamboo' a name also adopted in the west. *Nandina domestica* forms an evergreen unbranched shrub to 2 metres tall bearing doubly and trebly pinnate leaves to 45 cm long, red on emergence. White flowers are carried in erect panicles to 35 cm followed by red or reddish-purple fruits. A native of Central China and cultivated for centuries in Japan from where it was introduced to the West in 1804. According to Henry the sacred bamboo grew wild in the glens above the gorges and was cultivated around temples.

Nanhaia speciosa (Champ. ex Benth.) J. Compton & Schrire (FABACEAE)

Millettia speciosa Champ. ex Benth. Dunn in Journ. Linn. Soc. Bot. 41: 155. (1912).

Hainan: Environs of Haikou, 8341. Without locality, 8541. Ling-men, 8620.

Nanocnide japonica Blume (URTICACEAE)

Forbes & Hemsley in Journ. Linn. Soc. Bot. 26: 473. (1889).

Hubei: Liantuo, 3841. South Badong, 5280.

Narenga porphyrocoma (Hance) Bor (POACEAE)

Saccharum narenga (Nees ex Steud.) Wall. ex Hack. Rendle in Journ. Linn. Soc. Bot. 36: 349. (1904).

Taiwan: Kaohsiung Plain, on Lake Hill, 2036.

A reed to 2 metres tall, widely distributed in the tropics of south-east Asia.

Neanotis calycina (Wall. ex Hook. f.) W. H. Lewis (RUBIACEAE)

Anotis calycina Wall. ex Hook. f. Anonymous in Kew List of Determinations (volume 34).

Yunnan: Mengzi, 11293.

Nebularia wushanensis (F. N. Williams) M. T. Sharples & E. A. Tripp (CARYOPHYLLACEAE) New species

Stellaria wushanensis F. N. Williams in Journ. Linn. Soc. Bot. 34: 434. (1900), Diels in Bot. Jahrb. 29: 319. (1900).

Sichuan: North Wushan, 7047.

Neillia affinis Hemsley (ROSACEAE) New species

in Journ. Linn. Soc. Bot. 29: 304. (1892), Bretschneider in Hist. Eur. Bot. Disc. China 2: 782. (1898), Rehder in Sargent Pl. Wilson. 1: 434. (1913), J. Vidal in Adansonia 3(1): 156. pl. 1: 11. (1963), Cullen in Journ. Arnold Arb. 52: 146. (1971), Morley in Glasra 3: 79. (1979), Nelson in A Heritage of Beauty 315. (2000), O' Brien in Augustine

Henry, An Irish Plant Collector in China 26. (2002).

Sichuan: Kangding, Henry's Chinese collector with Pratt, 8968 (type).

Seeds of Wilson's 916a., collected on Wa Shan in Sichuan in September 1908 arrived at Glasnevin the following spring. A Wilson plant is still growing in the gardens.

Neillia affinis Hemsl. var. ***pauciflora*** (Rehder) J. Vidal (ROSACEAE) New variety

in Adansonia 3: 156. fig. 3. (1963), Cullen in Journ. Arnold Arb. 147. (1971).

Neillia pauciflora Rehder in Sargent Pl. Wilson. 1: 437. (1913).

Yunnan: Mountains to the north of Mengzi at 2,300 metres, 10231, 10231a., (type of *Neillia pauciflora* Rehder).

A slender shrub to 1 metre with reddish fruits. This variety was also later collected in Yunnan by George Forrest at Tengchong (formerly Tengyueh). Endemic to Sichuan and Yunnan.

Neillia sinensis Oliver (ROSACEAE) New species

in Hooker's. Icon. Pl. 16: t. 1540. (1886), Forbes & Hemsley in Journ. Linn. Soc. Bot. 23: 228. (1887), Focke in Pflanzenr. 3(3): 14. f. 4. (1888), Bretschneider in Hist. Eur. Bot. Disc. China 2: 782. (1898), J. H. Veitch in Journ. Roy. Hort. Soc. 28: 61. (1903—1904); in Hortus Veitchii 371. (1906), Rehder in Sargent Pl. Wilson. 1: 436. (1913), Handel-Mazzetti in Symb. Sin. 7: 449. (1933), Cullen in Journ. Arnold Arb. 52: 151. (1971), Morley in Glasra 3: 80. (1979), Zheng in Journ. Wuhan. Bot Res. 2.(1): 72. (1984), Bean in Trees & Shrubs 3: 2. (1992), Nelson in A Heritage of Beauty 316. (2000), O' Brien in Augustine Henry, An Irish Plant Collector in China 26. (2002).

Neillia sinensis Oliv. var. *caudata* Rehder in Sargent Pl. Wilson. 1: 436. (1913), Cullen in Journ. Arnold Arb. 52: 151. (1971). *Neillia thibetica* Bur. & Franch. var. *caudata* (Rehd.)Vidal in Adansonia 3: 164. pl. 3, figs. 5, 6. (1963).

Hubei: Badong, 605 (July 1885), 1733, 4055. Without locality, 5554a., 5554b. **Sichuan:** South Wushan, 5585, 5695, 5554. **Yunnan:** Mountains to the east of Mengzi at 2,000 metres, 9669 (type of *Neillia sinensis* Oliv. var. *caudata* Rehder).

A deciduous shrub to 1.5 metres tall, producing terminal racemes of pink flowers in summer. Introduced to cultivation by Wilson in 1901 through Messrs Veitch's Coombe Wood nursery where it first flowered in the summer of 1905. There are plants at the National Botanic Gardens, Glasnevin. It is an elegant shrub with distinctly attractive foliage. The genus commemorates the Scottish naturalist Patrick Neil (1776—1851).

Neillia thyrsiflora G. Don (ROSACEAE)

Rehder in Sargent Pl. Wilson. 1: 436. (1913), Cullen in Journ. Arnold Arb. 52: 145. (1971).

Neillia thyrsiflora G. Don var. *tunkinensis* (Vidal) Vidal Cullen in Journ. Arnold. Arb. 52: 145. (1971).

Yunnan: Mengzi at 1,800 metres, 9419. In forests near Simao at 1,500 metres, 12275. South of the Red River, 13653.

Distributed in Nepal, India, Bhutan, western China and Vietnam.

Nekemias megalophylla (Diels & Gilg) J. Wen & Z. L. Nie (VITACEAE) New species

Ampelopsis megalophylla Diels & Gilg in Bot. Jahrb. Syst. 29: 466. (1900), Schneider in Illus. Handb. Laubholzk. 2: 322. (1912), Zheng in Journ. Wuhan Bot. Res. 2.(1): 143. (1984), O' Brien in Augustine Henry, An Irish Plant Collector in China 12. (2002).

Hubei: Jianshi, 5850 (March 1888). Fang Xian, 6642. Donghu, 5850b. Without locality, 5850a., (isotype of *Ampelopsis megalophylla* Diels & Gilg.). **Sichuan:** South Wushan, 7294.

A vigorous climber with doubly pinnate leaves to 60 cm long and almost as wide. First grown in France in the nursery of Maurice de Vilmorin in 1894, it was later reintroduced by Wilson for Messrs Veitch in 1900. W. J. Bean stated it made 3 metre long growths in one season in Veitch's Coombe Wood nursery. In Ireland this vine is cultivated at the Talbot Botanic Gardens, Malahide Castle in County Dublin.

Nelumbo nucifera Gaertn. (NELUMBONACEAE)

Nelumbium speciosum Willd. Henry in Trans. Asiat. Soc. Jap. 24 suppl: 17. (1896) (List Pl. Formosa).

Hainan: Environs of Kiungchow, 8712. **Taiwan:** Kaohsiung Plain, without number.

The *lian xia* or sacred lotus, an aquatic perennial with thick fleshy rhizomes which are a valuable food crop in China and is also a favourite flower. The sacred lotus has been cultivated in China for over 14,000 years and is native from Iran to China, southwards to Australia. It was introduced to the Nile in Egypt in ancient times but no longer grows there (the lotus of the ancient Egyptians is a *Nymphaea*). An exceptionally beautiful plant, it has Buddhist and Hindu associations and is a symbol of eternal life. It is also regarded as a symbol of purity and the goddess of mercy (Kwan Yin) is often represented seated in the centre of a lotus flower. The sacred lotus is extensively planted in China especially on the plains of eastern Hubei where in flooded fields plants are grown in millions and make a spectacular show in summer when a succession of large fragrant, rose-pink flowers soar on slender stalks above large dark-green peltate leaves. All parts of the lotus are useful, the young leaves are either eaten raw or cooked or used to wrap small portions of food before cooking. The seeds, which are imbedded in a large flat-topped receptacle, are roasted and eaten and the long elongated tubers are edible and a common part of Chinese cuisine. The sacred lotus needs a temperature of 20 to 30° Celsius during the growing season to thrive. In Chinese herbal medicine this species is called the *he ye* and is used to control fever, diarrhoea and to stop bleeding. Wuhan Botanical Gardens in the eastern plains of Hubei province has the largest collection of cultivars in the world. By crossing a Japanese plant raised from seed unearthed from a 2000-year-old archaeological site with a plant raised from 1000 year old seed unearthed in Manchuria the beautiful cultivar *Nelumbo* 'Sino-Japanese Friendship' was raised. The longevity of the seeds of the sacred lotus is well known. In the mid-19th century, the Scottish botanist Robert Brown (1773—1858) used 150 year old seeds stored in a herbarium to test their germination. Of 16 seeds, 14 germinated and 2 failed. In 1925 the Japanese paleobotanist I. Ohga found *Nelumbo* seeds in a peat bed in Manchuria and on sowing the seed gained 100 percent germination. Using a Carbon-14 test these were later determined to be approximately 1,000 years old.

Nemosenecio solenoides (Dunn) B. Nord. (ASTERACEAE)
New species

in Op. Bot. 44: 46. (1978), Jeffrey & Chen in Kew Bull. 39: 265. (1985).

Senecio solenoides Dunn in Journ. Linn. Soc. Bot. 35: 508. (1903), S. Y. Hu in Quart. Journ. Taiwan Mus. 21: 145. (1968), H. Koyama in Mem. Fac. Sc. Kyoto. Univ. Ser. Biol. 2: 138. (1969).

Yunnan: Mountains to the east of Mengzi at 1,900 metres, 9678. Milê, 9678a.

Endemic to Yunnan.

Neoalsomitra clavigera (M. Roem.) Hutch. (CUCURBITACEAE)

Alsomitra clavigera (Wall.) M. Roem. Henry in Trans. Asiat. Soc. Jap. 24 suppl: 46. (1896) (List Pl. Formosa). *Alsomitra integrifolia* Hayata in Mater. Fl. Formos. in Journ. Coll. Sc. Imp. Univers. Tokyo 30: 121. (1911), Cogniaux in Engler, Pflanzenr. iv. 275. i: 17. (1916). *Hemsleya henryi* Cogniaux in Pflanzenr. (Engler) 4. Fam. 275. 26. (1916).

Taiwan: Wanchin, 183, 1556. **Yunnan:** Simao, 13420 (isotype of *Hemsleya henryi* Cogniaux).

Neocinnamomum caudatum (Nees) Merr. (LAURACEAE)

Neocinnamomum yunnanense Liou Ho in Laurac. Chine & Indochine 90. fig. 8. (1934).

Yunnan: Simao at 1,300 metres, 12401 (isotype of *Neocinnamomum yunnanensis* H. Liu).

Distributed in Nepal, India, Bhutan, western China and Vietnam.

Neocinnamomum delavayi (H. Lec.) H. Liu (LAURACEAE)
Liou Ho in Laurac. Chine & Indochine 90. (1934).

Yunnan: Mengzi, 10796, 10796b.

China [Tibet (Xizang), Sichuan and Yunnan] and Vietnam.

Neodicladiella pendula (Sull.) W. R. Buck. (METEORIACEAE)

Meteorium pendulum Sulliv. Salmon in Journ. Linn. Soc. Bot. 51. (1901).

Hubei: On the branches of a tree in a glen near Yichang, March 1888, 7910. (Moss).

Neohymenopogon parasiticus (Wall.) Bennet (RUBIACEAE)

Hymenopogon parasiticus Wallich Hutchinson in Sargent Pl. Wilson. 3: 407. (1916).

Yunnan: In Pingbian on the Daweishan Range at 1,930 metres, 9813. Mountains to the north of Mengzi at 2,300 metres, 9813a. On mountain cliffs to the east of Mengzi at 2,300 metres, 9813b. Near Pu'er on cliffs, 9813c. Mengzi, 9813d.

Native to India, the Himalaya, Myanmar, China [Yunnan and Tibet (Xizang)], Thailand and Vietnam.

Neolepisorus ovata (Wall. ex Bedd.) Ching (POLYPODIACEAE)

Polypodium ovatum Wall. ex Hook. & Grev. var. *populnea* Christ in Bull. Herb. Boiss. 6: 873. (1898). *Polypodium ovatum* Wall. ex Hook. & Grev. Christ in Bull. Herb. Boiss. 6: 873. (1898), Takeda in Notes Roy. Bot. Gard. Edinburgh viii. 310. (1913—1915). *Polypodium deltoideum* (Sw.) Baker in Journ. Bot. 26: 230. (1888); in Ann. Bot. 5: 478. (1890), Bretschneider in Hist. Eur. Bot. Disc. China 2: 794. (1898), Diels in Bot. Jahrb. 29: 205. (1900). *Polypodium phyllomanes* Christ. Christ in Bull. Soc. Bot. France 5: 17. (1905). *Neolepisorus ovatus* (Bedd.) Ching f. *deltoideus* (Bak.) Ching in Acta Phytotax. Sin. 21: 275. (1983).

Hubei: Antelope Glen, 3123, 7879. Yichang, 3279 (type of *Polypodium deltoideum* Baker). **Yunnan:** South of the Red River from Manmei at 1,600 metres, 10078. In Pingbian on the Daweishan Range in mountain forest at 1,650 metres, 10078a. In mountain forest near Mengzi at 1,700 metres, 10078b. In forests at Fengchunling at 2,2,50 metres, 10078c. Simao, 10078e.

Native to India, the Himalaya, Myanmar, China and Vietnam.

02

Neolitsea confertifolia (Hemsl.) Merrill (LAURACEAE) New species

in Lignan Sci. Journ. 15: 419. (1936), Allen in Ann. Missouri Bot. Gard. 25: 416. (1938).

Litsea confertifolia Hemsley in Journ. Linn. Soc. Bot. 26: 379. fig. 7. (1891). *Actinodaphne confertifolia* (Hemsl.) Gamble in Sargent Pl. Wilson 2: 74. (1914), Chun in Cont. Biol. Lab. Sci. Soc. China 1: 54. (1925), Liou in Laurac. Chine 7 Indoch. 159. (1932).

Hubei: Yichang, 1247. Liantuo, 2202, 2203, 3054. Changleping, 7829. Jianshi, 6007. Without locality, 7829a. **Sichuan:** South Wushan, 7197.

Along with other species of *Actinodaphne* and *Phoebe* Henry stated that this medium-sized tree was known as *nanmu shu* in Hubei.

Neolitsea sericea (Blume) Koidz. (LAURACEAE)

Neolitsea glauca (Thunb.) Koidz. H. Liu in Laurac. Chine & Indochine 148. (1934).

Zhejiang: Pootoo Island (at Shanghai, in the Yangtze delta at the mouth of the river), 13741.

A small evergreen dioecious tree. *Neolitsea sericea* is one of the most beautiful members of the Lauraceae and has formed a 7 metre tall tree near the alpine yard at Glasnevin. There it forms a beautiful picture each spring when its young fawn-brown leaves unfold. Rare in Britain and Ireland though perfectly hardy.

Neolitsea wushanica (Chun) Merrill (LAURACEAE) New species

in Sunyatsenia 3: 250. (1937), Allen in Ann. Missouri Bot. Gard. 25: 419. (1938).

Litsea wushanica Chun (nom. nov.) in Journ. Arnold Arb. 9: 153. (1928). *Litsea gracilipes* Hemsley in Journ. Linn. Soc. Bot. 26: 381. (1891), Bretschneider in Hist. Eur. Bot. Disc. China 2: 790. (1898). *Neolitsea gracilipes* (Hemsl.) H. Liu in Laurac. Chine & Indoch. 143. (1932).

Hubei: Liantuo, 2999. **Sichuan:** North Wushan, 7113 (type) 7114 [syntype of *Neolitsea wushanica* (Chun) Merrill, syntype of *Neolitsea gracilipes* (Hemsl.) H. Liu].

Neonatis hirsuta (L. f.) W. H. Lewis (RUBIACEAE)

Hedyotis stipulata R. Br. ex Hook. f. Anonymous in Kew List of Determinations (volume 34).

Yunnan: Mengzi, 10043. Simao, 12469.

Native to India, China, Korea and Japan.

Neonauclea griffithii (Hook. f.) Merr. (RUBIACEAE)

Nauclea griffithii (Hook. f.) Haviland Hutchinson in Sargent Pl. Wilson. 3: 406. (1916).

Yunnan: Forests to the south of Simao at 1,330 metres, 12676. Mountains to the west of Simao at 1,600 metres, 12880.

Distributed in India, Bhutan, Myanmar and western China.

Neonauclea reticulata (Havil.) Merr. (RUBIACEAE)

Li in Woody Flora of Taiwan 866. fig. 349. (1963).

Nauclea sp. Henry in Trans. Asiat. Soc. Jap. 24 suppl: 49. (1896) (List Pl. Formosa).

Taiwan: Oluanpi, 928.

A large evergreen tree, native to the Philippines and China (Taiwan). In Taiwan it is found in the southern part of the island where it is common on coral rocks.

Neoshirakia japonica (Sieb. & Zucc.) Esser (EUPHORBIACEAE)

Excoecaria japonica (Sieb. & Zucc.) Muell.-Arg. Forbes & Hemsley in Journ. Linn. Soc. Bot. 26: 446. (1894). *Sapium japonicum* (Sieb. & Zucc.) Pax & Hoffm. Zheng in Journ. Wuhan Bot. Res. 2.(1): 113. (1984).

Hubei: Yichang, 473, 558, 1128, 2302, 3602. On Tsui Fu Shan near Yichang (May 1888), 1615. Liantuo, 1970, 3927. Badong, 3769, 7317.

A deciduous shrub or small tree, native to Japan, the Ryukus, China and Korea. The leaves turn crimson in autumn. Grows well at Exbury, the Rothschild family's garden, in the New Forest in the south of England.

Neottia acuminata Schltr. (ORCHIDACEAE)

Neottia micrantha Rolfe non Lindl. in Journ. Linn. Soc. Bot. 36: 40. (1903).

Hubei: Fang Xian, 6877. Without locality, 6877a.

Neottia wardii (Rolfe) Szlach. (ORCHIDACEAE) New species

Listera grandiflora Rolfe in Kew Bull. Misc. Inf. 200. (1896), Bretschneider in Hist. Eur. Bot. Disc. China 2: 792. (1898), Rolfe in Journ. Linn. Soc. Bot. 36: 40. (1903).

Hubei: Fang Xian, 6876.

Based on a collection made by Captain Frank Kingdon Ward on the Mekong-Salween divide, Yunnan in June 1911, though the species had earlier been found in the late 1880s by the Rev. Dr. Ernst Faber on Emei Shan in Sichuan and by Augustine Henry in western Hubei.

Nepeta cataria L. (LAMIACEAE)

Forbes & Hemsley in Journ. Linn. Soc. Bot. 26: 288. (1890), Dunn in Notes Roy. Bot. Gard. Edinburgh 6: 167. (1915).

Hubei: Badong, 6156 (cultivated).

Nepeta fordii Hemsl. (LAMIACEAE)

Dunn in Notes Roy. Bot. Gard. Edinburgh 6: 167. (1911—1917).

Nepeta everardii Forbes & Hemsley non S. Moore in Journ. Linn. Soc. Bot. 26: 289. (1890).

Hubei: Yichang, 1132, 3978. Liantuo, 1986.

Discovered by Charles Ford, Superintendent of Hong Kong Botanic Gardens while he was travelling in

Guangdong province in the early 1880s.

Nephrodium intermedium (Muhl. ex Willd.) Baker (POLYPODIACEAE)

Henry in Trans. Asiat. Soc. Jap. 24 suppl: 113. (1896) (List Pl. Formosa).

Hubei: Yichang, 1110. Xiling (Yichang) Gorge, 7877. Antelope Glen, 4318. Liantuo, 1945, 2989. Badong, 2509. **Taiwan:** Tamsui, 1426. **Yunnan:** Mengzi, 9977

Nephrodium sp. (POLYPODIACEAE)

Anonymous in Kew List of Determinations (volume 34).

Yunnan: Simao, 13101.

Nephroia orbiculata (L.) L. Lian & Wei Wang (MENISPERMACEAE)

Cocculus orbiculatus (L.) DC Zheng in Journ. Wuhan Bot. Res. 2.(1): 44. (1984). *Cocculus thunbergii* DC. Henry in Journ. China Br. Roy. Asiat. Soc. 22: 250. (1887) (Chinese Names of Plants), Henry in Trans. Asiat. Soc. Jap. 24 suppl: 16. (1896) (List Pl. Formosa). *Cocculus trilobus* (Thunb) DC. Diels in Engler, Pflanzenr. iv. 94: 232. (1910), Rehder & E. H. Wilson in Sargent Pl. Wilson. 1: 388. (1913). *Cocculus cuneatus* Benth. Henry in Trans. Asiat. Soc. Jap. 24 suppl: 16. (1896) (List Pl. Formosa). *Cocculus sarmentosus* (Lour.) Diels Diels in Engler, Pflanzenr. iv. 94: 233. (1910), Li in Woody Flora of Taiwan 177. (1963). *Cocculus mollis* (Miers) Wall. ex Hook. f. & Thoms. Diels in Engler, Pflanzenr. iv. 94: 235. (1910).

Hubei: In glens near Yichang, 225, 240, 697, 1488, 1667, 1688, 2944, 2976. On the banks of the Yangtze opposite Yichang, 4226. San You Dong Glen near Yichang (July 1886), 1495. Antelope Glen, 3255, 3640. On Tsui Fu Shan near Yichang (May 1888), 1612. In the Monastery Valley near Yichang (July 1886), 1655. Liantuo, 2092. Badong, 6124. **Hainan:** Environs of Kiungchow, 8693 (collected November 1889). **Taiwan:** Kaohsiung, 1925, without number. **Yunnan:** In Pingbian in mountain forest on the Daweishan Range at 1,350 to 1,500 metres, 9500, 9500a., 9500b., 9500c., 9500d., 10656, 10656c., 10656d., 10656e. Simao, 11798. On the mountains to the south-west of Simao at 1,500 metres, 11902a.

A scandent shrub, native to the eastern Himalaya to China, Japan, Malesia and Hawaii. According to Wilson this climber was abundant in Hubei in the early 20th century and was known locally as *hsiao ching-teng.* Henry's Badong collector called it *ku erh t'eng.*

Nephrolepis biserrata (Sw.) Schott. (POLYPODIACEAE)

Nephrolepis acuta (Schkuhr) C. Presl. Henry in Trans. Asiat. Soc. Jap. 24 suppl: 114. (1896) (List Pl. Formosa).

Taiwan: Kaohsiung Plain, 1010, 2021. Wanchin, 1640.

Nephrolepis cordifolia (L.) C. Presl. (POLYPODIACEAE)

Henry in Trans. Asiat. Soc. Jap. 24 suppl: 112. (1896) (List Pl. Formosa).

Aspidium auriculatum (L.) Sw. Anonymous in Kew List of Determinations (volume 24 & 34).

Hubei: In a ravine on Moji Shan, 4175. Yichang 4292, 4293, 4319. Liantuo, 4490. **Taiwan:** Wanshoushan, 1154. Tamsui, 1424. Wanchin, 1638. **Yunnan:** Mengzi, 9239.

A widespread fern distributed through tropical Asia.

Nephrolepis exaltata (L.) Schott. (POLYPODIACEAE)

Henry in Trans. Asiat. Soc. Jap. 24 suppl: 115 (1896) (List Pl. Formosa).

Taiwan: Wanshoushan, 1154a.

Nephroma tropicum (Muell. Arg.) A. Zahlbruckner (NEPHROMATACEAE)

Wei in An Enumeration of Lichens in China 157. (1991).

Nephromium tropicum Muell.-Arg. Muell.-Arg. in Bull. Herb. Boissier 1: 236. (1893).

Hubei: Fang Xian, 6916. (Lichen).

Nerium oleander L. (APOCYNACEAE)

Nerium odorum Soland. Henry in Trans. Asiat. Soc. Jap. 24 suppl: 60. (1896) (List Pl. Formosa).

Taiwan: Kaohsiung, without number. **Yunnan:** Mengzi, 9730.

Oleander, a toxic glabrous, evergreen shrub to 4 metres tall, native to the Mediterranean area. This species has been cultivated in China since the middle ages and was thought to have been first cultivated in Fujian province on the east coast. Cultivated and called *chia-chu-t'ao.*

Neyraudia arundinacea (L.) Henrard (POACEAE)

Rendle in Journ. Linn. Soc. Bot. 36: 409. (1904).

Arundo madagascariensis Kunth Henry in Journ. China Br. Roy. Asiat. Soc. 22: 255. (1887) (Chinese Names of Plants).

Hubei: Yichang, 94. Antelope Glen, 4317. **Yunnan:** Mengzi at 1,500 metres, a plant 2 metres tall, 10063. Simao at 1,300 metres, 13544.

According to Henry the stems of this reed were used as bobbins for silk and the plant was known colloquially at Yichang as *liao chien kan tzu.*

Nicotiana rustica L. (SOLANACEAE)

Anonymous in Kew List of Determinations (volume 24).

Hubei: Badong, 2520, 5130.

An annual greenish-yellow flowered annual, native to Mexico and the South American Andes. This species was the original source of tobacco cultivated by the Indians of Virginia. According to Henry *Nicotiana rustica* was cultivated on a limited scale in mountainous districts of

Hubei and Sichuan.

Nicotiana tabacum L. (SOLANACEAE)

Henry in Trans. Asiat. Soc. Jap. 24 suppl: 65. (1896) (List Pl. Formosa).

Hubei: Yichang, 1091. Badong, 1074 (seeds only), 1075, (seeds only). **Hainan:** 32 km (20 miles) west of Haikou, 8133, 8418. Without locality, 8484. **Taiwan:** Wanchin, without number.

The *yen* or tobacco, a tall annual native to north-east Argentina and Bolivia. Introduced to China according to Wilson, about the same time as maize (c. 1530 A.D.). The commercial tobacco plant is thought to be an ancient hybrid derived from crossing several Andean species. Within the rice-belt in China it is a spring crop, the seeds being sown in late October and the crop harvested the following June. In the colder maize belts it is a summer crop. Tobacco is still widely grown all over China and is a common crop in Hubei and Yunnan. The Hani (an ethnic group) of Xishuangbanna in southern Yunnan call tobacco *yavhaoq* and use it to treat boils and snakebite. Tobacco was introduced to to Europe via France by Andre Thevet who brought seeds from Brazil. Linnaeus named the genus for Jean Nicot, the French ambassador to Portugal who made a fortune dealing tobacco in Paris. Introduced to Chinese Taiwan by the early Portugese settlers.

Nidorella aegyptiaca (L.) J. C. Manning & Goldblatt (ASTERACEAE)

Conyza aegyptiaca (L.) Aiton Henry in Trans. Asiat. Soc. Jap. 24 suppl: 52. (1896) (List Pl. Formosa).

Taiwan: Kaohsiung, 1772, 1779.

Niphobolus clavatus (Bak.) Christ (POLYPODIACEAE)

Diels in Bot. Jahrb. 29: 207. (1900) Christ in Bull. Soc. Bot. France 5: 24. (1905).

Polypodium clavatum Bak. Baker in Ann. Bot. 5: 472. (1890).

Hubei: Liantuo, 4624. Sichuan: South Wushan, 5570.

Notoseris henryi (Dunn) Shih (ASTERACEAE) New species

in Acta Phytotax. Sin. 25: 202. (1987).

Prenanthes henryi Dunn in Journ. Linn. Soc. Bot. 35: 514. (1903), Morley in Glasra 3: 80. (1979).

Sichuan: North Wushan, 7022 (isosyntype of *Prenanthes henryi* Dunn), 7022a., (syntype of *Prenanthes henryi* Dunn). **Yunnan:** In mountain forest on Fengchunling at 2,300 metres, 11214.

Endemic to China and distributed in the provinces of Guizhou Hubei, Hunan, Sichuan, and Yunnan.

Nuestanthus phaseoloides (Roxb.) Benth. (FABACEAE)

Pueraria phaseoloides (Roxb.) Benth. Henry in Trans.

Asiat. Soc. Jap. 24 suppl: 36. (1896) (List Pl. Formosa), Henry in Notes Econ. Bot. China 58. (1893).

Taiwan: Wanchin, 1503. **Yunnan:** Mengzi, 10931.

A twining perennial herb, native to Asia and common on the southern and central parts of Taiwan at low altitudes. Native to India, China, Cambodia, Laos and Vietnam.

Nyctocalos pinnatum Steenis (BIGNONIACEAE) New species

in Acta Bot. Neerl. 2: 306. (1953).

Yunnan: On Ma-an Shan near Simao, 13408 (type). Endemic to southern Yunnan.

Nymphaea nouchali Burm. f. (NYMPHAEACEAE)

Nymphaea stellata Willd. Anonymous in Kew List of Determinations (volume 24).

Hainan: Environs of Haikou, 8382.

An aquatic perennial with stout fleshy rhizomes, a beautiful blue water lily found in ponds and lakes in tropical and subtropical Africa and India to China.

Nymphoides hydrophyllum (Lour.) Kuntze (MENYANTHACEAE)

Limnanthemum cristatum (Roxb.) Gris. Henry in Trans. Asiat. Soc. Jap. 24 suppl: 61. (1896) (List Pl. Formosa).

Hainan: Environs of Haikou, 8436. **Taiwan:** Kaohsiung Plain, in pools 1183.

Nymphoides indica (L.) Kuntze (MENYANTHACEAE)

Limnanthemum indicum (L.) Griseb. Henry in Trans. Asiat. Soc. Jap. 24 suppl: 62. (1896) (List Pl. Formosa).

Taiwan: Kaohsiung Plain, in lakes, 1622, 1830.

The water snowflake, an aquatic rhizomatous perennial herb. Cosmopolitan in the tropics.

Nymphoides peltata (S. G. Gmel.) Kuntze (MENYANTHACEAE)

Limnanthemum nymphoides (L.) Hoffmanns. & Link. Henry in Journ. China Br. Roy. Asiat. Soc. 22: 256. (1887) (Chinese Names of Plants), Forbes & Hemsley in Journ. Linn. Soc. Bot. 142. (1890).

Hubei: Yichang, 760 (in part), 768, 2800, 3665.

According to Henry this species was known colloquially in the Three Gorges region as *ling chio ts'ao.*

Nyssa sinensis Oliver (NYSSACEAE) New species

in Hooker's Icon. Pl. 20: t. 1964. (1891), Bretschneider in Hist. Eur. Bot. Disc. China 2: 784. (1898), J. H. Veitch in Hortus Veitchii 371. (1906), Henry in Elwes and Henry The Trees Gt. Brit. & Irel. 3: 515. (1908), Morley in Glasra 3: 80. (1979), Zheng in Journ. Wuhan Bot. Res. 2.(1): 158. (1984), Chen in Acta Phytotax. Sin. 29: 504. (1991), Bean in Trees & Shrubs 3: 22. (1992), Lancaster in the Garden 117. 526. (1992), Nelson in A Heritage of Beauty 316. (2000), O' Brien in Augustine Henry, An Irish Plant Collector in China 26. (2002).

Hubei: Jianshi, 5832 (syntype). Changleping, 6273 (type).

A large deciduous shrub or a tree occasionally to 20 metres tall, rare in Hubei and introduced to cultivation by Wilson in 1901 through the nurserymen, Messrs Veitch. Only one tree was ever raised from this seed consignment, Wilson made his collections in the same region as Henry near Jianshi to the south of the Yangtze and on the Lu Shan mountain range where he described it as being common. According to Roy Lancaster the only introduction of this tree to Europe before 1982 was the Wilson collection and a pre-war collection from Lushan Botanical Garden. In Ireland, Henry's tupelo grows in several collections and there is a fine tree at Mount Usher in County Wicklow.

Oberonia caulescens Lindl. (ORCHIDACEAE)

Oberonia yunnanensis Rolfe in Journ. Linn. Soc. Bot. 36: 6. (1903).

Yunnan: At Manme, in the mountains south of the Red River at 2,300 metres, 9152 (type of *Oberonia yunnanensis* Rolfe).

Oberonia spp. (ORCHIDACEAE)

Anonymous in Kew List of Determinations (volume 34).

Yunnan: Mengzi, 11846. Simao, 12960, 13051, 13132, 13548.

Ocimum basilicum L. (LAMIACEAE)

Dunn in Notes Roy. Bot. Gard. Ebinburgh 6: 133. (1915).

Yunnan: Mengzi at 1,650 metres, 11378.

Common or sweet basil, an annual or short-lived perennial to 60 cm. Native to tropical Asia.

Ocimum tenuiflorum L. (LAMIACEAE)

Ocimum sanctum L. Henry in Trans. Asiat. Soc. Jap. 24 suppl: 72. (1896) (List Pl. Formosa), Forbes & Hemsley in Journ. Linn. Soc. Bot. 26: 266. (1890), Dunn in Notes Roy. Bot. Gard. Edinburgh 6: 133. (1915).

Hainan: Environs of Kiungchow, 8709. **Taiwan:** Kaohsiung, without number.

Sacred basil, a native to the Indian subcontinent. Henry stated it was known as *chiu-ts'eng-t'a* in China and in Sanskrit, the *tulsi* or *tulasi*. Widely dispersed over south-east Asia, this annual or short-lived perennial is sacred to the Hindu gods Krishna and Vishnu (Hindu god of preservation) and is frequently grown by Hindus near their homes where it is thought to provide devine protection to the household. Rosaries are made from its cut stems. It is also grown as a herb and is widely known for its beneficial medicinal properties.

Odontochilus chinensis (Rolfe) T. Yukawa (ORCHIDACEAE) New species

Myrmechis chinensis Rolfe in Journ. Linn. Soc. Bot. 36: 44. (1903).

Hubei: Fang Xian, 6841.

Odontochilus lanceolatus (Lindl.) Blume (ORCHIDACEAE)

Odontochilus yunnanensis Rolfe in Journ. Linn. Soc. Bot. 36: 43. (1903).

Yunnan: In mountain forest near Mengzi at 1,600 metres, 11110 (type of *Odontochilus yunnanensis* Rolfe).

Odontosoria chinensis (L.) J. Sm. supbsp. *tenuifolia* (Lam.) Fraser.-Jenk. & Kandel (LINDSAEACEAE)

Davallia tenuifolia (Lam.) Sw. Henry in Trans. Asiat. Soc. Jap. 24 suppl: 110. (1896) (List Pl. Formosa).

Hubei: Yichang, 2181, 4373, 7875. Changleping, 7816. **Taiwan:** Wanchin, 40. Oluanpi, 666. Tamsui, 1434. **Yunnan:** Mengzi, 9018.

A widespread fern, found over much of India, Sri Lanka, Bangladesh, Bhutan, Nepal, Myanmar, China, Korea, Vietnam, Thailand, Japan and the Philippines.

Oenanthe javanica (Blume) DC. (APIACEAE)

Oenanthe stolonifera (Roxb.) Wall ex DC. Forbes & Hemsley in Journ. Linn. Soc. Bot. 23: 331. (1887), Henry in Journ. China Br. Roy. Asiat. Soc. 22: 239. (1887) (Chinese Names of Plants).

Hubei: Liantuo, 6451. South Badong, 4925, 7282. **Sichuan:** South Wushan, 7458. **Hainan:** Environs of Haikou, 8353. **Yunnan:** Mengzi, 9498.

An aquatic perennial, this is the *shui ch'in ts'ai* and according to Henry it occured as a weed in edges of rice fields. In northern parts of China it is the *shui k'an* or water celery and is cultivated there as a vegetable. The leaves are cooked or eaten raw with rice. A very variable species, frequently inhabits ponds and paddy fields. A perennial herb to 80 cm tall, native to India, Malaysia, China, Japan and Australia.

Oenanthe linearis Wall. ex DC. (APIACEAE)

Oenanthe sp. Forbes & Hemsley in Journ. Linn. Soc. Bot. 332. (1887), Henry in Journ. China Br. Roy. Asiat. Soc. 22: 239. (1887) (Chinese Names of Plants). *Oenanthe sinense* Dunn in Journ. Linn. Soc. Bot. 35: 496. (1903).

Hubei: In rice-fields near Yichang, 1663, 4089 (syntype of *Oenanthe sinense* Dunn).

Henry stated this species was common around Yichang and was known locally as *shui ch'in ts'ai* or water-celery. It was later collected by Faber on the banks of the Min River in Sichuan.

Oenanthe linearis Wall. ex DC. subsp. *rivularis* (Dunn) C. F. Wu & F. T. Pu (APIACEAE)

New subspecies. *Oenanthe rivularis* Dunn in Journ. Linn. Soc. Bot. 35: 496. (1903).

Yunnan: Mengzi, in moist places, often in small rills, 10822 (isotype of *Oenanthe rivularis* Dunn).

Distributed in western China.

Oenanthe sp. (APIACEAE)

Anonymous in Kew List of Determinations (volume 24).

Hubei: Liantuo, 2071.

Oenanthe thomsonii C. B. Clarke (APIACEAE)

Forbes & Hemsley in Journ. Linn. Soc. Bot. 23: 332. (1887).

Oenanthe caudata C. Norman in Journ. of Bot. 147. (1929).

Hubei: Badong, 1709, 4874, 4874a., 4874b., 6051. Jianshi, 5924. Without locality, 5924a., 6051a. **Sichuan:** South Wushan, 7193. North Wushan, on rocks at 1,900 to 2,600 metres, 7152 (type of *Oenanthe caudata* C. Norman). **Hunan:** Shimen, 7549.

Ohwia caudata (Thunb.) H. Ohashi (FABACEAE)

Desmodium laburnifolium (Poir.) DC. Forbes & Hemsley in Journ. Linn. Soc. Bot. 23: 173. (1887) Henry in Trans. Asiat. Soc. Jap. 24 suppl: 33. (1896) (List Pl. Formosa). *Desmodium caudatum* (Thunb.) DC. Zheng in Journ. Wuhan Bot. Res. 2.(1): 99. (1984).

Hubei: Yichang, 1683 (in part), 2276, 2712, 6188. South Donghu, 7700. **Taiwan:** Wanchin, 1637.

Olax imbricata Roxb. (OLACACEAE)

Anonymous in Kew List of Determinations (volume 24).

Hainan: Without locality, 8589.

A scandent, almost glabrous shrub. Widespread across India, Java, China (Hainan and Taiwan) and the Philippines.

Oldenlandia corymbosa L. (RUBIACEAE)

Henry in Trans. Asiat. Soc. Jap. 24 suppl: 49. (1896) (List Pl. Formosa).

Hubei: In dry rice fields near Yichang, 4298. **Taiwan:** Kaohsiung, 1911. Wanchin, 396, 1725. **Yunnan:** Mengzi, 10146.

The dried stems and leaves of this species are currently used in Chinese herbal medicine to treat cases of cancer, appendicitis, hepatitis, urinary infection, bronchitis, tonsillitis and snake-bite.

Oldenlandia paniculata L. (RUBIACEAE)

Henry in Trans. Asiat. Soc. Jap. 24 suppl: 49. (1896) (List Pl. Formosa).

Hainan: At Nodoa in the interior of the island, 8465. **Taiwan:** Kaohsiung, 777. Wanchin, 869. Oluanpi, 1274.

Olea paniculata R. Br. (OLEACEAE)

Olea glandulifera Wall. ex G. Don Anonymous in Kew List of Determinations (volume 34).

Yunnan: Simao, 12500.

Native to India, Sri Lanka, Pakistan, Nepal and China (Yunnan).

Olea sp. (OLEACEAE)

Anonymous in Kew List of Determinations (volume 34).

Yunnan: Simao, 12012, 12643.

Oleandra cumingii J. Sm. (POLYPODACEAE)

Oleandra intermedia R. C. Ching in Bull. Fan. Mem. Inst. Biol. 2: 187. (1931).

Yunnan: Mengzi, 9484c., (type of *Oleandra intermedia* R. C. Ching).

Nepal to southern China and Malesia.

Onoclea orientalis (Hook) Hook. (ASPLENIACEAE)

Onoclea (*Struthiopteris*) *orientalis* (Hook) Hook. Henry in Journ. China Br. Roy. Asiat. Soc. 22: 251. (1887) (Chinese Names of Plants). *Matteuccia orientalis* (Hooker) Trev. Henry in Journ. China Br. Roy. Asiat. Soc. 22: 251. (1887) (Chinese Names of Plants). *Struthiopteris orientalis* Hook. var. *brevis* Christ in Bull. Soc. Bot. France 5: 44. (1905). *Struthiopteris orientalis* Hook. var. *incisa* Christ in Bull. Soc. Bot. France 5: 44. (1905).

Hubei: Yichang, 181. Badong, 1427 (type of *Struthiopteris orientalis* Hook. var. *brevis* Christ), 5037, 5062, 5113 (type of *Struthiopteris orientalis* Hook. var. *incisa* Christ). **Hunan:** Shimen, 7941.

Distributed through the Himalaya, China, Korea and Japan. According to Henry it was colloquially known in the mountains above the Three Gorges as *mao kuan chung*. This species was collected by the Glasnevin Central China Expedition (GCCE 251) on the hills above Badong in September 2002.

Onosma exsertum Hemsley (BORAGINACEAE) New species

in Hooker's Icon. Pl. 27: t. 2639. (1901).

Yunnan: On grassy hills near Mengzi at 1,950 metres, 9334 (type).

Endemic to China and distributed in the provinces of Guizhou, Sichuan and Yunnan.

Onychium japonicum (Thunb.) Kunze (PTERIDACEAE)

Henry in Trans. Asiat. Soc. Jap. 24 suppl: 111. (1896) (List Pl. Formosa), Christ in Bull. Soc. Bot. France 5: 60. (1905).

Trichomanes japonicum Thunb. Christ in Bull. Herb. Boiss. 6: 866. (1898).

Hubei: Yichang, 434, 2169. South Badon,g 6140, 7534. **Taiwan:** Wanchin, 530, 531, 1510. Kaohsiung, 1097. Tamsui, 1435. **Yunnan:** On mountains to the south of Mengzi at 1,650 metres, 9371. **Yunnan:** Simao, 13135, 13467.

A widespread species, native to Pakistan, India, the Himalaya, Myanmar, China (including Taiwan), Korea, Japan, the Philippines and Java. A handsome fern cultivated

02

in a number of Irish gardens. This fern was collected by the Sino-American Expedition in Shennongjia in 1980 and in nearby Xingshan by the Glasnevin Central China Expedition (GCCE 481).

Onychium siliculosum (Desv.) C. Chr. (PTERIDACEAE)
Onychium auratum Kaulf. Henry in Trans. Asiat. Soc. Jap. 24 suppl: 111. (1896) (List Pl. Formosa).

Taiwan: Wanchin, 71. **Yunnan:** Mengzi, 9510.
Widespread in Asia.

Operculina turpethum (L.) Silva Manso (CONVOLVULACEAE)
Ipomoea turpethum (L.) R. Br. Henry in Trans. Asiat. Soc. Jap. 24 suppl: 65. (1896) (List Pl. Formosa).

Taiwan: Wanchin, 500. Kaohsiung, 1624.

Ophioglossum pendulum L. (OPHIOGLOSSACEAE)
Ophioglossum pendulum L. Henry in Trans. Asiat. Soc. Jap. 24 suppl: 117. (1896) (List. Pl. Formosa).

Taiwan: Oluanpi, 1348.

An epiphytic fern, growing on tree trunks, palms and tree ferns or in clumps of the bird's nest fern (*Asplenium nidus*).

Ophioglossum reticulatum L. (OPHIOGLOSSACEAE)
Baker in Journ. Bot. 27: 178. (1889) Henry in Notes Econ. Bot. China 60. (1893), Diels in Bot. Jahrb. 29: 209. (1900).

Hubei: Jianshi, 5953.

Henry called this the *i-chih-chien*.

Ophiopogon clavatus C.H.Wright ex Oliver (ASPARAGACEAE) New species
in Kew Bull. Misc. Inf. 116. (1895), Wright in Hooker's Icon. Pl. t. 2382. (1894), Bretschneider in Hist. Eur. Bot. Disc. China 2: 793. (1898), Wright in Journ. Linn. Soc. Bot. 77. (1903).

Hubei: Badong, 6065. Zigui, 6065a.

Ophiopogon dracaenoides (Baker) Hook. f. (ASPARAGACEAE)
Henry in Journ. China Br. Roy. Asiat. Soc. 271. (1887) (Chinese Names of Plants).

Hubei: Liantuo, 2606, 3038.

According to Henry this species was known colloquially in the Three Gorges region as *shui chu ts'ao*. Cultivated in many parts of China for its medicinal roots.

Ophiopogon intermedius D. Don (ASPARAGACEAE)
Wright in Journ. Linn. Soc. Bot. 36: 78. (1903).
Ophiopogon japonicus Wright non (L. f.) Ker-Gawl. in Journ. Linn. Soc. Bot. 36: 78. (1903).

Taiwan: Wanshoushan, 1098a. Oluanpi, 1703a.

Widespread across Pakistan, India, Sri Lanka, Bangladesh, Nepal, Bhutan, Myanmar, Thailand, Cambodia, Vietnam, Indonesia (Sumatra and Java), the Philippines and China including Taiwan.

Ophiopogon japonicus (L. f.) Ker-Gawl. (ASPARAGACEAE)
Wright in Journ. Linn. Soc. Bot. 36: 78. (1903).

Hubei: Antelope Glen, 3352. Yichang, 3981, 6473. Badong, 3705, 4735, 4686. Liantuo, 4412, 4451. Donghu, 6473.

Ophiopogon spp. (ASPARAGACEAE)
Anonymous in Kew List of Determinations (volume 34).

Yunnan: In Pingbian, in forests on the Daweishan Range at 1,950 metres, 11478. Mengzi, 13747. Simao, 12171a.

Ophiopogon umbraticola C. H. Wright (ASPARAGACEAE)
Ophiopogon japonicus (L. f.) Ker-Gawl. var. *umbraticola* (Hance) C. H. Wright in Journ. Linn. Soc. Bot. 36: 78. (1903).

Hubei: Antelope Glen, 4344. Yichang, 4344a.

Ophiorrhiza caespitosa (Blume) Razafim & Rydin (RUBIACEAE)
Spiradiclis caespitosa Blume Anonymous in Kew List of Determinations (volume 34).

Yunnan: Simao, 12294.

Native to India, Myanmar, western China and Vietnam.

Ophiorrhiza cantonensis Hance (RUBIACEAE)
Forbes & Hemsley in Journ. Linn. Soc. Bot. 23: 378. (1888).

Hubei: Yichang, 806, 809, 1255, 3248. Antelope Glen, 3577. On tops of cliffs in the Antelope Glen, 3300. Liantuo, 2996. **Hainan:** Ling-men, 8618, 8624.

Discovered near the south-east coast of China by Theo. Sampson in 1861.

Ophiorrhiza japonica Blume (RUBIACEAE)
Forbes & Hemsley in Journ. Linn. Soc. Bot. 23: 378. (1888), Henry in Trans. Asiat. Soc. Jap. 24 suppl: 50. (1896) (List Pl. Formosa).
Ophiorrhiza japonica Blume var. Forbes & Hemsley in Journ. Linn. Soc. Bot. 23: 378. (1888).

Hubei: Badong, 3692, 5424. Jianshi, 7441. Changyang, 5424a. **Taiwan:** Tamsui, 1481.

Ophiorrhiza kingiana Watt. (RUBIACEAE)
Anonymous in Kew List of Determinations (volume 34).

Yunnan: Mengzi, 11072, 10366, 10902.

Ophiorrhiza lurida Hook. f. (RUBIACEAE)
Anonymous in Kew List of Determinations (volume 34).

Yunnan: Mengzi, 10757. Simao, 11715.

Ophiorrhiza pumila Champ. ex Benth. (RUBIACEAE)
Henry in Trans. Asiat. Soc. Jap. 24 suppl: 50. (1896)

(List Pl. Formosa).

Taiwan: Tamsui, 1466.

Ophiorrhiza rosea Hook. f. (RUBIACEAE)

Anonymous in Kew List of Determinations (volume 34).

Yunnan: Mengzi, 10883. Simao, 12702.

Ophiorrhiza sp. No. 1. (RUBIACEAE)

Anonymous in Kew List of Determinations (volume 24).

Hubei: Yichang, 2718. **Sichuan:** South Wushan, 7293.

Ophiorrhiza sp. No. 2. (RUBIACEAE)

Anonymous in Kew List of Determinations (volume 34).

Yunnan: Mengzi, 10635, 10894, 11427, 11142. Simao, 11674, 12203, 12462, 12600, 12601, 12710.

Oplismenus compositus (L.) P. Beauv. (POACEAE)

Henry in Trans. Asiat. Soc. Jap. 24 suppl: 106. (1896) (List Pl. Formosa), Rendle in Journ. Linn. Soc. Bot. 36: 337. (1904).

Taiwan: Oluanpi, 232. Wanchin, 1557. **Yunnan:** In woods near Mengzi at 1,750 metres, 10971.

An annual species with culms to 1 metre tall, widespread in the tropics and very common in shady places in Taiwan.

Oplismenus undulatifolius (Arduino) Roem. & Schult. (POACEAE)

Rendler in Journ. Linn. Soc. Bot. 36: 338. (1904).

Hubei: Yichang, 4349. Badong, 4782, 7277. Fang Xian, 6661.

Oreocharis amabilis Dunn (GESNERIACEAE) New species

in Journ. Linn. Soc. Bot. 38: 362. (1908), Pan in Acta Phytotax. Sin. 25: 275. (1987).

Yunnan: Mengzi, 13758 (phototype).

Endemic to Yunnan.

Oreocharis aurea Dunn (GESNERIACEAE) New species

in Kew Bull. Misc. Inf. 19. (1911), Pan in Acta Phytotax. Sin. 25: 281. (1987).

Yunnan: South of the Red River from Manmei at 2,000 metres, 9713 (syntype). On trees in mountain forest at Fengchunling at 1,500 metres, 9713a. In Pingbian, in forests on the Daweishan Range at 1,500 metres, 9713b., 9713c.

Distributed in China (Yunnan) and Vietnam.

Oreocharis henryana Oliver (GESNERIACEAE) New species

in Hooker's Icon. Pl. 20: t. 1944. (1890), Bretschneider in Hist. Eur. Bot. Disc. China 2: 789. (1898), Hand.-Mazz. in Symb. Sin. 7: 877. (1936), Morley in Glasra 3: 80. (1979), Lauener & Burtt in Notes Roy. Bot. Gard. Edinburgh 38:

469. (1980), H. W. Li in Bull. Bot. Res. 3: 11. (1983), Pan in Acta Phytotax. Sin. 25: 281. (1987).

Sichuan: Henry's Chinese collector with Pratt, no locality stated but may be Leshan, Emei Shan, Wa Shan or Kangding, 8999 (holotype).

Oreocharis saxatilis (Hemsl.) Mich.Möller & A. Weber (GESNERIACEAE) New genus

in Notes Roy. Bot. Gard. Edinburgh 11: 266. (1918—1919), in part.

Didissandra saxatilis Hemsley in Journ. Linn. Soc. Bot. 26: 227. (1890), Bretschneider in Hist. Eur. Bot. Disc. China 2: 789. (1898). *Didymocarpus saxitilis* Levl. in Compte Rend. Ass. Fr. 427. (1906).

Hubei: Fang Xian, 7615 (isotype of *Didissandra saxatilis* Hemsley), 6603. South Badong, 6162, 7346. **Sichuan:** South Wushan, on dry rocks, 5704. North Wushan at 2650 metres, 7346 (type of *Didissandra saxatilis* Hemsley).

Oreocharis speciosa (Hemsl.) Mich.Möller & W. C. Chen (GESNERIACEAE) New species

Didissandra speciosa Hemsley in Journ. Linn. Soc. Bot. 26: 228. (1890). *Didymocarpus speciosa* (Hemsl.) Levl. in Compte Rend. Ass. Fr. 428. (1906). *Briggsia speciosa* (Hemsl.) Craib in Notes Roy. Bot. Gard. Edinburgh 11: 264. (1918—1919).

Hubei: Liantuo, 3951, 6356. Xingshan, 6411a. Donghu, 6411 (type of *Didissandra speciosa* Hemsley). South Badong, mostly on the face of perpendicular cliffs, 7668.

A very handsome little plant with beautifully spotted flowers similar in size and shape to the common foxglove (*Digitalis purpurea* L.).

Oreocnide frutescens (Thunb.) Miq. (URTICACEAE)

Villebrunea frutescens (Thunb.) Bl. Henry in Journ. China Br. Roy. Asiat. Soc.22: 242. (1887) (Chinese Names of Plants) Forbes & Hemsley in Journ. Linn. Soc. Bot. 26: 491. (1899)

Hubei: Liantuo, 3078, 3212. Antelope Glen, 3114. **Yunnan:** Mengzi, 9343b., 9809.

Distributed in India, Bhutan, Myanmar, China, Thailand, Laos, Vietnam and Japan. According to Henry this species was known colloquially in the Three Gorges region as *ch'ai ch'u ma*.

Oreocnide obovata (C.H.Wright) Merrill (URTICACEAE) New species

Debregeasia ovata C. H. Wright in Journ. Linn. Soc. Bot. 26: 492. (1899).

Yunnan: In Pingbian, in forests on the Daweishan Range at 1,600 metres, 11254 (type of *Debregeasia ovata* C. H. Wright), 11254a.

Native to China and Vietnam.

Oreocnide sp. (URTICACEAE)

Villebrunea sp. Anonymous in Kew List of Determinations (volume 34).

Yunnan: Mengzi, 10855. Simao, 11654, 12188, 12672.

Oreocnide sylvatica (Blume) Miq. (URTICACEAE)

Villebrunea scabra (Blume) Wedd. Forbes & Hemsley in Journ. Linn. Soc. Bot. 26: 491. (1899).

Yunnan: Mengzi, 10682a., 11496, 11497.

Oreoseris henryi (Dunn) W. Zheng & J. Wen (ASTERACEAE) New species

Gerbera henryi Dunn in Journ. Linn. Soc. Bot. 35: 511. (1903).

Yunnan: In grassy mountains near Mengzi, in exposed rather barren spots, 9111 (isotype of *Gerbera henryi* Dunn).

Endemic to China and distributed in the provinces of Guizhou and Yunnan.

Oresitrophe sp. (SAXIFRAGACEAE)

Anonymous in Kew List of Determinations (volume 24).

Hubei: Badong, 1436.

Origanum vulgare L. (LAMIACEAE)

Henry in Journ. China Br. Roy. Asiat. Soc. 22: 266. (1887) (Chinese Names of Plants), Forbes and Hemsley in Journ. Linn. Soc. Bot. 26: 282. (1890), Dunn in Notes Roy. Bot. Gard. Edinburgh 6: 157. (1915).

Hubei: Yichang, 394, 457, 458, 623, 853, 2140, 2414. **Yunnan:** Mengzi, 11137.

Oregano, an aromatic perennial herb to 70 cm, native to Europe and central Asia. Known colloquially in the Three Gorges region as *yeh po ho* according to Henry.

Orixa japonica Thunb. (RUTACEAE)

Rehder & E. H. Wilson in Sargent Pl. Wilson. 2: 135. (1914) ; 3: 449. (1917), Zheng in Journ. Wuhan Bot. Res. 2.(1): 105. (1984).

Hubei: Liantuo, 1985, 3847, 3887. Yichang, 3477, 6477, 7773. Changyang, 5247.

A handsome deciduous shrub to 2.5 metres tall, native to both Japan and China. The leaves are aromatically scented and according to W. J. Bean have a pleasant spicy odour when crushed and turn pale yellow in autumn. The flowers are unisexual, female flowers are carried on different plants and fruits are only formed when both sexes are grown. According to Wilson when the fruits are ripe the seeds are shot out from the parent plant in the same manner as *Impatiens*. A common cliff-side inhabitant of western Hubei. This species was unknown in China previous to Henry's collections. It is used as a hedge plant in Japan.

Ormosia henryi Prain (FABACEAE) New species

in J. Asiat. Soc. Bengal. Pt. 2, Nat. Hist. 69: 180. (1900), Hemsley & E. H. Wilson in Kew Bull. Misc. Inf. 156. (1906), Craib in Sargent Pl. Wilson 2: 93. (1916), Gagnep. in Lecompte, Fl. Gen. Indo-Chine 2: 511. (1920), Hu in Contr. Biol. Lab. Sci. China 2: 9. (1926), Chun in Sci. Journ. Coll. Sci. Sunyatsen. Univ. 2: 46. (1930), Merr. & Chun in Sunyatsenia 1: 60. (1930); 532. t. 428. (1937), Merr. & L. Chen in sargentia 3: 109. (1943); 100. (1955), Chun in Chin. Econ. trees 187. (1922), Chang in Acta Phytotax. Sin. 22: 112. (1984).

Ormosia sp. Henry in Notes Econ. Bot. China 56. (1893). *Fedorovia henryi* (Prain) Yakovlev in Bot. Zhurn. (Moscow & Leningrad) 56: 656. (1971).

Hubei: South Donghu, 7577 (type).

The *ch'ing-t'an,* (according to Henry), a small ornamental tree to about 9 metres tall and now extremly rare in Hubei. Wilson who collected this species at Badong stated it was rare by the turn of the 20th century and that its finely grained yellow wood, once popular for turning probably led to its near extinction. Wilson found it near Changleping (in Wufeng County, south of Changyang) in 1907 and near Badong about 1900. *Ormosia henry* is cultivated in the rare and endangered garden at Wuhan Botanical Gardens, one of a number of gardens administered by the Chinese Academy of Sciences. The related *Ormosia hosei,* was discovered by Wilson in Changyang and he later collected it at Liantuo and on the Chengdu plain and by a temple in Chengdu in Sichuan. Like Henry's *Ormosia* this too is a rare tree. In September 2002 before travelling south through the Chengdu plain to climb Emei Shan, members of the Glasnevin Central China Expedition stayed for a brief time in the city of Chengdu and by our hotel was a large 20 metre tall tree of *Ormosia hosei.*

Ormosia longipes H. Y. Chen (FABACEAE) New species

in Sargentia 3: 100. (1943).

Yunnan: In Pingbian in forests on the Daweishan Range at 1,640 metres, 11854 (type).

Endemic to Yunnan.

Ormosia sp. (FABACEAE)

Anonymous in Kew List of Determinations (volume 34).

Yunnan: Simao, 12885.

Ormosia sp. nov. (FABACEAE)

Anonymous in Kew List of Determinations (volume 24).

Hainan: At Nodoa in the interior of the island, 8440.

Ormosia striata Dunn (FABACEAE) New species

in Journ. Linn. Soc. Bot. 35: 192. (1903), Chang in Acta Phytotax. Sin. 22: 114. (1984).

Fedorovia striata (Dunn) Yakolovlev in Nov. Syst. Pl.

Vasc. 10: 196. (1973)

Yunnan: Forests to the west of Simao at 1,650 metres, 11886 (isosyntype), 12843 (isosyntype). In forests near Simao at 1,300 metres, 12979 (isosyntype), 12979a., (isosyntype) 12979b., (isosyntype).

Distributed in Myanmar, China (Yunnan) and Thailand.

Ormosia yunnanensis Prain (FABACEAE) New species

in J. Asiat. Soc. Bengal. Pt. 2, Nat. Hist. 69: 183. (1900).

Yunnan: Mountains to the west of Simao at 1,300 metres, 11967 (isotype).

Endemic to Yunnan.

Ornithoboea henryi Craib (GESNERIACEAE) New species

in Kew Bull. Misc. Inf. 115. (1913).

Yunnan: Pu'er at 1,350 metres, 13378 (type).

Orobanche coerulescens Stephan (OROBANCHACEAE)

Orobanche ammophila C. A. Mey. Anonymous in Kew List of Determinations (volume 24).

Hubei: Badong, 668 (July 1885). Fang Xian, 6742, 7935.

Parasitic on various species of *Artemisia*, *Achillea* and *Lotus corniculatus*.

Orostachys fimbriata (Turcz.) Berger (CRASSULACEAE)

Cotyledon fimbriata Turcz. var. *ramosissima* (Maxim.) Maxim. Henry in Journ. China Br. Roy. Asiat. Soc. 22: 279. (1887) (Chinese Names of Plants).

Hubei: On roof tiles at Yichang, 893.

Known colloquially in the Three Gorges region as *wa sung* according to Henry. Seeds were collected from A. Henry 893 in the Kew herbarium on May 7th 1888 and sent for propagation.

Oroxylum indicum (L.) Kurz. (BIGNONIACEAE)

Oroxylum flavum Rehder in Sargent, Trees & Shrubs. 1: 193. fig. xcii. (1905).

Hainan: Environs of Kiungchow, 8740. **Yunnan:** Mengzi, without number.

The midnight horror or *ki tong tokang*, a wide-spreading semi-deciduous tree bearing foetid noturnal flowers in erect racemes. A widespread species distributed through the Himalaya, Sri Lanka, China and south-east Asia.

Oroxylum flavum was described by Alfred Rehder from plants that had been raised at the Arnold Arboretum from seeds sent from Yunnan by Henry, where it flowered for the first time in the spring of 1903. By then it formed trees to 3 metres high and flowered in spring, exhaling, according to Rehder, a rather disagreeable odour. Rehder went on to state that the tree would probably be suited to the Californian climate where he predicted it would be a highly desirable ornamental tree. There is a fine illustration of this species in Sargent's *Trees & Shrubs*.

Orthosiphon aristatus (Blume) Miq. (LAMIACEAE)

Orthosiphon stamineus Benth. Forbes & Hemsley in Journ. Linn. Soc. Bot. 26: 268. (1890), Dunn in Notes Roy. Bot. Gard. Edinburgh vi. 135. (1915).

Hainan: Without locality, 8533.

A very beautiful species, distributed from India to Myanmar, the Philippines and northern Australia. In China it is spread through the provinces of Fujian, Guangxi, Hainan, Taiwan and Yunnan.

Orthosiphon rubicundus (D. Don) Benth. (LAMIACEAE)

Dunn in Notes Roy. Bot. Gard. Edinburgh 6: 135. (1915).

Yunnan: Mengzi, 9800.

Orychophragmus violaceus (L.) O. E. Schulz (BRASSICACEAE)

Schulz in Engler, Pflanzenr. iv. 105: 74. (1923), Al-Shehbaz & Yang Guang in Novon 10: 350. (2000).

Orychophragmus violaceus (L.) Schulz var. *subintegrifolius* Pamp. Schulz. Schulz in Engler, Pflanzenr. iv. 105: 76. (1923).

Hubei: San You Dong Glen, 3411. Liantuo, 1998, 3838.

One of the most variable species of Chinese Brassicaceae, especially in leaf morphology. In A. Henry 3411 all the leaves are simple, whereas in most collections the leaves have lateral, leaflet-like lobes. *Orychophragmus violaceus* is also variable in flower size, fruit and style length. The shortest styles (ca. 3 mm) were observed in A. Henry 1998. Because of its rich oil contents, attempts are being made to grow this species as a crop plant. Native to China and North Korea. A highly ornamental annual or biennial herb. For an illustration of this species see Phillips & Rix in Annuals and Biennials, Macmillan, 1996.

Oryza sativa L. (POACEAE)

Henry in Trans. Asiat. Soc. Jap. 24 suppl: 107. (1896) (List Pl. Formosa), Rendle in Journ. Linn. Soc. Bot. 36: 344. (1904).

Taiwan: Kaohsiung Plain, without number.

Rice, the world's most important crop. An estimated 1.7 billion people depend on rice and it has been cultivated in Asia for at least 7,000 years. Many kinds of rice are cultivated in China, according to Henry the main devisions are; ordinary rice which is *keng* and glutinous rice is *no-mi*. The general term for all kinds of rice is *ku-tze*. Rice sprouts, which are used to aid digestion are known as *gu ya*. A tropical, aquatic crop, in the most favoured places such as the Chengdu plain in Sichuan (sometimes termed the granary of China) three crops of rice are harvested each year. Rice is generally grown on terraces in the low lying valley and above this maize is grown on terraces on higher ground. In China as in many other warm-temperate and subtropical parts of Asia it is grown in paddy fields, that is standing in level fields of shallow water. This still water prevents the growth of terrestrial weeds and rots the plant material left behind after harvesting and holds blue-green algae that acts as a green fertilizer.

Oryzopsis sp. (POACEAE)

Anonymous in Kew List of Determinations (volume 24).

Hubei: Yichang, 3975.

Osbeckia chinensis L. (MELASTOMATACEAE)

Forbes & Hemsley in Journ. Linn. Soc. Bot. 26: 298. (1887), Henry in Trans. Asiat. Soc. Jap. 24 suppl: 43. (1896) (List Pl. Formosa), Li in Journ. Arnold Arb. 25: 4. (1944) Li in Woody Flora of Taiwan 660. fig. 267. (1963).

Hubei: Yichang, 138, 2124, 2320. **Hainan:** Without locality, 8516. Ling-men (in the interior of the island), 8643. **Taiwan:** Oluanpi, 989. Wanchin, 1555.

A subshrub to 60 cm tall, distributed throughout Indo-Malaysia, southern China and Japan. According to Henry this was said to be the *chin-shih-liu* and used as a drug. Wilson stated this small shrub was rare in western Hubei. The genus *Osbeckia* commerates Peter Osbeck (1723—1805), a Swede and a pupil of Linnaeus who collected 155 plants in south-east China in 1751.

Osbeckia nepalensis Hook. (MELASTOMATACEAE)

Li in Journ. Arnold Arb. 25: 5. (1944).

Yunnan: Mengzi, 11026. Simao, 12309.

An erect subshrub to 1.5 metres tall, a widespread species distributed from the eastern Himalaya and western China to Thailand.

Osbeckia sp. (MELASTOMATACEAE)

Anonymous in Kew List of Determinations (volume 34).

Yunnan: Mengzi, 9442.

Osbeckia stellata Bech.-Ham. ex D. Don var. *crinita* (Benth. ex Naudin) C. Hansen (MELASTOMATACEAE)

Osbeckia crinita Benth. ex Naudin Rehder & E. H. Wilson in Sargent Pl. Wilson. 2: 421. (1915), Li in Journ. Arnold Arb. 25: 4. (1944).

Yunnan: Mengzi, 9978. On hills near Simao at 1,500 metres, 12458.

Osmanthus armatus Diels (OLEACEAE) New species

E. H. Wilson in Gard. Chron. ser. 3. 50: 113. (1913), Rehder & E. H. Wilson in Sargent Pl. Wilson. 2: 611. (1916), Green in Notes Roy. Bot. Gard. Edinburgh 22: 518. (1958).

Osmanthus fragrans Forbes & Hemsley non (Thunb.) Lour. in Journ. Linn. Soc. Bot. 26: 88. (1889), in part.

Hubei: Liantuo, 4597. Fang Xian, 6730. Changyang, 7587.

A shrub or small tree to 5 metres tall. Native to Hubei and Sichuan. The species was first found by Henry at Liantuo in 1888 (it was named from material later collected by Bock and von Rosthorn in 1891), a small hamlet in the Xiling (Yichang) Gorge and was later introduced to cultivation in 1902 by Wilson who collected a single living plant for Messrs Veitch. The Veitch catalogue, *New Hardy Plants from Western China* (autumn 1912) had plants available for 7 shillings and 6 pence. This species was collected by the Glasnevin Central China Expedition (GCCE 662) in the Red Dragon Reserve, Xingshan in October 2004. It is by no means common in its native habitat and in the Red Dragon Reserve it formed fine trees to 5 metres tall and grew in groves at the base of limestone cliffs.

Osmanthus cooperi Hemsl. (OLEACEAE)

Green in Notes Roy. Bot. Gard. Edinburgh 22: 525. (1958).

Yunnan: In Pingbian, in forests on the Daweishan Range at 1,650 metres, 11048.

Discovered by the Irish plant collector, George Playfair, endemic to China.

Osmanthus fragrans (Thunb.) Lour. (OLEACEAE)

Henry in Journ. China Br. Roy. Asiat. Soc. 22: 252. (1887) (Chinese Names of Plants), Forbes & Hemsley in Journ. Linn. Soc. Bot. 26: 88. (1889), Rehder & E. H. Wilson in Sargent Pl. Wilson 2: 609. (1916), Green in Notes Roy. Bot. Gard. Edinburgh 22: 484. (1958), Zheng in Journ. Wuhan Bot. Res. 2.(1): 182. (1984).

Hubei: Yichang, 2288, 7784 (in part). Liantuo, 2638. Badong, 2497, 2498, 5155, 5182. Changleping (cultivated) 7722. **Yunnan:** Cultivated at Mengzi, 11008. Yuanjiang, 13213.

A tree to 5 metres tall, bearing masses of sweetly scented creamy-yellow flowers in autumn The same flowers are used in China to perfume tea. According to Henry this tree was known colloquially in the Three Gorges region as *kui hua*. This is one of the most popular autumn flowering trees in China and the creamy-white flowers are produced in such lavish profusion that they virtually smother the entire tree and scent the air for a considerable distance. Because of its time of blooming during the mid-autumn or moon-festival this species has long been associated with Chinese lunar legends. It is believed that the figure on the moon's surface is an osmanthus tree. There is a double flowered form of this species in cultivation in China though the most beautiful variation must be *Osmanthus fragrans* Lour. f. *aurantiacus* (Mak.) P. S. Green which bears orange coloured flowers. There are several good trees of this very beautiful form in the grounds of the Chengdu Bamboo Garden where it grows alongside the type. A red flowered cultivar was recorded during the Ming Dynasty (1368—1644) though this is now extinct. *Osmanthus fragrans* was collected by the Glasnevin Central China Expedition in New Xingshan in October 2004.

Osmanthus henryi P. S. Green (OLEACEAE) New species

in Notes Roy. Bot. Gard. Edinburgh 22: 499. (1958), O' Brien in Augustine Henry, An Irish Plant Collector in China 27. (2002).

Yunnan: In mountain forest near Milê, 10020 (isotype, collected November 1st 1897).

Henry, in his field notes described this species as a 6 metre tall tree with white flowers. Endemic to China and distributed in Guizhou and Yunnan.

Osmanthus sp. (OLEACEAE)

Anonymous in Kew List of Determinations (volume 34).

Yunnan: Mengzi, 9764.

Osmorhiza aristata (Thunb.) Ryd. (APIACEAE)

Osmorhiza longistylis Diels non (Torr.) DC. in Bot. Jahrb. 29: 492. (1900).

Hubei: Jianshi, 5789. **Yunnan:** Mengzi, 10223.

Osmunda japonica Thunb. (OSMUNDACEAE)

Osmunda regalis Henry non L. in Journ. China Br. Roy. Asiat. Soc. 22: 258. (1887) (Chinese Names of Plants); in Notes Econ. Bot. China 60. (1893), Diels in Bot. Jahrb. 29: 208. (1900), Christ in Bull. Soc. Bot. France 5: 67. (1905). *Osmunda javanica* Henry non L. in Trans. Asiat. Soc. Jap. 24 suppl: 116. (1896) (List Pl. Formosa).

Hubei: Badong, 609 (July 1885), 1401. **Taiwan:** Tamsui, 1436.

According to Henry this species was known colloquially in the Three Gorges region as *mao erh t'ou*, in China it was generally called the *lao-hu-t'ai*. He also stated it was a common fern in the mountains of Hubei at altitudes above 1,000 metres. Native to Pakistan to China, Korea and Japan.

Osteomeles schweriniae C.K.Schneider (ROSACEAE) New species

In Illus. Handb. Laubholzk. 1: 762. fig. 430. (1906), Schneider in Fedde, Rep. Spec. Nov. Regni. Veg. 3: 222. (1906), E. H. Wilson in Sargent Pl. Wilson. 1: 184. (1913), O' Brien in Augustine Henry, An Irish Plant Collector in China 27. (2002).

Yunnan: Mengzi at 1,500 to 1,800 metres, 9315, 9315a., (type).

An evergreen shrub to 2.5 metres tall, leaves pinnate to 10 cm long, flowers white and produced in branching corymbs in June. Although described from Henry's material, Bean states it had been raised in the Jardin des Plantes in Paris from seeds sent there by Delavay in 1888, it reached Kew in 1892. *Osteomeles schweriniae* is native to China (Gansu, Guizhou, Shaanxi, Sichuan and Yunnan) and Japan. In Ireland this species is cultivated at the John F. Kennedy Arboretum and at the Talbot Botanic Gardens, Malahide Castle.

Ostodes paniculata Blume var. *katharinae* (Pax) Chakrab. & N. P. Balakr. (EUPHORBIACEAE) New variety

Ostodes katharinae Pax in Pflanzenr. (Engler) 4. Fam. 147(3): 19. (1911).

Yunnan: Simao at 1,300 metres, 13003 (isotype of *Ostodes katharinae* Pax). In forests near Simao at 1,940 metres, 13062 (syntype of *Ostodes katharinae* Pax).

A very beautiful evergreen tree to 15 metres tall, carrying axillary many-flowered racemes of white staminate flowers in April and May. Native to China [Yunnan and Tibet (Xizang)] and Thailand.

Ostodes sp. (EUPHORBIACEAE)

Anonymous in Kew List of Determinations (volume 34).

Yunnan: Simao, 12055, 12321, 12394, 12565, 12989, 13549.

Ostrya japonica Sarg. (BETULACEAE)

Schneider in Sargent Pl. Wilson. 2: 424. (1916), Elwes & Henry in Trees of Gr. Brit. & Irl. 3: 544. t. 201. fig. 10. (1908).

Ostrya virginica Willd. var. *japonica* Maxim. ex Sarg. Burkill in Journ. Linn. Soc. Bot. 26: 503. (1899). *Ostrya italica* Scop. ssp. *virginiana* Winkler non (Mill.) H. Wilk. in Engler, Pflanzenr. 61: 22. (1904).

Hubei: Fang Xian, 6581.

Osyris lanceolata Hochst. & Steud. (SANTALACEAE)

Osyris wightiana Wall. ex Wight Schneider in Sargent Pl. Wilson. 3: 320. (1916).

Yunnan: On barren hills near Mengzi at 1,600 metres, 9906, 9906a.

A widespread species in Asia, distributed in the western provinces of China.

Ototropis amethystina (Dunn) H. Ohashi & K. Ohashi (FABACEAE) New species

Desmodium amethystinum Dunn in Gard. Chron. ser. 3, 32: 210. (1902), Nelson in Pim, The Wood & the Trees 221. (1984); A Heritage of Beauty 312. (2000).

Yunnan: Simao, 12614 (isotype of *Desmodium amethystinum* Dunn). Mountains near Simao at 1,940 metres, 12614a., (isotype of *Desmodium amethystinum* Dunn).

A shrubby plant to 1.5 metres tall, introduced to cultivation by Henry through the Royal Botanic Gardens, Kew where after two years of growth it produced a panicle of amethyst-coloured flowers. This species is also found in Thailand on the Doi Angka range of the Pa Ngem.

Ototropis multiflora (DC.) H. Oshasi & K. Ohashi (FABACEAE)

Forbes & Hemsley in Journ. Linn. Soc. Bot. 23: 172. (1887), Henry in Journ. China Br. Roy. Asiat. Soc. 22: 273. (1887) (Chinese Names of Plants).

Desmodium multiflorum DC. Anonymous in Kew List of Determinations (volume 34). *Desmodium sambuense* (D. Don) DC. Zheng in Journ. Wuhan Bot. Res. 2.(1): 99. (1984).

Hubei: Badong, 2502. **Yunnan:** Mengzi, 9802, 9802a. Simao, 10003b.

Henry's Badong collector called this species the *yeh huang tou*. Native from the Himalayan region to mainland China, Thailand and Laos.

Ottelia alismoides (L.) Pers. (HYDROCHARITACEAE)

Wight in Journ. Linn. Soc. Bot. 36: 3. (1903).

Hubei: Yichang, 165. **Hainan:** Environs of Kiungchow, 8291. Ling-men (in the interior of the island), 8662.

A submersed freshwater perennial, widespread across Asia and south to Australia.

Oxalis corniculata L. (OXALIDACEAE)

Henry in Journ. China Br. Roy. Asiat. Soc. 22: 268. (1887) (Chinese Names of Plants); in Trans. Asiat. Soc. Jap. 24 suppl: 24. (1896) (List Pl. Formosa).

Hubei: Yichang, 265, 741. **Taiwan:** Wanchin, without number.

Creeping wood sorrel, which Henry stated was known colloquially in the Three Gorges region as *suan mi ts'ao*. In Chinese herbal medicine this species (in its dried state) is used as a diuretic. Widely distributed through temperate and warm regions of the world.

Oxalis debilis Kunth (OXALIDACEAE)

Oxalis acetosella Diels non L. in Bot. Jahrb. 29: 420. (1900).

Hainan: 32 km (20 miles) west of Haikou, 8157. **Hubei:** Yichang, 781. Badong, 1462. Fang Xian at 2,900 metres, 5288a.

A perennial acaulescent herb, native to South America and widely naturalised throughout the world.

Oxalis griffithii Edgew. & Hook. f. (OXALIDACEAE)

Knuth in Engler, Pflanzenr. 10: 234. (1930).

Hubei: South Badong at 1,950 metres, 5288. Without locality, 5288c. **Sichuan:** South Wushan, 5288b. **Yunnan:** Mengzi, 10623.

Native to India, Nepal and China.

Oxyspora howellii Jeffrey & W. W. Sm. (MELASTOMATACEAE) New variety

Allomorphia setosa Craib in Kew Bull. Misc. Inf. 68. (1913), Li in Journ. Arnold Arb. 25: 11. (1944).

Yunnan: Simao at 1,300 metres, 12993 (isotype of *Allomorphia setosa* Craib).

Oxyspora paniculata (D. Don) DC. (MELASTOMATACEAE)

Li in Journ. Arnold Arb. 25: 12. (1944).

Yunnan: Mengzi, 9010, 9010b., 11284. Simao, 12508, 12508a.

A handsome evergreen shrub to 2 metres tall, the small pink flowers are carried in terminal panicles composed of corymbs with two leaf-like involucres. Distributed in Nepal, Myanmar, China [Guizhou, Guangxi, Sichuan, Tibet (Xizang) and Yunnan] and northern Vietnam.

Oxyspora urophylla (Diels) Y. M. Shui (MELASTOMATACEAE) New species

Allomorphia urophylla Diels in Bot. Jahrb. Syst. 65: 102. (1932). *Oxyspora balansae* Li non (Cogn.) J. F. Maxwell, in Journ. Arnold Arb. 25: 10. (1944).

Yunnan: Mengzi, 9769 (isotype of *Allomorphia urophylla* Diels), 11448 (isoparatype of *Allomorphia urophylla* Diels), 11448a. In Pingbian, in forests on the Daweishan Range at 1,650 metres 9769a., (isotype of *Allomorphia urophylla* Diels).

Distributed in China (Guangdong, Guangxi and Yunnan) and Vietnam.

Pachygone laurifolia (DC.) L. Lian & Wei Wang (MENISPERMACEAE)

Cocculus laurifolius DC. Henry in Trans. Asiat. Soc. Jap. 24 suppl: 16. (1896) (List Pl. Formosa), Diels in Engler, Pflanzenr. iv. 94: 239. (1910), Li in Woody Flora of Taiwan 176. (1963), Huang in Flora of Taiwan 2: 592. (1996).

Taiwan: Wanchin, 56, 171. **Yunnan:** Mountains to the south-west of Mengzi at 1,200 metres, 12940a.

A widely distributed dioecious shrub or small tree with a range from the India, across the Himalaya, southern China to southern Japan and Malaysia.

Pachygone valida Diels (MENISPERMACEAE) New species

in Engler. Pflanzenr. 46. (iv. 94.): 243. (1910).

Yunnan: Mengzi, 13632 (isotype).

Endemic to China, distributed in Guangxi, Guizhou and Yunnan.

Pachyrhizus erosus (L.) Urban (FABACEAE)

Pachyrhizus angulatus Rich. ex DC. Henry in Trans. Asiat. Soc. Jap. 24 suppl: 37. (1896) (List Pl. Formosa). Henry in Notes Econ. Bot. China 58. (1893).

Taiwan: Kaohsiung, 1093. **Yunnan:** Mengzi, 11005. Simao, 12529, 12539a.

The yam bean, an herbaceous climber to 4.5 metres tall. Native to tropical Central America and naturalised at Kaohsiung, Henry stated the yam bean was known in eastern China as *fan-ko* and in Yunnan as *ti-kua*. Cultivated for its turnip-shaped root.

Pachysandra axillaris Franch. (BUXACEAE)

Forbes & Hemsley in Journ. Linn. Soc. Bot. 26: 419. (1894), Rehder & E. H. Wilson in Sargent Pl. Wilson. 2: 164. (1916).

Sichuan: South Wushan, 5589, 5709, 7529. **Yunnan:** Lunan, 9959.

An evergreen semi-woody shrub to 25 cm tall, flowers white, held in erect spikes in April. Discovered by Delavay and introduced to cultivation by Wilson in 1901.

Pachysandra axillaris Franch. var. *stylosa* (Dunn) Boufford & Q. Y. Xiang (BUXACEAE)

Pachysandra stylosa Dunn var. *tomentosa* H. C. Robbins in Sida 3: 237. (1968).

Yunnan: Lunan, 9959a., (holotype of *Pachysandra stylosa* Dunn var. *tomentosa* H. C. Robbins).

Pachysandra terminalis Sieb. & Zucc. (BUXACEAE)

Forbes & Hemsley in Journ. Linn. Soc. Bot. 26: 419. (1894), Zheng in Journ. Wuhan Bot. Res.2. (1): 115. (1984).

Hubei: Badong, 4904. Fang Xian at 2,500 to 2,900 metres, 6802. Changleping, 5331. **Sichuan:** South Wushan, 7836.

An evergreen procumbent semi-woody shrub to 25 cm tall. Wilson who later collected this species in Fang Xian (to the north of Yichang) stated it was a very common plant on rocks in the moist woods of western Hubei and Sichuan.

Pachystoma pubescens Blume (ORCHIDACEAE)

Pachystoma chinense (Lindl.) Reichb. f. Henry in Trans. Asiat. Soc. Jap. 24 suppl: 92. (1896) (List Pl. Formosa), Forbes & Hemsley in Journ. Linn. Soc. Bot. 17. (1903).

Taiwan: Wanchin, 384.

A terrestrial orchid to 50 cm tall, native to the Himalaya, southern China and southeast Asia.

Padbruggea filipes (Dunn) Craib (FABACEAE) New species

Adinobotrys filipes Dunn in Kew Bull. Misc. Inf. 195. (1911). *Whitfordiodendron filipes* (Dunn) Dunn in Kew Bull. Misc. Inf. 8: 364. (1912).

Yunnan: Simao, 11610 (type of *Adinobotrys filipes* Dunn).

Distributed in China (Guangxi and Yunnan) and Indochina.

Paederia foetida L. (RUBIACEAE)

Paederia tomentosa Blume Henry in Journ. China Br. Roy. Asiat. Soc. 22: 236. (1887) (Chinese Names of Plants), Forbes & Hemsley in Journ. Linn. Soc. Bot. 23: 389. (1888), Henry in Trans. Asiat. Soc. Jap. 24 suppl: 51. (1896) (List Pl. Formosa).

Hubei: Yichang, 173, 291, 898, 1031, 2771, 3084. San You Dong Glen near Yichang (July 1886), 1498. Badong, 1817, 1857, 2811, 4840. Liantuo, 2026. Changyang, 7503. **Hainan:** Ling-men, 8543. Environs of Kiungchow, 8710. Environs of Kiungchow, 8724. **Taiwan:** Oluanpi, 937. Wanchin, 1660.

According to Henry this species was known colloquially around Yichang as *chi shih t'eng*. It was collected by the Glasnevin Central China Expedition (GCCE 172) between Yichang and Badong in September 2002.

Paederia spp. (RUBIACEAE)

Anonymous in Kew List of Determinations (volume 34).

Yunnan: Mengzi, 9126, 9463. Simao, 12442.

Paeonia anomala L. (PAEONIACEAE)

Anonymous in Kew List of Determinations (volume 24).

Sichuan: Henry's Chinese collector with Pratt, no locality stated but may be any of the following, Leshan, Emei Shan, Wa Shan or Kangding, 8801.

Paeonia lactiflora Pall. (PAEONIACEAE)

Paeonia albiflora Pall. Henry in Journ. China Br. Roy. Asiat. Soc. 22: 268. (1887) (Chinese Names of Plants); in Notes Econ. Bot. China 7. (1893), Diels in Bot. Jahrb. 29: 324. (1900).

Hubei: Badong, 847, 1014. Liantuo, 3884.

Known as the *shaoyao* according to Henry, this species has a large geographical range from Siberia, Mongolia to central and northern China to Korea. *Shaoyao* means 'charming and beautiful' and has been the name of this species since ancient times. It was domesticated in China thousands of years ago and this has produced hundreds of distinct garden cultivars. It grows in thickets and forest margins between altitudes of 2,500 and 3,000 metres. In early summer two or more sweetly scented white flowers bloom on each stem and in later months the foliage takes on wonderful autumnal tints. Henry also stated that *Paeonia lactiflora* was known colloquially in the Three Gorges region as *so yo* and the root provides a drug called *chi-shao* which is used to invigorate the blood. Other Chinese herbals called it the *bai shao*, stating it is a tonic and that it is used in cases of menstrual disfunction. Along with the moutan (*Paoenia suffruticosa*) this species was introduced from China into Japan in medieval times and became a popular garden plant there. It was not until the beginning of the 19th century that *Paeonia lactiflora* reached Europe.

Paeonia obovata Maxim. subsp. *willmottiae* (Stapf.) D. Y. Hong & K. Y. Pan (PAEONIACEAE)

New subspecies. in Journ. Roy. Hort. Soc. 68: 28. (1943).

Paeonia obovata Diels non Maxim. in Bot. Jahrb. 29: 324. (1900).

Hubei: South Badong, 5365.

Paeonia obovata subsp. *willmottiae* was described from plants grown in the garden of Miss Ellen Willmott at Warley Place, having been raised from seeds collected by Wilson in Hubei, it had however been discovered by Henry in May of 1888. The Italian missionary Giraldi also later collected it in neighbouring Shanxi. This subspecies grows in thickets or forest edges in moist situations at altitudes between 1,600 and 2,400 metres.

Palaquium formosanum Hayata (SAPOTACEAE)

Li in Woody Flora of Taiwan 725. fig. 296. (1963).

Dichopsis sp. Henry in Trans. Asiat. Soc. Jap. 24 suppl: 58. (1896) (List Pl. Formosa)

Taiwan: Wanchin, 431, 842. Oluanpi, 341, 1983.

A large evergreen tree to 20 metres tall, native to the Batan and Babuyan islands in the Philippines and China (Taiwan) where it grows in tidal forests. According to Henry this was said to be the *t'u-kan-lan*, a tree 20 feet (6 metres) tall bearing edible fruits.

Paliurus hemsleyanus Rehder (RHAMNACEAE)
New species.

in Journ. Arnold Arb. 12: 74. (1931), Zheng in Journ. Wuhan Bot. Res. 2.(1): 138. (1984).

Hubei: Liantuo, 6379. **Sichuan:** South Wushan, 7205 (type).

A tree to 15 metres tall, its bark is covered with numerous spines. The fruits are white changing to brown at maturity.

Paliurus orientalis (Franch.) Hemsley (RHAMNACEAE)

Rehder in Journ. Arnold Arb. 12: 75. (1931).

Paliurus sinicus Schneider in Sargent Pl. Wilson. 2: 211. (1914).

Yunnan: In woods near Mengzi at 1,800 metres, 9427 (type of *Paliurus sinicus* Schneider), 9427b.

Discovered by Delavay near Dali in June 1883, endemic to China and distributed in Sichuan and Yunnan.

Paliurus ramosissimus (Lour.) Poir. (RHAMNACEAE)

Henry in Trans. Asiat. Soc. Jap. 24 suppl: 27. (1896) (List Pl. Formosa), Schneider in Sargent Pl. Wilson. 2: 210. (1916), Li in Woody Flora of Taiwan 511. fig. 198. (1963).

Taiwan: Tamsui, 1382.

A spiny shrub, native to China, Korea and Japan. In Chinese Taiwan it grows at low altitude and is often planted as a hedge.

Paliurus sp. (RHAMNACEAE)

Anonymous in Kew List of Determinations (volume 24).

Sichuan: South Wushan, 5528.

May be *Paliurus ramosissimus* (Lour.) Poir.

Panax bipinnatifidus Seem. (ARALIACEAE)

Aralia bipinnatifida (Seem.) C. B. Clarke Henry in Notes Econ. Bot. China 22. (1893). *Aralia quinquefolia* (L.) Decne & Planch. var. *major* Burkill in Kew Bull. Misc. Inf. 1: 7. (1902).

Hubei: Fang Xian, 6834, 6835. Zigui, 7728. **Sichuan:** North Wushan, 7121 (type of *Aralia quinquefolia* Decne & Planch. var. *major* Burkill).

Panax ginseng C. A. Mey. (ARALIACEAE)

Aralia quinquefolia (L.) Decne. & Planch. var. *ginseng* (C. A. Mey.) Anon. Anonymous in Kew List of Determinations (volume 24). *Aralia quinquefolia* Henry non (L.) Decne. & Planch. in Notes Econ. Bot. China 22. (1893).

Hubei: Badong, 932 (November 1885).

Ginseng, a small perennial native to the mountains of eastern Asia. The most famous of all Chinese medicinal plants, *ginseng* enhances the body's defense against stress and disease, it has been used in China for thousands of years to counteract weakness, fatigue and to enhance convalescence.

Panax japonicus (T. Nees) C. A. Mey (ARALIACEAE)

Aralia (*Panax*) *repens* (Steud. ex Maxim.) Matsum. Henry in Notes Econ. Bot. China 22. (1893).

Hubei: Badong, 3786.

The leaves of this species, according to Henry, were the source of a drug called *shen-yeh*, the drug extracted from the rhizome was called *chu-ken-ch'i*.

Panax notoginseng (Burkill) F. H. Chen (ARALIACEAE)

New variety. *Aralia quinquefolia* Decne. & Planch. var. *notoginseng* Burkill in Kew Bull. Misc. Inform. 7. (1902). *Panax* sp. Henry in Notes on Economic Botany of China 21. (1893).

Yunnan: Mengzi, 11407 (type of *Aralia quinquefolia* Decne. & Planch. var. *notoginseng* Burkill). Simao, 12259.

This is the famous Chinese drug *san ch'i*, Henry sent seeds to the Royal Botanic Gardens, Kew in 1896 instructing staff there to send a share of the seeds to the National Botanic Gardens, Glasnevin in Dublin.

Pandanus odorifer (Forssk.) Kuntze (PANDANACEAE)

Pandanus odoratissimus L. f. Henry in Trans. Asiat. Soc. Jap. 24 suppl: 100. (1896) (List Pl. Formosa), Wright in Journ. Linn. Soc. Bot. 36: 171. (1903). *Pandanus tectorius* Sol. var. *sinensis* Warburg in Engler, Pflanzenr. 9: 48. (1900).

Hainan: Environs of Kiungchow, 8289. **Taiwan:** Kaohsiung, without number.

02

A shrubby such branched species with numerous aerial roots. Widely distributed from the Indo-Malaysia region to tropical Australia and Polynesia. In Chinese Taiwan this species grows along the coast throughout the island and was known colloquially as *lin-ch'a* according to Henry. *Pandanus odoratissimus* was collected by the Glasnevin Taiwan Expedition (GTE 43) by the lighthouse at Oluanpi (South Cape of Henry and Wilson) in October 2004.

Pandanus sp. (PANDANACEAE)
Anonymous in Kew List of Determinations (volume 34).

Yunnan: Simao, 13604.

Panicum bisulcatum Thunb. (POACEAE)
Panicum acroanthum Steud. Rendle in Journ. Linn. Soc. Bot. 36: 327. (1904).

Sichuan: North Wushan, 7051.

Panicum brevifolium L. (POACEAE)
Rendle in Journ. Linn. Soc. Bot. 36: 328. (1904).

Hainan: Environs of Haikou, 8076.

An annual to 50 cm tall, widespread across tropical Africa and Asia.

Panicum decompositum R. Br. (POACEAE)
Rendle in Journ. Linn. Soc. Bot. 36: 330. (1904).
Isachne griffithii Henry non Munro in Trans. Asiat. Soc. Jap. 24 suppl: 106. (1896) (List Pl. Formosa).

Taiwan: Wanshoushan, 255, 1080.

Panicum khasianum Munro ex Hook. f. (POACEAE)
Rendle in Journ. Linn. Soc. Bot. 36: 331. (1904).

Yunnan: In woods near Mengzi at 1,700 metres, 9379.

Panicum miliaceum L. (POACEAE)
Rendle in Journ. Linn. Soc. Bot. 36: 331. (1904).

Hubei: Fang Xian, 6611.

Millet, cultivated in China since about 5,000 BC. Originally from Central Asia.

Panicum repens L. (POACEAE)
Rendle in Journ. Linn. Soc. Bot. 36: 332. (1904).

Taiwan: Wanshoushan, on the sea face in wet places 724. Kaohsiung, in moist places, 1171.

A perennial to 50 cm tall, pantropic in distribution and common in Taiwan.

Panicum sp. (POACEAE)
Anonymous in Kew List of Determinations (volume 34).

Yunnan: Mengzi, 13734.

Parabaena sagittata Miers. (MENISPERMACEAE)
Diels in Engler, Pflanzenr. iv. 94: 149. (1910), Forman in Kew Bull. 39: 106. (1984).

Yunnan: Mountains to the west of Simao at 1,650 metres, 11866. Simao at 1,200 metres, 12165.

A slender woody climber, native to north-east India, Bangladesh, Nepal, Sikkim, Bhutan, Myanmar, Thailand, Laos, Vietnam and in China, Yunnan.

Paraboea birmanica (Craib) C. Puglisi (GESNERIACEAE)
Dichiloboea birmanica (Craib) Stapf in Kew Bull. Misc. Inf. 357. (1913).

Yunnan: On cliffs and in forests near Simao between 1,500 and 1,800 metres, 12305a., 13112. Pu'er at 1,500 metres, 13396.

Distributed in Myanmar, western China and Thailand.

Paraboea crassifolia (Hemsley) B. L. Burtt (GESNERIACEAE) New species
Boea crassifolia Hemsley in Journ. Linn. Soc. Bot. 26: 232. (1890), Bretschneider in Hist. Eur. Bot. Disc. China 2: 789. (1898).

Hubei: Yichang, 3960. Liantuo, 6363.

Paraboea rufescens (Franch.) B. L. Burtt. (GESNERIACEAE)
Phylloboea henryi J. F. Duthie ex R. H. Beddome in Journ. Roy. Hort. Soc. 33: 96. (1908).

Yunnan: Mengzi at 1,200 metres, 9318 (isotype of *Phylloboea henryi* J. F. Duthie ex R. H. Beddome).

Distributed in western China, Thailand and Vietnam.

Paraboea sinensis (Oliv.) B. L. Burtt (GESNERIACEAE)
Phylloboea sinensis Oliver in Hooker's. Icon. Pl. xviii. t 1721. (1887), Forbes & Hemsley in Journ. Linn. Soc. Bot. 26: 225. (1890), Bretschneider in Hist. Eur. Bot. Disc. China 2: 789. (1898), Morley in Glasra 3: 80. (1979). *Chlamydoboea sinensis* (Oliv.) Stapf in Kew Bull. Misc. Inf. 345. (1913). *Chlamydoboea sinensis* (Oliv.) Stapf f. *macra* O. Stapf in Kew Bull. Misc. Inf. 355. (1913).

Hubei: In the Lung Wang Tung Glen near Yichang (June 1886), 1572. Yichang, 3958. Jianshi, 6017. In a ravine on Moji Shan, 4158. **Yunnan:** South of the Red River near Manmei at 1,433 metres, 9630 [isotype of *Chlamydoboea sinensis* (Oliv.) Stapf f. *macra* O. Stapf]. Mengzi at 1,800 metres, 9836. On wooded cliffs near Mengzi at 1,800 metres, 11223a. In mountain forest near Simao between 1,200 and 1,500 metres, 12305 [type of *Chlamydoboea sinensis* (Oliv.) Stapf f. *macrophylla* Stapf]. In forests near Simao between 1,200 and 1,500 metres 12161a., 12162b.

Paraboea sp. (GESNERIACEAE)
Phylloboea sp. Anonymous in Kew List of Determinations (volume 34).

Yunnan: Mengzi, 9385.

Paraboea swinhoei (Hance) B. L. Burtt (GESNERIACEAE)
Boea swinhoei Hance Henry in Trans. Asiat. Soc. Jap. 24 suppl: 68. (1896) (List Pl. Formosa).

Taiwan: Oluanpi, without number. Wanchin, 897.

Named *huo-ai*, as the dried leaves are used as tinder.

Paralamium griffithii (Hook. f.) Suddee & A. J. Paton (LAMIACEAE)

Paralamium gracile Dunn in Notes Roy. Bot. Gard. Edinburgh 8: 168. (1913), Dunn in Notes Roy. Bot. Gard. Edinburgh 6: 183. (1915).

Yunnan: In Pingbian, on the Daweishan Range at 1,640 metres, 10636 (isotype).

India (Assam) to China (Yunnan).

Paramignya confertifolia Swingle (RUTACEAE)

Paramignya citrifolia Anonymous non Oliver in Kew List of Determinations (volume 24).

Hainan: 32 km (20 miles) west of Haikou, 8153. Environs of Kiungchow, 8300, 8701.

Paraphlomis albiflora (Hemsl.) Hand.-Mazz. (LAMIACEAE) New species

Phlomis albiflora Hemsley in Forbes & Hemsley in Journ. Linn. Soc. Bot. 26: 304. (1890), Bretschneider in Hist. Eur. Bot. Disc. China 2: 789. (1898), Dunn in Notes Roy. Bot. Gard. Edinburgh 6: 185. (1915).

Hubei: In grass in a glen near Pingshanba (near Yichang, May 1886), 1575. Antelope Glen, 3576 (isosyntype of *Phlomis albiflora* Hemsley). Badong, 720. Liantuo, 1910.

Paraphlomis formosana (Hayata) T. H. Hsieh & T. C. Huang (LAMIACEAE)

Phlomis gracilis Hemsley in Journ. Linn. Soc. Bot. 26: 305. (1890), Bretschneider in Hist. Eur. Bot. Disc. China 2: 789. (1898), Dunn in Notes Roy. Bot. Gard. Edinburgh 6: 186. (1915). *Paraphlomis gracilis* (Hemsl.) Kudo in Mem. Fac. Sci. Taihoku Imp. Univ. 2: 210 (1929).

Hubei: Liantuo, 2089. Yichang, 4111 (isosyntype of *Phlomis gracilis* Hemsley).

Paraphlomis javanica (Blume) Prain (LAMIACEAE)

Phlomis rugosa Benth. Dunn in Notes Roy. Bot. Gard. Edinburgh 6: 186. (1915).

Yunnan: Mengzi at 1,950 metres, 10076, 11445. Fengchunling, 10076a. Simao at 1,300 metres, 12599.

Widespread in tropical and subtropical Asia.

Paraphlomis rugosa (Benth) Prain var. ***henryi*** Yamamoto (LAMIACEAE)

in Journ. Soc. Trop. Agric. 6: 555. (1934).

Paraphlomis javanica (Blume) Prain var. *henryi* (Yamamoto) C. Y. Wu ex H. W. Li C. Y. Wu in Flora Yunnanica 1: 633. (1977).

Taiwan: Wanchin, 1631 [isotype of *Paraphlomis rugosa* (Benth) Prain var. *henryi* Yamamoto].

Paraprenanthes polypodiifolia (Franch.) C. C. Chang ex C. Shih (ASTERACEAE)

Lactuca polypodiifolia Franch. Anonymous in Kew List of Determinations (volume 34).

Yunnan: Mengzi, 10621.

Endemic to China and distributed in the western provinces.

Paraprenanthes sororia (Miq.) C. Shih (ASTERACEAE)

Lactuca sororia Miq. Forbes & Hemsley in Journ. Linn. Soc. Bot. 23: 484. (1888).

Hubei: Antelope Glen, 3586. Fang Xian, 7573. **Sichuan:** South Wushan, 7390.

Paraprenanthes umbrosa (Dunn) Sennikov (ASTERACEAE) New species

Lactuca umbrosa Dunn in Journ. Linn. Soc. Bot. 35: 513. (1903).

Yunnan: In shade on dry clay walls on ravines in forests near Simao, 11694 (type of *Lactuca umbrosa* Dunn).

Distributed in China (Yunnan) and Indochina.

Pararuellia flagelliformis (Roxb.) Bremek. (ACANTHACEAE)

Ruellia flagelliformis Roxb. Forbes & Hemsley in Journ. Linn. Soc. Bot. 26: 237. (1890).

Hubei: San You Dong Glen, 2346. Liantuo, 4641, 6342.

Parasenecio ainsliaeiflorus (Franch.) Y. L. Chen (ASTERACEAE) New species

Senecio leucanthemus Dunn in Journ. Linn. Soc. Bot. 35: 506. (1903).

Hubei: Badong, 4667 (syntype of *Senecio leucanthemus* Dunn). South Badong, 7332 (syntype of *Senecio leucanthemus* Dunn). Fang Xian, 7571 (type of *Senecio leucanthemus* Dunn). **Hunan:** Shimen, 7556 (syntype of *Senecio leucanthemus* Dunn).

This species was discovered by Henry in 1887, Franchet based his *Senecio ainsliaefolius* on material later collected by Farges in Sichuan during the 1890s.

Parasenecio auriculatus (DC.) J. R. Grant (ASTERACEAE)

Senecio dahuricus Schultz-Bip. Forbes & Hemsley in Journ. Linn. Soc. Bot. 23: 451. (1888).

Hubei: Badong, 2437. Fang Xian at 1,900 to 3,000 metres, 6806.

Parasenecio begoniifolius (Franch.) Y. L. Chen (ASTERACEAE) New species

Senecio begoniifolius Franchet in J. Bot. (Morot) 8: 358. (1894), Bretschneider in Hist. Eur. Bot. Disc. China 2: 785. (1898).

Sichuan: North Wushan, 7116 (lectotype of *Senecio begoniifolius* Franchet).

Discovered by Henry in the autumn of 1888 while traveling with his Chinese collectors on his north of the Yangtze expedition.

Parasenecio bulbiferoides (Hand.-Mazz.) Y. L. Chen (ASTERACEAE)

In the Harvard University Herbarium (GH.).

Sichuan: North Wushan, 7214 (autumn 1888).

Parasenecio profundorum (Dunn) Y. L. Chen (ASTERACEAE) New species

Senecio profundorum Dunn in Journ. Linn. Soc. Bot. 35: 507. (1903).

Hubei: On cliffs in South Badong, 5434 (type of *Senecio profundorum* Dunn). Xingshan, 7612, (syntype of *Senecio profundorum* Dunn).

Parasenecio rubescens (S. Moore) Y. L. Chen (ASTERACEAE)

Senecio rubescens S. Moore Anonymous in Kew List of Determinations (volume 24).

Hubei: Fang Xian, 6572.

Senecio rubescens was discovered by Dr. George Shearer of the London Missionary Society near Jiujing in Jiangxi province in 1873.

Parasenecio rufipilis (Franch.) Y. L. Chen (ASTERACEAE)

Senecio bulbiferus Forbes & Hemsley non Maxim. in Journ. Linn. Soc. Bot. 23: 450. (1888).

Hubei: Badong, 2541. Jianshi, 7467. **Sichuan:** North Wushan, 7087.

Paratinospora dentata (Diels) Wei Wang (MENISPERMACEAE) New species

Limacia sp. Henry in Trans. Asiat. Soc. Jap. 24 suppl: 17. (1896) (List Pl. Formosa). *Tinospora dentata* Diels in Engler, Pflanzenr. 46(iv. 94) 139. (1910), Yamamoto in Journ. Soc. Trop. Agr. 12: 243. (1940); in Trans. Nat. Hist. Soc. Taiwan 34: 4. (1944), Li in Woody Flora of Taiwan. 186. (1963); in Fl. Taiwan 2: 540. (1976), Forman in Kew Bull. 36: 382. (1982).

Taiwan: Wanchin, 152 (type).

According to Henry the tuberous root of this climbing shrub was used as a medicine and was called *chin-sheng*. Endemic to Taiwan.

Paratinospora sagittata (Oliv.) Wei Wang (MENISPERMACEAE) New species

Limacia sagittata Oliver in Hooker's Icon. Pl. t. 1749. (1888), Henry in Notes Econ. Bot. China 51. (1893), Bretschneider in Hist. Eur. Bot. Disc. China 2: 779. (1898), Diels in Bot. Jahrb. 29: 345. (1900), Dunn & Tutcher in Kew Bull. Misc. Inf. 31. (1912). *Tinospora sagittata* (Oliv.) Gagnepain in Bull. Soc. Bot. France. ser. 4, viii. 45. (1908), Diels in Engler, Pflantzenr. iv.-94., 138. (1910), Rehder & E. H. Wilson in Sargent Pl. Wilson. 1: 390. (1913), Yamamoto in Journ. Soc. Trop. Agr. 13: 40. (1941); in Taiwania 1: 31. (1948), Lien in Acta Phytotax. Sinica 13: 36. t. 1-6. (1975), Forman in Kew. Bull. 36: 383. (1982), Zheng in Journ. Wuhan Bot. Res. 2.(1): 45. (1984).

Hubei: Antelope Glen, 3431 (holotype of *Limacia sagittata* Oliver), 4137. Changyang, 5227.

According to Henry the tuberous roots of this climber were used in Chinese herbal medicine in the Three Gorges region where it was known as *ch'ing niu tan*. Native to China [Hubei, Sichuan, Guangxi and south-east Tibet (Xizang)] and northern Vietnam. The Hani (an ethnic minority) in southern Yunnan call this species *alsiqhaq* and use it to treat cases of gastric ulcers and laryngitis.

Paris chinensis Franch. (MELANTHIACEAE)

Wright in Journ. Linn. Soc. Bot. 36: 144. (1903).

Hubei: Changyang, 5230.

Paris delavayi Franch. (MELANTHIACEAE)

Paris henryi Diels in Bot. Jahrb. Syst. 29: 252. (1900).

Hubei: South Badong, 5380 (type of *Paris henryi* Diels).

Paris fargesii Franchet (MELANTHIACEAE) New species

Paris petiolata Baker var. *membranacea* Wight in Journ. Linn. Soc. Bot. 36: 145. (1903).

Hubei: South Badong, 5385 (type *Paris petiolata* Baker var. *membranacea* Wight), 5385a.

Paris fargesii was discovered by Henry in May 1888 while on the first part of his six month plant hunting expedition to the south of the Yangtze River, Franchet based this species on material later collected in the early 1890s by the French missionary, Père Paul Farges in Sichuan.

Paris polyphylla Sm. (MELANTHIACEAE)

Henry in Journ. China Br. Roy. Asiat. Soc. 22: 249. (1887) (Chinese Names of Plants), Diels in Bot. Jahrb. 29: 252. (1900), Wright in Journ. Linn. Soc. Bot. 36: 145. (1903).

Hubei: Badong, 206 (June 1885), 2432, 2831, 4083, 4088. South Badong, 5287, 5373, 5387, 5409. Changleping, 6337. Changyang, 7515. Without locality, 5380a. **Sichuan:** South Wushan, 5387a. Henry's Chinese collector with Pratt, no locality stated but may be Leshan, Emei Shan, Wa Shan or Kangding, 8827.

Paris polyphylla was discovered by Francis Buchnan Hamilton in Nepal (he was the first westerner to collect there at Narainhetty in March 1803. This species was gave its present binomial by James Edward Smith some ten years later. According to Henry it was known colloquially in the Three Gorges region as *i chih hua* or *ch'i yeh i chih hua*. The dried rhizome is used in Chinese herbal medicine to allieviate pain and to treat cases of snake bite.

Paris sp. No. 1. (MELANTHIACEAE)

Anonymous in Kew List of Determinations (volume 24).

Hubei: Badong, 1405. Liantuo, 2151. Changleping, 7840.

Paris sp. No. 2. (MELANTHIACEAE)

Anonymous in Kew List of Determinations (volume 34).

Hubei: Badong, 1405. Liantuo, 2151. Changleping, 7840. **Yunnan:** Mengzi, 10752.

Parmotrema perlatum (Huds.) Ach. (PARMELIACEAE)

Parmelia perlata (Huds.) Ach. Anonymous in Kew List of Determinations (volume 24).

Hubei: Badong, 5184. (Lichen).

Parnassia delavayi Franch. (CELASTRACEAE)

Diels in Bot. Jahrb. 29: 370. (1900).

Hubei: Xingshan at 1,400 metres, 6542. **Sichuan:** South Wushan, 7260.

A handsome perennial to 45 cm tall, bearing during solitary white flowers June to September, distributed through Gansu, Henan, Hubei, Hunan, Shaanxi and Sichuan. The roots are used medicinally.

Parnassia sp. (CELASTRACEAE)

Anonymous in Kew List of Determinations (volume 24).

Hubei: On the boundary of Xingshan and Fang Xian, 6542a.

Parnassia wightiana Wall. ex Wight & Arn. (CELASTRACEAE)

Forbes & Hemsley in Journ. Linn. Soc. Bot. 23: 272. (1887).

Hubei: Yichang, 2734. **Yunnan:** Mengzi, 9332.

Native to Bhutan, Nepal, China and Thailand.

Parochetus communis Buch.-Ham. ex D. Don (FABACEAE)

Anonymous in Kew List of Determinations (volume 34).

Yunnan: In Pingbian on the Daweishan Range at 1,950 metres, 9053.

The shamrock pea, a prostrate clover-like herb, distributed through Africa and Asia to Java.

Parogonum cynanchoides (Hemsley) Desjardins & J. P. Bailey (POLYGONACEAE) New species

Polygonum cynanchoides Hemsley in Journ. Linn. Soc. Bot. 26: 338. (1891).

Hubei: Zigui, without number.

Parsonsia alboflavescens (Dennst.) Mabb. (APOCYNACEAE)

Parsonsia spiralis Henry non Wallich ex G. Don in Trans. Asiat. Soc. Jap. 24 suppl: 60. (1896) (List Pl. Formosa). *Parsonsia helicandra* Hook. & Arn. Li in Woody Flora of Taiwan 790. (1963).

Taiwan: Oluanpi, 306, 615.

A twining shrub, native to tropical Asia. In Taiwan this species is often found climbing over rocks near the seashore.

Parthenocissus henryana (Hemsl.) Graebn. ex Diels & Gilg. (VITACEAE) New species

in Bot. Jahrb. 24: 464. (1900), Gagnepain in Sargent Pl. Wilson. 1: 101. (1913), E. H. Wilson in A Naturalist in Western China 1: 29. (1913), Morley in Glasra 3: 80. (1979), Zheng in Journ. Wuhan Bot. Res. 2.(1): 143. (1984), Nelson in Pim, The Wood & the Trees 233. (1984), Bean in Trees & Shrubs 3: 93. (1992), Nelson in A Heritage of Beauty 316. (2000), O' Brien in Augustine Henry, An Irish Plant collector in China 27. (2002).

Vitis henryana Hemsley in Journ. Linn. Soc. Bot. 23: 132. (1886), Henry in Journ. China Br. Roy. Asiat. Soc. 22: 280. (1887) (Chinese Names of Plants), J. H. Veitch in Hortus Veitchii 384. (1906), Besant in Gard. Chron. ser. 3, 98: 334. (1935).

Hubei: Yichang, 599, 730, 4094, 4094a., 4094d., (isosyntype of *Vitis henryana* Hemsley). Donghu, 6397 (isosyntype of *Vitis henryana* Hemsley).

One of Henry's finest discoveries. A vigorous deciduous climber to 10 or more metres tall. Leaves composed of three to five leaflets beautifully marked with silver and pink variegations. In autumn the leaves turn a fiery red shade. *Parthenocissus henryana* was known colloquially in the Three Gorges region as *wu chao lung* and was discovered by Henry in 1885 and introduced to cultivation by Wilson when collecting for Messrs Veitch in 1900. Wilson stated it was common in the San You Dong Glen to the rear of the cave of the three pilgrims, five miles to the north of Yichang. In their *Novelties for 1908—1909* Messrs Veitch had plants available for three shillings and six pence each. *Parthenocissus henryana* was collected by the Glasnevin Central China Expedition (GCCE 622) in Dragon Gate River Forest Park in Xingshan in October 2004.

Parthenocissus semicordata (Wall.) Planch. (VITACEAE)

Vitis himalayana (Royle) Brandis Anonymous in Kew List of Determinations (volume 34).

Yunnan: Mengzi, 9285, 9438, 9507. Simao, 12591.

Parthenocissus tricuspidata (Sieb. & Zucc.) Planch. (VITACEAE)

Diels & Gilg. in Bot. Jahrb. 29: 464. (1900).

Vitis inconstans Miq. Henry in Trans. Asiat. Soc. Jap. 24 suppl: 28: (1896) (List Pl. Formosa).

Hubei: On the way to Lung Wang Tung (near Yichang), 4275. Lung Wang Tung valley, 3572. Yichang, 3572a. Liantuo, 3026, 3923, 4557, 4617. Xingshan, 6500. **Sichuan:** North Wushan, 7125. **Taiwan:** Wanshoushan, 1836.

A vigorous deciduous climber, scrambling its way through trees to 20 metres tall. Native to Japan and China and introduced to cultivation by John Gould Veitch in 1862. It is one of the most popular of all temperate climbers and also one of the most beautiful when carrying its autumnal cloak of scarlet.

Paspalum distichum L. (POACEAE)

Rendle in Journ. Linn. Soc. Bot. 36: 319. (1904).

Paspalum sp. Henry in Trans. Asiat. Soc. Jap. 24 suppl: 106. (1896) (List Pl. Formosa).

Taiwan: Kaohsiung Lagoon, growing in a few inches of salt water, 1036. Kaohsiung, growing in wet places, 1062. Kaohsiung, growing in water, 1900.

A perennial species to 50 cm tall, native to tropical and subtropical regions worldwide.

Paspalum scrobiculatum L. (POACEAE)

Rendle in Journ. Linn. Soc. Bot. 36: 320. (1904).

Paspalum sp. Henry in Trans. Asiat. Soc. Jap. 24 suppl: 106. (1896) (List Pl. Formosa).

Taiwan: Kaohsiung, 736, 1164.

Perennial to 70 cm tall, native to tropical Asia and the Pacific Islands. A common weed of tea plantations in Taiwan.

Paspalum thunbergii Kunth ex Steud. (POACEAE)

Paspalum scrobiculatum Rendle non L. in Journ. Linn. Soc. Bot. 36: 320. (1904), in part.

Hubei: Yichang, 3986a. Badong, 4816.

Native to China and Japan.

Passiflora altebilobata Hemsley (PASSIFLORACEAE) New species

in Kew Bull. Misc. Inf. 17. (1908).

Yunnan: Simao, 12987 (type), 12987a., (isotype).

Distributed in China (Yunnan) and Vietnam, this species was later collected in Simao by Joseph Rock.

Passiflora cochinchinensis Spreng. (PASSIFLORACEAE)

Passiflora hainanensis Hance Anonymous in Kew List of Determinations (volume 24).

Hainan: Environs of Kiungchow, 8251. At Nodoa in the interior of the island, 8448.

Passiflora cupiformis Masters (PASSIFLORACEAE) New species

in Hooker's Icon. Pl. xviii. t. 1768. (1988), Rehder & E. H. Wilson in Sargent Pl. Wilson. 2: 408. (1916), de Wilde in Blumea 20: 237. (1972), Bao in Acta Phytotax. Sin. 58. (1984).

Passiflora franchetiana Hemsley in Hooker's Icon Pl. xxvii. tt. 2623. (1899).

Yunnan: In woods near Mengzi at 1,800 metres, 10282 (type of *Passiflora franchetiana* Hemsley), 11192 (syntype of *Passiflora franchetiana* Hemsley).

Distributed in China and Vietnam.

Passiflora henryi Hemsley (PASSIFLORACEAE) New species

in Hooker's Icon. Pl. 27: t. 2622. (1899), Rehder & E. H. Wilson in Sargent Pl. Wilson. 2: 408. (1916), de Wilde in Blumea 20: 238. (1972), Bao in acta Phytotax. Sin. 22: 58. (1984).

Yunnan: In woods near the plains of Mengzi at 1,800 metres, 10252 (type).

Endemic to Yunnan. Hemsley cited A. Henry 10282 as the type of his *Passiflora henryi* the correct number is A. Henry 10252.

Passiflora wilsonii Hemsley (PASSIFLORACEAE) New species

in Kew Bull. Misc. Inf. 17. (1908), de Wilde in Blumea 20: 235. (1972), Bao in Acta Phytotax. Sin. 21: 59. (1983).

Yunnan: Mengzi, 9820. Mountain forest to the south of Simao at 1,650 metres, 11791 (isotype).

A handsome evergreen climber to 5 metres tall, the Hani (an ethnic group) of Xishuangbanna in southern Yunnan call this species *pavqcaiv* and use it to treat rheumatism and malaria. The distribution of *Passiflora wilsonii* extends from north-east India to Myanmar, Thailand (where it was collected by Kerr), Laos and Vietnam. In China Wilson's passion flower is distributed in Guizhou, Tibet (Xizang) and Yunnan.

Patis obtusa (Stapf ex Oliver) Romasch., P. M. Peterson & Soreng (POACEAE) New species

Oryzopsis obtusa Stapf ex Oliver in Hooker's. Icon. Pl. t. 2393. (1895), Bretschneider in Hist. Eur. Bot. Disc. China 2: 793. (1898), Pilger in Engl. Bot. Jahrb. xxix. 224. (1900), Rendle in Journ. Linn. Soc. Bot. 36: 383. (1904).

Hubei: Antelope Glen, 3507. Liantuo, 3896.

Distributed through China and Japan (south).

Patrinia heterophylla Bunge (CAPRIFOLIACEAE) New species

Patrinia angustifolia Hemsley in Journ. Linn. Soc. Bot. 23: 396. (1888), Bretschneider in Hist. Eur. Bot. Disc. China 2: 785. (1898).

Hubei: By the monastery in the Monastery Valley near Yichang (July 1886), 1657 (type of *Patrinia angustifolia* Hemsley). Yichang, 4281 (isotype of *Patrinia angustifolia* Hemsley). Liantuo, 4578.

This species was collected by the Sino-American Expedition to the Shennongjia Forest District in 1980.

Patrinia scabiosifolia Fisch. ex Trevir. (CAPRIFOLIACEAE)

Forbes & Hemsley in Journ. Linn. Soc. Bot. 23: 397. (1888).

Hubei: Badong, 8, 9 (collected September 1885, received Kew March 12th 1886), 1410, 2530, 2575, 3144, 4921, 5003. Yichang, 454, 2057.

A handsome perennial to 2 metres tall, native to Siberia, China, Indochina and Japan and is abundant in the Three Gorges region. This species is cultivated in the order beds at the National Botanic Gardens, Glasnevin from collections made by the Glasnevin Central China Expedition in Xingshan (GCCE 466) in September 2004 and on Tianthu Shan in Changyang (GCCE 818) in October of the same year.

Patrinia sp. No. 1. (CAPRIFOLIACEAE)

Anonymous in Kew List of Determinations (volume 24).

Hubei: Badong, 2399. **Sichuan:** South Wushan, 7222.

Patrinia sp. No. 2. (CAPRIFOLIACEAE)

Anonymous in Kew List of Determinations (volume 34).

Hubei: Badong, 2399. **Sichuan:** South Wushan, 7222. **Yunnan:** Mengzi, 9097, 9772.

Patrinia villosa (Thunb.) Dufr. (CAPRIFOLIACEAE)

Henry in Journ. China Br. Roy. Asiat. Soc. 22: 239. 1887 (Chinese Names of Plants), Forbes & Hemsley in Journ. Linn. Soc. Bot. 23: 398. (1888), Henry in Trans. Asiat. Soc. Jap. 24 suppl: 51. (1896) (List Pl. Formosa).

Patrinia dielsii Graebner in Bot. Jahrb. 29: 597. (1900).

Hubei: Yichang, 117, 117c., 118a., 149, 355, 928, 986, 2091. Antelope Glen, 3258. Badong, 535, 2426, 2532, 2557, 4837, 5017, 6147. Changyang, 7767. Donghu, 6147a. **Sichuan:** South Wushan, 7223, 7264. Fang Xian, 6831 (syntype of *Patrinia dielsii* Graebner). **Taiwan:** Wanchin, 1633.

Known colloquially in the districts of Changyang, Badong and Yichang as *yen chih ma*, according to Henry. It is also called the *bai jiang cao* and is currently used to treat cases of mumps. Distributed in China and Japan.

Pauldopia ghorta (Buch.-Ham ex G. Don) Steenis (BIGNONIACEAE)

Tecoma bipinnata Coll. & Hemsl. Anonymous in Kew List of Determinations (volume 34).

Yunnan: Simao, 12447, 13020.

Distributed in Nepal, India, Sri Lanka, Myanmar, China (Yunnan), Thailand, Laos and Vietnam.

Paulownia fargesii Dode (PAULOWNIACEAE)

Rehder in Sargent Pl. Wilson. 1: 575. (1913).

Paulownia fortunei Henry non (Seem.) Hemsl. in Flora & Sylva 1: 217. (1903). *Paulownia tomentosa* (Thunb.) Steud. var. *fargesii* (Dode) Henry in Elwes and Henry The Trees Gt. Brit. & Irel. 6: 1493. (1912). *Paulownia tomentosa* Walsh non (Thunb.) Steud. in Aug. Henry Forst. Herb. 111. (1957).

Hubei: Wild on cliffs at Jianshi, 5346a. **Yunnan:** Mengzi, 10831.

Paulownia fargesii was discovered by Augustine Henry while travelling on the first part of his six month plant hunting expedition in June of 1888 and was described by Dode from material later collected in north-east Sichuan by the French missionary, Père Paul Farges during the 1890s. It first flowered in cultivation in the nursery of M. Boucher in Paris during the summer of 1905 from seeds sent by Farges to Maurice de Vilmorin about 1896. From Vilmorin a tree was received at Kew in the autumn of 1908, by 1929 it was about 13 metres tall and flowered in the spring of that year. This species was collected by the Glasnevin Central China Expedition (GCCE 631) in Dragon Gate River Forest Park, Xingshan in October 2004. There are young seedlings from this collection at Glasnevin at present and there are fine young vigorous trees at the Royal Botanic Gardens, Kew from one of their Sichuan expeditions. Distributed in central and western China and Vietnam.

Paulownia tomentosa (Thunb.) Steud. (PAULOWNIACEAE)

Henry in Elwes and Henry The Trees Gt. Brit. & Irel. 6: 1493. (1912).

Paulownia imperialis Sieb. & Zucc. Henry in Journ. China Br. Roy. Asiat. Soc. 22: 276. (1887) (Chinese Names of Plants); in Notes Econ. Bot. China 60. (1893), Forbes & Hemsley in Journ. Linn. Soc. Bot. 26: 180. (1890).

Hubei: Wild on cliffs at Jianshi, 5346.

Henry's favourite tree, see *The Garden* (1902). A tree from Wilson's 769, grown from seeds collected in Fang Xian, Hubei in October 1907 still grows in the old Chinese shrubbery at the National Botanic Gardens, Glasnevin where it puts on a spectacular display each spring if the winter before has been a mild one. In the Three Gorges region *Paulownia tomentosa* was known colloquially as the *p'ao t'ung* according to Henry. This tree was introduced to Japan by Buddhist monks sometime after the 8th century.

Pavetta hongkongensis Bremekamp (RUBIACEAE) New species

in Repert Spec. Nov. Regni Veg. 37: 104. (1934).

Hainan: Environs of Haikou, 7962 (type). Without locality, 8508.

Pavetta scabrifolia Bremekamp (RUBIACEAE) New species

in Repert Spec. Nov. Regni Veg. 37: 100. (1934).

Yunnan: Simao, 11934 (type), 12416, 12701 (isotype).

Pavetta tomentosa Roxb. ex Sm. var. *glabrescens* (Kurz.) Bremek. (RUBIACEAE)

Pavetta indica L. var. *polyantha* (Wall.) Hook. f. Hutchinson in Sargent Pl. Wilson. 3: 412. (1916).

Yunnan: In Pingbian on the Daweishan Range at 1,640 metres, 10777. On the tea hills to the east of Simao 11934. Mountains to the south of Simao at 1,480 metres 11934a.

02

Pecteilis henryi Schlecter (ORCHIDACEAE) New species

in Repert. Spec. Nov. Regni Veg. Beih. 4: 76. (1919).

Yunnan: Simao, 12534 (type), 12534a., (isotype).

Distributed in Myanmar, China (Yunnan), Laos, Thailand and Cambodia.

Pecteilis susannae (L.) Raf.. (ORCHIDACEAE)

Platanthera susannae (L.) Lindl. Forbes & Hemsley in Journ. Linn. Soc. Bot. 57. (1903).

Yunnan: On grassy slopes near Mengzi at 1,800 metres, 11095. Simao, 13555.

Pedicularis artselaeri Maxim. (OROBANCHACEAE)

Forbes & Hemsley in Journ. Linn. Soc. Bot. 26: 205. (1890), Diels in Bot. Jahrb. 29: 572. (1900), W. Limpricht in Repert. Spec. Nov. Regni Veg. 20: 198. (1924).

Hubei: South Badong, 5326.

Pedicularis comptoniifolia Franch. ex Maxim. (OROBANCHACEAE)

Anonymous in Kew List of Determinations (volume 34).

Yunnan: Mengzi, 9708.

Native to China (Sichuan and Yunnan) and Myanmar.

Pedicularis conifera Maximowicz ex Hemsley (OROBANCHACEAE) New species

in Bull. Acad. St. Petersb. 32: f. 193. (1888); in Mel. Biol. 12: 193. (1888); in Journ. Linn. Soc. Bot. 26: 206. (1890); in Journ. Linn. Soc. Bot. 26: 206. (1890), Bretschneider in Hist. Bot. Disc. China 2: 788. (1898), Diels in Bot. Jahrb. 29: 572. (1900), W. Limprich in Repert Spec. Nov. Regni Veg. 20: 243. (1924), Li in Proc. Acad. Nat. Sci. Philad. 278. (1948).

Hubei: South Badong, 7625 (type).

Pedicularis filicifolia Hemsley (OROBANCHACEAE) New species

in Journ. Linn. Soc. Bot. 26: 208. (1890), Bretschneider in Hist. Bot. Disc. China 2: 789. (1898), Diels in Bot. Jahrb. 29: 571. (1900), Bonati in Bull. Soc. Bot. France liv. 185. (1907), W. Limprich in Repert. Spec. Nov. Regni Veg. 20: 257. (1924).

Hubei: Xingshan, 6980. South Badong, 6105 (type).

Pedicularis gracilis Wall. ex Benth. (OROBANCHACEAE)

Diels in Bot. Jahrb. 29: 571. (1900), W. Limprich in Repert. Spec. Nov. Regni. Veg. 20: 259. (1924).

Hubei: South Badong, without number. **Yunnan:** Mengzi, 9710.

Distributed in Afghanistan, Pakistan, Nepal India, Bhutan, and western China.

Pedicularis henryi Maximowicz (OROBANCHACEAE) New species

in Bull. Acad. St. Petersb. 32: 560. f. 54. (1888),

Maxim in Mel. Biol. 12: 833. n. 73. fig. 54. (1888), Forbes & Hemsley in Journ. Linn. Soc. Bot. 26: 209. (1890), Bretschneider in Hist. Eur. Bot. Disc. China 2: 789. (1898), Diels in Bot. Jahrb. 29: 572. (1900), W. Limprich in Repert. Spec. Nov. Regni Veg. 20: 240. (1924), Morley in Glasra 3: 80. (1979), Pai in Contr. Inst. Bot. Nat. Acad. Peiping 2: 212. (1934), Li in Proc. Acad. Nat. Sci. Philad. cl. 27. f. 105. (1949).

Hubei: Liantuo, 2216. South Badong, 4687, 6155 (type). Changleping, 6304.

Pedicularis oxycarpa Franch. ex Maxim. (OROBANCHACEAE)

Diels in Bot. Jahrb. 29: 572. (1900).

Sichuan: North Wushan, without number.

Pedicularis resupinata L. (OROBANCHACEAE)

Forbes & Hemsley in J. Linn.Soc. Bot. 26: 214. (1890), Diels in Bot. Jahrb. 29: 572. (1900).

Hubei: Badong, 112, 1048 (December 1885), 2547, 4940, 6523, 6523a. Fang Xian, 6866.

Pedicularis rudis Maxim. (OROBANCHACEAE)

Forbes & Hemsley in Journ. Linn. Soc. Bot. 26: 215. (1890), W. Limprich in Repert. Spec. Nov. Regni Veg. 20: 228. (1924), Diels in Bot. Jahrb. 29: 572. (1900).

Hubei: Fang Xian, 6741.

Pedicularis salviiflora Franch. (OROBANCHACEAE)

W. Limprich in Repert. Spec. Nov. Regni Veg. 20: 212. (1924).

Yunnan: Mengzi, 9282, 9822, 9822a.

Discovered by Père Delavay in north-west Yunnan. Endemic to China and distributed in Sichuan and Yunnan.

Pedicularis sp. No. 1. (OROBANCHACEAE)

Anonymous in Kew List of Determinations (volume 24).

Hubei: Badong, 1758. **Sichuan:** Henry's Chinese collector with Pratt, no locality stated but may be Lshan, Emei Shan, Wa Shan or Kangding, 8868.

Pedicularis sp. No. 2. (OROBANCHACEAE)

Anonymous in Kew List of Determinations (volume 34).

Yunnan: Mengzi, 9597, 9707, 9821, 10731, 10732. Simao, 12613, 13413.

Pedicularis spicata Pall. (OROBANCHACEAE)

Forbes & Hemsley in Journ. Linn. Soc. Bot. 26: 216. (1890), Diels in Bot. Jahrb. 29: 572. (1900), W. Limprich in Repert. Spec. Nov. Regni Veg. 20: 205. (1924).

Hubei: Fang Xian, 6744.

Pedicularis superba Franch. ex Maxim. (OROBANCHACEAE)

Anonymous in Kew List of Determinations (volume 24).

Sichuan: Henry's Chinese collector with Pratt, no locality stated but may be Leshan, Emei Shan, Wa Shan or Kangding, 8908.

Pedicularis tenuisecta Franch. ex Maxim. (OROBANCHACEAE)

W. Limprich in Repert. Spec. Nov. Regni Veg. 20: 240. (1924).

Yunnan: Yuanjiang, 13230.

Discovered by Delavay in north-west Yunnan. Native to western China and Laos.

Pedicularis torta Maxim. (OROBANCHACEAE)

Forbes & Hemsley in Journ. Linn. Soc. Bot. 26: 218. (1890), Diels in Bot. Jahrb. 29: 571. (1900).

Sichuan: North Wushan, 7093.

Peliosanthes spp. (ASPARAGACEAE)

Anonymous in Kew List of Determinations (volume 34).

Yunnan: Mengzi, 9402, 11818. Simao, 12323, 12649, 12950, 13362.

Peliosanthes teta Andr. (ASPARAGACEAE)

Henry in Trans. Asiat. Soc. Jap. 24 suppl: 95. (1896) (List Pl. Formosa).

Taiwan: Wanchin, 1591.

Pellionia griffithiana Wedd. (URTICACEAE)

Forbes & Hemsley in Journ. Linn. Soc. Bot. 26: 481. (1899).

Yunnan: Mengzi, 9163b., 9163c., 9163d.

Distributed in India, Bhutan, Myanmar, China, Laos and Vietnam.

Pellionia pellucida (Raf.) Merr. (URTICACEAE)

Pellionia scabra Benth. Forbes & Hemsley in Journ. Linn. Soc. Bot. 26: 481. (1899).

Yunnan: Mengzi, 10958.

A monoecious perennial, found on moist forest floors in China, Vietnam and Japan.

Pellionia spp. (URTICACEAE)

Anonymous in Kew List of Determinations (volume 34).

Yunnan: Mengzi, 13701, 11261, 13779. Simao, 11706, 12112, 12351, 12678, 12866, 12873.

Pellionia viridis C. H. Wright (URTICACEAE) New species

in Journ. Linn. Soc. Bot. 26: 481. (1899).

Hubei: Yichang, 4098 (isosyntype).

Peltigera aphlhosa Ach. (PELTIGERACEAE)

Wei in An Enumeration of Lichens in China 181. (1991).

Peltidea aphthosa Ach. Meller in Bull. Herb. Boiss. 1: 235. (1893).

Hubei: Fang Xian, 6920.

Peltigera membranacea (Ach.) Nyl. (PELTIGERACEAE)

Peltegera canina Hoffm. var. *membranacea* Ach. Muell.-Arg. in Bull. Herb. Boissier 1: 235. (1893).

Hubei: Fang Xian at 2,500 to 2, 900 metres, 6801.

Peltigera polydactyla (Necker) Hoffm. f. *lophyra* (Ach.) Nyl. (PELTIGERACEAE)

Wei in An enumeration of Lichens in China 185. (1991).

Peltigera polydactyla Hoffm. var. *dissecta* Mull.-Arg. Muell.-Arg. in Bull. Herb. Boissier 1: 236. (1898)

Hubei: Xingshan, 6472.

Peltigera praetexta (Flk. ex Sommerf.) Zopf. (PELTIGERACEAE)

Wei in An enumeration of Lichens in China 186. (1991).

Peltigera rufescens Hoffm. var. *praetexta* (Flk.) Nyl. Muell.-Arg. in Bull. Herb. Boissier 1: 235. (1893).

Hubei: Fang Xian, 6582.

Pemphis acidula J. R. & G. Forst. (LYTHRACEAE)

Henry in Trans. Asiat. Soc. Jap. 24 suppl: 44. (1896) (List Pl. Formosa), Li in Woody Flora of Taiwan 628. (1963).

Taiwan: Kaohsiung, 1633.

A maritime shrub or small tree, widely dispersed on tropical coasts of the Old World and disatributed on the south coast of Taiwan. This species was collected by the Glasnevin Taiwan Expedition (GTE 4) in Banana Bay Coastal Forest near Kenting in October 2004.

Pentapanax henryi Harms (ARALIACEAE) New species

in Bot. Jahrb. Syst. 23: .21. (1896) 29: 489. (1900), Bretschneider in Hist. Eur. Bot. Disc. China 2: 784. (1898), Morley in Glasra 3: 80. (1979), Nelson in A Heritage of Beauty 316. (2000), O' Brien in Augustine Henry, An Irish Plant Collector in China 27. (2002).

Sichuan: North Wushan, 7035 (type).

A large shrub to 4 metres tall, collected once only by Henry in Sichuan, Wilson, who had more time on his hands made several collection in the same province at Baoxing in August 1908, Kangding in the same year and Pan lan Shan in October 1910. This very handsome deciduous foliage plant was shown at the Chelsea Flower Show in an exhibit based on Augustine Henry's botanical activities in China in May of 2002. The same plant appears to be hardy on the Chinese slope at the National Botanic Gardens, Glasnevin.

Pentaphylacaceae (PENTAPHYLACEAE)

Ternstroemiaceae Anonymous in Kew List of Determinations (volume 34).

Yunnan: Fengchunling, 11853. Simao, 12547.

Penthorum sedoides L. (CRASSULACEAE)

Forbes & Hemsley in Journ. Linn. Soc. Bot. 23: 288. (1887).

Hubei: Yichang, 75, 451, 2283, 2724. Liantuo, 2692,

4598. Badong, 3186. **Yunnan:** Mengzi, 9647.

Peperomia japonica Makino (PIPERACEAE)

Peperomia dindyulensis Henry non Miq. in Trans. Asiat. Soc. Jap. 24 suppl: 77. (1896) (List Pl. Formosa).

Taiwan: Oluanpi, 580. Wanchin, 481. Kaohsiung, 1027.

A succulent herb to 30 cm tall, native to Japan and China (Taiwan).

Peperomia reflexa A. Dietr. (PIPERACEAE)

Forbes & Hemsley in Journ. Linn. Soc. Bot. 26: 366. (1891).

Sichuan: Henry's Chinese collector with Pratt, no locality stated but may be any of the following, Leshan, Emei Shan, Wa Shan or Kangding, 8889.

Peperomia spp. (PIPERACEAE)

Anonymous in Kew List of Determinations (volume 34).

Yunnan: Mengzi, 9150, 9776, 9838. Simao, 12642.

Peracarpa carnosa (Wall.) Hook. f. & Thoms. (CAMPANULACEAE)

Barnesky & Lammers in Bot. Bull. Acad. Sin. 38: 53. (1997).

Campanula circaeoides F. Schmidt Forbes & Hemsley in Journ. Linn. Soc. Bot. 26: 9. (1889).

Hubei: Antelope Glen, 3461. **Sichuan:** South Wushan, 5642, 5656. **Yunnan:** Mengzi, 9449.

A small perennial, bearing blue flowers from April to early August. Distributed through the Pacific rim from southern Sakhalin, Japan, South Korea, the Philippines, Papua New Guinea, China (including Taiwan) and northern Myanmar.

Peranema aspidioides (Blume) Mett. (DRYOPTERIDACEAE)

Diacalpe aspidioides Blume Anonymous in Kew List of Determinations (volume 34).

Yunnan: Mengzi, 11534. Simao, 12824.

A handsome fern, which is found over much of northern India, Sri Lanka, southern China, Indochina, the Philippines and Malaysia.

Pericampylus formosanus Diels (MENISPERMACEAE)

Pericampylus incanus Henry non (Colebr.) Miers. in Trans. Asiat. Soc. Jap. 24 suppl: 16. (1896) (List Pl. Formosa), Diels in Engler, Pflanzenr. iv. 94: 217. (1910).

Taiwan: Wanchin, 559.

A scandent shrub, native to southern China (including Taiwan) and Japan (south).

Pericampylus glaucus (Lam.) Merr. (MENISPERMACEAE)

Pericampylus incanus (Colebr.) Miers. Diels in Engler, Pflanzenr. iv. 94: 217. (1910).

Yunnan: South of the Red River on Fenchenglin at 2,100 metres, 10656. In woods near Mengzi at 1,500 metres

10656a., 10656b. South-west of Simao at 1,350 metres, 10656f.

Distributed in western and southern China.

Perilla frutescens (L.) Britt. (LAMIACEAE)

Perilla ocymoides L. Forbes & Hemsley in Journ. Linn. Soc. Bot. 26: 279. (1890), Dunn in Notes Roy. Bot. Gard. Edinburgh 6: 154. (1915).

Hubei: Yichang, 459, 2348, 2781. Badong, 4971.

Colloquially known as *su ma,* according to Henry it was cultivated at Badong for oil obtained from its seeds.

Perilla frutescens (L.) Britt. var. *crispa* (Thunb.) Hand.-Mazz. (LAMIACEAE)

Perilla nankinensis Decne. Forbes & Hemsley in Journ. Linn. Soc. Bot. 26: 279. (1890), Dunn in Notes Roy. Bot. Gard. Edinburgh 6: 154. (1915).

Hubei: Yichang, 692, 2258, 2258a. On the banks of the Yangtze at Yichang, 4356. Badong, 1050 (December 1885). Liantuo (October 1886), 3023.

An aromatic purple-leaved annual grown as a herb in both China and Japan and as an annual summer bedding plant in Europe. In China this is the *zi su ye* and it is used to cure symptoms like fever, chills, headache, nasal congestion or coughs. It also alleviates seafood poisoning. It is often used with other herbs eg. *sheng jiang* (*Zingiber officinalis*) and when mixed with *huang lian* (*Coptis chinensis*) it is used for morning sickness and irritability during pregnancy.

Perilla sp. (LAMIACEAE)

Anonymous in Kew List of Determinations (volume 34).

Yunnan: Mengzi, 11182.

Periploca calophylla (Wight) Falc. (APOCYNACEAE)

Forbes & Hemsley in Journ. Linn. Soc. Bot.26: 101. (1889), Schneider in Sargent Pl. Wilson. 3: 343. (1917), Zheng in Journ. Wuhan Bot. Res. 2.(1): 183. (1984).

Hubei: Antelope Glen, 3419. Yichang, 4119. **Yunnan:** Mengzi, 9130. Climbing on trees in a wood near Mengzi at 1,800 metres, 11311.

Native to India, the Himalaya, western and central China and Vietnam.

Peristrophe floribunda (Hemsl.) C. Y. Wu & H. S. Lo (ACANTHACEAE) New species

in Flora Hainanica 3: 595. (1974).

Dicliptera crinata (Thunb.) Nees var. *floribunda* Hemsley in Journ. Linn. Soc. Bot. 26: 248. (1890).

Hainan: On Liang Shan, 16 km (10 miles) east of Haikou, 8200.

Peristrophe japonica (Thunb.) Bremek. (ACANTHACEAE)

Dicliptera crinata (Thunb.) Nees. Forbes & Hemsley in Journ. Linn. Soc. Bot. 26: 248. (1890).

Hubei: Yichang, 2191, 4126. At Shih Pan Shan [off the

02

Xiling (Yichang) Gorge], 4191.

Peristrophe tinctoria Nees (ACANTHACEAE)

Henry in Journ. China Br. Roy. Asiat. Soc. 22: 251. 1887 (Chinese Names of Plants), Forbes & Hemsley in Journ. Linn. Soc. Bot. 26: 248. (1890).

Hubei: San You Dong Glen, 1695. Yichang, 3991, 4121, 4153.

Peristylus constrictus (Lindl.) Lindl. (ORCHIDACEAE)

K. Y. Lang in Acta Phytotax. Sin. 25: 450. (1987).

Yunnan: Simao, 13126.

Native to India, the Himalaya, Myanmar, Yunnan, Thailand, Cambodia and Vietnam.

Peristylus densus (Lindl.) Santapau & Kapadia (ORCHIDACEAE)

Habenaria buchneroides Schlecter in Repert. Spec. Nov. Regni Veg. Beih. 4: 46. (1919).

Yunnan: Simao, 13557 (isotype *Habenaria buchneroides* Schlecter).

Peristylus goodyeroides (D. Don) Lindl. (ORCHIDACEAE)

Rolfe in Journ. Linn. Soc. Bot. 36: 54. (1903).

Habenaria goodyeroides D. Don Henry in Trans. Asiat. Soc. Jap. 24 suppl: 93. (1896) (List Pl. Formosa).

Taiwan: Wanshoushan, 1126. Oluanpi, 1211. Wanchin, 1572. **Yunnan:** Mengzi, 9212. In mountains near Mengzi at 1,500 to 1,750 metres, 11127. Simao, 12963.

A terrestrial orchid to 60 cm tall, widespread in tropical and subtropical Asia.

Peristylus lacertiferus (Lindl.) J. J. Smith (ORCHIDACEAE)

K. Y. Lang in Acta Phytotax. Sin. 25: 453. (1987).

Peristylus chloranthus Lindl. Rolfe in Journ. Linn. Soc. Bot. 36: 53. (1903).

Taiwan: Wanchin, 895, 1572a. **Yunnan:** Simao, 12409.

A terrestrial orchid to 30 cm tall, a widespread species with a range across Sikkim, India, Myanmar, southern China, Thailand, Indochina, Japan, the Philippines and Malaysia.

Peristylus viridis (L.) Lindl. (ORCHIDACEAE)

Rolfe in Journ. Linn. Soc. Bot. 36: 54. (1903).

Hubei: Xingshan at 2,600 to 2,900 metres, 6996. Fang Xian, 6874.

Perotis latifolia Ait. (POACEAE)

Henry in Trans. Asiat. Soc. Jap. 24 suppl: 107. (1896) (List Pl. Formosa), Rendle in Journ. Linn. Soc. Bot. 36: 343. (1904).

Taiwan: Kaohsiung Spit, 1043, 1110.

Perrottetia racemosa (Oliv.) Loesener (CELASTRACEAE) New species

in Bot. Jahrb. xxiv. 201. (1893), Rehder & E. H. Wilson

in Sargent Pl. Wilson 2: 359. (1916), Hu in Journ. Arnold Arb. 31: 263. (1950), Zheng in Journ. Wuhan Bot. Res. 2.(1): 126. (1984).

Ilex racemosa Oliver in Hooker's Icon. Pl. 19: t. 1863. (1889).

Hubei: Antelope Glen, 3309. Yichang, 3527 (type of *Ilex racemosa* Oliver), 4117, 7189 (isosyntype of *Ilex racemosa* Oliver). Badong, 1863. Without locality, 4117a.

Discovered by Henry in the Antelope Glen near Yichang, the natives of this region called it the *t'ung-ts-ao*.

Persea ichangensis (Rehd. & Wils.) Kostermans (LAURACEAE) New species

in Reinwardtia 6: 192. (1962).

Machilus thunbergii Forbes & Hemsley non Sieb. & Zucc. in Journ. Linn. Soc. Bot. 26: 377. (1891), Henry in Notes Econ. Bot. China 43. (1893), Zheng in Journ. Wuhan Bot. Res. 2.(1): 55. (1984) .*Machilus ichangensis* Rehder & E. H. Wilson in Sargent Pl. Wilson 2: 621. (1916).

Hubei: Yichang, 1112, 2194, 3627, 7782. Yichang at the Lung-wang cave, 3427. Changyang, 5232. Antelope Glen, 7783. Without locality, 5232a. **Sichuan:** South Wushan, 5503.

A small evergreen tree, native to south-east Tibet (Xizang) and western and central China. *Persea ichangensis* was described from material collected by Wilson though it had been earlier (1885) discovered by Henry. It was also later introduced to cultivation by Wilson from central China and reintroduced by Forrest from Yunnan. It is common at low altitudes around Yichang where according to Henry it was known as *hsiao nanmu* (little nanmu) and makes a small tree to about 15 metres tall.

Persea robusta (W. W. Smith) Kosterm. (LAURACEAE) New species

Machilus robusta W. W. Smith in Notes Roy. Bot. Gard. Edinburgh 13: 169. (1921), Liou Ho in Laurac. Chine & Indoch. 52. (1934).

Yunnan: In woods near Mengzi at 1,800 metres, 10809 (syntype of *Machilus robusta* W. W. Smith).

An evergreen tree to 15 metres tall, distributed in southern Yunnan and also found in the provinces of Guizhou, Guangxi and Guangdong and in Myanmar .

Persea yunnanensis (H. Lec.) Kosterm. (LAURACEAE)

Machilus bracteata H. Lecompte Liou Ho in Laurac. Chine & Indoch. 58. (1934).

Yunnan: Simao, 11960, 11960c., 11960d.

Endemic to China and distributed in Sichuan and Yunnan.

Persicaria barbata (L.) H. Hara (POLYGONACEAE)

Polygonum barbatum L. Forbes & Hemsley in Journ. Linn. Soc. Bot. 26: 334. (1891).

Hubei: Yichang, 601, 1010.

Widely spread in tropical and subtropical Africa and Asia.

Persicaria bungeana (Turcz.) Nakai (POLYGONACEAE)

Polygonum pensylvanicum Bunge Anonymous in Kew List of Determinations (volume 24).

Hubei: Yichang, 31.

Persicaria chinense (L.) H. Gross (POLYGONACEAE)

Polygonum chinense L. Forbes & Hemsley in Journ. Linn. Soc. Bot. 26: 335. (1891), Henry in Trans. Asiat. Soc. Jap. 24 suppl: 76. (1896) (List Pl. Formosa).

Hubei: Yichang, 155, 1063, 3203. Xiling (Yichang) Gorge, 3535. In the Lung Tung Chi Glen near Yichang (October 1886), 2904. **Hainan:** Without locality, 8355. **Taiwan:** Tamsui, 1452. Oluanpi, 299, 628. Wanchin, 412. Kagee, 809. **Yunnan:** Mengzi, 9078.

India to China and Indonesia.

Persicaria criopolitanana (Hance) Migo (POLYGONACEAE)

Polygonum criopolitanum Hance Forbes & Hemsley in Journ. Linn. Soc. Bot. 26: 336. (1891).

Hubei: Yichang, 885, 3232.

Persicaria decipiens (R. Br.) K. L. Wilson (POLYGONACEAE)

Polygonum serrulatum Lagasca Henry in Trans. Asiat. Soc. Jap. 24 suppl: 76. (1896) (List Pl. Formosa).

Taiwan: Kaohsiung, 1121. Oluanpi, 1286.

Persicaria dichotoma (Blume) Masam. (POLYGONACEAE)

Polygonum pedunculare Wall. ex Meisn. Forbes & Hemsley in Journ. Linn. Soc. Bot. 26: 344. (1891), Henry in Trans. Asiat. Soc. Jap. 24 suppl: 76. (1896) (List Pl. Formosa).

Hubei: Yichang, 2247, 2793. **Taiwan:** Kaohsiung, 2027.

Distributed from northern India to China and south through south-east Asia to north-eastern Australia.

Persicaria filiformis (Thunb.) Nakai (POLYGONACEAE)

Polygonum filiforme Thunb. Henry in Journ. China Br. Roy. Asiat. Soc. 22: 261. (1887) (Chinese Names of Plants). *Polygonum virginianum* Forbes & Hemsley non L. in Journ. Linn. Soc. Bot. 26: 352. (1891).

Hubei: Yichang, 4123, 3244 (in part). Liantuo, 2175. Badong, 4784.

According to Henry this species was known as *niu hsi* in the Three Gorges region.

Persicaria glabra (Willd.) M. Gómez (POLYGONACEAE)

Polygonum glabrum Willd. Forbes & Hemsley in Journ. Linn. Soc. Bot. 26: 340. (1891).

Hubei: Yichang, 178. Badong, 366, 367 (June 1885). **Hainan:** 8 km (5 miles) south of Kuingchow, 8182.

Distributed throughout tropical and subtropical Asia, Africa and America.

Persicaria hydropiper (L.) Delarbre (POLYGONACEAE)

Polygonum hydropiper L. Forbes & Hemsley in Journ. Linn. Soc. Bot. 26: 340. (1891), Henry in Trans. Asiat. Soc. Jap. 24 suppl: 76. (1896) (List Pl. Formosa).

Hubei: Yichang, 62, 67. Badong, 1036 (December 1885). **Taiwan:** Wanchin, 88.

Native to temperate and subtropical regions of the Northern Hemisphere to Australia. In Chinese herbal medicine the dried aerial parts of this plant are used to treat rheumatic arthritis. The Hani (an ethnic group) of southern Yunnan call this species *anljilpavqqai* and is used in the treatment of gasteroenteritis, dysentry and rheumatism.

Persicaria japonica (Meissn.) Nakai (POLYGONACEAE)

Polygonum japonicum Meissn. Forbes & Hemsley in Journ. Linn. Soc. Bot. 26: 341. (1891).

Hubei: Liantuo, 3057.

Distributed through China, Korea and Japan.

Persicaria lapathifolia (L.) Delarbre (POLYGONACEAE)

Polygonum lapathifolium L. Forbes & Hemsley in Journ. Linn. Soc. Bot. 26: 342. (1891), Henry in Trans. Asiat. Soc. Jap. 24 suppl: 76. (1896) (List Pl. Formosa). *Polygonum lanigerum* R. Br. Henry in Trans. Asiat. Soc. Jap. 24 suppl: 76. (1896) (List Pl. Formosa). *Polygonum nodosum* Pers. Forbes & Hemsley in Journ. Linn. Soc. Bot. 26: 343. (1891).

Hubei: Yichang, 53, 299. In grass in the San You Dong Glen, 1529 (in part). Badong, without number. **Taiwan:** Kaohsiung, 271, 1192. Oluanpi, 349, 349a., 583.

An annual species to almost 1.5 metres tall. Native to temperate areas of the Northern Hemisphere, Malesia to Australia.

Persicaria longiseta (Bruijin) Kitag. (POLYGONACEAE)

Polygonum blumei Meisn. Anonymous in Kew List of Determinations (volume 24).

Hubei: Yichang, 1672, 2261, 2372.

Native to the Himalaya, China, Korea, Japan and Malaysia.

Persicaria minor (Huds.) Opiz (POLYGONACEAE)

Polygonum minus Huds. Anonymous in Kew List of Determinations (volume 24).

Hubei: Yichang, 196, 2784. Liantuo, 2215. Badong, 2407.

Persicaria muricata (Meisn.) Nemoto (POLYGONACEAE)

Polygonum muricatum Meisn. Forbes & Hemsley in Journ. Linn. Soc. Bot. 26: 343. (1891).

Hubei: Jianshi, 7408.

Distributed through northern India, China, Korea and Japan.

Persicaria nepalensis (Meisn.) H. Gross (POLYGONACEAE)

Polygonum nepalense Henry non Meiss. in Journ.

China Br. Roy. Asiat. Soc. 22: 238. (1887) (Chinese Names of Plants). *Polygonum alatum* (D. Don) Buch.-Ham. ex Spreng Forbes & Hemsley in Journ. Linn. Soc. Bot. 26: 332. (1891).

Hubei: Badong, 1870, 4996, 5133, 6117. At Shih Pan Shan (off the Xiling Gorge), 4187. Liantuo, 1913. Jianshi, 5816. Xingshan, 6953.

According to Henry this species was colloquially known as *ch'iao me han* at Yichang, Badong and Changyang.

Persicaria orientalis (L.) Spach (POLYGONACEAE)

Polygonum orientale L. Henry in Journ. China Br. Roy. Asiat. Soc. 22: 255. (1887) (Chinese Names of Plants), Forbes & Hemsley in Journ. Linn. Soc. Bot. 26: 343. (1891), Henry in Trans. Asiat. Soc. Jap. 24 suppl: 76. (1896) (List Pl. Formosa).

Hainan: Environs of Haikou, 8040. **Hubei:** Yichang, 43, 1668, 2189. Badong, 5004. **Taiwan:** Kaohsiung, Plain, A. Henry, without number.

An attractive annual once popular in European gardens during the 18th and 19th century. It is now rarely grown but is naturalised in central and southern Europe. According to Henry, in the Three Gorges region, cultivated plants of this species were used in the preparation of a wine ferment and were known as *liao-tzu*.

Persicaria perfoliata (L.) H. Gross (POLYGONACEAE)

Polygonum perfoliatum L. Henry in Journ. China Br. Roy. Asiat. Soc. 22: 235. (1887) (Chinese Names of Plants), Forbes & Hemsley in Journ. Linn. Soc. Bot. 26: 344. (1891).

Hubei: Yichang, 298, 2332. In the San You Dong Glen (June 1886), 1635.

An annual twining or climbing species, distributed through eastern Asia and eastern North America (where it was introduced) and was known colloquially at Yichang as *chi kai so* according to Henry. A handsome plant when seen carrying masses of beautiful blue fruits, this species is currently used in Chinese herbal medicine to treat various respiratory complaints. *Persicaria perfoliata* is cultivated in the order beds at the National Botanic Gardens, Glasnevin from collections made by the Glasnevin Central China Expedition near Yichang (GCCE 451) and near Tianthu Shan, Changyang (GCCE 761) in the autumn of 2004.

Persicaria pinetorum (Hemsley) H. Gross (POLYGONACEAE) New species

Polygonum pinetorum Hemsley in Journ. Linn. Soc. Bot. 26: 345. (1891).

Hubei: Fang Xian, 6848.

According to Hemsley who presumably supplied the information from Henry's field notes, this species is extremely common in conifer woods on the summits of mountains at 2,900 to 3,300 metres. Evidently this specimen was collected in Shennongjia Forest Reserve, Hubei's last great remaining tract of primary forest. The species was also based on material gathered on the summit of Emei Shan (Mount Omei) by Ernst Faber.

Persicaria posumbu (Buch.-Ham ex D. Don) H. Gross (POLYGONACEAE)

Polygonum posumbu Buch.-Ham. ex D. Don Forbes & Hemsley in Journ. Linn. Soc. Bot. 26: 346. (1891).

Hubei: Badong, 4797, 4939.

Persicaria praetermissa (Hook. f.) H. Hara (POLYGONACEAE)

Polygonum praetermissum Hook. f. Forbes & Hemsley in Journ. Linn. Soc. Bot. 26: 347. (1891).

Hubei: Yichang, 567, 962, 1669, 3659.

Native to northern India, China, Korea, Japan, the Philippines and Australia.

Persicaria runcinata (Buch.-Ham. ex D. Don) H. Gross (POLYGONACEAE)

Polygonum runcinatum Buch.-Ham. ex D. Don Forbes & Hemsley in Journ. Linn. Soc. Bot. 26: 347. (1891).

Hubei: Yichang, 1344. Liantuo, 3889. Jianshi, 6001. Badong, 6075. Xingshan, 6954. Changyang, 6075a.

A widespread species, distributed through the Himalaya, northern India, Myanmar, China, Thailand, Malaysia and Java.

Persicaria sagittata (L.) H. Gross (POLYGONACEAE)

Polygonum sagittatum L. Forbes & Hemsley in Journ. Linn. Soc. Bot. 26: 348. (1891).

Hubei: Yichang, 427. Badong, 519 (September 1885), 2514, 3172, 4951. **Hainan:** Environs of Kiungchow, 8309.

An annual to a metre tall, native to North America, eastern Siberia, China, Japan, Korea and India.

Persicaria senticosa (Meisn.) H. Gross (POLYGONACEAE)

Polygonum senticosum (Meisn.) Franch. & Sav. Forbes & Hemsley in Journ. Linn. Soc. Bot. 26: 349. (1891), Henry in Trans. Asiat. Soc. Jap. 24 suppl: 76. (1896) (List Pl. Formosa).

Hubei: Badong, 5080. **Taiwan:** Wanchin, 457.

An annual decumbent or climbing herb to 70 cm tall, native to Russia, China, Vietnam, Korea and Japan.

Persicaria strindbergii (Schust.) Galasso (POLYGONACEAE) New species

Polygonum strindbergii Schuster in Bull. Herb. Boiss. ser. 2. 8: 712. (1908).

Yunnan: Mengzi, 9245, 11361.

Persicaria thunbergii (Sieb. & Zucc.) H. Gross (POLYGONACEAE)

Polygonum thunbergii Sieb. & Zucc. Forbes & Hemsley in Journ. Linn. Soc. Bot. 26: 351. (1891).

Hubei: Badong, 2846, 4883. Jianshi, 7414. **Sichuan:** South Wushan, 7258.

Distributed from the Caucasus to northern India, China, Indochina, Korea, Japan and Sumatra.

Persicaria viscofera (Makino) H. Gross (POLYGONACEAE)

Polygonum excurrens Steward in Cont. Gray. Herb. n. s. 88: 65. (1930).

Hubei: Badong, 4790 (type of *Polygonum excurrens* Steward).

Persicaria viscosa (Buch.-Ham. ex D. Don) H. Gross ex T. Mori (POLYGONACEAE)

Polygonum viscosum Buch.-Ham. ex D. Don Forbes & Hemsley in Journ. Linn. Soc. Bot. 26: 352. (1891).

Hubei: Yichang, 2785.

Native from Russia to India, China, Korea and Japan.

Pertya sinensis Oliver (ASTERACEAE) New species

in Hooker. Icon. Pl. 23: t. 2214. (1894), Bretschneider in Hist. Eur. Bot. Disc. China 2: 786. (1898), J. H. Veitch in Hortus Veitchii 371. (1906), Bean in Kew Bull. Misc. Inf. 174. (1910), Schneider Illus. Handb. Laubholzk. 2: 767. (1912); in Sargent Pl. Wilson. 3: 419. (1917), Bean in trees & Shrubs 3: 123. (1992), Nelson in A Heritage of Beauty 316. (2000), O' Brien in Augustine Henry, An Irish Plant Collector in China 27. (2002).

Hubei: Xingshan, 6982 (type).

A deciduous shrub to 1.5 metres tall, producing pinkish-purple flowers in June and July. A genus new to China, previously to Henry's collection *Pertya* was known only in Japan and the Kurran valley, Afghanistan. Introduced to cultivation by Wilson through Veitch's Coombe Wood nursery in 1901. From this introduction Harry Veitch presented a plant to Kew, (the Veitch nursery had a policy of sending plants that were botanically interesting, though unlikely to be commercially successful to the Royal Botanic Gardens, Kew and Edinburgh and the National Botanic Gardens, Glasnevin, Dublin). At Kew it formed a shrub almost 2 metres tall and flowered annually in June. In Ireland *Pertya sinensis* is cultivated at Mount Congreve Gardens, County Waterford.

Petasites japonicus (Sieb. & Zucc.) F. Schmidt. (ASTERACEAE)

Henry in Journ. China Br. Roy. Asiat. Soc. 22: 244. 1887 (Chinese Names of Plants), Forbes & Hemsley in Journ. Linn. Soc. Bot. 23: 446. (1888), Henry in Notes Econ. Bot. China 29. (1893).

Hubei: Yichang, 1248. Antelope Glen, 3379. Liantuo, 6390, 6369. South Badong, 5482.

According to Henry, at Yichang *Petasites japonicus* was called *hu lu pao yeh* and was the source of the Chinese drug *tung hua*. In Japan the same plant is called *k'uan tung*

hua, a name applied at Yichang to the dried flowers of *Eriobotrya japonica*.

Petrocodon dealbatus Hance (GESNERACEAE)

Forbes & Hemsley in Journ. Linn. Soc. Bot. 26: 229. (1890).

Hubei: South Badong, on rocks, 7314. Changyang, on rocks, 7603.

Petrocosmea iodioides Hemsley (GESNERIACEAE) New species

in Hooker's Icon. Pl. xxvi. t. 2599. (1899), Craib in Notes Roy. Bot. Gard. Edinburgh 11: 275. (1918—1919).

Yunnan: On shaded cliffs on the mountains to the north of Mengzi at 2,300 metres, 10259 (holotype).

Petrocosmea kerrii Craib var. ***crinita*** W. T. Wang (GESNERIACEAE) New variety

in Acta Bot. Yunnan 7: 66. (1985).

Yunnan: Simao, 13120 (isotype).

Petrocosmea minor Hemsley (GESNERIACEAE) New species

in Hooker's Icon. Pl. t. xxvi. t. 2600. (1899).

Petrocosmea henryi Craib in Notes Roy. Bot. Gard. Edinburgh 10: 216. (1918).

Yunnan: On shaded cliffs near Mengzi at 1,640 metres, 9154 (syntype of *Petrocosmea minor* Hemsley, holotype of *Petrocosmea henryi* Craib).

Petrocosmea sinensis Oliver (GESNERACEAE) New genus

in Hooker's. Icon. Pl.xviii. t 1716. (1887), Forbes & Hemsley in Journ. Linn. Soc. Bot. 26: 229. (1890), Bretschneider in Hist. Eur. Bot. Disc. China 2: 789. (1898), Henry in Flora & Sylva 1: 218. (1903), Craib in Notes Roy. Bot. Gard. Edinburgh. 11: 272. (1918—1919), Morley in Glasra 3: 80. (1979).

Hubei: On Pingshanba near Yichang, on the surface of a rock in the bottom of a small cave, 2321. Xiling (Yichang) Gorge, in a cave, 2921. Yichang, 4256.

Discovered by Henry who found it growing on the surface of rocks in caves around the Xiling Gorge. It was later introduced by Père Delavay from Yunnan.

Petrodoxa argentea J. Anthony (GESNERIACEAE) New genus

in Notes Roy. Bot. Gard. Edinburgh 18: 203. (1934).

Yunnan: In Pingbian, in forests on the Daweishan Range at 1,650 metres, 10960 (type).

Peucedanum decursivum (Miq.) Maxim. (APIACEAE)

Forbes & Hemsley in Journ. Linn. Soc. Bot. 23: 335. (1887), Henry in Journ. China Br. Roy. Asiat. Soc. 22: 237. (1887) (Chinese Names of Plants); in Notes Econ. Bot. China 24. (1893).

Hubei: Badong, 7 (collected September 1885, received

at Kew on March 12th 1886), 285 (July 1885), 2429, 5195, 5203. Yichang, 74, 2741.

An important Chinese drug according to Henry who also stated it was colloquially known as *ch'iang huo* and *tu-huo* around Yichang.

Peucedanum dielsianum H. Wolff (APIACEAE) New species

in Repert. Spec. Nov. Regni. Veg. 33: 246. (1934).

Selinum monnieri Forbes & Hemsley non L. in Journ. Linn. Soc. Bot. 23: 332. (1887).

Hubei: Antelope Glen, 4381 (type).

Introduced to cultivation by Henry through the Royal Botanic Gardens, Kew. Seeds arrived there from Yichang on May 7th 1888 and plants were grown in the herb garden.

Peucedanum henryi H. Wolff (APIACEAE) New species

in Repert. Spec. Nov. Regni Veg. 33: 248. (1934).

Hubei: Yichang, 3604 (type).

Peucedanum medicum Dunn (APIACEAE) New species

in Journ. Linn. Soc. Bot. 35: 496. (1903).

Angelica sp. Forbes & Hemsley in Journ. Linn. Soc. Bot. 334. (1887), Henry in Journ. China Br. Roy. Asiat. Soc. 22: 239. (1887) (Chinese Names of Plants).

Hubei: Liantuo, 1906, 2006. On the road to the Dome, (June 1886), 1546 (type). Jianshi, 5868. Fang Xian, 5868a. **Sichuan:** South Wushan, 7473.

According to Henry this species was common in the countryside around Yichang and in the mountains to the north and south of the city. He also stated that the roots were collected by the Chinese for the drug *ch'ien-hu*.

Peucedanum praeruptorum Dunn (APIACEAE) New species

in Journ. Linn. Soc. Bot. 35: 497. (1903).

Hubei: In the Kan Chi Glen near Yichang (October 1886), 2911. Changyang, 7505. **Sichuan:** North Wushan, 7475.

A perennial herb to 120 cm tall. The root of this species was used in Chinese medicine according to Henry and was known as *fang-feng*. He stated it was also the *qian hu*, literally; 'before barbarians', and is used to control and stop coughing. It continues to be a valuable medicinal plant in China.

Peucedanum sp. (APIACEAE)

Anonymous in Kew List of Determinations (volume 24).

Hubei: Liantuo, 4219. Fang Xian, 6688. Badong, 4876. **Sichuan:** South Wushan, 6034.

Peucedanum terebinthaceum Fisch. (APIACEAE)

Forbes & Hemsley in Journ. Linn. Soc. Bot. 23: 335.

(1887), Henry in Notes Econ. Bot. China 4. (1893).

Hubei: Badong, 2421, 2526. Yichang, 4353. Without locality, 4353a. **Sichuan:** South Wushan, 7474.

The *fang-feng* of Hubei, the root of this species was used as a drug according to Henry.

Phacellanthus tubiflorus Sieb. & Zucc. (OROBANCHACEAE)

G. Beck in Engler, Pflanzenr. 261: 332. (1930).

Phacellanthus sp. Forbes & Hemsley in Journ. Linn. Soc. Bot. 26: 221. (1890).

Hubei: Jianshi, 5905.

Previous to this single collection made by Henry in the first part of his 1888 expedition this interesting parasite was known only in Japan.

Phaenosperma globosa Munro. (POACEAE)

in Hooker's. Icon. Pl. t. 1991. (1891), Rendle in Journ. Linn. Soc. Bot. 36: 340. (1904).

Hubei: Yichang, 626, 3966, 3968. Liantuo, 1943, 2073. **Sichuan:** Near Kangding, Henry's Chinese collector with Pratt, 8911. **Yunnan:** On wooded cliffs near Mengzi at 1,900 metres, 11299.

Phaenosperma globosa was discovered (as a new genus) by Père Armand David in July near Jiujing in Jiangxi province and introduced by him through the garden of the Paris Museum. Distributed across southern China, Korea and Japan.

Phaius tankervilleae (Banks ex L'Her.) Blume (ORCHIDACEAE)

Phaius grandifolius Lour. Anonymous in Kew List of Determinations (volume 34).

Yunnan: Simao, 13045.

A terrestrial orchid, distributed across India, Sri Lanka, southern China, Thailand, Malaysia to Indonesia, Australia and the Pacific Islands. A beautiful species bearing racemes of large, dark purple flowers between April and August.

Phalaenopsis difformis (Wall. ex Lindl.) Kocyan & Schuit. (ORCHIDACEAE)

Ornithochilus fuscus Wall. ex Heynh. Anonymous in Kew List of Determinations (volume 34).

Yunnan: Simao, 11812, 12199.

Distributed in Myanmar, China, Laos, Thailand and Vietnam.

Phalaenopsis formosana (Christenson) J. M. H. Shaw (ORCHIDACEAE) New species

Phalaenopsis aphrodite Rchb. f. Henry in Trans. Asiat. Soc. Jap. 24 suppl: 92. (1896) (List Pl. Formosa), Rolfe in Journ. Linn. Soc. Bot. 36: 34. (1903). *Phalaenopsis aphrodite* Reichb. f. ssp. *formosana* Christenson. in Phalaenopsis 197. (2001).

Taiwan: Oluanpi, 1705 (holotype of *Phalaenopsis*

aphrodite Reichb. f. ssp. *formosana* Christenson).

A tropical epiphytic orchid, endemic to Taiwan where it was once abundant in lowland tropical forest but now is very rare. Henry had the following to say about this species, "The most beautiful plants (in Taiwan) are the various tree-orchids, of which perhaps *Phalaenopsis* … is the most striking."

Phanera chalcophylla (H. Y. Chen) Mackinder (FABACEAE) New species

Bauhinia chalcophylla H. Y. Chen in Journ. Arnold Arb. 19: 130. (1938).

Yunnan: Tonguan, 13240 (type, collected 20th October 1898).

Endemic to Yunnan.

Phanera championii Benth. (FABACEAE)

Bauhinia championii (Benth.) Benth. Henry in Trans. Asiat. Soc. Jap. 24 suppl: 38. (1896) (List Pl. Formosa), Li in Woody Flora of Taiwan 337. fig. 118. (1963), Zheng in Journ. Wuhan Bot. Res. 2.(1): 96. (1984).

Hubei: Yichang, 4355, without number (May 1888). Antelope Glen, 7729. Without locality, 4355a. **Taiwan:** Wanchin, 1528.

A large woody climber, first described by Bentham in his *Flora of Hong Kong*. Native to southern China, including Taiwan where it grows in thickets and forests at low altitude. This species was collected by the Glasnevin Taiwan Expedition (GTE 108) on Wanshoushan (Ape's Hill of Henry, Wilson and Swinhoe) in October 2004.

Phanera comosa (Craib) Bandyop. & Ghoshal (FABACEAE) New species

Bauhinia comosa Craib in Kew Bull. Misc. Inf. 352. (1913). *Bauhinia saxatilis* Craib in Kew Bull. Misc. Inf. 354. (1913). *Bauhinia henryi* Craib in Kew Bull. Misc. Inf. 353. (1913), Harms in Fedde, Repert. Spec. Nov. Regni Veg. 17: 134. (1921).

Yunnan: On the plains to the north of Mengzi, between Tachay and the So-Pa Plains, 10193 (isotype of *Bauhinia saxatilis* Craib, collected 26th July 1896). Linan at 1,350 metres, 13358 (isotype of *Bauhinia saxatilis* Craib, collected November 1st 1898). South of the Red River from Manpan at 480 metres, 10175, (isotype of *Bauhinia henryi* Craib).

Endemic to China and distributed in the provinces of Sichuan and Yunnan.

Phanera yunnanensis (Franch.) Wunderlin (FABACEAE)

Bauhinia yunnanensis Franchet Anonymous in Kew List of Determinations (volume 34).

Yunnan: Mengzi, 9670.

A handsome climbing shrub, producing terminal racemes of 10~20 light-red, lilac or white flowers during July and August. Distributed through the provinces of Guizhou, Sichuan and Yunnan in China and also found in Myanmar and Thailand where it favours sun-baked mountain slopes.

Pharbitis nil (L.) Choisy (CONVOLVULACEAE)

Pharbitis hederacea Choisy Henry in Journ. China Br. Roy. Asiat. Soc. 22: 160. (1890) (Chinese Names of Plants). *Ipomoea hederacea* Jacq. Forbes & Hemsley in Journ. Linn. Soc. Bot. 26: 160. (1890), Henry in Notes Econ. Bot. China 49. (1893).

Hubei: Yichang, 1662. Liantuo, 4613.

Found in tropical and subtropical regions of both hemispheres. According to Henry this is the *ch'ien niu* (*qian niu zi*) and it occured as a garden weed around Yichang and Liantuo. The seeds of this species are used as a drug in China as well as India where they were called *kaladana*. In China it is used to control coughs and wheezing and to control internal parasites like roundworm and tapeworm infestations.

Pharbitis purpurea (L.) Voigt. (CONVOLVULACEAE)

Ipomoea purpurea Lam Henry in Annotation in the *Index. Flora Sinensis* (copy at NBG, Glasnevin).

Hubei: Cultivated in a garden at Yichang, 3599.

A native of tropical America, widely colonised in Asia and cultivated in China.

Phaseolus chrysanthus Savi. (FABACEAE)

Forbes & Hemsley in Journ. Linn. Soc. Bot. 23:193. (1887), Henry in Journ. China Br. Roy. Asiat. Soc. 22: 274. (1887) (Chinese Names of Plants); in Notes Econ. Bot. China 14. (1893).

Hubei: Yichang, 870, 2209.

Cultivated in the Three Gorges region where according to Henry this bean was known as *man tou.*

Phaseolus lunatus L. (FABACEAE)

Anonymous in Kew List of Determinations (volume 34).

Yunnan: Mengzi, 11376. Simao, 12685.

The Lima bean, a twining annual, native to tropical South America.

Phaseolus mungo L. (FABACEAE)

Henry in Trans. Asiat. Soc. Jap. 24 suppl: 36. (1896) (List Pl. Formosa).

Hainan: Environs of Haikou, 8351. **Taiwan:** Oluanpi, 361. Wanchin, 392.

Cultivated, the mung bean or black gram, which Henry stated was known in Taiwan as *fen-t'iao.* Also widely cultivated in India where it is the most valuable of all pulses.

Phaseolus sp. No. 1. (FABACEAE)

Anonymous in Kew List of Determinations (volume 24).

Hainan: 32 km (20 miles) west of Haikou, 8135.

Environs of Haikou, 8359.

Phaseolus sp. No. 2. (FABACEAE)

Anonymous in Kew List of Determinations (volume 34).

Yunnan: Mengzi, 11227.

Phaseolus trilobrus Ait. (FABACEAE)

Henry in Trans. Asiat. Soc. Jap. 24 suppl: 36. (1896) (List Pl. Formosa).

Taiwan: Kaohsiung, a common wild plant 1123.

Phedimus hybridus (L.) 't. Hart (CRASSULACEAE)

Sedum hybridum L. Anonymous in Kew List of Determinations (volume 24).

Hubei: Jianshi, 5857. Without locality, 5857a. **Sichuan:** South Wushan, 5728.

Phedimus kamtschaticus (Fisch.) t' Hart (CRASSULACEAE)

Sedum kamtschaticum Fisch. Forbes & Hemsley in Journ. Linn. Soc. Bot. 23: 285. (1887).

Hubei: Badong, 4658, 7393.

Phegopteris amaurophylla Christ (THELYPTERIDACEAE)
New species

Christ in Bull. Herb. Boiss. 7: 14. (1899).

Yunnan: In mountain forest to the east of Mengzi at 1,900 metres, 11536.

Phegopteris decursive-pinnata (van Hall) Fee (THELYPERIDACEAE)

Nephrodium decursivo-pinnatum (van Hall.) Bak. Henry in Journ. China Br. Roy. Asiat. Soc. 22: 263. (1887) (Chinese Names of Plants), Baker in Journ. Bot. 26: 228. (1888), Henry in Trans. Asiat. Soc. Jap. 24 suppl: 113. (1896) (List Pl. Formosa).

Hubei: Yichang, 2051. Badong, 3679, 5139, 5150. **Taiwan:** Tamsui, 1378, 1381.

Henry's Badong collector called this species the *pan chiu wang tzu*. Native to India, central and southeastern China, Vietnam, South Korea and Japan.

Phegopteris microstegia (Hook.) Christenh. (ASPLENIACEAE)

Aspidium distans (Mett.) Christ Diels in Bull. Soc. Bot. France 5: 35. (1905).

Hubei: Yichang, 315, 1861. **Yunnan:** Simao, 13178.

Phegopteris sp. (THELYPTERIDACEAE)

Anonymous in Kew List of Determinations (volume 34).

Yunnan: Mengzi, 11584.

Phellodendron chinense Schneid. var. *glabriusculum* Schneid. (RUTACEAE)

Rehder & E. H. Wilson in Sargent Pl. Wilson 2: 137. (1914); 3: 449. (1917), Zheng in Journ. Wuhan Bot. Res. 2.(1): 106. (1984).

Phellodendron japonicum Rehder non Maxim. in Sargent, Trees & Shrubs. 1: 198. & 201. (1905). *Phellodendron chinense* Schneider Sprague in Kew Bull. Misc. Inf. 235. (1920).

Hubei: Badong, 4003, 5202.

A tree to 10 metres tall, native to central China, it differs from *Phellodendron chinense* C. K. Schneid. in having the down on the under-sides of leaves confined to the mid-rib and veins, not spread throughout. The bark of this tree, according to Wilson was the *huang-po* or *huang peh*, a poor man's cure-all and used to treat almost every ailment known to the Chinese.

Philadelphus henryi Koehne (HYDRANGEACEAE)
New species

in Repert. Spec. Nov. Regni Veg. 10: 126. (1911), Levl. in Cat. Pl. Yunnan. 255. (1917), Rehder in Journ. Arnold Arb. 1: 197. (1920), Handel-Mazzetti in Symb. Sin. 7: 437. (1931), O' Brien in Augustine Henry, An Irish Plant collector in China 27. (2002).

Yunnan: Mountains to the south-west of Mengzi at 1,450 metres, 10749 (lectotype), 10749a., 10749b.

A very beautiful shrub to 3 metres tall with white, delightfully scented flowers in early summer. Native to Sichuan, Guizhou and Yunnan. In Ireland this species is cultivated at Glenveagh National Park, County Donegal having been sent there from Glasnevin during the 1990s.

Philadelphus incanus Koehne (HYDRANGACEAE)
New species

in Gartenflora. 45: 562. (1896), Diels in Bot. Jahrb. 29: 371. (1900), Koehne in Mitt. Deutsche. Dendr. Ges. 13: 84. (1904), Schneider in Illus. Handb. Laubholzk. 1: 370. (1905), Smith in Journ. Linn. Soc. Bot. 36: 500. (1905), Koehne in Sargent Pl. Wilson. 1: 5. (1913), Bean in Trees & Shrubs 2: 135. (1914), Hers in Journ. N. China Branch Roy. As. Soc. 53: 113. (1922), Rehder in Journ. Arnold Arb. 5: 153. (1924); in Man. Cult. Trees & Shrubs 274. (1927), Chen in Ill. Man. Chinese Trees & Shrubs 371. fig. 278. (1937), S. Y. Hu in Joun. Arnold Arb. 36: 338. (1955), Zheng in Journ. Wuhan Bot. Res. 2.(1): 60. (1984), Nelson in Pim, The Wood & the Trees 233. (1984); in A Heritage of Beauty 316. (2000), O' Brien in Augustine Henry, An Irish Plant Collector in China 27. (2002), O' Brien in Ir. Garden 13(5): 53. (2004).

Philadelphus coronarius L. var. *tomentosus* Forbes & Hemsley non (Wall. ex G. Don) Hook. f. & Thoms. in Journ. Linn. Soc. Bot. 23: 277. (1887).

Hubei: Liantuo, 1969, 2143. Xiling (Yichang) Gorge, 3949, 3489. Yichang, 3969, 4206. South Badong, 6093 (lectotype).

A shrub to 2.5 metres tall bearing racemes of five to nine fragrant flowers in July and August. Discovered by Henry in 1887, this is one of the very finest and most sweetly scented of all the mock oranges. *Philadelphus incanus* was introduced to Irish gardens through Wilson 574 (Xingshan, Hubei) and Wilson 583 (Yichang) sent to Glasnevin by Professor Sargent (Wilson had first introduced it in 1904). The same Wilson collection still grows at Glasnevin and in Ireland this exceptionally beautiful shrub is also grown at Birr Castle and Mount Usher.

Philadelphus sericanthus Koehne (HYDRANGACEAE)
New species

in Gartenflora. 45: 561. (1896), Koehne in Bot. Jahrb. 29: 371. (1900); in Mitt. Deutsch. Dendr. Ges. 13: 84. (1904); in Sargent Pl. Wilson. 1: 145. (1913), Schneider in Illus. Handb. Laubh. 1: 370. fig. 236. (1905), Smith in Journ. Linn. Soc. Bot. 36: 500. (1905), Stapf in Bot. Mag. 148. t. 8941. (1922), Rehder in Man. Cult. Trees & Shrubs 274. (1927), Chen in Ill. Man. Chin. Trees & Shrubs 372. (1937), in part., S. Y. Hu in Journ. Arnold Arbor. 36: 343. (1955), Zheng in Journ. Wuhan Bot. Res. 2. (1): 61. (1984), Bean in Trees & Shrubs 3: 140. (1992), O' Brien in Augustine Henry, An Irish Plant Collector in China 27. (2002).

Philadelphus coronarius L. var. *tomentosus* Forbes & Hemsley non (Wall. ex G. Don) Hook. f. & Thoms. in Journ. Linn. Soc. Bot. 23: 277. (1887). *Philadelphus incanus* Koehne in Gartenflora. 45: 562. (1896), in part. *Philadelphus hupehensis* (Koehne) S.Y. Hu Diels in Bot. Jahrb. 29: 371. (1900), Koehne in Mitt. Deutsche. Dendr. Ges. 13: 84. (1904), Schneider in Illus. Handb. Laubholzk. 1. 370. (1905), Smith in Journ. Linn. Soc. Bot. 36: 500. (1905), Koehne in Sargent Pl. Wilson. 1: 5. (1913), Bean in Trees & Shrubs 2: 135. (1914), Hers in Journ. N. China Branch Roy. As. Soc. 53: 113. (1922), Rehder in Journ. Arnold Arb. 5: 153. (1924); in Man. Cult. Trees & Shrubs 274. (1927), Chen in Ill. Man. Chinese Trees & Shrubs 371. fig. 278. (1937), S. Y. Hu in Joun. Arnold Arb. 36: 338. (1955), Zheng in Journ. Wuhan Bot. Res. 2.(1): 60. (1984), Nelson in Pim, The Wood & the Trees 233. (1984); in A Heritage of Beauty 316. (2000), O' Brien in Augustine Henry, An Irish Plant Collector in China 27. (2002).

Hubei: Xiling (Yichang) Gorge, 3625. Jianshi, 6013, 6015 (lectotype), 7428. South Badong, 5344 (isosyntype).

A deciduous shrub to 3.5 metres tall. *Philadelphus sericanthus* is endemic to central and southern China, the centre of its range, according to Hu, is the region immediately around the Wushan Gorge. It was introduced to cultivation by Père Farges who sent seeds to M. L. de Vilmorin from Sichuan in 1897. One of these seedlings had reached the Arnold Arboretum by 1902. E. H. Wilson later reintroduced it through Messrs Veitch in 1900. In Irish gardens *Philadelphus sericanthus* cultivated at the John F.

Kennedy Arboretum and at the National Botanic Gardens, Glasnevin.

Philadelphus subcanus Koehne (HYDRANGACEAE)
New species

in Mitt. Deutsch. Dendrol. Ges. 13: 83. (1904), Koehne in Illus. Handb. Laubholzk. 1: 369. (1906); in Sargent Pl. Wilson. 1: 5. (1913), Dunn in Journ. Linn. Soc. Bot. 39: 475. (1911), Rehder in Man. Cult. Trees & Shrubs 274. (1927), Chen in Ill. Man. Chin. Trees & Shrubs 372. (1937).

Philadelphus coronarius L. var. *tomentosus* Forbes & Hemsley non (Wall. ex G. Don) Hook. f. & Thoms. in Journ. Linn. Soc. Bot. 23: 277. (1887).

Sichuan: Henry's Chinese collector with Pratt, no locality stated but may be Leshan, Emei Shan, Wa Shan or Kangding, 8823 (isotype).

A species closely allied to *Philadelphus incanus* Koehne, though differing in the more sparsely hairy leaf underside and calyx, the disc and lower part of the style are downy whereas in *Philadelphus incanus* they are glabrous. Introduced to cultivation by E. H. Wilson.

Philonotis falcatus (Hook.) Mitt. (BARTRAMIACEAE)

Philonotis palustris Mitt. Anonymous in Kew List of Determinations (volume 24).

Hubei: Yichang, 4310. **Sichuan:** Henry's Chinese collector with Pratt, no locality stated but may be any of the following, Leshan, Emei Shan, Wa Shan or Kangding, 8837. (Moss).

Phlogacanthus curviflorus (Wall.) Nees (ACANTHACEAE)

Anonymous in Kew List of Determinations (volume 34).

Yunnan: Simao, 12785.

Phlogacanthus sp. (ACANTHACEAE)

Anonymous in Kew List of Determinations (volume 34).

Yunnan: Simao, 11622.

Phlomoides bracteosa (Royle ex Benth.) Kamelin & Makhm. (LAMIACEAE)

Phlomis bracteosa Royle ex Benth. Dunn in Notes Roy. Bot. Gard. Edinburgh 6: 187. (1915).

Yunnan: Forests to the north of Mengzi at 2,650 metres, 10216. Forests to the west of Mengzi at 2,300 metres, 10216a.

Phlomoides umbrosa (Turcz.) Kamelin & Makhm. LAMIACEAE

Phlomis umbrosa Turcz Dunn in Notes Roy. Bot. Gard. Edinburgh 6: 186. (1915). *Phlomis umbrosa* Turcz. var. *australis* Hemsley in Journ. Linn. Soc. Bot. 26: 306. (1890), J. H. Veitch in Hortus Veitchii 428. (1906) Nelson in A Heritage of Beauty 316. (2000).

Hubei: Badong, 875, 1872, 2442, 6029, 4749 (isosyntype of *Phlomis umbrosa* Turcz. var. *australis* Hemsley), 6179, 7360 (isosyntype of *Phlomis umbrosa* Turcz. var. *australis* Hemsley). Donghu, 6423 (isosyntype of *Phlomis umbrosa* Turcz. var. *australis* Hemsley).

An herbaceous perennial with two-lipped cream-white flowers, red at the base, held in false whorls in the axils of the uppermost leaves. Introduced to cultivation by E. H. Wilson through Veitch's Coombe Wood nursery from Hubei in 1901.

Phoebe chinensis Chun (LAURACEAE) New Species
Machilus macrophyllus Hemsley in Journ. Linn. Soc. Bot. 26: 375. (1891) non Blume (1851), non Nees (1836), Henry in Notes Econ. Bot. China 43. (1893), Bretschneider in Hist. Eur. Bot. Disc. China 2: 790. (1898), Rehder & E. H. Wilson in Sargent Pl. Wilson. 1: 107. (1913), Morley in Glasra 3: 79. (1979).

Sichuan: South Wushan, 5756, 5666 (isotype of *Machilus macrophyllus* Hemsley), 5699 (type of *Machilus macrophyllus* Hemsley), 7194. South Badong, 7323.

Phoebe faberi (Hemsl.) Chun (LAURACEAE) New species
in Cont. Biol. Labor. Sc. Soc. China 1: 31. (1925), H. Liu in Laurac. Chine & Indochine 72. (1934), Zheng in Journ. Wuhan Bot. Res. 2.(1): 56. (1984).

Litsea sp. Henry in Journ. China Br. Roy. Asiat. Soc.22: 260. (1887) (Chinese Names of Plants). *Machilus faberi* Hemsley in Journ. Linn. Soc. Bot. 26: 374. (1891).

Hubei: Antelope Glen, 3438. Liantuo, 3898, 4569. South Badong, 5507 (type of *Machilus faberi* Hemsley), 7297.

The species was based on material gathered in Hubei by Henry and in the same year on Emei Shan at 1,150 metres by Ernst Faber.

Phoebe lanceolata (Nees) Nees (LAURACEAE)
H. Liu in Laurac. Chine & Indochine 73. (1934).

Yunnan: Simao, 11734, 11734a., 11734b., 11734c., 11734d., 11734e., 11734f.

Native to India, Nepal, China (Yunnan) and Thailand.

Phoebe legendrei Lecompte (LAURACEAE)
H. Liu in Laurac. Chine & Indoch. 68. (1934).

Yunnan: Simao, 12652.

Endemic to China and distributed in Sichuan and Yunnan.

Phoebe neurantha (Hemsl.) Gamble (LAURACEAE) New species
Machilus neurantha Hemsley in Journ. Linn. Soc. Bot. 26: 376. (1891), Henry in Notes Econ. Bot. China 43. (1893), Bretschneider in Hist. Eur. Bot. Disc. China 2: 790. (1898).

Hubei: Liantuo, 4483, 6386 (type). Xingshan, 4540.

Jianshi, 6006. South Donghu, 7590. **Sichuan:** South Wushan, 5499, 7229. North Wushan, 7128.

Phoebe puwenensis Cheng (LAURACEAE) New species
Phoebe sheareri (Hemsl.) Gamble var. *longepaniculata* H. Liu in Laurac. Chine & Indochine 69. (1934).

Yunnan: Simao, 12922 [syntype of *Phoebe sheareri* (Hemsl.) Gamble var. *longepaniculata* H. Liu], 13016, 13016a., [syntype of *Phoebe sheareri* (Hemsl.) Gamble var. *longepaniculata* H. Liu].

Endemic to China.

Phoebe sheareri (Hemsl.) Gamble (LAURACEAE)
H. Liu in Laurac. Chine & Indoch. 69. (1934).

Jiangxi: Jiujiang, 219.

Phoebe sp. No. 1. (LAURACEAE)
Anonymous in Kew List of Determinations (volume 24).

Sichuan: South Wushan, 5516.

Phoebe sp. No. 2. (LAURACEAE)
Anonymous in Kew List of Determinations (volume 34).

Yunnan: Simao, 13033.

Phoebe tavoyana (Meisn.) Hook. f. (LAURACEAE)
Machilus henryi Hemsley in Journ. Linn. Soc. Bot. 26: 375. (1891). *Phoebe henryi* (Hemsley) Merr. in Lingnan Sci. Journ. 11: 43. (1932), Liou Ho in Laurac. Chine & Indoch. 65. (1934).

Hainan: Without locality, 8501.

Phoenix loureiroi Kunth (ARACEAE)
Phoenix humilis L. var. *hanceana* (Naud.) Becc. Henry in Trans. Asiat. Soc. Jap. 24 suppl: 99. (1896) (List Pl. Formosa). *Phoenix hanceana* Naudin. Wright in Journ. Linn. Soc. Bot. 36: 168. (1903).

Hainan: Near the Kiungchow Pagoda, 7973, 8253. Without locality, 8595. **Taiwan:** Kaohsiung, without number. Oluanpi, without number.

A very common palm to 4 metres tall, native to southeast China (including Hainan, Hong Kong and Taiwan). According to Henry this species bears edible fruits and was known coloquially in Taiwan as the *kuang-lang*, or *k'eng-lang*. This species was collected by the Glasnevin Taiwan Expedition (GTE 109) on Wanshoushan (Ape's Hill of Henry and Wilson) in October 2004.

Phoenix sp. (ARECACEAE)
Anonymous in Kew List of Determinations (volume 34).

Yunnan: Simao, 12924.

Photinia davidiana (Decne.) Cardot (ROSACEAE)
Stranvaesia henryi Diels in Bot. Jahrb. Syst. 36, Biebl. 82: 52. (1905), Schneider in Illus. Handb. Laubh. 1: 713.

(1906). *Stranvaesia davidiana* Decne. in Bot. Jahrb. Syst. 36, Biebl. 82: 52. (1905).

Sichuan: without number or locality, (type of *Stranvaesia henryi* Diels). **Yunnan:** Mengzi, 11325.

An evergreen shrub or occasionally a small tree, native to western, central and southern China and Vietnam south to Borneo. Discovered by Père Armand David near Baoxing in western Sichuan in 1869 and introduced by E. H. Wilson from Emei Shan in 1903 through Veitch's Coombe Wood nursery.

Photinia glabra (Thunb.) Pépin (ROSACEAE) New species

Photinia beckii C. K. Schneid. in Illus. Handb. Laubholzk. 1: 707. (1906), Schneider in Repert. Nov. Spec. Regni Veg. 3: 153. (1907).

Yunnan: Mengzi at 1,800 metres, 9575. In woods near Mengzi at 2,600 metres, 9575a., (isotype of *Photinia beckii* C. K. Schneid.).

China, Indochina and Japan.

Photinia griffithii Decne. (ROSACEAE) New species

in Sargent Pl. Wilson. 1: 190. (1912), O' Brien in Augustine Henry, An Irish Plant Collector in China 29. (2002).

Yunnan: Forests around Simao at 1,500 to 1,600 metres, 11716 (type of *Photinia glomerata* Rehder & E. H. Wilson). Forests near Simao at 1,500 metres, 11716a.

Himalaya to China (distributed in the provinces of Hubei, Sichuan and Yunnan), cultivated in several Irish gardens.

Photinia pustulata Lindl. (ROSACEAE)

Photinia lancifolia Rehder & E. H. Wilson in Sargent Pl. Wilson. 1: 191. (1912).

Yunnan: Near Mingheh at 1,100 metres, 12833 (holotype of *Photinia lancifolia* Rehder & E. H. Wilson). Simao at 1,300 metres, 13412.

Native to India, Myanmar, western China, Thailand and Vietnam.

Photinia serratifolia (Desf.) Kalkman (ROSACEAE)

Photinia serrulata Henry non Lindl. in Journ. China Br. Roy. Asiat. Soc. 22: 252. (1887) (Chinese Names of Plants), Forbes & Hemsley in Journ. Linn. Soc. Bot. 23: 263. (1887), Henry in Notes Econ. Bot. China 55. (1893), Rehder & E. H. Wilson in Sargent Pl. Wilson. 1: 184. (1913).

Hubei: Near Yichang in the Monastery Valley, 1490. Liantuo, 6353. **Sichuan:** South Wushan, 5490. **Yunnan:** Lunan, 10576.

A tree to 13 metres tall, native to China and introduced to cultivation by Captain Kirkpatrick of the East India Company in 1804. This is the *yu la shu* or *shui-hung-shu*, described by Henry as 'a beautiful, large evergreen tree,

occurring in the Monastery valley (the monastery valley lies to the approach of Moji Shan, a domed shape hill on the opposite banks of the Yangtze to where the foreign compound once lay. It is still a very visible landmark near the city of Yichang) There is a very fine tree by the scree beds near the rock garden at the National Botanic Gardens, Glasnevin and is a beautiful sight in spring when carrying its coppery-red new growths and corymbs of white flowers.

Photinia sp. (ROSACEAE)

Anonymous in Kew List of Determinations (volume 34).

Yunnan: Mengzi, 10526. Simao, 12120, 12139, 12842.

Photinia undulata (Decne.) Cardot (ROSACEAE)

Stranvaesia undulata Decne. Forbes & Hemsley in Journ. Linn. Soc. Bot. 23: 264. (1887), Schneider in Illus. Handb. Laubholzk.1. ii: 713. (1906). *Stranvaesia davidiana* Decne. var. *undulata* (Decne.) Rehder & E. H. Wilson in Sargent Pl. Wilson. 1: 107. (1912), Zheng in Journ. Wuhan Bot. Res. 2.(1): 95. (1984).

Hubei: Badong, 1760, 2807, 3179, 3703, 4080, 5132. Fang Xian, 6613. Xingshan, 5698a. Without locality, 5698b. **Sichuan:** South Wushan, 5698. Henry's Chinese collector with Pratt, no locality stated but may be any of the following, Leshan, Emei Shan, Wa Shan or Kangding, 8953.

A common and variable shrub in Hubei, it was introduced to cultivation by Wilson in 1901. Wilson's 382 from Baokang grew at Glasnevin during the early 20th century.

Phragmites australis (Cav.) Trin. ex Steud. (POACEAE)

Phragmites roxburghii Henry non Steud. in Journ. China Br. Roy. Asiat. Soc. 22: 257. (1887) (Chinese Names of Plants). *Phragmites communis* (L.) Trin. Rendle in Journ. Linn. Soc. Bot. 36: 409. (1904).

Hubei: Yichang, 79. **Taiwan:** Kaohsiung, on the side of a creek, 1149.

The *lu ch'ai*, at Yichang *p'ao* means 'hollow' and the second name signifies 'hollow reed stem', the rhizome is used in herbal medicine to reduce high fever. Widely distributed in the northern hemisphere.

Phragmites karka (Retz.) Trin. ex Steud. (POACEAE)

Rendle in Journ. Linn. Soc. Bot. 36: 410. (1904).

Taiwan: Kaohsiung, 1205. Kaohsiung Spit 1208. **Yunnan:** Mengzi, in moist places at 1,450 metres, 9209. Mengzi, in a ravine at 1,500 metres, 9209a.

A robust reed to 4 metres tall, widely spread from western Africa to Japan and south to Polynesia and Australia.

Phreatia elegans Lindl. (ORCHIDACEAE)

Rolfe in Journ. Linn. Soc. Bot. 36: 17. (1903).

Yunnan: Epiphytic on trees in forests on the

Daweishan Range in Pingbian at 1,650 metres, 11093.

Phreatia formosana Rolfe (ORCHIDACEAE) New species

in Ann. Bot. 9: 156. (1895), Henry in Trans. Asiat. Soc. Jap. 24 suppl: 91. (1896) (List Pl. Formosa), Bretschneider in Hist. Eur. Bot. Disc. China 2: 792. (1898), Rolfe in Journ. Linn. Soc. Bot. 36: 17. (1903), Kranzlin in Engler, Pflanzenr. 50: 34. (1911).

Taiwan: Oluanpi, 1349 (type).

An epiphytic orchid without pseudobulbs, native to China (Taiwan and Yunnan), Vietnam and Thailand.

Phryma nana Koidz. (PHRYMACEAE)

Phryma leptostachya Forbes & Hemsley non L. in Journ. Linn. Soc. Bot. 26: 251. (1890).

Hubei: Yichang, 2048. Badong, 925 (November 1885), 1842 (July 1886). Xingshan, 4533. Donghu, 6403, 6445. Yunnan: Mengzi, 10967.

This species has been used for a long time in Asian countries as a medicine and as an insecticide. Henry stated that in China it was known as *tou gu cao* or 'bone-penetrating herb', this name first appeared in Zhen Yi Yao Ping (Valuable and Grotesque Medicines) by Gao Lian in 1591. *Phryma nana* with a range extending from the Russia into the Himalaya to Japan, i.e. Russia, Pakistan, Kashmir, northern India, Nepal, China, Korea, Vietnam and Japan.

Phrynium pubinerve Blume (MARANTACEAE)

Phrynium capitatum Willd. Schumann in Engler, Pflanzenr. 56: 53. (1902), Wright in Journ. Linn. Soc. Bot. 36: 73. (1903).

Hainan: Ling-men, 8677. Yunnan: Simao at 1,400 metres, 11710.

A handsome tuberous perennial to 1 metre tall. The Hani (an ethnic group) of Xishuangbanna in southern Yunnan call this species *beepavq* and use it to treat cases of flu with high fever, mouth ulcers and dysentry. Distributed in India, China and Vietnam.

Phtheirospermum japonicum (Thunb.) Kanitz (OROBANCHACEAE)

Phtheirospermum chinense Bunge ex Fisch. & C. A. Mey. Forbes & Hemsley in Journ. Linn. Soc. Bot. 26: 204. (1890).

Hubei: Yichang, 132, 2761. Liantuo, 2676. Badong, 944 (November 1885), 1039 (December 1885), 2803, 5014. Without locality, 132a.

Phtheirospermum sp. (OROBANCHACEAE)

Anonymous in Kew List of Determinations (volume 34).

Yunnan: Mengzi, 9705.

Phyla nodiflora (L.) Greene (VERBENACEAE)

Lippia nodiflora (L.) Michx. Forbes & Hemsley in Journ. Linn. Soc. Bot. 26: 251. (1890), Henry in Trans. Asiat. Soc. Jap. 24 suppl: 70. (1896) (List Pl. Formosa).

Hubei: Yichang, 168, 543. Near Yichang, on the hills behind the San You Dong Glen (July 1886), 1516. Taiwan: Kaohsiung, without number. Oluanpi, without number. Wanchin, without number.

Phylacium majus Collett & Hemsley (FABACEAE)

Anonymous in Kew List of Determinations (volume 34).

Yunnan: Simao, 12484, 13502.

Native to Myanmar, western China, Thailand and Laos.

Phyllagathis ovalifolia H. L. Li (MELASTOMATACEAE) New species

in Journ. Arnold Arb. 25: 31. (1944).

Yunnan: Mengzi, 11035.

An undershrub to about a metre tall. A handsome species carrying up to 30 lavender-pink flowers per cyme on branch tips. Distributed in moist forests between 1,200 and 1,700 metres in southern Yunnan, Guangxi of China, Laos and Vietnam.

Phyllagathis tetrandra Diels (MELASTOMATACEAE) New species

in Bot. Jahrb. 65: 117. (1932).

Stapfiophyton tetrandrum (Diels) Li in Journ. Arnold Arb. 25: 29. (1944).

Yunnan: Mengzi, 10539 (isotype), 10539a.

Native to China (Yunnan) and Vietnam.

Phyllagathis velutina (Diels) C. Chen (MELASTOMATACEAE) New species

in Bull. Bot. Res. Harbin 4(3): 51. (1984).

Bredia velutina Diels in Bot. Jahrb. Syst. 65: 109. (1932) Li in Journ. Arnold Arb. 25: 22. (1944).

Yunnan: In Pingbian in forests on the Daweishan Range at 1,640 metres, 13479 (isotype of *Bredia velutina* Diels). Simao, 13497.

Endemic to Yunnan.

Phyllanthus amarus Schumach. & Thonn. (PHYLLANTHACEAE)

Phyllanthus niruri Henry non L. in Trans. Asiat. Soc. Jap. 24 suppl: 82. (1896) (List Pl. Formosa).

Taiwan: Oluanpi, 211, 235. Wanchin, 887.

An annual herb to 50 cm tall, thought to be a native to America but now a circumtropical weed.

Phyllanthus chekiangensis Croizat & F. P. Metcalf (PHYLLANTHACEAE)

Li in Acta Phytotax. Sin. 25: 379. (1987).

Phyllanthus leptoclados Forbes & Hemsley non (Hance) Benth. in Journ. Linn. Soc. Bot. 26: 422. (1894). *Phyllanthus glabricapsulus* Zheng non F. P. Metcalf in Journ. Wuhan Bot. Res. 2.(1): 113. (1984).

Hubei: Yichang, 432, 2722.

Phyllanthus emblica L. (PHYLLANTHACEAE)

Forbes & Hemsley in Journ. Linn. Soc. Bot. 26: 421. (1894).

Hainan: Environs of Kiungchow, 8762.

Phyllanthus flexuosus (Sieb. & Zucc.) Muell.-Arg. (PHYLLANTHACEAE)

Forbes & Hemsley in Journ. Linn. Soc. Bot. 26: 421. (1894), Zheng in Journ. Wuhan Bot. Res. 2.(1): 113. (1984).

Hubei: Yichang, 1206, 3420a. Antelope Glen, 3420.

A shrub to about 3 metres tall, common in the glens around Yichang at low altitudes.

Phyllanthus reticulatus Poir. (PHYLLANTHACEAE)

Henry in Trans. Asiat. Soc. Jap. 24 suppl: 82. (1896) (List Pl. Formosa), Li in Woody Flora of Taiwan 437. fig. 153. (1963).

Phyllanthus sp. Henry in Trans. Asiat. Soc. Jap. 24 suppl: 82. (1896) (List Pl. Formosa).

Taiwan: Kaohsiung, 750, 751. Wanshoushan, 1090.

A scandent shrub, native to tropical Africa, India, southern China to Malaysia and Australia. In Chinese Taiwan this species is distributed at low altitude up to 1,700 metres in the central and southern parts of the island.

Phyllanthus sootepensis Craib (PHYLLANTHACEAE)

Phyllanthus subpulchellus Croizat in J. Jap. Bot. 26: 652. (1940).

Yunnan: Simao at 1,300 metres, 12118a., (holotype of *Phyllanthus subpulchellus* Croizat).

Phyllanthus sp. No. 1. (PHYLLANTHACEAE)

Anonymous in Kew List of Determinations (volume 24).

Hainan: 32 km (20 miles) west of Haikou, 8132.

Phyllanthus sp. No. 2. (PHYLLANTHACEAE)

Anonymous in Kew List of Determinations (volume 34).

Yunnan: Mengzi, 9301, 10924, 13704. Simao, 12096, 12118.

Phyllanthus urinaria L. (PHYLLANTHACEAE)

Forbes & Hemsley in Journ. Linn. Soc. Bot. 26: 423. (1894), Zheng in Journ. Wuhan Bot. Res.2. (1): 113. (1984).

Hubei: Yichang, 2788.

This species is currently used in Chinese herbal medicine as a diuretic and to treat cases of snake-bite.

Phyllanthus virgatus G. Forst. (PHYLLANTHACEAE)

Phyllanthus simplex Retz. Forbes & Hemsley in Journ. Linn. Soc. Bot. 26: 423. (1894), Henry in Trans. Asiat. Soc. Jap. 24 suppl: 82. (1896) (List Pl. Formosa), Zheng in Journ. Wuhan Bot. Res.2. (1): 113. (1984).

Hubei: San You Dong Glen, 4279. Jianshi, 7409. **Hainan:** Environs of Haikou, 8421. **Taiwan:** Wanchin, 868.

Phyllodium elegans (Lour.) Desv. (FABACEAE)

Desmodium elegans (Lour.) Benth. Anonymous in Kew List of Determinations (volume 34).

Yunnan: Manpan, 11210. Simao, 12488.

Distributed in Afghanistan, India, Bhutan, western China and Indochina.

Phyllodium pulchellum (L.) Desv. (FABACEAE)

Desmodium pulchellum (L.) Benth. Anonymous in Kew List of Determinations (volume 24), Henry in Trans. Asiat. Soc. Jap. 24 suppl: 33. (1896) (List Pl. Formosa).

Hainan: Ling-men, 8786. **Taiwan:** Kaohsiung, without number. Wanchin, 3. Oluanpi, 1000, 1236.

Distributed in India, Sri Lanka, Myanmar, China, Thailand, Laos, Vietnam and Cambodia.

Phyllostachys heteroclada Oliver (POACEAE) New species

in Journ. Linn. Soc. Bot. 36: 441. (1904), Bretschneider in Hist. Eur. Bot. Disc. China 2: 793. (1898).

Phyllostachys congesta Rendle in Journ. Linn. Soc. Bot. 36: 438. (1904).

Hubei: Liantuo, 1932, 3828. San You Dong Glen, 1698 (syntype of *Phyllostachys congesta* Rendle). Badong, 4828 (syntype of *Phyllostachys congesta* Rendle). Changyang, 6238 (syntype of *Phyllostachys congesta* Rendle).

A handsome bamboo to about 5 metres tall, the stems of this species were used for paper-making according to Henry (Henry's 4828 was 'used for paper-making'). According to Wilson who also saw the process, the culms were cut into lengths, made into bundles and immersed into concrete pits, being weighed down under the water using heavy stones. After three months of soaking they were removed and washed and then stacked in layers, each layer was sprinkled with lime and water. After two months the stems were well retted. The fiberous mass was then washed to remove all traces of lime, steamed for 15 days, washed and placed back in the concrete tanks. There it was reduced to a fine pulp and was then ready for conversion to paper. A portion of the pulp was placed in troughs with a mucilage prepared from the roots of *Abelmoschus manihot*. From there it was placed in bamboo trays with a fine-meshed bottom to allow drainage. The film that collected formed a moist sheet of paper.

Phyllostachys nidularia Munro (POACEAE)

Rendle in Journ. Linn. Soc. Bot. 36: 442. (1904).

Phyllostachys nigra Diels non (Lodd. ex Lindl.) Munro in Bot. Jahrb. 29: 227. (1900).

Hubei: Yichang, 450, 1322, 7868, 7871. Liantuo, 3823. Badong, 3671. Changyang, 7809.

An abundant species in western Hubei. Henry stated that at Changyang its tall, thin culms were once used for paper making and was known as *twei-chu*.

Phyllostachys nigra (Lodd. ex Lindl.) Munro (POACEAE)

Phyllostachys nana Rendle in Journ. Linn. Soc. Bot. 36: 441. (1904).

Hubei: Yichang, 3278 (type of *Phyllostachys nana* Rendle).

Phyllostachys nigra (Lodd. ex Lindl.) Munro var. *henonis* (Mitford) Rendle (POACEAE)

Phyllostachys henryi Rendle in Journ. Linn. Soc. Bot. 36: 440. (1904), Morley in Glasra 3: 80. (1979). *Phyllostachys nevinii* Hance var. *hupehensis* Rendle in Journ. Linn. Soc. Bot. 36: 442. (1904).

Hubei: Liantuo, 6338. San You Dong Glen, 3318 (type of *Phyllostachys nevinii* Hance var. *hupehensis* Rendle).

Phyllostachys reticulata (Rupr.) K. Koch (POACEAE)

Phyllostachys bambusoides Sieb. & Zucc. Henry in Notes Econ. Bot. China 50. (1893), Rendle in Journ. Linn. Soc. Bot. 36: 438. (1904).

Hubei: Liantuo, 3867, 4405, 4420. Zigui, 7900.

A vigorous bamboo to 20 metres tall, native to China and Japan. Henry stated that this tall bamboo was known as the *kui-chu* at Yichang. One of the most important and useful species of bamboo in Asia.

Phyllostachys sp. (POACEAE)

Anonymous in Kew List of Determinations (volume 24).

Hubei: Liantuo, 6348. **Sichuan:** South Wushan, 5674.

Phymatosorus cuspidatus (D. Don) Pic. Serm. (POLYPODIACEAE)

Polypodium leiorhizum Wall. ex Mett. Christ in Bull. Herb. Boiss. 6: 876. (1898).

Yunnan: Mengzi at 1,600 metres, 9219.

Phymatosorus scolopendria (Burm. f.) Pic. Serm. (POLYPODIACEAE)

Polypodium phymatodes L. Henry in Trans. Asiat. Soc. Jap. 24 suppl: 115. (1896) (List Pl. Formosa).

Taiwan: Oluanpi, 912, 1213. Lambay Isle, 1133 (collected August 15th 1893. Tamsui, 1430.

A creeping epilithic fern, native to Africa, India, southern China (including Taiwan), Japan (south), the Philippines, and Polynesia. This species is cultivated out of doors at Glanleam, on Valentia Island, Co Kerry, a garden established by the Knights of Kerry and was collected by the Glasnevin Taiwan Expedition (GTE 5) at Banana Bay Coastal Forest near Oluanpi (South Cape of Henry and Wilson) in October 2004.

Physalis angulata L. (SOLANACEAE)

Henry in Trans. Asiat. Soc. Jap. 24 suppl: 66. (1896) (List Pl. Formosa).

Physalis minima L. Forbes & Hemsley in Journ. Linn. Soc. Bot. 26: 174. (1890), Henry in Journ. China Br. Roy. Asiat. Soc. 22: 273. (1887) (Chinese Names of Plants), Anonymous in Kew List of Determinations (volume 24 & 34).

Hubei: Yichang, 2935. On the banks of the Yangtze at Yichang, 4216. South Donghu, 7712. **Hainan:** Environs of Haikou, 8022. **Taiwan:** Kaohsiung, 280, 1146. **Yunnan:** Mengzi, 9734.

This species was collected by the Glasnevin Central China Expedition (GCCE 726) near Yichang in October 2004.

Physalis sp. (SOLANACEAE)

Anonymous in Kew List of Determinations (volume 24).

Hubei: Yichang, 1081 (in part).

Phytolacca acinosa Roxb. (PHYTOLACCACEAE)

Henry in Journ. China Br. Roy. Asiat. Soc. 22: 256. (1887) (Chinese Names of Plants), Forbes & Hemsley in Journ. Linn. Soc. Bot. 26: 331. (1891), Walter in Engler, Pflanzenr. 83: 40. (1909).

Phytolacca esculenta van Houtte Walter in Engler, Pflanzenr. 83: 40. (1909).

Hubei: Yichang, 327, 4351. Antelope Glen, 3647. Liantuo, 2045. South Badong, 327 (June 1885), 5511. Without locality, 5511b., 5511c. **Sichuan:** South Wushan, 5511a., Henry's Chinese collector with Pratt, no locality stated but may be Leshan, Emei Shan, Wa Shan or Kangding, 8895.

According to Henry this species was known colloquially in the Three Gorges region as *shan lo po* and *t'ien lo po*. From its roots the drug *shang lu* was extracted, *shang lu* was used to treat cases of severe constipation. This species was collected by the Glasnevin Central China Expedition (GCCE 590) near Zhenzi town in Xingshan in September 2004.

Phytolacca sp. (PHYTOLLACACEAE)

Anonymous in Kew List of Determinations (volume 34).

Yunnan: Mengzi, 10705.

Picea brachytyla (Franch.) E. Pritz. (PINACEAE)

Rehder & E. H. Wilson in Sargent Pl. Wilson. 2: 33. (1916), Zheng in Journ. Wuhan Bot. Res. 2.(1): 9. (1984).

Picea ajanensis Masters non Fisch. in Journ. Linn. Soc. Bot. 26: 553. (1902).

Hubei: Fang Xian, 6908. **Sichuan:** North Wushan, 7074, 7074a., 7157.

A tree to 27 metres tall, native to central and western China, it is one of the most stately of all Chinese spruces. Wilson who later collected this species at Xingshan in Hubei stated it was once very common in north-western Hubei but

02

by 1900 so many trees had been felled for lumber that it was rare. Wilson also stated that it formed medium-sized open branched trees to 25 metres tall. Based on collections made by Delavay and Farges.

Picea wilsonii Masters (PINACEAE) New species

Zheng in Journ. Wuhan Bot. Res. 2.(1): 9. (1984), Bean in Trees & Shrubs 3: 198. (2002), O' Brien in Augustine Henry, An Irish Plant Collector in China 29. (2002).

Picea watsoniana Mast. Masters in Journ. Linn. Soc. Bot. 37: 419. 1906, Rehder & E. H. Wilson in Sargent Pl. Wilson 2: 27. (1913), Walsh in Aug. Henry Forst. Herb. 24. (1957).

Hubei: Fang Xian, 6763.

In its native habitat in north-west Hubei and Sichuan this species makes a large pyramidal tree. Discovered by Henry in 1888, Wilson later collected it in the mountains to the north of Yichang where it grew on old graves and made trees to 22 metres tall and almost 2 metres in girth. *Picea wilsonii* was introduced by Wilson in 1901 through Veitch's Coombe Wood nursery and is cultivated in several Irish gardens. The tallest known specimen in Britain and Ireland grows in the garden of the Earl and Countess of Rosse at Birr Castle, County Offaly. Planted in 1916, it was found to be 21 metres tall with a trunk diameter of 56 cm when last measured in 2000.

Picrasma quassioides (D. Don) Benn. (SIMAROUBACEAE)

Henry in Journ. China Br. Roy. Asiat. Soc. 22: 255. (1887) (Chinese Names of Plants), Henry in Notes Econ. Bot. China 37. (1893), Rehder & E. H. Wilson in Sargent Pl. Wilson 2: 152. (1916), Zheng in Journ. Wuhan Bot. Res. 2.(1): 109. (1984).

Hubei: Liantuo, 2013, 3821, 3901. Xingshan, 7570. Jianshi, 7570a. Changleping, 7970b. Changyang, 7970c. **Sichuan:** South Wushan, 5491, 5534, 5534a., 5649.

A small deciduous tree with an extensive geographical range from the Himalaya, across China, Korea and into Japan. Common throughout central and western China at low altitudes. According to Henry it was known colloquially as *ku-lien-shu* (*shu* meaning tree) and its bark, known as *ku lien tzu*, supplies a valuable drug that was sometimes used as an insecticide or for treating colic, stomach pains and as a febrifuge. It is rare in cultivation but a wonderful sight in autumn when it turns a rich scarlet hue.

Picris hieracioides L. (ASTERACEAE)

Forbes & Hemsley in Journ. Linn. Soc. Bot. 23: 474. (1888), Anonymous in Kew List of Determinations (volume 34).

Hubei: Yichang, 2164. Badong, 950 (November 1885), 1798, 1832, 1833, 2467, 3137, 4710, 4714, 6074, 7368. Liantuo, 3940. Xingshan at 1,400 metres, 6549. Fang Xian,

6586. **Yunnan:** In Pingbian on the Daweishan Range at 1,950 metres, 9051. Mengzi, 10227.

Distributed across temperate Eurasia.

Pieris formosa (Wall.) D. Don (ERICACEAE)

Judd in Journ. Arnold Arb. 63: 122. (1982), Zheng in Journ. Wuhan Bot. Res. 2.(1): 167. (1984).

Hubei: Fang Xian, 6770. South Badong, 6081. **Yunnan:** Mengzi, 11295.

A large evergreen shrub or small tree to 6 metres tall, distributed across the Himalaya from Nepal to India (Assam) and eastwards to Myanmar and south-west and central China and Vietnam. An extremly variable species on account of its wide geographical range. *Pieris formosa* was collected by the Glasnevin Central China Expedition (GCCE 153) in the *Metasequoia* valley in September 2002. *Pieris formosa* (Wall.) D. Don var. *forrestii* (Harrow) Airy-Shaw is distinguished by its erect inflorescences, Lord Barrymore from Fota near Cobh in County Cork (a sponsor of the Scottish plant hunter, George Forrest) raised several forms of this variety and many of these still grow in his arboretum today. Some have recently been named as cultivars.

Pigea enneasperma (L.) P. I. Forst. (VIOLACEAE)

Ionidium suffruticosum (L.) Roth. ex Schult. Anonymous in Kew List of Determinations (volume 24).

Hainan: Hummocks near Haikou, 8101. On Liang Shan, 16 km (10 miles) east of Haikou, 8212. Environs of Haikou, 8435.

A small herb to about 20 cm tall, a widespread species native to Africa, Madagascar, south-east China, the Philippines, Borneo, Java, New Guinea and Australia.

Pilea angulata (Blume) Blume (URTICACEAE)

Pilea stipulosa (Miq.) Miq. Forbes & Hemsley in Journ. Linn. Soc. Bot. 26: 478. (1899).

Hubei: South Badong, 5352.

Pilea anisophylla Wedd. (URTICACEAE)

Forbes & Hemsley in Journ. Linn. Soc. Bot. 26: 475. (1899).

Taiwan: Wanchin, 1634. **Yunnan:** Mengzi, 11287.

Native to India, Bhutan, Nepal, Myanmar and China [Tibet (Xizang) and Yunnan].

Pilea gracilis Handel-Mazzetti. (URTICACEAE)

in Symb. Sin. 7: 148. (1929).

Yunnan: Simao, 12556 (type of *Pilea gracilis* Handel-Mazzetti), 13167 (isosyntype of *Pilea gracilis* Handel-Mazzetti), 13167a., (syntype of *Pilea gracilis* Handel-Mazzetti).

Pilea hookeriana Wedd. (URTICACEAE)

Forbes & Hemsley in Journ. Linn. Soc. Bot. 26; 476. (1899).

Yunnan: Mengzi, 11339.

Distributed in Nepal, India, Bhutan, to China (Yunnan) and Vietnam.

Pilea lomatogramma Handel-Mazzetti (URTICACEAE)
Pilea oxyodon Forbes & Hemsley non Wedd. in Journ. Linn. Soc. Bot. 26: 477. (1899).

Hubei: Jianshi, 5826.

Pilea notata C. H. Wright (URTICACEAE) New species
Forbes & Hemsley in Journ. Linn. Soc. Bot. 26: 476. (1899).

Hubei: Yichang, 2233, 4229a., 2706 (isosyntype). Antelope Glen, 4429 (isosyntype). Liantuo, 2592. South Badong, 7288 (isosyntype).

Pilea plataniflora C. H. Wright (URTICACEAE) New species
in Journ. Linn. Soc. Bot. 26: 477. (1899).

Yunnan: In mountain forest on the ascent of Mount Benvenu at 2,000 metres, 9781 (type). Mengzi, 9781a.

A widespread erect perennial distributed through northern India, Sri Lanka, China (including Hainan and Taiwan), Thailand, Vietnam, Japan and Java.

Pilea rubriflora C. H. Wright. (URTICACEAE) New species
in Journ. Linn. Soc. Bot. 26: 478. (1889).

Hubei: Xiling (Yichang), Gorge, 3333.

Pilea scripta (Buch.-Ham. ex D. Don) Wedd. (URTICACEAE)
Forbes & Hemsley in Journ. Linn. Soc. Bot. 478. (1899).

Yunnan: Mengzi, 10414, 11393.

Distributed in India, the Himalaya, Myanmar and China [Tibet (Xizang) and Yunnan].

Pilea semisessilis Handel-Mazzetti (URTICACEAE) New species
in Symb. Sin. 7: 137. (1929).
Pilea symmeria Forbes & Hemsley non Wedd. in Journ. Linn. Soc. Bot. 26: 479. (1899).

Yunnan: In woods on the mountains to the south-west of Mengzi at 1,829 metres, 9790 (type). Milê District, 10453.

Native to China and Thailand.

Pilea sp. (intermediate between *Pilea umbrosa* & *Pilea wightii*.) (URTICACEAE)
Forbes & Hemsley in Journ. Linn. Soc. Bot. 26: 479. (1899).

Yunnan: Mengzi, 11230.

Pilea spp. (URTICACEAE)
Forbes & Hemsley in Journ. Linn. Soc. Bot. 26: 479. (1889), Henry in Trans. Asiat. Soc. Jap. 24 suppl: 89. (1896) (List Pl. Formosa).

Hubei: Yichang, 2046, 4352a., 4352b., 4352c. San You Dong Glen, 4352. **Hainan:** Environs of Kiungchow, 8749. **Taiwan:** Oluanpi, 1223. **Yunnan:** Simao, 12007, 12243, 12460, 13139, 13185, 13352, 13534.

Pilea swinglei Merr. (URTICACEAE)
Pilea sp. Hemsley & Forbes in Journ. Linn. Soc. Bot. 480. (1899). *Pilea henryana* C. H. Wright in Journ. Bot. 20. (1920).

Hubei: South Badong, 7295 (type of *Pilea henryana* C. H. Wright).

A small perennial herb to 20 cm tall, distributed in Guangdong, Guangxi, Guizhou, Hubei, Jiangxi and Zhejiang.

Pilea symmeria Wedd. (URTICACEAE)
Forbes & Hemsley in Journ. Linn. Soc. Bot. 26: 479. (1899).

Hubei: Badong, 1815, 4043, 4989, 4752, 6178, 7340.

Pilea umbrosa Wedd. ex Blume (URTICACEAE)
Forbes & Hemsley in Journ. Linn. Soc. Bot. 26: 479. (1899).

Yunnan: On the peak of Fengchunling at 2,300 metres, 10896.

Distributed in Nepal, India, Bhutan and China [Tibet (Xizang) and Yunnan].

Pilea villicaulis Handel-Mazzetti, (URTICACEAE) New species
in Symb. Sin. 7(1): 125. (1929).

Yunnan: Simao, 13134 (type), 13195 (syntype, collected 16th July 1899).

Endemic to Yunnan.

Pilophorus acicularis (Ach.) Th. Fr. (CLADONIACEAE)
Muell.-Arg. in Bull. Herb. Boissier 1: 235. (1893), Wei in An Enumeration of Lichens in China 205. (1991).

Hubei: Xingshan, 6965, (in part). (Lichen).

Pimpinella bialata Wolff (APIACEAE) New species
in Repert. Spec. Nov. Regni Veg. 27: 188. (1930).

Hubei: Liantuo, 2649 (type).

Pimpinella coriacea (Franch.) H. Boissieu (APIACEAE)
Wolff. in Engler, Pflanzenr. iv. 228: 268. (1927).
Pimpinella diversifolia Wollf non DC. in Engler, Pflanzenr. iv. 228: 269. (1927). *Pimpinella wolffiana* Fedde in Repert. Spec. Nov. Regni Veg. 27: 185. (1930). *Pimpinella pseudocandolleana* H. Wolff in Repert. Spec. Nov. Regni Veg. 27: 189. (1929).

Yunnan: Mengzi at 1,650 metres, in grass, 9588 (isotype *Pimpinella pseudocandolleana* H. Wolff). Mountains to the west of Simao at 1,650 metres, 12485 (type of *Pimpinella wolffiana* Fedde). Simao, 12517.

Endemic to China and distributed in the western provinces.

02

Pimpinella diversifolia DC. (APIACEAE)

Forbes & Hemsley in Journ. Linn. Soc. Bot. 23: 329. (1887), Wolff in Engler, Pflanzenr. iv. 228: 269. (1927).

Hubei: Yichang, 85, 452, 1057, 2363, 2363a. Antelope Glen, 3256. Xiling (Yichang) Gorge, 4354. Liantuo, 4447. Badong, 4783, 4931. **Sichuan:** South Wushan, 7208.

Pimpinella fargesii H. Boissieu (APIACEAE) New species

Wolff in Engler, Pflanzenr. iv. 228: 279. (1927).

Sichuan: Wushan, without number.

Pimpinella henryi Diels (APIACEAE) New species

in Engl. Bot. Jahrb. 29: 495. (1900), Wolff in Engler, Pflanzenr. iv. 228: 282.(1927).

Pimpinella sutchuensis H. Boissieu in Bull. Herb. Boiss. 2. ser. 2. 808. (1902), Wolff in Engler, Pflanzenr. iv. 228: 279. (1927).

Sichuan: Wushan, 7101.

Pimpinella kingdon-wardii H. Wolff (APIACEAE) New species

Pimpinella thyrsiflora H. Wolff in Rep. Spec. Nov. Regni. Veg. 27: 320. (1930).

Yunnan: Mountains near Mengzi at 2,300 metres, 10225 (type of *Pimpinella thyrsiflora* Wolff).

Discovered by Henry in 1897 and described from a later collection made by Kingdon Ward.

Pimpinella renifolia H. Wolff (APIACEAE) New species

in Repert. Spec. Nov. Regni Veg. 27: 191. (1930).

Pimpinella diversifolia Forbes & Hemsley non DC. in Journ. Linn. Soc. Bot. 23: 329. (1887).

Hubei: Xiling (Yichang) Gorge, 2923 (type).

Pimpinella spp. (APIACEAE)

Anonymous in Kew List of Determinations (volume 24).

Hubei: Badong, 2419, 2516, 3155, 4854, 4907, 4936, 4941, 4948, 4968, 5095, 7275, 7331, 7353. Donghu, 6404. Xingshan, 6516. North Badong, 7167.

Pinalia formosana (Rolfe) Ormerod (ORCHIDACEAE) New species

Eria formosana Rolfe in Kew Bull. Misc. Inf. 194. (1896), Henry in Trans. Asiat. Soc. Jap. 24 suppl: 91. (1896) (List Pl. Formosa), Bretschneider in Hist. Eur. Bot. Disc. China 2: 792. (1898), Rolfe in Journ. Linn. Soc. Bot. 36: 16. (1903), Kranzlin in Engler, Pflanzenr. 50: 47. (1911).

Taiwan: Oluanpi, 1978 (type of *Eria formosana* Rolfe).

Pinanga sylvestris (Lour.) Hodel. (ARECACEAE)

Pinanga chinensis Beccari in Webbia 1: 326. (1905).

Yunnan: Near Simao at 1,300 metres, 12874 (isotype of *Pinanga chinensis* Beccari).

India (Assam) to Indochina.

Pinellia integrifolia N. E. Brown (ARACEAE) New species

in Hooker's. Icon. Pl. t. 1875. (1889), Bretschneider in Hist. Eur. Bot. Disc. China 2: 793. (1898), Brown in Journ. Linn. Soc. Bot. 36: 174. (1903), Engler, in Engler, Pflanzenr. iv. 23F: 221. (1920), Morley in Glasra 3: 80. (1979), Nelson in A Heritage of Beauty 316. (2000).

Hubei: Yichang, 633 (isotype). Xiling (Yichang) Gorge, 4323 (isosyntype).

Pinellia ternata (Thunb.) Makino (ARACEAE)

Pinellia tuberifera Brown non Tenore in Journ. Linn. Soc. Bot. 36: 174. (1903), Henry in Notes Econ. Bot. China 34. (1893). *Pinellia ternata* (Thunb.) Makino var. *vulgaris* Engler in Engler, Pflanzenr. 73.(iv. 23f.): 224. (1920).

Hubei: San You Dong Glen, 3646 [syntype of *Pinellia ternata* (Thunb.) Makino var. *vulgaris* Engler].

According to Henry this species was known at Yichang as *san-pu-tiao*, the tubers which were used as a drug were the *pan-hsia* (*ban xia* or 'half summer'), the same drug is administered to stop vomiting.

Pinus armandii Franch. (PINACEAE)

Henry in Notes Econ. Bot. China 35. (1893), Stapf in Bot. Mag. t. 8347. (1910), Henry in Elwes & Henry in The Trees of Gt. Brit. & Irel. 5: 1043. (1910), Rehder & E. H. Wilson in Sargent Pl. Wilson. 2: 12. (1914), E. H. Wilson in Journ. Arnold Arb. 7: 45. (1926), Nelson in Pim, The Wood & the Trees 222. (1984); in A Heritage of Beauty 316. (2000), O' Brien in Augustine Henry, An Irish Plant Collector in China 29. (2002).

Pinus scipioniformis Masters in Bull. Herb. Boiss. vi. 270. (1898); in Journ. Linn. Soc. Bot. 35: 552. (1902), Bretschneider in Hist. Eur. Bot. Disc. China 2: 791. (1898), Masters in Journ. Linn. Soc. Bot. 37: 415. (1906), Patschke in Bot. Jahrb. xlviii. 657. (1912). *Pinus mandschurica* Masters non Rupr. in Journ. Linn. Soc. Bot. 26: 551. (1902). *Pinus densiflora* Masters non Sieb. & Zucc. in Journ. Linn. Soc. Bot. 26: 549. (1902).

Hubei: Yichang, without number, (type of *Pinus scipioniformis* Masters). **Yunnan:** Milê District, 9868. Near Mengzi at 2,300 metres, 10519.

A noble tree to 35 metres tall, native to western and central China and Myanmar. *Pinus armandii* was discovered by the French missionary, Père Armand David in 1873 in Shaanxi province and was introduced to cultivation by Père Farges who sent seeds to Maurice de Vilmorin at Les Barres in 1895. Two years later Augustine Henry introduced it to Britain when he sent seeds from Mengzi in southern Yunnan to the Royal Botanic Gardens, Kew. By 1910 these trees were between 3 and 5 metres tall and were said to be the finest trees in Europe. One of his trees still grows there. In parts of south-east Tibet (Xizang) there still exist great forests of this species, as yet untouched by man. It was known in China as *peh-sung* (white pine) though Henry's Chinese collectors from Hubei called it the *niu-sung*. The seeds are roasted and eaten in Yunnan. The tallest known specimen in Britain and Ireland grows at the Fota Arboretum in Co Cork, it was found to be 29 metres tall with a trunk diameter of 101 cm when last measured in 1999. This species was collected by the Glasnevin Central China Expedition (GCCE 538) near Zhenzi town in Xingshan.

Pinus bungeana Zucc. ex Endl. (PINACEAE)

Henry in Journ. China Br. Roy. Asiat. Soc. 22: 269. (1887) (Chinese Names of Plants), Henry in Notes Econ. Bot. China 35. (1893), Bretschneider in Hist. Eur. Bot. Disc. China 2: 832. (1898), Masters in Journ. Linn. Soc. Bot. 26: 549.(1902), Henry in Elwes and Henry The Trees Gt. Brit. & Irel. 5: 1050. (1910).

Hubei: Yichang, 3274.

A tree to 30 metres tall with smooth, peeling white bark. Originally found by Bunge near Beijing (Peking) in 1831, it was from these cultivated trees that the species was described. The Beijing trees were planted by Buddhist temples and courtyards (always in pairs) and in tomb enclosures of the imperial classes, hense the superstition that to thrive they should be planted on the grave of a dead Manchu. *Pinus bungeana* was introduced to cultivation by Robert Fortune in 1846 by means of living plants. Henry's collections were the first records of truly wild trees in China. Wilson later found wild trees at Badong where he stated it was rare though Henry stated the following in *The Trees of Great Britain and Ireland,* 'Wilson found it south-west of Ichang on precipitous mountains at 2,000 to 4,000 feet. He saw many hundreds of trees scattered for miles, evidently the remains of a conciderable forest. It was also collected in southern Shensi by Père Giraldi.'. It is held in high esteem by the Chinese as an ornamental and Henry stated it was known as *peh sung* (white pine). In cultivation the lace-bark pine has remained a great rarity. At the Royal Horticultural Society's garden, Wisley it has reached 14 metres tall with a trunk diameter of 31 cm (2000).

Pinus henryi Masters (PINACEAE) New species

in Journ. Linn. Soc. Bot. 26: 550. (1902), Henry in Elwes and Henry The Trees Gt. Brit. & Irel. 5: 1126. (1910), E. H. Wilson in A Naturalist in Western China 39. (1913), Morley in Glasra 3: 80. (1979), Nelson in A Heritage of Beauty 316. (2000).

Pinus densiflora Henry non Sieb. & Zucc. in Notes Econ. Bot. China 35. (1893), Zheng in Journ. Wuhan. Bot. Res. 2(1): 9. (1984). *Pinus sinensis* Rehder & E. H. Wilson non Lambert in Sargent Pl. Wilson. 2: 16. (1916). *Pinus tabuliformis* Carr. var. *henryi* (Masters) C. T. Kuang in Fl. Sichuanica 2: 113 (1983).

Sichuan: South Wushan, 5327a. **Hubei:** South Badong, 5327b. Fang at 2,300 metres, 6909 (type of *Pinus henryi* Masters).

Henry's pine, a tree to 27 metres tall in its native habitat. According to Henry this species formed woods above altitudes of 1,300 metres.

Pinus massoniana Lamb. (PINACEAE)

Henry in Journ. China Br. Roy. Asiat. Soc. 22: 268. (1887) (Chinese Names of Plants), Henry in Notes Econ. Bot. China 35. (1893), Masters in Journ. Linn. Soc. Bot. 26: 551. (1902).

Hubei: Yichang, 1212, 3275. **Yunnan:** Mengzi, 10380.

The most common pine in Hubei, it is especially common in the *Metasequoia* valley where it was observed by the Glasnevin Central China Expedition (2002) growing with *Keteleeria davidiana* (Bertr.) Beissn. Henry stated that this species and other Chinese pines were known by the general name *sung shu* (pine tree). According to Wilson, pine soot, obtained by burning the branches of this common pine was used in 19th century China as a black dye for cotton goods. Rare in cultivation, a tree planted in 1957 at the Royal Botanic Gardens, Kew measures 13 metres tall with a trunk diameter of 44 cm (2001). This species was collected by the Glasnevin Central China Expedition (GCCE 578) in Xingshan in September 2004.

Pinus sp. cembra sect. (PINACEAE)

Masters in Journ. Linn. Soc. Bot. 38: 551. (1902).

Hubei: Badong, 5327.

Piper boehmeriifolium (Miquel) C. DC. var. *glabricaule* (C. DC.) Gilbert & Xia (PIPERACEAE)

New variety. in Novon 9: 191. (1999).

Piper glabricaule C. de Candolle in Notizbl. Bot. Gart. Berlin-Dahlem. 6(62): 477. (1918).

Yunnan: Mengzi at 1,600 metres, 9482 (isotype of *Piper glabricaule* C. de Candolle), 9482a., (holotype of *Piper glabricaule* C. de Candolle).

India (Assam) to China (Yunnan) and Malesia.

Piper flaviflorum (Miq.) Hand.-Mazz. (PIPERACEAE) New species

Piper flaviflorum C. de Candolle in Notizbl. Bot. Gart. Berlin-Dahlem 7: 477. (1917).

Yunnan: In forests near Simao at 1,650 metres, 11767 (isosyntype of *Piper flaviflorum* C. DC). To the south of Simao at 1,300 metres, 12931 (isosyntype of *Piper flaviflorum* C. DC).

02

Piper hainanense Hemsley (PIPERACEAE) New species

in Journ. Linn. Soc. Bot. 26: 365. (1891).

Hainan: Without locality, 8528.

Hemsley based this species on material first collected by the American missionary, the Rev'd. B. C. Henry and from specimens later collected in 1889 by Henry. Endemic to China and native to the provinces of Guangdong, Guangxi and Hainan.

Piper cf. ***hancei*** Maxim. (PIPERACEAE)

Henry in Trans. Asiat. Soc. Jap. 24 suppl: 77. (1896) (List Pl. Formosa).

Taiwan: Oluanpi, 1997.

Piper kadsura (Choisy) Ohwi (PIPERACEAE)

Henry in Trans. Asiat. Soc. Jap. 24 suppl: 77. (1896) (List Pl. Formosa), Li in Woody Flora of Taiwan 67. fig. 17. (1963).

Piper subglaucescens C. DC. in Ann. Cons. Jard. Geneve. 21: 222. (1920).

Taiwan: Kaohsiung, 715 (type of *Piper subglaucescens* C. DC.), 715c., 1134, 1870, 1886. Lambay Isle, 803 (collected August 15th 1893. Wanchin, 1526, 1581.

Found in coastal regions in Taiwan.

Piper macropodum C. de Candolle (PIPERACEAE) New species

in Bull. Herb. Boissier ser. 2, 4: 1026. t. 4. (1904), Gilbert & Xia in Novon 9: 193. (1999).

Piper szemaoense C. DC. in Notizbl. Konigl. Bot. Gart. Berlin-Dahlem 6: 481. (1917).

Yunnan: Mengzi, 10726. Simao, 12210 (isosyntype of *Piper szemaoense* C. DC.), 12210a., (isosyntype of *Piper szemaoense* C. DC.). Simao at 1,450 metres, 12210d., (holotype). Simao at 1,370 metres, 12210b., (lectotype of *Piper szemaoense* C. DC.).

Endemic to Yunnan.

Piper nudibaccatum Y. T. Tsang (PIPERACEAE)

Piper betle L. var. *psilocarpum* C. de Candolle in Notizbl. Bot. Gart. Berlin-Dahlem 6(62): 477. (1917).

Yunnan: In the mountains near Simao, 11767a., (isotype of *Piper betle* L. var. *psilocarpum* C. de Candolle).

Endemic to Yunnan.

Piper pedicellatum C. de Candolle (PIPERACEAE)

Piper curtipedunculum C. de Candolle in Notizbl. Bot. Gart. Berlin-Dahlem 6: 481. (1917).

Yunnan: Mengzi at 1,950 metres, 10438, (holotype of *Piper curtipedunculum* C. DC., type of *Piper pedicellatum* C. de Candolle).

Native to India, Bangladesh, Bhutan, China (Yunnan) and Vietnam.

Piper puberulilimbum C. de Candolle (PIPERACEAE) New species

in Notizbl. Bot. Gart. Berlin-Dahlem 6(62): 479. (1917).

Yunnan: South of the Red River from Mengzi, 13680 (isotype).

Endemic to Yunnan.

Piper retrofractum Vahl (PIPERACEAE)

Piper chaba W. Hunter Henry in Trans. Asiat. Soc. Jap. 24 suppl: 77. (1896) (List Pl. Formosa).

Taiwan: Oluanpi, cultivated, 2006.

Piper sarmentosum Roxb. (PIPERACEAE)

Forbes & Hemsley in Journ. Linn. Soc. Bot. 26: 366. (1891).

Hainan: 32 km (20 miles) west of Haikou, 8139. Environs of Kiungchow, 8737.

Piper semiimmersum C. de Candolle (PIPERACEAE) New species

in Notizbl. Bot. Gart. Berlin-Dahlem 6(62): 479. (1917).

Yunnan: In Pingbian, in forests on the Daweishan Range at 1,650 metres, 11037 (isotype).

Piper spp. (PIPERACEAE)

Anonymous in Kew List of Determinations (volume 34).

Yunnan: Mengzi, 9400, 9577, 10393, 11231, 11408. Simao 11635, 12155, 12786, 13012, 13039, 13533.

Piper umbellatum L. (PIPERACEAE)

Piper subpeltatum Willd. Henry in Trans. Asiat. Soc. Jap. 24 suppl: 77. (1896) (List Pl. Formosa), Li in Woody Flora of Taiwan 68. (1963).

Taiwan: Wanchin, 480. Oluanpi, 700.

Native to the Americas, widely naturalised in tropical regions of the world, in Taiwan this species is distributed on the southern part of the island.

Piper wallichii (Miq.) Hand.-Mazz. (PIPERACEAE)

Gilbert & Xia in Novon 9: 196. (1999).

Piper futokadsura Henry non Sieb. in Journ. China Br. Roy. Asiat. Soc. 22: 263. (1887) (Chinese Names of Plants). *Piper aurantiacum* Wallich. Forbes & Hemsley in Journ. Linn. Soc. Bot. 26: 364. (1891), Henry in Notes Econ. Bot. China 50. (1893). *Piper aurantiacum* Wall. ex DC. var. *hupeense* C. DC. in Notizbl. Konigl. Bot. Gart. Berlin-Dahlem 6: 478. (1917). *Piper wallichii* (Miq.) H.-M. var. *hupeense* (C. DC.) Hand.-Mazz. Zheng in Journ. Wuhan. Bot. Res. 2.(1): 13. (1984).

Hubei: Yichang, in the Antelope Glen, 3120, 3311. Yichang, 3311a. Liantuo, 3893 (lectotype of *Piper aurantiacum* Wall. ex C. DC. var. *hupeense* C. DC.).

According to Henry this species was known

colloquially in the Three Gorges region as *p'a ai hsiang*. He stated it was a common liana on the cliffs of the antelope glen near Yichang.

Piptanthus nepalensis (Hook.) Sweet (FABACEAE)

Anonymous in Kew List of Determinations (volume 34).

Yunnan: Mengzi, 10230.

Native to India, Bhutan, Nepal, Myanmar and China .

Pisonia aculeata L. (NYCTAGINACEAE)

Forbes & Hemsley in Journ. Linn. Soc. Bot. 26: 317. (1891), Henry in Trans. Asiat. Soc. Jap. 24 suppl: 74. (1896) (List Pl. Formosa), Li in Woody Flora of Taiwan 153. fig. 55. (1963).

Hainan: Environs of Kiungchow, 8226, 8248. Environs of Haikou, 8387. Without locality, 8470. **Taiwan:** Kaohsiung, without number. Wanchin, 87.

A pantropic scandent, prickly shrub, widespread across India, Malaysia, Indonesia and southern China. According to Henry, the flower buds of this species were used for 'blackening the teeth'. This was presumably by the 19th century indigenous inhabitants of Chinese Taiwan.

Pisonia grandis R. Br. (NYCTAGINACEAE)

Yang & Lu in Flora of Taiwan 2: 324. (1996).

Pisonia inermis G. Forst. Henry in Trans. Asiat. Soc. Jap. 24 suppl: 74. (1896) (List Pl. Formosa).

Taiwan: Wanchin, 137, 1596.

Henry's Chinese collector named this, *shui-tung-kua*. In different parts of China this name is applied to different trees, see Henry in *Notes on Economic Botany of China* 55. (1893).

Pistacia chinensis Bunge (ANACARDIACEAE)

Henry in Journ. China Br. Roy. Asiat. Soc. 22: 246. (1887) (Chinese Names of Plants), Henry in Notes Econ. Bot. China 38. (1893), Rehder & E. H. Wilson in Sargent Pl. Wilson 2: 173. (1916), Zheng in Journ. Wuhan Bot. Res. 2.(1): 116. (1984).

Hubei: Yichang, 3402. South Donghu, 7702. San You Dong Glen near Yichang (July 1886), 1500. Liantuo, 3809, 4471. **Taiwan:** Wanchin, 488. **Yunnan:** In woods near Mengze at 1,500 metres, 10254, 10254a.

A large deciduous tree to 27 metres tall. According to Henry this tree was known colloquially in the Three Gorges region as *huang lien ya*. Wilson summed up this tree when he called it 'one of the noblest, most widely distributed and useful of Chinese trees.' A glance at Henry's localities shows its wide geographical distribution. Fr. Hugh Scallan, a missionary from Dublin found it in north-central China, William Purdom found it in Shaanxi, Faber and Macgregor (at different dates) collected it near Ningpo in Zhejiang and Frank Meyer collected material in Shanxi in August

1907, just to give a few examples of the tree's widespread localities in China. In Hubei it is found in valleys to an altitude of about 1,000 metres. The young leafy shoots are red on emergence and eaten in the same manner as *Toona sinensis* (A. Juss.) M. Roem. In autumn the tree assumes wonderful autumnal tints of orange and crimson. The close-grained wood is tough and easily worked and was much used in boat construction in 19th century China. A log having a natural fork at one end was once used to balance the rudders of larger ships plying the Yangtze. *Pistacia chinensis* was introduced to Kew in 1897 and the tallest specimen in Britain and Ireland grows at Kew. When measured in 2001 this tree was found to be 5 metres tall with a girth of 28 cm at 1 metre above ground level. Native from the Transcaucasus to Afghanistan to China and the Philippines.

Pistacia weinmanniifolia J. Piosson ex Franchet (ANACARDIACEAE)

Rehder & E. H. Wilson in Sargent Pl. Wilson. 2: 174. (1914).

Yunnan: Mengzi plain at 1,500 metres, 9600, 9600a. Lunan, 9600a. Mountains to the west of Simao at 1,600 metres, 11913.

A tree to 20 metres tall, very common in parts of Yunnan and one of the most handsome trees found there. Wilson also collected it in nearby Sichuan for Messrs Veitch. According to Bretschneider it was discovered by Delavay in either 1882 or 1883.

Pisum sativum L. (FABACEAE)

Henry in Trans. Asiat. Soc. Jap. 24 suppl: 34. (1896) (List Pl. Formosa).

Taiwan: Kaohsiung, cultivated without number.

The garden pea, cultivated and called *ho-lan-tou*. In the Three Gorges area Henry also stated that the pea was known was *me wan tzu*. In many parts of China the young green shoots of the pea are stir fried and are delicuous when cooked this way. The pea reached China from India during the Tang dynasty (AD. 618-906). The garden pea is not known in the wild but is thought to have originated in south-east Asia.

Pithecellobium dulce (Roxb.) Benth. (FABACEAE)

Henry in Trans. Asiat. Soc. Jap. 24 suppl: 40. (1896) (List Pl. Formosa).

Taiwan: Tainan, cultivated, 1804.

The Manila tamarind, a spiny shrub to 1.2 metres, native to Mexico and Central America. Cultivated and naturalised in Chinese Taiwan, Henry called it the *chin-kuei-shu*.

Pithecellobium sp. (FABACEAE)

Anonymous in Kew List of Determinations (volume 34).

Yunnan: Mengzi, 11505. Simao, 11659, 11682.

Pittosporopsis kerrii Craib (ICACINACEAE) New genus

Sleumer in Blumea 17: 241. (1969).

Pittosporum nervosum (Gagnep.) Gowda in Journ. Arnold Arb. 32: 327. (1951).

Yunnan: Simao at 1,525 metres, 11778, 11778a., 11778b.

This genus was first found by Augustine Henry near Simao though it was described by Craib from material collected in northern Thailand by Kerr. Distributed in Myanmar, China (Yunnan) and Thailand.

Pittosporum brevicalyx (Oliv.) Gagnep. (PITTOSPORACEAE) New species

Gowda in Journ. Arnold Arb. 32: 329. (1951), Zheng in Journ. Wuhan Bot. Res. 2.(1): 63. (1984).

Pittosporum pauciflorum Hook. & Arn. var. *brevicalyx* Oliver in Hooker's Icon. Pl. 16: t. 1579. (1887). *Pittosporum glabratum* Rehder & E. H. Wilson non Lindl. in Sargent Pl. Wilson. 3: 326. (1917).

Hubei: Xiling (Yichang) Gorge, 1144. Yichang, 2957. Liantuo, 4455, 4503. Jianshi, 5999 (type of *Pittosporum pauciflorum* Hook. & Arn. var. *brevicalyx* Oliver).

Pittosporum glabratum Lindl. (PITTOSPORACEAE)

Rehder & E. H. Wilson in Sargent Pl. Wilson. 3: 326. (1917).

Hubei: Without locality, 4369a., 4369b.

An evergreen shrub to 1.5 metres tall carrying racemes of yellow scented flowers. in May Widespread over warm-temperate and subtropical China and introduced to cultivation by Robert Fortune from Hong Kong in 1845. Reintroduced by Wilson from Yichang in 1908.

Pittosporum glabratum Lindl. var. *neriifolium* Rehder & E. H. Wilson (PITTOSPORACEAE)

in Sargent Pl. Wilson. 3: 328. (1916), Gowda in Journ. Arnold Arb. 32: 293. (1951), Zheng in Journ. Wuhan Bot. Res. 2.(1): 63. (1984).

Hubei: Badong, 5359, 5422. Jianshi, 5422a.

This variety was first collected in Assam by Hooker and Thomson and in eastern Bengal by Griffith. It was first found in China by Henry and was later collected in western Sichuan by Wilson and on Emei Shan by Fr. Hugh Scallan in 1899.

Pittosporum henryi Gowda (PITTOSPORACEAE) New species

in J. Arnold Arbor. 32: 319. (1951).

Pittosporum glabratum Rehder & E. H. Wilson non Lindl. in Sargent Pl. Wilson. 3: 326. (1916), in part.

Sichuan: North Wushan, 7042 (holotype).

Pittosporum kerrii Craib (PITTOSPORACEAE) New species

Gowda in Journ. Arnold Arb. 32: 333. (1951).

Yunnan: Mengzi at 1,450 to 1,6509 metres, 10041. Simao at 1,650 metres 12219, 12219a., 12219b., 12219c., 12219d., 12219e.

An evergreen shrub or small tree to 12 metres tall. *Pittosporum kerrii* was described by Craib from collections by the Irish collector, Arthur Francis George Kerr (1877—1942) in Thailand in June 1921, Augustine Henry had discovered the species in Yunnan over two decades previously. Henry described the trees he collected from as being between 6 to 9 metres tall with yellow flowers. Native to India (Assam), Myanmar, China (Yunnan), Laos and Thailand.

Pittosporum moluccanum (Lam.) Miquel. (PITTOSPORACEAE)

Li in Woody Flora of Taiwan 258. (1963), Li & Huang in Flora of Taiwan 3: 66. (1993).

Pittosporum sp. nova. Henry in Trans. Asiat. Soc. Jap. 24 suppl: 18. (1896) (List Pl. Formosa). *Pittosporum formosanum* Rehder & E. H. Wilson non Hayata in Sargent Pl. Wilson. 3: 330. (1916).

Taiwan: Wanchin, 822, 1533. Oluanpi, 1058a.

A small evergreen shrub, native to Malaysia and China (Taiwan) where it grows by the coast on the southern part of the island.

Pittosporum pauciflorum Hook. & Arn. (PITTOSPORACEAE)

Anonymous in Kew List of Determinations (volume 34).

Yunnan: Mengzi, 9452.

Pittosporum pauciflorum Hook. & Arn. var. (PITTOSPORACEAE)

Anonymous in Kew List of Determinations (volume 24).

Hainan: Environs of Haikou, 8165. Environs of Kiungchow, 8296. 8187, 8226.

Pittosporum pentandrum (Blanco) Merr. var. *formosanum* (Hayata) Zhi Y. Zhang & Turland (PITTOSPORACEAE) New variety

in Novon 12: 153. (2002).

Pittosporum sp. nova. Henry in Trans. Asiat. Soc. Jap. 24 suppl: 18. (1896) (List Pl. Formosa). *Pittosporum pentandrum* Rehder & E. H. Wilson non (Blanco) Merr. in Sargent Pl. Wilson. 3: 330. (1916). *Pittosporum formosanum* Hayata in Journ. Coll. Sci. Tokyo 22: 32. (1906), Hayata in Mat. Formosa 34. (1911); in Icon. 1: 64. (1911), Rehder & E. H. Wilson in Sargent Pl. Wilson. 3: 330. (1917).

Taiwan: Kaohsiung, 256, 1888. Wanshoushan, 1070a.

Wanchin, 48, 52 (syntype of *Pittosporum formosanum* Hayata), 977, 1058, 1070. Kaohsiung, 1888.

A small tree native from China to Borneo and Sulawesi. In Taiwan this species is mainly distributed on the southern part of the island. Collected by the Glasnevin Taiwan Expedition (GTE 38) by the lighthouse at Oluanpi (South Cape of Henry and Wilson) in October 2004.

Pittosporum podocarpum Gagnep. (PITTOSPORACEAE)

Gowda in Journ. Arnold Arb. 32: 293. (1951), Zheng in Journ. Wuhan Bot. Res. 2.(1): 63. (1984).

Pittosporum glabratum Rehder & E. H. Wilson non Lindl. in Sargent Pl. Wilson. 3: 326. (1916), in part.

Hubei: Antelope Glen, 3414. Changleping, 7850. **Yunnan:** Mountains to the north of Mengtzi at 1,800 to 2,300 metres, 10191a., 10551, 10553. Yuanjiang, 13403.

Distributed in India, Myanmar, China and Vietnam.

Pittosporum podocarpum Gagnep. var. *angustatum* Gowda (PITTOSPORACEAE) New variety

in Journ. Arnold Arb. 32: 295. (1951), Zheng in Journ. Wuhan Bot. Res. 2.(1): 63. (1984).

Pittosporum glabratum Rehder & E. H. Wilson non Lindl. in Sargent Pl. Wilson. 3: 326. (1916), in part. *Pittosporum glabratum* Lindl. var. *neriifolium* Rehder & E. H. Wilson in Sargent Pl. Wilson. 3: 328. (1916), in part.

Hubei: Without locality, 7850a. **Yunnan:** In forests to the north of Mengzi at 1,950 metres, 10191b. In forests to the north of Mengzi at 2,300 metres, 10545 (holotype). Yuanjiang at 1,950 metres, 13299.

Distributed in India, Myanmar and China.

Pittosporum rehderianum Gowda (PITTOSPORACEAE) New species

in Journ. Arnold Arb. 32: 297. (1951), Zheng in Journ. Wuhan Bot. Res. 2.(1): 63. (1984).

Pittosporum glabratum Rehder & E. H. Wilson non Lindl. in Sargent Pl. Wilson. 3: 326. (1916), in part.

Hubei: Yichang, 3387, 3387a. Antelope Glen, 4369.

Discovered by Henry in 1887, *Pittosporum rehderianum* was later collected near Badong by Wilson and in neighbouring Shaanxi province by Giraldi on the 25th of September 1897.

Pittosporum sp. No. 1. (PITTOSPORACEAE)

Anonymous in Kew List of Determinations (volume 24).

Hubei: South Badong, 6164, Changyang, 7499. **Sichuan:** South Wushan, 7055.

Pittosporum sp. No. 2. (PITTOSPORACEAE)

Anonymous in Kew List of Determinations (volume 34).

Yunnan: Mengzi, 11059.

Pittosporum tobira (Thunb.) W. T. Aiton (PITTOSPORACEAE)

Li in Woody Flora of Taiwan 259. (1963), Li & Huang in Flora of Taiwan 3: 67. (1993).

Pittosporum makinoi Gowda non Nakai in Journ. Arnold Arb. 32: 311. (1951).

Taiwan: Oluanpi, 971. Tamsui, without number.

An evergreen shrub to 6 metres tall in its native habitat. This well known shrub has a wide distribution in east Asia and is a common coastal plant in Taiwan. Introduced to Kew in 1804.

Pittosporum truncatum E. Prtitz. (PITTOSPORACEAE) New species

Rehder & E. H. Wilson in Sargent Pl. Wilson. 3: 328. (1916), Gowda in Journ. Arnold Arb. 32: 339. (1951), Zheng in Journ. Wuhan Bot. Res. 2.(1): 64. (1984).

Pittosporum pauciflorum Henry non Hook. & Arn. in Journ. China Br. Roy. Asiat. Soc. 22: 234. (1887) (Chinese Names of Plants).

Hubei: Yichang, 1081. San You Dong Glen, (July 1886), 1524. South Badong, 7306. **Sichuan:** South Wushan, 5513.

A yellow-flowered shrub commonly found on the cliffs of glens and ravines around Yichang. According to Henry it was known colloquially as *ai hua tzu*. This species was discovered by Henry in 1886 and was described by Pritzel from material later collected by von Rosthorn in Sichuan between 1891 and 1892.

Plagiogyria euphlebia (Kunze) Mett. (CYATHEACEAE)

Lomariopsis sorbifolia Christ non (L.) Fée in Bull. Herb. Boiss. 6: 867. (1898). *Stenochlaena henryi* H. Christ in Notul. Syst. 1: 48. (1909).

Yunnan: In Pingbian on the Daweishan Range, in mountain forest at 2,200 metres, 9164 (type of *Stenochlaena henryi* H. Christ).

Widely distributed in Asia.

Plagiogyria stenoptera (Hance) Diels (CYATHEACEAE)

Plagiogyria henryi Christ in Bull. Herb. Boissier 7: 47. (1899), Christ in Bull. Soc. Bot. France 5: 64. (1905). *Lomaria decurrens* Baker in Kew Bull. Misc. Inf. 9. (1906).

Yunnan: Mountains to the east of Mengzi at 1,900 metres, 9036a., (isolectotype of *Plagiogyria henryi* Christ, collected 15th November 1896). In mountain forest to the east of Mengzi at 1,950 metres, 9036.

Distributed across China, Vietnam, Japan and the Philippines.

Plagiopetalum esquirolii (H. Lév.) Rehder (MELASTOMATACEAE) New genus

Li in Journ. Arnold Arb. 25: 10. (1944).

Sonerila henryi Kreanzlin in Vierteljahrsschr. Naturf. Ges. Zurich 76: 152. (1931). *Plagiopetalum henryi* (Kranzl.) Hu in Journ. Arnold. Arbor. 33: 172. (1952).

Yunnan: South of the Red River from Manhao, 9721. On the Daweishan Range in Pingbian at 1,500 metres, 9077 (isosyntype of *Sonerila henryi* Kreanzlin). In Pingbian, in forests on the Daweishan Range at 1,640 metres, 9077b., (isosyntype of *Sonerila henryi* Kreanzlin). On cliffs near Mengzi at 1,640 metres, 9077c., (isosyntype of *Sonerila henryi* Kreanzlin). Mengzi, 9077d., (isosyntype of *Sonerila henryi* Kreanzlin). Simao, 13520.

Rehder based this genus on *Plagiopetalum quadrangulum*, from Wilson's 3261 collection in western Sichuan on June 9th 1908. This particular species was discovered by Henry in 1896, Esquirol did not begin collecting until 1902. Distributed in western China and Vietnam.

Plantago major L. (PLANTAGINACEAE)

Henry in Journ. China Br. Roy. Asiat. Soc. 22: 235. (1887) (Chinese Names of Plants), Forbes & Hemsley in Journ. Linn. Soc. Bot. 26: 316. (1890), Henry in Notes Econ. Bot. China 7. (1893).

Hubei: Yichang, 199, 1362, 3139. Badong, 4707.

The greater plantain was known colloquially as *ch'e ch'ien ts'ao* at Yichang according to Henry who stated the seeds were added to a summer jelly called *liang fen*. The plant itself was used as a drug.

Plantago sp. (PLANTAGINACEAE)

Anonymous in Kew List of Determinations (volume 34).

Yunnan: Mengzi, 10860.

Platanthera japonica (Thunb.) Lindl. (ORCHIDACEAE)

Rolfe in Journ. Linn. Soc. Bot. 36: 56. (1903).

Hubei: Badong, 1835, 4082, 5041. Jianshi, 5889. Fang Xian, 5889a. **Sichuan:** South Wushan, 5713.

Platanthera minor (Miq.) Reichb. f. (ORCHIDACEAE)

Habenaria henryi Rolfe in Kew Bull. Misc. Inf. 202. (1896), Bretschneider in Hist. Eur. Bot. Disc. China 2: 792. (1898). *Platanthera henryi* (Rolfe) Rolfe in Journ. Linn. Soc. Bot. 36: 55. (1903), R. Schlechter in Fedde, Repert. Beihefte Bd. 4. 111. (1914), Kranzl in Orch. Gen & Spec. 1: 632. (1899), Morley in Glasra 3: 79. (1979).

Hubei: Badong, 4716, 6148. Zigui, 7663. **Sichuan:** South Wushan, 7453.

Platanthera ussuriensis (Regel.) Maxim. (ORCHIDACEAE)

Rolfe in Journ. Linn. Soc. Bot. 36: 57. (1903).

Hubei: Donghu, 6415. Xingshan, 6415a. Jianshi, 7469, 7470.

Platanus 'Augustine Henry' (PLATANACEAE)

Clarke in Bean, Trees & Shrubs 3: 270. (1976), Nelson in Pim, The Wood & the Trees 233. (1984); in A Heritage of Beauty 176. (2000).

Platanus hispanica Henry non Mill. ex Münchh. in Proc. Roy. Irish.Acad. 35: 18. (1919).

A tall, striking tree with lower pendelous branches and a well developed freely flaking trunk. Henry's plane was received at Kew from the Belgian nurseryman, Van Houtte in 1879 under the name *Platanus californica.* It was then widely taken to be *Platanus racemosa*, the Californian plane and was widely sold as such. It was Augustine Henry who first brought its merits to attention and so it is an appropriate cultivar to mark his great contribution to forestry and arboriculture. The type tree grows at Kew near the iron drinking fountain near the azalea garden. When measured in 2001 it was found to be 30 metres tall with a trunk diameter of 115 cm. A beautiful tree, though rare in cultivation.

Platostoma calcaratum (Hemsley) A. J. Paton (LAMIACEAE) New species

Plectranthus calcaratus Hemsley in Hooker's Icon. Pl. t. 2671. (1904), Forbes & Hemsley in Journ. Linn. Soc. Bot. 34: 477. (1900), Dunn in Notes Roy. Bot. Gard. Edinburgh 6: 144. (1915). *Ceratanthus calcaratus* (Hemsley) G. Taylor in Journ. Bot. Lxxiv. 40. (1936).

Yunnan: In mountain forest to the west of Simao at 1,450 metres, 12339 (type of *Plectranthus calcaratus* Hemsley, collected 7th August 1898).

Lilac flowers, distributed in Myanmar, western China and Indochina.

Platostoma coloratum (D. Don) A. J. Paton (LAMIACEAE)

Geniosporum strobiliferum Wall. ex Benth. Dunn in Notes Roy. Bot. Gard. Edinburgh 6: 134. (1915).

Yunnan: Pu'er, 13393.

A widespread species, found over much of India, Nepal, Bhutan, Myanmar, Chian (Yunnan) and Laos.

Platostoma hispidum (L.) A. J. Paton (LAMIACEAE)

Acrocephalus capitatus (Roth.) Benth. Dunn in Notes Roy. Bot. Gard. Edinburgh 6: 134. (1915).

Yunnan: Simao, 12610.

Platostoma palustre (Blume) A. J. Paton (LAMIACEAE)

Mesona procumbens Hemsley in Ann. of Bot. 9: 155. (1895), Henry in Trans. Asiat. Soc. Jap. 24 suppl: 72. (1896) (List Pl. Formosa), Forbes & Hemsley in Journ. Linn. Soc. Bot. 155. (1892).

Taiwan: Oluanpi, 1317.

Platycarya strobilacea Sieb. & Zucc. (JUGLANDACEAE)

Henry in Journ. China Br. Roy. Asiat. Soc. 22: 245. (1887) (Chinese Names of Plants), Henry in Notes Econ. Bot. China 62. (1893), Skan in Journ. Linn. Soc. Bot. 26: 495. (1899), Rehder & E. H. Wilson in Sargent Pl. Wilson. 3: 180. (1916), Zheng in Journ. Wuhan Bot. Res. 2.(1): 17.

(1984).

Hubei: Yichang, 69, 1008, 1358, 1606. Jianshi, 6014.
Yunnan: Milê district, 9937.

The *hua-kuo* according to Henry. A small tree to about 15 metres tall and also known colloquially in Hubei as *huan-hsiang-shu*. According to both Henry and Wilson the fruits were used to make a black dye for cotton goods. This species is cultivated at the National Botanic Gardens, Kilmacurragh from collections made by the Glasnevin Central China Expedition on Moji Shan (the Dome of Henry and Wilson) near Yichang in October 2004. Native to China, Korea, Vietnam and Japan.

Platycladus orientalis (L.) Franco (CUPRESSACEAE)

Biota not yet determined. Henry in Journ. China Br. Roy. Asiat. Soc. 22: 265. (1887) (Chinese Names of Plants). *Thuja orientalis* L. Henry in Notes Econ. Bot. China 37. (1893), Henry in Trans. Asiat. Soc. Jap. 24 suppl: 540. (1896) (List Pl. Formosa), Skan in Journ. Linn. Soc. Bot. 26: 540. (1902), Rehder & E. H. Wilson in Sargent Pl. Wilson. 2: 53. (1914), E. H. Wilson in Journ. Arnold Arb. 7: 62. (1926).

Hubei: Liantuo, 3910. **Taiwan:** Wanchin, 403. Oluanpi, 1975, 2076. **Yunnan:** Yuanjiang, planted by a temple at 2,000 metres, 13353.

A small tree to 20 metres tall, native to Russia, Korea and China and much planted there especially on graves and in the grounds of temples and monasteries. *Platycladus orientalis* was known colloquially in the Three Gorges region as *ai-peh* or *ai-po* according to Henry who also stated that the roots of this species were used in making incense. It was introduced to cultivation by the French Jesuit missionary, Père Pierre Nicholas le Chéron d'Incarville who sent seeds from the Beijing area to Paris. According to Wilson this tree has been cultivated in China since time immemorial and the tree was a favourite with Taoist, Buddhist and Confuscian priests. It was also planted on the tombs of Emperors and the fruits and leaves were much used in Chinese medicine. It is a long-lived tree, Wilson mentions meeting trees in Beijing some six centuries old. *Platycladus* differs from *Thuja* in the thicker cone-scales and wingless seeds. In cultivation it thrives best in hot dry areas on alkaline soils, a tree planted at the Royal Botanic Gardens, Kew in 1898 is 16 metres tall with a stem diameter of 74 cm at 0.4 metres above ground level (2001). At Lamorran House in Cornwall a tree is recorded having reached a height of 24 metres with a stem diameter of 90 cm in 1928. Collected by the Glasnevin Central China Expedition (GCCE 493) near Zhenzi town in Xingshan in September 2004.

Platycodon grandiflorus (Jacq.) A. DC. (CAMPANULACEAE)

Henry in Journ. China Br. Roy. Asiat. Soc. 22: 238.

(1887) (Chinese Names of Plants), Forbes & Hemsley in Journ. Linn. Soc. Bot. 26: 5. (1889), Henry in Notes Econ. Bot. China 4. (1893).

Hubei: Yichang, 307, 2329.

The Chinese bellflower, a pretty perennial to 60 cm, native to eastern Siberia, China, Korea and Japan. Henry stated this plant was common on barren hills around Yichang where it was known colloquially as *chieh keng* (*jie geng*). The root is used in traditional Chinese herbal medicine and according to Wilson were a cure for stomach chills. It also controls colds, bronchitis, tonsilitis, asthma, peptic ulcers, chronic inflammations, viral infections, high blood pressure, coughing, sore throat and loss of voice.

Pleione bulbocodioides (Franch.) Rolfe (ORCHIDACEAE)

Cribb & Butterfield in The Genus Pleione 67. (1988).

Pleione sp. nova Henry in Journ. China Br. Roy. Asiat. Soc. 22: 266. (1887) (Chinese Names of Plants). *Coelogyne bulbocodioides* Franch. Henry in Notes Econ. Bot. China 26. (1893). *Coelogyne pogonioides* Rolfe in Kew Bull. Misc. Inf. 196. (1896); in Journ. Linn. Soc. Bot. 36: 23. (1903). *Pleione pogonoides* (Rolfe) Rolfe in Orch. Rev. 11: 291. (1903), Kranzlin in Engler, Pflanzenr. 50: 124. (1907). *Coelogyne henryi* Rolfe in Kew Bull. Misc. Inf. 195. (1896), Bretschneider in Hist. Eur. Bot. Disc. China 2: 792. (1898), Rolfe in Journ. Linn. Soc. Bot. 36: 22. (1903), Morley in Glasra 3: 79. (1979). *Pleione henryi* (Rolfe) Schlecter in Repert. Spec. Nov. Regni Veg. 4: 186. (1919).

Hubei: Badong, in the high hills, 1473, 3785 (lectotype of *Pleione pogonoides* (Rolfe) Rolfe), 6068 (syntype of *Coelogyne pogonoides* Rolfe), 6068a. **Sichuan:** Henry's Chinese collector with Pratt, no locality stated but may be Leshan, Emei Shan, Wa Shan or Kangding, 8826 (syntype of *Coelogyne henryi* Rolfe).

A common orchid on wet, humus-clad rocks in the mountains of Hubei and Sichuan according to Henry. In Badong according to Henry, the pseudobulbs of this species were sold as a drug, the drug being called *pei mu* and *mao-ku*. E. H. Wilson in *A Naturalist in Western China* 2: 40. (1913) stated it was used as a cure for tuberculosis and asthma. *Pleione bulbocioides* was discovered by Père Armand David in Baoxing, Sichuan, in April 1869. Henry stated it was the *shan li gu* or 'benevolent aunt of the mountains' and is currently used in China to treat (externally) sores, ulcers, poisonous swellings, tuberculosis and snake-bite.

Pleione grandiflora (Rolfe) Rolfe (ORCHIDACEAE)

New species.

in Orch. Rev. 11: 291. (1903), Kranzlin in Engler, Pflanzenr. 50: 124. (1907), Hunt & Vosa in Kew Bull. 25:

428. (1971), Cribb et al. in Cutis's Bot Mag.184: 129. (1983), Cribb & Butterfield in the Genus Pleione 62. (1988).

Coelogyne grandiflora Rolfe in Journ. Linn. Soc. Bot. 36: 22. (1903).

Yunnan: In mountain forest to the north of Mengzi at 2,600 metres, 11116 (holotype of *Coelogyne grandiflora* Rolfe).

India to China (Yunnan) to Indochina.

Pleione aff. *yunnaneticus* Rolfe (ORCHIDACEAE)
Rolfe in Journ. Linn. Soc. Bot. 36: 23. (1903).

Yunnan: On rocks south of the Red River at 2,900 metres, 11115.

Pleione yunnanensis (Rolfe) Rolfe (ORCHIDACEAE)
New species.

in Orch. Rev. 11: 292. (1903), Rolfe in Gard. Chron. ser. 3, 115. (1906); in Journ. Hort. 1: 251. (1906); in Bot. Mag. t. 8106. (1906), Fedde in Repert Nov. Spec. Regni Veg. 4: 169. (1907), Kranzlin in Engler, Pflanzenr. 50: 122. (1907), Hunt & Vosa in Kew Bull. 25: 428. (1971), Lancaster in the Garden 107: 428. (1982), Cribb et al. in Curtis's Bot. Mag. 184: 127, t. 865. (1983), Nelson in Pim, The Wood & the Trees 233. (1984), Cribb & Butterfield in the Genus Pleione 61. (1988), O' Brien in Augustine Henry, An Irish Plant Collector in China 29. (2002).

Coelogyne (*Pleione*) *yunnanensis* Rolfe in Journ. Linn. Soc. Bot. 36: 23. (1903).

Yunnan: On grassy mountains to the north of Mengzi at 1,750 metres, 11113 (type of *Coelogyne yunnanensis* Rolfe).

A handsome orchid with flask-shaped pseudobulbs, producing slightly pendant red-purple flowers in May and June. This species was first collected by the Irish customs official, William Hancock near Mengzi, Rolfe based the species on material supplied by both Henry and Hancock. *Pleione yunnanensis* was introduced to cultivation almost a century later by British plantsman, Roy Lancaster. It is native to the northern extremity of Myanmar, China (Yunnan), Tibet (Xizang), southern Sichuan and western Guizhou where it grows on cliffs and in forests at 1,200 to 2,800 metres. The pseudobulbs are used in herbal medicine.

Pleurolobus gangeticus (L.) J. St.-Hil. Ex H. Ohashi & K. Ohashi (FABACEAE)
Desmodium gangeticum (L.) DC. Henry in Trans. Asiat. Soc. Jap. 24 suppl: 33. (1896) (List Pl. Formosa).

Hainan: Environs of Haikou, 8317. **Taiwan:** Kaohsiung, without number. Wanchin, 858. **Yunnan:** Mengzi, 9798. Yuanjiang at 1,600 metres, 13248.

An erect undershrub to 1.5 metres tall. Common in the central and southern parts of Taiwan at low altitudes.

Pleurospermum sp. (APIACEAE)
Anonymous in Kew List of Determinations (volume 24).

Hubei: Badong, 4950. Jianshi, 7430. Without locality, 4954a. North Badong, 7160. **Sichuan:** South Wushan, 5738.

Pluchea indica (L.) Less. (ASTERACEAE)
Henry in Trans. Asiat. Soc. Jap. 24 suppl: 55. (1896) (List Pl. Formosa).

Hainan: On Liang Shan, 16 km (10 miles) east of Haikou, 8210. Environs of Kiungchow, 8708. **Taiwan:** Kaohsiung, without number. Oluanpi, 302.

Plumbaginaceae (PLUMBAGINACEAE)
Statice sp. Anonymous in Kew List of Determinations (volume 34).

Yunnan: Mengzi, 13740.

Plumbago zeylanica L. (PLUMBAGINACEAE)
Henry in Trans. Asiat. Soc. Jap. 24 suppl: 57. (1896) (List Pl. Formosa).

Taiwan: Kaohsiung, without number. **Yunnan:** Simao, 12626.

Ceylon plumbago, the Hani (an ethnic group) of Xishuangbanna in southern Yunnan call this species *diaqdaqyaoq yaobajaiq* and use its roots to treat cases of rheumatism and sprains. Its leaves are used externally to treat bruising.

Plumeria rubra L. (APOCYNACEAE)
Plumeria acutifolia Poir. Henry in Trans. Asiat. Soc. Jap. 24 suppl: 60. (1896) (List Pl. Formosa).

Taiwan: Oluanpi, without number. Wanchin, 856. **Yunnan:** Simao, 12076.

The frangipangi or temple tree, native to tropical South America and widely cultivated in the tropics. According to Henry it was called the *fan-hua.*

Poa acroleuca Steud. (POACEAE)
Rendle in Journ. Linn. Soc. Bot. 36: 422. (1904).

Sichuan: South Wushan, 5685.

Poa annua L. (POACEAE)
Rendle in Journ. Linn. Soc. Bot. 36: 422. (1904).

Hubei: Yichang, 824. Badong, 3746, 5425.

Cosmopolitan in distribution.

Poa attenuata Trin. (POACEAE)
Poa nemoralis L. var. *ligulata* Stapf. Rendle in Journ. Linn. Soc. Bot. 26: 426. (1904).

Hubei: Fang Xian at 1,900 to 3,000 metres, 6807.

Poa faberi Rendle (POACEAE)
Poa prolixior Rendle in Journ. Linn. Soc. Bot. 36: 427. (1904).

Hubei: Badong at 1,900 metres, 6049.

Poa pratensis L. (POACEAE)
Rendle in Journ. Linn. Soc. Bot. 26: 426. (1904).

Hubei: Badong, 4026.

Poaceae unidentified (POACEAE)

Anonymous in Kew List of Determinations (volume 24).

Hubei: Liantuo, 4404.

Podocarpus neriifolius D. Don (PODOCARPACEAE)

E. H. Wilson in Journ. Arnold Arb. 7: 41. (1926).

Yunnan: South of Simao in Yulo forest (Xishuangbanna) at 1,300 metres, 12919.

A. Henry 12919 was quoted by Pilger as *Podocarpus neriifolius* D. Don var. *brevipes* (Blume) Pilger in Engler, Pflanzenr. 5: 81. (1903) in error.

Podochilus sp. (ORCHIDACEAE)

Anonymous in Kew List of Determinations (volume 34).

Yunnan: Simao, 12961.

Podophyllum pleianthum Hance (BERBERIDACEAE)

Dysosma pleiantha (Hance) Woodson in Ann. Missouri Bot. Gard. 15: 339. (1924).

Hubei: Liantuo, 3952.

A very beautiful woodland perennial, discovered by the Irish diplomat, Thomas Watters (in Herb Hance 21697) in Taiwan in April 1881. It was introduced to cultivation when Charles Ford, superintendent of Hong Kong Botanic Gardens sent living plants to Kew, where they flowered for the first time in 1889. The dried stem and rhizome of this species is currently used in Chinese herbal medicine in cases of traumatic injury and snake-bite.

Podophyllum sinense (H. L. Li) Christenh. & Byng (BERBERIDACEAE) New species

Diphylleia sinensis H. L. Li in Journ. Arnold Arb. 28: 442. (1947), Ying et al., in Journ. Arnold Arb. 65: 78. (1984). *Diphylleia cymosa* Henry non Michx. in Notes Econ. Bot. China 25. (1893); in Flora & Sylva 1: 217. (1903).

Hubei: Badong, 920 (November 1885). Fang Xian, in dark mountain forest at 2,300 metres, 6820.

Podophyllum versipelle Hance subsp. ***boreale*** Shaw (BERBERIDACEAE) New subspecies

in The New Plantsman 6(3): 160. (1999), Nelson in A Heritage of Beauty 316. (2000).

Diphylleia sp. nova Henry in Journ. China Br. Roy. Asiat. Soc. 22: 262. (1887) (Chinese Names of Plants). *Podophyllum versipelle* Henry non Hance in Notes Econ. Bot. China 25. (1893); in Flora & Sylva 1: 217. (1903). *Podophyllum pleianthum* Morley non Hance in The Garden 105(7): 287. (1980), in part quoad ref. to Mount Usher plant.

Hubei: South Badong, 5372 (holotype). Jianshi, 5372a., 5372e. Changyang, 5372b. Xingshan at 2,650 metres, 5372d. In a wood at Fang Xian, 5372c. **Sichuan:** South Wushan, 5372f., (syntype).

Henry stated this subspecies was known in Hubei and Sichuan as *pa-chio-lien*, its roots were used as a drug. Henry referred to this plant as *Diphylleia* in his *Chinese Names of Plants* (1887) but later updates this in *Notes on Economic Botany* (1893*). Podophyllum versipelle* subsp. *boreale* is in cultivation at Mount Usher in County Wicklow, the clone growing there is believed to have been an original Henry introduction. It is also cultivated at the National Botanic Gardens, Glasnevin.

Pogonatherum crinitum (Thunb.) Kunth (POACEAE)

Pogonatherum saccharoideum P. Beauv. var. *monandrum* (Roxb.) Hack. Rendle in Journ. Linn. Soc. Bot. 36: 357. (1904).

Taiwan: Wanchin, 382. Kaohsiung, 762.

Pogonatherum paniceum (Lam.) Hack. (POACEAE)

Pogonatherum saccharoideum P. Beauv. Henry in Trans. Asiat. Soc. Jap. 24 suppl: 108. (1896) (List Pl. Formosa), Rendle in Journ. Linn. Soc. Bot. 36: 357. (1904).

Hubei: Badong, 574, 3487. Yichang, 1142. **Yunnan:** Mengzi, 10360. On grassy hills near Mengzi at 1,600 metres, 11695.

Native to India, Sri Lanka, Myanmar, China, Vietnam, Malaysia and extending to Australia. Henry stated it was known in Chinese Taiwan as *pi-tze-ts'ao* and was used as a diuretic.

Pogonatum aloides (Hedw.) P. Beauv. (POLYTRICHACEAE)

Anonymous in Kew List of Determinations (volume 24).

Hubei: Jianshi, 5888 (in part). Changyang, 7752. (Moss).

Pogonatum cirratum (Sw.) Brid. (POLYTRICHACEAE)

Polytrichum convolutum L. var. *cirratum* (Sw.) Müll. Hal. Anonymous in Kew List of Determinations (volume 34).

Yunnan: Mengzi, 13715. (Moss).

Pogonatum fastigiatum Mitt. (POLYTRICHACEAE)

Polytrichum nudicaule C. H. Wright in Journ. of Bot. 106. (1891), Bretschneider in Hist. Eur. Bot. Disc. China 2: 794. (1898).

Hubei: Fang Xian, 6840 (type of *Polytrichum nudicaule* C. H. Wright, collected September 1888).

A very robust moss to 20 cm high.

Pogonatum proliferum (Griff.) Mitt. (POLYTRICHACEAE)

Polytrichum gymnophyllum Mitt. Salmon in Journ. Linn. Soc. Bot. 34: 460. (1900).

Yunnan: In mountain forest south of the Red River from Manmei, 10347. (Moss).

Pogonia japonica Reichb. f. (ORCHIDACEAE)

Pogonia ophioglossoides Rolfe non Ker. Gawl. in Journ. Linn. Soc. Bot. 36: 47. (1903).

Hubei: Badong, 1777. Yichang, 4044. Jianshi, 5811.

Pogostemon auricularia (L.) Hassk. (LAMIACEAE)

Dysophylla auricularia (L.) Blume Forbes & Hemsley in Journ. Linn. Soc. Bot. 26: 276. (1890). *Dysophylla auricularia* (L.) Blume Henry in Trans. Asiat. Soc. Jap. 24 suppl: 72. (1896) (List Pl. Formosa), Dunn in Notes Roy. Bot. Gard. 6: 147. & 202. (1915).

Hainan: Environs of Haikou, 8168. **Taiwan:** Wanchin, 541. **Yunnan:** In forests near Simao at 1,450 metres, 12311.

Pogostemon brachystachyus Benth. (LAMIACEAE)

Pogostemon nigrescens Dunn in Notes Roy. Bot. Gard. Edinburgh 8: 159. (1913), Dunn in Notes Roy. Bot. Gard. Edinburgh 6: 146. (1915).

Yunnan: In shaded woods near Mengzi, 9082. South of Red River; at Tachao Shan (at Fengchunling), 11174 (type of *Pogostemon nigrescens* Dunn).

Eastern Himalaya to China (Yunnan).

Pogostemon cruciatus (Benth.) Kuntze (LAMIACEAE)

Dysophylla cruciata Benth. Forbes & Hemsley in Journ. Linn. Soc. Bot. 26: 276. (1890), Dunn in Notes Roy. Bot. Gard. Edinburgh 6: 147. (1915).

Hainan: Environs of Haikou, without number.

Pogostemon formosanus Oliver (LAMIACEAE) New species

in Hooker's Icon. Pl. t. 2440. (1896), Henry in Trans. Asiat. Soc. Jap. 24 suppl: 72. (1896) (List Pl. Formosa), Bretschneider in Hist. Eur. Bot. Disc. China 2: 790. (1898).

Taiwan: Wanchin, 70. Wanshoushan, 1178 (isosyntype).

Pogostemon glaber Benth. (LAMIACEAE)

Dunn in Notes Roy. Bot. Gard. Edinburgh 6: 146. (1915).

Yunnan: Mengzi, 10418. In Pingbian, in forests on the Daweishan Range at 1,950 metres, 11487. Simao, 12832.

The Hani (an ethnic group) in southern Yunnan call this species *cmlciq* and the roots, flowers and leaves are used to treat cases of tuberculosis and gastroenteritis. A widespread species in Asia.

Pogostemon linearis (Benth.) Kuntze (LAMIACEAE)

Dysophylla linearis Benth. Dunn in Notes Roy. Bot. Gard. Edinburgh 6: 147. (1915).

Yunnan: Simao at 1,450 to 1,650 metres, 12226, 12226a., 12628.

Native to India, China (Yunnan) and Indochina.

Pogostemon menthoides Blume (LAMIACEAE)

Pogostemon fraternus Miq. Dunn in Notes Roy. Bot. Gard. Edinburgh 6: 146. (1915).

Yunnan: In forests near Simao at 1,450 metres, 11699.

Pogostemon sp. (LAMIACEAE)

Dysophylla sp. Anonymous in Kew List of Determinations (volume 34).

Yunnan: Simao, 12479. Yuanjiang, 13253.

Pogostemon stellatus (Lour.) Kuntze (LAMIACEAE)

Dysophylla verticillata Benth. ex Wall. Henry in Trans. Asiat. Soc. Jap. 24 suppl: 72. (1896) (List Pl. Formosa).

Taiwan: Wanchin, 1665.

Poikilospermum suaveolens (Blume) Merr. (URTICACEAE)

Conocephalus sinensis C. H. Wright in Journ. Linn. Soc. Bot. 26: 471. (1899).

Yunnan: Mengzi, 11074 (syntype of *Conocephalus sinensis* C. H. Wright).

Distributed in India, China (Yunnan), Thailand, Vietnam and the Philippines.

Polhillides velutina (Willd.) H. Ohashi & K. Ohashi (FABACEAE)

Desmodium latifolium (Roxb.) DC. Henry in Trans. Asiat. Soc. Jap. 24 suppl: 33. (1896) (List Pl. Formosa).

Taiwan: Wanchin, 1519.

Tropical and subtropical Old World.

Polhillides velutina (Willd.) H. Ohashi & K. Ohashi subsp. *longibracteata* (Schindl.) H. Ohashi & K. Ohashi (FABACEAE) New subspecies

Desmodium longibracteatum Schindler in Repert. Spec. Nov. Regni Veg. 21: 7. (1925).

Yunnan: Simao, 12433. Yuanjiang at 1,600 metres, 13250. Simao at 1,500 metres, 12557 (syntype of *Desmodium longibracteatum* Schindler).

India to China and Indonesia.

Poliothyrsis sinensis Oliver (SALICACEAE) New genus

in Hooker's Icon. Pl. 19: t. 1885. (1889), Henry in Notes Econ. Bot. China 55. (1893), Bretschneider in Hist. Eur. Bot. Disc. China 2: 780. (1898), Bean in Kew Bull. Misc. Inf. xxii. 355. (1909); in Trees & Shrubs 2: 206. (1921), Schneider in Illus. Handb. Laubholzk. 2: 361. (1912), Morley in Glasra 3: 80. (1979), Zheng in Journ. Wuhan Bot. Res. 2.(1): 153. (1984), Nelson in A Heritage of Beauty 316. (2000), O' Brien in Augustine Henry, An Irish Plant Collector In China 29. (2002).

Sichuan: South Wushan, 5522. North Wushan, 7140 (isosyntype). **Hubei:** Changyang, 7588. Xingshan at 1,400 metres, 6566 (isosyntype). Fang Xian, 6566a.

A deciduous tree to 15 metres tall, producing fragrant creamy-white flowers in a terminal inflorescence in late summer. Discovered by Henry in 1888, seeds of Wilson's 500 from Yichang reached Glasnevin from the Arnold Arboretum in 1908, from this consignment the tree currently grown in the old Chinese shrubbery at Glasnevin was raised. According to Wilson the colloquial for this tree around Yichang was *yu-kui-chou* (Henry gave *shan-kuai-*

tsao). Poliothyrsis sinensis is extremly rare in cultivation, at Caerhays Castle in Cornwall a tree was 15 metres tall with a girth of 36 cm in 1971. At the National Botanic Gardens, Glasnevin there is a superb tree raised from Wilson's 500 (collected Yichang, October 1907), when measured in 1992 this tree was found to be 9 metres tall with a girth of 29 cm. At the Royal Botanic Gardens, Kew near the side of the Great Palm House and the Victoria gate is a young vigorous, heavily-flowered tree. A beautiful medium-sized tree sporting masses of showy white flower racemes in early August, this species deserves to be better known.

Pollia secundiflora (Blume) Bakh. f. (COMMELINACEAE)

Pollia sorzogonensis (E. Mey. ex C. Presl.) Steud. Henry in Trans. Asiat. Soc. Jap. 24 suppl: 98. (1896) (List Pl. Formosa), Brown in Journ. Linn. Soc. Bot. 159. (1903). *Pollia sorzogonensis* (E. Mey. ex C. Presl.) Steud. var. *gigantea* (Hassk.) C. B. Clarke Brown in Journ. Linn. Soc. Bot. 36: 159. (1903).

Hubei: Antelope Glen, 4163. Changyang, 6231, 6231a. **Taiwan:** Wanchin, 133, 575.

Pollia sp. (COMMELINACEAE)

Anonymous in Kew List of Determinations (volume 34).

Yunnan: Mengzi, 10808, 11842.

Polyalthia suberosa (Roxb.) Thwaites (ANNONACEAE)

Anonymous in Kew List of Determinations (volume 24).

Hainan: Hummocks near Haikou, 8095. Environs of Kiungchow, 8252.

Polyblastidium hypoleucum (Ach.) Kalab. (PHYSCIACEAE)

Anaptychia speciosa Mass. var. *hypoleuca* Muehlbr f. *isidiifera* Muell.-Arg. Muell.-Arg in Bull. Herb. Boissier 1: 236. (1893). *Heterodermia hypoleuca* (Ach.) Trevis Wei in an Enumeration of Lichens in China 110. (1991).

Hubei: Fang Xian, 6718 (in part). (Lichen).

Polycarpaea corymbosa (L.) Lam. (CARYOPHYLLACEAE)

Anonymous in Kew List of Determinations (volume 24).

Hainan: Environs of Haikou, 7968.

Polygala arillata Buch.-Ham. ex D. Don (POLYGALACEAE)

Rehder & E. H. Wilson in Sargent Pl. Wilson. 1: 160. (1912), Zheng in Journ. Wuhan Bot. Res. 2.(1): 110. (1984), Chen in Acta Phytotax. Sin. 29: 204. (1991).

Hubei: South Badong, 7381. Jianshi, 5783. Xingshan, 5783a. Without locality, 5783b. **Sichuan:** Kangding, Henry's Chinese collector with Pratt, 8996. **Yunnan:** In the mountains near Mengzi at 1,600 metres, 9999, 9999c. Mountains near Simao at 1,600 metres, 12621, 12621a.

An erect deciduous shrub to 3 metres tall bearing terminal, pendelous racemes of yellow flowers to 10 cm long in June and July, sometimes tinged red at the apex. This beautiful shrub is common in thickets and on cliffs in Hubei and Sichuan The Hani (an ethnic group) in Xishuangbanna in southern Yunnan call this species *keepseq* and the roots are used to treat tuberculosis and post-delivery fatigue. Native to Nepal, India, Myanmar and China.

Polygala caudata Rehder & E. H. Wilson (POLYGALACEAE) New species

in Sargent Pl. Wilson. 2: 161. (1914), Chen in Acta Phytotax. Sin. 29: 208. (1991).

Hubei: South Donghu, 7714. **Yunnan:** On cliffs near Mengzi at 1,600 to 1,800 metres, 10901 (type).

A shrub to 2 metres tall with rose-coloured flowers. This species seems to have been first collected by the Rev'd. B. C. Henry in Guandong province in March 1881 and was described from his collections and from Augustine Henry's Yunnan and Hubei collections. Endemic to China.

Polygala chinensis L. (POLYGALACEAE)

Polygala glomerata Lour. Henry in Trans. Asiat. Soc. Jap. 24 suppl: 18. (1896) (List Pl. Formosa).

Hainan: Environs of Kiungchow, 8772. **Taiwan:** Wanchin, 1828.

A herb to 15 cm tall, native to the eastern Himalaya, India, China, Indochina and the Philippines.

Polygala furcata Royle (POLYGALACEAE)

Polygala triphyllos Buch.-Ham. ex D. Don Anonymous in Kew List of Determinations (volume 34).

Yunnan: Mengzi, 9839.

Polygala globulifera Dunn (POLYGALACEAE) New species

in J. Linn. Soc. Bot.35: 486. (1903), Chen in Acta Phytotax. Sin. 29: 206. (1991).

Yunnan: Simao, 12805, (syntype) 12805a., (syntype).

Distributed in India, Myanmar and China [Tibet (Xizang) and Yunnan].

Polygala isocarpa Chodat (POLYGALACEAE)

Chen in Acta Phytotax. Sin. 29: 216. (1991). *Polygala lacei* Craib non Craib in Notes Roy. Bot. Gard. Edinburgh 11: 187. (1918—1919).

Yunnan: On cliffs near Puer at 1,450 metres, 9303a.

Endemic to China and distributed in the western provinces of Guizhou, Sichuan and Yunnan.

Polygala japonica Houtt. (POLYGALACEAE)

Polygala sibirica L. var. *japonica* (Houtt.) Henry in Journ. China Br. Roy. Asiat. Soc. 22: 261. (1887) (Chinese Names of Plants).

Hubei: Yichang, 652, 777, 827, 905, 1133. Badong, 1756. Liantuo, 3036. Donghu, 6396.

At Badong this species was colloquially as *nu erh hung.*

02

Polygala karensium Kurz (POLYGALACEAE)

Polygala floribunda Dunn in Journ. Linn. Soc. Bot. 485. (1903) non Benth 1840. *Polygala congesta* Rehder & E. H. Wilson in Sargent Pl. Wilson. 2: 162. (1914). *Polygala tricornis* Gagnep. Chen in Acta Phytotax. Sin. 29: 210. (1991).

Yunnan: On grassy parts of mountains on the edge of forest near Mengzi at 1,600 to 2,000 metres, 9364, 11079, 10511 (syntype of *Polygala congesta* Rehder & E. H. Wilson), 10511, 10511a., 10511b., 10511c., 11416 (syntype of *Polygala floribunda* Dunn), 11472. In forests near Simao at 1,500 to 1,600 metres, 12272, 12272b., 12272d. Simao, 13519.

Bhutan to China and Indochina.

Polygala longifolia Poir. (POLYGALACEAE)

Chen in Acta Phytotax. Sin. 29: 224. (1991).

Yunnan: Mengzi, 11244.

A widespread species found over much of India, Sri Lanka, Nepal, western China, Laos and Cambodia to Australia.

Polygala persicariifolia DC. (POLYGALACEAE)

Chen in Acta Phytotax. Sin. 29: 225. (1991).

Yunnan: In forests near Mengzi at 2,300 metres, 9717. Mengzi, 9220, 10973. Simao, 12292.

Polygala saxicola Dunn (POLYGALACEAE) New species

in Journ. Linn. Soc. Bot. 35: 486. (1903), Chun in Acta Phytotax. Sin. 29: 213. (1991).

Yunnan: Near Mengzi on exposed dry rocks on the mountain-ridges at 1,900 to 2,600 metres, 9169 (syntype). Mengzi, 9169a.

The species was previously collected by the William Hancock near the Red River. Also native to Chian (Guangxi, Guangdong) and Vietnam.

Polygala sibirica L. (POLYGALACEAE)

Anonymous in Kew List of Determinations (volume 34).

Yunnan: Simao, 11731.

Polygala sp. (POLYGALACEAE)

Anonymous in Kew List of Determinations (volume 24).

Hubei: Changleping, 6272.

Polygala tatarinowii Regel (POLYGALACEAE)

Craib in Notes Roy. Bot. Gard. Edinburgh. 11: 187. (1918—1919), Chun in Acta Phytotax. Sin. 29: 214. (1991).

Hubei: Badong, 2486, 4822. Lung Wang Tung Valley near Yichang, 4165. Fang Xian, 6588. **Yunnan:** Mengzi, 9343.

An erect herb to about 10 cm tall, native to India, northern Myanmar, China, Japan, the Philippines and Papua New Guinea.

Polygala tenuifolia Willd. (POLYGALACEAE)

Anonymous in Kew List of Determinations (volume 34).

Yunnan: Milê, 10586.

Polygala wattersii Hance (POLYGALACEAE)

Rehder & E. H. Wilson in Sargent Pl. Wilson. 2: 161. (1914), Zheng in Journ. Wuhan Bot. Res. 2.(1): 110. (1984), Chen in Acta Phytotax. Sin. 29: 209. (1991).

Polygala mariesii Hemsley in Journ. Linn. Soc. Bot. 23: 61. (1886), Henry in Journ. China Br. Roy. Asiat. Soc. 22: 259. (1887) (Chinese Names of Plants).

Hubei: Xiling (Yichang) Gorge, 274, 808, 1104. Antelope Glen, 3260. In the Kan Chi Glen near Yichang (October 1886), 2912. Liantuo, 1973. **Yunnan:** Mengzi at 1,800 metres, 9395, 9770.

A small yellow flowered shrub to 1 metre tall, discovered by the Irish diplomat and plant collector Thomas Watters, in April 1880. Hemsley's *Polygala mariesii* was based on material collected by Charles Maries and Augustine Henry. *Polygala wattersii* is abundant on the cliffs of the glens and gorges around Yichang where Henry stated it was known colloquially as *mi hua tzu*. It was also collected by the Glasnevin Central China Expedition (GCCE 679) in the Red Dragon reserve, Xingshan in October 2004. Native to China and Vietnam.

Polygonatum cirrhifolium (Wall.) Royle (ASPARAGACEAE)

Wright in Journ. Linn. Soc. Bot. 36: 104. (1903).

Hubei: South Badong, 5501a.

Polygonatum cyrtonema Hua (ASPARAGACEAE) New species

Polygonatum henryi Diels in Bot. Jahrb. Syst. 29: 247. (1900). *Polygonatum multiflorum* Wright non (L.) All. in Journ. Linn. Soc. Bot. 36: 106. (1903).

Hubei: Yichang, 2020, 2343, 4087, 5460a., (isotype of *Polygonatum henryi* Diels), 5460b. South Badong, 5460. Liantuo, 2682, 3849. Xingshan at 1,400 metres, 6541.

Hua based the species on material collected by Père Farges during the early 1890s, Henry had discovered this species in the spring of 1887.

Polygonatum mengtzense F. T. Wang & Tang (ASPARAGACEAE) New species

in Bull. Fan. Mem. Inst. Biol. Peip. 7: 84. (1936).

Yunnan: In Pingbian in forests on the Daweishan Range at 1,640 metres, 11145 (isosyntype).

China (Yunnan and Guangxi) to Vietnam.

Polygonatum odoratum (Mill.) Druce (ASPARAGACEAE)

Polygonatum officinale All. Henry in Notes Econ. Bot. China 6. (1893), Wright in Journ. Linn. Soc. Bot. 36: 107. (1903).

Sichuan: South Wushan, 5566.

According to Henry the rhizomes of this species were the source of the drug *yu chu*.

Polygonatum sibiricum Delas ex Redouté (ASPARAGACEAE)

Henry in Journ. China Br. Roy. Asiat. Soc. 22: 237. (1887) (Chinese Names of Plants); in Notes Econ. Bot. China 60. (1893), Diels in Bot. Jahrb. 29: 248. (1900), Wright in Journ. Linn. Soc. Bot. 36: 109. (1903).

Hubei: Liantuo, 1984. **Sichuan:** South Wushan, 5501.

According to Henry this species was known colloquially at Yichang as *liang chiang* and from it the drug *huang ching* (*huang jing* or 'yellow essence', used as a tonic) is made.

Polygonatum spp. (ASPARAGACEAE)

Anonymous in Kew List of Determinations (volume 24 & 34).

Hubei: Yichang, 3290. Xingshan, 6497. Fang Xian, 6497a. **Yunnan:** Mengzi, 9387, 9548, 9655, 9866, 10812, 11378, 13750. Simao, 12225, 12255. South of the Red River from Manmei, 13668.

Polygonatum verticillatum (L.) All. (ASPARAGACEAE)

Polygonatum erythrocarpum Hua Diels in Bot. Jahrb. 29: 248. (1900). *Polygonatum roseum* Hook. Wright in Journ. Linn. Soc. Bot. 36: 108. (1903).

Hubei: Badong, 5205. Fang Xian, 6497.

Polygonum aviculare L. (POLYGONACEAE)

Forbes & Hemsley in Journ. Linn. Soc. Bot. 26: 334. (1891).

Hubei: Yichang, 1180, 3389. Badong, 4803, 4860, 4969, 5194.

A spreading, wiry annual weed, widely dispersed in the temperate regions of Europe and Asia, in China it supplies the drug *bian xi* which has been used in China since ancient times as a diuretic and to treat inflammation of the mouth and upper respiratory tract.

Polygonum plebeium R. Br. (POLYGONACEAE)

Forbes & Hemsley in Journ. Linn. Soc. Bot. 26: 346. (1891), Henry in Notes Econ. Bot. China 53. (1893); in Trans. Asiat. Soc. Jap. 24 suppl: 76. (1896) (List Pl. Formosa).

Hubei: Yichang, 830, 3601. Changyang, 7746. **Taiwan:** Oluanpi, 1756.

Distributed through tropical and subtropical Africa, Asia and Australia. According to Henry's notes this species is the *t'ieh-ma-ch'ih-han*.

Polygonum sp. (POLYGONACEAE)

Henry in Trans. Asiat. Soc. Jap. 24 suppl: 77. (1896) (List Pl. Formosa).

Taiwan: Kaohsiung, 1168, 1171. Wanshoushan, 2026.

Henry described his 2026 as a large climbing shrub seen only in a barren (unflowered) state.

Polygonum spp. (POLYGONACEAE)

Anonymous in Kew List of Determinations (volume 24 & 34).

Hubei: Yichang, 419, 649, 996, 2193, 2360, 2786, 4327, 6961. Badong, 332 (June 1885), 702 (July 1885), 1015, 4819, 4980, 5068, 5096. Zigui, 6196. Xingshan, 6961. Changyang, 7741. **Yunnan:** Mengzi, 9007, 9133, 9279, 9283, 9305, 9339, 9454, 9665, 9685, 9751, 10041, 10206, 10213, 10258, 10383, 10499, 10802, 10933, 11132, 11199, 11351, 13672. Simao 11918, 11970, 12071, 12132, 12187, 12217, 12220, 12348, 12519, 12655, 12662, 12667, 12695, 13201.

Polypodium spp. (POLYPODIACEAE)

Anonymous in Kew List of Determinations (volume 24 & 34).

Hubei: Liantuo, 4637. **Yunnan:** In woods near Mengzi at 1,650 metres, on trees, 9194b. Mengzi, 9557. Simao, 12554, 13071.

Polyporus sp. (Formes nigricans Fr.) (POLYPORACEAE)

Anonymous in Kew List of Determinations (volume 24).

Hubei: Yichang, 7948. (Fungus).

Polyscias fruticosa (L.) Harms (ARALIACEAE)

Panax fruticosus L. Henry in Notes Econ. Bot. China 67. (1893), Henry in Trans. Asiat. Soc. Jap. 24 suppl: 48. (1896) (List Pl. Formosa).

Taiwan: Wanchin, 494.

'Said by my native collector to be wild on the mountains; but this is doubtful. Often cultivated and named *t'u-san-ch'i*, in Formosa.' A. Henry. In his *Notes on Economic Botany in China* 67. (1893) Henry also calls this species *san-ch'i*, this belongs however to the medicinal *san-ch'i* of Yunnan (*Panax notoginseng*).

Polyspora axillaris (Roxb. ex Ker Gawl.) Sweet (THEACEAE)

Gordonia axillaris (Roxb. ex Ker Gawl.) Endl. Rehder & E. H. Wilson in Sargent Pl. Wilson. 2: 394. (1915), Li in Woody Flora of Taiwan 593. fig. 232. (1963). *Gordonia anomala* Spreng. Henry in Trans. Asiat. Soc. Jap. 24 suppl: 20. (1896) (List Pl. Formosa).

Taiwan: Oluanpi, 215, 685.

A small evergreen tree bearing handsome solitary white *Camellia*-like flowers on short stalks, native to Indochina and southern China. In Chinese Taiwan this species is distributed in broad-leaved forest at low and medium altitude. It is suited to the milder counties of the British Isles and Ireland and grows well at Glanleam on Valentia Island in Co Kerry.

02

Polyspora chrysandra (Cowan) Hu ex B. M. Barthol. & T. L. Ming (THEACEAE) New species

Gordonia axillaris Rehder & E. H. Wilson non (Roxb.) D. Dietr. in Sargent Pl. Wilson. 2: 394. (1915). *Gordonia chrysandra* Cowan in Notes Roy. Bot. Gard. Edinburgh 16: 184. (1931).

Yunnan: Mengzi, 10490. In forests near Mengzi at 1,800 to 2,800 metres, 10398 (syntype of *Gordonia chrysandra* Cowan), 11162.

Based on collections made by Henry at Mengzi in 1898 and on material later gathered at Tengchong in November 1912 by George Forrest (Forrest 9234 forms the type).

Polyspora sp. (THEACEAE)

Gordonia sp. Anonymous in Kew List of Determinations (volume 34).

Yunnan: Mengzi, 11398. Simao, 13380.

Polystachya sp. (ORCHIDACEAE)

Anonymous in Kew List of Determinations (volume 34).

Yunnan: Simao, 13590, 13611.

Polystichum aculeatum (L.) Roth. (POLYPODIACEAE)

Christ in Bull. Soc. Bot. France 5: 29. (1905).

Aspidium aculeatum (L.) Sw. Anonymous in Kew List of Determinations (volume 24 & 34). *Aspidium acuelatum* (L.) Sw. var. *angulare* Baker non (Kit. ex Willd.) D. C. Eaton in Journ. Bot. 26: 227. (1888).

Hubei: Yichang, 1121, 3043a. Antelope Glen, 3289. Jianshi, 5962, 5970. Liantuo, 2150, 2615, 2684, 2699, 3019, 3022, 4604, 7647. Badong, 1741, 2882 (October 1886), 2868, 2872, 3676, 5050, 5051, 5089, 5174, 5179. Changyang, 7640, 7643.Changleping, 7837. **Sichuan:** South Wushan, 7453 (in part). **Yunnan:** Mengzi, 10005, 11559. Simao, 12871, 12891, 12910.

Polystichum atkinsonii Bedd. (POLYPODIACEAE)

Aspidium atkinsonii (Bedd.) C. B. Clarke Baker in Journ. Bot. 27: 176. (1889).

Hubei: South Badong, 6171.

Polystichum braunii (Spenn.) Fée (POLYPODIACEAE)

Aspidium braunii Spenn. var. *clarkei* H. Christ in Bull. Herb. Boissier 7: 15. (1899).

Yunnan: Mengzi, 11549.

Polystichum craspedosorum (Maxim.) Diels (POLYPODIACEAE)

Aspidium craspedosorum Maxim. Anonymous in Kew List of Determinations (volume 24).

Hubei: Liantuo, 2651, 2987, 4607. Antelope Glen, 4321. South Badong, 6145 (in part).

This species was collected by the Sino-American Expedition in Shennongjia forest Reserve in 1980. Native to Russia, China, Korea and Japan.

Polystichum crinigerum (C. Chr.) Ching (POLYPODIACEAE E) New species

Polypodium crinitum Baker in Kew Bull. Misc. Inf. 12. (1906) non Poir 1804.

Yunnan: In a ravine in the mountains near Mengzi at 1,400 metres, 11557 (type of *Polypodium crinitum* Baker).

Polystichum deltodon (Diels) Diels (POLYPODIACEAE)

Christ in Bull. Soc. Bot. France 5: 26. (1905).

Hubei: Liantuo, 1948, 2009, 2010, 2011. Yichang, 3245. South Badong, 6145 (in part). Changleping, 7839. **Yunnan:** Mengzi, 9357.

Discovered by Charles Maries in the Yichang (Xiling) Gorge in 1879 while collecting for James Veitch & Sons, Chelsea. This species was collected by the Sino-American Expedition in Shennongjia forest Reserve in 1980, also native to Myanmar, Japan and the Philippines.

Polystichum fraxinellum (Christ) Diels (POLYPODIACEAE) New species

in Bull. Fan. Mem. Inst. Biol. 8: 329. pl. 7. (1938).

Aspidium fraxinellum Christ in Bull. Herb. Boissier 7: 15. (1899).

Yunnan: On wooded cliffs near Mengzi at 1,650 metres, 11550a., (isotype of *Aspidium fraxinellum* Christ, collected 6th May 1897).

Polystichum grandifrons C. Chr. (POLYPODIACEAE) New species

Polystichum grande R. C. Ching in Bull. Fan. Mem. Inst. Biol. 2: 189. (1931), nom. illeg, non Fee (1857).

Yunnan: Mengzi, 13686,(isotype of *Polystichum grande* R. C. Ching).

Polystichum ichangense Christ (POLYPODIACEAE) New species

in Bull. Soc. Bot. France 5: 28. (1905).

Hubei: Yichang, without number, (May 1888).

Endemic to China where it is distributed in the provinces of Guizhou, Hubei, Hunan and Sichuan. This species was named for the city of Ichang, now Yichang, once home to Augustine Henry, Ernest Wilson, Thomas Watters and visited at one stage or other by Père Armand David, Père Jean Marie Delavay, Reginald Farrer, Ernst Faber, Charles Maries, Pratt, Meyer and Frank Kingdon Ward. Yichang remains the most famous city in the history of botanical exploration in China.

Polystichum lanceolatum Diels (POLYPODIACEAE)

Aspidium lanceolatum Baker Baker in Journ. Bot. 27: 176. (1889).

Hubei: Changleping, 7835.

Discovered in the Yichang (Xiling) Gorge by Charles Maries while collecting for James Veitch & Sons, Chelsea. This beautiful fern is distributed through the provinces of

Hubei, Hunan, Guizhou and Jiangxi.

Polystichum mengziense Li Bing Zhang (POLYPODIACEAE) New species

Polystichum deltodon (Diels) Diels var. *henryi* Christ in Bull. Soc. Bot. France 5: 27. (1905).

Yunnan: Mengzi at 1,950 metres, 10346.

Endemic to China and distributed in the provinces of Gansu, Guizhou, Sichuan, and Yunnan.

Polystichum neolobatum Nakai (POLYPODIACEAE)

Polystichum lobatum (Sw.) Christ var. *chinense* Christ Christ in Bull. Soc. Bot. France 5: 29. (1905).

Hubei: Badong, 5086.

Polystichum subacutidens Ching ex L. L. Xiang (POLYPODIACEAE)

Polystichum deltodon (Diels) Diels var. *cultratum* Christ in Bull. Soc. Bot. France 5: 27. (1905).

Yunnan: Mengzi at 1,650 metres, 9773 (type of *Polystichum deltodon* (Diels) Diels var. *cultratum* Christ).

Distributed in China (Guizhou, Guangxi and Yunnan) and Vietnam.

Polystichum xiphophyllum (Bak.) Diels (POLYPODIACEAE)

Aspidium xiphophyllum Baker Baker in Journ. Bot. 27: 176. (1887).

Hubei: Changleping, 7838.

China to Malaysia.

Polystichum yunnanense Christ (POLYPODIACEAE) New species

in Not. Syst. 1: 34. (1911) Christ in Ind. Fil. Suppl. 1: 66. (1913), Zhang & Kung in Acta Phytotax. Sin. 34: 70. (1996).

Yunnan: Mengzi at 1,500 to 2,100 metres, 9101 (lectotype). Lunan, 9101a. Simao, 13423.

Native to Nepal and China [Tibet (Xizang), Yunnan and Sichuan] and Myanmar.

Polytrichum commune Hedw. (POLYTRICHACEAE)

Salmon in Journ. Linn. Soc. Bot. 34: 464. (1900).

Hubei: On the ground in a wood at Jianshi, 5817. (Moss).

Polytrichum sp. (POLYTRICHACEAE)

Anonymous in Kew List of Determinations (volume 24).

Hubei: Jianshi, 5693. South Badong, 7403. **Sichuan:** South Wushan, 5693.

Pometia pinnata J. R. Forst. & G. Forst. (SAPINDACEAE)

Anonymous in Kew List of Determinations (volume 34).

Yunnan: Simao, 12128.

Native to Sri Lanka, China (Yunnan) and the Philippines.

Pongamia pinnata (L.) Pierre (FABACEAE)

Li in Woody Flora of Taiwan 359. (1963).

Pongamia glabra Vent. in Trans. Asiat. Soc. Jap. 24 suppl: 38. (1896) (List Pl. Formosa).

Hainan: Environs of Kiungchow, 8277. **Taiwan:** Kaohsiung Spit, without number. Wanchin, 398.

A semi-deciduous tree, native to Indo-Malaysia, southern China and Japan (south) to northern Australia. Scattered along the sea-shore in Chinese Taiwan and often cultivated in gardens there.

Pontederia hastata L. (PONTEDERIACEAE)

Monochoria hastifolia C. Presl. Brown in Journ. Linn. Soc. Bot. 36: 148. (1903).

Hainan: Environs of Kiungchow, 8238.

Pontederia plantaginea Roxb. (PONTEDERIACEAE)

Henry in Journ. China Br. Roy. Asiat. Soc. 22: 242. (1887) (Chinese Names of Plants); in Trans. Asiat. Soc. Jap. 24 suppl: 98. (1896) (List Pl. Formosa), Brown in Journ. Linn. Soc. Bot. 36: 149. (1903).

Monochoria vaginalis (Burm. f.) C. Presl. ex Kunth. var. *plantaginea* Solms-Laubh. Brown in Journ. Linn. Soc. Bot. 36: 150. (1903).

Hubei: Yichang, 129, 2357. **Hainan:** Environs of Haikou, 8348, 8368, 8386. Environs of Kiungchow 8780. **Taiwan:** Oluanpi, without number. Kaohsiung Plain, in rice fields, 781.

A freshwater perennial, native to China (Taiwan), Japan, the Philippines, Malaysia and Java. In Taiwan this species is often found in ricefields and swamps. According to Henry this species was known colloquially at Yichang as *chu erh to.*

Populus adenopoda Maxim. (SALICACEAE)

Henry in Notes Econ. Bot. China 55. (1893), Schneider in Sargent Pl. Wilson. 3: 23. (1916).

Populus tremula Henry non L. in Notes Econ. Bot. China 55. (1893), Forbes & Hemsley in Journ. Linn. Soc. Bot. 26: 537. (1899). *Populus tremula* L. var. *adenopoda* (Maxim). Burkill. Burkill in Journ. Linn. Soc. Bot. 26: 537. (1899), Zheng in Journ. Wuhan Bot. Res. 2.(1): 14. (1984).

Hubei: Liantuo, 3798, 3816. Badong, 5281. Fang Xian, 6736. **Sichuan:** South Wushan (May 1888), 5281a.

The *pai-yang,* a tree to about 25 metres with a girth of almost 2 metres. The Chinese ally of *Populus tremula* L., this species is native to central and western China and was introduced to cultivation by Wilson in 1906. Propagation material from Wilson 4440, a living plant collected in Fang Xian in 1907 was sent from the Arnold Arboretum to Glasnevin soon after its collection. A common low-altitude poplar in Hubei and Sichuan. It was discovered about 1874 by the Russian explorer Pavel Jakovlevich Piasetski while travelling on an expedition sent to China by the Tsar to explore trade routes.

Populus × canadensis Moench. 'Henryana' (SALICACEAE)

Populus × henryana Dode in Mem. Soc. Hist. Nat. Autun. 18: 39. (1905), Henry in Elwes and Henry, The Trees Gt. Brit. & Irel. 7: 1829. (1913), Henry in Gard. Chron. ser. 3, 56: 47. fig. 17. (1914).

According to Henry this hybrid was of a much-branched habit. It formed a handsome tree when old such as the figure of the White Knights tree in the Gardeners' Chronicle which in 1914 was over 30 metres tall and 4 metres in girth. Its origin is unkown and it was named by Dode in recognition of Henry's work in bringing this group out of obscurity and sorting out enormous nomentlaclatural errors. A staminate tree with leaves cuneate at the broad base.

Populus × generosa Henry (SALICACEAE)

in Gard. Chron. ser. 3, 51: 258. fig. 102. (1914), Henry in Gard. Chron. ser. 3, 87: 24. (1930), Nelson in Pim, The Wood & the Trees 233. (1984), Bean in Trees & Shrubs 3: 312. (1992), O' Brien in Augustine Henry, An Irish Plant Collector in China 29. (2002).

Populus angulata × Populus trichocarpa Henry in Gard. Chron. ser. 3, 56: 67. (1914).

An artificial hybrid created by Augustine Henry at the Royal Botanic Gardens, Kew by crossing a female *Populus deltoides* 'Cordata' with the pollen of *Populus trichocarpa* in March 1912. By June of 1912, a few seeds had ripened and these germinated the following October at Cambridge (where Henry was Professor of Forestry). In 1913 Henry accepted a post as Professor of Forestry in the Royal College of Science for Ireland (now part of University College, Dublin) in Dublin and so had his seedlings (then 5 cm tall) planted in good soil in the enclosed garden at the National Botanic Gardens, Glasnevin. By the end of the 1913 growing season the tallest of Henry's poplar seedlings was over a metre tall and by September 1914 the tallest seedling was 3 metres tall. Henry repeated the cross in 1914 and in seven years seedlings from this crossing had reached a height of over 12 metres with a girth of 55 cm. The very first deliberate cross between a black and balsam popular this tree is prone to bacterial canker and has never become well known though Henry stated there were plantations of it in the south of France and northern Italy in 1930. In the UK *Populus × generosa* has reached 40 metres tall. In Ireland this tree is cultivated in the John f. Kennedy Arboretum near New Ross in Co Wexford and at the National Botanic Gardens, Glasnevin. There once grew a tree in Henry's garden at Ranelagh though this has long since disappeared having been swamped by *Populus × canadensis* Moech., formerly

Populus × vernirubens, another of Henry's other hybrid cottonwoods. *Populus × vernirubens* Henry in Gard. Chron. ser, 3. 87: 24. (1930) O' Brien in Augustine Henry, An Irish Plant Collector in China 29. (2002) Another forestry tree associated with Henry is *Larix × henryana* Rehder [in Journ. Arnold Arb. 1. 52. (1919)], a cross between the European larch *Larix decidua* and the Japanese larch *Larix kaempferi*. This hybrid larch is now referrable to *Larix × eurolepis* A. Henry in *The Gardeners' Chron*. Ser. 3., 36: 4. (1919).

Populus lasiocarpa Oliver (SALICACEAE) New species

in Hooker's. Icon. Pl. 20: t. 1943. (1890), Henry in Notes Econ. Bot. China 55. (1893), Bretschneider in Hist. Eur. Bot. Disc. China 2: 791. (1898), Burkill in Journ. Linn. Soc. Bot. 26: 536. (1899), J. H. Veitch in Journ. Roy. Hort. Soc. xxvii. 65. fig. 27. (1903); in Hortus Veitchii 372. (1906), Schneider in Illus. Handb. Laubholzk. 1: 17. (1904), Dode in Mem. Soc. Hist. Nat. Autun. xviii. (Extr. Monog. Ined. Populus, 66. (1905), Ascherson & Graebner in SynMitteleur. Fl. iv. 51. (1908), Gombocz in Math. Termesz. Kolz. 30: 120. (Monog. Gen. Populi) (1908), Henry in Elwes & Henry, Trees Gr. Brit. & Irel. 7: 1846. t. 408. fig. 9. (1913), Bean in Journ. Roy. Hort. Soc. 28: 65. fig. 27. (1903), Bean in Trees & shrubs. 2: 215. (1914), Skan in Bot. Mag. cxli. t. 8625. (1915), Schneider in Sargent Pl. Wilson. 3: 17. (1916), Morley in Glasra 3: 80. (1979), Chao in Journ. Wuhan Bot. Res. 9: 231. (1991), Nelson in A Heritage of Beauty 316. (2000), O' Brien in Augustine Henry, An Irish Plant Collector in China 29. (2002).

Hubei: Badong, 4013, 5423. Jianshi, 5423a.

A medium-sized tree to about 24 metres tall with a pyramidal crown, according to Henry it was known colloquially in Hubei as *tai-erh-po* or *ta-yeh-p'ao*. Discovered by Henry in 1887, Wilson introduced this magnificent tree by sending about a dozen living plants to Messrs Veitch in 1904. According to Henry's field notes at Kew, the wood of this tree was much used for planking and making ladles. This species was collected by the Glasnevin Central China Expedition (GCCE 286) on the hills above Badong in September 2002.

Populus sp. (SALICACEAE)

Anonymous in Kew List of Determinations (volume 24).

Hubei: Yichang, 1245.

Poranopsis discifera (Schneider) Staples (CONVOLVULACEAE) New species

in Novon 3: 200. (1993).

Porana discifera Schneider in Sargent Pl. Wilson. 3: 358. (1916).

Yunnan: Forests to the east of Simao, 12622. In forests

to the south of Simao at 1,600 metres, 12694 (holotype of *Porana discifera* Schneider).

Native to north-east India, Myanmar, China (Sichuan and Yunnan), northern Thailand and Vietnam.

Poranopsis sinensis (Handel-Mazzettii) Staples (CONVOLVULACEAE) New species

in Novon 3: 200. (1993).

Porana paniculata Schneider non Roxb. in Sargent Pl. Wilson. 3: 359. (1917). *Cardiochlamys sinensis* Hand.-Mazz. in Anz. Akad. Wiss. Wien. Math.-Nat. 57: 241. (1912); in Symb. Sinicae 7: 809. (1936). *Porana henryi* Verdcourt in Kew Bull. 26: 137. (1971).

Yunnan: Near Mengzi at 1,500 metres, 9489.

Discovered by Henry in about 1896, this species was later collected by Handel-Mazzetti near Kunming in Yunnan in March 1914. Henry described this particular plant as a large climber covering rocks and stated it had white flowers in masses. Endemic to China and distributed in Sichuan and Yunnan.

Porostereum spadiceum (Pers.) Hjortstam & Ryvarden (PHANEROCHAETACEAE)

Stereum spadiceum Fr. Anonymous in Kew List of Determinations (volume 24).

Hubei: Fang Xian, 6669. (Fungus).

Porotrichum makinoi Broth. f. forma (NECKERACEAE)

Salmons in Journ. Linn. Soc. Bot. 34: 467. (1900).

Taiwan: On the side of a cliff on Wanshoushan, 1841. (Moss).

Porphyra sp. (BANGIACEAE)

Henry in Trans. Asiat. Soc. Jap. 24 suppl: 118. (1896) (List Pl. Formosa).

Taiwan: Tamsui, 1938.

E. H. Wilson in *A Naturalist in Western China* 2: 63. (1913) states that the seaweed, *Porphyra vulgaris* was imported into China from Japan and from it a very nutritious jelly was made. Of the 160 species of algae used as a food source worldwide the bulk come fom three genera, *Porphyra*, *Laminaria* and *Undaria*. Species of *Porphyra* have been eaten throughout recorded history for their food value, flavour, colour and texture. The vitamin C content can be equal to that of citrus fruits. The consumption of this genus began in China between 533—544 A.D. and gifts of it were presented annually to the emperors of China during the Sung dynasty from Fujian province. Henry stated it was commonly known as *nori*.

Portulaca oleracea L. (PORTULACACEAE)

Henry in Trans. Asiat. Soc. Jap. 24 suppl: 19. (1896) (List Pl. Formosa); in Notes Econ. Bot. China 53. (1893).

Taiwan: Kaohsiung, 1790.

Purslane, an annual succulent herb Henry stated it was known in China by the name *ma-ch'ih-han* (*ma chi xian*) or 'horse's teeth amaranth'. A weed throughout the warmer parts of Europe, east Africa and Asia. Cultivated since ancient times and eaten like spinach. This plant has an enormous number of common names and in Malawi from the leaf shape it is called the 'buttocks of the wife of a chief'. In China it is used as an antidote to the pain and swelling of wasp stings and snake bites.

Potamogeton cristatus Regel & Maack. (POTAMOGETONACEAE)

Wright in Journ. Linn. Soc. Bot. 36: 194. (1903), Graebner in Engler, Pflanzenr. 13: 49. (1907).

Hubei: Yichang, 2377.

Potamogeton nodosus Poir. (POTAMOGETONACEAE)

Potamogeton malaianus Miq. Graebner in Engler, Pflanzenr. 13: 165. (1907). *Potamogeton mucronatus* Henry non C. Presl. in Trans. Asiat. Soc. Jap. 24 suppl: 101. (1896) (List Pl. Formosa). *Potamogeton gaudichaudii* Wright non Champ. & Schltdl. in Journ. Linn. Soc. Bot. 36: 194. (1903).

Taiwan: Kaohsiung, 1203, 1773, 1811. Wanchin, 1203a.

Potamogeton × orientalis Hagstr. (POTAMOGETONACEAE)

Potamogeton hillii Wright non Morong. in Journ. Linn. Soc. Bot. 36: 194. (1903). Graebner in Engler, Pflanzenr. 13: 165. (1907). *Potamogeton henryi* Fernald in Mem. Amer. Acad. Arts. 17: 73. (1932).

Hubei: Yichang, 4375 (holotype of *Potamogeton henryi* Fernald).

A natural hybrid between *Potamogeton berchtoldii* × *Potamogeton pulsillus*.

Potamogeton spp. (POTAMOGETONACEAE)

Anonymous in Kew List of Determinations (vol. 24 & 34), Henry in Trans. Asiat. Soc. Jap. 24 suppl: 102. (1896) (List Pl. Formosa).

Hebei: Anping, 1767. **Taiwan:** Kaohsiung Plain, 1813. **Yunnan:** Mengzi, 9985.

Potamogeton tepperi A. Benn. (POTAMOGETONACEAE)

Wright in Journ. Linn. Soc. Bot. 36: 196. (1903), Graebner in Engler, Pflanzenr. 13: 165. (1907).

Hubei: Yichang, 2366, 3603.

Potentilla centigrana Maxim. (ROSACEAE)

Diels in Bot. Jahrb. 29: 403. (1900).

Hubei: Badong, 4068.

Potentilla chinensis Ser. (ROSACEAE)

Forbes & Hemsley in Journ. Linn. Soc. Bot. 23: 241. (1887), Diels in Bot. Jahrb. 29: 403. (1900).

Potentilla multifida Henry non L. in Journ. China Br. Roy. Asiat. Soc. 22: 236. (1887) (Chinese Names of Plants).

Hubei: Yichang, 100, 101. Badong, 2416.

According to Henry this species was known

02

colloquially as *mao chi t'ui* at Yichang and the roots of this plant and those of *Potentilla discolor* were sometimes eaten by locals.

Potentilla cryptotaeniae Maxim. (ROSACEAE) New to China

Forbes & Hemsley in Journ. Linn. Soc. Bot. 23: 241. (1887), Diels in Bot. Jahrb. 29: 404. (1900).

Hubei: Badong, 1821, 4681.

Potentilla discolor Bunge (ROSACEAE)

Forbes & Hemsley in Journ. Linn. Soc. Bot. 23: 241. (1887), Henry in Journ. China Br. Roy. Asiat. Soc. 22: 236. (1887) (Chinese Names of Plants), Diels in Bot. Jahrb. 29: 403. (1900).

Hubei: Yichang, 211, 147. Liantuo, 1977.

Colloquially known at Yichang as *chi shih t'eng* according to Henry.

Potentilla fragarioides L. (ROSACEAE)

Anonymous in Kew List of Determinations (volume 24).

Hubei: Yichang, 3398.

Potentilla fragiformis Willd. ex D. F. K. Schltdl. (ROSACEAE)

Forbes & Hemsley in Journ. Linn. Soc. Bot. 242. (1887), Diels in Bot. Jahrb. 29: 403. (1900).

Hubei: Yichang, 7895. Badong, 349 (June 1885), 1441, 1278. **Yunnan:** Mengzi, 10598, 10905.

Potentilla indica (Andrews) Th.Wolf (ROSACEAE)

Fragaria indica Andrews Forbes & Hemsley in Journ. Linn. Soc. Bot. 23: 240. (1887), Henry in Journ. China Br. Roy. Asiat. Soc. 22: 264. (1887) (Chinese Names of Plants); in Notes Econ. Bot. China 46. (1893). *Duchesnea indica* (Andrews) Teschem. Diels in Bot. Jahrb. 29: 401. (1900).

Hubei: Yichang, 264 (in part), 319, 995, 1271, 3226. Xiling (Yichang) Gorge, 3296, 3453. **Yunnan:** Mengzi, 9295.

The fool's strawberry or Indian mock-strawberry, Henry stated this species was known colloquially in the Three Gorges region as the *she p'ao tzu*. In Chinese herbal medicine the dried aerial parts of this species are used in in cases of snake-bite. This species was collected by the Glasnevin Central China Expedition (GCCE 655) in Dragon Gate Forerst River Park in Xingshan in October 2004. Widespread across India, China, Korea, Japan, Indochina, the Philippines and Java. Also naturalised in Europe and North America.

Potentilla aff. *leschenaultiana* Ser. (ROSACEAE)

Anonymous in Kew List of Determinations (volume 34).

Yunnan: Mengzi, 9663.

Potentilla reptans L. (ROSACEAE)

Anonymous in Kew List of Determinations (volume 24).

Hubei: San You Dong Glen, 3410.

Potentilla reptans L. var. *sericophylla* Franch. (ROSACEAE)

Fragaria filipendula Hemsley in Journ. Linn. Soc. Bot. 23: 139. (1887), Henry in Journ. China Br. Roy. Asiat. Soc. 22: 264. (1887) (Chinese Names of Plants), Bretschneider in Hist. Eur. Bot. Disc. China 2: 783. (1898).

Hubei: Yichang, 264 (in part, type of *Fragaria filipendula* Hemsley).

According to Henry this species was known colloquially in the Three Gorges region as *she p'ao tzu*. Discovered by Père Armand David.

Potentilla sp. (ROSACEAE)

Anonymous in Kew List of Determinations (volume 24).

Hubei: Badong, 4825.

Potentilla sundaica (Blume) W. Theob. (ROSACEAE)

Potentilla kleiniana Wight & Arn. Diels in Bot. Jahrb. 29: 403. (1900).

Hubei: Yichang, 3852.

Pothos chinensis (Raf.) Merr. (ARACEAE)

Pothos seemannii Schott. Henry in Trans. Asiat. Soc. Jap. 24 suppl: 101. (1896) (List. Pl. Formosa), Brown in Journ. Linn. Soc. Bot. 36: 186. (1903), Pax & Knuth in Engler. Pflanzenr. iv. 23B: 29. (1905). *Pothos yunnanensis* Engler in Pflanzenr. (Engler) 4. Fam. 23B: 205. (1905).

Hubei: Liantuo, 4395. **Taiwan:** Wanchin, without number. Oluanpi, 689. **Yunnan:** In the mountains near Simao at 1,650 metres, 11779 (type of *Pothos yunnanensis* Engler).

A large root-climbing liana, distributed through India, Bangladesh, Bhutan, Myanmar, China, Cambodia, Laos, Vietnam and Thailand. Found growing over rocks and through trees at altitudes between 250 to 2,270 metres.

Pothos repens (Lour.) Druce (ARACEAE)

Pothos loureiroi Hook. & Arn. Brown in Journ. Linn. Soc. Bot. 36: 186. (1903), Engler in Engler, Pflanzenr. iv. 23B: 35. (1905).

Hainan: Without locality, 112, 7980, 7980a.

Pothos sp. (ARACEAE)

Anonymous in Kew List of Determinations (volume 34).

Yunnan: Mengzi, 10920.

Pottsia grandiflora Markgraf. (APOCYNACEAE) New species

Tsiang in Sunyatsenia 2: 170. (1934).

Yunnan: Mountains to the east of Simao at 1,650

metres, 12033a.

This large red flowered climber was discovered by Henry and described as a new species from material later collected by the Chinese botanist C. L. Tso in May 1929.

Pottsia laxiflora (Blume) Kuntze (APOCYNACEAE)

Pottsia cantonensis Hook. & Arn. Anonymous in Kew List of Determinations (volume 34).

Yunnan: Mengzi, 13674. **Hainan:** Without locality, 8571. Ling-men (in the interior of the island,) 8651.

Pourthiaea amphidoxa (C. K. Schneid.) Stapf. (ROSACEAE) New species

in Curtis's Bot. Mag. 155: t. 9275. (1932).

Photinia variabilis Forbes & Hemsley non F. B. Forbes & Hemsl. in Journ. Linn. Soc. Bot. 23: 263. (1887), Henry in Notes Econ. Bot. China 64. (1893). *Stranvaesia amphodoxa* Schneider in Bull. Herb. Boiss. ser. 2. 6: 319. (1906); in Illus. Handb. Laubholzk. I. 713, fig. k-l. (1906). *Photinia amphidoxa* (Schneid.) Rehder & E. H. Wilson in Sargent Pl. Wilson. 1: 190. (1912), Chun in Sunyatsenia 2: 65. (1924), Rehder in Journ. Arnold Arb. 13: 306. (1932), Metcalf in Journ. Arnold Arb. 20: 441. (1939), O' Brien in Augustine Henry, An Irish Plant Collector in China 27. (2002).

Sichuan: South Wushan, 7389, 5565 (type) 5565a., (topotype of *Stranvaesia amphodoxa* Schneider).

In Ireland this species is cultivated at the John F. Kennedy Arboretum, Co Wexford.

Pourthiaea arguta (Wall. ex Lindl.) Decne. (ROSACEAE)

Photinia variabilis Forbes & Hemsley non F. B. Forbes & Hemsl. in Journ. Linn. Soc. Bot. 23: 263. (1887), Henry in Notes Econ. Bot. China 64. (1893); in Trans. Asiat. Soc. Jap. 24 suppl: 41. (1896) (List Pl. Formosa). *Photinia beauverdiana* C. K. Schneid. in Bull. Herb. Boiss. ser. 2. 6: 319. (1906), Schneider in Illus. Handb. Laubholzk. 1: 710. (1906), Rehder & E. H. Wilson in Sargent Pl. Wilson. 1: 187. (1913), Bean in Trees & Shrubs 3: 275. (1936), Zheng in Journ. Wuhan Bot. Res. 2.(1): 72. (1984), Nelson in A Heritage of Beauty 316. (2000), O' Brien in Augustine Henry, An Irish Plant Collector in China 27. (2002). *Photinia lucida* (Decaisne) Schneider Li in Woody Flora of Taiwan 279. fig. 103. (1963).

Hubei: Badong, 3768. Liantuo, 4566. Jianshi, 7400. Changleping, 6268. **Sichuan:** South Wushan, 5599 (type of *Photinia beauverdiana* C. K. Schneid.), 5599a. North Wushan, 7095. **Taiwan:** Oluanpi, 675, 960, 1285.

According to Wilson this was a very common small, slender tree found in woodlands and copses in Hubei and Sichuan. In Ireland this species is cultivated at Dargle Glen (County Wicklow), the John F. Kennedy Arboretum, Mount

Congreve and at the National Botanic Gardens, Glasnevin.

Pourthiaea villosa (Thunb.) Decne (ROSACEAE)

Photinia variabilis Forbes & Hemsley non F. B. Forbes & Hemsl. in Journ. Linn. Soc. Bot. 23: 263. (1887), Henry in Notes Econ. Bot. China 64. (1893). *Pourthiaea parvifolia* Pritzel in Bot. Jahrb. 29: 389. (1900). *Photinia parvifolia* (E. Pritzl) C. K. Schneider in Illus. Handb. Laubholzk, 1: 711. fig. 392. (1906), Rehder & E. H. Wilson in Sargent Pl. Wilson. 1: 189. (1912). *Photinia komarovii* (H. Leveille & Vant.) L. T. Lu & C. L. Li in Illus. Handb. Laubholzk. 1. 711. fig 392. (1906), Rehder & E. H. Wilson in Sargent Pl. Wilson. 1: 189. (1912), Zheng in Journ. Wuhan Bot. Res. 2.(1): 73. (1984). *Photinia subumbellata* Rehder & E. H. Wilson in Sargent Pl. Wilson. 1: 189. (1912). *Photinia villosa* (Thunb.) DC. var. *sinica* Rehder & E. H. Wilson in Sargent Pl. Wilson. 1: 186. (1912), Bean in Trees & Shrubs 3: 276. (1936), Zheng in Journ. Wuhan Bot. Res. 2.(1): 74. (1984), Nelson in A Heritage of Beauty 316. (2000). *Photinia* sp. Zheng in Journ. Wuhan Bot. Res. 2.(1): 74. (1984).

Hubei: Yichang, 1108. Liantuo, 3002 (type of *Pourthiaea parvifolia* Pritzel), 6370, 6376, 7664. Xingshan, 4523. Badong, 234, 239 (June 1885), 4064, 5361, 5631. On the banks of the Yangtze at Badong, 7174. Changleping, 6295. Jianshi, 5830 (type of *Photinia parvifolia* (Pritzel) C. K. Schneider). Changleping, 7724. **Sichuan:** South Wushan, 5517, 5518 (syntype of *Photinia subumbellata* Rehder & E. H. Wilson), 7241. North Wushan, 7071. **Yunnan:** Simao, 11665.

Distributed in China, Korea and Japan.

Pouzolzia sanguinea (Blume) Merr. (URTICACEAE)

Pouzolzia ovalis Miq. Forbes & Hemsley in Journ. Linn. Soc. Bot. 26: 491. (1899).

Yunnan: Mengzi, 9344, 9344a.

Distributed in Nepal, India, Bhutan, Myanmar, China, Thailand and Vietnam.

Pouzolzia sanguinea (Blume) Merr. var. *elegans* (Wedd.) Friis, Wilmot-Dear & C. J. Chen (URTICACEAE)

Pouzolzia elegans Wedd. Forbes & Hemsley in Journ. Linn. Soc. Bot. 26: 489. (1899). *Boehmeria diffusa* Wedd. Henry in Trans. Asiat. Soc. Jap. 24 suppl: 89. (1896) (List Pl. Formosa).

Taiwan: Wanchin, 177.

A shrub or small tree, native to China (including Taiwan). According to Henry the pounded bark of this species was used by the early indigenous inhabitants of Taiwan as a medical application.

Pouzolzia sp. (URTICACEAE)

Forbes & Hemsley in Journ. Linn. Soc. Bot. 26: 491. (1899).

Hubei: Yichang, 80.

Pouzolzia zeylanica (L.) Benn. (URTICACEAE)

Pouzolzia indica (L.) Wight Henry in Trans. Asiat. Soc. Jap. suppl: 89. (1896) (List Pl. Formosa). *Pouzolzia indica* (L.) Wight var. *alienata* Weddell Forbes & Hemsley in Journ. Linn. Soc. Bot. 26: 490. (1899), Li in Woody Flora of Taiwan 138. (1963). *Pouzolzia hypericifolia* Henry non Blume in Trans. Asiat. Soc. Jap. 24 suppl: 90. (1896) (List Pl. Formosa). *Pouzolzia hispida* Henry non Benn. in Trans. Asiat. Soc. Jap. 24 suppl: 89. (1896) (List Pl. Formosa). *Pouzolzia hirta* Forbes & Hemsley non (Blume) Hassk. in Journ. Linn. Soc. Bot. 26: 489. (1899).

Hubei: At Shih Pan Shan [off the Xiling (Yichang) Gorge], 4188. **Hainan:** Environs of Haikou, 8032, 8366. **Taiwan:** Wanchin, 387, 437. Kaohsiung, 728, 784.

Pouzolzia zeylanica (L.) Benn. & R. Br. var. *angustifolia* (Wight.) J. Chen (URTICACEAE)

Pouzolzia indica Gaud. var. *angustifolia* (Wight.) Wedd. Forbes & Hemsley in Journ. Linn. Soc. Bot. 26: 490. (1899).

Hainan: Environs of Haikou, 8432.

Prasoxylon excelsum (Spreng.) Mabb. (MELIACEAE)

Dysoxylum procerum (Wall.) Hiern. Anonymous in Kew List of Determinations (volume 34).

Yunnan: Simao, 12046.

Native to India to western China and Papuasia.

Prasoxylon hongkongense (Tutch.) Mabb. (MELIACEAE)

Amoora rohituka Henry non (Roxb.) Wight & Arnott Henry in Trans. Asiat. Soc. Jap. 24 suppl: 26. (1896) (List Pl. Formosa). *Dysoxylum kusukusuense* (Hayata) Kanehira & Hatusima Li in Woody Flora of Taiwan 395. (1963).

Taiwan: Oluanpi, 1266.

A medium-sized tree, native to southern China, including Taiwan where it grows on the Hengchun Peninsula.

Premna bracteata Wall. ex C. B. Clarke (LAMIACEAE)

Anonymous in Kew List of Determinations (volume 34).

Yunnan: Simao, 11880a.

Distributed in India, Bangladesh, Bhutan and China [Tibet (Xizang) and Yunnan].

Premna fulva Craib (LAMIACEAE)

Premna longipila C. P'ei in Mem. Sci. Soc. China (The Verbenac. of China) 1(3): 75. (1932).

Yunnan: Mountains to the south of Simao at 1,500 metres, 12113 (type of *Premna longipila* C. P'ei).

Distributed in western China, Laos, China and Vietnam.

Premna henryana (Hand.-Mazz.) C. Y. Wu (LAMIACEAE)

New species
in Flora Yunnanica 1: 433. (1977).

Premna steppicola Hand.-Mazz. var. *henryana* H. von Handel-Mazzetti in Symb. Sin. 7: 902. (1936).

Yunnan: Mengzi, 10327 (type).

Endemic to China and distributed in Sichuan and Yunnan.

Premna interrupta Wall. ex Schauer. (LAMIACEAE)

Premna racemosa Wall. ex Schauer. C. Y. Wu in Flora Yunnanica 1: 439. (1977).

Yunnan: Mengzi, 11017.

Native to India, Bhutan, Nepal, Myanmar and China to Malaysia.

Premna microphylla Turcz. (LAMIACEAE)

Henry in Journ. China Br. Roy. Asiat. Soc. 22: 241. (1887) (Chinese Names of Plants), Forbes & Hemsley in Journ. Linn. Soc. Bot. 26: 256. (1890), Henry in Notes Econ. Bot. China 48. (1893), Zheng in Journ. Wuhan Bot. Res. 2.(1): 186. (1984).

Hubei: Liantuo, 2088, 3930.

According to Henry this species was known in the Three Gorges region as *ch'ou liang-tzu*.

Premna odorata Blanco (LAMIACEAE)

Li in Woody Flora of Taiwan 830. (1963), Sheng & Yang in Bull. Taiwan For. Inst. New Ser. 6(2): 112. (1991).

Premna vestita Schauer. Henry in Trans. Asiat. Soc. Jap. 24 suppl: 70. (1896) (List Pl. Formosa).

Taiwan: Wanchin, 381. Kaohsiung, 765.

A small tree, native to the Philippines to Australia and China (Taiwan) where it grows along the southern and eastern coast.

Premna puberula Pamp. (LAMIACEAE)

Rehder in Sargent Pl. Wilson. 3: 371. (1916).

Premna microphylla Forbes & Hemsley non Turcz. in Journ. Linn. Soc. Bot. 26: 256. (1890).

Hubei: Yichang, 3582, 4099. Badong, 6123. Liantuo, 6123a. Without locality, 4099a.

Premna serratifolia L. (LAMIACEAE)

Premna integrifolia L. Henry in Trans. Asiat. Soc. Jap. 24 suppl: 70. (1896) (List Pl. Formosa). *Premna* sp. Henry in Trans. Asiat. Soc. Jap. 24 supp: 71.(1896) (List Pl. Formosa). *Premna obtusifolia* R. Brown Li in Woody Flora of Taiwan 830. (1963), Lu & Yang in Bull. Taiwan For. Res. Inst. New Series 6(2): 108. (1991).

Taiwan: Oluanpi, 613. Kaohsiung, 791.

A small tree or shrub, distributed from tropical Asia to Australia. Common along coastal regions in Chinese Taiwan, this species was collected by the Glasnevin Taiwan Expedition (GTE 14) by the lighthouse at Oluanpi (South Cape of Henry and Wilson) in October 2004.

Premna spp. (LAMIACEAE)

Anonymous in Kew List of Determinations (volume

02

34).

Yunnan: Mengzi, 9521, 10337. Milê District, 9944a. Simao, 13205.

Premna szemaoensis C. P'ei (LAMIACEAE) New species

in Mem. Sci. Soc. China 1(3): 76. (1932).

Yunnan: Simao, 12105. In mountain forests near Simao at 1,650 metres, 12105a., (type).

Endemic to Yunnan.

Premna tomentosa Willd. (LAMIACEAE)

Anonymous in Kew List of Determinations (volume 34).

Yunnan: Mengzi, 9552.

Premna urticifolia Rehder (LAMIACEAE) New species

in Sargent Pl. Wilson. 3: 458. (1917).

Yunnan: Near Simao at 1,600 metres, 13389 (type).

Endemic to Yunnan.

Prenanthes sp. (ASTERACEAE)

Anonymous in Kew List of Determinations (volume 24).

Hubei: South Badong, 6132.

Primula barbicalyx C. H. Wright (PRIMULACEAE)

Pax & Knuth in Engler, Pflanzenr. iv. 237: 25. (1905).

Yunnan: On limestone crags near Mengzi at 2,900 metres, 10512, 11999.

Discovered by the Irish Plant collector and Imperial Maritime Customs officer, William Hancock near Mengzi in 1894. Similar to a short stemmed *Primula obconica*, not in cultivation and endemic to Yunnan.

Primula blinii H. Lév. (PRIMULACEAE)

Primula incisa Franch. Anonymous in Kew List of Determinations (volume 24).

Sichuan: Henry's Chinese collector with Pratt, no locality stated but may be Leshan, Emei Shan, Wa Shan or Kangding, 8869.

A widespread species in western Sichuan and extending southwards into Yunnan where it grows on limestone and sandstone in upper forested mountain slopes, on cliffs and in meadows between 3,200 to 4,200 metres. Introduced to cultivation by George Forrest in 1915, plants grew at the Royal Botanic Garden, Edinburgh until 1939.

Primula carolinehenryae S. O'Brien (PRIMULACEAE) New species

In Footsteps Augustine Henry 213. (2011).

Carolinella henryi Hemsley in Hooker's Ic. Pl. t. 2726. (1902). *Primula henryi* (Hemsley) Pax in Engler, Pflanzenr. iv. 236: 47. (1905), O'Brien in Augustine Henry, An Irish Plant Collector in China 7. (2002), non *Primula × henryi* T. Moore & Mast. in Gard. Chron. n. ser. 15: 404. (1881).

Yunnan: Mountains near Mengzi at 1,650 metres, 10735 (type of *Carolinella henryi* Hemsley, type of *Primula carolinehenryae* S. O'Brien).

A deciduous perennial bearing 2 to 3 heart-shaped leaves to 18cm long arising from a stout rhizome. Stems to 20 cm long, carrying up to 20 flowers in a short spikes. *Primula carolinehenryae* commerorates Henry's first wife Caroline Henry (nee Orridge d. 1894) who lived with him in Chinese Taiwan and collected plants (now housed in the Kew herbarium) in Japan and in Colorado, USA. There are nine species of *Primula* in the subgenus Carolinella (of which *Primula partschiana* is the type species). Native to China (south-east Yunnan), northern Vietnam and Thailand, they are usually found on steep slopes and cliffs under trees in warm temperate to subtropical regions at 1,500 to 2,600 metres. Little is known of these plants, many of the nine species have only ever been collected once. The earlier named *Primula × henryi* T. Moore & Mast. was raised by Mr. Anderson Henry as a *Primula denticulata* hybrid in the late 1870s and was published in *The Gardeners' Chronicle* in 1881, therefore Pax's *Primula henryi* from China (Yunnan) and northern Vietnam must be considered a homonym.

Primula denticulata Smith (PRIMULACEAE)

Pax & Knuth in Engler, Pflanzenr. iv. 236: 90. (1905).

Yunnan: Mengzi, 10918.

A beautiful clump forming deciduous perennial, overwintering by means of large ground-level buds. Widely known and grown for its large 'drumstick' heads of blue-violet, lavander or lilac flowers (several other shades have been developed in cultivation) produced between March and May. Native to the Sino-Himalaya from eastern Afghanistan, northern Pakistan, Kashmir, Sikkim, Nepal, Bhutan, northern Myanmar, China [Yunnan, south-east Tibet (Xizang), Sichuan and Guixhou], a range of over 3,000 km. It grows in in damp forest clearings, alpine meadows and by streams between 1,300 to 5,300 metres and can be abundant in certain regions so much so as to turn whole mountainsides into a sea of lilac-blue. It is extremly variable in its native habitat ranging from 30 cm to 4 cm tall when in flower. *Primula denticulata* was introduced from northern India by Dr. Royle of the East India Company to Messrs. Veitch who first exhibited plants before the Horticultural Society (later the Royal Horticultural Society) in March 1842.

Primula filipes G. Watt (PRIMULACEAE)

Henry in Gard. Chron. ser. 3, 31: 269. (1902), Pax & Knuth in Engler, Pflanzenr. iv. 236: 34. (1905).

Primula androsacea Pax in Engler, Pflanzenr. iv. 236: 34. (1905).

Yunnan: Mengzi at 1,500 metres, 9853, 10196, 10451

(type of *Primula androsace* Pax).

A very beautiful annual species bearing pale pink to white flowers, native from the Himalaya to China (Yunnan and Sichuan) where it is often found in marshes by ricefields and canals. According to Henry it appears in the great Chinese illustrated flora, the *Chih-Wu-Ming* (vol 29, folio 8) where it is called the *pao-ch'un* which translates as 'Herald of Spring'. Martyn Rix states it is cultivated in villages and temple gardens throughout the Lijiang valley, Yunnan. It was introduced from Yunnan by Père Delavay in 1889 through the garden of the Paris Museum.

Primula nutantiflora Hemsley (PRIMULACEAE) New species

in Journ. Linn. Soc. Bot. 29: 313. (1889), Bretschneider in Hist. Eur. Bot. Disc. China 2: 786. (1898), Pax & Knuth in Engler, Pflanzenr. iv. 236: 69. (1905), Nelson in Pim, The Wood & the Trees 227. (1984), O' Brien in Augustine Henry, An Irish Plant Collector in China 29. (2002).

Sichuan: South Wushan, 5584. On cliffs in North Wushan, 7073.

A dwarf perennial species, allied to *Primula yunnanensis* Franch. but with bell-shaped pendant lilac or rose flowers. Also native to Guizhou and Yunnan where it grows on damp limestone cliffs and rocks between 1,200 and 2,500 metres. Reginald Farrer wrote to Augustine Henry from Gansu (the letter is in the Glasnevin archives) seeking the location of this species. Farrer was planning to return to Beijing (then Peking) via the Three Gorges through Wushan that October and had hoped with further directions from Henry to obtain dormant crowns.

Primula obconica Hance (PRIMULACEAE)

Forbes & Hemsley in Journ. Linn. Soc. Bot. 26: 40. (1889), Pax & Knuth in Engler, Pflanzenr. iv. 237: 22. (1905).

Hubei: Yichang, 262, 1312. Liantuo, 3844. In a ravine on Moji Shan, 3366. Badong, 1435. **Sichuan:** Kangding, Henry's Chinese collector with Pratt 8899.

A short-lived perennial, well known in the West where it is treated as a spring flowered glasshouse annual. *Primula obconica* was described from a collection made by the Irish diplomat and plant collector, Thomas Watters near Yichang in the spring of 1879, though Bretschneider states it had been earlier found at Baoxing in western Sichuan by Père David in 1869. It was introduced to cultivation by Charles Maries from the Xiling (Yichang) Gorge for Messrs Veitch who first flowered it in September 1880. Both Henry and Wilson found it common in moist, grassy places on the banks of the Yangtze and in the glens of the Three Gorges. So common was this species on the steep slopes of the gorges that Henry concidered it a weed, he stated it grew in profusion throughout the grassy precipices of the gorges in places turning the slopes to a shade of violet-pink. It is also native to Guizhou, Guangdong, Sichuan, south-east Tibet (Xizang) and Yunnan, and is exceptionally variable over this range.

Primula odontocalyx (Franch.) Pax (PRIMULACEAE)

Primula hupehensis Craib in Notes Roy. Bot. Gard. Edinburgh 6: 252. (1911—1917).

Hubei: Fang Xian, without number (type of *Primula hupehensis* Craib).

A handsome perennial species allied to *Primula davidii* Franch., carrying on 20 cm tall stems 1 to 10 violet, nodding funnel-shaped flowers with a green and white eye in the centre. A widespread species in Hubei, Sichuan, Gansu and Shaanxi where it grows on damp cliffs and hills at the verge of *Abies delavayi* var. *fabri* forest between 1,400 to 2,800 metres. Flowering occurs during snow melt in April. Introduced to cultivation by Edward Needham from Emei Shan (Mount Omei) in Sichuan in 1990.

Primula ovalifolia Franch. (PRIMULACEAE)

Forbes & Hemsley in Journ. Linn. Soc. Bot. 26: 41. (1889), Pax & Knuth in Engler, Pflanzenr. iv. 236: 43. (1905).

Primula aequipila Craib. Craib in Notes Roy. Bot. Gard. Edinburgh 11: 182.(1918—1919). *Primula macropoda* Craib in Notes Roy. Bot. Gard. Edinburgh 11: 176. (1918—1919); in Notes Roy. Bot. Gard. Edinburgh 11: 183. (1918—1919).

Hubei: Badong at 1,950 metres, 707 (July 1885), 1456, 3731, 5286. Changyang, 5286a. **Sichuan:** North Wushan, 7148. Henry's Chinese collector with Pratt, no locality stated but may be Leshan, Emei Shan, Wa Shan or Kangding, 8860 (type of *Primula macropoda* Craib).

According to James Veitch (who obviously obtained his information from Ernest Wilson), in a wild state this species grows in masses in moist, shady woods, carpeting the ground with flowers as soon as the snow melts. Discovered by David in Baoxing in western Sichuan in March 1869. It is a remarkably beautiful plant with deep violet-purple (approaching blue) flowers and was introduced by means of living plants by Wilson in 1901. Plants first flowered in Veitch's Coombe Wood nursery in March 1905. Native to Yunnan, Sichuan (Baoxing and Emei Shan), and western Hubei where it grows on damp, shaded vertical cliffs with running water at 1,200 to 2,500 metres.

Primula partschiana Pax (PRIMULACEAE) New species

in Pflanzenr. (Engler) 4. Fam. 237: 45. (1905).

Carolinella cordifolia Hemsley in Hooker's Icon. Pl. 28: t. 2775. (1902), non *Primula cordifolia* Rupr. (1863).

Yunnan: South of Red River from Manmei (to the south-west of Mengzi, south of the Red River on the watershed with the Black River) in densely forested virgin country on shaded banks under trees at 2,286 metres, 10890 (type of *Carolinella cordifolia* Hemsley, isotype of *Primula partschiana* Pax).

A deciduous perennial bearing two heart-shaped leathery leaves. Flowering stem to 20 cm, bearing up to 15 rose-coloured flowers. Little is known of this species, it has only ever been collected once by Henry and may now be extinct. Endemic to Yunnan.

Primula pseudodenticulata Pax (PRIMULACEAE) New species

in Engler, Pflanzenr. iv. 236: 91. (1905).

Yunnan: Mengzi at 1,600 metres, 10579a.

Like a small form of *Primula denticulata* but lacking the persistent bud scales and sometimes stoloniferous. Endemic to China and distributed in Sichuan and Yunnan where it grows in damp places between 1,600 to 4,000 metres. The flower colour may be rose or lavander. Based on the collections of Delavay (1883) and those of Henry over a decade later.

Primula pulverulenta Duthie (PRIMULACEAE) New species

in Gard. Chron.ser. 3, 38: 259. (1905), Fedde in Repert. Nov. Spec. Regni Veg. 4: 174. (1907).

Primula japonica Hemsley non A. Gray in Journ. Linn. Soc. Bot. 29: 313. (1892). *Primula serratifolia* Pax & Knuth non Franchet in Engler, Pflanzenr. iv. 236: 126. (1905).

Sichuan: Kangding, Henry's Chinese collector with Pratt, growing abundantly in marshy ground at elevations between 2, 650 and 3,300 metres, 8879.

One of the largest of the candelabra primulas and probably the easiest to grow. A robust deciduous perennial bearing on 1 metre tall, white mealy stems up to 8 whorls of 5 to 10 carmine-red flowers on horizontal stalks during the month of May. Native to western Sichuan, especially abundant about Kangding where it grows in marshes and by streams in semi-shade at altitudes above 2,000 metres. Discovered by Henry's Chinese collector in 1889 and introduced to cultivation by Wilson through Messrs Veitch who first flowered it in their Coombe Wood nursery in May 1905. In their *Novelties for 1908—1909* Veitch had plants available for fifteen pence a dozen.

Primula rugosa N. P. Balakr. (PRIMULACEAE) New species

in J. Bombay Nat. Hist. Soc. 62: 63. (1973).

Carolinella obovata Hemsley in Hooker's Icon. Pl. t. 2726. (1902). *Primula obovata* (Hemsley) Pax in Engler, Pflanzenr. iv. 236: 47. (1905) non Duby (1844), nec Huter (1873).

Yunnan: In mountain forest to the south-west of Mengzi, 10626 (type of *Carolinella obovata* Hemsley).

A deciduous perennial bearing one to six leaves, rounded at the base and carrying on 14 cm long stems up to 12 pink to purple flowers on a short umbellate spike. Endemic to Yunnan.

Primula rupestris Balf. f. & Farrer (PRIMULACEAE)

Primula sinensis Forbes & Hemsley non Sabine ex Lindl. in Journ. Linn. Soc. Bot. 42. (1889), Hook. f. in Bot. Mag. t. 7559. (1889), Henry in Gard. Chron. ser. 3, 31: 269. fig. 84 et 85. (1902), Sutton in Journ. Roy. Hort. Soc. xiii. 99. (1891), Pax in Engl. Pflanzerr. Primul. iv. 237: 21. (1905), in part, E. H. Wilson in Gard. Chron. ser. 3, 40: 191. fig. 78. (1906). *Primula calciphila* Hutchinson in Kew Bull. Misc. Inf. 101. (1923); in Gard. Chron. ser. 3, 73: 283. (1923), Coutts in Gard. Chron. ser. 3, 73: 101. (1923), Stapf in Curtis's Bot. Mag. t. 8986. (1923), Pim in The Wood & the Trees 31. (1984).

Hubei: Yichang, by the cave of the three pilgrims at the mouth of the San You Dong Glen, 879 (syntype of *Primula calciphylla* Hutchinson), 1103, 1292.

Primula rupestris was introduced to cultivation by Antwerp Pratt who gathered seeds at Yichang. These he sent to Lady Hutt of Appleby Towers, Ryde (UK.), Lady Hutt in turn passed on seedlings to the taxonomist, Dr. Masters, who again passed on seedlings to a friend, Mr. Edmund Hyde of Ealing who first flowered the plant in 1892. It was from Lady Hutt's plant that Hooker figured the plant in *Curtis's Bot. Mag.* It is a limestone loving species hense Hutchinson's *Primula calciphila* (lover of lime) and Henry recorded seeing great ledges of is sweetly scented mauve-pink flowers in January and February on limestone shelves on the cliffs Three Gorges. There it grew where there was practically no soil or moisture, exposed to the sun and living in the decaying remains of former generations of the plants. Henry went on to state that these ledges were continuous for hundreds of feet and in December and January plants presented a scene of great beauty when in flower. The scent of the leaves was very strong and could be perceived at once on entering any of the ravines where the ledges occurred. The flowers, he said, were pinkish, with a yellow ring around the neck of the corolla. The Yichang population was initially taken to be *Primula sinensis* from which it differs in its smaller, neater appearance, it is also less fleshy and has persistent wiry leaf stalks, and the corolla tube is twice the size of the calyx. Native to Hubei and Sichuan, it was reintroduced to cultivation by Reginald Farrer from the borders of Shaanxi and Sichuan, plants from this collection flowered at the Royal Botanic Gardens, Edinburgh in 1916.

Farrer compared this species choice of habitat to *Primula allionii*, which also inhabits dry, hard cliff faces. In recent years was raised from seeds collected by Edward Needham on Emei Shan (Mount Omei).

Primula septemloba Franch. (PRIMULACEAE)

Anonymous in Kew List of Determinations (volume 24).

Sichuan: Henry's Chinese collector with Pratt, no locality stated but may be Leshan, Emei Shan, Wa Shan or Kangding, 8870.

A handsome perennial, allied to *Primula geraniifolia* Hook. f. Native to north-west Yunnan and south-west Sichuan where it grows in moist, mixed forest between 3,000 to 4,000 metres. In Yunnan it is said to be locally common at Lijiang and at Dali (on the nearby Cangshan Range) it was introduced to cultivation by George Forrest from Lijiang in 1906.

Primula spp. (PRIMULACEAE)

Anonymous in Kew List of Determinations (volume 24).

Hubei: South Badong, 5441. Xingshan at 2,300 to 2,900 metres, 6981, 6995. **Sichuan:** South Wushan, 7145. Henry's Chinese collector with Pratt, no locality stated but may be Leshan, Emei Shan, Wa Shan or Kangding, 8880.

Primula wilsonii S. T. Dunn (PRIMULACEAE) New species

in Gard. Chron. ser. 3, 31: 413. (1902), O' Brien in Augustine Henry, An Irish Plant Collector in China 29. (2002).

Primula angustidens Pax & Knuth non (Franch.) Pax in Engler, Pflanzenr. iv. 236: 128. (1905).

Yunnan: Simao, 12121 (lectotype), 12121a.

Discovered by Henry in 1898 and introduced to cultivation by Wilson in 1899 from Yunnan (while visiting Henry) through Messrs. Veitch. *Primula wilsonii* first flowered at Kew in the rock garden in 1902 and was named for Wilson on the day his daughter was born. In celebration of these two events Wilson named his daughter Muriel Primrose. *Primula wilsonii* was also later collected by Forrest (Forrest 16,791) in western Yunnan in August 1918. It is grown in the bog beds near the main pond at Glasnevin and is widely available in the trade. Endemic to China and distributed in Sichuan and Yunnan.

Primulina eburnea (Hance) Yin Z. Wang (GESNERACEAE)

Chirita eburnea Hance Forbes & Hemsley in Journ. Linn. Soc. Bot. 26: 231. (1890).

Hubei: Liantuo, 3794, 6364. **Sichuan:** South Wushan, 5531.

Introduced to cultivation by Wilson from Hubei through Veitch's Coombe Wood nursery where it first flowered in 1903.

Prinsepia utilis Royle (ROSACEAE)

Rehder in Sargent Pl. Wilson. 2: 345. (1915).

Yunnan: In mountains near Mengzi, 9281, 11343.

A very vigorous spiny shrub to 4 metres tall carrying creamy-white fragrant flowers on gracefully arching branches in January and February. A native to India, the Himalaya and western China. In south-east Tibet (Xizang) I have seen this plant (and in other areas *Rosa sericea* f. *pterocantha*) used for winter fencing. In the Tibetan highlands, stakes (of *Populus* spp.) are drove into the ground at regular intervals, between these the *Prinsepia* branches are weaved making a very effective barrier. It grows well at Glasnevin where there are old plants and should be more frequently planted in gardens on account of its handsome spring blossoms.

Prismatomeris tetrandra (Roxb.) K. Schumann (RUBIACEAE)

Hutchinson in Sargent Pl. Wilson. 3: 413. (1916)

Yunnan: Yiwu, near the Loas border in Xishuangbanna prefecture, 13573.

Procris crenata C. B. Rob. (ULMACEAE)

Procris laevigata Henry non Blume in Trans. Asiat. Soc. Jap. 24 suppl: 89. (1896) (List Pl. Formosa), Forbes & Hemsley in Journ. Linn. Soc. Bot. 26: 484. (1899).

Taiwan: Oluanpi, 1245. Wanchin, 1590. **Yunnan:** Mengzi, 11314.

A perennial herb or undershrub, epiphytic on tree trunks or rocks, native to Africa and tropical Asia.

Prunella vulgaris L. (LAMIACEAE)

Henry in Journ. China Br. Roy. Asiat. Soc. 22: 248. (1887) (Chinese Names of Plants), Forbes & Hemsley in Journ. Linn. Soc. Bot. 26: 299. (1890), Dunn in Notes Roy. Bot. Gard. 6: 177. (1916).

Hubei: Yichang, 198, 1592. Badong, 1779. Liantuo, 1923. **Yunnan:** Mengzi, 9804.

Selfheal, a common plant at low altitudes in many of China's provinces. Henry stated that in the Three Gorges region it was known as *hsia ku ts'ao* (*xia ku cao*), in China it is used with *Chrysanthemum × moriifolium* to treat headache, dizziness and vertigo.

Prunus arborea (Blume) Kalkm. var. *montana* (Hook. f.) Kalkman (ROSACEAE)

Kalkman in Blumea 13: 99 & 109. (1965).

Pygeum henryi Dunn in Journ. Linn. Soc. Bot. 35: 493. (1903), Koehne in Bot. Jahrb. li. 185. (1913), Rehder in Sargent Pl. Wilson. 2: 344. (1915).

Yunnan: In forests near Simao at 1,300 to 1,500 metres, 12313 (isosyntype of *Pygeum henryi* Dunn), 12313a., (holotype), 12313b., (isosyntype of *Pygeum henryi*

Dunn). Simao at 1,300 to 1,500 metres, 12708 (isosyntype of *Pygeum henryi* Dunn), 13447.

Himalaya to Indochina.

Prunus brachypoda Batalin (ROSACEAE) New species

in Act. Hort. Petrop. 12: 166. (1892), Batalin in Gartenflora xlii. 330. (1893).

Prunus ssiori Forbes & Hemsley non F. Schmidt. in Journ. Linn. Soc. Bot. 23: 221. (1887), Henry in Notes Econ. Bot. China 49. (1893). *Padus brachypoda* (Batal.) Schneider in Fedde, Repert. Nov. Spec. Regni. Veg. 1: 69. (1905); in Illus. Handb. Laubholzk. 1: 638. (1905). *Prunus brachypoda* Batalin var. *pseudossiori* Koehne in Sargent Pl. Wilson. 1: 65. (1911), Zheng in Journ. Wuhan Bot. Res. 2.(1): 74. (1984).

Hubei: Yichang, 1418. Jianshi, 5884, 5988 (syntype of *Prunus brachypoda* Batalin var. *pseudossiori* Koehne). Badong, 4723, 5324, 5395. Xingshan, 6484. **Sichuan:** South Wushan, 5591, 5739 (syntype of *Prunus brachypoda* Batalin var. *pseudossiori* Koehne), 5763 (syntype of *Prunus brachypoda* Batalin var. *pseudossiori* Koehne). Henry's Chinese collector with Pratt, no locality stated but may be Leshan, Emei Shan, Wa Shan or Kangding, 8956.

According to Henry this species was known colloquially in the Three Gorges region as *k'u t'ao*. This species was discovered about the same time (1885) by both Henry and Potanin.

Prunus buergeriana Miq. (ROSACEAE)

Anonymous in Kew List of Determinations (volume 24).

Sichuan: Henry's Chinese collector with Pratt, no locality stated but may be Leshan, Emei Shan, Wa Shan or Kangding, 8958.

Prunus cerasoides Buch.-Ham. ex D. Don (ROSACEAE)

Schneider in Fedde Repert. Nov. Spec. Regni Veg. 1: 54. (1905).

Prunus cerasoides D. Don var. *tibetica* (Batalin) Schneider in Fedde, Repert. Spec. Nov. Regni. Veg. 1: 54. (1905), except A. Henry 9411, 9411a., 9411b., 11469. *Prunus majestica* Koehne in Sargent Pl. Wilson. 1: 252. (1912).

Yunnan: In woods near Mengzi at 1,650 metres, 9411 (syntype of *Prunus majestica* Koehne). In woods near Mengzi at 1,500 to 1,600 metres, 9411a., (syntype of *Prunus majestica* Koehne), 11469 (syntype of *Prunus majestica* Koehne). Chu-yuan 9411b., (syntype of *Prunus majestica* Koehne).

Native to India, the Himalaya, Myanmar, China [Tibet (Xizang) and Yunnan], Thailand and Vietnam.

Prunus clarofolia C. K. Schneider (ROSACEAE) New species

Prunus litigiosa Schneider in Repert. Spec. Nov. Regni Veg. 1: 65. (1905), Schneider in Illus. Handb. Laubholzk. 1: 609. (1905), Koehne in Sargent Pl. Wilson. 1: 239. (1912), Bean in Trees & Shrubs 3: 311. (1936), Zheng in Journ. Wuhan. Bot. Res. 2.(1): 76. (1984), Nelson in A Heritage of Beauty 317. (2000), O'Brien in Augustine Henry, An Irish Plant Collector in China 30. (2002).

Prunus maximowiczii Henry non Rupr. in Notes Econ. Bot. China 49. (1893).

Hubei: Badong, 5295 (isotype of *Prunus litigiosa* Schneider), 6091. Changleping, 6275. **Sichuan:** South Wushan, 5721.

A very beautiful deciduous cherry to 7 metres tall carrying a profuse crop of pinkish-white flowers in April. Introduced to cultivation by Wilson in 1907 through the Arnold Arboretum. In Ireland there are trees at Dunloe Castle, the John F. Kennedy Arboretum and Mount Usher.

Prunus conradinae Koehne (ROSACEAE) New species

Zheng in Journ. Wuhan Bot. Res. 2.(1): 75. (1984).

Hubei: Yichang, 3378.

A beautiful and very graceful, early flowered cherry to 12 metres tall. Native to western Hubei this cherry was discovered by Henry in 1887 (and was based on collections later made by Wilson) and introduced to cultivation by Wilson who sent seeds to the Arnold Arboretum in 1907. The tree is at its best in February and early March when covered in pale pink flowers. The specific epithet commemorates Koehne's wife.

Prunus domestica L. (ROSACEAE)

Prunus communis Huds. Forbes & Hemsley in Journ. Linn. Soc. Bot. 23: 218. (1887).

Hubei: Liantuo, 3799.

Prunus glandulosa Thunb. (ROSACEAE)

Prunus japonica Forbes & Hemsley non Thunb. in Journ. Linn. Soc. Bot. 23: 219. (1887). *Prunus glandulosa* Thunb. var. *trichostyla* Koehne f. *paokangensis* (Schneid.) Koehne. Zheng in Journ. Wuhan Bot. Res. 2.(1): 76. (1984).

Hubei: South Badong, 5269. Liantuo, 6355. Cultivated in a garden at Yichang, 3598 [type of *Prunus glandulosa* Thunb. var. *trichostyla* Koehne f. *paokangensis* (Schneid.) Koehne], 7859.

Prunus grayana Maxim. (ROSACEAE)

Koehne in Sargent Pl. Wilson. 1: 69. (1911).

Padus acrophylla Schneider in Repert. Spec. Nov. Regni. Veg. 1: 70. (1905).

Hubei: Changleping, 6327. Badong, 4077 (isotype of *Padus acrophylla* Schneider).

Prunus henryi (Schneider) Koehne (ROSACEAE) New species

in Sarg. Pl. Wilson. 1: 240. (1912), Schneider in Ill. Handb. Laubholzk. 2: 987. (1912).

Prunus yunnanensis Franch. var. *henryi* Schneider in Repert. Spec. Nov. Regni. Veg. 1: 66. (1905), (in part.), Schneider in Ill. Handb. Laubholzk. 1: 609. (1905). *Prunus neglecta* Koehne in Sargent Pl. Wilson. 1: 241. (1913), Schneider in Ill. Handb. Laubholzk. 2: 987. (1912).

Yunnan: In woods near Mengzi at 1,800 metres, 10629 (type). In woods near Mengzi at 2,000 metres, 10629b., (in part, isotype of *Prunus yunnanensis* Franch. var. *henryi* Schneider, type of *Prunus neglecta* Koehne).

George Forrest later collected this species (Forrest 16,232) in the Mekong valley in Yunnan in February 1918 and described it as a shrub 6 to 9 metres tall bearing white flowers. Endemic to Yunnan.

Prunus himalayana J. Wen (ROSACEAE)

Maddenia himalaica Hook. f. & Thoms. Diels in Bot. Jahrb. 29: 406. (1900).

Sichuan: Henry's Chinese collector with Pratt, no locality stated but may be any of the following, Leshan, Emei Shan, Wa Shan or Kangding, 8952.

Prunus hypoxantha (Koehne) J. Wen (ROSACEAE) New species

Maddenia wilsonii Koehne Zheng in Journ. Wuhan Bot. Res. 2.(1): 71. (1984).

Hubei: Yichang, 3759.

Discovered by Henry in 1887 and based on collections made by Wilson in Hubei and Sichuan in 1908.

Prunus itosakura Siebold (ROSACEAE)

Prunus x *subhirtella* Miq. var. *ascendens* (Makino) E. H. Wilson in The Cherries of Japan 10. (1916), Zheng in Journ. Wuhan Bot. Res. 2.(1): 79. (1984).

Hubei: Changyang, 7804.

Wilson stated this variety was common wild at both Badong and Changyang in woods and thickets up to 1,000 metres in altitude. A popular tree around temples in Japan where large specimens occur.

Prunus japonica Thunb. (ROSACEAE)

Prunus consociiflora C. K. Schneider in Fedde, Repert. Spec. Nov. Regni Veg. 1: 54. (1905), Schneider in Illus. Handb. Laubholzk. 1: 612. (1905), Koehne in Sargent Pl. Wilson. 1: 279. (1912).

Forbes & Hemsley in Journ. Linn. Soc. Bot. 23: 219. (1887), Henry in Journ. China Br. Roy. Asiat. Soc. 22: 251. (1887) (Chinese Names of Plants); in Notes Econ. Bot. China 7 & 49. (1893).

Prunus japonica Thunb. var. *eujaponica* Koehne. f. *oldhamii* Koehne in Sargent Pl. Wilson. 1: 266. (1912),

Zheng in Wuhan Bot. Res.2. (1): 76. (1984).

Hubei: Yichang, 1039, 3958a. South Badong, 6313. Changyang, 7803.

A small deciduous cherry, allied to *Prunus glandulosa* Thubb., a native of China but described from a cultivated Japanese plant. Known colloquially in the Three Gorges region as *k'u li tzu* according to Henry.

Prunus leveilleana Koehne (ROSACEAE)

Prunus tenuiflora Koehne in Sargent Pl. Wilson. 1: 209. (1912).

Hubei: Jianshi, 5833.

Prunus persica (L.) Batsch. (ROSACEAE)

Forbes & Hemsley in Journ. Linn. Soc. Bot. 23: 220. (1887), Henry in Notes Econ. Bot. China 49. (1893); in Trans. Asiat. Soc. Jap. 24 suppl: 40. (1896) (List Pl. Formosa), Zheng in Journ. Wuhan Bot. Res. 2.(1): 77. (1984).

Hubei: Jianshi, 5815. Antelope Glen near Yichang, 5815a. **Taiwan:** Wanchin, 113, 408. **Yunnan:** Mengzi, 10789.

The origin of the peach has been a matter of great debate for centuries. It has been cultivated for so long and over such a large area that its place of origin has been confused though it is generally believed to have came from China and such was also the opinion of Augustine Henry. The emblem of long life, the peach figures predominently in the ancient literature of China and it was valued for both flower and fruit. Henry stated it grew in little orchards around peasant's cottages at Yichang (which lies at 30 metres above sea level) and into the mountainous gorges region to 1,300 metres, where the trees enjoyed a very hot spring and summer. He also goes on to say that he found it in the mountains near Yichang and had no doubts that these were truly wild trees. The wild plants produced almost inedible sour small fruits which Henry stated were known as *mao-t'ao-tzu*. The Hani (an ethnic group) of Xishuangbanna in southern Yunnan call the peach *siqymq* and use its roots to treat cases of gastroenteritis and gastritis.

Prunus phaeosticta (Hance) Maxim. (ROSACEAE)

Li in Woody Flora of Taiwan 285. (1963), Kalkman in Blumea 13: 40. (1965).

Prunus phaeosticta (Hance) Maxim. f. *lasioclada* Rehder in Journ. Arnold Arbor. 11: 163. (1930). *Prunus xerocarpa* Hemsley in Ann. of Bot. 9: 152. (1895), Henry in Trans. Asiat. Soc. Jap. 24 suppl: 40. (1896) (List Pl. Formosa), Bretschneider in Hist. Eur. Bot. Disc. China 2: 782. (1898), Forbes & Hemsley in Journ. Linn. Soc. Bot. 152. (1902).

Taiwan: Wanchin, 47, 1656 (type of *Prunus xerocarpa* Hemsley), 1658 (isosyntype of *Prunus xerocarpa* Hemsley).

471

Yunnan: Mengzi, 11333. In mountain forest to the east of Simao at 1,650 metres, 11666 [holotype of *Prunus phaeosticta* (Hance) Maxim. f. *lasioclada* Rehder].

An evergreen tree to 17 metres with a range across India (Assam) to southern China, including Taiwan where it is scattered at low altitude throughout the island.

Prunus pygeoides Koehne (ROSACEAE)

Kalkman in Blumea 13: 31. (1965).

Prunus semiarmillata Koehne in Bot. Jahrb. Syst. 52: 303. (1915).

Yunnan: Simao, 12887 (type of *Prunus semiarmillata* Koehne).

Distributed in India and China (Yunnan).

Prunus rufoides C. K. Schneider (ROSACEAE) New species

in Repert. Spec. Nov. Regni Veg. 1: 55. (1905), Koehne in Sargent Pl. Wilson. 1: 244. (1912).

Prunus dielsiana C. K. Schneider in Repert. Spec. Nov. Regni Veg. 1: 68. (1905), Koehne in Sargent Pl. Wilson. 1: 243. (1912), Zheng in Journ. Wuhan Bot. Res. 2.(1): 75. (1984), O' Brien in Augustine Henry, An Irish Plant Collector in China 30. (2002). *Prunus helenae* Koehne in Sargent Pl. Wilson. 1: 212. (1912), exclude A. Henry 5780. *Prunus rufoides* Schneider var. *glabrifolia* Schneider in Fedde. Rep. Spec. Regni. Veg. 1: 56. (1905).

Hubei: Jianshi, 5812 (isotype *Prunus dielsiana* C. K. Schneider). Sichuan: South Wushan, 5477 (type of *Prunus rufoides* Schneider, var. *glabrifolia* Schneider), 5780 (type of *Prunus rufoides* C. K. Schneider).

A charming deciduous cherry to 10 metres tall carrying masses of white or pink flowers before the leaves emerge in spring. Native to western Hubei, discovered by Henry in 1888 and introduced to cultivation by Wilson in 1907 through the Arnold Arboretum. In Ireland this tree is grown at the John F. Kennedy Arboretum, Co Wexford and at Dunloe Castle, Co Kerry. The Dunloe trees are the tallest in Britain and Ireland, one of these was 12 metres tall when measured in 1993.

Prunus aff. *serrula* Franchet (ROSACEAE)

Anonymous in Kew List of Determinations (volume 34).

Yunnan: Hoikow, 9589.

Prunus serrulata Lindl. (ROSACEAE)

Prunus wildeniana Koehne in Sargent Pl. Wilson. 1: 249. (1912), Zheng in Journ. Wuhan Bot. Res. 2.(1): 80. (1984).

Hubei: South Badong, 5308 (type *Prunus wildeniana* Koehne).

Prunus serrulata Lindl. var. *spontanea* (Maxim.) Makino (ROSACEAE)

E. H. Wilson in The Cherries of Japan 28. (1916),

Zheng in Journ. Wuhan Bot. Res.2. (1): 79. (1984).

Prunus cerasoides Schneider non Buch.-Ha. ex D. Don in Fedde, Repert Nov. Spec. Regni Veg. 1: 54. (1905).

Hubei: South Badong, 5300. Jianshi, 5833.

The hill cherry, a tree to 15 metres tall and native to China, Korea and Japan. In Japan it is the national tree and is much planted around shrines and temples. It was first found wild in China by Henry and introduced from Hubei by E. H. Wilson in 1900. It is a beautiful sight in spring when covered in white or pink flowers and copper-coloured emerging foliage.

Prunus sp. No. 1. (ROSACEAE)

Anonymous in Kew List of Determinations (volume 24).

Hubei: Badong, 3726. Jianshi 5901. Changyang, 5268. Changleping, 6324. Sichuan: South Wushan, 5546, 5703.

Prunus sp. No. 2. (ROSACEAE)

Anonymous in Kew List of Determinations (volume 34).

Yunnan: Mengzi, 10113, 10479, 11511. Simao, 11742.

Prunus spinulosa Sieb. & Zucc. (ROSACEAE)

Prunus spinulosa Sieb. & Zucc. var. *pubiflora* Koehne in Bot. Jahrb. Syst. 52: 300. (1915). *Prunus limbata* Cardot in Notul. Syst. (Paris) 4: 21. (1920).

Yunnan: In the mountains near Yuanjiang at 1,650 metres, 13228 (isotype of *Prunus limbata* Cardot, type of *Prunus spinulosa* Sieb. & Zucc. var. *pubiflora* Koehne).

Native to China and Japan.

Prunus tatsiensis Batalin (ROSACEAE)

Prunus tatsienensis Batal. var. *pilosiuscula* Schneider in Fedde. Rep. Sp. Nov. Regni. Veg. I. 66. (1905), *Prunus pilosiuscula* (Schneid.) Koehne. var. *media* Koehne in Sargent Pl. Wilson. 1: 204. (1912). *Prunus pilosiuscula* (C. K. Schneid.) Koehne in Sargent Pl. Wilson. 1: 204. (1912), Zheng in Journ. Wuhan Bot. Res. 2.(1): 77. (1984), Bean in Trees & Shrubs 3: 391. (1992).

Hubei: South Badong, 5325. Sichuan: South Wushan, 5604 (isotype of *Prunus tatsienensis* Batal. var. *pilosiuscula* Schneider).

Prunus undulata Buch.-Ham. ex D. Don (ROSACEAE)

Prunus acuminata (Wall.) Dietr. f. *elongata* Koehne in Bot. Jahrb. 52: 296. (1915). *Prunus wallichii* Steud. Kalkman in Blumea 13: 38. (1965).

Sichuan: South Wushan, 7188. Yunnan: Fengchunling, 11173 [syntype of *Prunus acuminata* (Wall.) Dietr. f. *elongata* Koehne]. Simao, 13187 (syntype of *Prunus acuminata* (Wall.) Dietr. f. *elongata* Koehne), 13187a., [syntype of *Prunus acuminata* (Wall.) Dietr. f. *elongata* Koehne].

Distributed in the Himalaya, India, Bangladesh,

Myanmar, China, Thailand and Vietnam.

Prunus veitchii Koehne (ROSACEAE)

Prunus concinna Koehne Zheng in Journ. Wuhan Bot. Res. 2.(1): 75. (1984).

Hubei: South Badong, 6308.

A shrubby cherry to 2.5 metres, tall bearing white or pale pink flowers in profusion during the months of March and April, carried in clusters of four before the leaves emerge. Discovered by Henry and later introduced to cultivation by Wilson while collecting for the Arnold Arboretum in 1907.

Prunus velutina Batalin (ROSACEAE) New species

in Act. hort. Petrop. 14: 168. (1895), Bretschneider in Hist. Eur. Bot. Disc. China 2: 782. (1898), Koehne in Sargent Pl. Wilson. 1: 69: (1913), Morley in Glasra 3: 80. (1979), Zheng in Journ. Wuhan Bot. Res. 2.(1): 79. (1984).

Padus velutina (Batalin) Schneider in Illus. Handb. Laubholzk. 1: 638. (1905).

Sichuan: South Wushan, 5592, 5774. **Hubei:** Jianshi, 5985.

Prunus wilsonii (Schneid.) Koehne (ROSACEAE) New species

in Sargent Pl. Wilson. 1: 63. (1911).

Padus nepaulensis (Ser.) Steud. forma Schneider in Fedde Rep. Sp. Regni. Veg. 1: 68. (1905); in Ill. Handb. Laubholzk. 1: 639. (1905). *Prunus nepaulensis* (Ser.) Steud. var. *sericea* Batalin in Act. Hort. Petrop. xiv. 169. (1895), Bretschneider in Hist. Eur. Bot. Disc. China 2: 782. (1898).

Sichuan: South Wushan, 5592a. 7050. **Yunnan:** Mengzi, 10547.

Wilson later collected this tree at Baokang (north of Yichang, near Fang Xian) in northern Hubei. Endemic to China.

Prunus zippeliana Miq. (ROSACEAE)

Kalkman in Blumea 13: 44 & 109. (1965).

Prunus macrophylla Sieb. & Zucc. Anonymous in Kew List of Determinations (volume 34).

Yunnan: Mengzi, 11070. Simao, 12943.

An evergreen tree, native to China (including Taiwan), Vietnam and Japan (south).

Pseudelephantopus spicatus (Juss. ex Aubl.) C. F. Baker (ASTERACEAE)

Elephantopus spicatus Juss. ex Aubl. Henry in Trans. Asiat. Soc. Jap. 24 suppl: 51. (1896) (List Pl. Formosa).

Hubei: On the banks of the Yangtze at Yichang, 4363. **Taiwan:** Kaohsiung, without number. Oluanpi, without locality.

Pseuderanthemum coudercii Benoist (ACANTHACEAE) New species

in Humbert, Not. Syst. 5: 111. (1935), R. Ben in Humbert, Fl. Gen. Indo-Chine 4: 720. (1935).

Eranthemum palatiferum (Nees) Nees Forbes & Hemsley in Journ. Linn. Soc. Bot. 26: 243. (1890).

Hainan: Ling-men, 8611. Without locality, 8749.

Pseuderanthemum haikangense C. Y. Wu & H. S. Lo (ACANTHACEAE) New species

in Flora Hainanica 3: 595. (1974).

Hainan: On hummocks near Haikou, 8111.

Pseuderanthemum polyanthum (C. B. Clarke ex Oliv.) Merr. (ACANTHACEAE)

Eranthemum polyanthum C. B. Clarke ex Oliv. Anonymous in Kew List of Determinations (volume 34).

Yunnan: Simao, 11655, 11933, 12933.

Pseuderanthemum shweliense (W. W. Smith) C. Y. Wu & C. C. Hu (ACANTHACEAE) New species

Eranthemum shweliense W. W. Smith in Notes Roy. Bot. Gard. Edinburgh 9: 176. (1916).

Yunnan: In the mountains to the south-west of Mengzi at 1,300 metres, 11010 (isosyntype of *Eranthemum shweliense* W. W. Smith).

Endemic to Yunnan. This species was also later collected in Yunnan by Forrest.

Pseudocaryopteris foetida (D. Don) P. D. Cantino (LAMIACEAE)

Caryopteris paniculata C. B. Clarke Anonymous in Kew List of Determinations (volume 34).

Yunnan: Mengzi, 10408, 10408a., 10408b., 10408c., 10408d.

Pseudocayratia oligocarpa (H. Levl. & Vantiot) J. Wen & L. M. Lu (VITACEAE)

Cayratia oligocarpa (H. Levl. & Vantiot) Gagn. Zheng in Journ. Wuhan Bot. Res. 2.(1): 143. (1984).

Hubei: Jianshi, 5841, 5952. Without locality 5952a.

Pseudocodon convolvulaceus (Kurz) D. Y. Hong & H. Sun (CAMPANULACEAE)

Codonopsis convolvulacea Kurz Anthony in Notes Roy. Bot. Gard. Edinburgh 15: 178. (1925—1927).

Yunnan: Mengzi, 9425.

An herbaceous climber to about 80 cm tall, a widespread species in western China, eastern Myanmar and northern Indochina, it was discovered by Dr. John Anderson in May 1868 and introduced by him to cultivation that same year (for a brief note on Dr. Anderson's collections in Yunnan see under *Dichotomanthes*). It was later collected by Forrest, Kingdon Ward and Henry from the region between Myanmar and Mengzi.

Pseudocydonia sinensis (Dum.-Cours.) Schneid. (ROSACEAE)

Pyrus sinensis Sprengel. Anonymous in Kew List of Determinations (volume 24).

Hubei: Liantuo, 4497. Badong, 5020.

A small, semi-evergreen tree to about 7 metres tall with brown-grey flaking bark. Solitary pink flowers are borne in spring, followed by large egg-shaped yellow fruits. This tree is occasionally planted in the grounds of Taoist temples and there is a very fine tree in one of the monastic courtyards on Wudang Shan in northern Hubei. Cultivated at the National Botanic Gardens, Glasnevin from collections made in Xingshan, Hubei by the 2004 Glasnevin Central China Expedition.

Pseudodissochaeta septentrionalis (W. W. Sm.) Nayar (MELASTOMATACEAE)

Anplectrum yunnanense Kraenzil in Vieteljahrsschr. Naturf. Ges. Zurich 76: 153. (1931).

Yunnan: In mountain forest to the east of Simao at 1,650 metres, 11705 (isotype of *Anplectrum yunnanense* Kraenzil). Simao, 11705a, 11705c., 11705d.

Native to Myanmar, western China, Thailand and Vietnam.

Pseudogalium paradoxum (Maxim.) L. E. Yang, Z. L. Nie, & H. Sun (RUBIACEAE)

Galium paradoxum Maxim. Anonymous in Kew List of Determinations (volume 24).

Hubei: Jianshi, 5851. South Badong, 6026.

Pseudognaphalium adnatum (DC.) Y. S. Chen (ASTERACEAE)

Anaphalis adnata DC. Anonymous in Kew List of Determinations (volume 34).

Yunnan: Mengzi, 9961.

Distributed from Pakistan to Myanmar, China, Thailand, Laos, Vietnam, Cambodia and the Philippines.

Pseudognaphalium affine (D. Don) Anderb. (ASTERACEAE)

Gnaphalium multiceps Wall. ex DC. Henry in Journ. China Br. Roy. Asiat. Soc. 22: 240. (1887) (Chinese Names of Plants), Forbes & Hemsley in Journ. Linn. Soc. Bot. 23: 427. (1888), Henry in Trans. Asiat. Soc. Jap. 24 suppl: 53. (1896) (List Pl. Formosa).

Hubei: Yichang, 481, 690, 776, 1579, 1961. Badong, 945 (November 1885), 1799, 3672, 4804, 5034, 5045. **Taiwan:** Wanchin, 482. Oluanpi, 637.

According to Henry, in Sichuan this species was known colloquially as *ch'ing ming ts'ao*. He also stated that this plant was sometimes pounded and made into pastry with flour In southern Yunnan the Hani (an ethnic group) call it *diqmevq* and is used to treat cases of influenza, cough, bronchitis and hypertension. External applications are used to treat cases of venomous snake bites.

Pseudognaphalium hypoleucum (DC.) Hilliard & B. L. Burtt (ASTERACEAE)

Gnaphalium hypoleucum DC. Forbes & Hemsley in Journ. Linn. Soc. Bot. 23: 426. (1888).

Hubei: Yichang, 1053, 2139. Badong, 2552, 2562. Jianshi, 7463. **Sichuan:** South Wushan, 7231 (in part). **Hunan:** Shimen, 7939.

Pseudognaphalium luteoalbum (L.) Hilliard & B. L. Burtt (ASTERACEAE)

Gnaphalium luteoalbum L. Henry in Trans. Asiat. Soc. Jap. 24 suppl: 53. (1896) (List Pl. Formosa).

Hubei: Yichang, 1170, 1218. **Taiwan:** Kaohsiung, without number.

Pseudopogonatherum quadrinerve (Hack.) Ohwi (POACEAE)

Pollinia quadrinervis Hack. Rendle in Journ. Linn. Soc. Bot. 36: 356. (1904).

Yunnan: On grassy mountains near Mengzi at 2,200 metres, 10052.

Pseudopogonatherum speciosum (Debeaux) Ohwi (POACEAE)

Pollinia speciosa (Debeaux) Hack. Rendle in Journ. Linn. Soc. Bot. 36: 357. (1904).

Hubei: Yichang, 2796, 4374.

A widespread species, native to India and extending into Thailand, China, Korea and Japan.

Pseudostachyum polymorphum Munro (POACEAE)

Rendle in Journ. Linn. Soc. Bot. 36: 448. (1904).

Hubei: Liantuo, 3857.

Pseudostellaria davidi (Franch.) Pax (CARYOPHYLLACEAE)

Krascheninnikovia davidii Franch. Williams in Journ. Linn. Soc. Bot. 34: 435. (1900), Diels in Bot. Jahrb. 29: 320. (1900).

Hubei: Antelope Glen, 3462. **Sichuan:** South Wushan, 5634.

Discovered by Père Armand David to the north of Beijing in May 1864.

Psidium guajava L. (MYRTACEAE)

Henry in Trans. Asiat. Soc. Jap. 24 suppl: 43. (1896) (List Pl. Formosa).

Hainan: Without locality, 7970. **Taiwan:** Kaohsiung, without number. **Yunnan:** Mengzi, 10631.

The common guava, a shrub or small tree to 4 metres tall, native to South American, the guava is widely cultivated and naturalised in the Old World tropics. The yellow pear-shaped fruits contain more vitamin C than ripe oranges.

Psilopeganum sinense Hemsley (RUTACEAE) New genus

in Journ. Linn. Soc. Bot. 23: 103. (1886).

Hubei: Xiling (Yichang) Gorge, 271, 318, 406, 737, 811, 1114 (type), 1274, 1303, 2413, 3427.

This new genus was first collected in the Xiling Gorge near Yichang by Charles Maries in 1879. Hemsley used material collected by both Maries and Henry on which to describe the species. Seeds were taken from Henry's herbarium specimens at Kew on October 29th 1887.

Psilotum nudum (L.) P. Beauv. (PSILOTACEAE)

Psilotum triquetrum Sw. Henry in Trans. Asiat. Soc. Jap. 24 suppl: 117. (1896) (List Pl. Formosa).

Taiwan: Oluanpi, 1282.

A terrestrial or often epiphytic fern, widely distributed in the tropics and subtropics.

Psychotria asiatica L. (RUBIACEAE)

Psychotria elliptica Henry non Ker. Gawl. in Trans. Asiat. Soc. Jap. 24 suppl: 50. (1896) (List Pl. Formosa). *Psychotria rubra* (Lour.) Poir. Li in Woody Flora of Taiwan 867. fig. 350. (1963).

Taiwan: Wanchin, 84, 476, 1512, 1676, 1680, 1907. Oluanpi, 385, 1003.

A small shrub, native from Indochina, southern China to Japan. This species was collected by the Glasnevin Taiwan Expedition (GTE 85) in the nature reserve of the Hong Chauw Tropical Botanic Gardens in Kenting National Park in October 2004.

Psychotria calocarpa Kurz. (RUBIACEAE)

Hutchinson in Sargent Pl. Wilson. 3: 415. (1916).

Yunnan: In forests near Simao at 1,640 metres, 12283b.

Psychotria elliptica Willd. ex Roem & Schult. (RUBIACEAE)

Anonymous in Kew List of Determinations (volume 24 & 34).

Hainan: Near the Kiungchow Pagoda, 7987. Without locality, 8517. **Yunnan:** Mengzi, 9260.

Psychotria henryi H. Léveille (RUBIACEAE) New species

in Repert. Spec. Nov. Regni Veg. 13: 179. (1914), Hutchinson in Sargent Pl. Wilson. 3: 415. (1917).

Yunnan: In forests near Simao at 1,320 metres, 12146 (type), 12146c. In forests to the south-east of Simao at 1,300 metres, 12146d., (isotype).

Psychotria morindoides Hutchinson (RUBIACEAE) New species

in Sargent Pl. Wilson. 3: 414. (1916).

Yunnan: Mountains to the east of Simao at 1,300 metres, 12069, 12069a., (type). In forests near Simao at 1,300 metres, 12069b., (isosyntype), 12069d., (isosyntype).

A slender shrub to 1 metre tall; flower white.

Psychotria pilifera Hutchinson (RUBIACEAE) New species

in Sargent Pl. Wilson. 3: 415. (1916).

Yunnan: In Pingbian in forests on the Daweishan Range at 1,640 metres, 11409, 13307. Simao, 13308.

Psychotria prainii (Craib) H. Lév. (RUBIACEAE)

Psychotria siamica (Craib) Hutchinson in Sargent Pl. Wilson. 3: 415. (1916).

Yunnan: In a ravine near Yuanjiang at 1,470 metres, 11589, 11589a. In forests to the south of Simao at 1,640 metres, 12368a.

Psychotria serpens L. (RUBIACEAE)

Henry in Trans. Asiat. Soc. Jap. 24 suppl: 50. (1896) (List Pl. Formosa).

Taiwan: Wanchin, 513. Oluanpi, 653.

A small shrub, native to southern China and Japan to Malaysia. Distributed in forests at low altitude in Chinese Taiwan where it is epiphytic on trees or creeping and climbing on rocks.

Psychotria spp. (RUBIACEAE)

Anonymous in Kew List of Determinations (volume 34).

Yunnan: Mengzi, 11384. Simao, 12726. Yuanjiang, 13321.

Psychotria symplocifolia Kurz. (RUBIACEAE)

Hutchinson in Sargent Pl. Wilson. 3: 415. (1916).

Yunnan: Mountains to the east of Simao at 1,460 metres, 12065, 12065a., 12065b.

Psychotria yunnanensis Hutchinson (RUBIACEAE) New species

in Sargent Pl. Wilson. 3: 414. (1916).

Yunnan: In forests to the south-west of Mengzi at 1,640 metres, 11447, 11447a. Mountains to the south of Simao 12032a., (type). Mountains to the east of Simao at 1,450 metres, 12032, 12032b. Forests to the south-east of Simao, 12806, 12806a.

Ptergocalyx volubilis Maxim. (GENTIANACEAE)

Crawfurdia pterygocalyx Forbes & Hemsley non Hemsl. in Journ. Linn. Soc. Bot. 26: 123. (1890).

Hubei: Xingshan, 6777a.

Pteridium aquilinum (L.) Kuhn (DENNSTAEDTIACEAE)

Christ in Bull. Soc. Bot. France 5: 57. (1905), Diels in Bot. Jahrb. 29: 202. (1900).

Pteris aquilina L. Henry in Journ. China Br. Roy. Asiat. Soc. 22: 241. (1887) (Chinese Names of Plants).

Hubei: Badong, 1865, 3146. Liantuo, 2642.

Bracken, Henry stated that this species was known colloquially aound Yichang and Badong as *chueh*, from the rhizomes an edible arrowroot-like substance called *chueh-fen* was made. The young unfurling fronds were eaten by the mountain peasantry according to Wilson, they continue to be used for this purpose today.

Pteridium rostratum (Burm. f.) Fraser-Jenk. (DENNSTAEDTIACEAE)

Polypodium rostratum Burm. f. Baker in Journ. Bot.

02

26: 230. (1888), Diels in Bot. Jahrb. 29: 204. (1900).

Hubei: Badong, 3685, 5963.

Pteridrys cnemidaria (H. Christ) C. Chr. & Ching (POLYPODIACEAE)

in Bull. Fan. Mem. Inst. Biol. 5: 136. pl. 12. (1934).

Dryopteris cnemidaria H. Christ in Bull. Acad. Geogr. Bot. 20: 140. (1910).

Yunnan: Mengzi, 10473 (type of *Dryopteris cnemidaria* H. Christ).

Distributed across India (including Sikkim, Assam), Bhutan, eastern Myanmar, China, northern Thailand and southern Vietnam.

Pteris actiniopteroides Christ (PTERIDACEAE) New species

in Bull. Herb. Boissier 7: 6. (1899), Christ in Bull. Soc. Bot. France 5: 56. (1905).

Hubei: Yichang, 1996. **Yunnan:** On the mountains to the east of Mengzi at 1,650 metres, 11833 (isotype).

NE India to China.

Pteris biaurita L. (PTERIDACEAE)

Christ in Bull. Herb. Boiss. 7: 6. (1899).

Yunnan: On the bank of a streamlet in the shade of mountain woods near Mengzi at 1,700 metres, 10334.

Distributed through the Himalaya, southern China, the Malay Peninsula, the Philippines, Borneo, Java, South Africa, the West Indies and Brazil.

Pteris cretica L. (PTERIDACEAE)

Henry in Trans. Asiat. Soc. Jap. 24 suppl: 112. (1896) (List Pl. Formosa), Christ in Bull. Soc. Bot. France 5: 56. (1905), Diels in Bot. Jahrb. 29: 202. (1900).

Hubei: Yichang, 439, 445, 2166, 2186. Liantuo, 4494. Badong, 5166. **Sichuan:** North Wushan, 7080. **Taiwan:** Tamsui, 1422. **Yunnan:** Mengzi, 10267. Simao, 13355.

A widespread species, distributed in tropical and temperate regions of the world.

Pteris dactylina Hook. (PTERIDACEAE)

Anonymous in Kew List of Determinations (volume 34).

Yunnan: Mengzi, 11518.

Distributed through the eastern Himalaya and western China.

Pteris deltodon Baker (PTERIDACEAE)

Pteris trifoliata Christ in Bull. Herb. Boissier 7: 9. (1899).

Yunnan: Near Mengzi on dry cliffs in shade at 1,900 metres, 11519 (isotype of *Pteris trifoliata* Christ, collected 27th October 1897).

Distributed through China, Laos, Vietnam and Japan.

Pteris ensiformis Burm. f. (PTERIDACEAE)

Henry in Trans. Asiat. Soc. Jap. 24 suppl: 111. (1896) (List Pl. Formosa).

Hubei: In the glens above Liantuo, 7874. **Taiwan:** Tamsui, 1412, 1417.

A widespread species distributed in the Himalaya, India, southern China (including Taiwan), Myanmar, the Malay Peninsula, southern Japan, the Philippines, tropical Australia, Polynesia and Micronesia.

Pteris esquirolii Christ (PTERIDACEAE) New species

in Notul. Syst. 1: 50. (1909).

Yunnan: Simao at 1,650 metres, 10349. Ouanchay at 1,650 metres, 13320. Simao at 1,650 metres, 13427.

Distributed in China and Vietnam.

Pteris fauriei Hieron. (PTERIDACEAE)

Pteris quadriaurita Henry non Retz. in Trans. Asiat. Soc. Jap. 24 suppl: 111. (1896) (List Pl. Formosa), Christ in Bull. Herb. Boiss. 7: 6. (1899) Diels in Bot. Jahrb. 29: 202. (1900).

Hubei: Xingshan, 4521. **Taiwan:** Tamsui, 1427. **Yunnan:** Mengzi, 11547. In forests near Mengzi near a cliff at 1,650 metres, 11520. Simao, 12957, 13002, 13199, 13200, 13622. Yuanjiang, 13222.

Distributed in China, Vietnam, Japan and the Philippines.

Pteris grevilleana Wall. ex J. Agardh. (PTERIDACEAE)

Henry in Trans. Asiat. Soc. Jap. 24 suppl: 111. (1896) (List Pl. Formosa).

Taiwan: Wanchin, 1495.

Native to northern India, southern China (including Taiwan), the Malay Peninsula, southern Japan, the Philippines and Borneo.

Pteris henryi H. Christ (PTERIDACEAE) New species

in Bull. Herb. Boissier 6: 957. (1898).

Yunnan: Mengzi, 9911 (isosyntype).

Endemic to China and distributed in the western provinces.

Pteris heteromorpha Fée (PTERIDACEAE)

Pteris inaequalis (Fée) Jenman Anonymous in Kew List of Determinations (volume 34).

Yunnan: Mengzi, 10152, 11546.

Pteris longifolia L. (PTERIDACEAE)

Henry in Journ. China Br. Roy. Asiat. Soc. 22: 280. (1887) (Chinese Names of Plants), Diels in Bot. Jahrb. 29: 202. (1900).

Hubei: Yichang, 1101, 1116, 2131. Badong, 2506, 5146, 5147. **Yunnan:** Simao, 13118.

According to Henry this species was known colloquially in the Three Gorges region as *wu kung ts'ao*. This species was collected by the Glasnevin Central China Expedition (GCCE 99) on Emei Shan (Mount Omei) in September 2002.

Pteris semipinnata L. (PTERIDACEAE)

Henry in Trans. Asiat. Soc. Jap. 24 suppl: 111. (1896) (List Pl. Formosa), Diels in Bot. Jahrb. 29: 202. (1900).

Hubei: Liantuo, 4631. **Taiwan:** Tamsui, 1418. Wanchin, 1418a.

Distributed through south-east China (including Taiwan), Japan, the Philippines, the Malay Peninsula and Borneo.

Pteris sp. (PTERIDACEAE)

Anonymous in Kew List of Determinations (volume 34).

Yunnan: Yuanjiang, 13325.

Pteris sp. **nov.** (PTERIDACEAE)

Anonymous in Kew List of Determinations (volume 34).

Yunnan: Simao, 13500.

Pteris terminalis Wall. ex J. Agardh. (PTERIDACEAE)

Pteris excelsa Gaudich. Baker in Journ. Bot. 27: 176. (1889), Diels in Bot. Jahrb. 29: 202. (1900).

Sichuan: North Wushan, 7109. **Yunnan:** Yuanjiang, 13319.

Distributed through Pakistan, the Himalaya, India, China, Korea, Japan, the Philippines, the Malay Peninsula and the Hawaiian Islands. This species was collected by the Sino-American Expedition in Shennongjia in 1980.

Pteris tripartita Sw. (PTERIDACEAE)

Pteris marginata Bory Henry in Trans. Asiat. Soc. Jap. 24 suppl: 111. (1896) (List Pl. Formosa).

Taiwan: Tamsui, 1413.

Pteris vittata L. (PTERIDACEAE)

Pteris longifolia Henry non L. in Trans. Asiat. Soc. Jap. 24 suppl: 111. (1896) (List Pl. Formosa).

Taiwan: Wanchin, without number. Kaohsiung, without number.

Found in the tropics and subtropics of the Old World.

Pteris wallichiana Agardh (PTERIDACEAE)

Pteris longipes D. Don Anonymous in Kew List of Determinations (volume 34).

Yunnan: Simao, 13143, 13524. Yuanjiang, 13338.

Native to the Himalaya, India, southern China, Japan and the Philippines.

Pternopetalum tanakae (Franch. & Savat.) Hand.-Mazz. (APIACEAE)

Pimpinella tanakae (Franch & Savat.) Diels Diels in Bot. Jahrb. 29: 494. (1900). *Cryptotaeniopsis tanakae* (Franch. & Savat.) de Boiss Wolff in Engler, Pflanzenr. iv. 228: 179. (1927). *Cryptotaeniopsis filicina* (Franch.) H. Boissieu Dunn in Journ. Linn. Soc. Bot. 35: 495. (1903), Wolff in Engler, Pflanzenr. iv. 228: 181. (1927).

Hubei: Jianshi, 5724a. Fang Xian, 6600, 6600a.

Sichuan: South Wushan, 5724.

Pternopetalum vulgare (Dunn) Hand.-Mazz. (APIACEAE) New species

in Symb. Sin. 7: 719. (1933).

Pternopetalum davidii Boissieu non Franch. in Bull. Herb. Boiss. 2. ser viii. 806. (1902). *Cryptotaeniopsis vulgaris* Dunn in Hooker's. Icon. Pl. t. 2723. (1902), Wolff in Engler, Pflanzenr. iv. 228: 176. (1927), Morley in Glasra 3: 79. (1979). *Cryptotaenopsis davidii* (Franch.) Wollf. Wollf. in Engler, Pflanzenr. iv. 228: 175. 1927).

Hubei: South Badong, 5384, 5444. Without locality, 5444a. **Sichuan:** South Wushan, 5406. **Yunnan:** Fengchunling, 10675.

Distributed in Nepal, India, Myanmar and China.

Pterocarya hupehensis Skan (JUGLANDACEAE) New species

in Journ. Linn. Soc. Bot. 26: 493. (1899), Rehder in Sargent Pl. Wilson. 3: 182. (1916), Bean in Trees & Shrubs 2: 262. (1921), Morley in Glasra 3: 80. (1980), Nelson in A Heritage of Beauty 317. (2000), O' Brien in Augustine Henry, An Irish Plant Collector in China 30. (2002).

Hubei: South Badong, 6158, (isotype).

A tree to about 20 metres tall, found by Henry in moist woods in Hubei in 1888. He stated it was known colloquially as *shan-liu-shu*. Introduced to cultivation by Wilson in 1901. It is cultivated in Ireland at the National Botanic Gardens, Glasnevin and was collected by the Glasnevin Central China Expedition on the border of Yichang and Changyang (GCCE 742) in October 2004.

Pterocarya stenoptera C. DC. (JUGLANDACEAE)

Henry in Journ. China Br. Roy. Asiat. Soc. 22: 256. (1887) (Chinese Names of Plants), Skan in Journ. Linn. Soc. Bot. 26: 494. (1899), Henry in Elwes and Henry The Trees Gt. Brit. & Irel. 2: 443. (1907), Rehder & E. H. Wilson in Sargent Pl. Wilson. 3: 181. (1916), Zheng in Journ. Wuhan. Bot. Res. 2.(1): 18. (1984).

Hubei: Yichang, 971, 1332. **Yunnan:** By a riverside at Lunan, 10537. Mengzi, 10573.

A tree to 30 metres tall, commonly found on river sides and associated sand beds at low altitudes in Hubei. According to Wilson it was always the first tree to colonise newly formed islands on the Yangtze. It is a tree of rapid growth attaining a height of 30 metres and Henry stated it was colloquially known at Yichang as *liu-shu*. Discovered by the Italian-French missionary Joseph-Marie Callery (1810—1862) who sent specimens to France in 1844, it was on this material that Casimir de Candolle based the species. Native to China, Korea and Japan.

Pteroceras asperatum (Schlecter) P. F. Hunt (ORCHIDACEAE) New species

Sarcochilus asperatus Schlecter in Repert. Spec. Nov. Regni. Veg. Beih. 4: 75. (1919).

Yunnan: Simao, 13581 (isotype of *Sarcochilus asperatus* Schlecter).

Endemic to Yunnan.

Pterolobium punctatum Hemsley (FABACEAE) New species

Forbes & Hemsley in Journ. Linn. Soc. Bot. 23: 207. (1887), Henry in Journ. China Br. Roy. Asiat. Soc. 22: 240. (1887) (Chinese Names of Plants), Bretschneider in Hist. Eur. Bot. Disc. China 2: 782. (1898), Zheng in Journ. Wuhan Bot. Res. 2.(1): 103. (1984).

Hubei: On the banks of a river in the San You Dong Glen, 1690. Yichang, 2940 (type). Liantuo, 4505. **Yunnan:** Mengzi, 9578.

A large, thorny rambling shrub with stems up to 15 metres long. A handsome climber bearing axillary racemes of small white flowrrs followed by masses of very attractive red flat seed pods with a prominent membranous wing. A common on the limestone cliffs of the glens and gorges throughout western Hubei where it was known colloquially as *chio-pu-t'a* according to Henry. Distributed through the western and central provinces of China and also in Laos.

Pterospermum proteus Burkill (MALVACEAE) New species

in Kew Bull. Misc. Inf. 138. (1901).

Yunnan: Mengzi, 10120 (type).

Endemic to Yunnan.

Pterospermum sp. **nov.** (MALVACEAE)

In 'Kew Inwards' book (archives of the Royal Botanic Gardens, Kew).

Yunnan: By the Red River at Manpan, 11354.

Pterospermum suberifolium (L.) Willd. (MALVACEAE)

Anonymous in Kew List of Determinations (volume 24).

Hainan: At Nodoa in the interior of the island, 8441.

Pterostyrax psilophyllus Diels ex Perkins (STYRACACEAE) New species

Halesia hispida Forbes & Hemsley non (Sieb. & Zucc.) N. E. Br. in Journ. Linn. Soc. Bot. 26: 76. (1889). *Pterostyrax hispidus* Perkins non Sieb. & Zucc. in Engler, Pflanzenr. 241: 102. (1907), Rehder in Sargent Pl. Wilson. 1: 295. (1913), Zheng in Journ. Wuhan Bot. Res. 2.(1): 175. (1984). *Pterostyrax corymbosus* Henry non Sieb. & Zucc. in Journ. China Br. Roy. Asiat. Soc. 22: 262. (1887) (Chinese Names of Plants).

Hubei: Badong, 611 (July 1885), 3176, 3774, 4884. Jianshi, 5595a. 5595c, 5595d. **Sichuan:** South Wushan,

5595. North Wushan, 5595b. Kangding, Henry's Chinese collector with Pratt, 8883.

Described from material collected near Liantuo in Hubei by Wilson in April 1900 for the Veitch nursery, it had been found fifteen years earlier at Badong by Henry.

Ptychomitrium gardneri Lesq. (GRIMMIACEAE)

Ptychomitrium polyphylloides (Müll. Hal.) Paris Salmon in Journ. Linn. Soc. Bot. 456. (1900).

Hubei: Changyang, 7482. In a glen near Yichang 7913 (February 1889). (Moss).

Puccinia henryana P. Syd. & Syd. (PUCCINIACEAE) New species

in Monogr. Ured. 1: 633. (1904).

Hubei: Yichang, without number, (isotype).

Puchiumazus lanceifolius (Hemsley) Bo Li, D. G. Zhang & C. L. Zhiang (MAZACEAE) New species

Mazus lanceifolius Hemsley in Journ. Linn. Soc. Bot. 26: 181. (1890), Bretschneider in Hist. Eur. Bot. Disc. China 2: 788. (1898).

Hubei: Jianshi, 5837. **Sichuan:** South Wushan, 7250 (type of *Mazus lanceifolius* Hemsley).

Pueraria montana (Lour.) Merr. (FABACEAE)

Shuteria vestita Benth. Anonymous in Kew List of Determinations (volume 34).

Yunnan: Mengzi, 9252, 9843, 10261. Simao, 11678.

Pueraria montana (Lour.) Merr. var. ***lobata*** (Willd.) Maesen & S. M. Almeida ex Sanjappa & Predeep (FABACEAE)

Pueraria thunbergiana (Sieb. & Zucc.) Benth. Forbes & Hemsley in Journ. Linn. Soc. Bot. 23: 191. (1887), Henry in Journ. China Br. Roy. Asiat. Soc. 22: 249. (1887) (Chinese Names of Plants); in Notes Econ. Bot. China 57. (1887); in Trans. Asiat. Soc. Jap. 24 suppl: 35. (1896) (List Pl. Formosa), Li in Woody Flora of Taiwan 359. fig. 128. (1963).

Hubei: Yichang, 17, 2120, 2287. Liantuo, (October 1886) 3045. Badong, 2427. **Taiwan:** Oluanpi, 510, 953. **Yunnan:** In Pingbian, in forests on the Daweishan Range at 1,650 metres, 11041. Simao, 12521.

The kudzu vine, a vigorous tuberous rooted climber to 10 metres tall, common at low altitudes in Hubei where Henry stated it was known as *ko t'eng (fan kot* in Cantonese). This plant, in Henry's time in China, furnished the *ko-pou* fibre, which was manufactured into a fabric used for underclothing in summer and as a slow burning fuse for explosives. Due to the costly and primitive process of preparation the manufacture of the fabric declined. From the the thickened rootstock (the tubers can weigh up to 30 kg) starch was once prepared as a food, though only used during famine times by poorer inhabitants. In places it is still cultivated as a source of starch for thickening soup, etc. The kudzu vine has been used in Chinese herbal medicine since ancient times as a remedy against colds and flu, muscle ache, dysentry, hypertension and migrine. It is naturalised in the south-east of the USA where it was introduced as cattle forage and where it is a pest. The costs of removing this rampant climber (which runs over entire houses, barns and telephone wires) are enormous and sets a good example for the care needed when introducing alien plants to a new locality. It has been recently found that this species contains a compound that reduces the cravings of alcohol addiction. Scientists at Harvard University have investigated extracts of the vine (which has been used in China since 200 AD to suppress the desire for alcohol) and have confirmed its effectiveness. The chemical has been synthesised and is still under trial.

In 19th century Taiwan, the root was made into an arrow-root like preparation named *ko-fen* and a cloth called *ko-pu* was made out of the fibre in its stems. The stems were cut into lengths and steeped in water with lime and wood-ashes for some days. They were then taken out and boiled, the bark was then stripped off and beaten with a mallet to cause the fibre to seperate, then washed and beaten and washed again. Then carded, spun and woven into a fabric by the hand-loom. Forty pounds of stem yielded one pound of fibre. A yellow colour was given to the cloth by soaking it in rice-water.

Pueraria sp. (FABACEAE)
Anonymous in Kew List of Determinations (volume 34).

Yunnan: At Yuanjiang, in a shady ravine at 1,100 metres, 11579. Simao, 13626.

Pueraria tuberosa (Roxb. ex Willd.) DC. (FABACEAE)
Henry in Notes Econ. Bot. China 58. (1893).

Sichuan: South Wushan, 7547.

According to Henry this species was cultivated for the sake of its flowers which were used as a drug. **Yunnan:** Mengzi, 9248, 10047.

Puhuaea sequax (Wall.) H. Ohashi & K. Ohashi (FABACEAE)
Desmodium sinuatum (Miq.) Blume ex Baker Henry in Trans. Asiat. Soc. Jap. 24 suppl: 33. (1896) (List Pl. Formosa).

Taiwan: Wanchin, 515. **Yunnan:** Simao, 12476.

Native to India, Nepal, Myanmar and China.

Punctularia strigosozonata (Schwerin.) P. H. B. Talbot (PUNCTULARIACEAE)
Phlebia reflexa Berk. Anonymous in Kew List of Determinations (volume 24).

Hubei: Fang Xian, 6685. (Fungus).

Punica granatum L. (LYTHRACEAE)
Henry in Trans. Asiat. Soc. Jap. 24 suppl: 44. (1896) (List Pl. Formosa).

Hubei: On the banks of the Yangtze near Yichang, an escape from cultivation, 3591. **Taiwan:** Kaohsiung, cultivated, without number.

The pomegranate, a subtropical fruit native to the near east and Africa and cultivated for its fruits in southern Europe, Iran, Palestine and India since ancient times. The pomegranate is an ancient cultigen and is suggested to be Eve's apple of the bible. Commercially cultivated for its fruits in China, it is naturalised in many of the warmer parts of the country. The globular fruits carry numerous juicy red seeds which in Asia are a symbol of fertility and abundance. In the certain countries in same region the fruits are an integral part of wedding cermonies, being offered to guests who throw them onto the floor of the honeymoon suite to shatter the fruits and cast the seeds to ensure fertility and and a large number of offspring for the newly weds. In China it is the *shi liu pi* and the husk of the fruit is used to kill and expel parasites like tapeworm and roundworm.

Pycnoporus cinnabarinus (Jacq.) Fr. (POLYPORACEAE)
Polyporus cinnabarinus (Jacq.) Fr. Anonymous in Kew List of Determinations (volume 24).

Hubei: Fang Xian, 6657. Changyang, on trees 7931. (Fungus).

Pycnospora lutescens (Poir.) Schldl. (FABACEAE)
Pycnospora hedysaroides R. Br. Henry in Trans. Asiat. Soc. Jap. 24 suppl: 34. (1896) (List Pl. Formosa).

Hainan: Environs of Haikou, 8357. Ling-men (in the interior of the island), 8665. **Taiwan:** Oluanpi, Wanchin, 1534.

A perennial herb, native to tropical Africa, India to the Philippines, New Guinea and Australia.

Pyracantha crenulata (D. Don) M. Roemer (ROSACEAE)
E. H. Wilson in Sargent Pl. Wilson. 1: 177. (1912).

Crataegus pyracantha Hemsley Henry in Journ. China Br. Roy. Asiat. Soc. 22: 234. (1887) (Chinese Names of Plants), Forbes & Hemsley in Journ. Linn. Soc. Bot. 23: 260. (1887). *Pyracantha coccinea* Zheng non M. Roem. in Journ. Wuhan. Bot. Res. 2.(1): 80. (1984).

Hubei: Yichang, 546, 979. Badong, 607 (July 1885), 2883 (October 1886). Liantuo, 2986. **Yunnan:** Mengzi at 1,800 metres, 10625.

According to Henry this shrub was common on the hills of western Hubei during his stay there (it was certainly locally common when the Glasnevin Central China Expedition visited Badong in the autumn of 2002.) and he stated it was known colloquially as *ch'a kou*. Seeds of Wilson's 662 from Yichang reached Glasnevin in 1908.

Native to Pakistan, India, Bhutan, Nepal, Myanmar and China.

Pyrenaria microcarpa (Dunn) H. Keng var. *ovalifolia* (H. L. Li) T. L. Ming & S. X. Yang (THEACEAE) New species

Tutcheria taiwanica Hung T. Chang & S. X. Ren in Acta Sci. Nat. Univ. Sunyatsenia 30(1): 71. (1991).

Taiwan: Wanchin, 123 (holotype of *Tutcheria taiwanica* Hung T. Chang & S. X. Ren, collected 25th February 1893).

Pyrola calliantha Andres. (ERICACEAE)

Pyrola rotundifolia Forbes & Hemsley non L. in Journ. Linn. Soc. Bot. 26: 32. (1889).

Hubei: Badong, 330 (June 1885), 4731, 5479.

This small perennial was collected by the Glasnevin Central China Expedition (GCCE 504) near Zhenzi town in Xingshan in September 2004.

Pyrola sp. (ERICACEAE)

Anonymous in Kew List of Determinations (volume 24).

Hubei: Badong, 1867, 3767.

Pyrrosia adnescens (Sw.) Ching (POLYPODIACEAE)

Polypodium adnescens Sw. Henry in Trans. Asiat. Soc. Jap. 24 suppl: 114. (1896) (List Pl. Formosa).

Sichuan: South Wushan, 5486. **Taiwan:** Wanshoushan, 747. Wanchin, 1532.

An epiphyte commonly found on hardwood trees. Distributed from India to Malaya, southern China (including Taiwan), Okinawa and the Philippines to Polynesia.

Pyrrosia angustissima (Gies) Tagawa & K. Iwats (POLYPODIACEAE) New species

Polypodium angustissimum Baker (non Fee 1869) in Ann. Bot. 5: 472. (1890), Bretschneider in Hist. Eur. Bot. Disc. China 2: 794. (1898). *Niphobolus angustissimus* Gies ex Diels in Engl. Jahrb. 29: 207. (1900), Christ in Bull. Soc. Bot. France 5: 23. (1905); legitimate name based on *Polypodium angustissimum* Baker.

Hubei: Badong, 5137. South Badong, 5447.

Distributed in China, Thailand and Japan.

Pyrrosia assimilis (Baker) Ching (POLYPODIACEAE)

Polypodium assimile Baker Baker in Ann. Bot. 5: 473. (1890).

Hubei: Liantuo, 1993.

Endemic to China.

Pyrrosia costata (Wall. ex C. Presl.) Tagawa & K. Iwats. (POLYPODIACEAE)

Nephrodium costatum Bedd. Christ in Bull. Soc. Bot. France 5: 34. (1905). *Niphobolus beddomeanus* Gies. f. *fallax* K. Giesenhagen in Farngatt. Niphob. 103. (1901).

Hubei: Yichang, 750, 2237. Liantuo, 2581, 3013.

Yunnan: Simao, 12704a., (isosyntype of *Niphobolus beddomeanus* Gies. f. *fallax* K. Giesenhagen).

Distributed in Nepal, India, Myanmar, China and Thailand.

Pyrrosia davidii (Giesenh.) Ching (POLYPODIACEAE)

Niphobolus gralla K. Giesenhagen in Farngatt. Niphob. 128. (1901), Christ in Bull. Soc. Bot. France 5: 24. (1905), *Pyrrosia gralla* (Gies.) Ching in Bull. Chin. Bot. Soc. 1: 50. (1935).

Yunnan: On mountains to the east of Mengzi at 1,650 metres, 9061a., (type of *Niphobolus gralla* K. Giesenhagen).

Distributed in China (including Taiwan).

Pyrrosia drakeana (Franch.) Ching (POLYPODIACEAE)

Polypodium drakeanus Franch. Christ in Bull. Herb. Boiss. 6: 871. (1898), Christ in Bull. Soc. Bot. France 5: 25. (1905).

Hubei: Badong, 3864. **Yunnan:** Mountains to the east of Mengzi at 1,600 metres, 9115. On mountains to the south-west of Mengzi at 1,650 metres, 9116.

This species was collected in the Shennongjia Forest District by the Sino-American Expedition in 1980. Northeast India to China.

Pyrrosia glabra (Desv.) Fraser-Jenk. (POLYPODIACEAE)

Niphobolus nudus K. Giesenhagen in Farngatt. Niphob. 149. (1901).

Yunnan: Simao, 12884a., (type of *Niphobolus nudus* K. Giesenhagen).

India to China and Malesia.

Pyrrosia linearifolia (Hook.) Ching. (POLYPODIACEAE)

Polypodium linearifolium (Hook.) Hook. Anonymous in Kew List of Determinations (volume 24).

Sichuan: South Wushan, 5488. **Hubei:** Without locality, 5488a.

Distributed in China, Korea and Japan.

Pyrrosia lingua (Thunb.) Farw. (POLYPODIACEAE)

Polypodium lingua (Thunb.) Sw. Baker in Journ. Bot. 26: 230. (1888), Christ in Bull. Herb. Boiss. 7: 5. (1899), Henry in Trans. Asiat. Soc. Jap. 24 suppl: 114. (1896) (List Pl. Formosa). *Niphobolus lingua* (Thunb.) J. Sm. Diels in Bot. Jahrb. 29: 205. (1900).

Hubei: Liantuo, 1996. Badong, 5028, 5128. Jianshi, 5810. **Taiwan:** Tamsui, 1429, 1431. **Yunnan:** In Pingbian on the Daweishan Range at 1,600 metres, on trees, 11816.

Distributed in Nepal, India, Bhutan, Myanmar, Thailand, Laos, Cambodia, Vietnam, China, Korea and Japan.

Pyrrosia polydactylos (Hance) Ching (POLYPODIACEAE)

Polypodium polydactylon Hance Henry in Trans. Asiat. Soc. Jap. 24 suppl: 114. (1896) (List Pl. Formosa).

Taiwan: Tamsui, 1390.

A fern to 30 cm., distributed in China (Taiwan) and Korea.

Pyrrosia porosa (C. Presl.) Hovenkamp (POLYPODIACEAE)

Polypodium mollissimum H. Christ in Bull. Herb. Boissier 7: 5. (1899), nom. illeg.

Yunnan: On mountains to the east of Mengzi at 1,900 metres, 9061b., (isotype of *Polypodium mollissimum* H. Christ).

Pyrrosia sheareri (Baker) Ching (POLYPODIACEAE)

Polypodium sheareri Bak. Baker in Ann. Bot. 5: 472. (1890). *Niphobolus sheareri* (Bak.) Christ Diels in Bot. Jahrb. 29: 205. (1900), Christ in Bull. Soc. Bot. France 5: 26. (1905). *Niphobolus inaequalis* Christ in Bull. Soc. Bot. France 5: 25. (1905).

Hubei: Badong, 1428 (syntype of *Niphobolus inaequalis* Christ), 2569, 2813.

The specific epithet commemorates Dr. George Shearer of the London Missionary Society who was appointed to take charge of their hospital at Hankow (now part of Wuhan), and made a large collection of plants in 1873 near Jiujing in Jiangxi province. This handsome fern was also collected by Prince Henry d'Orleans near Kangding in western Sichuan in 1889 and in March 1894 by Pére Delavay in Yunnan. *Pyrrosia sheareri* was collected in the Shennongjia Forest District by the Sino-American Expedition in 1980 and was met again by the Glasnevin Expedition near Badong in 2002 where it grew in association with *Keteeleria davidiana*, *Pinus massioniana*, *Quercus fabri* and *Sargentodoxa cuneata*.

Pyrrosia subfurfuracea (Hook.) Ching (POLYPODIACEAE)

Polypodium subfurfuraceum Hook. Christ in Bull. Herb. Boiss. 6: 871. (1898).

Yunnan: Mountains to the south of Mengzi at 1,650 metres, 9291. Mengzi at 1,500 metres, 9834.

Native to India, Bhutan, China [Tibet (Xizang) and Yunnan] and Vietnam.

Pyrularia edulis (Wall.) A. DC. (SANTALACEAE)

Anonymous in Kew List of Determinations (volume 34).

Yunnan: Mengzi, 10470.

Distributed in India, Bhutan, Myanmar and China.

Pyrus betulifolia Bunge (ROSACEAE)

Forbes & Hemsley in Journ. Linn. Soc. Bot. 23: 256. (1887), Henry in Journ. China Br. Roy. Asiat. Soc. 22: 272. (1887) (Chinese Names of Plants), Rehder in Proc. Amer. Acad. 50: 236. (1915), Zheng in Journ. Wuhan Bot. Res. 2.(1): 80. (1984).

Hubei: Yichang, 778. In the Monastery Valley near Yichang (July 1886), 1654.

A tree to 8 metres tall carrying corymbs of white flowers in spring. According to Henry this species was known in the Three Gorges region as *t'ang li*. It was introduced to cultivation by Emil Bretschneider who sent seeds to Arnold Arboretum in 1882. *Pyrus betulifolia* is used as an understock by the Chinese on which to graft pear cultivars. According to Bretschneider it was introduced to the garden of the Paris Museum by Gabriel Eugene Simon who collected it at Xiwanzi (Siwantze) near the Great Wall.

Pyrus pashia Buch.-Ham. ex. D. Don (ROSACEAE)

Rehder in Proc. Amer. Acad. 50: 239. (1915), Rehder in Sargent Pl. Wilson. 2: 264. (1915).

Pyrus wilhelmii C. K. Schneid in Ill. Handb. Laubholzk. 1: 665. fig. 363n. (1906); in Fedde Rep. Spec. Nov. Regni. Veg. 3: 120. (1907). *Pyrus pashia* Buch.-Hamilt. ex D. Don var. *kumaoni* (Decne.) Stapf. (Rehder) in Proc. Amer. Acad. 50: 239. (1915), Rehder in Sargent Pl. Wilson. 2: 265. (1915).

Yunnan: Mengzi at 1,400 to 1,500 metres, 10035, 10035c. In montane woodland near Mengzi at 1,400 metres, 10035b. In montane woodland near Mengzi at 1,600 metres, 10035a., (type of *Pyrus wilhelmii* C. K. Schneider).

A small tree of widespread distribution from Iran to Afghanistan across the Himalaya into Myanmar, western China, Thailand, Laos and Vietnam. Introduced to cultivation in 1825.

Pyrus pyrifolia (Burm. f.) Nakai (ROSACEAE)

Zheng in Journ. Wuhan Bot. Res. 2.(1): 82. (1984).

Pyrus serotina Rehder in Proc. Amer. Acad. 50: 232. (1915), Rehder in Sargent Pl. Wilson. 2: 263. (1915).

Hubei: South Badong, 5299. Jianshi, 5875 (cultivated).

A tree to 13 metres tall, native to western and central China and Indochina, introduced to cultivation by Wilson in 1909 while collecting for the Arnold Arboretum. It has remained rare in cultivation. *Pyrus pyrifolia* is locally common in thickets in Badong where it was collected by members of the Glasnevin Central China Expedition (GCCE 263) in October 2002 and again by the 2004 Glasnevin Expedition to Hubei in Xingshan (GCCE 507). The autumn colour (scarlet) is spectacular on this tree but this tends to vary in intensity from individual to individual.

Pyrus sp. No. 1. (ROSACEAE)

Anonymous in Kew List of Determinations (volume 24).

Hubei: Badong, 3743, 5301. Xingshan, 6499. Fang Xian, 6905.

Pyrus sp. No. 2. (ROSACEAE)

Anonymous in Kew List of Determinations (volume 34).

Yunnan: Mengzi, 10563.

Pyrus ussuriensis Maxim. (ROSACEAE)

Pyrus ovoidea Rehder in Proc. Amer. Acad. 50: 228. (1915).

Yunnan: Mengzi at 1,500 metres, cultivated, 11058.

This species had been cultivated at the Paris Museum prior to Henry's Mengzi collection.

Quercus acrodonta Seemen (FAGACEAE) New species

in Bot. Jahrb. Syst. 23. Biebl. 57: 48. (1897), Rehder & E. H. Wilson in Sargent Pl. Wilson. 3: 225. (1916), Zheng in Journ. Wuhan Bot. Res. 2.(1): 25. (1984), Li & Zhang in Wuhan Bot. Res. 16: 248. (1998).

Quercus phillyreoides Henry non A. Gray in Journ. China Br. Roy. Asiat. Soc. 22: 240. (1887) (Chinese Names of Plants); in Notes Econ. Bot. China 54. (1893). *Quercus ilex* L. var *acrodonta* (Seem.) Skan. Skan. in Journ. Linn. Soc. Bot. 26: 516. (1899).

Hubei: Yichang, 2945 (isosyntype), 2945a., (isosyntype), 3425. On cliffs near Yichang, 7619. Antelope Glen, 3425. Without locality 3425a., 3425b.

A small, evergreen densly branched tree to 5 metres tall or small bush. Common on the limestone cliffs around Yichang where according to Henry it was known as the *chiu-kang-shu* and was often planted by temples in the mountains.

Quercus aliena Blume (FAGACEAE)

Henry in Journ. China Br. Roy. Asiat. Soc. 22: 244. (1887) (Chinese Names of Plants); in Notes Econ. Bot. China 54. (1893), Skan in J. Linn.Soc. Bot. 26: 505. (1899).

Hubei: Yichang, 968, 1279. Badong, 553 (September 1885). Liantuo, 1899. **Yunnan:** Mengzi, 11499.

Known colloquially at Yichang as *hu li* according to Henry. This handsome oak was collected by the Glasnevin Central China Expedition (GCCE 594) in Xingshan in September 2004. Distributed in China, Korea and Japan.

Quercus aliena Blume var. acuteserrata Maxim. (FAGACEAE)

Rehder & E. H. Wilson in Sargent Pl. Wilson. 3. 215. (1916), Zheng in Journ. Wuhan Bot. Res. 2.(1): 25. (1984).

Quercus aliena Skan non Blume in Journ. Linn. Soc. Bot. 26: 505. (1889).

Hubei: Yichang, 68, 2292, 2293. Liantuo, 68a. **Hunan:** Shimen, 7945. **Yunnan:** Mengzi at 1,600 metres, 9394, 9394a., 11298

A tree to 25 metres tall with massive, widespreading branches, found at low altitudes to 2,000 metres in Hubei where it was known as *hu-li* according to Henry. Distributed through mainland China (including Taiwan), Korea and Japan.

Quercus augustinei Skan (FAGACEAE) New species

in Journ. Linn. Soc. Bot. 26: 507. (1899), Muir in The New Plantsman 3(4): 223. (1996).

Cyclobalanopsis augustinei (Skan) Schotty in Bot. Jahrb. Syst. 47: 656. (1912).

Yunnan: Mengzi at 1,650 metres, 11430 (isotype of *Quercus augustinei* Skan).

An evergreen tree of moderate size, discovered by Henry in 1897 and later collected by George Forrest and Joseph Rock on the Yunnan-Myanmar frontier. Forrest introduced it through his 30348 in 1925 and it has been reintroduced to cultivation from Fansipan in northern Vietnam in more recent times. Distributed in western China, Indochina and Myanmar.

Quercus championii Benth. (FAGACEAE)

Henry in Trans. Asiat. Soc. Jap. 24 suppl: 90. (1896) (List Pl. Formosa), Skan in Journ. Linn. Soc. Bot. 26: 509. (1899).

Cyclobalanopsis championii (Benth.) Oerst. Li in woody Flora of Taiwan 100. (1963).

Taiwan: Oluanpi, 1993, 1253.

An evergreen tree to 20 metres tall, native to southern China (Guangdong, Hainan, Taiwan and Hong Kong). In Taiwan this species is found in primary forest on the Hengchun Peninsula. It once grew in the collection at Caerhays Castle, Cornwall but is no longer known in cultivation.

Quercus ciliaris C. C. Huang & Y. T. Chang (FAGACEAE) New species

Quercus glauca Thunb. f. *gracilis* Rehder & E. H. Wilson in Sargent Pl. Wilson. 3: 228. (1916). *Quercus glauca* Thunb. var. *gracilis* (Rehd. & Wils.) A. Camus. Zheng in Journ. Wuhan Bot. Res. 2.(1): 27. (1984).

Hubei: Liantuo, 2992. Yichang, 3088. **Sichuan:** South Wushan, 7076.

Discovered by Henry in February 1887 and later collected by Wilson (W. 687).

Quercus delavayi Franchet (FAGACEAE)

Rehder & E. H. Wilson in Sargent Pl. Wilson. 3: 236. (1916).

Quercus gilva Skan non Blume in Journ. Linn. Soc. Bot. 26: 514. (1899).

Yunnan: Mengzi at 1,800 metres, 10504, 10504a.

Endemic to the western provinces of China.

Quercus dentata Thunb. (FAGACEAE)

Henry in Notes Econ. Bot. China 14. (1893), Skan in Journ. Linn. Soc. Bot. 26: 511. (1899), Rehder & E. H. Wilson in Sargent Pl. Wilson. 3: 210. (1917), Zheng in Journ. Wuhan Bot. Res. 2.(1): 26. (1894).

Quercus aliena Rehder & E. H. Wilson non Blume in

Sargent Pl. Wilson. 3: 214. (1916), in part.

Hubei: Liantuo, 3080. **Yunnan:** Mountains to the north of Mengzi at 2,000 metres, 9201.

A deciduous tree to 25 metres tall with trunks up to 3 metres in girth. Distributed through Mongolia, China, Korea and Japan.

Quercus dolicholepis A. Camus (FAGACEAE) New species

in Chenes 3: 1215. (1952—1954) nom nov. for *Quercus spathulata* Seem. non. *Quercus spathulata* Watelet in Pl. Foss. Basin. Paris. 136. (1886), Li & Zhang in Journ. Wuhan Bot. Res. 16: 249. (1998).

Quercus spathulata Seemen in Bot. Jahrb. xxiii. Beibl. 57: 49. (1897), Skan in Journ. Linn. Soc. Bot. 26: 521. (1899), Rehder & E. H. Wilson in Sargent Pl. Wilson. 3: 226. (1917), Zheng in Journ. Wuhan Bot. Res. 2.(1): 28. (1984).

Hubei: Liantuo, 6359, 6359c. Changyang, 6359a. Xingshan, 6359b., (holotype of *Quercus spathulata* Seemen)

A tree to 16 metres tall with a girth of 2 metres.

Quercus engleriana Seemen (FAGACEAE) New species

in Bot. Jahrb. Syst. 23 (Biebl. 57): 47. (1897), Skan in Journ. Linn. Soc. Bot. 26: 512. (1899), Koidzumi in Icon. Pl. Koisikav. 1: 111. t. 56. (1912), Rehder & E. H. Wilson in Sargent Pl. Wilson. 3: 220. (1917), Zheng in Journ. Wuhan Bot. Res. 2.(1): 26. (1984), Li & Zhang in Journ. Wuhan Bot. Res. 16: 249. (1998).

Quercus sp. Henry in Journ. China Br. Roy. Asiat. Soc. 22: 241. (1887) (Chinese Names of Plants). *Quercus obscura* Seemen in Bot. Jahrb. xxiii. Beibl. 57: 49. (1897), Skan in Journ. Linn. Soc. Bot. 519. (1899).

Hubei: Yichang, 967. Badong, 2827, 5682b., 6167 (co-type of *Quercus obscura* Seemen). **Sichuan:** South Wushan, 5682 (type), 5682a., (co-type of *Quercus obscura* Seemen), 5694.

A small tree to about 10 metres tall, common in rocky places in western Hubei and Sichuan. This is the 'white oak' of foreigners at Hankow, mentioned by Henry in his *Chinese Names of Plants*. Introduced to cultivation by Wilson through Veitch's Coombe Wood nursery in 1900.

Quercus fabri Hance (FAGACEAE)

Henry in Journ. China Br. Roy. Asiat. Soc. 22: 254. (1887) (Chinese Names of Plants), Henry in Notes Econ. Bot. China 53. (1893), Skan in Journ. Linn. Soc. Bot. 26: 512. (1899), Rehder & E. H. Wilson in Sargent Pl. Wilson. 3: 212. (1916), Zheng in Journ. Wuhan Bot. Res. 2.(1): 26. (1984).

Hubei: Yichang, 2294. Liantuo, 2636.

A tree to 20 metres tall though usually seen as scrub or coppiced growth in Hubei. At Yichang this oak was known as *hsiao pai-fan-li* according to Henry. Wilson made several collections in the same area. *Quercus fabri* was collected by the Glasnevin Central China Expedition (GCCE 554) in Xingshan in September 2004.

Quercus franchetii Skan (FAGACEAE) New species

in Journ. Linn. Soc. Bot. 26: 513. (1899).

Yunnan: Mengzi at 1,950 metres, 9298, 9298a., (type). Native to China (Sichuan and Yunnan) and Thailand.

Quercus glauca Thunb. (FAGACEAE)

Henry in Journ. China Br. Roy. Asiat. Soc. 22: 241. (1887) (Chinese Names of Plants); in Trans. Asiat. Soc. Jap. 24 suppl: 90. (1896) (List Pl. Formosa), Skan in Journ. Linn. Soc. Bot. 26: 515. (1899), Rehder & E. H. Wilson in Sargent Pl. Wilson. 3: 226. (1916), Zheng in Journ. Wuhan Bot. Res. 2.(1): 27. (1984).

Quercus glauca Thunb. var. *villosa* Skan in Journ. Linn. Soc. Bot. 26: 515. (1899).

Hubei: Yichang, 966, 3098, 342, 3422. Antelope Glen, 3421, 3426. In the Lung Wang Tung Glen (June 1886), 1561. Liantuo, 4473. South Badong, 5277b. Changleping, 6270. **Sichuan:** South Wushan, 5277a. North Wushan, 7064. **Taiwan:** Wanchin, 428. **Yunnan:** In woods near Mengzi at 1,800 metres, 9299 (type of *Quercus glauca* Thunb. var. *villosa* Skan), 9299a., 9299b.

The bamboo oak, a handsome evergreen tree to 20 metres tall with a dense rounded crown, widely distributed (this species has the greatest distribution of all the Asian oaks) through the Himalaya, China and Japan. In western Hubei this is the most common species of evergreen oak, and Henry stated it was known there as *tieh chou li*. Its wood, according to Henry was very hard and was used in making carrying poles. Introduced from Nepal in 1804, it was reintroduced by the Kew collector Richard Oldham from Japan in 1861 and by E. H. Wilson from Hubei in the first decade of the 20th century. This species was collected by the Glasnevin Central China Expedition (GCCE 664) in the Red Dragon Reserve in Xingshan in October 2004. It is rare in cultivation though there are fine trees to 8.5 metres tall at Caerhays Castle, Cornwall and at the Hillier Arboretum a tree reached 4 metres tall in less than 20 years. This species thrives best in the warmer, coastal regions of the British Isles and Ireland.

Quercus gomeziana A. Camus (FAGACEAE)

Quercus vestita Rehder & E. H. Wilson in Sargent Pl. Wilson. 3: 236. (1916), nom. illeg.

Yunnan: In forests near Simao at 1,500 to 1,600 metres, 11675, 11675b., 11675c., 11675d., 11675e.

Henry's collections of this spiese proved to be a new addition to the Chinese flora.

Quercus griffithii Hook. f. & Thoms. ex Miq. (FAGACEAE)

Rehder & E. H. Wilson in Sargent Pl. Wilson. 3: 233. (1916).

Yunnan: In forests near Simao at 1,600 to 1,800 metres, 11687, 12435, 12435a.

Native to India, Sri Lanka, Bhutan, Myanmar, western China and Thailand.

Quercus myrsinifolia Blume (FAGACEAE)

Rehder & E. H. Wilson in Sargent Pl. Wilson. 3: 236. (1916).

Yunnan: In forests near Simao at 1,500 metres, 11698.

A handsome evergreen tree to 25 metres tall in its native habitat, introduced to cultivation by Robert Fortune in 1845. Noted for its slender and elegant foliage this species may exceed the stated 25 metres when cultivated in the area of shrines and temples and in cultivation in the West it has reached 18 metres tall. Because of its wide geographical range (southern China, Laos, Korea and Japan) *Quercus myrsinifolia* is a variable species and thrives best in acid or neutral soil, it will not tolerate a high ph. A beautiful tree that deserves to be more widely planted.

Quercus oxyodon Miq. (FAGACEAE)

Skan in Journ. Linn. Soc. Bot. 26: 517. (1899), Rehder & E. H. Wilson in Sargent Pl. Wilson. 3: 228. (1917), Zheng in Journ. Wuhan Bot. Res. 2.(1): 228. (1984).

Quercus lineata Blume var. *grandiflora* Skan in Journ. Linn. Soc. Bot. 26: 517. (1899), A. Henry 5277a. *Quercus glauca* Skan non Thunb. in Journ. Linn. Soc. Bot. 26: 515. (1899). *Quercus lineata* Blume var. *oxyodon* (Miq.) Wenzig. Skan in Journ. Linn. Soc. Bot. 26: 517: (1899). *Quercus lineata* Blume var. *fargesii* (Franch.) Skan in Journ. Linn. Soc. Bot. 26: 517. (1899).

Hubei: Badong 554 (type of *Quercus lineata* Blume var. *grandiflora* Skan, collected September 1885). Donghu, 5632a. Jianshi, 5632c. Fang Xian, 5632e. **Sichuan:** Wushan, 5277a., 5632, 5632b., 5632d., 5632f.

A medium-sized tree to 15 metres tall with a broad flattened crown, composed of wide spreading branches and large leaves, common in Hubei and Sichuan between 600 and 1,200 metres above sea level. The leaves persist for two to three years and can be up to 22 cm long by 6 cm wide. Native to the eastern Himalaya, Myanmar and western and central China. Discovered in the Khasi hills, Meglalaya, NE India, *Quercus oxyodon* was introduced to cultivation nearly a century later by Wilson while collecting for Messrs Veitch in 1900. It is extremly rare in cultivation though there is a good tree at Caerhays Castle in Cornwall.

Quercus pachyloma O. Seemen (FAGACEAE)

Quercus sp. Henry in Trans. Asiat. Soc. Jap. 24 suppl:90. (1896) (List Pl. Formosa), Forbes & Hemsley in Journ. Linn. Soc. Bot. 519. (1899). *Cyclobalanopsis pachyloma* (O. Seemen) Schottky Forbes & Hemsley in Journ. Linn. Soc. Bot. 508. (1899).

Taiwan: Oluanpi, 1367.

A medium sized tree, found in forests in central and southern Taiwan. Also found in Fujian where it was collected by Warburg.

Quercus rex Hemsley (FAGACEAE) New species

in Hookers Icon. Pl. t. 2663. (1901), Hemsley in Journ. Linn. Soc. Bot. 34: 477. (1900).

Yunnan: Simao at 1,300 metres, 12665 (type of *Quercus rex* Hemsley).

Native to India, Myanmar, China (Yunnan), Laos and Vietnam. A very beautiful and aptly named species.

Quercus schottkyana Rehder & E. H. Wilson (FAGACEAE)

in Sargent Pl. Wilson. 3: 237. (1916).

Yunnan: In forests near Simao at 1,600 metres, 11698a., 11698b. Simao, 11770, 11770a.

Quercus serrata Murray (FAGACEAE)

Quercus glandulifera Blume Henry in Journ. China Br. Roy. Asiat. Soc. 22: 254. (1887) (Chinese Names of Plants); in Notes Econ. Bot. China 54. (1893), Skan in Journ. Linn. Soc. Bot. 26: 514. (1899), Rehder & E. H. Wilson in Sargent Pl. Wilson. 3: 212. (1916), Zheng in Journ. Wuhan Bot. Res. 2.(1): 26. (1984).

Hubei: Yichang, 157, 1227, 2760. Badong, 1419, 3906. Liantuo, 1890. **Sichuan:** South Wushan, 5496.

A deciduous tree to 15 metres tall, native to Japan, Korea and China, introduced to cultivation in 1893 by Professor Charles Sargent and later reintroduced by Wilson from Baoxing in Sichuan in 1909. A common tree throughout the Yangtze valley according to Wilson. On low hills and near centres of population it forms low scrub or coppice but when allowed to develop it forms trees to 25 metres with trunks to 3 metres in girth. In autumn the tree is particularly attractive when it takes on shades of orange-red and crimson. According to Henry in Hubei it was known as *peh fan li.* There are trees at Glasnevin and Kew.

Quercus sp. (FAGACEAE)

Henry in Trans. Asiat. Soc. Jap. 24 suppl: 90. (1896) (List Pl. Formosa).

Taiwan: Oluanpi, 1888.

Quercus sp. No. 1. (FAGACEAE)

Anonymous in Kew List of Determinations (volume 24).

Hubei: Yichang, 1323. Badong, 123. **Sichuan:** South Wushan, 5732.

Quercus sp. No. 2. (FAGACEAE)

Anonymous in Kew List of Determinations (volume 34).

Yunnan: Mengzi, 11841, 13645. Simao, 11597, 11613, 11727, 11936, 12850, 13022, 13042, 13055, 13056, 13345, 13348. Yuanjiang at 2,300 metres, 13249, 13341.

Probably several species.

Quercus sp. Specimen immaturum. (FAGACEAE)

Skan in Journ. Linn. Soc. Bot. 26: 522. (1899).

Yunnan: Mengzi, 9458, 9458a.

Quercus sp. With monstrous growth. (FAGACEAE)

Anonymous in Kew List of Determinations (volume 24).

Hubei: Liantuo, 4469.

Quercus spinosa David ex Franch. (FAGACEAE)

Henry in Notes Econ. Bot. China 54. (1893), Rehder & E. H. Wilson in Sargent Pl. Wilson. 3: 224. (1917), Zheng in Journ. Wuhan Bot. Res. 2.(1): 28. (1984).

Quercus ilex L. var. *spinosa* (David apud Franch.) Franch. Skan in Journ. Linn. Soc. Bot. 26: 516. (1899). *Quercus bullata* Seemen in Bot. Jahrb. xxiii. Beibl. 57: 48. (1897); xxix. 298. (1900). *Quercus semecarpifolia* Sm. var. *glabra* Skan non Franchet in Journ. Linn. Soc. Bot. 26: 151. (1899).

Hubei: Jianshi, 5981. Fang Xian, 5981a., 5981c. **Sichuan:** North Wushan, 5981b. **Yunnan:** On the summit of Ta hei Shan near Mengzi, 10217.

Under favourable conditions this oak forms a small tree to 10 metres tall but is usually seen in Hubei as a shrub from 1 to 5 metres tall and is common on limestone cliffs. In the mountains of Fang according to Henry it was known as the *t'ieh-chiang*, Henry also stated that Pére David (who discovered this species in February 1873 near Qingling in Shanxi) called it the *t'ieh-hsiang* or iron oak. *Quercus spinosa* was collected by the Glasnevin Central China Expedition (GCCE 518) in Xingshan in September 2004. Distributed in Myanmar and China.

Quercus variabilis Blume (FAGACEAE)

Rehder & E. H. Wilson in Sargent Pl. Wilson. 3: 219. (1916), Zheng in Journ. Wuhan Bot. Res. 2.(1): 28. (1984).

Quercus bungeana F. B. Forbes Skan in Journ. Linn. Soc. Bot. 26: 508. (1899).

Hubei: Yichang, 1956, 2291. Badong, 426 (June 1885). Liantuo, 2291a. Badong, 2877. **Yunnan:** In woods near Mengzi at 1,600 metres, 9913, 9913a. Yuanjiang, 13247.

A deciduous tree to 27 metres tall, native to China, Vietnam, Korea and Japan. *Quercus variabilis* was introduced to cultivation by Robert Fortune from Beijing in 1861 and reintroduced by Bretschneider from the same area in 1882. A common oak at low altitudes in Hubei and in southern Yunnan, Henry stated that in Hubei it was known as *hwa-k'o-li* and its tough durable wood was once much esteemed there for boat building, the cups of the acorns were used in dyeing silk-yarn black and the bark of the tree was used by the poorer classes to roof their houses. Young saplings were felled and on them the edible fungus, *Auricularia polytricha* (Mont.) Sacc. was cultivated. Cultivated at the National Botanic Gardens, Kilmacurragh from collections made by the Glasnevin Central China Expedition (GCCE 421) near the village of Bai Sha near Yesanguan in Badong.

Radermachera pentandra Hemsley (BIGNONIACEAE)

New species

in Hooker's Icon. Pl. 28: t. 2728. (1905).

Yunnan: Near Mengzi at 1,650 metres, 10909 (type, collected 13th November 1897).

Endemic to Yunnan.

Radermachera sinica (Hance) Hemsl. (BIGNONIACEAE)

Stereospermum sinicum Hance. Henry in Trans. Asiat. Soc. Jap. 24 suppl: 69. (1896) (List Pl. Formosa).

Taiwan: Oluanpi, 670. Wanchin, 1666. Tamsui, 1666a.

A very handsome deciduous tree named *shan-k'u-lien* according to Henry, as the leaves resemble those of *Melia*. Native to southern China and Japan.

Randia sp. (RUBIACEAE)

Anonymous in Kew List of Determinations (volume 34).

Yunnan: Mengzi, 9404, 9553. Simao, 11766, 12087.

Ranunculus cantoniensis DC. (RANUNCULACEAE)

Ranunculus pensylvanicus Anonymous non L. f. in Kew List of Determinations (volume 24).

Hubei: Yichang, 348, 656, 1263, 1315. Liantuo, 1931. Badong, 4838.

Ranunculus distans D. Don (RANUNCULACEAE)

Ranunculus laetus Wall. ex D. Don Anonymous in Kew List of Determinations (volume 34).

Yunnan: Mengzi, 10725.

Distributed in Afghanistan, Pakistan, Kyrgyzstan, Kazakstan, India, Bhutan, Nepal and China [Tibet (Xizang) and Yunnan].

Ranunculus japonicus Thunb. (RANUNCULACEAE)

Ranunculus acris Anonymous non L. in Kew List of Determinations (volume 24).

Hubei: Yichang, 589, 1232, 1272. Liantuo, 1988. Badong, 1442, 1844, 4718. Changyang, 7789.

In Chinese herbal medicine this species is currently used to subdue swelling, to relieve pain, to destroy parasites and prevent malaria.

Ranunculus sceleratus L. (RANUNCULACEAE)

Henry in Trans. Asiat. Soc. Jap. 24 suppl: 14. (1896)

(List Pl. Formosa).

Hubei: Xiling (Yichang) Gorge, 3328. **Taiwan:** Kaohsiung, without number.

Ranunculus sieboldii Miq. (RANUNCULACEAE)

Ranunculus arcuans S. S. Chien in Rhodora. 18: 190. (1916).

Hubei: Badong, 4039 (holotype of *Ranunculus arcuans* S. S. Chien).

Ranunculus sp. (RANUNCULACEAE)

Anonymous in Kew List of Determinations (volume 34).

Yunnan: Mengzi, 10565.

Ranunculus yunnanensis Franch. (RANUNCULACEAE)

Anonymous in Kew List of Determinations (volume 24).

Hubei: South Badong, 5310.

Raphanus raphanistrum L. (BRASSICACEAE)

Anonymous in Kew List of Determinations (volume 24).

Hubei: Badong, 3735.

Raphanus raphanistrum L. subsp. *sativus* (L.) Domin (BRASSICACEAE)

Raphanus sativus L. Henry in Journ. China Br. Roy. Asiat. Soc. 22: 256. (1887) (Chinese Names of Plants), Schulz in Engler, Pflanzenr. iv. 105: 205. (1919).

Hubei: Yichang, 810. Liantuo, 3813. **Hainan:** Environs of Haikou, cultivated 7969.

The radish, cultivated in the Three Gorges region where it is known colloquially as *hung-lo-po*. In Europe the radish plays only a minor role as a salad plant, whereas in China it is an important root, eaten raw, preserved or cooked. Chinese raised varieties are long and white and slow to mature, many will hold in the open ground throughout winter. It is also known as *lai fu zi* and the seeds are used in herbal medicine to treat abdominal pain or diarrhea. The radish is known to have grown in Egypt in 2780 B.C. and was rationed to the builders of the pyramids, it has grown in China since 500 B.C.

Raphiocarpus begoniifolius (H. Lévl.) B. L. Burtt (GESNERIACEAE) New species

Chirita chlamydata W. W. Smith in Notes Roy. Bot. Gard. Edinburgh 9: 170. (1916).

Yunnan: South of the Red River from Manmei at 1,950 metres, 9188a. In forests on Fengchunling at 2,300 metres, 9188a. In forests near Mengzi at 1,800 metres, 9413a. In a ravine near Mengzi at 1,950 metres, 9413b.

Rauvolfia sp. (APOCYNACEAE)

Anonymous in Kew List of Determinations (volume 34).

Yunnan: Simao, 11985.

Rauvolfia verticillata (Lour.) Baillon (APOCYNACEAE)

Li in Woody Flora of Taiwan 790. fig. 318. (1963).

Rauvolfia chinensis (Hance) Hemsley Henry in Trans. Asiat. Soc. Jap. 24 suppl: 59. (1896) (List. Pl. Formosa).

Hainan: Near the Kiungchow Pagoda, 7979. Environs of Kiungchow, 8250. Environs of Haikou, 8434. **Taiwan:** Oluanpi, without number. Wanchin, 529.

A small shrub, native to southern China, Indochina and Maleslia. A widespread spread species in Chinese Taiwan.

Reevesia pubescens Mast. (MALVACEAE)

Rehder & E. H. Wilson in Sargent Pl. Wilson. 2: 376. (1916), Anthony in Notes Roy. Bot. Gard. Edinburgh 15: 124. (1925—1927).

Reevesia wallichii Dunn non R. Br. ex Mast. in Journ. Linn. Soc. Bot. 484. (1911). *Reevesia sinica* E. H. Wilson in Journ. Arnold Arb. 5: 233. (1924).

Yunnan: Mengzi at 1,800 metres, 11510 (syntype of *Reevesia sinica* E. H. Wilson). Simao, 12062.

A handsome tree to 8 metres tall, native to India, Bhutan, Myanmar, China, Laos, Vietnam and Thailand (where it is rare). This species was introduced to cultivation by Wilson through the Arnold Arboretum from Sichuan and first flowered from this collection in the nursery of P. J. Berkmans Company, Augusta, Georgia in May 1907 and in England at Caerhays Castle, Cornwall in August 1924. The genus *Reevesia* commemorates John Reeves (died 1856) a tea inspector who lived in Macao and Guangzhou (then Canton). Reeves introduced many Chinese plants to cultivation.

Reevesia thyrsoidea Lindl. (MALVACEAE)

Reevesia thyrsoidea Lindl. Henry in Trans. Asiat. Soc. Jap. 24 suppl: 22. (1896) (List Pl. Formosa). *Reevesia taiwanensis* Chun & Hsue in Journ. Arnold Arb. 28: 330. (1920). *Reevesia formosana* Sprague in Kew Bull. Misc. Inf. 325. (1914), Anthony in Notes Roy. Bot. Gard. Edinburgh 15: 123. (1925—1927), Li in Woody Flora of Taiwan 565. fig. 222. (1963).

Taiwan: Oluanpi, 1970 (isotype of *Reevesia formosana* Sprague).

Rehmannia glutinosa (Gaert.) Libsch. (OROBANCHACEAE)

Anonymous in Kew List of Determinations (volume 24).

Hubei: Liantuo, 1989.

Rehmannia henryi N. E. Brown (OROBANCHACEAE) New species

in Kew Bull. Misc. Inf. 262. (1909), Prain in Curtis's Bot. Mag. 136: t. 8203. (1910), E. H. Wilson in A Naturalist in Western China 1: 35. (1913), Nelson in Pim, The Wood & the Trees 233. (1984), Rix in the Plantsman 8: 193-195. (1987), Morley in Glasra 3: 80. (1979), Nelson in A Heritage of Beauty 317. (2000).

Rehmannia piasezkii Henry non Maxim. in Journ. China Br. Roy. Asiat. Soc. 22: 242. (1887) (Chinese Names of Plants), Forbes & Hemsley in Journ. Linn. Soc. Bot. 23: 194. (1890).

Hubei: Yichang, 1157, 1376. Liantuo 3839.

A handsome species to about 50 cm tall, bearing violet-purple spikes of foxglove-like flowers. Described by Brown from living plants raised from seed at Kew which were gathered in Hubei by Wilson when collecting for the Arnold Arboretum, Mass. According to Henry it was known colloquially at Yichang as *feng t'ang kuan.* A hybrid between *R. henryi* and *R. glutinosa* is *R. × kewensis.*

Rehmannia piasezkii Maxim. (OROBANCHACEAE)

Forbes & Hemsley in Journ. Linn. Soc. Bot. 26: 194. (1890), Henry in Notes Econ. Bot. China 50. (1893).

Rehmannia glutinosa Lisbosch. var. *angulata* Oliver in Hooker's Icon. Pl. xvi. t. 1589. (1887). *Rehmannia angulata* (Oliv.) Hemsley in Journ. Linn. Soc. Bot. 26: 193. (1890), Henry in Notes Econ. Bot. China 50. (1893), Anon. in Gard. Chron. ser. 3, 33: 290. fig. (1903), J. H. Veitch in Hortus Veitchii 283. (1906), Rix in The Plantsman 8: 193-195. (1987), Nelson in A Heritage of Beauty 317. (2000).

Hubei: Yichang, 225, 1131. On a river bank at Yichang, 3600 (type).

According to Henry this was a common plant on the rocky banks of the gorges.

Reineckea carnea (Ander.) Kunth (ASPARAGACEAE)

Diels in Bot. Jahrb. 29: 249. (1900), Wright in Journ. Linn. Soc. Bot. 36: 113. (1903).

Hubei: Badong, 1000, 5933a. Antelope Glen, 3267, 4283. Jianshi, 5933. Xingshan, 6562.

A monotypic genus distributed through the eastern asiatic region. *Reineckia carnea* was collected by the Sino-Americam Expedition in the Shennongjia Forest District in 1980. Cultivated at the National Botanic Gardens, Glasnevin.

Reinwardtia indica Dumort. (LINACEAE)

Reinwardtia trigyna (Rchb.) Planch. Anonymous in Kew List of Determinations (volume 24 & 34).

Hubei: On the Banks of the Yangtze at Yichang, 1297. **Yunnan:** Chuyuan, 10607.

A glabrous subshrub to 90 cm tall, distributed in Pakistan, Nepal, northern India, China, Thailand and Vietnam.

Reissantia setulosa (A. C. Sm.) N. Hallé (CELASTRACEAE)

New species

Pristimera setulosa A. C. Sm. in Journ. Arnold Arbor. 26: 175. (1945).

Yunnan: By the Red River at Manhao, 9612 (type of *Pristimera setulosa* A. C. Sm., collected 19th June 1896).

Renanthera coccinea Lour. (ORCHIDACEAE)

Rolfe in Journ. Linn. Soc. Bot. 36: 35. (1903).

Hainan: 20 miles (32 km) north of Haikou, on rocks, 8147.

A beautiful epiphytic species found on trees at 800 to 1,400 metres above sea level.

Reynoutria japonica Houtt. (POLYGONACEAE)

Polygonum cuspidatum Sieb. & Zucc. Henry in Journ. China Br. Roy. Asiat. Soc. 22: 249. (1887) (Chinese Names of Plants), Forbes & Hemsley in Journ. Linn. Soc. Bot. 26: 336. (1891).

Hubei: Yichang, 14, 2747. Liantuo, 2649. Badong, 365 (June 1885), 2519, 4962.

Giant knotweed, according to Henry the roots of this species supplied a yellow dye at Badong, it was known there as *kan yen.* It is currently used in Chinese herbal medicine to relieve cases of cough.

Reynoutria multiflora (Thunb.) Moldenke (POLYGONACEAE)

Polygonum multiflorum Thunb. Henry in Journ. China Br. Roy. Asiat. Soc. 22: 243.(1887) (Chinese Names of Plants), Forbes & Hemsley in Journ. Linn. Soc. Bot. 26: 342. (1891), Henry in Trans. Asiat. Soc. Jap. 24 suppl: 76. (1896) (List Pl. Formosa), Schneider in Sargent Pl. Wilson. 3: 325. (1916).

Hubei: Yichang, 82. Badong, 2488. **Taiwan:** Wanshoushan, without number. Wanchin, without number. Tamsui, 1742.

According to Henry this species was known colloquially in the Three Gorges region as *ho shou wu* and when the root assumed a likeness to the human figure, it sold for a very large sum and was deemed an invaluable drug. It is currently used in Chinese herbal medicine to treat liver ailments, kidney ailments, premature ageing and infertility.

Rhamnaceae (RHAMNACEAE)

Anonymous in Kew List of Determinations (volume 34).

Yunnan: At Hsienkei on the Red River near the Vietnamese border, 9568.

Rhamnella martini (H. Lév.) C. K. Schneid. (RHAMNACEAE)

in Sargent Pl. Wilson. 2: 225. (1914).

Yunnan: In mountains near Mengzi at 2,800 metres, 10929.

Nepal to China.

Rhamnus bodinieri H. Léveille (RHAMNACEAE)

Rhamnus bodinieri H. Léveille f. *silvicola* Schneider in Sargent Pl. Wilson. 2: 247. (1916).

02

Yunnan: In woods near Mengzi at 2,200 metres, 10464 (syntype of *Rhamnus bodinieri* H. Léveille f. *silvicola* Schneider), 10464a., (syntype of *Rhamnus bodinieri* H. Léveille f. *silvicola* Schneider). On rocky mountains near Mengze at 2,600 to 2,800 metres, 10814, 10814a.

Nepal to China and Vietnam.

Rhamnus formosana Matsumura (RHAMNACEAE)

in Bot. Mag. Tokyo 12: 7: 22. (1898), Matsum. & Hayata in Journ. Coll. Sci. Univ. Tokyo 32: 88. t. 8. (1906), Kanehira, Form. Trees. Rev. Ed. f. 379. (1936), Li in Woody Flora of Taiwan 513. fig. 199. (1963).

Rhamnus sp. nova. near *Rhamnus javanica* Miq. Henry in Trans. Asiat. Soc. Jap. 24 suppl: 27. (1896) (List Pl. Formosa).

Taiwan: Kaohsiung, 298, 298a. Wanchin, 1172 (type).

A more or less scandent shrub to 10 metres, tall endemic to Taiwan where it grows in broad-leaved forest and is common on the southern part of the island.

Rhamnus globosa Bunge (RHAMNACEAE)

Zheng in Journ. Wuhan Bot. Res. 2.(1): 140. (1984).

Hubei: San You Dong Glen, 3612. Yichang, 3613. Xiling (Yichang) Gorge, 3593.

Rhamnus hemsleyana C. K. Schneider (RHAMNACEAE) New species

in Notizbl. Konigl. Bot. Gart. Berlin. 5: 78. (1908), Schneider in Sargent Pl. Wilson 2: 234. (1914).

Rhamnus napalensis Zheng non (Wall) Laws. in Journ. Wuhan. Bot. Res. 2. (1): 139. (1984).

Hubei: Jianshi, 5677a., 7462. **Sichuan:** South Wushan, 5677 (type).

Rhamnus heterophylla Oliver (RHAMNACEAE) New species

in Hooker's. Icon. Pl. 18: t. 1759. (1888), Zheng in Journ. Wuhan Bot. Res. 2.(1): 139. (1984).

Hubei: Yichang, 3083 (syntype), 3083a., (syntype). Antelope Glen, 3312 (type).

Rhamnus iteinophylla C. K. Schneider (RHAMNACEAE) New species

in Notizbl. Bot. Gart. Berlin. 5: 76. (1908), Schneider in Illus. Handb. Laubholzk. 2: 281. (1912); in Sargent Pl. Wilson 2: 239. (1914), Zheng in Journ. Wuhan Bot. Res. 2.(1): 139. (1984).

Hubei: Jianshi, 5915, 5915b. Fang Xian, 5915a. Without locality, 5915c., (type).

Rhamnus lamprophylla C. K. Schneider (RHAMNACEAE) New species

in Notizbl. Bot. Gart. Berlin. 5: 78. (1908), Schneider in Illus. Handb. Laubholzk. 2: 289. (1912); in Sargent Pl. Wilson 2: 252. (1914).

Hubei: Xingshan, 6504.

Rhamnus leptophylla C. K. Schneider (RHAMNACEAE) New species

in Notizbl. Bot. Gart. Berlin. 5: 77. (1908), Schneider in Illus. Handb. Laubholzk. 2: 285. (1912); in Sargent Pl. Wilson 2: 239. (1914), Zheng in Journ. Wuhan Bot. Res. 2.(1): 139. (1984).

Hubei: Yichang, 1493 (type), 3401 (October 1887), 3407, 3407a. Jianshi, 5968. South Badong, 6033a. Without locality, 3401a. **Sichuan:** South Wushan, 6033. North Wushan, 7048a.

A shrub to 2.5 metres tall, abundant around Yichang during Henry's time based there. Seeds of Wilson's 410 from Xingshan reached Glasnevin in 1908.

Rhamnus napalensis (Wall.) M. A. Lawson (RHAMNACEAE)

Zheng in Journ. Wuhan Bot. Res. 2.(1): 139. (1984).

Rhamnus paniculiflora C. K. Schneider in Sargent Pl. Wilson. 2: 223. (1914).

Sichuan: South Wushan, 5677. **Hubei:** Changyang, 7483. **Hainan:** Environs of Kiungchow, 8299. Environs of Kiungchow, 8707. **Yunnan:** Mountains to the west and east of Simao in forest at 1,800 to 2,000 metres, 1240, 12040a. Simao, 13364.

Rhamnus rugulosa Hemsley (RHAMNACEAE) New species

in Journ. Linn. Soc. Bot. 23: 129. (1886), Henry in Journ. China Br. Roy. Asiat. Soc. 22: 237. (1887) (Chinese Names of Plants), Bretschneider in Hist. Eur. Bot. Disc. China 2: 781. (1898), Schneider in Sargent Pl. Wilson 2: 238. (1916), Zheng in Journ. Wuhan Bot. Res. 2.(1): 140. (1984).

Hubei: In glens near Yichang, 448, 586, 1025, 1489, 2094, 3610, 3611. Glens off the Xiling (Yichang) Gorge, 3609. In the San You Dong Glen, 1583 (May 1886), 1689 (August 22nd 1886). In the Monastery Valley near Yichang, 1653 (July 1886). Liantuo, 1976, 2696 (type), 4560. Without locality 3593a.

Rhamnus sp. No. 1. (RHAMNACEAE)

Anonymous in Kew List of Determinations (volume 24).

Hubei: Yichang, 1169, 1395. Changyang, 7693.

Rhamnus sp. No. 2. (RHAMNACEAE)

Anonymous in Kew List of Determinations (volume 34).

Yunnan: Mengzi, 10378. Simao, 12845.

Rhamnus sp. No 3. (RHAMNACEAE)

Rhamnus parvifolius Bunge Henry in Journ. China Br. Roy. Asiat. Soc. 22: 237. (1887) (Chinese Names of Plants).

Hubei: Yichang, 1209. Liantuo, 1897.

Rhamnus utilis Decne. (RHAMNACEAE)

Forbes & Hemsley in Journ. Linn. Soc. Bot. 23: 128. (1886), Schneider in Illus. Handb. Laubholzk. 2: 289. (1912), Schneider in Sargent Pl. Wilson 2: 240. (1914), Zheng in Journ. Wuhan Bot. Res. 2.(1): 140. (1984).

Rhamnus davurica Forbes & Hemsley non Pall. in Journ. Linn. Soc. Bot. 23: 128. (1886), Zheng in Journ. Wuhan Bot. Res. 2.(1): 138. (1984).

Hubei: Yichang, 385, 991, 1162, 3605, 3606, 3607, 3614. Jianshi, 5990, 6018, 6020. San You Dong Glen, 3608. Antelope Glen, 4217. Badong, 2802, 2860, 5454. Liantuo, 4433. Changyang, 5224, 5240. Xingshan at 1,400 metres, 6553. **Sichuan:** South Wushan, 5895, 5601.

Green dyes were once obtained from the leaves of this species.

Rhamnus virgata Roxb. (RHAMNACEAE)

Rhamnus leptophylla Schneid. var. *milensis* C. K. Schneider in Sargent Pl. Wilson. 2: 250. (1914).

Yunnan: In mountain forest near Milê, 10021 (holotype of *Rhamnus leptophylla* C. K. Schneid. var. *milensis* C. K. Schneider). In mountain forest near Mengzi at 2,200 metres, 10021a. Mountains to the west of Simao at 2,000 metres, 11890.

Distributed in Afghanistan, Nepal, India, western China and Indochina.

Rhaphidophora hookeri Schott (ARACEAE)

Engler & Krause in Engler, Pflanzenr. iv. 23B: 32. (1908).

Yunnan: Simao at 1,600 metres, 11720.

Rhaphiolepis indica (L.) Lindl. (ROSACEAE)

Anonymous in Kew List of Determinations (volume 24).

Hainan: Without locality, 8481.

An evergreen shrub, widespread across India, Myanmar, southern China, Korea and Japan.

Rhaphiolepis indica (L.) Lindl. var. shilanensis (Kanehira) Yuen P. Yang & H. Y. Liu ROSACEAE)

Rhaphiolepis indica Lindl. var. Henry in Trans. Asiat. Soc. Jap. 24 suppl: 41. (1896) (List Pl. Formosa). *Rhaphiolepis indica* (L.) Lindl. var. *hiiranensis* (Kanehira) Li Li in Woody Flora of Taiwan 298. (1963).

Taiwan: Oluanpi, 643, 643a., 1323.

A shrub or small evergreen tree, in Taiwan this variety is distributed on the Hengchun Peninsula (on the island's southern tip).

Rhapis sp. (ARECACEAE)

Anonymous in Kew List of Determinations (volume 34).

Yunnan: Mengzi, 10173.

Rheum officinale Baill. (POLYGONACEAE)

Henry in Kew Bull. Misc. Inf. 226. (1889), Forbes & Hemsley in Journ. Linn. Soc. Bot. 26: 353. (1891), Henry in Notes Econ. Bot. China 33. (1893); in Flora & Sylva 1: 217. (1903).

Rheum sp. Henry in Journ. China Br. Roy. Asiat. Soc. 22: 271. (1887) (Chinese Names of Plants).

Hubei: Badong, 4086. Shennongjia at 2,300 to 3,250 metres 6850.

The *da huang* (according to Henry) or Chinese rhubarb, the root of which is a powerful laxative and has been used in Chinese herbal medicine since at least 2700 B.C. when it was mentioned in the Chinese herbal of *Pen-King*. It was first brought to Europe by Marco Polo and by the sixteenth century it was exported from China in enormous quantities to Europe as a cure for venereal disease. In present day China one of its uses is as a purgative, when used with the roots of *Aconitum carmichaelii* the preparation is used to treat constipation. Marco Polo (speaking of *Rheum tanguticum*) stated, 'All over the mountains of the province of Tangut, rhubarb is found in great abundance, and thither merchants come to buy it and carry it from thence all over the world.' According to Henry *Rheum officinale* was cultivated in Badong where it was known as *ta-huang*. The roots were sun-dried for the foreign market, sent down the gorges to Yichang and from there reached distant destinations such as London and Dublin. Henry's specimens from Shennongjia were the first wild specimens to reach the Kew herbarium and he described it as "a glorious plant having a remarkable rhizome 3 or 4 feet long, with a bright red cortex, but golden-yellow interiorly." It is known today that Shennongjia Forest District (where Henry collected his material) is the richest single home of medicinal plants in China. Henry was the first westerner to have visited the region. According to Henry, Antwerp Pratt found this rhubarb near Kangding on the Tibet (Xizang)-Sichuan frontier and at Sining in Gansu (once a collecting base for Farrer and Purdom). Seeds were taken from Henry's 6850 in the Kew herbarium on April 10th 1889 and sent for propagation.

Rhizophora mucronata Poir (RHIZOPHORACEAE)

Henry in Trans. Asiat. Soc. Jap. 24 suppl: 42. (1896) (List Pl. Formosa), Li in Woody Flora of Taiwan 634. fig. 251. (1963).

Taiwan: Kaohsiung Lagoon, in tidal mangrove swamps, without number.

A medium-sized tree found on the coasts of the paleotropics. An inhabitant of mangrove swamps, Henry stated it was known in Taiwan as *wu-chio-li*.

Rhizophora sp. (RHIZOPHORACEAE)

Anonymous in Kew List of Determinations (volume 34).

Yunnan: Mengzi, 9517.

Rhodiola yunnanensis (Franch.) S. H. Fu (CRASSULACEAE)

Sedum henryi Diels in Bot. Jahrb. Syst. 29: 361. (1900). *Sedum yunnanense* Franchet var. *henryi* (Diels) Hamet in Notes Roy. Bot. Gard. Edinburgh 8: 145. (1913). *Rhodiola henryi* (Diels) S. H. Fu in Acta Phytotax. Sin. Addit. 1: 126. (1965). *Sedum yunnanense* Franch. var. *valerianoides* Hamet Hamet in Notes Roy. Bot. Gard. Edinburgh 8: 145. (1913).

Sichuan: South Wushan, 5411 (isosyntype of *Sedum henryi* Diels), 5411b., 5411c., (type of *Sedum henryi* Diels). **Hubei:** South Badong, 5411a. **Yunnan:** Mountains to the north of Mengzi at 1,950 metres, 10204.

North-east India to Myanmar and China.

Rhodobryum giganteum (Schwägr.) Paris (BRYACEAE)

Bryum giganteum (Schwägr.) Arn. Salmon in Journ. Linn. Soc. Bot. 34: 457. (1900).

Hubei: Shennongjia at 2,200 to 3,100 metres, 6796 (August 1888), 6904. **Yunnan:** Mengzi, 13711. (Moss).

Rhododendron aliciae S. O'Brien nom. nov. (ERICACEAE) New species

Rhododendron excellens Hemsley & E. H. Wilson in Kew Bull. Misc. Inf. 113. (1910), Hutchinson in Notes Roy. Bot. Gard. Edinburgh 12: 29. (1919), Davidian in Rhodendron Spec. 1: 263. (1982), O'Brien in Augustine Henry, An Irish Plant Collector in China 30. (2002).

non *Rhododendron × excellens* Van Geert in Nursery Cat. (Auguste Van Geert) 26. (1854).

Yunnan: South of the Red River from Manmei, 13666.

Rhododendron excellens Hemsley & E. H. Wilson is an invalid name for this species since the specific epithet has been used earlier for an entirely different plant. *Rhododendron × excellens* Van Geert was first published by the eminent Ghent horticulturist Auguste van Geert in his nursery catalogue for 1854 for an existing *Rhododendron* hybrid. Hemsley & E. H. Wilson's later (1910) *Rhododendron excellens* must therefore be considered a homonym. In coining a new name I propose *Rhododendron aliciae* S. O'Brien, thus commemorating Augustine Henry's second wife Alice (Elsie) Henry (1882—1956), who, following her husband's death in March 1930, assembled and arranged his private herbarium collection of approximately ten thousand specimens, a task that took her eight years to complete. This she presented to the National Botanic Gardens, Glasnevin in Dublin where it forms The Augustine Henry Forestry Herbarium and contains material gathered by other famous collectors from across the globe.

She and Augustine lived in a large town house in Ranelagh, in the south-side suburbs of Dublin where in their garden they grew a wide range of Chinese plants named from Henry's many discoveries. She was also a noted plant authority in her own right.

Introduced to cultivation in recent years, *Rhododendron aliciae* forms a shrub of 3.5 metres tall, bearing 3 to 4 tubular funnel-shaped white flowers to 10 cm long in a terminal inflorescence in May. Henry's Chinese collector only ever made one collection of this magnificent plant and for a century this was the only representation of the species in herbaria. It is extremely rare in the wild and listed as Vulnerable in the *The Red List of Rhododendrons*. In Ireland it is grown at Mount Congreve, the John F. Kennedy Arboretum, the National Botanic Gardens, Kilmacurragh and at Glasnevin where there is a very good plant in the west-wing of the Curvilinear Range of glasshouses. A beautiful plant, native to China [Yunnan and (questionably) Guizhou] and Vietnam.

Rhododendron arboreum Sm. subsp. ***delavayi*** (Franch.) D. F. Chamb. (ERICACEAE)

Rhododendron delavayi Franch. Hemsley & E. H. Wilson in Kew Bull. Misc. Inf. 108. (1910), Watson in Gard. Chron. ser. 3, 35: 262. (1904).

Yunnan: Mengzi, 10983. Mountains to the east of Mengzi at 2,650 metres, 11330. Chuyuan, 10983a. Near Yuanjiang at 2,000 metres, 11060. Simao, 13208.

A shrub to 4 metres tall in cultivation but in its native range of the mountains of western Yunnan, Guizhou, India, Bhutan, Vietnam, Thailand and Myanmar it forms a tree to 12 metres tall. The Chinese counterpart of the Himalayan *Rhododendron arboreum*, it was described from a specimen collected by Delavay in Yunnan in April 1883. From *Rhododendron arboreum* it may be distinguished by its spongy (not thin and plastered) indumentum, the surface which is more or less fissured. It was first introduced to France in 1884 and reached Kew in 1889. It was reintroduced several times by Forrest, Kingdon Ward and Rock. The first flowering of this species in the British Isles and Ireland occured at Kilmacurragh (an annex garden of the National Botanic Gardens, Glasnevin), the garden of Thomas Acton in County Wicklow in 1904.

Rhododendron augustinii Hemsley (ERICACEAE) New species

in Journ. Linn. Soc. Bot. 26: 19. (1889), Bretschneider in Hist. Eur. Bot. Disc. China 2: 786. (1898), Bean in Flora and Sylva. 3: 162. (1905), Mottet in Rev. Hort. 18. fig. 16. (1909), Hemsley & E. H. Wilson in Kew Bull. Misc. Inf. 114. (1913), Raffill in Gard. Chron. ser. 3, 52: 4. fig. 3. (1912), Rehder & E. H. Wilson in Sargent Pl. Wilson. 1:

524. (1913), Hemsley in Curtis's Bot. Mag. 139: t. 8497. (1913), Besant in Gard. Chron. ser. 3, 98: 335. (1935), Morley in Glasra 3: 80. (1979), Davidian in Rhododend. Spec. 1: 335. (1982), Zheng in Journ. Wuhan Bot. Res. 2.(1): 168. (1984), Nelson in Pim, The Wood & the Trees 234. (1984), Bean in Trees & Shrubs 3: 602. (1992), Nelson in A Heritage of Beauty 317. (2000), O' Brien in Augustine Henry, An Irish Plant Collector in China 30. (2002), O' Brien in Ir. Garden 13(5): 53. (2004).

Hubei: Badong, 1420, 1421, 3736, 5414.

An evergreen shrub to 3 metres tall and occasionally up to 7 metres tall in its native habitat in central and western China. Flowers are carried in clusters of 2 to 6 in a terminal inflorescence, widely funnel shaped and varying in colour from any of the following, pink, rose, pale lavander-rose, lilac-purple, purple, lavander-purple, dark lavander-blue or violet. First found by Henry in 1886 in the mountains of Badong, this is one of his finest discoveries. This beautiful shrub first flowered in the collection of Maurice de Vilmorin at Les Barres from seeds sent there by Père Farges. It was introduced to Britain by Wilson in 1900 who sent seeds from Changyang, a region to the south-west of Yichang. It is a common species in Hubei, where it is extremly variable in colour. Wilson collected from rose-pink plants near Yichang and lilac-purple plants in Fang Xian. This species has the added advantage of being tolerant of soils with a high ph.

Rhododendron henryi Hance commemorates not Augustine Henry but the Rev'd B. C. Henry of the American Presbyterian Board of Missions who found it in Guangdong in March 1881. B. C. Henry was based at Guangzhou (formerly Canton) where he arrived in 1873. According to Bretschneider he made extensive travels in Guangdong and also collected in Hainan. These collections were carried out in the early 1880s. Other plants that commemorate this collector that may be confused with Augustine Henry are as follows, *Elaeocarpus henryi* Hance, *Eugenia henryi* Hance, *Machilus henryi* Hemsley, *Antidesma henryi* Hemsley while *Macaranga henricorum* Hemsley (now *Macaranga denticulata* (Blume) Muell.-Arg. was named to commemorate both Augustine Henry and B. C. Henry.

Rhododendron auriculatum Hemsley (ERICACEAE)
New species

in Journ. Linn. Soc. Bot. 26: 20. (1889), Bretschneider in Hist. Eur. Bot. Disc. China 2: 786. (1898), E. H. Wilson in Journ. Roy. Hort. Soc. 24: 657. (1905), Hemsley & E. H. Wilson in Kew Bull. Misc. Inf. 108. (1910), Schneider in Illus. Handb.Laubholzk. 2: 492. (1912), Hemsley in Curtis's Bot. Mag. 145: t. 8786. (1919), Morley in Glasra 3: 80. (1979), Zheng in Journ. Wuhan Bot. Res. 2.(1): 168. (1984), Nelson in Pim, The Wood & the Trees 220. (1984), Davidian

in Rhodo. Spec. 2: 103. (1989) Bean in Trees & Shrubs 3: 604. (1992), Nelson in A Heritage of Beauty 317. (2000), O' Brien in Augustine Henry, An Irish Plant Collector in China 30. (2002).

Hubei: Yichang, 513, 5029. South Badong, 7562. Changleping, 7725.

A small tree to 10 metres tall in its native habitat in central China where it was first found by Henry in 1885. The flowers are pure white, six to eight in a truss and 10 cm long. *Rhododendron auriculatum* is the latest of all Rhododendron species to flower in Hubei and was introduced to cultivation by Wilson for Messrs Veitch in 1901. Wilson stated this species was common on precipitous cliffs north of the Yangtze at elevations above 1,650 metres. *Rhododendron auriculatum* first flowered in cultivation at Caerhays Castle, Cornwall, (the garden of J. C. Williams) in 1912. Distributed in Guizhou, Hubei, Hunan, Jiangxi, Shaanxi and Sichuan.

Rhododendron ciliicalyx Franch. (ERICACEAE)
In Harvard University Herbarium (A.).

Yunnan: South of the Red River from Mengzi, 13681.

An evergreen shrub to 3 metres tall, sometimes epiphytic and producing teminal trusses of three to four white funnel shaped flowers stained yellow on the upper part of the tube in March and April. Discovered by Delavay in 1884 on Mt. Peechaho to the north of Lake Erhai at Dali in north west Yunnan and introduced by him to Paris. It then reached Kew in 1892 and first flowered there in 1900. *Rhododendron cilicalyx* was later reintroduced by Forrest from China (Yunnan) and from Myanmar by Kingdon Ward.

Rhododendron concinnum Hemsl. (ERICACEAE)
Hemsley & E. H. Wilson in Kew Bull. Misc. Inf. 115. (1910).

Sichuan: Henry's Chinese collector with Pratt, no locality stated but may be Leshan, Emei Shan, Wa Shan or Kangding, 8874.

A large shrub or small tree to 4.5 metres tall, flowers are carried in May in terminal trusses of up to eight purple or reddish-purple, funnel-shaped flowers. Discovered on Emei Shan, Sichuan by Faber in 1887 and introduced to cultivation by Wilson in 1904.

Rhododendron decorum Franch. (ERICACEAE)
Hemsley & E. H. Wilson in Kew Bull. Misc. Inf. 109. (1910), Rehder & E. H. Wilson in Sargent Pl. Wilson. 1: 541. (1913).

Yunnan: Grassy mountains near Mengzi at 2,000 metres, 9155. Mountains to the east of Mengzi at 1,900 metres, 9155a.

An evergreen shrub or small tree to 6 metres tall, producing terminal clusters of up to ten fragrant white or rose coloured flowers in early summer. Discovered in Baoxing, Sichuan by Père Armand David (though described from a collection later made by Père Delavay on the Cangshan Range above Dali), it was introduced to cultivation by E. H. Wilson from Kangdging in Sichuan in 1901. Farrer and Cox also later collected it on the Yunnan-Myanmar border. An adundant species in Yunnan province, particularly on the Cangshan Range above Dali. Also distributed in Guizhou, Sichuan and Tibet (Xizang). The flowers of this species (unlike many others which are poisonous) are edible and used as a vegetable in Yunnan.

Rhododendron dendrocharis Franch. (ERICACEAE)

Hemsley & E. H. Wilson in Kew Bull. Misc. Inf. 115. (1910).

Sichuan: Henry's Chinese collector with Pratt, no locality stated but may be any of the following, Leshan, Emei Shan, Wa Shan or Kangding, 8857.

An epiphytic (sometimes terrestrial) shrub to 70 cm tall, carrying pairs of bright rosy-red flowers in spring. These are widely funnel shaped, to 2.5 cm across and carried in the upper leaf axils. Discovered by Père Armand David at Baoxing in western Sichuan in May 1869, the specific epithet, *dendrocharis*, means tree-adorning and aptly describes the habit of this plant. In its native western Sichuan this species usually grows as an epiphyte on *Abies* and *Tsuga* at 2,600 to 3,000 metres. Recently introduced to cultivation.

Rhododendron emarginatum Hemsley & E. H. Wilson (ERICACEAE) New species

in Kew Bull. Misc. Inf. 118. (1910), Davidian in Rhododend. Spec. 1: 389. (1982).

Yunnan: On the mountains to the south-west of Mengzi at 680 metres, 9166 (type).

A shrub to 60 cm tall, flowers yellow, campanulate, to 1.3 cm long and carried in the axils of leaves in pairs or solitary in a terminal inflorescence. Known only from a single collection made by Henry in 1896 (not 1900 as stated by Davidian), it has recently been introduced to cultivation. Endemic to China and distributed in the western provinces of Guangxi, Guizhou and Yunnan.

Rhododendron formosanum Hemsley (ERICACEAE) New species

in Kew Bull. Misc. Inf. 185. (1895), Henry in Trans. Asiat. Soc. Jap. 24 suppl: 57. (1896) (List Pl. Formosa), Matsum. & Hayata in Journ. Coll. Sci. Univ. Tokyo. 22: 218. (1906), Hemsley & E. H. Wilson in Kew Bull. Misc. Inf. 108. (1910), Hayata in Icon. Pl. Form. 3: 132. (1913), Hutchinson in Millias, Rhodend. 168. (1917), E. H. Wilson in Journ. Arnold Arb. 6: 165. (1925), Kanehira in Form. Trees rev. ed. 538. f. 497. (1936), Li in Woody Flora of

Taiwan 689. fig. 288. (1963), Hsu in Proc. Nat. Sci. Counc. 6: 23. (1973), Ying in Quart. Journ. Chin. For. 9: 117. (1976), Li in Li et al. Fl. Taiwan 4: 25. (1978), Yamazaki in Journ. Jap. Bot. 56: 365. (1981), Davidian in Rhodo. Spec. 2: 92. (1989)

Taiwan: Oluanpi, 1976 (type).

A small tree to 10 metres tall with trunk girths of up to one metre. The funnel-shaped flowers which appear in April and May may be white or pink with purple-brown spots and are carried in a lax racemose umbel of 7 to 20. Endemic to Taiwan where it grows in mixed forest along streams between 800 and 2,000 metres. Li describes it as being the most beautiful species to inhabit the island. Introduced to cultivation by Patrick and Hsu in 1969 from near Chi Tou, Nantou.

Rhododendron fortunei Lindl. (ERICACEAE)

Forbes & Hemsley in Journ. Linn. Soc. Bot. 26: 23. (1889).

Hubei: Badong, 497. **Sichuan:** South Wushan, 6032.

An evergreen shrub or small tree to 9 metres tall, carrying terminal umbels of 6~12 broadly bell-shaped white or feint-pink flowers. in May and June the specific epithet commemorates the Scottish plant hunter, Robert Fortune.

Rhododendron fortunei Lindl. subsp. *discolor* (Franch.) D. F. Chamb. Houlstonii Group (ERICACEAE)

Rhododendron fortunei Henry non Lindl. in Journ. China Br. Roy. Asiat. Soc. 22: 266. (1887) (Chinese Names of Plants), Forbes & Hemsley in Journ. Linn. Soc. Bot. 26: 23. (1889), Hemsley & E. H. Wilson in Kew Bull. Misc. Inf. 109. (1910). *Rhododendron fortunei* Lindl. var. *houlstonii* (Rehd. & Wils.) Rehder & E. H. Wilson in Sargent Pl. Wilson. 1: 541. (1913). *Rhododendron houlstonii* Hemsley & E. H. Wilson Zheng in Journ. Wuhan Bot. Res. 2.(1): 168. (1984), Davidian in Rhodo. Spec. 2: 198. (1989), Bean in Trees & Shrubs 3: 685. (1992), O' Brien in Augustine Henry, An Irish plant collector in China 32. (2002).

Hubei: South Badong, 5354.

An evergreen shrub to 4 metres tall, flowers eight to a truss, flesh pink and 7 cm across. Discovered by Henry in 1888 on the borders of Changyang and Badong disctricts. It is native to western Hubei and Sichuan where it grows in forests between 1,400 and 2,100 metres. Introduced by E. H. Wilson in 1900 from Xingshan (a town to the north of Yichang), it first flowered in cultivation in Veitch's Coombe Wood nursery in 1913. It has remained rare in cultivation. The group epithet commemorates Mr. G. Houlston of the Chinese Imperial Maritime Customs at Yichang, who was a friend of Wilson. According to Henry, in the Three Gorges region it was known colloquially as *yeh p'i pa*.

02

Rhododendron hancockii Hemsley (ERICACEAE)

Hemsley & E. H. Wilson in Kew Bull. Misc. Inf. 116. (1910), Davidian in Rhodo. Spec. 3: 312. (1992).

Yunnan: Mountains to the north of Mengzi at 1,600 metres, 10523. Mengzi at 2,000 metres 10523a., 10523b.

A shrub or small tree to 4.5 metres tall, flowers white (sometimes pink) with a yellow blotch at the base, funnel-shaped to 7.5 cm long and solitary, in the uppermost axils of the leaves. Discovered by the Irish plant collector and Imperial Maritime Customs officer, William Hancock (1847—1914) near Mengzi in 1894 and found again by Henry two years later. Endemic to China (Guangxi and south-east Yunnan) where it grows on forest verges and open slopes between 1,500 and 2,000 metres. Recently introduced to cultivation.

Rhododendron hypoglaucum Hemsley (ERICACEAE)
New species

in Journ. Linn. Soc. Bot. 26: 25. (1889), Bretschneider in Hist. Eur. Bot. Disc. China 2: 786. (1898), Rehder & E. H. Wilson in Sargent Pl. Wilson. 1: 527. (1913), Bean in Trees & Shrubs 3: 361. (1936), Zheng in Journ. Wuhan Bot. Res. 2(1): 169. (1984), Davidian in Rhodo. Spec. 2: 95. (1989).

Rhododendron argyrophyllum Franch. ssp. *hypoglaucum* (Hemsl.) Chamberlain in Notes Roy. Bot. Gard. Edinburgh 2: 95. (1989), Nelson in A Heritage of Beauty 317. (2000), O' Brien in Augustine Henry, An Irish Plant Collector in China 30. (2002).

Hubei: Badong, 723 (type). Fang Xian, 6682.

A shrub to 6 metres tall, flowers carried in May, white or pink and spotted. According to Wilson who introduced this species in 1900, it was locally abundant in Hubei in thin woods and in open countryside amoungst rocks.

Rhododendron insigne Hemsley & E. H. Wilson (ERICACEAE) New species

in Kew Bull. Misc. Inf. 113. (1910), Bean in Trees & Shrubs 3: 690. (1992), O' Brien in Augustine Henry, An Irish Plant Collector in China 32. (2002).

Sichuan: Wa Shan, Henry's Chinese collector with Pratt, 8859.

An evergreen shrub to 5 metres tall in its native Sichuan, where it grows at elevations between 2,100 and 3,000 metres. A handsome sight in May when carrying trusses of up to 15 deep rosy pink bell-shaped flowers, maroon spotted on the upper lobes. Introduced to cultivation by Wilson in 1908 from Wa Shan.

Rhododendron irroratum Franch. subsp. *pogonostylum* (Balf. f. & W. W. Sm.) Chamberlain (ERICACEAE)

New subspecies. in Notes Roy. Bot. Gard. Edinburgh. 36: 117. (1978).

Rhododendron irroratum Hemsley & E. H. Wilson non Franch. in Kew Bull. Misc. Inf. 112. (1910). *Rhododendron gymnanthum* Rehder & E. H. Wilson non Diels in Sargent Pl. Wilson. 1: 539. (1913). *Rhododendron adenostemonum* I. B. Balfour & W. W. Smith in Trans. & Proc. Bot. Soc. Edinburgh. 27: 174. (1917). *Rhododendron pogonostylum* Balf. f. & W. W. Sm. in Trans. Bot. Soc. Edin. 27: 210. (1917), Davidian in Rhodo. Spec. 2: 273. (1989).

Yunnan: Mountains to the north of Mengzi at 2,300 metres, 11066 (type of *Rhododendron adenostemonum* Balf. f. & W. W. Sm.). In the mountains to the north of Mengzi at 2,600 metres, 11067, 11067b. In the mountains to the north of Mengzi at 2,800 metres, 11067a.

A shrub or small tree to 6 metres tall, carrying terminal, racemose umbels of pink or red tubular bell-shaped flowers in April and May. This subspecies is native to Vietnam and China (Guizhou and Yunnan) where it grows in woods with *Pinus yunnanensis* and *Albizia julibrissin* and on rocky slopes amoungst scrub between 2,100 and 3,000 metres. It was later collected by George Forrest and Joseph Rock in central and western Yunnan. Neither plant hunters appear to have collected seeds and it has only recently been introduced to cultivation. The specific epithet referrs to the bearded style.

Rhododendron lutescens Franch. (ERICACEAE)

Rehder & E. H. Wilson in Sargent Pl. Wilson. 1: 516. (1913).

Sichuan: Henry's Chinese collector with Pratt, no locality stated but may be Leshan, Emei Shan, Wa Shan or Kangding, 8862.

An evergreen shrub from to 2 to 6 metres tall, flowers appear in February and March, these are pale yellow with green spots and are produced singly or in carried in clusters of three in the axils of the uppermost leaves. Native to western Sichuan and north-east Yunnan and discovered by Père Armand David in Baoxing in Sichuan. Introduced to cultivation by E. H. Wilson through Messrs Veitch in 1904. One of the best and earliest flowered Subsection Triflora *Rhododendron*.

Rhododendron mariesii Hemsley & E. H. Wilson (ERICACEAE)

in Kew Bull. Misc. Inf. 244. (1907), Hutchinson in Bot. Mag. 134: t. 8206. (1908), Hemsley & E. H. Wilson in Kew Bull. Misc. Inf. 119. (1910), Rehder & E. H. Wilson in Sargent Pl. Wilson. 1: 548. (1913), Rehder & E. H. Wilson in A Monograph of Azaleas 80. (1921), Zheng in Journ. Wuhan. Bot. Res. 2.(1): 548. (1984), Nelson in Pim, The Wood & the Trees 222. (1984); in A Heritage of Beauty 317. (2000), O' Brien in Augustine Henry, An Irish Plant collector in China 32. (2002).

Rhododendron weyrichii Forbes & Hemsley non

Maxim. in Journ. Linn. Soc. Bot. 26: 32. (1889).

Hubei: Badong, 1422. Liantuo, 3829. Changyang, 5274. Jianshi, 5947.

A low growing deciduous azalea, flowers pink or rose, reddish purple on the upper lobes and carried on bare wood in April. Hemsley and Wilson based this species on collections made by Henry, Maries and Wilson. Robert Fortune had first found it in Zhejiang province and afterwards Maries had made a collection in the province of Guangxi. According to Wilson it was common on the cliffs around Yichang where it flowered in April and May before the leaves unfolded. It was introduced to cultivation by Henry who sent seeds to the Royal Botanic Gardens, Kew from Yichang in 1886. It first flowered there in the temperate house in April 1907.

Rhododendron mengtszense Balfour. f. & W. W. Smith (ERICACEAE)

in Trans. & Proc. Bot. Soc. Edinburgh 27: 206. (1917), Davidian in Rhododend. Spec. 2: 271. (1989).

Rhododendron irroratum Hemsley & E. H. Wilson non Franchet in Kew Bull. Misc. Inf. 112. (1910). *Rhododendron gymnanthum* Rehder & E. H. Wilson non Diels in Sargent Pl. Wilson. 1: 539. (1913).

Yunnan: In Pingbian, in mountain forest on the Daweishan Range at 2,300 metres. Tree, 6 metres tall. Flowers purple-red, 10275 (isotype).

A tree to 6 metres tall carrying up to 8 flowers in a racemose umbel, the flowers are campanulate, to 3.5 cm long and are purple-red with a crimson blotch at the base. This species is represented by a single gathering made by Henry in 1898 and it has recently been introduced to cultivation. Endemic to south-east Yunnan.

Rhododendron micranthum Turcz. (ERICACEAE)

Hemsley & E. H. Wilson in Kew Bull. Misc. Inf. 117. (1910).

Rhododendron pritzelianum Diels in Bot. Jahrb. 29: 510. (1900).

Hubei: Fang Xian, 6632 (type of *Rhododendron pritzelianum* Diels).

An evergreen shrub to 1.5 metre tall in cultivation (up to 2.5 metres in its native habitat), carrying white flowers in densely packed terminal racemes from May to July. Native to north and central China and Korea. Discovered by the French missionary Père d'Incarville in the mid 18th century in the mountains to the north of Beijing. It is widely distributed in north and central China and Korea. According to Wilson who introduced this species when collecting for Messrs Veitch in 1901, this *Rhododendron* was common on cliffs to the north of the Yangtze (in Hubei) at elevations above 1,650 metres. There it formed bushes

from 1 to 6 metres tall and produced erect racemes of small white flowers in great abundance. Wilson's introduction first flowered at Veitch's Coombe Wood nursery in May 1904 and at Caerhays Castle, Cornwall in the spring of 1905. In Veitch's catalogue *New Hardy Plants from Western China* (Autumn 1912) this species was available for two shillings and six pence.

Rhododendron microphyton Franch. (ERICACEAE)

Rehder & E. H. Wilson in A Monograph of Azaleas 57. (1921), Bean in Trees and Shrubs 3: 720. (1992).

Rhododendron indicum Hemsley & E. H. Wilson non (L.) Sweet in Kew Bull. Misc. Inf. 119. (1910).

Yunnan: Tongguan at 1,650 metres, 11596. In the mountains to the south-west of Simao at 1,650 metres, 12986.

An evergreen azalea from 30 cm to 90 cm tall, rosy-lilac flowers with crimson spots on the three upper lobes are produced in April in terminal clusters, four to six together. Discovered by Delavay in 1884, it was found by Forrest on the eastern flank of the Cangshan Range above the city of Dali (formerly Tali) in 1906 and was introduced by him in 1913. Distributed in Myanmar and western China, this species is hardy only in the milder counties of the British Isles and Ireland.

Rhododendron molle (Blume) G. Don (ERICACEAE)

Rehder & E. H. Wilson in A Monograph of Azaleas 95. (1921), Zheng in Journ. Wuhan Bot. Res. 2.(1): 169. (1984).

Rhododendron sinense Sweet Henry in Journ. China Br. Roy. Asiat. Soc. 22: 253. (1887) (Chinese Names of Plants), Rehder & E. H. Wilson in Sargent Pl. Wilson. 1: 549. (1913).

Hubei: Yichang, 268, 1220.

A deciduous azalea to 1 metre tall, native to eastern and central China and introduced to cultivation by Loddiges in 1823. Robert Fortune who reintroduced it in 1845 for the Horticultural Society (later the Royal Horticultural Society) frequently found it during his travels in eastern China and wrote, 'the yellow *Azalea sinensis* seems to paint the hillsides, so large were the flowers, so vivid were the colours'. According to Henry this species was known colloquially at Yichang as *lao hu hua*. Wilson described this as a rare species in Hubei, he collected it on the conglomerate hills on the opposite side of the Yangtze River to Yichang (Moji Shan) and it is likely Henry's specimen came from the same location as he frequently botanised there and was one of the sites where he found *Lilium henryi*.

Rhododendron moulmainense Hook. (ERICACEAE)

Rhododendron oxyphyllum Franch. Hemsley & E. H. Wilson in Kew Bull. Misc. Inf. 116. (1910).

Yunnan: In the mountains to the east of Simao, in forests at 1,650 metres, 11609. In the mountains to the west of Simao, in forests at 1,650 metres, 11609b. Mountains to the south of Simao at 1,100 metres, 11609c.

A shrub to 3 metres tall or a tree to 25 metres tall depending on the surrounding habitat. Flowers are tubular funnel-shaped, to 6 cm long, fragrant, white flushed yellow or flushed with any of the following, pink, rose, rose-red or lilac, with a pale green or yellow blotch at the base. These are produced in clusters of 2 to 8 in the axils of the uppermost 1 to 3 leaves. Discovered by Thomas Lobb at Moulmien on the Gerai Mountains in Myanmar and introduced by him through Messrs Veitch who first flowered it in January 1856. William Hooker described it in the same year. It was later reintroduced by George Forrest. This species ranges across south-east Tibet (Xizang), Yunnan, Guizhou and Guangxi in China and is also found in India, Myanmar and Thailand. Given this large geographical range this species is extremly variable in its morphological features. In the British Isles and Ireland it is best suited to the mildest coastal areas.

Rhododendron mucronatum (Blume) G. Don (ERICACEAE)

Rehder & E. H. Wilson in Monograph of Azaleas 68. (1921).

Rhododendron ledifolium G. Don Forbes & Hemsley in Journ. Linn. Soc. Bot. 26: 27. (1889).

Hubei: Cultivated in a garden at Yichang, 3503.

According to Wilson this species was widely cultivated in the gardens of the wealthy in many parts of China. Henry stated the Yichang plant was originally from Sichuan.

Rhododendron nivale Hook. f. subsp. *boreale* N. M. Philipson & Philipson (ERICACEAE)

Rhododendron nigropunctatum Bureau & Franch. Hemsley & E. H. Wilson in Kew Bull. Misc. Inf. 118. (1910).

Sichuan: Henry's Chinese collector with Pratt, no locality stated but probably Kangding, 8897.

Introduced to cultivation by Wilson from western Sichuan.

Rhododendron oldhamii Maxim. (ERICACEAE)

Henry in Trans. Asiat. Soc. Jap. 24 suppl: 57. (1896) (List Pl. Formosa), Rehder & E. H. Wilson in A Monograph of Azaleas 66. (1921), Li in Woody Flora of Taiwan 696. (1963).

Taiwan: Oluanpi, 588 (in part).

An evergreen much branched shrub to 4 metres tall, flowers carried in terminal clusters of two to four and are orange-red and stained pink on the upper lobe. Endemic to Taiwan and distributed there from sea-level to 2,800 metres, said to be most abundant on the northern part of the island. *Rhododendron oldhamii* was discovered by the Kew collector Richard Oldham in 1864 and was introduced to cultivation by Charles Maries who sent seeds to Messrs.

Veitch in 1878. It was reintroduced by E. H. Wilson in 1918 and is suited only to the mildest most sheltered areas of the British Isles and Ireland.

Rhododendron orbiculare Decne. (ERICACEAE)

Hemsley & E. H. Wilson in Kew Bull. Misc. Inf. 108. (1910).

Sichuan: Wa Shan, Henry's Chinese collector with Pratt, 8873.

An evergreen shrub to 3 metres tall, flowers produced in April, up to 10 in a terminal truss, these are bell-shaped and pale magenta-pink. Native to Sichuan, *Rhododendron orbiculare* was discovered by Père Armand David in 1870 and was introduced to cultivation by E. H. Wilson in 1904 through Messrs Veitch.

Rhododendron ovatum (Lindley) Planch. ex Maxim. (ERICACEAE)

Forbes & Hemsley in Journ. Linn. Soc. Bot. 26: 28. (1889), Rehder & E. H. Wilson in Sargent Pl. Wilson. 1: 546. (1913), Zheng in Journ. Wuhan Bot. Res. 2.(1): 169. (1984).

Hubei: Yichang, 734. Badong, 5278. Changleping, 6283, 7828.

An evergreen shrub to 2.5 metres tall, producing solitary pale purple, pink or white flowers from axillary buds in late May. Native to eastern and central China and discovered by Robert Fortune in 1843, introduced by him the following year. Wilson reintroduced this species from Changyang in Hubei in 1901 and again in 1907 from the same region.

Rhododendron pachypodum I. B. Balfour & W. W. Smith (ERICACEAE) New species

Rhododendron ciliicalyx Hemsley & E. H. Wilson non Franchet in Kew Bull. Misc. Inf. 113. (1910). *Rhododendron pilicalyx* Hutchinson in Notes Roy. Bot. Gard. Edinburgh. 12: 66. (1919), Davidian in Rhodo. Spec. 1: 248. (1982). *Rhododendron rufosquamosum* Hutchinson in Notes Roy. Bot. Gard. Edinburgh 12: 63. (1919), Davidian in Rhodo. Spec. 1: 250. (1982).

Yunnan: Mountains to the north of Mengzi at 2,650 metres, 10524 (isotype of *Rhododendron pilicalyx* Hutchinson). On hills near Simao at 1,464 metres, 11983 (isotype of *Rhododendron rufosquamosum* Hutchinson).

A shrub of 1 or 2 metres tall, allied to *Rhododendron burmanicum* and producing clusters of two to three fragrant white or yellow flowers. Though the discovery of this species is generally credited to George Forrest, it had in fact been first found by Augustine Henry in 1897. It was introduced to cultivation by Forrest from the western flank of the Cangshan Range above Dali and is a handsome plant though not as hardy as *Rhododendron burmanicum* (which

02

was introduced to cultivation by Lady Charlotte Wheeler Cuffe by means of a living plant through the National Botanic Gadens, Glasnevin). *Rhododendron pachypodum* is distributed throughout Yunnan province from Dali in the northwest to Simao, Mengzi and Pingbian in the exteme south. In Ireland this charming shrub grows well in several major collections including Glanleam on Valentia Island in County Kerry.

Rhododendron praevernum Hutchinson (ERICACEAE) New species

in Gard. Chron. ser. 3, 67: 127. (1920), O' Brien in Augustine Henry, an Irish Plant Collector in China 32. (2002).

Rhododendron fortunei Forbes & Hemsley non Lindl. in Journ. Linn. Soc. Bot. 23. (1889). *Rhododendron sutchuenense* Rehder & E. H. Wilson non Franch. in Sargent Pl. Wilson. 1: 544. (1913), Zheng in Journ. Wuhan Bot. Res. 2(1): 170. (1984).

Hubei: South Badong at 1,950 metres, 5285 (syntype).

A shrub or small tree to 4.5 metres tall, producing umbels of 8 to 10 white flowers, tinged rose with a large crimson blotch at the base, in February and March. A close ally of *Rhododendron sutchuenense* it was introduced to cultivation by Wilson in 1900. Wilson stated it was common in woods between 1,600 and 2,500 metres in the mountainous districts of north-west Hubei but was rare south of the Yangtze (i.e. Changyang and South Badong). The specific epithet referrs to the plants very early flowering season. Cultivated in Ireland at Ilnacullin, the John F. Kennedy Arboretum, Mount Congreve and Glasnevin.

Rhododendron scabrifolium Franch. var. *spiciferum* (Franch.) Cullen (ERICACEAE)

Rhododendron spiciferum Franch. Hemsley & E. H. Wilson in Kew Bull. Misc. Inf. 120. (1910).

Yunnan: On grassy hills near Mengzi at 1,650 metres, 9369. Chuyuan, 9369a.

A shrub to 1.5 metres tall, flowers produced in March and April in clusters of two or three in the upper most leaf axils. Flower colour varies from almost white to deep pink and rose. Native to south-west China and discovered by in Yunnan by Delavay in March 1891.

Rhododendron siderophyllum Franch. (ERICACEAE)

Hemsley & E. H. Wilson in Kew Bull. Misc. Inf. 115. (1910).

Yunnan: Mengzi, 9110a. Lunan, 9110b.

An evergreen shrub to 2.5 metres tall, pale pink flowers are produced in May in terminal and axillary clusters. A member of the Subsection Triflora, this species naturally hybridises occasionally with *Rhododendron yunnanense* where both species come in contact and I have seen such hybrids on the Cangshan Range above Dali in north-west Yunnan. *Rhododendron siderophyllum* was introduced to cultivation by Forrest from Yunnan. Endemic to China and distributed in the western provinces of Gansu, Sichuan and Yunnan.

Rhododendron simsii Planch. (ERICACEAE)

Rehder & E. H. Wilson in Monograph of Azaleas 45. (1921).

Rhododendron indicum Sw. Forbes and Hemsley in Journ. Linn. Soc. Bot. 26: 25. (1889). *Rhododendron indicum* (L.) Sweet var. *ignescens* Sweet. Rehder & E. H. Wilson in Sargent Pl. Wilson. 1: 547. (1913). *Rhododendron indicum* (L.) Sweet var. *formosanum* Hayata. Rehder & E. H. Wilson in A Monograph of Azaleas 45. (1921).

Hubei: Yichang, 782, 1160. Badong, 1416. Liantuo, 3194. **Taiwan:** Oluanpi, 588 (in part). **Yunnan:** Mengzi, 9900. In forests near Milê, 9900a. Near Mengzi, on grassy mountains at 1,500 to 2,000 metres, 9900b., 9900c. Mountains to the west of Simao at 1,600 metres, 9900d.

An evergreen shrub from 1 to 2.5 metres tall, blooms appearing in May in terminal clusters of up to six red, white, orange or rose-pink to red flowers. A widespread species, native to Myanmar, Thailand and in China (from Yunnan and Sichuan in the west to Hong Kong in the east). Kingdon Ward compared this shrub when in flower to 'the glow from an active volcano at night'. Cultivated at Kew as early as 1810 having been introduced from China in 1808 by the East India Company on the ship *Cuffnels*. According to Wilson it was abundant in the Yangtze valley during the early 20th century, in places whole hillsides in May were red with its flowers. It is one of the parents of the indoor azalea, used as a pot plant at Christmas.

Rhododendron sinofalconeri Balf. f. (ERICACEAE) New species

in Notes Roy. Bot. Gard. Edinburgh. 12: 63. (1916), Davidian in Rhodo. Spec. 2: 177. (1989), O' Brien in Augustine Henry, An Irish Plant Collector in China 32. (2002).

Rhododendron falconeri Hemsley & E. H. Wilson non Hook. f. in Kew Bull. Misc. Inf. 107. (1910).

Yunnan: On the summits of mountains in forest to the north of Mengzi at 2,745 metres, 9448 (type).

Discovered by William Hancock on the Great Black Mountain Range (Dahei Shan) at 950 metres in about 1893, this species was based on a collection later made by Augustine Henry on the mountains to the north of Mengzi, which is most likely the Great Black Mountain range where Hancock had previously collected. A tree to 12 metres tall with leaves to 36 cm long and carrying large clusters of 12~20 pale yellow or primrose flowers. It is the Chinese counterpart to the Himalayan, *Rhododendron falconeri*, itself a native of Nepal, Sikkim, Bhutan, India (northeat India and Assam). *Rhododendron sinofalconeri* was introduced to cultivation in recent decades. Native to China (Yunnan) and Vietnam. Grows well at the National Botanic Gardens, Kilmacurragh.

Rhododendron sp. (ERICACEAE)

Zheng in Journ. Wuhan Bot. Res. 2.(1): 170. (1924).

Hubei: Fang Xian, 6949.

Rhododendron sp. Subsection Irrorata (ERICACEAE)

Bayley Balfour in Trans. Bot. Soc. Edinburgh 27: 173. (1917).

Rhododendron irroratum Hemsley & E. H. Wilson non Franch. in Kew Bull. Misc. Inf. 112. (1910).

Yunnan: Mountains to the north of Mengzi at 2,650 metres, 10301.

This species was introduced to Kew by seeds sent there by Henry from the Mengzi area, according to Bayley Balfour the same plant at Kew had died by 1917.

Rhododendron sp. **nov**. (ERICACEAE)

Forbes & Hemsley in Journ. Linn. Soc. Bot. 26: 32. (1889).

Hubei: San You Dong Glen, 1700.

Rhododendron spanotrichum Balf. f. & W. W. Sm. (ERICACEAE) New species

in Trans. Bot. Soc. Edinburgh 27: 214. (1917), Davidian in Rhodo. Spec. 2: 275. (1989).

Rhododendron irroratum Hemsley & E. H. Wilson non Franch. in Kew Bull. Misc. Inf. 112. (1910). *Rhododendron gymnanthum* Rehder & E. H. Wilson non Diels in Sargent Pl. Wilson. 1: 539. (1913).

Yunnan: Fengchunling at 2,400 metres, 10853.

A tree to 6 metres tall, flowers crimson with a dark crimson blotch at the base, to 2 cm long and up to 10 carried in a racemose umbel. China (Sichuan and Yunnan) to northern Vietnam.

Rhododendron spinuliferum Franch. (ERICACEAE)

Hemsley & E. H. Wilson in Kew Bull. Misc. Inf. 120. (1917).

Yunnan: In the mountains near Mengzi at 1,700 metres, 10572.

An evergreen shrub to 2 metres tall, the young branches are covered with grey soft hairs and bristles, later shedding. Brick-red tubular flowers to 2.5 cm are carried in terminal clusters from the upper leaf axils in April. This species was founded on Delavay's 4883 and first flowered in the collection of M. L. de Vilmorin at des Barres in 1907 and at Kew in 1910. Endemic to China and distributed in Guizhou, Sichuan and Yunnan.

Rhododendron stamineum Franch. (ERICACEAE)

Hemsley & E. H. Wilson in Kew Bull. Misc. Inf. 116. (1910), Rehder & E. H. Wilson in Sargent Pl. Wilson. 1: 546. (1913), Zheng in Journ. Wuhan Bot. Res. 2.(1): 170. (1984).

Rhododendron aucubifolium Hemsley in Journ. Linn. Soc. Bot. 26: 29. (1889), quoad flores; folia *Daphniphyllum macropodum*, Bretschneider in Hist. Eur. Bot. Disc. China 2: 786. (1898), Diels in Bot. Jahrb. xxix. 515. (1900), Bean in Flora and Sylva 3: 164. (1905). *Rhododendron pittosporifolium* Hemsley in Journ. Linn. Soc. Bot. 26: 29. (1889), Bretschneider in Hist. Eur. Bot. Disc. China 2: 786. (1898).

Hubei: Jianshi, 5787 (with foliage of *Daphniphyllum macropodum*). In the mountains of South Badong, 4025 (syntype of *Rhododendron pittosporifolium* Hemsley), 4031 (type of *Rhododendron pittosporifolium* Hemsley), 4081 (type of *Rhododendron aucubifolium* Hemsley). Donghu, 6432 (with foliage of *Daphniphyllum macropodum*).

An evergreen shrub to 4 metres tall. White fragrant flowers are carried in April and May in clusters of four in the axils of the uppermost leaves. Native to central and western China and Myanmar, this species was discovered by Delavay in 1882 in north-east Yunnan and was introduced to cultivation by Wilson from Hubei in 1900 through Veitch's Coombe Wood nursery. Wilson reintroduced it in 1910 from Emei Shan in Sichuan through the Arnold Arboretum. *Rhododendron stamineum* first flowered in the collection of J. C. Williams at Caerhays Castle, Cornwall in 1914. *Rhododendron aucubifolium* was based on a false specimen constructed by Henry's Badong collector using the foliage of *Daphniphyllum macropodum* and the flowers of *Rhododendron stamineum*.

Rhododendron strigillosum Franch. (ERICACEAE)

Hemsley & E. H. Wilson in Kew Bull. Misc. Inf. 107. (1910).

Sichuan: Wa Shan, Henry's Chinese collector with Pratt, 8872.

An evergreen shrub or small tree to 7 metres tall, rich red (though sometimes white) flowers are carried in terminal trusses in March and April. A very rare species native to western Sichuan where it is known only from Baoxing, Emei Shan, Wa Shan and Pao-hsing-hsien. It was discovered by Père Armand David in Baoxing in western Sichuan in May 1869 and was introduced to cultivation by E. H. Wilson in 1904.

Rhododendron sutchuenense Franch. (ERICACEAE) New species

Rehder & E. H. Wilson in Sargent Pl. Wilson. 1: 544. (1913), Zheng in Journ. Wuhan Bot. Res. 2.(1): 170.

(1984), Davidian in Rhodo. Spec. 2: 187. (1989), O' Brien in Augustine Henry, An Irish Plant collector in China 33. (2002).

Rhododendron fortunei Forbes & Hemsley non Lindl. in Journ. Linn. Soc. Bot. 23. (1889).

Hubei: Fang Xian, 6914.

An evergreen shrub to 6 metres tall, up to ten rosy-lilac, purple-spotted bell-shaped flowers are produced in terminal clusters in March. Discovered by Augustine Henry in the autumn of 1888 and based on material later collected in 1891 by the French missionary, Père Paul Farges. It was introduced to cultivation by E. H. Wilson in 1900 through Veitch's Coombe Wood nursery where it first flowered in 1910 when only 60 cm tall. In Veitch's catalogue *New Hardy Plants from Western China* (Autumn 1912) this species was available for the rather hefty sum of ten shillings and six pence each. Wilson stated this was a common species in Hubei at 1,150 to 2,300 metres, in mixed woods, often in the shade of evergreen oaks and in the company of bamboo.

Rhododendron tutcherae Hemsley & E. H. Wilson (ERICACEAE) New species

in Kew Bull. Misc. Inf. 116. (1910), Davidian in Rhodo. Spec. 3: 319. (1992).

Yunnan: In Pingbian, in forests on the Daweishan Range at 2,000 metres, 10630. In mountain forest to the south of Mengzi at 660 metres, 10636.

A tree to 13 metre tall, bearing tubular, funnel-shaped violet flowers carried in pairs or singly in the axils of the uppermost 1 to 3 leaves. Discovered by Henry in 1898, it is closely allied to *Rhododendron hancockii* (q.v.) and the specific epithet commemorates Mrs. Tutcher, wife of W. J. Tutcher of Hong Kong Botanic Gardens. Distributed in China (southern Yunnan) and Vietnam. This species has never been introduced to cultivation.

Rhododendron vialii Delavay & Franchet (ERICACEAE)

Hemsley & E. H. Wilson in Kew Bull. Misc. Inf. 111. (1910), Davidian in Rhododendr. Spec. 3: 300. (1992).

Yunnan: In mountains near Shiping at 1,600 metres, 11563. Mengzi at 1,300 metres, 11563a. Tonguan at 1,960 metres, 13271.

A shrub to 4.5 metres tall, flowers crimson or pink, held singly or in pairs and carried in the axils of the uppermost 1 to 4 leaves in February and March. Discovered by Père Paul Vial, the French missionary with the Missions Etrangeres in China (Yunnan). Père Vial found it to the south of Tonghai (Tonghai lies to the south of Kunming) in 1871. Distributed in southern Yunnan and the border area between Laos and northern Vietnam. Père Delavay also collected this species in 1891. *Rhododendron vialii* has recently been introduced to cultivation.

Rhodoleia henryi K. Y. Tong (HAMAMELIDACEAE) New species

in Bull. Dept. Biol. Sun Yatsen Univ. 2: 35. (1930), Exell in Sunyatsenia 1: 97. (1933).

Rhodoleia sp. Henry in Flora & Sylva 1: 217. (1903).

Yunnan: South of the Red River, on Fengchunling at 2,600 metres, 10131 (type).

An evergreen tree to 15 metres tall, bearing clusters of 5 to 8 nodding crimson flowers in March and April. Endemic to Yunnan where it grows in evergreen broad-leaved forest in the southy east of the province.

Rhodoleia parvipetala K. Y. Tong (HAMAMELIDACEAE) New species

in Bull. Dept. Biol. Sun Yatsen Univ. 2: 35. (1930), Exell in Sunyatsenia 1: 99. (1933).

Yunnan: Mengzi at 1,640 metres, 11425 (isotype).

A handsome tree varying in height from 10 to 30 metres and bearing crimson flowers from December to April of the following year. Of limited distribution through the western provinces of Guangxi, Guangdong and Yunnan. Its territory also extends into northern Vietnam. A very beautiful species.

Rhodomyrtus tomentosa (Aiton) Hassk. (MYRTACEAE)

Henry in Trans. Asiat. Soc. Jap. 24 suppl: 43. (1896) (List Pl. Formosa), Merrill & Perry in Journ. Arnold Arb. 19: 196. (1938), Li in Woody Flora of Taiwan 645. fig 256. (1963).

Hainan: Environs of Haikou, 8020. Environs of Kiungchow, 8267. Without locality, 8491. **Taiwan:** Oluanpi, 982, 1366.

A small evergreen shrub to 2 metres tall, producing axillary cymes of red-purple flowers to 2 cm across. Widely distributed from India to southern China, through Malaysia and Australia. In Chinese Taiwan this species is distributed at low altitude in secondary forest.

Rhus chinensis Mill. (ANACARDIACEAE)

Zheng in Journ. Wuhan Bot. Res. 2.(1): 116. (1984).

Rhus semialata Murr. Forbes & Hemsley in Journ. Linn. Soc. Bot. 23: 146. (1886), Henry in Journ. China Br. Roy. Asiat. Soc. 22: 242. (1887) (Chinese Names of Plants).

Hubei: Yichang, 447, 2142, 4890. Badong, 187, 533 (September 1885), 2534. **Hainan:** Hummocks near Haikou, 8098.

The Chinese sumach, a small deciduous tree to 7 metres tall, widespread in Asia and distributed through the Himalaya, Myanmar, Thailand, China (including Taiwan), Indochina, Korea, Japan and Sumatra. Henry designated this species as the 'the red gall nut tree' or 'Chinese nutgall tree' of Hubei, the galls of which are used in medicine to alleviate coughing and stop diarrhea. He also stated that it was known at Yichang as *fu-yang-shu*, the galls that occured on the leaves were called *wu-p'ei-tzu* and were used in medicine and were also used at one time to dye expensive silk fibres black. These galls are produced by the parasitic Chinese sumach aphid, *Melaphis chinensis* (Bell) Baker on both *Rhus chinensis* and *Rhus potaninii*. Wilson stated this small tree was abundant in the Yangtze valley and that it was also used for tanning purposes, in his opinion it was 'the finest tanning material in the world.' The demand in the west was at one time actually greater than the Chinese supply. Like many others members of the genus this species has brilliant orange-red autumn colour.

Rhus chinensis Mill. var. *roxburghii* (DC.) Rehder (ANACARDIACEAE)

Rhus semi-alata Murr. Henry in Trans. Asiat. Soc. Jap. 24 suppl: 29. (1896) (List Pl. Formosa), *Rhus javanica* L. var. *roxburghii* (DC.) Rehder & E. H. Wilson in Sargent Pl. Wilson. 2: 179. (1914).

Taiwan: Wanchin, and Kaohsiung, 348.

A small deciduous tree with a native range from India to Indochina and China (Taiwan) at low altitude.

Rhus potaninii Maxim. (ANACARDIACEAE)

Rehder & E. H. Wilson in Sargent Pl. Wilson. 2: 177. (1916), Zheng in Journ. Wuhan Bot. Res. 2.(1): 117. (1984).

Rhus sp. Henry in Journ. China Br. Roy. Asiat. Soc. 22: 242. (1887) (Chinese Names of Plants). *Rhus henryi* Diels in Bot. Jahrb. 29: 432. (1900), Schneider in Illus. Handb. Laubholzk. 2: 154. (1912).

Hubei: Without locality, 5529c., (type of *Rhus henryi* Diels). Jianshi, 5903. Sichuan: South Wushan, 5529a.

Potanin's sumach, the *ch'ing fu yang*, a tree of moderate size (10 metres) in western Hubei though Wilson states he occasionally found specimens to 25 metres tall. Discovered by the Russian explorer, Grigori Nicolaevich Potanin in Gansu in 1885, Henry found it three years later in Hubei. He stated that the galls produced on the leaves were used as a medicine and were known as *ch'i-pei-tzu*. Introduced to cultivation by Wilson in 1904 through Veitch's Coombe Wood nursery.

Rhus punjabensis J. L. Stewart ex Brandis var. *sinica* (Diels) Rehder & E. H. Wilson (ANACARDIACEAE)

New variety. in Sargent Pl. Wilson. 2: 176. (1914), Zheng in Journ. Wuhan Bot. Res. 2.(1): 117. (1984).

Rhus punjabensis Henry non J. L. Stewart ex Brandis in Notes Econ. Bot. China 42. (1893).

Hubei: Badong, 3157, 4755, 5074. Without locality, 5529b.

A small tree to about 12 metres, differing from the type by its fewer and more sessile leaflets and the slightly winged upper rachis of the leaf. Common on woodland margins in Hubei where Wilson stated it was known as *hung-fu-yang*.

Discovered by Henry in 1887, this variety was based on material later collected by Giraldi, Bock and von Rosthorn in the early 1890s.

Rhus sp. No. 1. (ANACARDIACEAE)

Rhus toxicodendron Henry non L. in Trans. Asiat. Soc. Jap. 24 suppl: 30. (1896) (List Pl. Formosa).

Taiwan: Oluanpi, 614, 2005.

Rhus sp. No 2. (ANACARDIACEAE)

Anonymous in Kew List of Determinations (volume 34).

Yunnan: Mengzi, 9514, 10614.

Rhynchoglossum obliquum Blume (GESNERIACEAE)

Anonymous in Kew List of Determinations (volume 34).

Yunnan: Mengzi, 9253.

Rhynchosia lutea Dunn (FABACEAE) New species in Journ. Linn. Soc. Bot. 35: 491. (1903).

Yunnan: Mengzi, in woods and on rocky mountains 1,150 to 1,200 metres, 9105, 9105a. Mengzi in woods and on rocky mountains 1,500 to 1,850 metres, 9994 (isosyntype).

Endemic to Yunnan.

Rhynchosia minima (L.) DC. (FABACEAE)

Anonymous in Kew List of Determinations (volume 34).

Yunnan: Mengzi, 9509.

Distributed in India, Myanmar, China and Vietnam.

Rhynchosia minima (L.) DC. var. *nuda* (DC.) Kuntze (FABACEAE)

Rhynchosia minima Henry non (L.) DC. in Trans. Asiat. Soc. Jap. 24 suppl: 37. (1896) (List Pl. Formosa).

Taiwan: Oluanpi, 1271.

A slender twining herb, this variety is confined to China (Taiwan) and Japan (South).

Rhynchosia rothii Benth. ex Aitch. (FABACEAE)

Rhynchosia sericea Span. Henry in Trans. Asiat. Soc. Jap. 24 suppl: 37. (1896) (List Pl. Formosa).

Taiwan: Oluanpi, 1270.

A woody climber, native to India, Malaysia and Java. At low altitude in western Taiwan of China.

Rhynchosia rufescens (Willd.) DC. (FABACEAE)

Anonymous in Kew List of Determinations (volume 34).

Yunnan: In the Red River valley, 9605.

Native to India, Sri Lanka, China (Yunnan and Guangxi) and Cambodia.

Rhynchosia spp. (FABACEAE)

Anonymous in Kew List of Determinations (volume 24).

Hubei: Yichang, 3635. San You Dong Glen (July

1886), 1528. At a cave on way to Moji Shan, (June 27th 1886), 1528a. **Sichuan:** South Wushan, 7190, 7237.

Rhynchosia striata G. Don (FABACEAE)
Anonymous in Kew List of Determinations (volume 34).

Yunnan: Mengzi, 10163.

Rhynchosia viscosa (Roth.) DC. (FABACEAE)
Anonymous in Kew List of Determinations (volume 34).

Yunnan: Mengzi, 10722.

Native to India and China (Yunnan).

Rhynchosia volubilis Lour. (FABACEAE)
Forbes & Hemsley in Journ. Linn. Soc. Bot. 23: 196. (1887), Henry in Journ. China Br. Roy. Asiat. Soc. 22: 274. (1887) (Chinese Names of Plants); in Trans. Asiat. Soc. Jap. 24 suppl: 37. (1896) (List Pl. Formosa).

Hubei: Yichang, 320, 1098, 3003. In the Monastery Valley near Yichang (July 1886), 1656. Liantuo (October 1886), 3003, 4626. **Taiwan:** Wanchin, without number. Wanshoushan, 450. Oluanpi, 665. Kaohsiung Plain, 1923

A twining sub-shrub found in thickets at low altitudes around Yichang, where according to Henry it was called *yeh mao pien tou*. Native to China, Vietnam, Korea and Japan.

Rhynchospora corymbosa (L.) Britton (CYPERACEAE)
Rhynchospora aurea Vahl. Henry in Trans. Asiat. Soc. Jap. 24 suppl: 105. (1896) (List Pl. Formosa), C. B. Clarke in Journ. Linn. Soc. Bot. 36: 259. (1903).

Hainan: Without locality, 8551. **Taiwan:** Kaohsiung, 1843.

Rhynchospora rubra (Lour.) Makino (CYPERACEAE)
Rhynchospora wallichiana Kunth C. B. Clarke in Journ. Linn. Soc. Bot. 36: 260. (1903).

Yunnan: Mengzi at 1,650 metres, 11150, (in part).

Rhynchotechum discolor (Maxim.) B. L. Burt. (GESNERIACEAE)
Isanthera discolor Maxim. Henry in Trans. Asiat. Soc. Jap. 24 suppl: 68. (1896) (List Pl. Formosa).

Taiwan: Wanchin, 845, 849, 1540.

Discovered in Taiwan by the Kew collector, Richard Oldham.

Ribes franchetii Janczewski (GROSSULARIACEAE) New species
in Bull. Ac. Sci. Cracovie. 64, fig. 3-4. (1909), Schneider in Illus. Handb. Laubholzk. 2: 948. (1912), Eduardus Janczewski in Fedde, Repert. Spec. Nov. Regni. Veg. 6: 333. (1908—1909); in 12: 377. (1913).

Ribes sp. Zheng in Wuhan Bot. Res. 2. (1): 62. (1984), *Ribes alpinum* L. Zheng in Wuhan Bot. Res. 2. (1): 61. (1984).

Hubei: South Badong, 3741, 5465.

This currant was also later collected by the French missionary, Farges in eastern Sichuan during the early 1890s.

Ribes glaciale Wall. (GROSSULARIACEAE)
Anonymous in Kew List of Determinations (volume 24).

Sichuan: Henry's Chinese collector with Pratt, no locality stated but may be any of the following, Leshan, Emei Shan, Wa Shan or Kangding, 8973.

Ribes henryi Franchet (GROSSULARIACEAE) New species
in Bull. Mens. Soc. Linn. Paris. n.s.i. 87. (1898), Schneider in Illus. Handb. Laubholzk. 2: 948. (1912), Harrow in Gard. Chron. 56: ser. 3, 29. (1914), Besant in Gard. Chron. ser. 3, 98: 335. (1935), Bean in Trees & Shrubs 3: 434. (1936), Nelson in Pim, The Wood & the Trees 234. (1984), Bean in Trees & Shrubs 4: 8. (1989), Nelson in A Heritage of Beauty 317. (2000).

Sichuan: Henry's Chinese collector with Pratt, no locality stated but may be any of the following, Leshan, Emei Shan, Wa Shan or Kangding, 8941 (type).

This species was inadvertently introduced to cultivation by Wilson in 1908, when seeds of Wilson's (Wilson 584) collection of *Sinowilsonia henryi* Hemsl. (gathered on May 25th 1907 at Fang Xian in northern Hubei) were sown at the Royal Botanic Garden, Edinburgh. From this sowing a stray seedling of *Ribes henryi* mysteriously appeared. The Edinburgh plant proved to be male and flowered for the first time in the spring of 1912. Janczewski, the monographer of *Ribes*, who described the closely allied *Ribes laurifolium*, raised a single seedling of *Ribes henryi* in a case that almost parallels the Edinburgh plant. A single plant came up amoungst seedlings of *Ribes laurifolium*, this was female and produced fruits by crossing with *Ribes laurifolium*. This seedling undoubtedly was raised from *Ribes laurifolium* (Wilson 817) gathered on Wa Shan in western Sichuan in September 1908. A seed lot from the same Wilson collection was received at Glasnevin the following spring. Despite its horticultural merit *Ribes henryi* is extremely rare in cultivation. Henry grew it in his garden at Ranelagh in the south Dublin city suburbs.

Ribes heterotrichum C. A. Mey. (GROSSULARIACEAE)
Zheng in Journ. Wuhan Bot. Res. 2.(1): 62. (1984).
Hubei: 6602.

Ribes longeracemosum Franch (GROSSULARIACEAE)
Diels in Bot. Jahrb. 29: 377. (1900).
Hubei: Fang Xian at 1,400 to 3,000 metres, 6814.

Discovered by Père Armand David near Baoxing in Sichuan in March 1869.

02

Ribes nigrum L. (GROSSULARIACEAE)

Henry in Notes Econ. Bot. China 46. (1893), Diels in Bot. Jahrb. 29: 378. (1900), Zheng in Journ. Wuhan Bot. Res. 2.(1): 62. (1984).

Hubei: Fang Xian, 6781.

Ribes rubrum L. (GROSSULARIACEAE)

Diels in Bot. Jahrb. 29: 378. (1900).

Hubei: Badong, 4663.

Ribes spp. (GROSSULARIACEAE)

Zheng in Journ. Wuhan Bot. Res. 2.(1): 62. (1984). Anonymous in Kew List of Determinations (vol. 24 & 34).

Hubei: South Badong, 6097. **Sichuan:** South Wushan, 5700. Henry's Chinese collector with Pratt, no locality stated but may be any of the following, Leshan, Emei Shan, Wa Shan or Kangding, 8966. **Yunnan:** Mengzi, 10506, 10876.

Ribes tenue Janczewski (GROSSULARIACEAE)

Ribes alpinum Forbes & Hemsley non L. in Journ. Linn. Soc. Bot. 23: 279. (1887), Zheng in Journ. Wuhan Bot. Res. 2(1): 61. (1984).

Hubei: Badong, 1467, 5316, 5316a. Changleping, 7658. Jianshi, 5316b.

Ribes uva-crispa L. (GROSSULARIACEAE)

Ribes grossularia L. Henry in Notes Econ. Bot. China 46. (1893), Zheng in Journ. Wuhan Bot. Res. 2.(1): 62. (1984).

Hubei: Xingshan, 7000.

Ricinus communis L. (EUPHORBIACEAE)

Forbes & Hemsley in Journ. Linn. Soc. Bot. 26: 443. (1894), Henry in Trans. Asiat. Soc. Jap. 24 suppl: 84. (1896) (List Pl. Formosa).

Hainan: Ling-men, 8603. **Taiwan:** Oluanpi, without number.

A shrub to 13 metres tall, widely distributed in the tropics and more or less naturalised in the southern provinces of China, including Taiwan. *Ricinus communis* is the source of castor-oil and is used in paints, varnishes, resins and cosmetics. Hydrogenated castor oil is used as a lubricant for airplanes and rocket engines. Records come from Egypt where seeds have been unearthed in 6,000 year old tombs

Robiquetia succisa (Lindl.) Seidenf. & Garay (ORCHIDACEAE)

Sarcanthus henryi Schlecter in Repert. Spec. Nov. Regni. Veg. Beih. 4: 77. (1919).

Yunnan: Simao, 13582 (type of *Sarcanthus henryi* Schlecter).

Distributed in India, Bhutan, Myanmar, China, Thailand, Laos, Cambodia and Vietnam.

Rodgersia aesculifolia Batalin (SAXIFRAGACEAE)

Henry in Gard. Chron. ser. 3, 32: 132. (1902), J. H.

Veitch in Journ. Roy. Hort. Soc. 28: 62. (1903—1904).

Rodgersia podophylla Forbes & Hemsley non A. Gr. in Journ. Linn. Soc. Bot. 266. (1887). *Rodgersia* sp. nova Henry in Journ. China Br. Roy. Asiat. Soc. 22: 253. (1887) (Chinese Names of Plants).

Hubei: Badong, 849, 860 (September 1885), 4001. Xingshan, 5611a. South Badong, 5611b. **Sichuan:** South Wushan, 5711. Henry's Chinese collector with Pratt, no locality stated but may be any of the following, Leshan, Emei Shan, Wa Shan or Kangding, 8972.

According to Henry this *Rodgersia* was referred to in Chinese herbals as the *kuei teng ch'ing* or 'devil's lamp-stand' and was known colloquially in Hubei as *lao she p'an* or 'old serpent's dish', the rhizome was used as a drug. This species was described from collections made in Gansu by Potanin but according to Bretschneider it had been first found by Père Armand David near Baoxing in western Sichuan in 1869. It was introduced to cultivation by Ernest Wilson through Messrs Veitch who in their *Novelties for 1908—1909* were selling plants for ten shillings and six pence a dozen.

Rodgersia pinnata Franchet (SAXIFRAGACEAE)

Henry in Gard. Chron. ser. 3, 32: 131. fig. 44. (1902), Nelson in Pim, The Wood & the Trees 220. (1984); in A Heritage of Beauty 317. (2000), O' Brien in Augustine Henry, An Irish Plant Collector in China 33. (2002).

Yunnan: On cliffs on the mountains to the north of Mengzi at 2,650 metres, 11164.

One of a number of plants successfully introduced to cultivation by Henry through the Royal Botanic Gardens, Kew. *Rodgersia pinnata* was described by Franchet from a flowering specimen collected by Delavay to the north of Dali (probably the Cangshan Range) in north-west Yunnan. Henry's herbarium specimen is in fruit and it is from the same plant population that he sent seeds to Kew where plants first flowered in the summer of 1902. *Rodgersia pinnata* is native to Myanmar and China and is distributed in the western provinces of Guizhou, Sichuan and Yunnan. Not all the seeds Henry sent to Kew fared well. The 'Kew Inwards' (a volume recording donations to the gardens) lists large quantities of seeds sent directly from China by Henry or removed by herbarium staff from his dried specimens for propagation. For example, on April 17th 1886 Henry sent 97 packs of herbaceous seeds and 76 packets of trees and shrubs. On March 27th 1889, 34 packs of seeds were sent to Kew. A note for June 1891 records the death of 30 individuals seed lots sown on November 21st 1890, many of these, alas, were interesting new species. On March 21st 1898 Henry sent 20 packs of seeds for Kew, including 90 packs for A. K. Bulley. The amount for Bulley compared to

Kew is telling; Henry obviously felt a commercial nursery could do better.

Rohdea delavayi (Franchet) N. Tanaka (ASPARAGACEAE)

Tupistra delavayi Franchet in Bull. Soc. Bot. France 43: 40. (1896), Wright in Journ. Linn. Soc. Bot. 36: 114. (1903). *Campylandra delavayi* (Franchet) M. N. Tumara et al. in Novon. 10(2): 159. (2000).

Sichuan: South Wushan, 5231a., (syntype).

Discovered by Delavay in Yunnan, Franchet based this species on material collected by both Delavay and Henry.

Rohdea fargesii (Baill.) Y. F. Deng (ASPARAGACEAE)

Campylandra sp. nova. Henry in Journ. China Br. Roy. Asiat. Soc. 22: 249. (1887) (Chinese Names of Plants). *Tupistra chinensis* Baker in Hooker's. Icon. Pl. 19: t 1867. (1889), Henry in Notes Econ. Bot. China 21. (1893), Bretschneider in Hist. Eur. Bot. Disc. China 2: 793. (1898), Wright in Journ. Linn. Soc. Bot. 36: 114. (1903). *Campylandra chinensis* (Baker) M. N. Tumara et al. in Novon 10(2): 159. (2000).

Hubei: Yichang, 2044. Badong, 5023, 5023a., (type), 5231a., 5382. Changyang, 5231, 5692b. Jianshi, 5692a. Without locality, 5231c., 5692c. **Sichuan:** South Wushan, 5692.

According to Henry the roots of this species were used in treating mouth and throat diseases. It was known at Yichang as *k'ai kou chien* and at Badong as *ti liao yeh*. The Hani (an ethnic group) of Xishuangbanna in southern Yunnan call this species *lomanzanmanl* and the roots are used to treat flu, gastroenteritis, rheumatism and sprains.

Rohdea japonica (Thunb.) Roth (ASPARAGACEAE)

Henry in Notes Econ. Bot. China 21. (1893), Diels in Bot. Jahrb. 29: 250. (1900), Wright in Journ. Linn. Soc. Bot. 36: 115. (1903).

Campylandra sp. nova (No. 2.) Henry in Journ. China Br. Roy. Asiat. Soc. 22: 249. (1887) (Chinese Names of Plants).

Hubei: Liantuo, 5601. Badong 5023b.

According to Henry this species was called *wannien-ch'ing* and the root was used to treat cases of sore mouth. A handsome and distinctive evergreen perennial. Several cultivars are grown at the National Botanic Gardens, Glasnevin. In Japan over 70 cultivars have been named.

Rohdea spp. (ASPARAGACEAE)

Campylandra sp. Anonymous in Kew List of Determinations (volume 24). *Rohdea* sp. Anonymous in Kew List of Determinations (volume 34).

Hubei: Badong, 2873. **Yunnan:** Mengzi, 10871, 11292, 11825. Simao, 12205, 12900, 12956.

Rorippa globosa (Turcz. ex Fischer & C. A. Meyer) Hayek (BRASSICACEAE)

Nasturtium globosum Turcz. ex Fisch. ex C. A. Mey. Henry in Trans. Asiat. Soc. Jap. 24 suppl: 17. (1896) (List Pl. Formosa), Diels in Bot. Jahrb. 29: 357. (1900).

Hubei: Yichang, 194, 3225. **Taiwan:** Oluanpi, 265.

Rorippa indica (L.) Hiern. (BRASSICACEAE)

Nasturtium montanum Wall. ex Hook. f. & Thomson Henry in Trans. Asiat. Soc. Jap. 24 suppl: 17. (1896) (List Pl. Formosa), Diels in Bot. Jahrb. 29: 357. (1900).

Hubei: Yichang, 3993, 3534. Badong, 4066. **Taiwan:** Kaohsiung Plain, 1792.

Rosa banksiae W. T. Aiton (ROSACEAE)

Forbes & Hemsley in Journ. Linn. Soc. Bot. 248. 23: (1887), Anon in Gard. Chron. ser. 3. 24. (1887), Bretschneider in Hist. Eur. Bot. Disc. China 2: 783. (1898), Henry in Gard. Chron ser. 3. 33: 439. fig. 171. (1902), Nelson in Pim, The Wood & the Trees 219. (1984).

Rosa banksiae W. T. Aiton var. *normalis* Regel. Rehder & E. H. Wilson in Sargent Pl. Wilson. 2: 317. (1915), Zheng in Journ. Wuhan Bot. Res. 2.(1): 82. (1984), Nelson in A Heritage of Beauty 318. (2000).

Hubei: Hanging from rocks on the Xiling (Yichang) Gorge, 1153, 3198, 2922. **Sichuan:** South Wushan, in ravines and hedges between 650 and 1,000 metres in altitude, 5552.

Rosa banksiae was described from a cultivated plant with double white flowers, sent to England in 1807 by William Kerr from a garden in Guangzhou (Canton), this is the cultivar 'Alba Plena', which was later collected near Mengzi by Henry and in western Yunnan by Forrest. Wilson found a single plant of the double yellow cultivar 'Lutea' in Xingshan (north-west of Yichang) where it had been planted on a tomb. Both of these ancient cultivars derive from the single white wild plant (the former var. *normalis*), an abundant rose in the glens and gorges of western Hubei where it makes gigantic climbers to 15 metres tall and was known in the 19th century as *mu-hsiang* or "wood fragrance" (see Henry in Notes on Economic Botany of China 19. (1893) During the same period the root bark of this rose was used in dyeing and tanning fish-nets. Henry sent seeds to the Royal Botanic Gardens, Kew where they arrived on May 17th 1887. The wild Banksian rose was first found in China by Delavay on April 30th 1885, Henry found it a year later.

Wilson had the following to say about this rose:

The musk rose and Banksian roses often scale tall trees (in the Three Gorges) and a tree thus festooned with their branches laden with flowers is a sight to remember. To walk through a glen in the early morning or after a slight shower, when the air is laden with the soft delicious perfume from myriads of rose flowers is truly a walk through an earthly paradise.

E. H. Wilson in *A Naturalist in Western China* 1: 18. (1913)

Though no longer as common as it once was in the Three Gorges region, this variety was collected by the Glasnevin Central China Expedition (GCCE 732) on Moji Shan (the Dome of Henry and Wilson) in October 2004.

Rosa banksiae W. T. Aiton '**Alba Plena**' (ROSACEAE)

Henry in Gard. Chron. ser. 3, 31: 439. (1902).

Rosa banksiae Rehder & E. H. Wilson non W. T. Aiton. in Sargent Pl. Wilson. 2: 316. (1915).

Yunnan: Mountains to the north of Mengzi, 10508.

Evidently a cultivated plant, Henry describes this collection as a large climber with semi-double white flowers.

Rosa banksiopsis Baker (ROSACEAE) New species

Rehder & E. H. Wilson in Sargent Pl. Wilson. 2: 322. (1914).

Sichuan: South Wushan, 5746. **Hubei:** South Badong, 6071, 6071a.

Rosa bankiopsis was described from a plant flowered by Miss Ellen Wilmott, raised from seeds collected by Wilson in 1907, but had been discovered by Henry almost two decades previously.

Rosa chinensis Jacq. (ROSACEAE)

Rosa indica Lour. Forbes & Hemsley in Journ. Linn. Soc. Bot. 23: 249. (1887), Anon in Gard. Chron. ser. 3. Dec. 24th (1887), Henry in Gard. Chron. ser. 3. 33: 438. fig. 170. (1902); in Flora & Sylva 1: 217. (1903); in Trans. Asiat. Soc. Jap. 24 suppl: 40. (1896) (List Pl. Formosa), Rehder & E. H. Wilson in Sargent Pl. Wilson. 2: 320. (1914), Byhouwer in Journ. Arnold Arb. 10: 96. (1929).

Hubei: Liantuo, cultivated in a garden, 3018. **Taiwan:** Wanchin, 1609. **Yunnan:** Mountains to the south of Mengzi at 1,600 metres, 11272.

The *yueh chi* (according to Henry) or monthly rose, well known in European gardens and cultivated in China since ancient times. *Rosa chinensis* cultivars are the result of over a thousand years of hybridisation and selection in China. This rose is valued because of its perpetual blooming, its hardiness and adaptability to all locations and ease of culture. Because of its long flowering season. *Rosa chinensis* was crossed with European roses to create a race of exceptionally long flowered roses. Bean theorises that these hybrids may have been created in Yunnan where they are commonly grown even today and made their way from there, east to Guangzhou (formerly Canton) and from there to the gardens of Europe. Henry stated that *Rosa chinensis* was also known as the *yue ji hua* or *yue yue hong*

i.e. the 'moon-season flower', the partially opened flowers are used as a drug to invigorate the blood and to regulate menstruation. The Hani (an ethnic group) of Xishuangbanna in southern Yunnan call *Rosa chinensis alyaivneil* and the roots are used to treat sprains and contusions.

Rosa chinensis Jacq. f. *spontanea* Rehder & E. H. Wilson (ROSACEAE) New form

in Sargent Pl. Wilson. 2: 320. (1916), O' Brien in Augustine Henry, An Irish Plant Collector in China 33. (2002).

Rosa indica Forbes & Hemsley non Lour. in Journ. Linn. Soc. Bot. 23: 249. (1887), Henry in Gard. Chron. ser. 3, 33: 438. fig. 170. (1902); in Flora & Sylva 1: 217. (1903). *Rosa chinensis* Jacq. Nelson in Pim, The Wood & the Trees 234. (1984).

Hubei: Xiling (Yichang) Gorge, 738. Liantuo, 3018. Yichang, 4131. In the San You Dong (Three Pilgrim) Glen, 1151. **Sichuan:** Henry's Chinese collector with Pratt, no locality stated but may be any of the following, Leshan, Emei Shan, Wa Shan or Kangding, 8822.

The wild ancestor of the Chinese montly rose, itself one of the parent of our modern hybrid teas, to which it introduced important characteristics such as perpetual flowering and fragrance. Rose cultivation in China goes back for thousands of years and reached a peak during the late Ming Dynasty (1368—1644) when a great number of cultivars were raised and named. In the *Zhongguo Huajing* (China Floral Encyclopedia) it is stated that in the palace gardens of the Emperor Liang (502—547 A.D.) *Qiangwei* or rose culture flourished. During the Song Dynasty (960—1279 A.D.) perpetual flowered roses were widely cultivated in increasing variety. The Chinese generally favoured large double showy fragrant blooms and it is China rather than Europe that can claim to have improved the rose. The China rose was originally named *Rosa indica* by Linnaeus, according to Bretschneider, Linneaus had a very confused idea of the geographical position of China and often identified it with India.

Rosa corymbulosa Rolfe (ROSACEAE) New species

in Bot. Mag. cxl. t. 8566. (1914), Rehder & E. H. Wilson in Sargent Pl. Wilson 2: 323. (1914), Zheng in Journ. Wuhan Bot. Res. 2.(1): 82. (1984).

Hubei: Xingshan, 6491. Baokang, 6491a. Fang Xian, 6714.

Rosa cymosa Tratt. (ROSACEAE)

Zheng in Journ. Wuhan Bot. Res. 2.(1): 83. (1984), O' Brien in Augustine Henry, An Irish Plant Collector in China 33. (2002).

Rosa microcarpa Lindl. Forbes & Hemsley in Journ. Linn. Soc. Bot. 23: 251. (1887), Henry in Gard. Chron.

ser. 3, 31: 439. (1902), Rehder & E. H. Wilson in Sargent Pl. Wilson 2: 314. (1916), Nelson in Pim, The Wood & the Trees 219. (1984); in A Heritage of Beauty 318. (2000).

Hubei: Yichang, 248, 256, 495, 882, 1373, 3105 (on a grave), 3106. San You Dong Glen, (October 1886), 2887. Xiling (Yichang) Gorge, 3597. Changyang, 7589. **Fujian:** Xiamen (formerly Amoy), without number.

An abundant species in western Hubei and the most widely distributed of all Chinese roses. The cultivar 'Red Dragon' was selected at Glasnevin from seeds collected in western China by Keith Rushforth, the young emerging foliage is an intense fiery red and it grows at Glasnevin on the Chinese slope. *Rosa cymosa* was collected by the Glasnevin Central China Expedition in Xingshan (GCCE 489 & GCCE 568) in September 2004.

Rosa davidii Crepin (ROSACEAE)

Rehder & E. H. Wilson in Sargent Pl. Wilson. 2: 322. (1915).

Sichuan: Wa Shan, Henry's Chinese collector with Pratt, 8944.

A deciduous shrub to 3 metres tall, producing 5 cm wide rose-pink flowers in June and July followed in autumn by pendelous clusters of bright red fruits. Allied to the Himalayan *Rosa macrophylla*. Native to western and central China, *Rosa davidii* was discovered in Baoxing in western Sichuan in June 1869 and was introduced to cultivation by Wilson in 1903. Seeds of Wilson's 1238 from Baoxing and his 1060 and 1063 from Wenchuan, both in Sichuan were received at Glasnevin in 1909.

Rosa gigantea Collet ex Crép. (ROSACEAE)

Henry in Gard. Chron. ser. 3, 31: 438. (1902), Henry in Flora & Sylva 1: 217. (1903).

Rosa odorata Sweet. var *gigantea* (Collet ex Crép.) Rehder & E. H. Wilson in Sargent Pl. Wilson. 2: 338. (1915), Byhouwer in Journ. Arnold Arb. 10: 94. (1929).

Yunnan: Near Mengzi in ravines and on grassy hills at 1,500 to 1,600 metres, 9098a. Simao at 1,600 metres, 9098c.

A very vigorous semi-evergreen climber to 13 metres tall or in a suitable climate up to 27 metres tall. The stems, according to Kingdon Ward, could be as thick as a man's forearm (in France, at the Chateau Eleonore a plant gained a girth of 1.5 metres). Native to north-east India, Myanmar and Yunnan, this species was discovered by Sir George Watt in Manipur in 1882, but was described from a specimen later (1888) collected by Sir Henry Collett in Myanmar and was introduced to cultivation by him through Calcutta Botanic Gardens. One of the most rampant and largest flowered of all species roses. Kingdon Ward reintroduced it from Manipur and described it in his *Plant Hunting in Manipur* as follows:

... when the enormous flowers opened they were ivory white, borne singly all along the arching sprays, each petal faintly engraved with a network of veins like a watermark ... The globose hips look like crab apples. They are yellow with rosy cheeks when ripe, thick and iron hard.

George Forrest also met with it (Forrest 16,546) in the Mekong valley in June 1918.

Rosa helenae Rehder & E. H. Wilson (ROSACEAE)
New species

in Sargent Pl. Wilson. 2: 310. (1915), Zheng in Journ. Wuhan Bot. Res. 2.(1): 83. (1984), O' Brien in Augustine Henry, An Irish Plant Collector in China 33. (2002).

Hubei: Jianshi, 5973. **Sichuan:** North Wushan at 1,000 to 1, 500 metres, 7100.

Discovered by Henry and later collected by Wilson in 1900. Native to Sichuan and Hubei, this species commemorates Helen Wilson, wife of E. H Wilson, both of whom were tragically killed in a car crash in October 1930. Seeds of Wilson's 666 from Wushan in Sichuan were received at Glasnevin in 1908 from the Arnold Arboretum. It is a magnificent, vigorous, rambling species and at Glasnevin it is trained though an 8 metre tall tree of *Ilex aquifolium*, where in early summer it is covered in masses of fragrant white flowers.

Rosa laevigata Michx. (ROSACEAE)

Henry in Journ. China Br. Roy. Asiat. Soc. 22: 279. (1887) (Chinese Names of Plants), Diels in Bot. Jahrb. 29: 406. (1900), Rehder & E. H. Wilson in Sargent Pl. Wilson 2: 318. (1915), Zheng in Journ. Wuhan Bot. Res. 2.(1): 84. (1984).

Hubei: Yichang, 593, 1143.

A climbing shrub to 10 metres tall, producing fragrant, white flowers to 10 cm across in early summer. Native to western and central China, Myanmar and Indochina, this is a very beautiful rose when seen rambling through trees. Henry stated that it was known colloquially in the Three Gorges region as *tz'u pang t'ou*.

Rosa longicuspis A. Bertoloni (ROSACEAE)

Rehder & E. H. Wilson in Sargent Pl. Wilson. 2: 313. (1915), Byhouwer in Journ. Arnold Arb. 10: 88. (1929).

Yunnan: In rocky places near Mengzi at 1,525 metres, 9236a.

A large, semi-evergreen rambling shrub carrying lax panicles of up to 15 white flowers in summer. Described in 1861 from material collected by Hooker and Thomson in the Khasi hills of Meghalaya state, NE India, this species is native to the Himalaya, Myanmar and China (Yunnan, Guizhou and Sichuan). Wilson made several seed collections of this species in western Sichuan, those to be sent to Glasnevin included Wilson 1098, collected on Emei shan in October 1908, Wilson 1098a., from the neighbouring Wa Shan (November 1908) and south-east of Kanding in October 1908.

Rosa aff. *longicuspis* sp. nov. (ROSACEAE)

Rehder & E. H. Wilson in Sargent Pl. Wilson. 2: 336. (1916).

Rosa longicuspis Forbes & Hemsley non Bertol. in Journ. Linn. Soc. Bot. 486. (1911). *Rosa leschenaultiana* Willmott non (Thory) Wight & Arnott. in The Genus Rosa. 1: 52. (1910), Dunn in Journ. Linn. Soc. Bot. xxxix. 486. (1911).

Yunnan: In forests on Fengchunling, south of the Red River at 2, 250 metres, 10693.

Climber with white flowers.

Rosa multiflora Thunb. var. *cathayensis* Rehder & E. H. Wilson (ROSACEAE) New variety

in Sargent Pl. Wilson. 2: 313. (1915).

Rosa multiflora Forbes & Hemsley non Thunb. in Journ. Linn. Soc. Bot. 23: 253. (1887), Henry in Journ. China Br. Roy. Asiat. Soc. 22: 259. (1887) (Chinese Names of Plants), Zheng in Journ. Wuhan Bot. Res. 2.(1): 84. (1984). *Rosa multiflora* Thunb. var. Henry in Notes Econ. Bot. China 57. (1893).

Hubei: Yichang, 794 (in part), 826, 829, 1135, 1154. In the Antelope Glen near Yichang, 3129. Badong, 1780, 5289. Jianshi, 5786. Liantuo, 3854. San You Dong Glen, 7183. Changleping, 6267, 7639. **Yunnan:** Mengzi, 10385.

Described from material collected by Henry and Wilson, this rose had been previously gathered by other earlier collectors and is distributed from north-east India to China. The flowers corymbs are pink-tinged and larger than the type from Japan which is white and usually carries more flowers in the clusters. This variety is the parent of many good Chinese cultivars introduced to Europe in the early 19th century, the type was not introduced until 1860. Henry stated that in the Three Gorges region this variety was known as *mieh liang huang tz'u*. Several collections of this distinctive variety were made by the Glasnevin Central China Expedition in the autumn of 2002 including, the *Metasequoia* valley near Modaoqui (GCCE 150), between Yichang and Badong (GCCE 170 & GCCE 183) and in the mountains above Badong (GCCE 426). Seedlings raised from these collections flowered at Glasnevin for the first time in May of 2005. The resultant seedlings from these collections proved to be extremely variable in their vigour, leaf shape and leaf and stem colour. In the *Metasequoia* valley local farmers made low effective hedges of this variety.According to Henry, *hu-p'i* was the name given at Yichang to the root bark which was used for tanning fishing nets. Henry also recounts having seen enormous specimens

with stems of 15 cm. in diameter and over 30 metres long. (One can only guess at the luxuriance of the flora of the gorges in the late 19th century).

Rosa × *odorata* (Andr.) Sweet (ROSACEAE)

Byhouwer in Journ. Arnold Arb. 10: 93. (1929).

Yunnan: Mengzi, 10828.

An evergreen or semi-evergreen shrub belonging to a group of ancient and variable hybrids between *Rosa chinensis* and *Rosa gigantea*. In cultivation they are best grown against a warm, sheltered wall where they can reach 5 metres tall. This rose is commonly cultivated throughout Yunnan and it was probably from that province that it was sent to the east coast of China and from there to the gardens of Europe. According to Wilson, it could be seen in semi-double or fully double forms in colours of white, yellow, buff or pale rose-pink or a combination of these colours in Yunnan. The same may be said today, such roses are commonly grown around Lijiang and Dali. One of the best known of these hybrids is 'Fortune's Double Yellow' ('Pseudindica'), a beautiful old Chinese hybrid with semi-double, richly scented coppery yellow flushed scarlet flowers. In the Lijiang valley there is a rose cultivated that is almost identical to this cultivar. It was introduced by Robert Fortune in a mandarin's garden at Ningbo in eastern China in 1845. This rose flowers on two-year-old wood and should be pruned accordingly.

Rosa omeiensis Rolfe (ROSACEAE) New species

in Bot. Mag. cxxxviii. t. 8471. (1912), Rehder & E. H. Wilson in Sargent Pl. Wilson 2: 331. (1915), Zheng in Journ. Wuhan Bot. Res. 2.(1): 85. (1984).

Hubei: Shennongjia, 6782. **Sichuan:** Wa Shan, in woodlands at 2,300 to 3,300 m., Henry's Chinese collector with Pratt, 8947, 8961.

Rosa omiensis was described from collections made by Henry in Shennongjia in 1888 and by Faber on Emei Shan in 1887. Wilson's 4012 from Sungpan in Sichuan, collected on the 27th of August 1910 reached Glasnevin from the Arnold Arboretum that winter. A beautiful species, it is widely distributed through Bhutan and China [Hubei, Sichuan and Tibet (Xizang) and Yunnan].

Rosa roxburghii Tratt. f. *normalis* Rehder & E. H. Wilson. (ROSACEAE)

Rosa microphylla Anonymous non Roxb. in Kew List of Determinations (volume 24).

Sichuan: Henry's Chinese collector with Pratt, no locality stated but may be any of the following, Leshan, Emei Shan, Wa Shan or Kangding, 8949.

The species was originally described from plants cultivated in the Calcutta Botanic Gardens said to have been introduced from China. This is the wild, single flowered plant which is common in western Sichuan.

Rosa rubus H. Lévl. & Vant. (ROSACEAE) New species

Rehder & E. H. Wilson in Sargent Pl. Wilson. 2: 308. (1915), Zheng in Journ. Wuhan Bot. Res. 2.(1): 85. (1984), Bean in Trees & Shrubs 4: 133. (1989), Lancaster in Travels in China 316. (1993).

Sichuan: South Wushan, 5550.

A vigorous spreading shrub to 6 metres tall, the fragrant white flowers are held in dense clusters and are up to 3 cm across and appear during June and July. *Rosa rubus* was discovered by Augustine Henry in 1886, though the type was based on a collection made by L. Martin in Guizhou in May 12th 1899. It was introduced to cultivation by E. H. Wilson and was later reintroduced by Reginald Farrer from Gansu. This species has remained rare in gardens, there is a plant near the rose garden at Glasnevin. The bark of the roots of this species is used in China for extracting tannin.

Rosa sambucina Koidz. (ROSACEAE)

Rosa moschata Mill. var. *brunoni* Forbes & Hemsley non Lindl. in Journ. Linn. Soc. Bot. 23: 252. (1887). *Rosa cerasocarpa* R. A. Rolfe in Kew Bull. Misc. Inf. 89. (1915). *Rosa rubus* Rehder & E. H. Wilson non H. Lév. & Vanihot in Sargent Pl. Wilson 2: 308. (1915), (in part). *Rosa gentiliana* sens. Rehder & E. H. Wilson in Sargent Pl. Wilson 2: 312. (1915). *Rosa henryi* Boulenger in Ann. Soc. Bruxelles 53: 143. (1933), Zheng in Journ. Wuhan Bot. Res. 2.(1): 83. (1984), Bean in Trees & Shrubs. 4: 72. (1989), Lancaster in Travels in China 394. (1993), Nelson in A Heritage of Beauty 318. (2000).

Hubei: Yichang, 2952 (type of *Rosa cerasocarpa* R. A. Rolfe). North Badong, 7007 (syntype of *Rosa cerasocarpa* Rolfe). **Sichuan:** South Wushan, 5773.

Rosa saturata Baker (ROSACEAE) New species

Rehder & E. H. Wilson in Sargent Pl. Wilson. 2: 324. (1916), Zheng in Journ. Wuhan Bot. Res. 2.(1): 85. (1984).

Hubei: Xingshan, 6747.

Though Wilson is generally credited with the discovery of this rose it was originally found by Henry in August 1888. *Rosa saturata* is a close ally of *Rosa macrophylla* Lindl. and was described by Baker from a plant in Miss Willmott's garden raised from Wilson one of Wilson's collections. Wilson's 316 from Fang Xian, collected in September 1907 was received at Glasnevin from the Arnold Arboretum in 1908.

Rosa sertata Rolfe (ROSACEAE) New species

Rehder & E. H. Wilson in Sargent Pl. Wilson. 2: 327. (1916), Zheng in Journ. Wuhan Bot. Res. 2.(1): 85. (1984).

Sichuan: South Wushan 5679. **Hubei:** Xingshan at 2,600 to 2,900 metres, 6997.

Rosa sertata was introduced by one of the French

missionaries to Les Barres in 1897 and was distributed by Vilmorin as *Rosa webbiana* of which some authorities concider it a subspecies. Rolfe described the species from plants at Kew raised by Messrs Veitch from seeds collected by Wilson (probably Wilson-Veitch 1492, from Sichuan, collected October 1904). Henry, however had found this species previous to both introductions. A beautiful profusely flowered rose to 2 metres tall, carrying rose-red flowers to 5cm.

Rosa sp. No. 1. (ROSACEAE)

Anonymous in Kew List of Determinations (volume 34).

Yunnan: Mengzi, 13770.

Rosa sp. No. 2. (ROSACEAE)

Rosa moschata Herrm. Forbes & Hemsley in Journ. Linn. Soc. Bot. 23: 252. (1887).

Hubei: Yichang, 3082, 4308. Fang Xian, 7623. Xingshan, 7622.

Rosa sp. No. 3. (ROSACEAE)

Anonymous in Kew List of Determinations (volume 24).

Hubei: Yichang, 1327. Fang Xian, 6664, 6784, 7621. Badong, 4035. **Sichuan:** South Wushan 7255, 7528. North Wushan, 6664a.

Rosaceae (ROSACEAE)

Anonymous in Kew List of Determinations (volume 34).

Yunnan: Simao, 13272.

Roscoea debilis Gagnepain (ZINGIBERACEAE) New species

Cowley in Kew Bull. 36: 772. (1982).

Roscoea blanda K. Schumann in Engler, Pflanzenr. 20: 121. (1904), Cowan in New Flora and Silva 27. (1939).

Yunnan: Mengzi at 1,670 to 2,130 metres, 11102, 11102a. Mengzi at 1,830 metres, 11102c., (holotype of *Roscoea blanda* Schumann).

Gagnepain described this species from a collection made in Yunnan by Ducloux (Ducloux 688) in August 1899, Henry had found it two years previously. The flowers of this species are bluish-purple or sometime white and appear in groups of three or singly from June to August. In its native China (Yunnan) and Myanmar it grows in open grassland, in rough grass at the base of cliffs and on the margins of forests between 1,670 to 2,400 metres. In Yunnan it was later collected near Tengchong near the Myanmar border by Forrest in June 1912, he found it again on the eastern flank of the Cangshan range above Dali in July 1910. Ducloux collected the type near Luteng to the west of Kunming in 1899 and Henry made the first collection of this species at Mengzi in 1897. The genus commemorates William

Roscoe (1753—1831), one of the founders of Liverpool Botanic Gardens who grew a number of species that had been sent to him by Nathaniel Wallich of the Royal Botanic Gardens, Calcutta from a visit to Nepal. *Roscoea* is found in the Himalaya from Kashmir to north-east India and China [south-east Tibet (Xizang)] and from there into Sichuan, Yunnan and Myanmar.

Roscoea praecox K. Schumann (ZINGIBERACEAE)
New species

in Engler, Pflanzenr. 20: 122. (1984), Cowan in New Flora and Silva 27. (1939), Cowley in Kew Bull. 36: 756. (1982).

Yunnan: Mengzi at 1,525 metres, 11117, (isotype).

A low altitude species, endemic to Yunnan where it grows between the provincial capital of Kunming and Henry's former base at Mengzi. In its native habitat it grows along the banks of streams, in pastures, on shady limestone cliffs and in thickets between 500 to 2,300 metres. Flowers can be purple, violet or white and appear between the end of April to June. The only truly precociously flowering species, *Roscoea praecox* was later collected by Cavalerie near Kunming in 1916. It was discovered by the Irish plant collector William Hancock at Mengzi in April 1895 and was recollected by Henry about three years later.

Rostellularia diffusa (Willd.) Nees (ACANTHACEAE)
Justicia diffusa Willd. Forbes & Hemsley in Journ. Linn. Soc. Bot. 26: 245. (1890).

Hainan: Haikou, without number.

Rostellularia procumbens (L.) Nees (ACANTHACEAE)
Justicia procumbens L. Journ. Linn. Soc. Bot. 26: 246. (1890), Henry in Trans. Asiat. Soc. Jap. 24 suppl: 69. (1896) (List Pl. Formosa). *Justicia simplex* D. Don Henry in Trans. Asiat. Soc. Jap. 24 suppl: 69. (1896) (List Pl. Formosa).

Hubei: Yichang, 408, 983. Lung Tung Ch'i Glen (near Yichang), 4196. **Taiwan:** Wanchin, 136, 972, 1018. Oluanpi, 210. Kaohsiung Plain, 1169, 2045. **Yunnan:** Mengzi, 9388. Simao, 12408.

In Chinese herbal medicine this species is used in cases of cold and fever.

Rostellularia prostrata (Roxb. ex C. B. Clarke) R. B. Majumdar (ACANTHACEAE)
Rostellularia diffusa (Wiild.) Nees var. *prostrata* (Roxb. ex C. B. Clarke) J. L. Ellis Wu & Lo in Flora Hainanica 3: 598. (1974).

Hainan: Environs of Haikou, 8322.

Rostrinucula sinensis (Hemsl.) C. Y. Wu (LAMIACEAE) New species
Leucosceptrum sinense Hemsley in Journ. Linn. Soc. Bot. 26: 310. (1890), Bretschneider in Hist. Eur. Bot. Disc. China 2: 790. (1898), Dunn in Notes Roy. Bot. Gard.

Edinburgh 6: 192. (1915).

Hubei: On cliffs near Changyang, 7765.

Rotala indica (Willd.) Koehne (LYTHRACEAE)
Ammannia peploides Spreng. Henry in Trans. Asiat. Soc. Jap. 24 suppl: 44. 1896 (List Pl. Formosa).

Taiwan: Wanchin, 814.

Native to tropical and subtropical Asia, often in paddy fields at low altitude.

Rotala rotundifolia (Buch.-Ham. ex Roxb.) Koehne (LYTHRACEAE)
Ammannia rotundifolia Buch.-Ham. ex Roxb. Forbes & Hemsley in Journ. Linn. Soc. Bot. 23: 303. (1887), Henry in Trans. Asiat. Soc. Jap. 24 suppl: 44. 1896 (List Pl. Formosa).

Hubei: Yichang, 220. **Taiwan:** Kaohsiung Plain, 1788. **Yunnan:** Mengzi, 9355.

An erect herb to 30 cm tall, native to much of tropical Asia. In Taiwan this species is found in marshes, streamsides and paddy fields at low altitudes.

Rotheca serrata (L.) Steane & Mabb. (LAMIACEAE)
Clerodendrum serratum (L.) Moon var. *amplexifolium* Moldenke in Phytologia 4(1): 51. (1952).

Yunnan: Mengzi, 10077, 10077a., (type of *Clerodendrum serratum* (L.) Moon var. *amplexifolium* Moldenke). Simao, 12482.

An evergreen bush or small tree carrying spikes of small white or pale-pink flowers. In summer the Hani (an ethnic group) of Xishuangbanna in southern Yunnan call this species *smlpavqdovniov* and use it to treat tonsillitis, malaria and rheumatism. Native to China, India and Malaysia.

Rottboellia cochinchinensis (Lour.) Clayton (POACEAE)
Rottboellia exaltata L. f. Henry in Trans. Asiat. Soc. Jap. 24 suppl: 108. (1896) (List Pl. Formosa), Rendle in Journ. Linn. Soc. Bot. 36: 362. (1904).

Taiwan: Wanshoushan, 1127. Kaohsiung, 1127a.

A perennial to 30 cm tall, native to much of tropical Africa, Asia and Australia.

Rourea caudata Planch. (CONNARACEAE)
Anonymous in Kew List of Determinations (volume 34).

Yunnan: Simao, 11607.

Distributed in India and western China.

Rourea minor (Gaertn.) Leenh. (CONNARACEAE)
Santaloides roxburghii (Hook. & Arn.) O. Ktze. Schellenberg in Engler, Pflanzenr. iv. 127: 125. (1938).

Hainan: Without locality, 8561.

Rubia chinensis Regel & Maack (RUBIACEAE)
Anonymous in Kew List of Determinations (volume 24).

Sichuan: South Wushan, 5650.

02

Rubia cordifolia L. (RUBIACEAE)

Henry in Journ. China Br. Roy. Asiat. Soc. 22: 238. (1887) (Chinese Names of Plants).

Hubei: Yichang, 1096, 2032, 2226, 3252, 3954, 4174, 4305. San You Dong Glen (July 1886), 1525. Antelope Glen (November 1886), 3251, 3556. Badong, 2505, 1797, 3178, 4776. Liantuo, 2679. Changleping, 6264. **Yunnan:** Mengzi, 9981. Simao, 11998.

Ch'ien ts'ao (according to Henry), the source of 'Indian madder', though he stated that a dye was not extracted from the roots around Yichang. The Hani (an ethnic group) of Xishuangbanna in southern Yunnan call this species *qilxeel* and use its roots to treat diarrhea.

Rubiaceae (RUBIACEAE)

Anonymous in Kew List of Determinations (volume 24 & 34).

Hubei: South Badong, 7366. **Yunnan:** Simao, 12369, 13123, 13124, 13170, 13518.

Rubiteucris palmatum (Benth.) Kudo (LAMIACEAE)

Teucrium palmatum Benth. Forbes & Hemsley in Journ. Linn. Soc. Bot. 26: 313. (1890), Dunn in Notes Roy. Bot. Gard. Edinburgh 6: 191. (1915).

Hubei: Fang Xian at 2,600 metres, 6844.

Previous to Henry's collection this species was known only from Lachen in Sikkim.

Rubus alceifolius Poir. (ROSACEAE)

Rubus hainanensis Focke in Biblioth. Bot. 83. (1909). *Rubus gilvus* Focke in in Biblioth. Bot. 17: 79. (1910), Chun in Sunyatsenia 1: 251 (1934).

Hainan: Without locality, 8581 (holotype of *Rubus hainanensis* Focke). Ling-men, 8645 (isotype of *Rubus gilvus* Focke).

A scandent shrub, widely distributed across Myanmar, China, Thailand, Cambodia, Vietnam, Sumatra, Malaya, Java and Borneo.

Rubus assamensis Focke (ROSACEAE)

Anonymous in Kew List of Determinations (volume 34).

Yunnan: Mengzi, 10200.

Distributed in India, Myanmar and western and central China.

Rubus bambusarum Focke (ROSACEAE) New species

in Hooker's. Icon. Pl. 30: tt. 1952. (1891), Bretschneider in Hist. Eur. Bot. Disc. China 2: 783. (1898), J. H. Veitch in Hortus Veitchii 374. (1906), Focke in Bibl. Bot. 32: 44. (1910).

Rubus henryi Hemsl & Kuntze var. *bambusarum* (Focke) Rehder in Journ. Arnold Arb. 2: 179. (1922), O'Brien in Augustine Henry, An Irish Plant Collector in China 33. (2002).

Sichuan: South Wushan, 5618 (type). **Hubei:** Jianshi, 5618a.

This species differs from *Rubus henryi* in the 3-foliate, not 3~5 parted leaves. Introduced to cultivation by Wilson from Hubei in 1900, in Veitch's *Novelties for 1908—1909* plants were selling for three shillings and six pence each. There are good plants at Glasnevin, Kilmacurragh and Mount Congreve. This variety was collected by the Glasnevin Central China Expedition in Badong (GCCE 369) in October 2002 and again by the 2004 Glasnevin Expedition to nort-west Hubei (GCCE 788) on Tianthu Shan in Changyang in October of that year.

Rubus buergeri Miq. (ROSACEAE)

Zheng in Journ. Wuhan Bot. Res. 2.(1): 86. (1984).

Hubei: Changyang, 7759, 7684.

Rubus calycinus Wall. ex G. Don (ROSACEAE)

Anonymous in Kew List of Determinations (volume 34).

Yunnan: Fengchunling, 10641.

Rubus chiliadenus Focke (ROSACEAE) New species

in Hooker's Icon. Pl. 20: tt. 1952. (1891), Bretschneider in Hist. Eur. Bot. Disc. China 2: 783. (1898), Zheng in Journ. Wuhan Bot. Res. 2.(1): 86. (1984).

Hubei: Jianshi, 6009 (isotype).

Rubus chroosepalus Focke (ROSACEAE) New species

in Hooker's Icon. Pl. 20: t. 1952. (1891), Bretschneider in Hist. Eur. Bot. Disc. China 2: 783. (1898), J. H. Veitch in Hortus Veitchii 375. (1906), Bean in Kew Bull. Misc. Inf. 45. (1910); in Trees & Shrubs 2: 456. (1921), Zheng in Journ. Wuhan Bot. Res. 2.(1): 86. (1984), Nelson in A Heritage of Beauty 318. (2000).

Hubei: Badong, 5505 (type), 7291. Without locality, 5505a.

A large straggling semi-evergreen bush, introduced to cultivation by Wilson from Yichang in 1900 for Messrs Veitch and reintroduced a few years later when collecting for the Arnold Arboretum. Some of the seeds from this collection were received at the Royal Botanic Gardens, Kew in January 1908 from which it was first grown in England. According to Wilson, the fruits which are black, ripen in September but are of poor flavour. According to W. J. Bean, a colleague of Henry's and one-time curator of Kew, the large *Tilia*-like leaves of this *Rubus* with their white under-surfaces make it a distinct and striking species. In the Veitch catalogue *New Hardy Plants from Western China* (Autumn 1912) it was available for two shillings and six pence.

Rubus clinocephalus Focke (ROSACEAE) New species

in Biblioth. Bot. 72: 102. (1910).

Yunnan: Mengzi at 1,950 metres, 10293 (holotype).

Distributed in southern China and Vietnam.

Rubus cochinchinensis Tratt. (ROSACEAE)

Zheng in Journ. Wuhan Bot. Res. 2.(1): 86. (1984).

Rubus playfairianus Rehder & E. H. Wilson, non Hemsl. & Focke, in Sargent Pl. Wilson. 1: 49. (1911).

Hubei: Yichang 3417a., 3417b.

Rubus conduplicatus Duthie and J. H. Veitch (ROSACEAE)

Rubus trianthus Focke in Biblioth. Bot. 72: 140, fig. 59. (Spec. Rub.) (1910), Zheng in Journ. Wuhan Bot. Res. 2.(1): 90. (1984).

Hubei: South Badong, 6045 (type of *Rubus trianthus* Focke). Changyang, 6045a.

Rubus corchorifolius L.f. (ROSACEAE)

Forbes & Hemsley in Journ. Linn. Soc. Bot. 23: 230. (1887), Henry in Journ. China Br. Roy. Asiat. Soc. 22: 264. (1887) (Chinese Names of Plants), Bean in Kew Bull. Misc. Inf. 46. (1910), Zheng in Journ. Wuhan Bot. Res. 2.(1): 86. (1984).

Rubus involucratus Focke in Biblioth. Bot. 17: 132. (1909).

Hubei: Yichang, 1305. On the mountains above Jianshi at 1,650 metres, 5801, 6040. Zigui at 1,950 metres, 6045b. **Yunnan:** South of Red River from Manpan at 1,950 metres, 10358 (isotype of *Rubus involucratus* Focke).

As pointed out by W. J. Bean, although this species was named by the younger Linnaeus in 1781 from a Japanese specimen, it was not introduced to cultivation until collected by Wilson in 1907. A widespread species in central and western China, in the Three Gorges region Henry stated it was known colloquially as *shan p'ao-tzu*. This species has great potential as a fruit crop, according to Li et al. [Journ. Wuhan Bot. Res. 18(3): 240. 2000] the fruits are large, firm, early ripening with excellent flavour, aroma and good colour. The plant is vigorous, of erect habit, nearly spine free, resistant to pests and diseases and adapts to various situations well. A widespread species found over much of Myanmar, China, Korea, Vietnam and Japan.

Rubus coreanus Miq. (ROSACEAE)

Forbes & Hemsley in Journ. Linn. Soc. Bot. 23: 230. (1887), Henry in Journ. China Br. Roy. Asiat. Soc. 22: 264. (1887) (Chinese Names of Plants), J. H. Veitch in Hortus Veitchii 375. (1906), Bean in Kew Bull. Misc. Inf. 46. (1910), Zheng in Journ. Wuhan Bot. Res. 2.(1): 86. (1984).

Hubei: On the floor of a cave on way to Moji Shan, 1540 (June 1886),. On Tsui Fu Shan near Yichang 1625 (May 1888). Yichang, 3482.

A deciduous shrub to 3 metres tall with arching biennial stems covered in a blue-white bloom. According to Henry the fruits of this species are edible and of various colours, red, yellow and black. Henry stated that in the Three Gorges area it was known colloquially as *ch'aio me*. Brambles have a long history in Chinese traditional medicine, the unripe fruits of *Rubus coreanus* have been prescribed to invigorate the liver and kidney and to improve visual acuity. In most cases it is the roots of the bramble that are used in Chinese medicine, sometimes the whole plant as in *Rubus xanthocarpus*. The main uses of these medicinal rubi are to invigorate the kidneys, stop bleeding, and due to their rich tannin content, to act as an anti-inflammatory and anti-bacterial agent. Forty-five species and four varieties are claimed to have medicinal values (Weilin, He & Gu in Journ. Wuhan Bot. Res. 18(3): 238. (2000). Some of those used are *R. palmatus* Thunb., *R. corchorifolius* L. f., *R. parvifolius* L., *R. parkeri* Hance, *R. irenaeus* Focke, *R. innominatus* S. Moore, *R. clinocephalus* Focke, *R. peltatus* Maxim., *R. pungens* Camb., *R. ellipticus* Smith, *R. lambertianus* Ser. and its variety *glaber* Hemsl., *R. eucalyptus* Focke, *R. ichangensis* Hemsl. & Kuntze, *R. idaeus* L., and *R. pungens* Camb. var. *oldhamii* (Miq.) Maxim., for example.

W. J. Bean stated that the most attractive features of this species are its erect, stout stems covered with a beautiful blue-white bloom, and its lustrous, handsomely cut, pinnate leaves. Discovered in Korea by the Kew collector Richard Oldham, it was introduced to cultivation by E. H. Wilson in 1907. Seeds of *Rubus coreanus*, Wilson 31 from Yichang were received at Glasnevin in 1909 from the Arnold Arboretum.

Rubus coreanus Miq. var. *tomentosus* Cardot (ROSACEAE)

Zheng in Journ. Wuhan Bot. Res. 2.(1): 87. (1984).

Rubus coreanus Forbes & Hemsley non Miq. in Journ. Linn. Soc. Bot. 23: 230. (1887).

Hubei: Yichang, 755. Jianshi, 5802.

Rubus efferatus Craib (ROSACEAE)

Rubus laxus Focke in Biblioth. Bot. 72: 68. (1909), nom. illeg.

Yunnan: In Pingbian, on the Daweishan Range at 1,650 metres, 11030 (isotype *Rubus laxus* Focke).

Himalaya to China (Yunan) and Indochina.

Rubus ellipticus Sm. f. forma (ROSACEAE)

Anonymous in Kew List of Determinations (volume 34).

Yunnan: Mengzi, 9393.

Rubus eucalyptus Focke (ROSACEAE) New species

in Biblioth. Bot. 72: 102. (1911), Zheng in Journ. Wuhan Bot. Res. 2.(1): 87. (1984).

Hubei: South Badong, 5427 (syntype). Jianshi, 5872 (isosyntype, June 12th 1888).

02

Rubus eustephanos Focke (ROSACEAE) New species
in Bot. Jahrb. Syst. 36: 54. (1905).

Hubei: Changyang, 5237, 5237d. Liantuo, 5237a., (type). Jianshi, 5237c. By the Tatung rapid above the entrance to the Xiling (Yichang) Gorge, 5237e. **Sichuan:** South Wushan, 5237b.

Rubus evadens Focke (ROSACEAE) New species
in Bibl. Bot. 72(1): 75. (1909), Flora of China 9: 252. (2003).

Rubus viburnifolius Focke in Biblioth. Bot. 72: 75. (1910), non Franchet 1895.

Yunnan: Simao, 11714, (isotype of *Rubus evadens* Focke, isotype of *Rubus viburnifolius* Focke).

Endemic to southern Yunnan.

Rubus flagelliflorus Focke (ROSACEAE) New species
in Engler's Bot. Jahrb. 29: 393. (1900), Focke in Engler Bot. Jahrb. 29: 393. (1900); in Biblioth. Bot. Heft 72-74 17: 111. (1911), Bean in Trees & Shrubs 2: 458. (1921), Nelson in A Heritage of Beauty 318. (2000).

Rubus acerifolius Henry non Wallich in Notes Econ. Bot. China 45. (1893), Zheng in Journ. Wuhan Bot. Res. 2.(1): 85. (1984).

Hubei: Badong, 5416, 5416e. Jianshi, 5416c. **Sichuan:** South Wushan, 5416a., 5416g.

A graceful climbing evergreen shrub with white felted stems covered with small decurved prickles. This was the *pao-ku* according to Henry, the fruits of this species are black and of good flavour. Native to central and western China, it was iintroduced to cultivation by Wilson when collecting for Messrs Veitch in 1901. In their *Novelties for 1908—1909* Veitch were selling plants for three shillings and six pence each.

Rubus flosculosus Focke (ROSACEAE) New species
in Hooker's Icon. Pl. 20. tt. 1952. (1891), Bretschneider in Hist. Eur. Bot. Disc. China 2: 783. (1898), Zheng in Journ. Wuhan Bot. Res. 2.(1): 87. (1984).

Hubei: Jianshi, 5853 (type), 5921. South Badong, 7321. Xingshan, 6495 (isosyntype). Zigui, 5921a. South Badong, 5921b, 5921c.Without locality, 5854a.

A deciduous shrub to 4 metres tall with stout arching brown-coloured biennial stems and pinnate leaves to 17 cm long. In a paper on the future potential of Chinese rubi, Li, He & Gu in Journ. Wuhan Bot. Res. 18(3): 239. (2000) suggest that this species has potential for breeding towards productivity. With rich reserves and wide variability, Chinese rubi provide an abundant source of previously untapped germplasm for hybridization. Studies on this subject have been carried out at Nanjing Botanical Gardens since 1989, 27 species were identified to be important germplasm resources for raspberry and blackberry breeding because of their resistance to pests and diseases, tolerance to low and high temperatures, adaptability, good habit and fruit quality. *Rubus flosculosus* was introduced to cultivation by Wilson in 1907, seeds of his 145a., from Fang Xian were received at Glasnevin in 1908.

Rubus fockeanus S. Kurz (ROSACEAE)
Anonymous in Kew List of Determinations (volume 24).

Hubei: Fang Xian, 6839. Xingshan, 6839a.

A prostrate shrub to ahout 3 cm tall, discovered in Baoxing in western Sichuan by Père Armand David in Baoxing in western Sichuan in 1869. Though the locality given by Henry is Fang Xian, by his discription in *The Garden* 61: 4. (1902) it is clear that his 6839 is from present day Shennongjia. This species was collected by the Glasnevin Central China Expedition (GCCE 53) on Emei Shan (Mount Omei) in September 2002.

Rubus fraxinifolius Poir. (ROSACEAE)
Rubus sp. Henry in Trans. Asiat. Soc. Jap. 24 suppl: 40. (1896) (List Pl. Formosa). *Rubus alnifoliolatus* H. Lévl. & Vaniot Li in Woody Flora of Taiwan 320. (1963).

Taiwan: Wanchin, 517, 517a. Oluanpi, 1256.

Rubus henryi Hemsl & Kuntze (ROSACEAE) New species
in Journ. Linn. Soc. Bot. 23: 231. (1887), Henry in Notes Econ. Bot. China 45. (1893), Bretschneider in Hist. Eur. Bot. Disc. China 2: 782. (1887), Henry in Flora & Sylva 1: 217-218. (1902), Schneider in Illus. Handb. Laubholzk. 2: 965. (1912), Morley in Glasra 3: 80. (1979), Zheng in Journ. Wuhan Bot. Res. 2.(1): 87. (1984), Nelson in Pim, The Wood & the Trees 234. (1984); in A Heritage of Beauty 318. (2000), O' Brien in Augustine Henry, An Irish plant Collector in China 33. (2002).

Hubei: Badong, 1728. Without locality, 5665a. **Sichuan:** South Wushan, 5665 (type). **Hunan:** Shimen, 7551.

An elegant evergreen scandent shrub, scrambling its way through trees and shrubs to 7 metres tall. A remarkably distinct species, readily distinguished by handsome leaves of three to five narrow diverging lobes and covered on the underside with a fine white felt. Henry stated that in the Three Gorges area it was called *chi-chao-ch'a* as the leaves were once dried and used as a tea substitute in the 19th century. In the Veitch catalogue *New Hardy Plants from Western China* (Autumn 1912) this species was available for two shillings and six pence.

Rubus henryi Hemsl & Kuntze var. ***sozostylus*** (Focke) T. T. Yu & L. T. Lu (ROSACEAE) New variety
Rubus sozostylus Focke in Hookers Icon. Pl. tt. 1952. (1891).

Hubei: Badong, 5005 (in part).

Rubus hirsutus (ROSACEAE)

Rubus thunbergii Anonymous non Sieb. & Zucc. in Kew List of Determinations (volume 24).

Sichuan: Henry's Chinese collector with Pratt, no locality stated but may be any of the following, Leshan, Emei Shan, Wa Shan or Kangding, 8962, 8974.

Rubus ichangensis Hemsley & Kuntze (ROSACEAE) New species

in Journ. Linn. Soc. Bot. 23: 231. (1887), Henry in Journ. China Br. Roy. Asiat. Soc. 22: 264. (1887) (Chinese Names of Plants); in Notes Econ. Bot. China 45. (1893), Bretschneider in Hist. Eur. Bot. Disc. China 2: 782. (1898), J. H. Veitch in Hortus Veitchii 376. (1906), Schneider in Illus. Handb. Laubholzk. 2: 963. (1912), Bean in Trees & Shrubs 2: 376. (1921), Zheng in Journ. Wuhan Bot. Res. 2.(1):91. (1984), Nelson in A Heritage of Beauty 318. (2000), O' Brien in Augustine Henry, An Irish Plant Collector in China 33. (2002).

Hubei: Yichang, 3242, 4112. South Badong, 7305. Liantuo, 6339 (isotype).

A deciduous rambling shrub to about 7 metres tall with many slender stems carrying very handsome narrowly ovate-cordate, sparsely toothed leaves. The small red fruits are held in panicles of up to 60 cm long and are particularly beautiful in autumn. Henry stated that in the Three Gorges region it was known colloquially as *tung p'ao tzu* or huang *p'ao-tzu.* Writing in *The Garden,* p.3. 1902, Henry observed that each new valley and range yielded to him some new species of *Rubus* and that many of these brambles in a wild state have exquisetly flavoured fruits, thus having great potential in hybridization. In *Flora & Sylva* 1: 218. (1903) he stated;

I have collected of *Rubus* probably sixty or seventy distinct species; not meaning by *species* the trivial distinction which is imported into the word when British brambles are concerned, but meaning really different plants which no one would unite together.

Rubus ichangensis was collected by the Glasnevin Central China Expedition (GCCE 645) in the Dragon Gate River Forest Reserve in Xingshan in October 2004.

Rubus idaeopsis Focke (ROSACEAE) New species

in Biblioth. Bot. 72: 103. (1911).

Rubus innominatus Focke non S. Moore in Biblioth. Bot. Heft 72-74, 17: 195. (1911), Rehder & E. H. Wilson in Sargent Pl. Wilson. 3: 424. (1917).

Yunnan: Near Mengzi at 2,000 metres, 10922 (isotype).

Endemic to China.

Rubus idaeus L. f. forma (ROSACEAE)

Anonymous in Kew List of Determinations (volume 34).

Yunnan: Mengzi, 10199.

Rubus innominatus S. Moore (ROSACEAE)

Zheng in Journ. Wuhan Bot. Res. 2.(1): 88. (1984).

Hubei: Xingshan at 1,400 metres, 6536. South Badong, 7286.

This handsome bramble was discovered by Dr. George Shearer near Jiujing in Jiangxi province in 1873. It was introduced to cultivation by Wilson from Hubei while collecting for Messrs Veitch. According to James Veitch the main feature of this species are the large panicles of bright orange-scarlet panicles of edible fruits carried in September. In their *Novelties for 1908—1909*, Veitch had plants available for two shillings and six pence each.

Rubus innominatus S. Moore var. *kuntzeanus* (Hemsl.) L. H. Bailey (ROSACEAE) New variety

Rubus kuntzeanus Hemsley in Journ. Linn. Soc. Bot. 23: 232. (1887), Henry in Journ. China Br. Roy. Asiat. Soc. 22: 264. (1887) (Chinese Names of Plants), Bretschneider in Hist. Eur. Bot. Disc. China 2: 782. (1898), Bean in Trees & Shrubs 4: 226. (1989).

Hubei: Near Moji Shan, 1536 (June 1886). In a ravine on Moji Shan, 4178. Yichang, 3633. Badong, 1840, 6096 (type of *Rubus kuntzeanus* Hemsley). Liantuo, 6096a. South Badong, 6096b. North Badong, 6096d. Without locality, 3633a. **Sichuan:** South Wushan, 6096b.

A deciduous shrub carrying silvery-grey biennial stems to 3 metres tall. Native to central and western China and introduced to cultivation by Henry through the Royal Botanic Gardens, Kew from Yichang in 1886. It was reintroduced by Wilson in 1900 through Veitch's Coombe Wood nursery and again in 1907 through the Arnold Arboretum. Henry stated that *Rubus innominatus* var. *kuntzeanus* was known colloquially in the Three Gorges region as *tsao ku.* W. B. Hemsley named this shrub for the 19th century botanist, Dr. Otto Kuntze who had an unrivalled knowledge of the simple-leaved rubi.

Rubus inopertus (Focke) Focke ex Diels (ROSACEAE) New species

in Engl. Bot. Jahrb. 29: 400. (1901), Focke in Boblioth. Bot. Heft 72-74. 17: 182. (1911).

Rubus niveus Henry non Thunb. in Notes Econ. Bot. China 45. (1893).

Sichuan: South Wushan, 5523, 5623, 5772, 7336. Without locality, 8772. **Yunnan:** Mengzi, 10624. South of the Red River in forest on Fengchunlin at 2,300 metres, 10642.

According to Henry, in the Three Gorges region this

species was called *kou-shih*. Native to China and Vietnam.

Rubus irenaeus Focke (ROSACEAE) New species

in Bot. Jahrb. 29: 394. Bean in Kew Bull. Misc. Inf. 47. (1910), Zheng in Journ. Wuhan Bot. Res. 2.(1): 88. (1984).

Hubei: Badong, 4740, 6152 (isosyntype). **Hunan:** Shimen, 7541.

A low prostrate shrub with creeping stems and leaves suggestive of a small *Tussilago*, the plant is evergreen and bears large red fruits. Introduced to cultivation by Wilson for Messrs Veitch in 1900 and again in 1907 when collecting for the Arnold Arboretum.

Rubus lambertianus Ser. (ROSACEAE)

Rubus lambertianus Ser. Henry in Journ. China Br. Roy. Asiat. Soc. 22: 272. (1887) (Chinese Names of Plants); Ibid., 22: 264. (1887) (Chinese Names of Plants). *Rubus pycnanthus* Focke Zheng in Journ. Wuhan Bot. Res. 2.(1): 88. (1984). *Rubus davidianus* Kurze Zheng in Journ. Wuhan Bot. Res. 2.(1): 89. (1984).

Hubei: Yichang, 1120, 2704, 4152. Liantuo, 2092 (in part), 3046. Badong, 4241, 4786, 5022, 5129, 7359. **Sichuan:** North Wushan, 7359.

A semi-evergreen scandent shrub with four-angled stems. Henry records the fruits as being yellow and rather sour and stated it was known in the Three Gorges region as *tung p'ao tzu* (his Liantou collector called it the *tao pan lung*). Native to central China and introduced to cultivation by Wilson in 1907, seeds of his 482 collected near Yichang in September 1907 were received at Glasnevin in 1908 from the Arnold Arboretum. Li et al. in Journ. Wuhan Bot. Res. 18(3): 241. (2000) state that based on experiments carried out at Nanjing Botanical Gardens, this species may be an excellent source of pigment. With the wide use of food additives nowadays, there is a growing tendency to use natural and edible pigments instead of artificial products. These pigments are better extracted using alcohol than with water. The pigment extracted from *R. lambertianus* showed good colour and stability when used as a natural additive for juices, wines and jelly products.

Rubus lambertianus Ser. var. *glaber* Hemsley (ROSACEAE) New variety

in Journ. Linn. Soc. Bot. 23: 233. (1887).

Sichuan: North Wushan, 7032.

Rubus lambertianus Ser. var. *paykouangensis* (H. Lév.) Hand.-Mazz. (ROSACEAE)

Rubus viscidus Focke in Biblioth. Bot. 72: 108. (1910).

Yunnan: In the mountains near Mengzi at 1,950 metres, 9143 (isosyntype of *Rubus viscidus* Focke). In the mountains to the north of Mengzi at 2,800 metres 9143a., (type of *Rubus viscidus* Focke).

Distributed in western China and Thailand.

Rubus lasiostylus Focke (ROSACEAE) New species

in Hooker's. Icon. Pl. 20: t. 1951. (1891), Henry in Notes Econ. Bot. China 45. (1893), Hemsley in Curtis's Bot. Mag. 121: t. 7426. (1895), Bretschneider in Hist. Eur. Bot. Disc. China 2: 782. (1898), Besant in Gard. Chron. ser. 3, 98: 335. (1935), Zheng in Journ. Wuhan Bot. Res. 2.(1): 89. (1984), Nelson in Pim, The Wood & the Trees 222. (1984), Bean in Trees & Shrubs 4: 227. (1989), Nelson in A Heritage of Beauty 318. (2000).

Rubus lasiostylus Focke f. *tomentosa* Focke in Hooker's. Icon. Pl. 20: t. 1951. (1891).

Hubei: Jianshi, 5788 (type). South Badong, 5788a., (isotype of *Rubus lasiostylus* Focke f. *tomentosa* Focke). Zigui, 5788b. Fang Xian, 5788d., 5788e., 5788f. Xingshan, 5788g.

An erect deciduous shrub to 1.5 metres tall with biennial stems covered in a blue-white, waxy bloom. Introduced to cultivation by Augustine Henry through the Royal Botanic Gardens, Kew in 1889 where it first flowered in June 1894. Henry stated the fruits of this species have a delicious flavour and it was known in the Three Gorges region as *ch'an-mao-p'ao*. Seeds of E. H. Wilson's 279, collected in Fang in September 1907 were received at the National Botanic Gardens, Glasnevin from the Arnold Arboretum in 1908. Henry and his wife Alice grew this species in their garden at Ranelagh in the south Dublin suburbs. One of the most beautiful of all the white-stemmed rubi. Augustine

Rubus lasiostylus Focke var. *glabratus* (Focke) Chand. Gupta & S. S. Dash (ROSACEAE) New variety

Rubus lasiostylus Focke f. *glabratus* Focke in Hooker's. Icon. Pl. 20: t. 1951. (1891).

Hubei: Badong, 5788c., (isotype of *Rubus lasiostylus* Focke f. *glabrata* Focke, August 27th 1888).

Rubus lasiotrichos Focke (ROSACEAE) New species

in Biblioth. Bot. 72: 109. (1909).

Yunnan: In Pingbian, on the Daweishan Range at 2,300 metres, 9175 (type).

Native to China (Guizhou, Sichuan and Yunnan), Thailand and Vietnam.

Rubus lineatus Reinw. ex Blume (ROSACEAE)

Bean in Trees & Shrubs 4: 229. (1909), O' Brien in Augustine Henry, An Irish Plant Collector in China 33. (2002).

Yunnan: Mengzi, 13795.

A semi-evergreen shrub to 3 metres tall, native to the India, Bhutan, Nepal, Myanmar, China [Tibet (Xizang) and Yunnan] and Malaysia. Leaves are five-foliate with beautiful silver undersides. Most plants in cultivation derive from seeds collected by George Forrest.

02

Rubus lucens Focke (ROSACEAE)

Anonymous in Kew List of Determinations (volume 34).

Yunnan: Simao, 12731, collected 27th November 1898.

Distributed in India, China (Yunnan) and the Philippines.

Rubus malifolius Focke (ROSACEAE) New species

in Hooker's. Icon. Pl. 20: t. 1947. (1890), Bretschneider in Hist. Eur. Bot. Disc. China 2: 782. (1898).

Hubei: Jianshi, 5794 (isotype).

A deciduous shrub which may be prostrate or climbing depending on the availability of nearby host plants for support. One of the most beautiful of the Chinese species, particularly when seen in flower, this a common plant in the western provinces but is rare in Hubei. The specific epithet refers to the apple (*Malus*) like foliage.

Rubus mesogaeus Focke (ROSACEAE)

Focke in Sargent Pl. Wilson. 1: 56. (1911), Zheng in Journ. Wuhan Bot. Res. 2.(1): 89. (1984).

Hubei: Fang Xian, 6892. Jianshi, 5623b., 5623c.

A deciduous shrub to 3 metres tall with stems covered in a velvety down. *Rubus mesogaeus* was introduced to cultivation by Wilson in 1907, seeds of W. 71 from Yichang were received at Glasnevin in 1908.

Rubus moluccanus L. f. forma (ROSACEAE)

Anonymous in Kew List of Determinations (volume 24 & 34).

Sichuan: Henry's Chinese collector with Pratt, no locality stated but may be any of the following, Leshan, Emei Shan, Wa Shan or Kangding, 8940. **Yunnan:** Mengzi, 9546.

Rubus nigricaulis Proch. (ROSACEAE) New species

in Notul. Syst. Herb. Hort. Petropol. 5: 54. (1924).

Yunnan: Fengchunling, 9067 (isotype).

Rubus niveus Thunb. (ROSACEAE)

Rubus lasiocarpus Sw. Anonymous in Kew List of Determinations (volume 34).

Hubei: South Badong, 6090, 6090a., 7356. **Yunnan:** Mengzi, 9499.

Widespread in Asia.

Rubus pacificus Hance (ROSACEAE)

Anonymous in Kew List of Determinations (volume 24).

Hubei: Changleping, 6208.

Discovered in Jiujing in Jiangxi province in August 1873 by the German naturalist, Otto F. von Moellendorff who travelled to China in 1873 to collect zoological and botanical specimens.

Rubus parkeri Hance (ROSACEAE)

Henry in Notes Econ. Bot. China 45. (1893), Zheng in Journ. Wuhan Bot. Res. 2.(1): 89. (1984).

Rubus parkeri Hance var. *brevisetosus* Focke Zheng in Journ. Wuhan Bot. Res. 2.(1): 89. (1984).

Hubei: Yichang, 576, 3566. Badong, 604 (July 1885). Without locality, 3566a.

A shrub of scandent habit to 2 metres tall but up to 6 metres tall when cultivated in a cool greenhouse. Native to China and named for Mr. E. H. Parker (of the British Consular Service) who first collected it in eastern Sichuan in 1881. Henry stated this species was common on the banks of the Three Gorges where natives called it *wu-p'ao-tzu*. *Rubus parkeri* was introduced to cultivation from Yichang by Wilson through Messrs Veitch in 1900 and again in 1907 while collecting for the Arnold Arboretum. Plants were raised at Kew from this same collection. W. J. Bean had high praise for its distinct and striking foliage, stating it promised to make an eleagant climber.

Rubus parvifolius L. (ROSACEAE)

Henry in Journ. China Br. Roy. Asiat. Soc. 22: 264. (1887) (Chinese Names of Plants); in Trans. Asiat. Soc. Jap. 24 suppl: 40. (1896) (List Pl. Formosa), Li in Woody Flora of Taiwan 316. (1963), Zheng in Journ. Wuhan Bot. Res.2. (1): 89. (1984).

Rubus triphyllus Thunb. Focke. in Sargent Pl. Wilson. 3: 424. (1917). *Rubus triphyllus* Thunb. var. *eglandulosus* H. Bailey in Gentes Herb. 1: 30. (1920).

Hubei: Yichang, 581, 1134. Liantuo, 6375 (isosyntype of *Rubus triphyllus* Thunb. var. *eglandulosus* H. Bailey). Yichang, 6375a. **Taiwan:** Wanchin, 452. **Yunnan:** Mengzi, 10533.

A deciduous shrub to 1 metre tall, native to Japan and China and introduced to cultivation by the Horticultural Society (now the Royal Horticultural Society) in 1818. Henry called this raspberry the *ts'ai yuang,* though at Yichang it was the *p'ao-tzu* and at Badong the *mao p'ao-tzu.* Native to China, Vietnam, Korea and Japan. In Chinese herbal medicine this species is currently used as a diuretic and to promote blood circulation.

Rubus pedunculosus D. Don (ROSACEAE)

Zheng in Journ. Wuhan Bot. Res. 2.(1): 87. (1984).

Rubus foliolosus D. Don Anonymous in Kew List of Determinations (volume 24). *Rubus hypargyrus* Edgew. Zheng in Journ. Wuhan Bot Res. 2.(1): 87. (1984).

Hubei: Jianshi, 5996. Xingshan, 6494. **Hunan:** Shimen, 7558. **Yunnan:** Mengzi, 10198.

A robust deciduous shrub with biennial stems to 2 metres tall. Native to the Himalaya and western and central China, introduced to cultivation by Wilson in 1901 through

Veitch's Coombe Wood nursery. Native to India, Nepal, Bhutan and China [Tibet (Xizang) and Yunnan].

Rubus peltatus Maxim. (ROSACEAE)

Zheng in Journ. Wuhan Bot. Res. 2.(1): 89. (1984).

Hubei: Jianshi, 5835.

A deciduous shrub to 2 metres tall. Distributed in Anhui, Fujian, Guizhou, Hubei, Jiangxi, Sichuan and Zhejiang. Also native to Japan.

Rubus pentagonus Wall. ex Focke (ROSACEAE)

Rubus tridactylus Focke in Bibl. Bot. 72(2): 146. (1911).

Yunnan: In mountain forests to north of Mengzi at 2,800 metres, 10233 (type of *Rubus tridactylis* Focke).

Native to India, Bhutan, Nepal, western China and Vietnam.

Rubus pileatus Focke (ROSACEAE) New species

in Hooker's Icon. Pl. 20: tt. 1952. (1891), Bretschneider in Hist. Eur. Bot. Disc. China 2: 783. (1898), Focke in Bibl. Bot. LXXII. 167. (Spec. Rub) (1911).

Hubei: Badong at 1,900 metres, 6049 (in part).

Rubus piptopetalus Hayata ex Koidz. (ROSACEAE)

Rubus rosaefolius Henry non Sm. in Trans. Asiat. Soc. Jap. 24 suppl: 40.(1896) (List Pl. Formosa). *Rubus rosaefolius* Sm. var. *formosanus* Cardot. Li in Woody Flora of Taiwan 314. (1963).

Taiwan: Wanchin, 65. Kaohsiung, 65a. Oluanpi, 285.

Native to China (Taiwan) and Japan.

Rubus playfairianus Hemsl. Ex Focke (ROSACEAE)

Anonymous in Kew List of Determinations (volume 24).

Hubei: Antelope Glen, 3381, 3417. **Hainan:** 32 km (20 miles) west of Haikou, 8131. Ling-men (in the interior of the island), 8640.

A rambling shrub with slender, whip-like, dark-green stems. The habit of this shrub is very graceful and while its foliage is dintinctly handsome its flowers are horticulturally insignificant. Discovered by G. M. H. Playfair in Guangdong province, eastern China and introduced to cultivation by Wilson in 1907.

Rubus poliophyllus Kuntze (ROSACEAE)

Rubus distentus Focke in Bibl. Bot. 72(1): 68. (1909).

Yunnan: Simao at 1,650 metres, 11639 (isotype of *Rubus distentus* Focke).

Endemic to Yunnan.

Rubus preptanthus Focke (ROSACEAE) New species

in Biblioth. Bot. 72; 42. (1910).

Yunnan: Mountains to the north of Mengzi at 2,438 metres, 9598a., (isotype).

Endemic to Sichuan, Yunnan.

Rubus pungens Camb. (ROSACEAE)

Zheng in Journ. Wuhan Bot. Res. 2.(1): 90. (1984).

Hubei: South Badong, 5469. Fang Xian, 6856. Changleping, 5469a. **Sichuan:** South Wushan, 5469b.

Rubus aff. rosaefolius Smith (ROSACEAE)

Anonymous in Kew List of Determinations (volume 24).

Sichuan: Henry's Chinese collector with Pratt, no locality stated but may be any of the following, Leshan, Emei Shan, Wa Shan or Kangding, 8964.

Rubus rosifolius Sm. (ROSACEAE)

Rubus rosaefolius Smith Anonymous in Kew List of Determinations (volume 34).

Yunnan: Fengchunling, 10842.

Rubus rufus Focke (ROSACEAE) New species

in Biblioth. Bot. 72: 108. (1909).

Yunnan: Simao at 2,100 metres, 13076 (type).

Distributed in China, Thailand and Vietnam.

Rubus setchuenensis Bureau & Franchet (ROSACEAE)

Rubus singulifolius Focke in Bibl. Bot. 72(1): 77. f. 28. (1909). *Rubus clemens* Focke in Bilb. Bot. 72(1): 105. f. 46. (1909).

Yunnan: Milê district, 9945. Mengzi, 10926 (isotype of *Rubus singulifolius* Focke). Techen Shan near Mengzi, 10197 (type of *Rubus clemens* Focke).

Southern China and Vietnam.

Rubus simplex Focke (ROSACEAE) New species

in Hooker's Icon. Pl. 20: t. 1948. (1890), Bretschneider in Hist. Eur. Bot. Disc. China 2: 782. (1898), Zheng in Journ. Wuhan Bot. Res. 2.(1): 90. (1984).

Hubei: Jianshi, 5982 (isosyntype). South Badong, 7333 (isosyntype). Fang Xian, 5982b. Without locality, 5982b.

Introduced to Irish gardens through the National Botanic Gardens, Glasnevin from Wilson's 282 collected in Fang Xian (September 1907) and received from the Arnold Arboretum.

Rubus sp. No. 1. (ROSACEAE)

Anonymous in Kew List of Determinations (volume 34).

Yunnan: Simao, 11941.

Rubus sp. No. 2. (ROSACEAE)

Rubus illecebrosus Focke Zheng in Journ. Wuhan Bot. Res. 2.(1): 88. (1984).

Hubei: South Badong, 6310.

Rubus illecebrosus is a Japanese species.

Rubus sp. No. 3. (ROSACEAE)

Anonymous in Kew List of Determinations (volume 24).

Hubei: Yichang, 1371. Badong, 231 (June 1885), 4023, 4041, 5210, 5448. Zigui, 7667. **Sichuan:** Henry's Chinese collector with Pratt, no locality stated but may be any of the following, Leshan, Emei Shan, Wa Shan or Kangding, 8965.

Rubus sp. No. 4. (ROSACEAE)

Henry in Trans. Asiat. Soc. Jap. 24 suppl: 40. (1896) (List Pl. Formosa).

Taiwan: Wanchin, 1517.

Rubus sp. No. 5. near *Rubus innominatus* & *Rubus niveus*. (ROSACEAE)

Anonymous in Kew List of Determinations (volume 24).

Sichuan: Henry's Chinese collector with Pratt, no locality stated but may be any of the following, Leshan, Emei Shan, Wa Shan or Kangding, 8967.

Rubus sp. No. 6. (ROSACEAE)

Rubus (illeg.) Focke Anonymous in Kew List of Determinations (volume 34).

Yunnan: On the Daweishan Range in Pingbian at 1,500 metres, 9075.

Rubus swinhoei Hance (ROSACEAE)

Zheng in Journ. Wuhan Bot. Res. 2.(1): 87. (1984).

Rubus hupehensis Oliver in Hooker's Icon. Pl. t. 1816. (1889), Bretschneider in Hist. Eur. Bot. Disc. China 2: 782. (1898), Bean in Kew Bull. Misc. Inf. 46. (1910).

Hubei: Liantuo, 3931 (type of *Rubus hupehensis* Oliver), 6116a., 6116b, 6377. Badong, 6116.

A vigorous semi-evergreen bramble, leaves ovate-lanceolate on graceful, arching stems to 4.5 metres in length. In the Veitch catalogue *New Hardy Plants from Western China* (Autumn 1912) this species was available for two shillings and six pence.

Rubus wallichianus Wight & Arnott (ROSACEAE)

Rubus erythrolasius Focke in Biblioth. Bot. 72: 197. (1911). *Rubus pinfaensis* H. Lévl. & Vaniot Li in Journ. Arnold Arb. 25: 421. (1944), Zheng in Journ. Wuhan Bot. Res. 2.(1): 90. (1984).

Hubei: Xiling (Yichang) Gorge, 3336, 3336b., 3336c.. San You Dong Glen, 3336a. Without locality, 3336d. **Yunnan:** Lunan, 10583 (isotype of *Rubus erythrolasius* Focke), 10583a.

A climbing shrub, native to the Himalayan region, China and Vietnam.

Rubus xanthocarpus Bureau & Franch. (ROSACEAE)

Rubus spinipes Hemsley in Journ. Linn. Soc. Bot. 29: 306. (1892).

Sichuan: Henry's Chinese collector with Pratt, no locality stated but may be any of the following, Leshan, Emei Shan, Wa Shan or Kangding, 8969.

Rubus xanthoneurus Focke ex Diels (ROSACEAE)

Rubus dielsianus Focke in Bibl. Bot. 72(1): 53. f. 17. (1909).

Yunnan: Mengzi at 2,600 metres, 10201 (type of *Rubus dielsianus* Focke).

Native to China and Thailand.

Ruellia repens L. (ACANTHACEAE)

Henry in Trans. Asiat. Soc. Jap. 24 suppl: 69. (1896) (List Pl. Formosa).

Taiwan: Kaohsiung, 770, 1156. Wanchin, 880.

Rumex acetosa L. (POLYGONACEAE)

Henry in Journ. China Br. Roy. Asiat. Soc. 22: 261. (1887) (Chinese Names of Plants), Forbes & Hemsley in Journ. Linn. Soc. Bot. 26: 355. (1891).

Hubei: Yichang, 1175. Badong, 4024, 5293. Xingshan, 7923. Changleping, 6303. Fang Xian, 6852.

Sorrel, known colloquially in the Three Gorges region as *niu she t'ou* according to Henry.

Rumex aquaticus L. (POLYGONACEAE)

Forbes & Hemsley in Journ. Linn. Soc. Bot. 26: 355. (1891), Henry in Notes Econ. Bot. China 33. (1893).

Hubei: Liantuo, 5765, 6385.

According to Henry *Rumex aquaticus* was cultivated for the sake of its roots which were used as a drug. He stated that in Hubei it was generally called *t'u-ta-huang,* while at Liantuo it was called *chin-pu-huan.*

Rumex crispus L. (POLYGONACEAE)

Forbes & Hemsley in Journ. Linn. Soc. Bot. 26: 356. (1891).

Hubei: Yichang, 508, 508a.

The yellow dock, a leafy, vigorous plant to 1.5 metres tall, native to Europe and Asia, it has become a weed in other temperate areas of the world.

Rumex dentatus L. (POLYGONACEAE)

Forbes & Hemsley in Journ. Linn. Soc. Bot. 26: 356. (1891).

Hubei: Yichang, 584, 1001, 1364. Badong, 4070.

Henry's Badong collector called this species *ta huang* (rhubarb) and stated it was used as a cooling medicine.

Rumex pulcher L. (POLYGONACEAE)

Anonymous in Kew List of Determinations (volume 24).

Hubei: Yichang, 4110.

Rumex spp. (POLYGONACEAE)

Anonymous in Kew List of Determinations (volume 24 & 34).

Hubei: Yichang, 880. **Yunnan:** Mengzi, 11087. Simao, 12015, 13,446.

Rungia khasiana T. Anderss. (ACANTHACEAE)

Anonymous in Kew List of Determinations (volume 34).

Yunnan: Simao, 11784.

Rungia pectinata (L.) Nees (ACANTHACEAE)

Rungia parviflora (Retz.) Nees Forbes & Hemsley non Nees in Journ. Linn. Soc. Bot. 26: 247. (1890).

Hainan: Environs of Haikou, without number. **Yunnan:** In the Red River valley near Manhao, 9533. Simao, 13601.

Rungia spp. (ACANTHACEAE)

Henry in Trans. Asiat. Soc. Jap. 24 suppl: 69. (1896) (List Pl. Formosa), Kew List of Determinations (vol. 34).

Taiwan: Wanchin, 848. **Yunnan:** Simao, 11601.

Rungia stolonifera C. B. Clarke (ACANTHACEAE)

Anonymous in Kew List of Determinations (volume 34).

Yunnan: Simao, 13191.

Ruppia maritima L. (RUPPIACEAE)

Henry in Trans. Asiat. Soc. Jap. 24 suppl: 102. (1896) (List Pl. Formosa), Wright in Journ. Linn. Soc. Bot. 36: 197. (1903), Graebner in Engler, Pflanzenr. 13: 165. (1907).

Taiwan: Kaohsiung Lagoon, 1099.

A submerged herb with elongated many branched stems. Native to North America, Africa and Eurasia, most common in southern Taiwan of China where it grows in brackish waters.

Ruta chalepensis L. (RUTACEAE)

Ruta graveolens L. var. *chalepensis* (L.) Weston Anonymous in Kew List of Determinations (volume 24).

Hubei: Cultivated in a garden at Yichang, 3476.

Sabia campanulata Wall. (SABIACEAE)

Van de Water in Blumea 26: 25. (1980).

Sabia discolor H. H. Chun non Dunn in Mem. Sci. Soc. China 1: 152. (1924). *Sabia acutisepala* Stapf ex L. Chen in Acta Phytotax. Beobot. Kyoto 5: 77: (1936), nom. inval., Chen in Sargentia 3: 67. (1943).

Hubei: Changyang, 5265, (type of *Sabia acutisepala* Stapf). **Yunnan:** Mengzi, 10496.

Distributed in Nepal, Bhutan and China.

Sabia campanulata Wall. subsp. ***richieae*** (Rehder & E. H. Wilson) Y. F. Wu (SABIACEAE) New subspecies

Sabia richieae Rehder & E. H. Wilson in Sargent Pl. Wilson. 2: 195. (1914), Zheng in Journ. Wuhan Bot. Res. 2.(1): 136. (1984). *Sabia gaultheriifolia* Stapf. ex L. Chen in Sargentia. 3: 26. (1943).

Hubei: Changyang, 6227 (type of *Sabia gaultheriifolia* Stapf. ex Chen). Fang Xian, 6780.

A scandent shrub to 2 metres tall, discovered by Henry in 1888. The specific epithet commemorates Mrs. W. Ritchie, the wife of the postal commisioner at Chengdu who helped Wilson following his accident in 1909.

Sabia dielsii H. Léveillé (SABIACEAE) New species

Van de Water in Blumea 26: 33. (1980).

Sabia olacifolia Stapf ex L. Chen in Acta Phytotax. Geobot. 5: 78. (1936).

Yunnan: Mengzi at 1,650 metres, 10250 (holotype of *Sabia olacifolia* Stapf ex L. Chen). Mengzi 10400, 10400a., 10400b., 10400c., 10400d.

Léveillé based his type on a collection made by J. Esquirol who did not begin collecting until 1902 in the adjacent province of Guizhou. Henry's specimen dates from 1896, the earliest recorded. Endemic to China and distributed in Guizhou, Guangxi and Yunnan.

Sabia emarginata Lecomte (SABIACEAE) New species

in Bull. Soc. Bot. France. 54: 673. (1907), Rehder & E. H. Wilson in Sargent Pl. Wilson 2: 196. (1914), Handel-Mazzettii in Symb. Sin. 7: 644. (1933), Chen in Sargentia 20: 3. (1943), Zheng in Journ. Wuhan Bot. Res. 2.(1): 136. (1984).

Sabia campanulata Van de Water non Wallich in Blumea 26: 25. (1980).

Hubei: South of Badong, 5314 (type).

Sabia fasciculata Lecomte ex L. Chen (SABIACEAE) New species

in Sargentia 3: 42. (1943) Van de Water in Blumea 26: 36. (1980).

Yunnan: On the Daweishan Range in Pingbian at 1,650 metres, 10487 (holotype).

An evergreen climber or a scandent shrub, native to China, Myanmar and Vietnam.

Sabia japonica Maxim. (SABIACEAE)

Anonymous in Kew List of Determinations (volume 24).

Hubei: Without locality, 5421c.

Sabia schumanniana Diels subsp. ***pluriflora*** (Rehder & E. H. Wilson) Y. F. Wu (SABIACEAE)

New subspecies. in Acta Phytotax. Sin. 20: 427. (1982).

Sabia schumanniana Diels var. *plurifolia* Rehder & E. H. Wilson Zheng in Journ. Wuhan Bot. Res. 2.(1): 137. (1984).

Hubei: Badong, 5421, 6114. **Sichuan:** South Wushan, 5421b.

Discovered by Henry in 1888, this subspecies was based on material later collected by Wilson in May 1907.

Sabia swinhoei Hemsl. (SABIACEAE)

Van de Water in Blumea 26: 57. (1980), Zheng in Journ. Wuhan Bot. Res. 2.(1): 137. (1984).

Hubei: Yichang, 2762, 6626. South Badong, 7298. Fang Xian, 6626. Lung Tung Ch'i Glen (near Yichang), 4181. Near the entrance to the Xiling (Yichang) Gorge, 3460. Antelope Glen, 3126. Without locality, 3460a. **Sichuan:** Emei Shan, Henry's Chinese collector with Pratt,

8988. **Yunnan:** Simao, 11987.

An evergreen woody climber or scandent shrub to 6 metres tall. Widespread over southern and central China and Vietnam.

Sabia yunnanensis Franch. (SABIACEAE)

Sabia sp. nova Henry in Journ. China Br. Roy. Asiat. Soc. 22: 255. (1887) (Chinese Names of Plants). *Sabia puberula* Rehder & E. H. Wilson in Sargent Pl. Wilson. 2: 197. (1914). *Sabia pallida* Stapf & L. Chen in Sargentia 3: 33. (1943). *Sabia puberula* Rehder & E. H. Wilson var. *hupehensis* L. Chen in Sargentia 3: 23. (1943). *Sabia campanulata* Van de Water non Wallich in Blumea 26: 25. (1980).

Hubei: Badong, 4025. Jianshi 6022. Changleping, 6290, (type of *Sabia puberula* Rehder & E. H. Wilson var. *hupehensis* L. Chen). **Sichuan:** South Wushan, 7240. **Yunnan:** Mengzi at 1,950 metres, 10529, 10529a., (type *Sabia pallida* Stapf & L. Chen).

Henry stated that it was known colloquially in the Three Gorges region as *ling erh ch'ai.*

Saccharum formosanum (Stapf.) Ohwi (POACEAE)

Erianthus pollinioides Rendle in Journ. Linn. Soc. Bot. 36: 350. (1904).

Taiwan: Kaohsiung, 1143 (type of *Erianthus pollinioides* Rendle).

Southern China (including Hainan and Taiwan).

Saccharum longesetosum (Andersson) V. Naray. ex Bor. (POACEAE)

Erianthus hookeri Hack. Rendle in Journ. Linn. Soc. Bot. 36: 350. (1904).

Yunnan: Mengzi at 1,500 metres, 9660. Simao, 12870.

Saccharum officinarum L. (POACEAE)

Rendle in Journ. Linn. Soc. Bot. 36: 349. (1904).

Taiwan: Wanchin, on a plain, 1688.

Sugar-cane, this particular collection represents a selected form that was generally cultivated in southern Taiwan in the late nineteenth century. Sugar accounted for 90% of Taiwan's exports at this time. A crop that is still extensively cultivated on that island.

Saccharum × *sinense* Roxb. (POACEAE)

Henry in annotation in Journ. China Br. Roy. Asiat. Soc. 22: 263. (1887) (Chinese Names of Plants) (copy at Glasnevin).

Saccharum narenga Henry non (Nees ex Steud.) Hack. in Journ. China Br. Roy. Asiat. Soc. 22: 263. (1887) (Chinese Names of Plants), Rendle in Journ. Linn. Soc. Bot. 36: 349. (1904).

Hubei: Yichang, 154.

S. officinarum × *S. spontaneum.* Chinese sugarcane, which Henry stated was known colloquially in the Three

Gorges region as *hi pa wang.* The Hani (an ethnic group) of Xishuangbanna in southern Yunnan call this species *paoqqyul* and use it to reduce fever and vomiting and to treat constipation.

Saccharum sp. (POACEAE)

Anonymous in Kew List of Determinations (volume 34).

Yunnan: Simao, 12742.

Saccharum spontaneum L. (POACEAE)

Henry in Journ. China Br. Roy. Asiat. Soc. 22: 273. (1887) (Chinese Names of Plants).

Saccharum spontaneum L. ssp. *indicum* Hack. Rendle in Journ. Linn. Soc. Bot. 36: 349. (1904).

Hubei: Yichang, 34. On Pingshanba near Yichang, 2266. On the side of the San You Dong stream, 4222. **Taiwan:** Kaohsiung, 1077, 1077a.

According to Henry this species was known colloquially in the Three Gorges region as *t'ien ken tzu ts'ao.* Widely distributed in the warmer regions of the Old World.

Sacciolepis indica (L.) A. Chase (POACEAE)

Panicum indicum (L.) L. Rendle in Journ. Linn. Soc. Bot. 36: 330. (1904).

Hubei: Yichang, 2750. Xiling (Yichang) Gorge, 4312. Liantuo, 4423. **Taiwan:** Wanchin, 1547.

Native to tropical Asia and Australia. In Chinese Taiwan this species inhabits swampy places.

Saccolabium sp. (ORCHIDACEAE)

Anonymous in Kew List of Determinations (volume 34).

Yunnan: Simao, 12257, 12730, 12972, 12958, 12970, 13006, 13059, 13066, 13067, 13585,

Sageretia gracilis J. R. Drummond & Sprague (RHAMNACEAE) New species

in Kew Bull. Misc. Inform. 15. (1909).

Sageretia apiculata C. K. Schneider in Sargent Pl. Wilson. 2: 231. (1914).

Yunnan: Mengzi at 1,800 metres, 10144 (type of *Sageretia apiculata* Schneider, isotype of *Sageretia gracilis* J. R. Drummond & Sprague). 10144a., (type of *Sageretia gracilis* J. R. Drummond & Sprague).

Endemic to China and distributed in Guangxi, Yunnan and Tibet (Xizang).

Sageretia hamosa (Wall.) Brongn. (RHAMNACEAE)

Sageretia filiformis Zheng non (Roth.) G. Don in Journ. Wuhan Bot. Res. 2.(1): 141. (1984).

Hubei: Xingshan, 6364. Liantuo, 7184. Changyang, 7676. **Hainan:** Environs of Haikou, 8050.

Sageretia henryi J. R. Drummond & Sprague (RHAMNACEAE) New species

in Kew Bull. Misc. Inform. 14. (1908); Ibid., 178.

(1914), Schneider in Sargent Pl. Wilson 2: 623. (1916), Zheng in Journ. Wuhan Bot. Res. 2(1):141. (1984).

Sageretia cavaleriei (H. Lév.) C. K. Schneid. Schneider in Sargent Pl. Wilson 2: 228. (1914).

Hubei: South Badong, 5354. **Sichuan:** North Wushan, 7118 (syntype). **Yunnan:** On wooded cliffs near Mengzi at 2,200 metres, 11240 (isotype).

Endemic to China.

Sageretia rugosa Hance (RHAMNACEAE)

Zheng in Journ. Wuhan Bot. Res. 2.(1): 141. (1984).

Sageretia ferruginea Oliver in Hooker's Icon. Pl. 18: t. 1710. (1887), Bretschneider in Hist. Eur. Bot. Disc. China 2: 781. (1898).

Hubei: Yichang, 2701 (type of *Sageretia ferruginea* Oliver).

Sageretia sp. (RHAMNACEAE)

Anonymous in Kew List of Determinations (volume 24).

Hubei: Jianshi, 5967.

Sageretia thea (Osbeck.) M. C. Johnst. (RHAMNACEAE)

Zheng in Journ. Wuhan Bot. Res. 2.(1): 141. (1984).

Sageretia theezans (L.) Brong. Henry in Trans. Asiat. Soc. Jap. 24 suppl: 27. (1896) (List Pl. Formosa), Rehder & E. H. Wilson in Sargent Pl. Wilson. 2: 227. (1914), Li in Woody Flora of Taiwan 513. (1963).

Hubei: Yichang, 2710. **Taiwan:** Oluanpi, 227.

A widely distributed spiny shrub to 2 metres tall, native to India, Myanmar, China and the Philippines. In Chinese Taiwan this species grows from sea level to 800 metres and in Hubei it grows on the cliffs of the Three Gorges at low altitudes.

Sagina japonica (Swartz) Ohwi (CARYOPHYLLACEAE)

Sagina linnaei Anonymous non Presl. in Kew List of Determinations (volume 24). *Sagina maxima* auct. non A. Gray Anonymous in Kew List of Determinations (volume 24).

Hubei: Yichang, 3652. Badong, 3764, 7369. **Sichuan:** South Wushan, 5684.

Sagittaria pygmaea Miq. (ALISMATACEAE)

Sagittaria sagittifolia L. var. *oligocarpa* Micheli. Wright in Journ. Linn. Soc. Bot. 36: 191. (1903).

Hubei: Liantuo, 1919. In dry rice fields near Yichang, 4299.

Sagittaria sagittifolia L. (ALISMATACEAE)

Henry in Journ. China Br. Roy. Asiat. Soc. 22: 278. (1887) (Chinese Names of Plants); in Notes Econ. Bot. China 27. (1893), Wright in Journ. Linn. Soc. Bot. 36: 190. (1903).

Hubei: Yichang, 105, 300, 2324. Liantuo, 2664. South Badong, 7387. **Hainan:** Environs of Haikou, 8064, 8390. Ling-men (in the interior of the island), 8664.

Arrow-root, cultivated all over China for its edible roots. Several cultivars are grown, these tend to have larger roots than the wild type. In the Three Gorges region this species occurs wild in ponds and ditches and according to Henry was known as *tz'u ku* (or arrow-head).

Salix babylonica L. (SALICACEAE)

Burkill in Journ. Linn. Soc. Bot. 26: 526. (1899), Schneider in Sargent Pl. Wilson. 3: 42. (1917), Zheng in Journ. Wuhan. Bot. Res. 2.(1): 14. (1984).

Hubei: Yichang, 1328, 3355. Changleping, 6325.

The Chinese weeping willow, a tree to 15 metres tall with a low, wide-spreading head of branches. The weeping forms of this tree are mutations of the species. Long cultivated in the warmer parts of Europe, *Salix babylonica* probably traveled by one of the ancient trade routes from northern China. Linnaeus believed the tree to be a native of Babylon and coined the misleading specific epithet, *Salix babylonica* is native to northern China and Korea. The willow of Babylon (a region to the south of present day Baghdad) however, is *Populus euphratica*. The most common species of willow to be planted on the banks of the Yangtze, from its mouth at Shanghai westwards for 2,000 miles (3,200 km). Passing Yichang the tree becomes less and less common as pointed out by Wilson. This is probably due to the rapid increase in altitude on entering the Three Gorges, it is in this region that the great plains of eastern China end and the landscape becomes mountainous and continues to do so until the massive bulk of the Himalaya is met with above the Tibetan (Xizang) Plateau. This willow is a great favourite of the Chinese people and is often depicted in their drawings and on porcelains. In Japan where it is an introduced tree, it is often planted around Buddhist temples *Salix babylonica* is too tender to be cultivated successfully in Britain or Ireland and is replaced instead with *Salix × pendulina* nothof. *tristis* (Gaudin). I. V. Belyaeva.

Salix disperma Roxb. ex D. Don (SALICACEAE)

Salix wallichiana Anderson. var. *grisea* Anderss. Burkill in Journ. Linn. Soc. Bot. 26: 534. (1899). *Salix wallichiana* Anderss. C. K. Schneider in Sargent Pl. Wilson. 3: 64. (1916), Zheng in Journ. Wuhan Bot. Res. 2.(1): 16. (1984).

Hubei: Yichang, 1379, 3399. Badong, 1466. South Badong, 5355. Liantuo, 7653.

A small tree in western Hubei, where it is common and grows in moist woods along streams between 1,000 and 2,700 metres in altitude. Distributed from Afghanistan to Myanmar and China.

Salix fargesii Burkill (SALICACEAE) New species

in Journ. Linn. Soc. Bot. 26: 529. (1899), Seemen in Bot. Jahrb. xxix. 277. t. 3. fig. a-f. (1900); l.c. xxxvi. biebl.

lxxxii. 30. (1905), Leveille in Bull. Soc. Bot. France. lvi. 299. (1909), O' Brien in Augustine Henry, An Irish Plant Collector in China 33. (2002).

Sichuan: South Wushan, 5678 (type).

A deciduous shrub to 3 metres tall. One of the largest leaved Chinese willows and a handsome garden plant. *Salix fargesii* was first found by Augustine Henry and was later collected near Kangding in Sichuan by Père Paul Farges. It was introduced to cultivation by E. H. Wilson by means of cuttings from Fang in north-western Hubei in November 1910. Wilson found it in its typical state but also reported that it grew as a prostrate or procumbent shrub to 60 cm tall. This form would certainly be popular if ever introduced to western gardens. *Salix fargesii* was collected by the Glasnevin Central China Expedition (GCCE 773) on Tianthu Shan, Changyang in October 2004.

Salix heterochoma Seemen (SALICACEAE) New species

in Bot. Jahrb. xxi. Biebl. liii. 56. (1896), C. K. Schneider in Sargent Pl. Wilson. 3: 61. (1916), Zheng in Journ. Wuhan Bot. Res. 2.(1): 15. (1984).

Salix henryi Burkhill in Journ. Linn. Soc. Bot. 26: 530. (1899), Morley in Glasra 3: 80. (1979).

Hubei: South Badong, 5349. Jianshi, 5843 (fruiting co-type of *Salix heterochoma*, Seem., co-type of *Salix henryi* Burkill). **Sichuan:** South Wushan, 5671 (fruiting type of *Salix heterochoma* Seem., co-type of *Salix henryi* Burkhill). Jianshi, 5349a.

Salix hylonoma C. K. Schneid. (SALICACEAE) New species

Zheng in Journ. Wuhan Bot. Res. 2.(1): 15. (1984).

Salix longiflora Burkill non Wall. ex Anderrs. in Journ. Linn. Soc. Bot. 26: 530. (1899). *Salix wallichiana* C. K. Schneider non Anderss. in Sargent Pl. Wilson. 3: 65. (1916), in part. *Salix hylonoma* Schneider or *Salix* species nov. C. K. Schneider in Sargent Pl. Wilson. 3: 68. (1916).

Hubei: South Badong at 1,600 metres, 5296.

The type specimen was collected by Wilson on Wa-Shan, Sichuan in June 1908, two decades after Henry's initial collection.

Salix kusanoi (Hayata) C. K. Schneid. (SALICACEAE) New species

in Sargent Pl. Wilson. 3: 100. (1916), Li in Woody Flora of Taiwan 73. (1963).

Salix oldhamiana Henry non Miq. in Trans. Asiat. Soc. Jap. 24 suppl: 90. (1896). *Salix tetrasperma* Burkill non Roxb. in Journ. Linn. Soc. Bot. 26: 533. (1899).

Taiwan: Wanchin, without number. Kaohsiung, 1068, 1068a. Hengchun, 1068b. Tamsui, 1403.

A handsome tree to 8 metres tall, Kusano's type

specimen was collected in 1909, Henry had discovered it some sixteen years previously. Endemic to Taiwan where it forms small trees in wetlands and along the banks of creeks.

Salix moupinensis Franch. (SALICACEAE)

Burkill in Journ. Linn. Soc. Bot. 26: 531. (1899), C. K. Schneider in Sargent Pl. Wilson. 3: 346. (1916), Bean in Trees & Shrubs 4: 287. (1989).

Sichuan: Wa shan, Henry's Chinese collector with Pratt, 8891.

A deciduous shrub or small tree to 5 metres tall, similar to *Salix fargesii* (q.v.). A handsome willow, common, common in Baoxing (formerly Moupin) where it was first found by Père Armand David in March 1869.

Salix psilostigma Andersson (SALICACEAE)

C. K. Schneider in Sargent Pl. Wilson. 3: 116. (1916), Zheng in Journ. Wuhan Bot. Res. 2.(1): 16. (1984).

Salix tetrasperma Burkill non Roxb. in Journ. Linn. Soc. Bot. 26: 533. (1899). *Salix eriophylla* Anderss. Burkill in Journ. Linn. Soc. Bot. 26: 528. (1899).

Hubei: Changleping, 6267. **Yunnan:** Mengzi at 2,500 metres, 9338b. Mengzi at 2,000 metres, 10209. Mountains near Mengzi at 2,900 metres, 10493. Mountains to the north of Mengzi at 2,000 metres, 10493a.

A small tree to 5 metres tall, this species was also later collected near Dali in Yunnan by George Forrest (in May 1906). Distributed in Nepal, India, Bhutan, Myanmar and China [Tibet (Xizang), Sichuan and Yunnan].

Salix sp. (SALICACEAE)

Anonymous in Kew List of Determinations (volume 24).

Hubei: On Pingshanba near Yichang, 2271 (August 1886). Liantuo, 2980.

Salix tetrasperma Roxb. (SALICACEAE)

Burkill in Journ. Linn. Soc. Bot. 26: 533. (1899).

Salix mesnyi Hance Forbes & Hemsley in Journ. Linn. Soc. Bot. 530. (1899). *Salix araeostachya* C. K. Schneider in Sargent Pl. Wilson. 3: 96. (1916). *Salix oldhamiana* Henry non Miq. in Trans. Asiat. Soc. Jap. 24 suppl: 90. (1896). *Salix* sp. Henry in Trans. Asiat. Soc. Jap. 24 suppl: 90. (1896) (List Pl. Formosa). *Salix kusanoi* C. K. Schneider non (Hayata) C. K. Schneider, in part, in Sargent Pl. Wilson. 3: 100. (1916) (quoted as A. Henry 1404 & 1474). *Salix warburgii* O. Seem. Li in Woody Flora of Taiwan 73. (1963). *Salix wilsonii* Seemen in Bot. Jahrb. xxxvi. Beibl. lxxxii. 28. (1905), Leveille in Bull. Soc. Bot. France. lvi. 301. (1909), Schneider in Sargent Pl. Wilson.40. (1917), Zheng in Journ. Wuhan Bot. Res. 2.(1): 16. (1984). *Salix glandulosa* Seem. var. *wilsonii* (Seem.) R. Gorz in Repert. Spec. Nov. Regni Veg. 36: 21. (1934).

02

Hubei: Yichang, 246, 1277 (type of *Salix wilsonii* Seemen), 3538. Antelope Glen, 3442. Changyang, 5267. Without locality, 3538a. **Taiwan:** Tamsui, 1404, 1473, 1474, 1477. **Yunnan:** Mengzi at 1,800 metres, 9338, 9338a. Near a water-course at 1,600 metres at Mengzi, 9338c., (co-type of *Salix araeostachya* of C. K. Schneider), 9338d., (co-type of *Salix araeostachya* C. K. Schneider). Mountains to the south-east of Mengzi at 1,800 metres, 11250.

A tree from to 6 to 13 metres tall, this is the most common willow in the mountains of western Hubei, where according to Wilson, it was abundant on the banks of streams and mountain torrents between 1,300 and 2,000 metres altitude. Henry stated it was equally common on riverbanks in Taiwan and Yunnan.

Salix variegata Franch. (SALICACEAE)

Burkill in Journ. Linn. Soc. Bot. 26: 534. (1899), C. K. Schneider in Sargent Pl. Wilson. 3: 70. (1916), Zheng in Journ. Wuhan Bot. Res. 2.(1): 16. (1984).

Salix densifoliata Seemen in Bot. Jahrb. Syst. 21. Biebl. 53: 57. (1896), XIX. 278. (1900).

Hubei: Xiling (Yichang) Gorge, on banks of the Yangtze and its tributaries, 46, 48, 957, 3974. On the banks of the Yangtze at Badong, 7175 (type of *Salix densifoliata* Seemen), 7182.

A creeping willow to less than a metre tall, first found in the Three Gorges area by Père Armand David on the banks of the Yangtze in December 1868 while he was travelling to eastern Sichuan. During the summer floods this plant is totally submerged beneath the waters of the Yangtze for several weeks.

Salomonia cantoniensis Lour. (POLYGALACEAE)

Anonymous in Kew List of Determinations (volume 34).

Yunnan: Simao, 12364.

Distributed in Nepal, India, China, Thailand, Vietnam and the Philippines.

Salomonia ciliata (L.) DC. (POLYGALACEAE)

Salomonia oblongifolia DC. Anonymous in Kew List of Determinations (volume 24).

Hainan: Environs of Haikou, 8410.

An erect annual herb to 20 cm tall. Distributed across India, China, South Korea, Japan, the Philippines and Malaysia to tropical Australia.

Salvia cavaleriei H. Lévl. var. *erythrophylla* (Hemsl.) E. Peter (LAMIACEAE) New variety

Salvia japonica Thunb. var. *erythrophylla* Hemsley in Journ. Linn. Soc. Bot. 26: 285. (1890).

Hubei: Badong, 4014 (syntype of *Salvia japonica* Thunb. var. *erythrophylla* Hemsley). **Sichuan:** South Wushan, 5415 (type of *Salvia japonica* Thunb. var.

erythrophylla Hemsley).

Salvia japonica Thunb. (LAMIACEAE)

Forbes & Hemsley in Journ. Linn. Soc. Bot. 26: 284. (1890), Dunn in Kew Bull. Misc. Inf. 200. (1917).

Hubei: Yichang, 38, 382, 556. Badong, 5104.

Salvia maximowicziana Hemsley (LAMIACEAE) New species

in Journ. Linn. Soc. Bot. 26: 285. (1890), Bretschneider in Hist. Eur. Bot. Disc. China 2: 789. (1898), Dunn in Notes Roy. Bot. Gard. Edinburgh 6: 163. (1915).

Hubei: Fang Xian, 6822, 6864 (type).

The specific epithet commemorates Carl J. Maximowicz, who spent a period of over 40 years studying the flora of the far east.

Salvia miltiorrhiza Bunge (LAMIACEAE)

Forbes & Hemsley in Journ. Linn. Soc. Bot. 26: 286. (1890), Dunn in Notes Roy. Bot. Gard. Edinburgh 6: 162. (1915).

Hubei: Yichang, 488, 833, 1347. Liantuo, 833a., 1997.

In Chinese herbal medicine this species is used to treat menstrual disorders, insomnia and arthritis.

Salvia plebeia R. Br. (LAMIACEAE)

Henry in Trans. Asiat. Soc. Jap. 24 suppl: 72. (1896) (List Pl. Formosa), Forbes & Hemsley in Journ. Linn. Soc. Bot. 26: 287. (1890), Dunn in Kew Bull. Misc. Inf. 201. (1917).

Hubei: Yichang, 345, 559, 756, 1338, 3091. Liantuo, 1952. **Taiwan:** Wanchin, 547. Kaohsiung, 1786. **Yunnan:** In Pingbian, on the Daweishan Range at 1,640 metres, 10637

In current Chinese herbal medicine the dried aerial parts of this species are used to treat cases of sore throat and external skin diseases.

Salvia plectranthoides Griff. (LAMIACEAE)

Salvia japonica Thunb. var. *parvifoliola* Hemsley in Journ. Linn. Soc. Bot. 26: 285. (1890), exclude A. Henry 3772.

Hubei: Yichang, 340 (syntype of *Salvia japonica* Thunb. var. *parvifoliola* Hemsley), 813, 1183, 1203 (syntype of *Salvia japonica* Thunb. var. *parvifoliola* Hemsley). Changleping, 6284. Fang Xian, 6621. **Yunnan:** Mengzi, 9849.

Native to India, Bhutan and China.

Salvia sp. (LAMIACEAE)

Anonymous in Kew List of Determinations (volume 24).

Hubei: Badong, 916 (November 1885).

Salvia yunnanensis C. H. Wright (LAMIACEAE)

Dunn in Notes Roy. Bot. Gard. Edinburgh 6: 162. (1915).

Yunnan: Mountains to the south-west of Mengzi at

1,500 to 2,150 metres, 10053.

A perennial herb to 30 cm tall, producing spikes of blue-purple flowers from April to August. Discovered by the Irish plant collector and Imperial Maritime Customs official, William Hancock near Mengzi in 1894. Endemic to China and distributed in the provinces of Gansu, Sichuan and Yunnan.

Salvinia natans (L.) All. (SALVINIACEAE)
Diels in Bot. Jahrb. 29: 209. (1900).

Hubei: Yichang, 4376.

A small aquatic floating fern, distributed in Europe, Africa, Asia and North America. Often floating on ponds and paddy fields.

Sambucus javanica Reinw. ex Blume (VIBURNACEAE)
New species

Sambucus adnata Rehder non Wallich apud De Candolle in Sargent Pl. Wilson. 1: 308. (1912). *Sambucus henriana* Samutina in Bot. Zhurn. (Moscow & Leningrad) 71: 1121. (1986).

Yunnan: In forests in the mountains to the north of Mengzi at 2,300 metres, 10772 (isotype of *Sambucus henriana* Samutina).

Sambucus javanica Reinw. ex Blume subsp. *chinensis* (Lindl.) Fukuoka (VIBURNACEAE)
Sambucus javanica Henry non Reinw. ex Blume in Notes Econ. Bot. China 56. (1893); in Trans. Asiat. Soc. Jap. 24 suppl: 48. (1896) (List Pl. Formosa), Rehder in Sargent Pl. Wilson. 1: 106. & 307. (1911-12). *Sambucus formosana* Nakai Li in Woody Flora of Taiwan 889. fig. 360. (1963). *Sambucus chinensis* Lindl. Henry in Journ. China Br. Roy. Asiat. Soc. 22: 241. (1887) (Chinese Names of Plants), Li in Woody Flora of Taiwan 889. fig. 360. (1963), Zheng in Journ. Wuhan Bot. Res. 2.(1): 195. (1984).

Hubei: Yichang, 144, 331, 728, 1390. San You Dong Glen, 1694. Liantuo, 2007, 2537. Badong, 2388, 3183. **Taiwan:** Oluanpi, 214, 921. Wanchin, 553, 921. Tamsui, 1449, 1747. **Yunnan:** Milê district, 9924. Mengzi, 10616. In mountains to the south of Simao at 1,500 metres, 12340, 12809.

A low spreading shrub, or a suffrutescent herb which Henry stated was known around Yichang as *ch'o ch'o miao* or *ts'o-ts'o-miao*. *Sambucus javanica* subsp. *chinensis* is currently used in Chinese herbal medicine as a diuretic and as a painkiller in cases of traumatic injury.

Sambucus williamsii Hance (VIBURNACEAE)
Zheng in Journ. Wuhan Bot. Res. 2.(1): 195. (1984).

Sambucus racemosa Forbes & Hemsley non L. in Journ. Linn. Soc. Bot. 349. (1888), Henry in Notes Econ. Bot. China 56. (1893). *Sambucus sieboldiana* Henry non (Miq.) Graebner. in Journ. China Br. Roy. Asiat. Soc. 22: 241. (1887) (Chinese Names of Plants), Rehder in Sargent Pl. Wilson. 1: 106. (1913).

Hubei: South Badong, 331 (June 1885), 3755, 5309, 5438, 6101. Liantuo, 1947, 3835, 3950. **Sichuan:** South Wushan, 5532, 5533, 5775.

A small tree or large shrub, which Henry stated was known as the *ch'o shu* or *ts'o-chu* in the areas of Badong, Changyang and Yichang. Discovered near Beijing by S. Williams, an American Sinologist who lived in Beijing from 1862 to 1873 as secretary and interpreter to the United States Legation. He was also a friend of the renowned Latvian botanist, Emil Bretschneider and both men first met in Beijing in 1866. In Chinese herbal medicine the dried stems and leaves of this species is currently used to treat fractures, traumatic injuries and rheumatic arthritis.

Sanguisorba officinalis L. (ROSACEAE)
Poterium officinale (L.) A. Gray Forbes & Hemsley in Journ. Linn. Soc. Bot. 23: 247. (1887), Henry in Journ. China Br. Roy. Asiat. Soc. 22: 270. (1887) (Chinese Names of Plants).

Hubei: Badong, 369 (June 1885), 2404, 3159, 5094. Without locality, 369a., 3159a.

Greater burnet, an erect perennial to 90 cm tall, native to Europe and Asia. In Chinese herbal medicine this species is used to stop bleeding. This is reflected in the generic name *sanguis* (blood) *sorbeo* (I absorb).

Sanicula chinensis Bunge (APIACEAE)
Sanicula europaea Forbes & Hemsley non L. in Journ. Linn. Soc. Bot. 23: 326. (1887), Henry in Journ. China Br. Roy. Asiat. Soc. 22: 280. (1887) (Chinese Names of Plants).

Hubei: Badong, 1828, 4219, 4831. Liantuo, 2079. Yichang, 3631.

Henry stated that this species was known colloquially in the Three Gorges region as *ya chio pan*.

Sanicula lamelligera Hance (APIACEAE)
Forbes & Hemsley in Journ. Linn. Soc. Bot. 23: 326. (1887).

Sanicula yunnanensis Franch. Wolff in Engler, Pflanzenr. iv. 228: 53. (1913). *Sanicula ichangensis* Wolff in Engler, Pflanzenr. iv. 228: 54. (1913).

Hubei: Yichang, 1249, 1304, 3525a., (syntype of *Sanicula ichangensis* Wolff), 3526, 3526a., (syntype of *Sanicula ichangensis* Wolff).

Discovered in October 1876 by a Miss Whilden of the Methodist Episcopal Mission who arrived in Guangzhou (Canton) in 1872.

Sanicula orthacantha S. Moore (APIACEAE)
Forbes & Hemsley in Journ. Linn. Soc. Bot. 23: 326. (1887).

Sanicula henryi H. Wolff in Engler, Pflanzenr. 61. (iv.

228.): 55. (1913).

Hubei: Badong, 3525, 5627, 6311. **Yunnan:** Mengzi, 10015. Mengzi at 2,650 metres, 10500 (type of *Sanicula henryi* H. Wolff). Simao, 12073.

Sanicula orthacantha was described from a collection made by Dr. George Shearer near Jiujing in Jiangxi province in 1873. Distributed in India, China, Laos, Cambodia and Vietnam.

Sanicula sp. (APIACEAE)

Anonymous in Kew List of Determinations (volume 24).

Hubei: Badong, 908 (May 1885). Yichang, 1463.

Sapindaceae (SAPINDACEAE)

Cupania sp. Anonymous in Kew List of Determinations (volume 24 & 34).

Hainan: On hummocks near Haikou, 8123. Environs of Haikou, 8391, 8395. Without locality, 8547. **Yunnan:** Mengzi, 13799.

Sapindus mukorossi Gaert. (SAPINDACEAE)

Henry in Trans. Asiat. Soc. Jap. 24 suppl: 28. (1896) (List Pl. Formosa), Forbes & Hemsley in Journ. Linn. Soc. Bot. 23: 139. (1886), Henry in Journ. Chinese Br. Roy. Asiat. Soc. 22: 244. (1887) (Chinese Names of Plants), Rehder & E. H. Wilson in Sargent Pl. Wilson. 2: 191. (1914), Li in Woody Flora of Taiwan 871. fig. 193. (1963), Zheng in Journ. Wuhan. Bot. Res. 2.(1): 134. (1984).

Sapindus mukorossi Gaertn. var. *carinatus* Radlk. Radlkofer in Engler, Pflanzenr. iv. 165: 653. (1931-32).

Hubei: Yichang, 1032, 7960 (in part). In the Monastery Valley near Yichang (July 1886), 1652. Liantuo, 6367. **Hainan:** Environs of Haikou, 7958. Environs of Kiungchow, 8770. **Taiwan:** Wanchin, 552, without number. **Yunnan:** Simao, 12101.

The soap tree or *hou erh tsao*, a large, noble tree with a wide spreading head of branches, also known in Chinese books as *wu-huan-tzu* according to Henry and Wilson. The fruits of this tree are used as soap and are now available in Dublin and other European cities as an eco-friendly alternative to washing powder. *Sapindus mukorossi* was collected by the Glasnevin Taiwan Expedition (GTE 116) on Wanshoushan (Ape's Hill of Henry and Wilson) in October 2004. Native to India, China and Japan.

Sapindus sp. (SAPINDACEAE)

Anonymous in Kew List of Determinations (volume 34).

Yunnan: Mengzi, 10807.

Sapindus tomentosus Kurz (SAPINDACEAE)

Radlkofer in Engler, Pflanzenr. iv. 165: 667. (1931-32).

Yunnan: Mengzi, 10819.

Native to Myanmar and China (Gansu and Yunnan).

Sapium sp. (EUPHORBIACEAE)

Anonymous in Kew List of Determinations (volume 34).

Yunnan: Mengzi, 9513, 13705. Simao, 11569, 11942, 11946.

Saprosma henryi Hutchinson (RUBIACEAE) New species

in Sargent Pl. Wilson. 3: 417. (1916).

Yunnan: Mountains to the south of Simao at 1,320 metres, 12145 (type). In forests near Simao at 1,640 metres, 12646.

Saprosma ternata (Wall.) Hook. f. (RUBIACEAE)

Hutchinson in Sargent Pl. Wilson. 3: 416. (1916).

Yunnan: In a ravine near Simao at 1,320 metres, 11965. Mountains to the south of Simao at 1,320 metres, 11965a. Mountains to the south of Simao at 1,640 metres, 11965b.

Saraca sp. (FABACEAE)

Anonymous in Kew List of Determinations (volume 34).

Yunnan: Mengzi, 11089.

Sarcandra glabra (Thunb.) Nakai (CHLORANTHACEAE)

Li in Woody Flora of Taiwan 68. fig. 18. (1963).

Chloranthus brachystachys Henry non Blume in Trans. Asiat. Soc. Jap. 24 suppl: 78. (1896) (List Pl. Formosa).

Taiwan: Oluanpi, 1965. Wanchin, 1580.

An evergreen shrub to about a metre tall. Native to India, Sri Lanka, southern China, Korea, Japan, the Philippines and Java. In Chinese Taiwan this species is most common in thickets on the northern part of the island.

Sarcandra glabra (Thunb.) Nakai subsp. *brachystachys* (Blume) Verdc. (CHLORANTHACEAE)

Chloranthus brachystachys Blume Forbes & Hemsley in Journ. Linn. Soc. Bot. 26: 367. (1891), Rehder in Sargent Pl. Wilson. 3: 15. (1916).

Hainan: Ling-men, 8629. **Yunnan:** Mountains to the south-west of Mengzi at 1,200 metres, 9361a. On the Daweishan Range in Pingbian at 1,500 metres, 9361c. Simao at 1,300 to 1,500 metres 12341, 12341b., 12341c.

Distributed in India, China, Thailand, Laos and Vietnam to New Guinea.

Sarcochilus luniferus (Rchb. f.) Benth. ex Hook. f. (ORCHIDACEAE)

Rolfe in Journ. Linn. Soc. Bot. 36: 35. (1903).

Yunnan: In mountain forest in Pingbian on the Daweishan Range at 1,600 metres, epiphytic on trees, 11129.

Sarcochilus spp. (ORCHIDACEAE)

Henry in Trans. Asiat. Soc. Jap. 24 suppl: 92. (1896) (List Pl. Formosa). Anonymous in Kew List of

Determinations (volume 34).

Taiwan: Oluanpi, 1971. **Yunnan:** Mengzi, 11851. Simao, 12081, 12976, 13406, 13617.

Sarcococca confusa Sealy (BUXACEAE) New species

Nelson in A Heritage of Beauty 319. (2000), O' Brien in Augustine Henry, An Irish Plant Collector in China 33. (2002).

Sichuan: Wushan, without number.

The origins of this species are unclear, it is thought to have been first collected by Henry in central China but was described from cultivated plants. *Sarcococca confusa* was growing at Glasnevin and in Miss Willmott's garden at Warley Place during the 1930s.

Sarcococca hookeriana Baill. var. *digyna* Franch. (BUXACEAE)

Sarcococca pruniformis Forbes & Hemsley non Lindl. in Journ. Linn. Soc. Bot. 26: 418. (1894). *Sarcococca hookeriana* Baillon var. *humilis* (Stapf ex Sealy) Rehder & E. H. Wilson in Sargent Pl. Wilson 2: 164. (1914), O' Brien in Augustine Henry, An Irish Plant Collector in China 33. (2002). *Sarcococca humilis* Stapf ex Sealy in Journ. Royal Hort. Soc. Xiv. 302. (1949), Zheng in Journ. Wuhan Bot. Res. 2.(1): 115. (1984), Nelson in Pim, The Wood & the Trees 234. (1984), Bean in Trees & Shrubs 4: 328. (1989), Nelson in A Heritage of Beauty 319. (2000).

Hubei: Changyang, 7834. **Sichuan:** North Wushan, 7065 (type of *Sarcococca hookeriana* Baillon var. *humilis* (Stapf ex Sealy) Rehder & E. H. Wilson, type of *Sarcococca humilis* Stapf ex Sealy). **Yunnan:** Mountains to the north of Mengzi at 2,000 metres, 9859. Milê District, 9859a.

An evergreen shrub to 1 m tall, producing white, fragrant flowers in short axillary racemes in early spring, followed in autumn by blue-black fruits. Introduced to cultivation by Wilson from Wa-Shan, Sichuan in 1908. In the Veitch catalogue *New Hardy Plants from Western China* (Autumn 1912) this species was available for one shilling and six pence.

Sarcococca ruscifolia Stapf (BUXACEAE) New species

in Kew Bull. Misc. Inf. 394. (1910), Rehder & E. H. Wilson in Sargent Pl. Wilson 2: 163. (1914), Zheng in Journ. Wuhan Bot. Res. 2.(1): 115. (1984), Bean in Trees & Shrubs 4: 328. (1989), O' Brien in Augustine Henry, An Irish Plant Collector in China 33. (2002).

Sarcococca pruniformis Forbes & Hemsley non Lindl. in Journ. Linn. Soc. Bot. 26: 418. (1894), *Sarcococca ruscifolia* Stapf. var. *chinensis* (Franch) Rehder & E. H. Wilson in Sargent Pl. Wilson. 2: 163. (1914).

Hubei: Liantuo, 2588, 2589 (type), 2993, 3077, 3287, 4498, 3832. Yichang, 3077a. **Yunnan:** On cliffs near Mengzi at 1,600 metres, 9859b.

An evergreen shrub to 1 metre tall, the white, very fragrant flowers are produced in the axils of leaves in late winter. *Sarcococca ruscifolia* was introduced to cultivation by Wilson from Yichang in 1901 for Messrs Veitch who raised it at their Coombe Wood nursery. In the Veitch catalogue *New Hardy Plants from Western China* (Autumn 1912) this species was available for one shilling and six pence.

Sarcopyramis bodinieri H. Lévl. (MELASTOMATACEAE)

Sarcopyramis delicata C. B. Robinson Li in Journ. Arnold Arb. 25: 26. (1944).

Yunnan: In Pingbian in woods on the Daweishan Range at 2,600 metres, 9030.

Sarcopyramis nepalensis Wall. (MELASTOMATACEAE)

Li in Journ. Arnold Arb. 25: 25. (1944).

Yunnan: Mengzi, 9725, 10298, 10990. Simao, 13562b., 13562c.

A monotypic genus with a number of varieties, native to India, Nepal, China, the Philippines, Borneo, Java and Malaysia.

Sarcosperma arboreum Hook. f. (SAPOTACEAE)

Anonymous in Kew List of Determinations (volume 34).

Yunnan: In the mountains to the west of Simao at 1,640 metres, 11588. Simao, 12017, 12837, 13291.

Sarcosperma sp. (SAPOTACEAE)

Anonymous in Kew List of Determinations (volume 34).

Yunnan: Mengzi, 11424.

Sargentodoxa cuneata (Oliv.) Rehder & E. H. Wilson (LARDIZABALACEAE) New genus & species

in Sargent Pl. Wilson. 1: 350. (1913), Stapf in Curtis's Bot. Mag. 151: t. 9111 & 9112. (1925), Bean in Trees & Shrubs 4: 330. (1989), Nelson in A Heritage of Beauty 319. (2000), O' Brien in Augustine Henry, An Irish Plant Collector in China 35. (2002); in Ir. Garden 13(5): 55. (2004).

Holboellia cuneata Oliv. in Hooker's. Icon. Pl. t. 1817. (1888), Henry in Notes Econ. Bot. China 48. (1893), Bretschneider in Hist. Eur. Bot. Disc. China 779. (1898), Diels in Bot. Jahrb. 29: 343. (1900).

Hubei: Liantuo, 3830.

A vigorous, twining, deciduous climber to 8 metres tall in its native habitat, where it is often seen scrambling through nearby trees and shrubs (in Badong its support is often *Quercus fabri,* in Changyang it trails through *Sassafras tzumu,* for example). This monotypic genus was named in compliment to Professor Charles Sprague Sargent, director of the Arnold Arboretum and a friend and colleague of Augustine Henry. One of the most brilliant American botanists of the late 19th and early 20th century, it is mainly due to Sargent's abilities that the Arnold Arboretum reached worldwide acclaim.

The foliage of *Sargentodoxa* resembles *Sinofranchetia* so closely that Daniel Oliver desribed the flowers of *Sargentodoxa* and the fruits of *Sinofranchetia* from Henry's specimens as belonging to the same species (*Holboellia cuneata*). The fruits of *Sargentodoxa*, however are quite different to other members of Lardizabilaceae from central China, hense it was once placed in a new family, Sargentodoxaceae. Though the original description states that plants are dioecious the Kilmacurragh plant bears both male and female flowers in separate racemes, though it has never set fruit here. The plant is extremely rare in cultivation, with just a handful of plants cultivated in Europe including the National Botanic Gardens, Kilmacurragh.

In China, *Sargentodoxa cuneata* is commonly known as Sargent's glory vine (Henry gives the Chinese name *hung-t'eng*) and in 2002 while climbing Emei Shan in Sichuan, members of the Glasnevin Central China Expedition were able to purchase the dried stems of this species which are used in traditional Chinese medicine. In Chinese herbal medicine the dried stems of this species are used to relieve rheumatic conditions and to relieve the pain of traumatic swelling and acute appendicitis. Later on the same expedition (2002) we met with it in Badong in Hubei where it scaled though young trees of *Quercus fabri*. Given the amount of stems sold annually in China for medicinal purposes it must be still quite common in the wild. The plant is particularly handsome in autumn when its foliage takes on an autumnal gown of brilliant claret.

Sargentodoxa cuneata was collected by the Glasnevin Central China Expedition in Badong (GCCE 275) in October 2002 and further material was gathered by the 2004 Glasnevin Expedition to Hubei on Tianthu Shan, Changyang (GCCE 775) in October of that year.

Saruma henryi Oliver (ARISTOLOCHIACEAE) New genus

in Hooker's Icon. Pl. 19: t. 1895. (1889), Forbes & Hemsley in Journ. Linn. Soc. Bot. 26: 360. (1891), Hemsley in Gard. Chron. ser. 1, 422. (1890), Bretschneider in Hist. Eur. Bot. Disc. China 2: 790. (1898), Morley in Glasra 3: 80. (1979), Lancaster in The Garden 123: 238. (1998), Hinkley in The Explorers Garden 285. (1999), O' Brien in Augustine Henry, An Irish Plant Collector in China 35. (2002).

Hubei: Fang Xian, 6676 (isosyntype), 6683 (type).

This monotypic genus was discovered by Henry and introduced by him through the Royal Botanic Gardens, Kew. In naming this genus, Professor Daniel Oliver took the first letter from *Asarum* and by placing it at the end coined the generic epithet, *Saruma*. (*Saruma* bears petals unlike *Asarum*, on which they are absent or vestigial within a cup-like calyx). Lancaster (1998) states Henry did not collect or send home any seed of this species. He did, it was in still in cultivation by 1898 but appears to have been lost to gardens until the Japanese plant hunter, Mikinori Ogisu re-introduced it to England (Wilson also collected *Saruma* but it appears he did not collect seeds) from seeds of Japanese-grown plants that had been seed-raised from material from Wuhan in Hubei. A perennial to about 45 cm tall, the foliage of this species is extremely handsome and emerges purple-flushed in early spring. In China, *Saruma henryi* is found in the provinces of Jiangxi, Hubei, Henan, Shaanxi, Gansu, Sichuan and Guizhou, where it grows in damp deciduous woodland between 600 and 1,600 metres. Lancaster and Ogisu point out that in some of these areas the rootstock is used as a medicine for stomach ailments and the leaves are commonly eaten by the larvae of a butterfly, *Luehdorsia longicauda*. The newly emerged shoots are boiled and eaten as greens in other parts of its range.

Sassafras tzumu (Hemsl.) Hemsley (LAURACEAE) New species

in Kew Bull. Misc. Inf. 55. (1907), Hemsley in Hooker's Icon. Pl. xxix. t. 2833. (1907), Henry in Elwes and Henry, The Trees Gt. Brit. & Irel. 3: 515. (1908), Zheng in Journ. Wuhan Bot. Res. 2.(1): 56. (1984), Bean in Trees & Shrubs 4: 334. (1989).

Lindera aff. *triloba* Bl. Henry in Journ. China Br. Roy. Asiat. Soc. 22: 278. (1887) (Chinese Names of Plants). *Litsea laxiflora* Hemsley in Journ. Linn. Soc. Bot. 26: 383. (1891), Bretschneider in Hist. Eur. Bot. Disc. China 2: 790. (1898). *Lindera tzumu* Hemsley in Journ. Linn. Soc. Bot. 392. (1891), Bretschneider in Hist. Eur. Bot. Disc. China 2: 791. (1898).

Hubei: South Badong, 1465 (type of *Litsea laxiflora* Hemsley), 2856 (type of *Lindera tzumu* Hemsley), 5363. Zigui, 5363a.

During Henry's time in central China this was an abundant tree in the woods around Yichang where it made large trees to 35 metres tall with girths of up to 5 metres or more at the base. He stated it was known colloquially in the Three Gorges region as *tzu mu*, hense the specific epithet. It was discovered by Irishman, William Hancock, an employee of the Chinese Imperial Maritime Customs, near Ningbo in eastern Zheijiang on the 25th of May 1877 and was collected later again by Charles Maries on his rather unsuccessful trip to China. It was from Augustine Henry's material that Hemsley based his *Lindera tzumu* (though he also based his *Litsea laxiflora* on Henry's fruiting specimens of *Sassafras tzumu*). Previous to these collections *Sassafras* was thought to be monotypic and peculiar to the flora of eastern north America. We now know there are three species, *Sassafras albidum* (Nutt.) Nees, from the eastern United States, *Sassafras randaiense* (Hayata) Rehder from Chinese Taiwan (cultivated at Mount Usher in County Wicklow) and *Sassafras tzumu* (Hemsl.) Hemsley from mainland China.

The bark on old trees is longitudinally fissured and rugged and copious amounts of yellow flowers appear on the tree in early April before the leaves unfold. In autumn the leaves turn orange and red. The wood of this tree was once used to make packing cases. According to Bretschneider it is also the *t'ze* tree of the Chinese classics and was much valued by the ancient Chinese for its timber. *Sassafras tzumu* was introduced to cultivation by E. H. Wilson through Messrs Veitch in 1900 (Wilson-Veitch 64). Of rapid growth, four-year-old seedlings were over two metres tall in the Coombe Wood nursery from Wilson's seeds. At Wakehurst Place, an extension of the Royal Botanic Gardens, Kew, this species reached 18 metres tall until blown down by that memorable storm of 1987. *Sassafras tzumu* was also collected by the Glasnevin Central China Expedition (GCCE 792) on a steep ridge on Tianthu Shan in Changyang in October 2004.

Satyrium nepalense D. Don (ORCHIDACEAE)

Rolfe in Journ. Linn. Soc. Bot. 36: 63. (1903).

Yunnan: In Pingbian on the Daweishan Range amoungst grass at 1,900 metres, 9192, 9192b. South of the Red River from Manmei at 1,900 metres, 9192a. Simao, 12673.

Distributed in Nepal, India, Sri Lanka, Myanmar and China [Guizhou, Yunnan and Tibet (Xizang)].

Saurauia fasciculata Wall. (ACTINIDIACEAE)

Anonymous in Kew List of Determinations (volume 34).

Yunnan: Mengzi, 9592. Simao, 12126.

Saurauia napaulensis DC. (ACTINIDIACEAE)

Anonymous in Kew List of Determinations (volume 34).

Yunnan: Mengzi, 9142. Simao, 12178.

An evergreen tree to 10 metres tall, a highly ornamental species bearing bold, pleated leaves and large panicles of bell-shaped purple flowers. Distributed from Nepal to Laos, in China this species is found in the provinces of Guizhou, Guangxi, Sichuan and Yunnan.

Saurauia sp. (ACTINIDIACEAE)

Anonymous in Kew List of Determinations (volume 34).

Yunnan: Simao, 11732.

Saurauia tristyla DC. (ACTINIDIACEAE)

Saurauia oldhamii Hemsley Henry in Trans. Asiat. Soc. Jap. 24 suppl: 20. (1896) (List Pl. Formosa), Li in Woody Flora of Taiwan 575. fig. 226. (1963).

Taiwan: Wanchin, 507. Oluanpi, 1244. **Yunnan:** In the Red River valley near Manhao, 9540. Simao, 12124.

Sauromatum giganteum (Engl.) Cusimano & Hett. (ARACEAE)

Brown in Journ. Linn. Soc. Bot. 36: 180. (1903), Henry in Notes Econ. Bot. China 6 & 26. (1893).

Typhonium giraldii (Baroni) Engl. Engler in Engler, Pflanzenr. iv. 23F: 110. (1920).

Hubei: Xingshan at 1,400 metres, 6537.

According to Henry this species was known colloquially at Yichang as *sheng-king*, while at Fang it was called *tu-chio-lien*. Introduced to cultivation by Emil Bretschneider from the mountains near Beijing through the botanic gardens at St. Petersburg where it first flowered in 1883. The rhizome is the source of the drug *bai fu zi* which is used in cases of facial pain, headaches and dizziness.

Saururus chinensis (Lour.) Baill. (SAURURACEAE)

Henry in Journ. China Br. Roy. Asiat. Soc. 22: 262. (1887) (Chinese Names of Plants).

Saururus loureiri Decne. Forbes & Hemsley in Journ. Linn. Soc. Bot. 26: 363. (1891), Henry in Trans. Asiat. Soc. Jap. 24 suppl: 77. (1896) (List Pl. Formosa).

Hubei: Yichang, 501. Liantuo, 1901. **Hainan:** On hummocks near Haikou, 8110. **Taiwan:** Wanchin, without number. Kaohsiung, without number.

The Chinese lizardtail, a perennial herb to 80 cm tall and stated by Henry to be known colloquially in the Three Gorges region as *peh chieh ou*. The dried rhizome is used as a diuretic in Chinese herbal medicine.

Saussurea alatipes Hemsley (ASTERACEAE) New species

in Journ. Linn. Soc. Bot. 29: 308. (1892), Bretschneider in Hist. Eur. Bot. Disc. China 2: 786. (1898).

Sichuan: North Wushan, 7141, 7066.

Saussurea bullockii Dunn (ASTERACEAE) New species

in Journ. Linn. Soc. Bot. 35: 509. (1903).

Hubei: Fang Xian, 6692 (isotype).

Also collected by Shearer and Bullock in Jiangxi province on the east coast of China. Bullock's was the only specimen to reach Kew with perfect leaves and flowers, hence this species commemorates him rather than Henry.

Saussurea conyzoides Hemsley (ASTERACEAE) New species

in Journ. Linn. Soc. Bot. 29: 309. (1892), Bretschneider in Hist. Eur. Bot. Disc. China 2: 786. (1898).

02

Hubei: Fang Xian, 7575 (isotype).

Saussurea cordifolia Hemsley (ASTERACEAE) New species

in Journ. Linn. Soc. Bot. 29: 310. (1892), Bretschneider in Hist. Eur. Bot. Disc. China 2: 786. (1898), S. J. Lipschitz in Saussurea 186. (1979).

Saussurea aff. *triangulata* Trautv. & C. A. Mey. Forbes & Hemsley in Journ. Linn. Soc. Bot. 23: 468. (1888).

Hubei: Badong, 414, 5075. Fang Xian, 6640 (isolectotype). **Sichuan:** South Wushan, 7460.

This species was collected by the Sino-American Expedition to the Shennongjia Forest District in 1980. Henry's 6640 is from the same region.

Saussurea dolichopoda Diels (ASTERACEAE) New species

in Bot. Jahrb. Syst. 29: 623. (1901).

Hubei: South Badong, 7338. Xingshan, 6481 (isotype).

Saussurea henryi Hemsley (ASTERACEAE) New species

in Journ. Linn. Soc. Bot. 29: 311. (1892), Bretschneider in Hist. Eur. Bot. Disc. China 2: 786. (1898), Morley in Glasra 3: 80. (1979).

Sichuan: North Wushan, 7068 (type). **Hubei:** North Badong, 7068a.

Saussurea hieracioides Hook. f. (ASTERACEAE)

Saussurea villosa Franchet Hemsley in Journ. Linn. Soc. Bot. 29: 312. (1892), M. Smith in Journ. Linn. Soc. Bot. 36: 515. (1904—1906).

Hubei: Fang Xian, 6762.

Saussurea japonica (Thunb.) DC. (ASTERACEAE)

Forbes & Hemsley in Journ. Linn. Soc. Bot. 23: 464. (1888).

Hubei: Yichang, 136, 896. Badong, 1044 (December 1885), 3180, 5189. Liantuo, 2693. Antelope Glen, 3222. **Sichuan:** South Wushan, 7225.

Saussurea lyrata (Bunge) Franch. (ASTERACEAE)

Saussurea affinis Spreng. ex DC. Forbes & Hemsley in Journ. Linn. Soc. Bot. 23: 463. (1888), Henry in Trans. Asiat. Soc. Jap. 24 suppl: 55. (1896) (List Pl. Formosa).

Hubei: Yichang, 1174, 1336. Antelope Glen, 3639. South Badong, 7370. **Taiwan:** Wanshoushan, 156. Oluanpi, 619.

Saussurea macrota Franch. (ASTERACEAE)

Saussurea auriculata Hemsley in Journ. Linn. Soc. Bot. 29: 308. (1892), Bretschneider in Hist. Eur. Bot. Disc. China 2: 786. (1898).

Hubei: Fang at 2,200 to 3,100 metres, without number.

Saussurea nivea Turcz. (ASTERACEAE)

Saussurea eriolepis Bunge ex DC. Forbes & Hemsley in Journ. Linn. Soc. Bot. 23: 464. (1888).

Hubei: Badong, 4914.

Saussurea oligantha Franchet (ASTERACEAE) New species

in Journ. de Bot. 421. (1896), Bretschneider in Hist. Eur. Bot. Disc. China 2: 786. (1898).

Hubei: Yichang, without number.

Later collected by Père Paul Farges in Sichuan.

Saussurea populifolia Hemsley (ASTERACEAE) New species

in Journ. Linn. Soc. Bot. 29: 311. (1892), Bretschneider in Hist. Eur. Bot. Disc. China 2: 786. (1898).

Hubei: Shennongjia, common on mountain summits at 3,100 metres, 6942 (type).

A handsome biennial, this species was collected by the Glasnevin Central China Expedition in Badong (GCCE 287) in October 2002 and further collections were made by the 2004 Glasnevin Expedition to Hubei in Xingshan (GCCE 551) and on Tianthu Shan in Changyang (GCCE 816)

Saussurea salicifolia DC. forma (ASTERACEAE)

Forbes & Hemsley in Journ. Linn. Soc. Bot. 23: 467. (1888).

Hubei: Badong, 153, 931, 2533. Jianshi, 7465.

Saussurea spp. (ASTERACEAE)

Anonymous in Kew List of Determinations (volume 24 & 34). *Bennettia* sp. Anonymous in Kew List of Determinations (volume 34).

Hubei: Yichang, 789. **Sichuan:** South Wushan, 7153. **Yunnan:** On the Daweishan Range in Pingbian at 1,500 metres, 9074. Mengzi, 9324. Simao, 11911, 12583.

Saussurea tanakae Franch. & Savat. ex Maxim. (ASTERACEAE)

Anonymous in Kew List of Determinations (volume 24).

Hubei: Xingshan at 2,600 to 2,900 metres, 6998.

Saussurea vestita Franchet (ASTERACEAE)

Anonymous in Kew List of Determinations (volume 34).

Yunnan: Mengzi, 11322.

Saxifraga gemmipara Franchet (SAXIFRAGACEAE)

Engler & Irmscher in Engler, Pflanzenr. 67: 138. (1916).

Yunnan: Mengzi, 10380.

Native to China (Sichuan and Yunnan) and Thailand.

Saxifraga giraldiana Engl. (SAXIFRAGACEAE)

Saxifraga giraldiana Engl. var. *hupehensis* Engler in Bot. Jahrb. Syst. 29; 366. (1900), Engler & Irmscher in Engler, Pflanzenr. 67: 102. (1916).

Hubei: Fang Xian, 6861 (type of *Saxifraga giraldiana* Engl. var. *hupehensis* Engler).

02

Saxifraga imparilis I. B. Balfour (SAXIFRAGACEAE)
New species

in Trans. Bot. Soc. Edinburgh 27: 73. (1916).

Yunnan: Mengzi, 9917 (isosyntype).

Saxifraga mengtzeana Engler & Irmscher (SAXIFRAGACEAE) New species

in Notizbl. Konigl. Bot. Gart. Berlin 6: 37. (1913), Tebbitt in The New Plantsman 6(4): 210. (1999).

Saxifraga henryi I. B. Balfour in Trans. Bot. Soc. Edinburgh 27: 72. (1916). *Saxifraga mengtzeana* Engl. & Irmsch. var. *peltifolia* Engler in Notizbl. Konigl. Bot. Gart. Berlin 6: 37. (1913). Engler & Irmscher in Engler, Pflanzenr. 69: 648. (1919), Gornall et all. in Novon 10: 375. (2000). *Saxifraga mengtzeana* Engler & Irmscher var. *cordatifolia* Engler & Irmsch in Notizblatt Bot. Gart. u. Mus. Dahlem 6: 36. (1913), Engler & Irmscher in Engler, Pflanzenr. 69: 647. (1919), Gornall et all. in Novon 375. (2000).

Yunnan: Mengzi at 1,800 metres, 9118 (isotype of *Saxifraga henryi* I. B. Balfour, isotype of *Saxifraga mengtzeana* Engl. & Irmsch. var. *peltifolia* Engler), 10316 (syntype of *Saxifraga mengtzeana* Engler & Irmscher var. *cordatifolia* Engler & Irmsch), 10316b.

Saxifraga mengtzeana is one of ten species in the section of the Irregulares distributed through China, Japan and Korea, the best known species in the group being *Saxifraga fortunei* Hook. and *Saxifraga stolonifera* Curtis. *Saxifraga fortunei* was introduced to cultivation in 1862 from Japan. The name Irregulares reflects the asymetric flowers of the group, this unusual floral arrangement is said to aid the wind dispersal of seeds. Henry's *Saxifraga mengtzeana* was introduced to cultivation during the 1990s, it is said to be the most beautiful member of the section though coming from a subtropical region of China it needs winter protection. Around Mengzi it grows on clefts on limestone cliffs.

Saxifraga rufescens I. B. Balfour (SAXIFRAGACEAE)

Saxifraga sinensis Engl. & Irmsch. Engler & Irmscher in Engler, Pflanzenr. 69: 651. (1919), non *Saxifraga chinensis* Loureiro (1790).

Yunnan: Milê at 1,900 metres, without number.

Distributed in Hubei, Sichuan, Tibet (Xizang) and Yunnan.

Saxifraga rufescens Balf. f. var. *flabellifolia* C. Y. Wu & J. T. Pan (SAXIFRAGACEAE) New variety

in Acta Phytotax. Sin. 29: 7. (1991).

Saxifraga flabellifolia Franchet in Journ, de Bot. (Morot) 8: 295. (1894), non R. Brown (1840), Bretschneider in Hist. Eur. Bot. Disc. China 2: 783. (1898), Engler & Irmscher in Engler, Pflanzenr. 69: 652. (1919).

Sichuan: North Wushan, 7136.

Saxifraga stolonifera Meerb. (SAXIFRAGACEAE)

Saxifraga sarmentosa L. f. Forbes & Hemsley in Journ. Linn. Soc. Bot. 23: 268. (1887), Henry in Journ. China Br. Roy. Asiat. Soc. 22: 244. (1887) (Chinese Names of Plants), Engler & Irmscher in Engler, Pflanzenr. 69: 652. (1919). *Saxifraga dumetorum* I. B. Balfour in Trans. Bot. Soc. Edinburgh 27: 71. (1916).

Hubei: Yichang, 684. Yichang, 1129 (isosyntype of *Saxifraga dumetorum* I. B. Balfour). Badong, 4699.

According to Henry this species was known colloquially at Yichang as *hu erh ts'ao*. A handsome plant, well known in European gardens. In Chinese herbal medicine this species is currently used to relieve inflammation and it is also used as a detoxicant.

Scaevola hainanensis Hance (GOODENIACEAE)

Krause in Engler, Pflanzenr. iv. 277: 134. (1912).

Hainan: Environs of Haikou, 8159, collected November 1889.

Discovered in 1877 by Thomas Lowndes Bullock, British Consul on Hainan.

Scaevola taccada (Gaertn.) Roxb. (GOODENIACEAE)

Scaevola koenigii Henry non Vahl. in Trans. Asiat. Soc. Jap. 24 suppl: 56. (1896) (List Pl. Formosa). *Scaevola frutescens* (Mill.) Krause in Engler, Pflanzenr. iv. 277: 125. (1912). *Scaevola sericea* Forst. f. ex Valh. Li in Woody Flora of Taiwan 899. fig. 362. (1963).

Taiwan: Kaohsiung, without number. Oluanpi, 908.

A small tree or shrub to 7 metres tall, widely dispersed from Kenya, southeast Asia, Malaysia, tropical Australia, Micronesia, Melanesia and Hawaii. One of the most common shrubs of tropical beaches and coastal plant throughout Taiwan. This species was collected by the Glasnevin Taiwan Expedition (GTE 2) at Banana Bay Coastal Forest near Kenting in October 2004.

Schima argentea E. Pritz. (THEACEAE) New species

in Bot. Jahrb. Syst. 29: 473. (1900), Bean in Trees & Shrubs 4: 337. (1989), O' Brien in Augustine Henry, An Irish Plant Collector in China 35. (2002).

Yunnan: South of the Red River from Manmei, 9759. Mengzi, 10229.

An evergreen tree to 20 metres tall in its native habitat. Collected by Henry in 1898, introduced to cultivation by Forrest in November 1917 and distributed by the Royal Horticultural Society's garden at Wisley. Native to China and Indochina.

Schima superba Gardn. & Champ. (THEACEAE)

Schima noronhae Henry non Reinw. ex Blume in Trans. Asiat. Soc. Jap. 24 suppl: 20. (1896) (List Pl. Formosa).

Taiwan: Oluanpi, 366, 659.

Schima wallichii (DC.) Korth. (THEACEAE)

Anonymous in Kew List of Determinations (volume 34).

Yunnan: Mengzi, 9215. Simao, 11763.

An evergreen tree to 30 metres tall, bearing clusters of creamy-white fragrant flowers in April and May. Distributed in the Himalaya, Myanmar, western China, Thailand, Laos and Vietnam. The Hani (an ethnic group) of Xishuangbanna in southern Yunnan call this species *sivslav* and use the young leaves to treat burns, scalds and diarrhea.

Schisandra chinensis (Turcz.) Baill. (SCHISANDRACEAE)

Henry in Notes Econ. Bot. China 7. (1893).

Schisandra chinensis (Turcz.) Baill. var *glauca* or *Schisandra elongata* (Blume) Baill. var. Henry in Journ. China Br. Roy. Asiat. Soc. 22: 279. (1887) (Chinese Names of Plants).

Hubei: Badong, 1827, 4040. Liantuo, 4609. **Hunan:** Shimen, 7943.

A deciduous climbing shrub to 10 metres tall, producing clusters of two or three pale rose-coloured fragrant flowers in April and May followed by pendelous spikes of scarlet fruits. Native to China, Japan and Korea, introduced to cultivation in 1860. Source of the *wu wei tzu*, a Badong drug according to Henry.

Schisandra glaucescens Diels (SCHISANDRACEAE)
New species

Rehder in Sargent Pl. Wilson. 1: 413. (1913), Zheng in Journ. Wuhan Bot. Res. 2.(1): 46. (1984).

Hubei: South Badong, 5478. Jianshi, 5931. Liantuo, 6383. **Sichuan:** South Wushan, 5725.

A deciduous climber to 12 metres tall, allied to *Schisandra henryi*. The dioecious flowers of this species vary from pink to orange-red and are followed by scarlet fruits. Discovered in 1888 by Henry though described by Diels from material later collected by Bock and von Rosthorn in 1891 in Sichuan. Wilson stated it was exceedingly common in rocky places in western Hubei and seeds of his 164a., collected in Fang Xian, Hubei in October 1907 reached Glasnevin from the Arnold Arboretum in the spring of the following year.

Schisandra aff. *grandiflora* (Wall.) Hook. f. & Thoms. (SCHISANDRACEAE)

Anonymous in Kew List of Determinations (volume 24).

Sichuan: Henry's Chinese collector with Pratt, no locality stated but may be any of the following, Leshan, Emei Shan, Wa Shan or Kangding, 8798.

Schisandra henryi C. B. Clarke (SCHISANDRACEAE)
New species

in Gard. Chron. i. 162. fig. 55. (1905), Fedde in Repert.

Spec. Nov. Regni. Veg. 4: 172. (1907), Rehder & E. H. Wilson in Sargent Pl. Wilson. 1: 413. (1913), Zheng in Journ. Wuhan Bot. Res. 2.(1): 47. (1984), Bean in Trees & Shrubs 4: 341. (1989), Whiteley in The New Plantsman 4(2): 95. (1997), Nelson in A Heritage of Beauty 319. (2000).

Schisandra elongata Hook. f. & Thoms. var. *longissima* Dunn in Journ. Linn. Soc. Bot. 354. (1908).

Hubei: Badong, 1785. Changyang, 6226. **Yunnan:** Mengzi, 9193. South of Red River, from Manmei at 1,950 metres, 9193a., (isosyntype of *Schisandra elongata* Hook. f. & Thoms. var. *longissima* Dunn). Near Yuanjiang on the slopes of Fengchunling at 2,300 metres, 9193b., (isosyntype of *Schisandra elongata* Hook. f. & Thoms. var. *longissima* Dunn).

A deciduous, dioecious climbing shrub with twining stems, flowers are white followed by pendant spikes of fleshy fruit spikes. Wilson (who introduced this species to cultivation when collecting for Messrs Veitch in 1900) described this species as being rather rare and stated the fruits were eaten in the rural mountainous parts of Hubei where according to Henry the local people designated the plant as the *tieh-ku-san*. Wilson also collected *Schisandra henryi* on Emei Shan. It is likely that the plant grown by Alice and Augustine Henry their garden in Ranelagh, Dublin came from this source. Endemic to China.

Schisandra henryi C. B. Clarke subsp. *yunnanensis* (A. C. Smith) R. M. K. Saunders (SCHISANDRACEAE)

New variety. in Syst. Bot. Monogr. 58: 89. (2000)

Schisandra henryi C. B. Clarke var. *yunnanensis* A. C. Smith in Sargentia 7: 116. (1947). *Schisandra henryi* Li non C. B. Clarke in Acta Phytotax. Sin. 38: 538. (2000).

Yunnan: East of Simao at 1,650 metres, 12022 (holotype, collected May 30th 1899).

Schisandra plena A. C. Smith (SCHISANDRACEAE)
New species

in Sargentia 7: 154. (1947).

Yunnan: Simao at 1,300 metres, 10853 (holotype).

Distributed in India and China (Yunnan).

Schisandra propinqua (Wall.) Baill. (SCHISANDRACEAE)

Rehder & E. H. Wilson in Sargent Pl. Wilson. 1: 416. (1913).

Yunnan: Scrambling on rocks near Mengzi at 1,600 metres, 10719. Mountains to the west of Simao at 1,500 to 1,600 metres, 11893. Mountains to the south of Simao at 1,500 to 1,600 metres, 12192.

Distributed in from Nepal to Bhutan and China to Bali.

Schisandra propinqua (Wall.) Baill. subsp. *sinensis* (Oliver) R. M. K. Saunders (SCHISANDRACEAE)

Schisandra propinqua (Wall.) Baill. var. *sinensis* Oliver in Hooker's Icon. Pl. 18: t. 1715. (1887), Henry in Journ.

02

China Br. Roy. Asiat. Soc. 22: 248. (1887) (Chinese Names of Plants), Rehder & E. H. Wilson in Sargent Pl. Wilson. 1: 416. (1913), Zheng in Journ. Wuhan Bot. Res. 2.(1): 416. (1985), O' Brien in Augustine Henry, An Irish Plant Collector in China 35. (2002). *Schisandra propinqua* (Wall.) Hook.f. et. Thoms. var. *linearis* Finet & Gagnepain in Bull. Soc. Bot. France 52, Mem. 4: 48.(1905); in Contrib. Fl. As. Or. 2: 51. (1908).

Hubei: On the road to Moji Shan (June 1886), 1544 [lectotype of *Schisandra propinqua* (Wall.) Hook.f. et. Thoms. var. *linearis* Finet & Gagnepain, isolectotype of *Schisandra propinqua* (Wall.) Hook. f. & Thoms. var. *sinensis* Oliver]. San You Dong Glen, 1693. Yichang, 2028, 3243, 3354, 3699, 3961. Antelope Glen, 3434. Liantuo, 4418. Changleping, 6219.

A climber to 3 metres tall with narrow lanceolate leaves marbled white on the uppersurface, yellow flowers are produced in late summer, followed by yellow fruits. A handsome plant with attractive marbled leaves, known in central and western China as *hsueh-hu-teng*, Henry stated his Badong collector knew it as *t'ieh ku san*. Introduced to cultivation by Wilson in 1907. It is hardier than the type which is a Himalayan native and is cultivated at the National Botanic Gardens, Glasnevin.

Schisandra pubescens Hemsley & E. H. Wilson (SCHISANDRACEAE) New species

in Kew. Bull. Misc. Inform. 150. (1906).

Hubei: Badong, 1785. Chienshih, 5907.

Schisandra pubinervis (Rehder & E. H. Wilson) R. M. K. Saunders (SCHISANDRACEAE)

Schisandra sphenanthera Rehder & E. H. Wilson var. *pubinervis* Rehder & E. H. Wilson in Sargent Pl. Wilson. 1: 415. (1913).

Hubei: Jianshi, 5907 (syntype *Schisandra sphenanthera* Rehder & E. H. Wilson var. *pubinervis* Rehder & E. H. Wilson). Donghu, 6447.

An attractive yellow flowered species bearing pendelous spikes of orange-red fruits in autumn.

Schisandra sphenanthera Rehder & E. H. Wilson (SCHISANDRACEAE) New species

in Sargent Pl. Wilson. 1: 414. (1913), Zheng in Journ. Wuhan Bot. Res. 2.(1): 47. (1984).

Hubei: San You Dong Glen, 3446. Antelope Glen, 3469. Badong, 4059. Without locality, 5527a. Changleping, 6292. Liantuo, 6384. **Sichuan:** South Wushan, 5527. **Yunnan:** On Fengchunling in forests south of the Red River at 2,300 metres, 10697. In woods near Mengzi at 1,800 metres, 11211. Mountains near Simao at 1,600 to 1,800 metres, 12022a., 12022b.

A deciduous dioecious climber to 5 metres tall, carrying solitary flowers in the lower leaf axils (the individual flowers are greenish on the outside and orange within). Introduced to cultivation by Wilson in 1907, seeds of his 886 (from Wa Shan, Sichuan) 869 (Baoxing, Sichuan) 869a. (Wenchuan, Sichuan) were received at Glasnevin in 1909. *Schisandra sphenanthera* is rare in cultivation though there is a particularly good specimen on the *Clematis* tent at Glasnevin.

Schizomussaenda henryi (Hutchinson) X. F. Deng & D. X. Zhang (RUBIACEAE) New species

Mussaenda henryi Hutchinson in Sargent Pl. Wilson. 3: 397. (1916), *Mussaenda elongata* Hutchinson in Sargent Pl. Wilson. 3: 398. (1916). *Mussaenda dehiscens* Craib Anonymous in Kew List of Determinations (volume 34).

Yunnan: In forests to the south of Simao at 1,500 metres., 12363. Simao, 12825. South of the Red River from Manmei, 13660 (type of *Mussaenda henryi* Hutchinson). Mengzi, 13642.

A small evergreen tree to 3 metres tall, producing terminal cymes of golden yellow flowers surrounded by large white leaf-like bracts from May to October. The Hani (an ethnic group) of Xishuangbanna in southern Yunnan call this species *neev'qyul'albao!* and use it to treat cases of bronchitis and laryngiti. Distributed in China (Guangdong, Guangxi and Yunnan) and northern Indochina.

Schizopepon dioicus Cogniaux (CUCURBITACEAE) New species

in Hooker's Icon. Pl. 23: t. 2224. (1892).

Hubei: Badong, 4862, 5991c. Jianshi, 5991 (syntype). Fang Xian, 5991b. **Sichuan:** South Wushan, 5991a.

Schizotechium paniculatum (Edgew.) Pusalkar & S. K. Srivast. (CARYOPHYLLACEAE)

Stellaria paniculata Edgew. Anonymous in Kew List of Determinations (volume 34).

Yunnan: Mengzi, 11172.

Schnabelia terniflora (Maxim.) P. D. Cantino (LAMIACEAE)

Caryopteris terniflora Maxim. Forbes & Hemsley in Journ. Linn. Soc. Bot. 26: 265. (1890).

Hubei: Yichang, 336, 336a., 816, 822, 832, 855, 1230, 2038, 2726, 7376, 7776.

Discovered in Gansu about 1874 by the Russian explorer Pavel Jakovlevich Piasetski.

Schoenoplectiella erecta (Poir.) Lye (CYPERACEAE)

Scirpus erectus Poir. Henry in Trans. Asiat. Soc. Jap. 24 suppl: 105. (1896) (List Pl. Formosa), C. B. Clarke in Journ. Linn. Soc. Bot. 36: 248. (1903).

Taiwan: Kaohsiung, 1088.

Schoenoplectiella juncoides (Roxb.) Lye (CYPERACEAE)

Scirpus supinus L. var. *juncoides* Roxb. Henry in

Journ. China Br. Roy. Asiat. Soc. 22: 272. (1887) (Chinese Names of Plants). *Scirpus erectus* C. B. Clarke non Poir. in Journ. Linn. Soc. Bot. 36: 248. (1903).

Hubei: Yichang, 78, 2740, 4212. In dry rice fields near Yichang, 4296.

Henry stated that this species was known colloquially in the Three Gorges region as *shui teng ts'ao*.

Schoenoplectiella triangulata (Roxb.) J. Jung & H. K. Choi (CYPERACEAE)

Scirpus mucronatus Henry non L. in Trans. Asiat. Soc. Jap. 24 suppl: 105. (1896) (List Pl. Formosa), C. B. Clarke in Journ. Linn. Soc. Bot. 36: 252. (1903).

Hubei: Yichang, 2295, 2930, 4213, 4255. **Taiwan:** Kaohsiung, used to make mats, 1190.

Native to temperate and tropical Asia, growing in shallow water in marshes and by rivers and in rice fields.

Schoenoplectus lacustris (L.) Palla (CYPERACEAE)

Scirpus lacustris L. Henry in Trans. Asiat. Soc. Jap. 24 suppl: 105. (1896) (List Pl. Formosa), C. B. Clarke in Journ. Linn. Soc. Bot. 36: 250. (1903).

Taiwan: Kaohsiung, 754, 1057, 1059, 1775, 1777. **Yunnan:** Mengzi at 1,600 metres, 11135. Simao, 12194.

According to Henry this was the *pu chih ts'ao* and was cultivated and used for making mats.

Schoenoplectus triqueter (L.) Palla (CYPERACEAE)

Scirpus triqueter L. Henry in Trans. Asiat. Soc. Jap. 24 suppl: 105. (1896) (List Pl. Formosa). C. B. Clarke in Journ. Linn. Soc. Bot. 36: 255. (1903).

Taiwan: Tamsui, used for making mats, 1761.

Schoenus falcatus R. Br. (CYPERACEAE)

Henry in Trans. Asiat. Soc. Jap. 24 suppl: 105. (1896) (List Pl. Formosa), C. B. Clarke in Journ. Linn. Soc. Bot. 36: 261. (1903).

Taiwan: Kaohsiung, 1943.

A perennial to 1 metre tall, native to Indochina, China, Japan (south), Malesia, northern Australia and the Solomon Islands.

Schoepfia fragrans Wall. (SCHOEPFIACEAE)

Schneider in Sargent Pl. Wilson. 3: 322. (1916).

Yunnan: Mengzi, 11086. In the forests to the east of Simao on hills at 1,600 metres, 12274, 12660. In mountains to the south of Simao at 1,600 metres, 12274a., 12274b., 12274c.

A shrub to 1.8 metres tall carrying fragrant white flowers. Distributed in Nepal, India, Bangladesh, Bhutan, China [Tibet (Xizang) and Yunnan], Thailand, Laos, Cambodia and Vietnam.

Schoepfia jasminodora Sieb. & Zucc. (SCHOEPFIACEAE)

Forbes & Hemsley in Journ. Linn. Soc. Bot. 23: 114. (1886), Schneider in Sargent Pl. Wilson. 3: 321. (1916)

Zheng in Journ. Wuhan Bot. Res. 2.(1): 35. (1984).

Hubei: Jianshi, 5975. South Badong, 5975a. Donghu, 5597b. **Sichuan:** South Wushan, 5597. **Yunnan:** Milê District, 10605.

A small deciduous tree, native from north-east India to China (including Taiwan), Thailand, Vietnam and Japan (south).

Schoepfia sp. (fruit) (SCHOEPFIACEAE)

Anonymous in Kew List of Determinations (volume 24).

Sichuan: Henry's Chinese collector with Pratt, no locality stated but may be any of the following, Leshan, Emei Shan, Wa Shan or Kangding, 8814.

Schoepfia spp. (SCHOEPFIACEAE)

Anonymous in Kew List of Determinations (volume 34).

Yunnan: Simao, 12063, 12456, 12804, 13166. Mengzi, 13803.

Scirpus asiaticus Beetle (CYPERACEAE) New species in Amer. Journ. Bot. 33: 662. (1946).

Scirpus eriophorum C. B. Clarke non Michx. in Journ. Linn. Soc. Bot. 36: 249. (1903).

Sichuan: Kangding at 2,900 to 4,000 metres, Henry's Chinese collector with Pratt, 8836, 8912. **Yunnan:** Simao, 12527 (type).

Distributed from Russia to India, China, Korea and Japan.

Scirpus rosthornii Diels (CYPERACEAE)

Scirpus ternatensis C. B. Clarke in Journ. Linn. Soc. Bot. 36: 254. (1903).

Hubei: Near Yichang at the entrance to a cave in a moist place, 1400. Yichang, 1854, 3983. **Yunnan:** Mengzi at 1,900 metres, 10557. Simao, 12920.

Widely distributed from Nepal to Japan.

Scirpus sp. (CYPERACEAE)

Anonymous in Kew List of Determinations (volume 24).

Hubei: Badong, 303 (June 1885).

Scirpus ternatanus Reinw. ex Miq. (CYPERACEAE)

Henry in Trans. Asiat. Soc. Jap. 24 suppl: 105. (1896) (List Pl. Formosa), C. B. Clarke in Journ. Linn. Soc. Bot. 36: 254. (1903).

Scirpus mucronatus Henry non L. in Trans. Asiat. Soc. Jap. 24 suppl: 105. (1896) (List Pl. Formosa).

Taiwan: Oluanpi, 244, 593, 609.

A robust perennial, native to much of Asia and growing in wet places often on the slopes of low mountains near the coast.

Scleria annularis Nees ex Steud. (CYPERACEAE)

C. B. Clarke in Journ. Linn. Soc. Bot. 36: 263. (1903).

02

Hubei: Yichang, 4253.

Scleria levis Retz. (CYPERACEAE)

Scleria hebecarpa Nees Forbes & Hemsley in Journ. Linn. Soc. Bot. 36: 264. (1903). *Scleria hebecarpa* Nees var. *pubescens* (Steud.) C. B. Clarke C. B. Clarke in Journ. Linn. Soc. Bot. 36: 265. (1903).

Hubei: Yichang, 4095. **Hainan:** Environs of Haikou, 8009. **Yunnan:** Simao at 1,600 metres, 12230.

A perennial species to 90 cm tall, widely distributed across India, Sri Lanka, southern China, Japan, south-east Asia, Australia and the Pacific Islands.

Scleria scrobiculata Nees & Mey (CYPERACEAE)

C. B. Clarke in Journ. Linn. Soc. Bot. 36: 266. (1903).

Yunnan: Mengzi at 1,950 metres, 10484.

A widespread species found over much of Asia. Henry's 10484 was collected from a plant that was 2.5 metres tall and was used for making brooms.

Scleria terrestris (L.) Fassett (CYPERACEAE)

Scleria elata Thwaites var. *decolorans* Henry in Trans. Asiat. Soc. Jap. 24 suppl: 105. (1896) (List Pl. Formosa). *Scleria elata* Thwaites C. B. Clarke in Journ. Linn. Soc. Bot. 36: 264. (1903).

Taiwan: Oluanpi, 608.

Scleria terrestris (L.) Fassett var. ***hookeriana*** (Boeck.) D. M. Verma (CYPERACEAE)

Scleria hookeriana Boeck. C. B. Clarke in Journ. Linn. Soc. Bot. 36: 265. (1903).

Yunnan: Simao at 1,600 metres, 12230a.

Distributed from north-east India to China and Vietnam.

Scleromitrion diffusum (Willd.) R. J. Wang (RUBIACEAE)

Oldenlandia diffusa (Willd.) Roxb. Anonymous in Kew List of Determinations (volume 24 & 34).

Hainan: On Liang Shan, 16 km (10 miles) east of Haikou, 8203. **Yunnan:** Mengzi, 11020.

Native from Nepal to China and Japan.

Scleromitrion verticillatum (L.) R. J. Wang (RUBIACEAE)

Hedyotis hispida Retz. Anonymous in Kew List of Determinations (volume 34).

Yunnan: Simao, 12350.

Distributed in Nepal, India and China.

Scolopia buxifolia Gagnep. (SALICACEAE)

Sleumer in Blumea 20: 43. (1972).

Scolopia hainanensis Sleumer in Repert. Spec. Nov. Regni. Veg. 41: 123. (1936), Merrill & Chun in Sunyatsenia 5: 137. (1940), Chang & Chen in Fl. Hainan 1: 453. fig. 250. (1964).

Hainan: 32 km (20 miles) west of Haikou, 8151 (type

of *Scolopia hainanensis* Sleumer, collected November 1889).

A tropical tree to 8 metres tall, native to Thailand, southern Vietnam and China (Hainan).

Scolopia oldhamii Hance (SALICACEAE)

Li in Woody Flora of Taiwan 607. fig. 240.(1963), Sleumer in Blumea 20: 45. (1972).

Scolopia crenata Henry non (Wight. & Arn.) Clos. in Trans. Asiat. Soc. Jap. 24 suppl: 18. (1896) (List Pl. Formosa).

Taiwan: Oluanpi, 269, 2059. Wanchin, 478, 534, 976.

A small tree to 5 metres tall, native to China and the Philippines. In Chinese Taiwan this species is distributed at low altitude along the coast. *Scolopia oldhamii* was collected by the Glasnevin Taiwan Expedition (GTE 48) by the lighthouse at Oluanpi (South Cape of Henry and Wilson) in October 2004.

Scolopia saeva (Hance) Hance (SALICACEAE)

Sleumer in Blumea 20: 45. (1972).

Scolopia henryi Sleumer in Repert. Spec. Nov. Regni. Veg. 41: 123. (1936), Merr. & Chun in Sunyatsenia 5: 137. (1940).

Hainan: Environs of Haikou, 8346 (type of *Scolopia henryi* Sleumer, collected November 1889).

Scolopia spp. (SALICACEAE)

Anonymous in Kew List of Determinations (volume 34).

Yunnan: Mengzi, 11083. Simao, 12944.

Scoparia dulcis L. (PLANTAGINACEAE)

Anonymous in Kew List of Determinations (volume 24).

Hainan: Environs of Haikou, 8350.

Scorzonera albicaulis Bunge (ASTERACEAE)

Scorzonera macrosperma Turcz. ex DC. Henry in Journ. China Br. Roy. Asiat. Soc. 22: 266. (1887) (Chinese Names of Plants), Forbes & Hemsley in Journ. Linn. Soc. Bot. 23: 488. (1888)

Hubei: Liantuo, 2085. Yichang, 3645. Jianshi, 5994. Fang Xian, 6666. Without locality, 6666a.

According to Henry this species was known colloquially in the Three Gorges region as *pi kuan ts'ao*.

Scrophularia henryi Hemsley (SCROPHULARIACEAE) New species

in Journ. Linn. Soc. Bot. 26: 178. (1890), Bretschneider in Hist. Eur. Bot. Disc. China 2: 788. (1898).

Scrophularia henryi Hemsley var. *glabrescens* Hemsley in Journ. Linn. Soc. Bot. 26: 178. (1890).

Hubei: South Badong, 6180 (type of *Scrophularia henryi* Hemsley). Xingshan, 6946 (type of *Scrophularia henryi* Hemsley var. *glabrescens* Hemsley).

This species was collected by the Sino-American Expedition in Shennongjia Forest District in 1980. It is confined to that district and it is likely Henry made his collections in present day Shennongjia.

Scurrula atropurpurea (Blume) Dancer (LORANTHACEAE)

Loranthus philippensis Chamissso & Schlechtendal Lecompte in Sargent Pl. Wilson. 3: 317. (1917).

Yunnan: Mengzi at 1,500 metres, (on *Ehretia corylifolia*), 9991.

Scurrula ferruginea (Roxb. ex Jack) Dancer (LORANTHACEAE)

Loranthus ferrugineus Roxb. ex Jack Lecompte in Sargent Pl. Wilson. 3: 317. (1917). *Loranthus sootepensis* Craib Lecompte in Sargent Pl. Wilson. 3: 317. (1917).

Yunnan: Milê district, 9947. Mengzi at 1,400 to 2,000 metres, 10315.

Native to Myanmar, China (Yunnan), Laos, Thailand, Vietnam and the Philippines.

Scutellaria amoena Wright (LAMIACEAE)

Dunn in Notes Roy. Bot. Gard. Edinburgh 6: 176. (1915), C. Y. Wu in Flora Yunnanica 1: 560. (1977).

Yunnan: Mengzi at 1,950 to 2,300 metres, 10560.

Endemic to China and distributed in Guizhou, Sichuan and Yunnan, a beautiful species. For an illustration of this skullcap see Lancaster in *Travels in China* 158. (1993).

Scutellaria baicalensis Georgi (LAMIACEAE)

Scutellaria macrantha Fisch. Dunn in Notes Roy. Bot. Gard. Edinburgh 6: 177. (1915).

Yunnan: Mengzi at 1,650 to 1,950 metres, 13772.

Distributed across Russia, Mongolia, China, Korea and Japan.

Scutellaria barbata D. Don (LAMIACEAE)

Scutellaria rivularis Wall. ex Benth. Forbes & Hemsley in Journ. Linn. Soc. Bot. 26: 296. (1890), Henry in Trans. Asiat. Soc. Jap. 24 suppl: 73. (1896) (List Pl. Formosa), Dunn in Notes Roy. Bot. Gard. Edinburgh 6: 173. (1911-17).

Hubei: Yichang, 273, 480, 886, 1127, 1387. **Taiwan:** Wanchin, on the plain, 1728. **Yunnan:** In mountain forest to the west of Simao at 1,640 metres, 11919, 11919a.

A handsome skullcup, used in Chinese herbal medicine as a diuretic, to treat boils, sores, swelling and pain of the throat, venomous snake-bite and jaundice. Native to India, Nepal, Myanmar, China, Korea, Thailand, Vietnam and Japan.

Scutellaria cyrtopoda Miq. (LAMIACEAE)

Dunn in Notes Roy. Bot. Gard. Edinburgh 6: 176. (1915).

Yunnan: Mountain forests to the north of Mengzi at 2,650 metres, 10240.

Scutellaria discolor Colebr. (LAMIACEAE)

Dunn in Notes Roy. Bot. Gard. Edinburgh 6: 175. (1915).

Yunnan: In forests near Milê at 1,650 metres, 10458. Simao, 12386 (in part).

The Hani (an ethnic group) of Xishuangbanna in southern Yunnan call this species *naqhanq'alyaiv* and use it to treat cases of influenza, fever and sore throat.

Scutellaria franchetiana Leveille (LAMIACEAE)

Dunn in Notes Roy. Bot. Gard. 6: 174. (1915).

Scutellaria angulosa Forbes & Hemsley non Benth. in Journ. Linn. Soc. Bot. 26: 293. (1890).

Hubei: Antelope Glen, 3559. Yichang, 3559a. South Badong, 6087.

Scutellaria indica L. (LAMIACEAE)

Forbes & Hemsley in Journ. Linn. Soc. Bot. 26: 295. (1890), Henry in Trans. Asiat. Soc. Jap. 24 suppl: 73. (1896) (List Pl. Formosa), Dunn in Notes Roy. Bot. Gard. Edinburgh 6: 176. (1915).

Hubei: Yichang, 771, 825, 1189, 1345. Liantuo, 1974. Without locality, 825a. Jianshi, 5883, 5974. North Badong, 7162. **Taiwan:** Oluanpi, 2073. **Yunnan:** Mengzi, 11007.

Common on grassy hill-sides at low to medium altitudes in many of China's provinces, flowers varying from blue to purple and sometimes scented of violets.

Scutellaria laxa Dunn (LAMIACEAE) New species

in Notes Roy. Bot. Gard. Edinburgh 8: 166. (1913), Dunn in Notes Roy. Bot. Gard. Edinburgh 6: 176. (1915).

Yunnan: On Fengchunling at 2,650 metres, 13771 (type).

Endemic to Yunnan.

Scutellaria obtusifolia Hemsley (LAMIACEAE) New species

in Journ. Linn. Soc. Bot. 26: 296. (1890), Dunn in Notes Roy. Bot. Gard. Edinburgh 6: 175. (1915).

Hubei: At Shih Pan Shan (off the Xiling (Yichang) Gorge), 4208 (type).

Also later collected by Ernst Faber on Emei Shan.

Scutellaria playfairii Kudô (LAMIACEAE)

Scutellaria luzonica Henry non Rolfe in Trans. Asiat. Soc. Jap. 24 suppl: 73. (1896) (List Pl. Formosa).

Taiwan: Kaohsiung, without number.

Scutellaria sp. (LAMIACEAE)

Anonymous in Kew List of Determinations (volume 34).

Yunnan: Mengzi, 9639, 9794.

Scutellaria violacea B. Heyne ex Benth. (LAMIACEAE)

Scutellaria discolor Dunn non Colebr. in Notes Roy. Bot. Gard. Edinburgh 6: 175. (1915) in part. *Scutellaria violacea* B. Heyne ex Benth. var. *sikkimensis* Hook. f. C. Y.

Wu in Flora Yunnanica 1: 545. (1977).

Yunnan: To the south of the Red River on Fengchunling at 2,300 metres, 10657.

Distributed in India and China (Sichuan and Yunnan).

Searsia paniculata (Wall. ex G. Don) Moffett (ANACARDIACEAE)

Rhus paniculata Wall. ex G. Don Rehder & E. H. Wilson in Sargent Pl. Wilson. 2: 184. (1914).

Yunnan: In a ravine near Shiping at 1,150 metres, 11578. Yuanjiang at 1,300 metres, 11578a.

Distributed in India, Bhutan, Myanmar and China (Yunnan).

Secamone elliptica R. Br. (APOCYNACEAE)

Secamone micrantha (Decne.) Decne. Anonymous in Kew List of Determinations (volume 34).

Hainan: Environs of Haikou, 8046, 8007. **Yunnan:** Mengzi, 9875, 10815a.

Distributed from China to Vietnam and Cambodia to Australia and New Caledonia. This species is currently used in Chinese herbal medicine to treat cases of ulcers.

Sedum alfredi Hance (CRASSULACEAE)

Henry in Journ. China Br. Roy. Asiat. Soc. 22: 270. (1887) (Chinese Names of Plants), Forbes & Hemsley in Journ. Linn. Soc. Bot. 23: 283. (1887).

Hubei: Badong, 4652. Jianshi, 5876. **Sichuan:** Henry's Chinese collector with Pratt, no locality stated but may be any of the following, Leshan, Emei Shan, Wa Shan or Kangding, 8942.

Sedum alfredii was desribed by the China based, renowned British botanist, Dr. Henry Fletcher Hance, from a collection made by Hance's third son (Alfred C. Hance 1858—1890) who collected it on these east coast of China between 1868—1870. According to Henry it was known colloquially in the Three Gorges region as *shih pan ts'ai.*

Sedum barbeyi Raym.-Hamet (CRASSULACEAE) New species

in Notes Roy. Bot. Gard. Edinburgh 8: 143. (1913).

Hubei: Xingshan, 7002 (type).

Sedum constantini Raym.-Hamet. (CRASSULACEAE)

Sedum filipes Hemsl. var. *major* Hemsley in Journ. Linn. Soc. Bot. 23: 284. (1887), Hamet in Notes Roy. Bot. Gard. Edinburgh 8: 144. (1913). *Sedum majus* (Hemsl.) Migo in Bull. Shanghai Sci. Inst. 14: 293. (1944).

Hubei: Liantuo, 2667, 6989. **Sichuan:** South Wushan, 6989a.

Sedum dielsii Raym.-Hamet (CRASSULACEAE) New species

Hamet in Notes Roy. Bot. Gard. Edinburgh 8: 143. (1913).

Hubei: On mountains near Yichang at 750 metres, 961.

Sedum drymarioides Hance (CRASSULACEAE)

Forbes & Hemsley in Journ. Linn. Soc. Bot. 23: 283. (1887), Henry in Journ. China Br. Roy. Asiat. Soc. 22: 246. (1887) (Chinese Names of Plants).

Hubei: Yichang, 602, 736, 3997. Liantuo, 2070. Badong, 3709.

According to Bretschneider this species was first found by Tatarinov near Beijing but was described by Hance from a collection made by Theo. Sampson near Macao in June 1865. According to Henry this species was known colloquially in the Three Gorges region as *huo yen ts'ao.*

Sedum elatinoides Franch. (CRASSULACEAE)

Forbes & Hemsley in Journ. Linn. Soc. Bot. 23: 284. (1887), Hamet in Notes Roy. Bot. Gard. Edinburgh 8: 144. (1913).

Hubei: Yichang, 3655, 3994. Badong, 4730, 6025. Jianshi, 5960, 6025.

Sedum filipes Hemsley (CRASSULACEAE) New species

in Journ. Linn. Soc. Bot. 23: 284. fig. 7. (1887), Bretschneider in Hist. Eur. Bot. Disc. China 2: 783. (1898), Hamet in Notes Roy. Bot. Gard. Edinburgh 8: 144. (1913).

Sedum filipes Hemsley var. *genuinum* R. Hamet in Candollea 4: 31. (1929).

Hubei: Under a dripping rock in the San You Dong Glen (October 17th 1886), 3230. Yichang, 3230a. South Donghu, 7593. Fang Xian, 6756. **Sichuan:** South Wushan, 6989a.

Sedum multicaule Wall. ex Lindl. (CRASSULACEAE)

Hamet in Notes Roy. Bot. Gard. Edinburgh 8: 145. (1913).

Yunnan: On mountains near Mengzi at 1,950 metres, 9157. On a rock on the mountains near Mengzi at 1,950 metres, 10317. On cliffs near Pu'er at 1,950 metres, 13197.

Native to India, Pakistan, Bhutan, Nepal, Myanmar and western China.

Sedum oligospermum Maire (CRASSULACEAE) New species

Sedum bracteatum Diels non Viv. Hamet in Notes Roy. Bot. Gard. Edinburgh 8: 143. (1913). *Sedum amplibracteatum* K. T. Fu in Fl. Tsinlingensis 1(2): 425. (1974) (a nomen novum for *Sedum bracteatum* Diels, non *Sedum bracteatum* Viv.).

Hubei: Badong, 4888. North Badong, 5946b. Jianshi, 4946. Fang Xian, 5946a., 5946c.

Sedum rosthornianum Diels (CRASSULACEAE) New species

Hamet in Notes Roy. Bot. Gard. Edinburgh 8: 145. (1913).

Sichuan: South Wushan, 5727.

Discovered by Henry in the spring of 1888 and based on a collection made by the Austrian collector, A. von Rosthorn in 1891 in the same province (Sichuan).

Sedum sarmentosum Bunge (CRASSULACEAE)

Forbes & Hemsley in Journ. Linn. Soc. Bot. 23: 286. (1887), Henry in Journ. China Br. Roy. Asiat. Soc. 22: 247. (1887) (Chinese Names of Plants); in Notes Econ. Bot. China 53. (1893).

Hubei: Yichang, 487, 7772. Liantuo, 1902.

According to Henry this species was known colloquially in the Three Gorges region as *huo lien ts'ao* or *shui-ma-ch'ih-han*. This stonecrop is currently used in Chinese herbal medicine in cases of jaundice.

Sedum spp. (CRASSULACEAE)

Henry in Trans. Asiat. Soc. Jap. 24 suppl: 42. (1896) (List Pl. Formosa).

Hubei: Yichang, 57, 671, 940, 1389, 3224, 3654, 4224. Badong, 305 (June 1885), 858 (September 1885), 1471, 1703 (July 1886), 4654, 4655, 4796. Donghu, 6401. South Donghu, 7579. Changyang, 7718. Without locality, 3230b. **Sichuan:** South Wushan. 7025. **Taiwan:** Oluanpi, 354. Kaohsiung, 1186.

Sedum stellariifolium Franch. (CRASSULACEAE)

Sedum drymarioides Forbes & Hemsley non Hance in Journ. Linn. Soc. Bot.23: 283. (1887). *Sedum drymarioides* Hance var. *stellariifolium* (Franch.) Hamet Hamet in Notes Roy Bot. Gard. Edinburgh 8: 143. (1913).

Hubei: Badong, 4908, 7395. Yichang, 6240.

Sedum yvesii Hamet (CRASSULACEAE) New species

in Fedde, Repert. Sp. Nov. Regni Veg. 8: 27. (1910).

Sedum phyllanthum Rao non H. Lévl. & Vaniot in Acta Phytotax. Sin. 34: 624. (1996).

Hubei: Antelope Glen, 3643, (holotype).

Sehima nervosa (Rotter ex Roem.) Stapf (POACEAE)

Ischaemum laxum R. Br. Rendle in Journ. Linn. Soc. Bot. 36: 365. (1904).

Yunnan: Mengzi at 1,600 metres, 9618, 9618a.

Distributed from Africa to India, China (Guangdong and Yunnan) and the Philippines.

Selaginella braunii Baker (SELAGINELLACEAE)

Diels in Bot. Jahrb. 29: 211. (1900).

Hubei: Yichang, 4103. Badong, 4773.

Selaginella delicatula (Desv.) Alston (SELAGINELLACEAE)

Selaginella canaliculata Henry non (L.) Spring in Trans. Asiat. Soc. Jap. 24 suppl: 117. (1896) (List Pl. Formosa). *Selaginella flabellata* Henry non (L.) Spring in Trans. Asiat. Soc. Jap. 24 suppl: 117. (1896) (List Pl. Formosa).

Hubei: Yichang, 620. Changyang, 7602. Liantuo, 2624. **Sichuan:** North Wushan, 7129. **Taiwan:** Wanchin, 27.

Oluanpi, 596, 601, 610. Tamsui, 1469. Wanshoushan, 1095. **Yunnan:** Simao, 12410.

This species was collected by the Glasnevin Central China Expedition (GCCE 11) on Emei Shan (Mount Omei) in September 2002. Native to India, China (including Taiwan), Myanmar, the Philippines, Malaya, New Guinea and Polynesia.

Selaginella doederleinii Hieron (SELAGINELLACEAE)

Selaginella plumosa Baker non (L.) C. Presl. in Journ. Bot. 26: 231. (1888), Henry in Trans. Asiat. Soc. Jap. 24 suppl: 117. (1896) (List Pl. Formosa), Diels in Bot. Jahrb. 29: 211. (1900).

Hubei: Xiling (Yichang) Gorge, 3488. Badong, 7916. **Taiwan:** Wanshoushan, 61. Wanchin, 1560.

Distributed through India, China, Vietnam, Japan and Malaya.

Selaginella intermedia (Blume) Spring (SELAGINELLACEAE)

Selaginella atroviridis (Wall. ex Hook. & Grev.) Spring Anonymous in Kew List of Determinations (volume 34).

Yunnan: Mengzi, 11834, 13657.

Selaginella involvens (Sw.) Spring (SELAGINELLACEAE)

Selaginella caulescens (Wall. ex Hook. & Grev.) Spring Baker in Journ. Bot. 26: 231. (1888), Henry in Trans. Asiat. Soc. Jap. 24 suppl: 117. (1896) (List Pl. Formosa), Diels in Bot. Jahrb. 29: 211. (1900).

Hubei: Yichang, 437. Xiling (Yichang) Gorge, 3595. **Taiwan:** Wanshoushan, 797, 1096. **Yunnan:** Mengzi, 11552, 13658.

Widely distributed from India, Sri Lanka, Myanmar, China (including Taiwan), Vietnam, Korea and Japan where it is found growing on tree trunks and banks of streams in shade. This species was collected by the Glasnevin Taiwan Expedition (GTE 71) in Kenting National Park in October 2004.

Selaginella leptophylla Baker (SELAGINELLACEAE)

Selaginella proniflora Henry non (Lam.) Baker in Trans. Asiat. Soc. Jap. 24 suppl: 117. (1896) (List Pl. Formosa).

Taiwan: Wanshoushan, 1917.

Distributed through India, China, the Philippines, New Guinea and northern Australia.

Selaginella nipponica Franch. & Sav. (SELAGINELLACEAE)

Selaginella savatieri Baker Baker in Journ. Bot. 26: 231. (1888), Diels in Bot. Jahrb. 29: 211. (1900).

Hubei: Xiling (Yichang) Gorge, 3596.

Native to China (including Taiwan), Vietnam and Japan.

02

Selaginella repanda (Desv. ex Poir.) Spring (SELAGINELLACEAE)

Selaginella mongholica Henry non Rupr. in Trans. Asiat. Soc. Jap. 24 suppl: 117. (1896) (List Pl. Formosa). *Selaginella henryi* Koidz. in Fl. Symb. Orient.-Asiat. 85. (1930).

Taiwan: Kaohsiung, 1963 (isotype of *Selaginella henryi* Koidz).

Selaginella sp. No. 1. (SELAGINELLACEAE)

Henry in Trans. Asiat. Soc. Jap. 24 suppl: 117. (1896) (List Pl. Formosa).

Taiwan: Oluanpi, 606..

Selaginella sp. No. 2. (SELAGINELLACEAE)

Anonymous in Kew List of Determinations (volume 34).

Yunnan: Mengzi, 11835. Simao, 13595.

Selaginella tenera (Hook. & Grev.) Spring. (SELAGINELLACEAE)

Baker in Journ. Bot. 27: 178. (1889), Diels in Bot. Jahrb. 29: 211. (1900).

Hunan: Shimen, 7561.

Selaginella uncinata (Desv.) Spring (SELAGINELLACEAE)

Anonymous in Kew List of Determinations (volume 24).

Hubei: Liantuo, 1884. At Shih Pan Shan [off the Xiling (Yichang) Gorge], 4207.

This species was collected by the Sino-American expedition in Shennongjia to the north of Yichang in 1980.

Selaginella wallichii (Hook. & Grev.) Spring (SELAGINELLACEAE) New to China

Baker in Journ. Bot. 25: 171. (1887).

Hubei: In cave on way to Moji Shan, 2229.

Selenicereus triangularis (L.) D. R. Hunt (CACTACEAE)

Cereus triangularis (L.) Haw. Anonymous in Kew List of Determinations (volume 24).

Hainan: Environs of Haikou, 8336. Ling-men (in the interior of the island), 8623.

Naturalised, a New World species from the Carribbean.

Selinum vaginatum (Edgw.) C. B. Clarke (APIACEAE)

Anonymous in Kew List of Determinations (volume 34).

Yunnan: Mengzi, 10004.

Selliguea capitellata (Mett.) X. C. Zhang & L. J. He (POLYPODIACEAE)

Polypodium wallichianum Spreng. Takeda in Notes Roy. Bot. Gard. Edinburgh viii. 311. (1913—1915).

Yunnan: On mountains to the south of Mengzi at 1,600 metres, 10079, 10079a., 11514.

Selliguea crenatopinnata (C. B. Clarke) S. G. Lu, Hovenkamp & M. G. Gilbert (POLYPODIACEAE)

Polypodium pseudoserratum Christ in Bull. Herb. Boissier 6: 871. (1898), Takeda in Notes Roy. Bot. Gard. Edinburgh viii. 297. (1913—1915).

Yunnan: Milê at 1,600 metres, 9895 (isotype of *Polypodium pseudoserratum* H. Christ, collected 1st November 1896). In woods near Mengzi at 1,650 metres, 9895a., 10284

Native to the Himalaya, Myanmar and China.

Selliguea griffithiana (Hook.) Fraser-Jenk. (POLYPODIACEAE)

Polypodium griffithianum Hook. Takeda in Notes Roy. Bot. Gard. Edinburgh viii. 308. (1913—1915). *Polypodium simplex* Christ non (Hook.) E. J. Lowe Christ in Bull. Herb. Boiss. 6: 875. (1898).

Yunnan: On a tree in mountain forest to the north of Mengzi at 2,600 metres, 10272. On a rock on mountains to the east of Mengzi at 1,650 metres, 10272a.

Selliguea hastata (Thunb.) Fraser-Jenk. (POLYPODIACEAE)

Polypodium hastatum Thunb. Henry in Trans. Asiat. Soc. Jap. 24 suppl: 115.(1896) (List Pl. Formosa), Diels in Bot. Jahrb. 29: 205. (1900), Takeda in Notes Roy. Bot. Gard. Edinburgh 8: 301. (1913—1915). *Polypodium hastatum* Thunb. var. *simplex* Christ in Bull. Acad. Int. Georg. Bot. 105. (1906).

Hubei: Liantuo, 3025, 4436. Jianshi, 5969. **Taiwan:** Tamsui, 1410. **Yunnan:** Milê, 9897. Simao, 12633, 12633b., (type of *Polypodium hastatum* Thunb. var. *simplex* Christ).

Distributed through Russia, Manchuria, China (including Taiwan), Korea, Japan and the Philippines.

Selliguea lehmannii (Mett.) Christenh. (POLYPODIACEAE)

Polypodium lehmannii Mett. Christ in Bull. Herb. Boiss. 6: 876. (1898).

Yunnan: South of the Red River from Manmei at 2,200 metres, 9747. Simao, 13104.

Distributedfrom the Himalaya to China and Indochina.

Selliguea oxyloba (Wall.ex Kuntze) Fraser-Jenk. (POLYPODIACEAE)

Polypodium trifidum D. Don Henry in Trans. Asiat. Soc. Jap. 24 suppl: 115. (1896) (List Pl. Formosa), Christ in Bull. Herb. Boiss. 6: 875. (1898). *Polypodium oxylobum* Wall. ex Kunze Takeda in Notes Roy. Bot. Gard. viii. 299. & 310. (1913—1915).

Taiwan: Oluanpi, 1241. **Yunnan:** South of the Red River from Manmei at 1,950 metres, 10080, 10080a. On the mountains to the west of Simao at 1,950 metres, 10080b. On the mountains near Simao at 2,150 metres, 13074.

Distributed in Nepal, India, Myanmar, western China,

Thailand and Vietnam.

Selliguea stewartii (Bedd.) S. G. Lu, Hovenkamp & M. G. Gilbert (POLYPODIACEAE)

Polypodium malacodon Baker non Hook. in Journ. Bot. xvii. 177. (1889). *Polypodium shensiense* auct non Christ. Christ in Bull. Soc. Bot. France 5: 18. (1905). *Polypodium stewartii* (Bedd.) Baker Takeda in Notes Roy. Bot. Gard. Edinburgh 8: 298. & 311. (1913—1915).

Hubei: Fang Xian, 6170a., (in part). Without locality, 6170e.

Selliguea trisecta (Baker) Fraser-Jenk. (POLYPODIACEAE) New species

Polypodium trisectum Baker in Kew Bull. Misc. Inf. 232. (1898), Takeda in Notes Roy. Bot. Gard. Edinburgh viii. 295. & 311. (1913—1915). *Polypodium podobasis* H. Christ in Bull. Acad. Int. Georg. Bot. 215. (1902).

Yunnan: Milê, in woods, 9891 (type of *Polypodium trisectum* Baker). On the mountains to the west of Simao at 1,650 metres, 9891a., 13121 (type of *Polypodium podobasis* H. Christ), 13121a., 13121b.

Native to India (Assam), Myanmar, China (Sichuan and Yunnan) and Thailand.

Selliguea veitchii (Baker) H. Ohashi & K. Ohashi (POLYPODIACEAE)

Polypodium veitchii Baker Takeda in Notes Roy. Bot. Gard. Edinburgh 8: 296. (1913—1915). *Polypodium shensiense* Diels non Christ in Bot. Jahrb. 29: 206. (1900).

Hubei: South Badong, 6170. Fang Xian, 6170a.(in part). Without locality, 6170b.

Discovered by Delavay near Dali in Yunnan and later collected by Henry and Wilson in Hubei.

Semiaquilegia adoxoides (DC.) Makino (RANUNCULACEAE)

Isopyrum adoxoides DC. Diels in Bot. Jahrb. 29: 325. (1900).

Hubei: Yichang, 1253. Changyang, 7806. Without locality, 1253a.

The dried tubers of this pretty little plant are used in Chinese herbal medicine to treat cases of venomous snake-bite.

Senecio asperifolius Franch. (ASTERACEAE)

Jeffrey & Chen in Kew Bull. 39: 415. (1985).

Senecio luticola Dunn in Journ. Linn. Soc. Bot. 35: 507. (1903).

Yunnan: Common on barren clay-hills near Mengzi, in exposed arid situations, 9916 (syntype of *Senecio luticola* Dunn).

A perennial herb with woody tuberous rhizomes to 90 cm tall. Discovered near Dali by Delavay.

Senecio exul Hance (ASTERACEAE)

Forbes & Hemsley in Journ. Linn. Soc. Bot. 23: 451. (1888), Jeffrey & Chen in Kew Bull. 39: 429. (1985).

Hubei: Yichang, 1314, 7778.

An annual herb to 40 cm tall, discovered by Theo. Sampson near the south-east coast of China on February 22nd 1867. Also native to Thailand.

Senecio nemorensis L. (ASTERACEAE)

Forbes & Hemsley in Journ. Linn. Soc. Bot. 23: 455. (1888), Jeffrey & Chen in Kew Bull. 39: 362. (1985).

Hubei: Badong, 2430, 2559, 4849, 5011. Changyang, 7514. **Sichuan:** South Wushan, 7224.

Senecio nudicaulis Buch.-Ham ex D. Don (ASTERACEAE)

Jeffrey & Chen in Kew Bull. 39: 410. (1985).

Yunnan: Mengzi, 10603, 10603a. Yuanjiang, 13292.

Senecio sagittatus Schultz-Bip. var. *involucri bractiis* Hemsl. (ASTERACEAE)

Forbes & Hemsley in Journ. Linn. Soc. Bot. 23: 457. (1888).

Hubei: Badong, 4917. Fang Xian, 7569, 7569a.

Senecio scandens Buch.-Ham ex D. Don (ASTERACEAE)

Henry in Journ. China Br. Roy. Asiat. Soc. 22: 239. (1887) (Chinese Names of Plants), Forbes & Hemsley in Journ. Linn. Soc. Bot. 23: 457. (1888), Henry in Trans. Asiat. Soc. Jap. 24 suppl: 55. (1896) (List Pl. Formosa), Jeffrey & Chen in Kew Bull. 39: 419. (1985).

Hubei: Yichang, 385, 948, 2951, 3270. Liantuo, 2686, 2967. Antelope Glen, 3306. **Taiwan:** Wanchin, 1643. **Yunnan:** Mengzi at 1,150 to 2,600 metres, 9287, 9287a., 9287b. Simao, 9287c.

A perennial, scandent rhizomatous herb, becoming woody with age. Widespread in China as can be seen from the localities of Henry's collections, but also native to Nepal, Bhutan, Myanmar, Thailand, Indochina, Japan and the Philippines. According to Henry, around the districts of Yichang, Badong and Changyang this sprawling subshrub was known as either *ch'ien li kuang* or *huang hua chih ts'ao*. The Hani (an ethnic group) of Xishuangbanna in southern Yunnan call this species *seqnal'aqhhyu* and use it to treat cases of rheumatism, sprains and skin infections. *Senecio scandens* was collected by the Glasnevin Central China Expedition (GCCE 195) between Yichang and Badong in September 2002 and again in September 2004 (GCCE 531) in Xingshan.

Senecio spp. (ASTERACEAE)

Anonymous in Kew List of Determinations (volume 24 & 34).

Hubei: Yichang, 342. Badong, 284 (July 1885).

Changyang, 7586. Donghu, 6443. Fang Xian at 1,900 to 3,000 metres, 6813, 6925. **Sichuan:** South Wushan, 7142. **Yunnan:** Mengzi, 10923, 10923a., 10141. Simao, 12691, 12691a., 12668, 12578.

Senecio wightii (DC. ex Wight) Benth ex C. B. Cl. (ASTERACEAE)

Jeffrey & Chen in Kew Bull. 39: 416. (1985).

Yunnan: Mengzi at 1,500 metres, 9140, 9140a., 10154.

A widespread species, found in damp places by streams and ponds in India, Bhutan, Myanmar and China.

Senegalia caesia (L.) Maslin, Seigler & Ebinger (FABACEAE)

Acacia caesia (L.) Willd. Anonymous in Kew List of Determinations (volume 34).

Yunnan: Mengzi, 9127.

A climbing shrub, native to the Indo-Malaya region. Also found in the Kaohsiung area of Taiwan.

Senegalia intsia (L.) Maslin, Seigler & Ebinger (FABACEAE)

Acacia sp. Henry in Trans. Asiat. Soc. Jap. 24 suppl: 39.(1896) (List Pl. Formosa). *Acacia intsia* (L.) Willd. Li in Woody Flora of Taiwan 411. (1963).

Taiwan: Wanchin, 1571.

A climbing shrub, native to Indo-Malaysia and China (Taiwan) where Li describes it as rather scarce. Henry's local collector gathered this specimen in the mountains beyond the village of Wanchin.

Senegalia teniana (Harms) Maslin, Seigler & Ebinger (FABACEAE)

Acacia teniana Harms in Fedde, Repert. Spec. Nov. Regni Veg. 17: 133. (1921).

Yunnan: On the Mengzi Plain, 9601a. Simao, 11899a.

Endemic to China and distributed in the provinces of Sichuan and Yunnan.

Senna occidentalis (L.) Link (FABACEAE)

Cassia occidentalis L. Henry in Trans. Asiat. Soc. Jap. 24 suppl: 38. (1896) (List Pl. Formosa).

Taiwan: Kaohsiung, without number. Oluanpi, 301.

A erect, suffrutescent herb to 1.5 metres high. Native to South America and naturalised in Chinese Taiwan where according to Henry it was the *yang-chio-tou*

Senna siamea (Lam.) Irwin & Barn. (FABACEAE)

Cassia siamea Lam. Anonymous in Kew List of Determinations (volume 34).

Yunnan: Simao, 13172.

The kassod tree, a tropical tree to 13 metres tall, native from Myanmar to Malaysia. *Senna siamea* is used in parts of Asia to shade coffee and has escaped cultivation. The seed pods and leaves are highly toxic.

Senna sophera (L.) Roxb. (FABACEAE)

Cassia sophera L. Zheng in Journ. Wuhan Bot. Res. 2.(1): 97. (1984).

Hubei: Yichang, 441, 4625a. Liantuo, 4625. On the banks of the Yangtze at Badong, 7178. **Yunnan:** Mengzi, 9964.

Introduced to China, very widely spread in tropical and subtropical regions, but usually as a weed of cultivation.

Senna sulfurea (DC. ex Collad) H. S. Irwin & Barneby (FABACEAE)

Cassia glauca Lam. Anonymous in Kew List of Determinations (volume 24).

Hainan: Environs of Haikou, 7966. On Liang Shan, 16 km (10 miles) east of Haikou, 8209.

Senna tora (L.) Roxb. (FABACEAE)

Cassia tora L. Henry in Trans. Asiat. Soc. Jap. 24 suppl: 38. (1896) (List Pl. Formosa).

Hainan: Environs of Haikou, 8067. **Taiwan:** Kaohsiung, without number. Oluanpi, 360. **Yunnan:** Mengzi, 11277. Simao, 11793.

A suffrutescent herb to 50 cm tall, native to tropical America, widely distributed in tropical and subtropical regions of the world.

Serissa japonica (Thunb.) Thunb. (RUBIACEAE)

Serissa foetida (L. f.) Comm. Forbes & Hemsley in Journ. Linn. Soc. Bot. 23: 391. (1888), Zheng in Journ. Wuhan Bot. Res. 2.(1): 190. (1984). *Serissa democritea* Baill. ex Franch. Forbes & Hemsley in Journ. Linn. Soc. Bot. 23: 391. (1888). *Serissa serissoides* (DC.) Druce Zheng in Wuhan Bot. Res. 2.(1): 190. (1984).

Hubei: On Tsui Fu Shan near Yichang (May 1888), 1613. Yichang, 616, 5383.

Sesamum indicum L. (PEDALIACEAE)

Forbes & Hemsley in Journ. Linn. Soc. Bot. 26: 236. (1890), Henry in Trans. Asiat. Soc. Jap. 24 suppl: 69. (1896) (List Pl. Formosa).

Hubei: Yichang, without locality. **Taiwan:** Oluanpi, without number. Kaohsiung, cultivated, 370.

Sesame, grown for the oil extracted from its seeds and still a major seed oil source in some Asian countries.

Sesbania bispinosa (Jacq.) W. Wight (FABACEAE)

Sesbania aculeata (Willd.) Pers. Anonymous in Kew List of Determinations (volume 34).

Yunnan: Mengzi, 10985.

Sesbania sesban (L.) Merr. (FABACEAE)

Sesbania aegyptiaca (Poir.) Pers. Henry in Trans. Asiat. Soc. Jap. 24 suppl: 32. (1896) (List Pl. Formosa).

Taiwan: Kaohsiung, 1802.

According to Henry this was the *shan-ch'ing-tze*, a leguminous plant that was commonly cultivated in 19th century Taiwan. When plants were about a foot high, the ground was ploughed and the plants turned into the soil as green manure. When allowed to grow to maturity it was used for fuel.

Sesuvium portulacastrum (L.) L. (AIZOACEAE)

Henry in Trans. Asiat. Soc. Jap. 24 suppl: 46. (1896) (List Pl. Formosa).

Taiwan: Kaohsiung Spit, without number.

A decumbent perennial with stems to 60 cm long, pantropical in distribution and a common weed on the seashores of Taiwan.

Setaria barbata (Lam.) Kunth (POACEAE)

Setaria mauritiana Spreng. Rendle in Journ. Linn. Soc. Bot. 36: 336. (1904).

Yunnan: Mengzi at 1,500 to 1,800 metres, 9496, 9496a. Mengzi at 1,500 to 1,650 metres, 10999. Mengsi at 1,650 to 1,750 metres, 10013, 10013a. Mengzi at 1,500 to 1,750 metres, 10972.

Setaria forbesiana (Nees ex Steud) Hook. f. (POACEAE)

Rendle in Journ. Linn. Soc. Bot. 36: 334. (1904).

Hubei: At Shih Pan Shan [off the Xiling (Yichang) Gorge], 4210. Antelope Glen, 4246. Liantuo, 4642.

Setaria glauca (L.) P. Beauv. (POACEAE)

Henry in Journ. China Br. Roy. Asiat. Soc. 22: 250. (1887) (Chinese Names of Plants); in Trans. Asiat. Soc. Jap. 24 suppl: 106. (1896) (List Pl. Formosa), Rendle in Journ. Linn. Soc. Bot 36: 335. (1904).

Hubei: Yichang, 18, 2263. **Hainan:** Environs of Kiungchow, cultivated, 8285, 8687. **Taiwan:** Kaohsiung, in a ditch, 1109. Wanshoushan, 1904. **Yunnan:** Mengzi at 1,500 metres, 9613. On grassy mountains near Mengzi at 1,750 metres, 10994.

Native to the temperate regions of the Old World and introduced elsewhere. Henry stated that *Setaria glauca* was known colloquially at Yichang as *kou wei tzu*.

Setaria italica (L.) P. Beauv. (POACEAE)

Henry in Trans. Asiat. Soc. Jap. 24 suppl: 106. (1896) (List Pl. Formosa), Rendle in Journ. Linn. Soc. Bot. 36: 356. (1904).

Hubei: Badong, 2565a., 4885, 5054. Yichang, 4240. **Taiwan:** Wanchin, cultivated in the mountains, 571.

Foxtail millet, grown in China for over 7,000 years. Cultivated in the mountainous districts near Yichang, according to Henry it was known colloquially as *hsin ku*. This species continues to be cultivated by the indigenous people of Taiwan who grow numerous different cultigens. Henry's Taiwan collections were made at Wanchin, then the only safe trading post on Taiwan between the local Chinese settlers and the native headhunting indigenous peoples.

Setaria plicata (Lam.) T. Cooke (POACEAE)

Panicum plicatum Lam. Henry in Trans. Asiat. Soc. Jap. 24 suppl: 106. (1896) (List Pl. Formosa). *Setaria mauritiana* Rendle non Spreng. in Journ. Linn. Soc. Bot. 36: 336. (1904).

Hubei: Antelope Glen, 4244. San You Dong Glen, 4320. Antelope Glen, 4316, 4264, 4265. Liantuo, 4450, 4630. **Taiwan:** Oluanpi, 1238. Siko Hills, Tamsui, 1400. Wanchin, 1570.

Native to Nepal, India and Sri Lanka to China and Japan to Malaysia.

Setaria sp. (POACEAE)

Anonymous in Kew List of Determinations (volume 24).

Hubei: Yichang, 598, 695.

Setaria viridis (L.) P. Beauv. (POACEAE)

Henry in Journ. China Br. Roy. Asiat. Soc. 22: 250. (1887) (Chinese Names of Plants); in Trans. Asiat. Soc. Jap. 24 suppl: 106. (1896) (List Pl. Formosa), Rendle in Journ. Linn. Soc. Bot. 36: 336. (1904).

Hubei: Yichang, 191, 2565a. Badong, 361 (June 1885), 5054a. **Taiwan:** Oluanpi, 969. Kaohsiung, 1104

Widely distributed in the Old World. A coastal species in Taiwan.

Sheareria nana S. Moore (ASTERACEAE)

Forbes & Hemsley in Journ. Linn. Soc. Bot. 23: 432. (1888).

Hubei: Yichang, 2223, 2751, 4228.

Discovered by Dr. George Shearer of the London Missionary Society near Jiujing in Jiangxi province in 1873.

Shortia sinensis Hemsley (DIAPENSIACEAE) New species

in Hooker's Icon. Pl. t. 2624. (1901).

Yunnan: In Pingbian on the Daweishan Range at 1,650 metres, 11490 (isotype, collected January 10th 1898).

Distributed in China (Yunnan) and Vietnam.

Shuteria involucrata (Wall.) Wight & Arn. (FABACEAE)

Shuteria sinensis Hemsley in Hooker's Icon. Pl. t. 2626. (1901).

Yunnan: Mengzi at 1,650 metres, 12432.

Widely distributed in Asia, India to the Philippines.

Shuteria suffulta Benth. (FABACEAE)

Anonymous in Kew List of Determinations (volume 34).

Yunnan: Mengzi, 9216.

Sibirotrisetum henryi (Rendle) Barberá (POACEAE) New species

Trisetum henryi Rendle in Journ. Linn. Soc. Bot. 36: 400. (1904)

Hubei: Fang Xian, 6706, 6643 (type *Trisetum henryi* Rendle).

Sida acuta Burm. f. (MALVACEAE)

Henry in Trans. Asiat. Soc. Jap. 24 suppl: 20. (1896) (List Pl. Formosa), Li in Woody Flora of Taiwan 549. (1963).

Hainan: 32 km (20 miles) west of Haikou, 8145. **Taiwan:** Kaohsiung, without number. Oluanpi, 228.

A pantropic weed, grows as an undershrub from 0.5 to 3 metres tall.

Sida cordata (Burm. f.) Borss. Waalk. (MALVACEAE)

Sida humilis Cavan. Henry in Trans. Asiat. Soc. Jap. 24 suppl: 20. (1896) (List Pl. Formosa). *Sida veronicaefolia* Lamk. Li in Woody Flora of Taiwan 551. (1963).

Hainan: Environs of Haikou, 8354. **Taiwan:** Kaohsiung, without number. Oluanpi, 273.

A common pantropical suffruticose herb to 1 metre tall.

Sida cordifolia L. (MALVACEAE)

Henry in Trans. Asiat. Soc. Jap. 24 suppl: 20. (1896) (List Pl. Formosa), Li in Woody Flora of Taiwan 549. fig. 213.(1963).

Taiwan: Kaohsiung, without number.

A suffruticose weed to one metre tall. A common weed of tropical Africa and Asia.

Sida rhombifolia L. (MALVACEAE)

Henry in Trans. Asiat. Soc. Jap. 24 suppl: 20. (1896) (List Pl. Formosa).

Sida insularis Hatusima Li in Woody Flora of Taiwan 551. (1963).

Hainan: Environs of Kiungchow, 8702. **Taiwan:** Oluanpi, 1325. **Yunnan:** Mengzi, 11193.

A pantropical erect undershrub. This species was collected by the Glasnevin Taiwan Expedition (GTE 49) by the lighthouse at Oluanpi (South Cape of Henry and Wilson) in October 2004.

Sigesbeckia orientalis L. (ASTERACEAE)

Henry in Journ. China Br. Roy. Asiat. Soc. 22: 259. (1887) (Chinese Names of Plants); in Trans. Asiat. Soc. Jap. 24 suppl: 53. (1896) (List Pl. Formosa).

Hubei: Badong, 12, 2384, 5012. Yichang, 139, 871, 2748, 4306. **Hainan:** Without locality, 8502. **Taiwan:** Kaohsiung, without number. Oluanpi, 363, 618, 678. **Yunnan:** Mengzi, 10160. In forests on the slopes of Fengchunling near Yuanjiang at 2,300 metres, 11225.

Cosmopolitan in warm counteries and extending into some temperate regions. According to Henry in the Three Gorges region this species was known colloquially as *mu chu yu* and was also the *xi xian cao* and in herbal medicine it is used to treat cases of irritability, insomnia and forgetfulness.

Silene aprica Turcz. ex Fisch. & C. A. Mey. (CARYOPHYLLACEAE)

Anonymous in Kew List of Determinations (volume 24).

Hubei: Liantuo, 2115. Badong, 2862. Fang Xian, 6665, 6790. Jianshi, 7407.

Silene baccifera (L.) Durande (CARYOPHYLLACEAE)

Cucubalus baccifer L. F. N. Williams in Journ. Linn. Soc. Bot. 34: 428. (1900), Diels in Bot. Jahrb. 29: 319. (1900).

Hubei: Badong, 421 (June 1885), 2386, 4688. Jianshi, 5840. **Sichuan:** South Wushan, 5840b., 5731. Without locality, 6256a. Henry's Chinese collector with Pratt, no locality stated but may be Leshan, Emei Shan, Wa Shan or Kangding, 8805. **Yunnan:** South of the Red River from Manmei, 9732.

A scrambling perennial, widespread across much of Europe, north-west Africa and Asia as far east as Japan. This species was collected by the Sino-American Expedition in the Shennongjia Forest District in 1980. It was also collected in Badong by the Glasnevin Central China Expedition in October 2002 and again by the 2004 Glasnevin Expedition (GCCE 587) near Henji town in Xingshan.

Silene banksia (Meerb.) Mabb. (CARYOPHYLLACEAE)

Lychnis senno Sieb. & Zucc. Anonymous in Kew List of Determinations (volume 24).

Hubei: Xingshan, 6526. Fang Xian, 6574.

Silene conoidea L. (CARYOPHYLLACEAE)

Anonymous in Kew List of Determinations (volume 24).

Hubei: Liantuo, 3894.

Silene firma Sieb. & Zucc. (CARYOPHYLLACEAE)

Silene aprica Turcz. var. *firma* (Sieb. & Zucc.) F. N. Williams Anonymous in Kew List of Determinations (volume 24).

Hubei: Badong, 1045 (December 1885), 4922.

Silene fissipetala Turcz. (CARYOPHYLLACEAE)

Silene fortunei Vis. ex Rohrb. Henry in Trans. Asiat. Soc. Jap. 24 suppl: 19. (1896) (List Pl. Formosa).

Hubei: Yichang, 162, 2034, 2136. Badong, 529 (September 1885), 2563, 2863, 3701, 5073, 5177. Liantuo, 4549. Fang Xian, 6604. On the banks of the Yangtze at Badong, 7176. **Taiwan:** Tamsui, 1383

A very beautiful perennial to 60 cm tall, native to China (including Taiwan) and Japan. *Silene fissipetala* was discovered in Zhejiang province in eastern China by Irishman, Sir George Staunton (1737—1801) when travelling on Lord Macartney's embassy to the Emperor of China. It was later collected, according to Bretschneider by Sir Everard Home on the island of Zhousan (Chusan) and it was from the same region that Robert Fortune collected his material. Several collections of this species were made by the Glasnevin Central China Expedition in Xingshan (GCCE 458, GCC E834, GCCE 864).

Silene otodonta Franchet (CARYOPHYLLACEAE)

Anonymous in Kew List of Determinations (volume 34).

Yunnan: Simao, 13580.

Discovered by Père Delavay in Yunnan in September 1882.

Silene spp. (CARYOPHYLLACEAE)

Anonymous in Kew List of Determinations (volume 24 & 34).

Hubei: Fang Xian, 6631, 6631a., 6631b. **Yunnan:** Simao, 13503.

Silene tatarinowii Regel (CARYOPHYLLACEAE)

Anonymous in Kew List of Determinations (volume 24).

Hubei: Jianshi, 5914.

Silene viscidula Franchet (CARYOPHYLLACEAE)

Silene lankongensis Franchet Anonymous in Kew List of Determinations (volume 34).

Yunnan: Mengzi, 9625, 9698.

Discovered by Père Delavay in Yunnan in July 1883. Endemic to China and distributed in Guizhou, Sichuan, Tibet (Xizang) and Yunnan.

Silene cf. *viscidula* Franch. (CARYOPHYLLACEAE)

Anonymous in Kew List of Determinations (volume 34).

Yunnan: Mengzi, 9934.

Sinacalia davidii (Franch.) H. Koyama (ASTERACEAE)

Jeffrey & Chen in Kew Bull. 39: 219. (1984).

Senecio didymantha Dunn in Journ. Linn. Soc. Bot. 35: 305. (1903).

Sichuan: Near Kangding, Henry's Chinese collector with Pratt, 8920 (type of *Senecio didymanthus* Dunn).

A robust perennial to 150 cm tall, first found near Baoxing in Sichuan by Armand David.

Sinacalia macrocephala (H. Robins & Brettell) C. Jeffrey & Y. L. Chen (ASTERACEAE) New species

Jeffrey & Chen in Kew Bull. 39: 217. (1984).

Hubei: Antelope Glen, Yichang, 7638.

Handel-Mazzetti first based his *Cacalia macrocephala* (a nom. illegit.) on Wilson's 2644 from western Hubei, Henry had discovered this species in 1888.

Sinacalia tangutica (Maxim.) B. Nord. (ASTERACEAE)

Senecio (*Cacalia*) sp. nova Henry in Journ. China Br. Roy. Asiat. Soc. 22: 244. (1887) (Chinese Names of Plants). *Senecio henryi* Hemsley in Journ. Linn. Soc. Bot. 23: 452. (1888), Bretschneider in Hist. Eur. Bot. Disc. China 2: 785.

(1898), Morley in Glasra 3: 80. (1979). *Sinacalia henryi* (Hemsl.) H. Robins & Brettell in Act. Phytotax. Geobot. 27: 275. (1973) nom. non. rite public. *Senecio tangutica* Maxim. J. H. Veitch in Hortus Veitchii 436. (1906), Nelson in A Heritage of Beauty 319. (2000).

Hubei: Yichang, 180 (syntype of *Senecio henryi* Hemsley). Badong on cliffs, 2454, 4979, 5040. Liantuo, 4485, 4535, 4587. Xingshan at 1,400 metres, 6545.

A well known plant in cultivation bearing deeply cut leaves on stems to 2 metres tall, the bright yellow flowers are held in dense panicles at the apices of the stems. *Sinacalia tangutica* was discovered by the Russian botanical explorer Nicolai Mikhailovich Przewalski in Gansu in 1872 and was introduced to cultivation by Wilson through Messrs Veitch who first flowered it in their Coombe Wood nursery in the autumn of 1902. According to Henry it was known by his Badong collector as *shan hu lo po*.

Sindechites henryi Oliver (APOCYNACEAE) New genus

in Hooker's. Icon. Pl. 18: t.1772. (1888), Bretschneider in Hist. Eur. Bot. Disc. China 2: 787. (1898), Forbes & Hemsley in Journ. Linn. Soc. Bot. 26: 100. (1889), Schneider in Sargent Pl. Wilson. 3: 342. (1917), Morley in Glasra 3: 80. (1979).

Cleghornia henryi (Oliv.) P. T. Li in Guihaia 4: 192. (1984).

Hubei: Yichang, 3636 (isotype).

A climbing yellow flowered shrub to 3 metres tall.

Sinocarum vaginatum H. Wolff (APIACEAE) New species

Sinocarum pseudocruciatum H. Wolff in Repert Spec. Nov. Regni Veg. 28: 182. (1930).

Sichuan: North Wushan, 7067.

Sinocrassula indica (Decne.) A. Berger (CRASSULACEAE)

Sedum indicum (Decne.) Raym.-Hamet. ex Diels var. *genuinum* Hamet in Notes Roy. Bot. Gard. Edinburgh 8: 144. (1913).

Hubei: Yichang, 2733. **Yunnan:** Near Mengzi on rocks at 1,650 metres, 9151a. On the mountains to the south of Mengzi at 1,650 metres, 9151b.

Distributed in India, Pakistan, Nepal, Bhutan and China.

Sinodolichos lagopus (Dunn) Verdc. (FABACEAE) New species

Dolichos lagopus Dunn in Journ. Linn. Soc. Bot. 35: 490. (1903).

Yunnan: Simao, 12378 (syntype of *Dolichos lagopus* Dunn), 13554. Manpan, 11220.

China (Guangxi, Hainan and Yunnan,) to Indochina and Borneo.

Sinofranchetia chinensis (Franch.) Hemsley (LARDIZABILACEAE) New genus

in Hooker's Icon. Pl. 29: t. 2842. (1907), Rehder & E. H. Wilson in Sargent Pl. Wilson. 1: 349. (1913), Fedde in Repert. Spec. Nov. Regni Veg. 5: 343. (1908), Schneider in Illus. Handb. Laubholzk. 2: 912. (1912), Zheng in Journ. Wuhan Bot. Res. 2.(1): 42. (1984).

Holboellia cuneata Oliver in Hooker's Icon. Pl. 19: t. 1817. (1889), (in part, without fruits), Henry in Notes Econ. Bot. China 48. (1893), Bretschneider in Hist. Eur. Bot. Disc. China 2: 779. (1898). *Holboellia*, subgenus Sinofranchetia Diels in Engl. Bot. Jahrb. 29: 343. partim. (1900). *Parvatia chinensis* Franchet in Journ. de Bot. 8: 281. (1894). *Holboellia chinensis* Diels in Engl. Bot. Jahrb. 29: 343. (1900), Reauborg in Bull. Soc. Bot. France. liii. 455. (1906).

Hubei: Badong, 4887. Xinghan, 6480.

A deciduous climbing shrub to 13 metres tall. This genus was first found by Henry in the mountains of Badong. His collections however, were initially confused by Professor Daniel Oliver when describing his *Holboellia cuneata*. Henry's fruiting specimens did not belong to this plant, a fact first noticed by Diels, but to Franchet's *Parvatia chinensis* which was described from Paul Farges' 792 from Sichuan. Diels provisionally referred the two plants to *Holboellia* under the subgeneric name of *Sinofranchetia*. It was only when E. H. Wilson procured good specimens of both sexes that Diel's subgenus *Sinofranchetia* was raised to generic level, excluding male specimens of Oliver's *Holboellia cuneata* which was later placed in a new genus, *Sargentodoxa* (q.v.) by Rehder and E. H. Wilson. Introduced to cultivation by Wilson through the Arnold Arboretum in 1907.

Sinojackia henryi (Dummer) Merrill (STYRACACEAE) New genus

in Sunyatsenia 3: 257. (1937).

Pterostyrax henryi Dummer in Gard. Chron. ser. 3. 53: 19. (1913).

Sichuan: Kangding, between 2,900 and 4,200 metres, Henry's Chinese collector with Pratt, 8856 (isotype of *Pterostyrax henryi* Dummer), 8865 (type of *Pterostyrax henryi* Dummer).

Sinojohnstonia moupinensis (Franch.) W. T. Wang (BORAGINACEAE)

Omphalodes cordata Hemsley in Journ. Linn. Soc. Bot. 26: 148. (1890), Bretschneider in Hist. Bot. Disc. China 2: 788. (1898). *Omphalodes moupinensis* Franch. Brand in Engler, Pflanzenr. iv. 252: 105. (1921).

Hubei: Badong, 1445, 5329, 5412, 4029 (type of *Omphalodes cordata* Hemsley). **Sichuan:** South Wushan, 5610. Henry's Chinese collector with Pratt, no locality stated but may be any of the following, Leshan, Emei Shan, Wa Shan or Kangding, 8875.

Discovered by Père Armand David in Baoxing (formerly Moupin) in April 1869.

Sinomenium acutum (Thunb.) Rehder & E. H. Wilson (MENISPERMACEAE)

Zheng in Journ. Wuhan Bot. Res. 2.(1): 44. (1984).

Sinomenium diversifolium (Miq.) Diels var. *cinereum* Diels in Engler, Pflanzenr. iv. 94. 255. (1910). *Sinomenium acutum* (Thunb.) Rehder & E. H. Wilson var. *cinereum* (Diels) Rehder & E. H. Wilson Zheng in Journ. Wuhan Bot. Res. 2.(1): 45. (1984).

Hubei: Liantuo, 2014, 2590. Yichang, 4105.

Sinosenecio dryas (Dunn) C. Jeffrey & Y. L. Chen (ASTERACEAE) New species

in Kew Bull. 39: 231. (1984).

Senecio dryas Dunn in Journ. Linn. Soc. Bot. 35: 504. (1903).

Sichuan: South Wushan in forests, growing on rocks, 5697.

Sinosenecio globigerus (C. C. Chang) B. Nord. (ASTERACEAE) New species

Jeffrey & Chen in Kew Bull. 39: 239. (1984).

Senecio phalacrocarpus Forbes & Hemsley non Hance in Journ. Linn. Soc. Bot. 23: 456. (1888). *Senecio globigerus* C. C. Chang in Sunyatsenia 6: 21. (1941).

Hubei: Badong, 1402, 1406 (holotype). **Sichuan:** South Wushan, 5523, 5664. **Hunan:** Shimen, 7560.

An herbaceous perennial to 70 cm. tall, distributed in woods by streamsides in central and western China, flowering occurs from April to June.

Sinosenecio oldhamianus (Maxim.) B. Nord. (ASTERACEAE)

Jeffrey & Chen in Kew Bull. 39: 259. (1984).

Senecio oldhamianus Maxim. Forbes & Hemsley in Journ. Linn. Soc. Bot. 23: 455. (1888).

Hubei: Yichang, 630, 675. In a cave near Yichang, 1184. On the floor of a cave on the way to Moji Shan (June 1886) 1539. Badong, 1708, 4009, 4711. Liantuo, 1980, 3802. Antelope Glen, 3386. **Yunnan:** Mengzi, 9254.

A weedy species, widely distributed through China, Myanmar, Thailand and Vietnam. Discovered in Zhejiang in 1861 by the Kew collector, Richard Oldham.

Sinowilsonia henryi Hemsley (HAMAMELIDACEAE) New genus

in Hooker's Icon. Pl. 29: tt. 2817. (1906), Fedde in Repert. Spec. Nov. Regni. Veg. 5: 265. (1908), Bean in Kew Bull. Misc. Inf. xxii. 355. (1909), Pampanini in Nuov. Giorn. Bot. Ital. n. ser. xviii. 120. (1911), Schneider, Illus. Handb. Laubholzk. 2: 597. fig. 589. (1912), E. H. Wilson in

A Naturalist in Western China 1: 56. (1913), Bean in Trees & Shrubs 2: 512. (1921), Rehder in Journ. Arnold Arb. 4: 246. (1923), Zheng in Journ. Wuhan Bot. Res. 2.(1): 66. (1984), Lancaster in Travels in China 395. (1993), Nelson in A Heritage of Beauty 319. (2000), O' Brien in Augustine Henry an Irish Plant collector in China 35. (2002).

Hubei: Xingshan at 1,400 metres, 6559 (isotype).

A deciduous shrub or small tree to 8 metres tall, bearing foliage like that of *Hamamelis* and greenish flowers in terminal, pendelous racemes to 20 cm long. Discovered in north-west Hubei by Henry in 1888 and introduced by Wilson through the Arnold Arboretum in 1908. This monotypic genus first flowered at the Arnold Arboretum in 1923. Alice Henry grew this plant as a 'Henry plant' in her Ranelagh (Dublin) garden during the 1920s. In Ireland there are plants at Mount Usher and Kilmacurragh. A monotypic genus of botanical interest only, though autumn colour can be beautiful. Plants currently cultivated at the National Botanic Gardens, Kilmacurragh were raised from a collection made by the Glasnevin Central China Expedition (GCCE 885) in October 2004.

Siphocranion nudipes (Hemsl.) Kudô (LAMIACEAE) New species

Plectranthus (*Isodon*) *nudipes* Hemsley in Journ. Linn. Soc. Bot. 26: 272. (1890), Bretschneider in Hist. Eur. Bot. Disc. China 2: 789. (1898). *Hancea nudipes* (Hemsl.) Dunn in Notes Roy. Bot. Gard. Edinburgh vi. 153. (1911—1917).

Sichuan: North Wushan, 7037 (syntype of *Plectranthus nudipes* Hemsley).

Siphonostegia chinensis Benth. (OROBANCHACEAE)

Henry in Journ. China Br. Roy. Asiat. Soc. 22: 266. (1887) (Chinese Names of Plants), Forbes & Hemsley in Journ. Linn. Soc. Bot. 26: 202. (1890).

Hubei: Badong, 359 (June 1885). Yichang, 2141, 2160, 2310. **Yunnan:** Mengzi, 9990.

Henry stated that this species was known colloquially at Badong as *ta p'o chen*. Distributed in Russia, China, Korea and Japan.

Sirindhornia monophylla (Coll. & Hemsl.) H. A. Petersen & Suksathan (ORCHIDACEAE)

Orchis monophylla (Coll. & Hemsl.) Rolfe Rolfe in Journ. Linn. Soc. Bot. 36: 50. (1903).

Yunnan: On mountains near Mengzi at 1,900 to 2,300 metres, 11118, 11118a.

Native to Myanmar and China (Yunnan).

Skimmia melanocarpa Rehder & E. H. Wilson (RUTACEAE) New species

in Sargent Pl. Wilson. 2: 138. (1914), Zheng in Journ. Wuhan Bot. Res. 2.(1): 106. (1984).

Hubei: Fang Xian, 6888. **Sichuan:** South Wushan,

5608. North Wushan, 7146. **Yunnan:** In mountains to the south and east of Mengzi at 1,600 to 2,000 metres, 10469, 11069. On Fengchunling, in forests at 2,300 metres, 11200. Yuanjiang, 13328.

This species was based on material collected in the Sikkim Himalaya by Sir Joseph Dalton Hooker and in China by Henry and Wilson. In central and western China this black-fruited species is a common undershrub and is often growing in dense shade. In Sichuan and Hubei it makes a small shrub to 1 metre, Henry described the plants he saw in Yunnan as being esmall trees to 5 metres tall, agreeing well with Hooker's Sikkim material.

Skimmia sp. (RUTACEAE)

Anonymous in Kew List of Determinations (volume 34).

Yunnan: Mengzi, 10546, 11426.

Sladenia celastrifolia Kurz (SLADENIACEAE)

Anonymous in Kew List of Determinations (volume 34).

Yunnan: Simao, 11884a., 11884b., 11884d., 13491.

Discovered by John Anderson in Yunnan in May 1868. Distributed in Myanmar, western China and Thailand.

Sloanea dasycarpa (Benth.) Hemsley (ELAEOCARPACEAE)

Echinocarpus dasycarpus Benth. Henry in Trans. Asiat. Soc. Jap. 24 suppl: 24. (1896) (List Pl. Formosa). *Sloanea formosana* Li in Woody Flora of Taiwan 538. (1963).

Taiwan: Wanchin, 1654.

Sloanea hemsleyana (Ito) Rehder & E. H. Wilson (ECHINOCARPACEAE) New species

in Sargent Pl. Wilson. 2: 361. (1915), Zheng in Journ. Wuhan Bot. Res. 2.(1): 146. (1984), Coode in Kew Bull. 38: 395. (1984).

Echinocarpus sinensis Hemsley non Hance in Ann. of Bot. 9: 147. (1895), Bretschneider in Hist. Eur. Bot. Disc. China 2: 780. (1898), Diels in Bot. Jahrb. xxix. 467. (1900). *Echinocarpus hemsleyanus* Ito in Journ. Sci. Coll. Tokyo. xii. 349. (1899) nom. nov. for *Echinocarpus sinensis* Hemsley non Hance. *Sloanea hanceana* Hemsley in Hooker's Icon. Pl. xxv. tt. 2628. (1900) (a superflous nom. nov. for *Echinocarpus sinensis* Hemsley non Hance). *Sloanea assamica* (Benth.) Rehder and E. H. Wilson in Sargent Pl. Wilson. 2: 362. (1916) exclude A. Henry 13654. *Sloanea hemsleyana* (Ito) Rehder & E. H. Wilson var. *yunnanica* Coode in Kew Bull. 38: 397. (1984).

Hubei: South Donghu, 7488 (type of *Echinocarpus sinensis* Hemsley, holotype of *Sloanea hemsleyana* (Ito) Rehder & E. H. Wilson). **Yunnan:** South of the Red River near the Laos border, 13654.

A tall, very beautiful tree with stout spreading branches, covered in summer with masses of pure white, fragrant flowers. The fruits, too are handsome. A rare tree in Hubei, said to be more common in Sichuan.

Sloanea spp. (ELAEOCARPACEAE)

Echinocarpus sp. Anonymous in Kew List of Determinations (volume 24 & 34).

Sichuan: North Wushan, 7088. **Yunnan:** Simao, 11881.

Sloanea sterculiacea (Benth.) Rehder & E. H. Wilson (ELAEOCARPACEAE)

in Sargent Pl. Wilson. 2: 362. (1915), in part.

Sloanea forrestii W. W. Smith Coode in Kew Bull. 38: 399. (1984).

Yunnan: On wooded cliffs near Mengzi at 2,000 metres, 11501.

A tree to 15 metres tall, native to India. Bhutan and China (Yunnan). Both Joseph Rock and George Forrest later collected material in Yunnan.

Sloanea tomentosa (Benth) Rehder & E. H. Wilson (ELAEOCARPACEAE)

in Sargent Pl. Wilson. 2: 362. (1915), Coode in Kew Bull. 38: 403. (1984).

Yunnan: In forests to the south of Simao at 1,300 to 1,600 metres, 11745, 11745b., 12110.

A large deciduous tree, distributed in Nepal, India (including Assam), Bhutan, China (Yunnan) and Thailand.

Smilax bockii Warb. (SMILACACEAE.)

Heterosmilax japonica Kunth Norton in Sargent Pl. Wilson. 3: 13. (1916).

Taiwan: Wanchin, 892. Oluanpi, 1302.

A scandent shrub with stems reaching 4 metres long. Distributed from Nepal to Japan and China (Taiwan).

Smilax bracteata Presl. (SMILACACEAE)

Li in Woody Flora of Taiwan 925. (1963).

Smilax stenopetala A. Gray. Henry in Trans. Asiat. Soc. Jap. 24 suppl: 96. (1896) (List. Pl. Formosa), Norton in Sargent Pl. Wilson. 3: 12. (1917).

Taiwan: Wanchin, 52, 55, 115, 144.

A scandent prickly shrub, distributed in China (Taiwan), Japan and the Philippines.

Smilax chapaensis Gagn. (SMILACACEAE)

Smilax micropoda A. DC. var. *reflexa* Norton in Sargent Pl. Wilson. 3: 6. (1916).

Hubei: Yichang, 3327 (type of *Smilax micropoda* A. DC. var. *reflexa* Norton).

Smilax china L. (SMILACACEAE)

Henry in Journ. China Br. Roy. Asiat. Soc. 22: 239. (1887) (Chinese Names of Plants); in Trans. Asiat. Soc. Jap. 24 suppl: 96. (1896) (List Pl. Formosa), Wright in Journ. Linn. Soc. Bot. 36: 96. (1903).

Smilax stenopetala Norton non A. Gray in Sargent Pl. Wilson. 3: 12. (1916). *Smilax china* L. var. *taiheiensis* (Hayata) Koyama Li in Woody Flora of Taiwan 926. (1963).

Hubei: Yichang, 267, 897, 1177, 1197, 1208. Badong, 466 (June 1885), 1417, 1776, 5016, 5342. San You Dong Glen (July 1886), 1522. Changleping, 6329. **Hainan:** Without locality, 8520, 8538. **Taiwan:** Oluanpi, 284.

China root, a deciduous rambling shrub with stems armed with recurved prickles. Native to Myanmar, Thailand, Indochina, China, Japan, Korea and the Philippines. *Smilax china* was introduced to cultivation by Philip Miller from China before 1759 and was reintroduced by Wilson in 1907. According to Henry it was known colloquially around Yichang as the *chin pa tou* and yielded the drug 'China root', used to combat gout. *Smilax china* is currently used in Chinese herbal medicine in the treatment of cancer. The Hani (an ethnic group) of Xishuangbanna in southern Yunnan call this species is *hmqqiqhmqhav* and the roots, stems and young leaves are used to treat rheumatism, joint pain, dysentry and cancer.

Smilax cocculoides Warburg (SMILACACEAE) New species

in Bot. Jahrb. Syst. 29: 257. (1900), Norton in Sargent Pl. Wilson. 3: 7. (1916).

Smilax stans Forbes & Hemsley non Maxim. in Journ. Linn. Soc. Bot. 36: 101. (1903).

Hubei: South Badong 5436, (isotype). **Yunnan:** In forests near Mengzi at 2,000 metres, 11239.

Endemic to China.

Smilax corbularia Kunth (SMILACACEAE)

Smilax hypoglauca Wright non Bentham in Journ. Linn. Soc. Bot. 36: 98. (1903).

Hainan: At Nodoa in the interior of the island, 8443. Without locality, 8544.

A climbing vine with stems up to 5 metres long, native to southern China, northern Myanmar, Thailand and Vietnam south to Malaysia.

Smilax discotis Warburg (SMILACACEAE) New species

in Bot. Jahrb. Syst. 29: 256. (1900), Norton in Sargent Pl. Wilson. 3: 6. (1916).

Hubei: Without locality, 2943a. **Yunnan:** Mountains to the north of Mengzi at 2,600 metres, 10566.

A deciduous climber to 4 metres tall, native to mainland China from the Yangtze valley in Sichuan southwards to Fujian and Taiwan. Introduced to cultivation by Wilson in 1908 through the Arnold Arboretum. Henry stated the colloquial Chinese name for this species was *hsao-chin-pa-tao*. Forrest also collected *Smilax discotis* on the eastern flank of the Cangshan Range above Dali in Yunnan.

Smilax glabra Roxb. (SMILACACEAE)

Smilax trigona Warburg Norton in Sargent Pl. Wilson. 3: 10. (1916).

Yunnan: On exposed mountains near Mengzi at 1,800 metres, 9330. On rocky mountains near Mengzi at 1,800 metres, 9330a.

Native to India, Myanmar, China, Thailand and Vietnam.

Smilax glaucochina Warburg (SMILACACEAE) New species

in Bot. Jahrb. Syst. 29: 255. (1900).

Hubei: Yichang, 2943. Badong 4078.

Smilax hemsleyana Craib (SMILACACEAE)

Smilax indica Norton non Burm. f. in Sargent Pl. Wilson. 3: 12. (1916).

Yunnan: In forests near Mengzi at 1,500 metres, 11238.

Smilax hypoglauca Benth. (SMILACACEAE)

Norton in Sargent Pl. Wilson. 3: 10. (1916).

Yunnan: In the mountains to the south of Simao at 1,500 metres, 12115, 12115a.

Smilax lanceifolia Roxburgh var. **lanceolata** (J. B. Norton) T. Koyama (SMILACACEAE) New variety

Smilax cocculoides Warburg var. *lanceolata* Norton in Sargent Pl. Wilson. 3: 11. (1916).

Yunnan: In forests to the south-west of Mengzi at 1,600 metres, 10911. In mountain forest on the Daweishan Range in Pingbian at 2,000 metres, 10397. Mountains to the east of Simao at 1,600 metres 12577 (holotype of *Smilax cocculoides* Warburg var. *lanceolata* Norton). In forests on the Daweishan Range in Pingbian at 1,600 metres, 11489. In forests near Simao at 1,600 metres, 12799. In forests near Simao at 1,600 metres, 12902, 12902a., 12902b.

Endemic to China and distributed in southern Yunnan.

Smilax longipes Warburg (SMILACACEAE)

Norton in Sargent Pl. Wilson. 3: 5. (1916).

Smilax china Wight non L. in Journ. Linn. Soc. Bot. 36: 96. (1903).

Hubei: Yichang, 1219, 1222.

Smilax megalantha C. H. Wright (SMILACACEAE)

Norton in Sargent Pl. Wilson. 3: 4. (1916).

Yunnan: On the Daweishan Range in Pingbian at 2,000 metres, 9867.

An evergreen climber to 7 metres tall, endemic to China and distributed in Hubei, Sichuan and Yunnan. Discovered on Emei Shan (Mount Omei) in Sichuan in 1887 by Faber, it was introduced to cultivation by Wilson in 1907 when collecting for the Arnold Arboretum. A very handsome evergreen foliage plant.

Smilax menispermoidea A. DC. (SMILACACEAE)

C. H. Wright in Journ. Linn. Soc. Bot. 36: 99. (1903).

Hubei: Badong, 4679.

Smilax microphylla C. H. Wright (SMILACACEAE) New species

in Kew Bull. Misc. Inf. 117. (1895), Bretschneider in Hist. Eur. Bot. Disc. China 2: 793. (1898), Wright in Journ. Linn. Soc. Bot. 36: 99. (1903), Warburg in Bot. Jahrb. xxix. 259. (1900).

Hubei: San You Dong Glen (July 1886), 1521 (syntype). On way to Moji Shan, 3089. Yichang 3089a., 3980, 3996 (syntype). Liantuo, 4410.

Smilax nipponica Miq. (SMILACACEAE)

Smilax herbacea Henry non L. in Notes Econ. Bot. China 57. (1893). *Smilax herbacea* L. var. *oblonga* C. H. Wright in Journ. Linn. Soc. Bot. 36: 98. (1903). *Smilax herbacea* L. var. *nipponica* (Miq.) Maxim. Wright in Journ. Linn. Soc. Bot. 36: 98. (1903). *Smilax herbacea* L. var. *intermedia* C. H. Wright in Journ. Linn. Soc. 36: 97. (1903).

Hubei: Badong, 5216 (syntype of *Smilax herbacea* L. var. *oblonga* C. H. Wright), 5217 (syntype of *Smilax herbacea* L. var. *oblonga* C. H. Wright), 5600e., 5600g., 5600j. Jianshi, 5600c., 5600d. Changyang, 5600h., 5600i. Xingshan, 4509. **Sichuan:** South Wushan, 5600.

According to Henry this was the source of the drug *t'u-fu-ling.*

Smilax ocreata A. DC. (SMILACACEAE)

Norton in Sargent Pl. Wilson. 3: 11. (1916).

Yunnan: In forests near Mengzi at 1,800 metres, 11237. In forests to the south of Simao at 1,300 metres, 12902c. In forests near Simao at 1,600 metres, 12903.

Native to India, Bhutan, Nepal, Myanmar, China and Vietnam.

Smilax ovalifolia Roxb. ex D. Don (SMILACACEAE)

Norton in Sargent Pl. Wilson. 3: 12. (1916).

Yunnan: On the Daweishan Range in Pingbian at 2,300 metres, 9225. In woods near Mengzi at 1,800 metres, 9415a. In forests near Simao at 1,500 metres, 12719.

Smilax perfoliata Lour. (SMILACACEAE)

Li in Woody Flora of Taiwan 929. (1963).

Smilax china Henry non L. in Trans. Asiat. Soc. Jap. 24 suppl: 96. (1896) (List Pl. Formosa), in part. **Taiwan:** Oluanpi, 901.

A widespread species, native from India to southern China (including Taiwan).

Smilax riparia A. DC. (SMILACACEAE)

Smilax flaccida C. H. Wright in Kew Bull. Misc. Inf. 118. (1895), Bretschneider in Hist. Eur. Bot. Disc. China 2: 793. (1898), C. H. Wright in Journ. Linn. Soc. Bot. 36: 97. (1903). *Smilax herbacea* L. var. *angusta* C. H. Wright in

Journ. Linn. Soc. Bot. 36: 97. (1903).

Hubei: Yichang, 3630 (syntype of *Smilax flaccida* Wright), 3630a. Xiling (Yichang) Gorge, 3630b. Liantuo, 5600f., (type of *Smilax herbacea* L. var. *angusta* C. H. Wright). **Yunnan:** Mengzi at 1,600 metres, 13649.

Smilax riparia A. DC. var. *acuminata* (C. H. Wright.) Wang & Tang (SMILACACEAE) New variety

Smilax sp. Henry in Journ. China Br. Roy. Asiat. Soc. 22: 274. (1887) (Chinese Names of Plants). *Smilax herbacea* L. var. *acuminata* C. H. Wright in Journ. Linn. Soc. Bot. 36: 97. (1903).

Hubei: Badong, 4002. Jianshi, 5600b.

Henry stated that in the Three Gorges region this variety is known as the *t'u fu ling.*

Smilax riparia A. DC. var. *pubescens* (C. H. Wright.) Wang & Tang (SMILACACEAE) New variety

Smilax herbacea L. var. *pubescens* C. H. wright in Journ. Linn. Soc. Bot. 36: 98. (1903).

Hubei: Donghu, 5600k.

Smilax scobinicaulis C. H. Wright (SMILACACEAE) New species

in Kew Bull. Misc. Inform. 117. (1895), Bretschneider in Hist. Eur. Bot. Disc. China 2: 793. (1898), Wright in Journ. Linn. Soc. Bot. 36: 100. (1903), Warburg in Bot. Jahrb. 29: 259. (1900), Bean in Trees & Shrubs 3: 459. (1936), Nelson in A Heritage of Beauty 219. (2000).

Hubei: Xingshan at 1,400 metres, 6554 (isotype).

A deciduous climber to 4.5 metres tall, introduced to cultivation by Wilson through the Arnold Arboretum in 1907.

Smilax seisuiensis (Hayata) P. Li & C. F. Fu (SMILACACEAE.) New species

Smilax oldhamii Henry non Miq. in Trans. Asiat. Soc. Jap. 24 suppl: 96. (1896) (List Pl. Formosa). *Heterosmilax japonica* Norton non Kunth. (in part) in Sargent Pl. Wilson. 3: 13. (1917). *Heterosmilax seisuiensis* (Hayata) F. T. Wang & Tang, Li in Woody Flora of Taiwan 924. (1963).

Taiwan: Wanchin, 74.

A scandent shrub to 3 metres tall, distributed across southern China (including Taiwan) and Indochina, and known as *t'u-fu-ling* according to Henry (who first collected this species).

Smilax spp. (SMILACACEAE)

Anonymous in Kew List of Determinations (volume 24 & 34).

Hubei: Yichang, 929, 939, 3481. Liantuo, 3009. Badong, 1711 (in part), 1713, 2810, 2842, 4004. Jianshi, 6011. Changyang, 7675. **Sichuan:** South Wushan, 5575, 5726. **Yunnan:** Mengzi, 9363, 10130. Simao, 12579, 13458.

Smilax stans Maxim. (SMILACACEAE)

C. H. Wright in Journ. Linn. Soc. Bot. 36: 101. (1903).

Hubei: Jianshi, 5941, 5941a. Changleping, 6276, 6334. South Badong, 6436 (in part). **Sichuan:** South Wushan, 7906.

Smilax vaginata Decne. (SMILACACEAE)

C. H. Wright in Journ. Linn. Soc. Bot. 36: 101. (1903).

Hubei: South Badong, 5436 (in part).

Smithia ciliata Royle (FABACEAE)

Smithia sensitiva Henry non Aiton in Trans. Asiat. Soc. Jap. 24 suppl: 32. (1896) (List Pl. Formosa).

Taiwan: Wanchin, 1521. **Yunnan:** Yuanjiang at 1,600 metres, 13246.

An annual herb to 50 cm tall, native to India, Nepal, China and Japan.

Smithia sensitiva Aiton (FABACEAE)

Anonymous in Kew List of Determinations (volume 34).

Yunnan: Mengzi, 9079.

A widespread species found over much of tropical Africa, Asia and Australia.

Sohmaea gracillima (Hemsley) H. Ohashi & K. Ohashi (FABACEAE) New species

Desmodium gracillimum Hemsley in Ann. of Bot. 9: 152. (1895), Henry in Trans. Asiat. Soc. Jap. 24 suppl: 33. (1896) (List Pl. Formosa).

Taiwan: Summit of Wanshoushan, 1160 (type of *Desmodium gracillimum* Hemsley).

A prostrate herb, endemic to southern Taiwan and found in forested areas at low altitudes in the southern part of the island.

Sohmaea hispida (Franch.) H. Ohashi & K. Ohashi (FABACEAE)

Uraria henryi Schindler in Repert. Spec. Nov. Regni Veg. 21: 15. (1925).

Yunnan: Mengzi, 9342a., (type of *Uraria henryi* Schindler). Mengzi at 1,500 to 2,000 metres, 9624. In the mountains to the west of Simao at 1,450 metres, 12440 (isotype of *Uraria henryi* Schindler). In forests near Simao at 1,460 metres 12440b., (isotype of *Uraria henryi* Schindler).

Sohmaea laxiflora (DC.) H. Ohashi & K. Ohashi (FABACEAE)

Desmodium laxiflorum DC. Henry in Trans. Asiat. Soc. Jap. 24 suppl: 33. (1896) (List Pl. Formosa).

Taiwan: Wanchin, 891. Kaohsiung, 1176. **Yunnan:** Simao, 13181.

An erect undershrub to 1 metre tall, native to tropical Asia, very rare in Chian (Taiwan).

02

Solanum americanum Mill. (SOLANACEAE)

Solanum nigrum Anonymous non L.in Kew List of Determinations (volume 24).

Hainan: At Liang Shan, 16 km (10 miles) east of Haikou, 8209. Environs of Haikou, 8348. Without locality, 8511.

Solanum dulcamara L. (SOLANACEAE)

Anonymous in Kew List of Determinations (volume 34).

Yunnan: Mengzi, 11855.

Widely distributed in western China.

Solanum erianthum D. Don (SOLANACEAE)

Solanum verbascifolium Henry L. in Trans. Asiat. Soc. Jap. 24 suppl: 66. (1896) (List Pl. Formosa).

Hainan: Environs of Haikou, 8010, 8277. Ling-men (in the interior of the island), 8785. **Taiwan:** Kaohsiung, without number. **Yunnan:** Mengzi, 10869.

The mullein nightshade, the Hani (an ethnic group) of Xishuangbanna in southern Yunnan call this species *miqzal'albaol* and use its roots and leaves to treat diarrhoea.

Solanum incanum L. (SOLANACEAE)

Henry in Trans. Asiat. Soc. Jap. 24 suppl: 66. (1896) (List Pl. Formosa).

Hainan: Without locality, 8068. **Taiwan:** Wanshoushan, 795.

Solanum lasiocarpum L. (SOLANACEAE)

Solanum ferox L. Henry in Trans. Asiat. Soc. Jap. 24 suppl: 66. (1896) (List Pl. Formosa). *Solanum indicum* L. Henry in Trans. Asiat. Soc. Jap. 24 suppl: 66. (1896) (List Pl. Formosa).

Hainan: Environs of Haikou, 8001. 32 km (20 miles) west of Haikou, 8140. Ling-men (in the interior of the island), 8654. **Taiwan:** Oluanpi, 358. Wanchin, 447, 449. Kaohsiung, without number. **Yunnan:** Simao, 12158.

Indian nightshade, a subshrub to 2 metres tall, known in Taiwan as *huang-shui-ch'iao* according to Henry. A common nightshade in tropical and subtropical Asia. The Hani (an ethnic group) of Xishuangbanna in southern Yunnan call this species *siqhaq* and use its thorny stems, leaves and roots to treat cases of gastritis, toothache, stings and bites. Used in sauces and curries in several parts of Asia.

Solanum lyratum Thunb. (SOLANACEAE)

Henry in Journ. China Br. Roy. Asiat. Soc. 22: 258. (1887) (Chinese Names of Plants).

Solanum dulcamara Forbes & Hemsley non L. in Journ. Linn. Soc. Bot. 26: (1890).

Hubei: Badong, 110. Creeping on floor of a cave on way to Moji Shan (June 1886), 1537. Yichang, 635, 699, 2098, 2198. Liantuo, 4391. Jianshi, 5824. South Badong, 7476.

According to Henry this species was known colloquially in the Three Gorges region as *mao ho shang ts'ao*. In Chinese herbal medicine the dried aerial parts of this plant are used as a diuretic and in the treatment of cancer.

Solanum melongena L. (SOLANACEAE)

Forbes & Hemsley in Journ. Linn. Soc. Bot. 26: 169. (1890), Henry in Notes Econ. Bot. China 50. (1893).

Hubei: Yichang, 2792. **Hainan:** Environs of Haikou, 8403.

The *hung-ch'ieh* (according to Henry), aubergine or egg plant. A native of India and cultivated and naturalised in China. The earliest records of cultivation are from fifth century China, it was not known in Europe until the 13th century. China is now the world's largest producer of this fruit.

Solanum nigrum L. (SOLANACEAE)

Henry in Journ. China Br. Roy. Asiat. Soc. 22: 273. (1887) (Chinese Names of Plants), Forbes & Hemsley in Journ. Linn. Soc. Bot. 26: 171. (1890), Henry in Trans. Asiat. Soc. Jap. 24 suppl: 66. (1896) (List Pl. Formosa).

Hubei: Yichang, 625, 1024. Liantuo, 1903, 3210. Badong, 2485. **Taiwan:** Kaohsiung, without number. Wanchin, without number. Oluanpi, 365. **Yunnan:** Mengzi, 9870.

In Chinese herbal medicine this species is currently used to treat patients suffering from urinary infection, skin diseases and cancer. A common plant throughout China.

Solanum procumbens Lour. (SOLANACEAE)

Solanum hainanense Hance Anonymous in Kew List of Determinations (volume 24).

Hainan: Environs of Haikou, 8056, 8166. Environs of Kiungchow, 8733.

Solanum spirale Roxb. (SOLANACEAE)

Anonymous in Kew List of Determinations (volume 34).

Yunnan: Simao, 12254.

The spiral nightshade, the Hani (an ethnic group) of Xishuangbanna in southern Yunnan call this species *yavsailpavqhaq and* use it to treat cases of diarrhoea, malaria, flu and sore throat. Distributed in India, Myanmnar, China, Thailand and Vietnam to Java.

Solanum spp. (SOLANACEAE)

Anonymous in Kew List of Determinations (volume 24 & 34).

Hubei: Yichang, 628. Liantuo, 2680. **Hainan:** Ling-men, 8606. **Yunnan:** Mengzi, 9160, 10353. In Pingbian, in forests on the Daweishan Range at 1,650 metres, 11038. Simao, 12273, 13592.

Solanum torvum Sw. (SOLANACEAE)

Henry in Trans. Asiat. Soc. Jap. 24 suppl: 66. (1896) (List Pl. Formosa).

Taiwan: Oluanpi, 259. Kaohsiung, 746. **Yunnan:** Simao, 12552.

The Hani (an ethnic group) of Xishuangbanna in southern Yunnan call this species *siqhaqgaoq* and use ikts roots to treat toothache, bruising, menstrual pain and external wounds and cuts. Introduced and naturalised over much of China.

Solanum tuberosum L. (SOLANACEAE)

Anonymous in Kew List of Determinations (volume 34).

Yunnan: Mengzi, 9476.

The common potato, a South American native.

Solanum virginianum L. (SOLANACEAE)

Solanum xanthocarpum Schrad. Anonymous in Kew List of Determinations (volume 24).

Hainan: Environs of Haikou, 8380.

Solena amplexicaulis (Lam.) Gandhi (CUCURBITACEAE)

Zehneria umbellata (J. G. Klein ex Willd.) Thwaites Henry in Trans. Asiat. Soc. Jap. 24 suppl: 46. (1896) (List Pl. Formosa). *Melothria heterophylla* (Lour.) Cogn. Cogniaux in Engler, Pflanzenr. iv. 275. i: 121. (1916).

Taiwan: Wanshoushan, 1161. Kaohsiung, without number. **Yunnan:** Mengzi, 10135, 10135a.

A glabrescent perennial vine with a large tuberous roots. A widespread species found over much of India, southwestern China, Indochina and Malesia. The Hani (an ethnic group) of Xishuangbanna in southern Yunnan call this species *ho'qavqbiavyaq* and use its tuber to treat cases of poisonous snake bite, gastritis and diarrhoea.

Solidago decurrens Lour. (ASTERACEAE)

Solidago virgaurea L. var. *leiocarpa* (Benth) Miq. Journ. Linn. Soc. Bot. 23: 406. (1888). *Solidago virgaurea* Henry non L. in Trans. Asiat. Soc. Jap. 24 suppl: 52. (1896) (List Pl. Formosa).

Hubei: Yichang, 104, 383, 2736, 2961. Badong, 2383, 4933, 5009. Liantuo, 2697. **Taiwan:** Oluanpi, 1708. **Yunnan:** Mengzi, 10248.

The common goldenrod of China, in its dried state this species is used in Chinese herbal medicine to relieve inflammation.

Sonchus oleraceus L. (ASTERACEAE)

Forbes & Hemsley in Journ. Linn. Soc. Bot. 23: 487. (1888), Henry in Journ. China Br. Roy. Asiat. Soc. 22: 250. (1887) (Chinese Names of Plants); in Trans. Asiat. Soc. Jap. 24 suppl: 56. (1896) (List Pl. Formosa).

Hubei: Yichang, 1363. Badong, 1720, 1826, 3707, 4684. **Taiwan:** Kaohsiung, without number. **Yunnan:** Mengzi, 9935, 10594.

Henry stated that at Badong both *Sonchus arvensis* and *Sonchus oleraceus* were colloquially known as *k'u wo ma.*

Sonchus spp. (ASTERACEAE)

Anonymous in Kew List of Determinations (volume 24).

Hubei: Yichang, 468. **Hainan:** Environs of Kiungchow, 8774.

Sonchus wightianus DC. (ASTERACEAE)

Sonchus arvensis Forbes & Hemsley non L. in Journ. Linn. Soc. Bot. 23: 487. (1888), Henry in Journ. China Br. Roy. Asiat. Soc. 22: 250. (1887) (Chinese Names of Plants); in Trans. Asiat. Soc. Jap. 24 suppl: 56. (1896) (List Pl. Formosa)

Hubei: On way to Moji Shan (May 1886), 1595. **Taiwan:** Oluanpi, 1340. Wanchin, 1589, 1770. **Yunnan:** Mengzi, 10829.

Sonerila cantonensis Stapf. (MELASTOMATACEAE)

Sonerila yunnanensis Jeffrey ex W. W. Sm. in Notes Roy. Bot. Gard. Edinburgh 8: 207. (1914), Li in Journ. Arnold Arb. 25: 36. (1944), Hu in Journ. Arnold Arb. 33: 174. (1952).

Yunnan: In a forested ravine near Simao at 1,450 metres, 12337 (isotype of *Sonerila yunnanensis* Jeffrey ex W. W. Sm, collected August 18th 1898).

Distributed in southern China and Vietnam.

Sonerila maculata Roxb. (MELASTOMATACEAE)

Sonerila picta Korth. Li in Journ. Arnold Arb. 25: 36. (1944).

Yunnan: Mengzi, 9005, 9005a. Simao 12293a., 12293b.

Sonerila plagiocardia Diels (MELASTOMATACEAE)

New species
in Bot. Jahrb. Syst. 65: 117. (1932).

Yunnan: Simao, 13152 (isotype, collected 27th August 1899).

Distributed in southern China and Indochina.

Sonerila spp. (MELASTOMATACEAE)

Anonymous in Kew List of Determinations (volume 34).

Yunnan: Mengzi, 10753, 11036, 11438. Simao, 12562.

Sonerila tenera Royle (MELASTOMATACEAE)

Li in Journ. Arnold Arb. 25: 36. (1944).

Yunnan: Simao, 12564.

Sophora albescens (Rehder) C. Y. Ma (FABACEAE)

New species
in Acta Phytotax. Sin. 20: 468. (1982).

Sophora velutina Lind. var. *albescens* (Rehd.) Tsoong in Acta Phytotax. Sin. 19: 16. (1981).

Sichuan: Emei Shan, Henry's Chinese collector with

02

Pratt, 8987.

Rehder based his *Sophora glauca* Lesch. var. *albescens* on Wilson's 1179 from Wa Shan in Sichuan (collected in June 1908). Henry's Chinese collector had previously collected the plant on nearby Emei Shan in the autumn of 1889 while travelling with Pratt.

Sophora davidii (Franch.)Skeels (FABACEAE)

Zheng in Journ. Wuhan Bot. Res. 2.(1): 104. (1984), Nelson and Pim in The Wood and the Trees 220. (1984), O' Brien in Augustine Henry, An Irish Plant Collector in China 35. (2002).

Sophora moorcroftiana Forbes & Hemsley non (Benth.) Baker in Journ. Linn. Soc. Bot. 203. (1887). *Sophora viciifolia* Hance Forbes & Hemsl in Journ. Linn. Soc. Bot. 23: 203. (1887), Henry in Journ. China Br. Roy. Asiat. Soc. 22: 257. (1887), Tsoong in Acta Phytotax. Sin. 19: 21. (1981).

Hubei: Yichang, 505, 583, 762, 762a., 1237. Liantuo, 3825. Changyang, 7735. **Sichuan:** Emei Shan, Henry's Chinese collector with Pratt, 8980. **Yunnan:** Mengzi, 9316.

A deciduous, spiny shrub to 3 metres tall carrying pinnate leaves to 6 cm long. The pea-like flowers are borne in great profusion in early summer and are bluish-white, though pure white flowered plants sometimes occur in the wild. A widespread shrub in western and central China where it grows on open barren hillsides, it was known in the Three Gorges region as *tieh-ma-hu-tsao* according to Henry. Introduced to cultivation from southern Yunnan by Henry in 1897 through the Royal Botanic Gardens, Kew. Forrest later collected this species near the Zhongdian Plateau, northwest Yunnan in May 1917 (Forrest 13750). There are Forrest and Wilson plants at Glasnevin.

Sophora flavescens Aiton (FABACEAE)

Forbes & Hemsley in Journ. Linn. Soc. Bot. 23: 202. (1887), Henry in Journ. China Br. Roy. Asiat. Soc. 22: 251. (1887), Li in Woody Flora of Taiwan 361. (1963), Tsoong & Ma in Acta Phytotax. Sin. 19: 145. (1981), Zheng in Journ. Wuhan Bot. Res. 2.(1): 103. (1984), Bean in Trees & Shrubs 4: 388. (1989).

Sophora sp. Henry in Trans. Asiat. Soc. Jap. 24 suppl: 38. (1896) (List Pl. Formosa).

Hubei: Yichang, 507, 587, 743, 753, 1350, 3987. Badong, 6153, 7354. **Taiwan:** Wanchin, 1825.

A yellow-flowered subshrub to 2 metres tall, common in Hubei where it was colloquially known as *k'u-shen* according to Henry. The seeds and roots of this species are used to produce medicine for vetinary purposes and for human use as a diuretic and to treat cases of scabies and leprosy. Native to Siberia, China (including Taiwan) and Korea.

Sophora prazeri Prain (FABACEAE)

Tsoong & Ma in Acta Phytotax. Sin. 19: 155. (1981).

Sophora wightii Dunn in Journ. Linn. Soc. Bot. 39: 496. (1911), non Baker 1878.

Yunnan: Mengzi, 9673, 9673a., (type of *Sophora wightii* Dunn (non Baker), 9673c.

Native to Myanmar and China (Yunnan, Guizhou and Guangxi).

Sophora sp. (FABACEAE)

Anonymous in Kew List of Determinations (volume 24).

Sichuan: Emei Shan, Henry's Chinese collector with Pratt, 8985.

Sophora tomentosa L. (FABACEAE)

Henry in Trans. Asiat. Soc. Jap. 24 suppl: 38. (1896) (List Pl. Formosa), Tsoong & Ma in Acta Phytotax. Sin. 19: 153. (1981), Li in Woody Flora of Taiwan 361. fig. 129. (1963).

Taiwan: Oluanpi, 692.

A small tree, cosmopolitan in the tropics and scattered along the coast of the southern part of Taiwan where it grows on uplifted coral reef. This species was collected by the Glasnevin Taiwan Expedition (GTE 3) in Banana Bay Coastal Forest near Kenting in October 2004.

Sophora velutina Lindl. (FABACEAE)

Sophora glauca DC. Anonymous in Kew List of Determinations (volume 34).

Yunnan: Mengzi, 10720. Simao, 13094.

Sorbaria kirilowii (Regel) Maxim. var. *arborea* (C. K. Schneid.) J. H. Song & S. P. Hong (ROSACEAE) New variety

Spiraea sorbifolia Forbes & Hemsley non L. in Journ. Linn. Soc. Bot. 23: 227. (1887), Henry in Journ. China Br. Roy. Asiat. Soc. 22: 237. (1887). *Sorbaria arborea* Schneider in Illus. Handb. Laubholzk. 1: 490. fig. 297. (1905), Zheng in Journ. Wuhan Bot. Res. 2(1): 91. (1984). *Sorbaria arborea* Schneider var. *glabrata* Rehder in Sargent Pl. Wilson 1: 48. (1913).

Hubei: Yichang, 164, 6458. Badong 378 (June 1885), 1813 (syntype of *Sorbaria arborea* Schneider), 2435, 2828, 4705 (syntype of *Sorbaria arborea* Schneider), 4991, 5081. **Sichuan:** South Wushan, 6245 (syntype of *Sorbaria arborea* Schneider var. *glabrata* Rehder).

A deciduous shrub to 7 metres tall bearing large pinnate leaves and carrying upright pyramidal panicles (to 50 cm long) of small white flowers in July. Known colloquially around Yichang as *mao ch'i* according to Henry.

Sorbus discolor (Maxim) Maxim. (ROSACEAE)

in Journ. Linn. Soc. Bot. 23: 255. (1887).

Pyrus aucuparia Henry non Gaertn. in Notes Econ.

Bot. China 47. (1893).

Hubei: Badong, 2824.

Henry's Badong collections of *Sorbus discolor* represent the group which until recently was distinguished as *Sorbus hupehensis* Schneider. He described seeing a single tree in Badong as being three feet (0.9 metres) in diameter and fifty feet (15.2 metres) high. *Sorbus discolor* is abundant on the upper slopes of the Shennongjia Forest District. It was collected in neighbouring Xingshan (GCCE 471) by the Glasnevin Central China Expedition in September 2004.

Sorbus koehneana Schneider (ROSACEAE) New species

in Bull. Herb. Boissier. ser 2, 6: 316. (1906), Schneider in Illus. Handb. Laubholzk. I: 681. fig. 474. (1906), Koehne in Sargent Pl. Wilson. 1: 471. (1913), Bean in Trees & Shrubs 3: 324. (1936), Nelson in A Heritage of Beauty 319. (2000), O' Brien in Augustine Henry, An Irish Plant Collector in China 36. (2002).

Hubei: Fang Xian, 6766 (type).

A small tree to 4 metres tall, native to central and western China. Introduced to cultivation by Wilson when collecting for Messrs Veitch. This fine rowan is cultivated in several Irish gardens including the National Botanic Gardens, Glasnevin.

Sorbus scalaris Koehne (ROSACEAE) New species

Sorbus foliolosa (Wall.) Spach. var. *pluripinnata* Schneider in Bull. Herb. Boissier, ser. 2. vi. 315. (1906); in Illus. Handb. Laubholzk. 1: 680. (1906). *Sorbus pluripinnata* (Schneid.) Koehne in Sargent Pl. Wilson. 1: 481. (1913).

Sichuan: Henry's Chinese collector with Pratt, no locality stated but may be any of the following, Leshan, Emei Shan, Wa Shan or Kangding, 8960 [type of *Sorbus foliolosa* (Wall.) Spach. var. *pluripinnata* Schneider, type of *Sorbus pluripinnata* (Schneid.) Koehne].

A tree to 10 metres tall, native to western Sichuan. *Sorbus scalaris* was originally described from Wilson's Wa Shan collection (Wilson 922, June and October 1908), Henry's specimens had been collected 19 years previously.

Sorbus setschwanensis (Schneid) Koehne (ROSACEAE) New species

in Sargent Pl. Wilson. 1: 475. (1913).

Sorbus vilmorinii Schneid. var. *setschwanensis* Schneider in Bull. Herb. Boissier ser. 2. 6: 318. (1906); in Illus. Handb. Laubholzk. 1: 683. fig. 374. (1906).

Sichuan: Emei Shan, Henry's Chinese collector with Pratt, 8975.

A small tree to about 6 metres tall. This handsome little white fruited rowan is still very common on Emei Shan (Mount Omei) where it was collected by the Glasnevin Central China Expedition in September 2002 and where it grows with *Abies fabri* near the summit.

Sorbus sp. No. 1. (ROSACEAE)

Pyrus aff. *foliosae* Wall. (= Faber 538) Anonymous in Kew List of Determinations (volume 24).

Sichuan: Henry's Chinese collector with Pratt, no locality stated but may be Leshan, Emei Shan, Wa Shan or Kangding, 8963.

Sorbus sp. No. 2. (ROSACEAE)

Pyrus wallichii Hook. f. Anonymous in Kew List of Determinations (volume 24).

Hubei: Badong, 4062.

Sorbus sp. No. 3. (ROSACEAE)

Pyrus aria Forbes & Hemsley non L. in Journ. Linn. Soc. Bot. 23: 254. (1887)

Hubei: Badong on high hills, 5024. Changyang, 7905.

Sorbus wilsoniana Schneider (ROSACEAE) New species

Sorbus wilsoniana Schneid. var. (a) nova. Schneider in Bull. Herb. Boissier ser. 2, 4: 313. (1906), Koehne in Sargent Pl. Wilson. 1: 458. (1913).

Hubei: Yichang, 3757.

Sorghum bicolor (L.) Moench. (POACEAE)

Sorghum vulgare Pers. Henry in Trans. Asiat. Soc. Jap. 24 suppl: 108. (1896) (List Pl. Formosa), Rendle in Journ. Linn. Soc. Bot. 36: 368. (1904). *Sorghum vulgare* (L.) Pers. var. *nervosum* Hack. Rendle in Journ. Linn. Soc. Bot. 36: 368. (1904).

Hainan: Environs of Kiungchow, 8287, 8757, 8785. **Taiwan:** Oluanpi, 257. Wanchin, 807.

Millet or Guinea corn, widely cultivated as a cereal. Widely cultivated in warm countries. Known colloquially in the Three Gorges region as *lu-shu* or kao-*liang* according to Henry.

Sorghum halepense (L.) Pers. (POACEAE)

Rendle in Journ. Linn. Soc. Bot. 36: 367. (1904).

Hainan: Nartaisee near Kiungchow, 8295.

Widely cultivated as a fodder crop, also naturalised in parts of China.

Sorghum nitidum (Vahl.) Pers. (POACEAE)

Sorghum fulvum P. Beauv. var. *nitidum* (Vahl.) Hack. Rendle in Journ. Linn. Soc. Bot. 36: 367. (1904). *Sorghum fulvum* (R. Br.) P. Beauv. Rendle in Journ. Linn. Soc. Bot. 36: 367. (1904).

Hubei: Yichang, 28, 4234. **Taiwan:** Kaohsiung Plain, 1895.

Native to India, Sri Lanka, Myanmar, Thailand, Vietnam, China and extending to Australia.

Spathoglottis pubescens Lindl. (ORCHIDACEAE)

Rolfe in Journ. Linn. Soc. Bot. 36: 18. (1903).

Yunnan: Mengzi at 1,650 metres, 11096.

Native to India, Myanmar, China, Laos, Thailand, Cambodia and Vietnam.

Spatholobus pulcher Dunn (FABACEAE) New species in Journ. Linn. Soc. Bot. 35: 489. (1903).

Yunnan: In forests to the south-east of Simao at 1,600 metres, 12780 (isotype), 12780b., (type).

Endemic to Yunnan.

Spatholobus suberectus Dunn (FABACEAE) New species

in Journ. Linn. Soc. Bot. 35: 489. (1903).

Yunnan: In mountains to the west of Simao at 1,650 metres, 11977 (isotype). In the hills near Simao at 1,500 metres, 11977a., (isolectotype). In the mountains to the east of Simao at 1,450 metres, 11977b., (isosyntype). Mengzi, 13698.

Distributed from the eastern Himalaya to China (Fujian, Guangdong, Guangxi and Yunnan) and Indochina.

Spatholobus varians Dunn (FABACEAE) New species in Journ. Linn. Soc. Bot. 35: 490. (1903), Riddler-Numan & Wiriadinata in Reinwardtia 10: 195. (1985).

Yunnan: In mountain forest to the south of Simao at 1,600 metres, 11771 (type), 11771a., (isolectotype). In Yulo mountains near Simao at 1,600 metres, 11771b., (isosyntype). In mountain forest near Simao at 1,600 metres 11771c., (isosyntype).

Endemic to Yunnan.

Speranskia cantonensis (Hance) Pax. & Hoffm. (EUPHORBIACEAE)

Zheng in Journ. Wuhan Bot. Res. 2.(1): 114. (1984).

Speranskia henryi Oliver in Hooker's Icon. Pl. t. 1577. (1887), Henry in Journ. China Br. Roy. Asiat. Soc. 22: 273. (1887) (Chinese Names of Plants), Forbes & Hemsley in Journ. Linn. Soc. Bot. 26: 435. (1894), Bretschneider in Hist. Eur. Bot. Disc. China 2: 791. (1898).

Hubei: Yichang, 1273, 1372 (syntype of *Speranskia henryi* Oliver). Liantuo, 1971. In a cave in the San You Dong Glen, 2891.

Spermacoce hispida L. (RUBIACEAE)

Henry in Trans. Asiat. Soc. Jap. 24 suppl. 51. (1896) (List Pl. Formosa).

Hainan: Without locality, 8243. **Taiwan:** Wanchin, 49. Kaohsiung Spit, 1928.

Spermacoce pusilla Wall. (RUBIACEAE)

Spermacoce stricta Henry non L. f. in Trans. Asiat. Soc. Jap. 24 suppl: 51. (1896) (List Pl. Formosa).

Taiwan: Wanchin, 1550. **Yunnan:** Mengzi, 11233.

Spermacoce sp. (RUBIACEAE)

Anonymous in Kew List of Determinations (volume 34).

Yunnan: Simao, 12415, 12422.

Spermadictyon sp. (RUBIACEAE)

Hamiltonia sp. Anonymous in Kew List of Determinations (volume 34).

Yunnan: Mengzi, 9579, 11263. South of the Red River from Manmei at 1,900 metres, 9643.

Sphaeranthus africanus L. (ASTERACEAE)

Anonymous in Kew List of Determinations (volume 24).

Hainan: 8 km (5 miles) south of Kuingchow, 8179.

Native to Africa, India, Myanmar and China to Australia, a weed of paddy fields in the tropics.

Sphagneticola calendulacea (L.) Pruski (ASTERACEAE)

Wedelia calendulacea Less. Henry in Trans. Asiat. Soc. Jap. 24 suppl: 54. (1896) (List Pl. Formosa).

Hainan: Environs of Kiungchow, 8307. **Taiwan:** Kaohsiung, without number. Wanchin, 857.

Sphagnum capillifolium (Ehrh.) Hedw. (SPHAGNACEAE)

Sphagnum acutifolium Schrad. Salmon in Journ. Linn. Soc. Bot. 34: 472. (1900).

Hubei: Fang Xian, 6855 (September 1888).

Sphenoclea zeylanica Gaert. (SPHENOCLEACEAE)

Henry in Trans. Asiat. Soc. Jap. 24 suppl: 56. (1896) (List Pl. Formosa).

Hainan: Environs of Haikou, 8362. **Taiwan:** Oluanpi, 578.

Spinifex littoreus (Burm. f.) Merr. (POACEAE)

Spinifex squarrosus L. Henry in Trans. Asiat. Soc. Jap. 24 suppl: 107. (1896) (List Pl. Formosa).

Taiwan: Kaohsiung, in sea sand, without number.

A perennial species to 80 cm tall, native to Sri Lanka, Myanmar, Indochina, Malesia and southern China (including Taiwan) to Australia. Often found in littoral sand dunes.

Spiraea anomala Batalin (ROSACEAE) New species in Trudy Galvn. Bot. Sada. 13: 92. (1893).

Hubei: South Badong, 5305 (type).

Spiraea dasyantha Bunge (ROSACEAE)

Forbes & Hemsley in Journ. Linn. Soc. Bot. 23: 224. (1887).

Hubei: Yichang, 7781.

Spiraea henryi Hemsley (ROSACEAE) New species in Journ. Linn. Soc. Bot. 23: 225. fig. 6. (1887), Henry in Journ. China Br. Roy. Asiat. Soc. 22: 277. (1887) (Chinese Names of Plants), Bretschneider in Hist. Eur. Bot. Disc. China 2: 782. (1898), Veitch in Journ. Royal Hort. Soc. xxviii. 61. fig. 20. (1903); in Hortus Veitchii 379. (1906); The Garden, lxv. 44. (1904), Schneider in Illus. Handb. Laubholzk. I. 469. fig. 292. (1905), Bean in Bot. Mag. cxxxv. t. 8270. (1909), Rehder in Sargent Pl. Wilson. 1:

447. (1913), Morley in Glasra 3: 80. (1979), Bean in Trees & Shrubs 4: 480. (1989), Nelson in A Heritage of Beauty 319. (2000), O' Brien in Augustine Henry, An Irish Plant Collector in China 36. (2002).

Hubei: Badong, 1729 (type, collected July 1886), 5645. South Badong, 7335. **Sichuan:** South Wushan, 5645a., 5750.

A lax arching shrub to 1.5 metres tall, producing great masses of small white flowers in rounded corymbs along the stems in June. According to Henry the leaves were used as a substitute for tea in the Three Gorges region and was known colloquially there as *tsui lan ch'a*. *Spiraea henryi* was introduced to cultivation by Wilson in 1900 from Hubei through Messrs Veitch's Coombe Wood nursery where it first flowered in the summer of 1904. In the Veitch catalogue *New Hardy Plants from Western China* (Autumn 1912) this species was available for three shillings and six pence. Three of Wilson's seed collections reached Glasnevin fron the Arnold Arboretum during the first decade of the 20th century - Wilson 1318 (Baoxing, Sichuan, October 1908,) Wilson 1172, (Wa Shan, Sichuan, October 1908) and Wilson 4327 (Panlan Shan, Sichuan, October 1910). Augustine Henry grew this species in his garden at Ranelagh in the south Dublin suburbs. One of the best of the Chinese spiraeas.

Spiraea hingshanensis T. T. Yu & L. T. Lu (ROSACEAE)

Spiraea miyabei Koidz. var. *pilosula* Rehder in Sargent Pl. Wilson. 1: 445. (1913).

Sichuan: Badong, 5628.

Spiraea hirsuta (Hemsl.) C. K. Schneider (ROSACEAE) New species

in Bull. Herb. Boissier ser. 2. 5: 342. (1905), Schneider in Illus. Handb. Laubholzk. I. 463. fig. 292. (1905), Rehder in Sargent Pl. Wilson. 1: 444. (1913), Zheng in Journ. Wuhan Bot. Res. 2.(1): 93. (1984).

Spiraea blumei G. Don. var. *hirsuta* Hemsley in Journ. Linn. Soc. Bot. 23: 224. (1887).

Hubei: Yichang, 1109, 1181, 1229, 4115. On road to Moji Shan (June 1886), 1549. Liantuo, 3903. Without locality, 4115a.

Spiraea hirsuta (Hemsl.) C. K. Schneider var. *rotundifolia* (Hemsl.) Rehd. (ROSACEAE)

in Sargent Pl. Wilson. 1: 445. (1913), Zheng in Journ. Wuhan Bot. Res. 2.(1): 93. (1984).

Spiraea blumei G. Don. var. *rotundifolia* Hemsley in Journ. Linn. Soc. Bot. 23: 224. (1887). *Spiraea maximowicziana* Schneider in Illus. Handb. Laubholzk 1: 461. (1905). *Spiraea blumei* G. Don var. *maximowiczii* (C. K. Schneid.) Dunn in Journ. Linn. Soc. Bot. 38: 359. (1908), Zheng in Journ. Wuhan Bot. Res. 2.(1): 93. (1984).

Hubei: Yichang, 716, 3506a. Antelope Glen, 3506 (isotype of *Spiraea maximowicziana* Schneider). Badong, 501 (June 1885). On Moji Shan, 3570.

Spiraea japonica L. f. (ROSACEAE)

Spiraea japonica L.f. Forbes & Hemsley in Journ. Linn. Soc. Bot. 23: 225. (1887).

Hubei: Badong, 235, 363 (June 1885), 1003, 1843 (July 1886), 2835. Changyang, 5236.

May be any of the following varieties, var. *acuminata* Franch., var. *fortunei* (Planch.) Rehd. or var. *glabra* (Regel) Koidz.

Spiraea japonica L. f. var. *acuminata* Franch., (ROSACEAE) New variety

Spiraea japonica L. f. var. *stellaris* Rehder in Sargent Pl. Wilson. 1: 452. (1913).

Yunnan: In Pingbian, on the Daweishan Range at 1,950 metres, 9280 (type of *Spiraea japonica* L. f. var. *stellaris* Rehder).

Spiraea japonica L. f. var. *fortunei* (Planch.) Rehd. (ROSACEAE)

Rehder in Sargent Pl. Wilson. 1: 451. (1913).

Spiraea japonica Henry non L. f. in Journ. China Br. Roy. Asiat. Soc. 22: 276. (1887) (Chinese Names of Plants).

Hubei: Badong, 644 (July 1885). Jianshi, 5993.

Henry's Badong collector called this shrub the *tsa kan shu*. Native to eastern and central China and introduced to cultivation by Robert Fortune in 1849. This variety is cultivated on the Chinese slope at the National Botanic Gardens, Glasnevin from a collection made by the 2002 Glasnevin Central China Expedition (GCCE 253) on the mountains behind Badong.

Spiraea longigemmis Maxim. (ROSACEAE)

Zheng in Journ. Wuhan Bot. Res. 2.(1): 93. (1984).

Hubei: Changleping, 6333.

Spiraea martini H. Léveillé (ROSACEAE)

Rehder in Journ. Arnold Arb. 1: 258. (1920).

Spiraea fulvescens Rehder in Sargent Pl. Wilson. 1: 439. (1913), nom. illeg.

Yunnan: On barren dry hills near Mengzi at 1,600 metres, 10662 (type of *Spiraea fulvescens* Rehder).

Endemic to China and distributed in Guizhou, Guangxi, Sichuan and Yunnan.

Spiraea myrtilloides Rehder (ROSACEAE) New species

in Sargent Pl. Wilson. 1: 440. (1913), Zheng in Journ. Wuhan Bot. Res. 2.(1): 93. (1984).

Hubei: Xingshan, 6968.

Seeds of Wilson's 989 (collected at 3,300 metres near Kangding in September 1908) reached Glasnevin in the spring of 1909.

Spiraea prunifolia Sieb. & Zucc. (ROSACEAE)

Zheng in Journ. Wuhan. Bot. Res. 2.(1): 105. (1984).

Spiraea prunifolia Sieb. & Zucc. var. *plena* Schneid. Rehder in Sargent Pl. Wilson. 1: 438. (1913).

Hubei: Changyang, 5254.

In central China *Spiraea prunifolia* occurs only in its double flowered form and it is this form that is illustrated in the current *Flora Hubiensis*. This double form was introduced from Japan by Siebold in 1845 and was first collected in central China by Henry and later in May 1900 at Xingshan by Wilson where it was planted on graves. The single-flowered plant is distinguished as *Spiraea prunifolia* f. *simplicifolia* Nakai.

Spiraea spp. (ROSACEAE)

Anonymous in Kew List of Determinations (volume 24).

Hubei: Yichang, 257, 357, 989, 1023. Badong, 1431. On the banks of the Yangtze at Liantuo, 7851.

Spiranthes sinensis (Pers.) Ames. (ORCHIDACEAE)

Spiranthes australis Henry non (R. Br.) Lindl. in Trans. Asiat. Soc. Jap. 24 suppl: 93. (1896) (List Pl. Formosa).

Hubei: Yichang, 221. Badong, 4986, 7361. **Taiwan:** Kaohsiung, without number. Wanchin, without number.

A very beautiful and interesting terrestrial orchid, found on forest verges, thickets and in grass between 400 to 1,500 metres above sea level over most parts of China. The small white, pink or purple flowers appear in the typical manner of this genus in June and July and in China. This floral formation is compared to that of a coiled dragon. Native to Siberia, India, China, Indochina, the Malay Peninsula and the Philippines southwards to New Zealand and Tasmania.

Spodiopogon cotulifer (Thunb.) Hack. (POACEAE)

Rendle in Journ. Linn. Soc. Bot. 36: 351. (1904).

Miscanthus cotulifer (Thunb.) Benth. Anonymous in Kew List of Determinations (volume 24).

Hubei: Badong, 864 (September 1885). **Taiwan:** Wanshoushan, 1022. Kaohsiung, 1045. Wanchin, 1558.

Spodiopogon formosanus Rendle (POACEAE) New species

in Journ. Linn. Soc. Bot. 36: 351. (1904).

Taiwan: Wanchin, 76 (type).

A perennial to 1 metre tall, endemic to Taiwan. According to Henry this species was planted by indigenous people in mountains near Wanchin for harvesting as a food crop.

Spodiopogon sagittifolius Rendle (POACEAE) New species

in Journ. Linn. Soc. Bot. 36: 352. (1904).

Yunnan: On grassy mountains near Mengzi at 1,750 metres, 10997. In a wood near Mengzi at 1,500 metres,

11256 (type).

Endemic to China (Sichuan and Yunnan).

Spodiopogon sibiricus Trin. (POACEAE)

Henry in Trans. Asiat. Soc. Jap. 24 suppl: 107. (1896) (List Pl. Formosa), Rendle in Journ. Linn. Soc. Bot. 36: 353. (1904).

Hubei: Xingshan, 6520. Fang Xian, 6717. **Taiwan:** Kaohsiung, without number.

Sporobolus diandrus (Retz.) P. Beauv. (POACEAE)

Rendle in Journ. Linn. Soc. Bot. 36: 387. (1904).

Taiwan: Oluanpi, 625. Kaohsiung, 1044. On the summit of Wanshoushan, at 1,100 metres, 1106.

Native from Socotra to India, Myanmar, Sri Lanka and extending from China to Australia. One of the most common grasses growing on the low hills and plains of southern Taiwan, China.

Sporobolus indicus (L.) R. Br. (POACEAE)

Rendle in Journ. Linn. Soc. Bot. 36: 388. (1904).

Sporobolus elongatus Henry non R. Br. in Trans. Asiat. Soc. Jap. 24 suppl: 108. (1896) (List Pl. Formosa).

Hubei: Yichang, 889, 4287. Badong, 2510. **Taiwan:** Wanshoushan, 1030.

Sporobolus virginicus (L.) Kunth. (POACEAE)

Henry in Trans. Asiat. Soc. Jap. 24 suppl: 108. (1896) (List Pl. Formosa), Rendle in Journ. Linn. Soc. Bot. 36: 389. (1904).

Poa trivialis Henry non L. in Trans. Asiat. Soc. Jap. 24 suppl: 109. (1896) (List Pl. Formosa).

Taiwan: Kaohsiung, 1055, 1067, 1844.

Native to tropical Asia, Africa and America. A coastal species in delta regions of Chinese Taiwan.

Sporoxeia sciadophila W. W. Sm. (MELASTOMATACEAE)

Blastus latifolius H. L. Li in Journ. Arnold Arbor. 25: 15. (1944).

Yunnan: In Pingbian, in forests on the Daweishan Range at 1,640 metres, 9058a., (holotype of *Blastus latifolius* H. L. Li).

Distributed in China (Yunnan), Myanmar and Indochina.

Spuriopimpinella arguta (Diels) X. J. He & Z. X. Wang (APIACEAE) New species

Pimpinella arguta Diels in Bot. Jahrb. Syst. 29: 496. (1900), Wolff in Engler, Pflanzenr. iv. 228: 281. (1927).

Sichuan: North Wushan, 7086 (isotype of *Pimpinella arguta* Diels).

Stachyphrynium placentarium (Lour.) Clausager & Borsch. (MARANTACEAE)

Phrynium parviflorum Roxb. Wright in Journ. Linn. Soc. Bot. 36: 73. (1903), Schumann in Engler, Pflanzenr. 56: 55. (1903).

Hainan: Environs of Kiungchow, 8732.

Stachys adulterina Hemsley (LAMIACEAE) New species

in Journ. Linn. Soc. Bot. 26: 300. (1890), Bretschneider in Hist. Eur. Bot. Disc. China 2: 789. (1898), Dunn in Notes Roy. Bot. Gard. Edinburgh 6: 180. (1915).

Hubei: Badong, 2459, 4676. Zigui, 6192.

Stachys kouyangensis (Vaniot) Dunn (LAMIACEAE)

Dunn in Notes Roy. Bot. Gard. Edinburgh 6: 180. (1915).

Stachys sieboldii Forbes & Hemsley non Miq. in Journ. Linn. Soc. Bot. 171. (1890).

Hubei: Liantuo, cultivated, 4615. Jianshi (wild), 5913. Badong (wild), 7316. **Yunnan:** Mengzi at 2,300 metres, 10074, 10074a., 10074b.

According to Henry this species was cultivated in Hubei under the name *tsanyungtzu.*

Stachys oblongifolia Benth. ex Benth. (LAMIACEAE)

Henry in Journ. China Br. Roy. Asiat. Soc. 22: 276. (1887) (Chinese Names of Plants), Forbes & Hemsley in Journ. Linn. Soc. Bot. 26: 301. (1890), Henry in Trans. Asiat. Soc. Jap. 24 suppl: 73. (1896) (List Pl. Formosa), Dunn in Notes Roy. Bot. Gard. Edinburgh 6: 181. (1915).

Hubei: Yichang, 560. Liantuo, 2003. Badong, 1770 (July 1886), 2545, 4717. San You Dong Glen, 3514. Jianshi, 5943. **Taiwan:** Wanchin, 540.

According to Henry this species was known colloquially in the Three Gorges region as the *yeh ts'an yung tzu.*

Stachys palustris L. (LAMIACEAE)

Anonymous in Kew List of Determinations (volume 24).

Hubei: Badong, 918 (November 1885).

Stachyurus chinensis Franch. (STACHYURACEAE)

Rehder in Sargent Pl. Wilson. 1: 287. (1912), Li in Bull. Torrey Bot. Club 70: 625. (1943), Zheng in Journ. Wuhan Bot. Res. 2.(1): 154. (1984).

Stachyurus praecox Henry non Sieb. & Zucc. in Journ. China Br. Roy. Asiat. Soc. 22: 275. (1887) Chinese Names of Plants); in Notes Econ. Bot. China 63. (1893), Zheng in Journ. Wuhan Bot. Res. 2.(1): 154. (1984).

Hubei: Yichang, 1186, 3449. Liantuo, 2002. In a cave on Moji Shan, 3136. Badong, 3770. Fang Xian, 6917. Without locality, 3449b. **Sichuan:** South Wushan, 5744, 5744a.

A shrub to 5 metres tall, discovered by Père Delavay in Yunnan. In the Three Gorges region it was known colloquially as *t'ung tiao yeh* according to Henry who also stated the pith from the stems was utilised in the same was as that of *Tetrapanax papyrifer. Stachyurus chinensis* was

introduced to cultivation by E. H. Wilson in 1908. A close ally of the Japanese *Stachyurus praecox* though an earlier and more attractive plant than the latter. This species was collected by the Glasnevin Central China Expedition in Badong (GCCE 272) in October 2002 and two years later by the 2004 Glasnevin Expedition to Hubei in the Dragon Gate River Forest Park in Xingshan (GCCE 683).

Stachyurus himalaicus Hook. f. & Thoms. ex Benth. (STACHYURACEAE)

Henry in Notes Econ. Bot. China 63. (1893), Rehder in Sargent Pl. Wilson. 1: 287. (1912), Li in Bull. Torrey Bot. Club 70: 621. (1943); in Woody Flora of Taiwan 612. fig. 242. (1963).

Stachyurus himalaicus Hook.f. & Thoms. Henry in Trans. Asiat. Soc. Jap. 24 suppl: 20. (1896) (List Pl. Formosa).

Hubei: Yichang, 1287, 3449a. In the Kan Chi Glen near Yichang (October 1886), 2913. Liantuo, 3040. In a ravine on Moji Shan, 3362. Antelope Glen, 3349. Jianshi, 5819. Without locality, 1287a., 3362a., 3362b., 3362c., 3362d. **Sichuan:** North Wushan, 7139. **Taiwan:** Wanchin, 35. **Yunnan:** Mengzi at 1,700 to 2,000 metres, 10138, 10138a., 10543.

According to Henry this was the *t'ung-t'iao-shu* or *t'ung-ts'ao-shu,* a small tree, native to the eastern Himalaya, Myanmar and China (including Taiwan). Flowers may be yellow or pinkish. Henry's Taiwan specimen may be more accurately referred to as *Stachyurus chinensis* Franch.

Stachyurus sp. (STACHYURACEAE)

Anonymous in Kew List of Determinations (volume 24).

Hubei: Liantuo, 3800, 3900. **Sichuan:** South Wushan, 5555.

Staphylea bumalda (Thunb.) DC. (STAPHYLEACEAE)

Rehder & E. H. Wilson in Sargent Pl. Wilson. 2: 185. (1914), Zheng in Journ. Wuhan Bot. Res. 2.(1): 127. (1984).

Hubei: Yichang, 1241. Jianshi, 5948. Changyang, 5948a. Without locality, 5948b.

A deciduous shrub to 1.5 metres tall, greenish-white flowers appearing in May and June and carried in a terminal cymose cluster. Native to central and western China and Japan.

Staphylea cochinchinensis (Lour.) Byng & Christenh (STAPHYLEACEAE)

Turpinia nepalensis Wall Rehder & E. H. Wilson in Sargent Pl. Wilson. 2: 187. (1914).

Yunnan: On Fengchunling at 2,000 metres, 10694. In forests near Simao at 1,300 to 2,000 metres, 11612, 11612a., 11612b., 11612c., 11612d., 11612e., 11612f., 11612g., 11612h., 11612i.

Staphylea formosana (Nakai) Byng & Christenh (STAPHYLEACEAE) New species

Turpinia arguta Henry non (Lindl.) Seem. in Trans. Asiat. Soc. Jap. 24 suppl: 29. (1896) (List Pl. Formosa). *Turpinia formosana* Nakai in Journ. Arnold Arb. 5: 80. (1924), Li in Woody Flora of Taiwan 478. fig. 184. (1963).

Taiwan: Wanchin, 128, 434, 564.

A small evergreen tree, endemic to Taiwan where it is very common in forests from sea level to 1,500 metres. Nakai based this species on material gathered by Henry and later by Wilson (Wilson's 10130 from Nantou forms the type specimen).

Staphylea holocarpa Hemsley (STAPHYLEACEAE) New species

in Kew Bull. Misc. Inf. 15. (1895), Bretschneider in Hist. Eur. Bot. Disc. China 2: 781. (1898), Diels in Bot. Jahrb. xxix. 447. (1900), Schneider in Illus. Handb. Laubholzk. 2: 189. fig. 120. (1907), Bean in Kew Bull. Misc. Inf. 175. (1910), Rehder & E. H. Wilson in Sargent Pl. Wilson 2: 185. (1914), Bean in Trees & Shrubs 2: 548. (1921), Morley in Glasra 3: 80. (1979), Zheng in Journ. Wuhan Bot. Res. 2.(1): 127. (1984), Nelson in Pim, The Wood & the Trees 235. (1984); in A Heritage of Beauty 319. (2000), O' Brien in Augustine Henry, An Irish Plant Collector in China 36. (2002).

Hubei: Liantuo, 3017, 4536, 7651, (isosyntype). Badong, 5468. Changyang, 5751b. South Badong, 5751a., (type). Xingshan, 7651a.

A large shrub or small tree to 9 metres tall in its native habitat in central China, rarely reaching the same dimensions in cultivation. The white (rarely rose) flowers are carried on bare wood in panicles up to 10 cm long. This handsome bladder-nut was first found by Henry in 1887. According to that great authority on temperate trees and shrubs, W. J. Bean, Wilson collected 'an abundance of seeds' for the Arnold Arboretum in 1907 some of which were donated to the Royal Botanic Gardens, Kew in the spring of the following year. Wilson concidered this to be the most beautiful member in the genus and few who know this species would disagree with him. Distributed in the provinces of Gansu, Hubei, Henan, Shaanxi and Sichuan.

Staphylea holocarpa Hemsl. var. ***rosea*** Rehder & E. H. Wilson (STAPHYLEACEAE) New variety

in Sargent Pl. Wilson. 2: 186. (1916).

Yunnan: In forests near Mengzi at 2,300 metres, 10220a.

In this variety the flowers are larger and are pink and the leaves are woolly on the undersides when young. First collected by Henry in Yunnan in 1898, later by Wilson in Hubei and Sichuan.

Staphylea sp. (STAPHYLEACEAE)

Anonymous in Kew List of Determinations (volume 24).

Hubei: Changyang, 5239. **Sichuan:** South Wushan, 5659.

Stauntonia angustifolia (Wall.) R. Br. ex Wall. (LARDIZABALACEAE)

Holboellia fargesii Reaub. Rehder & E. H. Wilson in Sargent Pl. Wilson. 1: 346. (1913), Zheng in Journ. Wuhan Bot. Res. 2.(1): 41. (1984).

Hubei: Changyang, 5256. Without locality, 5256a.

These collections were made by Henry in the spring of 1888 while travelling on his south of the Yangtze expedition.

Stauntonia coriacea (Diels) Christenh. (LARDIZABALACEAE) New species

Holboellia latifolia Henry non Wallich in Notes. Econ. Bot. China 47. (1893); in Journ. China Br. Roy. Asiatic. Soc. 22: 262. (1887) (Chinese Names of Plants). *Stauntonia brevipes* Hemsley in Hooker's Icon. Pl. 29: t. adnot. add tab. 2849. (1907), Fedde in Repert. Nov. Spec. Regni Veg. 5: 346. (1908). *Holboellia coriacea* Diels Rehder & E. H. Wilson in Sargent Pl. Wilson. 1: 345. (1913), Zheng in Journ. Wuhan Bot. Res. 2.(1): 41. (1984).

Hubei: Yichang, 1178, 1276 (type of *Stauntonia brevipes* Hemsley). Changyang, 5225 (April 1888), 7788 (April 1889). Changleping, 6336.

According to Henry this was the *pa yueh cha*, a vigorous evergreen climber to 7 metres tall, climbing by means of twining stems. Corymbs of white, purple-tinged flowers are carried in late spring followed by fleshy purple fruits. Introduced to cultivation by Wilson in 1907, this beautiful climber should be grown wherever possible. Henry stated the fruits were eaten in the Three Gorges region. Discovered by Henry in the spring of 1887 and described as a new species by Diels from material later collected in Sichuan by Bock and von Rosthorn (c. 1891). This species was collected by the Glasnevin Central China Expedition (GCCE 656) in the Dragon Gate River Forest Park in Xingshan in October 2004.

Stauntonia latifolia (Wall.) R. Br. ex Wall. (LARDIZABALACEAE)

Holboellia latifolia Wall. Rehder & E. H. Wilson in Sargent Pl. Wilson. 1: 347. (1913).

Yunnan: Mengzi, 10297, 10527, 10527a. South of the Red River from Manmei at 1,900 metres, 9472.

Distributed in Nepal, India, Bhutan and China [Guizhou, Sichuan, Tibet (Xizang) and Yunnan].

Stauntonia parviflora Hemsley (LARDIZABALACEAE) New species

in Hooker's Icon. Pl. 29: t. 2849. (1907), Fedde in

Repert. Nov. Spec. Regni Veg. 5: 346. (1908).

Holboellia parviflora (Hemsley) Gagnepain in Bull. Mus. Hist. Nat. (Paris) 14: 68. (1908).

Yunnan: Mengzi at 1,500 metres, 10462, 10462a., (isotype).

Endemic to China and distributed in the western provinces.

Stauntonia sp. (LARDIZABALACEAE)

Holboellia sp. Anonymous in Kew List of Determinations (volume 34).

Yunnan: Simao, 12548.

Staurogyne sp. (ACANTHACEAE)

Ebermaiera sp. Anonymous in Kew List of Determinations (volume 34).

Yunnan: Simao, 11627.

Stellaria alsine Grimm. (CARYOPHYLLACEAE)

Stellaria uliginosa Murr. Anonymous in Kew List of Determinations (volume 24).

Hubei: Badong, 4708.

Stellaria aquatica (L.) Scop. (CARYOPHYLLACEAE)

Henry in Journ. China Br. Roy. Asiat. Soc. 22: 279. (1887) (Chinese Names of Plants), Diels in Bot. Jahrb. 29: 319. (1900).

Hubei: Yichang, 460, 761, 1214, 3540. Badong, 3763, 4674.

According to Henry this species was known in the Three Gorges region as *wo erh ch'ang*.

Stellaria graminea L. (CARYOPHYLLACEAE)

Diels in Bot. Jahrb. 29: 319. (1900).

Hubei: Badong, 1863 (in part).

Stellaria henryi Williams (CARYOPHYLLACEAE)

New species

in Journ. Linn. Soc. Bot. 34: 434. (1900), Diels in Bot. Jahrb. 29: 319. (1900).

Hubei: Xingshan, 6970 (type).

Stellaria media (L.) Vill. (CARYOPHYLLACEAE)

Diels in Bot. Jahrb. 29: 319. (1900).

Hubei: Yichang, 1269, 1320. Changyang, 5251.

Chickweed, a small creeping annual, native to Europe, north Africa and northern Asia and now a cosmopolitan weed. Chickweed is edible and may be cooked as a vegetable or may be eaten raw in salads.

Stellaria nipponica Ohwi (CARYOPHYLLACEAE)

Stellaria florida Fisch. var. *angustifolia* Maxim. Anonymous in Kew List of Determinations (volume 24).

Hubei: Badong, 614 (July 1885), 4670, 6070.

Stellaria spp. (CARYOPHYLLACEAE)

Anonymous in Kew List of Determinations (volume 34).

Yunnan: Milê District, 9932. Mengzi, 10858. Simao, 12434, 13501, 13552.

Stellaria vestita Kurz (CARYOPHYLLACEAE)

Stellaria saxatilis Buch.-Ham ex D. Don Diels in Bot. Jahrb. 29: 319. (1900).

Hubei: Badong, 1778 (July 1886), 4729.

Stellaria yunnanensis Franch. (CARYOPHYLLACEAE)

Anonymous in Kew List of Determinations (volume 34).

Yunnan: Mengzi, 10270.

Endemic to China and distributed in Sichuan and Yunnan.

Stellera chamaejasme L. var. *chrysantha* (H. C. Huang) Grey-Wilson (THYMELAEACEAE)

Stellera chamaejasme Rehder non L. in Sargent Pl. Wilson. 2: 551. (1916).

Yunnan: Mengzi, 10126.

A long-lived sub-shrub with a thick woody crown bearing numerous unbranched stems to 40 cm tall. The tough root stock can be up to 15 cm long on older plants. The inflorescence is made up of a rounded head of up to 50 flowers. These flowers may be white or yellow and are reddish-pink in bud. *Stellera chamaejasme* has a wide distribution from the Altai in the former eastern USSR, the Himalaya, the Tibetan (Xizang) Plateau into south-western China in the provinces of Yunnan and Sichuan and north into northern China and Mongolia. Both the type i.e. *Stellera chamaejasme* var. *chamaejasme* and the var. *chrysantha* are represented in China. In Tibet (Xizang) this plant grows on high plains where trees are absent and its woody roots are used by Buddhist Monks to make paper. The plant is poisonous (according to Henry it was the *lang du* or "wolf poison"), is graze-proof and forms large populations on the moorland it inhabits. It is particularly plentiful on the Zhongdian Plateau where, when in flower it makes a stunning display. The plant is extremely rare in cultivation and has been reintroduced several times in recent decades from China. The Royal Botanic Garden, Edinburgh has grown good plants for quite a number of years now, though the joy of seeing a sub-alpine plain covered in this wonderful plant can never be recreated in Europe. From Christopher Grey-Wilson's distribution map of this plant (see The New Plantsman 2(1): 45. (1995) I take it that Henry's plant is the yellow flowered var. *chrysantha*.

Stemona parviflora C. H. Wright (STEMONACEAE)

New species

in Journ. Linn. Soc. Bot. 32: 496. (1896); Journ. Linn. Soc. Bot. 36: 95. (1903).

Hainan: Environs of Kiungchow, 8698 (type).

Stemona sp. (STEMONACEAE)

Anonymous in Kew List of Determinations (volume 34).

Yunnan: Mengzi, 10773.

02

Stemona tuberosa Lour. (STEMONACEAE)

Henry in Journ. China Br. Roy. Asiat. Soc. 22: 265. (1887) (Chinese Names of Plants); in Trans. Asiat. Soc. Jap. 24 suppl: 96. (1896) (List Pl. Formosa), Wright in Journ. Linn. Soc. Bot. 32: 494. (1896); 36: 95. (1903).

Hubei: Yichang, 566. **Taiwan:** Wanchin, 816.

According to Henry the roots of this species were used as a medicine and in the Three Gorges region it was known colloquially as *peh pu ken*. Native to northern India, Myanmar, Vietnam, Cambodia, Thailand and southern China.

Stephania brachyandra Diels (MENISPERMACEAE) New species

in Pflanzenr. (Engler) 4. Fam. 94: 275. (1910).

Yunnan: Mengzi, 10776. Mengzi at 1,650 metres, 10776a., (isosyntype), 19776b., (isosyntype).

Distributed in Myanmar and China (Yunnan).

Stephania cephalantha Hayata (MENISPERMACEAE)

Stephania tetrandra S. Moore var. *glabra* Maxim. Diels in Engler, Pflanzenr. iv. 94: 282. (1910), Zheng in Journ. Wuhan Bot. Res. 2.(1): 45. (1984).

Hubei: Yichang, 4114. Without locality, 4114a.

Stephania delavayi Diels (MENISPERMACEAE)

in Engler, Pflanzenr. iv. 94: 275. (1910), Rehder & E. H. Wilson in Sargent Pl. Wilson. 1: 389. (1913), Zheng in Journ. Wuhan Bot. Res. 2.(1): 45. (1984).

Hubei: Yichang, 1664 (January 1887). **Yunnan:** Mengzi at 1,400 metres, 10312. Mountains to the east of Simao 10312b., 10312c. Mengzi at 1,400 metres, 13677.

A handsome vine with large wood flat tuber-like roots, the Hani (an ethnic group) of Xishuangbanna in southern Yunnan call this species *qilni'laolbail* and use the roots to treat gastric ulcers, fatigue, rheumatism and joint pain. Myanmar and China (Yunnan, Sichuan and Guizhou).

Stephania dolichopoda Diels (MENISPERMACEAE) New species

in Engler, Pflanzenr. 46. (iv. 94): 282. (1910).

Yunnan: Simao at 1,300 metres, 12008b., (isotype).

Distributed in India and China (Guangxi and Yunnan).

Stephania herbacea Gagnep. (MENISPERMACEAE)

Diels in Engler, Pflanzenr. iv. 94: 272. (1910).

Hubei: South Badong, 6089.

Stephania japonica (Thunb.) Miers (MENISPERMACEAE)

Diels in Engler, Pflanzenr. iv. 94: 277. (1910).

Stephania hernandiifolia Henry non (Willd.) Walp. in Trans. Asiat. Soc. Jap. 24 suppl: 16. (1896) (List Pl. Formosa).

Hubei: Yichang, 1564. Badong, 6004a. **Taiwan:** Wanchin, without number. Wanshoushan, without number. Oluanpi, 339.

A glabrous climber, native to India, China, Japan and the Philippines to Japan. A common inhabitant of forests throughout Taiwan of China.

Stephania japonica (Thunb.) Miers var. ***discolor*** (Blume) Forman (MENISPERMACEAE)

Stephania hernandiifolia (Willd.) Walp. Henry in Journ. China Br. Roy. Asiat. Soc. 22: 280. (1887) (Chinese Names of Plants), Zheng in Journ. Wuhan Bot. Res. 2.(1): 45. (1984).

Hubei: Jianshi, 6004. **Yunnan:** Mengzi, 9830. Simao, 12201, 12317

According to Henry this species was known colloquially in the Three Gorges region as *wu kuei shao*, he also stated that the root resemble a tortoise and was used as a drug.

Stephania longa Lour. (MENISPERMACEAE)

Diels in Engler, Pflanzenr. iv. 94: 278. (1910).

Hainan: 32 km (20 miles) west of Haikou, 8146, 8279.

Stephania sinica Diels (MENISPERMACEAE) New species

in Engler, Pflanzenr. iv. 94: 272. (1910), Zheng in Journ. Wuhan Bot. Res. 2.(1): 45. (1984).

Hubei: Badong, 4693. Fang Xian, 6662 (type).

Stephania tetrandra S. Moore (MENISPERMACEAE)

Henry in Journ. China Br. Roy. Asiat. Soc. 22: 280. (1887) (Chinese Names of Plants), Diels in Engler, Pflanzenr. iv. 94: 282. (1910), Li in Woody Flora of Taiwan 185. (1963).

Hubei: South Badong, 6061. **Taiwan:** Wanchin, 466.

A scandent shrub, native to Vietnam, southern China, including Taiwan (where it is rare). According to Henry the roots of this species supplied the drug *han fang ji*. Bretschneider stated this species was first found by the Kew collector Richard Oldham in 1864 (but was described as a new species from material collected by Shearer). Born in 1867 Oldham made extensive collections for Kew in Japan, Korea, China (Taiwan) and along the east coast of China. Oldham suffered from heart disease however and this cut short his exploration of Taiwan (1864).

Stephanotis volubilis (L. f.) S. Reuss, Liede & Meve (APOCYNACEAE)

Dregea volubilis (L. f.) Benth. ex Hook. f. Henry in Trans. Asiat. Soc. Jap. 24 suppl: 61. (1896) (List Pl. Formosa), Anonymous in Kew List of Determinations (volume 24). *Wattakaka volubilis* (Linn. f.) Stapf Li in Woody Flora of Taiwan 807. (1963).

Hainan: Environs of Haikou, 8026. Environs of Kiungchow, 8293. **Taiwan:** Kaohsiung, 945. **Yunnan:** Mengzi, 10976. Yiwu (Xishuangbanna), 13571.

A tall climber of widespread distribution from Pakistan

and India to Malaysia. In Chinese Taiwan this species is found on the southern part of the island, where it climbs through trees and is uncommon.

Sterculia henryi Hemsley (MALVACEAE) New species
in Kew Bull. Misc. Inf. 179. (1908).

Yunnan: In Pingbian, in forests on the Daweishan Range at 1,500 metres, 11016 (isotype).

Distributed in China (Yunnan) and Vietnam.

Sterculia lanceolata Cav. (MALVACEAE)
Anonymous in Kew List of Determinations (volume 24).

Hainan: Environs of Haikou, 8053, 8428. On hummocks near Haikou, 8118.

Sterculia lanceifolia Roxb. (MALVACEAE)
Sterculia roxburghii Wallich Anonymous in Kew List of Determinations (volume 34).

Yunnan: Simao, 11751.

Sterculia lanceolata Cav. var. ***coccinea*** (Jack) Phengklai (MALVACEAE)
Sterculia coccinea Roxb. ex G. Don Anonymous in Kew List of Determinations (volume 34).

Yunnan: Simao, 12318.

Sterculia sp. (MALVACEAE)
Anonymous in Kew List of Determinations (volume 34).

Yunnan: Mengzi, 11081, 11856.

Stereocaulon dactylophyllum Flöke (STEREOCAULACEAE)
Wei in An Enumeration of Lichens in China 234. (1991).

Stereocaulon coraloides Fr. Muell.-Arg. in Bull. Herb. Boissier 1: 235. (1893).

Hubei: Fang Xian at 1,900 to 3,000 metres, 6810 (in part). Xingshan, 6940.

Stereocaulon nesaeum Nyl. (STEREOCAULACEAE)
Anonymous in Kew List of Determinations (volume 34).

Yunnan: Mengzi, 9477. (Lichen).

Stereocaulon paschale (L.) Hoffm. (STEREOCAULACEAE)
Muell.-Arg. in Bull. Herb. Boissier 1: 235. (1893).

Hubei: Fang Xian at 1,900 to 3,000 metres, 6810 (in part). (Lichen).

Stereospermum colais (Buch.-Ham. ex Dillwyn.) Mabb. (BIGNONIACEAE)
Stereospermum personatum (Hassk.) Chatterjee Santisuk in Kew Bull. 28: 178. (1973).

Yunnan: Simao, 11976, 13153.

Native to the Himalaya, India, Bangladesh, China and Indochina.

Stereospermum spp. (BIGNONIACEAE)
Anonymous in Kew List of Determinations (volume 24 & 34).

Hainan: Environs of Kiungchow, 8288. **Yunnan:** Mengzi, 13763.

Steudnera henryana Engler (ARACEAE) New species
in Engler, Pflanzenr. iv. 23Da: 13. (1912).

Yunnan: Simao, 11986 (isotype).

Distributed in China (Yunnan), northern Laos and Vietnam.

Stewartia pteropetiolata W. C. Cheng (THEACEAE) New species
in Contr. Lab. Sci. Soc. China. Bot. Ser. 9: 202. (1934), Li in Acta Phytotax. Sin. 34: 66. (1996), Nelson in Pim, The Wood and the Trees 235. (1984), Bean in Trees & Shrubs 4: 512. (1989), Nelson in A Heritage of Beauty 319. (2000), (1989) O' Brien in Augustine Henry, An Irish Plant Collector in China 36. (2002).

Hartia sinensis Dunn in Hooker's Icon. Pl. t. 2727. (1902) non *Stewartia sinensis* Rehder & E. H. Wilson (1915)

Yunnan: South of the Red River, 10465 (syntype of *Hartia sinensis* Dunn). Mengzi, 10465a., (type).

A tree to 10 metres tall, originally the type of a new genus, *Hartia*. At Augustine Henry's request this species was named *Hartia sinensis* in honour of the Irishman, Sir Robert Hart, Head of the Chinese Imperial Maritime Customs Service, and at the time, the highest ranking foreigner in the pay of the Imperial Chinese goverment. George Forrest introduced *Stewartia pteropetiolata* from an area to the west of Tengchong (formerly Tengyueh, an area Henry himself had predicted to be botanically rich) near the Myanmar-Yunnan border. Endemic to Yunnan, this species is cultivated in Ireland at Mount Congreve, County Waterford. The tallest known specimens in Britain and Ireland grow at Caerhays Castle, Cornwall, by 1984 the tallest of these trees was 24 metres.

Stewartia sinensis Rehder & E. H. Wilson (THEACEAE) New species
in Sarg. Pl. Wilson. 2: 295. (1915), Zheng in Journ. Wuhan Bot. Res. 2.(1): 152. (1984), Li in Acta Phytotax. Sin. 34: 59. (1996), O' Brien in Augustine Henry, An Irish Plant Collector in China 36. (2002).

Stewartia monadelpha Henry non Sieb. & Zucc. in Notes Econ. Bot. China 44. (1893), Bean in Trees & Shrubs. 2: 553. (1914).

Hubei: South Badong, 6166 (paratype). **Sichuan:** North Wushan, 7392, 7392a., (paratype).

A medium-sized tree with beautiful peeling bark and white fragrant flowers. This was the *ma-liu-kuang* according to Henry. It is rare both in its native habitat and cultivation and at Trewithen in Cornwall a tree it has reached 14.5 metres tall with a girth of 63 cm (at ground level, 1995).

Sticta henryana Muell. Arg. (PELIGERACEAE) New species

in Flora 74: 374. (1891), Muell.-Arg. in Bull. Herb. Boissier 1: 236. (1893): in Stzbgr. in Flora 81: 124. (1895), A. Zahlbruckner in Handel-Mazzeti in Symb. Sin. 3: 85. (1930), Wei in An Enumeration of Lichens in China 237. (1991)

Hubei: Fang Xian, 6932. (Lichen)

Sticta nylanderiana A. Zahlbruckner (STICTACEAE)

Wei in An Enumeration of Lichens in China 238. (1991).

Sticta platyphylla Nyl. Muell.-Arg. in Bull. Herb. Boissier 1: 236. (1893).

Hubei: Fang Xian, 6635a. (Lichen).

Stictina retigera (Bory) Müell. Arg. (LOBARIACEAE)

Anonymous in Kew List of Determinations (volume 34).

Yunnan: Mengzi, 10781. (Lichen).

Stipa henryi Rendle (POACEAE) New species

in Journ. Linn. Soc. Bot. 36: 382. (1904), Morley in Glasra 3: 80. (1979).

Oryzopsis henryi (Rendle) Keng f. Fl. Tsinling 1(1): 145. (1976).

Hubei: Antelope Glen, 3444. **Sichuan:** South Wushan, 5530.

Stixis suaveolens (Roxb.) Pierre (RESEDACEAE)

Roydsia suaveolens Roxb. Anonymous in Kew List of Determinations (volume 34).

Yunnan: Mengzi, 13820.

Native to Nepal, Bhutan, Bangladesh, Myanmar and China.

Stranvaesia oblanceolata (Rehder & E. H. Wilson) Stapf (ROSACEAE) New species

in Curtis's Bot. Mag. 149: tt. 9008. (1924).

Stranvaesia nussia (Buch.-Ham. ex D. Don) Decne. var. *oblanceolata* Rehder & E. H. Wilson in Sargent Pl. Wilson. 1: 193. (1912), Hand.-Mazz. in Symb. Sin. 7: 483. (1933), Vidal in Adansonia 5: 232. (1965); in Fl. Camb., Laos & Vietn. 6: 56. pl. 8. (1968): in Fl. Thailand 2(1): 42. (1970).

Yunnan: Forests around Simao at 1,500 to 1,600 metres, 11615 [syntype of *Stranvaesia nussia* (Buch.-Ham.) Dcne. var. *oblanceolata* Rehder & E. H. Wilson], 11615a., (syntype of *Stranvaesia nussia* (Buch.-Ham.) Dcne. var. *oblanceolata* Rehder & E. H. Wilson), 11615f., (syntype of *Stranvaesia nussia* (Buch.-Ham.) Dcne. var. *oblanceolata* Rehder & E. H. Wilson. In woods near Mengzi, 11615b., [type of *Stranvaesia nussia* (Buch.-Ham.) Dcne. var. *oblanceolata* Rehder & E. H. Wilson], 11615e. [syntype of *Stranvaesia nussia* (Buch.-Ham.) Dcne. var. *oblanceolata* Rehder & E. H. Wilson].

Native to Myanmar, China (Yunnan), Laos and Thailand.

Stranvaesia sp. (ROSACEAE)

Anonymous in Kew List of Determinations (volume 24).

Hubei: Badong, 610 (July 1885), 873, 1016.

Streblus asper Lour. (MORACEAE)

Forbes & Hemsley in Journ. Linn. Soc. Bot. 26: 454. (1894).

Cudrania crenata C. H. Wright in Journ. Linn. Soc. Bot. 26: 469. (1899), Schneider in Sargent Pl. Wilson. 3: 309. (1917).

Hainan: Environs of Haikou, 7957, 8389. Without locality, 8565.

Streptocaulon juventas (Lour.) Merr. (APOCYNACEAE)

Streptocaulon griffithii Hook. f. Anonymous in Kew List of Determinations (volume 34).

Yunnan: Simao, 12851.

Widely distributed in Asia.

Streptocaulon sp. (APOCYNACEAE)

Anonymous in Kew List of Determinations (volume 34).

Yunnan: Simao, 12761.

Streptolirion sp. (COMMELINACEAE)

Anonymous in Kew List of Determinations (volume 34).

Yunnan: Mengzi, 9240.

Streptolirion volubile Edgew. (COMMELINACEAE)

Brown in Journ. Linn. Soc. Bot. 36: 159. (1903).

Hubei: Badong, 2428, 2482, 4170, 4926. Liantuo, 2587.

Striga lutea Lour. (OROBANCHACEAE)

Henry in Trans. Asiat. Soc. Jap. 24 suppl: 68. (1896) (List Pl. Formosa).

Hainan: Environs of Kiungchow, 8256. Environs of Haikou, 8326. Ling-men (in the interior of the island), 8671. **Taiwan:** Kaohsiung Plain, 1903.

Striga masuria (Buch.-Ham. ex Benth.) Benth. (OROBANCHACEAE)

Henry in Trans. Asiat. Soc. Jap. 24 suppl: 68. (1896) (List Pl. Formosa).

Taiwan: Kaohsiung, 1141. **Yunnan:** Simao, 13093.

Strobilanthes aprica (Hance) T. Anderson ex Benth. (ACANTHACEAE)

Anonymous in Kew List of Determinations (volume 34).

Yunnan: Simao, 13415.

Strobilanthes cusia (Nees) Kuntze (ACANTHACEAE)

Strobilanthes flaccidifolia Nees Anonymous in Kew List of Determinations (volume 34).

Yunnan: Mengzi, 10404, 11288, 11446. Simao, 12530, 12716.

Strobilanthes cycla C. B. Clarke ex W. W. Sm. (ACANTHACEAE) New species

in Notes Roy. Bot. Gard. Edinburgh 9: 192. (1916).

02

Yunnan: On grassy mountains near Mengzi at 2,300 metres, 10159 (type, collected 16th August 1897).

Endemic to China and distributed in Guangxi and Yunnan.

Strobilanthes dimorpotricha Hance (ACANTHACEAE)

Strobilanthes psilostachys C. B. Clarke ex W. W. Smith in Notes Roy. Bot. Gard. Edinburgh 9: 198. (1916).

Yunnan: In forests near Simao at 1,450 metres, 12914 (type of *Strobilanthes psilostachys* C. B. Clarke ex W. W. Smith).

Strobilanthes extensa (Nees) Nees (ACANTHACEAE)

Strobilanthes claviculata C. B. Clarke ex W. W. Sm. in Notes Roy. Bot. Gard. Edinburgh 9: 191. (1916).

Yunnan: In mountain forest near Milê, 10319 (type of *Strobilanthes claviculata* C. B. Clarke ex W. W. Sm).

Strobilanthes fluviatilis (C. B. Clarke) Moylan & Y. F. Deng (ACANTHACEAE) New species

Hemigraphis fluviatilis C. B. Clarke in Notes Roy. Bot. Gard. Edinburgh 10: 182. (1918).

Yunnan: In the Red River valley near Manhao, 9535 (isotype of *Hemigraphis fluviatilis* C. B. Clarke, collected 13th June 1896).

Strobilanthes glomerata (Nees) T. Anders. (ACANTHACEAE)

Anonymous in Kew List of Determinations (volume 34).

Yunnan: Simao, 12531.

Strobilanthes henryi Hemsley (ACANTHACEAE) New species

in Journ. Linn. Soc. Bot. 26: 240. (1890), Bretschneider in Hist. Eur. Bot. Disc. China 2: 789. (1898), Morley in Glasra 3: 80. (1979).

Gutzlaffia henryi (Hemsl.) C. B. Clarke ex S. Moore in J. Bot. 63: 167 (1925).

Hubei: In the Hsiao Pingshanba Glen (October 1886), 2918. In a ravine on Moji Shan, 4269.

Strobilanthes hossei C. B. Clarke (ACANTHACEAE)

Strobilanthes rufohirta C. B. Clarke ex W. W. Smith in Notes Roy. Bot. Gard. Edinburgh 9: 199. (1916).

Yunnan: In Pingbian, on the Daweishan Range at 1,640 metres, 11133 (type of *Strobilanthes rufohirta* C. B. Clarke ex W. W. Smith).

Strobilanthes hupehensis W. W. Smith (ACANTHACEAE) New species

in Notes Roy. Bot. Gard. Edinburgh 9: 193. (1916).

Strobilanthes penstemonoides Forbes & Hemsley non (Nees.) T. Anders. in Journ. Linn. Soc. Bot. 26: 241. (1890).

Hubei: Yichang, 1100, 1123, 2171, 2709. Liantuo, 2074, 3052. South Badong, 7289. Without locality, 2709a.

Strobilanthes ladbordei H. Lév. (ACANTHACEAE)

Strobilanthes radicans Forbes & Hemsley non T.

Anders. ex Benth. in Journ. Linn. Soc. Bot. 26: 242. (1890), *Strobilanthes debilis* Hemsley in Journ. Linn. Soc. Bot. 26: 239. (1890).

Hubei: Yichang, 87b., 324, 1007, 2795. On road to Moji Shan (June 1886), 1543. Jianshi, 7421 (type of *Strobilanthes debilis* Hemsley).

Strobilanthes lamiifolia (Nees) T. Anderson (ACANTHACEAE)

Strobilanthes austini C. B. Clarke ex W. W. Sm. in Notes Roy. Bot. Gard. Edinburgh 9: 190. (1916). *Strobilanthes hancockii* C. B. Clarke ex W. W. Sm. in Notes Roy. Bot. Gard. Edinburgh 9: 193. (1916).

Yunnan: Mengzi, in shaded woodland at 1,450 metres, 9159a. In mountain forest near Milê, 9956. On grassy mountains near Mengzi at 1,950 metres, 10027 (type of *Strobilanthes hancockii* C. B. Clarke ex W. W. Sm.). On the mountains to the south of Mengzi at 980 metres, 10027a. Near Yuanjiang at 1,650 metres, 13382 (type of *Strobilanthes austini* C. B. Clarke ex W. W. Sm.).

Strobilanthes pluriformis C. B. Clarke (ACANTHACEAE) New species

in Publ. Bur. Sci. Gov. Lab. 35: 93. (1900).

Acanthaceae, undetermined. Henry in Trans. Asiat. Soc. Jap. 24 suppl: 70. (1896) (List Pl. Formosa).

Taiwan: Wanchin, 1608 (type).

Strobilanthes polyneuros C. B. Clarke ex W. W. Smith (ACANTHACEAE) New species

in Notes Roy. Bot. Gard. Edinburgh 9: 198. (1916).

Yunnan: In the mountains to the south of Mengzi at 1,650 metres, 10352 (isotype).

Strobilanthes sp. (ACANTHACEAE)

Henry in Trans. Asiat. Soc. Jap. 24 suppl: 69. (1896) (List Pl. Formosa).

Taiwan: Tamsui, 1395. Kaohsiung, 1765.

Strobilanthes spp. (ACANTHACEAE)

Anonymous in Kew List of Determinations (vol. 34).

Yunnan: In Pingbian, in woods on the Daweishan Range at 2,600 metres, 9029. Mengzi, 11167. Simao, 12570, 12677. South of Red River from Manmei, 11183.

Strobilanthes triflorus Y. C. Tang (ACANTHACEAE)

Strobilanthes wallichii Forbes & Hemsley non Nees in Journ. Linn. Soc. Bot. 26: 242. (1890). *Strobilanthes latisepala* Hemsley in Journ. Linn. Soc. Bot. 26: 241. (1890), Bretschneider in Hist. Eur. Bot. Disc. China 2: 789. (1898).

Hubei: On cliffs in woods at Xingshan, 6502 (type of *Strobilanthes latisepala* Hemsley). Jianshi, 7413, 7420. Without locality, 7053a. **Sichuan:** North Wushan, 7053. South Wushan, 7218.

Strobocalyx esculenta (Hemsl.) H. Rob, S. C. Keeley, Skvarla & R. Chan (ASTERACEAE)

Vernonia papillosa Franch. Anonymous in Kew List of Determinations (volume 34).

Yunnan: Mengzi, 9502. Simao, 12439.

Native to Vietnam and China (Guizhou, Guangxi Sichuan and Yunnan). This species was discovered by Frederick Bourne of the British Consular Service during the 1880s. In his original description Hemsley states only 'south-west China' as the location of the type material, however, it is most likely Bourne collected it in Yunnan. Hemsley coined the specific epithet because Bourne stated that local children ate the pith of the stems of this species.

Strobocalyx solanifolia (Benth.) Sch.Bip. (ASTERACEAE)

Vernonia solanifolia Benth. Anonymous in Kew List of Determinations (volume 34).

Yunnan: Simao, 12863.

The Hani (an ethnic group) of Xishuangbanna in southern Yunnan call this perennial *sanq'anlpavqssaq* and use its roots and leaves to treat cases of sore throat. Native to India, Myanmar, China, Laos, Cambodia and Vietnam.

Strobocalyx sylvatica (Dunn) H. Rob., S. C. Keeley, Skvarla & R. Chan (ASTERACEAE) New species

Vernonia sylvatica Dunn in Journ. Linn. Soc. Bot. 35: 501. (1903).

Yunnan: In forests near Mengzi at 1,450 to 1,650 metres, 11051. In forests near Simao at 1,450 to 1,650 metres, 11697, 11687a.

Endemic to China and distributed in Guangxi and Yunnan. This species was based on collections made in Mengzi by Augustine Henry and William Hancock. Hancock, like Henry was an Irishman employed by the Chinese Imperial Maratime Customs Service.

Strongyleria pannea (Lindl.) Schuit, Y. P. Ng., & H. A. Pederson (ORCHIDACEAE)

Eria pannea Lindl. Kranzlin in Engler, Pflanzenr. 50: 48. (1911).

Yunnan: Simao, 12198, 13584.

Native to India, Bhutan, Myanmar, China and Indochina.

Strophanthus divaricatus (Lour.) Hook. & Arn. (APOCYNACEAE)

Strophanthus divergens Graham Anonymous in Kew List of Determinations (volume 24).

Hainan: Environs of Haikou, 8011, 8162. 8 km (5 miles) south of Kuingchow, 8185. Without locality, 8597.

Strychnos cathayensis Merr. (LOGANIACEAE)

Strychnos sp. Henry in Trans. Asiat. Soc. Jap. 24 suppl:

61. (1896) (List Pl. Formosa). *Strychnos henryi* Merr & Yamamoto ex Yamamoto in Jour. Soc. Trop. Agr. 7: 145. (1935), Li in Woody Flora of Taiwan 778. (1963).

Taiwan: Wanchin, 1662 (type of *Strychnos henryi* Merr & Yamamoto).

Various species of *Strychnos* are used in south-east Asia as arrow poison.

Stylophorum lasiocarpum (Oliv.) Fedde (PAPAVERACEAE) New species

in Das Pflanzenreich 104: 209. (1909), O' Brien in Augustine Henry, An Irish Plant Collector in China 36. (2002).

Chelidonium lasiocarpum Oliver in Hooker's Icon. Pl. 18: t. 1739. (1888), Henry in Notes on Economic Botany of China 49. (1893), Bretschneider in Hist. Eur. Bot. Disc. China 2: 779. (1889), J. H. Veitch in Hortus Veitchii 417. (1906), Nelson in A Heritage of Beauty 311. (2000).

Hubei: Xingshan, 5567a., 5567d. Fang Xian, 5567b., 5567c. Liantuo (cultivated), 3885 (type of *Chelidonium lasiocarpum* Oliver). **Sichuan:** South Wushan, 5567.

A perennial herb to 50 cm tall, distributed through the provinces of Hubei, Sichuan and Shaanxi where it grows on mountain slopes and forest ravines between altitudes of 800 to 1,800 metres. Henry stated that the Chinese name for this plant (*jen-hsueh-ts'ao*) signifies 'mans-blood herb' from the red juice in the root and stem which is used as a drug in Chinese herbal medicine. According to Bretschneider, the plants Henry collected from were cultivated (Henry himself, in his field notes at Kew states his 3885 from Liantuo was cultivated and had originally came from Sichuan), though the plant is native to the Three Gorges region. *Stylophorum lasiocarpum* was introduced to cultivation by Wilson through Veitch's Coombe Wood nursery.

Styphnolobium japonica (L.) Schott (FABACEAE)

Sophora japonica L. Forbes & Hemsley in Journ. Linn. Soc. Bot. 23: 202. (1887), Henry in Journ. China Br. Roy. Asiat. Soc. 22: 245. (1887) (Chinese Names of Plants), Tsoong & Ma in Acta Phytotax. Sin. 162. (1981), Zheng in Journ. Wuhan Bot. Res. 2.(1): 103. (1984). *Sophora japonica* L. var. *vestita* Rehder Tsoong & Ma in Acta Phytotax. Sin. 19: 164. (1981).

Hubei: Yichang, 2130, 2323. **Yunnan:** Mengzi, 9871.

Distributed in China, Korea, Vietnam and Japan, known colloquially in the Three Gorges region as *huai shu* according to Henry. Once cultivated in China for the sake of the 'Imperial yellow dye' obtained from its flowers and this was used to colour cotton and silk.

Styrax faberi Perkins (STYRACACEAE)

Styrax serrulatus Forbes & Hemsley non Roxb. in Journ. Linn. Soc. Bot. 26: 77. (1889). *Styrax confusus*

Hemsl. var. *microphyllus* Perkins in Engler, Pflanzenr. 241: 34. (1907). *Styrax confusa* Zheng non Hemsl. in Journ. Wuhan Bot. Res. 2.(1): 175. (1984).

Hubei: Yichang, 1115. Xiling (Yichang) Gorge, 3450 (type of *Styrax confusa* Hemsl. var. *microphyllus* Perkins). Xingshan, 4524. Without locality, 3450a.

Styrax formosana Matsum. (STYRACACEAE)

Styrax serrulatum Henry non Roxb. in Trans. Asiat. Soc. Jap. 24 suppl: 59. (1896) (List Pl. Formosa). *Styrax henryi* Perkins in Engler, Pflanzenr. 4(241): 33. (1907). *Styrax henryi* Perkins var. *microcalyx* Perkins in Engler, Pflanzenr. 4(241): 33. (1907).

Taiwan: Wanchin, 394 (isosyntype of *Styrax henryi* Perkins), 554. Oluanpi, 913, 2063 (holotype of *Styrax henryi* Perkins, var. *microcalyx* Perkins), 2064.

A shrub or small tree, endemic to Taiwan where it grows in forests to 1,300 metres throughout the island.

Styrax fortunei Hance (STYRACACEAE)

Styrax calvescens Perkins in Pflanzenr. (Engler) 4. Fam. 241: 32. (1907), Rehder in Sargent Pl. Wilson. 1: 290. (1913), O' Brien in Augustine Henry, An Irish Plant collector in China 36. (2002). *Styrax dasyanthus* Perkins in Bot. Jahrb. 29: 485. (1901). Perkins in Engl. Pflanzenr. iv. - 241: 31. (1907), Rehder in Sargent Pl. Wilson. 1: 289. (1913), Zheng in Journ. Wuhan Bot. Res. 2.(1): 175. (1984), Bean in Trees & Shrubs 4: 515. (1989), O' Brien in Augustine Henry, An Irish Plant Collector in China 36. (2002). *Styrax serrulatum* Forbes & Hemsley non Roxb. in Journ. Linn. Soc. Bot. 26: 77. (1889).

Hubei: Yichang, 721, (holotype of *Styrax calvescens* Perkins). Jianshi, 5977 (type of *Styrax dasyanthus* Perkins).

In Ireland this species is cultivated at Mount Congreve in County Waterford.

Styrax hemsleyanus Diels (STYRACACEAE) New species

in Bot. Jahrb. 29: 485. (1901), Perkins in Engl. Pflanzenr. iv. - 241: 31. (1907), Rehder in Sargent Pl. Wilson. 1: 289. (1913), Zheng in Journ. Wuhan Bot. Res. 2.(1): 175. (1984), Nelson in A Heritage of Beauty 319. (2000), O' Brien in Augustine Henry, An Irish Plant collector in China 36. (2002).

Hubei: Fang Xian, 6895 (type). **Sichuan:** South Wushan, 5676, 5676a.

A beautiful deciduous tree to 7 metres tall, producing terminal racemes of pure white flowers in June. Introduced to cultivation by Wilson in 1900 through Messrs Veitch's Coombe Wood nursery. It is grown in several Irish gardens.

Styrax hookeri C. B. Clarke (STYRACACEAE)

Styrax macranthus J. R. Perkins in Bot. Jahrb. Syst. 31: 487. (1902), Perkins in Engler, Pflanzenr. 241: 69. (1907).

Yunnan: Near Yuanjiang, on the slopes of Fengchunling at 2,300 metres, 10644 (type *Styrax macranthus* J. R. Perkins).

Styrax japonicus Sieb. & Zucc. (STYRACACEAE)

Henry in Journ. China Br. Roy. Asiat. Soc. 22: 259. (1887) (Chinese Names of Plants), Forbes & Hemsley in Journ. Linn. Soc. Bot. 26: 76. (1889), Henry in Notes Econ. Bot. China 42. (1893), Rehder in Sargent Pl. Wilson. 1: 291. (1913), Zheng in Journ. Wuhan Bot. Res. 2.(1):175. (1984).

Styrax serrulatus Forbes & Hemsley non Roxb. in Journ. Linn. Soc. Bot. 26: 77. (1889).

Hubei: Yichang, 1155. Badong, 1430, 1855, 2815, 5106, 6120, 7318. Liantuo, 1918, 2116, 3876, 3926, 3928, 3943, 4591. Without locality, 5639a. **Sichuan:** South Wushan, 5495, 5639, 5769, 5779, 7039. Henry's Chinese collector with Pratt, no locality stated but may be any of the following, Leshan, Emei Shan, Wa Shan or Kangding, 8882.

A small deciduous tree to 7 metres tall, producing clusters of 3 to 6 pure white flowers from lateral shoots in June. Native to Japan, China and Korea and introduced to cultivation by the Kew collector, Richard Oldham in 1862. According to Henry at Badong this species is colloquially known as *yeh mo li.* A widespread species in central China as can be seen by the number of collections made by Henry.

Styrax spp. (STYRACACEAE)

Anonymous in Kew List of Determinations (volume 24 & 34).

Hubei: Badong, 227 (June 1885). Xingshan, 4510, 4564. Jianshi, 7427. **Yunnan:** Mengzi, 10055, 10710, 13693.

Styrax suberifolius Hook. & Arn. (STYRACACEAE)

Henry in Trans. Asiat. Soc. Jap. 24 suppl: 59. (1896) (List Pl. Formosa), Rehder in Sargent Pl. Wilson. 1: 290. (1913), Li in Woody Flora of Taiwan 750. (1963), Zheng in Journ. Wuhan Bot. Res. 2.(1): 176. (1984).

Styrax caloneurus Perkins in Bot. Jahrb. Syst. 31: 484. (1902). *Styrax suberifolius* Hook. & Arn. var. *calaneurus* (Perk.) Perkins in Engler, Pflanzenr. iv.-241, 61. (1907).

Hubei: South Donghu, 7704. **Taiwan:** Wanchin, 536, 1579. Oluanpi, 592. **Yunnan:** Simao at 1,700 metres, 11885 (isosyntype of *Styrax suberifolius* Hook. & Arn. var. *calaneurus* (Perk.) Perkins. Simao at 1,400 metres, 11885a., (isosyntype of *Styrax suberifolius* Hook. & Arn. var. *calaneurus* (Perk.) Perkins, 11885b., (isotype of *Styrax caloneurus* Perkins).

Distributed in Myanmar, Vietnam and southern China including Taiwan where it is very common at low altitude.

Styrax suberifolius Hook. & Arn. var. *hayataianus* (Perkins) K. Mori (STYRACACEAE) New variety

Styrax suberifolius Henry non Hook. & Arn. in Trans.

Asiat. Soc. Jap. 24 suppl: 59. (1896) (List. Pl. Formosa). *Styrax hayataianus* Perkins in Repert. Spec. Nov. Regni Veg. 8: 82. (1910). *Styrax formosanus* Matsum. var. *hayataianaus* (Perkins) H. L. Li in Woody Flora of Taiwan 753. (1963).

Taiwan: Oluanpi, 1369 (isotype of *Styrax hayataianus* Perkins).

Endemic to Taiwan where it is scattered at low altitudes and is rare.

Styrax tonkinensis (Pierre) Craib ex Hartwich (STYRACACEAE)

Styrax hypoglaucus J. R. Perkins in Engler's Bot. Jahrb. Syst. 486. (1902), Perkins in Engler, Pflanzenr. 241: 82. (1907).

Yunnan: Mountains to the east of Simao at 1,600 metres, 12006 (type of *Styrax hypoglaucus* J. R. Perkins).

Native to China, Thailand, Laos, Cambodia and Vietnam.

Styrophyton caudatum (Diels) Hu (MELASTOMATACEAE) New genus

in Journ. Arnold Arb. 33: 174. (1952) .

Anerincleistus caudatus Diels in Bot. Jahrb. Syst. 65: 101. (1932). *Allomorphia caudata* (Diels) Li in Journ. Arnold Arb. 25: 11. (1944). *Oxyspora spicata* J. F. Maxwell in Gard. Bull. Singapore 35: 218. (1983).

Yunnan: In Pingbian, in forests on the Daweishan Range at 1,650 metres, 10761 (isotype of *Anerincleistus caudatus* Diels).

Styrophyton caudatum is a nom. nov. for *Anerincleistus caudatus* Diels, non *Oxyspora caudata* Geddes. Diels assigned this collection to *Anerincleistus* as fruits were missing on Henry's specimen. Endemic to China and distributed in Guangxi and Yunnan.

Suaeda maritima (L.) Dumort. (CHENOPODIACEAE)

Suaeda nudiflora (Willd.) Miq. Henry in Trans. Asiat. Soc. Jap. 24 suppl: 76. (1896) (List Pl. Formosa).

Taiwan: Kaohsiung, 1025.

A perennial, much-branched herb to 40 cm tall, cosmopolitan in distribution.

Suregada aequorea (Hance) Seem. (EUPHORBIACEAE)

Gelonium aequoreum Hance Henry in Trans. Asiat. Soc. Jap. 24 suppl: 84. (1896) (List Pl. Formosa).

Taiwan: Kaohsiung, without number. Oluanpi, 604, 691.

A small tree, distributed from the Philippines to China (Taiwan) and found on the seashore on the southern part of the island. This species is the *peh-shu*, according to Henry.

Suregada multiflora (A. Juss.) Baill. (EUPHORBIACEAE)

Gelonium aequoreum Hance var. *hainanense* Hemsley in Journ. Linn. Soc. Bot. 26: 444. (1894).

Hainan: Hummocks near Haikou, 8082. Environs of Kiungchow 8222.

Swertia angustifolia Buch.-Ham. ex D. Don (GENTIANACEAE)

Forbes & Hemsley in Journ. Linn. Soc. Bot. 26: 138. (1890).

Hubei: Yichang, 42, 2243, 4343 (in part). Badong, 527 (September 1885).

Swertia angustifolia Buch.-Ham. ex D. Don var. *pulchella* (D. Don) Burkill (GENTIANACEAE)

Swertia pulchella D. Don Anonymous in Kew List of Determinations (volume 34).

Yunnan: Simao, 12449 (in part).

Swertia bimaculata (Sieb. & Zucc.) Hook.f. & Thoms. ex C. B. Clarke (GENTIANACEAE)

Forbes & Hemsley in Journ. Linn. Soc. Bot. 26: 139. (1890), Hemsley in Hooker's Icon. Pl. t. 2462. (1896), Bretschneider in Hist. Eur. Bot. Disc. China 2: 780. (1898).

Hubei: Badong, 11 (collected September 1885, received at Kew on March 11th 1886), 1046 (December 1885), 2546, 3171, 4818, 4834, 4944, 5059, 5927, 7362. Fang Xian, 6591. Xingshan, 6591a. Without locality, 5059a.

A handsome plant to about 75 cm tall with greenish-yellow flowers speckled with small black spots. While travelling with the Glasnevin Central China Expedition in the autumn of 2002 we found this species locally common in Badong (GCCE 211, GCCE 261, GCCE 264) to the south of the gorges and again in Shennongjia and Xingshan in 2004 (GCCE 553 & GCCE 547) to the north of the Three Gorges. Plants raised from seeds collected by Wilson from this area flowered in Veitch's Coombe Wood nursery in the summer of 1905.

Swertia cincta Burkill (GENTIANACEAE) New species

in Journ. Asiatic Soc. Bengal 11: 319. (1906).

Yunnan: Yuanjiang at 2,200 metres, 13216.

Based on collections made in Yunnan by Delavay, Ducloux and Henry. Endemic to China and distributed in Guizhou, Sichuan and Yunnan.

Swertia davidii Franch. (GENTIANACEAE)

Forbes & Hemsley in Journ. Linn. Soc. Bot. 26: 140. (1890).

Hubei: Yichang, 142, 973, 1060, 3263. Liantuo, on the banks of the Yangtze, 4444.

Discovered by Père Armand David in the Yangtze valley in Sichuan in December 1868.

Swertia erythrosticta Maxim. (GENTIANACEAE)

Forbes & Hemsley in Journ. Linn. Soc. Bot. 26: 140. (1890).

Hubei: Xingshan, 6983.

Discovered by the Russian explorer, Nicolai

Mikhailovich Przewalski in the mountains of western Gansu in 1872.

Swertia kouitchensis Franch. (GENTIANACEAE)

In Harvard University Herbarium (GH.).

Sichuan: South Wushan, 7200.

Swertia oculata Hemsley (GENTIANACEAE) New species

in Journ. Linn. Soc. Bot. 26: 140. (1890), Bretschneider in Hist. Eur. Bot. Disc. China 2: 788. (1898).

Sichuan: North Wushan, 7106.

Swertia punicea Hemsley (GENTIANACEAE) New species

in Journ. Linn. Soc. Bot. 26: 140. (1890), Bretschneider in Hist. Eur. Bot. Disc. China 2: 788. (1898).

Hubei: Badong, 2823, 4879, 5112. In the Lung Tung Chi Glen near Yichang (October 1886), 2898 (type). San You Dong Glen, 4366. **Yunnan:** Mengzi, 11312.

This species was collected by the Sino-American Expedition in the Shennongjia Forest District in 1980 and was also collected in neighbouring Xingshan in September 2004 by the Glasnevin Central China Expedition (GCCE 547 & GCCE 573). Endemic to China and distributed in the provinces of Guizhou, Hubei, Hunan, Sichuan and Yunnan.

Swertia spp. (GENTIANACEAE)

Anonymous in Kew List of Determinations (volume 24 & 34).

Hubei: Badong, 3695. **Yunnan:** Mengzi, 9276, 9329.

At the Royal Botanic Gardens, Kew seeds from Henry's *Swertia* sp. 3695 (as 'probably an *Ophelia*') removed and sent for propagation in October 1887.

Swertia tetragona Edgew. (GENTIANACEAE)

Forbes & Hemsley in Journ. Linn. Soc. Bot. 26: 141. (1890).

Hubei: Badong, 3154. **Sichuan:** South Wushan, 7206.

Swertia yunnanensis Burkill (GENTIANACEAE) New species

in J. Asiat. Soc. Bengal 2: 230. (1906).

Yunnan: Milê at 1,829 metres, 9293a., (syntype).

Endemic to China and distributed in the provinces of Guizhou, Sichuan and Yunnan.

Sycopsis sinensis Oliver (HAMAMELIDACEAE) New species

in Hooker's Icon. Pl. 20. tt. 1931. (1890), Henry in Notes Econ. Bot. China 55. (1893), Bretschneider in Hist. Eur. Bot. Disc. China 2: 783. (1898), Bean in Trees & Shrubs 2: 561. (1921), Walker in Journ. Arnold Arb. 25: 337. (1944), Morley in Glasra 3: 80. (1979), Zheng in Journ. Wuhan Bot. Res. 2.(1): 66. (1984), O' Brien in Augustine Henry, An Irish Plant Collector in China 36. (2002).

Corylus sp. Henry in Journ. China Br. Roy. Asiat. Soc. 22: 271. (1887) (Chinese Names of Plants).

Hubei: Jianshi, 6019 (isosyntype). North Donghu, 7574 (isosyntype). Changleping, 7825. **Sichuan:** South Wushan, 7574a.

A small evergreen tree to 7 metres tall producing short dense clusters of yellow flowers in February. This species was common in ravines and rocky places near streams in Hubei at altitudes of up to 1,300 metres in the early 20th century according to Wilson who later introduced it for Messrs Veitch in 1901. In the Veitch catalogue *New Hardy Plants from Western China* (Autumn 1912) this species was available for two shillings and six pence. Henry's Chinese collector from Badong called it the *shui-ssu-li*. There is a fine 7 metre tall tree at the west end of the pond at the National Botanic Gardens, Glasnevin that originates from this introduction. It is capable of growing to twice that height in its native habitat and was collected and reintroduced by the Glasnevin Central China Expedition (GCCE 459 & GCCE 643) from Xingshan in the autumn of 2004.

Symplocos anomala Brand (SYMPLOCACEAE) New species

in Bot. Jahrb. 29: 529. (1900), Rehder in Sargent Pl. Wilson. 2: 596. (1916), Zheng in Journ. Wuhan Bot. Res. 2.(1): 174. (1984).

Symplocos alata Brand Brand in Engler, Pflanzenr. 242: 67. (1901). *Symplocos argentea* Brand in Pflanzenr. (Engler) 4. Fam. 242: 67. (1901).

Hubei: Fang Xian, 6691. Jianshi, 7440. **Yunnan:** In Pingbian, in woods on the Daweishan Range at 1,650 metres, 9034 (isosyntype of *Symplocos argentea* Brand). Mengzi at 2,300 metres, 11340.

Native to Myanmar, China, Thailand, Vietnam and Japan to western Malesia.

Symplocos caudata Wallich ex G. Don (SYMPLOCACEAE)

Rehder in Sargent Pl. Wilson. 2: 595. (1916).

Symplocos prunifolia Forbes & Hemsley non Sieb. & Zucc. in Journ. Linn. Soc. Bot. 26: 74. (1889).

Hubei: Yichang, 1244. Badong, 5119, 5292.

Symplocos cochinchinensis (Lour.) S. Moore (SYMPLOCACEAE)

Symplocos javanica (Blume) Kurz Rehder in Sargent Pl. Wilson. 2: 597. (1916).

Hainan: Without locality, 8492. Ling-men (in the interior of the island), 8621. Environs of Kiungchow, 8713.

Symplocos dryophila C. B. Clarke (SYMPLOCACEAE)

Symplocos longipetiolata Rehder in Sargent Pl. Wilson. 2: 599. (1916).

Yunnan: In forests on the mountains to the north

of Mengzi at 2,600 metres, 10874 (isotype of *Symplocos longipetiolata* Rehder).

Distributed in Nepal, India, Myanmar, western China, Thailand and Vietnam.

Symplocos glandulifera Brand (SYMPLOCACEAE) New species

in Pflanzenr. (Engler) 4. Fam. 242: 68. (1901), Hand.-Mazz. & Peter-Strbal in Beih. Bot. Centralbl. 62 (B): 28. (1943), H. P. Nooteboom in Leiden Bot. Ser. 1: 224. (1975), Wu in Acta Phytotax. Sin. 24: 284. (1986).

Yunnan: In mountain forest to the south of Mengzi, 10699 (isosyntype). In Pingbian, on the Daweishan Range at 1,650 metres, 11260.

Distributed in northern Vietnam and China (Guangxi, Hunan and Yunnan).

Symplocos glauca (Thunb.) Koidz. (SYMPLOCACEAE)

Litsea glauca (Thunb.) Sieb. Henry in Trans. Asiat. Soc. Jap. 24 suppl: 79. (1896) (List Pl. Formosa).

Taiwan: Wanchin, 391, 1585.

Symplocos glomerata King ex C. B. Clarke (SYMPLOCACEAE)

Anonymous in Kew List of Determinations (volume 34).

Yunnan: Mengzi, 10460, 11456.

Native to India, Bhutan and China.

Symplocos lancifolia Sieb. & Zucc. (SYMPLOCACEAE)

Henry in Trans. Asiat. Soc. Jap. 24. suppl: 58. (1896) (List Pl. Formosa), Wu in Acta Phytotax. Sin. 24: 281. (1986).

Symplocos formosana Brand in Engler, Pflanzenr. 6 (iv. 242.): 41. (1901), Li in Woody Flora of Taiwan 740. (1963).

Taiwan: Wanchin, 32, 127 (syntype of *Symplocos formosana* Brand), 516 (syntype of *Symplocos formosana* Brand), 1508 (syntype of *Symplocos formosana* Brand), 1564.

Symplocos lucida (Thunb.) Sieb. & Zucc. (SYMPLOCACEAE)

Symplocos japonica Forbes & Hemsley non A. DC. in Journ. Linn. Soc. Bot. 26: 73. (1889), Brand in Engler, Pflanzenr. 242: 31. (1901). *Symplocos henryi* Brand in Pflanzenr. (Engler) 4. Fam. 6 (iv. 242): 67. (1901), Hand.-Mazz. & Peter-Stibal in Beih. Bot. Centralbl. 62 (B): 16. (1943), Wu in Acta Phytotax. Sin. 24: 277. (1986). *Symplocos sinuata* Brand in Repert. spec. Nov. Regni. Veg. 14: 326. (1916).

Hubei: Badong, 2843, 3730, 5290. **Yunnan:** In Pingbian, in forests on the Daweishan Range at 1,650 metres, 11415 (lectotype of *Symplocos sinuata* Brand). Near Yuanjiang at 1,650 metres 13372 (type of *Symplocos sinuata* Brand), 13401.

Widely dispersed in Asia.

Symplocos macrophylla Wall. ex DC. subsp. *sulcata* (Kurz.) Noot. (SYMPLOCACEAE)

Symplocos yunnanensis Brand in Pflanzenr. (Engler) 6. (iv. 242.): 68. (1901), Hand.-Mazz. & Peter Stibal in Beih. Bot. Centralbl. 62. (B.) 28. (1943), Wu in Acta Phytotax. Sin. 24: 280. (1986).

Yunnan: In the mountains to the east of Simao at 1,650 metres, 12014 (isotype of *Symplocos yunnanensis* Brand). Simao, 12619.

Distributed in China (Yunnan) and Indochina.

Symplocos paniculata (Thunb.) Miq. (SYMPLOCACEAE)

Rehder in Sargent Pl. Wilson. 2: 593. (1916), Zheng in Journ. Wuhan Bot. Res. 2.(1): 174. (1984).

Symplocos crataegoides Buch.-Ham Henry in Journ. China Br. Roy. Asiat. Soc. 22: 243. (1887), Forbes & Hemsley in Journ. Linn. Soc. Bot. 26: 72. (1889), Henry in Notes Econ. Bot. Econ. China 55. (1893); in Trans. Asiat. Soc. Jap. 24 suppl: 58. (1896) (List Pl. Formosa), Brand in Engler, Pflanzenr. 242: 33. (1901). *Symplocos sinica* Ker. Forbes & Hemsley in Journ. Linn. Soc. Bot. 26: 74. (1889).

Hubei: Yichang, 175, 270, 270a., 1125, 1149, 1226, 2301, 2301a. Badong, 732 (June 1885), 1762, 2825, 2853, 3145, 6307. Liantuo, 1920, 2998, 4592. On Tsui Fu Shan near Yichang (May 1888), 1599. San You Dong Glen near Yichang (July 1886), 1499. Changleping, 6297, 7723. South Badong, 5755a., 5755c. Changyang, 5755c. North Badong, 7397. **Sichuan:** South Wushan, 5492, 5755, 5760. Henry's Chinese collector with Pratt, no locality stated but may be any of the following, Leshan, Emei Shan, Wa Shan or Kangding, 8864. **Taiwan:** Oluanpi, 635. **Yunnan:** South of the Red River from Manmei at 2,000 metres, 9466. Mengzi at 1,600 to 2,300 metres, 9466b., 9948b., 10554 In mountain forest in Milê District, 9948. Simao at 1,600 metres, 11671.

A deciduous shrub or a small tree bearing small white scented flowers in terminal panicles during May and early June. The chief glory of this shrub however, are the turquoise-blue fruits carried in autumn and it is necessary to grow at least two clones for these to form. Native to the Himalaya, China (including Taiwan) and Japan, according to Henry it was known colloquially in the Three Gorges region as *he-tzu-yeh*. Of very widespread distribution in China as can be seen from the localities of Henry's specimens. This species was collected by the Glasnevin Central China Expedition (GCCE 511) from Xingshan in September 2004.

Symplocos pilosa Rehder (SYMPLOCACEAE) New species

in Sargent Pl. Wilson. 2: 598. (1916).

Yunnan: In mountain forest to the south of Mengzi at 2,600 metres, 10698 (type, collected 15 March 1897).

Endemic to Yunnan.

02

Symplocos pseudobarberina Gontscharow (SYMPLOCACEAE) New species

in Bot. Mater. Gerb. Glavn. Bot. Sada RSFSR 5: 133. (1924).

Yunnan: Near Yuanjiang in forests on Fengchunlin at 2,300 metres, 11204 (isotype).

Native to China, Cambodia and Vietnam.

Symplocos racemosa Roxb. (SYMPLOCACEAE)

Symplocos intermedia Brand in Fedde, Repert. Nov. Spec. Regni Veg. 3: 217. (1907), Brand in Sinensia 5: 5. (1934).

Yunnan: Simao at 1,600 metres, 12503 (isotype of *Symplocos intermedia* Brand), 12503a., (type of *Symplocos intermedia* Brand). Mengzi, 9147.

Native to Nepal, India, Myanmar, western China, Thailand and Vietnam.

Symplocos spectabilis Brand (SYMPLOCACEAE) New species

in Pflanzenr. (Engler) 4. Fam. 242: 69. (1901).

Yunnan: Near Yuanjiang, in forests on Fengchunling at 2,400 metres, 10844 (isotype).

Native to Myanmar and China (Yunnan).

Symplocos spp. (SYMPLOCACEAE)

Henry in Trans. Asiat. Soc. Jap. 24 suppl: 58. (1896) (List Pl. Formosa), Anonymous in Kew List of Determinations (vol. 24. & 34.).

Hubei: Fang Xian, 6911. **Sichuan:** South Wushan, 7094. **Taiwan:** Oluanpi, 1315 (in part). **Yunnan:** Mengzi, 10671, 11271, 11422. Simao, 12412, 13280, 13480, 13525.

Symplocos sumuntia Buch.-Ham. ex D. Don (SYMPLOCACEAE)

Symplocos leucophylla Brand in Englers. Pflanzenr. iv. 242. (Heft 6): 60. (1901). *Symplocos botryantha* Franch. Brand in Engler, Pflanzenr. 242: 60. (1901), Rehder in Sargent Pl. Wilson 2: 596.(1916); 3: 455 (1917), Wu in Acta Phytotax. Sin. 24: 278. (1986). *Symplocos caudata* Wallich var. *maculata* Brand in Engler, Pflanzenr. 242: 42. (1901).

Hubei: Changyang, 5242, 7486 (isotype of *Symplocos leucophylla* Brand). Badong, 5319. South Badong 5357. **Sichuan:** South Wushan 5242a., 5242b., 5272b. **Yunnan:** Simao, 12841.

Native to the Himalaya, India, China, Korea, Thailand and Vietnam.

Symplocos theophrastifolia Sieb. & Zucc. (SYMPLOCACEAE)

Li in Woody Flora of Taiwan 738. (1963).

Symplocos spicata Henry non Roxb. in Trans. Asiat. Soc. Jap. 24 suppl: 58. (1896) (List Pl. Formosa).

Taiwan: Oluanpi, 1315 (in part).

Syneilesis aconitifolia (Bunge) Maxim. (ASTERACEAE)

Senecio aconitifolius Turcz. Anonymous in Kew List of Determinations (volume 24).

Hubei: Fang Xian, 6575.

A very beautiful woodland perennial, bearing remarkable deeply divided peltate leaves, which are covered in a dense white pubescence on emergence.

Synotis alata (Wall. ex DC.) C. Jeffrey & Y. L. Chen (ASTERACEAE)

Senecio alatus Wall. ex DC. Anonymous in Kew List of Determinations (volume 34).

Yunnan: Mengzi, 11159.

Synotis cappa (Buch.-Ham. ex D. Don) C. Jeffrey & Y. L. Chen (ASTERACEAE)

Jeffrey & Chen in Kew Bull. 39: 319. (1985).

Senecio ionodasys Hand.-Mazz. in Notizbl. Bot. Gart. Berlin 13: 637. (1937), S. Y. Hu in Quart. Journ. Taiwan Mus. 21: 51. (1968). *Synotis ionodasys* (Hand.-Mazz.) C. Jeffrey & Y. L. Chen in Kew Bull. 39: 320. (1984).

Yunnan: Mengzi, 9810. South of the Red River near Manpan at 1,830 metres, 10332, 10332a., (isotype of *Senecio ionodasys* Hand.-Mazz.). In Pingbian, in forests on the Daweishan Range at 1,640 metres, 11429.

Native to Nepal, India, Bhutan, Myanmar, western China and Thailand.

Synotis erythroppa (Bur. & Fr.) C. Jeffrey & Y. L. Chen (ASTERACEAE)

Jeffrey & Chen in Kew Bull. 39: 324. (1984).

Senecio glumaceus Dunn in Journ. Linn. Soc. Bot. 35: 505. (1903).

Sichuan: Near Kangding, Henry's Chinese collector with Pratt, 8919 (syntype of *Senecio glumaceus* Dunn), 8921 (syntype of *Senecio glumaceus* Dunn).

Discovered in western Sichuan in 1889 by Gabriel Bonvalot and Prince Henri d' Orleon. Henry's specimens were collected at the same time (1889).

Synotis nagensium (C. B. Clarke) C. Jeffrey & Y. L. Chen (ASTERACEAE)

Jeffrey & Chen in Kew Bull. 39: 320. (1985) .

Yunnan: Milê, 9810a.

Native to India, Myanmar and China.

Synotis rufinervis (DC.) C. Jeffrey & Y. L. Chen (ASTERACEAE)

Senecio rufinervis DC. Anonymous in Kew List of Determinations (volume 34).

Yunnan: Simao, 12759.

Synotis saluenensis (Diels) C. Jeffrey & Y. L. Chen (ASTERACEAE)

Jeffrey & Chen in Kew Bull. 39: 330. (1985).

Yunnan: Mengzi, 9086a.

A scrambling sub-shrub to 3 metres tall. The species was based on material gathered by George Forrest near Lijiang in Yunnan, but appears to have been first collected by Père Delavay (Delavay 477). Distributed in Myanmar, China [Guangxi, Guizhou, Tibet (Xizang) and Yunnan] and Vietnam.

Synotis triligulata (Buch.-Ham ex D. Don) C. Jeffrey & Y. L. Chen (ASTERACEAE)

Jeffrey & Chen in Kew Bull. 39: 329. (1985).

Yunnan: In Pingbian on the Daweishan Range at 1,950 metres, 9086. In the mountains to the east of Simao at 1,650 metres, 12779. In forests to the west of Simao at 1,300 metres, 12779a.

A widespread shrubby herb or sub-shrub to 1.5 metres tall, native to Nepal, north-east India, Bhutan, Nepal, Myanmar, China, [Tibet (Xizang) and Yunnan] and Thailand.

Synurus deltoides (Aiton) Nakai (ASTERACEAE)

Serratula pungens Henry non Franch. & Sav. in Journ. China Br. Roy. Asiat. Soc. 22: 266. (1887) (Chinese Names of Plants). *Serratula atriplicifolia* (Fisch. ex Trevir.) Sch. Bip. Forbes & Hemsley in Journ. Linn. Soc. Bot. 23: 468. (1888).

Hubei: Badong, 430 (June 1885), 2455, 5111. Liantuo, 4618.

A striking herbaceous plant with cordate leaves, white on the undersides and producing numerous globular purple flowers on upright stems to 1.5 metres tall. According to Henry this species was known colloquially in the Three Gorges region as *peh yeh tzu*. *Synurus deltoides* was introduced to cultivation by Wilson through Veitch's Coombe Wood nursery in 1901, in their *Novelties for 1908—1909* plants were available for a shilling each or ten shillings and six pence per dozen.

Syringa komarowii C. K. Schneid. subsp. *reflexa* (C. K. Schneid.) P. S. Green & M. C. Chang (OLEACEAE) New subspecies

Syringa reflexa Schneider in Fedde, Repert. Spec. Nov. Regni Veg. 9: 81. (1911), Schneider in Illus. Handb. Laubholzk. 2: 779. (1911), Zheng in Journ. Wuhan Bot. Res. 2.(1): 182. (1984), Nelson in Pim, The Wood & the Trees 235. (1984), Bean in Trees & Shrubs 4: 539. (1989), O' Brien in Augustine Henry, An Irish Plant Collector in China 36. (2002).

Hubei: Fang Xian at 2,650 to 2,950 metres, 6819.

A deciduous shrub to 4 metres tall, flowers purplish-pink, produced in June in a densely packed pendelous pyramidal panicle to 25 cm long. Introduced to cultivation by Wilson in 1910. One of the most beautiful of lilacs but unfortunately not scented.

Syringa pubescens Turcz. subsp. *microphylla* (Diels) M. C. Chang & Y. L. Chen (OLEACEAE)

New subspecies. Zheng in Journ. Wuhan Bot. Res. 2.(1): 182. (1984).

Syringa dielsiana C. K. Schneid. C. K.Schneider in Illus. Handb. Laubholzk. 2: 778. (1912). *Syringa microphylla* Diels Zheng in Journ. Wuhan Bot. Res. 2.(1): 182. (1984).

Hubei: North Badong, 6985.

A deciduous shrub to 1.5 metres tall, producing panicles of fragrant pinkish-lilac flowers in June. Diels based his *Syringa microphylla* on a collection made by the Italian missionary Giraldi in 1893. It had infact, been discovered by Henry in 1888, some five years previously. *Syringa pubescens* subsp. *microphylla* was introduced to Veitch's Coombe Wood nursery by William Purdom in 1910.

Syzygium balsameum (Wight.) Wall. ex Walp. (MYRTACEAE)

Merrill & Perry in Journ. Arnold Arb. 19: 229. (1938).
Yunnan: Simao, 12682, 12798.
Distributed in Yunnan.

Syzygium buxifolium Hook. & Arn. (MYRTACEAE)

Merrill & Perry in Journ. Arnold Arb. 19: 234. (1938).
Eugenia sinensis Hemsl. Forbes & Hemsley in Journ. Linn. Soc. Bot. 23: 298. (1887), Diels in Bot. Jahrb. 29: 484. (1900).

Hubei: Yichang, 4149. Changyang, 7605, 7758.

Syzygium claviflorum (Roxb.) Wall. ex Steud. (MYRTACEAE)

Syzygium leptanthum (Wight) Niedenzu Merrill & Perry in Journ. Arnold Arb. 19: 222. (1938).
Yunnan: Simao, 12860, 12921, 12921a.

Syzygium cumini (L.) Skeels (MYRTACEAE)

Merrill & Perry in Journ. Arnold. Arbor. 19: 231. (1938).

Eugenia jambolana Lam. Anonymous in Kew List of Determinations (volume 24). *Syzygium fruticosum* (Roxb.) DC. Merrill & Perry in Journ. Arnold Arb. 19: 231. (1938).

Hainan: Environs of Haikou, 8061. Environs of Kiungchow, 8695. **Yunnan:** South of Mengzi on the banks of the Red River at Hsinkai, 9644. In the Red River valley near Manpan, 10666. Simao, 11782a., 11782b., 11782c.

The Jambolan or Malabar plum, a tropical tree to about 30 metres tall. Native to the region from India to Java, this species has become a weed tree in many parts of the humid tropics.

Syzygium fluviatile (Hemsl.) Merr. & L. M. Perry (MYRTACEAE)

Eugenia fluviatilis Hemsley Anonymous in Kew List

of Determinations (volume 24).

Hainan: Without locality, 8478. Ling-men (in the interior of the island), 8604.

Discovered by the American missionary, the Rev'd B. C. Henry in Hainan in 1881.

Syzygium formosanum (Hayata) Mori (MYRTACEAE) New species

Eugenia sp. Henry in Trans. Asiat. Soc. Jap. 24 suppl: 43. (1896) (List Pl. Formosa). *Syzygium acutisepalum* (Hayata) Mori Li in Woody Flora of Taiwan 649. (1963).

Taiwan: Oluanpi, 959, 1312.

A small tree, endemic to Taiwan where it was first collected by Henry.

Syzygium forrestii Merrill & L. M. Perry (MYRTACEAE) New species

in Journ. Arnold Arb. 19: 238. (1938).

Yunnan: Simao, 11764, 11764a.

The specific epithet commemorates George Forrest who also later collected this species in Yunnan. Henry described the trees he collected from as being between 6 and 12 metres tall with creamy-yellow flowers. Endemic to Yunnan.

Syzygium globiflorum (Craib) Chantar. & J. Parn. (MYRTACEAE)

Syzygium brachyantherum Merrill & Perry in Journ. Arnold. Arb. 19: 218. (1938).

Yunnan: Simao, 12091, 12091a., 12091b., 12651.

Syzygium hancei Merrill & L. M. Perry (MYRTACEAE)

Syzygium szemaoense Merrill & L. M. Perry in Journ. Arnold Arb. 19: 105. (1938).

Yunnan: Simao, 12138 (type *Syzygium szemaoense* Merrill & L. M. Perry), 12895.

Endemic to China and distributed in Guangxi and Yunnan.

Syzygium handelii Merrill & L. M. Perry (MYRTACEAE) New species

in Journ. Arnold Arb. 19: 233. (1938), Zheng in Journ. Wuhan Bot. Res. 2.(1): 159. (1984).

Hubei: San You Dong Glen (October 1886), 2886.

Syzygium jambos (L.) Alston (MYRTACEAE)

Eugenia jambos L. Henry in Trans. Asiat. Soc. Jap. 24 suppl: 43. (1896) (List Pl. Formosa).

Taiwan: Kaohsiung, without number. Wanchin, 399a.

The rose apple, an evergreen tree to 12 metres tall. Naturalised in Taiwan and over much of the tropics in both the Old and the New World where it has in some parts became a major pest and threatens the survival of native vegetation. According to Henry this species was known as *lien-pu,* which is said to be a corruption of *na-mo.*

Syzygium aff. *latilimbum* (Merr.) Merr. & L. M. Perr. (MYRTACEAE)

Merrill & Perry in Journ. Arnold Arb. 29: 216. (1938).

Yunnan: Simao, 11945.

Syzygium malaccense (L.) Merr. & L. M. Perry (MYRTACEAE)

Eugenia malaccensis L. Henry in Trans. Asiat. Soc. Jap. 24 suppl: 43. 1896 (List Pl. Formosa).

Taiwan: Wanchin, 399. Kaohsiung Plain, 1821.

The Malay apple, a tropical tree to 15 metres tall and native from the Malay Archipelago to Australia. Introduced and cultivated in Taiwan for its edible fruits. *Syzygium malaccense* was introduced to Jamaica by Captain Bligh.

Syzygium polypetaloideum Merrill & L. M. Perry (MYRTACEAE) New species

in Journ. Arnold. Arbor. 19: 217. (1938).

Yunnan: On the banks of the Red River at Manpan, 10716, 10716a., (type).

Endemic to China and distributed in Guangxi and Yunnan.

Syzygium sp. Hainan No. 1. (MYRTACEAE)

Eugenia cf. *operculata* Roxb. Anonymous in Kew List of Determinations (volume 24).

Hainan: Ling-men, 8646.

Syzygium sp. Hainan No. 2. (MYRTACEAE)

Eugenia cf. *polyantha* WightAnonymous in Kew List of Determinations (volume 24).

Hainan: Environs of Kiungchow, 8717.

Syzygium sp. Hainan No. 3. (MYRTACEAE)

Eugenia cf. Ford. 286. Anonymous in Kew List of Determinations (volume 24).

Hainan: Without locality, 8487.

Syzygium sp. Hainan No. 4. (MYRTACEAE)

Eugenia sp. Anonymous in Kew List of Determinations (volume 24).

Hainan: Hummocks near Haikou, 8083, 8086. Environs of Kiungchow, 8218. Without locality, 8477. Ling-men (in the interior of the island), 8647.

Syzygium sp. No. 1. (MYRTACEAE)

Anonymous in Kew List of Determinations (volume 24).

Hubei: San You Dong Glen, 4160.

Syzygium sp. No. 2. (MYRTACEAE)

Anonymous in Kew List of Determinations (volume 34).

Yunnan: In the Red River valley, 11064. Simao, 12764.

Syzygium sp. No. 3. (MYRTACEAE)

Eugenia sp. Henry in Trans. Asiat. Soc. Jap. 24 suppl: 43. (1896) (List Pl. Formosa).

02

Taiwan: Oluanpi, 1711.

Syzygium tephrodes (Hance) Merr. & L. M. Perry (MYRTACEAE)

Merrill & Perry in Journ. Arnold Arb. 29: 223. (1938).

Hainan: Environs of Kiungchow, 8258.

Discovered by the Rev'd B. C. Henry in Hainan in November 1882.

Syzygium tetragonum (Wight.) Wall. ex Walp. (MYRTACEAE)

Merrill & Perry in Journ. Arnold Arb. 19: 218. (1938).

Yunnan: Simao, 12650, 12650a., 12650c.

Native to Nepal, India, Bhutan and China (Guangxi, Hainan and Yunnan), and Indochina.

Syzygium toddalioides (Wight) Walp. (MYRTACEAE)

in Journ. Arnold. Arbor. 19: 231. (1938).

Yunnan: In the mountains to the south of Simao at 1,650 metres, 11782 (type of *Syzygium angustinei* Merrill & L. M. Perry).

Syzygium yunnanense Merrill & L. M. Perry (MYRTACEAE) New species

in Journ. Arnold Arbor. 19: 227. (1938).

Yunnan: Simao, 12938 (isotype).

Endemic to Yunnan.

Tabernaemontana corymbosa Roxb. ex Wall. (APOCYNACEAE)

Ervatamia yunnanensis Tsiang in Acta Phytotax. Sin. 8(3): 242. (1963).

Yunnan: Mountains to the east of Simao at 1,450 metres, 12026 (isotype of *Ervatamia yunnanensis* Tsiang).

Native to Myanmar, western China, Thailand, Laos and Vietnam.

Tabernaemontana divaricata (L.) R. Br. ex Roem ex Schult. (APOCYNACEAE)

Tabernaemontana coronaria (L.) R. Br. Henry in Trans. Asiat. Soc. Jap. 24 suppl: 60. (1896) (List Pl. Formosa).

Hainan: Without locality, 8497. **Taiwan:** Wanchin, 545. Kaohsiung, 1876.

The crepe jasmine, a shrub to 1.5 metres tall bearing cymes of white fragrant tubular flowers. Distributed from northern India to China (Yunnan) and northern Thailand.

Tabernaemontana pandacaqui Poir. (APOCYNACEAE)

Tabernaemontana cumingiana A. DC. Henry in Trans. Asiat. Soc. Jap. 24 suppl: 60. (1896) (List Pl. Formosa).

Taiwan: Wanchin, without number. Oluanpi, 355.

A shrub or an evergreen tree to 14 metres tall carrying cymes of slightly scented white tubular flowers. Distributed from south-east Asia to Australia. According to Henry this species was known as *shan-ma-ti-hua* at Wanchin.

Tabernaemontana sp. (APOCYNACEAE)

Anonymous in Kew List of Determinations (volume 24).

Hainan: Ling-men, 8628.

Tacca chantrieri Andre (DIOSCOREACEAE)

Tacca paxiana H. Limpricht in Engler, Pflanzenr. 42: 16. (1928).

Yunnan: Simao at 1,220 to 1,525 metres, 12174 (type of *Tacca paxiana* Limpricht), 12592, 12592a.

The devil flower, bat plant or cat's whiskers, a spectacular perennial herb carrying a terminal inflorescence of purple black tubular flowers held within a purple-black bract in. Distributed in Cambodia, Laos, Thailand, Vietnam and southern China. In China the bat plant is known as 'tiger's paws' because of the flowers slender pendelous whisker-like bracteoles. The poisonous rhizomes are used in traditional Chinese herbal medicine. The Hani (an ethnic group) of Xishuangbanna in southern Yunnan call this species *beepavqnav* and the roots, leaves and stem are used in cases of gastroenteritis, dysentry, indegestion, hepatitis and lung infections. Still abundant in the tropical rainforests of Xishuangbanna in southern Yunnan.

Tadehagi triquetrum (L.) H. Ohashi (FABACEAE)

Desmodium triquetrum (L.) DC. Henry in Trans. Asiat. Soc. Jap. 24 suppl: 34. (1896) (List Pl. Formosa).

Hainan: Ling-men, 8660. **Taiwan:** Oluanpi, 336. Wanchin, 873. **Yunnan:** On the Daweishan Range in Pingbian at 1,500 metres, 9072.

Distributed in India, Sri Lanka, Myanmar, China and Indochina.

Tagetes erecta L. (ASTERACEAE)

Tagetes patula L. Henry in Trans. Asiat. Soc. Jap. 24 suppl: 54. (1896) (List Pl. Formosa).

Taiwan: Kaohsiung, without number. Oluanpi, 577. **Yunnan:** Mengzi, 11221. Simao, 12560.

The Aztec marigold, a native of Mexico and Central America. Cultivated throughout the world and naturalised in parts of Asia.

Tainia sp. (ORCHIDACEAE)

Anonymous in Kew List of Determinations (volume 34).

Yunnan: Simao, 12078, 13614, 13615.

Tamarindus indica L. (FABACEAE)

Anonymous in Kew List of Determinations (volume 34).

Yunnan: In the Red River valley, 9674.

The tamarind tree, a tropical xerophytic tree of unknown origin but thought to be from tropical Africa. The tamarind has many domestic and culinary uses, the pulp from its fruits is a rich sourse of vitamin C and is also used medicinally. Worcester sauce is made from its fruit pulp and when overripe, the pulp is used to clean copper and brass.

Tapiscia sinensis Oliver (TAPISCIACEAE) New genus

in Hooker's Icon. Pl. 20: t. 1928. (1890), Bretschneider in Hist. Eur. Bot. Disc. China 2: 781. (1898), Diels in Bot.

Jahrb. xxix. 448. (1900), Bean in Kew Bull. Misc. Inf. 356. (1909), Schneider in Illus. Handb. Laubholzk. 2: 1026. fig. 607. (1912), Rehder & E. H. Wilson in Sargent Pl. Wilson 2: 188. (1916).

Sichuan: Emei Shan, Henry's Chinese collector with Pratt, 8990 (isotype). **Yunnan:** Simao at 1,600 metres, 13151. On the bank of the Red River near Manpan, 10849.

A monotypic tree from 15 to 30 metres tall with a girth of 3.5 metres. Rare in western Hubei but according to Wilson it is common on Emei Shan and the largest trees occur there. The minute yellow flowers are carried on axillary panicles and emit a pleasant honey-like fragrance. The pinnate leaves turn golden yellow in autumn. Augustine Henry stated that *Tapiscia sinensis* was known as *yin-chi shu* in Hubei and it was later introduced to cultivation by E. H. Wilson through the Arnold Arboretum in 1908. The generic name is an anagram of *Pistacia* to which this species bears a resemblance, it is tender in Britain and Ireland and as a consequence is rare in cultivation.

Taraxacum mongolicum Hand.-Mazz. (ASTERACEAE)
Taraxacum officinale Henry non (L.) Weber ex F. H. Wigg. in Journ. China Br. Roy. Asiat. Soc. 22: 250. (1887) (Chinese Names of Plants), Forbes & Hemsley in Journ. Linn. Soc. Bot. 23: 478. (1888).

Hubei: Yichang, 147, 1267, 1268. Badong 4712.

According to Henry, at Yichang the Chinese dandelion was known colloquially as *k'u ts'ai* and was also known as the *pu cong ying*. In Chinese herbal medicine this species is currently used to treat boils and sores, inflammation of the eye, sore-throat, lung abcess and appendicitis.

Tarenna attenuata (Hook. f..) Hutch. (RUBIACEAE)
Tarenna sylvestris Hutchinson in Sargent Pl. Wilson. 3: 411. (1916).

Hainan: Environs of Haikou, 8060. **Yunnan:** Mengzi, 10116 (isotype of *Tarenna sylvestris* Hutchinson). In woods near Mengzi at 1,520 metres, 10006, 10006a., (type of *Tarenna sylvestris* Hutchinson), 10006c.

Native to India, China and Vietnam.

Tarenna depauperata Hutchinson (RUBIACEAE) New species
in Sargent Pl. Wilson. 3: 411. (1916).

Yunnan: In woods near Mengzi at 1,640 metres, 10816 (type).

Native to China and Vietnam.

Tarenna gracilipes (Hayata) Ohwi (RUBIACEAE) New species
Li in Woody Flora of Taiwan 871. fig. 352. (1963).

Webera attenuata Henry non Hook. f. in Trans. Asiat. Soc. Jap. 24 suppl: 50. (1896) (List Pl. Formosa).

Taiwan: Wanchin, 29, 89, 92, 519, 551. Oluanpi, 943,

991, 1217, 2061.

Native to Japan and China (Taiwan), first collected southern Taiwan by Henry.

Tarenna pubinervis Hutchinson (RUBIACEAE) New species
in Sargent Pl. Wilson. 3: 411. (1916).

Yunnan: In Pingbian, in forests on the Daweishan Range at 1,640 metres, 10059. At Fengchunling to the south of the Red River at 2,640 metres, 10678 (type).

Native to western China and Vietnam.

Tarenna spp. (RUBIACEAE)
Webera sp. Anonymous in Kew List of Determinations (volume 34).

Yunnan: Mengzi, 10724, 10910. Simao, 13435.

Tarennoidea wallichii (Hook. f.) Tirveng & Sastre (RUBIACEAE)
Tarenna pallida (Franch. ex Brandis) Hutch. Hutchinson in Sargent Pl. Wilson. 3: 410. (1916).

Yunnan: Near Manpan in the Red River valley at 820 metres, 10686. Simao, 11923c., 11923f.

Widely distributed in Asia.

Tateishia concinna (DC.) H. Oshashi & K. Oshashi (FABACEAE)
Desmodium concinnum DC. Anonymous in Kew List of Determinations (volume 34).

Yunnan: Simao, 12436.

A widespread species found over much of the Himalaya, India, Myanmar and south-west China.

Taxillus chinensis (DC.) Dancer (LORANTHACEAE)
Loranthus estipitatus Stapf. Forbes & Hemsley in Journ. Linn. Soc. Bot. 26: 405. (1894).

Hainan: Environs of Kiungchow, 8261.

Taxillus limprichtii (Grunning) H. S. Kiu var. *longiflorus* (Lecompte) H. S. Kiu (LORANTHACEAE)
New variety in Acta Phytotax. Sin. 21: 178. (1983).

Loranthus estipitatus Stapf var. *longiflorus* Lecompte in Sargent Pl. Wilson. 3: 316. (1916).

Yunnan: Mengzi at 2,150 metres, 10057 (type of *Loranthus estipitatus* Stapf var. *longiflorus* Lecompte).

Taxillus pseudochinensis (Yamam.) Danser (LORANTHACEAE)
Loranthus sp. Henry in Trans. Asiat. Soc. Jap. 24 suppl: 80. (1896) (List Pl. Formosa). *Loranthus chinensis* A. DC. var. *formosanus* Lecompte in Sargent Pl. Wilson. 3: 316. (1916).

Taiwan: Oluanpi, 979, 1303 (type of *Loranthus chinensis* A. DC. var. *formosanus* Lecompte).

Endemic to southern Taiwan where it was first collected by Henry.

Taxillus sutchuenensis (Lecompte) Dans. (LORANTHACEAE) New species

Danser in Blumea 2: 53. (1936).

Loranthus yadoriki Forbes & Hemsley non Sieb. ex Maxim. in Journ. Linn. Soc. Bot. 26: 407. (1894).

Hubei: Yichang, 1066, 1176 (on a mulberry tree), 2190. Liantuo, 3200, 3905. Antelope Glen, 3496. Badong, 2496. Jianshi, 5902, 5902a. Changyang, 5902b. South Donghu, 7701.

A semi-parasitic shrub. This species was collected by the Glasnevin Central China Expedition in Xingshan in September 2004. There it grew on trees of *Castanea seguinii*, the undersides of the leaves are beautifully fawn-brown and velvety. Originally described by Lecompte from Farges 444, this species was first found in Central China by Henry who stated the host trees were oak, elm, crab and mulberry. Henry stated it was coloquially known at Yichang as *sang chi-sheng*.

Taxillus sutchuenensis (Lecompte) Dans. var. duclouxii (Lecompte) H. S. Kiu (LORANTHACEAE)

Taxillus yadoriki (Sieb. ex Maxim.) Danser Zheng in Wuhan Bot. Res. 2.(1): 35. (1984).

Hubei: Yichang, without number.

Taxillus vestitus (Wall.) Danser (LORANTHACEAE)

Loranthus vestitus Wall. Zheng in Journ. Wuhan Bot. Res. 2.(1): 35. (1984). *Loranthus yadoriki* Sieb. & Zucc. Forbes & Hemsley non Sieb. ex Maxim. in Journ. Linn. Soc. Bot. 407. (1894).

Hubei: South Donghu, 7597. Yichang, 7876.

On oak, elm, crab and mulberry trees.

Taxotrophis sp. (MORACEAEE)

Forbes & Hemsley in Journ. Linn. Soc. Bot. 26: 454. (1894).

Hainan: Environs of Kiungchow, 8703.

Taxus chinensis (Pilg.) Rehder (TAXACEAE) New species

in Journ. Arnold Arb. 1: 51. (1920), Rehder in Journ. Arnold Arb. iv. 119. (1923), Rehder in Bailey, Cult. Evergreens. 187. (1923), Dallimore & Jackson in Handb. Conif. 71. (1923).

Taxus baccata Masters non L. in Journ. Linn. Soc. Bot. 26: 546. (1902). *Taxus baccata* L. ssp. *cuspidata* (Sieb. & Zucc.) Pilg. var. *chinensis* Pilger in Engler, Pflanzenr 4. Fam. 5: 112. (Taxaceae) (1903), Patschke in Bot. Jahrb. xlviii. 630. (1913), Henry in Elwes & Henry, Trees Gr. Brit. & Irel. i. 108. (1906). *Taxus cuspidata* Sieb. & Zucc. var. *chinensis* (Pilg.) Rehder & E. H. Wilson in Sargent Pl. Wilson 2: 8. (1916). *Taxus wallichiana* Zucc. var. *chinensis* (Pilger) Florin Bean in Trees and Shrubs 4: 570. (1989).

Hubei: Fang Xian, 6913. **Sichuan:** North Wushan

7097, 7155 (type of *Taxus baccata* L. ssp. *cuspidata* Sieb. & Zucc. var. *chinensis* Pilger).

The Chinese name of this tree is *houngtocha*, Henry described the trees he encountered as being 3 to 7 metres tall but of large diameter and with good timber. In the *Trees of Great Britain and Ireland* Henry went on to state that this species was very rare in the mountains of Hubei and Sichuan and it grew there on wooded cliffs between 1,800 and 2,500 metres. The largest tree he encountered on his travels was 6 metres tall with a girth of 2.4 metres and he stated it was called the *kuan-yin-sha* or 'the fir of the goddess of mercy'. The Chinese yew, according to Wilson occured scattered through western Hubei and Sichuan and was especially common where carboniferous limestone prevailed. It is interesting to note and compare that in Ireland our native yew, *Taxus baccata* L. grows abundantly on limestone outcrops at Muckross House near Killarney, County Kerry. *Taxus chinensis* was collected by the Glasnevin Central China Expedition (GCCE 799) on Tianthu Shan in Changyang in October 2004.

Tectaria cicutaria (L.) Copel. (POLYPODIACEAE)

Nephrodium cicutarium (L.) Baker Henry in Trans. Asiat. Soc. Jap. 24 suppl: 113. (1896) (List Pl. Formosa).

Taiwan: Wanchin, 192.

Tectaria decurrens (C. Presl.) Copel. (POLYPODIACEAE)

Nephrodium decurrens (C. Presl.) Baker Henry in Trans. Asiat. Soc. Jap. 24 suppl: 113. (1896) (List Pl. Formosa).

Taiwan: Wanchin, 194.

Tectaria dissecta (G. Forst.) Lellinger (POLYPODIACEAE)

Aspidium dissectum (G. Forst.) Christ Christ in Bull. Herb. Boiss. 7: 17. (1899).

Yunnan: In woods to the south of Mengzi near the Red River, 11530.

Tectaria membranacea (Hook.) Fraser-Jenk. & Kholia (POLYPODIACEAE)

Aspidium membranaceum Hook. Henry in Trans. Asiat. Soc. Jap. 24 suppl: 113. (1896) (List Pl. Formosa).

Taiwan: Kaohsiung, 1934.

Distributed in Sri Lanka, China (including Taiwan), Myanmar, Vietnam, Thailand and the Philippines.

Tectaria polymorpha (Wall. ex Hook.) Copel. (POLYPODIACEAE)

Nephrodium polymorphum (Wall. ex Hook.) Baker Henry in Trans. Asiat. Soc. Jap. 24 suppl: 113. (1896) (List Pl. Formosa).

Taiwan: Wanchin, 1616. **Yunnan:** Common along rocky banks of streamlets in shade of forests near Simao, 12654.

02

Tectaria subtriphylla (Hook. & Arn.) Copel. (POLYPODIACEAE)

Nephrodium subtriphylum (Hook. & Arn.) Baker Henry in Trans. Asiat. Soc. Jap. 24 suppl: 113. (1896) (List Pl. Formosa).

Taiwan: Wanchin, 193.

Telosma cordata (Burm. f.) Merr. (APOCYNACEAE)

Pergularia minor And. Anonymous in Kew List of Determinations (volume 24).

Hainan: Environs of Kiungchow, 8304.

Telosma pallida (Roxb.) Craib (APOCYNACEAE)

Pergularia pallida (Roxb.) Wight & Arn. Henry in Trans. Asiat. Soc. Jap. 24 suppl: 61. (1896) (List Pl. Formosa).

Taiwan: Wanchin, 861.

Tephroseris kirilowii (Turz. ex DC.) Holub. (ASTERACEAE)

Jeffrey & Chen in Kew Bull. 39: 281. (1924).

Senecio campestris Forbes & Hemsley non (Retz.) DC. in Journ. Linn. Soc. Bot. 23: 450. (1888).

Hubei: Yichang, 1210, 1403.

Tephroseris pseudosonchus (Vaniot) C. Jeffrey & Y. L. Chen (ASTERACEAE)

in Kew Bull. 39: 272. (1984).

Senecio campestris Forbes & Hemsley non (Retz.) DC. in Journ. Linn. Soc. Bot. 23: 450. (1888).

Hubei: Yichang, 1211, 1217.

Tephrosia purpurea (L.) Pers. (FABACEAE)

Henry in Trans. Asiat. Soc. Jap. 24 suppl: 32. (1896) (List Pl. Formosa). *Tephrosia tinctoria* Graham Anonymous in Kew List of Determinations (volume 34).

Hainan: Environs of Haikou, 7993, 8164. Without locality, 8535. **Taiwan:** Kaohsiung, without number. **Yunnan:** Simao, 12715.

An annual herb to 60 cm tall, pantropical in distribution.

Tephrosia vestita Vogel (FABACEAE)

Anonymous in Kew List of Determinations (volume 24).

Hainan: At Nodoa in the interior of the island, 8453. Ling-men (in the interior of the island), 8666.

Terminalia catappa L. (COMBRETACEAE)

Henry in Trans. Asiat. Soc. Jap. 24 suppl: 43. (1896) (List Pl. Formosa), Exell in Sunyatsenia 1: 91. (1933), Li in Woody Flora of Taiwan 642. fig. 254. (1963), Li & Lo in Flora of Taiwan 3: 933. (1993).

Hainan: Without locality, 48. **Taiwan:** Oluanpi, 917. Kaohsiung Spit, 1111.

A large semi-deciduous, dichotomously branched tree whose fruits are dispersed by ocean currents. A widespread littoral species, native to the tropics of the Old World and

found on the southern part of Taiwan. This species was collected by the Glasnevin Taiwan Expedition (GTE 17) by the lighthouse at Oluanpi (South Cape of Henry and Wilson) in October 2004.

Terminalia franchetii Gagnep. (COMBRETACEAE)

Terminalia franchetii Gagnep. var. *genuina* Exell in Sunyatsenia 1: 91. (1933).

Yunnan: In woods near Mengzi, 9300, 9300a. Simao, 9300b.

Discovered by Delavay in Yunnan, this species was also later collected in the same province by Forrest in the Lijiang valley. China (Guangxi, Sichuan and Yunnan) and Thailand.

Terminalia myriocarpa Van Heurck & Müell.-Arg. (COMBRETACEAE)

Exell in Sunyatsenia 1: 91. (1933).

Yunnan: Simao at 1,300 metres, 12523. Simao at 1,150 metres, 12523a. 12523b. Simao, 13436.

An evergreen tree to 35 metres tall bearing large panicles of small red flowers during the months of August and September. Distributed through India (including Sikkim), Myanmar, Malaysia, Thailand, Laos, northern Vietnam and western China. In Yunnan this very ornamental tree grows in subtropical regions and in tropical rainforest in rivervalleys in the extreme south.

Ternstroemia gymnanthera (Wight & Arn.) Bedd. (PENTAPHYLACEAE)

Ternstroemia japonica Henry non (Thunb.) Thunb. in Trans. Asiat. Soc. Jap. 24 suppl: 19. (1896) (List Pl. Formosa). *Ternstroemia japonica* Thunb. var. *wightii* (Choisy) Dyer Rehder & E. H. Wilson in Sargent Pl. Wilson 2: 397. (1913). *Ternstroemia gymnanthera* (Wight & Arn.) Bedd. var. *wightii* (Choisy) Hand.-Mazz. Zheng in Journ. Wuhan Bot. Res. 2.(1): 152. (1984).

Hubei: Yichang, 3301. Jianshi, 5987. **Taiwan:** Oluanpi, 1365. **Yunnan:** In woods near Simao at 1,500 to 1,800 metres, 12108, 12108b., 12108c., 12108d., 12108e., 12108f.

A widely distributed evergreen shrub or small tree, native to Indo-Malaysia, southern China to Japan and the Philippines. In Chinese Taiwan this species is distributed in broad-leaved forest.

Ternstroemia sp. (PENTAPHYLACEAE)

Anonymous in Kew List of Determinations (volume 24).

Hubei: Badong, 5159.

Tetracentron sinense Oliver (TROCHODENDRACEAE)
New genus

in Hooker's Icon. Pl. 19: t. 1892. (1889), Bretschneider in Hist. Eur. Bot. Disc. China 2: 779. (1898), Diels in Bot. Jahrb. 29: 323. (1900), Henry in Flora & Sylva 1: 218.

(1903), Finet & Gagnepain in Bull. Soc. Bot. France. LII. Mem. iv. 26. (1905); in Contrib. Fl. As. Or. 2: 26. (1907), J. H. Veitch in Hortus Veitchii 381. (1906), Bean in Kew Bull. Misc. Inf. 356. (1909), Schneider in Illus. Handb. Laubholzk. 2: 927. (1912), Rehder & E. H. Wilson in Sargent Pl. Wilson. 1: 417. (1913), Morley in Glasra 3: 80. (1979), Zheng in Journ. Wuhan Bot. Res. 2.(1): 36. (1984), Bean in Trees & Shrubs 4: 573. (1989), Nelson in A Heritage of Beauty 320. (2000), O' Brien in Augustine Henry, An Irish Plant Collector in China 36. (2002).

Hubei: Fang Xian, 6243 (type), 6690 (syntype). Jianshi, 7417 (syntype). **Yunnan:** South of the Red River from Manmei at 2,300 metres, 9744. In mountain forest on Fengchunling at 2,000 to 2,300 metres, 9744a.

A deciduous tree to 30 metres tall, native to central and western China and Myanmar. In the forests of central and western China, this tree is only surpassed in size by *Cercidiphyllum* according to Wilson, who later collected *Tetracentron* (and introduced it through Messrs Veitch in 1901) in several parts of Hubei and Sichuan. There are magnificent trees on the middle forested slopes of Emei Shan in Sichuan. The tallest specimen in Britain and Ireland grows at Cambridge Botanic Gardens, when measured in 2002 it was found to be 13 metres tall, it grows on two stems, each 34 cm and 27 cm in diameter. There are young trees at the National Botanic Gardens, Kilmacurragh and at Mount Stewart in County Down. Plants from Nepal, Bhutan and India differs by having longer-tipped leaves with finer tips and laxer fruiting catkins.

Tetracera scandens (L.) Merr. (DILLENIACEAE)

Tetracera sarmentosa Anonymous non (L.) Vahl in Kew List of Determinations (volume 24).

Hainan: Environs of Haikou, 8340. Without locality, 8537. Ling-men (in the interior of the island), 8641.

Tetradium daniellii (Benn.) T. G. Hartley (RUTACEAE)

O' Brien in Augustine Henry, An Irish Plant Collector in China 37. (2002).

Euodia daniellii (Benn.) F. B. Forbes & Hemsl. Henry in Journ. China Br. Roy. Asiat. Soc. 22: 253.(1887) (Chinese Names of Plants). *Euodia hupehensis* Dode in Bull. Soc. France 55: 707. (1908), Rehder & E. H. Wilson in Sargent Pl. Wilson. 1: 133. (1911). *Euodia daniellii* (Benn) Hemsl. var. *hupehensis* (Dode) Huang Zheng in Journ. Wuhan Bot. Res. 2.(1): 105. (1984).

Hubei: Badong, 2555. Liantuo, 2982, 4482, 4534. Yichang, 2939 (isotype of *Euodia hupehensis* Dode). Fang Xian, 6712 (type of *Euodia hupehensis* Dode). Without locality, 4534a.

The tallest growing species of *Tetradium* in western Hubei according to Wilson who later collected it at Xingshan, Changleping (in Wufeng to the south of Yichang), and in Changyang. It was also collected by the Irish missionary, Fr. Hugh Scallan (Pater Hugo) in Shaanxi province in 1899. Wilson stated it was known colloquially as the *ch'ou-la-shu* or stinking ash-tree and was also called *p'ao-la-tzu*, signifying that the fruit is worthless as a drug unlike *Tetradium ruticarpum*. Introduced to cultivation by Wilson while collecting for the Arnold Arboretum in 1907. *Tetradium daniellii* commemorates William Freemam Daniell, M.S., a Liverpool native (1818—1865) who travelled to China in 1860 as a medical officer to a division of a British army expedition to China. He collected plants in China and presented these to the herbarium at the British Museum. In Ireland *Tetradium daniellii* is sometimes planted by bee keepers as it is one of the very last trees to flower in the year.

Tetradium glabrifolium (Champ. ex Benth.) T. Hartley (RUTACEAE)

Euodia meliifolia (Hance ex Walpers) Benth. Henry in Trans. Asiat. Soc. Jap. 24 suppl: 24. (1896) (List Pl. Formosa). *Euodia glauca* Miq. Dode in Bull. Soc. Bot. France 55: 703. (1908), Rehder & E. H. Wilson in Sargent Pl. Wilson 2: 129. (1914).

Hubei: Yichang, 4577. **Hainan:** Environs of Kiungchow, 8706. **Taiwan:** Wanchin, 831, 934, 1562. Oluanpi, 932, 1296. **Yunnan:** Yiwu, 13577.

A large shrub or a small to medium-sized tree, native over a wide range from the Himalaya eastward to southern Japan and south to Sumatra and the Philippines.

Tetradium ruticarpum (Juss.) T. G. Hartley (RUTACEAE)

in Gard. Bull. Sing. 34: 91-131. (1981).

Euodia sp. nova. aff. *E. ruticarpa* Henry in Journ. China Br. Roy. Asiat. Soc. 22: 253. (1887) (Chinese Names of Plants). *Euodia officinalis* Dode in Bull. Soc. Bot. France. lv. 703. (1908), Rehder & E. H. Wilson in Sargent Pl. Wilson 2: 130. (1914). *Euodia ruticarpa* (Juss.) Bentham var. *officinalis* (Dode) Huang in Acta Phytotax. Sin. 6: 114. (1957). *Euodia rugosa* Rehder & E. H. Wilson in Sargent Pl. Wilson. 2: 132. (1914).

Hubei: Badong, 924 (November 1885), 6136 (syntype of *Evodia officinali*s Dode). Yichang, 1676, 2259, 4134 (syntype of *Evodia officinalis* Dode). Liantuo, 2043, 2077. Changyang, 6199 (syntype of *Evodia officinalis* Dode). Xingshan at 1,400 metres, 4525. Fang Xian, 6569. **Yunnan:** On the left summit of Taluchon in forests to the north of Mengzi at 2,600 metres, 10245 (holotype of *Evodia rugosa* Rehder & E. H. Wilson, collected 25th July 1897).

According to Farges (who later collected specimens for the Natural History Museum in Paris) this shrub was called *houang pei chou* or the 'yellow bark tree'. The fruits are used as a drug (to alleviate pain and to stop vomiting). Wilson also listed the colloquial names *chu-yu*, *wu-chu-yu* and *la-tzu-shu* for this species, it seems to be currently known as the *wu zhu ya*.

Tetradium sp. No. 1. (RUTACEAE)

Euodia sp. Anonymous in Kew List of Determinations (volume 24).

Hubei: Badong, 1802, 2816, 4719, 5116.

Tetradium sp. No. 2. (RUTACEAE)

Euodia meliifolia (Hance ex Walpole) Benth. var. Henry in Journ. China Br. Roy. Asiat. Soc. 22: 238. (1887) (Chinese Names of Plants).

Hubei: Badong 13, 6157. Xingshan, 6485.

According to Henry this species was known colloquially at Yichang as *chien feng kan*.

Tetradium sp. No. 3. (RUTACEAE)

Euodia sp. Anonymous in Kew List of Determinations (volume 24).

Hainan: Ling-men, 8605.

Tetradium trichotomum Lour. (RUTACEAE)

Euodia colorata Dunn in Kew Bull. Misc. Inf. 2. (1906). *Euodia trichotoma* (Lour.) Pierre Rehder & E. H. Wilson in Sargent Pl. Wilson. 2: 132. (1914).

Yunnan: Mountains to the south-west of Mengzi at 1,300 metres, 10951. Vicinity of Simao at 1,300 to 1,600 metres, 12137 (syntype of *Euodia colorata* Dunn), 12137a., (syntype of *Euodia colorata* Dunn), 12137b., 12137c.

A bush or small tree from 2 to 6 metres tall producing umbels of yellow flowers followed by reddish fruits. The Hani (an ethnic group) of Xishuangbanna in southern Yunnan call this species *naivqzaovqpievq* and use the fruits to treat gastritis, abdominal pain, dysentry and flu. The leaves are used externally to treat rheumatism and boils. Distributed from India (Assam) to China and Indochina.

Tetrapanax papyrifer (Hook.) K. Koch (ARALIACEAE)

Zheng in Journ. Wuhan Bot. Res. 2.(1): 162. (1984).

Fatsia papyrifera (Hook.) Miq. ex Witte Forbes & Hemsley in Journ. Linn. Soc. Bot. 23: 341. (1887), Henry in Journ. China Br. Roy. Asiat. Soc. 22: 275. (1887) (Chinese Names of Plants); in Notes Econ. Bot. China 63. (1893).

Hubei: Badong, 895, 5070. Changyang, 7763.

A deciduous shrub to 3 metres tall with stout stems covered in thick soft bristles. Henry's Badong collector called the rice-paper plant the *t'ung ts'ao* (*tong cao*, literally; 'facilitates urination', this species is used in Chinese herbal medicine for the same purpose. The stems of this handsome shrub are filled with a white pith once much used in the production of rice-paper used by Chinese artists. Distributed in the provinces of Guangdong, Guangxi, Guizhou, Hubei, Jiangxi, Sichuan and Taiwan. This species was collected by the Glasnevin Central China Expedition (GCCE 311) on the hills above Badong in September 2002.

Tetrapilus roseus (Craib ex Hosseus) L. A. S. Johnson (OLEACEAE) New species

Olea rosea Craib in Kew Bull. Misc. Inf. 411. (1910). *Olea densiflora* H. L. Li in Journ. Arnold Arb. 25: 315. (1944).

Yunnan: Simao, 11661 (type of *Olea densiflora* Li), 111661b., 11661e. Simao, at 1,500 metres, 11661a., (in part). In forests near Simao at 1,300 metres, 12598.

Distributed in Thailand, Cambodia, Laos, Vietnam and in China in southern Yunnan.

Tetrastigma dubium (M. A. Lawson) Gagnep. (VITACEAE) New species

Tetrastigma henryi Gagnepain in Notul. Syst. (Paris) 1: 264. (1911), O' Brien in Augustine Henry, An Irish Plant Collector in china 37. (2002).

Yunnan: Mengzi at 1,650 metres, 9992. Mengzi at 1,500 metres, 9992a., (syntype of *Tetrastigma henryi* Gagnepain). Near Simao at 1,300 metres, 11756 (syntype of *Tetrastigma henryi* Gagnepain). Simao, 11756c., (syntype of *Tetrastigma henryi* Gagnepain). Chuyuan, 10530c.

Distributed from Nepal to China and Borneo.

Tetrastigma formosanum (Hemsl.) Gagnep. (VITACEAE) New species

in Lecompte, Not. Syst. 1: 321. (1911), Li in Woody Flora of Taiwan 529. fig. 205. (1963).

Vitis sp. Henry in Trans. Asiat. Soc. Jap. 24 suppl: 28. (1896) (List Pl. Formosa). *Vitis formosana* Hemsley in Ann. of Bot. 9: 151. (1895), Henry in Trans. Asiat. Soc. Jap. 24 suppl: 28. (1896) (List Pl. Formosa), Bretschneider in Hist. Eur. Bot. Disc. China 2: 781. (1898).

Taiwan: Wanchin, 174, without number. Kaohsiung, 745 (syntype of *Vitis formosana* Hemsley).

A scandent shrub found in the southern part of China (Taiwan) and in Japan (south). This species was collected by the Glasnevin Taiwan Expedition (GTE 45) by the lighthouse at Oluanpi (South Cape of Henry and Wilson) and on Wanshoushan, (GTE 101) in October 2004.

Tetrastigma hemsleyanum Diels & Gilg (VITACEAE) New species

in Engler, Bot. Jahrb. 29: 443. (1900), Lecompte in Notul. Syst. (Paris) 1: 317. (1911).

Vitis sp. Henry in Trans. Asiat. Soc. Jap. 24 suppl: 28. (1896) (List Pl. Formosa). *Tetrastigma dentatum* (Hayata) H. L. Li in Woody Flora of Taiwan 857. (1963).

Hubei: Yichang, 3548 (isotype). **Taiwan:** Wanchin, 104.

Tetrastigma leucostaphylum (Dennst.) Alston (VITACEAE)

Vitis lanceolaria (Roxb.) Wall. Anonymous in Kew

02

List of Determinations (volume 34).

Yunnan: Simao, 12151, 12817.

Tetrastigma obovatum (Laws.) Gagnep. (VITACEAE)

Gagnepain in Notul. Syst. (Paris) 1: 266. (1911).

Yunnan: Simao, 12050b., 12051b.

Native to India, China (Yunnan), Thailand, Laos and Vietnam.

Tetrastigma obtectum (Wall. ex M. A. Lawson) Planch. ex Franch. (VITACEAE)

Vitis obtecta Wall. ex M. A. Lawson Henry in Journ. China Br. Roy. Asiat. Soc. 22: 280. (1887) (Chinese Names of Plants). *Tetrastigma obtectum* (Wall. ex M. A. Lawson) Planch. var. *pilosum* Gagnepain in Notul. Syst. (Paris) 1: 324. (1911), Zheng in Journ. Wuhan Bot. Res. 2.(1): 144. (1984).

Hubei: Liantuo, 2591, 4456. Yichang, 2035, 3539 (isosyntype of *Tetrastigma obtectum* [(Wall. ex M. A. Lawson) Planch. var. *pilosum* Gagnepain]. Antelope Glen, 3315. **Sichuan:** Henry's Chinese collector with Pratt, no locality stated but may be any of the following, Leshan, Emei Shan, Wa Shan or Kangding, 8991. **Yunnan:** Mengzi, 11385.

Henry stated that at the village of Liantuo near the eastern mouth of the Xiling (Yichang) Gorge this species was known as *wu chao lung*. Introduced to cultivation by Wilson when collecting for Messrs Veitch. Distributed across the Himalaya, western and central China and Vietnam, in the tropical rainforests of Xishuangbanna in southern Yunnan this species forms enormous lianas.

Tetrastigma pachyphyllum (Hemsl.) Chun (VITACEAE)

Tetrastigma strumarum (Planch.) Gagnep. Gagnepain in Notul. Syst. (Paris) 1: 267. (1911).

Hainan: Environs of Kiungchow, 8273.

Tetrastigma planicaule (Hook. f.) Gagnep. (VITACEAE)

Vitis planicaulis Hook. f. Anonymous in Kew List of Determinations (volume 34).

Yunnan: Mengzi, 10622. Simao, 11956.

Native from the Sikkim Himalaya to Sri Lanka, China and Vietnam.

Tetrastigma serrulatum (Roxb.) Planch (VITACEAE)

Lecompte in Notul. Syst. 1: 322. (1909).

Yunnan: Mengzi, 9695, 9877a.

Distributed from the temperate Himalaya to western China and Indochina.

Tetrastigma triphyllum (Gagnep.) W. T. Wang (VITACEAE) New species

Tetrastigma yunnanense Gagnep. var. *triphyllum* Gagnepain in Notul. Syst. 1: 38. (1909).

Yunnan: Simao, 9881d., (type of *Tetrastigma yunnanense* Gagnep. var. *triphyllum* Gagnepain). Simao, 11647b.

China (Sichuan and Yunnan) to northern Thailand.

Tetrataenium birmanicum (Kurz) Manden. (APIACEAE)

Heracleum birmanicum Kurz Anonymous in Kew List of Determinations (volume 34).

Yunnan: Mengzi, 9325, 9883.

Teucrium alborubrum Hemsley (LAMIACEAE) New species

in Journ. Linn. Soc. Bot. 26: 311. (1890), Bretschneider in Hist. Eur. Bot. Disc. China 2: 790. (1898), J. H. Veitch in Hortus Veitchii 437. (1906), Nelson in A Heritage of Beauty 320. (2000), *Teucrium pernyi* Dunn non Franch. in Notes Roy. Bot. Gard. Edinburgh 6: 192. (1915).

Hubei: In the Hsiao Pingshanba Glen (October 1886), 2919 (isosyntype). Antelope Glen, 4257 (isosyntype). Liantuo, 4600. **Sichuan:** South Wushan, 7227.

A hardy herbaceous perennial bearing lanceolate leaves and small rose-coloured flowers held in erect racemes in the axils of the uppermost leaves. Introduced to cultivation by Wilson through Veitch's Coombe Wood nursery where it first flowered during the summer of 1904.

Teucrium bidentatum Hemsley (LAMIACEAE) New species

in Journ. Linn. Soc. Bot. 26: 312. (1890), Bretschneider in Hist. Eur. Bot. Disc. China 2: 790. (1898), Dunn in Notes Roy. Bot. Gard. Edinburgh 6: 192. (1915).

Hubei: Antelope Glen near Yichang, 3119, 4150 (isosyntype).

Teucrium japonicum Houtt. (LAMIACEAE)

Teucrium japonicum Willd. (nom. illeg.) Dunn in Notes Roy. Bot. Gard. Edinburgh. 6: 191. (1915).

Hainan: Environs of Kiungchow, 8735.

Teucrium ornatum Hemsley (LAMIACEAE) New species

in Journ. Linn. Soc. Bot. 26: 313. (1890), Bretschneider in Hist. Eur. Bot. Disc. China 2: 790. (1898), Dunn in Notes Roy. Bot. Gard. 6: 192. (1915), Nelson in A Heritage of Beauty 320. (2000).

Hubei: In woods near Badong, 5141, 7309. Donghu, 6437. Xingshan, 6471, Fang Xian, 6700.

Introduced to cultivation by Wilson through Veitch's Coombe Wood nursery where it first flowered in 1904.

Teucrium pampaninii C. Du (LAMIACEAE)

Teucrium japonicum Forbes & Hemsley non Houtt. in Journ. Linn. Soc. Bot. 26: 312. (1890), Dunn in Notes Roy. Bot. Gard. Edinburgh 6: 191. (1915).

Hubei: Yichang, 941, 6673. Near Yichang, on the hills

behind the San You Dong Glen (July 1886), 1515. Liantuo, 4639. Badong, 5191. Jianshi, 7328. Donghu, 6406. Fang Xian, 6672.

Teucrium pernyi Franch. (LAMIACEAE)

Dunn in Notes Roy. Bot. Gard. Edinburgh 6: 192. (1915).

Yunnan: Mengzi at 1,950 metres, 10986.

Endemic to China.

Teucrium quadrifarium Buch.-Ham. ex D. Don (LAMIACEAE)

Dunn in Notes Roy. Bot. Gard. Edinburgh 6: 191. (1915).

Hunan: Shimen, 7938. **Yunnan:** Mengzi, 9845. In Pingbian, in forests on the Daweishan Range at 1,650 metres, 11047.

Native to Nepal, India, Myanmar and China.

Teucrium simplex Vaniot (LAMIACEAE)

Dunn in Notes Roy. Bot. Gard. Edinburgh 6: 191. (1915).

Yunnan: Mengzi at 1,750 metres, 10068.

Endemic to China and distributed in Guizhou and Yunnan.

Teucrium sp. No. 1. (LAMIACEAE)

Teucrium quadrifarium Buch.-Ham. ex D. Don Forbes & Hemsley in Journ. Linn. Soc. Bot. 26: 314. (1890), Dunn in Notes Roy. Bot. Gard. Edinburgh 6: 191. (1915).

Hubei: Badong, 1422, 5126. Jianshi, 7422.

Teucrium quadrifarium is not native to Hubei province, this entry belongs to another taxon.

Teucrium sp. No. 2. (LAMIACEAE)

Anonymous in Kew List of Determinations (volume 34).

Yunnan: Mengzi, 9549.

Teucrium viscidum Blume (LAMIACEAE)

Teucrium stoloniferum Roxb. Henry in Trans. Asiat. Soc. Jap. 24 suppl: 73. (1896) (List Pl. Formosa).

Taiwan: Tamsui, 1753.

Teyleria stricta (Kurz) A. N. Egan & B. Pan (FABACEAE)

Pueraria collettii Prain Anonymous in Kew List of Determinations (volume 34).

Yunnan: Yuanjiang, 13254. Simao, 13431.

Distributed in Myanmar, China (Yunnan) and Thailand.

Thalictrum angustifolium Jacq. (RANUNCULACEAE)

Diels in Bot. Jahrb. 29: 335. (1900).

Hubei: Yichang, 544, 561, 662, 1331.

Thalictrum delavayi Franchet (RANUNCULACEAE)

Thalictrum dipterocarpum Franch. Anonymous in Kew List of Determinations (volume 34).

Yunnan: Mengzi, 10001.

Distributed in Myanmar and China [Guizhou, Sichuan Tibet (Xizang) and Yunnan]. One of the most beautiful of all Chinese perennials, this species is common on the Cangshan range above Dali in Yunnan and in south-east Tibet (Xizang) in forest glades.

Thalictrum fortunei S. Moore (RANUNCULACEAE)

Anonymous in Kew List of Determinations (volume 24).

Hubei: Yichang, 339, 812, 1235. Badong, 913 (July 1885), 1437, 3749, 4076. Without locality, 5407a. **Sichuan:** South Wushan, 5407, Henry's Chinese collector with Pratt, no locality stated but may be any of the following, Leshan, Emei Shan, Wa Shan or Kangding, 8812.

Thalictrum ichangense Lecoyer ex Oliv. (RANUNCULACEAE) New species

in Hooker's. Icon. Pl. 18: t. 1765. (1888), Bretschneider in Hist. Eur. Bot. Disc. China 2: 778. (1898).

Hubei: Liantuo, 1949. Antelope Glen, 3583. Changyang, 6223. Fang Xian, 6223a., 6223b.

Thalictrum microgynum Lecoyer ex Oliv. (RANUNCULACEAE) New species

in Hooker's Icon. Pl. 18: t. 1766. (1888), Bretschneider in Hist. Eur. Bot. Disc. China 2: 778. (1898).

Hubei: Liantuo, 3932 (type).

A perennial species to 45 cm tall carrying crowded terminal corymbs of small white flowers from April to July. Distributed through Hubei, Hunan, Sichuan and Yunnan, where it grows between altitudes of 700 to 3,300 metres. A very beautiful plant both in terms of foliage and flowers.

Thalictrum minus L. (RANUNCULACEAE)

Anonymous in Kew List of Determinations (volume 24).

Hubei: Yichang, 1666. Badong 1869, 2424, 3142, 4747, 4870, 5208.

Thalictrum petaloideum L. (RANUNCULACEAE)

Anonymous in Kew List of Determinations (volume 34).

Yunnan: Mengzi, 10026.

Thalictrum reniforme Wall. (RANUNCULACEAE)

Anonymous in Kew List of Determinations (volume 34).

Yunnan: Milê, 9969.

Thalictrum robustum Maxim. (RANUNCULACEAE)

Thalictrum clematidifolium Franchet in Journ. de Bot. 273. (1894), Bretschneider in Hist. Eur. Bot. Disc. China 2: 778. (1898). *Thalictrum laxum* E. Ulbrich in Notizbl. Bot. Gart. Berlin-Dahlem 9: 225. (1925).

Hubei: Yichang, without number. Donghu, 6424 (isotype of *Thalictrum laxum* E. Ulbrich).

Thalictrum simaoense W. T. Wang & G. Zhu (RANUNCULACEAE) New species

in Phytologia 79: 385. (1996).

Yunnan: Simao, 13096 (isotype).

Endemic to south-west Yunnan and known only from the type locality at Simao.

Thalictrum spp. (RANUNCULACEAE)

Anonymous in Kew List of Determinations (volume 24 & 34).

Hubei: Yichang, 209. Badong, 333 (in part, June 1885), 6084, 7310, 7344. **Sichuan:** South Wushan, 7216, 7455. **Yunnan:** Mengzi, 11153.

Thamnolia vermicularis (Swartz) Ach. ex Schaerer (ICMADOPHILACEAE)

Muell.-Arg. in Bull. Herb. Boissier 1: 235. (1893).

Hubei: Xingshan, 6964. (Lichen).

Theligonum macranthum Franch. (RUBIACEAE)

Forbes & Hemsley in Journ. Linn. Soc. Bot. 26: 492. (1899).

Sichuan: South Wushan, 5588, 5652.

Thelypteris acuminata (Houtt.) C. V. Morton (ASPLENIACEAE)

Nephrodium sophoroides (Thunb.) Desv. Henry in Journ. China Br. Roy. Asiat. Soc. 22: 263. (1887) (Chinese Names of Plants), Henry in Trans. Asiat. Soc. Jap. 24 suppl: 113. (1896) (List Pl. Formosa), Christ in Bull. Soc. Bot. France 5: 34. (1905). *Dryopteris sinica* Christ in Lecompte, Notul. Syst. 1: 38. (1909). *Cyclosorus acuminatus* Holttum non (Houtt.) Nakai in Kew Bull. 31: 333. (1977).

Hubei: Liantuo, 2152, 2153. Badong, 2508. Antelope Glen, 4238. Changleping, 7812, 7813, 7814. **Taiwan:** Oluanpi, 1221 (in part). Wanchin, 1509. **Yunnan:** In woods near Mengzi at 1,500 metres, 11537a., (type of *Dryopteris sinica* Christ).

At Badong, according to Henry, this species was known locally as *pan chiu wang tyzu*.

Thelypteris aspera (C. Presl.) K. Iwats. (ASPLENIACEAE)

Polypodium urophyllum Wall. ex Hook. Henry in Trans. Asiat. Soc. Jap. 24 suppl: 114. (1896) (List Pl. Formosa).

Taiwan: Oluanpi, 1268. **Yunnan:** Mengzi, 10092.

Thelypteris beddomei (Baker) Ching (ASPLENIACEAE) New to China

Nephrodium beddomei Bak. Baker in Journ. Bot. 25: 171. (1887), Henry in Journ. China Br. Roy. Asiat. Soc. 22: 263. (1887) (Chinese Names of Plants), Baker in Ann. Bot. 5: 319. (1890).

Hubei: Badong, 1862, 4881.

When collected by Henry this species proved new to China. Henry's Badong collector called this species the *pan pien lien*. Native to India, Sri Lanka, China, Korea, Japan, the Philippines, the Malay Peninsula, Java and New Guinea.

Thelypteris cana (Baker) Ching (ASPLENIACEAE) New to China

Polypodium appendiculatum Wall. ex Baker Baker in Journ. Bot. 25: 171. (1887). *Aspidium appendiculatum* (Wall. ex Baker) Christ in Bull. Soc. Bot. France 5: 36. (1905).

Hubei: Liantuo, 2583.

Thelypteris crinipes (Hook.) K. Iwats. (ASPLENIACEAE)

Nephrodium crinipes Hook. Anonymous in Kew List of Determinations (volume 34).

Yunnan: Simao, 12877.

Distributed from the Himalaya to Malaysia.

Thelypteris dentata (Forrsk.) E. P. St. John (ASPLENIACEAE)

Nephrodium molle (Sw.) R. Br. Anonymous in Kew List of Determinations (volume 34). Henry in Trans. Asiat. Soc. Jap. 24 suppl: 113. (1896) (List Pl. Formosa).

Taiwan: Oluanpi, 1221 (in part), 1269. Tamsui, 1439, 1443. **Yunnan:** Simao, 13079.

Distributed in India, Myanmar, China, Thailand, Vietnam and Japan.

Thelypteris francoana (E. Fourn.) C. F. Reed (POLYPODIACEAE)

Polypodium subintegrum Baker in Kew Bull. Misc. Inf. 231. (1898).

Yunnan: South of the Red River from Manmei at 1,950 metres, 9194 (in part, type of *Polypodium subintegrum* Baker).

Thelypteris gracilescens (Blume) Ching (ASPLENIACEAE)

Nephrodium gracilescens (Blume) Hook. Henry in Trans. Asiat. Soc. Jap. 24 suppl: 113. (1896) (List Pl. Formosa).

Hubei: Liantuo, 3060. **Taiwan:** Tamsui, 1392.

Thelypteris glanduligera (Kunze) Ching (ASPLENIACEAE)

Dryopteris repentula Christ ex Clarke in Notul. Syst. (Paris) 1: 39. (1909).

Yunnan: Simao at 1,500 metres, 13077.

Thelypteris griffithii (T. Moore) C. F. Reed (PTERIDACEAE)

Hemionitis griffithii (T. Moore) Hook. f. & Thoms. Christ in Bull. Herb. Boiss. 6: 868. (1898).

Yunnan: In Pingbian on the Daweishan Range in forests at 1, 600 metres, 10422.

Thelypteris hirsutipes (C. B. Clarke) Ching (ASPLENIACEAE)

Dryopteris gracilescens (Blume) Kuntze var. *chinensis* Christ in Notul. Syst. (Paris) 1: 40. (1909).

Yunnan: Mengzi at 2,200 metres, 10111.

Thelypteris nudata (Roxb.) C. V. Morton (ASPLENIACEAE)

Polypodium multilineatum Wall. ex Hook. Anonymous in Kew List of Determinations (volume 34).

Yunnan: Simao, 13168.

02

Thelypteris penangiana (Hook.) C. F. Reed (ASPLENIACEAE)

Dryopteris rampans (Baker) Christ in Index Filicum Fasc. 5: 287. (1905). *Nephrodium rampans* Baker in Journ. Bot. 177. (1889), Baker in Ann. Bot. 5: 327. (1890), Bretschneider in Hist. Eur. Bot. Disc. China 2: 794. (1898), Christ in Bull. Soc. Bot. France 5: 34. (1905).

Hubei: Changleping, 7844 (type of *Nephrodium rampans* Baker).

Thelypteris prolifera (Retz.) C. F. Reed (ASPLENIACEAE)

Polypodium proliferum (Retz.) Roxb. ex Wall. Henry in Trans. Asiat. Soc. Jap. 24 suppl: 114. (1896) (List Pl. Formosa).

Taiwan: Kaohsiung, without number.

Thelypteris subelata (Baker) K. Iwats. (ASPLENIACEAE) New species

Nephrodium subelatum Baker in Kew Bull. Misc. Inf. 11. (1906). *Dryopteris subelata* (Bak.) C. Chr. Ind. Fil. Suppl. 40. (1913); Contrib. U. S. Nat. Herb. 26: 276. (1931) *Thelypteris subelata* (Bak.) Reed in Phytologia 17: 317. (1968). *Christella subelata* (Bak.) Holttum in Kew Bull. 31: 331. (1976). *Cyclosorus subelatus* (Bak.) Ching in Bull. Fan. Mem. Inst. Biol. 8: 224. (1938).

Yunnan: Simao, 11809a., (isotype of *Nephrodium subelatum* Baker, collected February 5th 1897).

A handsome fern to 1 metre tall, rhizomes short and creeping, native to from north-east India to China.

Thelypteris triphyllum (Sw.) K. Iwats (ASPLENIACEAE)

Meniscium triphyllum Sw. Henry in Trans. Asiat. Soc. Jap. 24 suppl: 116. (1896) (List Pl. Formosa).

Taiwan: Wanchin, 1561.

A widespread fern, native from India to Myanmar, Thailand, Malaysia, China, Japan and Australia.

Thelypteris truncata (Poir.) K. Iwats. (ASPLENIACEAE)

Nephrodium truncatum (Poir.) C. Presl. Henry in Trans. Asiat. Soc. Jap. 24 suppl: 113. (1896) (List Pl. Formosa).

Taiwan: Wanchin, without number. Kaohsiung, 740.

Distrubuted through the tropics of the Old World.

Thelypteris valida (Christ) Tagawa & K. Iwats (ASPLENIACEAE)

Dryopteris hirticarpa Ching in Bull. Fan Mem. Inst. Biol. 2(10): 197. (1931).

Yunnan: On Nalo Shan near Simao, 11809b., (isotype of *Dryopteris hirticarpa* Ching).

Thelypteris xylodes (Kunze) Ching (ASPLENIACEAE)

Dryopteris xylodes (Kunze) Christ in Notul. Syst. (Paris) 1: 41. (1909).

Yunnan: In humid forests near Simao at 1,700 metres, 10094b., 13105. In a ravine near Mengzi at 1,500 metres, 10359.

Themeda caudata (Nees) A. Camus (POACEAE)

Anthistiria ciliata Henry non L. f. in Journ. China Br. Roy. Asiat. Soc. 22: 264. (1887) (Chinese Names of Plants), Henry in Trans. Asiat. Soc. Jap. 24 suppl: 108. (1896) (List Pl. Formosa). *Themeda gigantea* (Cav.) Hackel. ex Dutie ssp. *caudata* (Nees) Hackel. Rendle in Journ. Linn. Soc. Bot. 36: 377. (1904).

Hubei: Yichang, 27. **Taiwan:** Wanchin, 1518. **Yunnan:** In mountains near Mengzi at 1,750 metres, 9459.

Native to India, Myanmar, China and Vietnam. According to Henry this species was known colloquially in the Three Gorges region as *pao tzu ts'ao*.

Themeda triandra Forssk. (POACEAE)

Themeda triandra Forsk. var. *major* subvar. *japonica* (Willd.) Hack. Rendle in Journ. Linn. Soc. Bot. 36: 378. (1904).

Hubei: Yichang, 959, 4091. **Yunnan:** On grassy mountains near Mengzi at 1,750 metres, 10996.

Themeda villosa (Poir.) A. Camus (POACEAE)

Themeda gigantea (Cav.) Hack. ex Dutie var. *villosa* (Poir.) Hack. Rendle in Journ. Linn. Soc. Bot. 36: 377. (1904).

Hubei: Yichang, 980. South Donghu, 7696.

Thesium chinense Turez. (SANTALACEAE)

Forbes & Hemsley in Journ. Linn. Soc. Bot. 26: 408. (1894).

Hubei: Badong, 1443. Yichang, 1960, 7777. Liantuo, 4567. **Sichuan:** South Wushan, 7211.

Thesium sp. (SANTALACEAE)

Anonymous in Kew List of Determinations (volume 34).

Yunnan: Mengzi, 10940.

Thladiantha cordifolia (Blume) Cogn. (CUCURBITACEAE)

Thladiantha calcarata (Wall.) C. B. Clarke Cogniaux in Engler, Pflanzenr. iv. 275. i: 50. (1916).

Yunnan: Mengzi, 10037a.

Distributed in India, China, Laos and Vietnam.

Thladiantha dentata Cogniaux (CUCURBITACEAE) New species

in Pflanzenr. (Engler) 4. Fam. 275: 44. (1916).

Hubei: North Badong, 7010 (isotype). **Yunnan:** Mengzi, 10207 (isosyntype).

Thladiantha glabra Cogniaux ex Oliver (CUCURBITACEAE) New species

in Hooker's Icon. Pl. 23: tt. 2222. (1892), Henry in Notes Econ. Bot. China 52. (1893), Cogniaux in Engler, Pflanzenr. iv. 275. i: 48. (1916).

Hubei: Jianshi, 5821, 5937 (syntype). Without locality, 5893a. **Sichuan:** South Wushan, 5893 (isosyntype).

Thladiantha henryi Hemsley (CUCURBITACEAE) New species

in Journ. Linn. Soc. Bot. 23: 316. (1887), Henry in

Journ. China Br. Roy. Asiat. Soc. 22: 251. (1887) (Chinese Names of Plants); in Notes Econ. Bot. China 52. (1893), Oliver in Hooker's Icon. Pl. t. 2223. (1892), Bretschneider in Hist. Eur. Bot. Disc. China 2: 784. (1898), Harms in Engler's Bot. Jahrb. 29: 603. (1901), Cogniaux in Engler, Pflanzenr. iv. 275. i: 47. (1916).

Thladiantha verrucosa Cogniaux in Hooker's Icon. Pl. 23: tt. 2223. (1892), Henry in Notes Econ. Bot. China 51. (1893), Cogniaux in Engler, Pflanzenr. iv. 275. i: 49. (1916).

Hubei: Badong, 1757. Jianshi, 5900 (isosyntype), 5936, 5997. Liantuo, 5936. Changleping, 6206. Zigui, 5936a. Xingshan at 1,400 metres, 6563 (type). Without locality, 5900a., (type of *Thladiantha verrucosa* Cogniaux), 5900b., 5936b.

Henry stated this species was known colloquially in the Three Gorges area as *kua lou t'eng* or *lai-kua* from the appearance of the fruit which is transversely plicated.

Thladiantha hookeri C. B. Clarke (CUCURBITACEAE)

Chakravarty in Notes Roy. Bot. Gard. Edinburgh 20: 121. (1939—1949).

Hemsleya trifoliolata Cogniaux in Fedde, Repert. Nov. Spec. Regni Veg. 6: 304. (1909); in Engler, Pflanzenr. 275: 26. (1916). *Hemsleya yunnanensis* Cogniaux in Engler, Pflanzenr. 66.(iv. 275): 27. (1916). *Thladiantha hookeri* C. B. Clarke var. *palmatifolia* Chakrav. Chakravarty in Notes Roy. Bot. Gard. Edinburgh 20: 121. (1939—1949). *Thladiantha pentadactyla* Cogniaux Cogniaux in Engler, Pflanzenr. 275: 52. (1916).

Yunnan: In Pingbian on the Daweishan Range, in mountain forest at 1,650 metres, 9057. Simao, 12295d., (syntype of *Hemsleya trifoliolata* Cogniaux). Simao at 1,700 metres, 12295d., (in part).

Thladiantha longifolia Cogniaux (CUCURBITACEAE)
New species

in Hooker's. Icon. Pl. 23: t. 2222. (1892), Henry in Notes Econ. Bot. China 52. (1893), Bretschneider in Hist. Eur. Bot. Disc. China 2: 784. (1898), Harms in Engler's Bot. Jahrb. 29: 603. (1901), Cogniaux in Engler, Pflanzenr. iv. 275. i: 49. (1916).

Hubei: Badong, 4767. South Badong, 6055. Without locality, 6055a.

Augustine Henry stated that in the Three Gorges district this species was known as the *chien-kua* or *chien-yeh-kua* due to the long pointed leaves.

Thladiantha maculata Cogniaux (CUCURBITACEAE)
New species

in Hooker's. Icon. Pl. 23: tt. 2222. (1892), Henry in Notes Econ. Bot. China 51. (1893), Cogniaux in Engler, Pflanzenr. iv. 275. i: 49. (1916).

Hubei: Xingshan, 6465. 6465a.

Thladiantha nudiflora Hemsley (CUCURBITACEAE)
New species

in Journ. Linn. Soc. Bot. 23: 316. pl. 8. (1887), Henry in Journ. China Br. Roy. Asiat. Soc. 22: 251. (1887) (Chinese Names of Plants), Bretschneider in Hist. Eur. Bot. Disc. China 2: 784. (1898), Harms in Engler's Bot. Jahrb. 29: 603. (1901), Cogniaux in Engler, Pflanzenr. iv. 275. i: 45. (1916).

Hubei: Liantuo, 2937a. Kan Chi Glen near Yichang, 2937 (isosyntype). Xiling (Yichang) Gorge, 3626. Yichang, 3957. Xingshan at 1,400 metres, 6544 (type). North Badong, 7009. South Badong, 7472.

According to Henry this species was known colloquially at Yichang as *mao k'u kua* and its roots and fruit were used as a medicine.

Thladiantha oliveri Cogniaux ex Oliv. (CUCURBITACEAE)
New species

in Hooker's Icon. Pl. 23: tt. 2223. (1892), Henry in Notes Econ. Bot. China 51. (1893), Mottet in Rev. Hort. 472. fig. 194. (1903); in Le Jardin 330. (1903); in Journ. Soc. n. d'Hort. Fr. 739. (1903); in Hort. Vilmor. 146. fig. 38. (1906), Cogniaux in Engler, Pflanzenr. iv. 275.i: 45. (1916).

Hubei: Jianshi, 5867 (type). North Badong, 7014. Without locality, 5377a.

Thladiantha spp. (CUCURBITACEAE)

Anonymous in Kew List of Determinations (volume 24).

Hubei: Liantuo, 2005. Yichang, 2208, 4892. Antelope Glen, 3253. Badong, 4745, 5377. Jianshi, 5596a. Changleping, 6259. **Sichuan:** South Wushan, 5596.

Thladiantha villosula Cogniaux (CUCURBITACEAE)
New species

in Engler, Pflanzenr. iv. 275 i: 44. (1916), Henry in Notes Econ. Bot. China 51. (1893).

Hubei: South Badong, 6144 (type). **Yunnan:** Simao at 1,300 metres, 11803a.

Native to central and western China.

Thomsonaria caloneura (Stapf.) Rushforth (ROSACEAE) New species

Micromeles caloneura Stapf in Kew Bull. Misc. Inf. 192. (1910); in Bot. Mag. cxxxvi. t. 8335. (1910). *Pyrus caloneura* (Stapf.) Bean in Trees & Shrubs. 2: 279. (1914). *Sorbus caloneura* (Stapf.) Rehder in Sargent Pl. Wilson. 2: 269. (1915), Bean in Trees & Shrubs 2: 279. (1921), Nelson in A Heritage of Beauty 319. (2000), O' Brien in Augustine Henry, An Irish Plant Collector in China 36. (2002).

Sichuan: North Wushan, 7027.

A tree to 11 metres tall, native to central, western and southern China, Indochina and Malaysia. Introduced to cultivation by Wilson in 1904, it first flowered at Veitch's nursery in 1910. Cultivated in several Irish gardens including the National Botanic Gardens, Glasnevin.

02

Thomsonaria decaisneana (Lavalée) Rushforth (ROSACEAE)

Micromeles keissleri Schneider in Illus. Handb. Laubholzk. 1: 701. fig. 388. (1906). *Micromeles decaisneana* Schneid. var. *keissleri* Schneider in Fedde. Rep. Spec. Regni. Veg. Spec. Nov. 3: 151. (1906). *Pyrus keissleri* Leveille in Fl. Kouy-Tcheou. 351. (1915). *Sorbus keissleri* (Schneid) Rehder in Sargent Pl. Wilson. 2: 269. (1915).

Hubei: Jianshi, 5715a. North Badong, 7166 (type of *Micromeles keissleri* Schneider). **Sichuan:** South Wushan, 5715.

A deciduous tree to 13 metres tall, native to central and western China and Myanmar. Introduced to cultivation by Wilson in 1904 through Veitch's Coombe Wood nursery and later reintroduced by Forrest in 1931 from Yunnan.

Thomsonaria granulosa (Bertol.) Rushforth (ROSACEAE)

Sorbus granulosa (Bertol.) Rehder in Sargent Pl. Wilson. 2: 274. (1915).

Yunnan: On the Daweishan Range in Pingbian at 2,000 metres, 10136.

Thomsonaria sp. (ROSACEAE)

Pyrus aff. *granulosa* Bertol. Forbes & Hemsley in Journ. Linn. Soc. Bot. 23: 258. (1887).

Hubei: Badong, 541.

Thuidium cymbifolium (Dozy. & Molk.) Dozy & Molk. (THUIDIACEAE)

Salmon in Journ. Linn. Soc. Bot. 34: 470. (1900).

Hubei: Glens near Yichang, 7911. **Yunnan:** Simao, 13553. (Moss).

Thunbergia coccinea Wallich ex D. Don (ACANTHACEAE)

Anonymous in Kew List of Determinations (volume 34).

Yunnan: Simao, 12360.

A woody climber with stems to 8 metres long. Distributed in India, Myanmar and China (Yunnan).

Thunbergia fragrans Roxb. (ACANTHACEAE)

Forbes & Hemsley in Journ. Linn. Soc. Bot. 26: 237. (1890).

Thunbergia hainanensis C. Y. Wu & H. S. Lo in Flora Hainanica 3: 591. (1974).

Hainan: Without locality, without number. **Yunnan:** Mengzi, 9727. Simao, 12376.

Thunbergia grandiflora (Roxb. ex Rottl.) Roxb. (ACANTHACEAE)

Forbes & Hemsley in Journ. Linn. Soc. Bot. 26: 237. (1890).

Hainan: Near Haikou, without number. **Yunnan:** On the bank of the Red River near Manpan, 10850.

The Bengal clock vine, an evergreen woody climber with stems to 5 metres long and producing pendelous racemes of blue-purple trumpet-shaped flowers from May to October. Distributed in India, southern China and the Indochina Peninsula.

Thunbergia lutea T. Anderson (ACANTHACEAE)

Anonymous in Kew List of Determinations (volume 34).

Yunnan: Simao, 13138.

Thunia alba (Lindl.) Rchb. f. (ORCHIDACEAE)

Thunia marshalliana Rchb. f. Anonymous in Kew List of Determinations (volume 34).

Yunnan: Simao, 12967.

Thunia alba (Lindl.) Rchb. f. var. ***bracteata*** (Roxb.) N. Pearce & P. J. Cribb. (ORCHIDACEAE)

Thunia marshalliana Rolfe non Rchb. f. in Journ. Linn. Soc. Bot. 36: 18. (1903). *Thunia venosa* Rolfe in Orchid Rev. 13: 206. (1905).

Yunnan: On damp mountains near Mengzi at 1,750 metres, 11091.

Thyrocarpus sampsonii Hance (BORAGINACEAE)

Forbes & Hemsley in Journ. Linn. Soc. Bot. 26: 149. (1890), Brand in Engler, Pflanzenr. iv. 252. (112. (1921).

Hubei: Yichang, 772, 823, 842, 1168,3391. In a ravine on Moji Shan, 3367. Xiling (Yichang) Gorge, 3367a. Jianshi, 5489. **Sichuan:** South Wushan, 5489a. **Yunnan:** At Hsienkei on the Red River near the Vietnamese border, 9564. Mengzi, 10790, 10791.

Thyrocarpus sampsonii was described from a collection made near Macao by Theo. Sampson in 1861. Collected by the Glasnevin Central China Expedition (GCCE 596) in Xingshan in September 2004. Native to China and Vietnam.

Thysanolaena latifolia (Roxb. ex Hornem.) Honda (POACEAE)

Thysanolaena agrostis Nees Rendle in Journ. Linn. Soc. Bot. 36: 340. (1904).

Taiwan: Wanchin, 407. **Yunnan:** In the Red River valley, 10950.

A tall reed-like perennial, native to India, southern China and Indochina to Malesia.

Tiarella polyphylla D. Don (SAXIFRAGACEAE)

Forbes & Hemsley in Journ. Linn. Soc. Bot. 23: 270. (1887), Diels in Bot. Jahrb. 29: 367. (1900).

Hubei: Badong, 1735, 4030, 5332. Without locality, 5332a., 5332b. **Sichuan:** Emei Shan, Henry's Chinese collector with Pratt, 8978. **Yunnan:** Mengzi, 10494.

This species was collected by the Glasnevin Central China Expedition (GCCE 340) in a valley near Yesanguan in Badong in September 2002. Native to the Himalaya, Myanmar, China and Japan.

Ticanto crista (L.) R. Clark & Gagnon (FABACEAE)

Caesalpinia bonducella Henry non (L.) Fleming

in Trans. Asiat. Soc. Jap. 24 suppl: 38. (1896) (List Pl. Formosa). *Caesalpinia nuga* Ait. Henry in Trans. Asiat. Soc. Jap. 24 suppl: 38. (1896) (List Pl. Formosa), Li in Woody Flora of Taiwan 339. (1963), Zheng in Journ. Wuhan Bot. Res. 2.(1): 96. (1984). *Caesalpinia crista* L. Li in Woody Flora of Taiwan 339. fig. 119. (1963).

Taiwan: Kaohsiung, without number. Oluanpi, 300.

Grows as a scandent woody vine in coastal thickets in Taiwan.

Ticanto sinensis (Hemsl.) R. Clark & Gagnon (FABACEAE) New species

Mezoneuron sinense Hemsley in Journ. Linn. Soc. Bot. 23: 204. (1887), Henry in Journ. China Br. Roy. Asiat. Soc. 22: 238. (1887) (Chinese Names of Plants), Hooker's Ic. Pl. t. 1960. (1891), Bretschneider in Hist. Eur. Bot. Disc. China 2: 782. (1898). *Mezoneuron sinense* Hemsl. var. *parvifolium* Hemsley in Journ. Linn. Soc. Bot. 23: 205. (1887).

Hubei: Yichang, 1122 (syntype of *Mezoneuron sinense* Hemsl. var. *parvifolium* Hemsley*)*, 2238 (syntype of *Mezoneuron sinense* Hemsl.), 2319, 2612, 3416a., 3416b. Liantuo 2612, 4629. Antelope Glen, 3416.

Tilia henryana Szyszylowicz (MALVACEAE) New species

in Hooker's Icon. Pl. 20: t. 1927. (1890), Henry in Notes Econ. Bot. China 21. (1893), Bretschneider in Hist. Eur. Bot. Disc. China 2: 780. (1898), Henry in Flora & Sylva 1: 217-218. (1903), J. H. Veitch in Hortus Veitchii 381. (1906), Schneider in Illus. Handb. Laubholzk. 2: 388. fig. 259. (1909), V. Engler in Monog. Tilia. 125. (1909), Pampanini in Nouv. Giorn. Bot. Ital. n. ser. xvii. 431. (1910), E. H. Wilson in A Naturalist in Western China 32. (1913), Bean in Trees & Shrubs. 2: 591. (1914), Morley in Glasra 3: 80. (1979), Zheng in Journ. Wuhan Bot. Res. 2.(1): 147. (1984), Lancaster in Travels in China 410. (1993), Nelson in A Heritage of Beauty 320. (2000), O' Brien in Augustine Henry, An Irish Plant Collector in China 37. (2002).

Hubei: Xingshan, 7452a., (type).

The largest of all the lindens from central and western China and without a doubt, the most beautiful. Wilson found it later on the mountains between Yichang and Xingshan at a place called *t'an shy ya* (the lime tree pass) The same pass was named for a gigantic specimen of Henry's linden which was 26 metres tall and 8.5 metres in girth. The unfolding toothed leaves in spring are covered with a white tomentum, emerging to bronze before finally maturing to green, the bark is light gray, firm and fissured. *Tilia henryana* was introduced to cultivation by Wilson when collecting for Messrs Veitch. This species is found in several Irish gardens, the finest trees in Europe are found at Birr Castle, County Offaly, these were raised by the sixth Earl of Rosse from

seeds sent from Lushan Botanic Gardens in 1938. When measured in 2002 the best of these trees was 14.5 metres tall with a trunk diameter of 37 cm.

Tilia oliveri Szyszylowicz (MALVACEAE) New species

in Hooker's Icon. Pl. 20: t. 1927. (1890), Henry in Notes Econ. Bot. China 21. (1893), Bretschneider in Hist. Eur. Bot. Disc. China 2: 780. (1898), Schneider in Illus. Handb. Laubholzk. 2: 387. fig. 259. (1909), V. Engler in Monog. Tilia. 114. (1909), Elwes & Henry in The Trees of Gt. Brit. & Irel. 7: 1681. (1912), Bean in Kew Bull. Misc. Inf. 53. (1914); Trees & Shrubs. 2: 593. (1914), Rehder & E. H. Wilson in Sargent Pl. Wilson 2: 366. (1913), Morley in Glasra 3: 80. (1979), Muir in Gard. Chron. 183: 178. (1978), Zheng in Journ. Wuhan Bot. Res. 2.(1): 147. (1984), Nelson in A Heritage of Beauty 320. (2000), O' Brien in Augustine Henry, An Irish Plant Collector in China 37. (2002).

Hubei: Fang Xian, 7452b. **Sichuan:** North Wushan, 7089 (type).

This species [which commemorates Professor Daniel Oliver (1813—1916) Keeper of the Kew Herbarium] was said by Wilson to be very abundant in the moist woods of north-west Hubei (i.e. Fang Xian, Shennongjia and Xingshan). In these regions it formed a tree of moderate size (12 to 15 metres tall) with spreading, down swept branches, though large trees were occasionally met with. Introduced by Wilson from Hubei in 1900. Seeds of Wilson's 4411 from Fang Xian, collected in Novermber 1910 were received at Glasnevin from the Arnold Arboretum the following spring. In the Veitch catalogue *New Hardy Plants from Western China* (Autumn 1912) this species was available for two shillings and six pence. *Tilia oliveri* has reached 25 metres tall with a trunk diameter of 64 cm at Westonbirt in Gloucestershire (2002).

Tilia tuan Szyszylowicz (MALVACEAE) New species

in Hooker's Icon. Pl. 20: t. 1926. (1890), Henry in Notes Econ. Bot. China 21. (1893), Bretschneider in Hist. Eur. Bot. Disc. China 2: 780. (1898), Diels in Bot. Jahrb. 29: 468. (1900), Schneider in Illus. Handb. Laubholzk. 2: 389. fig. 259. (1909), Rehder in Sargent Pl. Wilson 2: 368. (1916), Nelson in Moorea 2: 6. (1983), Zheng in Journ. Wuhan Bot. Res. 2.(1): 147. (1984), Nelson in A Heritage of Beauty 320. (2000), O' Brien in Augustine Henry, An Irish Plant Collector in China 37. (2002).

Hubei: Jianshi, 5874 (type, June 12th 1888). **Sichuan:** South Wushan, 7452 (isosyntype).

A tree of medium height, common in the woods of western Hubei in the early 20th century according to Wilson. In the Three Gorges region, according to Henry, it was known colloqually as *tuan-shu*, (as are all other species of *Tilia*) hense Szysylowicz's adoption of this name for the specific epithet. The bark of this species was used in making shoes and it was in these shoes that Henry sometimes did his plant collecting (see Henry in *Flora & Sylva* 1: 217. (1903). There is a young tree in the *Tilia* class in the National Botanic Gardens. The best known example of this species in cultivation grows at Thorp Perrow, Yorkshire, planted in 1936, when measured in 1991 this tree was found to be 16 metres tall with a trunk diameter of 49 cm.

Tilia tuan Szyszylowicz var. ***chinensis*** (Szyszyl) Rehder & E. H. Wilson ((MALVACEAE) New variety

in Sargent Pl. Wilson 2: 369. (1915), Bean in Trees & Shrubs 4: 610. (1989).

Tilia miqueliana Henry non Maxim. in Notes Econ. Bot. China 21. (1893). *Tilia miqueliana* Max. var. *chinensis* Szyszylowicz in Hooker's. Icon. Pl. t. 1927. (1890), J. H. Veitch in Hortus Veitchii 381. (1906).

Hubei: Xingshan, 6474 (type of *Tilia miqueliana* Max. var. *chinensis* Szyszylowicz).

Tinomiscium petiolare Miers ex Hook. f. & Thoms. (MENISPERMACEAE)

Tinomiscium tonkinense Gagnep. Diels in Engler, Pflanzenr. iv. 94: 118. (1910).

Yunnan: On the mountains to the east of Simao, 12068.

Distributed in China (Guangxi and Yunnan) and Vietnam.

Tinospora glabra (Burm. f.) Merrill (MENISPERMACEAE) Forman in Kew. Bull. 36: 414. (1982).

Hainan: Hummocks near Haikou, 8085.

A woody climber, widely distributed through the tropics from China (Hainan), the Philippines, the Malay peninsula, Sumatra, Java, New Guinea and Timor.

Tirpitzia sinensis (Hemsl.) Hallier f. (LINACEAE) New species

Reinwardtia sinensis Hemsley in Hooker's Icon. Pl. t. 2594. (1899).

Yunnan: Mengzi, 9081. On rocky mountains near Mengzi at 1,650 metres, 9081b., (isotype of *Reinwardtia sinensis* Hemsley).

Native to China and Vietnam.

Titanotrichum oldhamii (Hemsl.) Soler. (GESNERIACEAE)

Rehmannia oldhamii Hemsley in Ann. of Bot. 9: 155. (1895), Henry in Trans. Asiat. Soc. Jap. 24 suppl: 68. (1896) (List Pl. Formosa). *Matsumuria oldhamii* (Hemsl.) Hemsley in Kew Bull. Misc. Inf. 361. (1909).

Taiwan: Wanshoushan, on sides of cliffs in dry shade, 311, 1052.

Tofieldia sp. (TOFIELDIACEAE)

Anonymous in Kew List of Determinations (volume 34).

Yunnan: Mengzi, 10286, 10542.

Tofieldia thibetica Franch. (TOFIELDIACEAE)

Tofieldia nuda Wright non Maxim. in Journ. Linn. Soc. Bot. 36: 141. (1903).

Hubei: Xingshan, 6561. Fang Xian at 2,950 metres, 7627a. **Sichuan:** North Wushan, 7627.

Tongoloa dunnii (H. Boissieu) H. Wolff (APIACEAE) New species

Pimpinella dunnii H. Boissieu in Bull. Herb. Boissier. ser. 2. 3: 841. (1903).

Hubei: Xingshan, 6955 (type of *Pimpinella dunnii* H. Boissieu).

Toona ciliata M. Roem. (MELIACEAE)

Cedrela toona Roxb ex Rottler & Willd. var. *yunnanense* C. de Candolle in Rec. Bot. Surv. India. 3(4): 366. (1908). *Toona ciliata* M. Roemer var. *yunnanensis* (C. DC.) C. Y. Wu in Flora Yunnanica 1: 207. (1977).

Yunnan: Mengzi at 1,500 metres, 9486 (isosyntype of *Cedrela toona* Roxb ex Rottler & Willd. var. *yunnanense* C. de Candolle). Simao, 12807. Simao at 1,400 metres, 13001 (isosyntype of *Cedrela toona* Roxb ex Rottler & Willd. var. *yunnanense* C. de Candolle).

Toona sinensis (A. Juss.) M. Roem. (MELIACEAE)

Cedrela sinensis A. Juss. var. *hupehana* C. DC in Rec. Bot. Surv. India. 3(4): 361. (1908). *Cedrela sinensis* A. Juss. Henry in Notes Econ. Bot. China 38. (1893), Henry in Elwes and Henry The Trees Gt. Brit. & Irel. 2: 433. (1907), Rehder & E. H. Wilson in Sargent Pl. Wilson 2: 156. (1914), Zheng in Journ. Wuhan Bot. Res. 2(1): 109. (1984).

Hubei: Yichang, 3657 (isotype of *Cedrela sinensis* A. Juss. var. *hupehana* C. DC). **Yunnan:** Mengzi at 1,800 metres, 11131.

A common tree in central China and often cultivated as the young shoots are highly esteemed as a vegetable. To enable the gathering of shoots the trees are kept stunted. Under normal conditions *Toona sinensis* forms a tree to 25 metres tall and according to Henry was known colloquially as *ch'uen-tien shu* or *hsiang-ch'un,* its wood is very valuable and is used in high grade carpentry. Discovered by the French Jesuit missionary, Père d'Incarville (1706—1757) and introduced to cultivation by Père Armand David through the garden of the Paris Museum. The tree in the old Chinese shrubbery at the National Botanic Gardens, Glasnevin is one of the finest specimens in Europe. Raised from Wilson's 585 (collected near Yichang, September 1907) it was 15 metres tall with a trunk diameter of 76 cm when last measured in 1992. It has now far exceeded these dimensions and continues to grow vigorously. Native from China and Korea to western Malesia.

02

Toona sureni (Blume) Merr. (MELIACEAE)

Cedrela toona (Roxb. ex Rottler & Willd.) var *henryi* C. de Candolle in Rec. Bot. Surv. India 3(4): 369. (1908). *Toona ciliata* M. Roemer var. *henryi* (C. DC.) C. Y. Wu in Flora Yunnanica 1: 209. (1977).

Yunnan: Mountains to the south of Simao at 1,650 metres, 11963 [isotype of *Cedrela toona* (Roxb. ex Rottler & Willd.) var. *henryi* C. de Candolle].

Torenia benthamiana Hance (LINDERNIACEAE)

Anonymous in Kew List of Determinations (volume 24).

Hainan: Environs of Kiungchow, 8315.

Torenia concolor Lindl. (LINDERNIACEAE)

Henry in Trans. Asiat. Soc. Jap. 24 suppl: 67. (1896) (List Pl. Formosa).

Taiwan: Wanchin, 811, 1576, 1697. Oluanpi, 645, 1697.

Torenia crustacea (L.) Cham. & Schldt. (LINDERNIACEAE)

Vandellia crustacea (L.) Benth. Forbes & Hemsley in Journ. Linn. Soc. Bot. 26: 189. (1890), Henry in Trans. Asiat. Soc. Jap. 24 suppl: 67. (1896) (List Pl. Formosa).

Hubei: Yichang, 171. **Taiwan:** Wanchin, 886. Kaohsiung, 1075.

Torenia diffusa D. Don (LINDERNIACEAE)

Torenia vagans Roxb. Forbes & Hemsley in Journ. Linn. Soc. Bot. 26: 189. (1890).

Hubei: At Shih Pan Shan [off the Xiling (Yichang) Gorge], 4189.

Torenia flava Buch.-Ham. ex Benth. (LINDERNIACEAE)

Henry in Trans. Asiat. Soc. Jap. 24 suppl: 67. (1896) (List Pl. Formosa).

Taiwan: Wanchin, 1577.

Torenia cf. ***oblonga*** (Benth.) Hance (LINDERNIACEAE)

Anonymous in Kew List of Determinations (volume 24).

Hainan: Environs of Kiungchow, 8303.

Torenia spp. (LINDERNIACEAE)

Anonymous in Kew List of Determinations (volume 24 & 34).

Taiwan: Tamsui, 1455. **Yunnan:** South of the Red River from Manmei at 1,650 metres, 9467. Mengzi, 10375.

Torenia violacea (Azaola) Pennell (LINDERNIACEAE)

Torenia peduncularis Benth. ex Hook. f. Forbes & Hemsley in Journ. Linn. Soc. Bot. 26: 188. (1890), Henry in Trans. Asiat. Soc. Jap. 24 suppl: 67. (1896) (List Pl. Formosa).

Hubei: Yichang, 158, 2183, 2787. Liantuo, 4426. **Taiwan:** Wanchin, 1506. **Yunnan:** Mengzi, 9227.

A beautiful little plant, common in southern Yunnan. Distributed in India, Bhutan, China, Indochina and the Philippines.

Torilis japonica (Houtt.) DC. (APIACEAE)

Torilis anthriscus Henry non (L.) Gmel. in Journ. China Br. Roy. Asiat. Soc. 22: 266. (1887) (Chinese Names of Plants).

Hubei: Yichang, 565, 747, 981, 988. Near Yichang, on the hills behind the San You Dong Glen, (July 1886), 1512. Badong, 1848, 4751, 4836.

Henry stated this species was known colloquially in the Three Gorges region as *p'o tzu ts'ao.*

Torilis scabra (Thunb.) DC. (APIACEAE)

Torilis henryi Norman in Journ. of Bot. 67: 147. (1929).

Hubei: Yichang, 1349 (type of *Torilis henryi* Norman).

Torreya fargesii Franch. (TAXACEAE) New species

Pilger in Engler, Pflanzenr. 5: 108. (1903).

Torreya nucifera Henry non (L.) Sieb. & Zucc. in Journ. China Br. Roy. Asiat. Soc. 22: 269. (1887) (Chinese Names of Plants); in Notes Econ. Bot. China 36. (1893), Masters in Journ. Linn. Soc. Bot. 26: 546. (1902). *Torreya grandis* Henry non Fortune ex Lindl. in Trees of Gr. Brit. & Irel. vi. 1464. (1912), Rehder & E. H. Wilson in Sargent Pl. Wilson 2: 7. (1914), Walsh in Aug. Henry Forst. Herb. 5. (1957).

Hubei: Liantuo, 5346. Xingshan, 6478. **Sichuan:** North Wushan, 7096.

When working on Taxaceae for Sargent's *Plantae Wilsonianae* neither Rehder or Wilson had seen ripe fruits of *Torreya grandis* and confused *Torreya fargesii* and *Torreya nucifera*. *Torreya fargesii* Franch. is native to central and western China. It was discovered in Hubei by Henry and was later collected in eastern Sichuan by Farges on whose material Franchet based the type. Henry stated it was known colloquially in the Three Gorges region as *niu wei sha*. This species was collected by the Glasnevin Central China Expedition (GCCE 649) in Dragon Gate River Forest Park in Xingshan in October 2004.

Torricellia angulata Oliver (TORRICELLIACEAE) New species

in Hookers Icon Pl. 19. t. 1893. (1889), Bretschneider in Hist. Eur. Bot. Disc. China 2: 784. (1898), Harms in Bot. Jahrb. xxix. 506. (1900), Wagnerin in Engler. Pflanzenr. iv.-229. 33. fig. 6. (1910), Leveille in Fl. Kouy-Tcheou. 117. (1914), Rehder in Sargent Pl. Wilson 2: 569. (1916).

Sichuan: South Wushan, 5524.

Torricellia sp. (TORRICELLIACEAE)

Anonymous in Kew List of Determinations (volume 34).

Yunnan: Mengzi, 10793.

Torricellia tiliifolia De Candolle (TORRICELLIACEAE)

Rehder in Sargent Pl. Wilson. 2: 569. (1916).

Yunnan: Mountains to the west of Simao at 1,600 metres, 11909. Pu'er at 1,800 metres, 13297.

Previously unknown in China prior to Henry's collections at Simao.

Toxicodendron delavayi (Franch.) F. A. Barkley (ANACARDIACEAE)

Rhus delavayi Franch. Rehder & E. H. Wilson in Sargent Pl. Wilson. 2: 183. (1916).

Yunnan: Grassy hills near Mengzi at 1,600 metres, 10283.

Endemic to China and distributed in Sichuan and Yunnan.

Toxicodendron hookeri (Sahni & Bahadar) C. Y. Wu & T. L. Ming (ANACARDIACEAE)

Rhus insignis Hook. f. Anonymous in Kew List of Determinations (volume 34).

Yunnan: Simao, 11907.

Native to India and China [Tibet (Xizang) and Yunnan].

Toxicodendron orientale Greene subsp. hispidum (Engl.) Yonek. (ANACARDIACEAE)

Rhus toxicodendron Henry non L. in Notes Econ. Bot. China 42. (1893). *Rhus orientalis* C. K. Schneid. non (Greene) C. K. Schneid. in Sargent Pl. Wilson 2: 179. (1914). *Rhus toxicodendron* L. var. *hispida* Zheng in Journ. Wuhan Bot. Res. 2(1): 117. (1984).

Hubei: Badong, 922 (November 1885), 4861. Xingshan, 6448. **Sichuan:** Henry's Chinese collector with Pratt, no locality stated but may be any of the following, Leshan, Emei Shan, Wa Shan or Kangding, 8791.

Toxicodendron succedaneum (L.) Kuntze (ANACARDIACEAE)

Rhus succedanea L. Henry in Trans. Asiat. Soc. Jap. 24 suppl: 29. (1896) (List Pl. Formosa), Rehder & E. H. Wilson in Sargent Pl. Wilson 2: 182. (1914), Zheng in Journ. Wuhan Bot. Res. 2.(1): 117. (1984).

Hubei: South Donghu, 7578. **Hainan:** Environs of Kiungchow, 8728. **Taiwan:** Wanchin, mountains, 62. **Yunnan:** In mountains near Simao at 1,500 to 1,600 metres, 11690a., 11955, 11955a.

The wax tree, a deciduous tree to 10 metres tall, widely distributed across the Himalaya, China (including Taiwan), Japan and Malaysia. It was once much cultivated in Japan for vegetable tallow to make candles until replaced by American and Russian petroleum. According to Bretschneider it was introduced by Gabriel Eugene Simon through the garden of the Paris Museum before 1863. This species was collected by the Glasnevin Central China Expedition (GCCE 246) on the hills above Badong in September 2002. It is rather rare in Hubei and Sichuan.

Toxicodendron sylvestre (Sieb. & Zucc.) Kuntze (ANACARDIACEAE)

Rhus sylvestris Sieb. & Zucc. Forbes & Hemsley in Journ. Linn. Soc. Bot. 23: 147. (1886), Henry in Journ. Chinese Br. Roy. Asiat. Soc. 22: 237. (1887) (Chinese Names of Plants); in Notes Econ. Bot. China 41. (1893), Rehder & E. H. Wilson in Sargent Pl. Wilson 2: 180. (1914), Zheng in Journ. Wuhan Bot. Res. 2.(1): 117. (1984).

Hubei: Yichang, 314a., 4693. Badong, 923 (November 1885). Liantuo, 3875, 4648, 6349.

A species allied to *Toxicodendron vernicifluum,* but smaller in stature and very often a shrub. Native to eastern and central China, Korea and Japan. According to Henry this tree was known colloquially around Yichang as the *shan ch'i* or *yeh-ch'i.* Wilson stated it meets its western limits in western Hubei where it forms a large shrub or small tree and has no economic value. Introduced to cultivation in 1881 but still a rare tree in gardens. The foliage in autumn turns fiery red.

Toxicodendron vernicifluum (Stokes) F. A. Barkley (ANACARDIACEAE)

Rhus vernicflua Stokes Forbes & Hemsley in Journ. Linn. Soc. Bot. 23: 148. (1886), Henry in Notes Econ. Bot. China 41. (1893), Rehder & E. H. Wilson in Sargent Pl. Wilson 2: 181. (1914), Zheng in Journ. Wuhan Bot. Res. 2.(1): 117. (1984), Nelson in Pim, The Wood & the Trees 218. (1984); in A Heritage of Beauty 317. (2000). *Rhus vernicifera* DC. Henry in Journ. Chinese Br. Roy. Asiat. Soc. 22: 236. (1887) (Chinese Names of Plants); in Notes Econ. Bot. China 41. (1893), Pim in the Wood & the Trees 28. (1984).

Hubei: Yichang, 314b., Badong, 314 (June 1885). Jianshi, 5899, 6035. Changyang (cultivated), 6035a. Without locality, 5899a.

The Chinese varnish tree, a deciduous tree to 20 metres with pinnate leaves to 60 cm long. Native to the Himalaya and China to Korea. *Toxicodendron vernicifluum* is still a very familiar sight in the mountainous countryside of western Hubei, where it grows both wild and cultivated. Henry stated this tree was known as *ch'i shu,* (literally 'varnish tree' and from its sap a high quality lacquer is made. The method of extraction is much the same as that for rubber, in late summer 'v' shaped incisions are made about 30 cm apart and in these incisions mussel shells or pieces of bamboo are placed, the sap flows out over-night and is collected the following morning. This sap takes on a black appearance on exposure to air, becoming the Chinese black varnish and was once much favoured for varnishing coffins, etc. One of the strange features of this varnish is that it needs humid, cloudy weather to dry properly, if applied in hot sunny weather it will not dry. The use of lacquer was developed into an art form in China before the Christian era. The lacquer made from *Toxicodendron vernicifluum* (being a member of the Anacardiaceae) is extremely poisonous, but some people are immune. The same method of extraction is still used in the Three Gorges region. Henry stated that from the fruits of the tree an excellent tallow called *ch'i-la* was made, which had a high melting point and was used for making candles by people living in the mountains. By the chain-tent shrubbery at the National Botanic Gardens, Glasnevin is a plant of Wilson's 123, the seeds of which he collected near Yichang in September 1907 from a tree 20 metres tall with a girth of 2 metres. Wilson's 1907 introduction grew faster than Japanese (cultivated in Japan) sourced-plants (from where all the existing introductions had been made). Henry sent seeds (collected on his first plant collecting expedition in November 1884) of the varnish tree to Kew (where it arrived on May 12th 1885) with his very first letter to Sir Joseph Hooker while seeking advice on the nomenclature of local Yichang plants.

Toxicodendron vernicifluum was collected by the Glasnevin Central China Expedition (GCCE 262) on the mountains in Badong in October 2002 and again in September 2004 in Xingshan (GCCE 501).

Toxicopueraria peduncularis (Benth.) A. N. Egan & B. Pan (FABACEAE)

Pueraria peduncularis (Benth.) Graham ex Benth. Anonymous in Kew List of Determinations (volume 34).

Yunnan: Mengzi, 9177, 10628. Simao, 12483.

Native from Pakistan to India, Nepal, Myanmar and western China.

Toxocarpus himalensis Falc. ex Hook. f. (APOCYNACEAE)

Anonymous in Kew List of Determinations (volume 34).

Yunnan: Mengzi, 9526, 10927, 13628. Simao, 12002, 12927, 13202.

Native to India and China (Guangxi, Guizhou and Yunnan).

Toxocarpus spp. (APOCYNACEAE)

Anonymous in Kew List of Determinations (volume 34).

Yunnan: Mengzi, 10110. Simao, 13418.

Toxocarpus villosus (Blume) Decne. (APOCYNACEAE)

C. K. Schneider in Sargent Pl. Wilson. 3: 349. (1916), Zheng in Journ. Wuhan Bot. Res. 2.(1): 184. (1984).

Toxocarpus wightianus Forbes & Hemsley non Hook. & Arn. in Journ. Linn. Soc. Bot. 26: 101. (1889).

Hubei: Antelope Glen, 3281. Yichang, 3515.

Previously unknown in China.

Toxocarpus wightianus Hook. & Arn. (APOCYNACEAE)

Anonymous in Kew List of Determinations (volume 24).

Hainan: On hummocks near Haikou, 8109.

Trachelospermum asiaticum (Sieb. & Zucc.) Nakai (APOCYNACEAE)

Trachelospermum divaricatum (Thunb.) Kanitz. var. *brevisepalum* Schneider in Sargent Pl. Wilson. 3: 338. (1916). *Trachelospermum gracilipes* Hook. f. Li in Woody Flora of Taiwan 794. fig. 320. (1967).

Taiwan: Oluanpi, 1301[co-type of *Trachelospermum divaricatum* (Thunb.) Kanitz. var. *brevisepalum* Schneider].

A tall climber, distributed across north-east India, Indochina, southern China (including Taiwan) to Borneo.

Trachelospermum axillare Hook. f. (APOCYNACEAE)

C. K. Schneider in Sargent Pl. Wilson. 3: 335. (1916).

Hubei: Xiling (Yichang) Gorge, 3618. **Yunnan:** Mountains to the south-west of Mengzi at 1,900 metres, 9854. In Pingbian on the Daweishan Range at 1,600 metres, 9854a. Mountains to the south-west of Mengzi at 1,600 metres, 9854b. Mountains to the east of Simao at 1,600 metres, 12050. Mountains to the north-west of Simao at 1,600 metres, 13115.

A widespread woody climber, distributed from Sikkim to Sri Lanka, China and Vietnam. The bark is a source of fibre used to make ropes.

Trachelospermum jasminoides (Lind.) Lem. (APOCYNACEAE)

Forbes & Hemsley in Journ. Linn. Soc. Bot. 26: 99. (1889), Henry in Trans. Asiat. Soc. Jap. 24 suppl: 60. (1896) (List Pl. Formosa), Schneider in Sargent Pl. Wilson. 3: 334. (1917).

Hubei: In glens near Yichang, 496, 504, 792, 1361, 1367. Liantuo, 3945, 4585. Changyang, 7495. **Taiwan:** Oluanpi, 948, 1994. **Yunnan:** Mengzi, 9092.

An evergreen climber to 4 metres tall, producing white fragrant flowers in July and August on slender-stalked cymes to 5 cm long. Henry stated it was the *luo shi teng* or 'wrap around stone vine', in Chinese herbal medicine and is used to lower blood pressure and to relieve the pain of rheumatic arthritis. *Trachelospermum jasminoides* was introduced to cultivation from a Shanghai nursery by Robert Fortune in 1844. A common climber in woods and thickets in north-west Hubei, it was collected by the Glasnevin Central China Expedition in Xingshan (GCCE 463 & GCCE 629) in the autumn of 2004.

Trachelospermum lucidum (D. Don) K. Schum. (APOCYNACEAE)

Trachelospermum jasminoides Forbes & Hemsley non (Lind.) Lem. in Journ. Linn. Soc. Bot. 26: 99. (1899), (in part). *Trachelospermum cathayanum* C. K. Schneider in Sargent Pl. Wilson. 3: 333. (1916), Tsiang in Sunyatsenia 2: 140. (1934). *Trachelospermum tetanocarpum* C. K. Schneider in Sargent Pl. Wilson. 3: 339. (1916) Tsiang in Sunyatsenia 2: 142. (1934).

Hubei: Jianshi, 5976. **Yunnan:** Near Fengchunling, on the mountains south of the Red River at 1,900 metres, 10651

02

(syntype of *Trachelospermum cathayanum* Schneider). In Pingbian, in woods on the Daweishan Range at 1,600 metres, 10651a. In forests near Simao at 1,300 metres, 11949 (type of *Trachelospermum tetanocarpum* Schneider). In forests to the south-east of Simao, 12800 (paratype of *Trachelospermum tetanocarpum* Schneider).

Distributed from Pakistan to China (Taiwan).

Trachelospermum sp. (APOCYNACEAE)

Anonymous in Kew List of Determinations (volume 34).

Yunnan: Mengzi, 11243.

Trachycystis ussuriensis (Maack & Regel) D. J. Kop (MNIACEAE)

Mnium curvulum Müll. Hal. Salmon in Journ. Linn. Soc. Bot. 34: 458. (1900).

Hubei: Fang Xian, on a stone, September 1888, 6933. (Moss).

Trachypus bicolor Reinw. & Hornsch. (METEORIACEAE)

Anonymous in Kew List of Determinations (volume 34).

Yunnan: Mengzi, 13713. (Moss).

Trachyspermum scaberulum (Franch.) H. Wolff. ex. Hand.-Mazz. (APIACEAE)

Pimpinella scaberula Franch. Anonymous in Kew List of Determinations (volume 34).

Yunnan: Simao, 12278.

Endemic to China and distributed in the western provinces.

Trametes versicolor (L.) Lyoyd (POLYPORACEAE)

Polystictus versicolor (L.) Fr. Anonymous in Kew List of Determinations (volume 24).

Hubei: Fang Xian, 6568, 6577. Changyang, on trees, 7925, 7529. **Sichuan:** South Wushan, 7046. (Fungus).

Trapa natans L. (LYTHRACEAE)

Forbes & Hemsley in Journ. Linn. Soc. Bot. 23: 311. (1887), Henry in Trans. Asiat. Soc. Jap. 24 suppl: 45. (1896) (List Pl. Formosa).

Trapa bispinosa Roxb. Henry in Journ. China Br. Roy. Asiat. Soc. 22: 255. (1887) (Chinese Names of Plants).

Hubei: Yichang, 2929, 4145. **Taiwan:** Kaohsiung Plain, without number.

Jesuit's nut, water caltrops or water chestnut, a short-lived perennial, native to southern Europe and south-east Asia. Henry stated that at Yichang this species was known colloquially as *ling-ko*. An aquatic vegetable mainly grown in China. Poisonous when raw, the seeds must be boiled for at least an hour before eating. Needs a water temperature of 18 degrees celcius to grow well.

Trapa sp. (LYTHRACEAE)

Anonymous in Kew List of Determinations (volume 24).

Hubei: Yichang, 2340.

Trapella sinensis Oliver (PLANTAGINACEAE) New genus

in Hooker's. Icon. Pl. 16: t. 1595. (1887), Henry in Journ. China Br. Roy. Asiat. Soc. 22: 255. (1887) (Chinese Names of Plants), F. W. Oliver in Ann. of Bot. 2: 75. fig. 1. (1888); in Ann. Bot. 3: 134. (1889), Forbes & Hemsley in Journ. Linn. Soc. Bot. 26: 236. (1890), Bretschneider in Hist. Eur. Bot. Disc. China 2: 789. (1898), E. H. Wilson in A Naturalist in Western China 1: 21. (1913), Morley in Glasra 3: 80. (1979), Pim in The Wood & the Trees 27. (1984), Nelson in Pim, The Wood & the Trees 220. (1984).

Hubei: In abundance in one small pond a quarter mile behind the consulate at Yichang, the water in the pond being 30 to 60 cm deep, 1671, 4243. In dry rice fields near Yichang, 4300.

A floating aquatic perennial with long straggling, sparingly branched stems. Henry discovered this new genus in ponds close to the foreign compound at Yichang in which the customs building was situated in 1886. He stated that at Yichang it was known colloquially as *t'ieh ling chio* or iron *Trapa* indicating the uselessness of the plant compared to *Trapa* which is an important crop. In July 1887 Henry sent a further supply of material preserved in alcohol and this allowed F. W. Oliver (son of Professor Daniel Oliver who described *Trapella*) to give a complete and detailed account of the plant. In that same account Oliver states,

That I am able now to give this monographic account of a plant unknown to science before 1887, speaks of Dr. Henry's prompt courtesy in obtaining and dispatching material. No botanist in China in recent times has sent home collections richer in entirely new forms than has Dr. Henry, who is now working on the flora of central China, hitherto an almost sealed book.

Trema cannabina Lour. (CANNABACEAE)

Trema timorense (Decne.) Blume Forbes & Hemsley in Journ. Linn. Soc. Bot. 26: 452. (1894). *Trema virgatum* (Planch.) Blume Schneider in Sargent Pl. Wilson. 3: 289. (1916). *Trema orientale* Henry non (L.) Blume in Trans. Asiat. Soc. Jap. 24 suppl: 86. (1896) (List Pl. Formosa), in part.

Hainan: Ling-men, 8559. **Taiwan:** Oluanpi, 624, 999. **Yunnan:** Mengzi at 1,300 metres, 10011.

A medium-sized deciduous tree, widespread tree over much of Myanmar, Thailand, Indochina, China, (including Hainan and Taiwan), Malesia, Australia, Melanesia and Micronesia.

Trema cannabina Lour. var. *dielsianum* (Hand.-Mazz.) C. J. Chen (CANNABACEAE) New variety

Trema timorense Forbes & Hemsley non Blume in Journ. Linn. Soc. Bot. 26: 452. (1894). *Trema* spec.(verisim spec. nova) C. K. Schneider in Sargent Pl. Wilson. 3: 289. (1916). *Trema dielsianum* Handel-Mazzetti in Symb. Sin. 7: 106. (1929), Zheng in Journ. Wuhan Bot. Res. 2.(1): 30. (1984).

Hubei: Changleping, 6210 (isotype of *Trema dielsianum* Handel-Mazzetti).

Trema levigatum Handel-Mazzetti (CANNABACEAE) New species

in Symb. Sin. 7: 106. (1929), Zheng in Journ. Wuhan Bot. Res. 2.(1): 30. (1984).

Trema timorense Forbes & Hemsley non Blume in Journ. Linn. Soc. Bot. 26: 452. (1894). *Trema virgatum* C. K. Schneider non (Planch.) Blume in Sargent Pl. Wilson. 3: 289. (1916).

Hubei: South Donghu, 7710.

Trema orientale (L.) Blume (CANNABACEAE)

Li in Woody Flora of Taiwan 107. fig. 32. (1963).

Trema amboinense Henry non (Willd.) Blume in Trans. Asiat. Soc. Jap. 24 suppl: 85. (1896) (List Pl. Formosa).

Taiwan: Wanchin, 30, 1769, 1910. Tamsui, 1464. **Yunnan**: Mengzi, 10650.

A large deciduous tree with soft wood, known as the *shan-yu-ma* and used, according to Henry, for making ladles, buckets, etc. Distributed in Nepal, India, Sri Lanka, Myanmar, China, Thailand, Vietnam and Japan to Australia.

Trema spp. (CANNABACEAE)

Anonymous in Kew List of Determinations (volume 24 & 34).

Hubei: Yichang, 573. **Yunnan**: Simao, 12189.

Trema tomentosum (Roxb.) H. Hara (CANNABACEAE)

Trema amboinense (Willd.) Blume Forbes & Hemsley in Journ. Linn. Soc. Bot. 26: 451. (1894).

Hainan: Without locality, 8490.

Trevesia palmata (Roxb. ex Lindl.) Vis. (ARALIACEAE)

Jebb in Glasra 3: 98. (1998).

Yunnan: Mengzi, 9603, 13805. Simao, 11757, 11904.

A little branched tree to 8 metres tall, found in evergreen forest in Nepal, India, Bhutan, Bangladesh, Myanmar, China, Thailand, Laos, Cambodia and Vietnam at altitudes between 250 to 1,500 metres. Still a common inhabitant of the warm-temperate forests on the mountains to the south of Mengzi.

Triadica cochinchinensis Lour. (EUPHORBIACEAE)

Sapium discolor (Champ. ex Benth.) Müell.- Arg. Henry in Trans. Asiat. Soc. Jap. 24 suppl: 85. (1896) (List Pl. Formosa), Li in Woody Flora of Taiwan 438. (1963).

Taiwan: Wanchin, 549.

A shrub or small tree, native to Malaysia, southern China and India and known in China (Taiwan) as *shan-chiung* according to Henry.

Triadica sebifera (L.) Small (EUPHORBIACEAE)

Sapium sebiferum (L.) Roxb. Henry in Journ. China Br. Roy. Asiat. Soc. 22: 260. (1887) (Chinese Names of Plants), Forbes & Hemsley in Journ. Linn. Soc. Bot. 26: 445. (1894), Henry in Trans. Asiat. Soc. Jap. 24 suppl: 85. (1896) (List Pl. Formosa), Li in Woody Flora of Taiwan 438. (1963).

Hubei: On Tsui Fu Shan near Yichang (May 1888), 1602. **Hainan**: Environs of Haikou, 7992. **Taiwan**: Kaohsiung, without number. Wanchin, without number. Oluanpi, 584.

The Chinese vegetable tallow tree, a long-lived deciduous tree to 19 metres tall and with girths of up to 2 metres. Henry stated that the colloquial name in Hubei for the tallow tree was *mu tzu shu*. According to Wilson, this valuable tree was widely cultivated at low altitudes in western Hubei, where in autumn the leaves assumed the most brilliant tints. The seeds are covered with a type of fat or tallow and this was removed by steaming or by rubbbing the seeds through a bamboo sieve. By crushing the left-over seeds, oil was abtained and the yield of fat and oil was about 30% by weight of seed. The tallow was then made into candles and coated with insect white-wax (see *Fraxinus chinensis*). *Triadica sebifera* has been naturalised in the USA since the 19th century and was collected by the Glasnevin Central China Expedition by the Xiang Xi River near the border of Xingshan and Yichang (GCCE 693) and in Dragon Gate River Forest Park in Xingshan (GCCE 829) in October 2004.

Triaenophora rupestris (Hemsl.) Soler. (PLANTAGINACEAE) New species

Rehmannia sp. nova Augustine Henry in Journ. China Br. Roy. Asiat. Soc. 22: 234. (1887) (Chinese Names of Plants). *Rehmannia rupestris* Hemsley in Journ. Linn. Soc. Bot. 26: 195. (1890), Hooker in Curtis's Bot. Mag. 117: t. 7191. (1891), Henry in Notes Econ. Bot. China 50. (1893), Bretschneider in Hist. Eur. Bot. Disc. China 2: 788. (1898), Morley in Glasra 3: 80. (1979), Nelson in Pim, The Wood & the Trees 222. (1984); in A Heritage of Beauty 317. (2000).

Hubei: Liantuo, 2604, 4458 (syntype). Fang Xian, 6615 (type).

According to Henry this species bears the Chinese name *ai-pai-ts'ai* or cliff cabbage and was much esteemed in traditional herbal medicine. He also stated it grew only in almost inaccessible places on the faces of cliffs within the gorges where it was a striking plant when in flower. It was introduced to cultivation by Henry through the Royal Botanic Gardens, Kew where it first flowered in 1888.

Tribulus terrestris L. (ZYGOPHYLLACEAE)

Henry in Trans. Asiat. Soc. Jap. 24 suppl: 24. (1896) (List Pl. Formosa).

Taiwan: Kaohsiung, without number.

According to Henry this plant occured on sandy banks near the sea and was known as *po-chi-li* (the *bai ji li* in other parts of China). Known as caltrops in Europe this is a small creeping annual bearing yellow flowers above pinnate leaves. Native to southern Europe and south-west Asia it is now found as a weed in most parts of the world. Poisonous to cattle and used as a drug in China to treat headache and dizziness.

Trichaptum fuscoviolaceum (Ehrenb.) Ryvarden (HYMENOCHAETACEAE)

Irpex fuscoviolaceus (Ehrenb.) Fr. Anonymous in Kew List of Determinations (volume 24).

Hubei: Xingshan at 1,400 metres, 6535. (Fungus).

Trichodesma calycosum Collett & Hemsley (BORAGINACEAE)

Johnston in Journ. Arnold Arb. 33: 75. (1952).

Trichodesma khasyana Clarke. Henry in Trans. Asiat. Soc. Jap. 24 suppl: 63. (1896) (List Pl. Formosa). *Trichodesma formosana* Matsumara in Tokyo Bot. Mag. 12: 108. (1898). *Trichodesma calycosum* Collett & Hemsley var. *formosanum* (Matsum.) Li in Journ. Arnold Arb. 33: 78. (1952). *Trichodesma sinicum* Brand in Fedde, Repert Spec. Nov. Regni. Veg. 12: 504. (1913); in Pflanzenr. Heft. 78: 43. (1921).

Taiwan: Wanchin, 23 (syntype of *Trichodesma formosana* Matsumara). Oluanpi, 286, 939, 1239. **Yunnan:** Mengzi at 1,650 metres, 10124. Lunan, 10124a. Chuyuan, 10124b. South of the Red River from Manpan at 1,300 metres, 10124c. Simao at 1,640 metres, 10124d., (type of *Trichodesma sinicum* Brand).

Distributed in India, Myanmar, China, Thailand and Laos.

Trichomanes bipunctatum Poir (HYMENOPHYLLACEAE)

Trichomanes filicula (Bory ex Willd.) Bory Christ in Bull. Herb. Boiss. 6: 866. (1898).

Yunnan: On trees in montane woodland to the south of Mengzi at 1,600 metres, 10836. Simao, 12942.

Trichomanes birmanicum Bedd. (HYMENOPHYLLACEAE)

Trichomanes speciosum Sw. var. *umbrosum* (Wall.) Christ Christ in Bull. Herb. Boiss. 6: 866. (1898).

Yunnan: On mountain forest to the south of Mengzi at 1,850 metres, 9372. In mountain forest on the Daweishan range in Pingbian at 1,650 metres, 9372a. On rocks in mountain forest to the south-west of Mengzi at 1,650 metres, 9372b.

Trichosanthes cordata Roxb. (CUCURBITACEAE)

Anonymous in Kew List of Determinations (volume 34).

Yunnan: Mengzi, 11071.

Trichosanthes costata Blume (CUCURBITACEAE)

Gymnopetalum cochinchinense (Lour.) Kurz Henry in Trans. Asiat. Soc. Jap. 24 suppl: 45. (1896) (List Pl. Formosa).

Taiwan: Kaohsiung, without number.

A slender glabrescent vine, native from India to China and Malesia.

Trichosanthes kirilowii Maxim. (CUCURBITACEAE)

Forbes & Hemsley in Journ. Linn. Soc. Bot. 23: 313. (1887), Henry in Journ. China Br. Roy. Asiat. Soc. 22: 251. (1887) (Chinese Names of Plants); in Notes Econ. Bot. China 52. (1893).

Trichosanthes multiloba Forbes & Hemsley non Miq. in Journ. Linn. Soc. Bot. 23: 314. (1887), Henry in Journ. China Br. Roy. Asiat. Soc. 22: 251. (1887) (Chinese Names of Plants).

Hubei: Yichang, 634, 4230, 4250. Antelope Glen, 3663. In grass in a glen near Pingshanba (near Yichang, May 1886), 1578. On Tsui Fu Shan near Yichang (May 1888), 1626. Badong, 1849 (July 1886), 4739, 7311. Liantuo, 2004, 2072. Changyang, 6224. **Yunnan:** Mengzi, 9494.

A tuberous rooted, climbing perennial which Henry stated was known colloquially at Yichang as *yo-kua* or *hua k'u kua* (*gua lou*). He went on to state that the fruits of this species according were used as a drug (the fruits are currently used in Chinese herbal medicine as a cure for cough with associated chest pain, the root is currently used to treat diabetes). In autumn the orange-coloured fruits, as big as a fist, hang down in long peduncles, from the branches of trees on which the plant climbs. Cultivated at the National Botanic Gardens, Glasnevin from collections made by the Glasnevin Central China Expedition (GCCE 469) in Xingshan in September 2004. Native to China, Korea, Laos, Vietnam and Japan.

Trichosanthes multiloba Miq. (CUCURBITACEAE)

Henry in Trans. Asiat. Soc. Jap. 24 suppl: 45. (1896) (List Pl. Formosa).

Taiwan: Kaohsiung, 1195, 1927, 1951. **Yunnan:** Mengzi, 9432.

Trichosanthes ovigera Blume (CUCURBITACEAE)

Trichosanthes cucumeroides Anonymous non (Ser.) Maxim. ex Franch. & Sav. in Kew List of Determinations (volume 24).

Hubei: Yichang, 3988. Badong, 6130. Without locality, 6130a.

Trichosanthes pilosa Lour. (CUCURBITACEAE)

Trichosanthes cucumeroides (Ser.) Maxim. ex Franch. & Sav. Henry in Trans. Asiat. Soc. Jap. 24 suppl: 45. (1896) (List Pl. Formosa).

Taiwan: Kaohsiung, 1593, 1645, 1913.

Trichosanthes scabra Lour. var. ***penicaudii*** (Gagnep.) H. J. de Boer (CUCURBITACEAE)

Gymnopetalum penicaudii Gagnep Cogniaux in Engler, Pflanzenr. iv. 275. ii: 183. (1924).

Hainan: Environs of Haikou, 8078, 8404, 8412.

Trichosanthes sp. No. 1. (CUCURBITACEAE)

Trichosanthes palmata Roxb. Anonymous in Kew List of Determinations (volume 24).

Hainan: Environs of Kiungchow, 8730.

Trichosanthes sp. No. 2. (CUCURBITACEAE)

Henry in Trans. Asiat. Soc. Jap. 24 suppl: 45. (1896) (List Pl. Formosa).

Taiwan: Wanchin, 1513. Kaohsiung, 1926. Wanshoushan, 1952.

Trichosanthes spp. (CUCURBITACEAE)

Anonymous in Kew List of Determinations (volume 24 & 34).

Hubei: Badong, 1850. **Sichuan:** North Wushan, 7450. **Yunnan:** Mengzi, 11317. Simao, 12233. On Fengchunling, in forests at 2,300 metres, 11203.

Tricyrtis formosana Baker (LILIACEAE)

Henry in Trans. Asiat. Soc. Jap. 24 suppl: 98. (1896) (List Pl. Formosa), Wright in Journ. Linn. Soc. Bot. 36: 141. (1903).

Taiwan: Oluanpi, 1267, 1393a. Tamsui, 1393.

A very beautiful perennial to one metre tall, Native to Japan and China (Taiwan) (where it is rather uncommon). Discovered in Taiwan by the Kew collector, Richard Oldham in 1864.

Tricyrtis lasiocarpa Matsum.) (LILIACEAE)

Tricyrtis macropoda Henry non Miq. in Trans. Asiat. Soc. Jap. 24 suppl: 90. (1896) (List Pl. Formosa). *Tricyrtis latifolia* Wright non Maxim. in Journ. Linn. Soc. Bot. 36: 141. (1903).

Taiwan: Wanchin, 1485.

Differers from *Tricyrtis formosana* by its densely hirsute ovary and fruits. Endemic to the mountainous regions of Taiwan where it is uncommon.

Tricyrtis latifolia Maxim. (LILIACEAE)

Wright in Journ. Linn. Soc. Bot. 36: 141. (1903).

Hubei: Badong, 953 (November 1885), 4690. Liantuo, 4690a. Jianshi, 5871.

Discovered by Carl Maximowicz in northern Japan, this handsome toad lily attains a height of about 75 cms and bears white flowers, spotted with purple. Introduced to cultivation by Wilson through Messrs Veitch in 1900.

Tricyrtis macropoda Miq. (LILIACEAE)

Diels in Bot. Jahrb. 29: 241. (1900).

Hubei: Liantuo, 4619. Badong, 5215.

Tricyrtis maculata (D. Don) J. F. Macbr. (LILIACEAE)

Tricyrtis pilosa Wall. Wright in Journ. Linn. Soc. Bot. 36: 142. (1903).

Hubei: Badong, 4780.

Tricyrtis sp. (LILIACEAE)

Anonymous in Kew List of Determinations (volume 34).

Yunnan: In mountains near Mengzi, 11350.

Tridynamia megalantha (Merr.) Staples (CONVOLVULACEAE)

Porana spectabilis Anonymous non Kurz. in Kew List of Determinations (volume 24).

Hainan: Environs of Haikou, 8377.

Tridynamia sinensis (Hemsl.) Staples (CONVOLVULACEAE)

Porana sinensis Hemsl. Schneider in Sargent Pl. Wilson. 3: 355. (1917).

Yunnan: In the mountains near Mengzi at 1,600 metres, 10715a., 10715b. In mountains to the west of Simao at 1,600 metres, 11892.

Native to China and Vietnam.

Trigastrotheca stricta (L.) Thulin. (MOLLUGINACEAE)

Mollugo stricta L. Forbes & Hemsley in Journ. Linn. Soc. Bot. 23: 324. (1887), Henry in Trans. Asiat. Soc. Jap. 24 suppl: 46. (1896) (List Pl. Formosa).

Hubei: Yichang, 1677, 2749. **Hainan:** Without locality, 8114. **Taiwan:** Oluanpi, without number. Wanchin, 883.

An annual glabrous herb with a slender tap root. Native to eastern Asia from India to Japan and Polynesia.

Trigonostemon philippinensis Stapf (EUPHORBIACEAE) New species

Trigonostemon thyrsoideus Stapf in Kew Bull. Misc. Inf. 264. (1909).

Yunnan: Simao at 1,500 metres, 11947 (isotype of *Trigonostemon thyrsoideus* Stapf).

Trigonotis mollis Hemsley (BORAGINACEAE) New species

in Journ. Linn. Soc. Bot. 26: 153. (1890), Bretschneider in Hist. Eur. Bot. Disc. China 2: 788. (1898).

Hubei: Yichang, 630a. In a cave in the San You Dong Glen, 1574 (isosyntype). Fang Xian, 6735 (type). Changyang, 7796.

Trigonotis peduncularis (Trev.) Benth. (BORAGINACEAE)

Forbes & Hemsley in Journ. Linn. Soc. Bot. 26: 153. (1890).

Hubei: Badong, 1281. Yichang, 651 (in part), 1317, 3357, 3359. **Yunnan:** Mengzi, 9755.

Distributed from Russia to China, Korea and Japan.

Trigonotis sp. (BORAGINACEAE)

Anonymous in Kew List of Determinations (volume 34).

Yunnan: Mengzi, 9354.

Trigonotis vestita (Hemsley) I. M. Johnst. (BORAGINACEAE)

Trigonotis peduncularis (Trev.) Benth. var. *vestita*

Hemsley in Journ. Linn. Soc. Bot. 26: 154. (1890).

Sichuan: North Wushan, 7072 [type of *Trigonotis peduncularis* (Trev.) Benth. var. *vestita* Hemsley].

Trigonotis vestita (Hemsley) I. M. Johnst. (BORAGINACEAE)

Trigonotis peduncularis (Trev.) Benth. var. *vestita* Hemsley in Journ. Linn. Soc. Bot. 26: 154. (1890).

Sichuan: North Wushan, 7072 [type of *Trigonotis peduncularis* (Trev.) Benth. var. *vestita* Hemsley].

Trillium tschonoskii Maxim. (TRILLIACEAE)

Trillium erectum L. var. *japonicum* Anonymous non A. Gray, in Kew List of Determinations (volume 24).

Hubei: South Badong, 6067. Zigui (July 6th 1888), 6067a. Without locality, 6067b.

A perennial herb to 50 cm tall, flowering in May and June. A charming little species, native from Sikkim to China and Japan.

Triosteum himalayanum Wall. (CAPRIFOLIACEAE)

Triosteum hirsutum Henry non Wall. in Notes Ecom. Bot. China 46. (1893). *Triosteum himalayanum* Wall. var. *chinense* Diels & Graebner in Bot. Jahrb. Syst. 29: 590. (1901).

Hubei: Xingshan, 6751 (isotype of *Triosteum himalayanum* Wall. var. *chinense* Diels & Graebner). Sichuan: Henry's Chinese collector with Pratt, no locality stated but may be any of the following, Leshan, Emei Shan, Wa Shan or Kangding, 8925.

Henry called this the *t'ing-tzu-p'ao*. A beautiful woodland perennial, native from the Himalayan region to central China. Wilson later collected this species while travelling on his 1907 and 1910 expeditions for the Arnold Arboretum.. A remarkably beautiful plant when seen carrying spikes of large, fleshy red berries in autumn.

Tripidium arundinaceum (Retz.) Welker, Voronts. & E. A. Kellogg (POACEAE)

Saccharum arundinaceum Retz. Rendle in Journ. Linn. Soc. Bot. 36: 349. (1984).

Hubei: On the side of the San You Dong stream, 4227.

A tall handsome grass carrying pink flower spikes to 2.5 metres tall. This species was collected by the Glasnevin Central China Expedition between Yichang and Badong (GCCE 188) in October 2002 and was again collected by the 2004 Glasnevin Expedition to Hubei at Yichang (GCCE 478 & GCCE 724) and in Xingshan (GCCE 605) in the autumn of that year. Abundant in the low hills around Yichang.

Tripidium rufipilum (Steud.) Welker, Voronts. & E. A. Kellogg (POACEAE)

Erianthus japonicus Henry non (Thunb). P. Beauv. in Journ. China Br. Roy. Asiat. Soc. 22: 263. (1887) (Chinese Names of Plants). *Erianthus fulvus* Nees. ex Hack. Stapf in

Kew Bull. Misc. Inf. 228. (1898), Rendle in Journ. Linn. Soc. Bot. 36: 350. (1904).

Hubei: Badong, 5115.

According to Henry this species was known colloquially in the Three Gorges region as *pa wang ts'ao*.

Triplostegia glandulifera Wallich ex DC. (CAPRIFOLIACEAE)

Forbes & Hemsley in Journ. Linn. Soc. Bot. 23: 399. (1888).

Triplostegia delavayi Franch. Diels in Notes Roy. Bot. Gard. Edinburgh 5: 209. (1912).

Hubei: Badong, 2425, 4833, 4973, 4974. Xingshan, 6749. Yunnan: Mountains to the east of Mengzi at 1,950 metres, 9441.

This species was collected by the Sino-American Expedition to the Shennongjia Forest Expedition in 1980.

Tripogon chinensis (Franch.) Hack. (POACEAE)

Rendle in Journ. Linn. Soc. Bot. 36: 404. (1904).

Taiwan: Wanshoushan, 1064.

Native from eastern Siberia to China and the Philippines. Distributed in the south of Chinese Taiwan where it is rare.

Tripora divaricata (Maxim.) P. D. Cantino (LAMIACEAE)

Caryopteris divaricata Maxim. Forbes & Hemsley in Journ. Linn. Soc. Bot. 26: 263. (1890).

Hubei: South Badong, 7319.

Tripterospermum cordatum (Marq.) H. Sm. (GENTIANACEAE) New species

Gentiana cordata Marquand in Kew Bull. Misc. Inf. 77. (1931).

Sichuan: Kangding, at 2,700 to 4,000 metres, Henry's Chinese collector with Pratt, 8881. Yunnan: In mountain forest on Fengchunling at 2,100 metres, 11186.

Endemic to China.

Tripterospermum discoideum (Mirq.) H. Sm. (GENTIANACEAE) New species

Crawfurdia fasciculata Forbes & Hemsley non Wall. in Journ. Linn. Soc. Bot. 23: 122. (1890). *Gentiana discoidea* Marquand in Kew Bull. Misc. Inf. 72. (1931). *Gentiana caudata* Marquand in Kew Bull. Misc. Inf. 78. (1931).

Hubei: Badong, 1038 (December 1885), 2848, 4877. Liantuo, 4463. Sichuan: North Wushan, 7091 (type of *Gentiana caudata* Marquand).

This handsome climbing gentian was collected by the Glasnevin Central China Expedition (GCCE 567) in Xingshan in September 2004.

Tripterospermum fasciculatum (Wall.) Chater (GENTIANACEAE)

Crawfurdia fasciculata Wall. Anonymous in Kew List of Determinations (volume 34).

Yunnan: South of the Red River from Manmei at 1,900 metres, 9474.

Tripterospermum filicaule (Hemsl.) Harry Smith (GENTIANACEAE) New species

Gentiana filicaulis Hemsley in Journ. Linn. Soc. Bot. 26: 127. (1890,) Bretschneider in Hist. Eur. Bot. Disc. China 2: 788. (1898).

Hubei: In the mountains of Shennongjia at 2,600 to 3,100 metres, 6842 (isotype).

Tripterospermum sp. No. 1. (GENTIANACEAE)

Crawfurdia sp. Anonymous in Kew List of Determinations (volume 24).

Hubei: Liantou, 2661.

Tripterospermum sp. No. 2. (GENTIANACEAE)

Crawfurdia fasciculata Henry non Wall. in Trans. Asiat. Soc. Jap. 24 suppl: 61. (1896) (List Pl. Formosa).

Taiwan: Wanchin, 1603.

Tripterospermum trinervium (Thunb.) H. Ohashi & H. Nakai (GENTIANACEAE)

Crawfurdia fasciculata Forbes & Hemsley non Wallich in Journ. Linn. Soc. Bot. 23: 122. (1890). *Gentiana golowninia* Marq. var. *oblonga* Marquand in Kew Bull. Misc. Inf. 79. (1931).

Hubei: Jianshi, 7416.

Tripterospermum volubile (D. Don) H. Sm. (GENTIANACEAE)

Crawfurdia fasciculata Forbes & Hemsley non Wall. in Journ. Linn. Soc. Bot. 23: 122. (1890).

Hubei: Yichang, 95. Liantuo, 4579. Fang Xian, 6654, 6777. Without locality, 6654a., 6654b., 6654c.

Tripterygium wilfordii Hook. f. (CELASTRACEAE)

Tripterygium wilfordii Hook. f. Takeda in Kew Bull. Misc. Inf. 222. (1912). *Tripterygium wilfordii* Hook. f. var. *exesum* Sprague & Takeda in Kew Bull. Misc. Inf. 222. (1912).

Yunnan: Mountains to the north of Mengzi at 1,520 metres (probably Dahei Shan or Great Black Mountain of Henry), 10203a., (isotype *Tripterygium wilfordii* Hook. f. var. *exesum* Sprague & Takeda). Simao, 12024.

A handsome climbing shrub, native to Myanmar, China, Korea and Japan, introduced to cultivation by J. G. Jack (for whom the genus *Sinojackia* was named) through the Arnold Arboretum from Korea.

Trisetum flavescens (L.) P. Beauv. (POACEAE)

Rendle in Journ. Linn. Soc. Bot. 36: 399. (1904).

Hubei: Badong, 2558, 4027, 4696.

Tristellateia australasiae A. Richard (MALPIGHIACEAE)

Henry in Trans. Asiat. Soc. Jap. 24 suppl: 24. (1896) (List Pl. Formosa), Li in Woody Flora of Taiwan 403. fig. 149. (1963).

Taiwan: Oluanpi, 321, 598.

A woody vine to 10 metres tall, widely distributed from Malaysia to tropical Australia and the Pacific Islands. In Chinese Taiwan this species is found in coastal forest on the southernmost part of the island.

Triumfetta annua L. (MALVACEAE)

Anonymous in Kew List of Determinations (volume 24 & 34).

Hubei: Changyang, 7760. **Yunnan:** Fengchunling, 9251.

Of widespread distribution in India and China.

Triumfetta pilosa Roth. (MALVACEAE)

Henry in Trans. Asiat. Soc. Jap. 24 suppl: 23. (1896) (List Pl. Formosa).

Taiwan: Tamsui, 1472. Wanchin, 1505.

An annual herb to a metre tall, widely distributed in the tropics. This species inhabits fields and hillsides in northern Taiwan.

Triumfetta rhomboidea Jacq. (MALVACEAE)

Henry in Trans. Asiat. Soc. Jap. 24 suppl: 28. (1896) (List Pl. Formosa).

Taiwan: Kaohsiung, without number. Wanchin, 419. Oluanpi, 1231.

An erect annual herb to 1.5 metres tall, native to tropical Africa and Asia.

Triumfetta sp. (MALVACEAE)

Anonymous in Kew List of Determinations (volume 34).

Yunnan: South of the Red River, 10653. Simao, 12345, 12426.

Trochodendron aralioides Sieb. & Zucc. (TROCHODENDRACEAE)

Henry in Trans. Asiat. Soc. Jap. 24 suppl: 16. (1896) (List Pl. Formosa), Li in Woody Flora of Taiwan 155. fig. 56. (1963).

Taiwan: Tamsui, 1398. Oluanpi, 1981.

An evergreen tree to 25 metres tall. Native to Japan, China (Taiwan) (where it often forms pure stands), and Korea on Quelpaert Island (Cheju Do). Introduced to cultivation by Messrs Veitch from Japan.

Tropidia formosana Rolfe (ORCHIDACEAE) New species

in Ann. of Bot. 9: 158. (1895), Bretschneider in Hist. Eur. Bot. Disc. China 2: 792. (1898), Rolfe in Journ. Linn. Soc. Bot. 36: 40. (1903).

Taiwan: Wanchin, 1573 (isotype).

Tropidia somae Hayata (ORCHIDACEAE)

Tropidia angulosa Henry non (Lindl.) Blume in Trans. Asiat. Soc. Jap. 24 suppl: 92. (1896) (List Pl. Formosa), Rolfe in Journ. Linn. Soc. Bot. 36: 39. (1903).

Taiwan: Wanshoushan, in dark situations, 1905.

A terrestrial orchid to 20 cm tall, native to China (Taiwan), Japan (south) and probably the Philippines. In forests at low altitude throughout Taiwan.

Tropidia sp. (ORCHIDACEAE)

Anonymous in Kew List of Determinations (volume 34).

Yunnan: Simao, 13150.

Tsuga chinensis (Franch.) Pritz. (PINACEAE)

Rehder & E. H. Wilson in Sargent Pl. Wilson. 2: 37: (1914), Downie in Notes Roy. Bot. Gard. Edinburgh14: 14. & 18. (1923—1924), Walsh in Aug. Henry Forst. Herb. Cat. 33. (1957), Zheng in Journ. Wuhan Bot. Res. 2.(1): 10. (1984).

Tsuga sieboldii Henry non Carr. in Notes Econ. Bot. China 36. (1893), Masters in Journ. Linn. Soc. Bot. 26: 556. (1902). *Tsuga yunnanensis* Masters non (Franch.) Mast. in Gard. Chron. ser. 3. 39: 236. (1906).

Hubei: Fang Xian, 6907. **Sichuan:** North Wushan, 7156. Emei Shan, Henry's Chinese collector with Pratt, 8896.

A tree to 30 metres tall in its native habitat, distributed in central and western China and first collected by Père David near Baoxing in Sichuan though based on a collection later made by Père Farges in north-east Sichuan. Introduced to cultivation by E. H. Wilson in 1902, several of his seed collections reached Glasnevin from various locations and dates, Wilson-Veitch 572 (Xingshan, Hubei. October 1901), Wilson 2007 (Sichuan, August 1908). Henry stated it was called the *tieh-sha* (iron fir) by the Chinese and according to Wilson was made into shingles and used for roofing purposes.

Tupistra sp. (CONVALLARIACEAE)

Anonymous in Kew List of Determinations (volume 34).

Yunnan: Mengzi, 10486.

Turczaninovia fastigiata (Fisch.) DC. (ASTERACEAE)

Aster fastigiatus Fisch. Forbes & Hemsley in Journ. Linn. Soc. Bot. 23: 410. (1888).

Hubei: Yichang, 41, 88, 2184, 2379. Liantuo, 3031.

Turpinia montana (Blume) Kurz (STAPHYLEACEAE)

Turpinia nepalensis Rehder & E. H. Wilson non Wall. in Sargent Pl. Wilson. 2: 187. (1914). *Turpinia gracilis* Nakai in Journ. Arnold Arbor. 5: 79. (1924).

Yunnan: In forests near Simao at 1,300 to 2,000 metres, 12039 (holotype of *Turpinia gracilis* Nakai).

Turraea pubescens Hellen. (MELIACEAE)

Anonymous in Kew List of Determinations (volume 24).

Hainan: Hummocks near Haikou, 8088. Environs of Kiungchow, 8298.

Tylophora insulicola Meve & Liede (APOCYNACEAE) New species

Tylophora hispida Henry non Decne in Trans. Asiat.

Soc. Jap. 24 suppl: 61. (1896) (List Pl. Formosa). *Tylophora insulana* Tsiang & P. T. Li in Acta Phytotax. Sin. 12(1): 134. (1974).

Taiwan: Wanchin, 469 (holotype of *Tylophora insulana* Tsiang & P. T. Li). Kaohsiung, 1162. Oluanpi, 1279.

Discovered by Augustine Henry in 1893.

Tylophora spp. (APOCYNACEAE)

Anonymous in Kew List of Determinations (volume 24 & 34).

Hubei: Fang Xian, 6680. Badong, 7378. **Hainan:** Environs of Kiungchow, 8275. Ling-men (in the interior of the island), 8616. **Yunnan:** Mengzi, 10010. Simao, 12195, 12304, 12339, 13125, 13405, 13547.

Typha angustifolia L. (TYPHACEAE)

Brown in Journ. Linn. Soc. Bot. 36: 172. (1903).

Typha sp. Henry in Trans. Asiat. Soc. Jap. 24 suppl: 100. (1896) (List Pl. Formosa).

Taiwan: Kaohsiung, 1815.

Widely dispersed through Malaysia, Indonesia, China, the Philippines and North America.

Typha orientalis C. Presl. (TYPHACEAE)

Brown in Journ. Linn. Soc. Bot. 36: 173. (1903).

Typha near *T. shuttleworthii* W. D. K. Koch & Sond. Henry in Journ. China Br. Roy. Asiat. Soc. 22: 258. (1887) (Chinese Names of Plants).

Hubei: Liantuo, 2105. Badong, 5118, 7396. Xingshan, 6501.

According to Henry this species was used as a styptic and was known colloquially in the Three Gorges region as *mao la chu*.

Ulmus bergmanniana C. K. Schneider (ULMACEAE) New species

Schneider in Sargent Pl. Wilson. 3: 241. (1916).

Ulmus montana Forbes & Hemsley non With. in Journ. Linn. Soc. Bot. 26: 448. (1894).

Sichuan: South Wushan, 5690.

A tree to 28 metres tall with trunks with a girth of up to 6 metres, though usually seen as a smaller tree. Schneider described this species from material collected in Hubei by Wilson in 1900, it had been previously discovered by Henry in 1888. Wilson stated that in Hubei, this and other species of *Ulmus* were known as *lang shu*.

Ulmus castaneifolia Hemsley(ULMACEAE) New species

in Journ. Linn. Soc. Bot. 26: 446. (1894), Bretschneider in Hist. Eur. Bot. Disc. China 2: 791. (1898), Schneider in Illus. Handb. Laubholzk. 2: 904. (1912); in Sargent Pl. Wilson. 3: 256. (1916), Zheng in Journ. Wuhan Bot. Res. 2(1): 30. (1984).

Sichuan: South Wushan, 5498 (co-type). Changyang, 7780 (type).

A tree *Ulmus castaneifolia* at the Royal Botanic Gardens, Kew has reached a height of 11 metres with a trunk diameter of 26 cm (2001).

Ulmus davidiana Planch. var. *japonica* (Rehder) Nakai (ULMACEAE)

Ulmus sp. nov. Forbes & Hemsley in Journ. Linn. Soc. Bot. 26: 448. (1894), E. Pritzel in Bot. Jahrb. xxix. 296. (1900). *Ulmus wilsoniana* Schneider Rehder & E. H. Wilson in Sargent Pl. Wilson. 3: 239. (1916).

Sichuan: South Wushan, 5537.

A deciduous tree found on the margins of moist woods in the mountains of western Hubei where it forms trees to 25 metres tall with girths of up to 3 metres. The species was introduced to cultivation by Wilson who sent scions to the Arnold Arboretum from Fang Xian in 1910.

Ulmus lanceifolia Roxb. apud. Wall. (ULMACEAE)

Forbes & Hemsley in Journ. Linn. Soc. Bot. 26: 447. (1894), Schneider in Sargent Pl. Wilson. 3: 263. (1916), Walsh in Aug. Henry Forst. Herb. 77. (1957), Zheng in Journ. Wuhan Bot. Res. 2.(1): 31. (1984).

Hubei: By a temple at Yichang, 3271. **Yunnan:** Linan District, 10571, collected March 1st 1897. In Yulo forest near Simao, 12864.

Native to India, Bhutan, Myanmar, China (Guangxi, Hainan and Yunnan), Thailand, Laos and Vietnam.

Ulmus parvifolia Jacq. (ULMACEAE)

Henry in Journ. China Br. Roy. Asiat. Soc. 22: 253. (1887) (Chinese Names of Plants), Forbes & Hemsley in Journ. Linn. Soc. Bot. 26: 448. (1894), Henry in Trans. Asiat. Soc. Jap. 24 suppl: 85. (1896) (List Pl. Formosa), Schneider in Sargent Pl. Wilson. 3: 244. (1916), Li in Woody Flora of Taiwan 109. fig. 33. (1963), Zheng in Journ. Wuhan Bot. Res. 2.(1): 31. (1984).

Hubei: Yichang, 1030, 3219. Liantuo, 2602, 3079. Jianshi, 7518. Changyang, 7681. **Taiwan:** Wanchin, 1529.

According to both Henry and Wilson this was a common elm near the city of Yichang and on rich soil near watercourses it reached 25 metres tall with a girth of 2.5 metres. On the cliffs of the gorges and glens in Hubei it usually formed a bush. Native to China, Korea and Japan. In Taiwan this species is found in forests in the central and southern part of the island at low altitude.

Ulva intestinalis L. (ULVACEAE)

Gasteromorpha intestinalis (L.) Link. Henry in Trans. Asiat. Soc. Jap. 24 suppl: 118. (1896) (List Pl. Formosa).

Taiwan: Tamsui, 1936.

A seaweed exported from Tamsui in the late nineteenth century.

Ulva linza L. (ULVACEAE)

Phycoseris linza (L.) Ktzg. Henry in Trans. Asiat. Soc.

Jap. 24 suppl: 118. (1896) (List Pl. Formosa).

Taiwan: Tamsui, 1935. (Seaweed).

Umbilicaria esculenta (Miyoshi) Minks (UMBILICARIACEAE)

Wei in An Enumeration of Lichens in China 247. (1991).

Gyrophora spodochroa Muell.-Arg. non Ehrh. ex Hoffm.) Ach. in Bull. Herb. Boissier 1: 235. (1893).

Hubei: South Badong, 6184. (Lichen).

Uncaria hirsuta Haviland (RUBIACEAE)

Li in Woody Flora of Taiwan 878. fig. 356. (1963).

Uncaria florida Henry non S. Vidal. in Trans. Asiat. Soc. Jap. 24 suppl: 49. (1896) (List Pl. Formosa).

Taiwan: Oluanpi, 983.

A climbing shrub, native to southern China, including Taiwan where it grows in forests throughout the island.

Uncaria lancifolia Hutchinson (RUBIACEAE) New species

in Sargent Pl. Wilson. 3: 407. (1916).

Yunnan: In Pingbian, in forests on the Daweishan Range at 1,640 metres, 11389 (isotype).

Distributed in China (Yunnan) and Vietnam.

Uncaria macrophylla Wall. (RUBIACEAE)

Hutchinson in Sargent Pl. Wilson. 3: 407. (1916).

Yunnan: At Simao near the Nantan River at 1,490 metres, 12820.

Native to India, Bangladesh, Bhutan, Myanmar, China, Laos, Thailand and Vietnam.

Uncaria scandens (Smith) Wall. (RUBIACEAE)

Hutchinson in Sargent Pl. Wilson. 3: 406. (1916).

Yunnan: Mountains to the west of Simao at 1,640 metres, 11868.

Previously unknown in China.

Uncaria sessilifructus Roxburgh (RUBIACEAE)

Hutchinson in Sargent Pl. Wilson. 3: 406. (1916).

Yunnan: Chienhung (in Xishuangbanna, ex M. Bons d'Anty), 11195. Simao at 1,320 metres, 11195a. Simao, 13449.

According to Henry this large climber was known as *kou teng*. Distributed in the Himalaya, India, Myanmar, western China and Indochina.

Uncaria sinensis (Oliv.) Havil. (RUBIACEAE) New species

in Journ. Linn. Soc. Bot. 33: 89. (1897), Zheng in Journ. Wuhan Bot. Res. 2.(1): 190. (1984).

Nauclea sinensis Oliver in Hooker's Icon. Pl. t. 1956. (1891), Henry in Notes Econ. Bot. China 21. (1893), Bretschneider in Hist. Eur. Bot. Disc. China 2: 785. (1898), Morley in Glasra 3: 79. (1979).

Hubei: Liantuo, 4501, 4501a.

According to Henry, the hooks of this climbing plant were used in medicine and the drug was called *kou-t'eng* (the *gou teng* or 'hook vine'). Native to central and western China.

Uraria crinita (L.) Desv. (FABACEAE)

Henry in Trans. Asiat. Soc. Jap. 24 suppl: 34. (1896) (List Pl. Formosa).

Hainan: 8 km (5 miles) south of Kuingchow, 8180. Without locality, 8529. **Taiwan:** Wanchin, 836.

An undershrub to 1.5 metres tall, native to India, China, south-east Asia and Australia.

Uraria lacei Craib (FABACEAE) New species

in Kew Bull. Misc. Inf. 276. (1910).

Yunnan: Mengzi, 9144. Pu'er at 1,350 metres, 9144a.

Native to India, Myanmar and China (Yunnan) to Indochina.

Uraria lagopodioides (L.) Desv. ex DC. (FABACEAE)

Henry in Trans. Asiat. Soc. Jap. 24 suppl: 34. (1896) (List Pl. Formosa).

Hainan: Environs of Kiungchow, 8224. **Taiwan:** Kaohsiung, without number. Wanchin, 837. **Yunnan:** Mengzi, 10256.

A small shrub, native to India, China, the Philippines and Australia.

Uraria oblonga (Wallich ex Bentham) H. Ohashi & K. Ohashi (FABACEAE)

Desmodium oblongum Wallich ex Bentham Anonymous in Kew List of Determinations (volume 34).

Yunnan: Simao, 12515.

Native to India (Assam), Bhutan, western China and Indochina.

Uraria picta (Jacq.) Desv. ex DC. (FABACEAE)

Henry in Trans. Asiat. Soc. Jap. 24 suppl: 34. (1896) (List Pl. Formosa).

Taiwan: Kaohsiung, without number.

A perennial to 1.5 metres tall, native to tropical Africa and Asia. In Chinese Taiwan this species is confined to the Hengchun Peninsula where it grows in open grassland.

Uraria rufescens (DC.) Schind. (FABACEAE)

Uraria paniculata Hassk. Anonymous in Kew List of Determinations (volume 34).

Yunnan: Mengzi, 9114.

Uraria sinensis (Hemsl.) Franchet (FABACEAE) New species

Uraria hamosa Wall. var. *sinense* Hemsley in Journ. Linn. Soc. Bot. 23: 177. (1887).

Hubei: Yichang, 2361, 3137. **Yunnan:** Mengzi, 9434. Fengchunling, 11187.

Distributed from the eastern Himalaya to China.

Uraria sp. (FABACEAE)

Uraria hamosa Wall. Henry in Trans. Asiat. Soc. Jap. 24 suppl: 34. (1896) (List Pl. Formosa).

Taiwan: Wanchin, 1504.

Urceola laevigata (Juss.) J. J. Middleton & Livsh. (APOCYNACEAE)

Parameria glandulifera (Wall. ex G. Don) Benth. & Hook. f. ex Kurz. Anonymous in Kew List of Determinations (volume 34). *Parameria barbata* (Bl.) Schum. Tsiang in Sunyatsenia 2: 115. (1934).

Yunnan: Simao, 12232. In forests near Simao at 1,650, 12681.

China (Guangxi and Yunnan) to Malesia.

Urceola rosea (Hook. & Arn.) D. J. Middleton (APOCYNACEAE)

Ecdysanthera rosea Hook. & Arn. Henry in Trans. Asiat. Soc. Jap. 24 suppl: 82. (1896) (List Pl. Formosa) Schneider in Sargent Pl. Wilson. 3: 342. (1917) Li in Woody Flora of Taiwan 786. fig. 316. (1963)

Taiwan: Oluanpi, 338. Wanchin, 388, 838. **Yunnan:** Mengzi, 10128. Simao, 12760.

A large climber, native to southern China and Japan (south). In Chinese Taiwan this species is found in low altitude forest.

Urceola sp. (APOCYNACEAE)

Parameria sp. Anonymous in Kew List of Determinations (volume 34).

Yunnan: Simao, 12116, 12269.

Urena lobata L. (MALVACEAE)

Henry in Trans. Asiat. Soc. Jap. 24 suppl: 21. (1896) (List Pl. Formosa), Rehder & E. H. Wilson in Sargent Pl. Wilson. 2: 373. (1916).

Urena lobata L. var. *tomentosa* (Blume.) Walp. Li in Woody Flora of Taiwan 553. (1963).

Hubei: Liantuo, 4430, 4644. On the banks of the Yangtze at Badong, 7180. Changyang, 7761. Without locality, 7180a. **Hainan:** Environs of Haikou, 8028, 8209. **Taiwan:** Kaohsiung, without number. **Yunnan:** In the Red River valley near Manhao, 9542

A subshrub to 0.5 metres tall with small pink flowers, native to the tropics of both hemispheres and a common weed in China (Taiwan). In China the fibre from the stems of this species is used as a substitute for hemp and the roots are used medicinally.

Urena lobata L. var. *henryi* S. Y. Hu (MALVACEAE) New variety

in Fl. China, Fam. 153. Malvaceae 75. (1955).

Urena lobata Rehder & E. H. Wilson non L. in Sargent Pl. Wilson 2: 273. (1915), in part.

Hubei: Yichang, 7180b., (holotype).

Urena procumbens L. (MALVACEAE)

Li in Woody Flora of Taiwan 553. (1963).

Urena sinuata Henry non L. in Trans. Asiat. Soc. Jap. 24 suppl: 20. (1896) (List Pl. Formosa).

Taiwan: Kaohsiung, without number. Wanchin, 69.

A small shrub, native to the tropics of both hemispheres and a common weed in Taiwan.

Urena repanda Roxb. (MALVACEAE)

Anonymous in Kew List of Determinations (volume 34).

Yunnan: Mengzi, 10300.

A perennial herb to 90 cm tall, the small pink flowers of this species are carried towards the top of branches and appear between August and November. In China this species is distributed through Guizhou, Guangxi and Yunnan where it grows in thickets on mountain slopes between 850 to 1,600 metres.

Urochloa polystachya (Kunth) Mabb. (POACEAE)

Panicum antidotale Henry non Retz. in Trans. Asiat. Soc. Jap. 24 suppl: 106. (1896) (List. Pl. Formosa). *Eriochloa polystachya* Kunth Rendle in Journ. Linn. Soc. Bot. 36: 320. (1904).

Taiwan: Kaohsiung, 1024. Kaohsiung, 1108.

Urochloa reptans (L.) Stapf. (POACEAE)

Panicum prostratum Lam. Rendle in Journ. Linn. Soc. Bot. 36: 332. (1904). *Panicum procumbens* Nees Henry in Trans. Asiat. Soc. Jap. 24 suppl: 106. (1896) (List Pl. Formosa).

Taiwan: Kaohsiung, 1020, 1020a., 1919.

Urophysa henryi (Oliv.) Ulbr. (RANUNCULACEAE) New species

in Notizbl. Bot. Gart. Berlin-Dahlem 10: 870. (1929).

Isopyrum henryi Oliver in Hooker's Icon. Pl. 18: t. 1745. (1888), Bretschneider in Hist. Eur. Bot. Disc. China 2: 779. (1898), Henry in Flora & Sylva 1: 218. (1903), Morley in Glasra 3: 79. (1979). *Aquilegia henryi* (Oliv.) Finet & Gagnepain in Bull. Soc. Bot. France 51: 411. (1904). *Semiaquilegia henryi* (Oliv.) J. R. Drumm. & Hutch. in Kew Bull. Misc. Inf. 166. (1920).

Hubei: Liantuo, growing out of clefts of rocky cliffs, 3820 (type of *Isopyrum henryi* Oliver).

Urtica dioica L. var. ***holosericea*** Fr. (URTICACEAE)

Forbes & Hemsley in Journ. Linn. Soc. Bot. 26: 472. (1899).

Urtica dioica Henry non L. in Notes Econ. Bot. China 65. (1893).

Hubei: Jianshi, 5859.

Urtica parviflora Roxb. (URTICACEAE)

Forbes & Hemsley in Journ. Linn. Soc. Bot. 472. (1899).

Yunnan: Fengchunling, 11179, 11197.

Native to Nepal, India, Bhutan, and China [Guangxi, Tibet (Xizang) and Yunnan].

Urtica thunbergiana Sieb. & Zucc. (URTICACEAE)

Henry in Journ. China Br. Roy. Asiat. Soc. 270. (1887) (Chinese Names of Plants), Henry in Notes Econ. Bot. China 65. (1893), Forbes & Hemsley in Journ. Linn. Soc. Bot. 26: 472. (1899).

Hubei: In the Lung Tung Chi Glen near Yichang (October 1886), 2900. **Yunnan:** Mengzi, 9065a., 9065b.

According to Henry this stinging nettle was a common plant in the glens of the Three Gorges region where it was known colloquially as *she ma ts'ao*. Native to Russia, China, Korea and Japan.

Urtica thunbergiana subsp. ***dentata*** (Handel-Mazzetti) K. Becker & Weigend (URTICACEAE) New subspecies

Urtica dentata Handel-Mazzetti in Symb. Sin. 7: 112. (1929).

Hubei: South Badong, 5401 (type of *Urtica dentata* Handel-Mazzetti).

Usnea trichodea Ach. (PARMELIACEAE)

Muell.-Arg. in Bull. Herb. Boissier 1: 235. (1893), Wei in An Enumeration of Lichens in China 260. (1991).

Sichuan: North Wushan, 7070.

Utricularia aurea Lour. (LENTIBULARIACEAE)

Utricularia flexuosa Vahl. Forbes & Hemsley in Journ. Linn. Soc. Bot. 26: 223. (1890), Henry in Trans. Asiat. Soc. Jap. 24 suppl: 68. (1896) (List Pl. Formosa).

Hubei: Yichang, 2338. **Taiwan:** Kaohsiung Plain, 1806, 1069, 1789, 1840.

Utricularia bifida L. (LENTIBULARIACEAE)

Forbes & Hemsley in Journ. Linn. Soc. Bot. 26: 222. (1890).

Hubei: In dry rice fields near Yichang, 7787.

Most of Henry's aquatic collections from Yichang were made in ponds about a quarter of a mile behind the foreign compound on the outskirts of Yichang.

Utricularia sp. (LENTIBULARIACEAE)

Anonymous in Kew List of Determinations (volume 34).

Yunnan: Simao, 12405, 12497.

Utricularia striatula Sm. (LENTIBULARIACEAE)

Utricularia orbiculata Wall. ex DC. Forbes & Hemsley in Journ. Linn. Soc. Bot. 26: 224. (1890).

Hubei: Yichang, 4133. South Donghu, 7585.

Uvaria grandiflora Roxb. ex Hornem. (ANNONIACEAE)

Uvaria purpurea Blume Anonymous in Kew List of Determinations (volume 24).

Hainan: Without locality, 8569.

Uvaria microcarpa Champ. ex Benth. (ANNONIACEAE)

Anonymous in Kew List of Determinations (volume 24).

Hainan: Hummocks near Haikou, 8102. Environs of Kiungchow, 8257. Environs of Haikou, 8370. Without locality, 8494. **Yunnan:** Simao, 12148.

Uvaria sp. (ANNONACEAE)

Melodorum sp. Anonymous in Kew List of Determinations (volume 34).

Yunnan: Mengzi, 11090.

Vaccinium bracteatum Thunb. (ERICACEAE)

Forbes & Hemsley in Journ. Linn. Soc. Bot. 26: 14. (1889), Rehder & E. H. Wilson in Sargent Pl. Wilson. 1: 558. (1913), Zheng in Journ. Wuhan Bot. Res. 2.(1): 170. (1984).

Hubei: Liantuo, 2041, 3067. Jianshi, 5807. Changyang, 5807a. **Hainan:** Without locality, 8515, 8531.

An evergreen shrub to 1.5 metres tall, producing racemes of white bell-shaped flowers in August and September. Native to China, Korea and Japan, introduced by John Reeves from Guangzhou (Canton) in 1829, it was reintroduced by Maurice de Vilmorin in 1914. This species was collected by the Glasnevin Central China Expedition (GCCE 490) in Xingshan in September 2004.

Vaccinium dunalianum Wight (ERICACEAE)

Rehder & E. H. Wilson in Sargent Pl. Wilson. 1: 560. (1913), Bean in Trees & Shrubs 4: 670. (1989).

Yunnan: Mengzi at 2,000 metres, 9170, 9170b., 9170e.

A evergreen shrub to 6 metres tall, flowers white, bell-shaped and carried in racemes to 7 cm long. Native to the eastern Himalaya, India, western China and Vietnam.

Vaccinium dunalianum Wight var. *megaphyllum* Sleumer (ERICACEAE)

In the Harvard University Herbarium (A.).

Yunnan: In forests on the Daweishan Range in Pingbian at 1,600 to 2,300 metres, 9170. In forests near Yuanjiang at 1,600 metres, 13404.

Distributed from north-east India to China (Guizhou and Yunnan).

Vaccinium dunalianum Wight var. *urophyllum* Rehder & E. H. Wilson (ERICACEAE) New variety

in Sargent Pl. Wilson. 1: 560. (1913).

Yunnan: In forests on the Daweishan Range in Pingbian, 9170c., (type).

Native to Myanmar, China [Guizhou, Sichuan, Yunnan and Tibet (Xizang)] and Vietnam.

Vaccinium dunnianum Sleumer (ERICACEAE)

Agapetes vaccinioides Dunn in Journ. Linn. Soc. Bot. 35: 515. (1903).

Yunnan: On the Daweishan Range in Pingbian at 1,650 metres, 10707 (isotype of *Agapetes vaccinioides* Dunn). South of the Red River, 13664.

Endemic to China and distributed in the provinces of Guangxi and Yunnan.

Vaccinium fragile Franch. (ERICACEAE)

Rehder & E. H. Wilson in Sargent Pl. Wilson. 1: 559. (1913).

Yunnan: On grassy mountains near Mengzi at 2,100 metres, 10904.

A tufted evergreen shrub to 80 cm tall carrying clustered racemes of white or rosy-red flowers in May and June. Discovered by Delavay in Yunnan and introduced to cultivation by Forrest who stated that the fruits of this shrub were the principal food of pheasants in Yunnan. Forrest's 14152 from the Mekong-Salween divide (July 1917) was described by him as a shrub to 60 cm tall and with white flowers flushed rose. Wilson also collected this species in Sichuan. Endemic to China and distributed in and Guizhou, Sichuan, Yunnan and Tibet (Xizang). A beautiful little shrub, now rare in cultivation.

Vaccinium henryi Hemsley (ERICACEAE) New species

in Journ. Linn. Soc. Bot. 26: 15. (1889), Bretschneider in Hist. Eur. Bot. Disc. China 2: 786. (1898), Diels in Bot. Jahrb. xxix. 516. (1900), Rehder & E. H. Wilson in Sargent Pl. Wilson. 1: 561. (1913), Morley in Glasra 3: 80. (1979), Zheng in Journ. Wuhan Bot. Res. 2.(1): 171. (1984).

Hubei: Badong, 2579, 4703, 4826, 4985. Jianshi, 5911. Fang Xian, 6623, 6829. **Sichuan:** South Wushan, 7388. **Hunan:** Shimen, 7937.

According to Wilson's field notes this species is common in oak woods in western Hubei.

Vaccinium iteophyllum Hance (ERICACEAE)

Forbes & Hemsley in Journ. Linn. Soc. Bot. 26: 15. (1889), Rehder & E. H. Wilson in Sargent Pl. Wilson. 1: 557. (1913).

Hubei: Xingshan, 4526. Badong, 5176. **Yunnan:** In forests near Simao at 1,600 metres, 11648, 11648a.

Distributed in China and Vietnam.

Vaccinium japonicum Miq. var. *sinicum* (Nakai) Rehder (ERICACEAE) New variety

Zheng in Journ. Wuhan Bot. Res. 2.(1): 171. (1984).

Vaccinium japonicum Forbes & Hemsley non Miq. in Journ. Linn. Soc. Bot. 26: 16. (1889), Rehder & E. H. Wilson in Sargent Pl. Wilson.1: 562. (1913).

Hubei: Badong, 2826, 4666, 4864, 4949, 7347. Jianshi, 6021. Donghu, 6431. Without locality, 6431a.

A deciduous shrub to 80 cm tall, carrying solitary pink, bell-shaped flowers in leaf axils in June and July. Discovered by Henry in 1886 and introduced to cultivation by Wilson in 1907 when collecting for the Arnold Arboretum.

Vaccinium moupinense Franch. (ERICACEAE)

Anonymous in Kew List of Determinations (volume 24).

Sichuan: Henry's Chinese collector with Pratt, no locality stated but may be any of the following, Leshan, Emei Shan, Wa Shan or Kangding, 8887.

A small evergreen shrub to 60 cm tall, the bell-shaped flowers appear in May and June and are chocolate-red and carried on racemes from the axils of terminal leaves. Discovered by Père Armand David in Baoxing (formerly Moupin) in July 1869 and introduced to cultivation by Wilson in 1909 while collecting for the Arnold Arboretum. Common as an epiphyte on old trees in western Sichuan and also found on humus clad rocks and cliffs.

Vaccinium mandarinorum Diels (ERICACEAE)
New species

in Bot Jahrb. xxix. 516. (1900).

Vaccinium donianum Rehder & E. H. Wilson non Wight in Sargent Pl. Wilson. 1: 557. (1913)

Hubei: Jianshi, 5807b., (type).

Vaccinium petelotii Merr. (ERICACEAE)
Agapetes parviflora Dunn in Journ. Linn. Soc. Bot. 35: 515. (1903).

Yunnan: On the Daweishan Range in Pingbian at 1,650 metres, 10488 (syntype of *Agapetes parviflora* Dunn). Mengzi, 10488a., 13306.

Native to China (Yunnan) and Vietnam.

Vaccinium sp. (ERICACEAE)
Anonymous in Kew List of Determinations (volume 34).

Yunnan: In mountains near Mengzi, 13346, 13348.

Vaccinium sprengelii (G. Don) Sleumer ex Rehder (ERICACEAE)
Vaccinium donianum Wight. Rehder & E. H. Wilson in Sargent Pl. Wilson. 1: 557. (1913).

Hubei: Liantuo, 3918, 7660 (in part). Badong, 6129. **Yunnan:** In the forests to the north of Mengzi at 2,600 metres, 10552, 10552a. In woods near Mengzi at 1,800 metres, 9847b. Near Yuanjiang at 1,950 metres, 13298. In mountain forest to the east of Simao at 1,640 metres, 11626. Mountains to the west of Simao at 1,640 metres, 11917. In forests to the south of Simao at 1,960 metres, 12683. In mountains near Simao, 12745.

Vaccinium wrightii A. Gray (ERICACEAE)
Li in Woody Flora of Taiwan 703. (1963).

Vaccinium bracteatum Thunb. Henry in Trans. Asiat. Soc. Jap. 24 suppl: 56. (1896). *Vaccinium bracteatum* Thunb. var. *wrightii* (Gray) Rehder & E. H. Wilson in Sargent Pl. Wilson. 1: 559. (1913).

Taiwan: Oluanpi, 591, 947, 636, 2067.

A small evergreen tree, native to Japan (south) and China (Taiwan) where it grows at low altitude near the coast.

Vachellia farnesiana (L.) Wight & Arn. (FABACEAE)
Acacia farnesiana (L.) Willd Henry in Trans. Asiat. Soc. Jap. 24 suppl: 39. (1896) (List Pl. Formosa).

Taiwan: Kaohsiung, without number. **Yunnan:** Mengzi, 9966.

A shrub or tree to 7 metres tall, from tropical America and now pantropic in distribution. According to Henry this species was known colloquially as *tz'e-ch'iw*. In the south of France *Vachellia farnesiana* is cultivated for the perfume industry. The violet scented perfume is known commercially as *Cassie Avcienne* and is used to fortify violet oil. A handsome tree when seen covered in masses of small fragrant pompom flowers, but viciously armed with stout pairs of long slender spines.

Valeriana flaccidissima Maxim. (VALERIANACEAE)
Anonymous in Kew List of Determinations (volume 24).

Hubei: Changyang, 5255. Liantuo, 7652. **Sichuan:** South Wushan, 5569.

Valeriana hardwickii Wall. (VALERIANACEAE)
Anonymous in Kew List of Determinations (volume 24).

Hubei: Badong, 4875. **Yunnan:** Mengzi, 9621.

Native to Nepal, India, Bhutan, Myanmar and China.

Valeriana jatamanii Jones ex Roxb. (VALERIANACEAE)
Valeriana harmsii Graebner in Engler. Bot. Jahrb. xxiv. Biebl. 59, 32. (1898), Bretschneider in Hist. Eur. Bot. Disc. China 2: 785. (1898).

Hubei: Liantuo, without number (type of *Valeriana harmsii* Graebner). Yichang, 1139. Xiling (Yichang) Gorge, 3322. South Badong, 5294. Without locality, 3322a., 5294a. **Yunnan:** Mengzi, 10119, 10262, 10948.

Native to India and China.

Valeriana officinalis L. (VALERIANACEAE)
Henry in Journ. China Br. Roy. Asiat. Soc. 22: 262. (1887) (Chinese Names of Plants), Forbes & Hemsley in Journ. Linn. Soc. Bot. 23: 398. (1888).

Hubei: Yichang, 901. Badong, 4022, 5333, 6111. Without locality, 5333a., 5333b. **Sichuan:** Henry's Chinese collector with Pratt, no locality stated but may be any of the following, Leshan, Emei Shan, Wa Shan or Kangding, 8926. **Yunnan:** Mengzi, 10032.

Common valerian, native to Europe and Asia and naturalised in North America. At Badong in the Three Gorges region this species was used as a drug and Henry stated it was known there as *pa ti ma*. Widely used in herbal medicine as a non-addictive tranquilliser that is recommended against restlessness, sleeplessness and as a treatment against the symptoms of menopause.

Valeriana sp. No. 1. (VALERIANACEAE)
Anonymous in Kew List of Determinations (volume 24).

Hubei: Liantuo, 2037. South Badong, 5461.

02

Valeriana sp. No 2. (VALERIANACEAE)

Anonymous in Kew List of Determinations (volume 34).

Yunnan: Simao, 12450, 12468.

Vanda brunnea Rchb. f. (ORCHIDACEAE)

Vanda henryi Schlecter in Fedde, Repert. Spec. Nov. Regni Veg. 17: 71. (1921).

Yunnan: Epiphytic on trees near Simao, 12102 (isotype of *Vanda henryi* Schlecter).

Distributed in Myanmar, Thailand and China (Yunnan).

Vanda concolor Blume (ORCHIDACEAE)

Rolfe in Journ. Linn. Soc. Bot. 36: 35. (1903).

Vanda esquirolii Schlecter in Repert. Spec. Nov. Regni Veg. 17: 71. (1921).

Yunnan: In woods near Mengzi at 1,250 to 1,600 metres, 11092, 13624.

A handsome epiphytic orchid, distributed in Guangxi, Guizhou, southern Yunnan and Sichuan.

Vanda sp. (ORCHIDACEAE)

Anonymous in Kew List of Determinations (volume 34).

Yunnan: Simao, 12085, 13618, 13638.

Vandellia micrantha (D. Don) Eb. Fisch., Schaferh. & Kai Müell. (LINDERNIACEAE)

Vandellia angustifolia Benth. Forbes & Hemsley in Journ. Linn. Soc. Bot. 26: 189. (1890).

Hubei: Yichang, 1643, 4146, 2367 (in part).

Vandellia montana (Blume) Benth. (LINDERNIACEAE)

Vandellia mollis Benth. Anonymous in Kew List of Determinations (volume 24 & 34).

Hainan: Environs of Haikou, 8329, 8339. **Yunnan:** Mengzi, 11140. Simao, 12411.

Distributed in Pakistan, India, China, Laos and Vietnam.

Vandenboschia auriculata (Blume) Copel. (HYMENOPHYLLACEAE)

Trichomanes auriculatum Blume Henry in Trans. Asiat. Soc. Jap. 24 suppl: 110. (1896) (List Pl. Formosa), Christ in Bull. Herb. Boiss. 6: 866. (1898).

Taiwan: Oluanpi, 1229. **Yunnan:** On mountains to the south of Mengzi at 1,650 metres, 10100. Mengzi, 11527.

A widespread fern, native to the Himalaya, India, southern China (including Taiwan), Indonesia, Micronesia, Malaya, Thailand and Japan.

Vanilla albida Blume (ORCHIDACEAE)

Vanilla sp. Henry in Trans. Asiat. Soc. Jap. 24 suppl: 92. (1896) (List Pl. Formosa), Rolfe in Journ. Linn. Soc. Bot. 32: 458. (1903), Rolfe in Journ. Linn. Soc. Bot. 36: 39. (1903).

Taiwan: Wanchin, 479.

A semi-epiphytic orchid with climbing stems to 3 metres long. Native to Malaysia, Thailand, Indochina and China (Taiwan) where it grows in broadleaved forest at low

altitudes.

Ventilago denticulata Willd. (RHAMNACEAE)

Ventilago calyculata Tulasne Schneider in Sargent Pl. Wilson. 2: 253. (1914).

Yunnan: On the bank of the Red River near Manpan, 10889. Simao, 13286.

A large climber, native to India, the Himalaya, western China, Thailand and Vietnam.

Ventilago elegans Hemsley (RHAMNACEAE) New species

in Ann. of Bot. 9: 151. (1895) Henry in Trans. Asiat. Soc. Jap. 24 suppl: 27. (1896) (List Pl.Formosa), Bretschneider in Hist. Eur. Bot. Disc. China 2: 781. (1898), Forbes & Hemsley in Journ. Linn. Soc. Bot. 151. (1903), Li in woody Flora of Taiwan 520. fig. 201. (1963).

Taiwan: Wanchin, 489 (type). Kaohsiung, without number.

A small climbing shrub, endemic to Taiwan where it grows in forests on the Hengchun Peninsula and along the eastern and northeastern coasts.

Ventilago leiocarpa Benth. (RHAMNACEAE)

Henry in Trans. Asiat. Soc. Jap. 24 suppl: 27. (1896) (List Pl. Formosa).

Taiwan: Wanchin, 441. **Yunnan:** Mengzi, 11027.

A scandent shrub, native to tropical Asia. In Taiwan this species grows in thickets at low altitude.

Ventilago sp. (RHAMNACEAE)

Anonymous in Kew List of Determinations (volume 34).

Yunnan: Mengzi at 1,800 metres, 9125.

Veratrum grandiflorum (Maxim. ex Miq.) O. Loes. (MELANTHIACEAE)

Veratrum album Diels non L. in Bot. Jahrb. 29: 240. (1900), Journ. Linn. Soc. Bot. 147. (1903). *Veratrum puberlum* Loes. f. in Feddes Repert. Spec. Nov. Regni. Veg. 24: 63. (1927).

Hubei: South Badong, 6060 (isotype of *Veratrum puberlum* Loes. f.), 6082.

Veratrum maackii Regel (MELANTHIACEAE)

Diels in Bot. Jahrb. 29: 240. (1900), Wright in Journ. Linn. Soc. Bot. 36: 147. (1903).

Hubei: Yichang, 930.

Veratrum maackii Regel var. *parviflorum* (Maxim. ex Miq.) H. Hara (MELANTHIACEAE)

Veratrum maximowiczii Baker Diels in Bot. Jahrb. 29: 240. (1900), Wright in Journ. Linn. Soc. Bot. 36: 147. (1903).

Hubei: Liantuo, 4610. Yichang, 4610a.

Veratrum mengtzeanum Loes. f. (MELANTHIACEAE) New species

in Repert. Spec. Nov. Regni Veg. 24: 6. (1928).

Yunnan: Mengzi at 2,000 metres, 9979 (isolectotype). Distributed from China (Guizhou and Yunnan) to northern Thailand.

Veratrum nigrum L. (MELANTHIACEAE)

Diels in Bot. Jahrb. 29: 240. (1900), Wright in Journ. Linn. Soc. Bot. 36: 147 (1903).

Hubei: Fang Xian, 6082a. Zigui at 1,950 to 2,650 metres, 6082b.

Veratrum sp. (MELANTHIACEAE)

Anonymous in Kew List of Determinations (volume 34).

Yunnan: Mengzi, 13663.

Verbascum coromandelianum (Vahl.) Hub.-Mor. (SCROPHULARIACEAE)

Celsia coromandeliana Vahl. Forbes & Hemsley in Journ. Linn. Soc. Bot. 26: 177. (1890).

Hubei: Yichang, 193.

Verbena officinalis L. (VERBENACEAE)

Henry in Journ. China Br. Roy. Asiat. Soc. 22: 273. (1887) (Chinese Names of Plants), Forbes & Hemsley in Journ. Linn. Soc. Bot. 26: 252. (1890), Henry in Trans. Asiat. Soc. Jap. 24 suppl: 70. (1896) (List Pl. Formosa).

Hubei: Yichang, 759, 788, 1130. **Taiwan:** Oluanpi, 284. Kaohsiung, without number. Wanchin, without number.

An perennial herb to 80 cm, thought to be native to the Mediterranean region but now widely distributed in Europe, Asia, North Africa and North America. Henry stated that in the Three Gorges region this medicinal plant was known colloquially as *t'ieh ma pien*.

Vernicia fordii (Hemsl.) Airy-Shaw (EUPHORBIACEAE)
New species

in Kew Bull. Misc. Inf. 20: 394. (1966).

Aleurites cordata Henry non (Thunb.) R. Br. ex Steud. in Journ. China Br. Roy. Asiat. Soc. 22: 276. (1887) (Chinese Names of Plants); in Notes Econ. Bot. China 60. (1893), Forbes & Hemsley in Journ. Linn. Soc. Bot. 26: 433. (1894). *Aleurites fordii* Hemsley in Hooker's Icon. Pl. 29: tt. 2801, 2802. (1906), Fedde in Repert. Nov. Spec. Regni. Veg. 5: 260. (1908), Zheng in Journ. Wuhan Bot. Res. 2.(1): 110. (1984).

Hubei: Yichang, 678, 1243. **Sichuan:** South Wushan, 5856, 5856a. **Hainan:** Ling-men, 8639. Environs of Kiungchow, 8756. **Yunnan:** Milê, 10587 (isotype, collected 15th March 1897).

According to Augustine this is the *t'ung-tze-shu*, *tung-shu*, *tung-yu* or *yu-tung*, the wood oil tree which is sometimes called Ningpo varnish. A small tree occurring in the mountains of the Three Gorges region, it bears seed capsules as large as apples and from these seeds 'wood-oil' was extracted. In Henry's time in China it was used extensively for caulking, painting, varnishing and preserving woodwork. The best sort of Chinese ink used in traditional calligraphy was said to be made from the soot obtained by burning this oil. Central China was the chief area of production and from there enormous quantities were exported. Wilson stated it was abundant in the Yangtze valley in the region from Yichang to Chongqing but was best seen in the area of the Three Gorges and the mountains above them to an altitude of about 800 metres. It is a beautiful (small) tree when in flower bearing myriads of white flowers, stained pink and yellow in the month of April.

Tung oil is still produced in great quantities in China (it was once a relatively important commercial crop in the USA. Teak oil, sold in Europe and the USA for fine furniture, windows and doors is usually made from refined tung oil. David Fairchild of the United States Department of Agriculture was responsible for the development of this tree as a commercial crop in the United States. Fairchild had been in contact with Henry while based in China during the 1880s but met him during a visit to Kew in 1903. He related the following in his biography, *The World was my Garden, Travels of a plant Explorer* (page 284). (1941).

Doctor Augustine Henry, the great authority on Chinese plants, had returned to England and was living near Kew Gardens. He was a fascinating person, rather small in stature, with the nervous manner, which I have often noticed, in scientific explorers who are more accustomed to plants than people. He received me most cordially and helped me to fill one of my notebooks with information, even writing in it many Chinese characters and notes of his own.

During the twenty years which he had served in the Imperial Chinese Maritime Customs, he had spent a large part of his salary on his scientific studies of the plants of China. He gave me a pamphlet which he had published and distributed in 1893 called *Notes on Economic Botany of China*. I was delighted to possess this article, which was out of print, and later it proved so useful that the Department of Agriculture had it reproduced photographically for me, so that the library could retain the original copy on its shelves. My main object in calling on Dr. Henry had been to discuss exploration in China. I asked him whether he would consider a proposition to return there. He replied that Professor Sargent had already offered him a thousand pounds a year and his expenses if he would go back and collect for the Arnold Arboretum, but he had written saying he was tired of China and did not want to return. He was, however, most enthusiastic about the possibilities of the western provinces, particularly Yunnan and Szechwan. He declared that these provinces are immense plains irrigated by many rivers; seven crops a year are raised; and no botanist has touched this region. Doctor Henry told me of wild pear and peach varieties, hardy bamboos, eighty species of the genus *Rubus* to which the blackberry and raspberry belong, and a persimmon in Peking, which the dealers ripened by puncturing the skin with a small stick to let the air in. I was particularly anxious to know if he had seen tung oil trees growing. He knew it as *Aleurites cordata* and I took down the following notes at his dictation: "The tung oil tree, (*Aleurites cordata*) is cultivated about Ichang as an orchard tree and grows on waste land where nothing else will grow. A strictly mountain plant, (it) stands the snow, but is generally grown in the citrus belt. Hardier plants of the species are found high up in the mountains, where it is much colder."

He called my special attention to a tree called *Eucommia ulmoides*, which he said grew like a poplar in England. He advocated planting it because of a gum which it contained in the leaves and the bark. Later we distributed this tree throughout America, but, although the gum was so abundant that you could pull the leaves apart and the individual pieces would remain attached by delicate strands of latex, the gum has not been found to have commercial value. The industry of this unusual man was evident by a manuscript of his which he showed me - a dictionary containing 20,000 Chinese Names of Plants, including references from the principal Chinese botanical books. Moreover, he told me that he was working on a dictionary of the Lolo language, the lolos being a simple, independent people living in the interior of China, about whom little was known. That afternoon spent with Doctor Henry had far reaching results. The information which he gave me made a great impression on me, and subsequently, when I directed the plans and policy of the Office, it determined me to send explorers to China.

Fairchild's first agricultural explorer to travel to China on behalf of the United States Department of Agriculture was Frank Meyer, who later introduced to cultivation numerous fruits, vegetables and cereals from China. Meyer later drowned in the Yangtze. The tung oil tree is still a common tree in the mid-Yangtze valley. It was collected by the Glasnevin Central China Expedition in October 2004 (GCCE 715) near the San You Dong Glen near Yichang in October 2004 and near Dragon Gate River Park in Xingshan (GCCE 836) also in October 2004. In both cases it formed small trees to about 3 metres tall. Native to China and Vietnam.

Vernonia sp. (ASTERACEAE)
Anonymous in Kew List of Determinations (volume 34).
Yunnan: Mengzi, 10852.

Veronica anagallis-aquatica L. (PLANTAGINACEAE)
Forbes & Hemsley in Journ. Linn. Soc. Bot. 26: 198. (1890).
Hubei: Yichang, 774, 3533. Liantuo, 1954. **Yunnan:** Mengzi, 10928. Simao, 13515.

Veronica arvensis L. (PLANTAGINACEAE)
Veronica agrestis Forbes & Hemsley non L. in Journ. Linn. Soc. Bot. 26: 197. (1890).
Hubei: Yichang, 1199, 3329.

Veronica cana Wall. ex Benth. (PLANTAGINACEAE)
Anonymous in Kew List of Determinations (volume 34).
Yunnan: Mengzi, 10555.
Native to Nepal, India, Bhutan and China [Yunnan and Tibet (Xizang)].

Veronica henryi T. Yamazaki (PLANTAGINACEAE)
in Journ. Jap. Bot. 31: 296. (1956).
Veronica cana Forbes & Hemsley non Walli. ex Benth. in Journ. Linn. Soc. Bot. 26: 198. (1890).
Hubei: Badong, 3750, 5417, 6927. Jianshi, 5688, 5829. Fang Xian, 6843. **Sichuan:** South Wushan, 5568, 5729 (type).

Veronica laxa Benth. (PLANTAGINACEAE)
Forbes & Hemsley in Journ. Linn. Soc. Bot. 26: 199. (1890).
Hubei: Badong, 4051. Changleping, 6253. **Sichuan:** South Wushan, 5705, 5706. **Yunnan:** Mengzi, 10711.
Distributed in Pakistan, India, China and Japan.

Veronica serpyllifolia L. (PLANTAGINACEAE)
Forbes & Hemsley in Journ. Linn. Soc. Bot. 26: 199. (1890).
Hubei: Badong, 708 (July 1885), 1866, 3738, 4074. **Sichuan:** South Wushan, 5302. Henry's Chinese collector with Pratt, no locality stated but may be any of the following, Leshan, Emei Shan, Wa Shan or Kangding, 8877.

Veronica spicata L. (PLANTAGINACEAE)
Forbes & Hemsley in Journ. Linn. Soc. Bot. 26: 200. (1890).
Hubei: Xingshan, 6525. Fang Xian, 6533.

Veronicastrum axillare (Sieb. & Zucc.) T. Yamazaki (PLANTAGINACEAE)
Calorhabdos axillaris (Sieb. & Zucc.) Benth & Hook.f. Forbes & Hemsley in Journ. Linn. Soc. Bot. 26: 195. (1890).
Hubei: Fang Xian, 6673.
In Chinese herbal medicine the dried aerial parts of this plant are used to treat oedema.

Veronicastrum brunonianum (Benth.) D. Y. Hong (PLANTAGINACEAE)
Calorhabdos brunoniana Benth. Anonymous in Kew List of Determinations (volume 34).

Yunnan: On the Daweishan Range in Pingbian at 1,500 metres, 9076.

Veronicastrum caulopterum (Hance) T. Yamazaki (PLANTAGINACEAE)

Calorhabdos cauloptera Hance Forbes & Hemsley in Journ. Linn. Soc. Bot. 26: 195. (1890), Hooker in Bot. Mag. 7800 (1901), Nelson in Pim, The Wood & the Trees 221. (1984), Nelson in A Heritage of Beauty 310. (2000).

Hubei: Yichang, 2949. Xiling (Yichang) Gorge, 4326. Liantuo, on the banks of the Yangtze, 4445.

Discovered by the Rev'd. J. C. Nevin of the American United Presbyterian Mission in 1866. Introduced to cultivation by Augustine Henry through the Royal Botanic Gardens, Kew in 1896, it first flowered there under glass in August 1900.

Veronicastrum latifolium (Hemsl.) T. Yamazaki (PLANTAGINACEAE) New species

Calorhabdos latifolia Hemsley in Journ. Linn. Soc. Bot. 26: 196. pl. t. 4. (1890), Bretschneider in Hist. Eur. Bot. Disc. China 2: 788. (1898).

Hubei: Antelope glen, near Yichang, 4386 (type).

Veronicastrum stenostachyum (Hemsl.) T. Yamazaki (PLANTAGINACEAE) New species

Calorhabdos stenostachya Hemsley in Journ. Linn. Soc. Bot. 26: 196. (1890), Bretschneider in Hist. Eur. Bot. Disc. China 2: 788. (1898). *Calorhabdos venosa* Hemsley in Journ. Linn. Soc. Bot. 26: 197. (1890),

Hubei: Yichang, 55, 589, 2187 (syntype of *Calorhabdos venosa* Hemsley). Liantuo, 3044, 4638.

Viburnum amplifolium Rehder (VIBURNACEAE) New species

in Sargent, Trees & Shrubs 2: 112. (1908), Rehder in Fedde, Repert. Spec. Nov. Regni Veg. 9: 182. (1911).

Yunnan: In Pingbian, in forests on the Daweishan Range at 1,640 metres, 13470 (holotype).

Endemic to south-east Yunnan.

Viburnum atrocyaneum C. B. Clarke (VIBURNACEAE)

Viburnum calvum Rehder in Sargent Pl. Wilson. 1: 310. (1912), Bean in Trees & Shrubs 4: 712. (1989), O' Brien in Augustine Henry, An Irish Plant Collector in China 37. (2002).

Yunnan: In mountain forests to the north of Mengzi at 2,700 metres, 10564 (holotype, collected 9th April 1897).

A shrub to about 1.5 metres, allied to *Viburnum tinus* and *Viburnum propinquum*, the creamy-white flowers are held in rounded terminal corymbs to 7 cm across followed by blue-black fruits.

Viburnum betulifolium Batal. (VIBURNACEAE)

Rehder in Sargent, Trees & Shrubs 2: 99. (1907), Zheng in Journ. Wuhan Bot. Res. 2.(1): 198. (1984).

Viburnum phlebotrichum Forbes & Hemsley non Sieb. & Zucc. in Journ. Linn. Soc. Bot. 354. (1888). *Viburnum ovatifolium* Rehder in Sargent, Trees & Shrubs 2: 115. (1907), Rehder in Fedde, Repert. Spec. Nov. Regni Veg. 9: 185. (1911).

Hubei: Badong, 2885. Changleping, 6250, 6262. **Yunnan:** In the mountains to the north of Mengzi at 2,300 metres, 10211a., (syntype of *Viburnum ovatifolium* Rehder), 10211b., (syntype of *Viburnum ovatifolium* Rehder).

A deciduous shrub to 4 metres tall, native to central and western China. Discovered in Gansu by the Russian explorer Grigori Nicolaevich Potanin on July 10th 1885, this species was introduced to cultivation by Wilson in 1901 (seeds of his 238a. from Yichang were received at Glasnevin in 1908 from the Arnold Arboretum). It is a handsome shrub in flower but is magnificent when seen carrying dense corymbs of fleshy, blood-red fruits and scarlet autumnal foliage. In China the bark fibre of this species is used to make rope and paper.

Viburnum brachybotryum Hemsley (VIBURNACEAE) New species

in Journ. Linn. Soc. Bot. 23: 349. (1888), Bretschneider in Hist. Eur. Bot. Disc. China 2: 784. (1898), Rehder in Sargent, Trees & Shrubs 2: 107. (1907), Rehder in Sargent Pl. Wilson. 1: 309. (1913), Zheng in Journ. Wuhan Bot. Res. 2.(1): 196. (1984).

Hubei: Antelope Glen, 3324 (type). Yichang, 3324a. South Badong, 7302. Changleping, 6218. **Yunnan:** Mengzi at 2,000 metres, 10065, 10065a. In forests to the south-east of Simao at 1,700 metres, 12790, 12790a.

Endemic to China.

Viburnum buddleifolium C. H. Wright (VIBURNACEAE) New species

Zheng in Journ. Wuhan Bot. Res. 2.(1): 196. (1984).

Hubei: Badong, 6305a.

A deciduous shrub to 2 metres tall, discovered by Henry in 1888 and described from material later collected in Hubei by Wilson. Introduced by Wilson through Veitch's Coombe Wood nursery in 1900.

Viburnum congestum Rehder (VIBURNACEAE) New species

in Sargent, Trees & Shrubs 2: 111. (1907).

Yunnan: In woods near Mengzi at 1,600 metres, 9683, 9638a., (type).

Endemic to China and distributed in the provinces of Gansu, Guizhou, Sichuan and Yunnan.

Viburnum corylifolium Hook. f. & Thoms. (VIBURNACEAE)

Rehder in Sargent, Trees & Shrubs 2: 115. (1907), Rehder in Sargent Pl. Wilson. 1: 112. (1911), Zheng in

Journ. Wuhan Bot. Res. 2.(1): 196. (1984).

Viburnum dilatatum Rehder non Thunb. in Sargent, Trees & Shrubs 2: 114. (1907).

Hubei: Yichang, 544. **Yunnan:** Mengzi, 11362.

Viburnum cylindricum Buch.-Ham. ex D. Don (VIBURNACEAE)

Rehder in Sargent, Trees & Shrubs 2: 91. (1907).

Viburnum crassifolium Rehder in Sargent, Trees & Shrubs 2: 112. (1916); in Fedde, Repert. Spec. Nov. Regni Veg. 9: 181. (1911). *Viburnum cylindricum* Buch.-Ham. ex D. Don var. *crassifolium* (Rehd.) Schneider in Botanical Gazette lxiv. 77. (1917).

Hubei: Badong, 6125, 7006. **Sichuan:** South Wushan, 7238. **Yunnan:** Mengzi at 1,700 to 2,000 metres, 9757, 9757c., 9757d. On grassy mountains near Mengzi at 2,000 metres 9797, (isotype of *Viburnum crassifolium* Rehder).

Widely distributed across Nepal, India, Myanmar, China and Thailand.

Viburnum dilatatum Thunb. (VIBURNACEAE)

Forbes & Hemsley in Journ. Linn. Soc. Bot. 23: 351. (1888).

Hubei: On Tsui Fu Shan near Yichang (May 1888), 1598. Zigui, 5720a. **Sichuan:** South Wushan, 5743.

A deciduous shrub to 3 metres tall, flowers white, appearing in June and produced in five-rayed cymes to 12 cm across. Native to Japan and China, it first flowered in cultivation with Messrs Veitch in 1875. A handsome plant and prolific in the production of flowers.

Viburnum erosum Thunb. (VIBURNACEAE)

Viburnum erosum Thunb. var. *ichangense* Hemsley in Journ. Linn. Soc. Bot. 23: 352. (1887). *Viburnum ichangense* (Hemsl.) Rehder in Sargent, Trees & Shrubs. 2: 116. (1907), Rehder in Fedde, Repert. Spec. Nov. Regni Veg. 9: 186. (1911), Schneider in Illus. Handb. Laubholzk. 2: 650. (1912), Bean in Trees & Shrubs 2: 648. (1921), Zheng in Journ. Wuhan Bot. Res. 2.(1): 198. (1984), Nelson in A Heritage of Beauty 320. (2000), O' Brien in Augustine Henry, An Irish Plant Collector in China 38. (2002).

Hubei: Badong, 232 (June 1885), 2841. Liantuo, 1888. Yichang, 2289. Changyang, 5271. Changleping, 6247. Fang Xian, 6594 (isosyntype). **Sichuan:** South Wushan, 5476. North Wushan, 7052. **Hainan:** Ling-men, 8625.

Viburnum erubescens Wallich ex DC. (VIBURNACEAE)

Rehder in Sargent, Trees & Shrubs 2: 107. (1907), Zheng in Journ. Wuhan Bot. Res. 2.(1): 197. (1984).

Hubei: Changleping, 6296. Yichang, 6488. Without locality, 5605a., 5784b. **Sichuan:** South Wushan, 5605. North Wushan, 7024. **Yunnan:** Mengzi, 10247.

A variable species widely distributed across the Himalaya to India, northern Myanmar and China.

Viburnum erubescens Wallich ex DC. var. *gracilipes* Rehder (VIBURNACEAE) New variety

in Sargent Pl. Wilson. 1: 107. (1911), O' Brien in Augustine Henry, An Irish Plant Collector in China 38. (2002).

Hubei: Xingshan at 1,400 metres, 6543. **Sichuan:** South Wushan, 5660, 5691. Henry's Chinese collector with Pratt, no locality stated but may be any of the following, Leshan, Emei Shan, Wa Shan or Kangding, 8931.

This variety is hardier than the type and bears longer flowering panicles, it was introduced to cultivation by Wilson in 1910, an old plant from this introduction grows in the Chinese shrubbery at the National Botanic Gardens, Glasnevin.

Viburnum foetidum Wall. (VIBURNACEAE)

Rehder in Sargent, Trees & Shrubs 2: 114. (1907).

Yunnan: Mengzi at 1,700 metres, 9244. On Fengchunlin at 1,700 metres, 9244a.

Native to Nepal, India, Bhutan, Myanmar, China, Thailand and Laos.

Viburnum formosanum (Hance) Hayata (VIBURNACEAE)

Viburnum erosum Henry non Thunb. in Trans. Asiat. Soc. Jap. 24 suppl: 49. (1896) (List Pl. Formosa). *Viburnum luzonicum* Rolfe var. *formosanum* (Hance) Rehd. Rehder in Sargent, Trees & Shrubs 2: 97. (1907), Li in Woody Flora of Taiwan 896. fig. 361. (1963).

Taiwan: Wanchin, 161, 569. Oluanpi, 607, 612, 652, 949, 1272.

A large shrub or a small deciduous tree. Endemic to Taiwan where according to Henry the indigenous inhabitants of the mountainous interior made their bows from its wood. Discovered by the Kew collector Richard Oldham.

Viburnum hainanense Merrill & Chun (VIBURNACEAE)

in Journ. Arnold Arb. 23: 378. (1942).

Viburnum sempervirens Rehder non K. Koch in Sargent, Trees & Shrubs. 2: 95. t. 145. (1908), Rehder in Journ. Arnold Arb. 16: 331. (1935). *Viburnum tsangii* Rehder in Journ. Arnold Arb. 23: 378. (1942).

Yunnan: Mountains to the south of Simao at 1,650 metres, 12753 (syntype of *Viburnum tsangii* Rehder).

Native to southern China and Vietnam.

Viburnum henryi Hemsley (VIBURNACEAE) New species

in Journ. Linn. Soc. Bot. 23: 353. (1888), Bretschneider in Hist. Eur. Bot. Disc. China 2: 784. (1898), Hemsley in Curtis's Bot. Mag. 137: t. 8393. (1911), Schneider in Illus. Handb. Laubholzk. 2: 668. (1912), Rehder in Sargent, Trees & Shrubs. 2: 35. fig. 116. (1913), Besant in Gard. Chron. ser. 3, 98: 335. (1935), Morley in Glasra 3: 80. (1979), Zheng in Journ.

Wuhan Bot. Res. 2.(1): 197. (1984), Nelson in A Heritage of Beauty 320. (2000), O' Brien in Augustine Henry, An Irish Plant Collector in China 38. (2002).

Hubei: Badong, 1705 (isosyntype), 1730, 4060 (isosyntype), 5350, 6092. Jianshi, 5784. South Badong, 6489. Jianshi, 7466. Fang Xian, 7608a. **Sichuan:** South Wushan, 5617. North Wushan, 7608. **Yunnan:** Fengchunling, 10645.

An erect evergreen shrub to 3 metres tall, carrying white flowers in erect pyramidal panicles in June followed by red berries that ripen to black. One of the most beautiful and little grown of Chinese viburnums, this species was exhibited by the Irish Garden Plant Society and Bord Glas at the Chelsea Flower Show, London in May 2002 in an exhibit based on Henry's botanical activities in China. Of the many plants on display this was the most popular and commented on. It is a beautiful plant in habit, fragrance, foliage, flower and fruit and deserves to be better known. Introduced to cultivation by Wilson through Veitch's Coombe Wood nursery in 1901. Endemic to China.

Viburnum hupehense Rehder (VIBURNACEAE) New species

in Sargent, Trees & Shrubs. 2: 116. (1913), Rehder in Fedde, Repert. Spec. Nov. Regni Veg. 9: 185. (1911), Bean in Trees & Shrubs 2: 650. (1921), Zheng in Journ. Wuhan Bot. Res. 2.(1): 198. (1984), Hemsley in Curtis's Bot. Mag. n. ser. 165: t. 41. (1948), Nelson in A Heritage of Beauty 320. (2000), O' Brien in Augustine Henry, An Irish Plant Collector in China 38. (2002).

Hubei: Shennongjia at 1,900 to 3,000 metres, 6805 (holotype).

A deciduous shrub carrying corymbs of white flowers to 5 cm across in early summer. Introduced to cultivation by Wilson in 1908 through the Arnold Arboretum.

Viburnum lobophyllum Graebner (VIBURNACEAE) New species

in Engler, Bot. Jahrb. xxix. 589. (1901).

Sichuan: Henry's Chinese collector with Pratt, no locality stated but may be any of the following, Leshan, Emei Shan, Wa Shan or Kangding, 8930.

Viburnum macrocephalum Fortune f. *keteleeri* (Carr.) Rehd. (VIBURNACEAE)

Zheng in Journ. Wuhan Bot. Res. 2.(1): 199. (1984).

Viburnum arborescens Hemsley in Journ. Linn. Soc. Bot. 349. (1888), Bretschneider in Hist. Eur. Bot. Disc. China 2: 784. (1898). *Viburnum macrocephalum* Forbes and Hemsley non Fortune in Journ. Linn. Soc. Bot. 23: 353. (1888), Rehder in Sargent, Trees & Shrubs 2: 110. (1907).

Hubei: Liantuo, 3852, 3180 (type of *Viburnum arborescens* Hemsley), 4643. Badong, 4756. Liantuo, 7661.

A deciduous shrub to 4 metres tall, *Viburnum macrocephalum* is a perfectly sterile garden selection and was introduced by Robert Fortune from China in 1844. The form *keteleeri* is the normal wild plant.

Viburnum odoratissimum Ker.-Gawl. (VIBURNACEAE)

Henry in Trans. Asiat. Soc. Jap. 24 suppl: 49. (1896) (List Pl. Formosa), Li in Woody Flora of Taiwan 891. (1963).

Taiwan: Wanchin, 67, 75, 86, 139. Oluanpi, 682, 2053.

An evergreen shrub to 6 metres tall with bold, leathery leaves to 15 cm long. The white flowers are fragrant and produced in pyramidal panicles to 15 cm high.

Viburnum oliganthum Batalin (VIBURNACEAE)

Rehder in Sargent, Trees & Shrubs 2: 108. (1907).

Sichuan: Henry's Chinese collector with Pratt, no locality stated but may be any of the following, Leshan, Emei Shan, Wa Shan or Kangding, 8934.

Discovered by the Russian explorer, Grigori Nicolaevich Potanin in eastern Gansu on August 15th 1885 and based on that collection and Henry's Sichuan collection made by his Chinese collector in 1889.

Viburnum plicatum Thunb. f. *tomentosum* (Thunb.) Rehd. (VIBURNACEAE)

Zheng in Journ. Wuhan Bot. Res. 2.(1): 199. (1984).

Viburnum tomentosum Thunb. Henry in Journ. China Br. Roy. Asiat. Soc. 22: 261. (1887) (Chinese Names of Plants), Forbes & Hemsley in Journ. Linn. Soc. Bot. 23: 356. (1888). *Viburnum tomentosum* Thunb. var. *parvifolium* (Miq.) Rehder in Sargent, Trees & Shrubs 2: 108. (1907).

Hubei: Badong, 228, 712 (June 1885), 711 (July 1885), 1761. Xingshan, 4516. Liantuo, 7654, 7654a. Jianshi, 7418. Changleping, 7654. Xingshan, 7654b. **Sichuan:** South Wushan, 5720.

A deciduous, wide spreading, horizontally branched shrub to 3 metres tall, new growths covered when young with minute down. Henry stated this handsome species was known colloquially in the Three Gorges region as *no mi shu*. Flowers are produced in June and consist of a flat umbel to 10 cm across, the centre filled with perfect flowers surrounded by large white sterile flowers. Native to Japan and China, introduced from China in 1865. It was reintroduced by Charles Maries and Ernest Wilson through Veitch's Coombe Wood nursery. A beautiful sight in early summmer when carrying its broad corymbs on horizontal branches. The two finest selections are 'Mariesii' and 'Rowallane'.

Viburnum propinquum Hemsley (VIBURNACEAE) New species

in Journ. Linn. Soc. Bot. 23: 355. (1888), Bretschneider in Hist. Eur. Bot. Disc. China 2: 784. (1898), Rehder in Sargent, Trees & Shrubs. 2: 33. (1913), Bean in Trees &

Shrubs 2: 645. (1921), Zheng in Journ. Wuhan Bot. Res. 2.(1): 199. (1984), Nelson in Pim, The Wood & the Trees 220. (1984); in A Heritage of Beauty 320. (2000), O' Brien in Augustine Henry, An Irish Plant Collector in China 38. (2002).

Hubei: Antelope Glen, 3415 (type). Liantuo, 3914. Yichang, 4237. Xiling (Yichang) Gorge, 4313 (isosyntype). Fang Xian, 6624, 6658. Changyang, 7745. Without locality, 3415a.

An evergreen shrub to 2 metres tall, flowers greenish-white and produced in small cymes in early summer. Native to central and western China and south China (Taiwan) and the Philippines, *Viburnum propinquum* was introduced to cultivation by Wilson in 1901 through Veitch's Coombe Wood nursery [seeds of his 498 (from Yichang) reached Glasnevin from the Arnold Arboretum in 1908]. Though its flowers leave little to recommend, it is a very handsome foliage and fruit plant, there is a good example in the rock garden at Glasnevin.

Viburnum propinquum Hemsley var. *mairei* W.W. Smith (VIBURNACEAE)

Zheng in Journ. Wuhan Bot. Res. 2.(1): 199. (1984).

Hubei: Changyang (April 1888), 5221.

This variety commemorates Charles Maries who made a brief collecting trip to the Three Gorges region for Messrs Veitch in 1879.

Viburnum pyramidatum Rehder (VIBURNACEAE) New species

in Sargent, Trees & Shrubs 2: 93. (1907), Rehder in Fedde, Repert. Spec. Nov. Regni Veg. 9: 181. (1911).

Yunnan: Near Mengzi at 1,700 metres, 11475 (holotype).

Native to China (Guangxi and Yunnan) and Vietnam.

Viburnum rhytidophyllum Hemsley (VIBURNACEAE) New species

in Journ. Linn. Soc. Bot. 23: 355. (1888), Bretschneider in Hist. Eur. Bot. Disc. China 2: 784. (1898), Hemsley in Curtis's Bot. Mag. 137: 8382. (1911), Schneider in Illus. Handb. Laubholzk. 2: 665. (1912), Rehder in Sargent, Trees & Shrubs. 2: 39. (1913), Besant in Gard. Chron. ser. 3, 98: 335. (1935), Morley in Glasra 3: 80. (1979), Zheng in Journ. Wuhan Bot. Res. 2.(1): 200. (1984), Nelson in A Heritage of Beauty 320. (2000), O' Brien in Augustine Henry, An Irish Plant Collector in China 38. (2002), O' Brien in Ir. Garden 13(5): 53. (2004).

Hubei: Badong, 613 (type, collected July 1885). South Badong, 6305. Jianshi, 7451. Without locality 5328b., 7451a. **Sichuan:** South Wushan, 5328.

An upright, evergreen shrub to 6 metres tall carrying ovate-oblong leaves to 18 cm long, the upper surface wrinkled, the undersides plastered with a thick felt of grey down. Dull creamy flowers are produced in large terminal trusses to 20 cm across in early summer. Introduced to cultivation by Wilson when collecting for Messrs Veitch in 1900, in their *Novelties for 1908—1909* plants were available for up to twenty-one shillings each. On his return to Dublin in 1913 to take a Chair in Forestry at the Royal College for Science (now part of University College, Dublin), Henry and his wife Alice established a fine garden at their house in Ranelagh in the south Dublin suburbs and this species and *Viburnum henryi* were grown there. Other 'Henry plants' listed as growing there during the 1920's include *Acer henryi, Buddleja davidii, Cotoneaster henryana, Lilium henryi, Ligustrum henryi, Lonicera henryi, Rubus lasiostylus, Spiraea henryi, Parthenocissus henryana, Rhododendron augustinii, Corydalis cheilanthifolia, Corydalis thalictrifolia, Iris wattii, Ribes henryi, Lysimachia henryi, Itea ilicifolia, Lonicera pileata, Actinidia chinensis, Jasminum mesnyi, Lonicera pileata, Androsace henryi, Deinanthe caerulea, Dipsacus asper, Actinidia henryi, Dipteronia sinensis, Schisandra henryi, Lilium centifolium* and *Sinowilsonia henryi*.

Viburnum sargentii Koehne (VIBURNACEAE)

Zheng in Journ. Wuhan Bot. Res. 2.(1): 200. (1984).

Hubei: Yichang, 6450.

The Asian ally of *Viburnum opulus*, a deciduous shrub to 3 metres tall. Introduced to cultivation by Professor Charles Sprague Sargent of the Arnold Arboretum in 1892.

Viburnum setigerum Hance (VIBURNACEAE)

Rehder in Journ. Arnold Arb. 12: 77. (1931).

Viburnum phlebotrichum Forbes & Hemsley non Sieb. & Zucc. in Journ. Linn. Soc. Bot. 26: 354. (1888). *Viburnum theiferum* Rehder in Sargent, Trees & Shrubs 2: 45. (1907); in Fedde, Repert. Spec. Nov. Regni Veg. 9: 183. (1911).

Hubei: Badong, 899 (October 1885), 2857. Liantuo, 2670, 3877, 3941. Changyang, 7513. **Sichuan:** South Wushan 5586 (isotype of *Viburnum theiferum* Rehder).

Viburnum setigerum was discovered by William Mesny near Chongqing in July 1880.

Viburnum spp. (VIBURNACEAE)

Anonymous in Kew List of Determinations (volume 24 & 34).

Hubei: Badong, 2840, 4822. Yichang, 3484, 5218. Xingshan, 4517, 4522. Jianshi, 5826a. **Yunnan:** Mengzi, 10211, 11262, 11276, 13647, 13708. Simao, 13523.

Viburnum sympodiale Graebn. (VIBURNACEAE) New species

in Bot. Jahrb. xxix. 587. (1901), Rehder in Sargent, Trees & Shrubs 2: 83. (1907), Schneider in Illus. Handb. Laubholzk. 2: 665. (1912), Zheng in Journ. Wuhan Bot. Res.

2.(1): 201. (1984), O' Brien in Augustine Henry, An Irish Plant Collector in China 38. (2002).

Viburnum furcatum Forbes & Hemsley non Blume ex Maxim. in Journ. Linn. Soc. Bot. 23: 352. (1888).

Hubei: Badong, 1470, 5312, 6312. Zigui, 5759a., (isotype). Xingshan, 7607. **Sichuan:** South Wushan, 5663, 5702, 5759. Henry's Chinese collector with Pratt, no locality stated but may be any of the following, Leshan, Emei Shan, Wa Shan or Kangding, 8929.

The Chinese ally of the Japanese *Viburnum furcatum*. Introduced by Wilson through Messrs Veitch's Coombe Wood nursery in 1900, it is cultivated at the National Botanic Gardens, Glasnevin.

Viburnum utile Hemsley (VIBURNACEAE) New species

in Journ. Linn. Soc. Bot. 23: 356. (1888), J. H. Veitch in Hotus Veitchii 410. (1906), Rehder in Sargent, Trees & Shrubs 2: 89. (1907), Schneider in Illus. Handb. Laubholzk. 2: 663. (1912), Zheng in Journ. Wuhan Bot. Res. 2.(1): 201. (1984), O' Brien in Augustine Henry, An Irish Plant Collector in China 38. (2002).

Viburnum sp. nova. Henry in Journ. China Br. Roy. Asiat. Soc. 22: 243. (1887).

Hubei: Yichang, 131, 260 (isotype). Liantuo, 2622.

A handsome evergreen shrub to two metres tall, producing rounded trusses of white flowers to 7 cm across in May. Discovered by the Irish diplomat, Thomas Watters near Yichang, Hemsley based this species on material collected by both Henry and Watters. Henry stated it was known colloquially at Yichang as *he kan t'iao* and *no mi t'iao tzu* and said the branches were used for making pipe-stems. *Viburnum utile* was introduced to cultivation by E. H. Wilson when collecting for Messrs Veitch in 1901. In the Veitch catalogue *New Hardy Plants from Western China* (Autumn 1912) this species was available for five shillings. George Forrest later collected this species (Forrest 14,792) in the Mekong-Salween divide in September 1917. Still a common plant in the hills around Yichang, it was collected by the Glasnevin Central China Expedition in the autumn of 2002.

Viburnum yunnanense Rehder (VIBURNACEAE) New species

in Sargent, Trees & Shrubs 2: 106. (1907) Rehder in Fedde, Repert. Spec. Nov. Regni Veg. 9: (1911).

Yunnan: In mountain forest near Mengzi at 2,300 metres, 11015 (isotype).

Endemic to Yunnan.

Vicia cracca L. (FABACEAE)

Forbes & Hemsley in Journ. Linn. Soc. Bot. 23: 184. (1887).

Hubei: On the cliffs of the Xiling (Yichang Gorge) and in glens, 1159. Badong, 667 (July 1885), 2418, 4017.

Vicia hirsuta (L.) S. T. Gray (FABACEAE)

Anonymous in Kew List of Determinations (volume 24).

Hubei: Yichang, 3393.

Vicia lens (L.) Coss. & Germ. (FABACEAE)

Ervum lens L. Henry in Journ. China Br. Roy. Asiat. Soc. 22: 258. (1887) (Chinese Names of Plants), Henry in Notes Econ. Bot. China 14. (1893).

Hubei: Yichang, 1152.

Lentils, cultivated in the glens and mountains near Yichang as a winter crop where in Henry's time it was colloquially known as *chin me wan tzu*.

Vicia sativa L. (FABACEAE)

Forbes & Hemsley in Journ. Linn. Soc. Bot. 23: 185. (1887), Henry in Journ. China Br. Roy. Asiat. Soc. 22: 258. (1887) (Chinese Names of Plants), Henry in Trans. Asiat. Soc. Jap. 24 suppl: 34. (1896) (List Pl. Formosa).

Hubei: Yichang, 596, 1216, 1366. **Taiwan:** On the plains near Wanchin, 186, 1832. **Yunnan:** Mengzi, 10426.

Native to the temperate northern hemisphere. Widely colonised through cultivation, in the Three Gorges region Henry stated it was known as *yeh me wan tzu*.

Vicia sp. (FABACEAE)

Anonymous in Kew List of Determinations (volume 24).

Hubei: Zigui, 6191.

Vicia tetrasperma (L.) Moench. (FABACEAE)

Anonymous in Kew List of Determinations (volume 24).

Hubei: Yichang, 3490.

Vicia unijuga A. Braun (FABACEAE)

Anonymous in Kew List of Determinations (volume 24 & 34).

Hubei: Fang Xian, 6596. **Yunnan:** Mengzi, 11184.

Native to Russia, China, Mongolia, Korea and Japan.

Vigna minima (Roxb.) Ohwi & H. Ohashi (FABACEAE)

Vigna lutea Henry non A. Gray in Trans. Asiat. Soc. Jap. 24 suppl: 36. (1896) (List Pl. Formosa).

Taiwan: Kaohsiung, without number.

An annual herb, native to China (Taiwan) and Japan (south).

Vigna radiata (L.) R. Wilczek (FABACEAE)

Phaseolus sp. Henry in Trans. Asiat. Soc. Jap. 24 suppl: 36. (1896) (List Pl. Formosa).

Taiwan: Oluanpi, 1273.

An erect herb to 30 cm tall, cultivated in Taiwan and escaped there to waste land and roadsides.

Vigna sp. No. 1. (FABACEAE)

Henry in Trans. Asiat. Soc. Jap. 24 suppl: 37. (1896) (List Pl. Formosa).

Taiwan: Wanchin, 542, 1783. Kaohsiung, in creeks, 1181.

Vigna sp. No. 2. (FABACEAE)

Anonymous in Kew List of Determinations (volume 34).

Yunnan: Simao, 12499.

Vigna trilobata (L.) Verdic (FABACEAE)

Dolichos trilobatus L. Henry in Trans. Asiat. Soc. Jap. 24 suppl: 37. (1896) (List Pl. Formosa).

Taiwan: Wanshoushan, 1082, 2012.

A twining vine to 120 cm long, native to tropical Asia. Found in the southern part of Taiwan at low altitude.

Vigna umbellata (Thunb.) Ohwi & H. Ohashi (FABACEAE)

Phaseolus calcaratus Roxb. Anonymous in Kew List of Determinations (volume 34).

Yunnan: Simao, 12554.

Vigna vexillata (L.) A. Rich. (FABACEAE)

Forbes & Hemsley in Journ. Linn. Soc. Bot. 23: 193. (1887), Henry in Journ. China Br. Roy. Asiat. Soc. 22: 274. (1887) (Chinese Names of Plants).

Hubei: Yichang, 381, 2197, 2286. Badong, 2548, 4970. Liantuo, 4614. Yunnan: Mengzi, 9703. In Pingbian, in forests on the Daweishan Range at 1,650 metres, 11042. Fengchunling, 11181. Simao, 12449.

Spread over the tropics, Henry stated that in the Three Gorges region this species was colloquially known as *yeh man tou*.

Vincetoxicum amplexicaule Sieb. & Zucc. (APOCYNACEAE)

Cynanchum amplexicaule (Sieb. & Zucc.) Hemsl. Henry in annotation in *Index Flora Sinensis* (copy at Glasnevin).

Hubei: Xingshan, 5624a. Jianshi, 5624b. Sichuan: South Wushan, 5624, 5624c.

Vincetoxicum atratum (Bunge) C. Morren & Decne. (APOCYNACEAE)

Cynanchum atratum Bunge Anonymous in Kew List of Determinations (volume 34).

Yunnan: Mengzi, 9424.

Vincetoxicum augustinianum (Hemsl.) Meve & Liede (APOCYNACEAE) New species

Henrya augustiniana Hemsley (genus novum) in Journ. Linn. Soc. Bot. 26: 111. (1889); in Hooker's Icon. Pl. t. 1971. (1891), Bretschneider in Hist. Eur. Bot. Disc. China 2: 788. (1898). *Henryastrum augustinianum* (Hemsl.) Happ in Ann. Missouri Bot. Gard. 24: 567. (1937). *Neohenrya augustiniana* Hemsley in Bull. Torrey Bot. Club 19: 97. (1892). *Tylophora angustiniana* (Hemsl.) Craib in Bull. Misc. Inf. Kew 417. (1911).

Hubei: Yichang, 4252 (type of *Henrya augustinianum*

Hemsley).

In dedicating his new genus *Henrya*, W. B. Hemsley stated the following;

This genus is named after the discoverer, Dr. Augustine Henry; and it may also serve to commemorate the Rev. B. C. Henry, a friend and correspondent of the late Dr. Hance, and a collector of plants who has discovered many novelties in the south of China.

Vincetoxicum auriculatum (Royle ex Wight) Kunze (APOCYNACEAE)

Cynanchum wilfordii Henry non (Maxim.) Hook. f. in Journ. China Br. Roy. Asiat. Soc. 22: 250. (1887) (Chinese Names of Plants). *Cynanchum auriculatum* Royle ex Wight Forbes & Hemsley in Journ. Linn. Soc. Bot. 26: 105. (1889), Schneider in Sargent Pl. Wilson. 3: 346. (1917).

Hubei: Yichang, 2196, 4271. Liantuo, 2659. Badong, 2481, 2845, 4910, 4938. Antelope Glen, 4282, 4379. In a ravine on Moji Shan, 4272. Xingshan at 1,400 metres, 6546. Without locality, 4379.

Henry stated that this species was known colloquially in the Three Gorges region as *ke shan hsiao*.

Vincetoxicum chinense S. Moore (APOCYNACEAE)

Cynanchum mooreanum Hemsl. Forbes & Hemsley in Journ. Linn. Soc. Bot. 26: 108. (1889).

Hubei: On Tsui Fu Shan near Yichang (May 1888), 1597.

Vincetoxicum hemsleyanum (Warb.) Meve & Liede (APOCYNACEAE) New species

Phytotaxa 369: 147. (2018).

Gongronema hemsleyanum Warburg in Fedde, Repert. Nov. Spec. Regni. Veg. 3: 341. (1907), *Biondia hemsleyana* (Warb.) Tsiang in Sunyatsenia 6: 124 (1941).

Sichuan: South Wushan, 5606 (isotype of *Gongronema hemsleyana* Warburg).

Vincetoxicum henryanum Meve & Liede (APOCYNACEAE) New species

Tylophora henryi Warburg in Fedde, Repert. Nov. Spec. Regni Veg. 3: 313. (1906).

Hubei: Jianshi, 5940. Without locality, 5940a.

Vincetoxicum henryi (Warb. ex Schltr. & Diels) Meve & Liede (APOCYNACEAE) New species

Cynanchum henryi Warburg ex Schlecter & Diels in Bot. Jahrb. 29: 542. (1900). *Biondia henryi* (Warb. ex Schltr. & Diels) Tsiang & P. T. Li in Acta Phytotax. Sin. 12: 114. (1974).

Sichuan: South Wushan, 5514 (holotype of *Cynanchum henryi* Warburg ex Schlecter & Diels).

Vincetoxicum hirsutum (Wall.) Kuntze (APOCYNACEAE)

Tylophora hispida Decne. Anonymous in Kew List of Determinations (volume 24).

02

Hainan: Environs of Kiungchow, 8758.

Vincetoxicum insulicola Meve & Liede (APOCYNACEAE)
New species

Tylophora hispida Henry non Decne in Trans. Asiat. Soc. Jap. 24 suppl: 61. (1896) (List Pl. Formosa). *Tylophora insulana* Tsiang & P. T. Li in Acta Phytotax. Sin. 12(1): 134. (1974).

Taiwan: Wanchin, 469 (holotype of *Tylophora insulana* Tsiang & P. T. Li). Kaohsiung, 1162. Oluanpi, 1279.

Discovered by Henry in 1893.

Vincetoxicum muckdenense Kitag. (APOCYNACEAE)
Pycnostelma chinensis Bunge ex Decne. Henry in Journ. China Br. Roy. Asiat. Soc. 22: 247. (1887) (Chinese Names of Plants), Forbes & Hemsley in Journ. Linn. Soc. Bot. 26: 102. (1889).

Hubei: Yichang, 597, 2137, 2308. Liantuo, 3897. South Badong, 7280.

According to Henry this species was known in the Three Gorges region as *chu yeh hsi hsin*.

Vincetoxicum stenophyllum (Hemsley) Kuntze (APOCYNACEAE) New species
Cynanchum stenophyllum Hemsley in Journ. Linn. Soc. Bot. 26: 108. (1889).

Hubei: Liantuo, 49 (isosyntype). On the banks of the Yangtze at Badong, 7181.

Later collected by Faber on the banks of the Yangtze in the Wu Gorge.

Vincetoxicum uncinatum (Gilbert & Li) Meve & Liede (APOCYNACEAE) New species
Tylophora uncinata Gilbert & Li in Novon 5: 16. (1995).

Hainan: Hummocks near Haikou, 8105 (collected November 1889).

Vincetoxicum verticillatum Hemsl. (APOCYNACEAE)
Anonymous in Kew List of Determinations (volume 24).

Sichuan: Henry's Chinese collector with Pratt, no locality stated but may be any of the following, Leshan, Emei Shan, Wa Shan or Kangding, 8852.

Viola acuminata Ledeb. (VIOLACEAE)
Viola canina Henry non L. in Journ. China Br. Roy. Asiat. Soc. 22: 251. (1887) (Chinese Names of Plants).

Hubei: Yichang, 1256. Badong, 1455, 1461. Liantuo, 2082. Jianshi, 7439.

Viola canina L. (VIOLACEAE)
Viola sylvestris Lam. Diels in Bot. Jahrb. 29: 477. (1900).

Hubei: Yichang, 1307. Liantuo, 3845. South Badong, 5390. Changyang, 7793.

Viola davidii Franch. (VIOLACEAE)
Diels in Bot. Jahrb. 29: 477. (1900)

Viola davidii Franch. var. *paucicrenata* W. Becker in Beih. Bot. Centralab. 34(2): 420. (1917).

Hubei: South Badong, 5362 (isotype of *Viola davidii* Franch. var. *paucicrenata* W. Becker).

Viola davidii was discovered by Père Armand David near Baoxing in western Sichuan in April 1869.

Viola delavayi Franch. (VIOLACEAE)
Anonymous in Kew List of Determinations (volume 34).

Yunnan: Mengzi, 9503.

Endemic to China and distributed in Guizhou, Sichuan and Yunnan.

Viola diffusa Ging. (VIOLACEAE)
Diels in Bot. Jahrb. 29: 477. (1900).

Hubei: Yichang, 1311. In a ravine on Moji Shan, 3361. Badong, 3754. Changyang, 7795. **Yunnan:** Mengzi, 10977.

Distributed across the Himalaya, India to China, Indochina, southern Japan, the Philippines and New Guinea.

Viola hamiltoniana D. Don (VIOLACEAE)
Viola verecunda A. Gray Diels in Bot. Jahrb. 29: 476. (1900).

Sichuan: South Wushan, 5625, 5761.

Viola henryi Boissieu (VIOLACEAE) New species
in Bull. Herb. Boiss. ser. 2, 1: 1075. (1901).

Hubei: Jianshi, 5607a., (type). Without locality, 5607b. **Sichuan:** South Wushan, 5607.

Viola hirta L. (VIOLACEAE)
Anonymous in Kew List of Determinations (volume 24).

Hubei: Badong, 3787. Liantuo, 6317.

Viola mirabilis L. (VIOLACEAE)
Viola mirabilis L. Henry in Notes Econ. Bot. China 52. (1893).

Hubei: Badong, 5358, 6028. Fang Xian, 6681. **Sichuan:** South Wushan, 5771, 5611.

Henry stated that this is the *chien-chung-hsiao* which is a name given to several other medicinal plants.

Viola cf. moupinensis Franch. (VIOLACEAE)
Anonymous in Kew List of Determinations (volume 24).

Sichuan: South Wushan, 7239.

Viola patrinii DC. (VIOLACEAE)
Henry in Journ. China Br. Roy. Asiat. Soc. 22: 251. (1887) (Chinese Names of Plants), Diels in Bot. Jahrb. 29: 476. (1900).

Hubei: Yichang, 263, 1325, 2723, 3360. Badong, 1460, 1446, 3737, 3753. **Yunnan:** Mengzi, 10949.

According to Henry this species was colloquially known in the Three Gorges region as *kuan t'ou chien*.

Viola phalacrocarpa Maxim. (VIOLACEAE)
Anonymous in Kew List of Determinations (volume 34).

Yunnan: Mengzi, 9351, 11226.

Native to Russia, China, Korea and Japan.

Viola pilosa Blume (VIOLACEAE)

Viola serpens Wall. ex Ging. Anonymous in Kew List of Determinations (volume 34).

Yunnan: Mengzi, 10444.

Distributed in India, Myanmar, western China and Thailand.

Viola principis H. Boissieu (VIOLACEAE)

Viola canescens Wall. subsp. *lanuginosa* W. Becker in Beih. Bot. Centralab. 34(2): 256. (1913).

Sichuan: South Wushan, 5626 (isosyntype of *Viola canescens* Wall. subsp. *lanuginosa* W. Becker).

Commemorates Prince Henri d'Orleans (1867—1901).

Viola spp. (VIOLACEAE)

Anonymous in Kew List of Determinations (volume 24 & 34).

Hubei: Yichang, 903. Badong, 1457, 4785. Antelope Glen, 3448. Changyang, 5272, 7799. Fang Xian, 6663, 6674. **Yunnan:** Lunan, 10577. Mengzi, 10949. Simao, 11768.

Viola sylvatica Fries. (VIOLACEAE)

Anonymous in Kew List of Determinations (volume 24).

Hubei: Xingshan.

Viola variegata Fisch. ex Link. (VIOLACEAE)

Henry in Journ. China Br. Roy. Asiat. Soc. 22: 251. (1887) (Chinese Names of Plants), Diels in Bot. Jahrb. 29: 477. (1900).

Hubei: Yichang, 802, 963.

Viola yunnanensis W. Becker (VIOLACEAE) New species

in Bull. Herb. Boissier ser. 2. 8: 740. (1908).

Yunnan: In Pingbian on the Daweishan Range at 1,650 metres, 10685 (isotype).

North-east India to China (Hainan, Sichuan and Yunnan) and Malesia.

Viscum album L. (LORANTHACEAE)

Henry in Journ. China Br. Roy. Asiat. Soc. 22: 237. (1887) (Chinese Names of Plants), Forbes & Hemsley in Journ. Linn. Soc. Bot. 26: 407. (1894), Lecompte in Sargent Pl. Wilson. 3: 318. (1915), Danser in Blumea 4: 268. (1941), Zheng in Journ. Wuhan Bot. Res. 2.(1): 36. (1984).

Hubei: Yichang, 7883.

An evergreen semi-parasitic shrub, native to Europe, North Africa and temperate Asia. According to Wilson mistletoe was parasitic on *Pterocarya stenoptera* C. DC. at Yichang.

Viscum articulatum Burm. f. (LORANTHACEAE)

Henry in Journ. China Br. Roy. Asiat. Soc. 22: 237. (1887) (Chinese Names of Plants), Forbes & Hemsley in Journ. Linn. Soc. Bot. 26: 407. (1894), Henry in Trans. Asiat. Soc. Jap. 24 suppl: 81. (1896) (List Pl. Formosa), Lecompte in Sargent Pl. Wilson 2: 318. (1915), Danser in Blumea 4: 289. (1941), Li in Woody Flora of Taiwan 148. fig. 53. (1963), Zheng in Journ. Wuhan Bot. Res. 2.(1): 36. (1984).

Hubei: Liantuo, 3206. Yichang, without number. **Taiwan:** Wanchin, 59.

A leafless parasitic shrub to 50 cm tall. Henry stated that at Yichang *Viscum articulatum* was known colloquially as *sang chi-sheng* and grew there on *Dalbergia hupeana* Hance, near Badong it grew on *Diospyros kaki* L. In Taiwan this species is parasitic on *Liquidamber formosana* Hance, *Trochodendron aralioides* Sieb. & Zucc., *Cyclobalanopsis glauca* (Thunb.) Oest., *Quercus dentata* Thunb., *Quercus variabilis* Blume and *Magnolia compressa* Maxim. In Taiwan it is most common on the southern half of the island. The Hani (an ethnic group) of Xishuangbanna in southern Yunnan call this species *xavtevqtevqlevq* and use it to treat rheumatism and joint pain. Native from southern China to tropical Asia and Australia.

Viscum liquidambaricola Hayata (VISCACEAE)

Danser in Blumea 4: 289. (1941).

Viscum articulatum Lecompte non Burm. f. in Sargent Pl. Wilson. 3: 318. (1915), in part.

Yunnan: Milê District, on *Zanthoxylum,* 9942. Milê at 1,300 metres, on *Castanea,* 10303. Mountains to the north-west of Simao at 1,650 metres, 10303a.

Distributed across Nepal, India, Bhutan, China, Thailand and Vietnam.

Viscum multinerve (Hayata) Hayata (VISCACEAE)

Danser in Blumea 4: 307. (1941).

Viscum stipitatum Lecompte in Sargent Pl. Wilson. 3: 319. (1915).

Yunnan: Simao at 1,600 metres, 12758 (isotype of *Viscum stipitatum* Lecompte).

Native from Nepal to China, Thailand and Vietnam.

Viscum ovalifolium Wall. ex DC. (LORANTHACEAE)

Danser in Blumea 4: 296. (1941).

Viscum orientale Forbes & Hemsley non Willd. in Journ. Linn. Soc. Bot. 26: 408. (1894).

Hainan: Environs of Haikou, 8420.

Vitex burmensis Moldenke (LAMIACEAE)

Vitex lanceolata C. P'ei in Mem. Sci. Soc. China 1: 114. (1932), non Turcz. (1863). *Vitex lanceifolia* S. C. Huang in Flora Yunnanica 1: 450. (1977).

Yunnan: Simao, 11883a., (holotype of *Vitex lanceolata* C. P'ei, type of *Vitex lanceifolia* S. C. Huang).

Native to Myanmar and western China.

Vitex canescens Kurz (LAMIACEAE)

Anonymous in Kew List of Determinations (volume 34).

02

607

Yunnan: Mengzi, 10654.

Distributed in India, Myanmar, China and Indochina.

Vitex henryi Moldenke (LAMIACEAE) New species
in Phytologia 3: 488. (1951).

Vitex quinata Rehder non (Lour.) F. N. Williams in
Sargent Pl. Wilson. 3: 374. (1916).

Yunnan: Simao at 1,500 metres, 12638 (holotype).

Vitex kwangsiensis C. P'ei (LAMIACEAE)
in Mem. Sci. Soc. China. 1(3): 93. (1932).

Guangxi: Longshen, 617 (holotype).

Vitex negundo L. (LAMIACEAE)

Henry in Journ. China Br. Roy. Asiat. Soc. 22: 245.
(1887) (Chinese Names of Plants), Forbes & Hemsley in
Journ. Linn. Soc. Bot. 26: 258. (1890), Henry in Trans.
Asiat. Soc. Jap. 24 suppl: 71. (1896) (List Pl. Formosa),
Rehder in Sargent Pl. Wilson. 3: 372. (1916), Li in Woody
Flora of Taiwan 832. fig. 334. (1963), Zheng in Journ.
Wuhan Bot. Res. 2.(1): 186. (1984).

Vitex negundo L. var. *incisa* (Lam.) Mill. Zheng in
Journ. Wuhan Bot. Res. 2.(1): 186. (1984).

Hubei: Yichang, 91, 92. **Hainan:** Near Haikou, without
number. **Taiwan:** Oluanpi, 905. Kaohsiung, 1142. **Yunnan:**
Kunming, 9750. Near Yuanjiang at 750 metres, 13210.

A deciduous shrub to 3 metres tall, well known in
cultivation on account of its handsome digitate foliage and
handsome loose panicles of violet flowers. Native from
tropical east Africa to Asia and Polynesia.. Henry stated it
was known colloquially at Yichang as *huang ching*. This
species was collected by the Glasnevin Central China
Expedition (GCCE 603) in Xingshan in September 2004).

Vitex quinata (Lour.) F. N. Williams (LAMIACEAE)

Rehder in Sargent Pl. Wilson. 3: 374. (1916), Li in
Woody Flora of Taiwan 834. (1963).

Vitex heterophylla Roxb. Forbes & Hemsley in Journ.
Linn. Soc. Bot. 26: 257. (1890), Henry. in Trans. Asiat. Soc.
Jap. 24 suppl: 71. (1896) (List Pl. Formosa).

Hainan: Near Haikou, without number. **Taiwan:**
Wanchin, without number. Kaohsiung, 1182. Oluanpi, 1258.
Yunnan: Mengzi at 1,500 metres, 9787.

A large tree to 14 metres, native to tropical Asia.
Common at low altitude in Taiwan where Henry stated it
was known as *pu-chiang-mu*. This species was collected
by the Glasnevin Taiwan Expedition (GTE 51) by the
lighthouse at Oluanpi (South Cape of Henry and Wilson) in
October 2004.

Vitex rotundifolia L.f. (LAMIACEAE)

Li in Woody Flora of Taiwan 834. (1963).

Vitex trifolia L. var. *unifoliolata* Schauer. Henry
in Trans. Asiat. Soc. Jap. 24 suppl: 71. (1896) (List Pl.
Formosa).

Taiwan: Kaohsiung, without number.

According to Henry this is common as a low creeper
on sea-shore sands, and was known as *peh-pu-chiang* during
his time in Chinese Taiwan. Native from China to Korea,
Japan and the Philippines.

Vitex spp. (LAMIACEAE)

Anonymous in Kew List of Determinations (volume 34).

Yunnan: At Hsienkei on the Red River near the
Vietnamese border, 9567. Simao, 11883. Mengzi, 13651.

Vitex trifolia L. (LAMIACEAE)

Forbes & Hemsley in Journ. Linn. Soc. Bot. 26: 258.
(1890).

Hainan: Near Haikou, without number. **Yunnan:**
Simao, 12302.

The Hani (an ethnic group) of southern Yunnan call this
species *diamabiaqxang* and use its edible seeds and leaves
to treat cases of flu, fever, headaches and eye pain. Widely
distributed in Asia and extending to Australia and the Pacific
Islands.

Vitex vestita Wall. ex Schaeur. (LAMIACEAE)

Anonymous in Kew List of Determinations (volume 34).

Yunnan: Simao, 12310.

Vitis betulifolia Diels & Gilg (VITACEAE) New species
Zheng in Journ. Wuhan Bot. Res. 2.(1): 144. (1984).

Vitis labrusca Forbes & Hemsley non L. in Journ. Linn.
Soc. Bot. 23: 134. (1886).

Hubei: Zigui, 6194.

An ally of *Vitis piasezkii*, but with simple leaves. The
fruits are blue black and with a slight bloom. Introduced
to cultivation by Wilson in 1907 while collecting for the
Arnold Arboretum. *Vitis betulifolia* was discovered in
1888 by Henry but was based on material later collected in
Sichuan by Bock and von Rosthorn (c. 1891).

Vitis davidii (Rom.Caill.) Foëx. (VITACEAE)

Henry in Notes Econ. Bot. China 47. (1893), Zheng in
Journ. Wuhan Bot. Res. 2.(1): 144. (1984).

Vitis armata Diels & Gilg in Bot. Jahrb. Syst. 29: 462.
(1900), Schneider in Illus. Handb. Laubholzk. 2:L 303.
(1912).

Hubei: Antelope Glen, 3521 (isosyntype of *Vitis
armata* Diels & Gilg.).

The spiny vine or *p'u-tao-tzu*, according to Wilson this
species was sometimes cultivated and produced good sized
grapes but with a harsh flavour. A luxuriant deciduous vine
with large heart shaped leaves to 25 cm long. Discovered
by Père Armand David in Shaanxi in 1872 and introduced
by him to the gardens of the Paris Museum that year. Plants
now in cultivation derive from Wilson's 1900 reintroduction
and it is likely that an old plant on the vine border wall at
Glasnevin is from this source.

Vitis ficifolia Bunge (VITACEAE)

Henry in Journ. China Br. Roy. Asiat. Soc. 22: 280. (1887) (Chinese Names of Plants).

Vitis thunbergii Sieb. & Zucc. Zheng in Journ. Wuhan Bot. Res. 2.(1): 145. (1984).

Hubei: Yichang, 290. San You Dong Glen (May 1886), 1573. **Yunnan:** Simao, 11870.

Distributed in China, Korea and Japan.

Vitis flexuosa Thunb. (VITACEAE)

Henry in Journ. China Br. Roy. Asiat. Soc. 22: 280. (1887) (Chinese Names of Plants); in Trans. Asiat. Soc. Jap. 24 suppl: 27. (1896) (List Pl. formosa), Zheng in Journ. Wuhan Bot. Res. 2.(1): 144. (1984).

Hubei: Badong, 1758. Liantuo, 1963, 4570. Antelope Glen, 3544. **Hainan:** Without locality, 8408. **Taiwan:** Wanchin, 400. **Yunnan:** Mengzi, 10549, 10818.

According to Wilson this was a common climber on cliffs near Yichang. An elegant deciduous climber, native to and China, Japan, Korea and Indochina.

Vitis heyneana Roemer & Schultes (VITACEAE)

Vitis pentagona Diels & Gilg. in Bot. Jahrb. 29: 460. (1900).

Hubei: In the San You Dong Glen (June 1886), 1632 (syntype of *Vitis pentagona* Diels & Gilg). On the floor of a cave on Moji Shan, 3109.

Henry stated this vine was known colloquially in the Three Gorges region as *wa wa tzu*. Distributed in Nepal, India, Bhutan and China.

Vitis labrusca L. (VITACEAE)

Forbes & Hemsley in Journ. Linn. Soc. Bot. 23: 134. (1886), Henry in Journ. China Br. Roy. Asiat. Soc. 22: 280. (1887) (Chinese Names of Plants).

Hubei: Yichang, 508 (in part), 734 (in part), 1167, 1202, 1357. Badong, 1722. Xiling (Yichang) Gorge, 3623. **Yunnan:** Mengzi, 9456.

Native to North America, this entry may belong to another taxon.

Vitis leeoides Maxim. (VITACEAE)

Anonymous in Kew List of Determinations (volume 24).

Hubei: South Donghu, 7489.

Vitis piasezkii Maxim. (VITACEAE)

Henry in Notes Econ. Bot. China 47. (1893), Schneider in Illus. Handb. Laubholzk. 2: 303. (1912), Rehder in Journ. Arnold. Arb. 14: 349. (1933), Zheng in Journ. Wuhan Bot. Res. 2.(1): 145. (1984).

Vitis pagnuccii Rom du Caill. Ex Planch. Gilg. & Diels in Bot. Jahrb. 29: 461. (1900).

Hubei: Badong, 1721, 5446. Yichang, 2242, 4307. On Pingshanba near Yichang, 2268 (August 1886). **Sichuan:** South Wushan, 5766.

A vigorous deciduous vine, native to western and central China and introduced to cultivation by Wilson in 1900 through Veitch's Coombe Wood nursery. Henry stated that this species was known in the Three Gorges region as the *tz'p'u-t'ao*.

Vitis romanetii Rom. Caill. (VITACEAE)

Henry in Notes Econ. Bot. China 47. (1893), Schneider in Illus. Handb. Laubholzk. 2: 303. (1912), Zheng in Journ. Wuhan Bot. Res. 2.(1): 145. (1984).

Hubei: Liantuo, 4543. Yichang, in the Antelope Gorge, 4754. South Donghu, 7592.

According to Henry this species bears excellent grapes in large bunches. In the late 19th century trials had been carried out in France to assess the potential of this species as a commercial crop and for hybridising. Discovered and introduced by Père Armand David in 1872 from Shaanxi through the garden of the Paris Museum.

Vitis sinocinerea W. T. Wang (VITACEAE)

Vitis labrusca Henry non L. in Trans. Asiat. Soc. Jap. 24 suppl: 28. (1896) (List Pl. Formosa). *Vitis thunbergii* Sieb. & Zucc. var. *adstricta* Li non (Hance) Gagnep. in Woody Flora of Taiwan 531. fig. 206. (1963).

Taiwan: Kaohsiung, without number.

Vitis sp. No. 1. (VITACEAE)

Henry in Trans. Asiat. Soc. Jap. 24 suppl: 28. (1896) (List Pl. Formosa).

Taiwan: Wanchin, 826.

Vitis sp. No. 2. (VITACEAE)

Henry in Notes Econ. Bot. China 52. (1893).

Cissus carnosa Diels & Gilg non Wall. in Bot. Jahrb. 29: 466. (1900).

Hubei: On road to Moji Shan (June 1886), 1545. Antelope Glen, 3553. Yichang, 4109.

According to Henry this was a common wild vine at Yichang and was known there as *wu-yueh-wu*.

Vitis spp. (VITACEAE)

Anonymous in Kew List of Determinations (volume 24 & 34).

Hubei: Yichang, 226, 681, 1814, 2156, 2772, 2947, 3246, 4129. Changyang, 7748. Antelope Glen, 3317. Badong, 1782. Liantuo, 1915, 2217, 2600, 2626, 2630, 3039. Xingshan, 4539. Changleping, 6301. Without locality, 7536. **Sichuan:** South Wushan, 7401. **Hainan:** Environs of Kiungchow, 8738. **Yunnan:** Mengzi, 9704, 10875, 11289, 11465, 13706, 13703. In the Red River Valley near Manhao, 9534. By the Red River at Hsinkai, 9593. Simao, 11805, 11806, 11899, 12141, 12342, 12674, 13485, 13514, 13527.

Vitis vinifera L. (VITACEAE)

Henry in Notes Econ. Bot. China 46. (1893).

Hubei: Liantuo, 3869.

The grape-vine, cultivated in Hubei. According to Bretschneider it was introduced to China in the 2nd century B.C.

Vittaria costata Kunze (PTERIDACEAE)

Christ in Bull. Herb. Boiss. 7: 3. (1899).

Yunnan: In Pingbian on the Daweishan Range at 1,650 metres, on trees, 9195a.

Vittaria lineata (L.) Sw. (PTERIDACEAE)

Diels in Bot. Jahrb. 29: 200. (1900).

Hubei: Changleping, 7726. **Yunnan:** Simao, 12197.

Widely distributed in Asia.

Volkameria inermis L. (LAMIACEAE)

Clerodendrum inerme (L.) Gaertn. Forbes & Hemsley in Journ. Linn. Soc. Bot. 26: 261. (1890), Henry in Trans. Asiat. Soc. Jap. 24 suppl: 71. (1896) (List Pl. Formosa), Li in Woody Flora of Taiwan 827. fig. 332. (1963).

Hainan: Haikou, without number. **Taiwan:** Kaohsiung, without number. Oluanpi, without number.

A prostrate shrub, widely dispersed from India to tropical Asia, Australia to the Pacific Islands.

Wahlenbergia marginata (Thunb.) A. DC. (CAMPANULACEAE)

Wahlenbergia gracilis Forbes & Hemsley non (G. Forst.) A. DC. in Journ. Linn. Soc. Bot. 26: 4. (1889).

Hubei: Yichang, 210, 659, 607. On Tsui Fu Shan near Yichang (May 1888), 1610. **Hainan:** Without locality, 7995.

Wallichia spp. (ARECACEAE)

Anonymous in Kew List of Determinations (volume 34).

Yunnan: Mengzi, 10411. Simao, 12331.

Walsura pinnata Hassk (MELIACEAE)

Walsura yunnanensis C. Y. Wu in Flora Yunnanica 1: 226. (1977).

Yunnan: Simao, 12929.

Waltheria indica L. (MALVACEAE)

Henry in Trans. Asiat. Soc. Jap. 24 suppl: 22. (1896) (List Pl. Formosa).

Hainan: Environs of Haikou, 7967, 8192, 8352, 8423. Without locality, 8477, 8539. **Taiwan:** Wanchin, 395.

An erect shrub to 1.5 metres tall bearing clusters of fragrant yellow flowers, a pantropical weed.

Wedelia sp. (ASTERACEAE)

Anonymous in Kew List of Determinations (volume 24).

Hainan: 32 km (20 miles) west of Haikou, 8138.

Weigela japonica Thunb. var. ***sinica*** (Rehd.) L. H. Bailey (CAPRIFOLIACEAE) New variety

Zheng in Journ. Wuhan Bot. Res. 2.(1): 202. (1984).

Diervilla floribunda Forbes & Hemsley non Sieb. & Zucc. in Journ. Linn. Soc. Bot. 23: 369. (1888). *Diervilla japonica* (Thunb.) DC. var. *sinica* Rehder in Mitt. Deutsch. Dendr. Ges. 22: 264. (1913), Rehder & E. H. Wilson in Sargent Pl. Wilson. 3: 430. (1917).

Hubei: Badong, 203 (June 1885), 1412, 5379. Yichang, 1164. Liantuo, 1928, 2022, 3831. Antelope Glen, 3435. **Sichuan:** South Wushan, 5485.

The Chinese counterpart of *Weigelia japonica*, a larger plant than the former (to 6 metres tall), leaves longer stalked and downy beneath. The flowers are pale pink. Henry's Badong collector named this variety as the *lu pien hua*. Collected by the Glasnevin Central China Expedition in Xingshan (GCCE 564) and on the lower slopes of Tianthu Shan in Changyang (GCCE 768) in the autumn of 2004.

Wendlandia augustinii Cowan (RUBIACEAE) New species

in Notes Roy. Bot. Gard. Edinburgh 16: 298. (1932).

Yunnan: On hills near Simao at 1,650 metres, 11773. Mountains to the west of Simao at 1,650 metres, 11773a., (type).

Endemic to Yunnan (Simao region).

Wendlandia bouvardioides Hutchinson (RUBIACEAE) New species

in Sargent Pl. Wilson. 3: 393. (1916), Cowan in Notes Roy. Bot. Gard. Edinburgh 16: 303. (1932).

Yunnan: In Pingbian in forests on the Daweishan Range at 1,640 metres, 10956 (type).

Endemic to Yunnan.

Wendlandia erythroxylon Cowan (RUBIACEAE)

in Notes Roy. Bot. Gard. Edinburgh 16: 299. (1932).

Wendlandia uvariifolia L. L. Li non Hance in Woody Flora of Taiwan 882. (1963). *Wendlandia paniculata* Henry non (Roxb.) DC. in Trans. Asiat. Soc. Jap. 24 suppl: 49. (1896) (List Pl. Formosa).

Taiwan: Wanchin, 125 (type).

Endemic to Taiwan. Henry's local collector stated this was a tree 6 metres tall with good wood and was known as *hung-mu*, i.e. red wood.

Wendlandia formosana Cowan (RUBIACEAE) New species

in Notes Roy. Bot. Gard. Edinburgh 16: 247. (1932), Li in Woody Flora of Taiwan 880. fig. 357. (1963).

Wendlandia glabrata Henry non DC. in Trans. Asiat. Soc. Jap. 24 suppl: 49. (1896) (List Pl. Formosa). *Wendlandia heyneana* Nakai non Wallich in Journ. Arnold Arb. 5: 83. (1924).

Taiwan: Oluanpi, 672, 924 (isotype), 931, 950.

A small, semi-deciduous tree, native to Vietnam, southern China and Japan (south). Scattered along the sea-shore in Chinese Taiwan and often cultivated in gardens.

Wendlandia longidens (Hance) Hutch. (RUBIACEAE)

Cowan in Notes Roy. Bot. Gard. Edinburgh 16: 301.

(1932), Chen in Acta Phytotax. Sin. 21: 401. (1983), Zheng in Journ. Wuhan Bot. Res. 2.(1): 202. (1984).

Wendlandia henryi Oliver in Hooker's Icon. Pl. 18: t. 1712. (1887), Henry in Journ. China Br. Roy. Asiat. Soc. 271. (1887) (Chinese Names of Plants), Forbes & Hemsley in Journ. Linn. Soc. Bot. 23: 372. (1888), Bretschneider in Hist. Eur. Bot. Disc. China 2: 785. (1898).

Hubei: Yichang, 317 (type of *Wendlandia henryi* Oliver), 603, 4101. On Tsui Fu Shan near Yichang (May 1888), 1619. On Pingshanba near Yichang, 2269 (August 1886).

A small bush to 1.5 metres tall, Henry stated that in the Three Gorges region this species was known as the *shui ching k'o tzu*.

Wendlandia pendula (Wall.) DC. (RUBIACEAE)

Cowan in Notes Roy. Bot. Gard. Edinburgh 16: 302. (1932).

Yunnan: Mountains to the west of Simao at 1,650 metres, 12878. In the mountains at Yuanjiang at 850 metres, 13227.

Distributed in Nepal, India, Bhutan, Myanmar and China (Yunnan).

Wendlandia salicifolia Franch. ex A. Paiva (RUBIACEAE)

Cowan in Notes Roy. Bot. Gard. Edinburgh 16: 244. (1932).

Yunnan: At Hsienkei on the Red River near the Vietnamese border, 9559.

Native to western China, Laos and Vietnam.

Wendlandia tinctoria (Roxb.) DC. subsp. *barbata* Cowan (RUBIACEAE) New subspecies

in Notes Roy. Bot. Gard. Edinburgh 16: 268. (1932), How in Sunyatsenia 7: 42. (1948), Chen in Acta Phytotax. Sin. 21: 390. (1983).

Wendlandia floribunda (Craib) Craib Hutchinson in Sargent Pl. Wilson. 3: 392. (1917).

Yunnan: In Pingbian, in forests on the Daweishan Range at 1,640 metres, 10176a., (type). Simao, 13014.

Native to China (Guangxi and Yunnan) and Vietnam.

Wendlandia tinctoria (Roxb.) DC. subsp. *callitricha* (Cowan) W. C. Chen (RUBIACEAE) New to China

in Acta Phytotax. Sin. 21: 389. (1983).

Wendlandia tinctoria (Roxb.) DC. Hutchinson in Sargent Pl. Wilson. 3: 392. (1916). *Wendlandia tinctoria* (Roxb.) DC. var. *callitricha* Cowan in Notes Roy. Bot. Gard. Edinburgh 16: 265. (1932).

Yunnan: In the Red River valley, on grassy hills near Manpan at 820 metres, 10176 (in part). Milê, 10568 (holotype).

This subspecies appears to have been first collected in Myanmar by Hugh Falconer on the 4th of February

1849. Augustine Henry's specimens were the first to have been collected in China and were used by Cowan in his description. Native to Myanmar and China (Guangxi and Yunnan).

Wendlandia tinctoria (Roxb.) DC. var. *orientalis* Cowan (RUBIACEAE)

in Notes Roy. Bot. Gard. Edinburgh 16: 268. (1932), Chen in Acta Phytotax. Sin. 21: 388. (1983).

Yunnan: In the Red River valley near Manhao at 850 metres, 10176 (in part). Simao at 1,500 metres, 10176b.

Henry's collections were used in describing this variety which was first found in Assam by Collett in March 1882 and later by Delavay and Henry in China. Distributed in India (Assam), Myanmar, China (Guangxi and Yunnan) and Thailand.

Wendlandia uvarifolia Hance (RUBIACEAE)

Wendlandia paniculata Hutchinson non (Roxb.) DC. in Sargent Pl. Wilson. 3: 392. (1916). *Wendlandia uvarifolia* Hance subsp. *yunnanensis* Cowan in Notes Roy. Bot. Gard. Edinburgh 16: 288. (1932).

Yunnan: In Pingbian, in forests on the Daweishan Range at 1,640 metres, 10953. In Pingbian, in forests on the Daweishan Range at 1,950 metres, 11479 (type). Mountains to the south-west of Simao at 1,330 metres, 12982.

Wendlandia wallichii Wight & Arn. (RUBIACEAE)

Anonymous in Kew List of Determinations (volume 34).

Yunnan: Mengzi, 11450.

Weniomeles bodinieri (H. Lév.) B. B. Liu (ROSACEAE)

Photinia davidsoniae Rehder & E. H. Wilson in Sargent Pl. Wilson. 1: 185. (1912), Zheng in Journ. Wuhan Bot. Res. 2.(1): 73. (1984), O'Brien in Augustine Henry, An Irish Plant Collector in China 29. (2002).

Hubei: At side of rice fields in the Monastery Valley, 1649. Changyang, 7604.

According to Wilson this evergreen tree was often planted around shrines and tombs in central China.

Whytockia chiritiflora (Oliv.) W. W. Smith (GESNERIACEAE) New species

in Trans. Bot. Soc. Edinburgh 27: 338. pl. 7. (1917).

Stauranthera chiritifolia Oliver in Hooker's Icon. Pl. 25: t. 2454. (1896)

Yunnan: In Pingbian, in forests on the Daweishan Range at 1,950 metres, 11232.

Wikstroemia angustifolia Hemsley (THYMELAEACEAE) New species

in Journ. Linn. Soc. Bot. 26: 396. (1894), Bretschneider in Hist. Eur. Bot. Disc. China 2: 791. (1898), Rehder in Sargent Pl. Wilson 2: 535. (1916), Zheng in Journ. Wuhan Bot. Res. 2.(1): 155. (1984).

Hubei: On the banks of the river passing through the San You Dong Glen near Yichang, 1519 (July 1886). Yichang, 3313b. Xiling Gorge, 3594, 4184. On Pingshanba near Yichang, 2270 (August 1886).

A pink flowered shrub to one metre tall, a common inhabitant of the cliffs and gorges near Yichang during Henry's time there.

Wikstroemia dolichantha Diels (THYMELAEACEAE)

Rehder in Sargent Pl. Wilson. 2: 537. (1916).

Wikstroemia effusa Rehder in Sargent Pl. Wilson. 2: 538. (1916). *Wikstroemia dolichantha* Diels var. *pubescens* Domke in Notizbl. Bot. Gard. Berlin-Dahlem 11: 358. (1932).

Yunnan: On grassy hills near Mengzi at 1,400 metres, 10277 (isotype of *Wikstroemia dolichantha* Diels var. *pubescens* Domke). Simao at 1,500 metres, 13367 (type of *Wikstroemia effusa* Rehder).

Described from collection made by George Forrest in Yunnan in September 1904, it had been found by Augustine Henry eight years earlier.

Wikstroemia gracilis Hemsley (THYMELAEACEAE)
New species

in Journ. Linn. Soc. Bot. 26: 397. (1894), Bretschneider in Hist. Eur. Bot. Disc. China 2: 791. (1898), Rehder in Sargent Pl. Wilson. 3: 455. (1917), Zheng in Journ. Wuhan Bot. Res. 2.(1): 155. (1984).

Hubei: Xingshan at 1, 400 metres, 6540.

Wikstroemia indica (L.) C. A. Meyer (THYMELIACEAE)

Forbes & Hemsley in Journ. Linn. Soc. Bot. 26: 398. (1894), Henry in Trans. Asiat. Soc. Jap. 24 suppl: 80. (1896) (List Pl. Formosa), Rehder in Sargent Pl. Wilson. 2: 534. (1916).

Hainan: 32 km (20 miles) west of Haikou, 8129. **Taiwan:** Kaohsiung, 10, 1197, 1888. Lambay Isle, 1131 (in part, collected August 15th 1893).

The Indian stringbush, a deciduous shrub. Widely distributed from India to China, south through southeast Asia, Malaysia to Australia and Melanesia. According to the Chinese Imperial Maritime Customs Service list of medicines, this species produced the *pu-lun-t'ou*, a drug exported from Xiamen in Fujian province. According to the Irish plant collector, George Playfair, this shrub was used for making paper and paper blankets in Guangxi.

Wikstroemia linoides Hemsley (THYMELAEACEAE)
New species

in Journ. Linn. Soc. Bot. 26: 398. (1984), Bretschneider in Hist. Eur. Bot. Disc. China 2: 791. (1898), Rehder in Sargent Pl. Wilson. 2: 535. (1916).

Hubei: Yichang, 615, 646, 1542, 3467, 3467a.

An herbaceous species from a mainly shrubby genus.

Wikstroemia micrantha Hemsley (THYMELAEACEAE)
New species

in Journ. Linn. Soc. Bot. 26: 399. (1894), Rehder in Sargent Pl. Wilson. 2: 530. (1916), Zheng in Journ. Wuhan Bot. Res. 2.(1): 155. (1984).

Hubei: On road to Moji Shan (June 1886), 1550. At the side of rice fields in the Monastery valley, 1648. In a ravine on Moji Shan, 4173 (co-type). Yichang, 4273 (type).

A handsome shrub to about one metre with fragrant yellow flowers.

Wikstroemia retusa A. Gray (THYMELIACEAE)

Rehder in Sargent Pl. Wilson. 2: 534. (1916).

Taiwan: Lambay Isle, 1131 (in part, collected August 15th 1893).

A small shrub, native to southern China (Taiwan), Japan (south) and the Philippines.

Wikstroemia sp. (THYMELAEACEAE)

Anonymous in Kew List of Determinations (volume 34).

Yunnan: Mengzi, 10276.

Wisteria sinensis (Sims) Sweet (FABACEAE)

Zheng in Journ. Wuhan Bot. Res. 2.(1): 104. (1984).

Wisteria chinensis (Sims) DC. Henry in Journ. China Br. Roy. Asiat. Soc. 22: 237. (1887) (Chinese Names of Plants).

Hubei: In glens and ravines around Yichang, 310, 1221. On Tsui Fu Shan near Yichang, 1603 (May 1888).

Henry stated this was the *tsu teng* or 'purple vine', this well known climbing shrub was once fairly common around Yichang on trees and on the cliffs of the gorges. It is also known as *chiao-t'eng*. According to Wilson albino forms occured around Yichang. A native of central China *Wisteria sinensis* was introduced to cultivation in May 1816 by John Reeves, Chief Inspector of tea at Canton. Despite the fact that it was a common plant around Yichang and albino forms occured there Wilson never collected seeds of this species. This handsome climber was collected by the Glasnevin Central China Expedition in the *Metasequoia* valley near Modaoqui (GCCE 215) in September 2002 and in the San You Dong Glen (GCCE 705) (San Yu Tung of Henry and Wilson) near Yichang in October 2004.

Wisteriopsis reticulata (Benth.) J. Compton & Schrire (FABACEAE)

Millettia reticulata Benth. Forbes & Hemsley in Journ. Linn. Soc. Bot. 159. (1886), Henry in Journ. China Br. Roy. Asiat. Soc. 22: 237. (1887) (Chinese Names of Plants); in Trans. Asiat. Soc. Jap. 24 suppl: 32. (1896) (List Pl. Formosa), Dunn in Journ. Linn. Soc. Bot. 41: 154. (1912).

Hubei: Yichang, 631, 2280, 2411, 2333, 2334. San You Dong Glen, 1554a., 1636 (June 1886). **Hainan:** Without locality, 8392. **Taiwan:** Wanshoushan, without number.

Wanchin, 68, 574, 894. Oluanpi, 1306.

A scandent shrub which Henry stated was colloquially known in Hubei as *ch'ai-chiao-t'eng*. The flowers are dark purple to dark red in the coastal regions, purple or rose coloured in the interior of China. The genus extends across China, Indochina and temperate east Asia. In China the stems of climbing species are used to make a kind of rough cord. The first species to reach Europe was *Wisteriopsis reticulata* and was collected by the Irish diplomat George Staunton in Guangdong province in 1793.

Withania heterophylla (Hemsl.) Hunz. (SOLANACEAE) New species

Chamaesaracha heterophylla Hemsley in Journ. Linn. Soc. Bot. 26: 174. (1890), Henry in Notes Econ. Bot. China 50. (1893), Bretschneider in Hist. Eur. Bot. Disc. China 2: 788. (1898).

Hubei: Fang Xian, 6702 (isotype of *Chamaesarascha heterophylla* Hemsley).

According to Henry this species was known as the *yeh-ch'ieh-tzu* in the Three Gorges region.

Withania sinensis (Hemsl.) Hunz. (SOLANACEAE) New species

Chamaesaracha sinensis Hemsley in Journ. Linn. Soc. Bot. 26: 174. (1890), Bretschneider in Hist. Eur. Bot. Disc. China 2: 788. (1898). *Archiphysalis sinensis* (Hemsley) Kuang in Acta Phytotax. Sin. 11: 62. (1966). *Physalis sinensis* (Hemsl.) Averett in Ann. Missouri Bot. Gard. 57: 380. (1970). *Physaliastrum sinense* (Hemsl.) D'Arcy & Zhang in Novon 2: 127. (1992).

Hubei: In the Lung Tung Chi Glen near Yichang (October 1886), 2902 (isotype of *Chamaesaracha sinense* Hemsley).

Wollastonia biflora (L.) DC. (ASTERACEAE)

Wedelia biflora (L.) DC. Henry in Trans. Asiat. Soc. Jap. 24 suppl: 54. (1896) (List Pl. Formosa).

Taiwan: Wanchin, 389, without number.

Woodsia polystichoides D. C. Eaton (ASPLENIACEAE)

Baker in Journ. Bot. 26: 225. (1888), Christ in Bull. Soc. Bot. France 5: 44. (1905).

Hubei: Liantuo, 2065, 3933, 4602. Badong, 4774, 6102.

Distributed in Russia, Mongolia, China, Korea and Japan. This species was collected by the Sino-American Expedition in Shennongjia Forest District in 1980.

Woodwardia japonica (L. f.) Sm. (ASPLENIACEAE)

Christ in Bull. Soc. Bot. France 5: 64. (1905).

Hubei: Liantuo, 2648.

Distributed in China, Vietnam, Korea and Japan. According to Henry the rhizome of this fern is known as *guan zhong* and was used in Chinese herbal medicine.

Woodwardia prolifera Hook. & Arn. (ASPLENIACEAE)

Woodwardia orientalis Henry non (Sw.) Sw. in Trans. Asiat. Soc. Jap. 24 suppl: 111. (1896) (List Pl. Formosa).

Taiwan: Kaohsiung, without number.

Native to China (including Taiwan) and Japan (south).

Woodwardia unigemmata (Makino) Nakai (ASPLENIACEAE)

Woodwardia radicans Henry non (L.) Sm. in Journ. China Br. Roy. Asiat. Soc. 22: 251. (1887) (Chinese Names of Plants), Diels in Bot. Jahrb. 29: 197. (1900), Christ in Bull. Soc. Bot. France 5: 64. (1905).

Hubei: Yichang, 1106, 1118, 2158.

Distributed in China, Japan and the Philippines. According to Henry the rhizomes of this fern were used as a drug and it was known colloquially in the Three Gorges region as *kuan chung*. This fern is still abundant in Shennongjia and Xingshan and was collected by the Glasnevin Central China Expedition (GCCE 537) in Xingshan in September 2004.

Wrightia coccinea (Lodd.) Sims. (APOCYNACEAE)

Ngan in Ann. Missouri Bot. Gard. 52: 171. (1965).

Yunnan: In forests near Yiwu (in Xishuangbanna), 13574.

A shrub or sometimes a tree to 20 metres tall, native to eastern Pakistan, Bengal, India (Assam) and Myanmar, China (Guangxi and Yunnan) and Thailand.

Wrightia laevis Hook. f. (APOCYNACEAE)

Ngan in Ann. Missouri Bot. Gard. 52: 136. (1965).

Yunnan: Simao, 12393, 12743.

A tree to 40 metres tall, of widespread distribution from India, Myanmar, southern China, Thailand, Indochina, Malaya, Indonesia, the Philippines, New Guinea and northern Australia. *Wrightia* commemorates Dr. William Wright, a Scottish physician and botanist who spent 18 years on the island of Jamaica. The genus is confined to the eastern hemisphere, from east Africa to the Solomon Islands and from India and southern China to northeastern Australia.

Wrightia pubescens R. Br. ssp. *laniti* (Blanco) P. T. Ngan Ngan (APOCYNACEAE) New subspecies

in Ann. Missouri Bot. Gard. 52: 153. (1965).

Hainan: Kien Cha Fu, 7956. Near the Kiungchow Pagoda, 7980. Environs of Kiungchow, 8751. **Yunnan:** Mengzi, 13724.

Xanthium strumarium L. (ASTERACEAE)

Henry in Journ. China Br. Roy. Asiat. Soc. 22: 259. (1887) (Chinese Names of Plants), Forbes & Hemsley in Journ. Linn. Soc. Bot. 23: 433. (1888), Henry in Trans. Asiat. Soc. Jap. 24 suppl: 54. (1896) (List Pl. Formosa).

Hubei: Yichang, 51, 1674, 2330. **Taiwan:** Kaohsiung, without number.

A common weed in the warmer parts of China, native to the New World and now a cosmopolitan weed. Henry stated it was known colloquially in the Three Gorges region as *mu chu lai*.

Xenostegia tridentata (L.) D. F. Austin & Staples subsp. *angustifolia* (Jacq.) Lejoly & Lisowski (CONVOLVULACEAE)

Ipomoea filicaulis (Vahl.) Blume Anonymous in Kew List of Determinations (volume 24). *Ipomoea angustifolia* Jacq. Henry in Trans. Asiat. Soc. Jap. 24 suppl: 64. (1896) (List Pl. Formosa).

Hainan: Near the Kiungchow Pagoda, 7984. Environs of Haikou, 7998, 8354. Without locality, 8582. **Taiwan:** Wanchin, 859.

Henry knew this as the *kang-tan* stating it was cultivated.

Xylanche himalaica (Hook.f. & Thoms.) G. Beck. (OROBANCHACEAE)

G. Beck in Engler, Pflanzenr. 261: 330. (1930).

Boschniakia himalaica Hook.f. & Thoms. Forbes & Hemsley in Journ. Linn. Soc. Bot. 221. (1890).

Hubei: Fang Xian, 6898.

First collected in China by Faber on the summit of Emei Shan, Sichuan in 1887 and in the following year by Henry in Hubei. In the Himalaya and China the host plant for this parasite is *Rhododendron*.

Xylosma congesta (Lour.) Merr. (SALICACEAE)

Xylosma racemosa Henry non (Sieb. & Zucc.) Miq. in Journ. China Br. Roy. Asiat. Soc. 22: 275. (1887) (Chinese Names of Plants). *Xylosma racemosa* (Sieb. & Zucc.) Miq. var. *pubescens* Rehder & E. H. Wilson in Sargent Pl. Wilson. 1: 283. (1912). *Xylosma congestum* (Lour.) Merr. var. *pubescens* (Rehd. & E. H. Wils.) Rehd. Zheng in Journ. Wuhan. Bot. Res. 2.(1): 154. (1984).

Hubei: Yichang, 444, 2312 (syntype of *Xylosma racemosa* (Sieb. & Zucc.) Miq. var. *pubescens* Rehder & E. H. Wilson), 3111, 3231. In the Monastery Glen near Yichang, 1492. Changyang, 7766. Without locality, 3231a. **Yunnan:** In a ravine near Mengzi at 1,500 metres, 10804.

According to both Henry and Wilson this is the *tung-ching shu* (winter-green tree) of the Chinese and one of the most handsome of their evergreen trees. In Hubei this tree was commonly planted over shrines and tombs where it often reached a height of 20 metres or more. It was planted by the many taoist temples that crowned the crags and peaks of the Three Gorges. Henry's 1492 from the Monastery Valley was obviously cultivated by a temple. The Monastery Valley lies to the approach of the Moji Shan, a pyamidal-shaped hill on the opposite bank of the Yangtze from Yichang where Henry carried out a lot of his collecting and

from where he collected *Lilium henryi*. Antwerp Pratt and Ernest Wilson also later collected there.

Xylosma controversa Clos. (SALICACEAE)

Rehder & E. H. Wilson in Sargent Pl. Wilson. 1: 284. (1912).

Yunnan: Simao at 1,300 to 2,000 metres, 11884c., 12757.

Xylosma longifolia Clos. (SALICACEAE)

Rehder & E. H. Wilson in Sargent Pl. Wilson. 1: 284. (1912).

Yunnan: Mengzi at 1,500 metres, 9901, 9901a., 9901b. Simao at 1,500 metres, 12635. Tonguan at 1,600 metres, 13334.

Native to India, China, Laos and Vietnam.

Xylosma sp. (SALICACEAE)

Anonymous in Kew List of Determinations (volume 34).

Yunnan: Yuanjiang at 1,600 metres, 13243.

Yamazakia pusilla (Willd.) W. R. Barker, Y. S. Liang & Wannan (LINDERNIACEAE)

Vandellia scabra Benth. Henry in Trans. Asiat. Soc. Jap. 24 suppl: 68. (1896) (List Pl. Formosa).

Taiwan: Wanchin, 885.

Yamazakia viscosa (Hornem.) W. R. Barker, Y. S. Liang & Wannan (LINDERNIACEAE)

Vandellia hirsuta Buch.-Ham. ex Benth. Henry in Trans. Asiat. Soc. Jap. 24 suppl: 67. (1896) (List Pl. Formosa).

Taiwan: Wanchin, 860.

Yinshania henryi (Oliv.) Y. H. Zhang (BRASSICACEAE) New genus

in Acta Phytotax. Sin. 25: 213. (1987).

Nasturtium henryi Oliver in Hooker's Icon Pl. 18: t. 1719. (1887), Bretschneider in Hist. Eur. Bot. Disc. China 2: 779. (1898), Diels in Bot. Jahrb. 29: 357. (1900). *Cochleria henryi* (Oliv.) O. E. Schulz in Notizbl. Bot. Gart. Berlin 8: 546. (1923).

Hubei: In the Lung Tung Chi Glen near Yichang (October 1886), 2899 (holotype). Yichang, 4125.

An annual to about 35 cm tall, endemic to China and found in the provinces of Hubei, Guizhou, Sichuan and Yunnan. Flowering and fruiting occurs from June through to September. This species inhabits rocky mountain slopes and valleys between altitudes of 800 to 1,200 metres.

Yinshania paradoxa (Hance) Y. Z. Zhao (BRASSICACEAE)

in Acta Sci. Nat. Univ. Intramongol 23: 567. (1992), Al-Shehbaz et al. in Harvard Papers in Botany 3(1): 79-94. (1998).

Cardamine paradoxa Hance Hemsley in Hooker's

Icon. Pl. t. 1818. (1889), Bretschneider in Hist. Eur. Bot. Disc. China 2: 779. (1898), Diels in Bot. Jahrb. 29: 357. (1900).

Hubei: Lung Tung Ch'i Glen (near Yichang), 4199.

An annual to 70 cm tall, distributed through the provinces of Guangdong, Guangxi, Hubei and Sichuan where it grows on mountain slopes and valleys between altitudes of 300 to 1,000 metres. According to Bretschneider it was discovered by Mr. Theo Sampson (a friend of the famous China-based botanist Dr. H. F. Hance) on the 27th of May 1867.

Youngia cineripappa (E. B. Babcock) Babc. & Stebbins (ASTERACEAE) New species

Crepis cineripappa E. B. Babcock in Univ. Calif. Publ. 14: 325. (1928).

Yunnan: Simao, 11997 (type of *Crepis cineripappa* E. B. Babcock).

Youngia fuscipappa (Benth. & Hook. f.) Thwaites (ASTERACEAE)

Crepis fuscipappa Benth. & Hook. f. Anonymous in Kew List of Determinations (volume 34).

Yunnan: Mengzi, 10680, 13678.

Youngia henryi (Diels) Babc. & Stebb. (ASTERACEAE) New species

Crepis henryi Diels in Bot. Jahrb. Syst. 29: 633. (1900).

Hubei: South Badong, 6069 (isotype of *Crepis henryi* Diels).

Youngia heterophylla (Hemsl.) Babc. & Stebb. (ASTERACEAE) New species

Crepis heterophylla Hemsley in Journ. Linn. Soc. Bot. 23: 475. (1888).

Hubei: Antelope Glen, 3440 (type of *Crepis heterophylla* Hemsley).

Youngia humifusa (Dunn) Y. L. Peng & X. F. Gao (ASTERACEAE) New species

Lactuca humifusa Dunn in Journ. Linn. Soc. Bot. 35: 512. (1903).

Hubei: Badong, 4760. **Sichuan:** South Wushan, 5762.

Henry stated that this species was a cliff plant, remarkable for its spreading stolons that sometimes completly covered the suruonding areas.

Youngia japonica (L.) DC. (ASTERACEAE)

Crepis japonica (L.) Benth. Henry in Journ. China Br. Roy. Asiat. Soc. 22: 245. (1887), Forbes & Hemsley in Journ. Linn. Soc. Bot. 23: 475. (1888), Henry in Notes Econ. Bot. China 53. (1893); in Trans. Asiat. Soc. Jap. 24 suppl: 55. (1896) (List Pl. Formosa).

Hubei: Yichang, 391, 694, 3131, 7870. Jianshi, 5877. Antelope Glen, 3470, 3664. In the Lung Tung Chi Glen near Yichang (October 1886), 2897. In the San You Dong Glen (May 1886), 1593. Badong, 1715, 1860, 4006, 4069. **Taiwan:** Wanchin, without number. Oluanpi, 293. **Yunnan:** Mengzi, 10681. Simao, 11772.

According to Henry the leaves of this species were eaten at Yichang and Badong where it was known as *huang kue tsa*. A pretty species, cultivated at Glasnevin in the rock garden. Distributed in India, China, Korea and the Philippines.

Youngia longipes (Hemsley) Babc. & Stebbins (ASTERACEAE) New species

Crepis longipes Hemsley in Journ. Linn. Soc. Bot. 23: 476. (1888), Bretschneider in Hist. Eur. Bot. Disc. China 2: 786. (1898).

Hubei: Yichang, 3964.

Yua thomsonii (Laws.) C. L. Li (VITACEAE)

Parthenocissus thomsonii (Laws.) Planch. Rehder & E. H. Wilson in Sargent Pl. Wilson. 3: 427. (1917).

Hubei: South Badong, 6137. Fang Xian, 6656. **Yunnan:** In ravines near Mengzi at 2,600 to 2,700 metres, 10754, 10754a.

A slender deciduous climber, native to India, the Himalaya and central and western China. Introduced to cultivation by Wilson when collecting for Messrs Veitch in 1900. In the Veitch catalogue, *Novelties for 1908—1909*, plants were available for two shillings and six pence each. Distributed in Nepal, India and China.

Yushania confusa (McClure.) Z. P. Wang & G. H. Ye (POACEAE) New species

Indocalamus confusus McClure in Lignan Univ. Sci. Bull. 9: 20. (1940).

Hubei: On cliffs in Shennongjia between 1,900 and 3,100 metres, 6832 (isotype of *Indocalamus confusus* McClure), in part.

Zabelia dielsii (Graebn. & Buchw.) Makino (CAPRIFOLIACEAE) New species

Abelia umbellata (Graebn. & Buchw.) Rehder in Sargent Pl. Wilson. 1: 122. (1911), O' Brien in Augustine Henry, An Irish Plant Collector in China 9. (2002).

Linnaea umbellata Graebn. & Buchw. in Bot. Jahrb. Syst. 29: 143. (1900).

Sichuan: South Wushan, 7083 [isotype of *Linnaea umbellata* Graebn. & Buchw., type of *Abelia umbellata* (Graebn. & Buchw.) Rehder].

Cultivated at the National Botanic Gardens, Glasnevin.

Zanthoxylum acanthopodium DC. (RUTACEAE)

Anonymous in Kew List of Determinations (volume 34).

Yunnan: South of the Red River from Manmei, 9875. Mengzi, 10150. Yuanjiang, 13336. Simao, 12451.

A shrub or small tree to 10 metres tall. George Forrest also collected this species (Forrest 15,914) in western

Yunnan in August 1917. The Hani (an ethnic group) of Xishuangbanna in southern Yunnan call this species *naivzaovq* and use its roots to treat rheumatism. Distributed in Nepal, India, Myanmar, China (Guangxi, Sichuan and Yunnan), Thailand, Laos and Vietnam.

Zanthoxylum ailanthoides Sieb. & Zucc. (RUTACEAE)

Henry in Trans. Asiat. Soc. Jap. 24 suppl: 25. (1896) (List Pl. Formosa).

Fagara ailanthoides (Sieb & Zucc.) Engler. Li in Woody Flora of Taiwan 372. (1963).

Taiwan: Oluanpi, 1353. Wanchin, 1630.

A medium-sized tree, native to China (including Taiwan), Korea, Japan (south) and the Philippines.

Zanthoxylum armatum DC. (RUTACEAE)

Zheng in Journ. Wuhan Bot. Res.2. (1): 106. (1984).

Zanthoxylum alatum Roxb. Henry in Journ. China Br. Roy. Asiat. Soc. 22: 238. (1887) (Chinese Names of Plants); in Notes Econ. Bot. China 38. (1893). *Zanthoxylum alatum* Roxb. var. *planispinum* (Sieb. & Zucc.) Rehder & E. H. Wilson in Sargent Pl. Wilson 2: 125. (1914). *Zanthoxylum planispinum* Sieb. & Zucc. Henry in Notes Econ. Bot. China 39. (1893).

Hubei: Yichang, 1072, 1225, 3247. Antelope Glen, 3584. Liantuo, 3021, 3811. Without locality, 3021a. Changyang, 7687. **Yunnan:** Mengzi, growing in rocky places at 1,500 to 1,600 metres, 9366, 9366a., 9366c., 9366d., 9366e., 9366f., 11413. Simao at 1,500 to 1,600 metres, 11908, 12249.

A small tree to about 4 metres tall, native to the Himalaya, Myanmar, Laos, Vietnam, Korea, Japan, China and the Philippines. In Nepal, toothbrushes are made from the branches of this tree. In the same country the fruits are used to relieve tooth-ache and to stupefy fish. Henry stated that this species was known colloquially as *kou hua chiao* around Yichang where it was common on the plains in hedges and on graves.

Zanthoxylum asiaticum (L.) Appelhans, Groppo & J. Wen (RUTACEAE)

Toddalia aculeata Pers. Henry in Trans. Asiat. Soc. Jap. 24 suppl: 25. (1896) (List Pl. Formosa). *Toddalia asiatica* (L.) Lam. Henry in Journ. China Br. Roy. Asiat. Soc. 22: 263. (1887) (Chinese Names of Plants), Rehder in Sargent Pl. Wilson 2: 137. (1914), Zheng in Journ. Wuhan Bot. Res. 2.(1): 106. (1984).

Hubei: Yichang, 1306, 3496. Liantuo, 4586. **Hainan:** Environs of Kiungchow, 8278. Without locality, 8473. 8787. **Taiwan:** Oluanpi, 378, without number. **Yunnan:** Milê District, 9928. Simao at 1,600 metres, 9928d. Mengzi at 1,600 metres, 13650.

A thorny, semi-scandent shrub to 6 metres, known as

pa-ta-wang according to Henry. This species inhabits the glens and gorges in western Hubei at low altitudes. A plant of widespread distribution as can be seen from the locations of Henry's specimens. Native from Ethiopia to India, Sri Lanka and southern China to the Philippines. A very common coastal plant in Chinese Taiwan.

Zanthoxylum avicennae (Lam.) DC. (RUTACEAE)

Anonymous in Kew List of Determinations (volume 24).

Hainan: Environs of Haikou, 8408. Ling-men (in the interior of the island), 8782.

Zanthoxylum bungeanum Maxim. (RUTACEAE)

Zanthoxylum bungei Planch. Henry in Journ. China Br. Roy. Asiat. Soc. 22: 238. (1887) (Chinese Names of Plants); in Notes Econ. Bot. China 38. (1893), Rehder & E. H. Wilson in Sargent Pl. Wilson 2: 121. (1914). *Zanthoxylum fraxinoides* Hemsley in Ann. of Bot. 9: 148. (1895), Bretschneider in Hist. Eur. Bot. Disc. China 2: 780. (1898), Hemsley in Journ. Linn. Soc. Bot. 148. (1903). *Zanthoxylum simulans* Hance var. *imperforatum* (Franch.) Reeder & S. Y. Cheo Zheng in Journ. Wuhan Bot. Res. 2.(1): 108. (1984).

Hubei: Yichang, 1068, 1117, 1205, 1291, 3110, 3956, 7852. Badong, 1447. San You Dong Glen (July 1886), 1527. Liantuo, 3156. In a ravine on Moji Shan, 3363. Fang Xian, 6653, 6905 (type of *Zanthoxylum fraxinoides* Hemsley). Changyang, 7852. Yichang (Xiling) Gorge, 7884. **Sichuan:** South Wushan, 5781.

The Chinese spice or pepper bush, a variable shrub to 4 metres tall which Henry stated was known at Yichang as *hua chiao*, introduced to Kew in 1869. Henry's 5781 from Wushan came from a cultivated plant. This species was known colloquially at Yichang as *yeh hua chiao* or 'wild pepper.'

Zanthoxylum dimorphophyllum Hemsley (RUTACEAE) New species

in Ann. of Bot. 9: 150. (1895), Bretschneider in Hist. Eur. Bot. Disc. China 2: 780. (1898), Zheng in Journ. Wuhan Bot. Res. 2.(1): 107. (1984).

Fagara dimorphophylla (Hemsl.) Engler in Engler & Prantl, Nat. Pflanzenfam. 3: Abt. iv. 118. (1896), Pritzel in Bot. Jahrb. xxix. 422. (1900).

Hubei: Liantuo, 3902, 4462. Xingshan, 4512. North Badong, 7003 (isosyntype). **Sichuan:** South Wushan, 5512 (isosyntype). **Yunnan:** Mengzi, 11337, 11367.

A common cliff shrub in Hubei and free from prickles according to Wilson, which is remarkable for a species of *Zanthoxylum*.

Zanthoxylum dimorphophyllum Hemsley var. *spinifolium* Rehder & E. H. Wilson (RUTACEAE)

New variety. in Sargent Pl. Wilson. 2: 126. (1914).

Sichuan: South Wushan, 5494.

Differs from the type by its spiny branches and petiole,

the 3 and occasionally 5 foliate leaves and the long setose spines on the margins of leaflets.

Zanthoxylum dissitum Hemsley (RUTACEAE)

Henry in Journ. China Br. Roy. Asiat. Soc. 22: 267. (1887) (Chinese Names of Plants), Rehder & E. H. Wilson in Sargent Pl. Wilson 2: 128. (1914), Zheng in Journ. Wuhan Bot. Res. 2.(1): 107. (1984).

Hubei: Yichang, 2728, 3325. Antelope Glen near Yichang, 3121. Badong, 3713. Liantuo, 4397. Fang Xian, 6732. **Sichuan:** South Wushan, 7137, 7138. **Yunnan:** On the Daweishan Range in Pingbian at 1,600 to 2,000 metres, 10429, 11437. Yuanjiang, 13326, 13326a.

A handsome scandent shrub to 4 metres tall, common in the glens and gorges of western Hubei. *Zanthoxylum dissitum* was discovered by Charles Maries in the Yichang (Xiling) Gorge in 1879 and was introduced by him through James Veitch & Sons, Chelsea. Known by Henry's Badong collector as *san peh pang*. Endemic to China.

Zanthoxylum echinocarpum Hemsley (RUTACEAE)
New species

in Ann. Bot. 9: 145. (1895), Bretschneider in Hist. Eur. Bot. Disc. China 2: 780. (1898), Rehder & E. H. Wilson in Sargent Pl. Wilson. 2: 128. (1914), Zheng in Journ. Wuhan Bot. Res. 2.(1): 107. (1984).

Fagara echinocarpa (Hemsl.) Engler in Engler & Prantl. Nat. Pflanzenfam. 3: Abt. iv. 118. (1896), Pritzel in Bot. Jahrb. xxix. 422. (1900).

Hubei: Yichang, 3416b., (type), 3416d. **Yunnan:** In mountains to the south-west of Mengzi at 1,600 metres, 10932.

A scandent bush to 2 metres tall, confined to warm temperate areas of Sichuan and Hubei. Endemic to China.

Zanthoxylum micranthum Hemsley (RUTACEAE)
New species

in Ann. Bot. 9: 147. (1895), Bretschneider in Hist. Eur. Bot. Disc. China 2: 780. (1898), Schneider in Illus. Handb. Laubholzk. 2: 120. (1912), Rehder & E. H. Wilson in Sargent Pl. Wilson 2: 127. (1914), Zheng in Journ. Wuhan. Bot. Res. 2.(1): 107. (1984).

Fagara micrantha (Hemsl.) Engler in Engler & Prantl. Nat. Pflanzenfam. 3: Abt. iv. 118. (1896), Pritzel in Bot. Jahrb. xxix. 422. (1900).

Hubei: Liantuo, 2690, 2095 (type). Yichang, 4127, 4127a., (syntype). Xingshan, 4528.

The only species of *Zanthoxylum* from central China to grow as a tree. Henry stated it was known as *tz'u-chin-shu*, all parts of the tree emit a pungent odour when crushed.

Zanthoxylum multijugum Franchet (RUTACEAE)

Rehder & E. H. Wilson in Sargent Pl. Wilson. 2: 129. (1914).

Zanthoxylum multifoliolatum Hemsley in Hooker's Icon. Pl. xxvi. t. 2595. (1899).

Yunnan: Mengzi, in woods and on cliffs at 1,500 to 1,600 metres, 9998 (type of *Zanthoxylum multifoliolatum* Hemsley), 9988a., 9988b., 9988c.

Discovered by Père Delavay in May 1886. Endemic to Yunnan.

Zanthoxylum nitidum (Roxb.) DC. (RUTACEAE)

Henry in Trans. Asiat. Soc. Jap. 24 suppl: 25. (1896) (List Pl. Formosa).

Fagara nitida Roxb. Li in Woody Flora of Taiwan 373. fig. 134. (1963).

Hainan: Environs of Haikou, 8399. **Taiwan:** Oluanpi, 205, 2050. Wanchin, 462, 1655, 1782. **Yunnan:** Mengzi, 10421, 11163.

A scandent shrub or small deciduous tree to 5 metres tall with branchlets covered in short recurved prickles. Distributed in southern China, including Taiwan where it grows at low altitude throughout the island. The Hani (an ethnic group) of Xishuangbanna in southern Yunnan call this species *xaqbeevjavhlaq* and use its bark and roots to treat cases of rheumatism, joint pain, toothache, gastritis, venomous snake bite, sprains and contusions.

Zanthoxylum scandens Blume (RUTACEAE)

Zanthoxylum cuspidatum Champ. ex Benth. Henry in Trans. Asiat. Soc. Jap. 24 suppl: 25. (1896) (List Pl. Formosa). *Fagara cuspidata* (Champ.) Engler. Li in Woody Flora of Taiwan 373. (1963).

Taiwan: Oluanpi, 334, 1969.

A scandent shrub, armed with small prickles. Native to India, China, Japan (south), Sumatra, Java and Borneo.

Zanthoxylum simulans Hance (RUTACEAE)

Zanthoxylum setosum Hemsl. Henry in Journ. China Br. Roy. Asiat. Soc. 22: 238. 1887 (Chinese Names of Plants), Rehder & E. H. Wilson in Sargent Pl. Wilson. 2: 124. (1914), Zheng in Journ. Wuhan Bot. Res. 2.(1): 108. (1984).

Hubei: Yichang, 1504, 1571.

Wilson called this species 'wild pepper', stating it was common on cliffs in Hubei and Sichuan.

Zanthoxylum spp. (RUTACEAE)

Anonymous in Kew List of Determinations (volume 24 & 34).

Hubei: Liantuo, 6319. **Yunnan:** Mengzi, 10689, 10984.

Zanthoxylum stenophyllum Hemsley (RUTACEAE)
New species

in Ann. of Bot. 9: 147. (1895), Bretschneider in Hist. Eur. Bot. Disc. China 2: 780. (1898), Rehder & E. H. Wilson in Sargent Pl. Wilson 2: 127. (1914), Zheng in Journ. Wuhan Bot. Res. 2.(1): 108. (1984), O' Brien in Augustine Henry,

02

An Irish Plant Collector in China 38. (2002).

Fagara stenophylla (Hemsl.) Engler in Engler & Prantl. Nat. Pflanzenfam. 3: Abt. iv. 118. (1896), Pritzel in Bot. Jahrb. xxix. 422. (1900).

Hubei: Xingshan, 6466 (type), 6555. **Sichuan:** South Wushan, 5560. **Yunnan:** Mengzi, 9874.

A common subscandent shrub at low altitudes in western and central China. Cultivated at the National Botanic Gardens, Glasnevin. Endemic to China and distributed in Gansu, Hubei, Shaanxi, Sichuan and Yunnan.

Zanthoxylum undulatifolium Hemsley (RUTACEAE)
New species

in Ann. Bot. 9: 148. (1895), Bretschneider in Hist. Eur. Bot. Disc. China 2: 780. (1898), Schneider in Illus. Handb. Laubholzk. 2: 119. (1912), Rehder & E. H. Wilson in Sargent Pl. Wilson. 2: 124. (1916), Zheng in Journ. Wuhan Bot. Res. 2.(1): 108. (1984).

Hubei: Liantuo, 3938 (type). **Sichuan:** South Wushan, 5646 (syntype).

Zehneria mucronata (Blume) Miq. (CUCURBITACEAE)

Zehneria maysorensis Henry non Arn. in Trans. Asiat. Soc. Jap. 24 suppl: 46. (1896) (List Pl. Formosa). *Melothria mucronata* (Blume) Cogn. Cogniaux in Engler, Pflanzenr. iv. 275. i: 108. (1916).

Hubei: Liantuo, 4416. **Taiwan:** Oluanpi, 209, 1712. Wanchin, 344.

A glabrescent herbaceous vine, native to China (including Taiwan), Japan (south), Malesia and the Pacific Islands.

Zehneria odorata (Hook. f. & Thoms. ex Benth.) M. D. Dwivedi, A. K. Pandey & H. Schaef. (CUCURBITACEAE)

Melothria odorata (Hook. f. & Thoms ex Benth.) Hook. f. & Thoms. ex C. B. Clarke Henry in Trans. Asiat. Soc. Jap. 24 suppl: 46. (1896) (List Pl. Formosa). *Melothria indica* (L.) Lour. Cogniaux in Engler, Pflanzenr. iv. 275. i: 98. (1916).

Hainan: Environs of Haikou, 8356. **Taiwan:** Wanchin, 1729. **Yunnan:** Mengzi, 10007.

A stout, climbing perennial herb, distributed in east Asia. The fruits of this species are edible.

Zehneria sp. (CUCURBITACEAE)

Anonymous in Kew List of Determinations (volume 34).

Yunnan: On the banks of the Red River at Manpan, 9611. Simao, 12551.

Zeuxine nervosa (Wall. ex Lindl.) Benth. ex Clarke (ORCHIDACEAE)

Zeuxine formosana Rolfe in Ann. of Bot. 9: 158. (1895), Henry in Trans. Asiat. Soc. Jap. 24 suppl: 92. (1896) (List Pl. Formosa), Bretschneider in Hist. Eur. Bot. Disc. China 2: 792. (1898), Rolfe in Journ. Linn. Soc. Bot. 36: 42. (1903).

Taiwan: Oluanpi, 644 (isotype of *Zeuxine formosana* Rolfe).

Zeuxine sp. (ORCHIDACEAE)

Anonymous in Kew List of Determinations (volume 34).

Yunnan: Simao, 12724.

Zeuxine strateumatica (L.) Schltr. (ORCHIDACEAE)

Zeuxine sulcata (Roxb.) Lindl. ex Wight Henry in Trans. Asiat. Soc. Jap. 24 suppl: 92. (1896) (List Pl. Formosa, Rolfe in Journ. Linn. Soc. Bot. 36: 42. (1903).

Hubei: Fang Xian, 6589. **Taiwan:** Wanchin, on the plain, without number.

A small terrestrial orchid, native to tropical and subtropical Asia and North America.

Zingiber officinale Roscoe (ZINGIBERACEAE)

Henry in Trans. Asiat. Soc. Jap. 24 suppl: 94. (1896) (List Pl. Formosa), Wright in Journ. Linn. Soc. Bot. 36: 70. (1903).

Hubei: Yichang, cultivated 4142. **Taiwan:** Wanchin, 491, 1575.

Ginger, widely cultivated in the tropics since ancient times but thought to be native to northeastern India. Ginger is an ancient sterile cultigen and has never been found in the wild. The Sanskritic name *singabera* is the origin of *Zingiber*. It has been extensively cultivated in China for centuries as a culinary spice and in Chinese herbal medicine fresh ginger has been traditionally used to treat fever, coughs and nausea, while dried ginger is used in cases of stomach pain and diarrhoea. One of the first eastern spices to be commercially cultivated in the West Indies (before 1547).

Zingiber oligophyllum Schumann (ZINGIBERACEAE)
New species

in Engler, Pflanzenr. 20. (iv. 46.): 185. (1904).

Zingiber (*Cryptanthium*) sp. Henry in Trans. Asiat. Soc. Jap. 24 suppl: 94. (1896) (List Pl. Formosa).

Taiwan: Kaohsiung, 1605 (type).

A perennial with pseudostems to 60 cm tall. Southeast China to southern China (Taiwan).

Zingiber pleiostachyum Schumann (ZINGIBERACEAE)
New species

in Engler, Pflanzenr. 20. (iv. 46.): 185. (1904), Wright in Journ. Linn. Soc. Bot. 36: 71. (1903), Wang in Flora of Taiwan 5: 724. (2000).

Zingiber (*Cryptanthium*) sp. Henry in Trans. Asiat. Soc. Jap. 24 suppl: 94. (1896) (List Pl. Formosa).

Taiwan: Wanchin, 147, 1659.

Known only from Henry's original collections made at Wanchin.

Zingiber sp. No. 1. (ZINGIBERACEAE)

Zingiber (*Cryptanthium*) sp. Henry in Trans. Asiat.

Soc. Jap. 24 suppl: 94. (1896) (List Pl. Formosa), Wright in Journ. Linn. Soc. Bot. 36: 71. (1903).

Taiwan: Oluanpi, 1966. Wanchin, 1607.

Zingiber sp. No. 2. indescript. (No. 6. of *Index Florae Sinensis*). (ZINGIBERACEAE)

Wright in Journ. Linn. Soc. Bot. 36: 71. (1903).

Sichuan: North Wushan, 7126. **Hubei:** South Donghu, 7594, 7671.

Zingiber sp. No. 3. indescript. (No. 7. of *Index Florae Sinensis*). (ZINGIBERACEAE)

Wright in Journ. Linn. Soc. Bot. 36: 71. (1903)

Hubei: Yichang, 7104a.

Zingiber spp. (ZINGIBERACEAE)

Anonymous in Kew List of Determinations (volume 24 & 34).

Hubei: Without locality, 7568. **Yunnan:** Mengzi, 9880, 13636, 13807. Simao, 12398, 13068, 13587.

Zingiber striolatum Diels (ZINGIBERACEAE) New species

in Engler's Bot. Jahrb. 29: 263. (1900) Schumann in Engler, Pflanzenr. 20: 182. (1904).

Zingiber didymoglossum K. M. Schumann in Engler. Pflanzenr. 20.(iv. 46): 186. (1904).

Sichuan: North Wushan, 7104.

Zingiberaceae No. 1. (ZINGIBERACEAE)

Scitamineae Anonymous in Kew List of Determinations (volume 34).

Yunnan: Simao, 12413.

Zingiberaceae No. 2. (leaves only) (ZINGIBERACEAE)

Anonymous in Kew List of Determinations (volume 24).

Hainan: Without locality, 8447.

Ziziphus incurva Roxb. (RHAMNACEAE)

Zizyphus sp. Schneider in Sargent Pl. Wilson. 2: 213. (1916). *Ziziphus yunnanensis* Schneider in Sargent Pl. Wilson. 2: 213. (1916).

Yunnan: Simao at 1,450 metres, 11726. Simao at 1,300 metres, 12086b. In mountains to the west of Simao at 1,800 metres, 12086a., (type of *Ziziphus yunnanensis* Schneider). Yiwu (in Xishuangbanna), 13572.

Native to India, Bhutan, Nepal and western China.

Ziziphus jujuba Mill. (RHAMNACEAE)

Henry in Trans. Asiat. Soc. Jap. 24 suppl: 27. (1896) (List Pl. Formosa).

Ziziyphus vulgaris Lam. Forbes & Hemsley in Journ. Linn. Soc. Bot. 23: 126. (1886), Henry in Journ. China Br. Roy. Asiat. Soc. 22: 276. (1887) (Chinese Names of Plants). *Zizyphus sativa* Gaert. var. *inermis* Schneid. Schneider in Sargent Pl. Wilson 2: 212. (1916). *Ziziphus jujuba* Mill. var. *inermis* (Bunge) Rehd. Zheng in Journ. Wuhan Bot. Res.

2.(1): 141. (1984).

Hubei: Yichang, 578, 2021. Changyang, 7512. **Hainan:** On Liang Shan, 16 km (10 miles) east of Haikou, 8206. **Taiwan:** Kaohsiung, without number. **Yunnan:** Yuanjiang, 13258.

The Chinese date-plum, a bush or small tree to 10 metres tall, known in western China as *tsao-tzu shu* (*da zao shu* i.e. 'big date tree' according to Henry. In this variety the branches are unarmed. The fruits (which are delicious) are used as a tonic and to treat disorders of the upper respiratory tract. Native to Russia, China and Korea where it is very widely distributed.

Ziziphus rugosa Lam. (RHAMNACEAE)

Anonymous in Kew List of Determinations (volume 34).

Yunnan: Mengzi, 10674.

Distributed in India, Sri Lanka, Myanmar and China (Guangxi, Hainan and Yunnan).

Zizyphus sp. (RHAMNACEAE)

Anonymous in Kew List of Determinations (volume 24).

Hainan: Environs of Haikou, 8189.

Zoysia matrella (L.) Merr. (POACEAE)

Zoysia pungens Willd. Henry in Trans. Asiat. Soc. Jap. 24 suppl: 107. (1896) (List Pl. Formosa), Rendle in Journ. Linn. Soc. Bot. 36: 344. (1904).

Taiwan: Kaohsiung, 1798, 1798a., 1906.

NO DATA. CENTRAL CHINA (HUBEI and SICHUAN): 29, 118, 395, 531, 539, 542, 564, 637, 661, 669, 673, 689, 793, 784, 821, 828, 845, 857, 881, 888, 915, 1012, 1019, 1029, 1074, 1075, 1076, 1077, 1078, 1079, 1080, 1165, 1283, 1476, 1477, 1478, 1479, 1480, 1481, 1482, 1483, 1485, 1486, 1487, 1577, 1594, 1629, 1784, 1796, 1820, 2053, 2066, 2097, 2112, 2254, 2313, 2335, 2362, 2408, 2438, 2542, 2565, 2468, 2479, 2484, 2492, 2512, 2524, 2553, 2570, 2601, 2625, 2644, 2652, 2677, 2708, 2711, 2758, 2764, 2847, 2850, 3027, 3149, 3305, 3340, 3667, 3708, 3880, 4407, 4417, 4419, 4449, 4484, 4664, 4777, 4593, 4805, 4809, 5007, 5013, 5042, 5100, 5209, 5258, 5351, 5384, 5449, 5657, 5776, 5777, 6606, 5979, 6135, 6161, 6309, 6893, 6974, 7057, 7391, 7540, 7543, 7552, 7665, 8909, 8815, 8816, 8817, 8818, 8819, 8820, 8821, 8824, 8825, 8834, 8835, 8838, 8839, 8840, 8841, 8842, 8844, 8845, 8846, 8847, 8848, 8849, 8850, 8886, 8900, 8901, 8902, 8904, 8905, 8906, 8906, 8907, 8914, 8903, 8910, 8928, 8932, 8938, 8959, 8998. **HAINAN:** 7964, 7965, 7971, 8005, 8012, 8033, 8039, 8071, 8097, 8113, 8115, 8141, 8143, 8155, 8170, 8173, 8191, 8196, 8214, 8215, 8216, 8227, 8649, 6260, 8276,

8301, 8311, 8314, 8330, 8342, 8344, 8394, 8401, 8405, 8425, 8446, 8449, 8493, 8507, 8513, 8545, 8546, 8549, 8568, 8591, 8592, 8593, 8596, 8598, 8599, 8600, 8607, 8608, 8635, 8648, 8659, 8661, 8683, 8704, 8720, 8723, 8727, 8750, 8753, 8764, 8766. **TAIWAN:** 7, 22, 46, 224, 226, 234, 253, 272, 296, 326, 343, 351, 372, 374, 376, 377, 423, 455, 475, 487, 518, 524, 525, 526, 528, 546, 579, 582, 621, 650, 681, 694, 718, 737, 738, 767, 916, 941, 974, 994, 1054, 1261, 1266, 1299, 1302, 1324, 1339, 1341, 1361, 1379, 1429, 1442, 1455, 1457, 1463, 1480, 1494, 1621, 1628, 1685, 1701, 1737, 1741, 1746, 1750, 1791, 1805, 1809, 1827, 1834, 1837, 1838, 1850, 1851, 1853, 1854, 1899, 1901, 1908, 1921, 1930, 1932, 1933, 1944, 1945, 1959, 1982, 1988, 1998, 2004, 2008, 2010, 2015, 2022, 2024, 2025, 2031, 2033, 2037, 2043, 2047, 2056, 2070, 2077. **YUNNAN:** 9003, 9015, 9064, 9095, 9096, 9113, 9141, 9146, 9256, 9165, 9167, 9232, 9247, 9296, 9307, 9314, 9358, 9378, 9382, 9392, 9396, 9405, 9420, 9437, 9446, 9447, 9462, 9481, 9483, 9488, 9512, 9515, 9520, 9523, 9571, 9585, 9587, 9591, 9594, 9599, 9620, 9664, 9709, 9719, 9724, 9752, 9756, 9763, 9778, 9779, 9788, 9879, 9806, 9816, 9823, 9826, 9827, 9861, 9903, 9905, 9921, 9922, 9923, 9925, 9967, 9995, 10019, 10022, 10033, 10058, 10066, 10084, 10085, 10091, 10102, 10112, 10139, 10156, 10157, 10166, 10172, 10219, 10221, 10234, 10263, 10280, 10296, 10335, 10336, 10338, 10345, 10354, 10365, 10373, 10374, 10379, 10388, 10389, 10399, 10402, 10403, 10410, 10412, 10419, 10430, 10448, 10466, 10474, 10489, 10491, 10509, 10550, 10558, 10562, 10620, 10663, 10664, 10750, 10751, 10783, 10795, 10827, 10837, 10865, 10886, 10898, 10942, 11004, 11013, 11031, 11049, 11120, 11136, 11155, 11190, 11224, 11241, 11242, 11249, 11278, 11310, 11323, 11373, 11380, 11418, 11432, 11433, 11453, 11455, 11457, 11500, 11509, 11553, 11555, 11817, 13673, 13683, 13685, 13687, 13700, 13707, 13709, 13712, 13714, 13721, 13722, 13736.

REFERENCES

HENRY A, 1893. *Notes on Economic Botany of China* 16 [M]. The Presbyterian Mission Press, Shanghai.

O'BRIEN S, 2011. *In the Footsteps of Augustine Henry and his Chinese Plant* [M]. Collectors 19. Garden Art Press.

SARGENT C S, 1911. *Plantae Wilsonianiae* [M]. Preface. V. Cambridge, The University Press.

WALSH T, 1957. *The Augustine Henry Forestry Herbarium – A Catalogue of Specimens* [M]. An Roinn Talmhaíochta (Department of Agriculture), Dublin.

AUTHOR

Seamus O'Brien is Head Gardener at the National Botanic Gardens, Kilmacurragh, County Wicklow, the country estate and rural annex of the National Botanic Gardens of Ireland, Glasnevin in Dublin. He is Ireland's leading authority on the flora of China and the history of botanical exploration. At Kilmacurragh he manages an important plant collection from temperate regions of the world, particularly from China, the Himalaya, South America, New Zealand, Australia and Tasmania. With a wide range of endangered and threatened taxa, conservation is an important feature of the Garden's work and this includes material sourced through botanical expeditions across the globe and from collaborative projects such as the Royal Botanic Garden, Edinburgh's International Conifer Conservation Programme.

Education and Career

Seamus was born in 1970 in Baltinglass, County Wicklow and completed his Secondary (High) School education at Scoil Conglais, Baltinglass in 1989. He received his formal horticultural training at the National Botanic Gardens of Ireland, Glasnevin, between 1990—1993 and also holds an International Diploma in Botanic Gardens Management from the Royal Botanic Gardens, Kew, London (2006).

In the past he has managed several notable Irish gardens with important historic plant collections, including Glanleam on Valentia Island, County Kerry (1993—1997), famed for its southern hemisphere trees and shrubs, and Beech Park in Clonsilla, Dublin (1997—2000), then renowned for one of the largest collections of herbaceous and alpine plants in Britain and Ireland. He returned to the National Botanic Gardens, Glasnevin as a staff member in February 2000 and from there moved to manage the National Botanic Gardens, Kilmacurragh in May 2006.

He has travelled extensively across the globe to study plants in their native habitats, most notably to Australia (including Tasmania), the Azores, Bhutan, China [Hong Kong, Hubei, Sichuan, Yunnan, Taiwan and Xizang (Tibet)], California, Chile, India (Assam, Ladakh, Manipur, Nagaland, Sikkim, West Bengal), Myanmar, Nepal, New Zealand, Kenya, Uganda and South Africa.

He has retraced the routes taken by Irish botanical explorer, Dr Augustine Henry across China, publishing the internationally acclaimed and award winning book *In the Footsteps of Augustine Henry and his Chinese collectors* (2011). From more recent travels in north-east India he has published *In the Footsteps of Joseph Dalton Hooker: A*

Sikkim Adventure (2018). He is currently writing a biography of the English plant collector and explorer Captain Frank Kingdon Ward (1885—1958). He also contributes to several publications including *Curtis's Botanical Magazine, The Irish Garden,* the Irish Garden Plant Society newsletter and the *International Dendrology Society Year Book.*

He lectures on garden history and botanical travel internationally and is a member of the Royal Horticultural Society's Woody Plant Committee, as well as being a judge of the Royal Horticultural Society's Early Spring Shows. Rhododendrons are a passion. He is Irish Branch Chairman of the Royal Horticultural Society's Rhododendron Camellia and Magnolia Group and is a trustee of the Exbury Garden Trust. Exbury is the Rothschild family's 200 acre garden near Southampton in England, which is home to a world-famous *Rhododendron* collection. Created by Lionel de Rothschild (1882—1942), the gardens contain the Chinese collections of George Forrest (1873—1932) and Captain Frank Kingdon Ward. He is also a member of the Advisory Group to the Annesley Garden and Castlewellan, the National Arboretum for Northern Ireland.

Awards and Honours

In 2018 Seamus was awarded the Royal Horticultural Society of Ireland's Gold Medal of Honour and was made an Honorary Member of the Society in 2019. In 2021 he was made a Fellow of the Explorer's Club (New York), in 2022 he was awarded the Royal Horticultural Society's (UK) Loder Rhododendron Cup 'for his work in conserving the famous *Rhododendron* collection at Kilmacurragh and for his willingness to share his knowledge through his books and publications'. In 2023 was made a Fellow of the Linnean Society (FLS).

Publications

O'BRIEN S, 1994. Glanleam opens to the Public *Irish Garden Plant Society Newsletter*, 52: 9. IGPS, Dublin.

O'BRIEN S, 1995. From Southern Shores *Borderlines - Journal of the Half Hardy Group of the Hardy Plant Society*, 9: 8-12. Cumbria, UK.

O'BRIEN S, 1995. Gold at Glanleam *The Irish Garden*, 4(6): 27. Mediateam Ltd., Dublin.

O'BRIEN S, 1996. From a Kerry Rainforest *Irish Garden Plant Society Newsletter*, 63: 2-5. IGPS, Dublin.

O'Brien, S, 1996. Serial Solanums *Borderlines - Journal of the Half Hardy Group of the Hardy Plant Society*, 10: 32-35. Cumbria, UK. (with Tim Longville *et al.*).

O'BRIEN S, 1996. Gossip from Glanleam *Borderlines- Journal of the Half Hardy Group of the Hardy Plant Society*, 9(2): 52-54. Cumbria, UK.

O'BRIEN S, 1997. Glasnevin's Californian Fuchsia *Irish Garden Plant Society Newsletter*, 64: 22. IGPS, Dublin.

O'BRIEN S, 1997. In Search of Chinese Exotics *Borderlines - Journal of the Half Hardy Group of the Hardy Plant Society*, 12 & 13: 7-11. Cumbria, UK.

O'BRIEN S, 1998. Autumn in Yunnan – An Expedition to south-west China *Moorea*, 13: 16-21. IGPS, Dublin.

O'BRIEN S, 1998. Beech Park SE Tibet Expedition *Irish Garden Plant Society Newsletter*, 67: 3-4. IGPS, Dublin.

O'BRIEN S, 1998. Beech Park – Alive and Well *Irish Garden Plant Society Newsletter*, 68: 3-6. IGPS, Dublin.

O'BRIEN S, 1998. Rambles on the Burren *Irish Garden Plant Society* Newsletter, 69: 6-9. IGPS, Dublin.

O'BRIEN S, 1998. Christmas Eve Storm on Valentia and Knight of Kerry Papers *Irish Garden Plant Society Newsletter*, 70: 2-6. IGPS, Dublin.

O'BRIEN S, 1998. On the Roof of the World *The Irish Garden*, 7(2): 48-49. Mediateam Ltd., Dublin.

O'BRIEN S, 1998. Quest for the Dove Tree *The Irish Garden*, 7(4): 9. Mediateam Ltd., Dublin.

O'BRIEN S, 1998. A Glimpse of the Emperor's Garden *The Irish Garden*, 7(9): 46-49. Mediateam Ltd., Dublin.

O'BRIEN S, 1999. Society News *Irish Garden Plant Society* Newsletter, 71: 19-21. IGPS, Dublin.

O'BRIEN S, 1999. Cornwall in February *Irish Garden Plant Society* Newsletter 72: 20-24. IGPS, Dublin.

O'BRIEN S, 1999. Mid-summer at Clonsilla *Irish Garden Plant Society Newsletter*, 73: 3-7. IGPS, Dublin.

O'BRIEN S, 1999. Fronded Giants *The Irish Garden*, 8(4): 38-41. Mediateam Ltd., Dublin.

O'BRIEN S, 1999. Creator of Glanleam *The Irish Garden*, 8(5): 38-41. Mediateam Ltd., Dublin.

O'BRIEN S, 2000. From China to Chelsea *Irish Garden Plant Society Newsletter*, 75: 2-8. IGPS, Dublin.

O'BRIEN S, 2000. Carolinella – The women in Henry's life *Irish Garden Plant Society Newsletter*, 76: 2-7. IGPS, Dublin.

O'BRIEN S, 2000. Chelsea Flower Show – May 2001 *Irish Garden Plant Society Newsletter* 76: 7-8. IGPS, Dublin.

O'BRIEN S, 2000. An Irishman in China *Irish Garden Plant Society Newsletter*, 78: 2-5. IGPS, Dublin.

O'BRIEN S, 2000. Richard Shackleton – An Appreciation *Irish Garden Plant Society Newsletter*, 78: 25. IGPS, Dublin.

O'BRIEN S, 2000. White Ladies of Spring *The Irish Garden*, 9(1): 16-19. Mediateam Ltd., Dublin.

O'BRIEN S, 2001. Chelsea Flower Show 2001 Update *Irish Garden Plant Society Newsletter*, 79: 22-24. IGPS, Dublin.

02

O'BRIEN S, 2001. Mid-summer in Glasnevin's woodland garden and Captain Madden's gargantuan lily *Irish Garden Plant Society Newsletter*, 81: 14-17. IGPS, Dublin.

O'BRIEN S, 2001. Trekking in south-east Tibet – scorpions, snakes, leeches and all! *Irish Garden Plant Society Newsletter*, 82: 3-6. IGPS, Dublin.

O'BRIEN S, 2002. New Plants at Glasnevin (*Helleborus × hybridus* 'Seamus O'Brien' and *Clematis tibetana* subsp. *vernayi* 'Glasnevin Dusk') *Irish Garden Plant Society Newsletter*, 84: 11-12. IGPS, Dublin.

O'BRIEN S, 2003 Glasnevin Central China Expedition *Irish Garden Plant Society Newsletter*, 87: 3-7. IGPS, Dublin.

O'BRIEN S, 2003. Notes from Glasnevin *Irish Garden Plant Society Newsletter*, 90: 4-8. IGPS, Dublin.

O'BRIEN S, 2003. A Tree from the Dawn of Time *Crann*, 60: 10. Imprinta, Maynooth.

O'BRIEN S, 2004. Once again to China *Irish Garden Plant Society Newsletter*, 93: 11. IGPS, Dublin.

O'BRIEN S, 2004. Woodland Gardening at Glasnevin *Irish Garden Plant Society Newsletter*, 91: 24-28. IGPS, Dublin.

O'BRIEN S, 2004. The Glasnevin Central China Expedition 2004 *Irish Garden Plant Society Newsletter*, 95: 3-10. IGPS, Dublin.

O'BRIEN S, et al., 2004. Looking for Chinese Rarities *The Garden* (Journal of the Royal Horticultural Society), 129 (2):103. RHS, London, UK.

O'BRIEN S, 2005. Report on the (National) Botanic Gardens China Expedition 2004 *Horticulture and Landscape Ireland*, 13: 39. SWP Publishers.

O'BRIEN S, 2006. A Journey to Western China (part 1) *Irish Garden Plant Society Newsletter*, 99: 3-8. IGPS, Dublin.

O'BRIEN S, 2006. A Journey to Western China (part 2) *Irish Garden Plant Society Newsletter*, 100: 28-35. IGPS, Dublin.

O'BRIEN S, 2007. Valerie Finnis (Lady Scott), An Appreciation *Irish Garden Plant Society Newsletter*, 103: 2-24. IGPS, Dublin.

O'BRIEN S, 2007. On the Shoulders of Giants *The Irish Garden*, 16(3): 96-98. Mediateam Ltd., Dublin.

O'BRIEN S, 2008. Thomas Acton, A Centennial Celebration at Kilmacurragh *Irish Garden Plant Society Newsletter*, 109: 8-16. IGPS, Dublin.

O'BRIEN S, 2008. Thomas Acton – A centennial celebration at Kilmacurragh *International Dendrology Society Yearbook*: 160-169. Dendrology Charitable Company, Herefordshire, UK.

O'BRIEN S, 2009. *Lilium henryi* – 120 years in cultivation *Irish Garden Plant Society Newsletter*, 112: 34-36. IGPS, Dublin.

O'BRIEN S, 2010. In Praise of the Monkey Puzzle *Journal of the Royal Horticultural Society of Ireland*, 9-21. RHSI, Dublin.

O'BRIEN S, 2010. The Chilean Fire Bush – Continuing a Tale *Irish Garden Plant Society Newsletter*, 115: 19-30. IGPS, Dublin.

O'BRIEN S, 2010. A tale of two Gardens *The Irish Garden*, 19(8): 58-61. Mediateam Ltd., Dublin.

O'BRIEN S, 2011. Two New Irish Cultivars *Irish Garden Plant Society Newsletter*, 121: 10-11. IGPS, Dublin.

O'BRIEN S, 2011. *In the Footsteps of Augustine Henry and his Chinese Plant Collectors* (Foreword by Roy Lancaster) 367 pp. Garden Art Press, Suffolk, England.

O'BRIEN S, 2011. Modern Plant Hunters *The Irish Garden*, 20(2): 50-53. Mediateam Ltd., Dublin

O'BRIEN S, 2011. Mr Hodgins and his Hollies *The Irish Garden*, 20(10): 46-49. Mediateam Ltd., Dublin.

O'BRIEN S, 2012. Spring at Kilmacurragh *Irish Garden Plant Society Newsletter*, 123: 28-36. IGPS, Dublin.

O'BRIEN S, 2012 Aextoxicon punctatum *Curtis's Bot. Mag.*, 29(2): t. 737. 182-193. Wiley.

O'BRIEN S, 2012. What's in a Name? *The Irish Garden*, 21(4): 56-58. Mediateam Ltd., Dublin.

O'BRIEN S, 2013. Visiting Roy Lancaster's Garden – A Plantsman's Paradise *Irish Garden Plant Society Newsletter*, 127: 35-41. IGPS, Dublin.

O'BRIEN S, 2013. A Sikkim Adventure *International Dendrology Society Yearbook*: 166-180. Dendrology Charitable Company, Herefordshire, UK.

O'BRIEN S, 2013. Irish Botanical Art Lives On. *Garden Heaven Annual*, 87-91. Image Publications, Dublin.

O'BRIEN S, 2013. In the Footsteps of Augustine Henry. *Garden Heaven Annual*: 152-157. Image Publications, Dublin.

O'BRIEN S, 2013. Filoli – The Californian Muckross *The Irish Garden*, 22(2): 50-53. Mediateam Ltd., Dublin.

O'BRIEN S, 2013. In Search of Rare Trees *The Irish Garden*, 22(6): 46-49. Mediateam Ltd., Dublin.

O'BRIEN S, 2013 Patrinia heterophylla *Curtis's Bot. Mag.*, 30(2): t. 761. 101-107. Wiley.

O'BRIEN S, 2013. Patrinia villosa *Curtis's Bot. Mag.*, 30(2): t. 762. 108-113. Wiley.

O'BRIEN S, 2013. Meeting with Magnolia campbellii *Journal of the Royal Horticultural Society of Ireland*: 13-16. RHSI, Dublin.

O'BRIEN S, 2014. In the Footsteps of Joseph Dalton Hooker *Moorea*, 16: 11-22. IGPS, Dublin.

O'BRIEN S, 2014. Travels in the Himalaya *Garden*

Heaven spring/summer: 102-107. Image Publications, Dublin.

O'BRIEN S, 2014. Tourin, County Waterford *The Irish Garden*, 23(8): 46-48. Mediateam Ltd., Dublin.

O'BRIEN S, 2015. Rhododendron magnificum 'Lady Rose' *Irish Garden Plant Society Newsletter*, 132: 3-10. IGPS, Dublin.

O'BRIEN S, 2016. Narcissus 'Countess of Annesley' *Irish Garden Plant Society Newsletter*, 135: 4. IGPS Dublin.

O'BRIEN S, 2016. Myrceugenia leptospermoides *Curtis's Bot. Mag.*, 33(4): t. 850. 310-319 Wiley (with Martin F. Gardner, Royal Botanic Garden, Edinburgh).

O'BRIEN S, 2016. Myrceugenia lanceolata *Curtis's Bot. Mag.*, 33(4): t. 851. 320-326 Wiley (with Martin F. Gardner, Royal Botanic Garden, Edinburgh).

O'BRIEN S, 2016. Woody Plants (in Heritage Irish Plants – Plandaí Oidhreacta) 92-101. ISBA & IGPS, Dublin.

O'BRIEN S, 2017. Retracing Joseph Hooker's routes in the Sikkim Himalaya *Curtis's Bot. Mag.*, 34(3): 175-189. Wiley.

O'BRIEN S, 2018. Commemorating Armistice Day *Irish Garden Plant Society Newsletter*, 143: 7-9. IGPS, Dublin.

O'BRIEN S, 2018. In the Footsteps of Joseph Dalton Hooker *Irish Garden Plant Society Newsletter*, 143: 18-21. IGPS, Dublin.

O'BRIEN S, 2018. *In the Footsteps of Joseph Dalton Hooker: A Sikkim Adventure* (Foreword by Lady Mary Keen). 324 pp. Kew Publishing, Richmond, London, UK.

O'BRIEN S, 2018. Trees and Shrubs of Tasmania *International Dendrology Society Yearbook*: 75-86 Dendrology Charitable Company, Herefordshire, UK.

O'BRIEN S, 2019. The British-Irish Botanical Expedition to Tasmania *Irish Garden Plant Society Newsletter*, 144: 13-19. IGPS, Dublin.

O'BRIEN S., 2019. David Jervois Thetford Gilliland *International Dendrology Society Yearbook*: 10-12 Dendrology Charitable Company, Herefordshire, UK.

O'BRIEN S, 2019. Castlewellan *International Dendrology Society Yearbook*: 166-172. Dendrology Charitable Company, Herefordshire, UK.

O'BRIEN S, 2019. Quercus lamellosa *Curtis's Bot. Mag.*, 36(4): t. 923. 355-364. Wiley.

O'BRIEN S, 2020. Kilmacurragh re-united *Irish Garden Plant Society* Newsletter, 151: 29-30. IGPS, Dublin.

O'BRIEN S, 2020. Notable magnolias and rhododendrons at Kilmacurragh *Irish Tree Society Newsletter*: 24-26. ITS, Dublin.

O'BRIEN S, 2020. Rhododendron barbatum *Curtis's Bot. Mag.*, 27(2): t. 945. 200-211. Wiley.

O'BRIEN S, 2020. Wuhan and the Plant Hunters *Journal of the Royal Horticultural Society of Ireland*: 5-11. RHSI, Dublin.

O'BRIEN S, 2021. *Luma apiculata* 'Glanleam Gold' *Irish Garden Plant Society Newsletter*, 151: 19. IGPS, Dublin.

O'BRIEN S, 2021. New Irish Rhododendrons *Irish Garden Plant Society Newsletter*, 152: 6-9. IGPS, Dublin.

O'BRIEN S, 2021. *Melanophylla dianeae* A new species of rainforest tree from Madagascar *Irish Garden Plant Society Newsletter*, 152: 38. IGPS, Dublin.

O'BRIEN S, 2021. *Rhododendron magnificum Curtis's Bot. Mag.*, 38(2): t. 984. 202-216. Wiley.

O'BRIEN S, 2021. Lady Franklin's whitey wood *International Dendrology Society Yearbook*: 50-53. Dendrology Charitable Company, Herefordshire, UK.

O'BRIEN S, 2021. In Search of McCabe's Rhododendron – an autumn expedition to north-east India *Rhododendrons, Camellias & Magnolias*: 21-34. The Royal Horticultural Society, Vincent Square, London.

O'BRIEN S, 2022. Members' Memories – Exhibiting at Chelsea *Irish Garden Plant Society Newsletter*, 155: 20. IGPS, Dublin.

O'BRIEN S, 2023. *Pinus armandii* var. *dabeishanensis Curtis's Bot. Mag.*, 40(3): t. 1067. 267-273. Wiley.

O'BRIEN S, 2023. Woody Plants of Ladakh *International Dendrology Society Yearbook*: 150-161. Dendrology Charitable Company, Herefordshire, UK.

O'BRIEN S, 2023. Foreword (in *Ireland's Native Trees – Crainn na h Éireann*): 11-15. ISBA, Dublin.

O'BRIEN S, 2023. Cornus capitata 'Kilmacurragh Rose' *Irish Garden Plant Society Newsletter*, 158: 21. IGPS, Dublin.

O'BRIEN S, 2024. Spring Flowers *Irish Arts Review*, 41(1): 86-89 Irish Arts Review Ltd., Dublin.

O'BRIEN S, 2024. Gigantic Plants of East Africa *Journal of the Royal Horticultural Society of Ireland*: 44-51. RHSI, Dublin.

O'BRIEN S, 2024. Winter at Kilmacurragh *Irish Garden Plant Society Newsletter*, 160: 5-8. IGPS, Dublin.

O'BRIEN S, 2024. A new Hydrangea at Tullynally Castle *Irish Garden Plant Society Newsletter*, 161: 11-12. IGPS, Dublin.

O'BRIEN S, 2024. To Uganda's Mountains of the Moon *Journal of the Royal Horticultural Society of Ireland*: 25-34. RHSI, Dublin.

Seamus O'Brien

National Botanic Gardens of Ireland, Kilmacurragh, Irelan

November 2024

植物中文名索引
Plant Names in Chinese

植物学名索引
Plant Names in Latin

中文人名索引
Persons Index in Chinese

西文人名索引
Persons Index